# STUDENT SOLUTIONS MANUAL

KEVIN BODDEN ▪ RANDY GALLAHER
JULIE MUNIZ ▪ KATHY KOPELOUSOS ▪ MIKE ZIEGLER
*Lewis & Clark Community College*

# INTERMEDIATE ALGEBRA
*for college students*



# *Essentials of* INTERMEDIATE ALGEBRA
*for college students*

# ALGEBRA
*for college students*

FIFTH EDITION

# Blitzer

PEARSON
Prentice Hall

Upper Saddle River, NJ 07458

Editor-in-Chief: Chris Hoag
Executive Editor: Paul Murphy
Supplements Editor: Christina Simoneau
Executive Managing Editor: Kathleen Schiaparelli
Assistant Managing Editor: Becca Richter
Production Editor: Allyson Kloss
Supplement Cover Manager: Paul Gourhan
Supplement Cover Designer: Joanne Alexandris
Manufacturing Buyer: Ilene Kahn


Pearson Prentice Hall
Pearson Education, Inc.
Upper Saddle River, NJ 07458



Printed in the United States of America

10 9 8 7 6 5 4 3 2

ISBN 0-13-192200-9 Standalone
0-13-185797-5 Student Study Pack Component

Pearson Education Ltd., *London*
Pearson Education Australia Pty. Ltd., *Sydney*
Pearson Education Singapore, Pte. Ltd.
Pearson Education North Asia Ltd., *Hong Kong*
Pearson Education Canada, Inc., *Toronto*
Pearson Educación de Mexico, S.A. de C.V.
Pearson Education—Japan, *Tokyo*
Pearson Education Malaysia, Pte. Ltd.

## TABLE OF CONTENTS for STUDENT SOLUTIONS

### *INTERMEDIATE ALGEBRA FOR COLLEGE STUDENTS 4E*


Chapter 1 Algebra, Mathematical Models, and Problem Solving ........ 1
Chapter 2 Functions and Linear Functions ........ 45
Chapter 3 Systems of Linear Equations ........ 87
Chapter 4 Inequalities and Problem Solving ........ 180
Chapter 5 Polynomials, Polynomial Functions, and Factoring ........ 250
Chapter 6 Rational Expressions, Functions, and Equations ........ 320
Chapter 7 Radicals, Radical Functions, and Rational Exponents ........ 443
Chapter 8 Quadratic Equations and Functions ........ 512
Chapter 9 Exponential and Logarithmic Functions ........ 619
Chapter 10 Conic Sections and Systems of Nonlinear Equations ........ 692
Chapter 11 Sequences, Series, and the Binomial Theorem ........ 784


### *ESSENTIALS OF INTERMEDIATE ALGEBRA FOR COLLEGE STUDENTS*


Chapter 1 Algebra, Mathematical Models, and Problem Solving ........ 1
Chapter 2 Functions and Linear Functions ........ 45
Chapter 3 Systems of Linear Equations ........ 87
Chapter 4 Inequalities and Problem Solving ........ 180
Chapter 5 Polynomials, Polynomial Functions, and Factoring ........ 250
Chapter 6 Rational Expressions, Functions, and Equations ........ 320
Chapter 7 Radicals, Radical Functions, and Rational Exponents ........ 443
Chapter 8 Quadratic Equations and Functions ........ 512


### *ALGEBRA FOR COLLEGE STUDENTS 5E*


Chapter 1 Algebra, Mathematical Models, and Problem Solving ........ 1
Chapter 2 Functions and Linear Functions ........ 45
Chapter 3 Systems of Linear Equations ........ 87
Chapter 4 Inequalities and Problem Solving ........ 180
Chapter 5 Polynomials, Polynomial Functions, and Factoring ........ 250
Chapter 6 Rational Expressions, Functions, and Equations ........ 320
Chapter 7 Radicals, Radical Functions, and Rational Exponents ........ 443
Chapter 8 Quadratic Equations and Functions ........ 512
Chapter 9 Exponential and Logarithmic Functions ........ 619
Chapter 10 Conic Sections and Systems of Nonlinear Equations ........ 692
Chapter 11 More on Polynomial and Rational Functions ........ 845
Chapter 12 Sequences, Induction, and Probability ........ 908
Appendix A Distance and Midpoint Formulas; Circles ........ 989
Appendix B Summation Notation and the Binomial Theorem ........ 993

# Chapter 1
# Algebra, Mathematical Models, and Problem Solving

## 1.1 Exercise Set

**1.** $x+5$

**3.** $x-4$

**5.** $4x$

**7.** $2x+10$

**9.** $6-\frac{1}{2}x$

**11.** $\frac{4}{x}-2$

**13.** $\frac{3}{5-x}$

**15.** $7+5(10)=7+50=57$

**17.** $6(3)-8=18-8=10$

**19.** $8^2+3(8)=64+24=88$

**21.** $7^2-6(7)+3=49-42+3=7+3=10$

**23.** $4+5(9-7)^3=4+5(2)^3$
$=4+5(8)=4+40=44$

**25.** $8^2-3(8-2)=64-3(6)$
$=64-18=46$

**27.** $\{1, 2, 3, 4\}$

**29.** $\{-7, -6, -5, -4\}$

**31.** $\{8, 9, 10, \ldots\}$

**33.** $\{1, 3, 5, 7, 9\}$

**35.** True. Seven is an integer.

**37.** True. Seven is a rational number.

**39.** False. Seven is a rational number.

**41.** True. Three is not an irrational number.

**43.** False. $\frac{1}{2}$ is a rational number.

**45.** True. $\sqrt{2}$ is not a rational number.

**47.** False. $\sqrt{2}$ is a real number.

**49.** –6 is less than –2. True.

**51.** 5 is greater than –7. True.

**53.** 0 is less than –4. False. 0 is greater than –4.

**55.** –4 is less than or equal to 1. True.

**57.** –2 is less than or equal to –6. False. –2 is greater than –6.

**59.** –2 is less than or equal to –2. True.

**61.** –2 is greater than or equal to –2. True.

**63.** 2 is less than or equal to $-\frac{1}{2}$. False.

2 is greater than $-\frac{1}{2}$.

**65.** True.

**66.** True.

**67.** False. $\{3\} \not\subseteq \{1,2,3,4\}$.

**68.** False. $\{4\} \not\subseteq \{1,2,3,4,5\}$.

**69.** True.

**70.** True.

**71.** False. The value of $\{x \mid x$ is an integer between $-3$ and $0\}$ = $\{-2,-1\}$, not $\{-3,-2,-1,0\}$.

**72.** False. The value of $\{x \mid x$ is an integer between $-4$ and $0\}$ = $\{-3,-2,-1\}$, not $\{-4,-3,-2,-1,0\}$.

**73.** False. Twice the sum of a number and three is represented by $2(x+3)$, not $2x+3$.

**74.** False. Three times the sum of a number and five is represented by $3(x+5)$, not $3x+5$.

**75.** The year 2000 is $x = 20$ years after 1980. $E = 183.6(20) + 992$
$$= 3672 + 992 = 4664$$
According to the formula, each American paid \$4664 for health care. This is an underestimate of \$8 from the actual amount of \$4672.

**77.** The year 2001 is $x = 4$ years after 1997. $N = 0.5(4)^2 - 3(4) + 36$
$$= 8 - 12 + 36 = 32$$
According to the formula, 32 million Americans were living below the poverty level in 2001. The formula models the data quite well.

**79.** $$C = \frac{5}{9}(50-32) = \frac{5}{9}(18) = 10$$
$10^\circ$C is equivalent to $50^\circ$F.

**81.** $h = 4 + 60t - 16t^2 = 4 + 60(2) - 16(2)^2$
$$= 4 + 120 - 16(4) = 4 + 120 - 64$$
$$= 124 - 64 = 60$$
Two seconds after it is kicked, the ball's height is 60 feet.

**83. – 99.** Answers will vary.

**101.** Statement ***c*** is true. Some rational numbers are not positive.

Statement **a.** is false. Not all rational numbers can be expressed as integers.

Statement **b.** is false. All whole numbers are integers.

Statements **d.** is false. Irrational numbers can be negative. $-\sqrt{3}$ is one example.

**103.** $(2 \cdot 3 + 3) \cdot 5 = 45$

**105.** 26 is not a perfect square and $\sqrt{26}$ cannot be simplified. Consider the numbers closest to 26, both smaller and larger, which are perfect squares. The first perfect square smaller than 26 is 25. The first perfect square larger than 26 is 36. We know that the square root of 26 will lie between these numbers. We have $-\sqrt{36} < -\sqrt{26} < -\sqrt{25}$. If we simplify, we have $-6 < -\sqrt{26} < -5$. Therefore, $-\sqrt{26}$ lies between –6 and –5.

## 1.2 Exercise Set

**1.** $|-7| = 7$

**3.** $|4| = 4$

**5.** $|-7.6| = 7.6$

**7.** $\left|\frac{\pi}{2}\right| = \frac{\pi}{2}$

**9.** $|-\sqrt{2}| = \sqrt{2}$

**11.** $-\left|-\frac{2}{5}\right| = -\frac{2}{5}$

**13.** $-3 + (-8) = -11$

**15.** $-14 + 10 = -4$

**17.** $-6.8 + 2.3 = -4.5$

**19.** $\frac{11}{15} + \left(-\frac{3}{5}\right) = \frac{11}{15} + \left(-\frac{9}{15}\right) = \frac{2}{15}$

**21.** $-\frac{2}{9} - \frac{3}{4} = -\frac{2}{9} + \left(-\frac{3}{4}\right)$
$= -\frac{8}{36} + \left(-\frac{27}{36}\right) = -\frac{35}{36}$

**23.** $-3.7 + (-4.5) = -8.2$

**25.** $0 + (-12.4) = -12.4$

**27.** $12.4 + (-12.4) = 0$

**29.** $x = 11$
$-x = -11$

**31.** $x = -5$
$-x = 5$

**33.** $x = 0$
$-x = 0$

**35.** $3 - 15 = 3 + (-15) = -12$

**37.** $8 - (-10) = 8 + 10 = 18$

**39.** $-20 - (-5) = -20 + 5 = -15$

**41.** $\frac{1}{4} - \frac{1}{2} = \frac{1}{4} + \left(-\frac{1}{2}\right) = \frac{1}{4} + \left(-\frac{2}{4}\right) = -\frac{1}{4}$

**43.** $-2.3 - (-7.8) = -2.3 + 7.8 = 5.5$

**45.** $0 - (-\sqrt{2}) = 0 + \sqrt{2} = \sqrt{2}$

**47.** $9(-10) = -90$

**49.** $(-3)(-11) = 33$

**51.** $\frac{15}{13}(-1) = -\frac{15}{13}$

**53.** $-\sqrt{2} \cdot 0 = 0$

**55.** $(-4)(-2)(-1) = (8)(-1) = -8$

**57.** $2(-3)(-1)(-2)(-4)$
$= (-6)(-1)(-2)(-4)$
$= (6)(-2)(-4) = (-12)(-4) = 48$

**59.** $(-10)^2 = (-10)(-10) = 100$

**61.** $-10^2 = -(10)(10) = -100$

**63.** $(-2)^3 = (-2)(-2)(-2) = -8$

**65.** $(-1)^4 = (-1)(-1)(-1)(-1) = 1$

**67.** Since a product with an odd number of negative factors is negative, $(-1)^{33} = -1$.

**69.** $-\left(-\frac{1}{2}\right)^3 = -\left(-\frac{1}{2}\right)\left(-\frac{1}{2}\right)\left(-\frac{1}{2}\right) = \frac{1}{8}$

**71.** $\frac{12}{-4} = -3$

**73.** $\frac{-90}{-2} = 45$

**75.** $\frac{0}{-4.6} = 0$

**77.** $-\frac{4.6}{0}$ is undefined.

**79.** $-\frac{1}{2} \div \left(-\frac{7}{9}\right) = -\frac{1}{2} \cdot \left(-\frac{9}{7}\right) = \frac{9}{14}$

**81.** $6 \div \left(-\frac{2}{5}\right) = \frac{6}{1} \cdot \left(-\frac{5}{2}\right) = -\frac{30}{2} = -15$

**83.** $4(-5) - 6(-3) = -20 - (-18)$
$= -20 + 18 = -2$

**85.** $3(-2)^2 - 4(-3)^2 = 3(4) - 4(9)$
$= 12 - 36 = -24$

**87.** $8^2 - 16 \div 2^2 \cdot 4 - 3 = 64 - 16 \div 4 \cdot 4 - 3$
$= 64 - 4 \cdot 4 - 3$
$= 64 - 16 - 3$
$= 48 - 3$
$= 45$

**89.** $\frac{5 \cdot 2 - 3^2}{\left[3^2 - (-2)\right]^2} = \frac{5 \cdot 2 - 9}{\left[9 - (-2)\right]^2}$
$= \frac{10 - 9}{(9+2)^2} = \frac{1}{11^2} = \frac{1}{121}$

**91.** $8 - 3[-2(2-5) - 4(8-6)]$
$= 8 - 3[-2(-3) - 4(2)]$
$= 8 - 3[6 - 8] = 8 - 3[-2] = 8 + 6 = 14$

**93.** $\frac{2(-2) - 4(-3)}{5-8} = \frac{-4+12}{-3} = \frac{8}{-3} = -\frac{8}{3}$

**95.** $\frac{(5-6)^2 - 2|3-7|}{89 - 3 \cdot 5^2} = \frac{(-1)^2 - 2|-4|}{89 - 3 \cdot 25}$
$= \frac{1 - 2(4)}{89 - 75}$
$= \frac{1-8}{14} = \frac{-7}{14} = -\frac{1}{2}$

**97.** $15 - \sqrt{3 - (-1)} + 12 \div 2 \cdot 3$
$= 15 - \sqrt{4} + 12 \div 2 \cdot 3$
$= 15 - 2 + 12 \div 2 \cdot 3$
$= 15 - 2 + 6 \cdot 3$
$= 15 - 2 + 18 = 13 + 18 = 31$

**99.** $20 + 1 - \sqrt{10^2 - (5+1)^2}\,(-2)$
$= 20 + 1 - \sqrt{10^2 - 6^2}\,(-2)$
$= 20 + 1 - \sqrt{100 - 36}\,(-2)$
$= 20 + 1 - \sqrt{64}\,(-2)$
$= 20 + 1 - 8(-2) = 20 + 1 + 16 = 37$

**101.** Commutative Property of Addition: $4x + 10 = 10 + 4x$

Commutative Property of Multiplication: $4x + 10 = x \cdot 4 + 10$

**103.** Commutative Property of Addition: $7x - 5 = -5 + 7x$

Commutative Property of Multiplication: $7x - 5 = x \cdot 7 - 5$

**105.** $4 + (6 + x) = (4 + 6) + x = 10 + x$

**107.** $-7(3x) = (-7 \cdot 3)x = -21x$

**109.** $-\frac{1}{3}(-3y) = \left(-\frac{1}{3} \cdot -3\right)y = y$

**111.** $3(2x + 5) = 3 \cdot 2x + 3 \cdot 5 = 6x + 15$

**113.** $-7(2x+3) = -7\cdot 2x + (-7)3$
$= -14x - 21$

**115.** $-(3x-6) = -1\cdot 3x + (-1)6 = -3x - 6$

**117.** $7x + 5x = (7+5)x = 12x$

**119.** $6x^2 - x^2 = (6-1)x^2 = 5x^2$

**121.** $6x + 10x^2 + 4x + 2x^2$
$= 6x + 4x + 10x^2 + 2x^2$
$= (6+4)x + (10+2)x^2 = 10x + 12x^2$

**123.** $8(3x-5) - 6x$
$= 8\cdot 3x - 8\cdot 5 - 6x$
$= 24x - 40 - 6x$
$= 24x - 6x - 40$
$= (24-6)x - 40 = 18x - 40$

**125.** $5(3y-2) - (7y+2)$
$= 5\cdot 3y - 5\cdot 2 - 1\cdot 7y + (-1)2$
$= 15y - 10 - 7y - 2$
$= 15y - 7y - 10 - 2$
$= (15-7)y - 12 = 8y - 12$

**127.** $7 - 4[3 - (4y-5)]$
$= 7 - 4[3 - 4y + 5]$
$= 7 - 12 + 16y - 20 = 16y - 25$

**129.** $18x^2 + 4 - [6(x^2-2) + 5]$
$= 18x^2 + 4 - [6x^2 - 12 + 5]$
$= 18x^2 + 4 - [6x^2 - 7]$
$= 18x^2 + 4 - 6x^2 + 7$
$= 18x^2 - 6x^2 + 4 + 7$
$= (18-6)x^2 + 11 = 12x^2 + 11$

**131.** $x - (x+4) = x - x - 4 = -4$

**132.** $x - (8-x) = x - 8 + x = 2x - 8$

**133.** $6(-5x) = -30x$

**134.** $10(-4x) = -40x$

**135.** $5x - 2x = 3x$

**136.** $6x - (-2x) = 6x + 2x = 8x$

**137.** $8x - (3x+6) = 8x - 3x - 6 = 5x - 6$

**138.** $8 - 3(x+6) = 8 - 3x - 18 = -3x - 10$

**139.** Hong Kong is at $+8$, and Los Angeles is at $-8$.
$8 - (-8) = 8 + 8 = 16$
Thus, Hong Kong is 16 hours ahead of Los Angeles.

**141.** Los Angeles is at $-8$, and New York is at $-5$. $-8 - (-5) = -8 + 5 = -3$
Thus, Los Angeles is 3 hours behind New York.

**143.** For the group whose highest level of parental education is an associates degree, the verbal score is $-20$, and the math score is $-28$.
$-20 - (-28) = -20 + 28 = 8$
Thus, the verbal score is 8 points above the math score.

**145.** For the group whose highest level of parental education is a graduate degree, the math score is 50. For the group whose parent does not have a high school diploma, the math score is $-76$. $50 - (-76) = 50 + 76 = 126$
Thus, there is a difference of 126 points.

**147. a.** $0.05x + 0.12(10{,}000 - x)$
$= 0.05x + 1200 - 0.12x$
$= 1200 - 0.07x$

**b.** $0.05(6000) + 0.12(10{,}000 - 6000)$
$= 0.05(6000) + 0.12(4000)$
$= 300 + 480 = 780$

$1200 - 0.07(6000) = 1200 - 420$
$= 780$

The total interest will be \$780.

**149. – 165.** Answers will vary.

**167.** Statement **d.** is correct.

Statement **a.** is false.
$16 \div 4 \cdot 2 = 4 \cdot 2 = 8$

Statement **b.** is false.
$6 - 2(4+3) = 6 - 2(7) = 6 - 14 = -8$

Statement **c.** is false. The expression cannot be simplified. The terms are not like terms and cannot be combined.

**169.** $\left(2 \cdot 5 - \frac{1}{2} \cdot 10\right) \cdot 9 = 45$

**171.** $\frac{10}{x} - 4x$

**172.** $10 + 2(x-5)^4$
$= 10 + 2(7-5)^4$
$= 10 + 2(2)^4$
$= 10 + 2(16) = 10 + 32 = 42$

**173.** True. $\frac{1}{2}$ is not an irrational number.

## 1.3 Exercise Set

**1. – 9.**

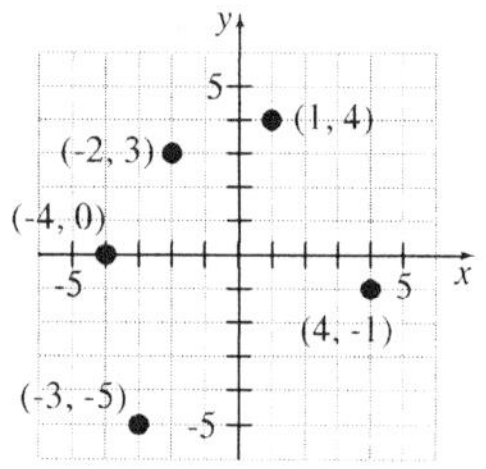


**11.**

| $x$ | $(x, y)$ |
|---|---|
| $-3$ | $(-3, 7)$ |
| $-2$ | $(-2, 2)$ |
| $-1$ | $(-1, -1)$ |
| 0 | $(0, -2)$ |
| 1 | $(1, -1)$ |
| 2 | $(2, 2)$ |
| 3 | $(3, 7)$ |

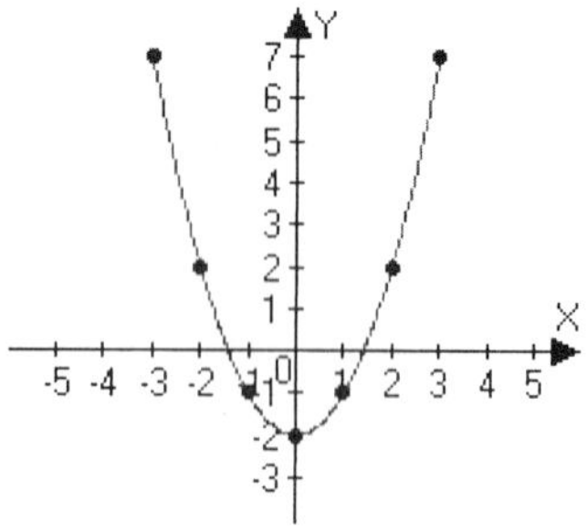


**13.**

| $x$ | $(x, y)$ |
|---|---|
| $-3$ | $(-3, -5)$ |
| $-2$ | $(-2, -4)$ |
| $-1$ | $(-1, -3)$ |
| 0 | $(0, -2)$ |
| 1 | $(1, -1)$ |
| 2 | $(2, 0)$ |
| 3 | $(3, 1)$ |

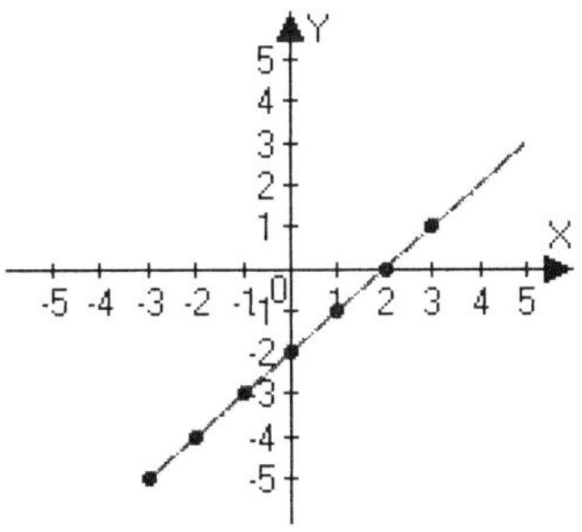


**15.**

| $x$ | $(x, y)$ |
|---|---|
| $-3$ | $(-3,-5)$ |
| $-2$ | $(-2,-3)$ |
| $-1$ | $(-1,-1)$ |
| $0$ | $(0,1)$ |
| $1$ | $(1,3)$ |
| $2$ | $(2,5)$ |
| $3$ | $(3,7)$ |

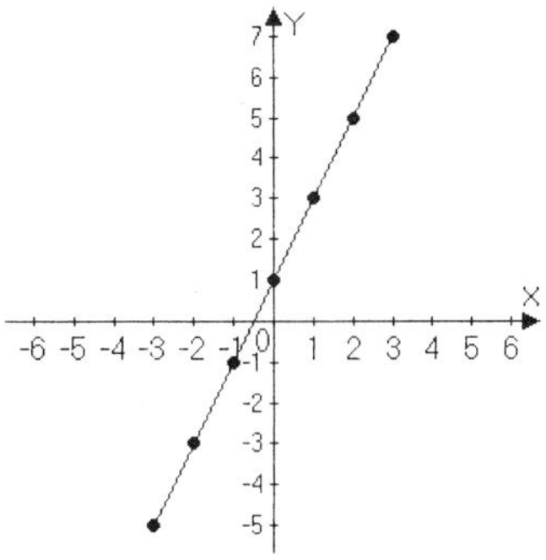


**17.**

| $x$ | $(x, y)$ |
|---|---|
| $-3$ | $\left(-3,\frac{3}{2}\right)$ |
| $-2$ | $(-2,1)$ |
| $-1$ | $\left(-1,\frac{1}{2}\right)$ |
| $0$ | $(0,0)$ |
| $1$ | $\left(1,-\frac{1}{2}\right)$ |
| $2$ | $(2,-1)$ |
| $3$ | $\left(3,-\frac{3}{2}\right)$ |

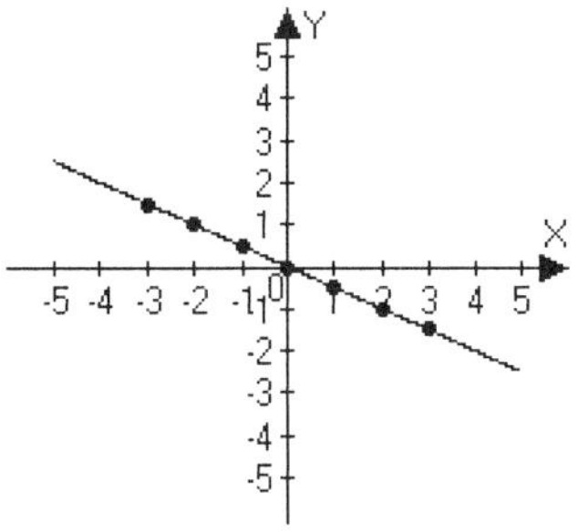


**19.**

| $x$ | $(x, y)$ |
|---|---|
| $-3$ | $(-3,4)$ |
| $-2$ | $(-2,3)$ |
| $-1$ | $(-1,2)$ |
| $0$ | $(0,1)$ |
| $1$ | $(1,2)$ |
| $2$ | $(2,3)$ |
| $3$ | $(3,4)$ |

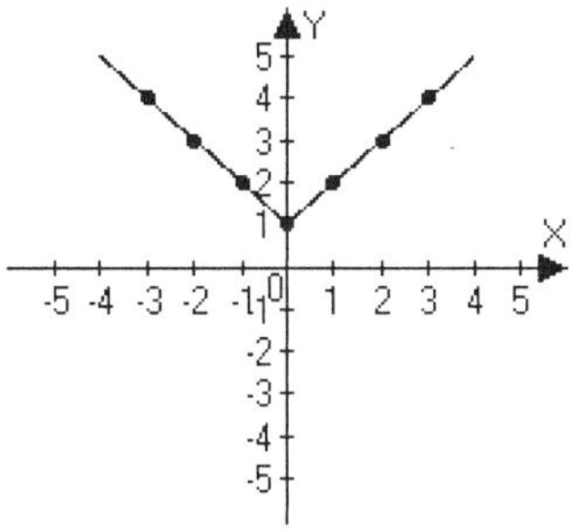


**21.**

| $x$ | $(x, y)$ |
|---|---|
| $-3$ | $(-3,6)$ |
| $-2$ | $(-2,4)$ |
| $-1$ | $(-1,2)$ |
| $0$ | $(0,0)$ |
| $1$ | $(1,2)$ |
| $2$ | $(2,4)$ |
| $3$ | $(3,6)$ |

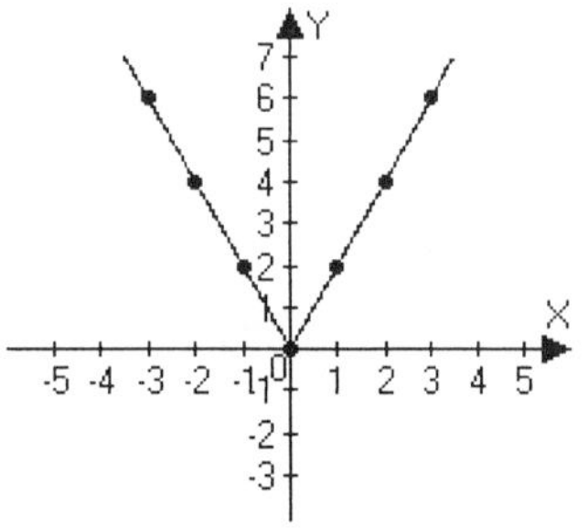

**23.**

| $x$ | $(x, y)$ |
|---|---|
| $-3$ | $(-3,-9)$ |
| $-2$ | $(-2,-4)$ |
| $-1$ | $(-1,-1)$ |
| $0$ | $(0,0)$ |
| $1$ | $(1,-1)$ |
| $2$ | $(2,-4)$ |
| $3$ | $(3,-9)$ |

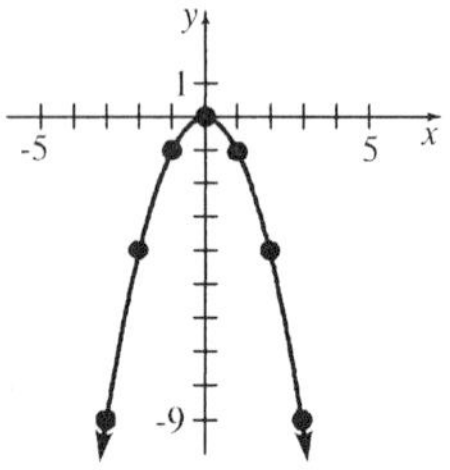

**25.**

| $x$ | $(x, y)$ |
|---|---|
| $-3$ | $(-3,-27)$ |
| $-2$ | $(-2,-8)$ |
| $-1$ | $(-1,-1)$ |
| $0$ | $(0,0)$ |
| $1$ | $(1,1)$ |
| $2$ | $(2,8)$ |
| $3$ | $(3,27)$ |

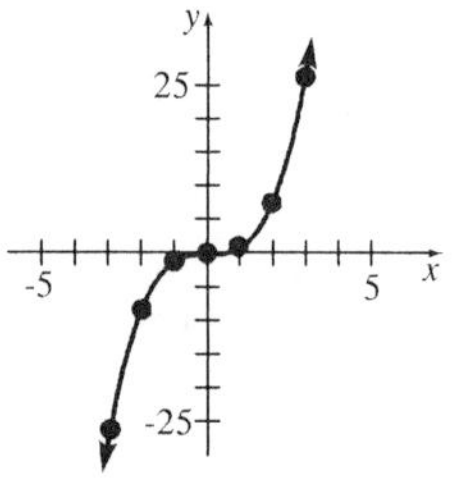

**27.** $[-5, 5, 1]$ by $[-5, 5, 1]$
This matches figure (c).

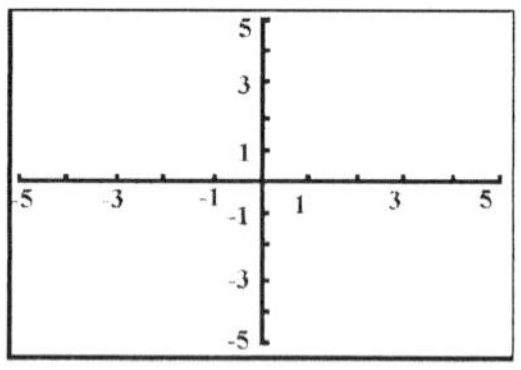

**29.** $[-20, 80, 10]$ by $[-30, 70, 10]$
This matches figure (b).

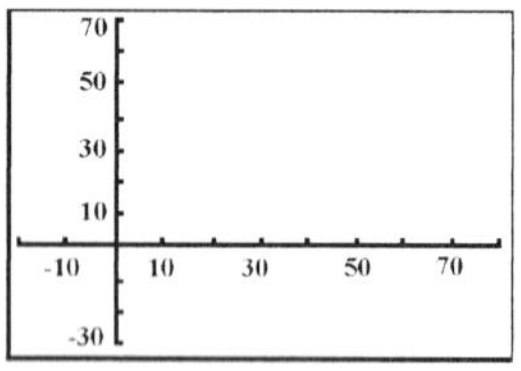

**31.** The equation that corresponds to $Y_2$ in the table is (c), $y_2 = 2 - x$. We can tell because all of the points $(-3,5)$, $(-2,4)$, $(-1,3)$, $(0,2)$, $(1,1)$, $(2,0)$, and $(3,-1)$ are on the line $y = 2 - x$, but all are not on any of the others.

**33.** No. It passes through the point $(0,2)$.

**35.** $(2,0)$

**37.** The graphs of $Y_1$ and $Y_2$ intersect at the points $(-2,4)$ and $(1,1)$.

**39.** $y = 2x + 4$

**40.** $y = 4 - 2x$

**41.** $y = 3 - x^2$

**42.** $y = x^2 + 2$

**43.**

| $x$ | $(x, y)$ |
|---|---|
| $-3$ | $(-3,5)$ |
| $-2$ | $(-2,5)$ |
| $-1$ | $(-1,5)$ |
| $0$ | $(0,5)$ |
| $1$ | $(1,5)$ |
| $2$ | $(2,5)$ |
| $3$ | $(3,5)$ |

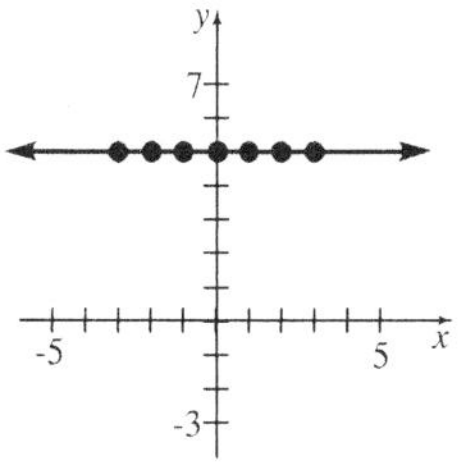


**44.**

| $x$ | $(x, y)$ |
|---|---|
| $-3$ | $(-3,-1)$ |
| $-2$ | $(-2,-1)$ |
| $-1$ | $(-1,-1)$ |
| $0$ | $(0,-1)$ |
| $1$ | $(1,-1)$ |
| $2$ | $(2,-1)$ |
| $3$ | $(3,-1)$ |

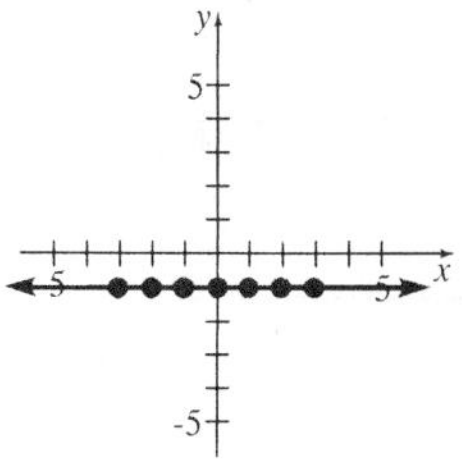


**45.**

| $x$ | $(x, y)$ |
|---|---|
| $-2$ | $\left(-2,-\frac{1}{2}\right)$ |
| $-1$ | $(-1,-1)$ |
| $-\frac{1}{2}$ | $\left(-\frac{1}{2},-2\right)$ |
| $-\frac{1}{3}$ | $\left(-\frac{1}{3},-3\right)$ |
| $\frac{1}{3}$ | $\left(\frac{1}{3},3\right)$ |
| $\frac{1}{2}$ | $\left(\frac{1}{2},2\right)$ |
| $1$ | $(1,1)$ |
| $2$ | $\left(2,\frac{1}{2}\right)$ |

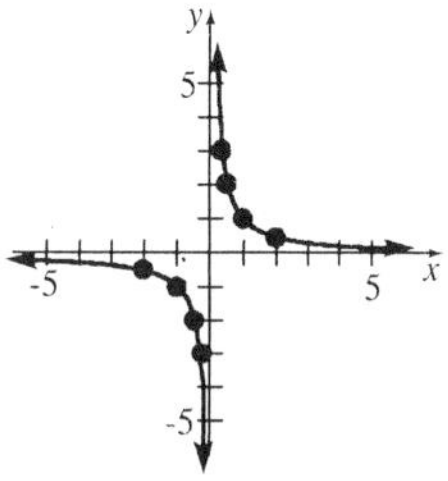


**46.**

| $x$ | $(x, y)$ |
|---|---|
| $-2$ | $\left(-2,\frac{1}{2}\right)$ |
| $-1$ | $(-1,1)$ |
| $-\frac{1}{2}$ | $\left(-\frac{1}{2},2\right)$ |
| $-\frac{1}{3}$ | $\left(-\frac{1}{3},3\right)$ |
| $\frac{1}{3}$ | $\left(\frac{1}{3},-3\right)$ |
| $\frac{1}{2}$ | $\left(\frac{1}{2},-2\right)$ |
| $1$ | $(1,-1)$ |
| $2$ | $\left(2,-\frac{1}{2}\right)$ |

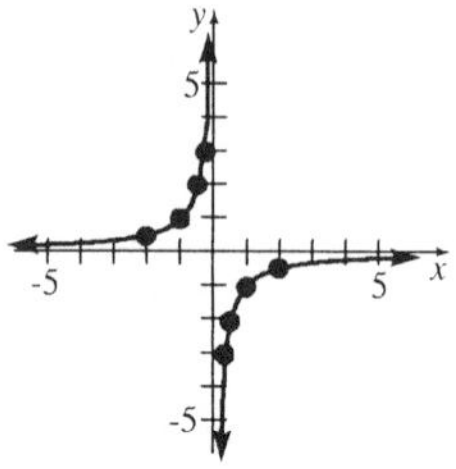


**47.** The chance of divorce increases from year 0 to year 4.

**49.** The chance of divorce is highest in the 4th year with chance of approximately 8.3%.

**51.** The greatest percentage of high school seniors used alcohol in 1980. That year, about 72% of seniors used alcohol.

**53.** In 1980, about 33% of high school seniors used marijuana, and about 30% used cigarettes. This is about a 3% difference.

**55.** (a)

**57.** (b)

**59.** (b)

**61.** The maximum healthy weight for a man who is 6 feet tall is about 181 pounds.

**63.** A man who is 70 inches tall and needs to gain 10 pounds weighs about 135 pounds.

**65. – 71.** Answers will vary.

**73.** Answers will vary. For example, consider Exercise 11. $y = x^2 - 2$

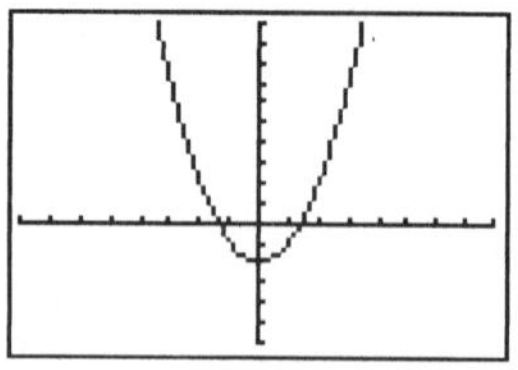

**75.** $y = x^2 + 10$

**a.**

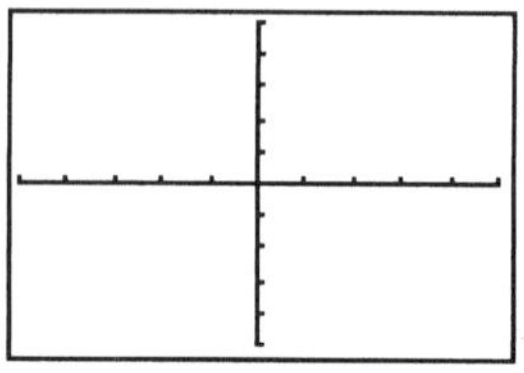

**b.**

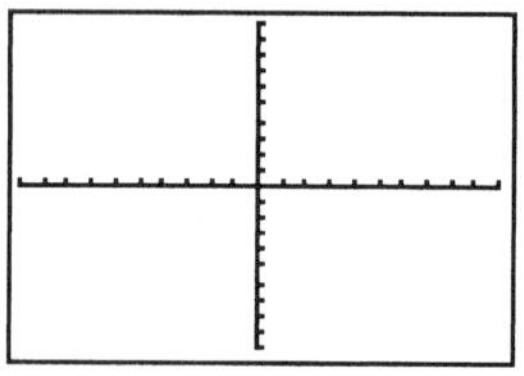

**c.**

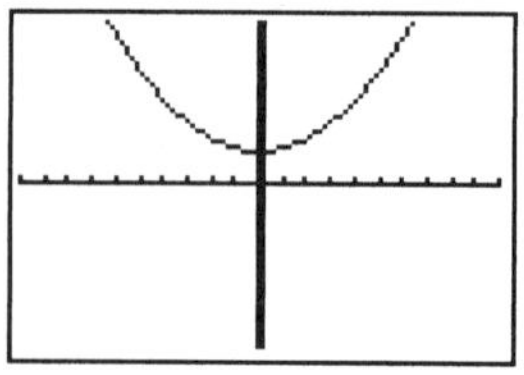

Graph **c.** gives a complete graph.

**77.** $y = \sqrt{x} + 18$

**a.**

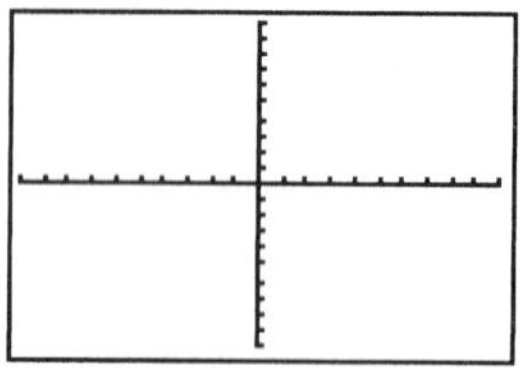

**b.**

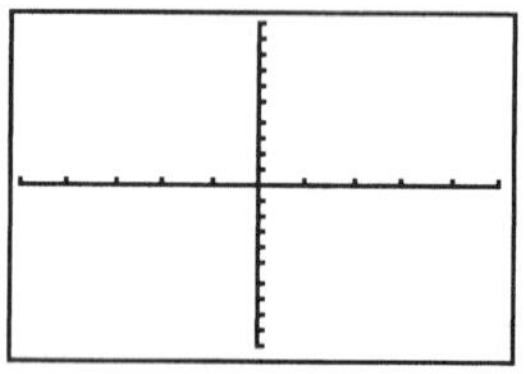

**c.**

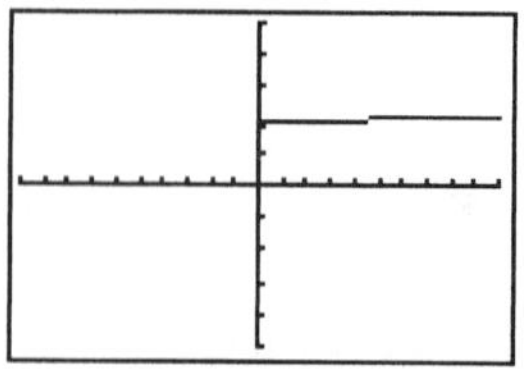

Graph **c.** gives a complete graph.

**79.** Statement **d.** is true.

Statement **a.** is false. If the product of a point's coordinates is positive, the point could be in quadrant I <u>or</u> III.

Statement **b.** is false. When a point lies on the *x*-axis, $y = 0$.

Statement **c.** is false. The majority of graphing utilities do not display numbers along the axes.

**81.** The ball will hit the ground approximately 8.8 seconds from when it is dropped.

**82.** $|-14.3| = 14.3$

**83.**
$$\begin{aligned} &[12-(13-17)]-[9-(6-10)] \\ &=[12-(-4)]-[9-(-4)] \\ &=[12+4]-[9+4]=16-13=3 \end{aligned}$$

**84.**
$$\begin{aligned} &6x-5(4x+3)-10 \\ &=6x-20x-15-10 \\ &=(6-20)x-(15+10)=-14x-25 \end{aligned}$$

### 1.4 Exercise Set

**1.**
$$\begin{aligned} 5x+3&=18 \\ 5x+3-3&=18-3 \\ 5x&=15 \\ \frac{5x}{5}&=\frac{15}{5} \\ x&=3 \end{aligned}$$

**3.**
$$\begin{aligned} 6x-3&=63 \\ 6x-3+3&=63+3 \\ 6x&=66 \\ \frac{6x}{6}&=\frac{66}{6} \\ x&=11 \end{aligned}$$

**5.**
$$\begin{aligned} 14-5x&=-41 \\ 14-5x-14&=-41-14 \\ -5x&=-55 \\ \frac{-5x}{-5}&=\frac{-55}{-5} \\ x&=11 \end{aligned}$$

**7.**
$$\begin{aligned} 11x-(6x-5)&=40 \\ 11x-6x+5&=40 \\ 5x+5&=40 \\ 5x+5-5&=40-5 \\ 5x&=35 \\ \frac{5x}{5}&=\frac{35}{5} \\ x&=7 \end{aligned}$$

**9.**
$$\begin{aligned} 2x-7&=6+x \\ 2x-x-7&=6+x-x \\ x-7&=6 \\ x-7+7&=6+7 \\ x&=13 \end{aligned}$$

**11.**
$$\begin{aligned} 7x+4&=x+16 \\ 7x-x+4&=x-x+16 \\ 6x+4&=16 \\ 6x+4-4&=16-4 \\ 6x&=12 \\ \frac{6x}{6}&=\frac{12}{6} \\ x&=2 \end{aligned}$$

**13.**
$$\begin{aligned} 8y-3&=11y+9 \\ 8y-8y-3&=11y-8y+9 \\ -3&=3y+9 \\ -3-9&=3y+9-9 \\ -12&=3y \\ \frac{-12}{3}&=\frac{3y}{3} \\ -4&=y \end{aligned}$$

**15.**
$$\begin{aligned} 3(x-2)+7&=2(x+5) \\ 3x-6+7&=2x+10 \\ 3x-2x-6+7&=2x-2x+10 \\ x-6+7&=10 \\ x+1&=10 \\ x+1-1&=10-1 \\ x&=9 \end{aligned}$$

**17.**
$$\begin{aligned}3(x-4)-4(x-3)&=x+3-(x-2)\\3x-12-4x+12&=x+3-x+2\\-x&=5\\x&=-5\end{aligned}$$

**19.**
$$\begin{aligned}16&=3(x-1)-(x-7)\\16&=3x-3-x+7\\16&=2x+4\\16-4&=2x+4-4\\12&=2x\\\frac{12}{2}&=\frac{2x}{2}\\6&=x\end{aligned}$$

**21.**
$$\begin{aligned}7(x+1)&=4[x-(3-x)]\\7x+7&=4[x-3+x]\\7x+7&=4[2x-3]\\7x+7&=8x-12\\7x-7x+7&=8x-7x-12\\7&=x-12\\7+12&=x-12+12\\19&=x\end{aligned}$$

**23.**
$$\begin{aligned}\frac{1}{2}(4z+8)-16&=-\frac{2}{3}(9z-12)\\2z+4-16&=-6z+8\\2z-12&=-6z+8\\8z-12&=8\\8z&=20\\z&=\frac{20}{8}=\frac{5}{2}\end{aligned}$$

**25.**
$$\begin{aligned}\frac{x}{3}&=\frac{x}{2}-2\\6\left(\frac{x}{3}\right)&=6\left(\frac{x}{2}-2\right)\\2x&=3x-12\\2x-3x&=3x-3x-12\\-x&=-12\\x&=12\end{aligned}$$

**27.**
$$\begin{aligned}20-\frac{x}{3}&=\frac{x}{2}\\6\left(20-\frac{x}{3}\right)&=6\left(\frac{x}{2}\right)\\120-2x&=3x\\120-2x+2x&=3x+2x\\120&=5x\\\frac{120}{5}&=\frac{5x}{5}\\24&=x\end{aligned}$$

**29.**
$$\begin{aligned}\frac{3x}{5}&=\frac{2x}{3}+1\\15\left(\frac{3x}{5}\right)&=15\left(\frac{2x}{3}+1\right)\\9x&=10x+15\\9x-10x&=10x-10x+15\\-x&=15\\x&=-15\end{aligned}$$

**31.**
$$\begin{aligned}\frac{3x}{5}-x&=\frac{x}{10}-\frac{5}{2}\\10\left(\frac{3x}{5}-x\right)&=10\left(\frac{x}{10}-\frac{5}{2}\right)\\6x-10x&=x-25\\-4x&=x-25\\-4x-x&=x-x-25\\-5x&=-25\\x&=5\end{aligned}$$

**33.**
$$\begin{aligned}\frac{x+3}{6}&=\frac{2}{3}+\frac{x-5}{4}\\12\left(\frac{x+3}{6}\right)&=12\left(\frac{2}{3}\right)+12\left(\frac{x-5}{4}\right)\\2(x+3)&=4(2)+3(x-5)\\2x+6&=8+3x-15\\2x+6&=3x-7\\-x+6&=-7\\-x&=-13\\x&=13\end{aligned}$$

**35.**
$$\begin{aligned}\frac{x}{4} &= 2+\frac{x-3}{3}\\ 12\left(\frac{x}{4}\right) &= 12\left(2+\frac{x-3}{3}\right)\\ 3x &= 24+4(x-3)\\ 3x &= 24+4x-12\\ 3x &= 12+4x\\ 3x-4x &= 12+4x-4x\\ -x &= 12\\ x &= -12\end{aligned}$$

**37.**
$$\begin{aligned}\frac{x+1}{3} &= 5-\frac{x+2}{7}\\ 21\left(\frac{x+1}{3}\right) &= 21\left(5-\frac{x+2}{7}\right)\\ 7(x+1) &= 105-3(x+2)\\ 7x+7 &= 105-3x-6\\ 7x+3x+7 &= 105-3x+3x-6\\ 10x+7 &= 99\\ 10x &= 92\\ x &= \frac{92}{10}=\frac{46}{5}\end{aligned}$$

**39.**
$$\begin{aligned}5x+9 &= 9(x+1)-4x\\ 5x+9 &= 9x+9-4x\\ 5x+9 &= 5x+9\end{aligned}$$
The solution set is $\{x|x$ is a real number$\}$ or $\mathbb{R}$. The equation is an identity.

**41.**
$$\begin{aligned}3(y+2) &= 7+3y\\ 3y+6 &= 7+3y\\ 3y-3y+6 &= 7+3y-3y\\ 6 &= 7\end{aligned}$$
There is no solution. The solution set is $\varnothing$. The equation is inconsistent.

**43.**
$$\begin{aligned}10x+3 &= 8x+3\\ 10x-8x+3 &= 8x-8x+3\\ 2x &= 0\\ x &= 0\end{aligned}$$
The solution set is $\{0\}$. The equation is conditional.

**45.**
$$\begin{aligned}\frac{1}{2}(6z+20)-8 &= 2(z-4)\\ 3z+10-8 &= 2z-8\\ 3z+2 &= 2z-8\\ z+2 &= -8\\ z &= -10\end{aligned}$$
The solution set is $\{-10\}$. The equation is conditional.

**47.**
$$\begin{aligned}-4x-3(2-2x) &= 7+2x\\ -4x-6+6x &= 7+2x\\ 2x-6 &= 7+2x\\ -6 &= 7\end{aligned}$$
There is no solution. The solution set is $\varnothing$. The equation is inconsistent.

**49.**
$$\begin{aligned}y+3(4y+2) &= 6(y+1)+5y\\ y+12y+6 &= 6y+6+5y\\ 13y+6 &= 11y+6\\ 2y+6 &= 6\\ 2y &= 0\\ y &= 0\end{aligned}$$
The solution set is $\{0\}$. The equation is conditional.

**51.** The equation is $3(x-4)=3(2-2x)$, and the solution is $x=2$.

**53.** The equation is $-3(x-3)=5(2-x)$, and the solution is $x=0.5$.

**55.** Solve: $4(x-2)+2=4x-2(2-x)$
$$\begin{aligned}4x-8+2 &= 4x-4+2x\\ 4x-6 &= 6x-4\\ -2x-6 &= -4\\ -2x &= 2\\ x &= -1\end{aligned}$$
Now, evaluate $x^2-x$ for $x=-1$:
$$\begin{aligned}x^2-x &= (-1)^2-(-1)\\ &= 1-(-1)=1+1=2\end{aligned}$$

**56.** Solve: $2(x-6)=3x+2(2x-1)$

$$\begin{aligned}2x-12&=3x+4x-2\\2x-12&=7x-2\\-5x-12&=-2\\-5x&=10\\x&=-2\end{aligned}$$

Now, evaluate $x^2-x$ for $x=-2$:

$$\begin{aligned}x^2-x&=(-2)^2-(-2)\\&=4-(-2)=4+2=6\end{aligned}$$

**57.** Solve for $x$: $\dfrac{3(x+3)}{5}=2x+6$

$$\begin{aligned}3(x+3)&=5(2x+6)\\3x+9&=10x+30\\-7x+9&=30\\-7x&=21\\x&=-3\end{aligned}$$

Solve for $y$: $-2y-10=5y+18$

$$\begin{aligned}-7y-10&=18\\-7y&=28\\y&=-4\end{aligned}$$

Now, evaluate $x^2-(xy-y)$ for $x=-3$ and $y=-4$:

$$\begin{aligned}&x^2-(xy-y)\\&=(-3)^2-[-3(-4)-(-4)]\\&=(-3)^2-[12-(-4)]\\&=9-(12+4)=9-16=-7\end{aligned}$$

**58.** Solve for $x$: $\dfrac{13x-6}{4}=5x+2$

$$\begin{aligned}13x-6&=4(5x+2)\\13x-6&=20x+8\\-7x-6&=8\\-7x&=14\\x&=-2\end{aligned}$$

Solve for $y$: $5-y=7(y+4)+1$

$$\begin{aligned}5-y&=7y+28+1\\5-y&=7y+29\\5-8y&=29\\-8y&=24\\y&=-3\end{aligned}$$

Now, evaluate $x^2-(xy-y)$ for $x=-2$ and $y=-3$:

$$\begin{aligned}&x^2-(xy-y)\\&=(-2)^2-[-2(-3)-(-3)]\\&=(-2)^2-[6-(-3)]\\&=4-(6+3)=4-9=-5\end{aligned}$$

**59.**

$$\begin{aligned}\left[(3+6)^2\div 3\right]\cdot 4&=-54x\\\left(9^2\div 3\right)\cdot 4&=-54x\\(81\div 3)\cdot 4&=-54x\\27\cdot 4&=-54x\\108&=-54x\\-2&=x\end{aligned}$$

The solution set is $\{-2\}$.

**60.**

$$\begin{aligned}2^3-\left[4(5-3)^3\right]&=-8x\\8-\left[4(2)^3\right]&=-8x\\8-4\cdot 8&=-8x\\8-32&=-8x\\-24&=-8x\\3&=x\end{aligned}$$

The solution set is $\{3\}$.

**61.** $5-12x=8-7x-\left[6\div 3\left(2+5^3\right)+5x\right]$

$5-12x=8-7x-\left[6\div 3\left(2+125\right)+5x\right]$

$5-12x=8-7x-\left[6\div 3\cdot 127+5x\right]$

$5-12x=8-7x-\left[2\cdot 127+5x\right]$

$5-12x=8-7x-\left[254+5x\right]$

$5-12x=8-7x-245-5x$

$5-12x=-12x-237$

$5=-237$

The final statement is a contradiction, so the equation has no solution. The solution set is $\varnothing$.

**62.** $2\left(5x+58\right)=10x+4\left(21\div 3.5-11\right)$

$10x+116=10x+4\left(6-11\right)$

$10x+116=10x+4\left(-5\right)$

$10x+116=10x-20$

$116=-20$

The final statement is a contradiction, so the equation has no solution. The solution set is $\varnothing$.

**63.** $0.7x+0.4(20)=0.5(x+20)$

$0.7x+8=0.5x+10$

$0.2x+8=10$

$0.2x=2$

$x=10$

The solution set is $\{10\}$.

**64.** $0.5(x+2)=0.1+3(0.1x+0.3)$

$0.5x+1=0.1+0.3x+0.9$

$0.5x+1=0.3x+1$

$0.2x+1=1$

$0.2x=0$

$x=0$

The solution set is $\{0\}$.

**65.** $4x+13-\left\{2x-\left[4\left(x-3\right)-5\right]\right\}=2(x-6)$

$4x+13-\left\{2x-\left[4x-12-5\right]\right\}=2x-12$

$4x+13-\left\{2x-\left[4x-17\right]\right\}=2x-12$

$4x+13-\left\{2x-4x+17\right\}=2x-12$

$4x+13-\left\{-2x+17\right\}=2x-12$

$4x+13+2x-17=2x-12$

$6x-4=2x-12$

$4x-4=-12$

$4x=-8$

$x=-2$

The solution set is $\{-2\}$.

**66.** $-2\left\{7-\left[4-2\left(1-x\right)+3\right]\right\}=10-\left[4x-2(x-3)\right]$

$-2\left\{7-\left[4-2+2x+3\right]\right\}=10-\left[4x-2x+6\right]$

$-2\left\{7-\left[2x+5\right]\right\}=10-\left[2x+6\right]$

$-2\left\{7-2x-5\right\}=10-2x-6$

$-2\left\{-2x+2\right\}=-2x+4$

$4x-4=-2x+4$

$6x-4=4$

$6x=8$

$x=\frac{8}{6}=\frac{4}{3}$

The solution set is $\left\{\frac{4}{3}\right\}$.

**67.** Let T = 4421. Then

$4421=165x+2771$

$1650=165x$

$10=x$

Tuition will be \$4421 ten years after 1996, which is the school year ending 2006.

**69.** If the desirable heart rate for a woman is 117 beats per minute, she is 40 years old. This solution is approximated by the point (40, 117) on the female graph.

**71. – 81.** Answers will vary.

**83.** $2x+3(x-4)=4x-7$

Let $y_1=2x+3(x-4)$ and let $y_2=4x-7$.

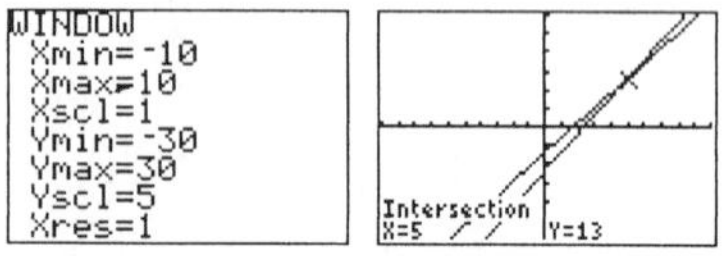


Since the graphs intersect at the point (5, 13), the solution is 5.

Verify by direct substitution:

$$2x+3(x-4)=4x-7$$
$$2(5)+3(5-4)=4(5)-7$$
$$10+3(1)=20-7$$
$$10+3=13$$
$$13=13$$

The equation is conditional and the solution set is $\{5\}$.

**85.** $\frac{2x-1}{3}-\frac{x-5}{6}=\frac{x-3}{4}$

Let $y_1=\frac{2x-1}{3}-\frac{x-5}{6}$ and let $y_2=\frac{x-3}{4}$.

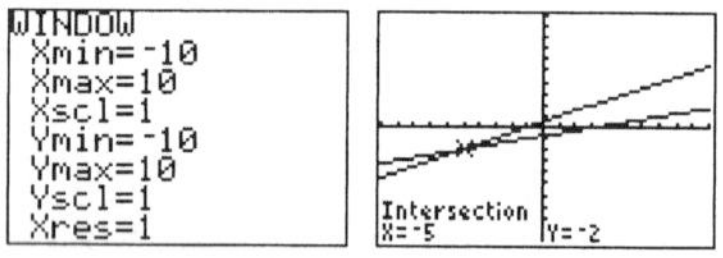


Since the graphs intersect at the point $(-5,-2)$, the solution is $-5$.

Verify by direct substitution:

$$\frac{2x-1}{3}-\frac{x-5}{6}=\frac{x-3}{4}$$
$$\frac{2(-5)-1}{3}-\frac{(-5)-5}{6}=\frac{(-5)-3}{4}$$
$$\frac{-10-1}{3}-\frac{-5-5}{6}=\frac{-5-3}{4}$$
$$\frac{-11}{3}-\frac{-10}{6}=\frac{-8}{4}$$
$$6\left(\frac{-11}{3}-\frac{-10}{6}\right)=6(-2)$$
$$-22-(-10)=-12$$
$$-22+10=-12$$
$$-12=-12$$

The equation is conditional and the solution set is $\{-5\}$.

**87.**
$$ax+b=c$$
$$ax+b-b=c-b$$
$$ax=c-b$$
$$x=\frac{c-b}{a}$$

**89.** Answers will vary.

**91.**
$$-\frac{1}{5}-\left(-\frac{1}{2}\right)=-\frac{1}{5}+\frac{1}{2}$$
$$=-\frac{1}{5}\cdot\frac{2}{2}+\frac{1}{2}\cdot\frac{5}{5}$$
$$=-\frac{2}{10}+\frac{5}{10}=\frac{3}{10}$$

**92.** $4(-3)(-1)(-5)=(-12)(5)=-60$

**93.**

| $x$ | $(x, y)$ |
|---|---|
| −3 | (−3,5) |
| −2 | (−2,0) |
| −1 | (−1,−3) |
| 0 | (0,−4) |
| 1 | (1,−3) |
| 2 | (2,0) |
| 3 | (3,5) |

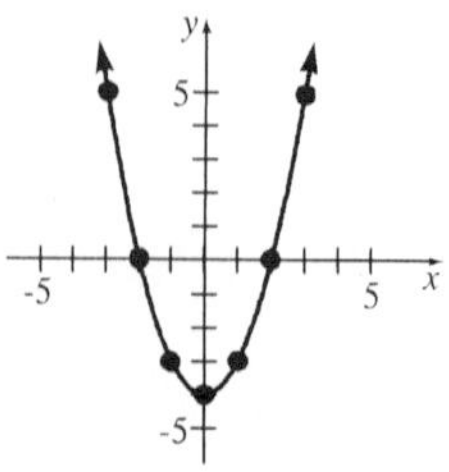

## Mid-Chapter Check Point – Chapter 1

**1.** $-5+3(x+5)=-5+3x+15$
$=3x+10$

**2.**
$$\begin{aligned}-5+3(x+5)&=2(3x-4)\\-5+3x+15&=6x-8\\3x+10&=6x-8\\-3x+10&=-8\\-3x&=-18\\x&=6\end{aligned}$$
The solution set is $\{6\}$.

**3.**
$$\begin{aligned}3[7-4(5-2)]&=3[7-4(3)]\\&=3[7-12]\\&=3(-5)\\&=-15\end{aligned}$$

**4.**
$$\begin{aligned}\frac{x-3}{5}-1&=\frac{x-5}{4}\\20\left(\frac{x-3}{5}-1\right)&=20\left(\frac{x-5}{4}\right)\\4(x-3)-20&=5(x-5)\\4x-12-20&=5x-25\\4x-32&=5x-25\\-x-32&=-25\\-x&=7\\x&=-7\end{aligned}$$
The solution set is $\{-7\}$.

**5.** $$\frac{-2^4+(-2)^2}{-4-(2-2)}=\frac{-16+4}{-4-0}=\frac{-12}{-4}=3$$

**6.**
$$\begin{aligned}&7x-[8-3(2x-5)]\\&=7x-[8-6x+15]\\&=7x-[-6x+23]\\&=7x+6x-23\\&=13x-23\end{aligned}$$

**7.**
$$\begin{aligned}3(2x-5)-2(4x+1)&=-5(x+3)-2\\6x-15-8x-2&=-5x-15-2\\-2x-17&=-5x-17\\3x-17&=-17\\3x&=0\\x&=0\end{aligned}$$
The solution set is $\{0\}$.

**8.**
$$\begin{aligned}&3(2x-5)-2(4x+1)-5(x+3)-2\\&=6x-15-8x-2-5x-15-2\\&=(6x-8x-5x)+(-15-2-15-2)\\&=-7x-34\end{aligned}$$

**9.**
$$\begin{aligned}&-4^2\div2+(-3)(-5)\\&=-16\div2+(-3)(-5)\\&=-8+15\\&=7\end{aligned}$$

**10.**
$$\begin{aligned}3x+1-(x-5)&=2x-4\\3x+1-x+5&=2x-4\\2x+6&=2x-4\\6&=-4\end{aligned}$$
This is a contradiction, so the equation has no solution. The solution set is $\varnothing$.

**11.**
$$\begin{aligned}\frac{3x}{4}-\frac{x}{3}+1&=\frac{4x}{5}-\frac{3}{20}\\60\left(\frac{3x}{4}-\frac{x}{3}+1\right)&=60\left(\frac{4x}{5}-\frac{3}{20}\right)\\45x-20x+60&=48x-9\\25x+60&=48x-9\\-23x+60&=-9\\-23x&=-69\\x&=3\end{aligned}$$
The solution set is $\{3\}$.

**12.**
$$(6-9)(8-12)\div\frac{5^2+4\div 2}{8^2-9^2+8}$$
$$=(-3)(-4)\div\frac{25+2}{64-81+8}$$
$$=(-3)(-4)\div\frac{27}{-9}$$
$$=(-3)(-4)\div(-3)$$
$$=12\div(-3)$$
$$=-4$$

**13.**
$$4x-2(1-x)=3(2x+1)-5$$
$$4x-2+2x=6x+3-5$$
$$6x-2=6x-2$$
The equation is an identity. The solution set is $\{x|x \text{ is a real number}\}$ or $\mathbb{R}$ .

**14.**
$$\frac{3\left[4-3(-2)^2\right]}{2^2-2^4}=\frac{3(4-3\cdot 4)}{4-16}$$
$$=\frac{3(4-12)}{-12}$$
$$=\frac{3(-8)}{-12}=\frac{-24}{-12}=2$$

**15.**

| $x$ | $(x, y)$ |
|---|---|
| −2 | −5 |
| −1 | −3 |
| 0 | −1 |
| 1 | 1 |
| 2 | 3 |

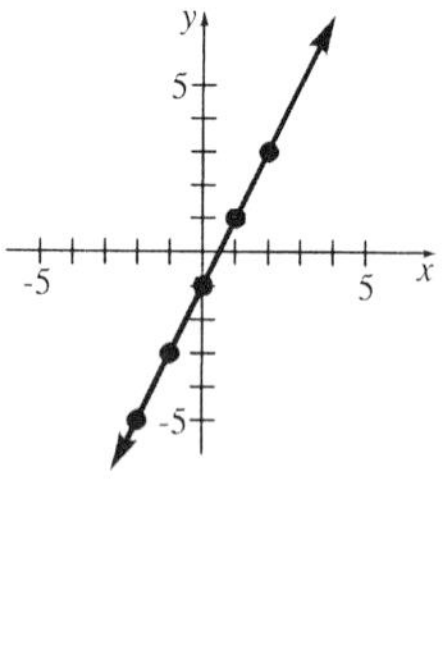


**16.**

| $x$ | $(x, y)$ |
|---|---|
| −3 | −2 |
| −2 | −1 |
| −1 | 0 |
| 0 | 1 |
| 1 | 0 |
| 2 | −1 |
| 3 | −2 |

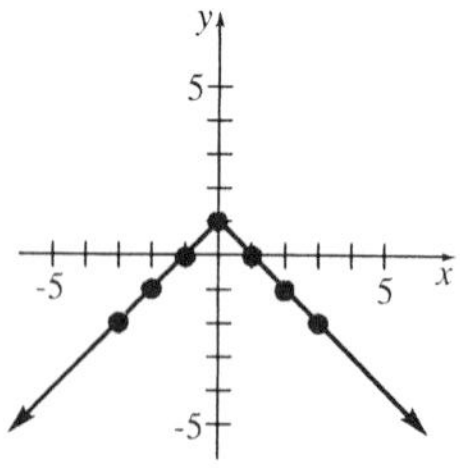


**17.**

| $x$ | $(x, y)$ |
|---|---|
| −2 | 6 |
| −1 | 3 |
| 0 | 2 |
| 1 | 3 |
| 2 | 6 |

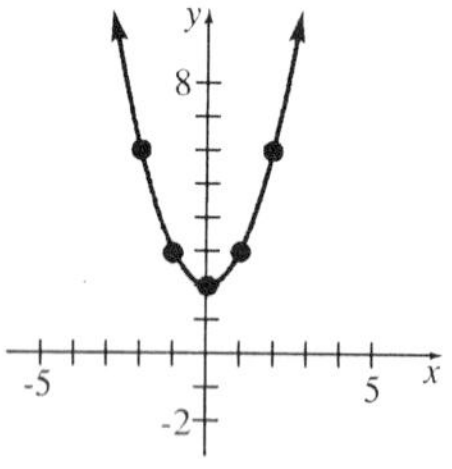


**18.** True.

**19.** False.
$\{x|x \text{ is a negative greater than } -4\}$ $= \{-3,-2,-1\}$, not $\{-4,-3,-2,-1\}$.

**20.** False. −17 does belong to the set of rational numbers.

**21.** True.

$$-128 \div (2 \cdot 4) > (-128 \div 2) \cdot 4$$
$$-128 \div 8 > -64 \cdot 4$$
$$-16 > -256$$

which is true because $-16$ is to the right of $-256$ on the number line.

## 1.5 Exercise Set

**1.** Let $x$ = a number

$$5x - 4 = 26$$
$$5x = 30$$
$$x = 6$$

The number is 6.

**3.** Let $x$ = a number

$$x - 0.20x = 20$$
$$0.80x = 20$$
$$x = 25$$

The number is 25.

**5.** Let $x$ = a number

$$0.60x + x = 192$$
$$1.6x = 192$$
$$x = 120$$

The number is 120.

**7.** Let $x$ = a number

$$0.70x = 224$$
$$x = 320$$

The number is 320.

**9.** Let $x$ = a number

$x + 26$ = another number

$$x + (x + 26) = 64$$
$$x + x + 26 = 64$$
$$2x + 26 = 64$$
$$2x = 38$$
$$x = 19$$

If $x = 19$, then $x + 26 = 45$.
The numbers are 19 and 45.

**11.** Let $x$ = the number of Internet users in China.

$x + 10$ = the number of Internet users in Japan.

$x + 123$ = the number of Internet users in the United States.

$$x + (x + 10) + (x + 123) = 271$$
$$3x + 133 = 271$$
$$3x = 138$$
$$x = 46$$

If $x = 46$, then $x + 10 = 56$ and $x + 123 = 169$. Thus, there are 46 million Internet users in China, 56 million Internet users in Japan, and 169 Internet users in the United States.

**13.** Let $x$ = the measure of the $2^{nd}$ angle.

$2x$ = the measure of the $1^{st}$ angle.

$x - 8$ = the measure of the $3^{rd}$ angle.

$$x + 2x + (x - 8) = 180$$
$$4x - 8 = 180$$
$$4x = 188$$
$$x = 47$$

If $x = 47$, then $2x = 94$ and $x - 8 = 39$. Thus, the measure of the $1^{st}$ angle is $47°$, the $2^{nd}$ angle is $94°$, and the $3^{rd}$ angle is $39°$.

**15.** Let $x$ = the measure of the $1^{st}$ angle.

$x + 1$ = the measure of the $2^{nd}$ angle.

$x + 2$ = the measure of the $3^{rd}$ angle.

$$x + (x + 1) + (x + 2) = 180$$
$$3x + 3 = 180$$
$$3x = 177$$
$$x = 59$$

If $x = 59$, then $x + 1 = 60$ and $x + 2 = 61$. Thus, the measures of the three angles are $59°$, $60°$, and $61°$.

**17.** Let $L$ = the life expectancy of an American man.
$y$ = the number of years after 1900.

$$L = 55 + 0.2y$$
$$85 = 55 + 0.2y$$
$$30 = 0.2y$$
$$150 = y$$

The life expectancy will be 85 years in the year $1900 + 150 = 2050$.

**19.** Let $p$ = the percentage of babies born to unmarried parents.
$y$ = the number of years after 1990.

$$p = 28 + 0.6y$$
$$40 = 28 + 0.6y$$
$$12 = 0.6y$$
$$20 = y$$

The percentage of babies born to unmarried parents will be 40% in the year 1990 + 20 = 2010.

**21.** Let $v$ = the car's value.
$y$ = the number of years (after 2003).

$$v = 80,500 - 8705y$$
$$19,565 = 80,500 - 8705y$$
$$-60,935 = -8705y$$
$$7 = y$$

The car's value will be $19,565 after 7 years.

**23.** Let $x$ = the number of months.
The cost for Club A: $25x + 40$
The cost for Club B: $30x + 15$

$$25x + 40 = 30x + 15$$
$$-5x + 40 = 15$$
$$-5x = -25$$
$$x = 5$$

The total cost for the clubs will be the same at 5 months. The cost will be $25(5) + 40 = 30(5) + 15 = \$165$

**25.** Let $x$ = the number of uses.
Cost without coupon book: $1.25x$
Cost with coupon book: $15 + 0.75x$

$$1.25x = 15 + 0.75x$$
$$0.50x = 15$$
$$x = 30$$

The bus must be used 30 times in a month for the costs to be equal.

**27.** **a.** Let $x$ = the number of years (after 2005).
College A's enrollment: $13,300 + 1000x$

College B's enrollment: $26,800 - 500x$

$$13,300 + 1000x = 26,800 - 500x$$
$$13,300 + 1500x = 26,800$$
$$1500x = 13,500$$
$$x = 9$$

The two colleges will have the same enrollment in the year $2005 + 9 = 2014$. That year the enrollments will be

$$13,300 + 1000(9)$$
$$= 26,800 - 500(9)$$
$$= 22,300 \text{ students}$$

**b.** Check some points to determine that $y_1 = 13,300 + 1000x$ and $y_2 = 26,800 - 500x$.

**29.** Let $x$ = the cost of the television set.

$$x - 0.20x = 336$$
$$0.80x = 336$$
$$x = 420$$

The television set's price is $420.

**31.** Let $x$ = the nightly cost

$$x + 0.08x = 162$$
$$1.08x = 162$$
$$x = 150$$

The nightly cost is $150.

**33.** Let $x$ = the annual salary for men whose highest educational attainment is a high school degree.

$$x + 0.22x = 44{,}000$$
$$1.22x = 44{,}000$$
$$x \approx 36{,}000$$

The annual salary for men whose highest educational attainment is a high school degree is about \$36,000.

**35.** Let c = the dealer's cost

$$584 = c + 0.25c$$
$$584 = 1.25c$$
$$467.20 = c$$

The dealer's cost is \$467.20.

**37.** Let $w$ = the width of the field
Let $2w$ = the length of the field

$$P = 2(\text{length}) + 2(\text{width})$$
$$300 = 2(2w) + 2(w)$$
$$300 = 4w + 2w$$
$$300 = 6w$$
$$50 = w$$

If $w = 50$, then $2w = 100$. Thus, the dimensions are 50 yards by 100 yards.

**39.** Let $w$ = the width of the field
Let $2w + 6$ = the length of the field

$$228 = 6w + 12$$
$$216 = 6w$$
$$36 = w$$

If $w = 36$, then $2w + 6 = 2(36) + 6 = 78$. Thus, the dimensions are 36 feet by 78 feet.

**41.** Let $x$ = the width of the frame.
Total length: $16 + 2x$
Total width: $12 + 2x$

$$P = 2(\text{length}) + 2(\text{width})$$
$$72 = 2(16 + 2x) + 2(12 + 2x)$$
$$72 = 32 + 4x + 24 + 4x$$
$$72 = 8x + 56$$
$$16 = 8x$$
$$2 = x$$

The width of the frame is 2 inches.

**43.** Let $x$ = the length of the call.

$$0.43 + 0.32(x - 1) + 2.10 = 5.73$$
$$0.43 + 0.32x - 0.32 + 2.10 = 5.73$$
$$0.32x + 2.21 = 5.73$$
$$0.32x = 3.52$$
$$x = 11$$

The person talked for 11 minutes.

**45.** $$A = lw$$
$$l = \frac{A}{w}$$

**47.** $$A = \frac{1}{2}bh$$
$$2A = bh$$
$$b = \frac{2A}{h}$$

**49.** $$I = Prt$$
$$P = \frac{I}{rt}$$

**51.** $$T = D + pm$$
$$T - D = pm$$
$$p = \frac{T - D}{m}$$

**53.**
$$A = \frac{1}{2}h(a+b)$$
$$2A = h(a+b)$$
$$\frac{2A}{h} = a+b$$
$$a = \frac{2A}{h} - b \text{ or } a = \frac{2A-hb}{h}$$

**55.**
$$V = \frac{1}{3}\pi r^2 h$$
$$3V = \pi r^2 h$$
$$h = \frac{3V}{\pi r^2}$$

**57.**
$$y - y_1 = m(x - x_1)$$
$$m = \frac{y - y_1}{x - x_1}$$

**59.**
$$V = \frac{d_1 - d_2}{t}$$
$$Vt = d_1 - d_2$$
$$d_1 = Vt + d_2$$

**61.**
$$Ax + By = C$$
$$Ax = C - By$$
$$x = \frac{C - By}{A}$$

**63.**
$$s = \frac{1}{2}at^2 + vt$$
$$2s = \cancel{2}\left(\frac{1}{\cancel{2}}at^2\right) + 2vt$$
$$2s = at^2 + 2vt$$
$$2s - at^2 = 2vt$$
$$\frac{2s - at^2}{2t} = \frac{2vt}{2t}$$
$$v = \frac{2s - at^2}{2t}$$

**65.**
$$L = a + (n-1)d$$
$$L - a = (n-1)d$$
$$\frac{L-a}{d} = n - 1$$
$$n = \frac{L-a}{d} + 1$$

**67.**
$$A = 2lw + 2lh + 2wh$$
$$A - 2wh = 2lw + 2lh$$
$$A - 2wh = l(2w + 2h)$$
$$l = \frac{A - 2wh}{2w + 2h}$$

**69.**
$$IR + Ir = E$$
$$I(R + r) = E$$
$$I = \frac{E}{R + r}$$

**71. – 75.** Answers will vary.

**77.** Answers will vary.

**79.** Exercise 19.

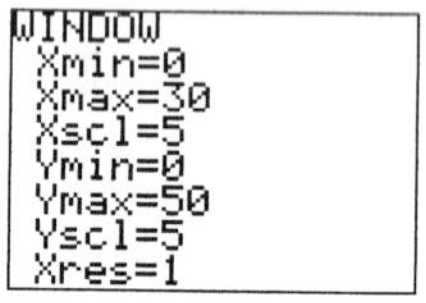


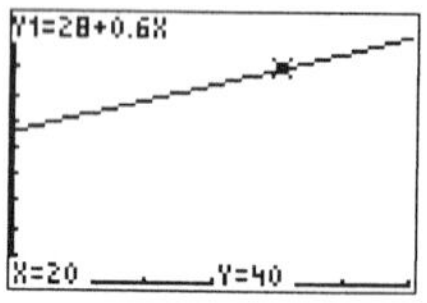


The percentage of babies born to unmarried parents will be 40% in the year 2010.

Exercise 20.

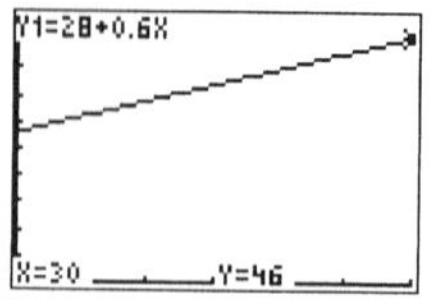


The percentage of babies born to unmarried parents will be 46% in the year 2020.

**81.** Let $x$ = the original price of the dress. If the reduction in price is 40%, the price paid is 60%.

$$\text{price paid} = 0.60(0.60x)$$
$$72 = 0.60(0.60x)$$
$$72 = 0.36x$$
$$200 = x$$

The original price is \$200.

**83.** Let $x$ = the amount a girl would receive

$2x$ = the amount Mrs. Ricardo would receive

$4x$ = the amount a boy would receive

$$\text{Total Savings} = x + 2x + 4x$$
$$14{,}000 = 7x$$
$$2{,}000 = x$$

Mrs. Ricardo received \$4000, the boy received \$8000, and the girl received \$2000.

**85.**
$$V = C - \frac{C-S}{L}N$$
$$V = C - \left(\frac{C-S}{L}\right)\frac{N}{1}$$
$$V = C - \frac{CN - SN}{L}$$
$$V = \frac{CL}{L} - \frac{CN - SN}{L}$$
$$V = \frac{CL - CN + SN}{L}$$
$$LV = CL - CN + SN$$
$$LV - SN = CL - CN$$
$$LV - SN = C(L - N)$$
$$C = \frac{LV - SN}{L - N}$$

**86.** $-6$ is less than or equal to $-6$. True.

**87.**
$$\frac{(2+4)^2 + (-1)^5}{12 \div 2 \cdot 3 - 3} = \frac{(6)^2 + (-1)}{6 \cdot 3 - 3}$$
$$= \frac{36 + (-1)}{18 - 3} = \frac{35}{15} = \frac{7}{3}$$

**88.**
$$\frac{2x}{3} - \frac{8}{3} = x$$
$$3\left(\frac{2x}{3} - \frac{8}{3}\right) = 3(x)$$
$$2x - 8 = 3x$$
$$-8 = x$$

The solution set is $\{-8\}$.

## 1.6 Exercise Set

**1.** $b^4 \cdot b^7 = b^{(4+7)} = b^{11}$

**3.** $x \cdot x^3 = x^{(1+3)} = x^4$

**5.** $2^3 \cdot 2^2 = 2^{(3+2)} = 2^5 = 32$

**7.** $3x^4 \cdot 2x^2 = 6x^{(4+2)} = 6x^6$

**9.** $(-2y^{10})(-10y^2) = 20y^{(10+2)} = 20y^{12}$

**11.**
$$(5x^3y^4)(20x^7y^8) = 100x^{(3+7)}y^{(4+8)}$$
$$= 100x^{10}y^{12}$$

**13.**
$$(-3x^4y^0z)(-7xyz^3y^8)$$
$$= 21x^{(4+1)}y^{(0+1)}z^{(1+8)}$$
$$= 21x^5y^1z^9 = 21x^5yz^9$$

**15.** $\frac{b^{12}}{b^3} = b^{(12-3)} = b^9$

**17.** $\frac{15x^9}{3x^4} = 5x^{(9-4)} = 5x^5$

**19.** $\dfrac{x^9y^7}{x^4y^2} = x^{(9-4)}y^{(7-2)} = x^5y^5$

**21.** $\dfrac{50x^2y^7}{5xy^4} = 10x^{(2-1)}y^{(7-4)} = 10xy^3$

**23.** $\dfrac{-56a^{12}b^{10}c^8}{7ab^2c^4} = -8a^{(12-1)}b^{(10-2)}c^{(8-4)}$
$= -8a^{11}b^8c^4$

**25.** $6^0 = 1$

**27.** $(-4)^0 = 1$

**29.** $-4^0 = -1$

**31.** $13y^0 = 13(1) = 13$

**33.** $(13y)^0 = 1$

**35.** $3^{-2} = \dfrac{1}{3^2} = \dfrac{1}{9}$

**37.** $(-5)^{-2} = \dfrac{1}{(-5)^2} = \dfrac{1}{25}$

**39.** $-5^{-2} = -(5^{-2}) = -\dfrac{1}{5^2} = -\dfrac{1}{25}$

**41.** $x^2y^{-3} = \dfrac{x^2}{y^3}$

**43.** $8x^{-7}y^3 = \dfrac{8y^3}{x^7}$

**45.** $\dfrac{1}{5^{-3}} = 5^3 = 125$

**47.** $\dfrac{1}{(-3)^{-4}} = (-3)^4 = 81$

**49.** $\dfrac{x^{-2}}{y^{-5}} = \dfrac{y^5}{x^2}$

**51.** $\dfrac{a^{-4}b^7}{c^{-3}} = \dfrac{b^7c^3}{a^4}$

**53.** $(x^6)^{10} = x^{(6 \cdot 10)} = x^{60}$

**55.** $(b^4)^{-3} = \dfrac{1}{(b^4)^3} = \dfrac{1}{b^{(4 \cdot 3)}} = \dfrac{1}{b^{12}}$

**57.** $(7^{-4})^{-5} = 7^{-4 \cdot (-5)} = 7^{20}$

**59.** $(4x)^3 = 4^3x^3 = 64x^3$

**61.** $(-3x^7)^2 = (-3)^2x^{7 \cdot 2} = 9x^{14}$

**63.** $(2xy^2)^3 = 8x^{(1 \cdot 3)}y^{(2 \cdot 3)} = 8x^3y^6$

**65.** $(-3x^2y^5)^2 = (-3)^2x^{(2 \cdot 2)}y^{(5 \cdot 2)} = 9x^4y^{10}$

**67.** $(-3x^{-2})^{-3} = (-3)^{-3}(x^{-2})^{-3}$
$= \dfrac{x^6}{(-3)^3} = \dfrac{x^6}{-27} = -\dfrac{x^6}{27}$

**69.** $(5x^3y^{-4})^{-2} = 5^{-2}(x^3)^{-2}(y^{-4})^{-2}$
$= 5^{-2}x^{-6}y^8 = \dfrac{y^8}{25x^6}$

**71.** $(-2x^{-5}y^4z^2)^{-4} = (-2)^{-4}x^{20}y^{-16}z^{-8}$
$= \dfrac{x^{20}}{(-2)^4y^{16}z^8}$
$= \dfrac{x^{20}}{16y^{16}z^8}$

**73.** $\left(\frac{2}{x}\right)^4 = \frac{2^4}{x^4} = \frac{16}{x^4}$

**75.** $\left(\frac{x^3}{5}\right)^2 = \frac{x^{(3\cdot 2)}}{5^2} = \frac{x^6}{25}$

**77.** $\left(-\frac{3x}{y}\right)^4 = \frac{(-3)^4 x^4}{y^4} = \frac{81x^4}{y^4}$

**79.** $\left(\frac{x^4}{y^2}\right)^6 = \frac{x^{(4\cdot 6)}}{y^{(2\cdot 6)}} = \frac{x^{24}}{y^{12}}$

**81.** $\left(\frac{x^3}{y^{-4}}\right)^3 = \frac{x^{(3\cdot 3)}}{y^{(-4\cdot 3)}} = \frac{x^9}{y^{-12}} = x^9 y^{12}$

**83.** $\left(\frac{a^{-2}}{b^3}\right)^{-4} = \frac{a^{(-2\cdot(-4))}}{b^{(3\cdot(-4))}} = \frac{a^8}{b^{-12}} = a^8 b^{12}$

**85.** $\frac{x^3}{x^9} = x^{3-9} = x^{-6} = \frac{1}{x^6}$

**87.** $\frac{20x^3}{-5x^4} = -4x^{3-4} = -4x^{-1} = -\frac{4}{x}$

**89.** $\frac{16x^3}{8x^{10}} = 2x^{3-10} = 2x^{-7} = \frac{2}{x^7}$

**91.** $\frac{20a^3b^8}{2ab^{13}} = 10a^{3-1}b^{8-13}$

$= 10a^2b^{-5} = \frac{10a^2}{b^5}$

**93.** $x^3 \cdot x^{-12} = x^{3+(-12)} = x^{-9} = \frac{1}{x^9}$

**95.** $(2a^5)(-3a^{-7}) = -6a^{5+(-7)}$

$= -6a^{-2} = -\frac{6}{a^2}$

**97.** $\left(-\frac{1}{4}x^{-4}y^5z^{-1}\right)\left(-12x^{-3}y^{-1}z^4\right)$

$= 3x^{-4+(-3)}y^{5+(-1)}z^{-1+4}$

$= 3x^{-7}y^4z^3 = \frac{3y^4z^3}{x^7}$

**99.** $\frac{6x^2}{2x^{-8}} = 3x^{2-(-8)} = 3x^{2+8} = 3x^{10}$

**101.** $\frac{x^{-7}}{x^3} = x^{-7-3} = x^{-10} = \frac{1}{x^{10}}$

**103.** $\frac{30x^2y^5}{-6x^8y^{-3}} = -5x^{2-8}y^{5-(-3)}$

$= -5x^{-6}y^8 = -\frac{5y^8}{x^6}$

**105.** $\frac{-24a^3b^{-5}c^5}{-3a^{-6}b^{-4}c^{-7}} = 8a^{3-(-6)}b^{-5-(-4)}c^{5-(-7)}$

$= 8a^9b^{-1}c^{12} = \frac{8a^9c^{12}}{b}$

**107.** $\left(\frac{x^3}{x^{-5}}\right)^2 = \left(x^{3-(-5)}\right)^2 = \left(x^8\right)^2 = x^{16}$

**109.** $\left(\frac{-15a^4b^2}{5a^{10}b^{-3}}\right)^3 = \left(-3a^{4-10}b^{2-(-3)}\right)^3$

$= \left(-3a^{-6}b^{2+3}\right)^3$

$= \left(-3a^{-6}b^5\right)^3$

$= (-3)^3\left(a^{-6}\right)^3\left(b^5\right)^3$

$= -27a^{-18}b^{15}$

$= -\frac{27b^{15}}{a^{18}}$

**111.** $\left(\dfrac{3a^{-5}b^2}{12a^3b^{-4}}\right)^0 = 1$

Recall the Zero Exponent Rule.

**113.** $\left(\dfrac{x^{-5}y^8}{3}\right)^{-4} = \dfrac{x^{(-5)(-4)}y^{8(-4)}}{3^{-4}}$

$= \dfrac{x^{20}y^{-32}}{3^{-4}} = \dfrac{3^4x^{20}}{y^{32}} = \dfrac{81x^{20}}{y^{32}}$

**115.** $\left(\dfrac{20a^{-3}b^4c^5}{-2a^{-5}b^{-2}c}\right)^{-2}$

$= \left(10a^{-3-(-5)}b^{4-(-2)}c^{5-1}\right)^{-2}$

$= \dfrac{1}{\left(10a^2b^6c^4\right)^2}$

$= \dfrac{1}{10^2a^{2(2)}b^{6(2)}c^{4(2)}} = \dfrac{1}{100a^4b^{12}c^8}$

**117.** $\dfrac{9y^4}{x^{-2}} + \left(\dfrac{x^{-1}}{y^2}\right)^{-2} = 9x^2y^4 + \dfrac{x^{(-1)(-2)}}{y^{2(-2)}}$

$= 9x^2y^4 + \dfrac{x^2}{y^{-4}}$

$= 9x^2y^4 + x^2y^4$

$= 10x^2y^4$

**118.** $\dfrac{7x^3}{y^{-9}} + \left(\dfrac{x^{-1}}{y^3}\right)^{-3} = 7x^3y^9 + \dfrac{x^{(-1)(-3)}}{y^{3(-3)}}$

$= 7x^3y^9 + \dfrac{x^3}{y^{-9}}$

$= 7x^3y^9 + x^3y^9$

$= 8x^3y^9$

**119.** $\left(\dfrac{3x^4}{y^{-4}}\right)^{-1}\left(\dfrac{2x}{y^2}\right)^3$

$= \dfrac{3^{-1}x^{4(-1)}}{y^{(-4)(-1)}} \cdot \dfrac{2^3x^{1\cdot3}}{y^{2\cdot3}}$

$= \dfrac{x^{-4}}{3y^4} \cdot \dfrac{8x^3}{y^6} = \dfrac{8x^{-4+3}}{3y^{4+6}} = \dfrac{8x^{-1}}{3y^{10}} = \dfrac{8}{3xy^{10}}$

**120.** $\left(\dfrac{2^{-1}x^{-2}y}{x^4y^{-1}}\right)^{-2}\left(\dfrac{xy^{-3}}{x^{-3}y}\right)^3$

$= \dfrac{2^{(-1)(-2)}x^{(-2)(-2)}y^{1(-2)}}{x^{4(-2)}y^{(-1)(-2)}} \cdot \dfrac{x^{1\cdot3}y^{-3\cdot3}}{x^{-3\cdot3}y^{1\cdot3}}$

$= \dfrac{2^2x^4y^{-2}}{x^{-8}y^2} \cdot \dfrac{x^3y^{-9}}{x^{-9}y^3}$

$= 4x^{4-(-8)}y^{-2-2} \cdot x^{3-(-9)}y^{-9-3}$

$= 4x^{12}y^{-4} \cdot x^{12}y^{-12}$

$= 4x^{12+12}y^{-4+(-12)} = 4x^{24}y^{-16} = \dfrac{4x^{24}}{y^{16}}$

**121.** $\left(-4x^3y^{-5}\right)^{-2}\left(2x^{-8}y^{-5}\right)$

$= \dfrac{2x^{-8}y^{-5}}{\left(-4x^3y^{-5}\right)^2}$

$= \dfrac{2x^{-8}y^{-5}}{(-4)^2x^{3\cdot2}y^{-5\cdot2}}$

$= \dfrac{2x^{-8}y^{-5}}{16x^6y^{-10}} = \dfrac{y^{-5-(-10)}}{8x^{6-(-8)}} = \dfrac{y^5}{8y^{14}}$

**122.** $\left(-4x^{-4}y^5\right)^{-2}\left(-2x^5y^{-6}\right)$

$= \dfrac{-2x^5y^{-6}}{\left(-4x^{-4}y^5\right)^2}$

$= \dfrac{-2x^5y^{-6}}{(-4)^2x^{-4\cdot2}y^{5\cdot2}}$

$= -\dfrac{2x^5y^{-6}}{16x^{-8}y^{10}} = -\dfrac{x^{5-(-8)}}{8y^{10-(-6)}} = -\dfrac{x^{13}}{8y^{16}}$

**123.** $\dfrac{(2x^2y^4)^{-1}(4xy^3)^{-3}}{(x^2y)^{-5}(x^3y^2)^4}$

$= \dfrac{(x^2y)^5}{(2x^2y^4)^1(4xy^3)^3(x^3y^2)^4}$

$= \dfrac{x^{2\cdot5}y^{1\cdot5}}{(2x^2y^4)(4^3x^{1\cdot3}y^{3\cdot3})(x^{3\cdot4}y^{2\cdot4})}$

$= \dfrac{x^{10}y^5}{(2x^2y^4)(64x^3y^9)(x^{12}y^8)}$

$= \dfrac{x^{10}y^5}{128x^{2+3+12}y^{4+9+8}}$

$= \dfrac{x^{10}y^5}{128x^{17}y^{21}}$

$= \dfrac{1}{128x^{17-10}y^{21-5}} = \dfrac{1}{128x^7y^{16}}$

**124.** $\dfrac{(3x^3y^2)^{-1}(2x^2y)^{-2}}{(xy^2)^{-5}(x^2y^3)^3}$

$= \dfrac{(xy^2)^5}{(3x^3y^2)^1(2x^2y)^2(x^2y^3)^3}$

$= \dfrac{x^{1\cdot5}y^{2\cdot5}}{(3x^3y^2)(2^2x^{2\cdot2}y^2)(x^{2\cdot3}y^{3\cdot3})}$

$= \dfrac{x^5y^{10}}{(3x^3y^2)(4x^4y^2)(x^6y^9)}$

$= \dfrac{x^5y^{10}}{12x^{3+4+6}y^{2+2+9}}$

$= \dfrac{x^5y^{10}}{12x^{13}y^{13}}$

$= \dfrac{1}{12x^{13-5}y^{13-10}} = \dfrac{1}{12x^8y^3}$

**125.** If $x = 90$,

$y = 1000\left(\dfrac{1}{2}\right)^{\frac{90}{30}}$

$= 1000\left(\dfrac{1}{2}\right)^3 = 1000\left(\dfrac{1}{8}\right) = 125$

The amount of cesium-137 in the atmosphere will be 125 kilograms. Chernobyl will be unsafe for human habitation.

**127.** If $n = 1$,

$d = \dfrac{3(2^{n-2})+4}{10}$

$= \dfrac{3(2^{1-2})+4}{10}$

$= \dfrac{3(2^{-1})+4}{10}$

$= \dfrac{3\left(\frac{1}{2}\right)+4}{10} = \dfrac{1.5+4}{10} = \dfrac{5.5}{10} = 0.55$

Mercury is 0.55 astronomical units from the sun.

**129.** If $n = 3$,

$d = \dfrac{3(2^{n-2})+4}{10}$

$= \dfrac{3(2^{3-2})+4}{10}$

$= \dfrac{3(2)+4}{10} = \dfrac{6+4}{10} = \dfrac{10}{10} = 1$

Earth is 1 astronomical unit from the sun.

**131. – 137.** Answers will vary.

**139.**

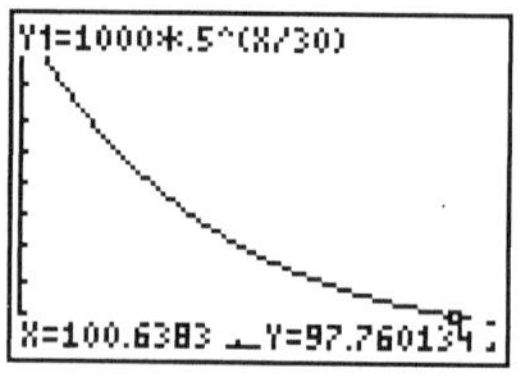


Approximately 100 years after 1986 in 2086, Chernobyl will be safe for human habitation.

**141.** Statement **d.** is true.

Statement **a.** is false.

$\frac{1}{(-2)^3} = \frac{1}{-8} = -\frac{1}{8}$, but $2^{-3} = \frac{1}{2^3} = \frac{1}{8}$.

Statement **b.** is false.

$\frac{2^8}{2^{-3}} = 2^{8-(-3)} = 2^{11}$, not $2^5$.

Statement **c.** is false.

$2^4 + 2^5 = 16 + 32 = 48$, but $2^9 = 512$.

**143.**

$$\left(x^{-4n} \cdot x^n\right)^{-3} = \left(x^{-4n+n}\right)^{-3} = \left(x^{-3n}\right)^{-3} = x^{(-3n)(-3)} = x^{9n}$$

**145.**

$$\left(\frac{x^n y^{3n+1}}{y^n}\right)^{-2} = \left(x^n y^{(3n+1)-n}\right)^3 = \left(x^n y^{2n+1}\right)^3 = x^{n \cdot 3} y^{(2n+1) \cdot 3} = x^{3n} y^{6n+3}$$

**146.**

| $x$ | $y = 2x - 1$ | $(x, y)$ |
|---|---|---|
| –3 | $y = 2(-3) - 1$ <br> $= -6 - 1 = -7$ | $(-3, -7)$ |
| –2 | $y = 2(-2) - 1$ <br> $= -4 - 1 = -5$ | $(-2, -5)$ |
| –1 | $y = 2(-1) - 1$ <br> $= -2 - 1 = -3$ | $(-1, -3)$ |
| 0 | $y = 2(0) - 1$ <br> $= 0 - 1 = -1$ | $(0, -1)$ |
| 1 | $y = 2(1) - 1$ <br> $= 2 - 1 = 1$ | $(1, 1)$ |
| 2 | $y = 2(2) - 1$ <br> $= 4 - 1 = 3$ | $(2, 3)$ |
| 3 | $y = 2(3) - 1$ <br> $= 6 - 1 = 5$ | $(3, 5)$ |

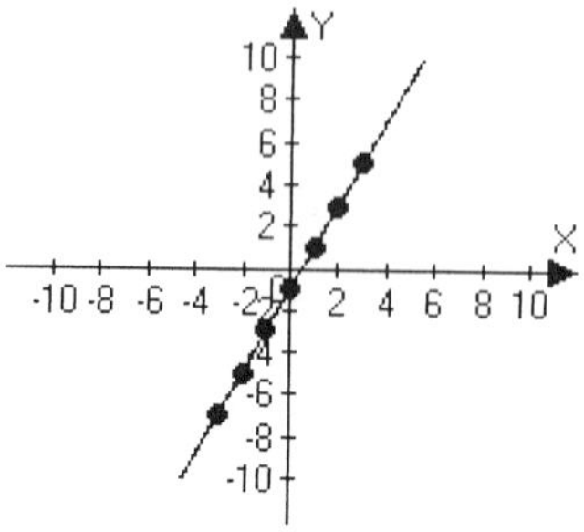


**147.**

$$Ax + By = C$$
$$By = C - Ax$$
$$y = \frac{C - Ax}{B}$$

**148.** Let $w$ = the width of the field.
$2w - 5$ = the length of the field.

$$P = 2(\text{length}) + 2(\text{width})$$
$$230 = 2(2w-5)+2w$$
$$230 = 4w-10+2w$$
$$230 = 6w-10$$
$$240 = 6w$$
$$40 = w$$

Find the length.

$$2w-5 = 2(40)-5 = 80-5 = 75$$

The playing field is 40 meters by 75 meters.

### 1.7 Exercise Set

**1.** $3.8\times10^2 = 380$

**3.** $6\times10^{-4} = 0.0006$

**5.** $-7.16\times10^6 = -7,160,000$

**7.** $1.4\times10^0 = 1.4\times1 = 1.4$

**9.** $7.9\times10^{-1} = 0.79$

**11.** $4.15\times10^{-3} = 0.00415$

**13.** $-6.00001\times10^{10} = -60,000,100,000$

**15.** $32,000 = 3.2\times10^4$

**17.** $638,000,000,000,000,000$
$= 6.38\times10^{17}$

**19.** $-317 = -3.17\times10^2$

**21.** $-5716 = -5.716\times10^3$

**23.** $0.0027 = 2.7\times10^{-3}$

**25.** $-0.00000000504 = -5.04\times10^{-9}$

**27.** $0.007 = 7\times10^{-3}$

**29.** $3.14159 = 3.14159\times10^0$

**31.** $(3\times10^4)(2.1\times10^3)$
$= (3\times2.1)(10^4\times10^3)$
$= 6.3\times10^{4+3} = 6.3\times10^7$

**33.** $(1.6\times10^{15})(4\times10^{-11})$
$= (1.6\times4)(10^{15}\times10^{-11})$
$= 6.4\times10^{15+(-11)} = 6.4\times10^4$

**35.** $(6.1\times10^{-8})(2\times10^{-4})$
$= (6.1\times2)(10^{-8}\times10^{-4})$
$= 12.2\times10^{-8+(-4)}$
$= 12.2\times10^{-12} = 1.22\times10^{-11}$

**37.** $(4.3\times10^8)(6.2\times10^4)$
$= (4.3\times6.2)(10^8\times10^4)$
$= 26.66\times10^{8+4}$
$= 26.66\times10^{12}$
$= 2.666\times10^{13} \approx 2.67\times10^{13}$

**39.** $\dfrac{8.4\times10^8}{4\times10^5} = \dfrac{8.4}{4}\times\dfrac{10^8}{10^5}$
$= 2.1\times10^{8-5} = 2.1\times10^3$

**41.** $\dfrac{3.6\times10^4}{9\times10^{-2}} = \dfrac{3.6}{9}\times\dfrac{10^4}{10^{-2}}$
$= 0.4\times10^{4-(-2)}$
$= 0.4\times10^6 = 4\times10^5$

**43.** $\dfrac{4.8\times10^{-2}}{2.4\times10^6} = \dfrac{4.8}{2.4}\times\dfrac{10^{-2}}{10^6}$
$= 2\times10^{-2-6} = 2\times10^{-8}$

**45.** $\frac{2.4\times10^{-2}}{4.8\times10^{-6}}=\frac{2.4}{4.8}\times\frac{10^{-2}}{10^{-6}}$
$=0.5\times10^{-2-(-6)}$
$=0.5\times10^{4}=5\times10^{3}$

**47.** $\frac{480,000,000,000}{0.00012}=\frac{4.8\times10^{11}}{1.2\times10^{-4}}$
$=\frac{4.8}{1.2}\times\frac{10^{11}}{10^{-4}}$
$=4\times10^{11-(-4)}$
$=4\times10^{15}$

**49.** $\frac{0.00072\times0.003}{0.00024}$
$=\frac{(7.2\times10^{-4})(3\times10^{-3})}{2.4\times10^{-4}}$
$=\frac{7.2\times3}{2.4}\times\frac{10^{-4}\cdot10^{-3}}{10^{-4}}=9\times10^{-3}$

**51.** $(2\times10^{-5})x=1.2\times10^{9}$
$x=\frac{1.2\times10^{9}}{2\times10^{-5}}$
$=\frac{1.2}{2}\times\frac{10^{9}}{10^{-5}}$
$=0.6\times10^{9-(-5)}$
$=0.6\times10^{14}=6\times10^{13}$

**52.** $(3\times10^{-2})x=1.2\times10^{4}$
$x=\frac{1.2\times10^{4}}{3\times10^{-2}}$
$=\frac{1.2}{3}\times\frac{10^{4}}{10^{-2}}$
$=0.4\times10^{4-(-2)}$
$=0.4\times10^{6}=4\times10^{5}$

**53.** $\frac{x}{2\times10^{8}}=-3.1\times10^{-5}$
$x=(2\times10^{8})(-3.1\times10^{-5})$
$=[2\cdot(-3.1)]\times(10^{8}\cdot10^{-5})$
$=-6.2\times10^{8+(-5)}=-6.2\times10^{3}$

**54.** $\frac{x}{5\times10^{11}}=-2.9\times10^{-3}$
$x=(5\times10^{11})(-2.9\times10^{-3})$
$=[5(-2.9)]\times(10^{11}\cdot10^{-3})$
$=-14.5\times10^{11+(-3)}$
$=-14.5\times10^{8}=-1.45\times10^{9}$

**55.** $x-(7.2\times10^{18})=9.1\times10^{18}$
$x=(9.1\times10^{18})+(7.2\times10^{18})$
$=(9.1+7.2)\times10^{18}$
$=16.3\times10^{18}=1.63\times10^{19}$

**56.** $x-(5.3\times10^{-16})=8.4\times10^{-16}$
$x=(8.4\times10^{-16})+(5.3\times10^{-16})$
$=(8.4+5.3)\times10^{-16}$
$=13.7\times10^{-16}=1.37\times10^{-15}$

**57.** $(-1.2\times10^{-3})x=(1.8\times10^{-4})(2.4\times10^{6})$
$x=\frac{(1.8\times10^{-4})(2.4\times10^{6})}{-1.2\times10^{-3}}$
$=\frac{1.8\cdot2.4}{-1.2}\times\frac{10^{-4}\cdot10^{6}}{10^{-3}}$
$=1.8(-2)\times10^{-4+6-(-3)}=-3.6\times10^{5}$

**58.** $\left(-7.8\times10^{-4}\right)x=\left(3.9\times10^{-7}\right)\left(6.8\times10^{5}\right)$

$$x=\frac{\left(3.9\times10^{-7}\right)\left(6.8\times10^{5}\right)}{-7.8\times10^{-4}}$$

$$=\frac{3.9\cdot6.8}{-7.8}\times\frac{10^{-7}\cdot10^{5}}{10^{-4}}$$

$$=\frac{6.8}{-2}\times10^{-7+5-(-4)}=-3.4\times10^{2}$$

**59.** $62.6 \text{ million} = 62.6\times10^{6} = 6.26\times10^{7}$

So, $6.26\times10^{7}$ people will be 65 and over in 2025.

**61.** $131.2-34.9=96.3$

$96.3 \text{ million} = 96.3\times10^{6} = 9.63\times10^{7}$

There will be $9.63\times10^{7}$ more people 65 and over in the year 2100 than in 2000.

**63.** $20 \text{ billion} = 2\times10^{10}$

$$\frac{2\times10^{10}}{2.88\times10^{8}}=\frac{2}{2.88}\times\frac{10^{10}}{10^{8}}$$

$$\approx0.694444\times10^{10-8}$$

$$=0.694444\times10^{2}$$

$$=6.94444\times10^{1}\approx69$$

The average American consumes about 69 hotdogs each year.

**65.** $8 \text{ billion} = 8\times10^{9}$

$$\frac{8\times10^{9}}{3.2\times10^{7}}=\frac{8}{3.2}\times\frac{10^{9}}{10^{7}}$$

$$=2.5\times10^{9-7}$$

$$=2.5\times10^{2}=250$$

$2.5\times10^{2}=250$ chickens are raised for food each second in the U.S.

**67.** $26\left(2\times10^{4}\right)=52\times10^{4}=5.2\times10^{5}$

The total distance covered by all of the runners is $5.2\times10^{5}$ miles.

**69.** $20{,}000\left(5.3\times10^{-23}\right)$

$$=\left(2\times10^{4}\right)\left(5.3\times10^{-23}\right)$$

$$=\left(2\cdot5.3\right)\times\left(10^{4}\cdot10^{-23}\right)$$

$$=10.6\times10^{4+(-23)}$$

$$=10.6\times10^{-19}$$

$$=1.06\times10^{-18}$$

The mass of 20,000 oxygen molecules is $1.06\times10^{-18}$ grams.

**71.** $\dfrac{365 \text{ days}}{1 \text{ year}}\cdot\dfrac{24 \text{ hours}}{1 \text{ day}}$

$$=8760 \text{ hours/year}$$

$$=8.76\times10^{3} \text{ hours/year}$$

$$\frac{8.76\times10^{3} \text{ hours}}{1 \text{ year}}\cdot\frac{60 \text{ minutes}}{1 \text{ hour}}$$

$$=525.6\times10^{3} \text{ minutes/year}$$

$$=5.256\times10^{5} \text{ minutes/year}$$

$$\frac{5.256\times10^{5} \text{ minutes}}{1 \text{ year}}\cdot\frac{60 \text{ seconds}}{1 \text{ minute}}$$

$$=315.36\times10^{5} \text{ seconds/year}$$

$$=3.1536\times10^{7} \text{ seconds/year}$$

There are $3.1536\times10^{7}$ seconds in a year.

**73. – 75.** Answers will vary.

**77.** Answers will vary. For example, consider Exercise 15.

$32{,}000=3.2\times10^{4}$

**79.** Statement **d.** is true.

$(4\times10^3)+(3\times10^2)$

$=(40\times10^2)+(3\times10^2)$

$=(40+3)\times10^2=43\times10^2$

Statement **a.** is false.

$534.7=5.347\times10^2$, not $5.347\times10^3$.

Statement **b.** is false.

$\frac{8\times10^{30}}{4\times10^{-5}}=2\times10^{30-(-5)}=2\times10^{35}$, not $2\times10^{25}$

Statement **c.** is false.

$(7\times10^5)+(2\times10^{-3})$

$=700,000+0.002=700,000.002,$

not $9\times10^2=900$.

**81.** $8.2\times10^{-16}+4.3\times10^{-16}$

$=(8.2+4.3)\times10^{-16}$

$=12.5\times10^{-16}=1.25\times10^{-15}$

**83.** Answers will vary.

**84.** $9(10x-4)-(5x-10)$

$=90x-36-5x+10$

$=90x-5x-36+10=85x-26$

**85.**

$$\frac{4x-1}{10}=\frac{5x+2}{4}-4$$

$$20\left(\frac{4x-1}{10}\right)=20\left(\frac{5x+2}{4}-4\right)$$

$$2(4x-1)=5(5x+2)-80$$

$$8x-2=25x+10-80$$

$$8x-2=25x-70$$

$$-2=17x-70$$

$$68=17x$$

$$4=x$$

The solution set is $\{4\}$.

**86.** $(8x^4y^{-3})^{-2}=8^{-2}(x^4)^{-2}(y^{-3})^{-2}$

$=8^{-2}x^{-8}y^6$

$=\frac{1}{64}\cdot\frac{1}{x^8}y^6$

$=\frac{y^6}{64x^8}$

## Chapter 1 Review Exercises

**1.** $2x-10$

**2.** $6x+4$

**3.** $\frac{9}{x}+\frac{1}{2}x$

**4.** $x^2-7x+4=(10)^2-7(10)+4$

$=100-70+4$

$=30+4$

$=34$

**5.** $6+2(x-8)^3=6+2(11-8)^3$

$=6+2(3)^3$

$=6+2(27)$

$=6+54$

$=60$

**6.** $x^4-(x-y)=(2)^4-(2-1)$

$=16-1$

$=15$

**7.** $\{1,2\}$

**8.** $\{-3,-2,-1,0,1\}$

**9.** False. Zero is not a natural number.

**10.** True. –2 is a rational number.

**11.** True. $\frac{1}{3}$ is not an irrational number.

**12.** Negative five is less than two. True.

**13.** Negative seven is greater than or equal to negative three. False.

**14.** Negative seven is less than or equal to negative seven. True.

**15.**
$$\begin{aligned} S &= 0.015x^2 + x + 10 \\ &= 0.015(60)^2 + 60 + 10 \\ &= 0.015(3600) + 70 \\ &= 54 + 70 \\ &= 124 \end{aligned}$$
The recommended safe distance is 124 feet.

**16.** $|-9.7| = 9.7$

**17.** $|5.003| = 5.003$

**18.** $|0| = 0$

**19.** $-2.4 + (-5.2) = -7.6$

**20.** $-6.8 + 2.4 = -4.4$

**21.** $-7 - (-20) = -7 + 20 = 13$

**22.** $(-3)(-20) = 60$

**23.**
$$\begin{aligned} -\frac{3}{5} - \left(-\frac{1}{2}\right) &= -\frac{3}{5} + \frac{1}{2} \\ &= -\frac{3}{5}\cdot\frac{2}{2} + \frac{1}{2}\cdot\frac{5}{5} \\ &= -\frac{6}{10} + \frac{5}{10} \\ &= -\frac{1}{10} \end{aligned}$$

**24.** $\left(\frac{2}{7}\right)\left(-\frac{3}{10}\right) = -\frac{6}{70} = -\frac{2\cdot 3}{2\cdot 35} = -\frac{3}{35}$

**25.**
$$\begin{aligned} 4(-3)(-2)(-10) &= -12(-2)(-10) \\ &= 24(-10) \\ &= -240 \end{aligned}$$

**26.** $(-2)^4 = 16$

**27.** $-2^5 = -32$

**28.**
$$\begin{aligned} -\frac{2}{3} \div \frac{8}{5} &= -\frac{2}{3}\cdot\frac{5}{8} \\ &= -\frac{2\cdot 5}{3\cdot 2\cdot 4} = -\frac{5}{3\cdot 4} = -\frac{5}{12} \end{aligned}$$

**29.** $\frac{-35}{-5} = \frac{-5\cdot 7}{-5} = 7$

**30.** $\frac{54.6}{-6} = -9.1$

**31.**
$$\begin{aligned} x &= -7 \\ -1(x) &= -1(-7) \\ -x &= 7 \end{aligned}$$

**32.**
$$\begin{aligned} -11 - [-17 + (-3)] &= -11 - [-20] \\ &= -11 + 20 \\ &= 9 \end{aligned}$$

**33.** $\left(-\frac{1}{2}\right)^3 \cdot 2^4 = -\frac{1}{8}\cdot 16 = -\frac{16}{8} = -2$

**34.**
$$\begin{aligned} -3[4 - (6 - 8)] &= -3[4 - (-2)] \\ &= -3[6] \\ &= -18 \end{aligned}$$

**35.**
$$8^2-36\div 3^2\cdot 4-(-7)$$
$$=64-36\div 9\cdot 4-(-7)$$
$$=64-4\cdot 4-(-7)$$
$$=64-16-(-7)$$
$$=48-(-7)$$
$$=48+7=55$$

**36.**
$$\frac{(-2)^4+(-3)^2}{2^2-(-21)}=\frac{16+9}{4-(-21)}$$
$$=\frac{25}{4+21}=\frac{25}{25}=1$$

**37.**
$$\frac{(7-9)^3-(-4)^2}{2+2(8)\div 4}=\frac{(-2)^3-16}{2+16\div 4}$$
$$=\frac{-8-16}{2+4}$$
$$=\frac{-24}{6}=-4$$

**38.**
$$4-(3-8)^2+3\div 6\cdot 4^2$$
$$=4-(-5)^2+3\div 6\cdot 4^2$$
$$=4-25+3\div 6\cdot 16$$
$$=4-25+\frac{1}{2}\cdot 16$$
$$=4-25+8$$
$$=-21+8=-13$$

**39.**
$$5(2x-3)+7x=10x-15+7x$$
$$=17x-15$$

**40.**
$$5x+7x^2-4x+2x^2=9x^2+x$$

**41.**
$$3(4y-5)-(7y+2)$$
$$=12y-15-7y-2=5y-17$$

**42.**
$$8-2[3-(5x-1)]=8-2[3-5x+1]$$
$$=8-6+10x-2$$
$$=10x$$

**43.**
$$6(2x-3)-5(3x-2)$$
$$=12x-18-15x+10$$
$$=-3x-8$$

**44. – 46.**

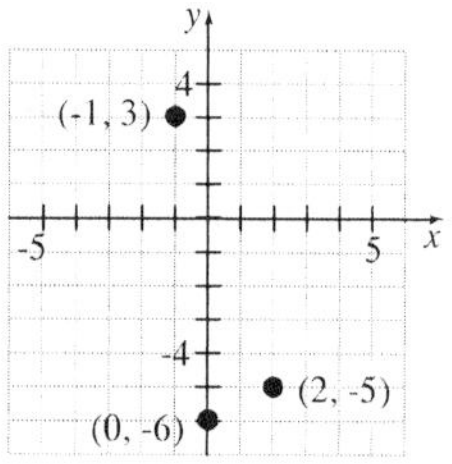


**47.**

| $x$ | $y=2x-2$ | $(x,y)$ |
|---|---|---|
| $-3$ | $y=2(-3)-2$ <br> $=-6-2=-8$ | $(-3,-8)$ |
| $-2$ | $y=2(-2)-2$ <br> $=-4-2=-6$ | $(-2,-6)$ |
| $-1$ | $y=2(-1)-2$ <br> $=-2-2=-4$ | $(-1,-4)$ |
| $0$ | $y=2(0)-2$ <br> $=0-2=-2$ | $(0,-2)$ |
| $1$ | $y=2(1)-2$ <br> $=2-2=0$ | $(1,0)$ |
| $2$ | $y=2(2)-2$ <br> $=4-2=2$ | $(2,2)$ |
| $3$ | $y=2(3)-2$ <br> $=6-2=4$ | $(3,4)$ |

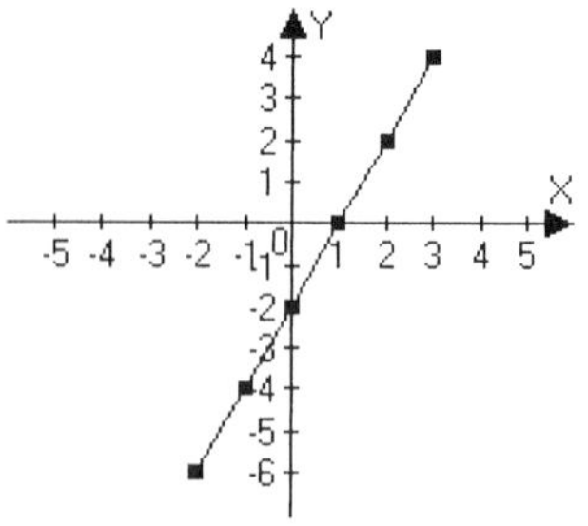

**48.**

| $x$ | $y = x^2 - 3$ | $(x, y)$ |
|---|---|---|
| $-3$ | $y = (-3)^2 - 3$ <br> $= 9 - 3 = 6$ | $(-3, 6)$ |
| $-2$ | $y = (-2)^2 - 3$ <br> $= 4 - 3 = 1$ | $(-2, 1)$ |
| $-1$ | $y = (-1)^2 - 3$ <br> $= 1 - 3 = -2$ | $(-1, -2)$ |
| $0$ | $y = (0)^2 - 3$ <br> $= 0 - 3 = -3$ | $(0, -3)$ |
| $1$ | $y = (1)^2 - 3$ <br> $= 1 - 3 = -2$ | $(1, -2)$ |
| $2$ | $y = (2)^2 - 3$ <br> $= 4 - 3 = 1$ | $(2, 1)$ |
| $3$ | $y = (3)^2 - 3$ <br> $= 9 - 3 = 6$ | $(3, 6)$ |

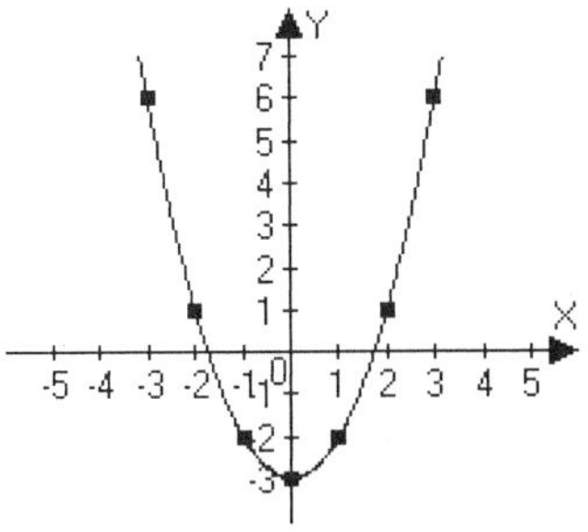

**49.**

| $x$ | $y = x$ | $(x, y)$ |
|---|---|---|
| $-3$ | $y = -3$ | $(-3, -3)$ |
| $-2$ | $y = -2$ | $(-2, -2)$ |
| $-1$ | $y = -1$ | $(-1, -1)$ |
| $0$ | $y = 0$ | $(0, 0)$ |
| $1$ | $y = 1$ | $(1, 1)$ |
| $2$ | $y = 2$ | $(2, 2)$ |
| $3$ | $y = 3$ | $(3, 3)$ |

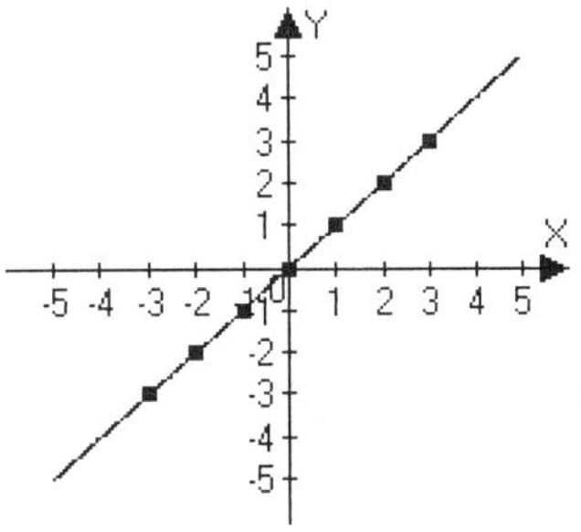

**50.**

| $x$ | $y = \|x\| - 2$ | $(x, y)$ |
|---|---|---|
| $-3$ | $y = \|-3\| - 2 = 3 - 2 = 1$ | $(-3, 1)$ |
| $-2$ | $y = \|-2\| - 2 = 2 - 2 = 0$ | $(-2, 0)$ |
| $-1$ | $y = \|-1\| - 2 = 1 - 2 = -1$ | $(-1, -1)$ |
| $0$ | $y = \|0\| - 2 = 0 - 2 = -2$ | $(0, -2)$ |
| $1$ | $y = \|1\| - 2 = 1 - 2 = -1$ | $(1, -1)$ |
| $2$ | $y = \|2\| - 2 = 2 - 2 = 0$ | $(2, 0)$ |
| $3$ | $y = \|3\| - 2 = 3 - 2 = 1$ | $(3, 1)$ |

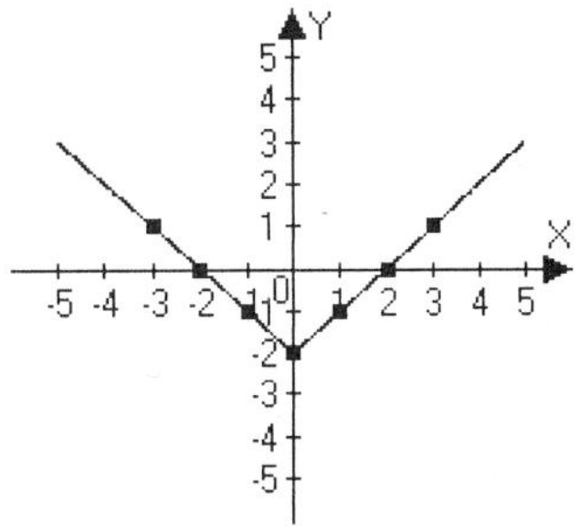

**51.** The minimum $x$-value is $-20$ and the maximum $x$-value is 40. The distance between tick marks is 10. The minimum $y$-value is $-5$ and the maximum $y$-value is 5. The distance between tick marks is 1.

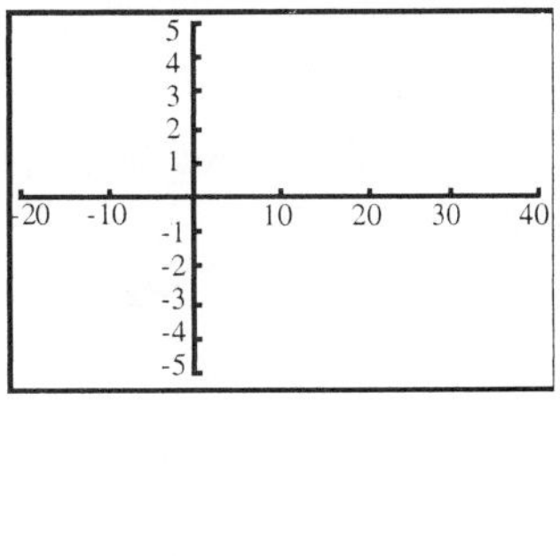

**52.** 20% of 75 year old Americans have Alzheimer's.

**53.** Age 85 represents a 50% prevalence.

**54.** Answers will vary.

**55.** Graph **c.** illustrates the description.

**56.**
$$\begin{aligned} 2x-5&=7 \\ 2x&=12 \\ x&=6 \end{aligned}$$
The solution is 6 and the solution set is $\{6\}$.

**57.**
$$\begin{aligned} 5x+20&=3x \\ 2x+20&=0 \\ 2x&=-20 \\ x&=-10 \end{aligned}$$
The solution is –10 and the solution set is $\{-10\}$.

**58.**
$$\begin{aligned} 7(x-4)&=x+2 \\ 7x-28&=x+2 \\ 6x-28&=2 \\ 6x&=30 \\ x&=5 \end{aligned}$$
The solution is 5 and the solution set is $\{5\}$.

**59.**
$$\begin{aligned} 1-2(6-x)&=3x+2 \\ 1-12+2x&=3x+2 \\ -11+2x&=3x+2 \\ -11&=x+2 \\ -13&=x \end{aligned}$$
The solution is –13 and the solution set is $\{-13\}$.

**60.**
$$\begin{aligned} 2(x-4)+3(x+5)&=2x-2 \\ 2x-8+3x+15&=2x-2 \\ 5x+7&=2x-2 \\ 3x+7&=-2 \\ 3x&=-9 \\ x&=-3 \end{aligned}$$
The solution is –3 and the solution set is $\{-3\}$.

**61.**
$$\begin{aligned} 2x-4(5x+1)&=3x+17 \\ 2x-20x-4&=3x+17 \\ -18x-4&=3x+17 \\ -4&=21x+17 \\ -21&=21x \\ -1&=x \end{aligned}$$
The solution is –1 and the solution set is $\{-1\}$.

**62.**
$$\begin{aligned} \frac{2x}{3}&=\frac{x}{6}+1 \\ 6\left(\frac{2x}{3}\right)&=6\left(\frac{x}{6}+1\right) \\ 2(2x)&=6\left(\frac{x}{6}\right)+6(1) \\ 4x&=x+6 \\ 3x&=6 \\ x&=2 \end{aligned}$$

The solution is 2 and the solution set is $\{2\}$.

**63.**
$$\frac{x}{2}-\frac{1}{10}=\frac{x}{5}+\frac{1}{2}$$
$$10\left(\frac{x}{2}-\frac{1}{10}\right)=10\left(\frac{x}{5}+\frac{1}{2}\right)$$
$$10\left(\frac{x}{2}\right)-10\left(\frac{1}{10}\right)=10\left(\frac{x}{5}\right)+10\left(\frac{1}{2}\right)$$
$$5x-1=2x+5$$
$$3x-1=5$$
$$3x=6$$
$$x=2$$
The solution is 2 and the solution set is $\{2\}$.

**64.**
$$\frac{2x}{3}=6-\frac{x}{4}$$
$$12\left(\frac{2x}{3}\right)=12\left(6-\frac{x}{4}\right)$$
$$4(2x)=12(6)-12\left(\frac{x}{4}\right)$$
$$8x=72-3x$$
$$11x=72$$
$$x=\frac{72}{11}$$
The solution is $\frac{72}{11}$ and the solution set is $\left\{\frac{72}{11}\right\}$.

**65.**
$$\frac{x}{4}=2+\frac{x-3}{3}$$
$$12\left(\frac{x}{4}\right)=12\left(2+\frac{x-3}{3}\right)$$
$$3x=12(2)+12\left(\frac{x-3}{3}\right)$$
$$3x=24+4(x-3)$$
$$3x=24+4x-12$$
$$3x=12+4x$$
$$-x=12$$
$$x=-12$$
The solution is –12 and the solution set is $\{-12\}$.

**66.**
$$\frac{3x+1}{3}-\frac{13}{2}=\frac{1-x}{4}$$
$$12\left(\frac{3x+1}{3}-\frac{13}{2}\right)=12\left(\frac{1-x}{4}\right)$$
$$12\left(\frac{3x+1}{3}\right)-12\left(\frac{13}{2}\right)=3(1-x)$$
$$4(3x+1)-6(13)=3(1-x)$$
$$12x+4-78=3-3x$$
$$12x-74=3-3x$$
$$15x-74=3$$
$$15x=77$$
$$x=\frac{77}{15}$$
The solution is $\frac{77}{15}$ and the solution set is $\left\{\frac{77}{15}\right\}$.

**67.**
$$7x+5=5(x+3)+2x$$
$$7x+5=5x+15+2x$$
$$7x+5=7x+15$$
$$5=15$$
There is no solution. The solution set is $\varnothing$. The equation is inconsistent.

**68.**
$$7x+13=4x-10+3x+23$$
$$7x+13=7x+13$$
The solution set is $\{x|x \text{ is a real number}\}$ or $\mathbb{R}$. The equation is an identity.

**69.** $7x+13=3x-10+2x+23$

$7x+13=5x-10+23$

$7x+13=5x+13$

$2x+13=13$

$2x=0$

$x=0$

The solution set is $\{0\}$. The equation is conditional.

**70.** $4(x-3)+5=x+5(x-2)$

$4x-12+5=x+5x-10$

$4x-7=6x-10$

$-2x-7=-10$

$-2x=-3$

$x=\frac{-3}{-2}=\frac{3}{2}$

The solution set is $\left\{\frac{3}{2}\right\}$. The equation is conditional.

**71.** $(2x-3)2-3(x+1)=(x-2)4-3(x+5)$

$4x-6-3x-3=4x-8-3x-15$

$x-9=x-23$

$-9=-23$

There is no solution. The solution set is $\varnothing$. The equation is inconsistent.

**72.** $E=10x+167$

$437=10x+167$

$270=10x$

$27=x$

There will be 437 endangered species in the United States in the year 1980 + 27 = 2007.

**73.** Let $x$ = the number of calories in Burger King's Chicken Caesar.
$x+125$ = the number of calories in Taco Bell's Express Taco Salad.
$x+95$ = the number of calories in Wendy's Mandarin Chicken Salad.

$x+(x+125)+(x+95)=1705$

$3x+220=1705$

$3x=1485$

$x=495$

$x+125=495+125=620$

$x+95=495+95=590$

There are 495 calories in the Chicken Caesar, 620 calories in the Express Taco Salad, and 590 calories in the Mandarin Chicken Salad.

**74.** Let $x$ = the measure of the 2nd angle.
$x+10$ = the measure of the 1st angle.
$2[x+(x+10)]$ = the measure of the 3rd angle.

$x+(x+10)+2[x+(x+10)]=180$

$x+x+10+2x+2x+20=180$

$6x+30=180$

$6x=150$

$x=25$

$x+10=25+10=35$

$2[x+(x+10)]=2[10+(10+10)]$

$=3(30)=60$

The angles measure $25°$, $35°$, and $60°$

**75.** Let $E$ = the number of endangered plant species, and let $x$ represent the number of years since 1998.

$E=6.4x+567$

$663=6.4x+567$

$96=6.4x$

$15=x$

There will be 663 endangered species in the United States in the year 1998 + 15 = 2013.

**76.** Let $x$ = the number of minutes of long distance
Plan A: $C = 15 + 0.05x$
Plan B: $C = 5 + 0.07x$
Set the costs equal.
$15 + 0.05x = 5 + 0.07x$
$15 = 5 + 0.02x$
$10 = 0.02x$
$500 = x$
The cost will be the same if 500 minutes of long distance are used.

**77.** Let $x$ = the original price of the phone.
$48 = x - 0.20x$
$48 = 0.80x$
$48 = 0.80x$
$60 = x$
The original price is \$60.

**78.** Let $x$ = the amount sold to earn \$800 in one week
$800 = 300 + 0.05x$
$500 = 0.05x$
$500 = 0.05x$
$10,000 = x$
Sales must be \$10,000 in one week to earn \$800.

**79.** Let $w$ = the width of the playing field.
Let $3w - 6$ = the length of the playing field.
$P = 2(\text{length}) + 2(\text{width})$
$340 = 2(3w - 6) + 2w$
$340 = 6w - 12 + 2w$
$340 = 8w - 12$
$352 = 8w$
$44 = w$
The dimensions are 44 yards by 126 yards.

**80.** **a.** Let $x$ = the number of years (after 2005).
College A's enrollment:
$14,100 + 1500x$
College B's enrollment:
$41,700 - 800x$
$14,100 + 1500x = 41,700 - 800x$

**b.** Check some points to determine that $y_1 = 14,100 + 1500x$ and $y_2 = 41,700 - 800x$. Since $y_1 = y_2 = 32,100$ when $x = 12$, the two colleges will have the same enrollment in the year $2005 + 12 = 2017$. That year the enrollments will be 32,100 students.

**81.**
$$V = \frac{1}{3}Bh$$
$$3V = Bh$$
$$h = \frac{3V}{B}$$

**82.**
$$y - y_1 = m(x - x_1)$$
$$\frac{y - y_1}{m} = \frac{m(x - x_1)}{m}$$
$$\frac{y - y_1}{m} = x - x_1$$
$$x = \frac{y - y_1}{m} + x_1$$

**83.**
$$E = I(R + r)$$
$$\frac{E}{I} = \frac{I(R + r)}{I}$$
$$\frac{E}{I} = R + r$$
$$R = \frac{E}{I} - r \text{ or } R = \frac{E - Ir}{I}$$

**84.**
$$C = \frac{5F - 160}{9}$$
$$9C = 5F - 160$$
$$9C + 160 = 5F$$
$$F = \frac{9C + 160}{5} \text{ or } F = \frac{9}{5}C + 32$$

**85.**
$$s = vt + gt^2$$
$$s - vt = gt^2$$
$$g = \frac{s - vt}{t^2}$$

**86.**
$$T = gr + gvt$$
$$T = g(r + vt)$$
$$g = \frac{T}{r + vt}$$

**87.** $$(-3x^7)(-5x^6) = 15x^{7+6} = 15x^{13}$$

**88.** $$x^2 y^{-5} = \frac{x^2}{y^5}$$

**89.** $$\frac{3^{-2}x^4}{y^{-7}} = \frac{x^4 y^7}{3^2} = \frac{x^4 y^7}{9}$$

**90.** $$(x^3)^{-6} = x^{3(-6)} = x^{-18} = \frac{1}{x^{18}}$$

**91.** $$(7x^3 y)^2 = 7^2 x^{3\cdot 2} y^{1\cdot 2} = 49x^6 y^2$$

**92.** $$\frac{16y^3}{-2y^{10}} = -8y^{3-10} = -8y^{-7} = -\frac{8}{y^7}$$

**93.**
$$(-3x^4)(4x^{-11}) = -12x^{4+(-11)}$$
$$= -12x^{-7} = -\frac{12}{x^7}$$

**94.** $$\frac{12x^7}{4x^{-3}} = 3x^{7-(-3)} = 3x^{10}$$

**95.**
$$\frac{-10a^5 b^6}{20a^{-3}b^{11}} = \frac{-1}{2}a^{5-(-3)}b^{6-11}$$
$$= \frac{-1}{2}a^8 b^{-5} = -\frac{a^8}{2b^5}$$

**96.**
$$(-3xy^4)(2x^2)^3 = (-3xy^4)(2^3 x^{2\cdot 3})$$
$$= (-3xy^4)(8x^6)$$
$$= -24x^{1+6}y^4$$
$$= -24x^7 y^4$$

**97.**
$$2^{-2} + \frac{1}{2}x^0 = \frac{1}{2^2} + \frac{1}{2}\cdot 1$$
$$= \frac{1}{4} + \frac{1}{2} = \frac{1}{4} + \frac{2}{4} = \frac{3}{4}$$

**98.**
$$(5x^2 y^{-4})^{-3} = \left(\frac{5x^2}{y^4}\right)^{-3}$$
$$= \left(\frac{y^4}{5x^2}\right)^3$$
$$= \frac{y^{4\cdot 3}}{5^3 x^{2\cdot 3}} = \frac{y^{12}}{125x^6}$$

**99.**
$$(3x^4 y^{-2})(-2x^5 y^{-3}) = \left(\frac{3x^4}{y^2}\right)\left(\frac{-2x^5}{y^3}\right)$$
$$= \left(\frac{-6x^{4+5}}{y^{2+3}}\right)$$
$$= -\frac{6x^9}{y^5}$$

**100.**
$$\left(\frac{3xy^3}{5x^{-3}y^{-4}}\right)^2 = \left(\frac{3x^{1-(-3)}y^{3-(-4)}}{5}\right)^2$$
$$= \left(\frac{3x^4 y^7}{5}\right)^2$$
$$= \frac{3^2 x^{4\cdot 2} y^{7\cdot 2}}{5^2} = \frac{9x^8 y^{14}}{25}$$

**101.** $\left(\frac{-20x^{-2}y^{3}}{10x^{5}y^{-6}}\right)^{-3} = \left(-2x^{-2-5}y^{3-(-6)}\right)^{-3}$

$= \left(-2x^{-7}y^{9}\right)^{-3}$

$= (-2)^{-3}x^{(-7)(-3)}y^{9(-3)}$

$= \frac{x^{21}y^{-27}}{(-2)^{3}}$

$= \frac{x^{21}}{-8y^{27}} = -\frac{x^{21}}{8y^{27}}$

**102.** $7.16\times10^{6} = 7{,}160{,}000$

**103.** $1.07\times10^{-4} = 0.000107$

**104.** $-41{,}000{,}000{,}000{,}000 = -4.1\times10^{13}$

**105.** $0.00809 = 8.09\times10^{-3}$

**106.** $\left(4.2\times10^{13}\right)\left(3\times10^{-6}\right)$

$= (4.2\times3)\left(10^{13}\times10^{-6}\right)$

$= (4.2\times3)\left(10^{13}\times10^{-6}\right)$

$= 12.6\times10^{13+(-6)}$

$= 12.6\times10^{7}$

$= 1.26\times10^{8}$

**107.** $\frac{5\times10^{-6}}{20\times10^{-8}} = \frac{5}{20}\times\frac{10^{-6}}{10^{-8}}$

$= 0.25\times10^{-6-(-8)}$

$= 0.25\times10^{-6+8}$

$= 0.25\times10^{2}$

$= 2.5\times10^{1}$

**108.** $150\left(2.9\times10^{8}\right) = \left(1.5\times10^{2}\right)\left(2.9\times10^{8}\right)$

$= (1.5\cdot2.9)\times\left(10^{2}\cdot10^{8}\right)$

$= 4.35\times10^{2+8}$

$= 4.35\times10^{10}$

## Chapter 1 Test

**1.** $4x-5$

**2.** $8+2(x-7)^{4} = 8+2(10-7)^{4}$

$= 8+2(3)^{4}$

$= 8+2(81)$

$= 8+162$

$= 170$

**3.** $\{-4,-3,-2,-1\}$

**4.** True. $\frac{1}{4}$ is not a natural number.

**5.** Negative three is greater than negative one. False.

**6.** $F = 24t^{2}-260t+816$

$= 24(10)^{2}-260(10)+816$

$= 24(100)-2600+816$

$= 2400-2600+816$

$= -200+816$

$= 616$

There were 616 convictions in 2000.

**7.** $|-17.9| = 17.9$

**8.** $-10.8+3.2 = -7.6$

**9.** $-\frac{1}{4}-\left(-\frac{1}{2}\right) = -\frac{1}{4}+\frac{1}{2} = -\frac{1}{4}+\frac{2}{4} = \frac{1}{4}$

**10.** $2(-3)(-1)(-10) = -6(-1)(-10)$

$= 6(-10)$

$= -60$

**11.** $-\frac{1}{4}\left(-\frac{1}{2}\right) = \frac{1}{8}$

**12.** $\frac{-27.9}{-9} = 3.1$

**13.** $24 - 36 \div 4 \cdot 3 = 24 - 9 \cdot 3$
$= 24 - 27 = -3$

**14.** $(5^2 - 2^4) + [9 \div (-3)]$
$= (25 - 16) + [-3] = (9) + [-3] = 6$

**15.** $\frac{(8-10)^3 - (-4)^2}{2 + 8(2) \div 4}$
$= \frac{(-2)^3 - 16}{2 + 16 \div 4} = \frac{-8 - 16}{2 + 4} = \frac{-24}{6} = -4$

**16.** $7x - 4(3x + 2) - 10 = 7x - 12x - 8 - 10$
$= -5x - 18$

**17.** $5(2y - 6) - (4y - 3)$
$= 10y - 30 - 4y + 3 = 6y - 27$

**18.** $9x - [10 - 4(2x - 3)]$
$= 9x - [10 - 8x + 12]$
$= 9x - 10 + 8x - 12 = 17x - 22$

**19.**

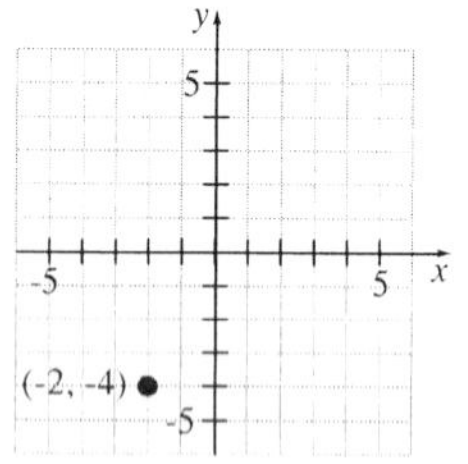


**20.**

| $x$ | $y = x^2 - 4$ | $(x, y)$ |
|---|---|---|
| -3 | $y = (-3)^2 - 4 = 9 - 4 = 5$ | $(-3, 5)$ |
| -2 | $y = (-2)^2 - 4 = 4 - 4 = 0$ | $(-2, 0)$ |
| -1 | $y = (-1)^2 - 4 = 1 - 4 = -3$ | $(-1, -3)$ |
| 0 | $y = (0)^2 - 4 = 0 - 4 = -4$ | $(0, -4)$ |
| 1 | $y = (1)^2 - 4 = 1 - 4 = -3$ | $(1, -3)$ |
| 2 | $y = (2)^2 - 4 = 4 - 4 = 0$ | $(2, 0)$ |
| 3 | $y = (3)^2 - 4 = 9 - 4 = 5$ | $(3, 5)$ |

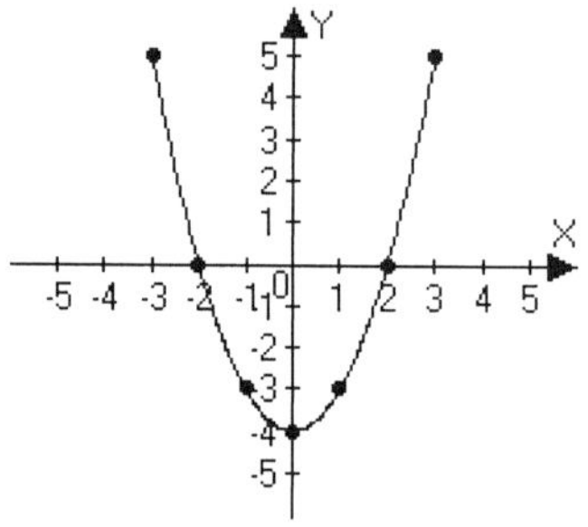


**21.** $3(2x - 4) = 9 - 3(x + 1)$
$6x - 12 = 9 - 3x - 3$
$6x - 12 = 6 - 3x$
$9x - 12 = 6$
$9x = 18$
$x = 2$
The solution is 2 and the solution set is $\{2\}$.

**22.** $\frac{2x - 3}{4} = \frac{x - 4}{2} - \frac{x + 1}{4}$
$2(2x - 3) = 8\left(\frac{x - 4}{2}\right) - 8\left(\frac{x + 1}{4}\right)$
$4x - 6 = 4(x - 4) - 2(x + 1)$
$4x - 6 = 4x - 16 - 2x - 2$
$4x - 6 = 2x - 18$
$2x - 6 = -18$
$2x = -12$
$x = -6$
The solution is –6 and the solution set is $\{-6\}$.

**23.** $3(x - 4) + x = 2(6 + 2x)$
$3x - 12 + x = 12 + 4x$
$4x - 12 = 12 + 4x$
$-12 = 12$
There is no solution. The solution set is $\varnothing$. The equation is inconsistent.

**24.** Let $x$ = the first number.
Let $2x + 3$ = the second number.
$$x+2x+3=72$$
$$3x+3=72$$
$$3x=69$$
$$x=23$$
Find the second number:
$$2x+3=2(23)+3=46+3=49$$
The numbers are 23 and 49.

**25.** Let $x$ = the number of years since the car was purchased.
$$\text{Value}=13,805-\$1820x$$
$$4705=13,805-\$1820x$$
$$-9100=-\$1820x$$
$$5=x$$
The car will have a value of \$4705 in 5 years.

**26.** Let $x$ = the number of prints.
Photo Shop A: $0.11x+1.60$
Photo Shop B: $0.13x+1.20$
$$0.13x+1.20=0.11x+1.60$$
$$0.02x+1.20=1.60$$
$$0.02x=0.40$$
$$x=20$$
The cost will be the same for 20 prints. That common price is
$$0.11(20)+1.60=0.13(20)+1.20$$
$$=\$3.80$$

**27.** Let $x$ = the original selling price
$$20=x-0.60x$$
$$20=0.40x$$
$$50=x$$
The original price is \$50.

**28.** Let $x$ = the width of the playing field.
Let $x + 260$ = the length of the playing field.
$$P=2(\text{length})+2(\text{width})$$
$$1000=2(x+260)+2x$$
$$1000=2x+520+2x$$
$$1000=4x+520$$
$$480=4x$$
$$x=120$$
The dimensions of the playing field are 120 yards by 380 yards.

**29.**
$$V=\frac{1}{3}lwh$$
$$3V=lwh$$
$$h=\frac{3V}{lw}$$

**30.**
$$Ax+By=C$$
$$By=C-Ax$$
$$y=\frac{C-Ax}{B}$$

**31.**
$$(-2x^5)(7x^{-10})=-14x^{5+(-10)}$$
$$=-14x^{-5}=-\frac{14}{x^5}$$

**32.**
$$(-8x^{-5}y^{-3})(-5x^2y^{-5})=40x^{-5+2}y^{-3+(-5)}$$
$$=40x^{-3}y^{-8}$$
$$=\frac{40}{x^3y^8}$$

**33.**
$$\frac{-10x^4y^3}{-40x^{-2}y^6}=\frac{1}{4}x^{4-(-2)}y^{3-6}$$
$$=\frac{1}{4}x^6y^{-3}=\frac{x^6}{4y^3}$$

**34.** $$\left(4x^{-5}y^2\right)^{-3} = \left(\frac{4y^2}{x^5}\right)^{-3}$$
$$= \left(\frac{x^5}{4y^2}\right)^3$$
$$= \frac{\left(x^5\right)^3}{(4)^3\left(y^2\right)^3}$$
$$= \frac{x^{15}}{64y^6}$$

**35.** $$\left(\frac{-6x^{-5}y}{2x^3y^{-4}}\right)^{-2} = \left(-3x^{-5-3}y^{1-(-4)}\right)^{-2}$$
$$= \left(-3x^{-8}y^5\right)^{-2}$$
$$= (-3)^{-2}x^{(-8)(-2)}y^{5(-2)}$$
$$= \frac{x^{16}y^{-10}}{(-3)^2}$$
$$= \frac{x^{16}}{9y^{10}}$$

**36.** $3.8\times10^{-6} = 0.0000038$

**37.** $407{,}000{,}000{,}000 = 4.07\times10^{11}$

**38.** $$\frac{4\times10^{-3}}{8\times10^{-7}} = 0.5\times10^{-3-(-7)}$$
$$= 0.5\times10^4 = 5.0\times10^3$$

**39.** $$2\left(6.08\times10^9\right) = 12.16\times10^9$$
$$= 1.216\times10^{10}$$
The population will be $1.216\times10^{10}$.

# Chapter 2
# Functions and Linear Functions

## 2.1 Exercise Set

**1.** The relation is a function.
The domain is {1, 3, 5}.
The range is {2, 4, 5}.

**3.** The relation is not a function.
The domain is {3, 4}.
The range is {4, 5}.

**5.** The relation is a function.
The domain is {-3, -2, -1, 0}.
The range is {-3, -2, -1, 0}.

**7.** The relation is not a function.
The domain is {1}.
The range is {4, 5, 6}.

**9.** **a.** $f(0) = 0 + 1 = 1$

**b.** $f(5) = 5 + 1 = 6$

**c.** $f(-8) = -8 + 1 = -7$

**d.** $f(2a) = 2a + 1$

**e.** $f(a+2) = (a+2) + 1$
$= a + 2 + 1 = a + 3$

**11.** **a.** $g(0) = 3(0) - 2 = 0 - 2 = -2$

**b.** $g(-5) = 3(-5) - 2$
$= -15 - 2 = -17$

**c.** $g\left(\frac{2}{3}\right) = 3\left(\frac{2}{3}\right) - 2 = 2 - 2 = 0$

**d.** $g(4b) = 3(4b) - 2 = 12b - 2$

**e.** $g(b+4) = 3(b+4) - 2$
$= 3b + 12 - 2 = 3b + 10$

**13.** **a.** $h(0) = 3(0)^2 + 5 = 3(0) + 5$
$= 0 + 5 = 5$

**b.** $h(-1) = 3(-1)^2 + 5 = 3(1) + 5$
$= 3 + 5 = 8$

**c.** $h(4) = 3(4)^2 + 5 = 3(16) + 5$
$= 48 + 5 = 53$

**d.** $h(-3) = 3(-3)^2 + 5 = 3(9) + 5$
$= 27 + 5 = 32$

**e.** $h(4b) = 3(4b)^2 + 5 = 3(16b^2) + 5$
$= 48b^2 + 5$

**15.** **a.** $f(0) = 2(0)^2 + 3(0) - 1$
$= 0 + 0 - 1 = -1$

**b.** $f(3) = 2(3)^2 + 3(3) - 1$
$= 2(9) + 9 - 1$
$= 18 + 9 - 1 = 26$

**c.** $f(-4) = 2(-4)^2 + 3(-4) - 1$
$= 2(16) - 12 - 1$
$= 32 - 12 - 1 = 19$

**d.** $f(b)=2(b)^2+3(b)-1$

$=2b^2+3b-1$

**e.** $f(5a)=2(5a)^2+3(5a)-1$

$=2(25a^2)+15a-1$

$=50a^2+15a-1$

**17. a.** $f(0)=\dfrac{2(0)-3}{(0)-4}=\dfrac{0-3}{0-4}$

$=\dfrac{-3}{-4}=\dfrac{3}{4}$

**b.** $f(3)=\dfrac{2(3)-3}{(3)-4}=\dfrac{6-3}{3-4}$

$=\dfrac{3}{-1}=-3$

**c.** $f(-4)=\dfrac{2(-4)-3}{(-4)-4}=\dfrac{-8-3}{-8}$

$=\dfrac{-11}{-8}=\dfrac{11}{8}$

**d.** $f(-5)=\dfrac{2(-5)-3}{(-5)-4}=\dfrac{-10-3}{-9}$

$=\dfrac{-13}{-9}=\dfrac{13}{9}$

**e.** $f(a+h)=\dfrac{2(a+h)-3}{(a+h)-4}$

$=\dfrac{2a+2h-3}{a+h-4}$

**f.** Four must be excluded from the domain, because four would make the denominator zero. Division by zero is undefined.

**19.**

| $x$ | $f(x)=x$ | $(x,y)$ |
|---|---|---|
| -2 | -2 | $(-2,-2)$ |
| -1 | -1 | $(-1,-1)$ |
| 0 | 0 | $(0,0)$ |
| 1 | 1 | $(1,1)$ |
| 2 | 2 | $(2,2)$ |

| $x$ | $g(x)=x+3$ | $(x,y)$ |
|---|---|---|
| -2 | $g(-2)=-2+3=1$ | $(-2,1)$ |
| -1 | $g(-1)=-1+3=2$ | $(-1,2)$ |
| 0 | $g(0)=0+3=3$ | $(0,3)$ |
| 1 | $g(1)=1+3=4$ | $(1,4)$ |
| 2 | $g(2)=2+3=5$ | $(2,5)$ |

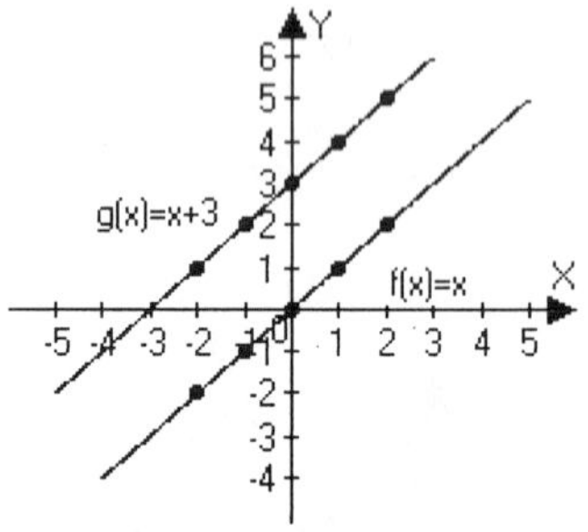


The graph of $g$ is the graph of $f$ shifted up 3 units.

**21.**

| $x$ | $f(x)=-2x$ | $(x,y)$ |
|---|---|---|
| -2 | $f(-2)=-2(-2)=4$ | $(-2,4)$ |
| -1 | $f(-1)=-2(-1)=2$ | $(-1,2)$ |
| 0 | $f(0)=-2(0)=0$ | $(0,0)$ |
| 1 | $f(1)=-2(1)=-2$ | $(1,-2)$ |
| 2 | $f(2)=-2(2)=-4$ | $(2,-4)$ |

| $x$ | $g(x)=-2x-1$ | $(x,y)$ |
|---|---|---|
| -2 | $g(-2)=-2(-2)-1$ $=4-1=3$ | $(-2,3)$ |
| -1 | $g(-1)=-2(-1)-1$ $=2-1=1$ | $(-1,1)$ |
| 0 | $g(0)=-2(0)-1$ $=0-1=-1$ | $(0,-1)$ |
| 1 | $g(1)=-2(1)-1$ $=-2-1=-3$ | $(1,-3)$ |
| 2 | $g(2)=-2(2)-1$ $=-4-1=-5$ | $(2,-5)$ |

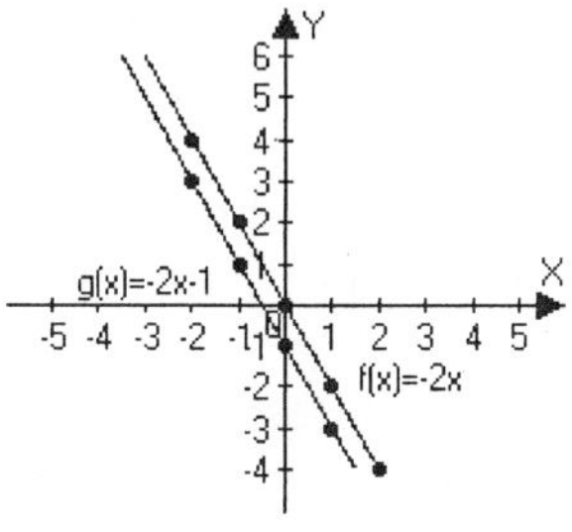


The graph of $g$ is the graph of $f$ shifted down 1 unit.

**23.**

| $x$ | $f(x)=x^2$ | $(x,y)$ |
|---|---|---|
| -2 | $f(-2)=(-2)^2=4$ | $(-2,4)$ |
| -1 | $f(-1)=(-1)^2=1$ | $(-1,1)$ |
| 0 | $f(0)=(0)^2=0$ | $(0,0)$ |
| 1 | $f(1)=(1)^2=1$ | $(1,1)$ |
| 2 | $f(2)=(2)^2=4$ | $(2,4)$ |

| $x$ | $g(x)=x^2+1$ | $(x,y)$ |
|---|---|---|
| -2 | $g(-2)=(-2)^2+1$ $=4+1=5$ | $(-2,5)$ |
| -1 | $g(-1)=(-1)^2+1$ $=1+1=2$ | $(-1,2)$ |
| 0 | $g(0)=(0)^2+1$ $=0+1=1$ | $(0,1)$ |
| 1 | $g(1)=(1)^2+1$ $=1+1=2$ | $(1,2)$ |
| 2 | $g(2)=(2)^2+1$ $=4+1=5$ | $(2,5)$ |

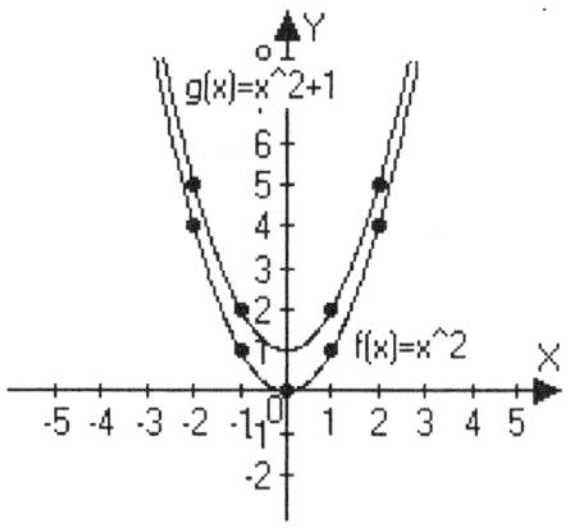


The graph of $g$ is the graph of $f$ shifted up 1 unit.

**25.**

| $x$ | $f(x)=\lvert x\rvert$ | $(x, y)$ |
|---|---|---|
| $-2$ | $f(-2)=\lvert -2\rvert=2$ | $(-2,2)$ |
| $-1$ | $f(-1)=\lvert -1\rvert=1$ | $(-1,1)$ |
| $0$ | $f(0)=\lvert 0\rvert=0$ | $(0,0)$ |
| $1$ | $f(1)=\lvert 1\rvert=1$ | $(1,1)$ |
| $2$ | $f(2)=\lvert 2\rvert=2$ | $(2,2)$ |

| $x$ | $g(x)=\lvert x\rvert-2$ | $(x, y)$ |
|---|---|---|
| $-2$ | $g(-2)=\lvert -2\rvert-2$ <br> $=2-2=0$ | $(-2,0)$ |
| $-1$ | $g(-1)=\lvert -1\rvert-2$ <br> $=1-2=-1$ | $(-1,-1)$ |
| $0$ | $g(0)=\lvert 0\rvert-2$ <br> $=0-2=-2$ | $(0,-2)$ |
| $1$ | $g(1)=\lvert 1\rvert-2$ <br> $=1-2=-1$ | $(1,-1)$ |
| $2$ | $g(2)=\lvert 2\rvert-2$ <br> $=2-2=0$ | $(2,0)$ |

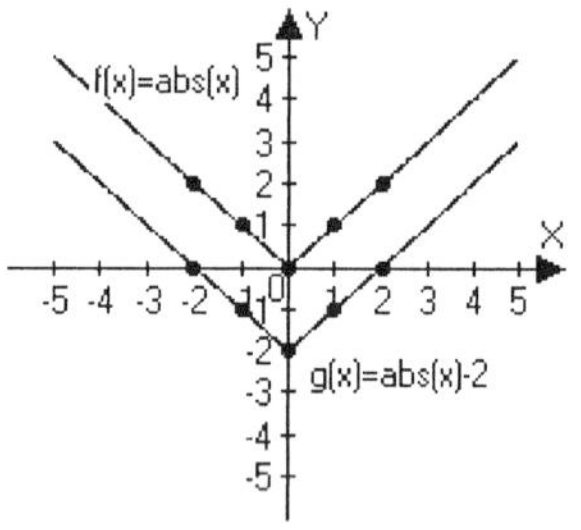


The graph of $g$ is the graph of $f$ shifted down 2 units.

**27.**

| $x$ | $f(x)=x^3$ | $(x, y)$ |
|---|---|---|
| $-2$ | $f(-2)=(-2)^3=-8$ | $(-2,-8)$ |
| $-1$ | $f(-1)=(-1)^3=-1$ | $(-1,-1)$ |
| $0$ | $f(0)=(0)^3=0$ | $(0,0)$ |
| $1$ | $f(1)=(1)^3=1$ | $(1,1)$ |
| $2$ | $f(2)=(2)^3=8$ | $(2,8)$ |

| $x$ | $g(x)=x^3+2$ | $(x, y)$ |
|---|---|---|
| $-2$ | $g(-2)=(-2)^3+2$ <br> $=-8+2=-6$ | $(-2,-6)$ |
| $-1$ | $g(-1)=(-1)^3+2$ <br> $=-1+2=1$ | $(-1,1)$ |
| $0$ | $g(0)=(0)^3+2$ <br> $=0+2=2$ | $(0,2)$ |
| $1$ | $g(1)=(1)^3+2$ <br> $=1+2=3$ | $(1,3)$ |
| $2$ | $g(2)=(2)^3+2$ <br> $=8+2=10$ | $(2,10)$ |

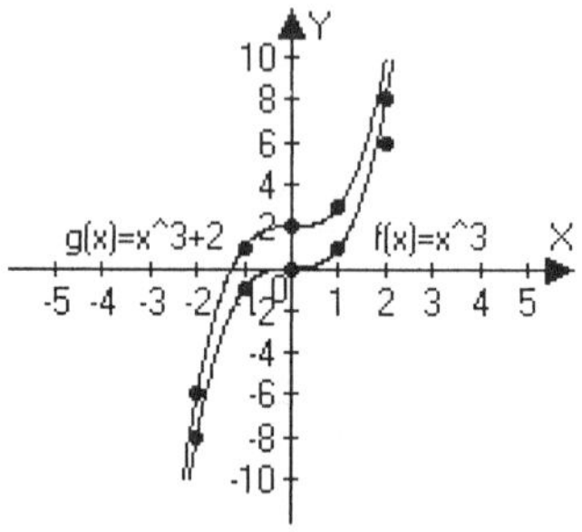


The graph of $g$ is the graph of $f$ shifted up 2 units.

**29.** The vertical line test shows that the graph represents a function.

**31.** The vertical line test shows that the graph does not represent a function.

**33.** The vertical line test shows that the graph represents a function.

**35.** The vertical line test shows that the graph does not represent a function.

**37.** $f(-2)=-4$

**39.** $f(4)=4$

**41.** $f(-3)=0$

**43.** $g(-4)=2$

**45.** $g(-10)=2$

**47.** When $x=-2$, $g(x)=1$.

**49.** The domain is $\{x\,|\,0\le x<5\}$.
The range is $\{y\,|\,-1\le y<5\}$.

**51.** The domain is $\{x\,|\,x\ge 0\}$.
The range is $\{y\,|\,y\ge 1\}$.

**53.** The domain is $\{x\,|\,-2\le x\le 6\}$.
The range is $\{y\,|\,-2\le y\le 6\}$.

**55.** The domain is
$\{x\,|\,x \text{ is a real number}\}$.
The range is $\{y\,|\,y\le -2\}$.

**57.** The domain is $\{x\,|\,x=-5,-2,0,1,3\}$.
The range is $\{y\,|\,y=2\}$.

**59.**
$$g(1)=3(1)-5=3-5=-2$$
$$f(g(1))=f(-2)=(-2)^2-(-2)+4$$
$$=4+2+4=10$$

**60.**
$$g(-1)=3(-1)-5=-3-5=-8$$
$$f(g(-1))=f(-8)=(-8)^2-(-8)+4$$
$$=64+8+4=76$$

**61.**
$$\sqrt{3-(-1)}-(-6)^2+6\div(-6)\cdot 4$$
$$=\sqrt{3+1}-36+6\div(-6)\cdot 4$$
$$=\sqrt{4}-36+-1\cdot 4$$
$$=2-36+-4$$
$$=-34+-4$$
$$=-38$$

**62.**
$$|-4-(-1)|-(-3)^2+-3\div 3\cdot -6$$
$$=|-4+1|-9+-3\div 3\cdot -6$$
$$=|-3|-9+-1\cdot -6$$
$$=3-9+6=-6+6=0$$

**63.**
$$f(-x)-f(x)$$
$$=(-x)^3+(-x)-5-(x^3+x-5)$$
$$=-x^3-x-5-x^3-x+5=-2x^3-2x$$

**64.**
$$f(-x)-f(x)$$
$$=(-x)^2-3(-x)+7-(x^2-3x+7)$$
$$=x^2+3x+7-x^2+3x-7$$
$$=6x$$

**65.** **a.** $f(-2)=3(-2)+5=-6+5=-1$

**b.** $f(0)=4(0)+7=0+7=7$

**c.** $f(3)=4(3)+7=12+7=19$

**d.**
$$f(-100)+f(100)$$
$$=3(-100)+5+4(100)+7$$
$$=-300+5+400+7$$
$$=100+12$$
$$=112$$

**66.** **a.** $f(-3)=6(-3)-1=-18-1=-19$

**b.** $f(0)=7(0)+3=0+3=3$

**c.** $f(4)=7(4)+3=28+3=31$

**d.** $f(-100)+f(100)$
$=6(-100)-1+7(100)+3$
$=-600-1+700+3$
$=100+2=102$

**67.** The domain is $\{x\,|\,x\neq 1\}$.
The range is $\{y\,|\,y\neq 0\}$.

**68.** The domain is
$\{x\,|\,x$ is a real number$\}$.
The range is $\{y\,|\,y>0\}$.

**69.** a. {(EL,1%),(L,7%),(SL,11%),(M,52%),(SC,13%),(C,13%),(EC,3%)}

b. Yes, the relation in part (a) is a function because each ideology corresponds to exactly one percentage.

c. {(1%,EL),(7%,L),(11%,SL),(52%,M),(13%,SC),(13%,C),(3%,EC)}

d. No, the relation in part (b) is not a function because 13% in the domain corresponds to two ideologies, SC and C, in the range.

**71.** $W(16)=0.07(16)+4.1$
$=1.12+4.1=5.22$
In 2000 there were 5.22 million women enrolled in U.S. colleges. (2000,5.22)

**73.** $W(20)=0.07(20)+4.1$
$=1.4+4.1=5.5$
$M(20)=0.01(20)+3.9$
$=0.2+3.9=4.1$
$W(20)-M(20)=5.5-4.1=1.4$
In 2004, there will be 1.4 million more women than men enrolled in U.S. colleges.

**75.** $f(20)=0.4(20)^2-36(20)+1000$
$=0.4(400)-720+1000$
$=160-720+1000$
$=-560+1000=440$

Twenty-year-old drivers have 440 accidents per 50 million miles driven. (20,440)

**77.** The graph reaches its lowest point at $x=45$.
$f(45)=0.4(45)^2-36(45)+1000$
$=0.4(2025)-1620+1000$
$=810-1620+1000$
$=-810+1000=190$
Drivers at age 45 have 190 accidents per 50 million miles driven. This is the least number of accidents for any driver between ages 16 and 74.

**79.** $f(60)\approx 3.1$
In 1960, 3.1% of the U.S. population was made up of Jewish Americans.

**81.** In 1919 and 1964, $f(x)=3$. This means that in 1919 and 1964, 3% of the U.S. population was made up of Jewish Americans.

**83.** The percentage of Jewish Americans in the U.S. population reached a maximum in 1940. Using the graph to estimate, approximately 3.7% of the U.S. population was Jewish American.

**85.** Each year is paired with exactly one percentage. This means that each member of the domain is paired with one member of the range.

**87.** $f(3) = 0.83$
The cost of mailing a first-class letter weighing 3 ounces is $0.83.

**89.** The cost to mail a letter weighing 1.5 ounces is $0.60.

**91. - 99.** Answers will vary.

**101.** Statement **d.** is true. A horizontal line can intersect the graph of a function in more than one point.

Statement **a.** is false. All relations are not functions.

Statement **b.** is false. Functions can have ordered pairs with the same second component. It is the first component that cannot be duplicated.

Statement **c.** is false. The graph of a vertical line is not a function.

**103.** Answers will vary. An example is $\{(3,3),(4,3)\}$

**105.**
$$\begin{aligned}&24 \div 4\left[2-(5-2)\right]^2 - 6\\&= 24 \div 4\left[2-(3)\right]^2 - 6\\&= 24 \div 4[-1]^2 - 6\\&= 6[1] - 6 = 6 - 6 = 0\end{aligned}$$

**106.** $$\left(\frac{3x^2y^{-2}}{y^3}\right)^{-2} = \left(\frac{3x^2}{y^5}\right)^{-2} = \left(\frac{y^5}{3x^2}\right)^2 = \frac{y^{10}}{9x^4}$$

**107.**
$$\begin{aligned}\frac{x}{3} &= \frac{3x}{5} + 4\\15\left(\frac{x}{3}\right) &= 15\left(\frac{3x}{5} + 4\right)\\15\left(\frac{x}{3}\right) &= 15\left(\frac{3x}{5}\right) + 15(4)\\5x &= 3(3x) + 60\\5x &= 9x + 60\\5x - 9x &= 9x - 9x + 60\\-4x &= 60\\\frac{-4x}{-4} &= \frac{60}{-4}\\x &= -15\end{aligned}$$
The solution is –15 and the solution set is $\{-15\}$.

## 2.2 Exercise Set

**1.** The domain of $f$ is $\{x|x \text{ is a real number}\}$.

**3.** The domain of $g$ is $\{x|x$ is a real number and $x \neq -4\}$.

**5.** The domain of $f$ is $\{x|x$ is a real number and $x \neq 3\}$.

**7.** The domain of $g$ is $\{x|x$ is a real number and $x \neq 5\}$.

**9.** The domain of $f$ is $\{x|x$ is a real number and $x \neq -7$ and $x \neq 9\}$.

**11. a.** $(f+g)(x) = (3x+1)+(2x-6)$
$= 3x+1+2x-6$
$= 5x-5$

**b.** $(f+g)(5) = 5(5)-5$
$= 25-5 = 20$

**13. a.** $(f+g)(x) = (x-5)+(3x^2)$
$= x-5+3x^2$
$= 3x^2+x-5$

**b.** $(f+g)(5) = 3(5)^2+5-5$
$= 3(25) = 75$

**15. a.** $(f+g)(x)$
$= (2x^2-x-3)+(x+1)$
$= 2x^2-x-3+x+1 = 2x^2-2$

**b.** $(f+g)(5) = 2(5)^2-2$
$= 2(25)-2$
$= 50-2 = 48$

**17.** The domain of $f+g$ is $\{x|x$ is a real number$\}$.

**19.** The domain of $f+g$ is $\{x|x$ is a real number and $x \neq 5\}$.

**21.** The domain of $f+g$ is $\{x|x$ is a real number and $x \neq 0$ and $x \neq 5\}$.

**23.** The domain of $f+g$ is $\{x|x$ is a real number and $x \neq 2$ and $x \neq -3\}$.

**25.** Domain of $f+g$ is $\{x|x$ is a real number and $x \neq 2\}$.

**27.** Domain of $f+g$ is $\{x|x$ is a real number$\}$.

**29.** $(f+g)(x) = f(x)+g(x)$
$= x^2+4x+2-x$
$= x^2+3x+2$
$(f+g)(3) = (3)^2+3(3)+2$
$= 9+9+2 = 20$

**31.** $f(-2) = (-2)^2+4(-2)$
$= 4+(-8) = -4$
$g(-2) = 2-(-2) = 2+2 = 4$
$f(-2)+g(-2) = -4+4 = 0$

**33.** $(f-g)(x) = f(x)-g(x)$
$= (x^2+4x)-(2-x)$
$= x^2+4x-2+x$
$= x^2+5x-2$
$(f-g)(5) = (5)^2+5(5)-2$
$= 25+25-2 = 48$

**35.** From Exercise 31, we know $f(-2) = -4$, and $g(-2) = 4$.
$f(-2)-g(-2) = -4-4 = -8$

**37.** From Exercise 31, we know $f(-2) = -4$, and $g(-2) = 4$.

$(fg)(-2) = f(-2) \cdot g(-2)$
$= -4(4) = -16$

**39.** $f(5) = (5)^2 + 4(5) = 25 + 20 = 45$
$g(5) = 2 - 5 = -3$
$(fg)(5) = f(5) \cdot g(5)$
$= 45(-3) = -135$

**41.** $\left(\frac{f}{g}\right)(x) = \frac{f(x)}{g(x)} = \frac{x^2 + 4x}{2 - x}$

$\left(\frac{f}{g}\right)(1) = \frac{(1)^2 + 4(1)}{2 - (1)} = \frac{1+4}{1} = \frac{5}{1} = 5$

**43.** $\left(\frac{f}{g}\right)(-1) = \frac{(-1)^2 + 4(-1)}{2 - (-1)}$
$= \frac{1-4}{3} = \frac{-3}{3} = -1$

**45.** The domain of $f + g$ is $\{x | x$ is a real number$\}$.

**47.** $\left(\frac{f}{g}\right)(x) = \frac{f(x)}{g(x)} = \frac{x^2 + 4x}{2 - x}$

The domain of $\frac{f}{g}$ is $\{x | x$ is a real number and $x \neq 2\}$.

**49.** $(f + g)(-3) = f(-3) + g(-3)$
$4 + 1 = 5$

**50.** $(g - f)(-2) = g(-2) - f(-2)$
$= 2 - 3 = -1$

**51.** $(fg)(2) = f(2)g(2) = (-1)(1) = -1$

**52.** $\left(\frac{g}{f}\right)(3) = \frac{g(3)}{f(3)} = \frac{0}{-3} = 0$

**53.** The domain of $f + g$ is $\{x | -4 \le x \le 3\}$.

**54.** The domain of $\frac{f}{g}$ is $\{x | -4 < x < 3\}$

**55.** The graph of $f + g$

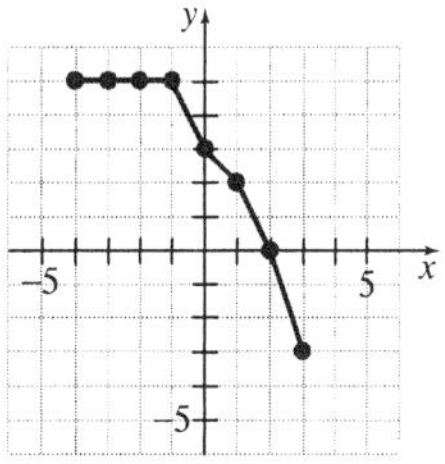


**56.** The graph of $f - g$

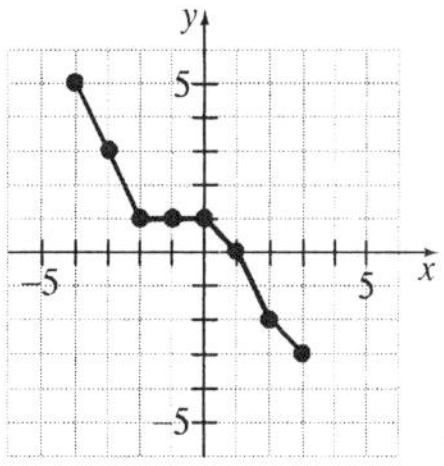


**57.** $(f + g)(1) - (g - f)(-1)$
$= f(1) + g(1) - [g(-1) - f(-1)]$
$= f(1) + g(1) - g(-1) + f(-1)$
$= -6 + -3 - (-2) + 3$
$= -6 + -3 + 2 + 3 = -4$

**58.** $(f + g)(-1) - (g - f)(0)$
$= f(-1) + g(-1) - [g(0) - f(0)]$
$= 3 + (-2) - [4 - (-2)]$
$= 3 + -2 - (4 + 2) = 3 + -2 - 6 = -5$

**59.** $(fg)(-2)-\left[\left(\frac{f}{g}\right)(1)\right]^2$

$= f(-2)g(-2)-\left[\frac{f(1)}{g(1)}\right]^2$

$= 5\cdot 0-\left[\frac{-6}{-3}\right]^2 = 0-2^2 = 0-4 = -4$

**60.** $(fg)(2)-\left[\left(\frac{g}{f}\right)(0)\right]^2$

$= f(2)g(2)-\left[\frac{g(0)}{f(0)}\right]^2$

$= 0(1)-\left(\frac{4}{-2}\right)^2 = 0-(-2)^2$

$= 0-4 = -4$

**61.** The domain is $[x \mid x = 0,1,2,...,7]$.

**63.** **a.** $(B-D)(x)$

$= 24,770x + 3,873,266$
$\quad - (20,205x + 2,294,970)$

$= 24,770x + 3,873,266$
$\quad - 20,205x - 2,294,970$

$= 4565x + 1,578,296$

This function represents the change in population; i.e. the difference between births and deaths in the U.S.

**b.** $(B-D)(6) = 4565(6) + 1,578,296$
$= 1,605,686$

This means there was an increase in the population of 1,605,686 in 2001.

**c.** Using the table,

$(B-D)(6) = 4,025,933 - 2,416,425$
$= 1,609,508$

This function models the data in the table fairly well since the difference in the answers to part (b) and part (c) is fairly small.

**65.** $(f+g)(x)$ represents the total world population, $h(x)$.

**67.** $(f+g)(2000) = f(2000) + g(2000)$
$= h(2000) = 7.5$

This means that the total world population is 7.5 billion.

**69.** First, find $(R-C)(x)$.

$(R-C)(x) = 65x - (600,000 + 45x)$
$= 65x - 600,000 - 45x$
$= 20x - 600,000$

$(R-C)(20,000)$
$= 20(20,000) - 600,000$
$= 400,000 - 600,000 = -200,000$

This means that if the company produces and sells 20,000 radios, it will lose \$200,000.

$(R-C)(30,000)$
$= 20(30,000) - 600,000$
$= 600,000 - 600,000 = 0$

If the company produces and sells 30,000 radios, it will break even with its costs equal to its revenue.

$(R-C)(40,000)$
$= 20(40,000) - 600,000$
$= 800,000 - 600,000 = 200,000$

This means that if the company produces and sells 40,000 radios, it will make a profit of \$200,000.

**71. – 75.** Answers will vary.

**77.** $y_1 = x - 4$ $\quad y_2 = 2x$

$y_3 = y_1 - y_2$

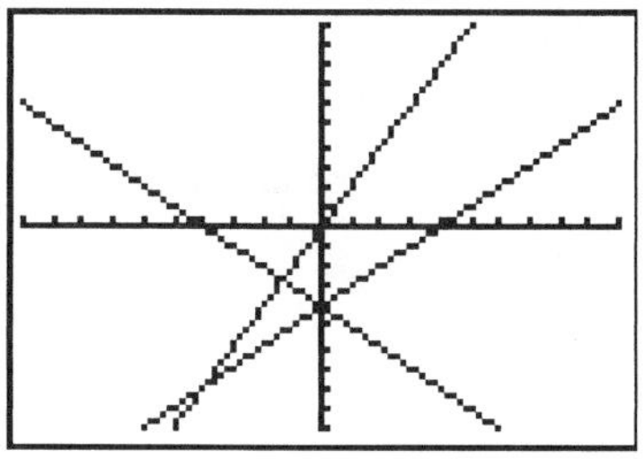

**79.** $y_1 = x^2 - 2x$ $\quad y_2 = x$

$y_3 = \dfrac{y_1}{y_2}$

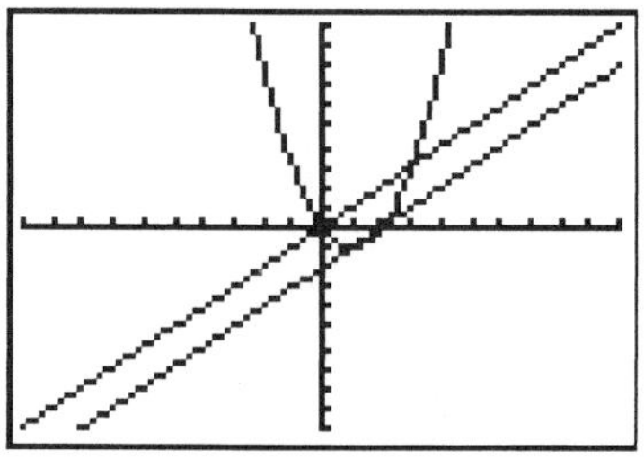

**81.** Statement **d.** is false. If $(fg)(a) = 0$, then $f(a) = 0$ or $g(a) = 0$. If $g(a) = 0$, $(fg)(a) = 0$ regardless of the value of $f(a)$.

Statement **a.** is true. If $(f+g)(a) = 0$, then $f(a) + g(a) = 0$ and $f(a) = -g(a)$. This means that $f(a)$ and $g(a)$ are additive inverses.

Statement **b.** is true. If $(f-g)(a) = 0$, then $f(a) - g(a) = 0$ and $f(a) = g(a)$.

Statement **c.** is true. If $\left(\dfrac{f}{g}\right)(a) = 0$, then $\dfrac{f(a)}{g(a)} = 0$. The only way that $\dfrac{f(a)}{g(a)}$ can equal zero is if $f(a)$ equals zero. (Recall: $g(a)$ cannot equal zero because division by zero is meaningless.)

**83.**

$$R = 3(a+b)$$
$$R = 3a + 3b$$
$$R - 3a = 3b$$
$$\frac{R-3a}{3} = \frac{3b}{3}$$
$$b = \frac{R-3a}{3} \text{ or } \frac{R}{3} - a$$

**84.**

$$3(6-x) = 3 - 2(x-4)$$
$$18 - 3x = 3 - 2x + 8$$
$$18 - 3x = 11 - 2x$$
$$18 = 11 + x$$
$$7 = x$$

The solution is 7, and the solution set is $\{7\}$.

**75.**

$$f(b+2) = 6(b+2) - 4$$
$$= 6b + 12 - 4 = 6b + 8$$

## Mid-Chapter Check Point

**1.** The relation is not a function.
The domain is $\{1,2\}$.
The range is $\{-6,4,6\}$.

**2.** The relation is a function.
The domain is $\{0,2,3\}$.
The range is $\{1,4\}$.

**3.** The relation is a function.
The domain is $\{x \mid -2 \le x < 2\}$.
The range is $\{y \mid 0 \le y \le 3\}$.

**4.** The relation is not a function.
The domain is $\{x \mid -3 < x \le 4\}$.
The range is $\{y \mid -1 \le y \le 2\}$.

**5.** The relation is not a function.
The domain is $\{-2,-1,0,1,2\}$.
The range is $\{-2,-1,1,3\}$.

**6.** The relation is a function.
The domain is $\{x \mid x \le 1\}$.
The range is $\{y \mid y \ge -1\}$.

**7.** The graph of $f$ represents the graph of a function because every element in the domain is corresponds to exactly one element in the range. It passes the vertical line test.

**8.** $f(-4)=3$

**9.** The function $f(x)=4$ when $x=-2$.

**10.** The function $f(x)=0$ when $x=2$ and $x=-6$.

**11.** The domain of $f$ is $\{x \mid x \text{ is a real number}\}$.

**12.** The range of $f$ is $\{y \mid y \le 4\}$.

**13.** The domain is $\{x \mid x \text{ is a real number}\}$.

**14.** The domain of $g$ is
$\{x \mid x \text{ is a real number}$
$\text{and } x \ne -2 \text{ and } x \ne 2\}$.

**15.**
$$f(0)=0^2-3(0)+8=8$$
$$g(-10)=-2(-10)-5=20-5=15$$
$$f(0)+g(-10)=8+15=23$$

**16.**
$$f(-1)=(-1)^2-3(-1)+8$$
$$=1+3+8=12$$
$$g(3)=-2(3)-5=-6-5=-11$$
$$f(-1)-g(3)=12-(-11)$$
$$=12+11=23$$

**17.**
$$f(a)=a^2-3a+8$$
$$g(a+3)=-2(a+3)-5$$
$$=-2a-6-5=-2a-11$$
$$f(a)+g(a+3)=a^2-3a+8+-2a-11$$
$$=a^2-5a-3$$

**18.**
$$(f+g)(x)=x^2-3x+8+-2x-5$$
$$=x^2-5x+3$$
$$(f+g)(-2)=(-2)^2-5(-2)+3$$
$$=4+10+3=17$$

**19.**
$$(f-g)(x)=x^2-3x+8-(-2x-5)$$
$$=x^2-3x+8+2x+5$$
$$=x^2-x+13$$
$$(f-g)(5)=(5)^2-5+13$$
$$=25-5+13=33$$

**20.** $f(-1) = (-1)^2 - 3(-1) + 8$
$= 1 + 3 + 8 = 12$
$g(-1) = -2(-1) - 5 = 2 - 5 = -3$
$(fg)(-1) = 12(-3) = -36$

**21.** $\left(\dfrac{f}{g}\right)(x) = \dfrac{x^2 - 3x + 8}{-2x - 5}$

$\left(\dfrac{f}{g}\right)(-4) = \dfrac{(-4)^2 - 3(-4) + 8}{-2(-4) - 5}$

$= \dfrac{16 + 12 + 8}{8 - 5} = \dfrac{36}{3} = 12$

**22.** The domain of $\dfrac{f}{g}$ is $\left\{x \mid x \neq -\dfrac{5}{2}\right\}$.

## 2.3 Exercise Set

**1.** $x + y = 4$
Find the $x$–intercept by setting $y = 0$.
$x + y = 4$
$x + 0 = 4$
$x = 4$
Find the $y$–intercept by setting $x = 0$.
$x + y = 4$
$0 + y = 4$
$y = 4$

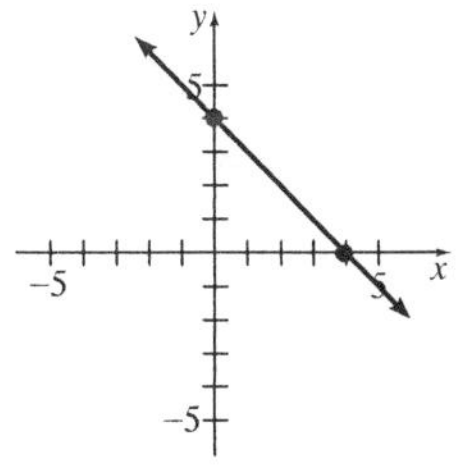


**3.** $x + 3y = 6$
Find the $x$–intercept by setting $y = 0$.
$x + 3(0) = 6$
$x = 6$
Find the $y$–intercept by setting $x = 0$.
$(0) + 3y = 6$
$y = 2$

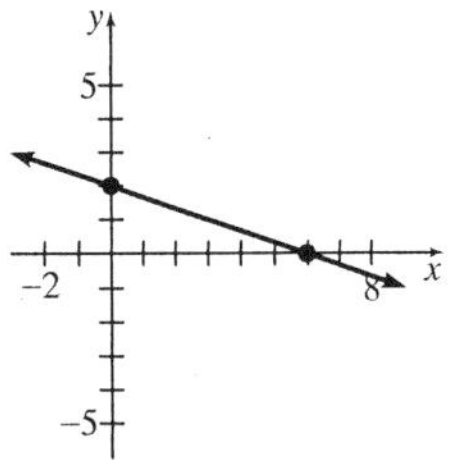


**5.** $6x - 2y = 12$
Find the $x$–intercept by setting $y = 0$.
$6x - 2(0) = 12$
$6x = 12$
$x = 2$
Find the $y$–intercept by setting $x = 0$.
$6(0) - 2y = 12$
$-2y = 12$
$y = -6$

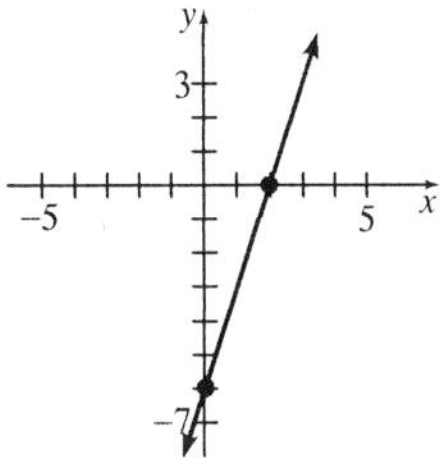

**7.** $3x - y = 6$

Find the $x$–intercept by setting $y = 0$.

$$3x - 0 = 6$$
$$3x = 6$$
$$x = 2$$

Find the $y$–intercept by setting $x = 0$.

$$3(0) - y = 6$$
$$-y = 6$$
$$y = -6$$

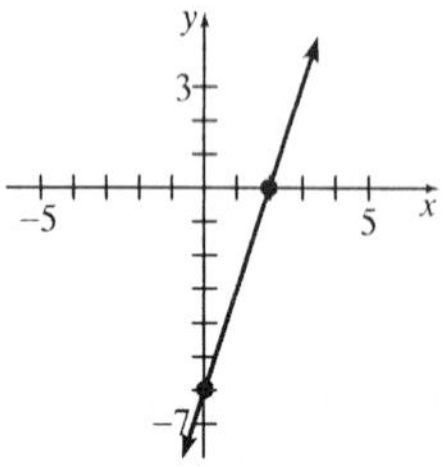


**9.** $x - 3y = 9$

Find the $x$–intercept by setting $y = 0$.

$$x - 3(0) = 9$$
$$x = 9$$

Find the $y$–intercept by setting $x = 0$.

$$(0) - 3y = 9$$
$$-3y = 9$$
$$y = -3$$

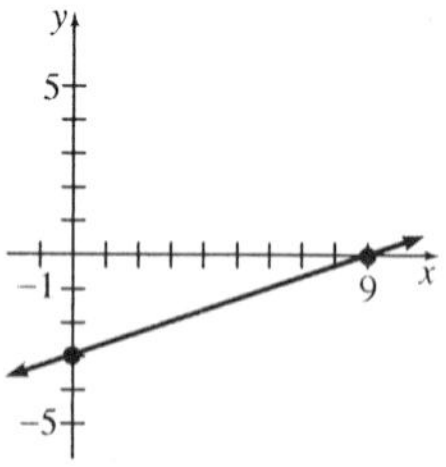


**11.** $2x = 3y + 6$

Find the $x$–intercept by setting $y = 0$.

$$2x = 3(0) + 6$$
$$2x = 6$$
$$x = 3$$

Find the $y$–intercept by setting $x = 0$.

$$2(0) = 3y + 6$$
$$0 = 3y + 6$$
$$-6 = 3y$$
$$-2 = y$$

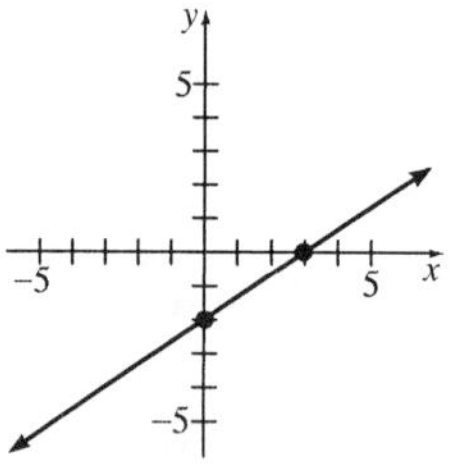


**13.** $6x - 3y = 15$

Find the $x$–intercept by setting $y = 0$.

$$6x - 3(0) = 15$$
$$6x = 15$$
$$x = \frac{15}{6} = \frac{5}{2}$$

Find the $y$–intercept by setting $x = 0$.

$$6(0) - 3y = 15$$
$$-3y = 15$$
$$y = -5$$

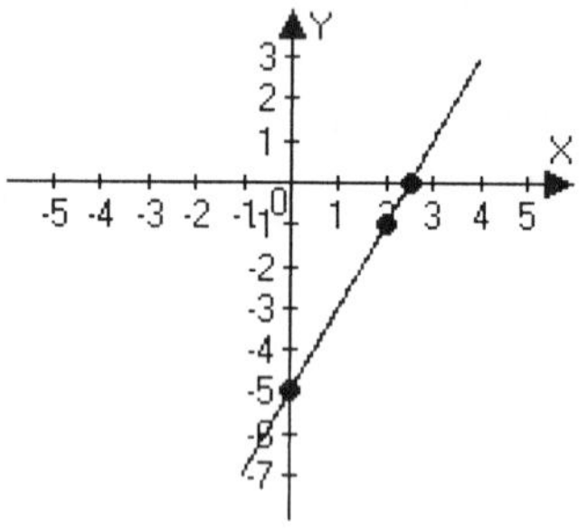


**15.** $m = \dfrac{8-4}{3-2} = \dfrac{4}{1} = 4$

The line rises.

**17.** $m = \dfrac{5-4}{2-(-1)} = \dfrac{1}{2+1} = \dfrac{1}{3}$

The line rises.

**19.** $m = \dfrac{5-5}{-1-2} = \dfrac{0}{-3} = 0$

The line is horizontal.

**21.** $m = \dfrac{-3-1}{-4-(-7)} = \dfrac{-4}{-4+7} = \dfrac{-4}{3} = -\dfrac{4}{3}$

The line falls.

**23.** $m = \dfrac{\frac{1}{4}-(-2)}{\frac{7}{2}-\frac{7}{2}} = \dfrac{\frac{1}{4}+2}{0} = \dfrac{\frac{9}{4}}{0}$

$m$ is undefined

The line is vertical.

**25.** Line 1 goes through $(-3,0)$ and $(0,2)$.

$m = \dfrac{2-0}{0-(-3)} = \dfrac{2}{3}$

Line 2 goes through $(2,0)$ and $(0,4)$.

$m = \dfrac{4-0}{0-2} = \dfrac{4}{-2} = -2$

Line 3 goes through $(0,-3)$ and $(2,-4)$.

$m = \dfrac{-4-(-3)}{2-0} = \dfrac{-4+3}{2} = \dfrac{-1}{2} = -\dfrac{1}{2}$

**27.** $y = 2x+1$

$m = 2$ $\quad$ $y-\text{intercept} = 1$

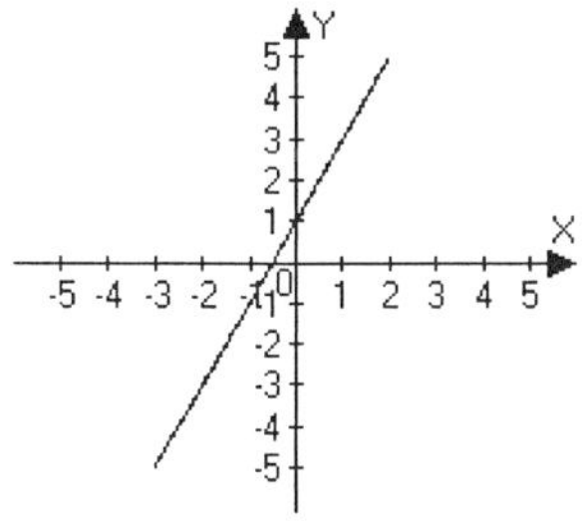


**29.** $y = -2x+1$

$m = -2$ $\quad$ $y-\text{intercept} = 1$

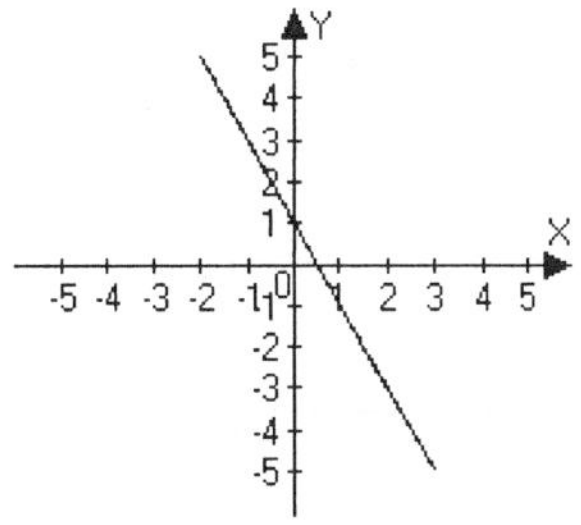


**31.** $f(x) = \dfrac{3}{4}x - 2$

$m = \dfrac{3}{4}$ $\quad$ $y-\text{intercept} = -2$

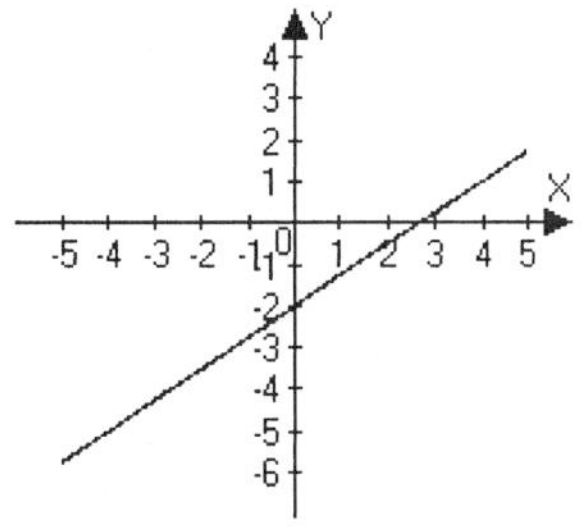


**33.** $f(x) = -\dfrac{3}{5}x + 7$

$m = -\dfrac{3}{5}$ $\quad$ $y-\text{intercept} = 7$

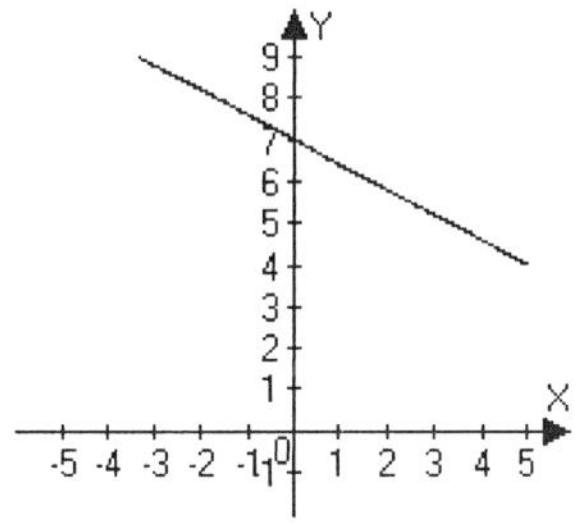

**35.** $y = -\frac{1}{2}x$

$m = -\frac{1}{2}$ $\qquad y - \text{intercept} = 0$

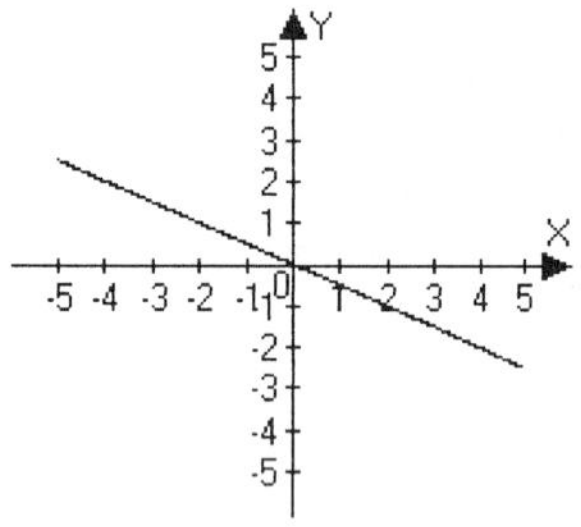


**37.** **a.** $2x + y = 0$

$y = -2x$

**b.** $m = -2$ $\qquad y - \text{intercept} = 0$

**c.**

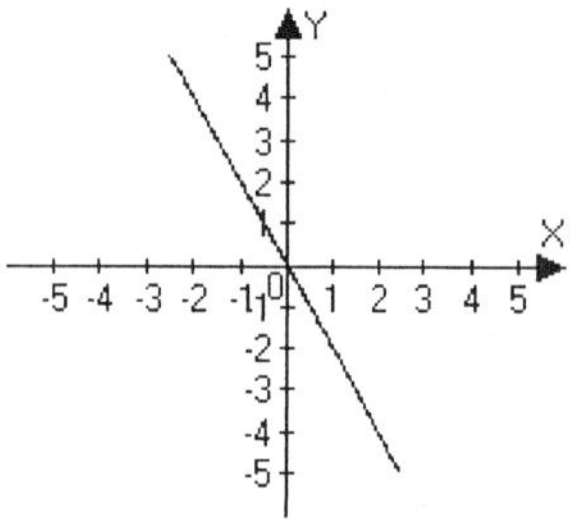


**39.** **a.** $5y = 4x$

$y = \frac{4}{5}x$

**b.** $m = \frac{4}{5}$ $\qquad y - \text{intercept} = 0$

**c.**

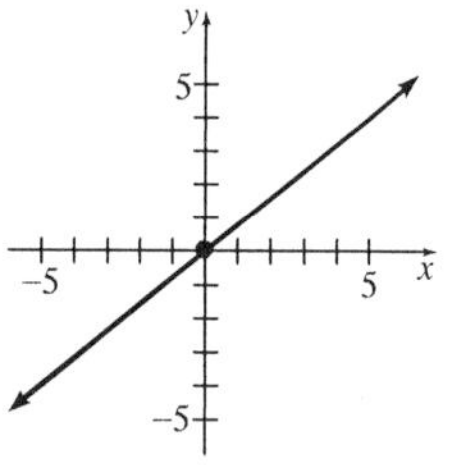


**41.** **a.** $3x + y = 2$

$y = -3x + 2$

**b.** $m = -3$ $\qquad y - \text{intercept} = 2$

**c.**

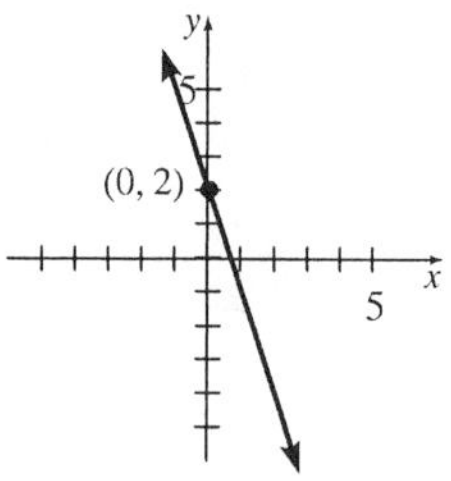


**43.** **a.** $5x + 3y = 15$

$3y = -5x + 15$

$y = -\frac{5}{3}x + 5$

**b.** $m = -\frac{5}{3}$ $\qquad y - \text{intercept} = 5$

**c.**

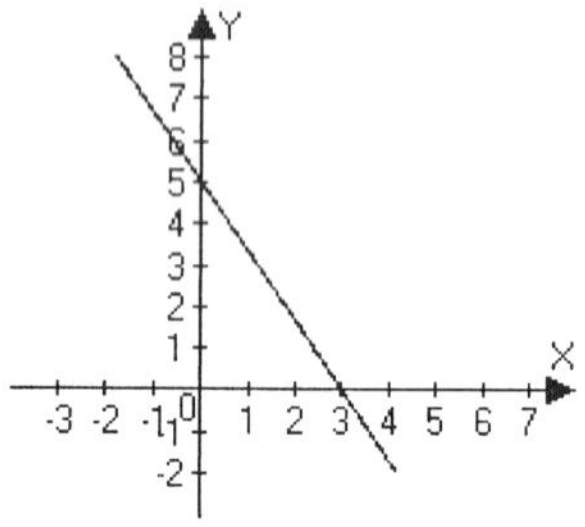


**45.** $y = 3$

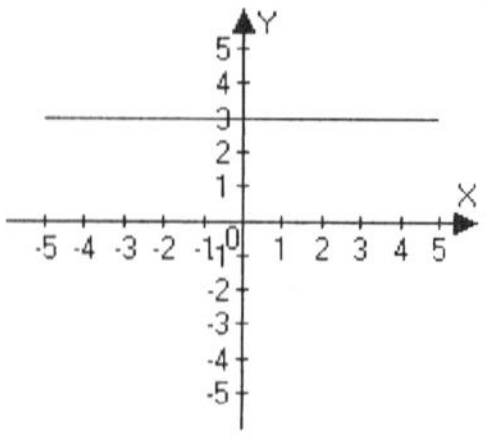

**47.** $f(x) = -2$

$y = -2$

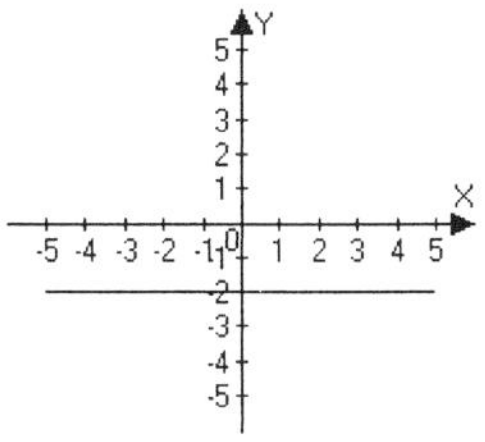


**49.** $3y = 18$

$y = 6$

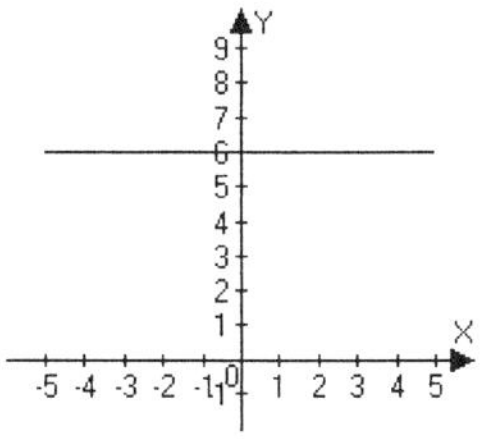


**51.** $f(x) = 2$

$y = 2$

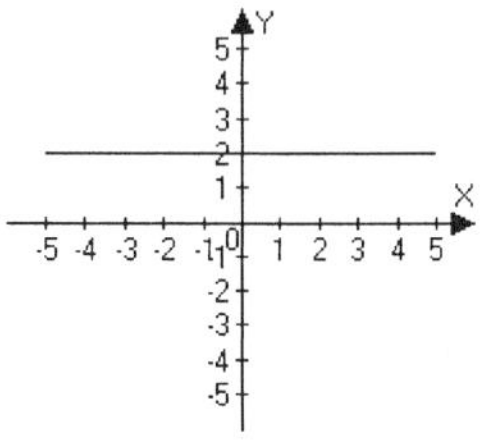


**53.** $x = 5$

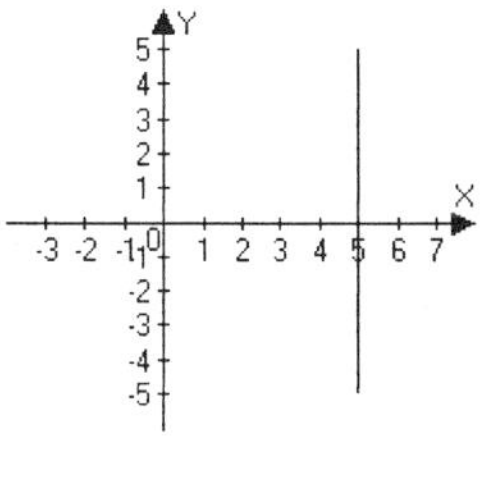


**55.** $3x = -12$

$x = -4$

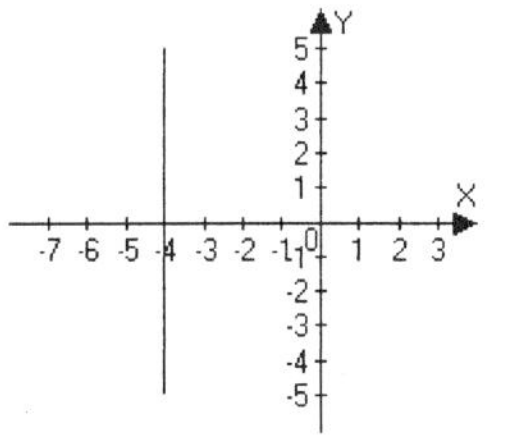


**57.** $x = 0$

This is the equation of the $y$–axis.

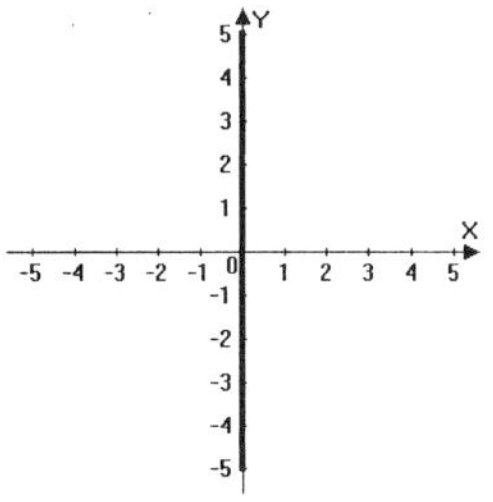


**59.** $$m = \frac{0-a}{b-0} = \frac{-a}{b} = -\frac{a}{b}$$

Since $a$ and $b$ are both positive, $-\frac{a}{b}$ is negative. Therefore, the line falls.

**60.** $$m = \frac{-b-0}{0-(-a)} = \frac{-b}{a} = -\frac{b}{a}$$

The line falls.

**61.** $$m = \frac{(b+c)-b}{a-a} = \frac{c}{0}$$

The slope is undefined.
The line is vertical.

**62.** $$m = \frac{(a+c)-c}{a-(a-b)} = \frac{a}{b}$$

The line rises.

**63.** $Ax + By = C$

$$By = -Ax + C$$

$$y = -\frac{A}{B}x + \frac{C}{B}$$

The slope is $-\frac{A}{B}$ and the $y-$intercept is $\frac{C}{B}$.

**64.** $Ax = By - C$

$$Ax + C = By$$

$$\frac{A}{B}x + \frac{C}{B} = y$$

The slope is $\frac{A}{B}$ and the $y-$intercept is $\frac{C}{B}$.

**65.** $-3 = \frac{4-y}{1-3}$

$$-3 = \frac{4-y}{-2}$$

$$6 = 4 - y$$

$$2 = -y$$

$$-2 = y$$

**66.** $\frac{1}{3} = \frac{-4-y}{4-(-2)}$

$$\frac{1}{3} = \frac{-4-y}{4+2}$$

$$\frac{1}{3} = \frac{-4-y}{6}$$

$$6 = 3(-4-y)$$

$$6 = -12 - 3y$$

$$18 = -3y$$

$$-6 = y$$

**67.** $3x - 4f(x) = 6$

$$-4f(x) = -3x + 6$$

$$f(x) = \frac{3}{4}x - \frac{3}{2}$$

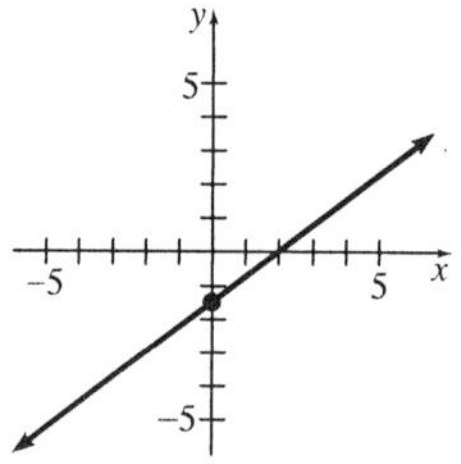


**68.** $6x - 5f(x) = 20$

$$-5f(x) = -6x + 20$$

$$f(x) = \frac{6}{5}x - 4$$

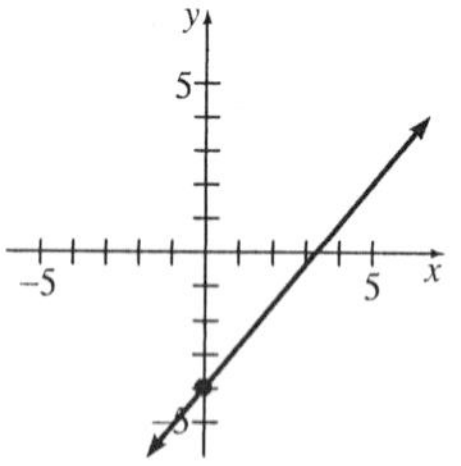


**69.** Using the slope-intercept form for the equation of a line:

$$-1 = -2(3) + b$$

$$-1 = -6 + b$$

$$5 = b$$

**70.** $-6 = -\frac{3}{2}(2) + b$

$$-6 = -3 + b$$

$$-3 = b$$

**71.** The line with slope $m_1$ is the steepest rising line so its slope is the biggest positive number. Then the line with slope $m_3$ is next because it is the only other line whose slope is positive. Since the line with slope $m_2$ is less steep but decreasing, it is next. The slope $m_4$ is the smallest because it is negative and the line with slope $m_4$ is steeper than the line with slope $m_3$, so its slope is more negative.
Decreasing order: $m_1, m_3, m_2, m_4$

**72.** Decreasing order: $b_2, b_1, b_4, b_3$

**73.** $m = -0.4$
This means that the percentage of U.S. men smoking has decreased at a rate of 0.4% per year after 1980.

**75.** $m = 0.01$
This means the global average temperature of Earth has increased at a rate of 0.01 °F per year since 1995.

**77.** $m = \dfrac{24.7 - 32.5}{2002 - 1997} = \dfrac{-7.8}{5} = -1.56$
The percentage of total sales of rock decreases on average 1.56% per year.

**79.**
a. $m = \dfrac{41 - 33}{4 - 0} = \dfrac{8}{4} = 2$

$b = 33$

$y = 2x + 33$

b. Since 2010 is 11 years after 1999,
$y = 2(11) + 33 = 55$
This model predicts there will be 55,000,000 HIV/AIDS cases in 2010.

**81.** Let $x$ be the number of years since 1995 and $S(x)$ be the average salary.
$S(x) = 1700x + 49{,}100$

**83.** Let $x$ be the number of years since 1996 and $A(x)$ be percentage of audited taxpayers.
$A(x) = -0.28x + 1.7$

**85.** Let $x$ be the number of years since 1980 and $f(x)$ be the percentage of U.S. households with televisions.
$f(x) = 98$

**87.**

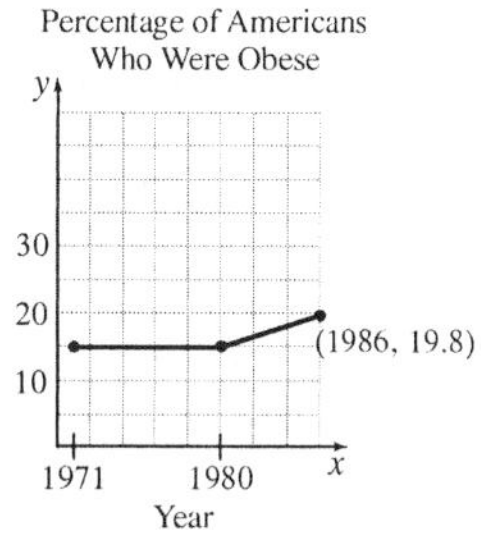


**89. – 105.** Answers will vary.

**107.** $y = -3x + 6$

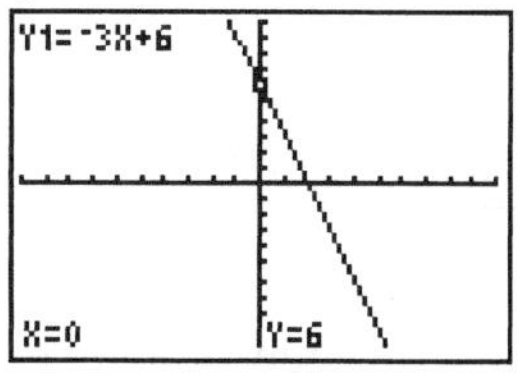


Two points found using [TRACE] are (0, 6) and (2, 0). Based on these points, the slope is:
$m = \dfrac{6 - 0}{0 - 2} = \dfrac{6}{-2} = -3.$
This is the same as the coefficient of $x$ in the line's equation.

**109.** $y = \frac{3}{4}x - 2$

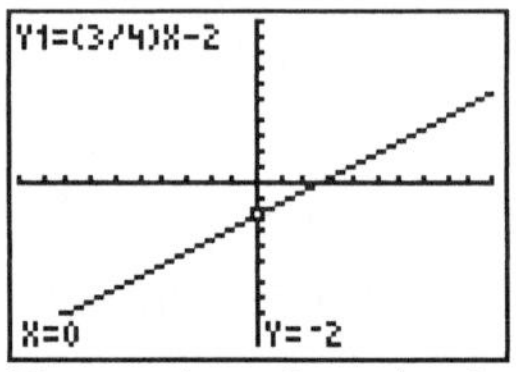


Two points found using [TRACE] are $(0,-2)$ and $(1,-1.25)$. Based on these points, the slope is:

$$m = \frac{-2-(-1.25)}{0-1} = \frac{-0.75}{-1} = 0.75.$$

This is equivalent to the coefficient of $x$ in the line's equation.

**111.** We are given that the $x$-intercept is $-2$ and the $y$-intercept is 4. We can use the points $(-2,0)$ and $(0,4)$ to find the slope.

$$m = \frac{4-0}{0-(-2)} = \frac{4}{0+2} = \frac{4}{2} = 2$$

Using the slope and one of the intercepts, we can write the line in point-slope form.

$$y - y_1 = m(x - x_1)$$
$$y - 0 = 2(x-(-2))$$
$$y = 2(x+2)$$
$$y = 2x + 4$$
$$-2x + y = 4$$

Find the $x$– and $y$–coefficients for the equation of the line with right-hand-side equal to 12. Multiply both sides of $-2x + y = 4$ by 3 to obtain 12 on the right-hand-side.

$$-2x + y = 4$$
$$3(-2x + y) = 3(4)$$
$$-6x + 3y = 12$$

The coefficients are –6 and 3.

**113.** **a.** $f(x_1 + x_2) = m(x_1 + x_2) + b$
$= mx_1 + mx_2 + b$

**b.** $f(x_1) + f(x_2)$
$= mx_1 + b + mx_2 + b$
$= mx_1 + mx_2 + 2b$

**c.** no

**114.** $$\left(\frac{4x^2}{y^{-3}}\right)^2 = (4x^2y^3)^2 = 4^2(x^2)^2(y^3)^2 = 16x^4y^6$$

**115.** $$(8\times10^{-7})(4\times10^3) = (8\times4)(10^{-7}\times10^3) = 32\times10^{-4} = (3.2\times10^1)\times10^{-4} = 3.2\times(10^1\times10^{-4}) = 3.2\times10^{-3}$$

**116.** $$5-[3(x-4)-6x] = 5-[3x-12-6x] = 5-3x+12+6x = 3x+17$$

## 2.4 Exercise Set

**1.** Slope $= 3$, passing through $(2,5)$

Point-Slope Form

$$y - y_1 = m(x - x_1)$$
$$y - 5 = 3(x-2)$$

Slope-Intercept Form

$$y - 5 = 3(x-2)$$
$$y - 5 = 3x - 6$$
$$y = 3x - 1$$
$$f(x) = 3x - 1$$

**3.** Slope $= 5$, passing through $(-2,6)$

Point-Slope Form

$$y - y_1 = m(x - x_1)$$
$$y - 6 = 5(x - (-2))$$
$$y - 6 = 5(x + 2)$$

Slope-Intercept Form

$$y - 6 = 5(x + 2)$$
$$y - 6 = 5x + 10$$
$$y = 5x + 16$$
$$f(x) = 5x + 16$$

**5.** Slope $= -4$, passing through $(-3,-2)$

Point-Slope Form

$$y - y_1 = m(x - x_1)$$
$$y - (-2) = -4(x - (-3))$$
$$y + 2 = -4(x + 3)$$

Slope-Intercept Form

$$y + 2 = -4(x + 3)$$
$$y + 2 = -4x - 12$$
$$y = -4x - 14$$
$$f(x) = -4x - 14$$

**7.** Slope $= -5$, passing through $(-2,0)$

Point-Slope Form

$$y - y_1 = m(x - x_1)$$
$$y - 0 = -5(x - (-2))$$
$$y - 0 = -5(x + 2)$$

Slope-Intercept Form

$$y - 0 = -5(x + 2)$$
$$y = -5(x + 2)$$
$$y = -5x - 10$$
$$f(x) = -5x - 10$$

**9.** Slope $= -1$, passing through $\left(-2, -\frac{1}{2}\right)$

Point-Slope Form

$$y - y_1 = m(x - x_1)$$
$$y - \left(-\frac{1}{2}\right) = -1(x - (-2))$$
$$y + \frac{1}{2} = -1(x + 2)$$

Slope-Intercept Form

$$y + \frac{1}{2} = -1(x + 2)$$
$$y + \frac{1}{2} = -x - 2$$
$$y = -x - \frac{5}{2}$$
$$f(x) = -x - \frac{5}{2}$$

**11.** Slope $= \frac{1}{4}$, passing through $(0,0)$

Point-Slope Form

$$y - y_1 = m(x - x_1)$$
$$y - 0 = \frac{1}{4}(x - 0)$$

Slope-Intercept Form

$$y - 0 = \frac{1}{4}(x - 0)$$
$$y = \frac{1}{4}x$$
$$f(x) = \frac{1}{4}x$$

**13.** Slope $= -\frac{2}{3}$, passing through $(6,-4)$

Point-Slope Form

$$y - y_1 = m(x - x_1)$$
$$y-(-4) = -\frac{2}{3}(x-6)$$
$$y+4 = -\frac{2}{3}(x-6)$$

Slope-Intercept Form

$$y+4 = -\frac{2}{3}(x-6)$$
$$y+4 = -\frac{2}{3}x+4$$
$$y = -\frac{2}{3}x$$
$$f(x) = -\frac{2}{3}x$$

**15.** Passing through $(6,3)$ and $(5,2)$

First, find the slope:

$$m = \frac{2-3}{5-6} = \frac{-1}{-1} = 1$$

Then use the slope and one of the points to write the equation in point-slope form.

$$y - y_1 = m(x - x_1)$$
$$y-3 = 1(x-6)$$

or

$$y-2 = 1(x-5)$$

Slope-Intercept Form

$$y-2 = 1(x-5)$$
$$y-2 = x-5$$
$$y = x-3$$
$$f(x) = x-3$$

**17.** Passing through $(-2,0)$ and $(0,4)$

First, find the slope:

$$m = \frac{4-0}{0-(-2)} = \frac{4}{2} = 2$$

Then use the slope and one of the points to write the equation in point-slope form.

$$y - y_1 = m(x - x_1)$$
$$y-4 = 2(x-0)$$

or

$$y-0 = 2(x-(-2))$$
$$y-0 = 2(x+2)$$

Slope-Intercept Form

$$y-0 = 2(x+2)$$
$$y = 2x+4$$
$$f(x) = 2x+4$$

**19.** Passing through $(-6,13)$ and $(-2,5)$

First, find the slope:

$$m = \frac{5-13}{-2-(-6)} = \frac{-8}{-2+6} = \frac{-8}{4} = -2$$

Then use the slope and one of the points to write the equation in point-slope form.

$$y - y_1 = m(x - x_1)$$
$$y-5 = -2(x-(-2))$$
$$y-5 = -2(x+2)$$

or

$$y-13 = -2(x-(-6))$$
$$y-13 = -2(x+6)$$

Slope-Intercept Form

$$y-13 = -2(x+6)$$
$$y-13 = -2x-12$$
$$y = -2x+1$$
$$f(x) = -2x+1$$

**21.** Passing through (1,9) and (4,−2)

First, find the slope:

$$m=\frac{-2-9}{4-1}=\frac{-11}{3}=-\frac{11}{3}$$

Then use the slope and one of the points to write the equation in point-slope form.

$$y-y_1=m(x-x_1)$$
$$y-(-2)=-\frac{11}{3}(x-4)$$
$$y+2=-\frac{11}{3}(x-4)$$

or

$$y-9=-\frac{11}{3}(x-1)$$

Slope-Intercept Form

$$y-9=-\frac{11}{3}(x-1)$$
$$y-9=-\frac{11}{3}x+\frac{11}{3}$$
$$y=-\frac{11}{3}x+\frac{38}{3}$$
$$f(x)=-\frac{11}{3}x+\frac{38}{3}$$

**23.** Passing through (−2,−5) and (3,−5)

First, find the slope:

$$m=\frac{-5-(-5)}{4-(-2)}=\frac{0}{6}=0$$

Then use the slope and one of the points to write the equation in point-slope form.

$$y-y_1=m(x-x_1)$$
$$y-(-5)=0(x-3)$$
$$y+5=0(x-3)$$

or

$$y-(-5)=0(x-(-2))$$
$$y+5=0(x+2)$$

Slope-Intercept Form

$$y+5=0(x+2)$$
$$y+5=0$$
$$y=-5$$
$$f(x)=-5$$

**25.** Passing through $(7,8)$ with $x-\text{intercept}=3$

If the line has an $x-\text{intercept}=3$, it passes through the point $(3,0)$.

First, find the slope:

$$m=\frac{8-0}{7-3}=\frac{8}{4}=2$$

Then use the slope and one of the points to write the equation in point-slope form.

$$y-y_1=m(x-x_1)$$
$$y-0=2(x-3)$$
$$y-0=2(x-3)$$

or

$$y-8=2(x-7)$$

Slope-Intercept Form

$$y-8=2(x-7)$$
$$y-8=2x-14$$
$$y=2x-6$$
$$f(x)=2x-6$$

**27.** $x-\text{intercept}=2$ and
$y-\text{intercept}=-1$
If the line has an $x-\text{intercept}=2,$ it passes through the point $(2,0)$. If the line has a $y-\text{intercept}=-1,$ it passes through $(0,-1)$.
First, find the slope:
$$m=\frac{-1-0}{0-2}=\frac{-1}{-2}=\frac{1}{2}$$
Then use the slope and one of the points to write the equation in point-slope form.
$$y-y_1=m(x-x_1)$$
$$y-0=\frac{1}{2}(x-2)$$
or
$$y-(-1)=\frac{1}{2}(x-0)$$
$$y+1=\frac{1}{2}(x-0)$$
Slope-Intercept Form
$$y-(-1)=\frac{1}{2}(x-0)$$
$$y+1=\frac{1}{2}x$$
$$y=\frac{1}{2}x-1$$
$$f(x)=\frac{1}{2}x-1$$

**29.** For $y=5x,\ m=5$. A line parallel to this line would have the same slope, $m=5$. A line perpendicular to it would have slope $m=-\frac{1}{5}$.

**31.** For $y=-7x,\ m=-7$. A line parallel to this line would have the same slope, $m=-7$. A line perpendicular to it would have slope $m=\frac{1}{7}$.

**33.** For $y=\frac{1}{2}x+3,\ m=\frac{1}{2}$. A line parallel to this line would have the same slope, $m=\frac{1}{2}$. A line perpendicular to it would have slope $m=-2$.

**35.** For $y=-\frac{2}{5}x-1,\ m=-\frac{2}{5}$. A line parallel to this line would have the same slope, $m=-\frac{2}{5}$. A line perpendicular to it would have slope $m=\frac{5}{2}$.

**37.** To find the slope, we rewrite the equation in slope-intercept form.
$$4x+y=7$$
$$y=-4x+7$$
So, $m=-4$. A line parallel to this line would have the same slope, $m=-4$. A line perpendicular to it would have slope $m=\frac{1}{4}$.

**39.** To find the slope, we rewrite the equation in slope-intercept form.
$$2x+4y-8=0$$
$$4y=-2x+8$$
$$y=-\frac{1}{2}x+2$$
So, $m=-\frac{1}{2}$. A line parallel to this line would have the same slope, $m=-\frac{1}{2}$. A line perpendicular to it would have slope $m=2$.

**41.** To find the slope, we rewrite the equation in slope-intercept form.

$$2x-3y-5=0$$
$$-3y=-2x+5$$
$$y=\frac{2}{3}x-\frac{5}{3}$$

So, $m=\frac{2}{3}$. A line parallel to this line would have the same slope, $m=\frac{2}{3}$. A line perpendicular to it would have slope $m=-\frac{3}{2}$.

**43.** We know that $x=6$ is a vertical line with undefined slope. A line parallel to it would also be vertical with undefined slope. A line perpendicular to it would be horizontal with slope $m=0$.

**45.** Since $L$ is parallel to $y=2x$, we know it will have slope $m=2$. We are given that it passes through (4, 2). We use the slope and point to write the equation in point-slope form.

$$y-y_1=m(x-x_1)$$
$$y-2=2(x-4)$$

Solve for $y$ to obtain slope-intercept form.

$$y-2=2(x-4)$$
$$y-2=2x-8$$
$$y=2x-6$$

In function notation, the equation of the line is $f(x)=2x-6$.

**47.** Since $L$ is perpendicular to $y=2x$, we know it will have slope $m=-\frac{1}{2}$.

We are given that it passes through (2, 4). We use the slope and point to write the equation in point-slope form.

$$y-y_1=m(x-x_1)$$
$$y-4=-\frac{1}{2}(x-2)$$

Solve for $y$ to obtain slope-intercept form.

$$y-4=-\frac{1}{2}(x-2)$$
$$y-4=-\frac{1}{2}x+1$$
$$y=-\frac{1}{2}x+5$$

In function notation, the equation of the line is $f(x)=-\frac{1}{2}x+5$.

**49.** Since the line is parallel to $y=-4x+3$, we know it will have slope $m=-4$. We are given that it passes through (−8, −10 ). We use the slope and point to write the equation in point-slope form.

$$y-y_1=m(x-x_1)$$
$$y-(-10)=-4(x-(-8))$$
$$y+10=-4(x+8)$$

Solve for $y$ to obtain slope-intercept form.

$$y+10=-4(x+8)$$
$$y+10=-4x-32$$
$$y=-4x-42$$

In function notation, the equation of the line is $f(x)=-4x-42$.

**51.** Since the line is perpendicular to $y=\frac{1}{5}x+6$, we know it will have slope $m=-5$. We are given that it passes through (2, –3). We use the slope and point to write the equation in point-slope form.

$$y-y_1=m(x-x_1)$$
$$y-(-3)=-5(x-2)$$
$$y+3=-5(x-2)$$

Solve for $y$ to obtain slope-intercept form.

$$y+3=-5(x-2)$$
$$y+3=-5x+10$$
$$y=-5x+7$$

In function notation, the equation of the line is $f(x)=-5x+7$.

**53.** To find the slope, we rewrite the equation in slope-intercept form.

$$2x-3y-7=0$$
$$-3y=-2x+7$$
$$y=\frac{2}{3}x-\frac{7}{3}$$

Since the line is parallel to $y=\frac{2}{3}x-\frac{7}{3}$, we know it will have slope $m=\frac{2}{3}$. We are given that it passes through (–2, 2). We use the slope and point to write the equation in point-slope form.

$$y-y_1=m(x-x_1)$$
$$y-2=\frac{2}{3}(x-(-2))$$
$$y-2=\frac{2}{3}(x+2)$$

Solve for $y$ to obtain slope-intercept form.

$$y-2=\frac{2}{3}(x+2)$$
$$y-2=\frac{2}{3}x+\frac{4}{3}$$
$$y=\frac{2}{3}x+\frac{10}{3}$$

In function notation, the equation of the line is $f(x)=\frac{2}{3}x+\frac{10}{3}$.

**55.** To find the slope, we rewrite the equation in slope-intercept form.

$$x-2y-3=0$$
$$-2y=-x+3$$
$$y=\frac{1}{2}x-\frac{3}{2}$$

Since the line is perpendicular to $y=\frac{1}{2}x-\frac{3}{2}$, we know it will have slope $m=-2$. We are given that it passes through (4, –7). We use the slope and point to write the equation in point-slope form.

$$y-y_1=m(x-x_1)$$
$$y-(-7)=-2(x-4)$$
$$y+7=-2(x-4)$$

Solve for $y$ to obtain slope-intercept form.

$$y+7=-2(x-4)$$
$$y+7=-2x+8$$
$$y=-2x+1$$

In function notation, the equation of the line is $f(x)=-2x+1$.

**57.** Since the line is perpendicular to $x=6$ which is a vertical line, we know the graph of $f$ is a horizontal line with 0 slope. The graph of $f$ passes through $(-1,5)$, so the equation of $f$ is $f(x)=5$.

**58.** Since the line is perpendicular to $x=-4$ which is a vertical line, we know the graph of $f$ is a horizontal line with 0 slope. The graph of $f$ passes through $(-2,6)$, so the equation of $f$ is $f(x)=6$.

**59.** First we need to find the equation of the line with $x-$intercept of 2 and $y-$intercept of $-4$. This line will pass through $(2,0)$ and $(0,-4)$. We use these points to find the slope.

$$m=\frac{-4-0}{0-2}=\frac{-4}{-2}=2$$

Since the graph of $f$ is perpendicular to this line, it will have slope $m=-\frac{1}{2}$.

Use the point $(-6,4)$ and the slope $-\frac{1}{2}$ to find the equation of the line.

$$y-y_1=m(x-x_1)$$
$$y-4=-\frac{1}{2}(x-(-6))$$
$$y-4=-\frac{1}{2}(x+6)$$
$$y-4=-\frac{1}{2}x-3$$
$$y=-\frac{1}{2}x+1$$
$$f(x)=-\frac{1}{2}x+1$$

**60.** First we need to find the equation of the line with $x-$intercept of 3 and $y-$intercept of $-9$. This line will pass through $(3,0)$ and $(0,-9)$. We use these points to find the slope.

$$m=\frac{-9-0}{0-3}=\frac{-9}{-3}=3$$

Since the graph of $f$ is perpendicular to this line, it will have slope $m=-\frac{1}{3}$.

Use the point $(-5,6)$ and the slope $-\frac{1}{3}$ to find the equation of the line.

$$y-y_1=m(x-x_1)$$
$$y-6=-\frac{1}{3}(x-(-5))$$
$$y-6=-\frac{1}{3}(x+5)$$
$$y-6=-\frac{1}{3}x-\frac{5}{3}$$
$$y=-\frac{1}{3}x+\frac{13}{3}$$
$$f(x)=-\frac{1}{3}x+\frac{13}{3}$$

**61.** First put the equation $3x-2y=4$ in slope-intercept form.

$$3x-2y=4$$
$$-2y=-3x+4$$
$$y=\frac{3}{2}x-2$$

The equation of $f$ will have slope $-\frac{2}{3}$ since it is perpendicular to the line above and the same $y-$intercept $-2$.

So the equation of $f$ is

$$f(x)=-\frac{2}{3}x-2.$$

**62.** First put the equation $4x - y = 6$ in slope-intercept form.

$$4x - y = 6$$
$$-y = -4x + 6$$
$$y = 4x - 6$$

The equation of $f$ will have slope $-\frac{1}{4}$ since it is perpendicular to the line above and the same $y$-intercept $-6$.

So the equation of $f$ is $f(x) = -\frac{1}{4}x - 6.$

**63.** The graph of $f$ is just the graph of $g$ shifted down 2 units. So subtract 2 from the equation of $g(x)$ to obtain the equation of $f(x)$.

$$f(x) = g(x) - 2 = 4x - 3 - 2 = 4x - 5$$

**64.** The graph of $f$ is just the graph of $g$ shifted up 3 units. So add 3 to the equation of $g(x)$ to obtain the equation of $f(x)$.

$$f(x) = g(x) + 3 = 2x - 5 + 3 = 2x - 2$$

**65.** To find the slope of the line whose equation is $Ax + By = C$, put this equation in slope-intercept form by solving for $y$.

$$Ax + By = C$$
$$By = -Ax + C$$
$$y = -\frac{A}{B}x + \frac{C}{B}$$

The slope of this line is $m = -\frac{A}{B}$ so the slope of the line that is parallel to it is the same, $-\frac{A}{B}$.

**66.** From exercise 65, we know the slope of the line is $-\frac{A}{B}$. So the slope of the line that is perpendicular would be $\frac{B}{A}$.

**67.** **a.** First we must find the slope using $(10,16)$ and $(16,12.7)$.

$$m = \frac{12.7 - 16}{16 - 10} = -\frac{3.3}{6} = -0.55$$

Then use the slope and one of the points to write the equation in point-slope form.

$$y - y_1 = m(x - x_1)$$
$$y - 16 = -0.55(x - 10)$$

or

$$y - 12.7 = -0.55(x - 16)$$

**b.**
$$y - 16 = -0.55(x - 10)$$
$$y - 16 = -0.55x + 5.5$$
$$y = -0.55x + 21.5$$
$$f(x) = -0.55x + 21.5$$

**c.** $f(20) = -0.55(20) + 21.5 = 10.5$

The linear function predicts 10.5% of adult women will be on weight-loss diets in 2007.

**69.** **a.**

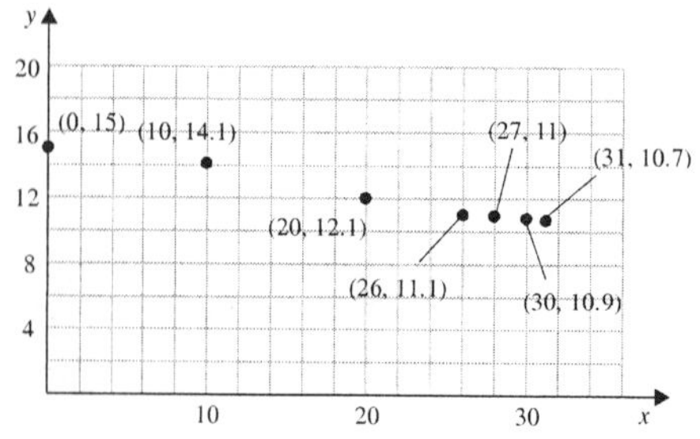

**b.**

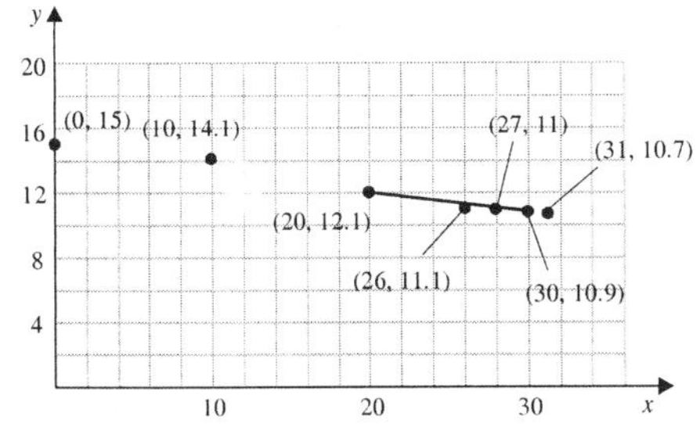


Use the two points $(20,12.1)$ and $(30,10.9)$ to find the slope.

$$m=\frac{12.1-10.9}{20-30}=\frac{1.2}{-10}=-0.12$$

Then use the slope and one of the points to write the equation in point-slope form.

$$y-y_1=m(x-x_1)$$
$$y-12.1=-0.12(x-20)$$

or

$$y-10.9=-0.12(x-30)$$
$$y-10.9=-0.12x+3.6$$
$$y=-0.12x+14.5$$
$$f(x)=-0.12x+14.5$$

**c.** $f(40)=-0.12(40)+14.5=9.7$

The equation predicts 9.7% of U.S. students will drop out of school in 2010.

**71.**

**a.** $m=\dfrac{34.2-24}{2001-1995}=\dfrac{10.2}{6}=1.7$

This means that stimulant use has increased at a rate of 1.7 per thousand children per year.

**b.** $m=\dfrac{16.4-8}{2001-1995}=\dfrac{8.4}{6}=1.4$

This means that antidepressant use has increased at a rate of 1.4 per thousand children per year.

**c.** The red and blue line segments do not lie on parallel lines; the slopes are not the same. This means stimulant use has increased at a faster rate than the use of antidepressant.

**73. – 79.** Answers will vary.

**81. a.**

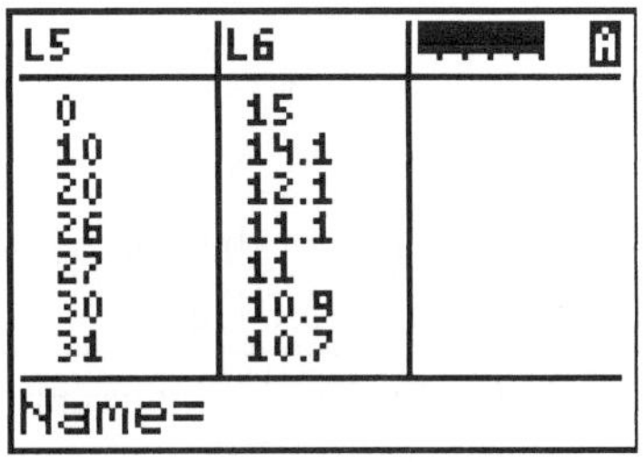

| L5 | L6 |
|---|---|
| 0 | 15 |
| 10 | 14.1 |
| 20 | 12.1 |
| 26 | 11.1 |
| 27 | 11 |
| 30 | 10.9 |
| 31 | 10.7 |

Name=

**b.**

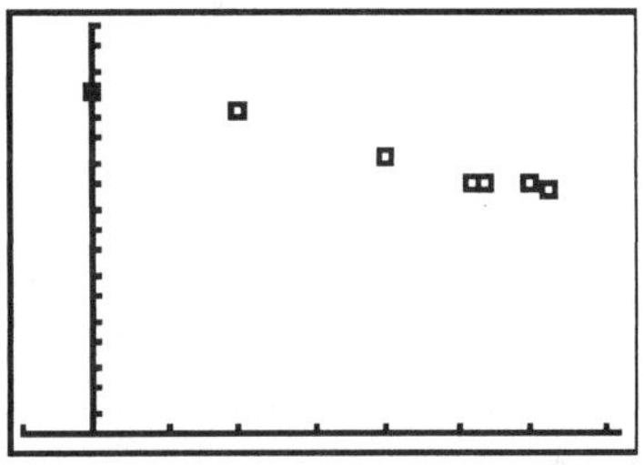

**c.**

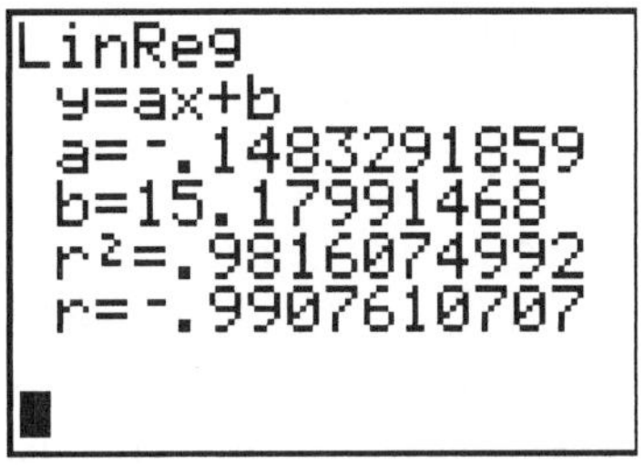


**d.**

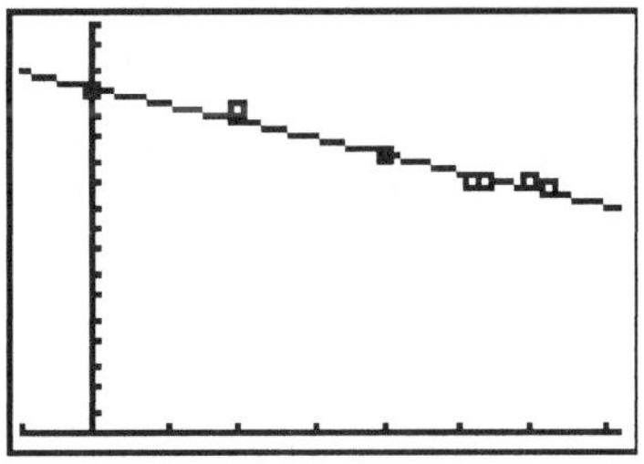

**83.** $By = 8x - 1$

$y = \frac{8}{B}x - 1$

Since $\frac{8}{B}$ is the slope, $\frac{8}{B}$ must equal -2.

$\frac{8}{B} = -2$

$8 = -2B$

$-4 = B$

**85.** Find the slope of the line by using the two points, $(-3,0)$, the $x-$intercept and $(0,-6)$, the $y-$intercept.

$m = \frac{-6-0}{0-(-3)} = \frac{-6}{3} = -2$

So the equation of the line is $y = -2x - 6$.

Substitute $-40$ for $x$:

$y = -2(-40) - 6 = 80 - 6 = 74$

This is the $y-$coordinate of the first ordered pair.

Substitute $-200$ for $y$:

$-200 = -2x - 6$

$-194 = -2x$

$97 = x$

This is the $x-$coordinate of the second ordered pair.

Therefore, the two ordered pairs are $(-40,74)$ and $(97,-200)$.

**78.** $f(-2) = 3(-2)^2 - 8(-2) + 5$

$= 3(4) + 16 + 5 = 12 + 16 + 5 = 33$

**79.** Since $(fg)(-1) = (f)(-1)\cdot(g)(-1)$, find $(f)(-1)$ and $(g)(-1)$.

$f(-1) = (-1)^2 - 3(-1) + 4$

$= 1 + 3 + 4 = 8$

$g(-1) = 2(-1) - 5 = -2 - 5 = -7$

$(fg)(-1) = (f)(-1)\cdot(g)(-1)$

$= 8(-7) = -56$

**80.** Let $x$ = the measure of the smallest angle

$x + 20$ = the measure of the second angle

$2x$ = the measure of the third angle

$x + (x+20) + 2x = 180$

$x + x + 20 + 2x = 180$

$4x + 20 = 180$

$4x = 160$

$x = 40$

Find the other angles.

$x + 20 = 40 + 20 = 60$

$2x = 2(40) = 80$

The angles are 40E, 60E and 80E.

## Chapter 2 Review Exercises

**1.** The relation is a function.
Domain {3, 4, 5}
Range {10}

**2.** The relation is a function.
Domain {1, 2, 3, 4}
Range {-6, B, 12, 100}

**3.** The relation is not a function.
Domain {13, 15}
Range {14, 16, 17}

**4.** **a.** $f(0)=7(0)-5=0-5=-5$

**b.** $f(3)=7(3)-5=21-5=16$

**c.** $f(-10)=7(-10)-5$
$=-70-5=-75$

**d.** $f(2a)=7(2a)-5=14a-5$

**e.** $f(a+2)=7(a+2)-5$
$=7a+14-5=7a+9$

**5.** **a.** $g(0)=3(0)^2-5(0)+2$
$=0-0+2=2$

**b.** $g(5)=3(5)^2-5(5)+2$
$=3(25)-25+2$
$=75-25+2=52$

**c.** $g(-4)=3(-4)^2-5(-4)+2$
$=3(16)+20+2$
$=48+20+2=70$

**d.** $g(b)=3(b)^2-5(b)+2$
$=3b^2-5b+2$

**e.** $g(4a)=3(4a)^2-5(4a)+2$
$=3(16a^2)-20a+2$
$=48a^2-20a+2$

**6.**

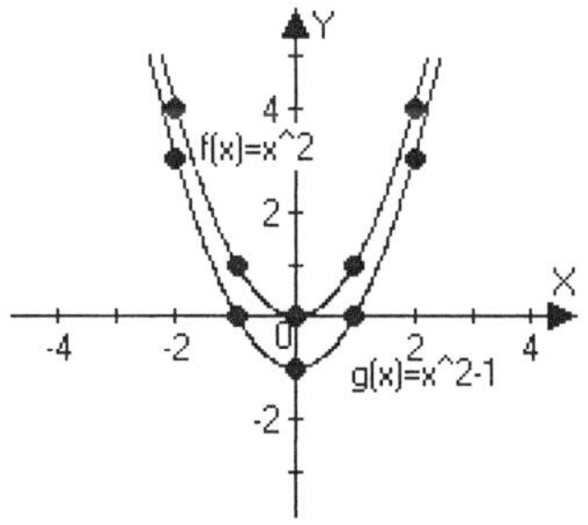


*g* shifts the graph of *f* down one unit

**7.**

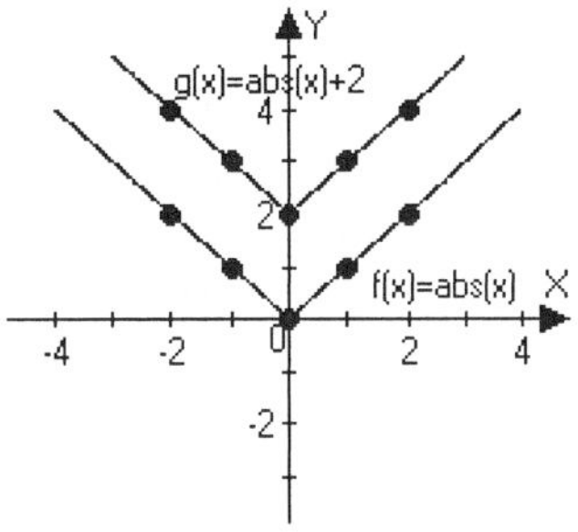


*g* shifts the graph of *f* up two units

**8.** The vertical line test shows that this is not the graph of a function.

**9.** The vertical line test shows that this is the graph of a function.

**10.** The vertical line test shows that this is the graph of a function.

**11.** The vertical line test shows that this is not the graph of a function.

**12.** The vertical line test shows that this is not the graph of a function.

**13.** The vertical line test shows that this is the graph of a function.

**14.** $f(-2)=-3$

**15.** $f(0)=-2$

**16.** When $x=3$, $f(x)=-5$.

**17.** The domain of $f$ is $\{x\mid -3\le x<5\}$.

**18.** The range of $f$ is $\{y\mid -5\le y\le 0\}$.

**19.** **a.** The vulture's height is a function of its time in flight because every time, *t*, is associated with at most one height.

**b.** $f(15) = 0$

At time $t$ = 15 seconds, the vulture is at height zero. This means that after 15 seconds, the vulture is on the ground.

**c.** The vulture's maximum height is 45 meters.

**d.** For $x = 7$ and 22, $f(x) = 20$. This means that at times 7 seconds and 22 seconds, the vulture is at a height of 20 meters.

**e.** The vulture began the flight at 45 meters and remained there for approximately 3 seconds. At that time, the vulture descended for 9 seconds. It landed on the ground and stayed there for 5 seconds. The vulture then began to climb back up to a height of 44 meters.

**20.** The domain of $f$ is $\{x|x \text{ is a real number}\}$.

**21.** The domain of $f$ is $\{x|x \text{ is a real number and } x \neq -8\}$.

**22.** The domain of $f$ is $\{x|x \text{ is a real number and } x \neq 5\}$.

**23. a.** $(f+g)(x) = (4x-5)+(2x+1)$
$= 4x-5+2x+1$
$= 6x-4$

**b.** $(f+g)(3) = 6(3)-4$
$= 18-4 = 14$

**24. a.** $(f+g)(x)$
$= (5x^2-x+4)+(x-3)$
$= 5x^2-x+4+x-3 = 5x^2+1$

**b.** $(f+g)(3) = 5(3)^2+1 = 5(9)+1$
$= 45+1 = 46$

**25.** The domain of $f+g$ is $\{x|x \text{ is a real number and } x \neq 4\}$

**26.** The domain of $f+g$ is $\{x\,|\,x \text{ is a real number and } x \neq -6 \text{ and } x \neq -1\}$.

**27.** $f(x) = x^2-2x, \quad g(x) = x-5$
$(f+g)(x) = (x^2-2x)+(x-5)$
$= x^2-2x+x-5$
$= x^2-x-5$
$(f+g)(-2) = (-2)^2-(-2)-5$
$= 4+2-5 = 1$

**28.** From Exercise 27 we know $(f+g)(x) = x^2-x-5$. We can use this to find $f(3)+g(3)$.
$f(3)+g(3) = (f+g)(3)$
$= (3)^2-(3)-5$
$= 9-3-5 = 1$

**29.** $f(x) = x^2-2x, \quad g(x) = x-5$
$(f-g)(x) = (x^2-2x)-(x-5)$
$= x^2-2x-x+5$
$= x^2-3x+5$
$(f-g)(x) = x^2-3x+5$
$(f-g)(1) = (1)^2-3(1)+5$
$= 1-3+5 = 3$

**30.** From Exercise 29 we know $(f-g)(x)=x^2-3x+5$. We can use this to find $f(4)-g(4)$.

$$f(4)-g(4)=(f-g)(4)$$
$$=(4)^2-3(4)+5$$
$$=16-12+5=9$$

**31.** Since $(fg)(-3)=f(-3)\cdot g(-3)$, find $f(-3)$ and $g(-3)$ first.

$$f(-3)=(-3)^2-2(-3)$$
$$=9+6=15$$
$$g(-3)=-3-5=-8$$
$$(fg)(-3)=f(-3)\cdot g(-3)$$
$$=15(-8)=-120$$

**32.** $f(x)=x^2-2x,\quad g(x)=x-5$

$$\left(\frac{f}{g}\right)(x)=\frac{x^2-2x}{x-5}$$
$$\left(\frac{f}{g}\right)(4)=\frac{(4)^2-2(4)}{4-5}=\frac{16-8}{-1}$$
$$=\frac{8}{-1}=-8$$

**33.** $(f-g)(x)=x^2-3x+5$

The domain of $f-g$ is $\{x|x \text{ is a real number}\}$.

**34.** $\left(\frac{f}{g}\right)(x)=\frac{x^2-2x}{x-5}$

The domain of $\frac{f}{g}$ is $\{x|x \text{ is a real number and } x\neq 5\}$.

**35.** $x+2y=4$

Find the $x$–intercept by setting $y=0$ and the $y$–intercept by setting $x=0$.

| | |
|---|---|
| $x+2(0)=4$ | $0+2y=4$ |
| $x+0=4$ | $2y=4$ |
| $x=4$ | $y=2$ |

Choose another point to use as a check. Let $x=1$.

$$1+2y=4$$
$$2y=3$$
$$y=\frac{3}{2}$$

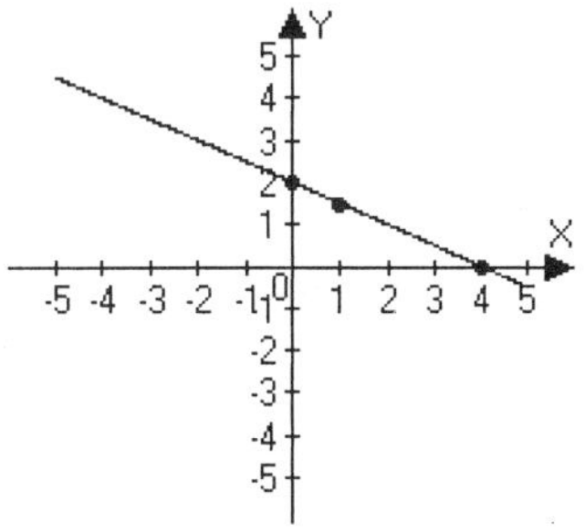


**36.** $2x-3y=12$

Find the $x$–intercept by setting $y=0$ and the $y$–intercept by setting $x=0$.

| | |
|---|---|
| $2x-3(0)=12$ | $2(0)-3y=12$ |
| $2x+0=12$ | $0-3y=12$ |
| $2x=12$ | $-3y=12$ |
| $x=6$ | $y=-4$ |

Choose another point to use as a check. Let $x=1$.

$$2(1)-3y=12$$
$$2-3y=12$$
$$-3y=10$$
$$y=-\frac{10}{3}$$

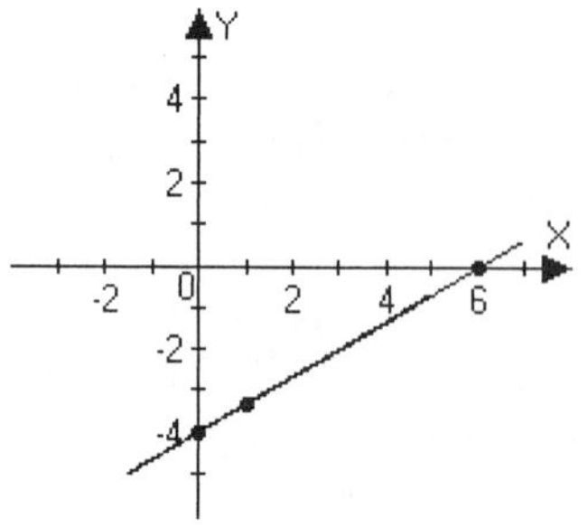


**37.** $4x = 8 - 2y$

Find the $x$–intercept by setting $y = 0$ and the $y$–intercept by setting $x = 0$.

$$4x = 8 - 2(0) \qquad 4(0) = 8 - 2y$$
$$4x = 8 - 0 \qquad 0 = 8 - 2y$$
$$4x = 8 \qquad 2y = 8$$
$$x = 2 \qquad y = 4$$

Choose another point to use as a check. Let $x = 1$.

$$4(1) = 8 - 2y$$
$$4 = 8 - 2y$$
$$-4 = -2y$$
$$2 = y$$

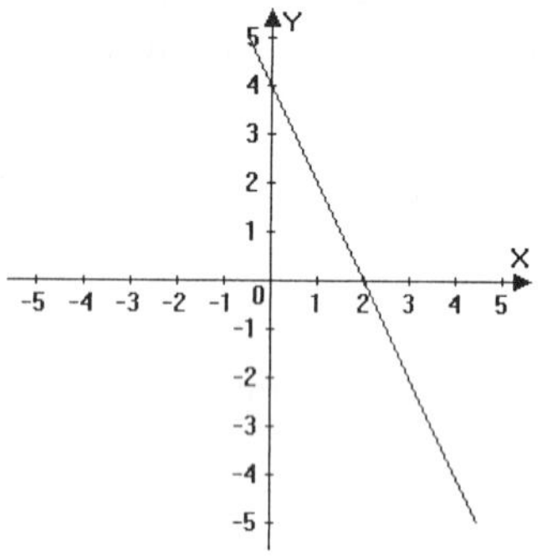


**38.** $m = \dfrac{2-(-4)}{5-2} = \dfrac{6}{3} = 2$

The line through the points rises.

**39.** $m = \dfrac{3-(-3)}{-2-7} = \dfrac{6}{-9} = -\dfrac{2}{3}$

The line through the points falls.

**40.** $m = \dfrac{2-(-1)}{3-3} = \dfrac{3}{0}$

$m$ is undefined. The line through the points is vertical.

**41.** $m = \dfrac{4-4}{-3-(-1)} = \dfrac{0}{-2} = 0$

The line through the points is horizontal.

**42.** $y = 2x - 1$

$m = 2 \qquad y - \text{intercept} = -1$

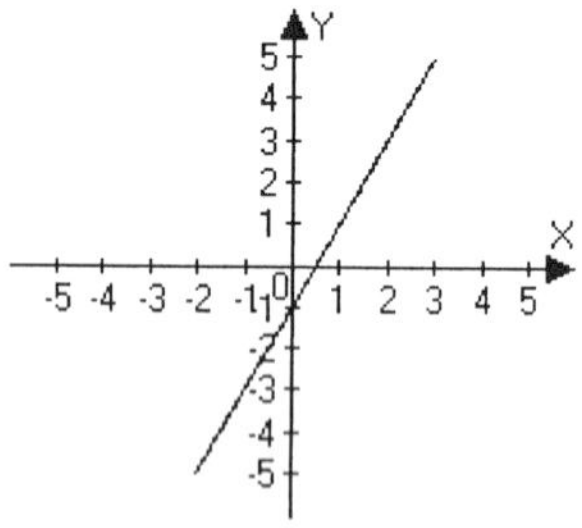


**43.** $f(x) = -\dfrac{1}{2}x + 4$

$m = -\dfrac{1}{2} \qquad y - \text{intercept} = 4$

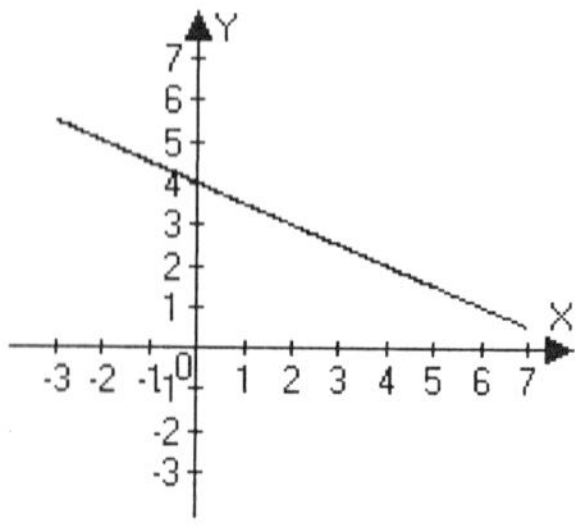

**44.** $y = \frac{2}{3}x$

$m = \frac{2}{3}$ $\quad y - \text{intercept} = 0$

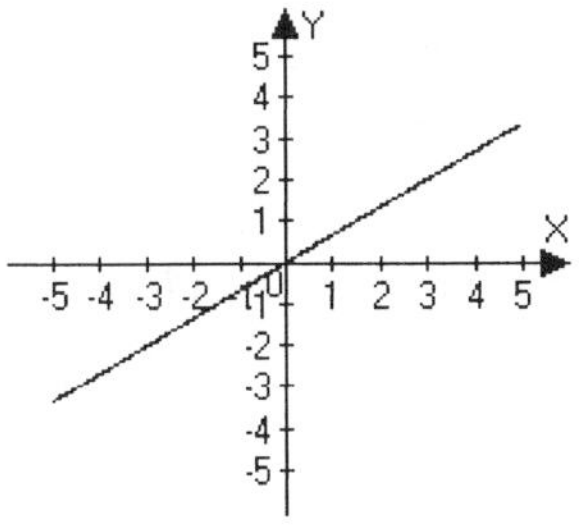

**45.** To rewrite the equation in slope-intercept form, solve for *y*.

$2x + y = 4$

$y = -2x + 4$

$m = -2$ $\quad y - \text{intercept} = 4$

**46.** $-3y = 5x$

$y = -\frac{5}{3}x$

$m = -\frac{5}{3}$ $\quad y - \text{intercept} = 0$

**47.** $5x + 3y = 6$

$3y = -5x + 6$

$y = -\frac{5}{3}x + 2$

$m = -\frac{5}{3}$ $\quad y - \text{intercept} = 2$

**48.** $y = 2$

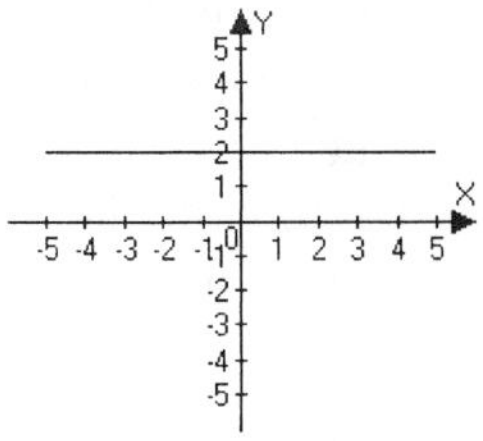

**49.** $7y = -21$

$y = -3$

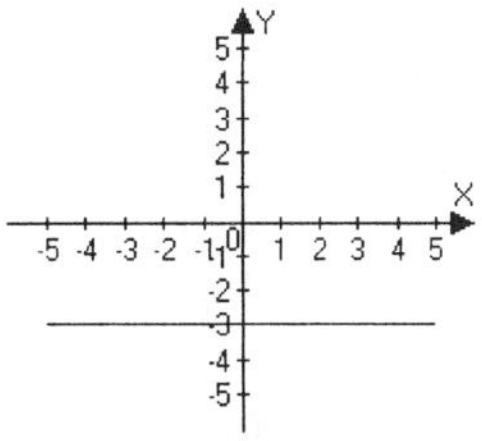

**50.** $f(x) = -4$

$y = -4$

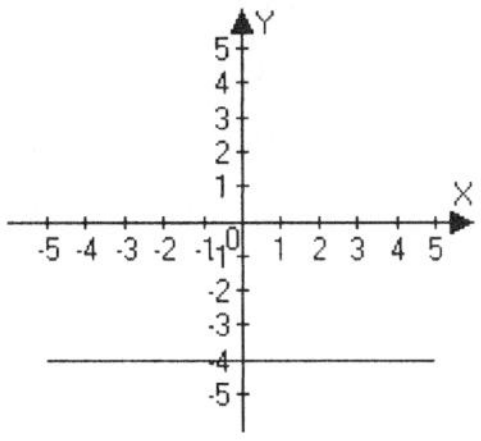

**51.** $x = 3$

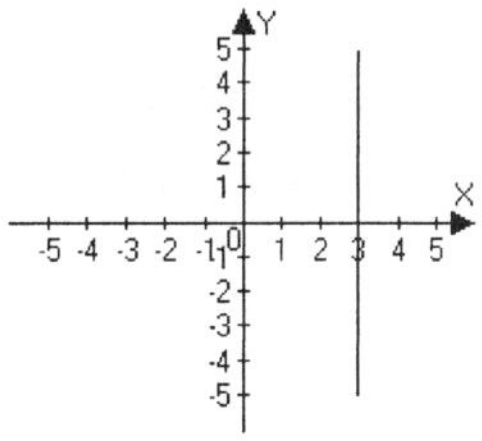

**52.** $2x = -10$

$x = -5$

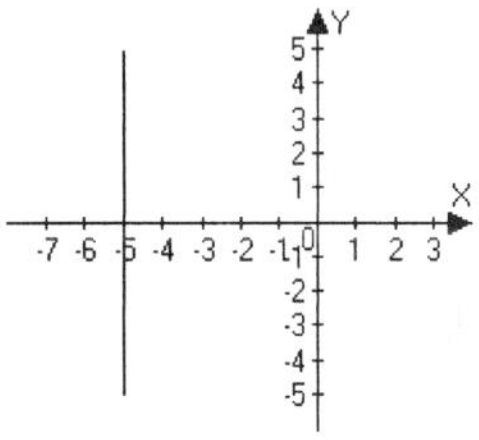

**53.** In $f(t) = -0.27 + 70.45$, the slope is $-0.27$. A slope of $-0.27$ indicates that the record time for the women's 400-meter has been decreasing by 0.27 seconds per year since 1900.

**54.** Use the points (1946,600) and (2004,400) to find the average rate of change in military spending per year. This would be the slope of the line passing through those two points.

$$m = \frac{600-400}{1946-2004} = \frac{200}{-58} \approx -3.4$$

This means that there has been an average decrease in military spending of \$3.4 billion per year from 1946 to 2004.

**55.** **a.** Find the slope of the line by using the two points (0,32) and (100,212).

$$m = \frac{212-32}{100-0} = \frac{180}{100} = \frac{9}{5}$$

We use the slope and one of the points to write the equation in point-slope form.

$$y - y_1 = m(x - x_1)$$
$$y - 32 = \frac{9}{5}(x - 0)$$
$$y - 32 = \frac{9}{5}x$$
$$y = \frac{9}{5}x + 32$$
$$F = \frac{9}{5}C + 32$$

**56.** Slope $= -6$, passing through $(-3, 2)$

Point-Slope Form

$$y - y_1 = m(x - x_1)$$
$$y - 2 = -6(x - (-3))$$
$$y - 2 = -6(x + 3)$$

Slope-Intercept Form

$$y - 2 = -6(x + 3)$$
$$y - 2 = -6x - 18$$
$$y = -6x - 16$$

In function notation, the equation of the line is $f(x) = -6x - 16$.

**57.** Passing through $(1, 6)$ and $(-1, 2)$

First, find the slope:

$$m = \frac{6-2}{1-(-1)} = \frac{4}{2} = 2$$

Then use the slope and one of the points to write the equation in point-slope form.

$$y - y_1 = m(x - x_1)$$
$$y - 6 = 2(x - 1)$$

or

$$y - 2 = 2(x - (-1))$$
$$y - 2 = 2(x + 1)$$

Slope-Intercept Form

$$y - 6 = 2(x - 1)$$
$$y - 6 = 2x - 2$$
$$y = 2x + 4$$

In function notation, the equation of the line is $f(x) = 2x + 4$.

**58.** Rewrite $3x + y - 9 = 0$ in slope-intercept form.

$$3x + y - 9 = 0$$
$$y = -3x + 9$$

Since the line we are concerned with is parallel to this line, we know it will have slope $m = -3$. We are given that it passes through (4, –7). We use the slope and point to write the equation in point-slope form.

$$y - y_1 = m(x - x_1)$$
$$y - (-7) = -3(x - 4)$$
$$y + 7 = -3(x - 4)$$

Solve for $y$ to obtain slope-intercept form.

$$y+7=-3(x-4)$$
$$y+7=-3x+12$$
$$y=-3x+5$$

In function notation, the equation of the line is $f(x)=-3x+5$.

**59.** The line is perpendicular to $y=\frac{1}{3}x+4$, so the slope is –3. We are given that it passes through (–2, 6). We use the slope and point to write the equation in point-slope form.

$$y-y_1=m(x-x_1)$$
$$y-6=-3(x-(-2))$$
$$y-6=-3(x+2)$$

Solve for $y$ to obtain slope-intercept form.

$$y-6=-3(x+2)$$
$$y-6=-3x-6$$
$$y=-3x$$

In function notation, the equation of the line is $f(x)=-3x$.

**60.** **a.** $(2, 4.85)$ and $(5, 4.49)$

First, find the slope using the points $(2, 4.85)$ and $(5, 4.49)$

$$m=\frac{4.49-4.85}{5-2}=\frac{-0.36}{3}=-0.12$$

Then use the slope and one of the points to write the equation in point-slope form.

$$y-y_1=m(x-x_1)$$
$$y-4.85=-0.12(x-2)$$

or

$$y-4.49=-0.12(x-5)$$

**b.** Solve for $y$ to obtain slope-intercept form.

$$y-4.85=-0.12(x-2)$$
$$y-4.85=-0.12x+0.24$$
$$y=-0.12x+5.09$$
$$f(x)=-0.12x+5.09$$

**c.** To predict the minimum hourly inflation-adjusted wages in 2007, let $x$ = 2007 – 1997 = 10.

$$f(10)=-0.12(10)+5.09=3.89$$

The linear function predicts the minimum hourly inflation-adjusted wage in 2007 to \$3.89.

## Chapter 2 Test

**1.** The relation is a function.
Domain {1, 3, 5, 6}
Range {2, 4, 6}

**2.** The relation is not a function.
Domain {2, 4, 6}
Range {1, 3, 5, 6}

**3.** $f(a+4)=3(a+4)-2$
$=3a+12-2=3a+10$

**4.** $f(-2)=4(-2)^2-3(-2)+6$
$=4(4)+6+6=16+6+6=28$

**5.**

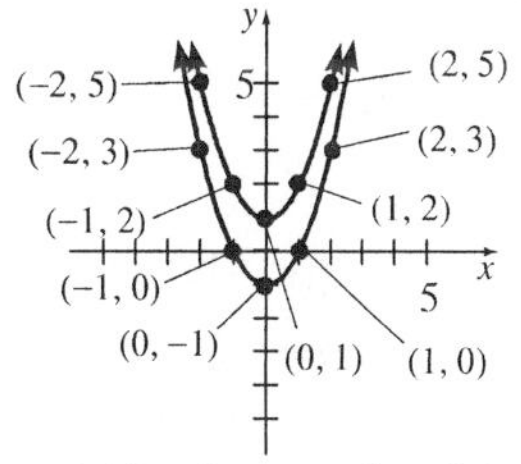


$g$ shifts the graph of $f$ up 2 units

**6.** The vertical line test shows that this is the graph of a function.

**7.** The vertical line test shows that this is not the graph of a function.

**8.** $f(6) = -3$

**9.** $f(x) = 0$ when $x = -2$ and $x = 3$.

**10.** The domain of $f$ is $\{x|x \text{ is a real number}\}$

**11.** The range of $f$ is $\{y \mid y \le 3\}$.

**12.** The domain of $f$ is $\{x \mid x \text{ is a real number and } x \ne 10\}$.

**13.** $f(x) = x^2 + 4x \qquad g(x) = x + 2$

$$\begin{aligned}(f+g)(x) &= f(x) + g(x)\\ &= (x^2+4x)+(x+2)\\ &= x^2+4x+x+2\\ &= x^2+5x+2\end{aligned}$$

$$\begin{aligned}(f+g)(3) &= (3)^2+5(3)+2\\ &= 9+15+2 = 26\end{aligned}$$

**14.** $f(x) = x^2 + 4x \qquad g(x) = x + 2$

$$\begin{aligned}(f-g)(x) &= f(x) - g(x)\\ &= (x^2+4x)-(x+2)\\ &= x^2+4x-x-2\\ &= x^2+3x-2\end{aligned}$$

$$\begin{aligned}(f-g)(-1) &= (-1)^2+3(-1)-2\\ &= 1-3-2 = -4\end{aligned}$$

**15.** We know that $(fg)(x) = f(x)\cdot g(x)$. So, to find $(fg)(-5)$, we use $f(-5)$ and $g(-5)$.

$$f(-5) = (-5)^2 + 4(-5) = 25 - 20 = 5$$

$$g(-5) = -5 + 2 = -3$$

$$\begin{aligned}(fg)(-5) &= f(-5)\cdot g(-5)\\ &= 5(-3) = -15\end{aligned}$$

**16.** $f(x) = x^2 + 4x \qquad g(x) = x + 2$

$$\left(\frac{f}{g}\right)(x) = \frac{x^2+4x}{x+2}$$

$$\left(\frac{f}{g}\right)(2) = \frac{(2)^2+4(2)}{2+2} = \frac{4+8}{4} = \frac{12}{4} = 3$$

**17.** Domain of $\frac{f}{g}$ is

$\{x|x \text{ is a real number and } x \ne -2\}$.

**18.** $4x - 3y = 12$

Find the $x$–intercept by setting $y = 0$.

$$\begin{aligned}4x - 3(0) &= 12\\ 4x &= 12\\ x &= 3\end{aligned}$$

Find the $y$–intercept by setting $x = 0$.

$$\begin{aligned}4(0) - 3y &= 12\\ -3y &= 12\\ y &= -4\end{aligned}$$

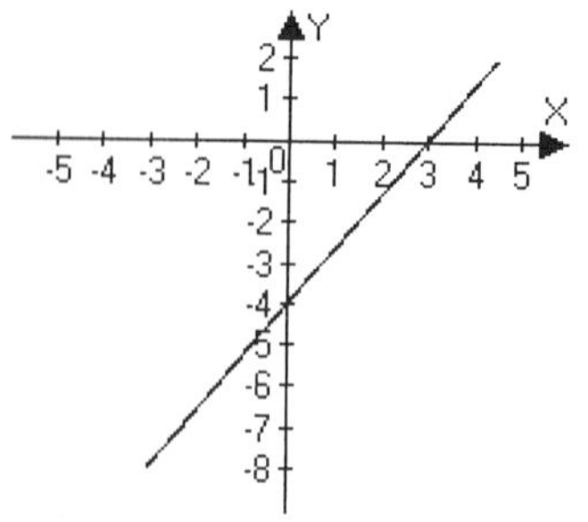

**19.** $f(x) = -\frac{1}{3}x + 2$

$m = -\frac{1}{3}$ $\quad$ $y - \text{intercept} = 2$

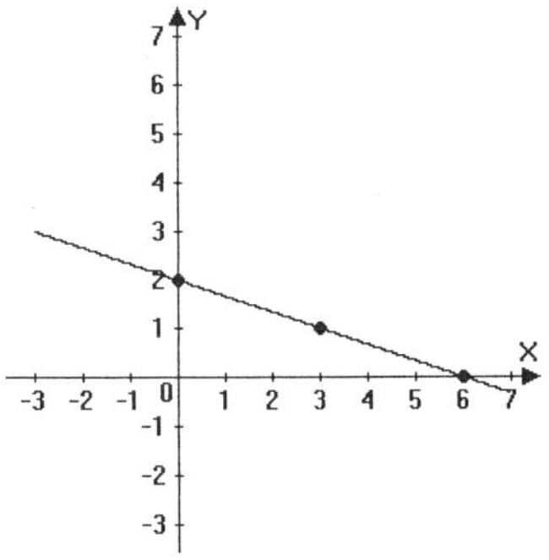


**20.** $f(x) = 4$

$y = 4$

An equation of the form $y = b$ is a horizontal line.

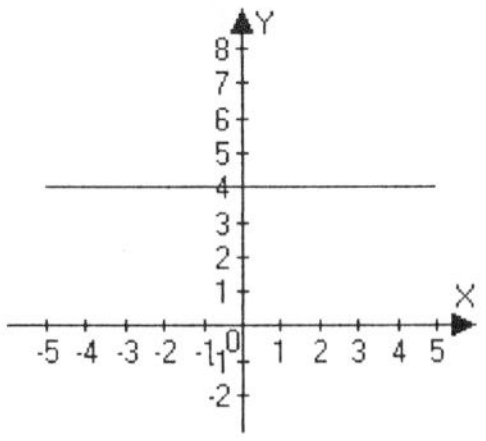


**21.** $m = \frac{4-2}{1-5} = \frac{2}{-4} = -\frac{1}{2}$

The line through the points falls.

**22.** $m = \frac{5-(-5)}{4-4} = \frac{10}{0}$

$m$ is undefined

The line through the points is vertical.

**23.** $V(10) = 3.6(10) + 140$

$= 36 + 140 = 176$

In the year 2005, there were 176 million Super Bowl viewers.

**24.** The slope is 3.6. This means the number of Super Bowl viewers is increasing at a rate of 3.6 million per year.

**25.** Passing through $(-1,-3)$ and $(4,2)$

First, find the slope:

$m = \frac{2-(-3)}{4-(-1)} = \frac{5}{5} = 1$

Then use the slope and one of the points to write the equation in point-slope form.

$y - y_1 = m(x - x_1)$

$y - (-3) = 1(x - (-1))$

$y + 3 = 1(x + 1)$

or

$y - 2 = 1(x - 4)$

$y - 2 = x - 4$

Slope-Intercept Form

$y - 2 = x - 4$

$y = x - 2$

In function notation, the equation of the line is $f(x) = x - 2$.

**26.** The line is perpendicular to $y = -\frac{1}{2}x - 4,$ so the slope is 2. We are given that it passes through $(-2,3)$. We use the slope and point to write the equation in point-slope form.

$y - y_1 = m(x - x_1)$

$y - 3 = 2(x - (-2))$

$y - 3 = 2(x + 2)$

Solve for $y$ to obtain slope-intercept form.

$y - 3 = 2(x + 2)$

$y - 3 = 2x + 4$

$y = 2x + 7$

In function notation, the equation of the line is $f(x) = 2x + 7$.

**27.** The line is parallel to $x+2y=5$. Put this equation in slope-intercept form by solving for $y$.

$x+2y=5$

$2y=-x+5$

$y=-\frac{1}{2}x+\frac{5}{2}$

Therefore the slopes are the same; $m=-\frac{1}{2}$.

We are given that it passes through $(6,-4)$. We use the slope and point to write the equation in point-slope form.

$y-y_1=m(x-x_1)$

$y-(-4)=-\frac{1}{2}(x-6)$

$y+4=-\frac{1}{2}(x-6)$

Solve for $y$ to obtain slope-intercept form.

$y+4=-\frac{1}{2}(x-6)$

$y+4=-\frac{1}{2}x+3$

$y=-\frac{1}{2}x-1$

In function notation, the equation of the line is $f(x)=-\frac{1}{2}x-1$.

**28. a.** First, find the slope using the points $(1, 2.3)$ and $(4, 3.2)$.

$m=\frac{3.2-2.3}{4-1}=\frac{0.9}{3}=0.3$

Then use the slope and a point to write the equation in point-slope form.

$y-y_1=m(x-x_1)$

$y-3.2=0.3(x-4)$

or

$y-2.3=0.3(x-1)$

**b.** Solve for $y$ to obtain slope-intercept form.

$y-3.2=0.3(x-4)$

$y-3.2=0.3x-1.2$

$f(x)=0.3x+2$

**c.** To predict the number of worldwide death due to AIDS in 2007, let $x = 2007 - 1999 = 8$.

$f(8)=0.3(8)+2=2.4+2=4.4$

If the current trend continues, the number of worldwide deaths due to AIDS in 2007 will be 4.4 million.

## Cumulative Review Exercises

**1.** {0, 1, 2, 3}

**2.** False. B is an irrational number.

**3.** $\frac{8-3^2\div 9}{|-5|-[5-(18\div 6)]^2}$

$=\frac{8-9\div 9}{5-[5-(3)]^2}=\frac{8-1}{5-[2]^2}$

$=\frac{7}{5-4}=\frac{7}{1}=7$

**4.** $4-(2-9)^0+3^2\div 1+3$

$=4-(-7)^0+9\div 1+3=4-1+9\div 1+3$

$=4-1+9+3=3+9+3=15$

**5.** $3-[2(x-2)-5x]$
$=3-[2x-4-5x]=3-[-3x-4]$
$=3+3x+4=3x+7$

**6.** $2+3x-4=2(x-3)$
$3x-2=2x-6$
$x-2=-6$
$x=-4$
The solution is –4 and the solution set is $\{-4\}$.

**7.** $4x+12-8x=-6(x-2)+2x$
$12-4x=-6x+12+2x$
$12-4x=-4x+12$
$12=12$
$0=0$
The solution set is $\{x|x$ is all real numbers$\}$ or $\mathbb{R}$. The equation is an identity.

**8.** $\frac{x-2}{4}=\frac{2x+6}{3}$
$4(2x+6)=3(x-2)$
$8x+24=3x-6$
$5x+24=-6$
$5x=-30$
$x=-6$
The solution is –6 and the solution set is $\{-6\}$.

**9.** Let $x$ = the price before reduction
$x-0.20x=1800$
$0.80x=1800$
$x=2250$
The price of the computer before the reduction was $2250.

**10.** $A=p+prt$
$A-p=prt$
$\frac{A-p}{pr}=t$

**11.** $(3x^4y^{-5})^{-2}=\left(\frac{3x^4}{y^5}\right)^{-2}=\left(\frac{y^5}{3x^4}\right)^2=\frac{y^{10}}{9x^8}$

**12.** $\left(\frac{3x^2y^{-4}}{x^{-3}y^2}\right)^2=\left(\frac{3x^2x^3}{y^2y^4}\right)^2$
$=\left(\frac{3x^5}{y^6}\right)^2=\frac{9x^{10}}{y^{12}}$

**13.** $(7\times10^{-8})(3\times10^2)$
$=(7\times3)(10^{-8}\times10^2)=21\times10^{-6}$
$=(2.1\times10)\times10^{-6}=2.1(10\times10^{-6})$
$=2.1\times10^{-5}$

**14.** The relation is a function.
Domain $\{1, 2, 3, 4, 6\}$
Range $\{5\}$

**15.**

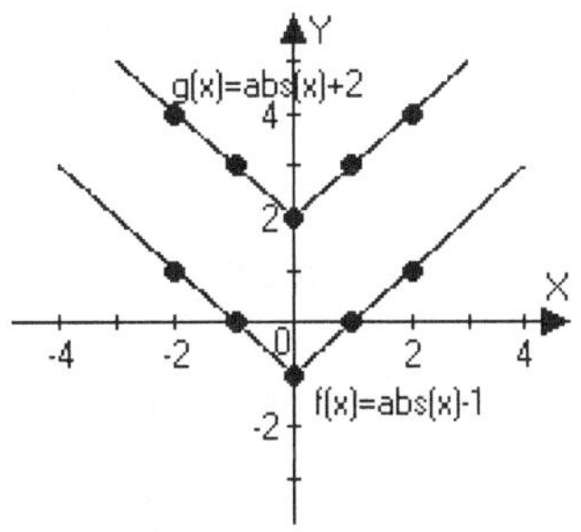


$g$ shifts the graph of $f$ up three units

**16.** The domain of $f$ is
$\{x|x$ is a real number and $x\neq15\}$.

**17.** $(f-g)(x)$
$= (3x^2-4x+2)-(x^2-5x-3)$
$= 3x^2-4x+2-x^2+5x+3$
$= 2x^2+x+5$
$(f-g)(-1) = 2(-1)^2+(-1)+5$
$= 2(1)-1+5 = 2-1+5 = 6$

**18.** $f(x) = -2x+4$
$y = -2x+4$
$m = -2 \qquad y-\text{intercept} = 4$

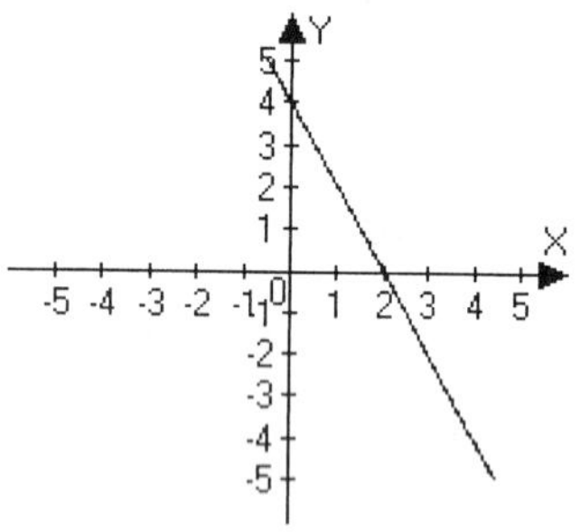

**19.** $x-2y=6$
Rewrite the equation of the line in slope intercept form.
$x-2y=6$
$-2y = -x+6$
$y = \frac{1}{2}x-3$
$m = \frac{1}{2} \qquad y-\text{intercept} = -3$

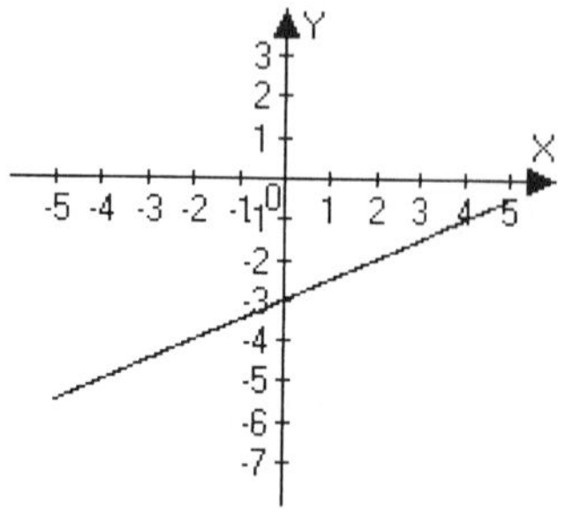

**20.** The line is parallel to $y = 4x+7$, so the slope is 4. We are given that it passes through (3, –5). We use the slope and point to write the equation in point-slope form.
$y-y_1 = m(x-x_1)$
$y-(-5) = 4(x-3)$
$y+5 = 4(x-3)$
Solve for $y$ to obtain slope-intercept form.
$y+5 = 4(x-3)$
$y+5 = 4x-12$
$y = 4x-17$
In function notation, the equation of the line is $f(x) = 4x-17$.

# Chapter 3
# Systems of Linear Equations

## 3.1 Exercise Set

**1.** $x-y=12$ $\quad$ $x+y=2$

$7-(-5)=12$ $\quad$ $7+(-5)=2$

$12=12$ $\quad$ $2=2$

The pair is a solution of the system.

**3.** $3x+4y=2$ $\quad$ $2x+5y=1$

$3(2)+4(-1)=2$ $\quad$ $2(2)+5(-1)=1$

$6-4=2$ $\quad$ $4-5=1$

$2=2$ $\quad$ $-1=1$

The pair is not a solution of the system.

**5.** $y=2x-13$ $\quad$ $4x+9y=-7$

$-3=2(5)-13$ $\quad$ $4(5)+9(-3)=-7$

$-3=10-13$ $\quad$ $20-27=-7$

$-3=-3$ $\quad$ $-7=-7$

The pair is a solution of the system.

**7.** $x+y=4$

Find the $x$–intercept by setting $y=0$.

$x+0=4$

$x=4$

Find the $y$–intercept by setting $x=0$.

$0+y=4$

$y=4$

$x-y=2$

Find the $x$–intercept by setting $y=0$.

$x-0=2$

$x=2$

Find the $y$–intercept by setting $x=0$.

$0-y=2$

$-y=2$

$y=-2$

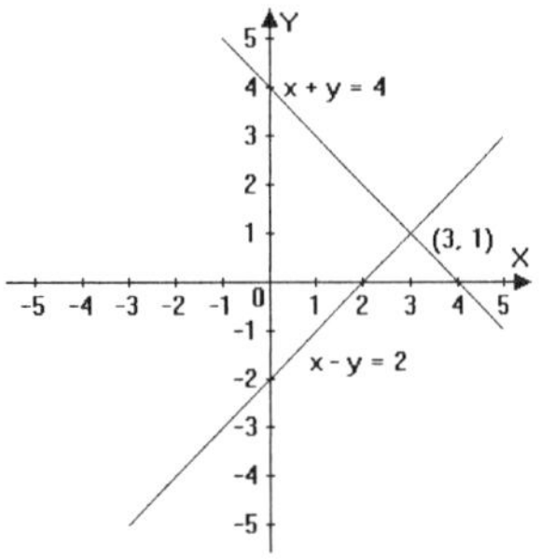


The solution is $(3,1)$.

**9.** $2x+y=4$

Find the $x$–intercept by setting $y=0$.

$2x+0=4$

$2x=4$

$x=2$

Find the $y$–intercept by setting $x=0$.

$2(0)+y=4$

$y=4$

$y=4x+1$

$y-\text{intercept}=1$

$\text{slope}=4$

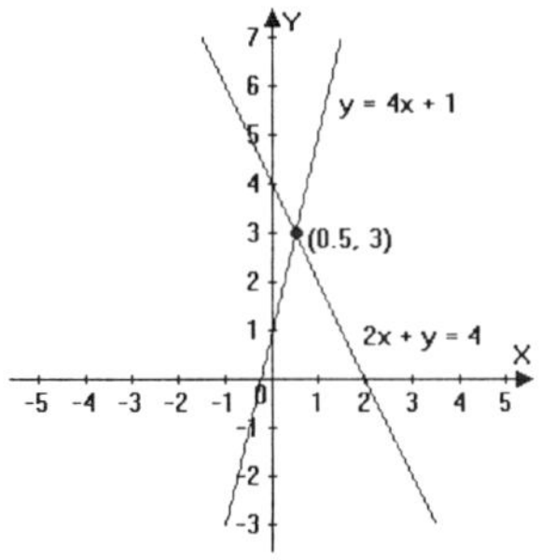


The solution is $\left(\frac{1}{2},3\right)$.

**11.** Solve both equations for $y$ to obtain slopes and intercepts.

$$3x-2y=6 \qquad\qquad x-4y=-8$$
$$-2y=-3x+6 \qquad\qquad -4y=-x-8$$
$$y=\frac{3}{2}x-3 \qquad\qquad y=\frac{1}{4}x+2$$
$$y-\text{intercept}=-3 \qquad\qquad y-\text{intercept}=2$$
$$\text{slope}=\frac{3}{2} \qquad\qquad \text{slope}=\frac{1}{4}$$

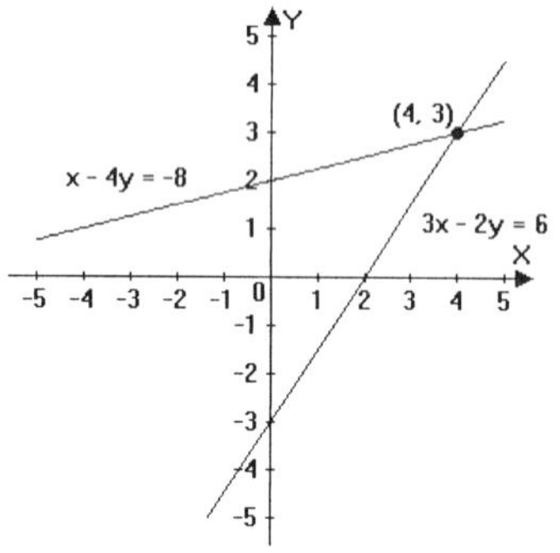


The solution is $(4,3)$.

**13.** Solve both equations for $y$ to obtain slopes and intercepts.

$$2x+3y=6 \qquad\qquad 4x=-6y+12$$
$$3y=-2x+6 \qquad\qquad 6y=-4x+12$$
$$y=-\frac{2}{3}x+2 \qquad\qquad y=-\frac{4}{6}x+2$$
$$y=-\frac{2}{3}x+2$$
$$y-\text{intercept}=2 \qquad\qquad y-\text{intercept}=2$$
$$\text{slope}=-\frac{2}{3} \qquad\qquad \text{slope}=-\frac{2}{3}$$

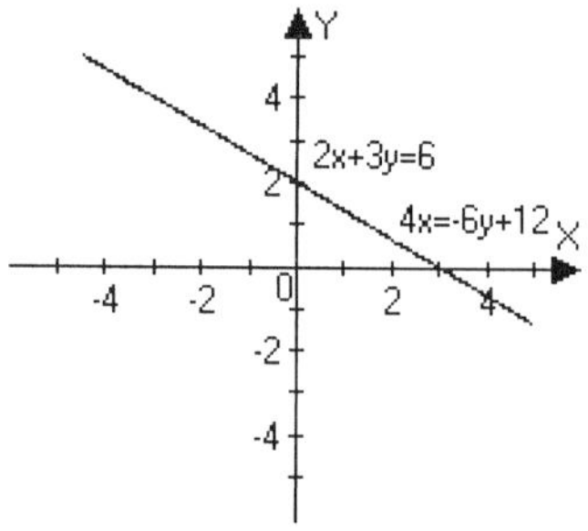


The lines coincide. The solution set is $\{(x,y)|2x+3y=6\}$ or $\{(x,y)|4x=-6y+12\}$.

**15.** Both equations are in slope–intercept form, so graph using slopes and intercepts.

$$y=2x-2 \qquad\qquad y=-5x+5$$
$$y-\text{intercept}=-2 \qquad\qquad y-\text{intercept}=5$$
$$\text{slope}=2 \qquad\qquad \text{slope}=-5$$

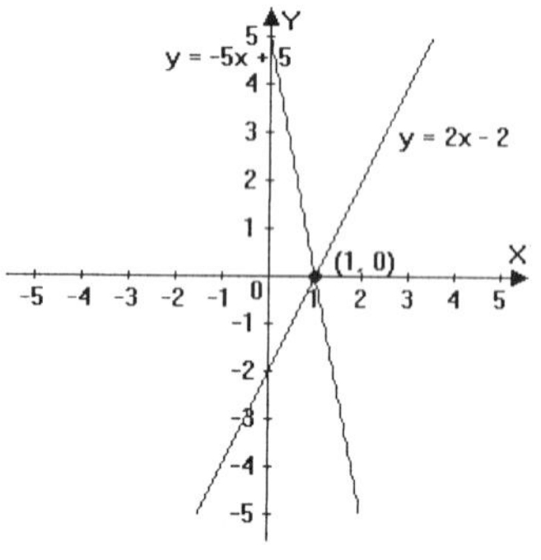


The solution is $(1,0)$.

**17.** Solve both equations for $y$ to obtain slopes and intercepts.

$$3x-y=4$$
$$-y=-3x+4$$
$$y=3x-4$$
$$y-\text{intercept}=-4$$
$$\text{slope}=3$$

$6x-2y=4$

$-2y=-6x+4$

$y=3x-2$

$y-\text{intercept}=-2$

$\text{slope}=3$

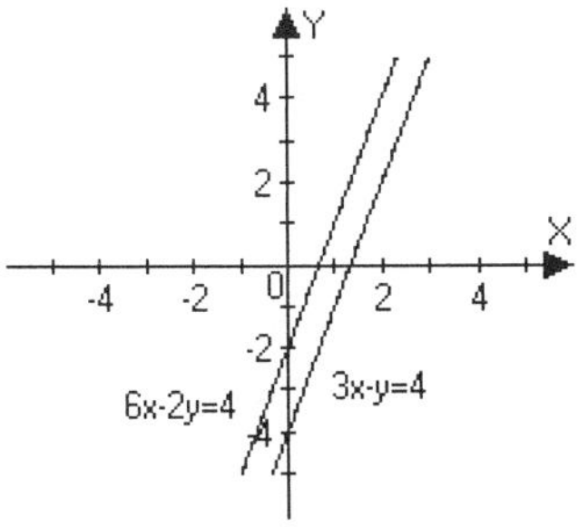


Since the lines do not intersect, there is no solution. The solution set is $\varnothing$ or $\{\ \}$.

**19.** Solve both equations for $y$ to obtain slopes and intercepts.

$2x+y=4$

$y=-2x+4$

$y-\text{intercept}=4$

$\text{slope}=-2$

$4x+3y=10$

$3y=-4x+10$

$y=-\frac{4}{3}x+\frac{10}{3}$

$y-\text{intercept}=\frac{10}{3}=3\frac{1}{3}$

$\text{slope}=-\frac{4}{3}$

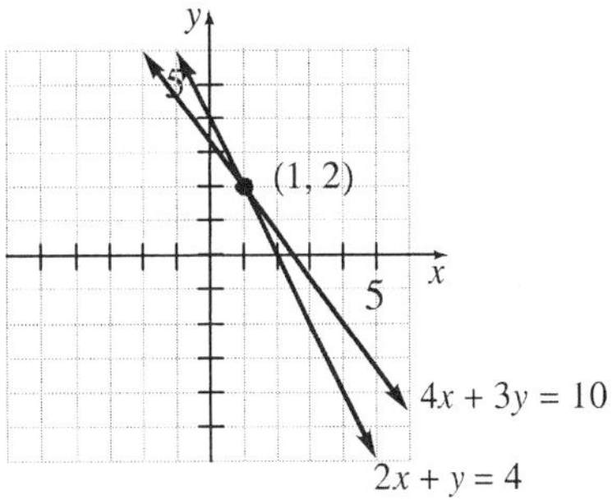


The solution is $(1,2)$.

**21.** Solve the first equation for $y$ to obtain the slope and intercept.

$x-y=2$

$-y=-x+2$

$y=x-2$

$y-\text{intercept}=-2$

$\text{slope}=1$

$y=1$

This is a horizontal line passing through $y$–intercept, $(0,1)$.

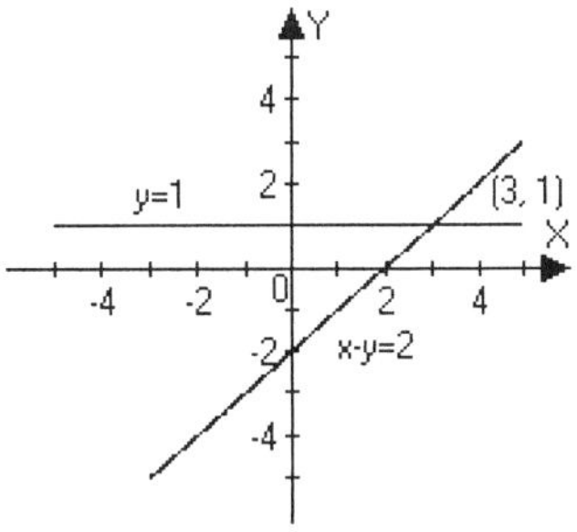


The solution is $(3,1)$.

**23.** Solve both equations for $y$ to obtain slopes and intercepts.

$3x+y=3$

$y=-3x+3$

$y-\text{intercept}=3$

$\text{slope}=-1$

$6x+2y=12$

$2y=-6x+12$

$y=-3x+6$

$y-\text{intercept}=6$

$\text{slope}=-3$

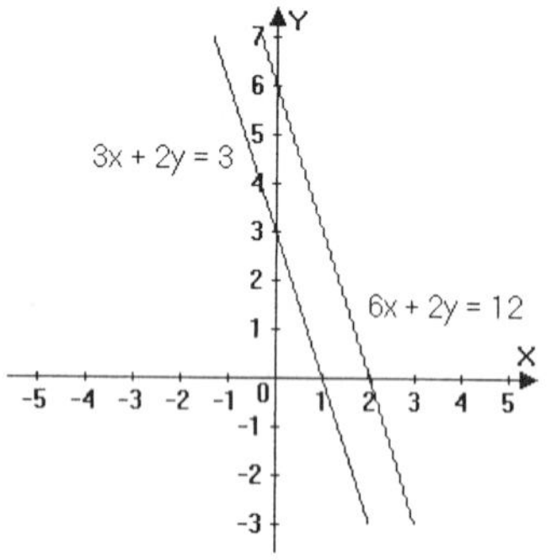


Since the lines do not intersect, there is no solution. The solution set is $\varnothing$ or $\{\ \}$.

**25.** $x+y=6$

$y=2x$

Substitute $2x$ for $y$ in the first equation.

$x+y=6$

$x+2x=6$

$3x=6$

$x=2$

Back-substitute to find $y$.

$2+y=6$

$y=4$

The solution is $(2,4)$.

**27.** $2x+3y=9$

$x=y+2$

Substitute $y+2$ for $x$ in the first equation.

$2x+3y=9$

$2(y+2)+3y=9$

$2y+4+3y=9$

$5y+4=9$

$5y=5$

$y=1$

Back-substitute to find $x$.

$x=y+2$

$x=1+2$

$x=3$

The solution is $(3,1)$.

**29.** $y=-3x+7$

$5x-2y=8$

Substitute $-3x+7$ for $y$ in the second equation.

$5x-2y=8$

$5x-2(-3x+7)=8$

$5x+6x-14=8$

$11x-14=8$

$11x=22$

$x=2$

Back-substitute to find $y$.

$y=-3(2)+7$

$y=-6+7$

$y=1$

The solution is $(2,1)$.

**31.** $4x + y = 5$

$2x - 3y = 13$

Solve for $y$ in the first equation.

$4x + y = 5$

$y = -4x + 5$

Substitute $-4x + 5$ for $y$ in the second equation.

$2x - 3(-4x + 5) = 13$

$2x + 12x - 15 = 13$

$14x - 15 = 13$

$14x = 28$

$x = 2$

Back-substitute to find $y$.

$4x + y = 5$

$4(2) + y = 5$

$8 + y = 5$

$y = -3$

The solution is $(2, -3)$.

**33.** $x - 2y = 4$

$2x - 4y = 5$

Solve for $x$ in the first equation.

$x - 2y = 4$

$x = 2y + 4$

Substitute $2y + 4$ for $x$ in the second equation.

$2x - 4y = 5$

$2(2y + 4) - 4y = 5$

$4y + 8 - 4y = 5$

$4y + 8 - 4y = 5$

$8 \neq 5$

The system is inconsistent. There are no values of $x$ and $y$ for which 8 will equal 5. The solution set is $\varnothing$ or $\{\ \}$.

**35.** $2x + 5y = -4$

$3x - y = 11$

Solve for $y$ in the second equation.

$3x - y = 11$

$-y = -3x + 11$

$y = 3x - 11$

Substitute $3x - 11$ for $y$ in the first equation.

$2x + 5y = -4$

$2x + 5(3x - 11) = -4$

$2x + 15x - 55 = -4$

$17x - 55 = -4$

$17x = 51$

$x = 3$

Back-substitute to find $y$.

$3x - y = 11$

$3(3) - y = 11$

$9 - y = 11$

$-y = 2$

$y = -2$

The solution is $(3, -2)$.

**37.** $2(x - 1) - y = -3$

$y = 2x + 3$

Substitute $2x + 3$ for $y$ in the first equation.

$2(x - 1) - y = -3$

$2(x - 1) - (2x + 3) = -3$

$2x - 2 - 2x - 3 = -3$

$-5 \neq -3$

Since there are no values of $x$ and $y$ for which –5 will equal –3, the system is inconsistent. The solution set is $\varnothing$ or $\{\ \}$.

**39.** $\frac{x}{4}-\frac{y}{4}=-1$

$x+4y=-9$

Solve for $x$ in the second equation.

$x+4y=-9$

$x=-4y-9$

Substitute $-4y-9$ for $x$ in the first equation.

$$\frac{x}{4}-\frac{y}{4}=-1$$
$$\frac{-4y-9}{4}-\frac{y}{4}=-1$$
$$4\left(\frac{-4y-9}{4}-\frac{y}{4}\right)=4(-1)$$
$$-4y-9-y=-4$$
$$-5y-9=-4$$
$$-5y=5$$
$$y=-1$$

Back-substitute to find $x$.

$$x+4y=-9$$
$$x+4(-1)=-9$$
$$x-4=-9$$
$$x=-5$$

The solution is $(-5,-1)$.

**41.** $y=\frac{2}{5}x-2$

$2x-5y=10$

Substitute $\frac{2}{5}x-2$ for $y$ in the second equation.

$$2x-5y=10$$
$$2x-5\left(\frac{2}{5}x-2\right)=10$$
$$2x-2x+10=10$$
$$10=10$$

Since 10 = 10 for all values of $x$ and $y$, the system is dependent. The solution set is $\left\{(x,y)\middle|\ y=\frac{2}{5}x-2\right\}$ or $\{(x,y)|\ 2x-5y=10\}$.

**43.** Solve by addition.

$$x+y=7$$
$$\underline{x-y=3}$$
$$2x=10$$
$$x=5$$

Back-substitute to find $y$.

$$x+y=7$$
$$5+y=7$$
$$y=2$$

The solution is $(5,2)$.

**45.** Solve by addition.

$$12x+3y=15$$
$$\underline{2x-3y=13}$$
$$14x=28$$
$$x=2$$

Back-substitute to find $y$.

$$12(2)+3y=15$$
$$24+3y=15$$
$$3y=-9$$
$$y=-3$$

The solution is $(2,-3)$.

**47.** $x+3y=2$

$4x+5y=1$

Multiply the first equation by –4.

$$-4x-12y=-8$$
$$\underline{4x+\ 5y=\ \ 1}$$
$$-7y=-7$$
$$y=1$$

Back-substitute to find $x$.

$x+3y=2$
$x+3(1)=2$
$x+3=2$
$x=-1$
The solution is $(-1,1)$.

**49.** $6x-y=-5$
$4x-2y=6$
Multiply the first equation by –2.
$-12x+2y=10$
$\underline{4x-2y=6}$
$-8x=16$
$x=-2$
Back-substitute to find $y$.
$6(-2)-y=-5$
$-12-y=-5$
$-y=7$
$y=-7$
The solution is $(-2,-7)$.

**51.** $3x-5y=11$
$2x-6y=2$
Multiply the first equation by –2 and the second equation by 3.
$-6x+10y=-22$
$\underline{6x-18y=\ \ 6}$
$-8y=-16$
$y=2$
Back-substitute to find $x$.
$2x-6(2)=2$
$2x-12=2$
$2x=14$
$x=7$
The solution is $(7,2)$.

**53.** $2x-5y=13$
$5x+3y=17$
Multiply the first equation by 3 and the second equation by 5.
$6x-15y=39$
$\underline{25x+15y=85}$
$31x=124$
$x=4$
Back-substitute to find $y$.
$5(4)+3y=17$
$20+3y=17$
$3y=-3$
$y=-1$
The solution is $(4,-1)$.

**55.** $2x+6y=8$
$3x+9y=12$
Multiply the first equation by –3 and the second equation by 2.
$-6x-18y=-24$
$\underline{6x+18y=\ 24}$
$0=0$
Since 0 = 0 for all values of $x$ and $y$, the system is dependent. The solution set is $\{(x,y)|\ 2x+6y=8\}$ or $\{(x,y)|\ 3x+9y=12\}$.

**57.** $2x-3y=4$
$4x+5y=3$
Multiply the first equation by –2.
$-4x+6y=-8$
$\underline{4x+5y=3}$
$11y=-5$
$y=-\frac{5}{11}$
Back-substitute to find $x$.

$$2x-3y=4$$
$$2x-3\left(-\frac{5}{11}\right)=4$$
$$2x+\frac{15}{11}=4$$
$$2x=\frac{29}{11}$$
$$x=\frac{29}{22}$$

The solution is $\left(\frac{29}{22},-\frac{5}{11}\right)$.

**59.** $3x-7y=1$

$2x-3y=-1$

Multiply the first equation by –2 and the second equation by 3.

$$\begin{aligned}-6x+14y&=-2\\ 6x-\ 9y&=-3\\ \hline 5y&=-5\\ y&=-1\end{aligned}$$

Back-substitute to find $x$.

$$3x-7(-1)=1$$
$$3x+7=1$$
$$3x=-6$$
$$x=-2$$

The solution is $(-2,-1)$.

**61.** $x=y+4$

$3x+7y=-18$

Substitute $y+4$ for $x$ in the second equation.

$$3x+7y=-18$$
$$3(y+4)+7y=-18$$
$$3y+12+7y=-18$$
$$10y+12=-18$$
$$10y=-30$$
$$y=-3$$

Back-substitute to find $x$.

$x=y+4$

$x=-3+4$

$x=1$

The solution is $(1,-3)$.

**63.** $9x+\frac{4y}{3}=5$

$4x-\frac{y}{3}=5$

Multiply the second equation by 4.

$$\begin{aligned}9x+\frac{4y}{3}&=\ 5\\ 16x-\frac{4y}{3}&=20\\ \hline 25x&=25\\ x&=1\end{aligned}$$

Back-substitute to find $y$.

$$4x-\frac{y}{3}=5$$
$$4(1)-\frac{y}{3}=5$$
$$4-\frac{y}{3}=5$$
$$-\frac{y}{3}=1$$
$$y=-3$$

The solution is $(1,-3)$.

**65.** $\frac{1}{4}x-\frac{1}{9}y=\frac{2}{3}$

$\frac{1}{2}x-\frac{1}{3}y=1$

Multiply the first equation by –2.

$$-2\left(\frac{1}{4}x\right)-(-2)\left(\frac{1}{9}y\right)=-2\left(\frac{2}{3}\right)$$
$$-\frac{1}{2}x+\frac{2}{9}y=-\frac{4}{3}$$

We now have a system of two equations in two variables.

$$-\frac{1}{2}x+\frac{2}{9}y=-\frac{4}{3}$$
$$\frac{1}{2}x-\frac{1}{3}y=1$$

Solve the addition.

$$-\frac{1}{2}x+\frac{2}{9}y=-\frac{4}{3}$$
$$\underline{\frac{1}{2}x-\frac{1}{3}y=1}$$
$$-\frac{1}{9}y=-\frac{1}{3}$$
$$y=-\frac{1}{3}(-9)$$
$$y=3$$

Back-substitute to find $x$.

$$\frac{1}{2}x-\frac{1}{3}y=1$$
$$\frac{1}{2}x-\frac{1}{3}(3)=1$$
$$\frac{1}{2}x-1=1$$
$$\frac{1}{2}x=2$$
$$x=4$$

The solution is $(4,3)$.

**67.** $x=3y-1$

$2x-6y=-2$

Substitute $3y-1$ for $x$ in the second equation.

$$2x-6y=-2$$
$$2(3y-1)-6y=-2$$
$$6y-2-6y=-2$$
$$-2=-2$$

Since –2 = –2 for all values of $x$ and $y$, the system is dependent. The solution set is $\{(x,y)|\ x=3y-1\}$ or $\{(x,y)|\ 2x-6y=-2\}$.

**69.** $y=2x+1$

$y=2x-3$

Multiply the first equation by –1.

$$-y=-2x-1$$
$$\underline{y=\ 2x-3}$$
$$0\neq-4$$

Since there are no values of $x$ and $y$ for which 0 = –4, the system is inconsistent. The solution set is $\varnothing$ or $\{\ \}$.

**71.** $0.4x+0.3y=2.3$

$0.2x-0.5y=0.5$

Multiply the second equation by –2.

$$0.4x+0.3y=\ 2.3$$
$$\underline{-0.4x+1.0y=-1.0}$$
$$1.3y=1.3$$
$$y=1$$

Back-substitute to find $x$.

$$0.2x-0.5y=0.5$$
$$0.2x-0.5(1)=0.5$$
$$0.2x-0.5=0.5$$
$$0.2x=1.0$$
$$x=5$$

The solution is $(5,1)$.

If you prefer to work with whole numbers and not decimals, you can multiple both of the original equations by 10. This will clear the equations of decimals. This is similar to the method we use to clear fractions.

**73.** $5x - 40 = 6y$

$2y = 8 - 3x$

Rewrite the equations in general form.

$5x - 6y = 40$

$3x + 2y = 8$

Multiply the second equation by 3.

$5x - 6y = 40$

$\underline{9x + 6y = 24}$

$14x = 64$

$x = \frac{32}{7}$

Back-substitute to find *y*.

$2y = 8 - 3x$

$2y = 8 - 3\left(\frac{32}{7}\right)$

$2y = 8 - \frac{96}{7}$

$2y = -\frac{40}{7}$

$y = -\frac{40}{14} = -\frac{20}{7}$

The solution is $\left(\frac{32}{7}, -\frac{20}{7}\right)$.

**75.** $3(x + y) = 6$

$3(x - y) = -36$

Divide both equations by 3.

$x + y = 2$

$x - y = -12$

Solve the system by addition.

$x + y = 2$

$\underline{x - y = -12}$

$2x = -10$

$x = -5$

Back-substitute to find *y*.

$x + y = 2$

$-5 + y = 2$

$y = 7$

The solution is $(-5, 7)$.

**77.** $3(x - 3) - 2y = 0$

$2(x - y) = -x - 3$

Rewrite the equations in general form.

$3(x-3) - 2y = 0 \quad 2(x - y) = -x - 3$

$3x - 9 - 2y = 0 \quad 2x - 2y = -x - 3$

$3x - 2y = 9 \quad 3x - 2y = -3$

We now have a system of two equations in two variables.

$3x - 2y = 9$

$3x - 2y = -3$

Multiply the second equation by -1.

$3x - 2y = 9$

$\underline{-3x + 2y = 3}$

$0 \neq 12$

Since there are no values or *x* and *y* for which 0 = 12, the system is inconsistent. The solution set is $\varnothing$ or $\{\ \}$.

**79.** $x + 2y - 3 = 0$

$12 = 8y + 4x$

Rewrite the equations in general form.

$x + 2y - 3 = 0 \quad 12 = 8y + 4x$

$x + 2y = 3 \quad 8y + 4x = 12$

$4x + 8y = 12$

We now have a system of two equations in two variables.

$x + 2y = 3$

$4x + 8y = 12$

Multiply the first equation by -4.

$$\begin{array}{r} -4x-8y=-12 \\ \underline{4x+8y=\ \ 12} \\ 0=0 \end{array}$$

Since 0 = 0 for all values of $x$ and $y$, the system is dependent. The solution set is $\{(x,y)|\ x+2y-3=0\}$ or $\{(x,y)|\ 12=8y+4x\}$.

**81.** $3x+4y=0$

$7x=3y$

After rewriting the second equation in general form, the system becomes:

$3x+4y=0$

$7x-3y=0$

Multiply the first equation by 3 and the second equation by 4.

$$\begin{array}{r} 9x+12y=0 \\ \underline{28x-12y=0} \\ 37x=0 \\ x=0 \end{array}$$

Back-substitute to find $y$.

$7(0)=3y$

$0=3y$

$0=y$

The solution is $(0,0)$.

**83.** $\dfrac{x+2}{2}-\dfrac{y+4}{3}=3$

$\dfrac{x+y}{5}=\dfrac{x-y}{2}-\dfrac{5}{2}$

Start by multiplying each equation by its LCD and simplifying to clear the fractions.

$\dfrac{x+2}{2}-\dfrac{y+4}{3}=3$

$\dfrac{x+y}{5}=\dfrac{x-y}{2}-\dfrac{5}{2}$

Start by multiplying each equation by its LCD and simplifying to clear the fractions.

$$6\left(\frac{x+2}{2}-\frac{y+4}{3}\right)=6(3)$$

$3(x+2)-2(y+4)=18$

$3x+6-2y-8=18$

$3x-2y=20$

$$10\left(\frac{x+y}{5}\right)=10\left(\frac{x-y}{2}-\frac{5}{2}\right)$$

$2(x+y)=5(x-y)-5(5)$

$2x+2y=5x-5y-25$

$3x-7y=25$

We now need to solve the equivalent system of equations:

$3x-2y=20$

$3x-7y=25$

Subtract the two equations:

$$\begin{array}{r} 3x-2y=20 \\ \underline{-(3x-7y=25)} \\ 5y=-5 \\ y=-1 \end{array}$$

Back-substitute this value for $y$ and solve for $x$.

$3x-2y=20$

$3x-2(-1)=20$

$3x+2=20$

$3x=18$

$x=6$

The solution is $(6,-1)$.

**84.** $\dfrac{x-y}{3}=\dfrac{x+y}{2}-\dfrac{1}{2}$

$\dfrac{x+2}{2}-4=\dfrac{y+4}{3}$

Start by multiplying each equation by its LCD and simplifying to clear the fractions.

$$6\left(\frac{x-y}{3}\right)=6\left(\frac{x+y}{2}-\frac{1}{2}\right)$$
$$2(x-y)=3(x+y)-3(1)$$
$$2x-2y=3x+3y-3$$
$$x+5y=3$$

$$6\left(\frac{x+2}{2}-4\right)=6\left(\frac{y+4}{3}\right)$$
$$3(x+2)-6(4)=2(y+4)$$
$$3x+6-24=2y+8$$
$$3x-2y=26$$

We now need to solve the equivalent system of equations:

$x+5y=3$

$3x-2y=26$

Multiply the first equation by -3 and then add the equations.

$$-3x-15y=-9$$
$$3x-\ 2y=26$$
$$-17y=17$$
$$y=-1$$

Back-substitute this value for $y$ into one of the above equations and solve for $x$.

$$x+5(-1)=3$$
$$x-5=3$$
$$x=8$$

The solution is $(8,-1)$.

**85.** $5ax+4y=17$

$ax+7y=22$

Multiply the second equation by $-5$ and add the equations.

$$5ax+4y=17$$
$$-5ax-35y=-110$$
$$-31y=-93$$
$$y=3$$

Back-substitute into one of the original equations to solve for $x$.

$$ax+7y=22$$
$$ax+7(3)=22$$
$$ax+21=22$$
$$ax=1$$
$$x=\frac{1}{a}$$

The solution is $\left(\frac{1}{a},3\right)$.

**86.** $4ax+by=3$

$6ax+5by=8$

Multiply the first equation by $-5$ and add the equations.

$$-20ax-5by=-15$$
$$6ax+5by=8$$
$$-14ax=-7$$
$$x=\frac{1}{2a}$$

Back-substitute into one of the original equations to solve for $y$.

$$4a\left(\frac{1}{2a}\right)+by=3$$
$$2+by=3$$
$$by=1$$
$$y=\frac{1}{b}$$

The solution is $\left(\dfrac{1}{2a},\dfrac{1}{b}\right)$.

**87.** $f(-2)=11 \rightarrow -2m+b=11$

$f(3)=-9 \rightarrow 3m+b=-9$

We need to solve the resulting system of equations:

$$-2m+b=11$$
$$3m+b=-9$$

Subtract the two equations:

$$-2m+b=11$$
$$\underline{3m+b=-9}$$
$$-5m=20$$
$$m=-4$$

Back-substitute into one of the original equations to solve for *b*.

$$-2m+b=11$$
$$-2(-4)+b=11$$
$$8+b=11$$
$$b=3$$

Therefore, $m=-4$ and $b=3$.

**88.** $f(-3)=23 \rightarrow -3m+b=23$

$f(2)=-7 \rightarrow 2m+b=-7$

We need to solve the resulting system of equations:

$$-3m+b=23$$
$$2m+b=-7$$

Subtract the two equations:

$$-3m+b=23$$
$$\underline{2m+b=-7}$$
$$-5m=30$$
$$m=-6$$

Back-substitute into one of the original equations to solve for *b*.

$$-3m+b=23$$
$$-3(-6)+b=23$$
$$18+b=23$$
$$b=5$$

Therefore, $m=-6$ and $b=5$.

**89.** The solution to a system of linear equations is the point of intersection of the graphs of the equations in the system. If $(6,2)$ is a solution, then we need to find the lines that intersect at that point.
Looking at the graph, we see that the graphs of $x+3y=12$ and $x-y=4$ intersect at the point $(6,2)$. Therefore, the desired system of equations is

$x+3y=12$ or $y=-\frac{1}{3}x+4$

$x-y=4$ $\quad$ $y=x-4$

**90.** A system whose solution set is the empty set consists of parallel lines (assuming there are only two equations in the system). Therefore, we check the graph for two parallel lines.
From the graph, the desired system is

$x-3y=-6$ or $y=\frac{1}{3}x+2$

$x-3y=6$ $\quad$ $y=\frac{1}{3}x-2$

**91.** We need to solve the system of equations

$$E=0.20x+70.0$$
$$E=0.11x+77.5$$

Since both equations are solved for *E*, we can solve this system by using substitution.

$$0.20x+70.0=0.11x+77.5$$
$$0.09x+70.0=77.5$$
$$0.09x=7.5$$
$$x=\frac{7.5}{0.09}\approx 83$$
$$E=0.20(83)+70.0=16.6+70.0=86.6$$

The two life expectancies will be the same in 2063 (83 years after 1980). At that time, the life expectancy for both men and women will be about 86.6 years. Explanations will vary.

**93.** **a.** $m = \dfrac{25.3-38}{17-0} = \dfrac{-12.7}{17} \approx -0.75$

From the point $(0,38)$ we have that the y-intercept is $b = 38$. Therefore, the equation of the line is $y = -0.75x + 38$.

**b.** $m = \dfrac{23-40}{17-0} = \dfrac{-17}{17} = -1$

From the point $(0,40)$ we have that the y-intercept is $b = 40$. Therefore, the equation of the line is $y = -x + 40$.

**c.** To find the year when cigarette use is the same, we set the two equations equal to each other and solve for $x$.

$$\begin{aligned} -0.75x + 38 &= -x + 40 \\ 0.25x + 38 &= 40 \\ 0.25x &= 2 \\ x &= \frac{2}{0.25} = 8 \end{aligned}$$

$y = -(8) + 40 = 32$

Cigarette use was the same for African Americans and Hispanics in 1993 (8 years after 1985). At that time, 32% of each group used cigarettes.

**95.** **a.**

$$\begin{aligned} N_d &= -5p + 750 \\ &= -5(120) + 750 \\ &= -600 + 750 \\ &= 150 \end{aligned}$$

$N_s = 2.5(120) = 300$

If the price of the televisions is \$120, 150 sets can be sold and 300 sets can be supplied.

**b.** To find the price at which supply and demand are equal, we set the two equations equal to each other and solve for $p$.

$$\begin{aligned} -5p + 750 &= 2.5p \\ 750 &= 7.5p \\ \frac{750}{7.5} &= p \\ 100 &= p \end{aligned}$$

$N = 2.5(100) = 250$.

Supply and demand will be equal if the price of the televisions is \$100. At that price, 250 sets can be supplied and sold.

**97.** Answers will vary.

**99.** Answers will vary.

**101.** Answers will vary.

**103.** Answers will vary.

**105.** Answers will vary.

**107.** $y = 0.94x + 5.64$

$0.74x + y = 146.76 \rightarrow y = -0.74x + 146.76$

Chicken: slope = 0.94

Red meat: slope = −0.74

Chicken consumption is increasing and red meat consumption is decreasing. The two lines are not parallel so there must be an intersection point. To the right of the intersection point, chicken consumption is greater than red meat consumption. This is indicated by the graph for chicken being above the graph for red meat.

**109.** Choice **d** is true. Since the two lines are already in slope-intercept form, we can see that they have the same slope, but different y-intercepts. This means that the lines are parallel and that the system has no solution.

Choice **a** is false because the addition method *can* be used to eliminate either variable.

Choice **b** is false because the solution set for the system contains all real numbers. The second equation is just twice the first equation.

Choice **c** is false because a system of linear equations can never have exactly to ordered pair solutions. There can be 0, 1, or an infinite number of solutions.

**111.** Answers will vary. One system of equations with solution set $\{(-2,7)\}$ is: $3x+4y=22$

$-x-y=-5.$

**113.** $6x=10+5(3x-4)$

$6x=10+15x-20$

$6x=15x-10$

$-9x=-10$

$x=\frac{10}{9}$

The solution is $\frac{10}{9}$.

**114.** $\left(4x^2y^4\right)^2\left(-2x^5y^0\right)^3$

$=4^2x^{2\cdot2}y^{4\cdot2}(-2)^3x^{5\cdot3}y^{0\cdot3}$

$=4^2x^4y^8(-2)^3x^{15}y^0$

$=16(-8)x^4x^{15}y^8y^0$

$=-128x^{19}y^8$

**115.** $f(x)=x^2-3x+7$

$f(-1)=(-1)^2-3(-1)+7$

$=1+3+7$

$=11$

### 3.2 Exercise Set

**1.** Let $x$ = the first number
Let $y$ = the second number

$x+y=7$

$x-y=-1$

Solve by addition:

$x+y=7$

$\underline{x-y=-1}$

$2x=6$

$x=3$

Back-substitute to find $y$.

$3+y=7$

$y=4$

The numbers are 3 and 4.

**3.** Let $x$ = the first number
Let $y$ = the second number

$3x-y=1$

$x+2y=12$

Multiply the first equation by 2.

$6x-2y=2$

$\underline{x+2y=12}$

$7x=14$

$x=2$

Back-substitute to find $y$.

$x+2y=12$

$2+2y=12$

$2y=10$

$y=5$

The numbers are 2 and 5.

**5.** **a.** At the break-even point, $R(x)=C(x)$.

$25500+15x=32x$

$25500=17x$

$1500=x$

$C(x)=32x$

$C(1500)=32(1500)$

$=48000$

Fifteen hundred units must be produced and sold to break-even. At this point, there will $48,000 in costs and revenue.

**b.** $P(x)=R(x)-C(x)$

$=(32x)-(25,500+15x)$

$=32x-25,500-15x$

$=17x-25,500$

**7.** **a.** At the break-even point, $R(x)=C(x)$.

$105x+70000=245x$

$70000=140x$

$500=x$

$C(x)=245x$

$C(500)=245(500)$

$=122500$

Five hundred units must be produced and sold to break-even. At this point, there will $122,500 in costs and revenue.

**b.** $P(x)=R(x)-C(x)$

$=(245x)-(105x+70,000)$

$=245x-105x-70,000$

$=140x-70,000$

**9.** $x+y=36$

$x-y=16$

Add the two equations.

$x+y=36$

$\underline{x-y=16}$

$2x=52$

$x=26$

Back-substitute to solve for $y$.

$x+y=36$

$26+y=36$

$y=10$

26% have a Bachelor's or higher and 10% have an Associate.

**11.** Let $x$ = the number of computers sold. Let $y$ = the number of hard drives sold.

$x+y=36$

$1180x+125y=27,710$

Multiply the first equation by $-125$ and add the two equations.

$-125x-125y=-4500$

$\underline{1180x+125y=27,710}$

$1055x=23,210$

$x=22$

Back-substitute to solve for $y$.

$x+y=36$

$22+y=36$

$y=14$

The store sold 22 computers and 14 hard drives.

**13.** Let $x$ = the amount invested at 6%.
Let $y$ = the amount invested at 8%.

$$x + y = 7000$$
$$0.06x + 0.08y = 520$$

Solve the first equation for $x$.

$$x = 7000 - y$$

Substitute this result for $x$ in the second equation.

$$0.06(7000 - y) + 0.08y = 520$$
$$420 - 0.06y + 0.08y = 520$$
$$0.02y = 100$$
$$y = \frac{100}{0.02} = 5000$$

Back-substitute to solve for $x$.

$$x + y = 7000$$
$$x + 5000 = 7000$$
$$x = 2000$$

\$2000 was invested at 6% and \$5000 was invested at 8%.

**15.** Let $x$ = the amount in the first fund
Let $y$ = the amount in the second fund.

$$0.09x + 0.03y = 900$$
$$0.10x + 0.01y = 860$$

Multiply the second equation by $-3$ and add the two equations.

$$0.09x + 0.03y = 900$$
$$\underline{-0.30x - 0.03y = -2580}$$
$$-0.21x = -1680$$
$$x = 8000$$

Back-substitute to solve for $y$.

$$0.10x + 0.01y = 860$$
$$0.10(8000) + 0.01y = 860$$
$$800 + 0.01y = 860$$
$$0.01y = 60$$
$$y = \frac{60}{0.01} = 6000$$

\$8000 was invested in the first fund and \$6000 was invested in the second fund.

**17.** Let $x$ = amount invested with 12% return.
Let $y$ = amount invested with the 5% loss.

$$x + y = 20{,}000$$
$$0.12x - 0.05y = 1890$$

Multiply the first equation by $0.05$ and add the two equations.

$$0.05x + 0.05y = 1000$$
$$\underline{0.12x - 0.05y = 1890}$$
$$0.17x = 2890$$
$$x = 17{,}000$$

Back-substitute to solve for $y$.

$$x + y = 20{,}000$$
$$17{,}000 + y = 20{,}000$$
$$y = 3{,}000$$

\$17,000 was invested at 12% interest and \$3,000 was invested at a 5% loss.

**19.** Let $x$ = gallons of 5% wine.
Let $y$ = gallons of 9% wine.

$$x + y = 200$$
$$0.05x + 0.09y = 0.07(200)$$

or

$$x + y = 200$$
$$0.05x + 0.09y = 14$$

Solve the first equation for $x$.

$$x + y = 200$$
$$x = 200 - y$$

Substitute this expression for $x$ in the second equation.

$$0.05(200 - y) + 0.09y = 14$$
$$10 - 0.05y + 0.09y = 14$$
$$0.04y = 4$$
$$y = 100$$

Back-substitute and solve for $x$.

$x = 200 - y$

$= 200 - 100$

$= 100$

The wine company should mix 100 gallons of the 5% California wine with 100 gallons of the 9% French wine.

**21.** Let $x$ = grams of 18-karat gold.
Let $y$ = grams of 12-karat gold.

$$x + y = 300$$
$$0.75x + 0.5y = 0.58(300)$$

or

$$x + y = 300$$
$$0.75x + 0.5y = 174$$

Solve the first equation for $x$.

$x = 300 - y$

Substitute this result for $x$ into the second equation and solve for $y$.

$$0.75(300 - y) + 0.5y = 174$$
$$225 - 0.75 + 0.5y = 174$$
$$-0.25y = -51$$
$$y = 204$$

Back-substitute to solve for $x$.

$x = 300 - y = 300 - 204 = 96$

You would need 96 grams of 18-karat gold and 204 grams of 12-karat gold.

**23.** Let $x$ = pounds of cheaper candy.
Let $y$ = pounds of more expensive candy.

$$x + y = 75$$
$$1.6x + 2.1y = 1.9(75)$$

or

$$x + y = 75$$
$$1.6x + 2.1y = 142.5$$

Multiply the first equation by $-1.6$ and add the two equations.

$$-1.6x - 1.6y = -120$$
$$\underline{1.6x + 2.1y = 142.5}$$
$$0.5y = 22.5$$
$$y = 45$$

Back-substitute to solve for $x$.

$$x + 45 = 75$$
$$x = 30$$

The manager should mix 30 pounds of the cheaper candy and 45 pounds of the more expensive candy.

**25.** Let $n$ = the number of nickels.
Let $d$ = the number of dimes.

$$n + d = 15$$
$$0.05n + 0.1d = 1.10$$

Solve the first equation for $n$.

$n = 15 - d$

$$0.05(15 - d) + 0.1d = 1.10$$
$$0.75 - 0.05d + 0.1d = 1.10$$
$$0.05d = 0.35$$
$$d = 7$$

Back-substitute to solve for $n$.

$$n + 7 = 15$$
$$n = 8$$

The purse has 8 nickels and 7 dimes.

**27.** Let $x$ = the speed of the plane in still air
Let $y$ = the speed of the wind

| | r × | t = | d |
|---|---|---|---|
| With Wind | $x + y$ | 5 | 800 |
| Against Wind | $x - y$ | 8 | 800 |

$$5(x + y) = 800$$
$$\underline{8(x - y) = 800}$$
$$5x + 5y = 800$$
$$8x - 8y = 800$$

Multiply the first equation by 8 and the second equation by 5.

$$40x + 40y = 6400$$
$$\underline{40x - 40y = 4000}$$
$$80x = 10400$$
$$x = 130$$

Back-substitute to find $y$.

$$5x + 5y = 800$$
$$5(130) + 5y = 800$$
$$650 + 5y = 800$$
$$5y = 150$$
$$y = 30$$

The speed of the plane in still air is 130 miles per hour and the speed of the wind is 30 miles per hour.

**29.** Let $x$ = the crew's rowing rate
Let $y$ = the rate of the current

| | r × | t = | d |
|---|---|---|---|
| With Current | $x + y$ | 2 | 16 |
| Against Current | $x - y$ | 4 | 16 |

$$2(x + y) = 16$$
$$4(x - y) = 16$$

Rewrite the system is $Ax + By = C$ form.

$$2x + 2y = 16$$
$$4x - 4y = 16$$

Multiply the first equation by –2.

$$-4x - 4y = -32$$
$$\underline{4x - 4y = \ \ 16}$$
$$-8y = -16$$
$$y = 2$$

Back-substitute to find $x$.

$$2x + 2(2) = 16$$
$$2x + 4 = 16$$
$$2x = 12$$
$$x = 6$$

The crew's rowing rate is 6 kilometers per hour and the rate of the current is 2 kilometers per hour.

**31.** Let $x$ = the speed in still water
Let $y$ = the speed of the current

| | r · | t = | d |
|---|---|---|---|
| With Current | $x + y$ | 4 | 24 |
| Against Current | $x - y$ | 6 | $\frac{3}{4}(24)$ |

$$4(x + y) = 24$$
$$6(x - y) = \frac{3}{4}(24)$$

Rewrite the system is $Ax + By = C$ form.

$$4x + 4y = 24$$
$$6x - 6y = 18$$

Multiply the first equation by –3 and the second equation by 2.

$$-12x - 12y = -72$$
$$\underline{12x - 12y = 36}$$
$$-24y = -36$$
$$y = 1.5$$

Back-substitute to find $x$.

$$4x + 4y = 24$$
$$4x + 4(1.5) = 24$$
$$4x + 6 = 24$$
$$4x = 18$$
$$x = 4.5$$

The speed in still water is 4.5 miles per hour and the speed of the current is 1.5 miles per hour.

**33.** Let $x$ = the larger score.
Let $y$ = the smaller score.

$$x - y = 12$$
$$\frac{x + y}{2} = 80$$

Rewrite the second equation in standard form.

$$x - y = 12$$
$$x + y = 160$$

Add the two equations.

$$\begin{aligned} x - y &= 12 \\ \underline{x + y} &\underline{= 160} \\ 2x \quad &= 172 \\ x \quad &= 86 \end{aligned}$$

Back-substitute to solve for $y$.

$$\begin{aligned} x - y &= 12 \\ 86 - y &= 12 \\ -y &= -74 \\ y &= 74 \end{aligned}$$

The two test scores are 74 and 86.

**35.** $$\begin{aligned} x + 2y &= 180 \\ (2x - 30) + y &= 180 \end{aligned}$$

Rewrite the second equation in standard form.

$$\begin{aligned} x + 2y &= 180 \\ 2x + y &= 210 \end{aligned}$$

Multiply the first equation by $-2$ and add the equations.

$$\begin{aligned} -2x - 4y &= -360 \\ \underline{2x + y} &\underline{= 210} \\ -3y &= -150 \\ y &= 50 \end{aligned}$$

Back-substitute to solve for $x$.

$$\begin{aligned} x + 2y &= 180 \\ x + 2(50) &= 180 \\ x + 100 &= 180 \\ x &= 80 \end{aligned}$$

The three interior angles measure $80°$, $50°$, and $50°$.

**37.** Let $x$ = the lot length.
Let $y$ = the lot width.

$$\begin{aligned} 2x + 2y &= 220 \\ 20x + 8(2y) &= 2040 \end{aligned}$$

Rewriting the second equation gives the following equivalent system:

$$\begin{aligned} 2x + 2y &= 220 \\ 20x + 16y &= 2040 \end{aligned}$$

Multiply the first equation by $-8$ and add the two equations.

$$\begin{aligned} -16x - 16y &= -1760 \\ \underline{20x + 16y} &\underline{= 2040} \\ 4x &= 280 \\ x &= 70 \end{aligned}$$

Back-substitute to find $y$.

$$\begin{aligned} 2x + 2y &= 220 \\ 2(70) + 2y &= 220 \\ 140 + 2y &= 220 \\ 2y &= 80 \\ y &= 40 \end{aligned}$$

The lot is 70 feet long and 40 feet wide.

**39.** Let $x$ = the number of two-seat tables.
Let $y$ = the number of four-seat tables.

$$\begin{aligned} 2x + 4y &= 56 \\ x + y &= 17 \end{aligned}$$

Multiply the second equation by $-2$ and add the two equations.

$$\begin{aligned} 2x + 4y &= 56 \\ \underline{-2x - 2y} &\underline{= -34} \\ 2y &= 22 \\ y &= 11 \end{aligned}$$

Back-substitute to find $x$.

$$\begin{aligned} x + y &= 17 \\ x + 11 &= 17 \\ x &= 6 \end{aligned}$$

The owners should buy 6 two-seat tables and 11 four-seat tables.

**41.** At the break-even point, $R(x)=C(x)$.

$10000+30x=50x$

$10000=20x$

$10000=20x$

$500=x$

Five hundred radios must be produced and sold to break-even.

**43.** $R(x)=50x$

$R(200)=50(200)=10000$

$C(x)=10000+30x$

$C(200)=10000+30(200)$

$=10000+6000=16000$

$R(200)-C(200)=10000-16000$

$=-6000$

This means that if 200 radios are produced and sold the company will lose $6,000.

**45.** **a.** $P(x)=R(x)-C(x)$

$=50x-(10000+30x)$

$=50x-10000-30x$

$=20x-10000$

$P(x)=20x-10000$

**b.** $P(10000)=20(10000)-10000$

$=200000-10000=190000$

If 10,0000 radios are produced and sold the profit will be $190,000.

**47.** **a.** The cost function is:

$C(x)=18,000+20x$

**b.** The revenue function is:

$R(x)=80x$

**c.** At the break-even point,

$R(x)=C(x)$.

$80x=18000+20x$

$60x=18000$

$x=300$

$R(x)=80x$

$R(300)=80(300)$

$=24,000$

When approximately 300 canoes are produced the company will break-even with cost and revenue at $24,000.

**49.** **a.** The cost function is:

$C(x)=30000+2500x$

**b.** The revenue function is:

$R(x)=3125x$

**c.** At the break-even point,

$R(x)=C(x)$.

$3125x=30000+2500x$

$625x=30000$

$x=48$

After 48 sold out performances, the investor will break-even. ($150,000)

**51.** Answers will vary.

**53.** Answers will vary.

**55.** Answers will vary.

**57.** Answers will vary.

**59.** $R(x)=92.5x;\quad C(x)=52x+1782$

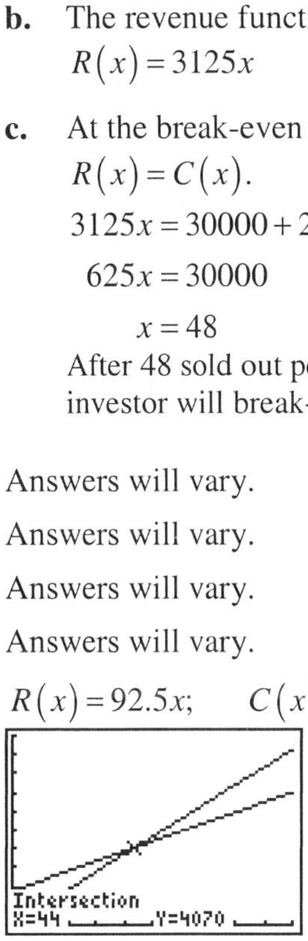


The break even point is 44 units. When 44 units are produced and sold, cost and revenue are the same at $4070.

**61.** Let $h$ = the number of hexagons.
Let $s$ = the number of squares.
We need to solve the following system of equations:

$6h+s=52$ (band)

$h+4s=24$ (pom-pom)

Multiply the first equation by $-4$ and add the equations.

$$\begin{aligned} -24h-4s&=-208 \\ h+4s&=24 \\ \hline -23h \quad &=-184 \\ h \quad &=8 \end{aligned}$$

Back-substitute to solve for $s$.

$$\begin{aligned} 6h+s&=52 \\ 6(8)+s&=52 \\ 48+s&=52 \\ s&=4 \end{aligned}$$

The students can form 8 hexagons and 4 squares.

**63.** Let $t$ = the original tens-place digit.
Let $u$ = the original ones-place digit.

$x+u=14$ (digits sum)

$(10x+u)-(10u+x)=36$ (score diff.)

Simplify the second equation so it is in standard form.

$$\begin{aligned} 10x+u-10u-x&=36 \\ 9x-9u&=36 \end{aligned}$$

Solve the following system:

$$\begin{aligned} x+u&=14 \\ 9x-9u&=36 \end{aligned}$$

Multiply the first equation by 9 and add the two equations.

$$\begin{aligned} 9x+9u&=126 \\ 9x-9u&=36 \\ \hline 18x \quad &=162 \\ x \quad &=9 \end{aligned}$$

Back-substitute to solve for $u$.

$$\begin{aligned} x+u&=14 \\ 9+u&=14 \\ u&=5 \end{aligned}$$

Therefore, your original score was 95.

**65.** Let $x$ = the cost of the mangos
Let $y$ = the cost of the avocados

$$\begin{aligned} x+y&=67 \\ 0.20x-0.02y&=8.56 \end{aligned}$$

Multiply the first equation by 0.02.

$$\begin{aligned} 0.02x+0.02y&=1.34 \\ 0.20x-0.02y&=8.56 \\ \hline 0.22x&=9.90 \\ x&=45 \end{aligned}$$

Back-substitute to find $y$.

$$\begin{aligned} x+y&=67 \\ 45+y&=67 \\ y&=22 \end{aligned}$$

The dealer paid $45 for the mangos and $22 for the avocados.

**66.** Passing through $(-2,5)$ and $(-6,13)$

First, find the slope:

$$m=\frac{y_2-y_1}{x_2-x_1}=\frac{13-5}{-6-(-2)}=\frac{8}{-4}=-2$$

Use the slope and one of the points to write the equation in point-slope form.

$$\begin{aligned} y-y_1&=m(x-x_1) \\ y-5&=-2(x-(-2)) \\ y-5&=-2(x+2) \end{aligned}$$

or

$$\begin{aligned} y-y_1&=m(x-x_1) \\ y-13&=-2(x-(-6)) \\ y-13&=-2(x+6) \end{aligned}$$

Rewrite the equation in slope-intercept form by solving for $y$.

$$y-13=-2(x+6)$$
$$y=-2x-12+13$$
$$y=-2x+1$$

In function notation, the equation of the line is $f(x)=-2x+1$.

**67.** Since the line is parallel to $-x+y=7$, we can use it to obtain the slope. Rewriting the equation in slope-intercept form, we obtain $y=x+7$. The slope is $m=1$. We are given that it passes through (–3, 0). We use the slope and point to write the equation in point-slope form.

$$y-y_1=m(x-x_1)$$
$$y-0=1(x-(-3))$$
$$y-0=1(x+3)$$

Rewrite the equation in slope-intercept form by solving for $y$.

$$y-0=1(x+3)$$
$$y=x+3$$

In function notation, the equation of the line is $f(x)=x+3$.

**68.** Since the denominator of a fraction cannot be zero, the domain of $g$ is $\{x \mid x \text{ is a real number and } x \neq 3\}$.

### 3.3 Exercise Set

**1.**
$$x+y+z=4$$
$$2-1+3=4$$
$$4=4$$
True

$$x-2y-z=1$$
$$2-2(-1)-3=1$$
$$2+2-3=1$$
$$1=1$$
True

$$2x-y-z=-1$$
$$2(2)-(-1)-3=-1$$
$$4+1-3=-1$$
$$1=-1$$
False

The ordered triple (2, –1, 3) does not make all three equations true, so it is not a solution.

**3.**
$$x-2y=2$$
$$4-2(1)=2$$
$$4-2=2$$
$$2=2$$
True

$$2x+3y=11$$
$$2(4)+3(1)=11$$
$$8+3=11$$
$$11=11$$
True

$$y-4z=-7$$
$$1-4(2)=-7$$
$$1-8=-7$$
$$-7=-7$$
True

The ordered triple (4, 1, 2) makes all three equations true, so it is a solution.

**5.** $x + y + 2z = 11$

$x + y + 3z = 14$

$x + 2y - z = 5$

Multiply the second equation by –1 and add to the first equation..

$x + y + 2z = 11$

$-x - y - 3z = -14$

$-z = -3$

$z = 3$

Back-substitute 3 for $z$ in the first and third equations:

$x + y + 2z = 11$ $\quad$ $x + 2y - z = 5$

$x + y + 2(3) = 11$ $\quad$ $x + 2y - 3 = 5$

$x + y + 6 = 11$ $\quad$ $x + 2y = 8$

$x + y = 5$

We now have two equations in two variables.

$x + y = 5$

$x + 2y = 8$

Multiply the first equation by –1 and solve by addition.

$-x - y = -5$

$x + 2y = 8$

$y = 3$

Back-substitute 3 for $y$ into one of the equations in two variables.

$x + y = 5$

$x + 3 = 5$

$x = 2$

The solution is $(2,3,3)$ and the solution set is $\{(2,3,3)\}$.

**7.** $4x - y + 2z = 11$

$x + 2y - z = -1$

$2x + 2y - 3z = -1$

Multiply the second equation by –4 and add to the first equation.

$4x - y + 2z = 11$

$-4x - 8y + 4z = 4$

$-9y + 6z = 15$

Multiply the second equation by –2 and add it to the third equation.

$-2x - 4y + 2z = 2$

$2x + 2y - 3z = -1$

$-2y - z = 1$

We now have two equations in two variables.

$-9y + 6z = 15$

$-2y - z = 1$

Multiply the second equation by 6 and solve by addition.

$-9y + 6z = 15$

$-12y - 6z = 6$

$-21y = 21$

$y = 21$

Back-substitute –1 for $y$ in one of the equations in two variables.

$-2y - z = 1$

$-2(-1) - z = 1$

$2 - z = 1$

$-z = -1$

$z = 1$

Back-substitute –1 for $y$ and 1 for $z$ in one of the original equations in three variables.

$x + 2y - z = -1$

$x + 2(-1) - 1 = -1$

$x - 2 - 1 = -1$

$x - 3 = -1$

$x = 2$

The solution is $(2,-1,1)$ and the solution set is $\{(2,-1,1)\}$.

**9.** $3x + 2y - 3z = -2$

$2x - 5y + 2z = -2$

$4x - 3y + 4z = 10$

Multiply the second equation by –2 and add to the third equation.

$$\begin{aligned} -4x + 10y - 4z &= 4 \\ 4x - 3y + 4z &= 10 \\ \hline 7y &= 14 \\ y &= 2 \end{aligned}$$

Back-substitute 2 for $y$ in the first and third equations to obtain two equations in two unknowns.

$$\begin{aligned} 3x + 2y - 3z &= -2 \\ 3x + 2(2) - 3z &= -2 \\ 3x + 4 - 3z &= -2 \\ 3x - 3z &= -6 \end{aligned}$$

$$\begin{aligned} 4x - 3y + 4z &= 10 \\ 4x - 3(2) + 4z &= 10 \\ 4x - 6 + 4z &= 10 \\ 4x + 4z &= 16 \end{aligned}$$

The system of two equations in two variables becomes:

$3x - 3z = -6$

$4x + 4z = 16$

Multiply the first equation by –4 and the second equation by 3.

$$\begin{aligned} -12x + 12z &= 24 \\ 12x + 12z &= 48 \\ \hline 24z &= 72 \\ z &= 3 \end{aligned}$$

Back-substitute 3 for $z$ to find $x$.

$$\begin{aligned} 3x - 3z &= -6 \\ 3x - 3(3) &= -6 \\ 3x - 9 &= -6 \\ 3x &= 3 \\ x &= 1 \end{aligned}$$

The solution is $(1, 2, 3)$ and the solution set is $\{(1, 2, 3)\}$.

**11.** $2x - 4y + 3z = 17$

$x + 2y - z = 0$

$4x - y - z = 6$

Multiply the second equation by –1 and add it to the third equation.

$$\begin{aligned} -x - 2y + z &= 0 \\ 4x - y - z &= 6 \\ \hline 3x - 3y &= 6 \end{aligned}$$

Multiply the second equation by 3 and add it to the first equation.

$$\begin{aligned} 2x - 4y + 3z &= 17 \\ 3x + 6y - 3z &= 0 \\ \hline 5x + 2y &= 17 \end{aligned}$$

The system in two variables becomes:

$3x - 3y = 6$

$5x + 2y = 17$

Multiply the first equation by 2 and the second equation by 3 and solve by addition.

$$\begin{aligned} 6x - 6y &= 12 \\ 15x + 6y &= 51 \\ \hline 21x &= 63 \\ x &= 3 \end{aligned}$$

Back-substitute 3 for $x$ in one of the equations in two variables.

$$\begin{aligned} 3x - 3y &= 6 \\ 3(3) - 3y &= 6 \\ 9 - 3y &= 6 \\ -3y &= -3 \\ y &= 1 \end{aligned}$$

Back-substitute 3 for $x$ and 1 for $y$ in one of the original equations in three variables.

$$x+2y-z=0$$
$$3+2(1)-z=0$$
$$3+2-z=0$$
$$5-z=0$$
$$5=z$$

The solution is $(3,1,5)$ and the solution set is $\{(3,1,5)\}$.

**13.** $2x+y=2$
$x+y-z=4$
$3x+2y+z=0$

Add the second and third equations together to obtain an equation in two variables.

$$x+y-z=4$$
$$3x+2y+z=0$$
$$\overline{4x+3y=4}$$

Use this equation and the first equation in the original system to write two equations in two variables.

$$2x+y=2$$
$$4x+3y=4$$

Multiply the first equation by –2 and solve by addition.

$$-4x-2y=-4$$
$$4x+3y=4$$
$$\overline{y=0}$$

Back-substitute 0 for $y$ in one of the equations in two unknowns.

$$2x+y=2$$
$$2x+0=2$$
$$2x=2$$
$$x=1$$

Back-substitute 1 for $x$ and 0 for $y$ in one of the equations in three unknowns.

$$x+y-z=4$$
$$1+0-z=4$$
$$1-z=4$$
$$-z=3$$
$$z=-3$$

The solution is $(1,0,-3)$ and the solution set is $\{(1,0,-3)\}$.

**15.** $x+y=-4$
$y-z=1$
$2x+y+3z=-21$

Multiply the first equation by –1 and add to the second equation.

$$-x-y=4$$
$$y-z=1$$
$$\overline{-x-z=5}$$

Multiply the second equation by –1 and add to the third equation.

$$-y+z=-1$$
$$2x+y+3z=-21$$
$$\overline{2x+4z=-22}$$

The system of two equations in two variables becomes.

$$-x-z=5$$
$$2x+4z=-22$$

Multiply the first equation by 2 and add to the second equation.

$$-2x-2z=10$$
$$2x+4z=-22$$
$$\overline{2z=-12}$$
$$z=-6$$

Back-substitute –6 for $z$ in one of the equations in two variables.

$-x - z = 5$

$-x - (-6) = 5$

$-x + 6 = 5$

$-x = -1$

$x = 1$

Back-substitute 1 for $x$ in the first equation of the original system.

$x + y = -4$

$1 + y = -4$

$y = -5$

The solution is $(1, -5, -6)$ and the solution set is $\{(1, -5, -6)\}$.

**17.** $2x + y + 2z = 1$

$3x - y + z = 2$

$x - 2y - z = 0$

Add the first and second equations to eliminate $y$.

$2x + y + 2z = 1$

$\underline{3x - y + z = 2}$

$5x \quad + 3z = 3$

Multiply the second equation by –2 and add to the third equation.

$-6x + 2y - 2z = -4$

$\underline{x - 2y - z = 0}$

$-5x \quad - 3z = -4$

We obtain two equations in two variables.

$5x + 3z = 3$

$-5x - 3z = -4$

Adding the two equations, we obtain:

$5x + 3z = 3$

$\underline{-5x - 3z = -4}$

$0 = -1$

The system is inconsistent. There are no values of $x$, $y$, and $z$ for which $0 = -1$.

**19.** $5x - 2y - 5z = 1$

$10x - 4y - 10z = 2$

$15x - 6y - 15z = 3$

Multiply the first equation by –2 and add to the second equation.

$-10x + 4y + 10z = -2$

$\underline{10x - 4y - 10z = 2}$

$0 = 0$

The system is dependent and has infinitely many solutions.

**21.** $3(2x + y) + 5z = -1$

$2(x - 3y + 4z) = -9$

$4(1 + x) = -3(z - 3y)$

Rewrite each equation and obtain the system of three equations in three variables.

$6x + 3y + 5z = -1$

$2x - 6y + 8z = -9$

$4x - 9y + 3z = -4$

Multiply the second equation by –3 and add to the first equation.

$6x + 3y + 5z = -1$

$\underline{-6x + 18y - 24z = 27}$

$21y - 19z = 26$

Multiply the second equation by –2 and add to the third equation.

$-4x + 12y - 16z = 18$

$\underline{4x - 9y + 3z = -4}$

$3y - 13z = 14$

The system of two variables in two equations is:

$21y - 19z = 26$

$3y - 13z = 14$

Multiply the second equation by –7 and add to the third equation.

$$\begin{aligned} 21y-19z &= 26 \\ -21y+91z &= -98 \\ \hline 72z &= -72 \\ z &= -1 \end{aligned}$$

Back-substitute –1 for $z$ in one of the equations in two variables to find $y$.

$$\begin{aligned} 3y-13z &= 14 \\ 3y-13(-1) &= 14 \\ 3y+13 &= 14 \\ 3y &= 1 \\ y &= \frac{1}{3} \end{aligned}$$

Back-substitute –1 for $z$ and $\frac{1}{3}$ for $y$ in one of the original equations in three variables.

$$\begin{aligned} 6x+3y+5z &= -1 \\ 6x+1-5 &= -1 \\ 6x-4 &= -1 \\ 6x &= 3 \\ x &= \frac{1}{2} \end{aligned}$$

The solution is $\left(\frac{1}{2},\frac{1}{3},-1\right)$ and the solution set is $\left\{\left(\frac{1}{2},\frac{1}{3},-1\right)\right\}$.

**23.** Use each ordered pair to write an equation as follows:

$(x, y) = (-1, 6)$

$$\begin{aligned} y &= ax^2+bx+c \\ 6 &= a(-1)^2+b(-1)+c \\ 6 &= a-b+c \end{aligned}$$

$(x, y) = (1, 4)$

$$\begin{aligned} y &= ax^2+bx+c \\ 4 &= a(1)^2+b(1)+c \\ 4 &= a+b+c \end{aligned}$$

$(x, y) = (2, 9)$

$$\begin{aligned} y &= ax^2+bx+c \\ 9 &= a(2)^2+b(2)+c \\ 9 &= a(4)+2b+c \\ 9 &= 4a+2b+c \end{aligned}$$

The system of three equations in three variables is:

$$\begin{aligned} a-b+c &= 6 \\ a+b+c &= 4 \\ 4a+2b+c &= 9 \end{aligned}$$

Add the first and second equations.

$$\begin{aligned} a-b+c &= 6 \\ a+b+c &= 4 \\ \hline 2a+2c &= 10 \end{aligned}$$

Multiply the first equation by 2 and add to the third equation.

$$\begin{aligned} 2a-2b+2c &= 12 \\ 4a+2b+c &= 9 \\ \hline 6a+3c &= 21 \end{aligned}$$

The system of two equations in two variables becomes:

$$\begin{aligned} 2a+2c &= 10 \\ 6a+3c &= 21 \end{aligned}$$

Multiply the first equation by –3 and add to the second equation.

$$\begin{aligned} -6a-6c &= -30 \\ 6a+3c &= 21 \\ \hline -3c &= -9 \\ c &= 3 \end{aligned}$$

Back-substitute 3 for $c$ in one of the equations in two variables.

$$2a+2c=10$$
$$2a+2(3)=10$$
$$2a+6=10$$
$$2a=4$$
$$a=2$$

Back-substitute 3 for $c$ and 2 for $a$ in one of the equations in three variables.

$$a+b+c=4$$
$$2+b+3=4$$
$$b+5=4$$
$$b=-1$$

The quadratic function is $y=2x^2-x+3.$

**25.** Use each ordered pair to write an equation.

$$(x,y)=(-1,-4)$$
$$y=ax^2+bx+c$$
$$-4=a(-1)^2+b(-1)+c$$
$$-4=a-b+c$$

$$(x,y)=(1,-2)$$
$$y=ax^2+bx+c$$
$$-2=a(1)^2+b(1)+c$$
$$-2=a+b+c$$

$$(x,y)=(2,5)$$
$$y=ax^2+bx+c$$
$$5=a(2)^2+b(2)+c$$
$$5=a(4)+2b+c$$
$$5=4a+2b+c$$

The system of three equations in three variables is:

$$a-b+c=-4$$
$$a+b+c=-2$$
$$4a+2b+c=5$$

Multiply the second equation by –1 and add to the first equation.

$$a-b+c=-4$$
$$\underline{-a-b-c=2}$$
$$-2b=-2$$
$$b=1$$

Back-substitute 4 for $b$ in first and third equations to obtain two equations in two variables.

$$a-b+c=-4 \qquad 4a+2b+c=5$$
$$a-1+c=-4 \qquad 4a+2(1)+c=5$$
$$a+c=-3 \qquad 4a+2+c=5$$
$$4a+c=3$$

The system of two equations in two variables becomes:

$$a+c=-3$$
$$4a+c=3$$

Multiply the first equation by –1 and add to the second equation.

$$-a-c=3$$
$$\underline{4a+c=3}$$
$$3a=6$$
$$a=2$$

Back-substitute 2 for a and 1for $b$ in one of the equations in three variables.

$$a-b+c=-4$$
$$2-1+c=-4$$
$$1+c=-4$$
$$c=-5$$

The quadratic function is $y=2x^2+x-5.$

**27.** Let $x$ = the first number
Let $y$ = the second number
Let $z$ = the third number

$$x + y + z = 16$$
$$2x + 3y + 4z = 46$$
$$5x - y = 31$$

Multiply the first equation by –4 and add to the second equation.

$$-4x - 4y - 4z = -64$$
$$\underline{2x + 3y + 4z = 46}$$
$$-2x - y = -18$$

The system of two equations in two variables becomes:

$$5x - y = 31$$
$$-2x - y = -18$$

Multiply the first equation by –1 and add to the second equation.

$$-5x + y = -31$$
$$\underline{-2x - y = -18}$$
$$-7x = -49$$
$$x = 7$$

Back-substitute 7 for $x$ in one of the equations in two variables.

$$5x - y = 31$$
$$5(7) - y = 31$$
$$35 - y = 31$$
$$-y = -4$$
$$y = 4$$

Back-substitute 7 for $x$ and 4 for $y$ in one of the equations in two variables.

$$x + y + z = 16$$
$$7 + 4 + z = 16$$
$$11 + z = 16$$
$$z = 5$$

The numbers are 7, 4 and 5.

**29.**

$$\frac{x+2}{6} - \frac{y+4}{3} + \frac{z}{2} = 0$$
$$6\left(\frac{x+2}{6} - \frac{y+4}{3} + \frac{z}{2}\right) = 6(0)$$
$$(x+2) - 2(y+4) + 3z = 0$$
$$x + 2 - 2y - 8 + 3z = 0$$
$$x - 2y + 3z = 6$$

$$\frac{x+1}{2} + \frac{y-1}{2} - \frac{z}{4} = \frac{9}{2}$$
$$4\left(\frac{x+1}{2} + \frac{y-1}{2} - \frac{z}{4}\right) = 4\left(\frac{9}{2}\right)$$
$$2(x+1) + 2(y-1) - z = 18$$
$$2x + 2 + 2y - 2 - z = 18$$
$$2x + 2y - z = 18$$

$$\frac{x-5}{4} + \frac{y+1}{3} + \frac{z-2}{2} = \frac{19}{4}$$
$$12\left(\frac{x-5}{4} + \frac{y+1}{3} + \frac{z-2}{2}\right) = 12\left(\frac{19}{4}\right)$$
$$3(x-5) + 4(y+1) + 6(z-2) = 57$$
$$3x - 15 + 4y + 4 + 6z - 12 = 57$$
$$3x + 4y + 6z = 80$$

We need to solve the equivalent system:

$$x - 2y + 3z = 6$$
$$2x + 2y - z = 18$$
$$3x + 4y + 6z = 80$$

Add the first two equations together.

$$x - 2y + 3z = 6$$
$$\underline{2x + 2y - z = 18}$$
$$3x + 2z = 24$$

Multiply the second equation by –2 and add it to the third equation.

$$\begin{aligned}-4x-4y+2z&=-36\\ 3x+4y+6z&=80\\ \hline -x+8z&=44\end{aligned}$$

Using the two reduced equations, we solve the system

$3x+2z=24$

$-x+8z=44$

Multiply the second equation by 3 and add the equations.

$$\begin{aligned}3x+2z&=24\\ -3x+24z&=132\\ \hline 26z&=156\\ z&=6\end{aligned}$$

Back-substitute to find $x$.

$$\begin{aligned}-x+8(6)&=44\\ -x+48&=44\\ -x&=-4\\ x&=4\end{aligned}$$

Back substitute to find $y$.

$$\begin{aligned}x-2y+3z&=6\\ 4-2y+3(6)&=6\\ -2y&=-16\\ y&=8\end{aligned}$$

The solution is $(4,8,6)$.

**30.** $$\frac{x+3}{2}-\frac{y-1}{2}+\frac{z+2}{4}=\frac{3}{2}$$

$$4\left(\frac{x+3}{2}-\frac{y-1}{2}+\frac{z+2}{4}\right)=4\left(\frac{3}{2}\right)$$

$$\begin{aligned}2(x+3)-2(y-1)+(z+2)&=6\\ 2x+6-2y+2+z+2&=6\\ 2x-2y+z&=-4\end{aligned}$$

$$\frac{x-5}{2}+\frac{y+1}{3}-\frac{z}{4}=-\frac{25}{6}$$

$$12\left(\frac{x-5}{2}+\frac{y+1}{3}-\frac{z}{4}\right)=12\left(-\frac{25}{6}\right)$$

$$\begin{aligned}6(x-5)+4(y+1)-3z&=-50\\ 6x-30+4y+4-3z&=-50\\ 6x+4y-3z&=-24\end{aligned}$$

$$\frac{x-3}{4}-\frac{y+1}{2}+\frac{z-3}{2}=-\frac{5}{2}$$

$$4\left(\frac{x-3}{4}-\frac{y+1}{2}+\frac{z-3}{2}\right)=4\left(-\frac{5}{2}\right)$$

$$\begin{aligned}(x-3)-2(y+1)+2(z-3)&=-10\\ x-3-2y-2+2z-6&=-10\\ x-2y+2z&=1\end{aligned}$$

We need to solve the equivalent system:

$2x-2y+z=-4$

$6x+4y-3z=-24$

$x-2y+2z=1$

Multiply the first equation by 2 and add it to the second equation.

$$\begin{aligned}4x-4y+2z&=-8\\ 6x+4y-3z&=-24\\ \hline 10x-z&=-32\end{aligned}$$

Multiply the third equation by $-1$ and add to the first equation.

$$\begin{aligned}2x-2y+z&=-4\\ -x+2y-2z&=-1\\ \hline x-z&=-5\end{aligned}$$

Using the two reduced equations, we solve the system

$10x-z=-32$

$x-z=-5$

Multiply the second equation by $-1$ and add it to the first.

$$\begin{aligned} 10x - z &= -32 \\ -x + z &= 5 \\ \hline 9x &= -27 \\ x &= -3 \end{aligned}$$

Back-substitute to solve for $z$.

$$\begin{aligned} x - z &= -5 \\ -3 - z &= -5 \\ -z &= -2 \\ z &= 2 \end{aligned}$$

Back-substitute to solve for $y$.

$$\begin{aligned} x - 2y + 2z &= 1 \\ -3 - 2y + 2(2) &= 1 \\ -3 - 2y + 4 &= 1 \\ -2y &= 0 \\ y &= 0 \end{aligned}$$

The solution is $(-3, 0, 2)$.

**31.** Selected points may vary, but the equation will be the same.

$y = ax^2 + bx + c$

Use the points $(2, -2)$, $(4, 1)$, and $(6, -2)$ to get the system

$$\begin{aligned} 4a + 2b + c &= -2 \\ 16a + 4b + c &= 1 \\ 36a + 6b + c &= -2 \end{aligned}$$

Multiply the first equation by $-1$ and add to the second equation.

$$\begin{aligned} -4a - 2b - c &= 2 \\ 16a + 4b + c &= 1 \\ \hline 12a + 2b &= 3 \end{aligned}$$

Multiply the first equation by $-1$ and add to the third equation.

$$\begin{aligned} -4a - 2b - c &= 2 \\ 36a + 6b + c &= -2 \\ \hline 32a + 4b &= 0 \end{aligned}$$

Using the two reduced equations, we get the system

$$\begin{aligned} 12a + 2b &= 3 \\ 32a + 4b &= 0 \end{aligned}$$

Multiply the first equation by $-2$ and add to the second equation.

$$\begin{aligned} -24a - 4b &= -6 \\ 32a + 4b &= 0 \\ \hline 8a &= -6 \\ a &= -\frac{3}{4} \end{aligned}$$

Back-substitute to solve for $b$.

$$\begin{aligned} 12a + 2b &= 3 \\ 12\left(-\frac{3}{4}\right) + 2b &= 3 \\ -9 + 2b &= 3 \\ 2b &= 12 \\ b &= 6 \end{aligned}$$

Back-substitute to solve for $c$.

$$\begin{aligned} 4a + 2b + c &= -2 \\ 4\left(-\frac{3}{4}\right) + 2(6) + c &= -2 \\ -3 + 12 + c &= -2 \\ c &= -11 \end{aligned}$$

The equation is:

$$y = -\frac{3}{4}x^2 + 6x - 11$$

**32.** Selected points may vary, but the equation will be the same.

$y = ax^2 + bx + c$

Use the points $(3, 4)$, $(4, 2)$, and $(5, 2)$ to get the system

$$\begin{aligned} 9a + 3b + c &= 4 \\ 16a + 4b + c &= 2 \\ 25a + 5b + c &= 2 \end{aligned}$$

Multiply the first equation by $-1$ and add to the second equation.

$$-9a-3b-c=-4$$
$$\underline{16a+4b+c=2}$$
$$7a+b=-2$$

Multiply the first equation by $-1$ and add to the third equation.

$$-9a-3b-c=-4$$
$$\underline{25a+5b+c=2}$$
$$16a+2b=-2$$

Use the two reduced equations to get the system

$$7a+b=-2$$
$$16a+2b=-2$$

Multiply the first equation by $-2$ and add to the second equation.

$$-14a-2b=4$$
$$\underline{16a+2b=-2}$$
$$2a=2$$
$$a=1$$

Back-substitute to solve for $b$.

$$7a+b=-2$$
$$7(1)+b=-2$$
$$7+b=-2$$
$$b=-9$$

Back-substitute to solve for $c$.

$$9a+3b+c=4$$
$$9(1)+3(-9)+c=4$$
$$9-27+c=4$$
$$c=22$$

The equation is:

$$y=x^2-9x+22$$

**33.** $ax-by-2cz=21$

$$ax+by+cz=0$$
$$2ax-by+cz=14$$

Add the first two equations.

$$ax-by-2cz=21$$
$$\underline{ax+by+cz=0}$$
$$2ax-cz=21$$

Multiply the first equation by $-1$ and add to the third equation.

$$-ax+by+2cz=-21$$
$$\underline{2ax-by+cz=14}$$
$$ax+3cz=-7$$

Use the two reduced equations to get the following system:

$$2ax-cz=21$$
$$ax+3cz=-7$$

Multiply the second equation by $-2$ and add the equations.

$$2ax-cz=21$$
$$\underline{-2ax-6cz=14}$$
$$-7cz=35$$
$$z=-\frac{5}{c}$$

Back-substitute to solve for $x$.

$$ax+3cz=-7$$
$$ax+3c\left(-\frac{5}{c}\right)=-7$$
$$ax-15=-7$$
$$ax=8$$
$$x=\frac{8}{a}$$

Back-substitute to solve for $y$.

$$ax+by+cz=0$$
$$a\left(\frac{8}{a}\right)+by+c\left(-\frac{5}{c}\right)=0$$
$$8+by-5=0$$
$$by=-3$$
$$y=-\frac{3}{b}$$

The solution is $\left(\frac{8}{a},-\frac{3}{b},-\frac{5}{c}\right)$.

**34.** $ax - by + 2cz = -4$

$ax + 3by - cz = 1$

$2ax + by + 3cz = 2$

Multiply the first equation by $-1$ and add to the second equation.

$-ax + by - 2cz = 4$

$\underline{ax + 3by - cz = 1}$

$4by - 3cz = 5$

Multiply the first equation by $-2$ and add to the third equation.

$-2ax + 2by - 4cz = 8$

$\underline{2ax + by + 3cz = 2}$

$3by - cz = 10$

Use the two reduced equations to get the following system:

$4by - 3cz = 5$

$3by - cz = 10$

Multiply the second equation by $-3$ and add to the first equation.

$4by - 3cz = 5$

$\underline{-9by + 3cz = -30}$

$-5by = -25$

$y = \frac{5}{b}$

Back-substitute to solve for $z$.

$4by - 3cz = 5$

$4b\left(\frac{5}{b}\right) - 3cz = 5$

$20 - 3cz = 5$

$-3cz = -15$

$z = \frac{5}{c}$

Back-substitute to solve for $x$.

$ax - by + 2cz = -4$

$ax - b\left(\frac{5}{b}\right) + 2c\left(\frac{5}{c}\right) = -4$

$ax - 5 + 10 = -4$

$ax = -9$

$x = -\frac{9}{a}$

The solution is $\left(-\frac{9}{a}, \frac{5}{b}, \frac{5}{c}\right)$.

**35.** **a.** $(0, 13.6)$, $(70, 4.7)$, $(102, 11.5)$

**b.**

$0a + 0b + 0c = 13.6$

$4900a + 70b + c = 4.7$

$10{,}404a + 102b + c = 11.5$

**37.** **a.** Using the three ordered pairs, $(1, 224)$, $(3, 176)$, and $(4, 104)$, we get the following system:

$a + b + c = 224$

$9a + 3b + c = 176$

$16a + 4b + c = 104$

Multiply the first equation by $-1$ and add to the second equation.

$-a - b - c = -224$

$\underline{9a + 3b + c = 176}$

$8a + 2b = -48$

Multiply the first equation by $-1$ and add to the third equation.

$-a - b - c = -224$

$\underline{16a + 4b + c = 104}$

$15a + 3b = -120$

Using the two reduced equations, we get the following system:

$8a + 2b = -48$

$15a + 3b = -120$

Multiply the first equation by $-3$ and multiply the second equation by 2, then add to the equations.

$$-24a-6b=144$$
$$30a+6b=-240$$
$$6a=-96$$
$$a=-16$$

Back-substitute to solve for $b$.

$$8a+2b=-48$$
$$8(-16)+2b=-48$$
$$-128+2b=-48$$
$$2b=80$$
$$b=40$$

Back-substitute to solve for $c$.

$$a+b+c=224$$
$$-16+40+c=224$$
$$c=200$$

The function is

$$y=-16x^2+40x+200$$

**b.** When $x=5$, we get

$$y=-16(5)^2+40(5)+200$$
$$=-16(25)+200+200$$
$$=-400+400$$
$$=0$$

After 5 seconds, the ball hits the ground.

**39.** Let $x$ = estimated wealth of Carnegie.
Let $y$ = estimated wealth of Vanderbilt.
Let $z$ = estimated wealth of Gates.

$$x+y+z=256$$
$$x-y=4$$
$$y-z=36$$

Solve the second equation for $x$.

$$x-y=4$$
$$x=y+4$$

Solve the third equation for $z$.

$$y-z=36$$
$$-z=-y+36$$
$$z=y-36$$

Substitute the expressions for $x$ and $z$ into the first equation and solve for $y$.

$$(y+4)+y+(y-36)=256$$
$$y+4+y+y-36=256$$
$$3y=288$$
$$y=96$$

Back-substitute to solve for $x$ and $z$.

$$x=y+4=96+4=100$$
$$z=y-36=96-36=60$$

The estimated wealth was Carnegie (100 billion), Vanderbilt (96 billion) and Gates (60 billion).

**41.** Let $x$ = the amount invested at 8%
Let $y$ = the amount invested at 10%
Let $z$ = the amount invested at 12%

$$x+y+z=6700$$
$$0.08x+0.10y+0.12z=716$$
$$z-x-y=300$$

Rewrite the system in $Ax+By+Cz=D$ form.

$$x+\quad y+\quad z=6700$$
$$0.08x+0.10y+0.12z=716$$
$$-x-\quad y+\quad z=300$$

Add the first and third equations to find $z$.

$$x+y+z=6700$$
$$-x-y+z=300$$
$$2z=7000$$
$$z=3500$$

Back-substitute 3500 for $z$ to obtain two equations in two variables.

$$x+y+z=6700$$
$$x+y+3500=6700$$
$$x+y=3200$$

$$0.08x + 0.10y + 0.12(3500) = 716$$
$$0.08x + 0.10y + 420 = 716$$
$$0.08x + 0.10y = 296$$

The system of two equations in two variables becomes:

$$x + y = 3200$$
$$0.08x + 0.10y = 296$$

Multiply the second equation by –10 and add it to the first equation.

$$\begin{array}{r} x + y = 3200 \\ \underline{-0.8x + -y = -2960} \\ 0.2x = 240 \\ x = 1200 \end{array}$$

Back-substitute 1200 for $x$ in one of the equations in two variables.

$$x + y = 3200$$
$$1200 + y = 3200$$
$$y = 2000$$

\$1200 was invested at 8%, \$2000 was invested at 10%, and \$3500 was invested at 12%.

**43.** Let $x$ = the number of \$8 tickets
Let $y$ = the number of \$10 tickets
Let $z$ = the number of \$12 tickets

$$x + y + z = 400$$
$$8x + 10y + 12z = 3700$$
$$x + y = 7z$$

Rewrite the system in $Ax + By + Cz = D$ form.

$$x + y + z = 400$$
$$8x + 10y + 12z = 3700$$
$$x + y - 7z = 0$$

Multiply the first equation by –1 and add to the third equation.

$$\begin{array}{r} -x - y - z = -400 \\ \underline{x + y - 7z = 0} \\ -8z = -400 \\ z = 50 \end{array}$$

Back-substitute 50 for $z$ in two of the original equations to obtain two of equations in two variables.

$$x + y + z = 400$$
$$x + y + 50 = 400$$
$$x + y = 350$$
$$8x + 10y + 12z = 3700$$
$$8x + 10y + 12(50) = 3700$$
$$8x + 10y + 600 = 3700$$
$$8x + 10y = 3100$$

The system of two equations in two variables becomes:

$$x + y = 350$$
$$8x + 10y = 3100$$

Multiply the first equation by –8 and add to the second equation.

$$\begin{array}{r} -8x - 8y = -2800 \\ \underline{8x + 10y = 3100} \\ 2y = 300 \\ y = 150 \end{array}$$

Back-substitute 50 for $z$ and 150 for $y$ in one of the original equations in three variables.

$$x + y + z = 400$$
$$x + 150 + 50 = 400$$
$$x + 200 = 400$$
$$x = 200$$

There were 200 \$8 tickets, 150 \$10 tickets, and 50 \$12 tickets sold.

**45.** Let $A$ = the number of servings of $A$
Let $B$ = the number of servings of $B$
Let $C$ = the number of servings of $C$

$$\begin{aligned}40A + 200B + 400C &= 660\\ 5A + 2B + 4C &= 25\\ 30A + 10B + 300C &= 425\end{aligned}$$

Multiply the second equation by –8 and add to the first equation to obtain an equation in two variables.

$$\begin{aligned}40A + 200B + 400C &= 660\\ -40A - 16B - 32C &= -200\\ \hline 184B + 368C &= 460\end{aligned}$$

Multiply the second equation by –6 and add to the third equation to obtain an equation in two variables.

$$\begin{aligned}-30A - 12B - 24C &= -150\\ 30A + 10B + 300C &= 425\\ \hline -2B + 276C &= 275\end{aligned}$$

The system of two equations in two variables becomes:

$$\begin{aligned}184B + 368C &= 460\\ -2B + 276C &= 275\end{aligned}$$

Multiply the second equation by 92 and eliminate $B$.

$$\begin{aligned}184B + 368C &= 460\\ -184B + 25392C &= 25300\\ \hline 25760C &= 25760\\ C &= 1\end{aligned}$$

Back-substitute 1 for $C$ in one of the equations in two variables.

$$\begin{aligned}-2B + 276C &= 275\\ -2B + 276(1) &= 275\\ -2B + 276 &= 275\\ -2B &= -1\\ B &= \frac{1}{2}\end{aligned}$$

Back-substitute 1 for $C$ and $\frac{1}{2}$ for $B$ in one of the original equations in three variables.

$$\begin{aligned}5A + 2B + 4C &= 25\\ 5A + 2\left(\frac{1}{2}\right) + 4(1) &= 25\\ 5A + 1 + 4 &= 25\\ 5A + 5 &= 25\\ 5A &= 20\\ A &= 4\end{aligned}$$

To meet the requirements, 4 ounces of Food $A$, $\frac{1}{2}$ ounce of Food $B$, and 1 ounce of Food $C$ should be used.

**47.-53.** Answers will vary.

**55.** Statement **c.** is true. The variable terms of the second equation are multiples of the variable terms in the first equation, but the constants are not. If we multiply the first equation by 2 and add to the second equation, we obtain:

$$\begin{aligned}-2x - 2y + 2z &= -20\\ 2x + 2y - 2z &= 7\\ \hline 0 &= -13\end{aligned}$$

This is a contradiction, so the system is inconsistent.
Statement **a.** is false. The ordered triple is one solution to the equation, but there are an infinite number of other ordered triples which satisfy the equation.
Statement **b.** is false.

$$\begin{aligned}x - y - z &= -6\\ 2-(-3)-5 &= -6\\ 2+3-5 &= -6\\ 0 &\neq -6\end{aligned}$$

Statement **d.** is false. An equation with four variables can be satisfied by real numbers.

**57.** $x+y+z=180$

$(2x+5)+y=180$

$(2x-5)+z=180$

Rewrite the system in standard form as

$x+y+z=180$

$2x+y=175$

$2x+z=185$

Multiply the first equation by $-1$ and add to the second equation to obtain an equation with two variables.

$$\begin{array}{r} -x-y-z=-180 \\ \underline{2x+y\quad\;\; =175} \\ x-z=-5 \end{array}$$

Combine this equation with the third equation to make a system of two equations.

$$\begin{array}{r} x-z=-5 \\ \underline{2x+z=185} \\ 3x=180 \\ x=60 \end{array}$$

Back-substitute to find $z$.

$x-z=-5$

$60-z=-5$

$-z=-65$

$z=65$

Back-substitute to find $y$.

$x+y+z=180$

$60+65+z=180$

$z=55$

The angles measure $55°$, $60°$, and $65°$.

**59.** $f(x)=-\dfrac{3}{4}x+3$

Use the slope and the $y$–intercept to graph the line.

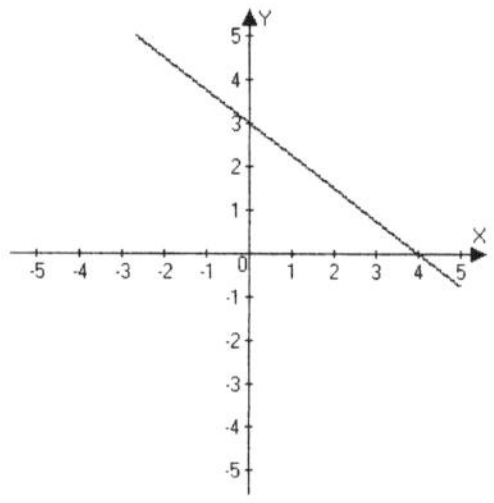

**60.** $-2x+y=6$

Rewrite the equation in slope-intercept form.

$-2x+y=6$

$y=2x+6$

Use the slope and the $y$–intercept to graph the line.

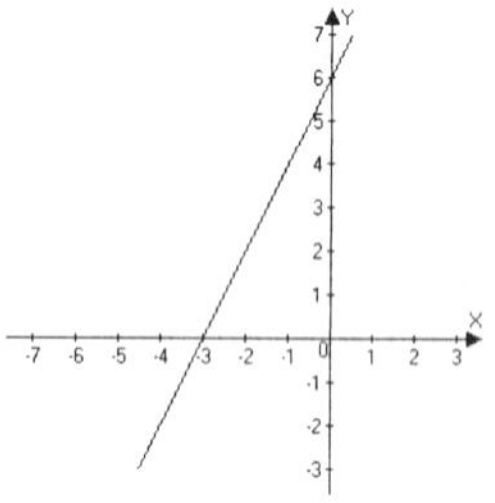

**61.** $f(x)=-5$

This line is the horizontal line, $y=-5$.

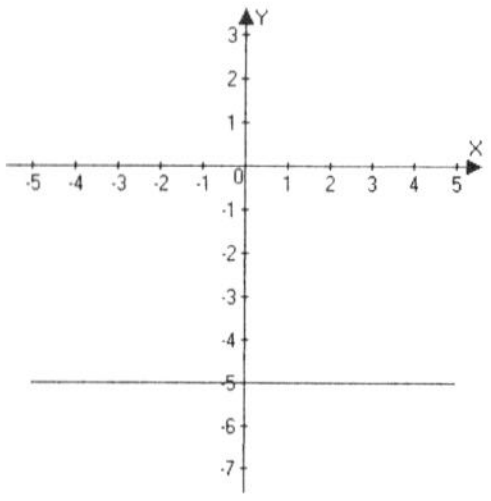

## Mid-Chapter 3 Check Point

**1.** $x = 3y - 7$

$4x + 3y = 2$

Since the first equation is solved for $x$ already, we will use substitution.

Let $x = 3y - 7$ in the second equation and solve for $y$.

$4(3y-7) + 3y = 2$

$12y - 28 + 3y = 2$

$15y = 30$

$y = 2$

Substitute this value for $y$ in the first equation.

$x = 3(2) - 7 = 6 - 7 = -1$

The solution is $(-1, 2)$.

**2.** $3x + 4y = -5$

$2x - 3y = 8$

Multiply the first equation by 3 and the second equation by 4, then add the equations.

$9x + 12y = -15$

$8x - 12y = 32$

$17x = 17$

$x = 1$

Back-substitute to solve for $y$.

$3x + 4y = -5$

$3(1) + 4y = -5$

$3 + 4y = -5$

$4y = -8$

$y = -2$

The solution is $(1, -2)$.

**3.** $\dfrac{2x}{3} + \dfrac{y}{5} = 6$

$\dfrac{x}{6} - \dfrac{y}{2} = -4$

Multiply the first equation by 15 and the second equation by 6 to eliminate the fractions.

$15\left(\dfrac{2x}{3} + \dfrac{y}{5}\right) = 15(6)$

$10x + 3y = 90$

$6\left(\dfrac{x}{6} - \dfrac{y}{2}\right) = 6(-4)$

$x - 3y = -24$

We now need to solve the equivalent system

$10x + 3y = 90$

$x - 3y = -24$

Add the two equations to eliminate $y$.

$10x + 3y = 90$

$x - 3y = -24$

$11x = 66$

$x = 6$

Back-substitute to solve for $y$.

$x - 3y = -24$

$6 - 3y = -24$

$-3y = -30$

$y = 10$

The solution is $(6, 10)$.

**4.** $y = 4x - 5$

$8x - 2y = 10$

Since the first equation is already solved for $y$, we will use substitution.

Let $y = 4x - 5$ in the second equation and solve for $x$.

$8x - 2(4x - 5) = 10$

$8x - 8x + 10 = 10$

$10 = 10$

This statement is an identity. The system is dependent so there are an infinite number of solutions. The solution set is $\{(x, y) \mid y = 4x - 5\}$.

**5.** $2x+5y=3$

$3x-2y=1$

Multiply the first equation by 3 and the second equation by $-2$, then add the equations.

$$\begin{array}{r} 6x+15y=9 \\ -6x+4y=-2 \\ \hline 19y=7 \end{array}$$

$$y=\frac{7}{19}$$

Back-substitute to solve for $x$.

$$2x+5y=3$$
$$2x+5\left(\frac{7}{19}\right)=3$$
$$2x+\frac{35}{19}=3$$
$$2x=\frac{22}{19}$$
$$x=\frac{11}{19}$$

The solution is $\left(\frac{11}{19},\frac{7}{19}\right)$.

**6.** $\frac{x}{12}-y=\frac{1}{4}$

$4x-48y=16$

Solve the first equation for $y$.

$$\frac{x}{12}-y=\frac{1}{4}$$
$$-y=-\frac{x}{12}+\frac{1}{4}$$
$$y=\frac{x}{12}-\frac{1}{4}$$

Let $y=\frac{x}{12}-\frac{1}{4}$ in the second equation and solve for $x$.

$$4x-48\left(\frac{x}{12}-\frac{1}{4}\right)=16$$
$$4x-4x+12=16$$
$$12=16$$

This statement is a contradiction. The system is inconsistent so there is no solution. The solution is $\{\ \}$ or $\varnothing$.

**7.** $2x-y+2z=-8$

$x+2y-3z=9$

$3x-y-4z=3$

Multiply the first equation by 2 and add to the second equation.

$$\begin{array}{r} 4x-2y+4z=-16 \\ x+2y-3z=9 \\ \hline 5x+z=-7 \end{array}$$

Multiply the first equation by $-1$ and add to the third equation.

$$\begin{array}{r} -2x+y-2z=8 \\ 3x-y-4z=3 \\ \hline x-6z=11 \end{array}$$

Use the two reduced equations to get the following system:

$5x+z=-7$

$x-6z=11$

Multiply the first equation by 6 and add to the second equation.

$$\begin{array}{r} 30x+6z=-42 \\ x-6z=11 \\ \hline 31x=-31 \end{array}$$

$$x=-1$$

Back-substitute to solve for $z$.

$$5x+z=-7$$
$$5(-1)+z=-7$$
$$-5+z=-7$$
$$z=-2$$

Back-substitute to solve for $y$.

$$2x - y + 2z = -8$$
$$2(-1) - y + 2(-2) = -8$$
$$-2 - y - 4 = -8$$
$$-y = -2$$
$$y = 2$$

The solution is $(-1, 2, -2)$.

**8.** $x \quad - 3z = -5$

$2x - y + 2z = 16$

$7x - 3y - 5z = 19$

Multiply the second equation by $-3$ and add to the third equation.

$$-6x + 3y - 6z = -48$$
$$\underline{7x - 3y - 5z = 19}$$
$$x - 11z = -29$$

Use this reduced equation and the original first equation to obtain the following system:

$x - 3z = -5$

$x - 11z = -29$

Multiply the second equation by $-1$ and add to the first equation.

$$x - 3z = -5$$
$$\underline{-x + 11z = 29}$$
$$8z = 24$$
$$z = 3$$

Back-substitute to solve for $x$.

$$x - 3z = -5$$
$$x - 3(3) = -5$$
$$x - 9 = -5$$
$$x = 4$$

Back-substitute to solve for $y$.

$$2x - y + 2z = 16$$
$$2(4) - y + 2(3) = 16$$
$$8 - y + 6 = 16$$
$$-y = 2$$
$$y = -2$$

The solution is $(4, -2, 3)$.

**9.** Graph the two lines by using the intercepts.

$2x - y = 4$

x-intercept: $2x - y = 4$
$$2x - 0 = 4$$
$$2x = 4$$
$$x = 2$$

y-intercept: $2x - y = 4$
$$2(0) - y = 4$$
$$-y = 4$$
$$y = -4$$

$x + y = 5$

x-intercept: $x + y = 5$
$$x + 0 = 5$$
$$x = 5$$

y-intercept: $x + y = 5$
$$0 + y = 5$$
$$y = 5$$

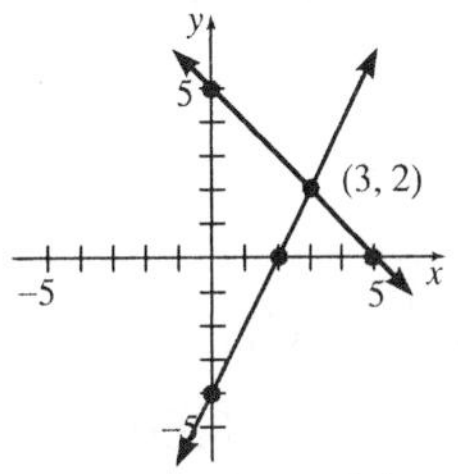


The solution of the system is the intersection point of the graphs. Therefore, the solution is $(3, 2)$.

**10.** Graph the two lines by using the slope and y-intercept.

$y = x + 3$

y-intercept: $b = 3$

slope: $m = 1 = \frac{1}{1}$

Plot the points $(0, 3)$ and

$(0+1, 3+1) = (1, 4)$

$y = -\frac{1}{2}x$

y-intercept: $b = 0$

slope: $m = -\frac{1}{2} = \frac{-1}{2}$

Plot the points $(0,0)$ and $(0+2, 0-1) = (2, -1)$.

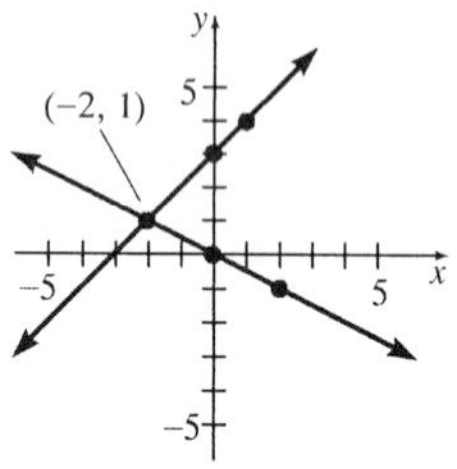


The solution of the system is the intersection point of the graphs. Therefore, the solution is $(-2, 1)$.

**11.** **a.** $C(x) = 400{,}000 + 20x$

**b.** $R(x) = 100x$

**c.** $P(x) = R(x) - C(x)$
$= 100x - (400{,}000 + 20x)$
$= 80x - 400{,}000$

**d.** The break even point is the point where cost and revenue are the same. We need to solve the system
$y = 400{,}000 + 20x$
$y = 100x$
Let $y = 400{,}000 + 20x$ in the second equation and solve for $x$.
$400{,}000 + 20x = 100x$
$400{,}000 = 80x$
$5000 = x$
Back-substitute to solve for $y$.
$y = 100x$
$= 100(5000)$
$= 500{,}000$
Thus, the break-even point is $(5000,\ 500{,}000)$. The company will break even when it produces and sells 5000 PDAs. At this level, the cost and revenue will both be $500,000.

**12.** Let $x$ = the number of roses.
Let $y$ = the number of carnations.
$x + y = 20$
$3x + 1.5y = 39$
Solve the first equation for $x$.
$x + y = 20$
$x = 20 - y$
Substitute this expression for $x$ in the second equation and solve for $y$.
$3(20 - y) + 1.5y = 39$
$60 - 3y + 1.5y = 39$
$-1.5y = -21$
$y = 14$
Back-substitute to solve for $x$.
$x = 20 - y = 20 - 14 = 6$
There are 6 roses and 14 carnations in the bouquet.

**13.** Let $x$ = the amount invested at 5%.
Let $y$ = the amount invested at 6%.
$x + y = 15{,}000$
$0.05x + 0.06y = 837$
Solve the first equation for $x$.
$x + y = 15{,}000$
$x = 15{,}000 - y$

Substitute this expression for $x$ in the second equation and solve for $y$.

$$0.05(15,000-y)+0.06y=837$$
$$750-0.05y+0.06y=837$$
$$0.01y=87$$
$$y=8700$$

Back-substitute to solve for $x$.

$$x=15,000-y$$
$$=15,000-8700$$
$$=6300$$

You invested \$6300 at 5% and \$8700 at 6%.

**14.** Let $x$ = gallons of 13% nitrogen.
Let $y$ = gallons of 18% nitrogen.

$$x+y=50$$
$$0.13x+0.18y=0.16(50)$$

or

$$x+y=50$$
$$0.13x+0.18y=8$$

Solve the first equation for $x$.

$$x+y=50$$
$$x=50-y$$

Substitute this expression for $x$ in the second equation and solve for $y$.

$$0.13(50-y)+0.18y=8$$
$$6.5-0.13y+0.18y=8$$
$$0.05y=1.5$$
$$y=30$$

Back-substitute to solve for $x$.

$$x=50-y=50-30=20$$

The manager should mix 20 gallons of the 13% nitrogen with 30 gallons of the 18% nitrogen.

**15.** Let $w$ = the rate of the water (current).
Let $r$ = your average rowing rate.
For this problem we will make use of the distance traveled formula: $d=r\cdot t$
In addition, remember that when you go *with* the current you add the rate of the current to your rowing rate. If you go *against* the current, you subtract the rate of the current you're your rowing rate.
With this in mind, we obtain the following system:

$$9=(r+w)(2)$$
$$9=(r-w)(6)$$

or

$$2r+2w=9$$
$$6r-6w=9$$

Multiply the first equation by 3 and add the two equations.

$$6r+6w=27$$
$$\underline{6r-6w=9}$$
$$12r=36$$
$$r=3$$

Back-substitute to solve for $w$.

$$2r+2w=9$$
$$2(3)+2w=9$$
$$6+2w=9$$
$$2w=3$$
$$w=1.5$$

Your rowing rate in still water is 3 miles per hour; the rate of the current is 1.5 miles per hour.

**16.** Let $x$ = the amount invested at 2%.
Let $y$ = the amount invested at 5%.

$$x + y = 8000$$
$$0.05y = 0.02x + 85$$

Multiply the second equation by 20.

$$x + y = 8000$$
$$y = 0.4x + 1700$$

Let $y = 0.4x + 1700$ in the first equation and solve for $x$.

$$x + (0.4x + 1700) = 8000$$
$$1.4x = 6300$$
$$x = 4500$$

Back-substitute to solve for $y$.

$$x + y = 8000$$
$$4500 + y = 8000$$
$$y = 3500$$

You invested \$4500 at 2% and \$3500 at 5%.

**17.** Using the points $(-1,0)$, $(1,4)$, and $(2,3)$ in the equation $y = ax^2 + bx + c$, we get the following system of equations:

$$a - b + c = 0$$
$$a + b + c = 4$$
$$4a + 2b + c = 3$$

Add the first two equations.

$$a - b + c = 0$$
$$\underline{a + b + c = 4}$$
$$2a + 2c = 4$$

Multiply the first equation by 2 and add to the third equation.

$$2a - 2b + 2c = 0$$
$$\underline{4a + 2b + c = 3}$$
$$6a + 3c = 3$$

Using the two reduced equations, we get the following system of equations:

$$2a + 2c = 4$$
$$6a + 3c = 3$$

Multiply the first equation by $-3$ and add to the second equation.

$$-6a - 6c = -12$$
$$\underline{6a + 3c = 3}$$
$$-3c = -9$$
$$c = 3$$

Back-substitute to solve for $a$.

$$2a + 2c = 4$$
$$2a + 2(3) = 4$$
$$2a + 6 = 4$$
$$2a = -2$$
$$a = -1$$

Back-substitute to solve for $b$.

$$a + b + c = 4$$
$$-1 + b + 3 = 4$$
$$b = 2$$

The equation is $y = -x^2 + 2x + 3$.

**18.** Let $n$ = the number of nickels.
Let $d$ = the number of dimes.
Let $q$ = the number of quarters.
From the problem statement, we have

$$n + d + q = 26$$
$$0.05n + 0.10d + 0.25q = 4.00$$
$$q = n + d - 2$$

If we multiply the second equation by 20 and rearrange the third equation, we get the following equivalent system:

$$n + d + q = 26$$
$$n + 2d + 5q = 80$$
$$n + d - q = 2$$

Multiply the first equation by $-1$ and add to the second equation.

$$-n - d - q = -26$$
$$\underline{n + 2d + 5q = 80}$$
$$d + 4q = 54$$

Multiply the third equation by $-1$ and add to the first equation.

$$\begin{aligned} n+d+q&=26\\ -n-d+q&=-2\\ \hline 2q&=24\\ q&=12 \end{aligned}$$

Back-substitute to solve for $d$.

$$\begin{aligned} d+4q&=54\\ d+4(12)&=54\\ d+48&=54\\ d&=6 \end{aligned}$$

Back-substitute to solve for $n$.

$$\begin{aligned} n+d+q&=26\\ n+6+12&=26\\ n&=8 \end{aligned}$$

The collection contains 8 nickels, 6 dimes, and 12 quarters.

## 3.4 Exercise Set

**1.** $x-3y=11$

$y=-3$

Substitute –3 for $y$ in the first equation.

$$\begin{aligned} x-3y&=11\\ x-3(-3)&=11\\ x+9&=11\\ x&=2 \end{aligned}$$

The solution is $(2,-3)$.

**3.** $x-3y=1$

$y=-1$

Substitute –1 for $y$ in the first equation.

$$\begin{aligned} x-3y&=1\\ x-3(-1)&=1\\ x+3&=1\\ x&=-2 \end{aligned}$$

The solution is $(-2,-1)$.

**5.** $x \quad -4z=5$

$y-12z=13$

$z=-\frac{1}{2}$

Substitute $-\frac{1}{2}$ for $z$ in the second equation to find $y$.

$$\begin{aligned} y-12z&=13\\ y-12\left(-\frac{1}{2}\right)&=13\\ y+6&=13\\ y&=7 \end{aligned}$$

Substitute 7 for $y$ in the first equation to find $x$.

$$\begin{aligned} x-4z&=5\\ x-4\left(-\frac{1}{2}\right)&=5\\ x+2&=5\\ x&=3 \end{aligned}$$

The solution is $\left(3,7,-\frac{1}{2}\right)$.

**7.** $x+\frac{1}{2}y+\quad z=\frac{11}{2}$

$y+\frac{3}{2}z=7$

$z=4$

Substitute 4 for $z$ in the second equation to find $y$.

$$\begin{aligned} y+\frac{3}{2}z&=7\\ y+\frac{3}{2}(4)&=7\\ y+6&=7\\ y&=1 \end{aligned}$$

Substitute 1 for $y$ and 4 for $z$ in the first equation to find $x$.

$$x+\frac{1}{2}y+z=\frac{11}{2}$$
$$x+\frac{1}{2}(1)+4=\frac{11}{2}$$
$$x+\frac{9}{2}=\frac{11}{2}$$
$$x=\frac{2}{2}=1$$

The solution is $(1,1,4)$.

**9.** $\left[\begin{array}{cc|c} 2 & 2 & 5 \\ 1 & -\frac{3}{2} & 5 \end{array}\right] \quad R_1 \leftrightarrow R_2$

$=\left[\begin{array}{cc|c} 1 & -\frac{3}{2} & 5 \\ 2 & 2 & 5 \end{array}\right]$

**11.** $\left[\begin{array}{cc|c} -6 & 8 & -12 \\ 3 & 5 & -2 \end{array}\right] \quad -\frac{1}{6}R_1$

$=\left[\begin{array}{cc|c} 1 & -\frac{4}{3} & 2 \\ 3 & 5 & -2 \end{array}\right]$

**13.** $\left[\begin{array}{cc|c} 1 & -3 & 5 \\ 2 & 6 & 4 \end{array}\right] \quad -2R_1+R_2$

$=\left[\begin{array}{cc|c} 1 & -3 & 5 \\ 0 & 12 & -6 \end{array}\right]$

**15.** $\left[\begin{array}{cc|c} 1 & -\frac{3}{2} & \frac{7}{2} \\ 3 & 4 & 2 \end{array}\right] \quad -3R_1+R_2$

$=\left[\begin{array}{cc|c} 1 & -\frac{3}{2} & \frac{7}{2} \\ 0 & \frac{17}{2} & -\frac{17}{2} \end{array}\right]$

**17.** $\left[\begin{array}{ccc|c} 2 & -6 & 4 & 10 \\ 1 & 5 & -5 & 0 \\ 3 & 0 & 4 & 7 \end{array}\right] \quad \frac{1}{2}R_1$

$=\left[\begin{array}{ccc|c} 1 & -3 & 2 & 5 \\ 1 & 5 & -5 & 0 \\ 3 & 0 & 4 & 7 \end{array}\right]$

**19.** $\left[\begin{array}{ccc|c} 1 & -3 & 2 & 0 \\ 3 & 1 & -1 & 7 \\ 2 & -2 & 1 & 3 \end{array}\right] \quad -3R_1+R_2$

$=\left[\begin{array}{ccc|c} 1 & -3 & 2 & 0 \\ 0 & 10 & -7 & 7 \\ 2 & -2 & 1 & 3 \end{array}\right]$

**21.** $\left[\begin{array}{ccc|c} 1 & 1 & -1 & 6 \\ 2 & -1 & 1 & -3 \\ 3 & -1 & -1 & 4 \end{array}\right] \quad \begin{array}{c} -2R_1+R_2 \\ \text{and} \\ -3R_1+R_3 \end{array}$

$=\left[\begin{array}{ccc|c} 1 & 1 & -1 & 6 \\ 0 & -3 & 3 & -15 \\ 0 & -4 & 2 & -14 \end{array}\right]$

**23.** $\left[\begin{array}{cc|c} 1 & 1 & 6 \\ 1 & -1 & 2 \end{array}\right] \quad -R_1+R_2$

$=\left[\begin{array}{cc|c} 1 & 1 & 6 \\ 0 & -2 & -4 \end{array}\right] \quad -\frac{1}{2}R_2$

$=\left[\begin{array}{cc|c} 1 & 1 & 6 \\ 0 & 1 & 2 \end{array}\right]$

The resulting system is:

$x+y=6$

$y=2$

Back-substitute 2 for $y$ in the first equation.

$x+2=6$

$x=4$

The solution is $(4,2)$.

**25.** $\left[\begin{array}{cc|c} 2 & 1 & 3 \\ 1 & -3 & 12 \end{array}\right] \quad R_1 \leftrightarrow R_2$

$= \left[\begin{array}{cc|c} 1 & -3 & 12 \\ 2 & 1 & 3 \end{array}\right] \quad -2R_1 + R_2$

$= \left[\begin{array}{cc|c} 1 & -3 & 12 \\ 0 & 7 & -21 \end{array}\right] \quad \frac{1}{7}R_2$

$= \left[\begin{array}{cc|c} 1 & -3 & 12 \\ 0 & 1 & -3 \end{array}\right]$

The system is:

$x - 3y = 12$

$y = -3$

Back-substitute –3 for $y$ in the first equation.

$x - 3y = 12$

$x - 3(-3) = 12$

$x + 9 = 12$

$x = 3$

The solution is $(3, -3)$.

**27.** $\left[\begin{array}{cc|c} 5 & 7 & -25 \\ 11 & 6 & -8 \end{array}\right] \quad \frac{1}{5}R_1$

$= \left[\begin{array}{cc|c} 1 & \frac{7}{5} & -5 \\ 11 & 6 & -8 \end{array}\right] \quad -11R_1 + R_2$

$= \left[\begin{array}{cc|c} 1 & \frac{7}{5} & -5 \\ 0 & -\frac{47}{5} & 47 \end{array}\right] \quad -\frac{5}{47}R_2$

$= \left[\begin{array}{cc|c} 1 & \frac{7}{5} & -5 \\ 0 & 1 & -5 \end{array}\right]$

The resulting system is:

$x + \frac{7}{5}y = -5$

$y = -5$

Back-substitute –5 for $y$ in the first equation.

$x + \frac{7}{5}y = -5$

$x + \frac{7}{5}(-5) = -5$

$x - 7 = -5$

$x = 2$

The solution is $(2, -5)$.

**29.** $\left[\begin{array}{cc|c} 4 & -2 & 5 \\ -2 & 1 & 6 \end{array}\right] \quad \frac{1}{4}R_1$

$= \left[\begin{array}{cc|c} 1 & -\frac{1}{2} & \frac{5}{2} \\ -2 & 1 & 6 \end{array}\right] \quad 2R_1 + R_2$

$= \left[\begin{array}{cc|c} 1 & -\frac{1}{2} & \frac{5}{2} \\ 0 & 0 & \frac{17}{2} \end{array}\right]$

The resulting system is:

$x - \frac{1}{2}y = \frac{5}{2}$

$0x + 0y = \frac{17}{2}$

This is a contradiction. The system is inconsistent.

**31.** $\left[\begin{array}{cc|c} 1 & -2 & 1 \\ -2 & 4 & -2 \end{array}\right] \quad 2R_1 + R_2$

$= \left[\begin{array}{cc|c} 1 & -2 & 1 \\ 0 & 0 & 0 \end{array}\right]$

The resulting system is:

$x - 2y = 1$

$0x + 0y = 0$

The system is dependent. There are infinitely many solutions.

**33.** $\left[\begin{array}{ccc|c} 1 & 1 & -1 & -2 \\ 2 & -1 & 1 & 5 \\ -1 & 2 & 2 & 1 \end{array}\right] \quad -2R_1 + R_2$

$= \left[\begin{array}{ccc|c} 1 & 1 & -1 & -2 \\ 0 & -3 & 3 & 9 \\ -1 & 2 & 2 & 1 \end{array}\right] \quad R_1 + R_3$

$= \left[\begin{array}{ccc|c} 1 & 1 & -1 & -2 \\ 0 & -3 & 3 & 9 \\ 0 & 3 & 1 & -1 \end{array}\right] \quad R_2 + R_3$

$= \left[\begin{array}{ccc|c} 1 & 1 & -1 & -2 \\ 0 & -3 & 3 & 9 \\ 0 & 0 & 4 & 8 \end{array}\right] \quad \frac{1}{4}R_3$

$= \left[\begin{array}{ccc|c} 1 & 1 & -1 & -2 \\ 0 & -3 & 3 & 9 \\ 0 & 0 & 1 & 2 \end{array}\right]$

The resulting system is:

$x + y - z = -2$

$y - z = -3$

$z = 2$

Back-substitute 2 for $z$ to find $y$.

$y - z = -3$

$y - 2 = -3$

$y = -1$

Back-substitute 2 for $z$ and $-1$ for $y$ to find $x$.

$x + y - z = -2$

$x - 1 - 2 = -2$

$x - 3 = -2$

$x = 1$

The solution is $(1, -1, 2)$.

**35.** $\left[\begin{array}{ccc|c} 1 & 3 & 0 & 0 \\ 1 & 1 & 1 & 1 \\ 3 & -1 & -1 & 11 \end{array}\right] \quad -R_1 + R_2$

$= \left[\begin{array}{ccc|c} 1 & 3 & 0 & 0 \\ 0 & -2 & 1 & 1 \\ 3 & -1 & -1 & 11 \end{array}\right] \quad -3R_1 + R_3$

$= \left[\begin{array}{ccc|c} 1 & 3 & 0 & 0 \\ 0 & -2 & 1 & 1 \\ 0 & -10 & -1 & 11 \end{array}\right] \quad -\frac{1}{2}R_2$

$= \left[\begin{array}{ccc|c} 1 & 3 & 0 & 0 \\ 0 & 1 & -\frac{1}{2} & -\frac{1}{2} \\ 0 & -10 & -1 & 11 \end{array}\right] \quad -\frac{1}{10}R_3$

$= \left[\begin{array}{ccc|c} 1 & 3 & 0 & 0 \\ 0 & 1 & -\frac{1}{2} & -\frac{1}{2} \\ 0 & 1 & \frac{1}{10} & -\frac{11}{10} \end{array}\right] \quad -R_2 + R_3$

$= \left[\begin{array}{ccc|c} 1 & 3 & 0 & 0 \\ 0 & 1 & -\frac{1}{2} & -\frac{1}{2} \\ 0 & 0 & \frac{3}{5} & -\frac{3}{5} \end{array}\right] \quad \frac{5}{3}R_3$

$= \left[\begin{array}{ccc|c} 1 & 3 & 0 & 0 \\ 0 & 1 & -\frac{1}{2} & -\frac{1}{2} \\ 0 & 0 & 1 & -1 \end{array}\right]$

The resulting system is:

$x + 3y = 0$

$y - \frac{1}{2}z = -\frac{1}{2}$

$z = -1$

Back-substitute $-1$ for $z$ and solve for $y$.

$$y-\frac{1}{2}z=-\frac{1}{2}$$
$$y-\frac{1}{2}(-1)=-\frac{1}{2}$$
$$y+\frac{1}{2}=-\frac{1}{2}$$
$$y=-1$$

Back-substitute $-1$ for $y$ to find $x$.

$$x+3y=0$$
$$x+3(-1)=0$$
$$x-3=0$$
$$x=3$$

The solution is $(3,-1,-1)$.

**37.** $\left[\begin{array}{ccc|c} 2 & 2 & 7 & -1 \\ 2 & 1 & 2 & 2 \\ 4 & 6 & 1 & 15 \end{array}\right]$ $\frac{1}{2}R_1$

$$=\left[\begin{array}{ccc|c} 1 & 1 & \frac{7}{2} & -\frac{1}{2} \\ 2 & 1 & 2 & 2 \\ 4 & 6 & 1 & 15 \end{array}\right] \quad -2R_1+R_2$$

$$=\left[\begin{array}{ccc|c} 1 & 1 & \frac{7}{2} & -\frac{1}{2} \\ 0 & -1 & -5 & 3 \\ 4 & 6 & 1 & 15 \end{array}\right] \quad -R_2$$

$$=\left[\begin{array}{ccc|c} 1 & 1 & \frac{7}{2} & -\frac{1}{2} \\ 0 & 1 & 5 & -3 \\ 4 & 6 & 1 & 15 \end{array}\right] \quad -4R_1+R_3$$

$$=\left[\begin{array}{ccc|c} 1 & 1 & \frac{7}{2} & -\frac{1}{2} \\ 0 & 1 & 5 & -3 \\ 0 & 2 & -13 & 17 \end{array}\right] \quad -2R_2+R_3$$

$$=\left[\begin{array}{ccc|c} 1 & 1 & \frac{7}{2} & -\frac{1}{2} \\ 0 & 1 & 5 & -3 \\ 0 & 0 & -23 & 23 \end{array}\right] \quad -R_3$$

$$=\left[\begin{array}{ccc|c} 1 & 1 & \frac{7}{2} & -\frac{1}{2} \\ 0 & 1 & 5 & -3 \\ 0 & 0 & 1 & -1 \end{array}\right]$$

The resulting system is:

$$x+y+\frac{7}{2}z=-\frac{1}{2}$$
$$y+5z=-3$$
$$z=-1$$

Back-substitute $-1$ for $z$ to find $y$.

$$y+5z=-3$$
$$y+5(-1)=-3$$
$$y-5=-3$$
$$y=2$$

Back-substitute $-1$ for $z$ and 2 for $y$ to find $x$.

$$x+y+\frac{7}{2}z=-\frac{1}{2}$$
$$x+2+\frac{7}{2}(-1)=-\frac{1}{2}$$
$$x+2-\frac{7}{2}=-\frac{1}{2}$$
$$x-\frac{3}{2}=-\frac{1}{2}$$
$$x=1$$

The solution is $(1,2,-1)$.

**39.** $\left[\begin{array}{ccc|c} 1 & 1 & 1 & 6 \\ 1 & 0 & -1 & -2 \\ 0 & 1 & 3 & 11 \end{array}\right] \quad R_2 \leftrightarrow R_3$

$= \left[\begin{array}{ccc|c} 1 & 1 & 1 & 6 \\ 0 & 1 & 3 & 11 \\ 1 & 0 & -1 & -2 \end{array}\right] \quad -R_1 + R_3$

$= \left[\begin{array}{ccc|c} 1 & 1 & 1 & 6 \\ 0 & 1 & 3 & 11 \\ 0 & -1 & -2 & -8 \end{array}\right] \quad R_2 + R_3$

$= \left[\begin{array}{ccc|c} 1 & 1 & 1 & 6 \\ 0 & 1 & 3 & 11 \\ 0 & 0 & 1 & 3 \end{array}\right]$

The resulting system is:

$x + y + z = 6$

$y + 3z = 11$

$z = 3$

Back-substitute 3 for $z$ to find $y$.

$y + 3z = 11$

$y + 3(3) = 11$

$y + 9 = 11$

$y = 2$

Back-substitute 3 for $z$ and 2 for $y$ to find $x$.

$x + y + z = 6$

$x + 2 + 3 = 6$

$x + 5 = 6$

$x = 1$

The solution is $(1, 2, 3)$.

**41.** $\left[\begin{array}{ccc|c} 1 & -1 & 3 & 4 \\ 2 & -2 & 6 & 7 \\ 3 & -1 & 5 & 14 \end{array}\right] \quad \begin{array}{c} -2R_1 + R_2 \\ \text{and} \\ -3R_1 + R_3 \end{array}$

$= \left[\begin{array}{ccc|c} 1 & -1 & 3 & 4 \\ 0 & 0 & 0 & -1 \\ 0 & 2 & -4 & 2 \end{array}\right]$

The resulting system is:

$x - y + 3z = 4$

$0x + 0y + 0z = -1$

$2y - 4z = 2$

The second row is a contradiction, since $0x + 0y + 0z$ cannot equal $-1$. We conclude that the system is inconsistent.

**43.** $\left[\begin{array}{ccc|c} 1 & -2 & 1 & 4 \\ 5 & -10 & 5 & 20 \\ -2 & 4 & -2 & -8 \end{array}\right] \quad \frac{1}{5}R_2$

$= \left[\begin{array}{ccc|c} 1 & -2 & 1 & 4 \\ 1 & -2 & 1 & 4 \\ -2 & 4 & -2 & -8 \end{array}\right]$

$R_1$ and $R_2$ are the same. The system is dependent and there are infinitely many solutions.

**45.** $\left[\begin{array}{ccc|c} 1 & 1 & 0 & 1 \\ 0 & 1 & 2 & -2 \\ 2 & 0 & -1 & 0 \end{array}\right] \quad -2R_1 + R_3$

$= \left[\begin{array}{ccc|c} 1 & 1 & 0 & 1 \\ 0 & 1 & 2 & -2 \\ 0 & -2 & -1 & -2 \end{array}\right] \quad 2R_2 + R_3$

$= \left[\begin{array}{ccc|c} 1 & 1 & 0 & 1 \\ 0 & 1 & 2 & -2 \\ 0 & 0 & 3 & -6 \end{array}\right] \quad \frac{1}{3}R_3$

$= \left[\begin{array}{ccc|c} 1 & 1 & 0 & 1 \\ 0 & 1 & 2 & -2 \\ 0 & 0 & 1 & -2 \end{array}\right]$

The resulting system is:

$x + y = 1$

$y + 2z = -2$

$z = -2$

Back-substitute –2 for $z$ to find $y$.

$$y+2z=-2$$
$$y+2(-2)=-2$$
$$y-4=-2$$
$$y=2$$

Back-substitute 2 for $y$ to find $x$.

$$x+y=1$$
$$x+2=1$$
$$x=-1$$

The solution is $(-1,2,-2)$.

**47.** The system is

$$w-x+y+z=3$$
$$x-2y-z=0$$
$$y+6z=17$$
$$z=3$$

Back-substitute $z=3$ to solve for $y$.

$$y+6(3)=17$$
$$y+18=17$$
$$y=-1$$

Back-substitute $z=3$ and $y=-1$ to solve for $x$.

$$x-2(-1)-(3)=0$$
$$x+2-3=0$$
$$x=1$$

Back-substitute $x=1$, $y=-1$ and $z=3$ to solve for $w$.

$$w-(1)+(-1)+(3)=3$$
$$w-1-1+3=3$$
$$w=2$$

The solution is $(2,1,-1,3)$.

**48.** The system is

$$w+2x-y=2$$
$$x+y-2z=0$$
$$2w+3y+4z=11$$
$$y-z=-2$$
$$z=3$$

Back-substitute $z=3$ to solve for $y$.

$$y-(3)=-2$$
$$y=1$$

Back-substitute $y=1$ and $z=3$ to solve for $x$.

$$x+(1)-2(3)=-3$$
$$x+1-6=-3$$
$$x=2$$

Back-substitute $x=2$ and $y=1$ to solve for $w$.

$$w+2(2)-(1)=2$$
$$w+4-1=2$$
$$w=-1$$

The solution is $(-1,2,1,3)$.

**49.**
$$\left[\begin{array}{cccc|c} 1 & -1 & 1 & 1 & 3 \\ 0 & 1 & -2 & -1 & 0 \\ 2 & 0 & 3 & 4 & 11 \\ 5 & 1 & 2 & 4 & 6 \end{array}\right]\begin{array}{l} \\ \\ -2R_1+R_3 \\ -5R_1+R_4 \end{array}$$
$$=\left[\begin{array}{cccc|c} 1 & -1 & 1 & 1 & 3 \\ 0 & 1 & -2 & -1 & 0 \\ 0 & 2 & 1 & 2 & 5 \\ 0 & 6 & -3 & -1 & -9 \end{array}\right]$$

**50.**
$$\left[\begin{array}{cccc|c} 1 & -5 & 2 & -2 & 4 \\ 0 & 1 & -3 & -1 & 0 \\ 3 & 0 & 2 & -1 & 6 \\ -4 & 1 & 4 & 2 & -3 \end{array}\right]\begin{array}{l} \\ \\ -3R_1+R_3 \\ 4R_1+R_4 \end{array}$$
$$=\left[\begin{array}{cccc|c} 1 & -5 & 2 & -2 & 4 \\ 0 & 1 & -3 & -1 & 0 \\ 0 & 15 & -4 & 5 & -6 \\ 0 & -19 & 12 & -6 & 13 \end{array}\right]$$

**51.** $\left[\begin{array}{cccc|c} 1 & 1 & 1 & 1 & 4 \\ 2 & 1 & -2 & -1 & 0 \\ 1 & -2 & -1 & -2 & -2 \\ 3 & 2 & 1 & 3 & 4 \end{array}\right] \begin{array}{l} \\ -2R_1+R_2 \\ -1R_1+R_3 \\ -3R_1+R_4 \end{array}$

$= \left[\begin{array}{cccc|c} 1 & 1 & 1 & 1 & 4 \\ 0 & -1 & -4 & -3 & -8 \\ 0 & -3 & -2 & -3 & -6 \\ 0 & -1 & -2 & 0 & -8 \end{array}\right] \begin{array}{l} \\ -1R_2 \\ \\ \\ \end{array}$

$= \left[\begin{array}{cccc|c} 1 & 1 & 1 & 1 & 4 \\ 0 & 1 & 4 & 3 & 8 \\ 0 & -3 & -2 & -3 & -6 \\ 0 & -1 & -2 & 0 & -8 \end{array}\right] \begin{array}{l} \\ \\ 3R_2+R_3 \\ R_2+R_4 \end{array}$

$= \left[\begin{array}{cccc|c} 1 & 1 & 1 & 1 & 4 \\ 0 & 1 & 4 & 3 & 8 \\ 0 & 0 & 10 & 6 & 18 \\ 0 & 0 & 2 & 3 & 0 \end{array}\right] \begin{array}{l} \\ \\ \frac{1}{2}R_4 \\ R_3 \end{array}$

$= \left[\begin{array}{cccc|c} 1 & 1 & 1 & 1 & 4 \\ 0 & 1 & 4 & 3 & 8 \\ 0 & 0 & 1 & \frac{3}{2} & 0 \\ 0 & 0 & 10 & 6 & 18 \end{array}\right] \begin{array}{l} \\ \\ \\ -10R_3+R_4 \end{array}$

$= \left[\begin{array}{cccc|c} 1 & 1 & 1 & 1 & 4 \\ 0 & 1 & 4 & 3 & 8 \\ 0 & 0 & 1 & \frac{3}{2} & 0 \\ 0 & 0 & 0 & -9 & 18 \end{array}\right] \begin{array}{l} \\ \\ \\ -\frac{1}{9}R_4 \end{array}$

$= \left[\begin{array}{cccc|c} 1 & 1 & 1 & 1 & 4 \\ 0 & 1 & 4 & 3 & 8 \\ 0 & 0 & 1 & \frac{3}{2} & 0 \\ 0 & 0 & 0 & 1 & -2 \end{array}\right]$

The resulting system is

$$w+x+y+z=4$$
$$x+4y+3z=8$$
$$y+\frac{3}{2}z=0$$
$$z=-2$$

Back-substitute $z=-2$ to solve for $y$.

$$y+\frac{3}{2}(-2)=0$$
$$y-3=0$$
$$y=3$$

Back-substitute $y=3$ and $z=-2$ to solve for $x$.

$$x+4(3)+3(-2)=8$$
$$x+12-6=8$$
$$x=2$$

Back-substitute $x=2$, $y=3$, and $z=-2$ to solve for $w$.

$$w+(2)+(3)+(-2)=4$$
$$w+5-2=4$$
$$w=1$$

The solution is $(1,2,3,-2)$

**52.** $\left[\begin{array}{cccc|c} 1 & 1 & 1 & 1 & 5 \\ 1 & 2 & -1 & -2 & -1 \\ 1 & -3 & -3 & -1 & -1 \\ 2 & -1 & 2 & -1 & -2 \end{array}\right]\begin{array}{l} \\ -1R_1+R_2 \\ -1R_1+R_3 \\ -2R_1+R_4 \end{array}$

$=\left[\begin{array}{cccc|c} 1 & 1 & 1 & 1 & 5 \\ 0 & 1 & -2 & -3 & -6 \\ 0 & -4 & -4 & -2 & -6 \\ 0 & -3 & 0 & -3 & -12 \end{array}\right]\begin{array}{l} \\ \\ 4R_2+R_3 \\ 3R_2+R_4 \end{array}$

$=\left[\begin{array}{cccc|c} 1 & 1 & 1 & 1 & 5 \\ 0 & 1 & -2 & -3 & -6 \\ 0 & 0 & -12 & -14 & -30 \\ 0 & 0 & -6 & -12 & -30 \end{array}\right]\begin{array}{l} \\ \\ -\frac{1}{6}R_4 \\ R_3 \end{array}$

$=\left[\begin{array}{cccc|c} 1 & 1 & 1 & 1 & 5 \\ 0 & 1 & -2 & -3 & -6 \\ 0 & 0 & 1 & 2 & 5 \\ 0 & 0 & -12 & -14 & -30 \end{array}\right]\begin{array}{l} \\ \\ \\ 12R_3+R_4 \end{array}$

$=\left[\begin{array}{cccc|c} 1 & 1 & 1 & 1 & 5 \\ 0 & 1 & -2 & -3 & -6 \\ 0 & 0 & 1 & 2 & 5 \\ 0 & 0 & 0 & 10 & 30 \end{array}\right]\begin{array}{l} \\ \\ \\ \frac{1}{10}R_4 \end{array}$

$=\left[\begin{array}{cccc|c} 1 & 1 & 1 & 1 & 5 \\ 0 & 1 & -2 & -3 & -6 \\ 0 & 0 & 1 & 2 & 5 \\ 0 & 0 & 0 & 1 & 3 \end{array}\right]$

The resulting system is

$$\begin{aligned} w+x+y+z&=5 \\ x-2y-3z&=-6 \\ y+2z&=5 \\ z&=3 \end{aligned}$$

Back-substitute $z=3$ to solve for $y$.

$$\begin{aligned} y+2(3)&=5 \\ y+6&=5 \\ y&=-1 \end{aligned}$$

Back-substitute $y=-1$ and $z=3$ to solve for $x$.

$$\begin{aligned} x-2(-1)-3(3)&=-6 \\ x+2-9&=-6 \\ x&=1 \end{aligned}$$

Back-substitute $x=1$, $y=-1$, and $z=3$ to solve for $w$.

$$\begin{aligned} w+(1)+(-1)+(3)&=5 \\ w+4-1&=5 \\ w&=2 \end{aligned}$$

The solution is $(2,1,-1,3)$.

**53.** **a.** Use each ordered pair to write an equation as follows:

$$\begin{aligned} (x, y)&=(1,344) \\ y&=ax^2+bx+c \\ 344&=a(1)^2+b(1)+c \\ 344&=a+b+c \end{aligned}$$

$$\begin{aligned} (x, y)&=(5,480) \\ y&=ax^2+bx+c \\ 480&=a(5)^2+b(5)+c \\ 480&=a(25)+5b+c \\ 480&=25a+5b+c \end{aligned}$$

$$\begin{aligned} (x, y)&=(10,740) \\ y&=ax^2+bx+c \\ 740&=a(10)^2+b(10)+c \\ 740&=a(100)+10b+c \\ 740&=100a+10b+c \end{aligned}$$

The system of three equations in three variables is:

$$\begin{aligned} a+b+c&=344 \\ 25a+5b+c&=480 \\ 100a+10b+c&=740 \end{aligned}$$

$$\left[\begin{array}{ccc|c} 1 & 1 & 1 & 344 \\ 25 & 5 & 1 & 480 \\ 100 & 10 & 1 & 740 \end{array}\right] \quad \begin{array}{c} -25R_1 + R_2 \\ \text{and} \\ -100R_1 + R_3 \end{array}$$

$$= \left[\begin{array}{ccc|c} 1 & 1 & 1 & 344 \\ 0 & -20 & -24 & -8120 \\ 0 & -90 & -99 & -33660 \end{array}\right] \quad -\frac{1}{20}R_2$$

$$= \left[\begin{array}{ccc|c} 1 & 1 & 1 & 344 \\ 0 & 1 & \frac{6}{5} & 406 \\ 0 & -90 & -99 & -33660 \end{array}\right] \quad -\frac{1}{90}R_3$$

$$= \left[\begin{array}{ccc|c} 1 & 1 & 1 & 344 \\ 0 & 1 & \frac{6}{5} & 406 \\ 0 & 1 & \frac{11}{10} & 374 \end{array}\right] \quad -R_2 + R_3$$

$$= \left[\begin{array}{ccc|c} 1 & 1 & 1 & 344 \\ 0 & 1 & \frac{6}{5} & 406 \\ 0 & 0 & -\frac{1}{10} & -32 \end{array}\right] \quad -10R_3$$

$$= \left[\begin{array}{ccc|c} 1 & 1 & 1 & 344 \\ 0 & 1 & \frac{6}{5} & 406 \\ 0 & 0 & 1 & 320 \end{array}\right]$$

The resulting system is:

$$x + y + z = 344$$
$$y + \frac{6}{5}z = 406$$
$$z = 320$$

Back-substitute 320 for $z$ to find $y$.

$$y + \frac{6}{5}z = 406$$
$$y + \frac{6}{5}(320) = 406$$
$$y + 384 = 406$$
$$y = 22$$

Back-substitute 22 for $y$ and 320 for $z$ to find $x$.

$$x + y + z = 344$$
$$x + 22 + 320 = 344$$
$$x + 342 = 344$$
$$x = 2$$

The solution set is $\{(2, 22, 320)\}$.

The quadratic function is $y = 2x^2 + 22x + 320.$

**b.** $f(x) = 2x^2 + 22x + 320$

$$f(30) = 2(30)^2 + 22(30) + 320$$
$$= 2(900) + 660 + 320$$
$$= 1800 + 660 + 320$$
$$= 2780$$

The model predicts that there will be 2,780,000 inmates in 2010.

**c.** Answers will vary. One possible answer is the number of prison cells available in the year 2010. If there isn't room to house 2,780,000 inmates, adjustments would have to be made.

**55.** From the problem statement, we have the following equations:

$$y = x + z + 22$$
$$2x = y + 7$$
$$x + y + z = 100$$

Writing the equations in standard form gives

$$x - y + z = -22$$
$$2x - y \quad = 7$$
$$x + y + z = 100$$

The corresponding augmented matrix is

$$\left[\begin{array}{ccc|c} 1 & -1 & 1 & -22 \\ 2 & -1 & 0 & 7 \\ 1 & 1 & 1 & 100 \end{array}\right] \begin{array}{l} \\ -2R_1 + R_2 \\ -1R_1 + R_3 \end{array}$$

$$= \left[\begin{array}{ccc|c} 1 & -1 & 1 & -22 \\ 0 & 1 & -2 & 51 \\ 0 & 2 & 0 & 122 \end{array}\right] \begin{array}{l} \\ \\ -2R_2 + R_3 \end{array}$$

$$= \left[\begin{array}{ccc|c} 1 & -1 & 1 & -22 \\ 0 & 1 & -2 & 51 \\ 0 & 0 & 4 & 20 \end{array}\right] \begin{array}{l} \\ \\ \frac{1}{4}R_4 \end{array}$$

$$= \left[\begin{array}{ccc|c} 1 & -1 & 1 & -22 \\ 0 & 1 & -2 & 51 \\ 0 & 0 & 1 & 5 \end{array}\right]$$

The resulting system of equations is

$$x - y + z = -22$$
$$y - 2z = 51$$
$$z = 5$$

Back-substitute $z = 5$ to solve for $y$.

$$y - 2(5) = 51$$
$$y - 10 = 51$$
$$y = 61$$

Back-substitute $y = 61$ and $z = 5$ to solve for $x$.

$$x + 61 + 5 = 100$$
$$x = 34$$

Therefore, 34% of the single women polled responded YES, 61% responded NO, and 5% responded NOT SURE.

**57.** Answers will vary.

**59.** Answers will vary.

**61.** Answers will vary. The system will have no solution if you obtain a row that contains all zeros except for the last entry.

**63.** Answers will vary. For example, verify Exercise 11.

```
[A]
     [[-6 8 -12]
      [3  5 -2 ]]
```

```
*row(-1/6,[A],1)
▸Frac
   [[1 -4/3 2 ]
    [3 5     -2]]
```

This matches the result obtained in Exercise 11.

**65.** Statement **d.** is true.

Statement **a.** is false. Multiplying a row by a negative fraction is permitted.

Statement **b.** is false because there are three variables in the system. The augmented matrix should be:

$$\left[\begin{array}{ccc|c} 1 & -3 & 0 & 5 \\ 0 & 1 & -2 & 7 \\ 2 & 0 & 1 & 4 \end{array}\right]$$

Statement **c.** is false. When solving a system of three equations in three variables, we use row operations to obtain ones along the diagonal and zeros below the ones.

**67.**

$$f(x) = -3x + 10$$
$$f(2a-1) = -3(2a-1) + 10$$
$$= -6a + 3 + 10$$
$$= -6a + 13$$

**68.** $f(x)=3x$ and $g(x)=2x-3$

$$\begin{aligned}(fg)(x)&=f(x)\cdot g(x)\\&=3x(2x-3)\\&=6x^2-9x\end{aligned}$$

$$\begin{aligned}(fg)(-1)&=6(-1)^2-9(-1)\\&=6(1)+9\\&=6+9=15\end{aligned}$$

**69.** $$\frac{-4x^8y^{-12}}{12x^{-3}y^{24}}=\frac{-4x^8x^3}{12y^{24}y^{12}}=\frac{-x^{11}}{3y^{36}}$$

## 3.5 Exercise Set

**1.** $$\begin{vmatrix}5 & 7\\ 2 & 3\end{vmatrix}=5(3)-2(7)=15-14=1$$

**3.** $$\begin{vmatrix}-4 & 1\\ 5 & 6\end{vmatrix}=-4(6)-5(1)=-24-5=-29$$

**5.** $$\begin{vmatrix}-7 & 14\\ 2 & -4\end{vmatrix}=-7(-4)-2(14)=28-28=0$$

**7.** $$\begin{vmatrix}-5 & -1\\ -2 & -7\end{vmatrix}=-5(-7)-(-2)(-1)=35-2=33$$

**9.** $$\begin{vmatrix}\frac{1}{2} & \frac{1}{2}\\ \frac{1}{8} & -\frac{3}{4}\end{vmatrix}=\frac{1}{2}\left(-\frac{3}{4}\right)-\frac{1}{8}\left(\frac{1}{2}\right)=-\frac{3}{8}-\frac{1}{16}=-\frac{6}{16}-\frac{1}{16}=-\frac{7}{16}$$

**11.** $$D=\begin{vmatrix}1 & 1\\ 1 & -1\end{vmatrix}=1(-1)-1(1)=-1-1=-2$$

$$D_x=\begin{vmatrix}7 & 1\\ 3 & -1\end{vmatrix}=7(-1)-3(1)=-7-3=-10$$

$$D_y=\begin{vmatrix}1 & 7\\ 1 & 3\end{vmatrix}=1(3)-1(7)=3-7=-4$$

$$x=\frac{D_x}{D}=\frac{-10}{-2}=5$$

$$y=\frac{D_y}{D}=\frac{-4}{-2}=2$$

The solution is $(5,2)$.

**13.** $$D=\begin{vmatrix}12 & 3\\ 2 & -3\end{vmatrix}=12(-3)-2(3)=-36-6=-42$$

$$D_x=\begin{vmatrix}15 & 3\\ 13 & -3\end{vmatrix}=15(-3)-13(3)=-45-39=-84$$

$$D_y=\begin{vmatrix}12 & 15\\ 2 & 13\end{vmatrix}=12(13)-2(15)=156-30=126$$

$$x=\frac{D_x}{D}=\frac{-84}{-42}=2$$

$$y=\frac{D_y}{D}=\frac{126}{-42}=-3$$

The solution is $(2,-3)$.

**15.** $D = \begin{vmatrix} 4 & -5 \\ 2 & 3 \end{vmatrix} = 4(3) - 2(-5)$

$= 12 + 10 = 22$

$D_x = \begin{vmatrix} 17 & -5 \\ 3 & 3 \end{vmatrix} = 17(3) - 3(-5)$

$= 51 + 15 = 66$

$D_y = \begin{vmatrix} 4 & 17 \\ 2 & 3 \end{vmatrix} = 4(3) - 2(17)$

$= 12 - 34 = -22$

$x = \frac{D_x}{D} = \frac{66}{22} = 3$

$y = \frac{D_y}{D} = \frac{-22}{22} = -1$

The solution is $(3, -1)$.

**17.** $D = \begin{vmatrix} 1 & -3 \\ 3 & -4 \end{vmatrix} = 1(-4) - 3(-3)$

$= -4 + 9 = 5$

$D_x = \begin{vmatrix} 4 & -3 \\ 12 & -4 \end{vmatrix} = 4(-4) - 12(-3)$

$= -16 + 36 = 20$

$D_y = \begin{vmatrix} 1 & 4 \\ 3 & 12 \end{vmatrix} = 1(12) - 3(4)$

$= 12 - 12 = 0$

$x = \frac{D_x}{D} = \frac{20}{5} = 4$

$y = \frac{D_y}{D} = \frac{0}{5} = 0$

The solution is $(4, 0)$.

**19.** $D = \begin{vmatrix} 3 & -4 \\ 2 & 2 \end{vmatrix} = 3(2) - 2(-4)$

$= 6 + 8 = 14$

$D_x = \begin{vmatrix} 4 & -4 \\ 12 & 2 \end{vmatrix} = 4(2) - 12(-4)$

$= 8 + 48 = 56$

$D_y = \begin{vmatrix} 3 & 4 \\ 2 & 12 \end{vmatrix} = 3(12) - 2(4)$

$= 36 - 8 = 28$

$x = \frac{D_x}{D} = \frac{56}{14} = 4; \quad y = \frac{D_y}{D} = \frac{28}{14} = 2$

The solution is $(4, 2)$.

**21.** First, rewrite the system in standard form.

$2x - 3y = 2$

$5x + 4y = 51$

$D = \begin{vmatrix} 2 & -3 \\ 5 & 4 \end{vmatrix} = 2(4) - 5(-3)$

$= 8 + 15 = 23$

$D_x = \begin{vmatrix} 2 & -3 \\ 51 & 4 \end{vmatrix} = 2(4) - 51(-3)$

$= 8 + 153 = 161$

$D_y = \begin{vmatrix} 2 & 2 \\ 5 & 51 \end{vmatrix} = 2(51) - 5(2)$

$= 102 - 10 = 92$

$x = \frac{D_x}{D} = \frac{161}{23} = 7; \quad y = \frac{D_y}{D} = \frac{92}{23} = 4$

The solution is $(7, 4)$.

**23.** $3x + 3y = 2$

$2x + 2y = 3$

First, rewrite the system in standard form.

$D = \begin{vmatrix} 3 & 3 \\ 2 & 2 \end{vmatrix} = 3(2) - 2(3)$

$= 6 - 6 = 0$

$D_x = \begin{vmatrix} 2 & 3 \\ 3 & 2 \end{vmatrix} = 2(2) - 3(3)$

$= 4 - 9 = -5$

$D_y = \begin{vmatrix} 2 & 3 \\ 3 & 2 \end{vmatrix} = 2(2) - 3(3)$

$= 4 - 9 = -5$

Because $D = 0$ but neither $D_x$ nor $D_y$ is zero, Cramer's Rule cannot be used to solve the system. Instead, use matrices. $\left[\begin{array}{cc|c} 3 & 3 & 2 \\ 2 & 2 & 3 \end{array}\right] \quad \frac{1}{3}R_1$

$= \left[\begin{array}{cc|c} 1 & 1 & 2/3 \\ 2 & 2 & 3 \end{array}\right] \quad -2R_1 + R_2$

$= \left[\begin{array}{cc|c} 1 & 1 & 2/3 \\ 0 & 0 & 5/3 \end{array}\right]$

This is a contradiction. There are no values for $x$ and $y$ for which $0 = \frac{5}{3}$.

The solution set is $\varnothing$ and the system is inconsistent.

**25.** Rewrite the system in standard form.

$3x + 4y = 16$

$6x + 8y = 32$

$D = \begin{vmatrix} 3 & 4 \\ 6 & 8 \end{vmatrix} = 3(8) - 6(4)$

$= 24 - 24 = 0$

$D_x = \begin{vmatrix} 16 & 4 \\ 32 & 8 \end{vmatrix} = 16(8) - 32(4)$

$= 128 - 128 = 128$

$D_y = \begin{vmatrix} 3 & 16 \\ 6 & 32 \end{vmatrix} = 3(32) - 6(16)$

$= 96 - 96 = 0$

Since $D = 0$ and all determinants in the numerators are 0, the equations in the system are dependent and there are infinitely many solutions.

**27.** $\begin{vmatrix} 3 & 0 & 0 \\ 2 & 1 & -5 \\ 2 & 5 & -1 \end{vmatrix} = 3\begin{vmatrix} 1 & -5 \\ 5 & -1 \end{vmatrix} - 2\begin{vmatrix} 0 & 0 \\ 5 & -1 \end{vmatrix} + 2\begin{vmatrix} 0 & 0 \\ 1 & -5 \end{vmatrix}$

$$= 3(1(-1)-5(-5)) - 2(0(-1)-5(0)) + 2(0(-5)-1(0))$$
$$= 3(-1+25) - 2(0-0) + 2(0-0)$$
$$= 3(24) - \cancel{2(0)} + \cancel{2(0)} = 72$$

**29.** $\begin{vmatrix} 3 & 1 & 0 \\ -3 & 4 & 0 \\ -1 & 3 & -5 \end{vmatrix} = 3\begin{vmatrix} 4 & 0 \\ 3 & -5 \end{vmatrix} - (-3)\begin{vmatrix} 1 & 0 \\ 3 & -5 \end{vmatrix} + (-1)\begin{vmatrix} 1 & 0 \\ 4 & 0 \end{vmatrix}$

$$= 3(4(-5)-3(0)) + 3(1(-5)-3(0)) - 1(1(0)-4(0))$$
$$= 3(-20-0) + 3(-5-0) - 1(0-0)$$
$$= 3(-20) + 3(-5) - \cancel{1(0)} = -60 - 15 = -75$$

**31.** $\begin{vmatrix} 1 & 1 & 1 \\ 2 & 2 & 2 \\ -3 & 4 & -5 \end{vmatrix} = 1\begin{vmatrix} 2 & 2 \\ 4 & -5 \end{vmatrix} - 2\begin{vmatrix} 1 & 1 \\ 4 & -5 \end{vmatrix} + (-3)\begin{vmatrix} 1 & 1 \\ 2 & 2 \end{vmatrix}$

$$= 1(2(-5)-4(2)) - 2(1(-5)-4(1)) - 3(1(2)-2(1))$$
$$= 1(-10-8) - 2(-5-4) - 3(2-2)$$
$$= 1(-18) - 2(-9) - \cancel{3(0)} = -18 + 18 = 0$$

**33.** $x + y + z = 0$
$2x - y + z = -1$
$-x + 3y - z = -8$

$D = \begin{vmatrix} 1 & 1 & 1 \\ 2 & -1 & 1 \\ -1 & 3 & -1 \end{vmatrix} = 1\begin{vmatrix} -1 & 1 \\ 3 & -1 \end{vmatrix} - 2\begin{vmatrix} 1 & 1 \\ 3 & -1 \end{vmatrix} - 1\begin{vmatrix} 1 & 1 \\ -1 & 1 \end{vmatrix}$

$$= 1(-1(-1)-3(1)) - 2(1(-1)-3(1)) - 1(1(1)-(-1)(1))$$
$$= 1(1-3) - 2(-1-3) - 1(1+1)$$
$$= 1(-2) - 2(-4) - 1(2) = -2 + 8 - 2 = 4$$

$$D_x = \begin{vmatrix} 0 & 1 & 1 \\ -1 & -1 & 1 \\ -8 & 3 & -1 \end{vmatrix} = 0\begin{vmatrix} -1 & 1 \\ 3 & -1 \end{vmatrix} - (-1)\begin{vmatrix} 1 & 1 \\ 3 & -1 \end{vmatrix} - 8\begin{vmatrix} 1 & 1 \\ -1 & 1 \end{vmatrix}$$
$$= 1(1(-1) - 3(1)) - 8(1(1) - (-1)(1))$$
$$= 1(-1-3) - 8(1+1)$$
$$= 1(-4) - 8(2) = -4 - 16 = -20$$

$$D_y = \begin{vmatrix} 1 & 0 & 1 \\ 2 & -1 & 1 \\ -1 & -8 & -1 \end{vmatrix} = 1\begin{vmatrix} -1 & 1 \\ -8 & -1 \end{vmatrix} - 2\begin{vmatrix} 0 & 1 \\ -8 & -1 \end{vmatrix} - 1\begin{vmatrix} 0 & 1 \\ -1 & 1 \end{vmatrix}$$
$$= 1(-1(-1) - (-8)1) - 2(0(-1) - (-8)1) - 1(0(1) - (-1)1)$$
$$= 1(1+8) - 2(0+8) - 1(0+1)$$
$$= 1(9) - 2(8) - 1(1) = 9 - 16 - 1 = -8$$

$$D_z = \begin{vmatrix} 1 & 1 & 0 \\ 2 & -1 & -1 \\ -1 & 3 & -8 \end{vmatrix} = 1\begin{vmatrix} -1 & -1 \\ 3 & -8 \end{vmatrix} - 2\begin{vmatrix} 1 & 0 \\ 3 & -8 \end{vmatrix} - 1\begin{vmatrix} 1 & 0 \\ -1 & -1 \end{vmatrix}$$
$$= 1(-1(-8) - 3(-1)) - 2(1(-8) - 3(0)) - 1(1(-1) - (-1)0)$$
$$= 1(8+3) - 2(-8-0) - 1(-1+0)$$
$$= 1(11) - 2(-8) - 1(-1) = 11 + 16 + 1 = 28$$

$$x = \frac{D_x}{D} = \frac{-20}{4} = -5 \qquad y = \frac{D_y}{D} = \frac{-8}{4} = -2 \qquad z = \frac{D_z}{D} = \frac{28}{4} = 7$$

The solution is $(-5, -2, 7)$.

**35.** 
$$\begin{aligned} 4x - 5y - 6z &= -1 \\ x - 2y - 5z &= -12 \\ 2x - y \phantom{- 5z} &= 7 \end{aligned}$$

$$D = \begin{vmatrix} 4 & -5 & -6 \\ 1 & -2 & -5 \\ 2 & -1 & 0 \end{vmatrix} = 4\begin{vmatrix} -2 & -5 \\ -1 & 0 \end{vmatrix} - 1\begin{vmatrix} -5 & -6 \\ -1 & 0 \end{vmatrix} + 2\begin{vmatrix} -5 & -6 \\ -2 & -5 \end{vmatrix}$$
$$= 4(-2(0) - (-1)(-5)) - 1(-5(0) - (-1)(-6)) + 2(-5(-5) - (-2)(-6))$$
$$= 4(-5) - 1(-6) + 2(25 - 12)$$
$$= -20 + 6 + 2(13) = -20 + 6 + 26 = 12$$

$$D_x = \begin{vmatrix} -1 & -5 & -6 \\ -12 & -2 & -5 \\ 7 & -1 & 0 \end{vmatrix} = -1\begin{vmatrix} -2 & -5 \\ -1 & 0 \end{vmatrix} - (-12)\begin{vmatrix} -5 & -6 \\ -1 & 0 \end{vmatrix} + 7\begin{vmatrix} -5 & -6 \\ -2 & -5 \end{vmatrix}$$

$$= -1\left(\cancel{-2(0)} - (-1)(-5)\right) - (-12)\left(\cancel{-5(0)} - (-1)(-6)\right) + 7\left(-5(-5) - (-2)(-6)\right)$$

$$= -1(-5) - (-12)(-6) + 7(25 - 12)$$

$$= 5 - 72 + 7(13) = 5 - 72 + 91 = 24$$

$$D_y = \begin{vmatrix} 4 & -1 & -6 \\ 1 & -12 & -5 \\ 2 & 7 & 0 \end{vmatrix} = 4\begin{vmatrix} -12 & -5 \\ 7 & 0 \end{vmatrix} - 1\begin{vmatrix} -1 & -6 \\ 7 & 0 \end{vmatrix} + 2\begin{vmatrix} -1 & -6 \\ -12 & -5 \end{vmatrix}$$

$$= 4\left(\cancel{-12(0)} - 7(-5)\right) - 1\left(\cancel{-1(0)} - 7(-6)\right) + 2\left(-1(-5) - (-12)(-6)\right)$$

$$= 4(35) - 1(42) + 2(5 - 72)$$

$$= 140 - 42 + 2(-67) = 140 - 42 - 134 = -36$$

$$D_z = \begin{vmatrix} 4 & -5 & -1 \\ 1 & -2 & -12 \\ 2 & -1 & 7 \end{vmatrix} = 4\begin{vmatrix} -2 & -12 \\ -1 & 7 \end{vmatrix} - 1\begin{vmatrix} -5 & -1 \\ -1 & 7 \end{vmatrix} + 2\begin{vmatrix} -5 & -1 \\ -2 & -12 \end{vmatrix}$$

$$= 4\left(-2(7) - (-1)(-12)\right) - 1\left(-5(7) - (-1)(-1)\right) + 2\left(-5(-12) - (-2)(-1)\right)$$

$$= 4(-14 - 12) - 1(-35 - 1) + 2(60 - 2)$$

$$= 4(-26) - 1(-36) + 2(58) = -104 + 36 + 116 = 48$$

$$x = \frac{D_x}{D} = \frac{24}{12} = 2 \qquad y = \frac{D_y}{D} = \frac{-36}{12} = -3 \qquad z = \frac{D_z}{D} = \frac{48}{12} = 4$$

The solution is $(2, -3, 4)$.

**37.** $x + y + z = 4$

$x - 2y + z = 7$

$x + 3y + 2z = 4$

$$D = \begin{vmatrix} 1 & 1 & 1 \\ 1 & -2 & 1 \\ 1 & 3 & 2 \end{vmatrix} = 1\begin{vmatrix} -2 & 1 \\ 3 & 2 \end{vmatrix} - 1\begin{vmatrix} 1 & 1 \\ 3 & 2 \end{vmatrix} + 1\begin{vmatrix} 1 & 1 \\ -2 & 1 \end{vmatrix}$$

$$= 1\left(-2(2) - 3(1)\right) - 1\left(1(2) - 3(1)\right) + 1\left(1(1) - (-2)1\right)$$

$$= 1(-4 - 3) - 1(2 - 3) + 1(1 + 2)$$

$$= 1(-7) - 1(-1) + 1(3) = -7 + 1 + 3 = -3$$

$$D_x = \begin{vmatrix} 4 & 1 & 1 \\ 7 & -2 & 1 \\ 4 & 3 & 2 \end{vmatrix} = 4\begin{vmatrix} -2 & 1 \\ 3 & 2 \end{vmatrix} - 7\begin{vmatrix} 1 & 1 \\ 3 & 2 \end{vmatrix} + 4\begin{vmatrix} 1 & 1 \\ -2 & 1 \end{vmatrix}$$

$$= 4(-2(2)-3(1)) - 7(1(2)-3(1)) + 4(1(1)-(-2)1)$$

$$= 4(-4-3) - 7(2-3) + 4(1+2)$$

$$= 4(-7) - 7(-1) + 4(3) = -28 + 7 + 12 = -9$$

$$D_y = \begin{vmatrix} 1 & 4 & 1 \\ 1 & 7 & 1 \\ 1 & 4 & 2 \end{vmatrix} = 1\begin{vmatrix} 7 & 1 \\ 4 & 2 \end{vmatrix} - 1\begin{vmatrix} 4 & 1 \\ 4 & 2 \end{vmatrix} + 1\begin{vmatrix} 4 & 1 \\ 7 & 1 \end{vmatrix}$$

$$= 1(7(2)-4(1)) - 1(4(2)-4(1)) + 1(4(1)-7(1))$$

$$= 1(14-4) - 1(8-4) + 1(4-7)$$

$$= 1(10) - 1(4) + 1(-3) = 10 - 4 - 3 = 3$$

$$D_z = \begin{vmatrix} 1 & 1 & 4 \\ 1 & -2 & 7 \\ 1 & 3 & 4 \end{vmatrix} = 1\begin{vmatrix} -2 & 7 \\ 3 & 4 \end{vmatrix} - 1\begin{vmatrix} 1 & 4 \\ 3 & 4 \end{vmatrix} + 1\begin{vmatrix} 1 & 4 \\ -2 & 7 \end{vmatrix}$$

$$= 1(-2(4)-3(7)) - 1(1(4)-3(4)) + 1(1(7)-(-2)4)$$

$$= 1(-8-21) - 1(4-12) + 1(7+8)$$

$$= 1(-29) - 1(-8) + 1(15) = -29 + 8 + 15 = -6$$

$$x = \frac{D_x}{D} = \frac{-9}{-3} = 3 \qquad y = \frac{D_y}{D} = \frac{3}{-3} = -1 \qquad z = \frac{D_z}{D} = \frac{-6}{-3} = 2$$

The solution is $(3, -1, 2)$.

**39.** $x \quad + 2z = 4$

$2y - z = 5$

$2x + 3y \quad = 13$

$$D = \begin{vmatrix} 1 & 0 & 2 \\ 0 & 2 & -1 \\ 2 & 3 & 0 \end{vmatrix} = 1\begin{vmatrix} 2 & -1 \\ 3 & 0 \end{vmatrix} - 0\begin{vmatrix} 0 & 2 \\ 3 & 0 \end{vmatrix} + 2\begin{vmatrix} 0 & 2 \\ 2 & -1 \end{vmatrix}$$

$$= 1(2(0) - 3(-1)) + 2(0(-1) - 2(2))$$

$$= 1(3) + 2(-4) = 3 - 8 = -5$$

$$D_x = \begin{vmatrix} 4 & 0 & 2 \\ 5 & 2 & -1 \\ 13 & 3 & 0 \end{vmatrix} = 4\begin{vmatrix} 2 & -1 \\ 3 & 0 \end{vmatrix} - 5\begin{vmatrix} 0 & 2 \\ 3 & 0 \end{vmatrix} + 13\begin{vmatrix} 0 & 2 \\ 2 & -1 \end{vmatrix}$$

$$= 4\left(\cancel{2(0)} - 3(-1)\right) - 5\left(\cancel{0(0)} - 3(2)\right) + 13\left(\cancel{0(-1)} - 2(2)\right)$$

$$= 4(3) - 5(-6) + 13(-4) = 12 + 30 - 52 = -10$$

$$D_y = \begin{vmatrix} 1 & 4 & 2 \\ 0 & 5 & -1 \\ 2 & 13 & 0 \end{vmatrix} = 1\begin{vmatrix} 5 & -1 \\ 13 & 0 \end{vmatrix} - \cancel{0\begin{vmatrix} 4 & 2 \\ 13 & 0 \end{vmatrix}} + 2\begin{vmatrix} 4 & 2 \\ 5 & -1 \end{vmatrix}$$

$$= 1\left(\cancel{5(0)} - 13(-1)\right) + 2\left(4(-1) - 5(2)\right)$$

$$= 1(13) + 2(-4 - 10)$$

$$= 1(13) + 2(-14) = 13 - 28 = -15$$

$$D_z = \begin{vmatrix} 1 & 0 & 4 \\ 0 & 2 & 5 \\ 2 & 3 & 13 \end{vmatrix} = 1\begin{vmatrix} 2 & 5 \\ 3 & 13 \end{vmatrix} - \cancel{0\begin{vmatrix} 0 & 4 \\ 3 & 13 \end{vmatrix}} + 2\begin{vmatrix} 0 & 4 \\ 2 & 5 \end{vmatrix}$$

$$= 1\left(2(13) - 3(5)\right) + 2\left(\cancel{0(5)} - 2(4)\right)$$

$$= 1(26 - 15) + 2(-8)$$

$$= 1(11) - 16 = 11 - 16 = -5$$

$$x = \frac{D_x}{D} = \frac{-10}{-5} = 2 \qquad y = \frac{D_y}{D} = \frac{-15}{-5} = 3 \qquad z = \frac{D_z}{D} = \frac{-5}{-5} = 1$$

The solution is $(2,3,1)$.

**41.** $$\begin{vmatrix} \begin{vmatrix} 3 & 1 \\ -2 & 3 \end{vmatrix} & \begin{vmatrix} 7 & 0 \\ 1 & 5 \end{vmatrix} \\ \begin{vmatrix} 3 & 0 \\ 0 & 7 \end{vmatrix} & \begin{vmatrix} 9 & -6 \\ 3 & 5 \end{vmatrix} \end{vmatrix} = \begin{vmatrix} 3(3)-(-2)(1) & 7(5)-1(0) \\ 3(7)-0(0) & 9(5)-3(-6) \end{vmatrix} = \begin{vmatrix} 9+2 & 35-0 \\ 21-0 & 45+18 \end{vmatrix} = \begin{vmatrix} 11 & 35 \\ 21 & 63 \end{vmatrix}$$

$$= 11(63) - 21(35)$$

$$= 693 - 735$$

$$= -42$$

**42.** $$\begin{vmatrix} \begin{vmatrix} 5 & 0 \\ 4 & -3 \end{vmatrix} & \begin{vmatrix} -1 & 0 \\ 0 & -1 \end{vmatrix} \\ \begin{vmatrix} 7 & -5 \\ 4 & 6 \end{vmatrix} & \begin{vmatrix} 4 & 1 \\ -3 & 5 \end{vmatrix} \end{vmatrix} = \begin{vmatrix} 5(-3)-4(0) & (-1)(-1)-0 \\ 7(6)-4(-5) & 4(5)-(-3)(1) \end{vmatrix} = \begin{vmatrix} -15-0 & 1+0 \\ 42+20 & 20+3 \end{vmatrix} = \begin{vmatrix} -15 & 1 \\ 62 & 23 \end{vmatrix}$$

$$= (-15)(23) - 62(1) = -345 - 62 = -407$$

**43.** From $D = \begin{vmatrix} 2 & -4 \\ 3 & 5 \end{vmatrix}$ we obtain the coefficients of the variables in our equations:

$2x - 4y = c_1$

$3x + 5y = c_2$

From $D_x = \begin{vmatrix} 8 & -4 \\ -10 & 5 \end{vmatrix}$ we obtain the constant coefficients: 8 and $-10$

$2x - 4y = 8$

$3x + 5y = -10$

**44.** From $D = \begin{vmatrix} 2 & -3 \\ 5 & 6 \end{vmatrix}$ we obtain the coefficients of the variables in our equations:

$2x - 3y = c_1$

$5x + 6y = c_2$

From $D_x = \begin{vmatrix} 8 & -3 \\ 11 & 6 \end{vmatrix}$ we obtain the constant coefficients: 8 and 11

$2x - 3y = 8$

$5x + 6y = 11$

**45.** $$\begin{vmatrix} -2 & x \\ 4 & 6 \end{vmatrix} = 32$$

$$-2(6) - 4(x) = 32$$

$$-12 - 4x = 32$$

$$-4x = 44$$

$$x = -11$$

The solution is $-11$.

**46.**

$$\begin{vmatrix} x+3 & -6 \\ x-2 & -4 \end{vmatrix} = 28$$

$$(x+3)(-4)-(x-2)(-6)=28$$

$$-4x-12+6x-12=28$$

$$2x-24=28$$

$$2x=52$$

$$x=26$$

The solution is 26.

**47.**

$$\begin{vmatrix} 1 & x & -2 \\ 3 & 1 & 1 \\ 0 & -2 & 2 \end{vmatrix} = -8$$

$$0\begin{vmatrix} x & -2 \\ 1 & 1 \end{vmatrix} - (-2)\begin{vmatrix} 1 & -2 \\ 3 & 1 \end{vmatrix} + 2\begin{vmatrix} 1 & x \\ 3 & 1 \end{vmatrix} = -8$$

$$2[1(1)-3(-2)]+2[1(1)-3(x)]=-8$$

$$2(1+6)+2(1-3x)=-8$$

$$2(7)+2(1-3x)=-8$$

$$14+2-6x=-8$$

$$-6x=-24$$

$$x=4$$

The solution is 4.

**48.**

$$\begin{vmatrix} 2 & x & 1 \\ -3 & 1 & 0 \\ 2 & 1 & 4 \end{vmatrix} = 39$$

$$-(-3)\begin{vmatrix} x & 1 \\ 1 & 4 \end{vmatrix} + \begin{vmatrix} 2 & 1 \\ 2 & 4 \end{vmatrix} - 0\begin{vmatrix} 2 & x \\ 2 & 1 \end{vmatrix} = 39$$

$$3(4x-1)+(8-2)=39$$

$$12x-3+6=39$$

$$12x=36$$

$$x=3$$

The solution is 3.

**49.** $\text{Area} = \pm\frac{1}{2}\begin{vmatrix} 3 & -5 & 1 \\ 2 & 6 & 1 \\ -3 & 5 & 1 \end{vmatrix} = \pm\frac{1}{2}\left[3\begin{vmatrix} 6 & 1 \\ 5 & 1 \end{vmatrix} - 2\begin{vmatrix} -5 & 1 \\ 5 & 1 \end{vmatrix} - 3\begin{vmatrix} -5 & 1 \\ 6 & 1 \end{vmatrix}\right]$

$= \pm\frac{1}{2}\left[3(6(1)-5(1))-2(-5(1)-5(1))-3(-5(1)-6(1))\right]$

$= \pm\frac{1}{2}\left[3(6-5)-2(-5-5)-3(-5-6)\right]$

$= \pm\frac{1}{2}\left[3(1)-2(-10)-3(-11)\right]$

$= \pm\frac{1}{2}[3+20+33] = \pm\frac{1}{2}[56] = \pm 28$

The area is 28 square units.

**51.** $\begin{vmatrix} 3 & -1 & 1 \\ 0 & -3 & 1 \\ 12 & 5 & 1 \end{vmatrix} = 3\begin{vmatrix} -3 & 1 \\ 5 & 1 \end{vmatrix} - 0\begin{vmatrix} -1 & 1 \\ 5 & 1 \end{vmatrix} + 12\begin{vmatrix} -1 & 1 \\ -3 & 1 \end{vmatrix}$

$= 3(-3(1)-5(1))+12(-1(1)-(-3)1)$

$= 3(-3-5)+12(-1+3)$

$= 3(-8)+12(2) = -24+24 = 0$

Because the determinant is equal to zero, the points are collinear.

**53.** $\begin{vmatrix} x & y & 1 \\ 3 & -5 & 1 \\ -2 & 6 & 1 \end{vmatrix} = x\begin{vmatrix} -5 & 1 \\ 6 & 1 \end{vmatrix} - 3\begin{vmatrix} y & 1 \\ 6 & 1 \end{vmatrix} - 2\begin{vmatrix} y & 1 \\ -5 & 1 \end{vmatrix}$

$= x(-5(1)-6(1))-3(y(1)-6(1))-2(y(1)-(-5)1)$

$= x(-5-6)-3(y-6)-2(y+5)$

$= x(-11)-3y+18-2y-10 = -11x-5y+8$

To find the equation of the line, set the determinant equal to zero.

$-11x-5y+8=0$

Solve for $y$ to obtain slope-intercept form.

$-11x-5y+8=0$

$-5y = 11x-8$

$y = -\frac{11}{5}x+\frac{8}{5}$

**55.** Answers will vary.

**57.** Answers will vary.

**59.** Answers will vary.

**61.** Answers will vary.

**63.** Answers will vary. For example, verify Exercise 7.

```
[A]
        [[-5  -1]
         [-2  -7]]
det([A])
              33
```

This is the same result obtained in Exercise 7.

**65.** Statement **d.** is true. Only one determinant is necessary to evaluate the determinant.

$$\begin{vmatrix} 2 & 3 & -2 \\ 0 & 1 & 3 \\ 0 & 4 & -1 \end{vmatrix} = 2\begin{vmatrix} 1 & 3 \\ 4 & -1 \end{vmatrix} - 0\cancel{\begin{vmatrix} 3 & -2 \\ 4 & -1 \end{vmatrix}} + 0\cancel{\begin{vmatrix} 3 & -2 \\ 1 & 3 \end{vmatrix}}$$

Two of the three $2\times2$ determinants are multiplied by zero so only one $2\times2$ is necessary to evaluate the expression.

Statement **a.** is false. For the determinant to equal zero, not every variable in the system must be zero.

Statement **b.** is false. Using Cramer's rule, we use $\frac{D_y}{D}$ to get the value of $y$.

Statement **c**. is false. Despite determinants being different (i.e., all entries are not identical), they can have the same value. This means that the numerators of the $x$ and $y$ when using Cramer's rule can have the same value, without being the same determinant. As a result, $x$ and $y$ can have the same value.

**67.** Original System:

$$a_1x + b_1y = c_1$$
$$a_2x + b_2y = c_2$$

$$D = \begin{vmatrix} a_1 & b_1 \\ a_2 & b_2 \end{vmatrix} = a_1b_2 - a_2b_1$$

$$D_x = \begin{vmatrix} c_1 & b_1 \\ c_2 & b_2 \end{vmatrix} = c_1b_2 - c_2b_1$$

$$D_y = \begin{vmatrix} a_1 & c_1 \\ a_2 & c_2 \end{vmatrix} = a_1c_2 - a_2c_1$$

$$x = \frac{D_x}{D} = \frac{c_1b_2 - c_2b_1}{a_1b_2 - a_2b_1}$$

$$y = \frac{D_y}{D} = \frac{a_1c_2 - a_2c_1}{a_1b_2 - a_2b_1}$$

Replace the first equation by the sum of the two equations:

$$(a_1 + a_2)x + (b_1 + b_2)y = (c_1 + c_2)$$
$$a_2x + b_2y = c_2$$

$$D = \begin{vmatrix} a_1 + a_2 & b_1 + b_2 \\ a_2 & b_2 \end{vmatrix}$$
$$= b_2(a_1 + a_2) - a_2(b_1 + b_2)$$
$$= a_1b_2 + \cancel{a_2b_2} - a_2b_1 - \cancel{a_2b_2}$$
$$= a_1b_2 - a_2b_1$$

$$D_x = \begin{vmatrix} c_1 + c_2 & b_1 + b_2 \\ c_2 & b_2 \end{vmatrix}$$
$$= b_2(c_1 + c_2) - c_2(b_1 + b_2)$$
$$= c_1b_2 + \cancel{c_2b_2} - c_2b_1 - \cancel{c_2b_2}$$
$$= c_1b_2 - c_2b_1$$

$$D_y = \begin{vmatrix} a_1 + a_2 & c_1 + c_2 \\ a_2 & c_2 \end{vmatrix}$$
$$= c_2(a_1 + a_2) - a_2(c_1 + c_2)$$
$$= a_1c_2 + \cancel{a_2c_2} - a_2c_1 - \cancel{a_2c_2}$$
$$= a_1c_2 - a_2c_1$$

$$x = \frac{D_x}{D} = \frac{c_1 b_2 - c_2 b_1}{a_1 b_2 - a_2 b_1}$$

$$y = \frac{D_y}{D} = \frac{a_1 c_2 - a_2 c_1}{a_1 b_2 - a_2 b_1}$$

These are the same solutions as for the original system.

**69.** $6x - 4 = 2 + 6(x-1)$

$6x - 4 = 2 + 6x - 6$

$6x - 4 = 6x - 4$

$0 = 0$

Since $0 = 0$ for all $x$, the solution set is $\{x | x \text{ is a real number}\}$ or $\mathbb{R}$.

**70.** $-2x + 3y = 7$

$3y = 2x + 7$

$y = \frac{2x+7}{3}$ or $y = \frac{2}{3}x + \frac{7}{3}$

**71.** $\frac{4x+1}{3} = \frac{x-3}{6} + \frac{x+5}{6}$

$6\left(\frac{4x+1}{3}\right) = 6\left(\frac{x-3}{6}\right) + 6\left(\frac{x+5}{6}\right)$

$2(4x+1) = x - 3 + x + 5$

$8x + 2 = 2x + 2$

$6x + 2 = 2$

$6x = 0$

$x = \frac{0}{6}$

$x = 0$

The solution is 0 and the solution set is $\{0\}$.

## Chapter 3 Review Exercises

**1.**

| | |
|---|---|
| $2x - 5y = -2$ | $3x + 4y = 4$ |
| $2(4) - 5(2) = -2$ | $3(4) + 4(2) = 4$ |
| $8 - 10 = -2$ | $12 + 8 = 4$ |
| $-2 = -2$ | $20 \neq 4$ |

The pair is not a solution of the system.

**2.**

$-x + 2y = 11$

$-(-5) + 2(3) = 11$

$5 + 6 = 11$

$11 = 11$

$y = -\frac{x}{3} + \frac{4}{3}$

$3 = -\frac{(-5)}{3} + \frac{4}{3}$

$3 = \frac{5}{3} + \frac{4}{3}$

$3 = \frac{9}{3}$

$3 = 3$

The pair is a solution of the system.

**3.** $x + y = 5$

$y = -x + 5$

$m = -1$

$y\text{-intercept} = 5$

$3x - y = 3$

$-y = -3x + 3$

$y = 3x - 3$

$m = 3$; $y\text{-intercept} = -3$

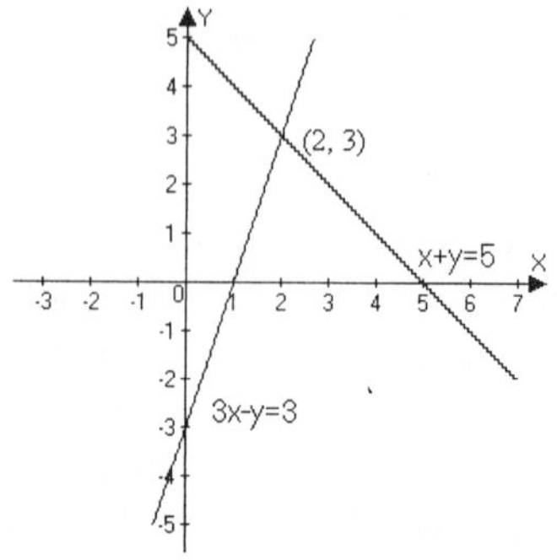


The solution is $(2,3)$.

**4.** $3x-2y=6$

$$-2y=-3x+6$$
$$y=\frac{3}{2}x-3$$
$$m=\frac{3}{2}$$
$y$-intercept $=-3$

$$6x-4y=12$$
$$-4y=-6x+12$$
$$y=\frac{6}{4}x-3$$
$$y=\frac{3}{2}x-3$$
$$m=\frac{3}{2}$$
$y$-intercept $=-3$

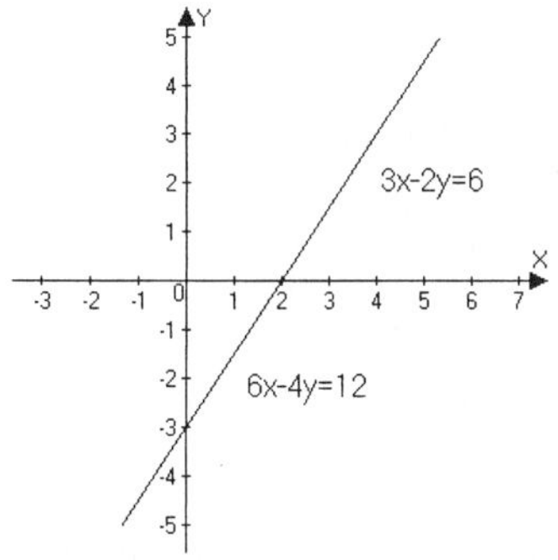


Since the lines coincide, the solution set is $\{(x,y)|3x-2y=6\}$ or $\{(x,y)|6x-4y=12\}$.

**5.** $y=\frac{3}{5}x-3$

$$m=\frac{3}{5}$$
$y$-intercept $=-3$

$$2x-y=-4$$
$$-y=-2x-4$$
$$y=2x+4$$
$$m=2$$
$y$-intercept $=4$

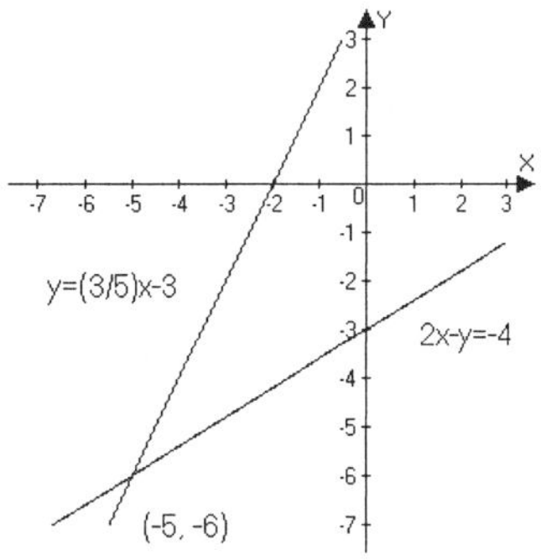


The solution is $(-5,-6)$.

**6.** $y=-x+4$

$$m=-1$$
$y$-intercept $=4$

$$3x+3y=-6$$
$$3y=-3x-6$$
$$y=-x-2$$
$$m=-1$$
$y$-intercept $=-2$

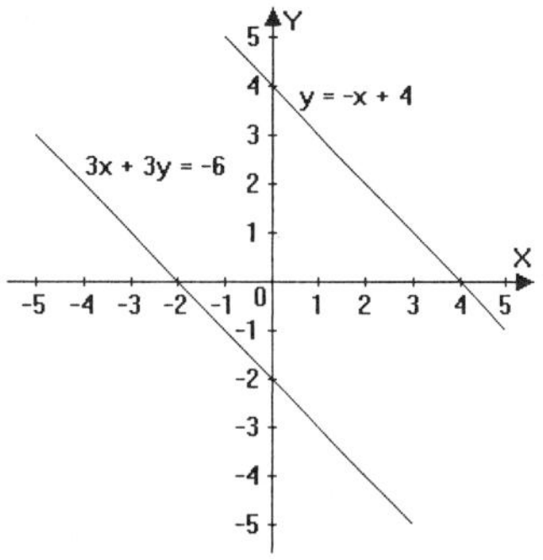


The lines do not intersect. The solution set is $\varnothing$ or $\{\ \}$.

**7.** $2x - y = 2$

$x + 2y = 11$

Solve the second equation for $x$.

$x + 2y = 11$

$x = 11 - 2y$

Substitute for $x$ in the first equation solve for $y$.

$2(11 - 2y) - y = 2$

$22 - 4y - y = 2$

$22 - 5y = 2$

$-5y = -20$

$y = 4$

Back-substitute 4 for $y$ to find $x$.

$2x - 4 = 2$

$2x = 6$

$x = 3$

The solution is $(3, 4)$.

**8.** $y = -2x + 3$

$3x + 2y = -17$

Substitute for $y$ in the second equation and solve for $x$.

$3x + 2y = -17$

$3x + 2(-2x + 3) = -17$

$3x - 4x + 6 = -17$

$-x + 6 = -17$

$-x = -23$

$x = 23$

Back-substitute 23 for $x$ to find $y$.

$y = -2(23) + 3$

$y = -46 + 3$

$y = -43$

The solution is $(23, -43)$.

**9.** $3x + 2y = -8$

$2x + 5y = 2$

Multiply the first equation by –2 and the second equation by 3 and solve by addition.

$-6x - 4y = 16$

$6x + 15y = 6$

$11y = 22$

$y = 2$

Back-substitute 2 for $y$ in one of the original equations.

$3x + 2(2) = -8$

$3x + 4 = -8$

$3x = -12$

$x = -4$

The solution is $(-4, 2)$.

**10.** $5x - 2y = 14$

$3x + 4y = 11$

Multiply the first equation by 2 and add to the second equation.

$10x - 4y = 28$

$3x + 4y = 11$

$13x = 39$

$x = 3$

Back-substitute 3 for $x$ in one of the original equations.

$5x - 2y = 14$

$5(3) - 2y = 14$

$15 - 2y = 14$

$-2y = -1$

$y = \frac{1}{2}$

The solution is $\left(3, \frac{1}{2}\right)$.

**11.** $y = 4 - x$

$3x + 3y = 12$

Substitute for $x$ in the first equation solve for $y$.

$$3x + 3y = 12$$
$$3x + 3(4 - x) = 12$$
$$3x + 12 - 3x = 12$$
$$12 = 12$$

The system has infinitely many solutions. The solution set is

$\{(x, y) | y = 4 - x\}$ or

$\{(x, y) | 3x + 3y = 12\}$.

**12.** $\frac{x}{8} + \frac{3y}{4} = \frac{19}{8}$

$-\frac{x}{2} + \frac{3y}{4} = \frac{1}{2}$

To clear fractions, multiply the first equation by 8 and the second equation by 4.

$$8\left(\frac{x}{8}\right) + 8\left(\frac{3y}{4}\right) = 8\left(\frac{19}{8}\right)$$
$$x + 2(3y) = 19$$
$$x + 6y = 19$$
$$4\left(-\frac{x}{2}\right) + 4\left(\frac{3y}{4}\right) = 4\left(\frac{1}{2}\right)$$
$$2(-x) + 3y = 2(1)$$
$$-2x + 3y = 2$$

The system becomes:

$x + 6y = 19$

$-2x + 3y = 2$

Multiply the first equation by 2 and add to the second equation.

$$2x + 12y = 38$$
$$-2x + 3y = 2$$
$$15y = 40$$
$$y = \frac{40}{15}$$
$$y = \frac{8}{3}$$

Back-substitute $\frac{8}{3}$ for $y$ in one of the equations to find $x$.

$$x + 6\left(\frac{8}{3}\right) = 19$$
$$x + 2(8) = 19$$
$$x + 16 = 19$$
$$x = 3$$

The solution is $\left(3, \frac{8}{3}\right)$.

**13.** $x - 2y + 3 = 0$

$2x - 4y + 7 = 0$

Rewrite the system is standard form.

$x - 2y = -3$

$2x - 4y = -7$

Multiply the first equation by –2 and add to the second equation.

$$-2x + 4y = 6$$
$$2x - 4y = -7$$
$$0 = -1$$

There are no values for $x$ and $y$ that will make 0 = -1, so the solution set is $\varnothing$ or $\{\ \}$.

**14.** Let $t$ = the cost of the televisions
Let $s$ = the cost of the stereos

$3t + 4s = 2530$

$4t + 3s = 2510$

Multiply the first equation by –4 and the second by 3 and solve by addition.

$$3t + 4s = 2530$$
$$-4(3t) + (-4)(4s) = -4(2530)$$
$$-12t - 16s = -10120$$

$$4t + 3s = 2510$$
$$3(4t) + 3(3s) = 3(2510)$$
$$12t + 9s = 7530$$

The system of two equations in two variables:

$$\begin{array}{r} -12t - 16s = -10120 \\ \underline{12t + \ \ 9s = \ \ \ 7530} \\ -7s = -2590 \\ s = 370 \end{array}$$

Back-substitute 370 for $s$ in one of the original equations to find $t$.

$$3t + 4s = 2530$$
$$3t + 4(370) = 2530$$
$$3t + 1480 = 2530$$
$$3t = 1050$$
$$t = 350$$

The televisions cost \$350 each and the stereos costs \$370 each.

**15.** Let $x$ = the amount invested at 4%
Let $y$ = the amount invested at 7%

$x + \quad y = 9000$

$0.04x + 0.07y = 555$

Multiply the first equation by –0.04 and add.

$$\begin{array}{r} -0.04x - 0.04y = -360 \\ \underline{0.04x + 0.07y = 555} \\ 0.03y = 195 \\ y = 6500 \end{array}$$

Back-substitute 6500 for $y$ in one of the original equations to find $x$.

$$x + y = 9000$$
$$x + 6500 = 9000$$
$$x = 2500$$

There was \$2500 invested at 4% and \$6500 invested at 7%.

**16.** Let $x$ = the amount of the 34% solution
Let $y$ = the amount of the 4% solution

$x + \quad y = 100$

$0.34x + 0.04y = 0.07(100)$

Simplified, the system becomes:

$x + \quad y = 100$

$0.34x + 0.04y = \ \ 7$

Multiply the first equation by –0.34 and add to the second equation.

$$\begin{array}{r} -0.34x - 0.34y = -34 \\ \underline{0.34x + 0.04y = \ \ \ 7} \\ -0.30 = -27 \\ y = 90 \end{array}$$

Back-substitute 90 for $y$ to find $x$.

$$x + y = 100$$
$$x + 90 = 100$$
$$x = 10$$

10 ml of the 34% solution and 90 ml of the 4% solution must be used.

**17.** Let $r$ = the speed of the plane in still air
Let w = the speed of the wind

| | r · | t = | d |
|---|---|---|---|
| With Wind | **r + w** | 3 | $3(r + w)$ |
| Against Wind | **r – w** | 4 | $3(r - w)$ |

$3(r + w) = 2160$

$4(r - w) = 2160$

Simplified, the system becomes:

$3r + 3w = 2160$

$4r - 4w = 2160$

Multiply the first equation by 4, the second equation by 3, and solve by addition.

$$\begin{aligned} 12r+12w&=8640 \\ 12r-12w&=6480 \\ \hline 24r&=15120 \\ r&=630 \end{aligned}$$

Back-substitute 630 for $r$ to find $w$.

$$\begin{aligned} 3r+3w&=2160 \\ 3(630)+3w&=2160 \\ 1890+3w&=2160 \\ 3w&=270 \\ w&=90 \end{aligned}$$

The speed of the plane in still air is 630 miles per hour and the speed of the wind is 90 miles per hour.

**18.** Let $l$ = the length of the table
Let $w$ = the width of the table

$$2l+2w=34$$
$$4l-3w=33$$

Multiply the first equation by –2 and solve by addition.

$$\begin{aligned} -4l-4w&=-68 \\ 4l-3w&=\ \ 33 \\ \hline -7w&=-35 \\ w&=5 \end{aligned}$$

Back-substitute 5 for $w$ to find $l$.

$$\begin{aligned} 2l+2w&=34 \\ 2l+2(5)&=34 \\ 2l+10&=34 \\ 2l&=24 \\ l&=12 \end{aligned}$$

The dimensions of the table are 12 feet by 5 feet.

**19.** $C(x)=22500+40x$

$$\begin{aligned} C(400)&=22500+40(400) \\ &=22500+16000 \\ &=38500 \end{aligned}$$

$$\begin{aligned} R(x)&=85x \\ R(400)&=85(400) \\ &=34000 \end{aligned}$$

$$\begin{aligned} P(x)&=R(x)-C(x) \\ P(400)&=R(400)-C(400) \\ &=34000-38500 \\ &=-4500 \end{aligned}$$

There is a \$4500 loss when 400 calculators are sold.

**20.** $R(x)=C(x)$

$$\begin{aligned} 85x&=22500+40x \\ 45x&=22500 \\ x&=500 \end{aligned}$$

$$\begin{aligned} R(x)&=85x \\ R(500)&=85(500) \\ &=42500 \end{aligned}$$

The break-even point is $(500,\ 42,500)$.

This means that when 500 calculators are produced and sold, the cost is the same as the revenue at \$42,500. At this point, there is no net loss or gain.

**21.**

$$\begin{aligned} P(x)&=R(x)-C(x) \\ &=85x-(22500+40x) \\ &=85x-22500-40x \\ &=45x-22,500 \end{aligned}$$

**22.** **a.** $C(x) = 60000 + 200x$

**b.** $R(x) = 450x$

**c.** $R(x) = C(x)$

$$450x = 60000 + 200x$$
$$250x = 60000$$
$$x = 240$$
$$R(x) = 450x$$
$$R(240) = 450(240)$$
$$= 108,000$$

The break-even point is $(240,\ 108,000)$. This means that when 240 desks are produced and sold, the cost is the same as the revenue at $108,000. At this point, there is no net loss or gain.

**23.**
$$x + y + z = 0$$
$$-3 + (-2) + 5 = 0$$
$$-5 + 5 = 0$$
$$0 = 0$$

True

$$2x - 3y + z = 5$$
$$2(-3) - 3(-2) + 5 = 5$$
$$-6 + 6 + 5 = 5$$
$$5 = 5$$

True

$$4x + 2y + 4z = 3$$
$$4(-3) + 2(-2) + 4(5) = 3$$
$$-12 - 4 + 20 = 3$$
$$4 = 3$$

False

The ordered triple (–3, –2, 5) does not satisfy all three equations, so it is not a solution.

**24.**
$$2x - y + z = 1$$
$$3x - 3y + 4z = 5$$
$$4x - 2y + 3z = 4$$

Multiply the first equation by –2 and add to the third.

$$-4x + 2y - 2z = -2$$
$$\underline{4x - 2y + 3z = 4}$$
$$z = 2$$

Back-substitute 2 for $z$ in two of the original equations to obtain a system of two equations in two variables.

$$2x - y + z = 1 \qquad 3x - 3y + 4z = 5$$
$$2x - y + 2 = 1 \qquad 3x - 3y + 4(2) = 5$$
$$2x - y = -1 \qquad 3x - 3y + 8 = 5$$
$$3x - 3y = -3$$

Th e system of two equations in two variables becomes:

$$2x - y = -1$$
$$3x - 3y = -3$$

Multiply the first equation by –3 and add solve by addition.

$$-6x + 3y = 3$$
$$\underline{3x - 3y = -3}$$
$$-3x = 0$$
$$x = 0$$

Back-substitute 0 for $x$ to find $y$.

$$2x - y = -1$$
$$2(0) - y = -1$$
$$-y = -1$$
$$y = 1$$

The solution is $(0,1,2)$.

**25.** $x+2y-z=5$

$2x-y+3z=0$

$2y+z=1$

Multiply the first equation by –2 and add to the second equation.

$-2x-4y+2z=-10$

$2x-y+3z=0$

$-5y+5z=-10$

We now have two equations in two variables.

$2y+z=1$

$-5y+5z=-10$

Multiply the first equation by –5 and solve by addition.

$-10y-5z=-5$

$-5y+5z=-10$

$-15y=-15$

$y=1$

Back-substitute 1 for $y$ to find $z$.

$2(1)+z=1$

$2+z=1$

$z=-1$

Back-substitute 1 for $y$ and –1 for $z$ to find $x$.

$x+2y-z=5$

$x+2(1)-(-1)=5$

$x+2+1=5$

$x+3=5$

$x=2$

The solution is $(2,1,-1)$.

**26.** $3x-4y+4z=7$

$x-y-2z=2$

$2x-3y+6z=5$

Multiply the second equation by –3 and add to the third equation.

$-3x+3y+6z=-6$

$2x-3y+6z=5$

$-x+12z=-1$

Multiply the second equation by –4 and add to the first equation.

$3x-4y+4z=7$

$-4x+4y+8z=-8$

$-x+12z=-1$

The system of two equations in two variables becomes:

$-x+12z=-1$

$-x+12z=-1$

The two equations in two variables are identical. The system is dependent. There are an infinite number of solutions to the system.

**27.** Use each ordered pair to write an equation as follows:

$(x, y)=(1,4)$

$y=ax^2+bx+c$

$4=a(1)^2+b(1)+c$

$4=a+b+c$

$(x, y)=(3,20)$

$y=ax^2+bx+c$

$20=a(3)^2+b(3)+c$

$20=a(9)+3b+c$

$20=9a+3b+c$

$(x, y)=(-2,25)$

$y=ax^2+bx+c$

$25=a(-2)^2+b(-2)+c$

$25=a(4)-2b+c$

$25=4a-2b+c$

The system of three equations in three variables is:

$$a + b + c = 4$$
$$9a + 3b + c = 20$$
$$4a - 2b + c = 25$$

Multiply the first equation by –1 and add to the second equation.

$$-a - b - c = -4$$
$$\underline{9a + 3b + c = 20}$$
$$8a + 2b = 16$$

Multiply the first equation by –1 and add to the third equation.

$$-a - b - c = -4$$
$$\underline{4a - 2b + c = 25}$$
$$3a - 3b = 21$$

The system of two equations in two variables becomes:

$$8a + 2b = 16$$
$$3a - 3b = 21$$

Multiply the first equation by 3, the second equation by 2 and solve by addition.

$$24a + 6b = 48$$
$$\underline{6a - 6b = 42}$$
$$30a = 90$$
$$a = 3$$

Back-substitute 3 for $a$ to find $b$.

$$3(3) - 3b = 21$$
$$9 - 3b = 21$$
$$-3b = 12$$
$$b = -4$$

Back-substitute 3 for $a$ and –4 for $b$ to find $c$.

$$a + b + c = 4$$
$$3 + (-4) + c = 4$$
$$-1 + c = 4$$
$$c = 5$$

The quadratic function is $y = 3x^2 - 4x + 5$.

**28.** $x + y + z = 307$

$$x - y = 32$$
$$y - z = 16$$

Add the first and third equations to eliminate $z$.

$$x + y + z = 307$$
$$\underline{y - z = 16}$$
$$x + 2y = 323$$

We obtain a system of two equations in two variables.

$$x - y = 32$$
$$x + 2y = 323$$

Multiply the first equation by 2 and solve by addition.

$$2x - 2y = 64$$
$$\underline{x + 2y = 323}$$
$$3x = 387$$
$$x = 129$$

Back-substitute 129 for $x$ to find $y$.

$$x - y = 32$$
$$129 - y = 32$$
$$-y = -97$$
$$y = 97$$

Back-substitute 97 for $y$ to find $z$.

$$y - z = 16$$
$$97 - z = 16$$
$$-z = -81$$
$$z = 81$$

The average yearly cost for veterinary care is \$129 per dog, \$97 per horse, and \$81 per cat.

**29.** $\left[\begin{array}{cc|c} 2 & 3 & -10 \\ 0 & 1 & -6 \end{array}\right]$

The system is:

$$2x+3y=-10$$
$$y=-6$$

Back-substitute –6 for $y$ to find $x$.

$$2x+3y=-10$$
$$2x+3(-6)=-10$$
$$2x-18=-10$$
$$2x=8$$
$$x=4$$

The solution is $(4,-6)$.

**30.** $\left[\begin{array}{ccc|c} 1 & 1 & 3 & 12 \\ 0 & 1 & -2 & -4 \\ 0 & 0 & 1 & 3 \end{array}\right]$

The system is:

$$x+y+3z=12$$
$$y-2z=-4$$
$$z=3$$

Back-substitute 3 for $z$ to find $y$.

$$y-2z=-4$$
$$y-2(3)=-4$$
$$y-6=-4$$
$$y=2$$

Back-substitute 2 for $y$ and 3 for $z$ to find $x$.

$$x+2+3(3)=12$$
$$x+2+9=12$$
$$x+11=12$$
$$x=1$$

The solution is $(1,2,3)$.

**31.** $\left[\begin{array}{cc|c} 1 & -8 & 3 \\ 0 & 7 & -14 \end{array}\right] \quad \frac{1}{7}R_2$

$=\left[\begin{array}{cc|c} 1 & -8 & 3 \\ 0 & 1 & -2 \end{array}\right]$

**32.** $\left[\begin{array}{cc|c} 1 & -3 & 1 \\ 2 & 1 & -5 \end{array}\right] \quad -2R_1+R_2$

$=\left[\begin{array}{cc|c} 1 & -3 & 1 \\ 0 & 7 & -7 \end{array}\right]$

**33.** $\left[\begin{array}{ccc|c} 2 & -2 & 1 & -1 \\ 1 & 2 & -1 & 2 \\ 6 & 4 & 3 & 5 \end{array}\right] \quad \frac{1}{2}R_1$

$=\left[\begin{array}{ccc|c} 1 & -1 & 1/2 & -1/2 \\ 1 & 2 & -1 & 2 \\ 6 & 4 & 3 & 5 \end{array}\right]$

**34.** $\left[\begin{array}{ccc|c} 1 & 2 & 2 & 2 \\ 0 & 1 & -1 & 2 \\ 0 & 5 & 4 & 1 \end{array}\right] \quad -5R_2+R_3$

$=\left[\begin{array}{ccc|c} 1 & 2 & 2 & 2 \\ 0 & 1 & -1 & 2 \\ 0 & 0 & 9 & -9 \end{array}\right]$

**35.** $\left[\begin{array}{cc|c} 1 & 4 & 7 \\ 3 & 5 & 0 \end{array}\right] \quad -3R_1+R_2$

$=\left[\begin{array}{cc|c} 1 & 4 & 7 \\ 0 & -7 & -21 \end{array}\right] \quad -\frac{1}{7}R_2$

$=\left[\begin{array}{cc|c} 1 & 4 & 7 \\ 0 & 1 & 3 \end{array}\right]$

The resulting system is:

$$x+4y=7$$
$$y=3$$

Back-substitute 3 for $y$ to find $x$.

$$x+4y=7$$
$$x+4(3)=7$$
$$x+12=7$$
$$x=-5$$

The solution is $(-5,3)$.

**36.** $\left[\begin{array}{cc|c} 2 & -3 & 8 \\ -6 & 9 & 4 \end{array}\right] \quad 3R_1+R_2$

$$=\left[\begin{array}{cc|c} 2 & -3 & 8 \\ 0 & 0 & 28 \end{array}\right]$$

This is a contradiction. $R_2$ states that

$$0x+0y=28$$
$$0=28$$

There are no values of $x$ and $y$ for which 0 = 28. The system is inconsistent and the solution set is $\varnothing$ or $\{\ \}$.

**37.** $\left[\begin{array}{ccc|c} 1 & 2 & 3 & -5 \\ 2 & 1 & 1 & 1 \\ 1 & 1 & -1 & 8 \end{array}\right] \quad -2R_1+R_2$

$$=\left[\begin{array}{ccc|c} 1 & 2 & 3 & -5 \\ 0 & -3 & -5 & 11 \\ 1 & 1 & -1 & 8 \end{array}\right] \quad -R_1+R_3$$

$$=\left[\begin{array}{ccc|c} 1 & 2 & 3 & -5 \\ 0 & -3 & -5 & 11 \\ 0 & -1 & -4 & 13 \end{array}\right] \quad -\frac{1}{3}R_2$$

$$=\left[\begin{array}{ccc|c} 1 & 2 & 3 & -5 \\ 0 & 1 & \frac{5}{3} & -\frac{11}{3} \\ 0 & -1 & -4 & 13 \end{array}\right] \quad R_2+R_3$$

$$=\left[\begin{array}{ccc|c} 1 & 2 & 3 & -5 \\ 0 & 1 & \frac{5}{3} & -\frac{11}{3} \\ 0 & 0 & -\frac{7}{3} & \frac{28}{3} \end{array}\right] \quad -\frac{3}{7}R_3$$

$$=\left[\begin{array}{ccc|c} 1 & 2 & 3 & -5 \\ 0 & 1 & \frac{5}{3} & -\frac{11}{3} \\ 0 & 0 & 1 & -4 \end{array}\right]$$

The resulting system is:

$$x+2y+3z=-5$$
$$y+\frac{5}{3}z=-\frac{11}{3}$$
$$z=-4$$

Back-substitute –4 for $z$ to find $y$.

$$y+\frac{5}{3}z=-\frac{11}{3}$$
$$y+\frac{5}{3}(-4)=-\frac{11}{3}$$
$$y-\frac{20}{3}=-\frac{11}{3}$$
$$y=\frac{9}{3}$$
$$y=3$$

Back-substitute 3 for $y$ and –4 for $z$ to find $x$.

$$x+2y+3z=-5$$
$$x+2(3)+3(-4)=-5$$
$$x+6-12=-5$$
$$x-6=-5$$
$$x=1$$

The solution is $(1,3,-4)$.

**38.** $\left[\begin{array}{ccc|c} 1 & -2 & 1 & 0 \\ 0 & 1 & -3 & -1 \\ 0 & 2 & 5 & -2 \end{array}\right] \quad -2R_2 + R_3$

$= \left[\begin{array}{ccc|c} 1 & -2 & 1 & 0 \\ 0 & 1 & -3 & -1 \\ 0 & 0 & 11 & 0 \end{array}\right] \frac{1}{11}R_3$

$= \left[\begin{array}{ccc|c} 1 & -2 & 1 & 0 \\ 0 & 1 & -3 & -1 \\ 0 & 0 & 1 & 0 \end{array}\right]$

The resulting system is:

$$\begin{aligned} x - 2y + z &= 0 \\ y - 3z &= -1 \\ z &= 0 \end{aligned}$$

Back-substitute 0 for $z$ to find $y$.

$$\begin{aligned} y - 3z &= -1 \\ y - \cancel{3(0)} &= -1 \\ y &= -1 \end{aligned}$$

Back-substitute 1 for $y$ and 0 for $z$ to find $x$.

$$\begin{aligned} x - 2y + z &= 0 \\ x - 2(-1) + 0 &= 0 \\ x + 2 &= 0 \\ x &= -2 \end{aligned}$$

The solution is $(-2, -1, 0)$.

**39.** $\begin{vmatrix} 3 & 2 \\ -1 & 5 \end{vmatrix} = 3(5) - (-1)2 = 15 + 2 = 17$

**40.** $\begin{vmatrix} -2 & -3 \\ -4 & -8 \end{vmatrix} = -2(-8) - (-4)(-3)$

$= 16 - 12 = 4$

**41.** $\begin{vmatrix} 2 & 4 & -3 \\ 1 & -1 & 5 \\ -2 & 4 & 0 \end{vmatrix} = 2\begin{vmatrix} -1 & 5 \\ 4 & 0 \end{vmatrix} - 1\begin{vmatrix} 4 & -3 \\ 4 & 0 \end{vmatrix} - 2\begin{vmatrix} 4 & -3 \\ -1 & 5 \end{vmatrix}$

$$\begin{aligned} &= 2\left(\cancel{-1(0)} - 4(5)\right) - 1\left(\cancel{4(0)} - 4(-3)\right) - 2\left(4(5) - (-1)(-3)\right) \\ &= 2(-20) - 1(12) - 2(20 - 3) \\ &= -40 - 12 - 2(17) \\ &= -40 - 12 - 34 = -86 \end{aligned}$$

**42.**
$$\begin{vmatrix} 4 & 7 & 0 \\ -5 & 6 & 0 \\ 3 & 2 & -4 \end{vmatrix} = 4\begin{vmatrix} 6 & 0 \\ 2 & -4 \end{vmatrix} - (-5)\begin{vmatrix} 7 & 0 \\ 2 & -4 \end{vmatrix} + 3\begin{vmatrix} 7 & 0 \\ 6 & 0 \end{vmatrix}$$
$$= 4\left(6(-4) - \cancel{2(0)}\right) + 5\left(7(-4) - \cancel{2(0)}\right) + 3\left(\cancel{7(0)} - \cancel{6(0)}\right)$$
$$= 4(-24) + 5(-28) + \cancel{3(0)} = -96 - 140 = -236$$

**43.**
$$D = \begin{vmatrix} 1 & -2 \\ 3 & 2 \end{vmatrix} = 1(2) - 3(-2) = 2 + 6 = 8$$
$$D_x = \begin{vmatrix} 8 & -2 \\ -1 & 2 \end{vmatrix} = 8(2) - (-1)(-2) = 16 - 2 = 14$$
$$D_y = \begin{vmatrix} 1 & 8 \\ 3 & -1 \end{vmatrix} = 1(-1) - 3(8) = -1 - 24 = -25$$
$$x = \frac{D_x}{D} = \frac{14}{8} = \frac{7}{4} \qquad y = \frac{D_y}{D} = \frac{-25}{8} = -\frac{25}{8}$$
The solution is $\left(\frac{7}{4}, -\frac{25}{8}\right)$.

**44.**
$$D = \begin{vmatrix} 7 & 2 \\ 2 & 1 \end{vmatrix} = 7(1) - 2(2) = 7 - 4 = 3$$
$$D_x = \begin{vmatrix} 0 & 2 \\ -3 & 1 \end{vmatrix} = \cancel{0(1)} - (-3)(2) = 6 \qquad D_y = \begin{vmatrix} 7 & 0 \\ 2 & -3 \end{vmatrix} = 7(-3) - \cancel{2(0)} = -21$$
$$x = \frac{D_x}{D} = \frac{6}{3} = 2 \qquad y = \frac{D_y}{D} = \frac{-21}{3} = -7$$
The solution is $(2, -7)$.

**45.**
$$D = \begin{vmatrix} 1 & 2 & 2 \\ 2 & 4 & 7 \\ -2 & -5 & -2 \end{vmatrix} = 1\begin{vmatrix} 4 & 7 \\ -5 & -2 \end{vmatrix} - 2\begin{vmatrix} 2 & 2 \\ -5 & -2 \end{vmatrix} - 2\begin{vmatrix} 2 & 2 \\ 4 & 7 \end{vmatrix}$$
$$= 1(4(-2) - (-5)7) - 2(2(-2) - (-5)2) - 2(2(7) - 4(2))$$
$$= 1(-8 + 35) - 2(-4 + 10) - 2(14 - 8)$$
$$= 1(27) - 2(6) - 2(6) = 27 - 12 - 12 = 3$$

$$D_x = \begin{vmatrix} 5 & 2 & 2 \\ 19 & 4 & 7 \\ 8 & -5 & -2 \end{vmatrix} = 5\begin{vmatrix} 4 & 7 \\ -5 & -2 \end{vmatrix} - 19\begin{vmatrix} 2 & 2 \\ -5 & -2 \end{vmatrix} + 8\begin{vmatrix} 2 & 2 \\ 4 & 7 \end{vmatrix}$$

$$= 5(4(-2)-(-5)7) - 19(2(-2)-(-5)2) + 8(2(7)-4(2))$$

$$= 5(-8+35) - 19(-4+10) + 8(14-8)$$

$$= 5(27) - 19(6) + 8(6) = 135 - 114 + 48 = 69$$

$$D_y = \begin{vmatrix} 1 & 5 & 2 \\ 2 & 19 & 7 \\ -2 & 8 & -2 \end{vmatrix} = 1\begin{vmatrix} 19 & 7 \\ 8 & -2 \end{vmatrix} - 2\begin{vmatrix} 5 & 2 \\ 8 & -2 \end{vmatrix} - 2\begin{vmatrix} 5 & 2 \\ 19 & 7 \end{vmatrix}$$

$$= 1(19(-2)-(8)7) - 2(5(-2)-8(2)) - 2(5(7)-19(2))$$

$$= 1(-38-56) - 2(-10-16) - 2(35-38)$$

$$= 1(-94) - 2(-26) - 2(-3) = -94 + 52 + 6 = -36$$

$$D_z = \begin{vmatrix} 1 & 2 & 5 \\ 2 & 4 & 19 \\ -2 & -5 & 8 \end{vmatrix} = 1\begin{vmatrix} 4 & 19 \\ -5 & 8 \end{vmatrix} - 2\begin{vmatrix} 2 & 5 \\ -5 & 8 \end{vmatrix} - 2\begin{vmatrix} 2 & 5 \\ 4 & 19 \end{vmatrix}$$

$$= 1(4(8)-(-5)19) - 2(2(8)-(-5)5) - 2(2(19)-4(5))$$

$$= 1(32+95) - 2(16+25) - 2(38-20)$$

$$= 1(127) - 2(41) - 2(18) = 127 - 82 - 36 = 9$$

$$x = \frac{D_x}{D} = \frac{69}{3} = 23 \qquad y = \frac{D_y}{D} = \frac{-36}{3} = -12 \qquad z = \frac{D_z}{D} = \frac{9}{3} = 3$$

The solution is $(23, -12, 3)$.

**46.** Rewrite the system in standard form.

$$\begin{aligned} 2x + y \quad\;\; &= -4 \\ y - 2z &= 0 \\ 3x \quad\;\; - 2z &= -11 \end{aligned}$$

$$D = \begin{vmatrix} 2 & 1 & 0 \\ 0 & 1 & -2 \\ 3 & 0 & -2 \end{vmatrix} = 2\begin{vmatrix} 1 & -2 \\ 0 & -2 \end{vmatrix} - 0\begin{vmatrix} 1 & 0 \\ 0 & -2 \end{vmatrix} + 3\begin{vmatrix} 1 & 0 \\ 1 & -2 \end{vmatrix}$$

$$= 2(1(-2) - 0(-2)) + 3(1(-2) - 1(0))$$

$$= 2(-2) + 3(-2) = -4 - 6 = -10$$

$$D_x = \begin{vmatrix} -4 & 1 & 0 \\ 0 & 1 & -2 \\ -11 & 0 & -2 \end{vmatrix} = -4\begin{vmatrix} 1 & -2 \\ 0 & -2 \end{vmatrix} - 0\begin{vmatrix} 1 & 0 \\ 0 & -2 \end{vmatrix} - 11\begin{vmatrix} 1 & 0 \\ 1 & -2 \end{vmatrix}$$

$$= -4(1(-2) - 0(-2)) - 11(1(-2) - 1(0))$$

$$= -4(-2) - 11(-2) = 8 + 22 = 30$$

$$D_y = \begin{vmatrix} 2 & -4 & 0 \\ 0 & 0 & -2 \\ 3 & -11 & -2 \end{vmatrix} = 2\begin{vmatrix} 0 & -2 \\ -11 & -2 \end{vmatrix} - 0\begin{vmatrix} -4 & 0 \\ -11 & -2 \end{vmatrix} + 3\begin{vmatrix} -4 & 0 \\ 0 & -2 \end{vmatrix}$$

$$= 2(0(-2) - (-11)(-2)) + 3(-4(-2) - 0(0))$$

$$= 2(-22) + 3(8) = -44 + 24 = -20$$

$$D_z = \begin{vmatrix} 2 & 1 & -4 \\ 0 & 1 & 0 \\ 3 & 0 & -11 \end{vmatrix} = 2\begin{vmatrix} 1 & 0 \\ 0 & -11 \end{vmatrix} - 0\begin{vmatrix} 1 & -4 \\ 0 & -11 \end{vmatrix} + 3\begin{vmatrix} 1 & -4 \\ 1 & 0 \end{vmatrix}$$

$$= 2(1(-11) - 0(0)) + 3(1(0) - 1(-4))$$

$$= 2(-11) + 3(4) = -22 + 12 = -10$$

$$x = \frac{D_x}{D} = \frac{30}{-10} = -3 \qquad y = \frac{D_y}{D} = \frac{-20}{-10} = 2 \qquad z = \frac{D_z}{D} = \frac{-10}{-10} = 1$$

The solution is $(-3, 2, 1)$.

**47.** Use each ordered pair to write an equation as follows:

| $(x, y) = (20, 400)$ | $(x, y) = (40, 150)$ | $(x, y) = (60, 400)$ |
|---|---|---|
| $y = ax^2 + bx + c$ | $y = ax^2 + bx + c$ | $y = ax^2 + bx + c$ |
| $400 = a(20)^2 + b(20) + c$ | $150 = a(40)^2 + b(40) + c$ | $400 = a(60)^2 + b(60) + c$ |
| $400 = a(400) + 20b + c$ | $150 = a(1600) + 40b + c$ | $400 = a(3600) + 60b + c$ |
| $400 = 400a + 20b + c$ | $150 = 1600a + 40b + c$ | $400 = 3600a + 60b + c$ |

The system of three equations in three variables is:

$400a + 20b + c = 400$

$1600a + 40b + c = 150$

$3600a + 60b + c = 400$

$$D = \begin{vmatrix} 400 & 20 & 1 \\ 1600 & 40 & 1 \\ 3600 & 60 & 1 \end{vmatrix} = 400\begin{vmatrix} 40 & 1 \\ 60 & 1 \end{vmatrix} - 1600\begin{vmatrix} 20 & 1 \\ 60 & 1 \end{vmatrix} + 3600\begin{vmatrix} 20 & 1 \\ 40 & 1 \end{vmatrix}$$

$$= 400(40(1)-60(1)) - 1600(20(1)-60(1)) + 3600(20(1)-40(1))$$

$$= 400(40-60) - 1600(20-60) + 3600(20-40)$$

$$= 400(-20) - 1600(-40) + 3600(-20)$$

$$= -8000 + 64000 - 72000 = -16000$$

$$D_x = \begin{vmatrix} 400 & 20 & 1 \\ 150 & 40 & 1 \\ 400 & 60 & 1 \end{vmatrix} = 400\begin{vmatrix} 40 & 1 \\ 60 & 1 \end{vmatrix} - 150\begin{vmatrix} 20 & 1 \\ 60 & 1 \end{vmatrix} + 400\begin{vmatrix} 20 & 1 \\ 40 & 1 \end{vmatrix}$$

$$= 400(40(1)-60(1)) - 150(20(1)-60(1)) + 400(20(1)-40(1))$$

$$= 400(40-60) - 150(20-60) + 400(20-40)$$

$$= 400(-20) - 150(-40) + 400(-20)$$

$$= -8000 + 6000 - 8000 = -10000$$

$$D_y = \begin{vmatrix} 400 & 400 & 1 \\ 1600 & 150 & 1 \\ 3600 & 400 & 1 \end{vmatrix} = 400\begin{vmatrix} 150 & 1 \\ 400 & 1 \end{vmatrix} - 1600\begin{vmatrix} 400 & 1 \\ 400 & 1 \end{vmatrix} + 3600\begin{vmatrix} 400 & 1 \\ 150 & 1 \end{vmatrix}$$

$$= 400(150(1)-400(1)) - 1600(400(1)-400(1)) + 3600(400(1)-150(1))$$

$$= 400(150-400) - 1600(400-400) + 3600(400-150)$$

$$= 400(-250) - \cancel{1600(0)} + 3600(250)$$

$$= 400(-250) + 3600(250)$$

$$= -100000 + 900000 = 800000$$

$$D_z = \begin{vmatrix} 400 & 20 & 400 \\ 1600 & 40 & 150 \\ 3600 & 60 & 400 \end{vmatrix} = 400\begin{vmatrix} 40 & 150 \\ 60 & 400 \end{vmatrix} - 1600\begin{vmatrix} 20 & 400 \\ 60 & 400 \end{vmatrix} + 3600\begin{vmatrix} 20 & 400 \\ 40 & 150 \end{vmatrix}$$

$$= 400(40(400)-60(150)) - 1600(20(400)-60(400)) + 3600(20(150)-40(400))$$

$$= 400(16000-9000) - 1600(8000-24000) + 3600(3000-16000)$$

$$= 400(7000) - 1600(-16000) + 3600(-13000)$$

$$= 2800000 + 25600000 - 46800000 = -18400000$$

$$x = \frac{D_x}{D} = \frac{-10000}{-16000} = \frac{5}{8} \qquad y = \frac{D_y}{D} = \frac{800000}{-16000} = -50 \qquad z = \frac{D_z}{D} = \frac{-18400000}{-16000} = 1150$$

The quadratic function is $f(x) = \frac{5}{8}x^2 - 50x + 1150.$

To predict the average number of automobile accidents in which 30-year-olds and 50-year-olds are involved, find $f(30)$ and $f(50)$.

$$\begin{aligned} f(30) &= \frac{5}{8}(30)^2 - 50(30) + 1150 \\ &= \frac{5}{8}(900) - 1500 + 1150 \\ &= 562.5 - 1500 + 1150 \\ &= 212.5 \end{aligned}$$

$$\begin{aligned} f(50) &= \frac{5}{8}(50)^2 - 50(50) + 1150 \\ &= \frac{5}{8}(2500) - 2500 + 1150 \\ &= 1562.5 - 2500 + 1150 \\ &= 212.5 \end{aligned}$$

Both 30-year-olds and 50-year-olds are involved in approximately 212.5 accidents per day in the United States.

## Chapter 3 Test

**1.**

$$\begin{aligned} x + y &= 6 \\ y &= -x + 6 \end{aligned}$$

$m = -1$

$y - \text{intercept} = 6$

$$\begin{aligned} 4x - y &= 4 \\ -y &= -4x + 4 \\ y &= 4x - 4 \end{aligned}$$

$m = 4$

$y - \text{intercept} = -4$

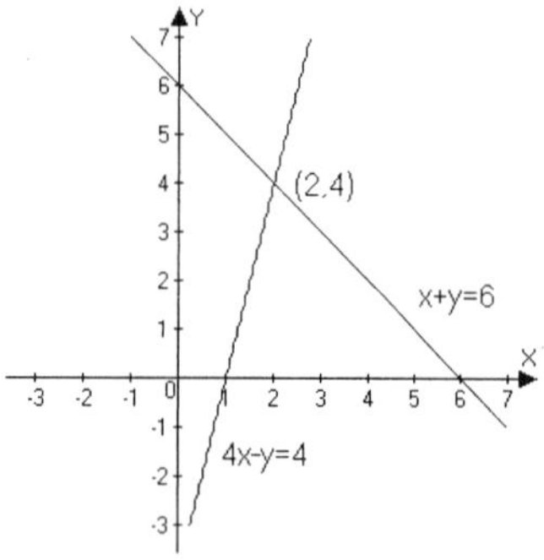


The solution is $(2, 4)$.

**2.**

$$\begin{aligned} 5x + 4y &= 10 \\ 3x + 5y &= -7 \end{aligned}$$

Multiply the first equation by –5, and the second equation by 4, and solve by addition.

$$\begin{aligned} -25x - 20y &= -50 \\ 12x + 20y &= -28 \\ \hline -13x &= -78 \\ x &= 6 \end{aligned}$$

Back-substitute 6 for $x$ to find $y$.

$$\begin{aligned} 5x + 4y &= 10 \\ 5(6) + 4y &= 10 \\ 30 + 4y &= 10 \\ 4y &= -20 \\ y &= -5 \end{aligned}$$

The solution is $(6, -5)$.

**3.** $x = y + 4$

$3x + 7y = -18$

Substitute $y + 4$ for $x$ to find $y$.

$3x + 7y = -18$

$3(y+4) + 7y = -18$

$3y + 12 + 7y = -18$

$10y + 12 = -18$

$10y = -30$

$y = -3$

Back-substitute $-3$ for $y$ to find $x$.

$x = y + 4$

$x = -3 + 4$

$x = 1$

The solution is $(1, -3)$.

**4.** $4x = 2y + 6$

$y = 2x - 3$

Substitute $2x - 3$ for $y$ to find $x$.

$4x = 2y + 6$

$4x = 2(2x - 3) + 6$

$4x = 4x - 6 + 6$

$0 = -6 + 6$

$0 = 0$

The system is dependent. There are infinitely many solutions. The solution set is $\{(x, y) | 4x = 2y + 6\}$ or $\{(x, y) | y = 2x - 3\}$.

**5.** Let $x$ = the number of one-bedroom condominiums.
Let $y$ = the number of two-bedroom condominiums.

$x + y = 50$

$120x + 150y = 7050$

Solve the first equation for $x$.

$x + y = 50$

$x = 50 - y$

Substitute $x = 50 - y$ into the second equation and solve for $y$.

$120(50 - y) + 150y = 7050$

$6000 - 120y + 150y = 7050$

$30y = 1050$

$y = 35$

Back-substitute $y = 35$ to solve for $x$.

$x + y = 50$

$x + 35 = 50$

$x = 15$

15 one-bedroom and 35 two-bedroom condominiums were sold.

**6.** Let $x$ = the amount invested at 6%
Let $y$ = the amount invested at 7%

$x + \quad y = 9000$

$0.06x + 0.07y = 610$

Multiply the first equation by –0.06 and add to the second equation.

$x + \quad y = 9000$

$0.06x + 0.07y = 610$

$$\begin{array}{r} -0.06x - 0.06y = -540 \\ 0.06x + 0.07y = 610 \\ \hline 0.01y = 70 \end{array}$$

$y = 7000$

Back-substitute 7000 for $y$ to find $x$.

$x + y = 9000$

$x + 7000 = 9000$

$x = 2000$

There is \$2000 invested at 6% and \$7000 invested at 7%.

**7.** Let $x$ = the number of ounces of 6% solution
Let $y$ = the number of ounces of 9% solution

$$x + y = 36$$
$$0.06x + 0.09y = 0.08(36)$$

Rewrite the system in standard form.

$$x + \quad y = 36$$
$$0.06x + 0.09y = 2.88$$

Multiply the first equation by –0.06 and add to the second equation.

$$-0.06x - 0.06y = -2.16$$
$$\underline{0.06x + 0.09y = 2.88}$$
$$0.03y = 0.72$$
$$y = 24$$

Back-substitute 24 for $y$ to find $x$.

$$x + y = 36$$
$$x + 24 = 36$$
$$x = 12$$

12 ounces of 6% peroxide solution and 24 ounces of 9% peroxide solution must be used.

**8.** Let $r$ = the speed of the paddleboat in still water.
Let $w$ = the speed of the current.

| | r · | t = | d |
|---|---|---|---|
| With current | $r + w$ | 3 | 48 |
| Against current | $r - w$ | 4 | 48 |

$$3(r + w) = 48$$
$$4(r - w) = 48$$

Simplified, the system becomes:

$$3r + 3w = 48$$
$$4r - 4w = 48$$

Multiply the first equation by 4, the second equation by 3, and solve by addition.

$$12r + 12w = 192$$
$$\underline{12r - 12w = 144}$$
$$24r = 336$$
$$r = 14$$

Back-substitute 14 for $r$ to find $w$.

$$3r + 3w = 48$$
$$3(14) + 3w = 48$$
$$42 + 3w = 48$$
$$3w = 6$$
$$w = 2$$

The speed of the paddleboat in still water is 14 miles per hour and the speed of the current is 2 miles per hour.

**9.** Let $x$ = the number of computers produced
$C(x) = 360{,}000 + 850x$

**10.** Let $x$ = the number of computers sold
$R(x) = 1150x$

**11.**
$$R(x) = C(x)$$
$$1150x = 360000 + 850x$$
$$300x = 360000$$
$$x = 1200$$
$$R(x) = 1150x$$
$$R(1200) = 1150(1200)$$
$$= 1{,}380{,}000$$

The break-even point is (1200, 1,380,000). When 1200 computers are produced and sold, the revenue will equal the cost at $1,380,000.

**12.** $P(x) = R(x) - C(x)$

$$= 125x - (40x + 350,000)$$
$$= 125x - 40x - 350,000$$
$$= 85x - 350,000$$

**13.**
$$x + y + z = 6$$
$$3x + 4y - 7z = 1$$
$$2x - y + 3z = 5$$

Multiply the first equation by 7 and add to the second equation.

$$7x + 7y + 7z = 42$$
$$3x + 4y - 7z = 1$$
$$10x + 11y = 43$$

Multiply the first equation by –3 and add to the third equation.

$$-3x - 3y - 3z = -18$$
$$2x - y + 3z = 5$$
$$-x - 4y = -13$$

The system of two equations in two variables.

$$10x + 11y = 43$$
$$-x - 4y = -13$$

Multiply the second equation by 10 and solve by addition.

$$10x + 11y = 43$$
$$-10x - 40y = -130$$
$$-29y = -87$$
$$y = 3$$

Back-substitute 3 for $y$ to find $x$.

$$-x - 4y = -13$$
$$-x - 4(3) = -13$$
$$-x - 12 = -13$$
$$-x = -1$$
$$x = 1$$

Back-substitute 1 for $x$ and 3 for $y$ to find z.

$$x + y + z = 6$$
$$1 + 3 + z = 6$$
$$4 + z = 6$$
$$z = 2$$

The solution is $(1,3,2)$ and the solution set is $\{(1,3,2)\}$.

**14.**
$$\left[\begin{array}{ccc|c} 1 & 0 & -4 & 5 \\ 6 & -1 & 2 & 10 \\ 2 & -1 & 4 & -3 \end{array}\right] \quad -6R_1 + R_2$$
$$= \left[\begin{array}{ccc|c} 1 & 0 & -4 & 5 \\ 0 & -1 & 26 & -20 \\ 2 & -1 & 4 & -3 \end{array}\right]$$

**15.**
$$\left[\begin{array}{cc|c} 2 & 1 & 6 \\ 3 & -2 & 16 \end{array}\right] \quad \frac{1}{2}R_1$$
$$= \left[\begin{array}{cc|c} 1 & 1/2 & 3 \\ 3 & -2 & 16 \end{array}\right] \quad -3R_1 + R_2$$
$$= \left[\begin{array}{cc|c} 1 & 1/2 & 3 \\ 0 & -7/2 & 7 \end{array}\right] \quad -\frac{2}{7}R_2$$
$$= \left[\begin{array}{cc|c} 1 & 1/2 & 3 \\ 0 & 1 & -2 \end{array}\right]$$

The resulting system is:

$$x + \frac{1}{2}y = 3$$
$$y = -2$$

Back-substitute –2 for $y$ to find $x$.

$$x + \frac{1}{2}y = 3$$
$$x + \frac{1}{2}(-2) = 3$$
$$x - 1 = 3$$
$$x = 4$$

The solution is $(4,-2)$.

**16.** $\left[\begin{array}{rrr|r} 1 & -4 & 4 & -1 \\ 2 & -1 & 5 & 6 \\ -1 & 3 & -1 & 5 \end{array}\right] \quad -2R_1 + R_2$

$= \left[\begin{array}{rrr|r} 1 & -4 & 4 & -1 \\ 0 & 7 & -3 & 8 \\ -1 & 3 & -1 & 5 \end{array}\right] \quad \frac{1}{7}R_2$

$= \left[\begin{array}{rrr|r} 1 & -4 & 4 & -1 \\ 0 & 1 & -3/7 & 8/7 \\ -1 & 3 & -1 & 5 \end{array}\right] \quad R_1 + R_2$

$= \left[\begin{array}{rrr|r} 1 & -4 & 4 & -1 \\ 0 & 1 & -3/7 & 8/7 \\ 0 & -1 & 3 & 4 \end{array}\right] \quad R_2 + R_3$

$= \left[\begin{array}{rrr|r} 1 & -4 & 4 & -1 \\ 0 & 1 & -3/7 & 8/7 \\ 0 & 0 & 18/7 & 36/7 \end{array}\right] \quad \frac{7}{18}R_3$

$= \left[\begin{array}{rrr|r} 1 & -4 & 4 & -1 \\ 0 & 1 & -3/7 & 8/7 \\ 0 & 0 & 1 & 2 \end{array}\right]$

The resulting system is:

$$\begin{aligned} x - 4y + 4z &= -1 \\ y - \frac{3}{7}z &= \frac{8}{7} \\ z &= 2 \end{aligned}$$

Back-substitute 2 for $z$ to find $y$.

$$\begin{aligned} y - \frac{3}{7}(2) &= \frac{8}{7} \\ y - \frac{6}{7} &= \frac{8}{7} \\ y &= \frac{14}{7} \\ y &= 2 \end{aligned}$$

Back-substitute 2 for $y$ and 2 for $z$ to find $x$.

$$\begin{aligned} x - 4y + 4z &= -1 \\ x - 4(2) + 4(2) &= -1 \\ x - 8 + 8 &= -1 \\ x &= -1 \end{aligned}$$

The solution is $(-1, 2, 2)$.

**17.** $\begin{vmatrix} -1 & -3 \\ 7 & 4 \end{vmatrix} = -1(4) - 7(-3) = -4 + 21 = 17$

**18.** $\begin{vmatrix} 3 & 4 & 0 \\ -1 & 0 & -3 \\ 4 & 2 & 5 \end{vmatrix} = 3\begin{vmatrix} 0 & -3 \\ 2 & 5 \end{vmatrix} - (-1)\begin{vmatrix} 4 & 0 \\ 2 & 5 \end{vmatrix} + 4\begin{vmatrix} 4 & 0 \\ 0 & -3 \end{vmatrix}$

$= 3\left(\cancel{0(5)} - 2(-3)\right) - (-1)\left(4(5) - \cancel{2(0)}\right) + 4\left(4(-3) - \cancel{0(0)}\right)$

$= 3(6) + 1(20) + 4(-12) = 18 + 20 - 48 = -10$

**19.** $D = \begin{vmatrix} 4 & -3 \\ 3 & -1 \end{vmatrix} = 4(-1) - 3(-3) = -4 + 9 = 5$

$D_x = \begin{vmatrix} 14 & -3 \\ 3 & -1 \end{vmatrix} = 14(-1) - 3(-3) = -14 + 9 = -5$

$$D_y = \begin{vmatrix} 4 & 14 \\ 3 & 3 \end{vmatrix} = 4(3) - 3(14) = 12 - 42 = -30$$

$$x = \frac{D_x}{D} = \frac{-5}{5} = -1 \qquad y = \frac{D_y}{D} = \frac{-30}{5} = -6$$

The solution is $(-1, -6)$.

**20.**

$$D = \begin{vmatrix} 2 & 3 & 1 \\ 3 & 3 & -1 \\ 1 & -2 & -3 \end{vmatrix} = 2\begin{vmatrix} 3 & -1 \\ -2 & -3 \end{vmatrix} - 3\begin{vmatrix} 3 & 1 \\ -2 & -3 \end{vmatrix} + 1\begin{vmatrix} 3 & 1 \\ 3 & -1 \end{vmatrix}$$

$$= 2(3(-3) - (-2)(-1)) - 3(3(-3) - (-2)1) + 1(3(-1) - 3(1))$$

$$= 2(-9 - 2) - 3(-9 + 2) + 1(-3 - 3)$$

$$= 2(-11) - 3(-7) + 1(-6) = -22 + 21 - 6 = -7$$

$$D_x = \begin{vmatrix} 2 & 3 & 1 \\ 0 & 3 & -1 \\ 1 & -2 & -3 \end{vmatrix} = 2\begin{vmatrix} 3 & -1 \\ -2 & -3 \end{vmatrix} - 0\cancel{\begin{vmatrix} 3 & 1 \\ -2 & -3 \end{vmatrix}} + 1\begin{vmatrix} 3 & 1 \\ 3 & -1 \end{vmatrix}$$

$$= 2(3(-3) - (-2)(-1)) + 1(3(-1) - 3(1))$$

$$= 2(-9 - 2) + 1(-3 - 3) = 2(-11) + 1(-6) = -22 - 6 = -28$$

$$D_y = \begin{vmatrix} 2 & 2 & 1 \\ 3 & 0 & -1 \\ 1 & 1 & -3 \end{vmatrix} = 2\begin{vmatrix} 0 & -1 \\ 1 & -3 \end{vmatrix} - 3\begin{vmatrix} 2 & 1 \\ 1 & -3 \end{vmatrix} + 1\begin{vmatrix} 2 & 1 \\ 0 & -1 \end{vmatrix}$$

$$= 2\left(\cancel{0(-3)} - 1(-1)\right) - 3(2(-3) - 1(1)) + 1\left(2(-1) - \cancel{0(1)}\right)$$

$$= 2(1) - 3(-6 - 1) + 1(-2) = 2 - 3(-7) - 2 = 2 + 21 - 2 = 21$$

$$D_z = \begin{vmatrix} 2 & 3 & 2 \\ 3 & 3 & 0 \\ 1 & -2 & 1 \end{vmatrix} = 2\begin{vmatrix} 3 & 0 \\ -2 & 1 \end{vmatrix} - 3\begin{vmatrix} 3 & 2 \\ -2 & 1 \end{vmatrix} + 1\begin{vmatrix} 3 & 2 \\ 3 & 0 \end{vmatrix}$$

$$= 2\left(3(1) - \cancel{(-2)(0)}\right) - 3(3(1) - (-2)2) + 1\left(\cancel{3(0)} - 3(2)\right)$$

$$= 2(3) - 3(3 + 4) + 1(-6)$$

$$= 6 - 3(7) - 6 = 6 - 21 - 6 = -21$$

$$x = \frac{D_x}{D} = \frac{-28}{-7} = 4 \qquad y = \frac{D_y}{D} = \frac{21}{-7} = -3 \qquad z = \frac{D_z}{D} = \frac{-21}{-7} = 3$$

The solution is $(4, -3, 3)$.

**Cumulative Review Exercises (Chapters 1-3)**

1. $$\frac{6(8-10)^3+(-2)}{(-5)^2(-2)}=\frac{6(-2)^3-2}{(25)(-2)}=\frac{6(-8)-2}{-50}=\frac{-48-2}{-50}=\frac{-50}{-50}=1$$

2. $$\begin{aligned}&7x-[5-2(4x-1)]\\&=7x-[5-8x+2]\\&=7x-[7-8x]\\&=7x-7+8x\\&=15x-7\end{aligned}$$

3. $$\begin{aligned}5-2(3-x)&=2(2x+5)+1\\5-6+2x&=4x+10+1\\-1+2x&=4x+11\\-1-2x&=11\\-2x&=12\\x&=-6\end{aligned}$$
The solution is –6.

4. $$\begin{aligned}\frac{3x}{5}+4&=\frac{x}{3}\\15\left(\frac{3x}{5}\right)+15(4)&=15\left(\frac{x}{3}\right)\\3(3x)+60&=5x\\9x+60&=5x\\4x+60&=0\\4x&=-60\\x&=-15\end{aligned}$$
The solution is –15.

5. $$\begin{aligned}3x-4&=2(3x+2)-3x\\3x-4&=6x+4-3x\\3x-4&=3x+4\\-4&=4\end{aligned}$$
This is a contradiction. There is no value of $x$ for which –4 equals 4. The solution set is $\varnothing$ or $\{\ \}$.

6. Let $x$ = the amount of the sales
$$\begin{aligned}200+0.05x&=0.15x\\200&=0.10x\\2000&=x\end{aligned}$$
For sales of \$2000, the earnings will be the same under either pay arrangement.

7. $$\frac{-5x^6y^{-10}}{20x^{-2}y^{20}}=\frac{-x^6x^2}{4y^{20}y^{10}}=-\frac{x^8}{4y^{30}}$$

8. $$\begin{aligned}f(x)&=-4x+5\\f(a+2)&=-4(a+2)+5\\&=-4a-8+5\\&=-4a-3\end{aligned}$$

9. To find the denominator, set it equal to zero.
$$\begin{aligned}x+3&=0\\x&=-3\end{aligned}$$
The domain is $\{x|x$ is a real number and $x\neq-3\}$.

**10.** $(f-g)(x)$
$= f(x)-g(x)$
$= 2x^2-5x+2-(x^2-2x+3)$
$= 2x^2-5x+2-x^2+2x-3$
$= x^2-3x-1$

$(f-g)(3)=(3)^2-3(3)-1$
$=9-9-1$
$=-1$

**11.** $f(x)=-\frac{2}{3}x+2$

$y=-\frac{2}{3}x+2$

$y-\text{intercept}=2$

$m=-\frac{2}{3}$

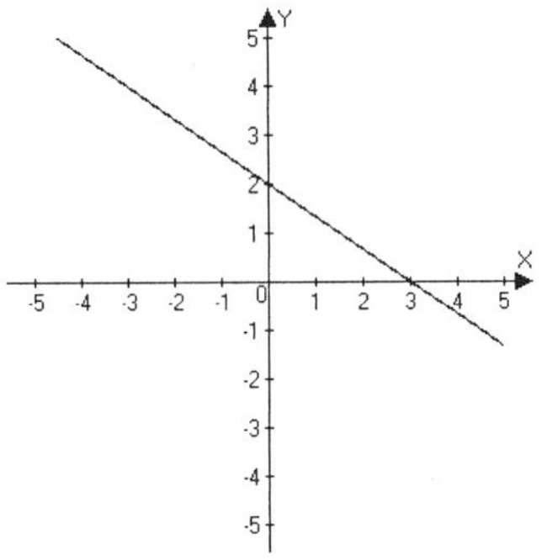

**12.** $2x-y=6$
$-y=-2x+6$
$y=2x-6$
$y-\text{intercept}=-6$
$m=2$

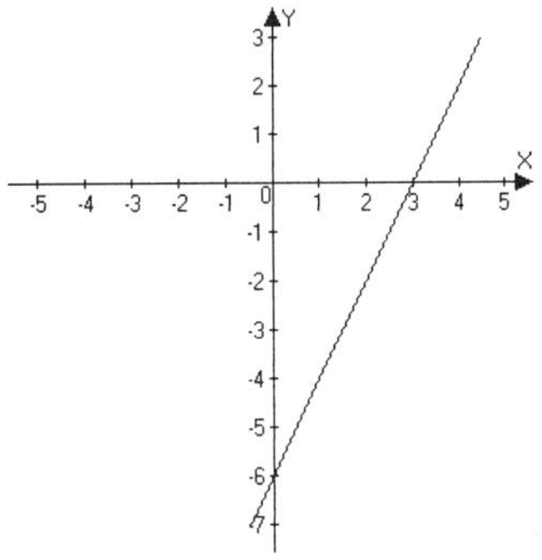

**13.** First, find the slope.
$m=\frac{4-(-2)}{2-4}=\frac{4+2}{-2}=\frac{6}{-2}=-3$
Use the slope and one of the points to write the line in point-slope form.
$y-y_1=m(x-x_1)$
$y-4=-3(x-2)$
Solve for $y$ to obtain slope-intercept form.
$y-4=-3(x-2)$
$y-4=-3x+6$
$y=-3x+10$

**14.** Parallel lines have the same slope. Find the slope of the given line.
$3x+y=6$
$y=-3x+6$
We obtain, $m=-3$.
Use the slope and the point to put the line in point-slope form.
$y-y_1=m(x-x_1)$
$y-0=-3(x-(-1))$
$y-0=-3(x+1)$
Solve for $y$ to obtain slope-intercept form.
$y-0=-3(x+1)$
$y=-3x-3$

**15.** $3x+12y=25$

$2x-\ 6y=12$

Multiply the second equation by 2 and solve by addition.

$$\begin{aligned} 3x+12y&=25 \\ 4x-12y&=24 \\ \hline 7x&=49 \\ x&=7 \end{aligned}$$

Back-substitute 7 for $x$ to find $y$.

$$\begin{aligned} 2x-6y&=12 \\ 2(7)-6y&=12 \\ 14-6y&=12 \\ -6y&=-2 \\ y&=\frac{-2}{-6} \\ y&=\frac{1}{3} \end{aligned}$$

The solution is $\left(7, \frac{1}{3}\right)$.

**16.** $x+3y-\ z=\ 5$

$-x+2y+3z=13$

$2x-5y-\ z=-8$

Add the first and second equations to eliminate $x$.

$$\begin{aligned} x+3y-z&=5 \\ -x+2y+3z&=13 \\ \hline 5y+2z&=18 \end{aligned}$$

Multiply the first equation by –2 and add to the third equation.

$$\begin{aligned} -2x-6y+2z&=-10 \\ 2x-5y-z&=-8 \\ \hline -11y+z&=-18 \end{aligned}$$

The system of two equations in two variables becomes:

$5y+2z=\ 18$

$-11y+\ z=-18$

Multiply the second equation by –2 and solve by addition.

$$\begin{aligned} 5y+2z&=18 \\ 22y-2z&=36 \\ \hline 27y&=54 \\ y&=2 \end{aligned}$$

Back-substitute 2 for $y$ to find $z$.

$$\begin{aligned} 5y+2z&=18 \\ 5(2)+2z&=18 \\ 10+2z&=18 \\ 2z&=8 \\ z&=4 \end{aligned}$$

Back-substitute 2 for $y$ and 4 for $z$ in one of the equations in three variables.

$$\begin{aligned} x+3y-z&=5 \\ x+3(2)-4&=5 \\ x+6-4&=5 \\ x+2&=5 \\ x&=3 \end{aligned}$$

The solution is $(3,2,4)$.

**17.** Let $x$ = the number of pads

Let $y$ = the number of pens

$2x+19y=5.40$

$7x+\ 4y=6.40$

Multiply the first equation by 7 and the second equation by –2 and solve by addition.

$$\begin{aligned} 14x+133y&=37.80 \\ -14x-8y&=-12.80 \\ \hline 125y&=25.00 \\ y&=.20 \end{aligned}$$

Back-substitute .20 for $y$ to find $x$.

$$7x+4y=6.40$$
$$7x+4(.20)=6.40$$
$$7x+.80=6.40$$
$$7x=5.60$$
$$x=.80$$

Pads cost \$0.80 each and pens cost \$0.20 each.

**18.** $\begin{vmatrix} 0 & 1 & -2 \\ -7 & 0 & -4 \\ 3 & 0 & 5 \end{vmatrix}$

$= 0\begin{vmatrix} 0 & -4 \\ 0 & 5 \end{vmatrix} - (-7)\begin{vmatrix} 1 & -2 \\ 0 & 5 \end{vmatrix} + 3\begin{vmatrix} 1 & -2 \\ 0 & -4 \end{vmatrix}$

$=7\left(1(5)-0(-2)\right)+3\left(1(-4)-0(-2)\right)$

$=7(5)+3(-4)=35-12=23$

**19.** $\left[\begin{array}{ccc|c} 2 & 3 & -1 & -1 \\ 1 & 2 & 3 & 2 \\ 3 & 5 & -2 & -3 \end{array}\right]$ $R_1 \leftrightarrow R_2$

$=\left[\begin{array}{ccc|c} 1 & 2 & 3 & 2 \\ 2 & 3 & -1 & -1 \\ 3 & 5 & -2 & -3 \end{array}\right]$ $-2R_1+R_2$ and $-3R_1+R_3$

$=\left[\begin{array}{ccc|c} 1 & 2 & 3 & 2 \\ 0 & -1 & -7 & -5 \\ 0 & -1 & -11 & -9 \end{array}\right]$ $-R_2$

$=\left[\begin{array}{ccc|c} 1 & 2 & 3 & 2 \\ 0 & 1 & 7 & 5 \\ 0 & -1 & -11 & -9 \end{array}\right]$ $R_2+R_3$

$=\left[\begin{array}{ccc|c} 1 & 2 & 3 & 2 \\ 0 & 1 & 7 & 5 \\ 0 & 0 & -4 & -4 \end{array}\right]$ $-\frac{1}{4}R_3$

$=\left[\begin{array}{ccc|c} 1 & 2 & 3 & 2 \\ 0 & 1 & 7 & 5 \\ 0 & 0 & 1 & 1 \end{array}\right]$

**20.** $D=\begin{vmatrix} 3 & 4 \\ -2 & 1 \end{vmatrix}=3(1)-(-2)4$

$=3+8=11$

$D_x=\begin{vmatrix} -1 & 4 \\ 8 & 1 \end{vmatrix}=-1(1)-8(4)$

$=-1-32=-33$

$D_y=\begin{vmatrix} 3 & -1 \\ -2 & 8 \end{vmatrix}=3(8)-(-2)(-1)$

$=24-2=22$

$x=\frac{D_x}{D}=\frac{-33}{11}=-3$

$y=\frac{D_y}{D}=\frac{22}{11}=2$

The solution is $(-3,2)$.

# Chapter 4
# Inequalities and Problem Solving

## 4.1 Exercise Set

**1.** $x > 5$

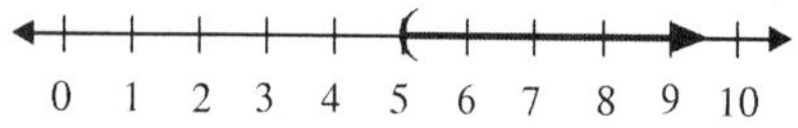

**3.** $x < -2$

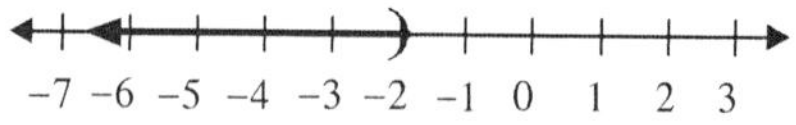

**5.** $x \geq -2.5$

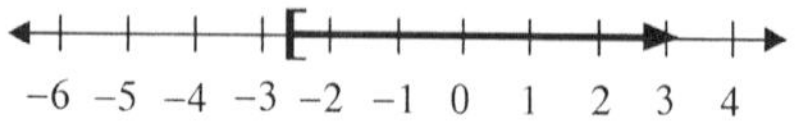

**7.** $x \leq 3$

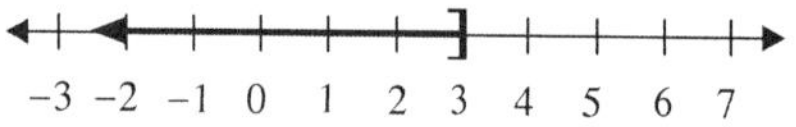

**9.** $-3 < x \leq 5$

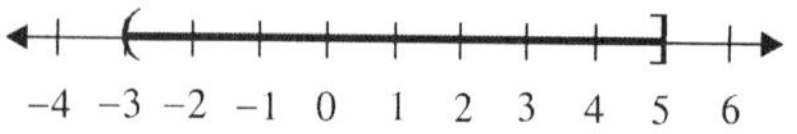

**11.** $-1 < x < 3$

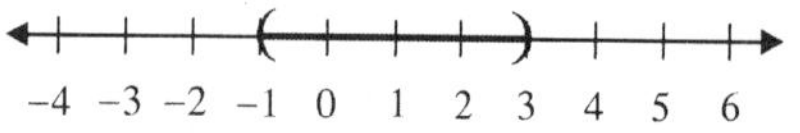

**13.** $(1,6]$

$\{x|1 < x \leq 6\}$

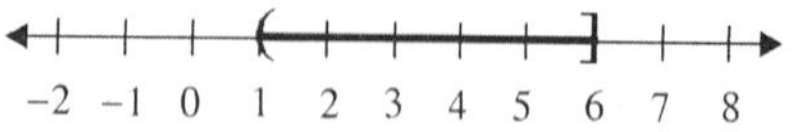

**15.** $[-5,2)$

$\{x|-5 \leq x < 2\}$

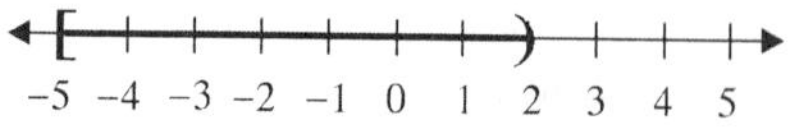

**17.** $[-3,1]$

$\{x|-3 \leq x \leq 1\}$

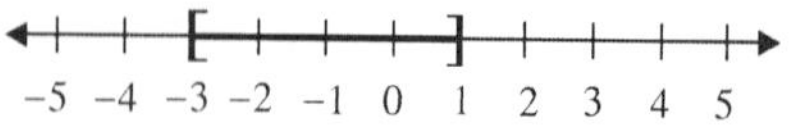

**19.** $(2,\infty)$

$\{x|x > 2\}$

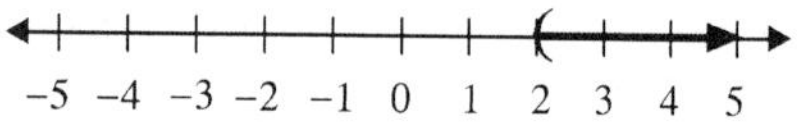

**21.** $[-3,\infty)$

$\{x|x \geq -3\}$

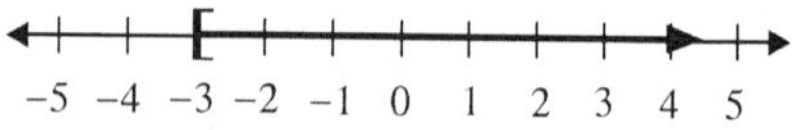

**23.** $(-\infty,3)$

$\{x|x < 3\}$

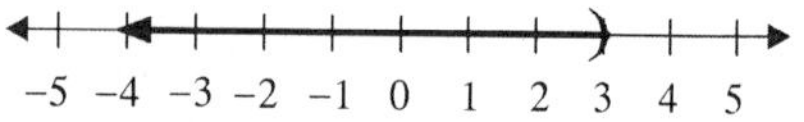

**25.** $(-\infty,5.5)$

$\{x|x < 5.5\}$

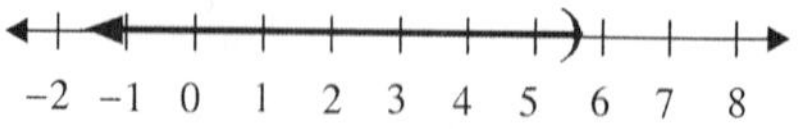

**27.** $5x + 11 < 26$

$5x < 15$

$x < 3$

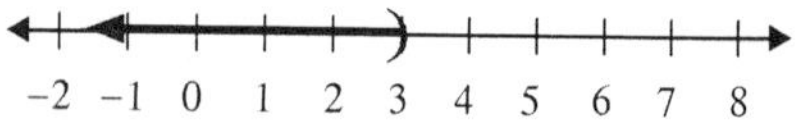

The solution set is $\{x|x < 3\}$ or $(-\infty,3)$.

**29.** $3x - 8 \geq 13$

$3x \geq 21$

$x \geq 7$

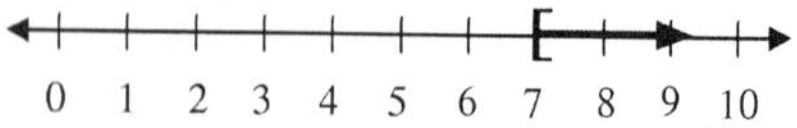

The solution set is $\{x|x \geq 7\}$ or $[7,\infty)$.

**31.** $-9x \geq 36$

$x \leq -4$

The solution set is $\{x \mid x \leq -4\}$ or $(-\infty, -4]$.

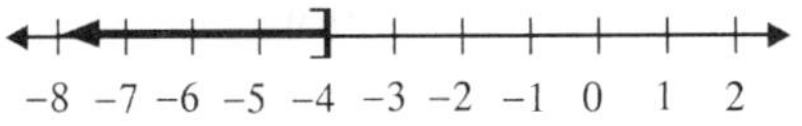

**33.** $8x - 11 \leq 3x - 13$

$5x - 11 \leq -13$

$5x \leq -2$

$x \leq -\frac{2}{5}$

The solution set is $\left\{x \mid x \leq -\frac{2}{5}\right\}$ or $\left(-\infty, -\frac{2}{5}\right]$.

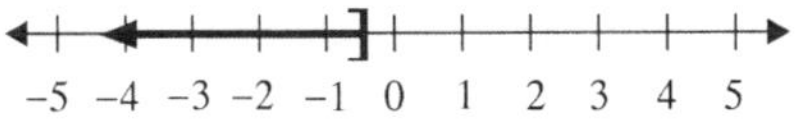

**35.** $4(x+1) + 2 \geq 3x + 6$

$4x + 4 + 2 \geq 3x + 6$

$4x + 6 \geq 3x + 6$

$x + 6 \geq 6$

$x \geq 0$

The solution set is $\{x \mid x \geq 0\}$ or $[0, \infty)$.

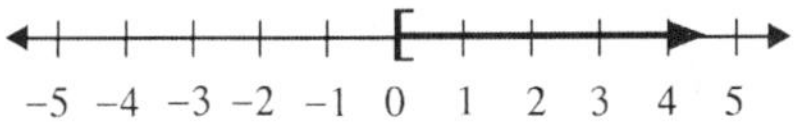

**37.** $2x - 11 < -3(x+2)$

$2x - 11 < -3x - 6$

$5x - 11 < -6$

$5x < 5$

$x < 1$

The solution set is $\{x \mid x < 1\}$ or $(-\infty, 1)$.

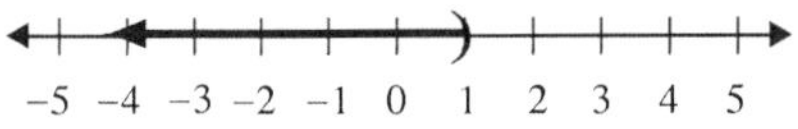

**39.** $1 - (x+3) \geq 4 - 2x$

$1 - x - 3 \geq 4 - 2x$

$-x - 2 \geq 4 - 2x$

$x - 2 \geq 4$

$x \geq 6$

The solution set is $\{x \mid x \geq 6\}$ or $[6, \infty)$.

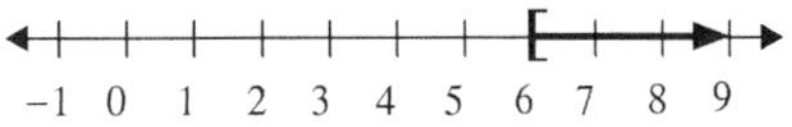

**41.** $\frac{x}{4} - \frac{1}{2} \leq \frac{x}{2} + 1$

$4\left(\frac{x}{4}\right) - 4\left(\frac{1}{2}\right) \leq 4\left(\frac{x}{2}\right) + 4(1)$

$x - 2 \leq 2x + 4$

$-x - 2 \leq 4$

$-x \leq 6$

$x \geq -6$

The solution set is $\{x \mid x \geq -6\}$ or $[-6, \infty)$.

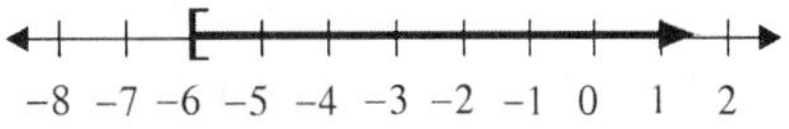

**43.** $1 - \frac{x}{2} > 4$

$2(1) - 2\left(\frac{x}{2}\right) > 2(4)$

$2 - x > 8$

$-x > 6$

$x < -6$

The solution set is $\{x \mid x < -6\}$ or $(-\infty, -6)$.

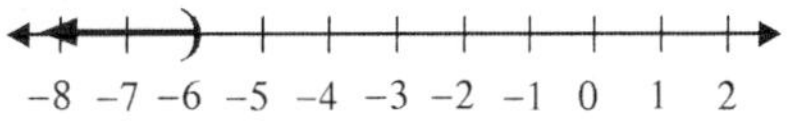

**45.** $\frac{x-4}{6} \ge \frac{x-2}{9} + \frac{5}{18}$

$18\left(\frac{x-4}{6}\right) \ge 18\left(\frac{x-2}{9}\right) + 18\left(\frac{5}{18}\right)$

$3(x-4) \ge 2(x-2) + 5$

$3x - 12 \ge 2x - 4 + 5$

$3x - 12 \ge 2x + 1$

$x - 12 \ge 1$

$x \ge 13$

The solution set is $\{x | x \ge 13\}$ or $[13, \infty)$.

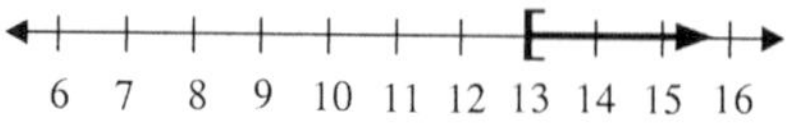

**47.** $4(3x-2) - 3x < 3(1+3x) - 7$

$12x - 8 - 3x < 3 + 9x - 7$

$9x - 8 < 9x - 4$

$-8 < -4$

Regardless of the value of $x$, $-8 < -4$ is always true. Therefore, the solution set is $\{x | x \text{ is a real number}\}$, $\mathbb{R}$ or $(-\infty, \infty)$.

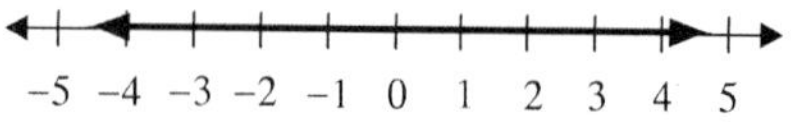

**49.** $8(x+1) \le 7(x+5) + x$

$8x + 8 \le 7x + 35 + x$

$8x + 8 \le 8x + 35$

$8 \le 35$

Regardless of the value of $x$, $8 \le 35$ is always true. Therefore, the solution set is $\{x | x \text{ is a real number}\}$, $\mathbb{R}$ or $(-\infty, \infty)$.

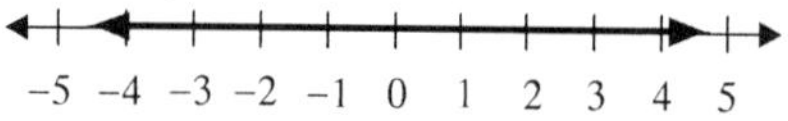

**51.** $3x < 3(x-2)$

$3x < 3x - 6$

$0 < -6$

There are no values of $x$ for which $0 < -6$. The solution set is $\varnothing$ or $\{\ \}$.

**53.** $7(y+4) - 13 < 12 + 13(3+y)$

$7y + 28 - 13 < 12 + 39 + 13y$

$7y + 15 < 13y + 51$

$-6y + 15 < 51$

$-6y < 36$

$\frac{-6y}{-6} > \frac{36}{-6}$

$y > -6$

$\{y | y > -6\}$ or $(-6, \infty)$

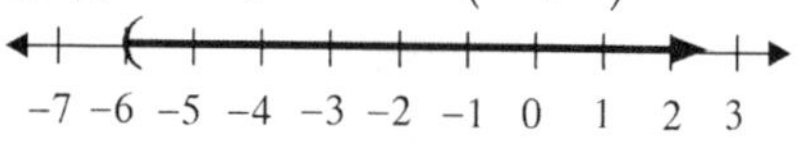

**55.** $6 - \frac{2}{3}(3x - 12) \le \frac{2}{5}(10x + 50)$

$6 - 2x + 8 \le 4x + 20$

$-2x + 14 \le 4x + 20$

$-6x + 14 \le 20$

$-6x \le 6$

$x \ge -1$

$\{x | x \ge -1\}$ or $[-1, \infty)$

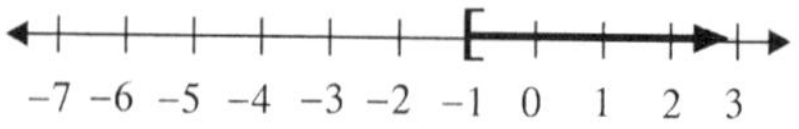

**57.**

$$3\left[3(y+5)+8y+7\right] + 5\left[3(y-6)-2(y-5)\right] < 2(4y+3)$$

$$3[3y+15+8y+7] + 5[3y-18-2y+10] < 8y+6$$

$$3[11y+22]+5[y-8]<8y+6$$

$$33y+66+5y-40<8y+6$$

$$28y+26<8y+6$$

$$20y+26<6$$

$$20y<-20$$

$$y<-2$$

$\{y \mid y \geq -2\}$ or $(-\infty,-2)$

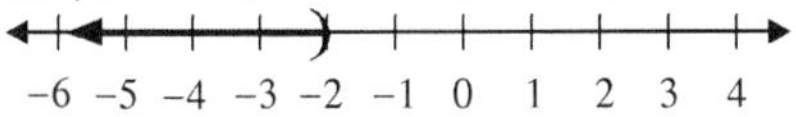

**59.** $(39\%, 55\%]$

Gardening, Amusement Parks, and Home Improvement

**61.**

$$f(x) > g(x)$$

$$3x+2>5x-8$$

$$-2x+2>-8$$

$$-2x>-10$$

$$\frac{-2x}{-2}<\frac{-10}{-2}$$

$$x<-5$$

$\{x \mid x<-5\}$ or $(-\infty,-5)$

**63.**

$$2(x+3)>6-\left\{4\left[x-(3x-4)-x\right]+4\right\}$$

$$2x+6>6-\{4[x-3x+4-x]+4\}$$

$$2x+6>6-\{4[-3x+4]+4\}$$

$$2x+6>6-\{-12x+16+4\}$$

$$2x+6>6-\{-12x+20\}$$

$$2x+6>6+12x-20$$

$$2x+6>12x-14$$

$$6>10x-14$$

$$20>10x$$

$$2>x$$

$$x<2$$

$\{x \mid x<2\}$ or $(-\infty, 2)$

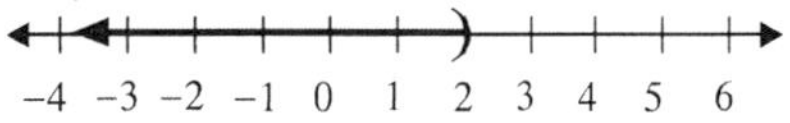

**64.**

$$3(4x-6)<4-\left\{5x-\left[6x-(4x-(3x+2))\right]\right\}$$

$$12x-18<4-\left\{5x-\left[6x-(4x-3x-2)\right]\right\}$$

$$12x-18<4-\left\{5x-\left[6x-(x-2)\right]\right\}$$

$$12x-18<4-\{5x-[6x-x+2]\}$$

$$12x-18<4-\{5x-[5x+2]\}$$

$$12x-18<4-\{5x-5x-2\}$$

$$12x-18<4-\{-2\}$$

$$12x-18<6$$

$$12x<24$$

$$x<2$$

$\{x \mid x<2\}$ or $(-\infty, 2)$

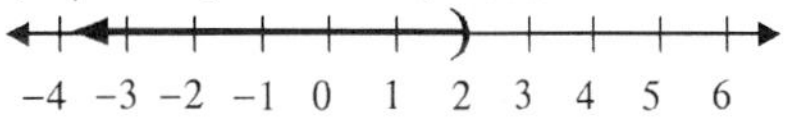

**65.**

$$ax + b > c, a < 0$$
$$ax + b - b > c - b$$
$$ax > c - b$$
$$\frac{ax}{a} < \frac{c-b}{a}, a < 0$$
$$x < \frac{c-b}{a}$$

**66.**

$$\frac{ax+b}{c} > b, c < 0$$
$$ax + b < bc$$
$$ax + b - b < bc - b$$
$$ax < bc - b$$
$$x < \frac{bc-b}{a}$$

**67.** $\{x \mid x \leq -3\}$ or $(-\infty, -3]$

**68.** $\{x \mid x \geq -3\}$ or $[-3, \infty)$

**69.** $\{x \mid x > -1.4\}$ or $(-1.4, \infty)$

**70.** $\{x \mid x < -1.4\}$ or $(-\infty, -1.4)$

**71.** $(0, 4)$

**73.** passion $\leq$ intimacy

**75.** passion<commitment

**77.** 9, after 3 years

**79.**

$$3.1x + 25.8 > 63$$
$$3.1x > 37.2$$
$$x > 12$$

Since x is the number of years after 1994, we calculate 1994+12=2006. 63% of voters will use electronic systems after 2006.

**81.**

$$W < M$$
$$-0.19t + 57 < -0.15t + 50$$
$$-0.04t + 57 < 50$$
$$-0.04t < -7$$
$$t > 175$$

The women's winning time will be less than the men's winning time after 1900 + 175 = 2075.

**83.** The daily rental cost to rent a truck from Basic Rental is $C_B = 50 + 0.20x$. The daily cost to rent a truck from Continental is $C_C = 20 + 0.50x$.

$$C_B < C_C$$
$$50 + 0.20x < 20 + 0.50x$$
$$50 - 0.30x < 20$$
$$-0.30x < -30$$
$$x > 100$$

Basic Rental is a better deal when the truck is driven more than 100 miles in a day.

**85.** The tax bill assessed under the first tax bill is $T_1 = 1800 + 0.03x$. The tax bill assessed under the second tax bill is $T_2 = 200 + 0.08x$.

$$T_1 < T_2$$
$$1800 + 0.03x < 200 + 0.08x$$
$$1800 - 0.05x < 200$$
$$-0.05x < -1600$$
$$x > 32000$$

The first tax bill is a better deal when the assessed value of the home is greater than \$32,000.

**87.** The cost is $C = 10{,}000 + 0.40x$.
The revenue is $R = 2x$.

$$C < R$$
$$10{,}000 + 0.40x < 2x$$
$$10{,}000 - 1.6x < 0$$
$$-1.6x < -10{,}000$$
$$x > 6250$$

The company will make a profit gain when more than 6,250 tapes are produced and sold each week.

**89.** Let $x$ = the number of bags of cement that can be lifted safely in the elevator.

$$W_{\text{operator}} + W_{\text{cement}} \le 3000$$
$$200 + 70x \le 3000$$
$$70x \le 2800$$
$$x \le 40$$

40 bags of cement can be lifted safely on the elevator per trip.

**91.** A parenthesis signifies that a number is not included in the solution set, bracket signifies that the number is included in the solution set.

**93.** You can use the exact same algebraic operations on inequalities that you use on equations.

**95.** If the solution set is $(-\infty, \infty)$, then every value of the variable is a solution of the inequality.

**99.** $-2(x+4) > 6x+16$

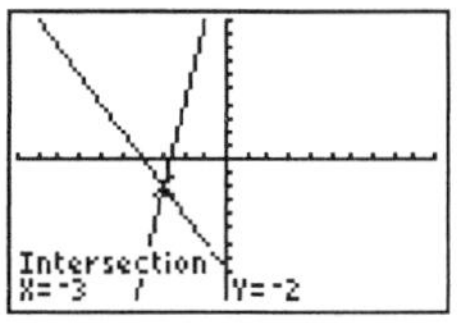


Moving from left to right on the graphing calculator screen, we see that the graph of $-2(x+4)$ is above the graph of $6x+16$ from $-\infty$ to $-3$.
The solution set is
$\{x \mid x < -3\}$ or $(-\infty, -3)$.

**101.** Graph $y = 12x - 10$ and $y = 2(x-4) + 10x$.

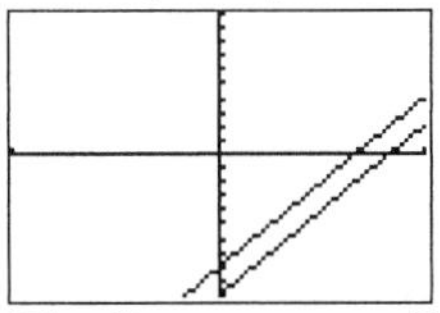

The lines are parallel. They do not intersect. There is no solution.
When solving the inequality algebraically, we arrive at a *false* statement:

$$12x - 10 > 2(x-4) + 10x$$
$$12x - 10 > 2x - 8 + 10x$$
$$12x - 10 > 12x - 8$$
$$-10 > -8 \text{ false}$$

There are no $x$-values that solve the inequality. The solution set is $\varnothing$.

**103.** **a.** Plan A: $4 + 0.10x$
Plan B: $2 + 0.15x$

**b.**

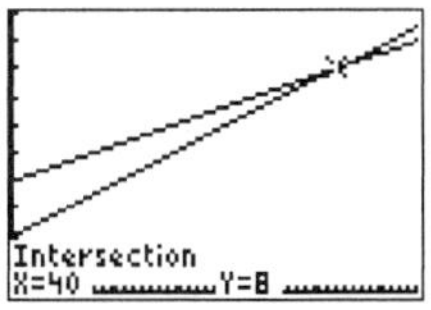


**c.** Plan A is better than Plan B for more than 40 checks per month.

**d.**

$$A < B$$
$$4 + 0.10x < 2 + 0.15x$$
$$4 < 2 + 0.05x$$
$$0.05x + 2 > 4$$
$$0.05x > 2$$
$$x > 40$$

**105.** Since $x < y \Rightarrow y - x < 0$. When multiplying both sides of the inequality by $(y-x)$, remember to *flip* the inequality.

$$2 > 1$$
$$2(y-x) < 1(y-x)$$
$$2y - 2x < y - x$$
$$y - 2x < -x$$
$$y < x$$

**107.** $f(x) = x^2 - 2x + 5$

$$f(-4) = (-4)^2 - 2(-4) + 5$$
$$= 16 + 8 + 5$$
$$= 29$$

**108.**
$$2x - y - z = -3$$
$$3x - 2y - 2z = -5$$
$$-x + y + 2z = 4$$

Add the first and third equations to eliminate $y$.

$$2x - y - z = -3$$
$$\underline{-x + y + 2z = 4}$$
$$x + z = 1$$

Multiply the third equation by 2 and add to the second equation.

$$3x - 2y - 2z = -5$$
$$\underline{-2x + 2y + 4z = 8}$$
$$x + 2z = 3$$

The system of two equations in two variables becomes:

$$x + z = 1$$
$$x + 2z = 3$$

Multiply the second equation by $-1$ and solve for $z$.

$$x + z = 1$$
$$\underline{-x - 2z = -3}$$
$$-z = -2$$
$$z = 2$$

Back-substitute 2 for $z$ to find $x$.

$$x + z = 1$$
$$x + 2 = 1$$
$$x = -1$$

Back-substitute 2 for $z$ and $-1$ and $x$ in one of the original equations in three variables to find $y$.

$$2x - y - z = -3$$
$$2(-1) - y - 2 = -3$$
$$-2 - y - 2 = -3$$
$$-y - 4 = -3$$
$$-y = 1$$
$$y = -1$$

The solution is $(-1, -1, 2)$ and the solution set is $\{(-1, -1, 2)\}$.

**109.**
$$\left(\frac{2x^4y^{-2}}{4xy^3}\right)^3 = \frac{2^3x^{12}y^{-6}}{4^3x^3y^9}$$
$$= \frac{8}{64}x^{12-3}y^{-6-9}$$
$$= \frac{1}{8}x^9y^{-15}$$
$$= \frac{x^9}{8y^{15}}$$

## 4.2 Exercise Set

**1.** $\{1, 2, 3, 4\} \cap \{2, 4, 5\} = \{2, 4\}$

**3.** $\{1, 3, 5, 7\} \cap \{2, 4, 6, 8, 10\} = \{\ \}$

The empty set is also denoted by $\varnothing$.

**5.** $\{a, b, c, d\} \cap \varnothing = \varnothing$

**7.** $x > 3$, $x > 6$, $x > 3$ and $x > 6$

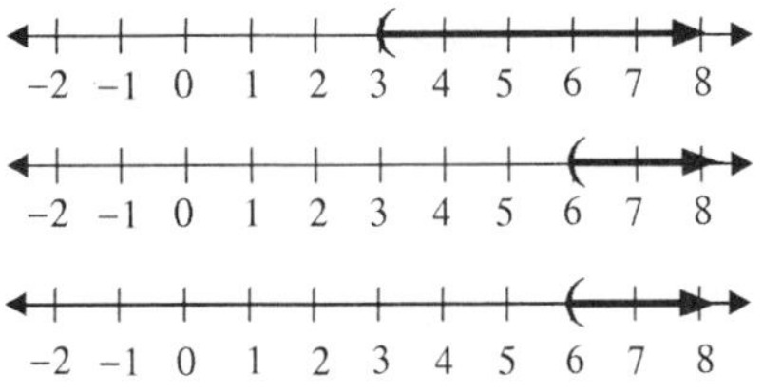


The solution set is $\{x|x > 6\}$ or $(6, \infty)$.

**9.** $x \le 5$, $x \le 1$, $x \le 5$ and $x \le 1$

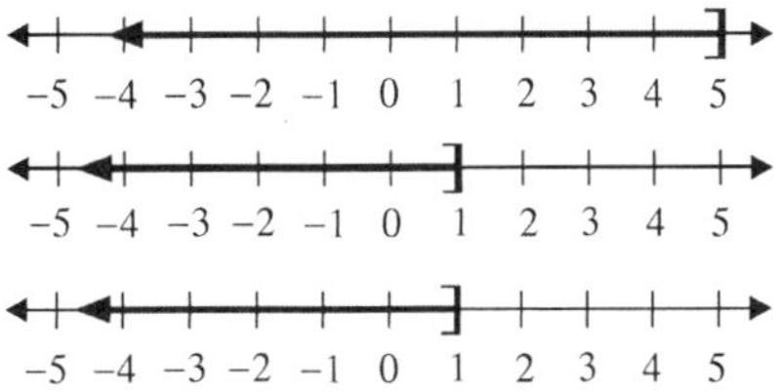


The solution set is $\{x|x \le 1\}$ or $(-\infty, 1]$.

**11.** $x < 2$, $x \ge -1$, $x < 2$ and $x \ge -1$

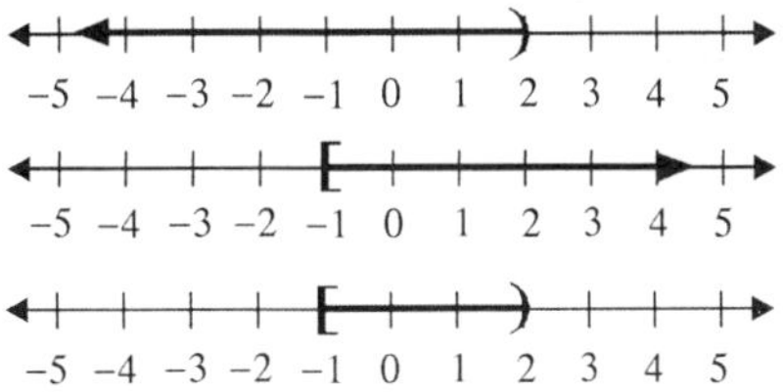


The solution set is $\{x|-1 \le x < 2\}$ or $[-1, 2)$.

**13.** $x > 2$, $x < -1$, $x > 2$ and $x < -1$

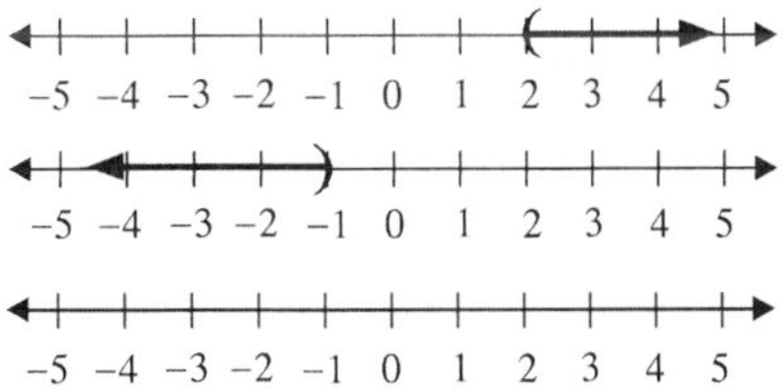


Since the two sets do not intersect, the solution set is $\varnothing$ or $\{\ \}$.

**15.** $5x < -20$, $3x > -18$,

$x < -4$ $\quad$ $x > -6$

$x < -4$ and $x > -6$

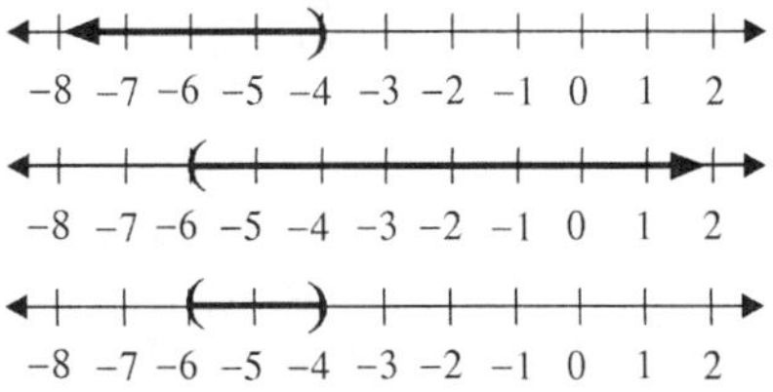


The solution set is $\{x|-6 < x < -4\}$ or $(-6, -4)$.

**17.** $x - 4 \le 2$ and $3x + 1 > -8$

$x \le 6$ $\quad$ $3x > -9$

$x > -3$

$x \le 6$, $x > -3$, $x \le 6$ and $x > -3$

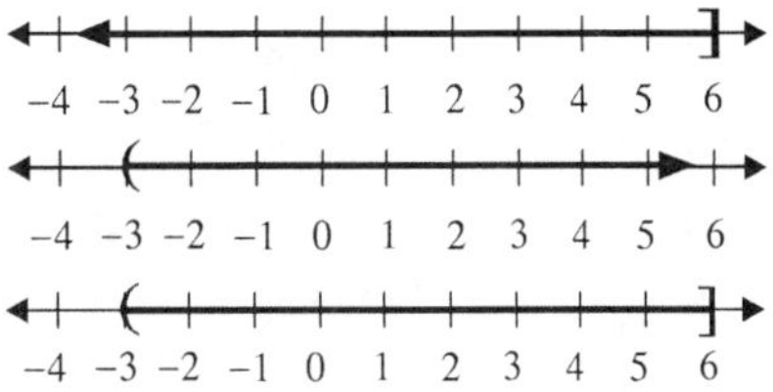


The solution set is $\{x|-3 < x \le 6\}$ or $(-3, 6]$.

**19.** $2x > 5x - 15$ and $7x > 2x + 10$

$-3x > -15$ $\quad$ $5x > 10$

$x < 5$ $\quad$ $x > 2$

$x < 5$, $x > 2$, $x < 5$ and $x > 2$

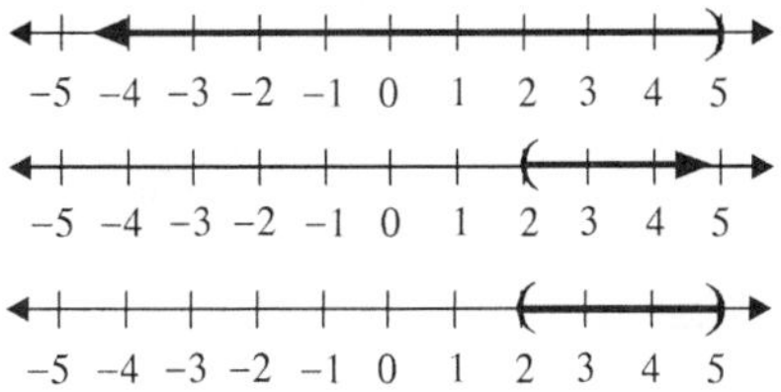


The solution set is $\{x|2 < x < 5\}$ or $(2, 5)$.

**21.** $4(1-x)<-6$

$4-4x<-6$

$-4x<-10$

$x>\dfrac{5}{2}$

$\dfrac{x-7}{5}\le -2$

$5\left(\dfrac{x-7}{5}\right)\le 5(-2)$

$x-7\le -10$

$x\le -3$

$x>\dfrac{5}{2}$, $x\le -3$, $x>\dfrac{5}{2}$ and $x\le -3$

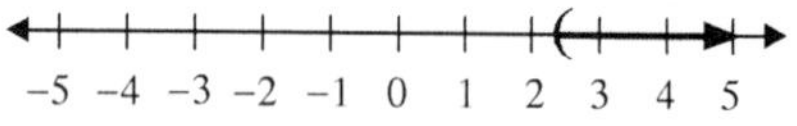


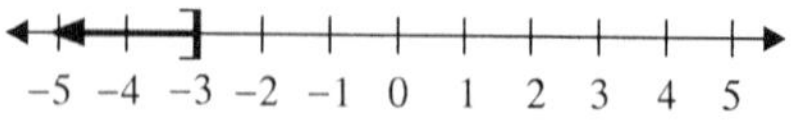


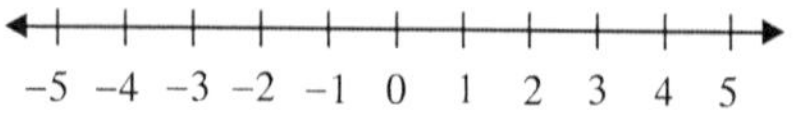


Since the two sets do not intersect, the solution set is $\varnothing$ or $\{\ \}$.

**23.** $x-1\le 7x-1$ and $4x-7<3-x$

$-1\le 6x-1$ $\quad$ $5x-7<3$

$0\le 6x$ $\quad$ $5x<10$

$0\le x$ $\quad$ $x<2$

$x\ge 0$

$x<2$, $x\ge 0$, $x<2$ and $x\ge 0$

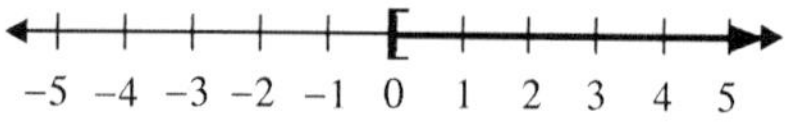


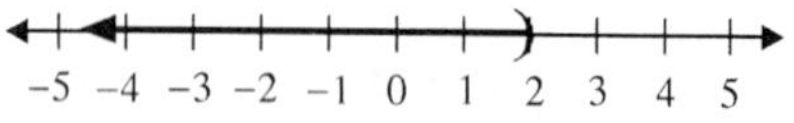

The solution set is $\{x|0\le x<2\}$ or $[0,2)$.

**25.** $6<x+3<8$

$6-3<x+3-3<8-3$

$3<x<5$

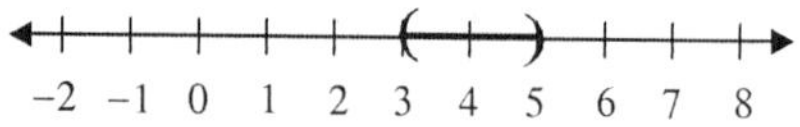


The solution set is $\{x|3<x<5\}$ or $(3,5)$.

**27.** $-3\le x-2<1$

$-3+2\le x-2+2<1+2$

$-1\le x<3$

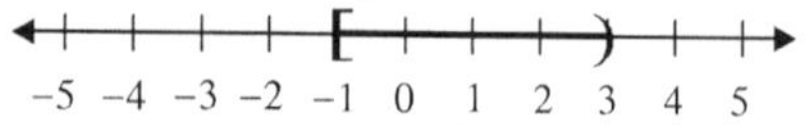


The solution set is $\{x|-1\le x<3\}$ or $[-1,3)$.

**29.** $-11<2x-1\le -5$

$-11+1<2x-1+1\le -5+1$

$-10<2x\le -4$

$-5<x\le -2$

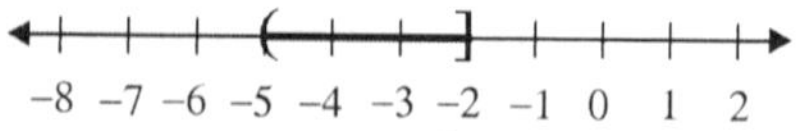


The solution set is $\{x|-5<x\le -2\}$ or $(-5,-2]$.

**31.** $-3\le \dfrac{2x}{3}-5<-1$

$-3+5\le \dfrac{2x}{3}-5+5<-1+5$

$2\le \dfrac{2x}{3}<4$

$3(2)\le 3\left(\dfrac{2x}{3}\right)<3(4)$

$6\le 2x<12$

$3\le x<6$

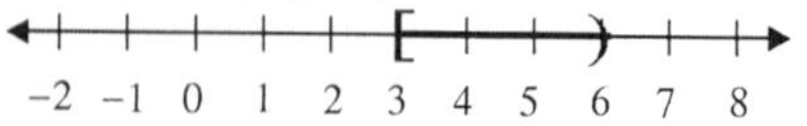


The solution set is $\{x|3\le x<6\}$ or $[3,6)$.

**33.** $\{1,2,3,4\} \cup \{2,4,5\} = \{1,2,3,4,5\}$

**35.** $\{1,3,5,7\} \cup \{2,4,6,8,10\}$
$= \{1,2,3,4,5,6,7,8,10\}$

**37.** $\{a,e,i,o,u\} \cup \varnothing = \{a,e,i,o,u\}$

**39.** $x > 3$, $x > 6$, $x > 3$ or $x > 6$

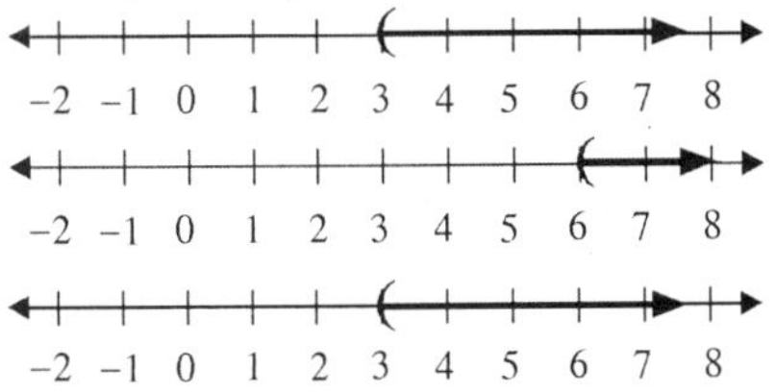


The solution set is $\{x \mid x > 3\}$ or $(3, \infty)$.

**41.** $x \le 5$, $x \le 1$, $x \le 5$ or $x \le 1$

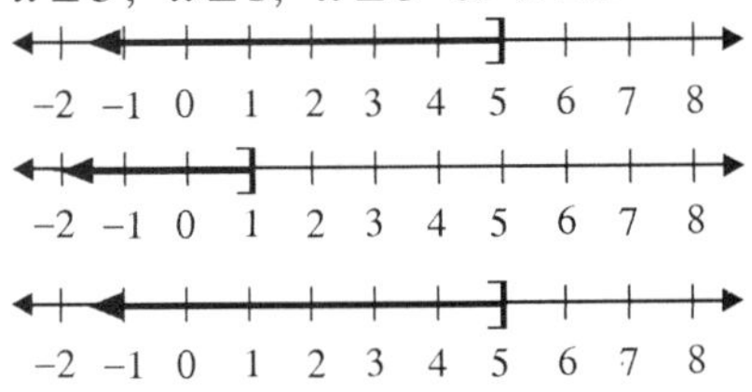


The solution set is $\{x \mid x \le 5\}$ or $(-\infty, 5]$.

**43.** $x < 2$, $x \ge -1$, $x < 2$ or $x \ge -1$

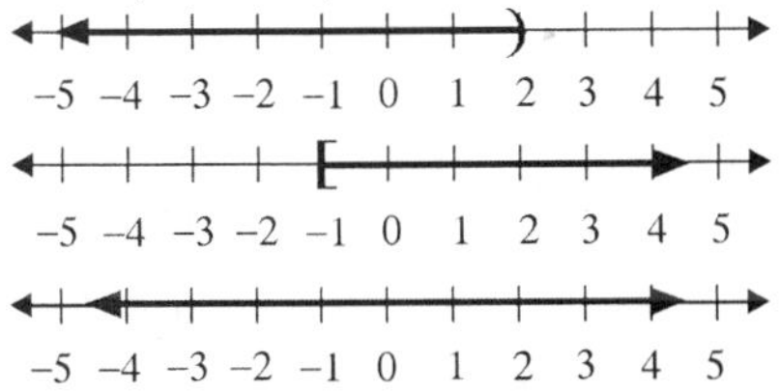


The solution set is $\mathbb{R}$, $(-\infty, \infty)$ or $\{x \mid x \text{ is a real number}\}$.

**45.** $x \ge 2$, $x < -1$, $x \ge 2$ or $x < -1$

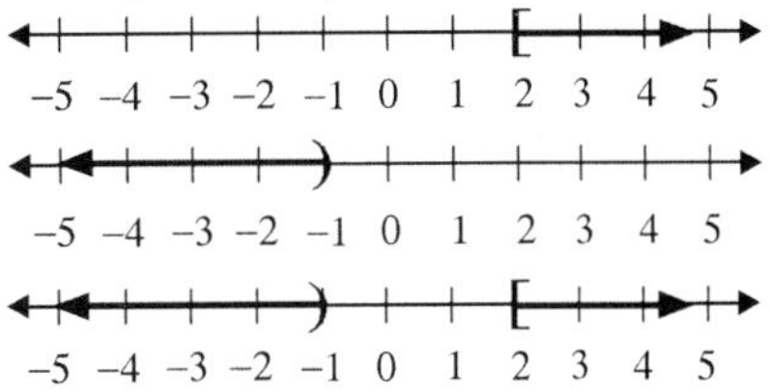


The solution set is $\{x \mid x < -1 \text{ or } x \ge 2\}$ or $(-\infty, -1) \cup [2, \infty)$.

**47.** $3x > 12$ or $2x < -6$

$x > 4$ $\qquad$ $x < -3$

$x > 4$, $x < -3$, $x > 4$ or $x < -3$

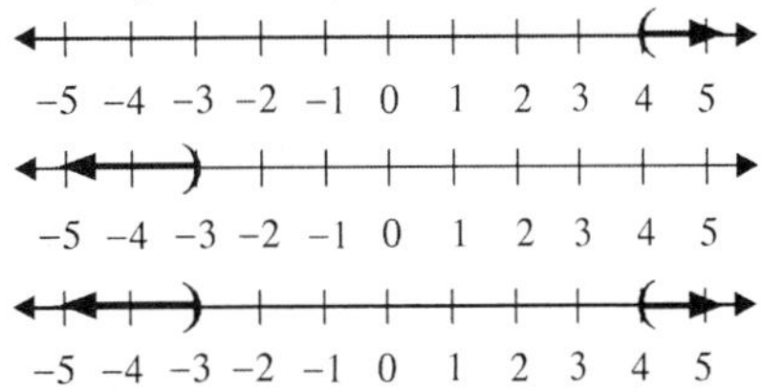


The solution set is $\{x \mid x < -3 \text{ or } x > 4\}$ or $(-\infty, -3) \cup (4, \infty)$.

**49.** $3x + 2 \le 5$ or $5x - 7 \ge 8$

$3x \le 3$ $\qquad$ $5x \ge 15$

$x \le 1$ $\qquad$ $x \ge 3$

$x \le 1$, $x \ge 3$, $x \le 1$ or $x \ge 3$

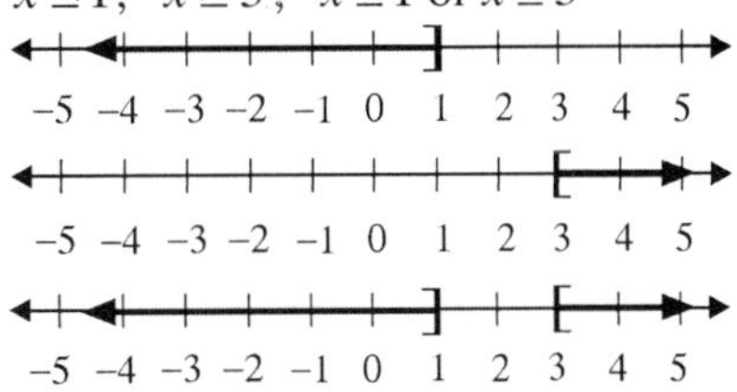


The solution set is $\{x \mid x \le 1 \text{ or } x \ge 3\}$ or $(-\infty, 1] \cup [3, \infty)$.

**51.** $4x+3<-1$ or $2x-3\geq -11$

$4x<-4$ $\quad$ $2x\geq -8$

$x<-1$ $\quad$ $x\geq -4$

$x<-1,\ x\geq -4,\ x<-1 \text{ or } x\geq -4$

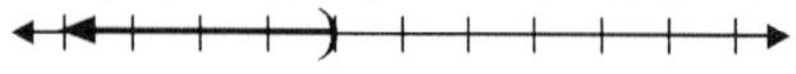

The solution set is $\mathbb{R}$, $(-\infty,\infty)$ or $\{x|x \text{ is a real number}\}$.

**53.** $-2x+5>7$ or $-3x+10>2x$

$-2x>2$ $\quad$ $-5x+10>0$

$x<-1$ $\quad$ $-5x>-10$

$x<2$

$x<-1,\ x<2,\ x<-1$ or $x<2$

The solution set is $\{x|x<2\}$ or $(-\infty,2)$.

**55.** $2x+3\geq 5$ and $3x-1>11$

$2x\geq 2$ $\quad$ $3x>12$

$x\geq 1$ $\quad$ $x>4$

The solution set is $\{x|x>4\}$ or $(4,\infty)$.

**57.** $3x-1<-1$ and $4-x<-2$

$3x<0$ $\quad$ $4<-2+x$

$x<0$ $\quad$ $6<x$

The solution set is $\{x|x<0 \text{ or } x>6\}$ or $(-\infty,0)\cup(6,\infty)$.

**59.** $a>0, b>0, c>0$

$-c<ax-b<c$

$b-c<ax<b+c$

$\frac{b-c}{a}<x<\frac{b+c}{a}$

**61.** $\{x|-1\leq x\leq 3\}$ or $[-1,3]$

**62.** $\{x|-1<x<3\}$ or $(-1,3)$

**63.** Solving in separate pieces:

$x-2<2x-1$ $\quad$ or $\quad$ $2x-1<x+2$

$-2<x-1$ $\quad$ $x-1<2$

$-1<x$ $\quad$ $x<3$

Since both exterior pieces of the inequality have the same number of $x$'s, we can solve this inequality all in one piece:

$x-2<2x-1<x+2$

$x-2-x<2x-1-x<x+2-x$

$-2<x-1<2$

$-1<x<3$

The solution set is $\{x|-1<x<3\}$ or $(-1,3)$.

**64.** Solving in separate pieces:

$x\leq 3x-10$ $\quad$ or $\quad$ $3x-10\leq 2x$

$0\leq 2x-10$ $\quad$ $x-10\leq 0$

$10\leq 2x$ $\quad$ $x\leq 10$

$5\leq x$

The solution set is $\{x|5\leq x\leq 10\}$ or $[5,10]$.

**65.** The solution set is $\{x|-1\leq x<2\}$ or $[-1,2)$.

**66.** The solution set is $\{x \mid 1 < x \leq 4\}$ or $(1,4]$.

**67.**

$$\begin{aligned} 5-4x &\geq 1 \\ -4x &\geq -4 \\ x &\leq 1 \end{aligned} \quad \text{and} \quad \begin{aligned} 3-7x &< 31 \\ 3 &< 7x+31 \\ -28 &< 7x \\ -4 &< x \end{aligned}$$

The solution set is $\{x \mid -4 < x \leq 1\}$ or $(-4,1]$. The set of negative integers that fall within this set are $\{-3,-2,-1\}$.

**68.**

$$-5 < 3x+4 \leq 16$$
$$-9 < 3x \leq 12$$
$$-3 < x \leq 4$$

The solution set is $\{x \mid -3 < x \leq 4\}$ or $(-3,4]$. The set of negative integers that fall within this set are $\{-2,-1\}$.

**69.** toys cars and trucks
sports equipment
spatial-temporal toys

**71.** toys cars and trucks
sports equipment
spatial-temporal toys
doll houses}

**73.** No toys were requested by more than 40% of the boys and more than 10% of the girls.

**75.**

$$28 \leq 20+0.40(x-60) \leq 40$$
$$28 \leq 20+0.40x-24 \leq 40$$
$$28 \leq 0.40x-4 \leq 40$$
$$32 \leq 0.40x \leq 44$$
$$80 \leq x \leq 110$$

Between 80 and 110 ten minutes, inclusive.

**77.** Let $x$ = the score on the fifth exam

$$80 \leq \frac{70+75+87+92+x}{5} < 90$$
$$80 \leq \frac{324+x}{5} < 90$$
$$5(80) \leq 5\left(\frac{324+x}{5}\right) < 5(90)$$
$$400 \leq 324+x < 450$$
$$400-324 \leq 324-324+x < 450-324$$
$$76 \leq x < 126$$

A grade between 76 and 125 is needed on the fifth exam. (Because the inequality states the score must be less than 126, we say 125 is the highest possible score. In interval notation, we can use parentheses to exclude the maximum value. The range of scores can be expressed as $[76,126)$.) If the highest grade is 100, the grade would need to be between 76 and 100.

**61.** Let $x$ = the number of times the bridge is crossed per three month period
The cost with the 3-month pass is $C_3 = 7.50+0.50x$.
The cost with the 6-month pass is $C_6 = 30$.

Because we need to buy two 3-month passes per 6-month pass, we multiply the cost with the 3-month pass by 2.

$$2(7.50+0.50x) < 30$$
$$15+x < 30$$
$$x < 15$$

We also must consider the cost without purchasing a pass. We need this cost to be less than the cost with a 3-month pass.

$$3x > 7.50 + 0.50x$$
$$2.50x > 7.50$$
$$x > 3$$

The 3-month pass is the best deal when making more than 3 but less than 15 crossings per 3-month period.

**81.-87.** Answers will vary.

**89.** $-1 < \dfrac{x+4}{2} < 3$

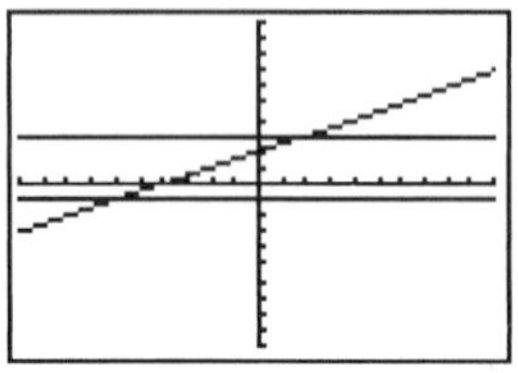

We need to find the range of the $x$-values of the points lying between the two constant functions. Using the intersection feature, we can determine the $x$-values of the endpoints of the range.

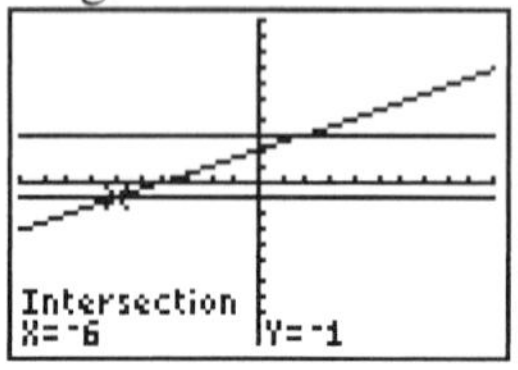


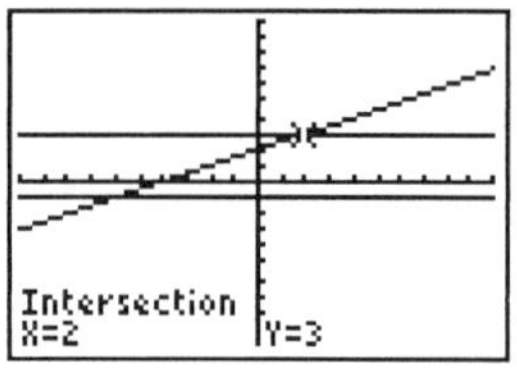


The solution set is $\{x|-6 < x < 2\}$ or $(-6, 2)$.

**91.** $2 \le 4 - x \le 7$

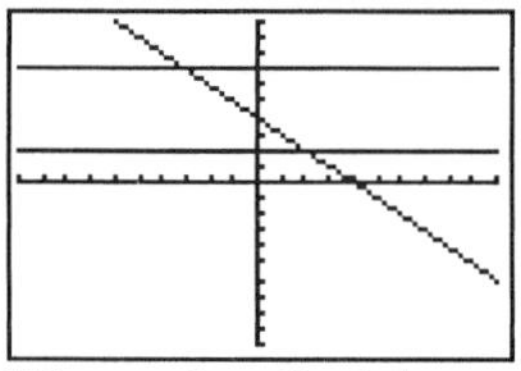

We need to find the range of the $x$-values of the points lying between the two constant functions. Using the intersection feature, we can determine the $x$-values of the endpoints of the range.

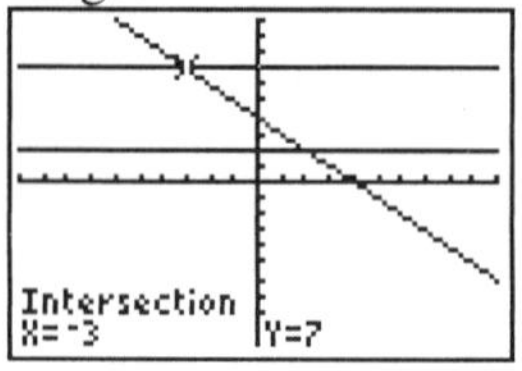


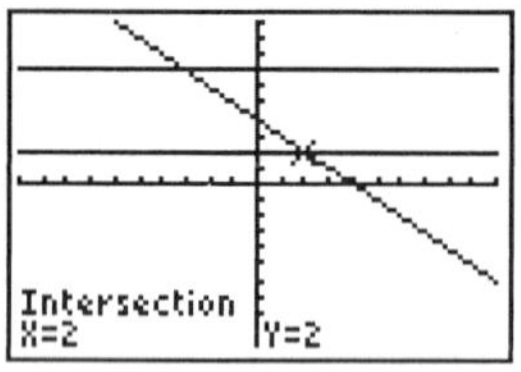


The solution set is $\{x|-3 \le x \le 2\}$ or $[-3, 2]$.

**93.**

$$-7 \le 8 - 3x \le 20$$
$$-15 \le -3x \le 12$$
$$5 \ge x \ge -4$$
$$-4 \le x \le 5$$

$$-7 < 6x - 1 < 41$$
$$-6 < 6x < 42$$
$$-1 < x < 7$$

The intersection of the above two sets is the solution set: $\{x|-1 < x \le 5\}$ or $(-1, 5]$.

**95.** $[-1, \infty)$

**97.** $(-1, 4]$

**99.** $f(x) = x^2 - 3x + 4 \qquad g(x) = 2x - 5$

$$\begin{aligned}(g-f)(x) &= g(x) - f(x)\\ &= (2x-5) - (x^2 - 3x + 4)\\ &= 2x - 5 - x^2 + 3x - 4\\ &= -x^2 + 5x - 9\end{aligned}$$

$$\begin{aligned}(g-f)(x) &= -x^2 + 5x - 9\\ (g-f)(-1) &= -(-1)^2 + 5(-1) - 9\\ &= -1 - 5 - 9\\ &= -15\end{aligned}$$

**100.** Passing through (4, 2) and perpendicular to the line $4x - 2y = 8$

The slope of the line $4x - 2y = 8$ can be found by rewriting the equation in slope-intercept form.

$$\begin{aligned}4x - 2y &= 8\\ -2y &= -4x + 8\\ y &= 2x - 4\end{aligned}$$

The slope of this line is 2. The slope of the line perpendicular to this line is $-\frac{1}{2}$.

Use the slope and the point to write the equation of the line in point-slope form. Then, solve for $y$ and rewrite the equation using function notation.

$$\begin{aligned}y - y_1 &= m(x - x_1)\\ y - 2 &= -\frac{1}{2}(x - 4)\\ y - 2 &= -\frac{1}{2}x + 2\\ y &= -\frac{1}{2}x + 4\\ f(x) &= -\frac{1}{2}x + 4\end{aligned}$$

**101.** $$\begin{aligned}4 - [2(x-4) - 5] &= 4 - [2x - 8 - 5]\\ &= 4 - [2x - 13]\\ &= 4 - 2x + 13\\ &= 17 - 2x\end{aligned}$$

## 4.3 Exercise Set

**1.** $|x| = 8$

$x = 8 \quad \text{or} \quad x = -8$

The solutions are –8 and 8 and the solution set is $\{-8, 8\}$.

**3.** $|x - 2| = 7$

$x - 2 = 7 \quad \text{or} \quad x - 2 = -7$

$x = 9 \qquad\qquad x = -5$

The solutions are –5 and 9 and the solution set is $\{-5, 9\}$.

**5.** $|2x - 1| = 7$

$2x - 1 = 7 \quad \text{or} \quad 2x - 1 = -7$

$2x = 8 \qquad\qquad 2x = -6$

$x = 4 \qquad\qquad x = -3$

The solutions are –3 and 4 and the solution set is $\{-3, 4\}$.

**7.** $\left|\frac{4x - 2}{3}\right| = 2$

$\frac{4x-2}{3} = 2 \quad \text{or} \quad \frac{4x-2}{3} = -2$

$4x - 2 = 3(2) \qquad 4x - 2 = 3(-2)$

$4x - 2 = 6 \qquad 4x - 2 = -6$

$4x = 8 \qquad 4x = -4$

$x = 2 \qquad x = -1$

The solutions are –1 and 2 and the solution set is $\{-1, 2\}$.

**9.** $|x| = -8$

The solution set is $\varnothing$ or $\{\ \}$. There are no values of $x$ for which the absolute value of $x$ is a negative number. By definition, absolute values are always zero or positive.

**11.** $|x+3| = 0$

Since the absolute value of the expression equals zero, we set the expression equal to zero and solve.

$x+3=0$

$x=-3$

The solution is –3 and the solution set is $\{-3\}$.

**13.** $2|y+6| = 10$

$|y+6| = 5$

$y+6=5$ or $y+6=-5$

$y=-1$ $\qquad$ $y=-11$

The solutions are –11 and –1 and the solution set is $\{-11,-1\}$.

**15.** $3|2x-1| = 21$

$|2x-1| = 7$

$2x-1=7$ or $2x-1=-7$

$2x=8$ $\qquad$ $2x=-6$

$x=4$ $\qquad$ $x=-3$

The solutions are –3 and 4 and the solution set is $\{-3,4\}$.

**17.** $|6y-2|+4 = 32$

$|6y-2| = 28$

$6y-2=28$ or $6y-2=-28$

$6y=30$ $\qquad$ $6y=-26$

$y=5$ $\qquad$ $y=-\frac{26}{6}$

$y=-\frac{13}{3}$

The solutions are $-\frac{13}{3}$ and 5 and the solution set is $\left\{-\frac{13}{3},5\right\}$.

**19.** $7|5x|+2 = 16$

$7|5x| = 14$

$|5x| = 2$

$5x=2$ or $5x=-2$

$x=\frac{2}{5}$ $\qquad$ $x=-\frac{2}{5}$

The solutions are $-\frac{2}{5}$ and $\frac{2}{5}$ and the solution set is $\left\{-\frac{2}{5},\frac{2}{5}\right\}$.

**21.** $|x+1|+5 = 3$

$|x+1| = -2$

The solution set is $\varnothing$ or $\{\ \}$. By definition, absolute values are always zero or positive.

**23.** $|4y+1|+10 = 4$

$|4y+1| = -6$

The solution set is $\varnothing$ or $\{\ \}$. By definition, absolute values are always zero or positive.

**25.** $|2x-1|+3=3$

$|2x-1|=0$

Since the absolute value of the expression equals zero, we set the expression equal to zero and solve.

$2x-1=0$

$2x=1$

$x=\frac{1}{2}$

The solution is $\frac{1}{2}$ and the solution set is $\left\{\frac{1}{2}\right\}$.

**27.** $|5x-8|=|3x+2|$

$5x-8=3x+2$ or $5x-8=-3x-2$

$2x-8=2$ $\quad$ $8x-8=-2$

$2x=10$ $\quad$ $8x=6$

$x=5$ $\quad$ $x=\frac{6}{8}=\frac{3}{4}$

The solutions are $\frac{3}{4}$ and 5 and the solution set is $\left\{\frac{3}{4},5\right\}$.

**29.** $|2x-4|=|x-1|$

$2x-4=x-1$ or $2x-4=-x+1$

$x-4=-1$ $\quad$ $3x-4=1$

$x=3$ $\quad$ $3x=5$

$x=\frac{5}{3}$

The solutions are $\frac{5}{3}$ and 3 and the solution set is $\left\{\frac{5}{3},3\right\}$.

**31.** $|2x-5|=|2x+5|$

$2x-5=2x+5$ or $2x-5=-2x-5$

$-5\neq 5$ $\quad$ $4x-5=-5$

$4x=0$

$x=0$

The solution is 0 and the solution set is $\{0\}$.

**33.** $|x-3|=|5-x|$

$x-3=5-x$ or $x-3=-(5-x)$

$2x-3=5$ $\quad$ $x-3=-5+x$

$2x=8$ $\quad$ $-3\neq -5$

$x=4$

The solution is 4 and the solution set is $\{4\}$.

**35.** $|2y-6|=|10-2y|$

$2y-6=10-2y$ or $2y-6=-10+2y$

$4y-6=10$ $\quad$ $-6\neq -10$

$4y=16$

$y=4$

The solution is 4 and the solution set is $\{4\}$.

**37.** $\left|\frac{2x}{3}-2\right|=\left|\frac{x}{3}+3\right|$

$\frac{2x}{3}-2=\frac{x}{3}+3$

$3\left(\frac{2x}{3}\right)-3(2)=3\left(\frac{x}{3}\right)+3(3)$

$2x-6=x+9$

$x-6=9$

$x=15$

or

$$\frac{2x}{3}-2=-\left(\frac{x}{3}+3\right)$$
$$\frac{2x}{3}-2=-\frac{x}{3}-3$$
$$3\left(\frac{2x}{3}\right)-3(2)=3\left(-\frac{x}{3}\right)-3(3)$$
$$2x-6=-x-9$$
$$3x-6=-9$$
$$3x=-3$$
$$x=-1$$

The solutions are $-1$ and 15 and the solution set is $\{-1,15\}$.

**39.** $|x|<3$

$-3<x<3$

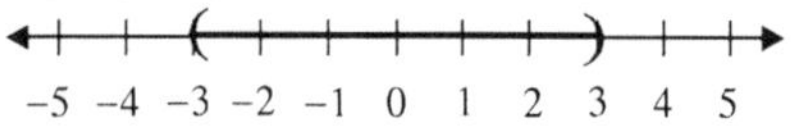


The solution set is $\{x|-3<x<3\}$ or $(-3,3)$.

**41.**
$$|x-2|<1$$
$$-1<x-2<1$$
$$-1+2<x-2+2<1+2$$
$$1<x<3$$

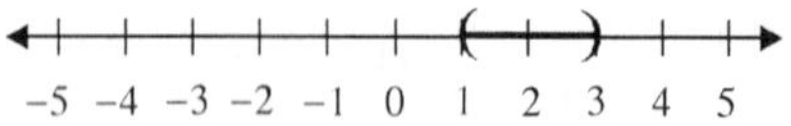


The solution set is $\{x|1<x<3\}$ or $(1,3)$.

**43.**
$$|x+2|\le 1$$
$$-1\le x+2\le 1$$
$$-1-2\le x+2-2\le 1-2$$
$$-3\le x\le -1$$

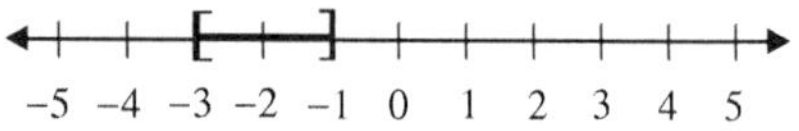


The solution set is $\{x|-3\le x\le -1\}$ or $[-3,-1]$.

**45.**
$$|2x-6|<8$$
$$-8<2x-6<8$$
$$-8+6<2x-6+6<8+6$$
$$-2<2x<14$$
$$-1<x<7$$

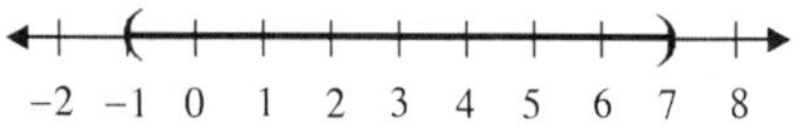


The solution set is $\{x|-1<x<7\}$ or $(-1,7)$.

**47.** $|x|>3$

$x<-3$ or $x>3$

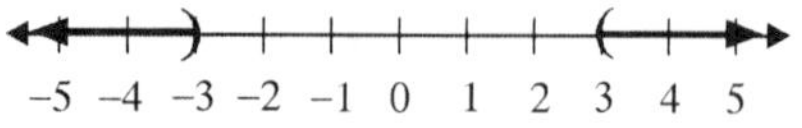


The solution set is $\{x|x<-3$ and $x>3\}$ or $(-\infty,-3)\cup(3,\infty)$.

**49.** $|x+3|>1$

$x+3<-1$ or $x+3>1$

$x<-4$ $\quad$ $x>-2$

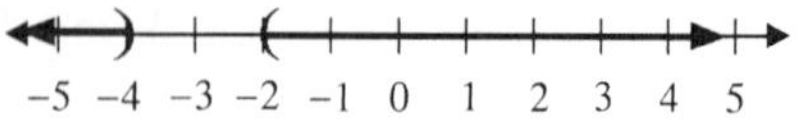


The solution set is $\{x|x<-4$ and $x>-2\}$ or $(-\infty,-4)\cup(-2,\infty)$.

**51.** $|x-4| \geq 2$

$x-4 \leq -2$ or $x-4 \geq 2$

$x \leq 2$ $\quad$ $x \geq 6$

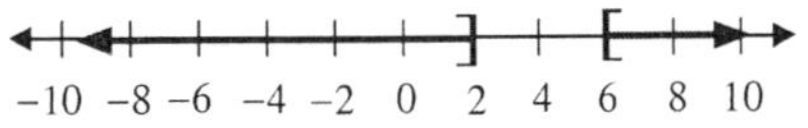

The solution set is $\{x | x \leq 2$ and $x \geq 6\}$ or $(-\infty, 2] \cup [6, \infty)$.

**53.** $|3x-8| > 7$

$3x-8 < -7$ or $3x-8 > 7$

$3x < 1$ $\quad$ $3x > 15$

$x < \frac{1}{3}$ $\quad$ $x > 5$

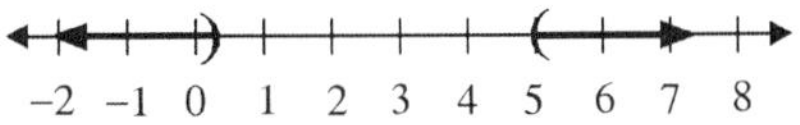

The solution set is $\left\{x \middle| x < \frac{1}{3}\right.$ and $x > 5\}$ or $\left(-\infty, \frac{1}{3}\right) \cup (5, \infty)$.

**55.**

$$|2(x-1)+4| \leq 8$$
$$|2x-2+4| \leq 8$$
$$|2x+2| \leq 8$$
$$-8 \leq 2x+2 \leq 8$$
$$-8-2 \leq 2x+2-2 \leq 8-2$$
$$-10 \leq 2x \leq 6$$
$$-5 \leq x \leq 3$$

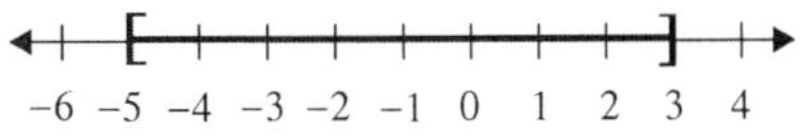

The solution set is $\{x | -5 \leq x \leq 3\}$ or $[-5, 3]$.

**57.**

$$\left|\frac{2y+6}{3}\right| < 2$$
$$-2 < \frac{2y+6}{3} < 2$$
$$3(-2) < 3\left(\frac{2y+6}{3}\right) < 3(2)$$
$$-6 < 2y+6 < 6$$
$$-6-6 < 2y+6-6 < 6-6$$
$$-12 < 2y < 0$$
$$-6 < y < 0$$

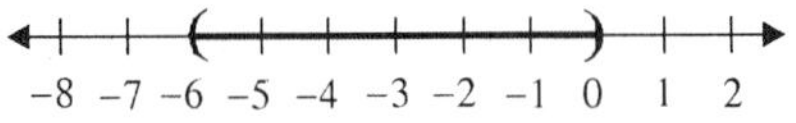

The solution set is $\{x | -6 < x < 0\}$ or $(-6, 0)$.

**59.** $\left|\frac{2x+2}{4}\right| \geq 2$

$\frac{2x+2}{4} \leq -2$ or $\frac{2x+2}{4} \geq 2$

$2x+2 \leq -8$ $\quad$ $2x+2 \geq 8$

$2x \leq -10$ $\quad$ $2x \geq 6$

$x \leq -5$ $\quad$ $x \geq 3$

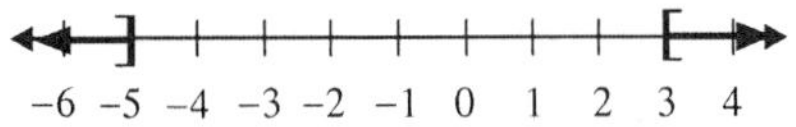

The solution set is $\{x | x \leq -5$ and $x \geq 3\}$ or $(-\infty, -5] \cup [3, \infty)$.

**61.** $\left|3-\frac{2x}{3}\right|>5$

$3-\frac{2x}{3}<-5$ or $3-\frac{2x}{3}>5$

$-\frac{2x}{3}<-8$ $\quad$ $-\frac{2x}{3}>2$

$-2x<-24$ $\quad$ $-2x>6$

$x>12$ $\quad$ $x<-3$

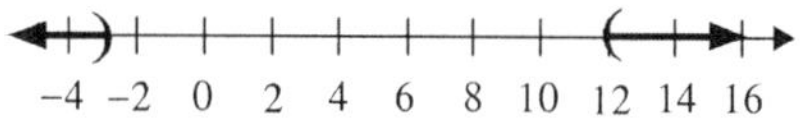

The solution set is $\{x|x<-3$ and $x>12\}$ or $(-\infty,-3)\cup(12,\infty)$.

**63.** $|x-2|<-1$

The solution set is $\varnothing$ or $\{\ \}$. Since all absolute values are zero or positive, there are no values of $x$ that will make the absolute value of the expression less than $-1$.

**65.** $|x+6|>-10$

Since all absolute values are zero or positive, we know that when simplified, the left hand side will be a positive number. We also know that any positive number is greater than any negative number. This means that regardless of the value of $x$, the left hand side will be greater than the right hand side of the inequality. The solution set is $\{x|x$ is a real number$\}$, $\mathbb{R}$ or $(-\infty,\infty)$.

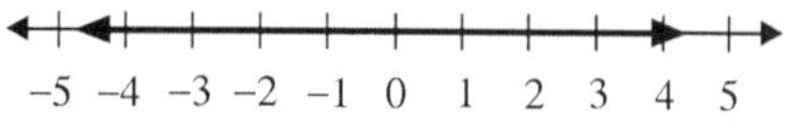

**67.**

$|x+2|+9\le16$

$|x+2|\le7$

$-7\le x+2\le7$

$-7-2\le x+2-2\le7-2$

$-9\le x\le5$

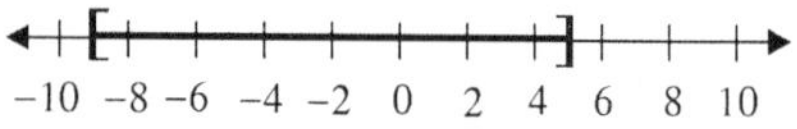

The solution set is $\{x|-9\le x\le5\}$ or $[-9,5]$.

**69.** $2|2x-3|+10>12$

$2|2x-3|>2$

$|2x-3|>1$

$2x-3<-1$ $\quad$ $2x-3>1$

$2x<2$ or $2x>4$

$x<1$ $\quad$ $x>2$

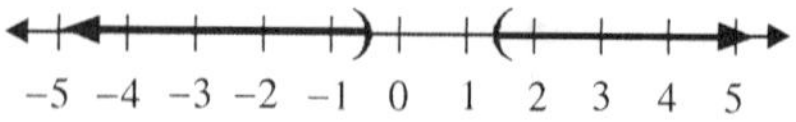

The solution set is $\{x|x<1$ and $x>2\}$ or $(-\infty,1)\cup(2,\infty)$.

**71.** $|5-4x|=11$

$5-4x=11$ or $5-4x=-11$

$-4x=6$ $\quad$ $-4x=-16$

$x=-\frac{3}{2}$ $\quad$ $y=4$

The solutions are $-\frac{3}{2}$ and 4 and the solution set is $\left\{-\frac{3}{2},4\right\}$.

**73.** $|3-x|=|3x+11|$

$3-x=3x+11$
$-4x+3=11$
$-4x=8$
$x=-2$
or
$3-x=-(3x+11)$
$3-x=-3x-11$
$2x+3=-11$
$2x=-14$
$x=-7$

The solutions are $-2$ and -7 and the solution set is $\{-7,-2\}$.

**75.** $|-1+3(x+1)|\le 5$
$-5\le -1+3x+3\le 5$
$-5\le 3x+2\le 5$
$-7\le 3x\le 3$
$-\frac{7}{3}\le x\le 1$

The solution set is $\left\{x \mid -\frac{7}{3}\le x\le 1\right\}$ or $\left[-\frac{7}{3},1\right]$.

**77.** $|2x-3|+1>6$
$|2x-3|>5$

$2x-3>5$ or $2x-3<-5$
$2x>8$ or $2x<-2$
$x>4$ or $x<-1$

The solution set is $\{x \mid x<-1 \text{ or } x>4\}$ or $(-\infty,-1)\cup(4,\infty)$.

**79.** Let $x$ be the number.
$|4-3x|\ge 5$ or $|3x-4|\ge 5$

$3x-4\ge 5$ or $3x-4\le -5$
$3x\ge 9$ or $3x\le -1$
$x\ge 3$ or $x\le -\frac{1}{3}$

The solution set is $\left\{x \mid x\le -\frac{1}{3} \text{ or } x\ge 3\right\}$ or $\left(-\infty,-\frac{1}{3}\right]\cup[3,\infty)$.

**80.** Let $x$ be the number.
$|5-4x|\le 13$ or $|4x-5|\le 13$

$-13\le 4x-5\le 13$
$-8\le 4x\le 18$
$-2\le x\le \frac{9}{2}$

The solution set is $\left\{x \mid -2\le x\le \frac{9}{2}\right\}$ or $\left[-2,-\frac{2}{9}\right]$.

**81.** $|ax+b|<c$
When solving, we do not "flip" the inequality symbol when dividing by $a$ since $a>0$.
$-c<ax+b<c$
$-c-b<ax<c-b$
$\frac{-c-b}{a}<x<\frac{c-b}{a}$

The solution set is $\left\{x \mid \frac{-c-b}{a}<x<\frac{c-b}{a}\right\}$.

**82.** $|ax+b| \geq c$

$ax+b \geq c$ or $ax+b \leq -c$

$ax \geq c-b$ or $ax \leq -c-b$

$x \geq \dfrac{c-b}{a}$ $\quad x \leq \dfrac{-c-b}{a}$

The solution set is

$\left\{x \mid x \leq \dfrac{-c-b}{a} \text{ or } x \geq \dfrac{c-b}{a}\right\}$.

**83.** $|4-x|=1$

$4-x=1$ or $4-x=-1$

$-x=-3$ or $-x=-5$

$x=3$ $\quad x=5$

**84.** $|4-x|<5$ or $|x-4|<5$

$-5<x-4<5$

$-1<x<9$

The solution set is $\{x \mid -1<x<9\}$.

**85.** The solution set is $\{x \mid -2 \leq x \leq 1\}$.

**86.** The solution set is $\{x \mid x \leq -2 \text{ or } x \geq 1\}$.

**87.**

$$|x-60.2| \leq 1.6$$
$$-1.6 \leq x-60.2 \leq 1.6$$
$$-1.6+60.2 \leq x-60.2+60.2 \leq 1.6+60.2$$
$$58.6 \leq x \leq 61.8$$

The percentage of the U.S. population that watched M*A*S*H is between 58.6% and 61.8%, inclusive. The margin of error is 1.6%.

**89.**

$$|T-57| \leq 7$$
$$-7 \leq T-57 \leq 7$$
$$-7+57 \leq T-57+57 \leq 7+57$$
$$50 \leq T \leq 64$$

The monthly average temperature for San Francisco, California ranges from 50°F to 64°F, inclusive.

**91.**

$$|x-8.6| \leq 0.01$$
$$-0.01 \leq x-8.6 \leq 0.01$$
$$-0.01+8.6 \leq x-8.6+8.6 \leq 0.01+8.6$$
$$8.59 \leq x \leq 8.61$$

The length of the machine part must be between 8.59 and 8.61 centimeters, inclusive.

**93.** $\left|\dfrac{h-50}{5}\right| \geq 1.645$

$\dfrac{h-50}{5} \leq -1.645$

$h-50 \leq 5(-1.645)$

$h-50 \leq -8.225$

$h \leq 41.775$

or

$\dfrac{h-50}{5} \geq 1.645$

$h-50 \geq 5(1.645)$

$h-50 \geq 8.225$

$h \geq 58.225$

The coin would be considered unfair if the tosses resulted in 41 or less heads, or 59 or more heads.

**94.-101.** Answers will vary.

**103.** $|3(x+4)|=12$

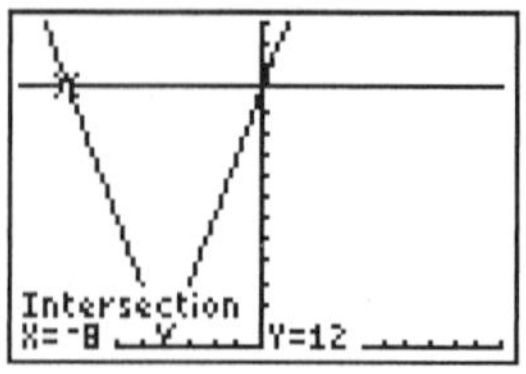

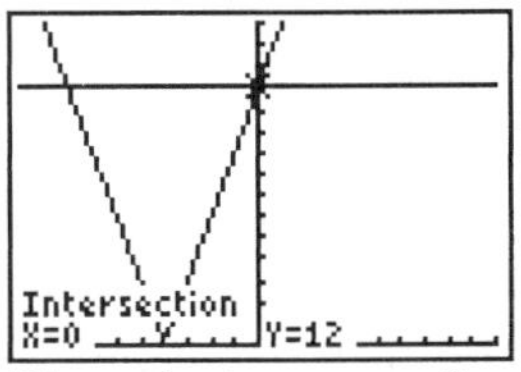


The solutions are –8 and 0 and the solution set is $\{-8, 0\}$.

**105.** $|2x+3|<5$

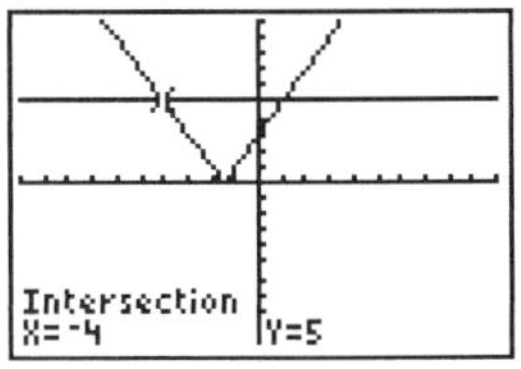


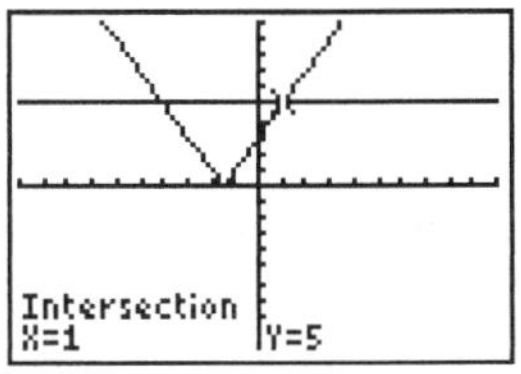


The solution set is $\{x|-4<x<1\}$ or $(-4,1)$.

**107.** $|x+4|<-1$

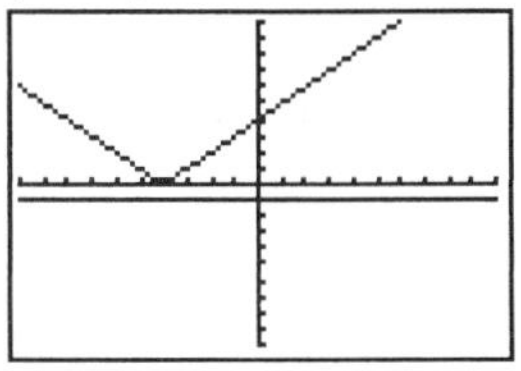

No part of the graph of the absolute value lies below the graph of the constant. The solution set is $\varnothing$ or $\{\ \}$.

**109.** $|0.1x-0.4|+0.4>0.6$

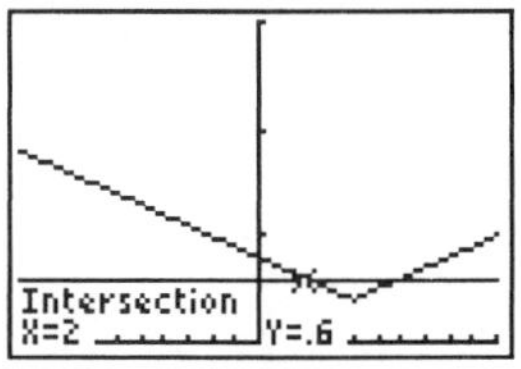


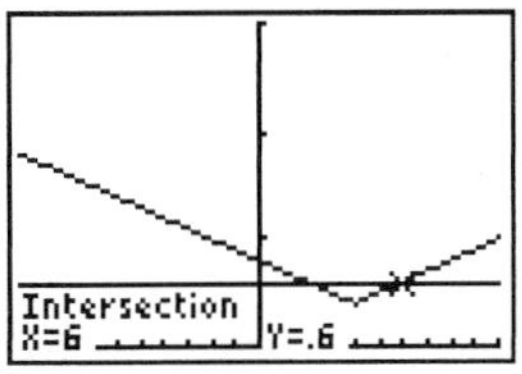


The solution set is $\{x|x<2 \text{ or } x>6\}$ or $(-\infty,2)\cup(6,\infty)$.

**111.** Answers will vary. For example, consider Exercise 5.

$|2x-1|=5$

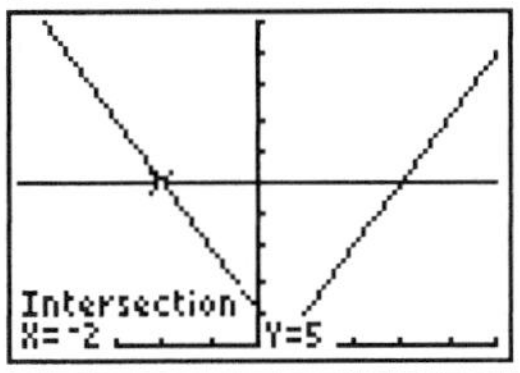


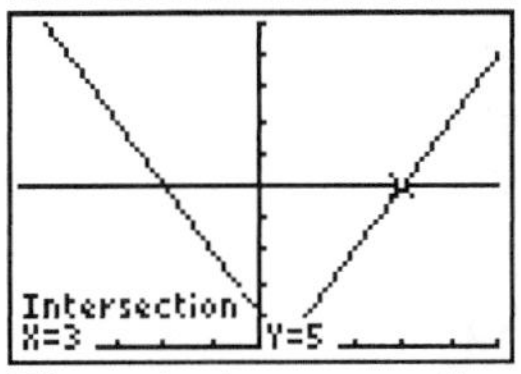


The solutions are –2 and 3 and the solution set is $\{-2,3\}$.

**113.** **a.** $|x-4|<3$

**b.** $|x-4|\geq 3$

**115.** $|2x+5|=3x+4$

$$\begin{aligned} 2x+5&=3x+4 \\ -x+5&=4 \\ -x&=-1 \\ x&=1 \end{aligned} \qquad \text{or} \qquad \begin{aligned} 2x+5&=-(3x+4) \\ 2x+5&=-3x-4 \\ 5x+5&=-4 \\ 5x&=-9 \\ x&=-\frac{9}{5} \end{aligned}$$

**116.** Solve for $y$ to obtain slope-intercept form.

$$3x-5y=15$$
$$-5y=-3x+15$$
$$y=\frac{3}{5}x-3$$

The $y$-intercept is –3 and the slope is $\frac{3}{5}$.

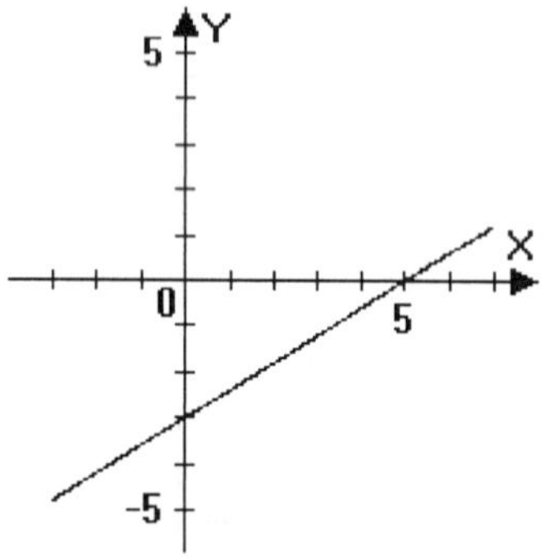


**117.** $f(x)=-\frac{2}{3}x$

$$y=-\frac{2}{3}x$$

The $y$-intercept is 0 and the slope is $-\frac{2}{3}$.

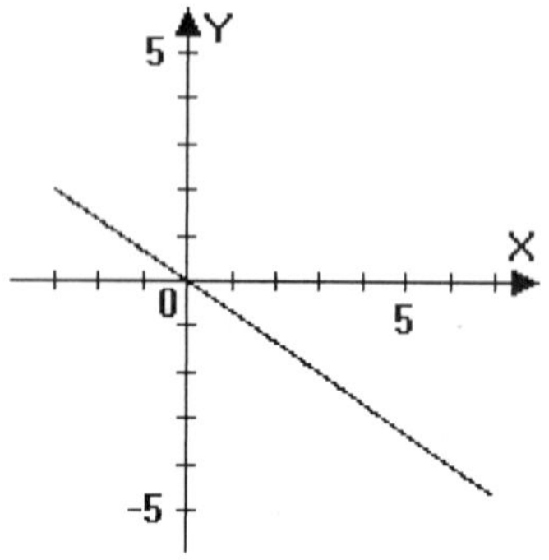


**118.** $f(x)=-2$

$$y=-2$$

When graphed, an equation of the form $y = b$ is a horizontal line. $f(x)=-2$ is the horizontal line positioned at $y=-2$.

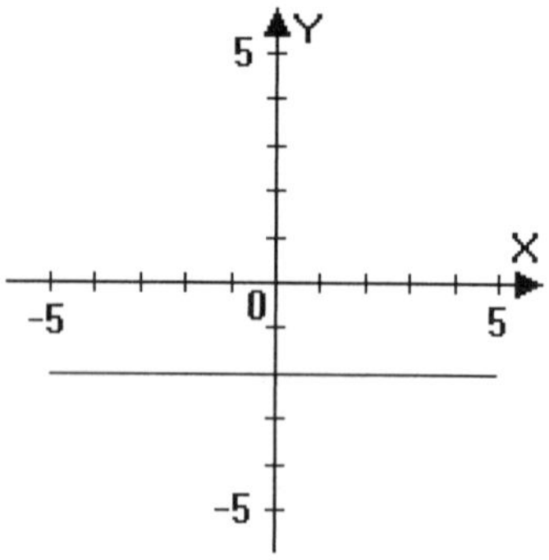


## Mid-Chapter 4 Check Point

**1.** 
$$4-3x\ge 12-x$$
$$4\ge 12+2x$$
$$-8\ge 2x$$
$$-4\ge x$$
$$x\le -4$$

$\{x\,|\,x\le -4\}$ or $(-\infty,4]$

**2.** 
$$5\le 2x-1<9$$
$$6\le 2x<10$$
$$3\le x<5$$

$\{x\,|\,3\le x<5\}$ or $[3,5)$

**3.** $|4x-7|=5$

$4x-7=5$ or $4x-7=-5$

$4x=12$ or $4x=2$

$x=3$ or $x=\frac{1}{2}$

The solution set is $\left\{\frac{1}{2},3\right\}$.

**4.** $-10-3(2x+1)>8x+1$

$$-10-6x-3>8x+1$$
$$-6x-13>8x+1$$
$$-13>14x+1$$
$$-14>14x$$
$$-1>x$$
$$x<-1$$

$\{x\,|\,x<-1\}$ or $(-\infty,-1)$

**5.** $2x+7<-11$ or $-3x-2<13$

$$2x<-18 \qquad -3x<15$$
$$x<-9 \qquad x>-5$$

$\{x\,|\,x<-9 \text{ or } x>-5\}$
or $(-\infty,-9)\cup(-5,\infty)$

**6.** $|3x-2|\le 4$

$$-4\le 3x-2\le 4$$
$$-2\le 3x\le 6$$
$$-\frac{2}{3}\le x\le 2$$

$\left\{x\,\middle|\,-\frac{2}{3}\le x\le 2\right\}$ or $\left[-\frac{2}{3},2\right]$

**7.** $|x+5|=|5x-8|$

$$x+5=5x-8 \qquad x+5=-(5x-8)$$
$$-4x+5=-8 \qquad x+5=-5x+8$$
$$-4x=-13 \quad \text{or} \quad 6x+5=8$$
$$x=\frac{13}{4} \qquad 6x=3$$
$$x=\frac{1}{2}$$

The solution set is $\left\{\frac{1}{2},\frac{13}{4}\right\}$.

**8.** $5-2x\ge 9$

$$5\ge 2x+9 \qquad 5x+3>-17$$
$$-4\ge 2x \quad \text{and} \quad 5x>-20$$
$$-2\ge x \qquad x>-4$$
$$x\le -2$$

$\{x\,|\,-4<x\le -2\}$ or $(-4,-2]$

**9.** $3x-2>-8$ or $2x+1<9$

$$3x>-6 \qquad 2x<8$$
$$x>-2 \qquad x<4$$

The union of these sets is the entire number line. The solution set is $\{x\,|\,x \text{ is a real number}\}$ or $(-\infty,\infty)$.

**10.**

$$\frac{x}{2}+3\le\frac{x}{3}+\frac{5}{2}$$
$$6\left(\frac{x}{2}+3\right)\le 6\left(\frac{x}{3}+\frac{5}{2}\right)$$
$$3x+18\le 2x+15$$
$$x+18\le 15$$
$$x\le -3$$

$\{x\,|\,x\le -3\}$ or $(-\infty,-3]$

**11.** $\frac{2}{3}(6x-9)+4>5x+1$

$$4x-6+4>5x+1$$
$$4x-2>5x+1$$
$$-2>x+1$$
$$-3>x$$
$$x<-3$$

$\{x\,|\,x<-3\}$ or $(-\infty,-3)$

**12.** $|5x+3|>2$

$5x+3>2$ or $5x+3<-2$

$5x>-1$ or $5x<-4$

$x>-\frac{1}{5}$ or $x<-1$

$\left\{x \mid x<-1 \text{ or } x>-\frac{1}{5}\right\}$ or $(-\infty,-1)\cup\left(-\frac{1}{5},\infty\right)$

**13.**
$$7-\left|\frac{x}{2}+2\right|\le 4$$
$$-\left|\frac{x}{2}+2\right|\le -3$$
$$\left|\frac{x}{2}+2\right|\ge 3$$
$$|x+4|\ge 6$$

$x+4\ge 6$ or $x+4\le -6$

$x\ge 2$ or $x\le -10$

$\{x \mid x\le -10 \text{ or } x\ge 2\}$ or $(-\infty,-10]\cup[2,\infty)$

**14.**
$$5(x-2)-3(x+4)\ge 2x-20$$
$$5x-10-3x-12\ge 2x-20$$
$$2x-22\ge 2x-20$$
$$-22\ge -20$$

This is a *false* statement. There are no solutions to the inequality. The solution set is { } or $\varnothing$.

**15.**
$$\frac{x+3}{4}<\frac{1}{3}$$
$$3x+9<4$$
$$3x<-5$$
$$x<-\frac{5}{3}$$

$\left\{x \mid x<-\frac{5}{3}\right\}$ or $\left(-\infty,-\frac{5}{3}\right)$

**16.** $5x+1\ge 4x-2$ and $2x-3>5$

$x+1\ge -2$ and $2x>8$

$x\ge -3$ and $x>4$

The solution set is $\{x \mid x>4\}$ or $(4,\infty)$.

**17.**
$$3-|2x-5|=-6$$
$$-|2x-5|=-9$$
$$|2x-5|=9$$

$2x-5=9$ or $2x-5=-9$

$2x=14$ or $2x=-4$

$x=7$ or $x=-2$

The solution set is $\{-2,7\}$.

**18.**
$$3+|2x-5|=-6$$
$$|2x-5|=-9$$

Since absolute values cannot be negative, there are no solutions. The solution set is $\varnothing$.

**19.** Let $x$ = number of miles.
$$24+0.20x\le 40$$
$$0.20x\le 16$$
$$x\le 80$$
No more than 80 miles per day.

**20.** Let $x$ = grade on the fifth exam.

$80 \le \frac{95+79+91+86+x}{5} < 90$

$80 \le \frac{351+x}{5} < 90$

$400 \le x+351 < 450$

$48 \le x < 98$

$[48,98)$

**21.** Let $x$ = amount invested.

$x(0.075) \ge 9000$

$x \ge 120,000$

The retiree should invest at least \$120,000.

**22.** Let $x$ = number of compact discs.
Cost: $60,000+0.18x$
Revenue: $0.30x$
Profit = Revenue – Cost
We need

$0.30x-(60,000+0.18x) \ge 30,000$

$0.30x-60,000-0.18x \ge 30,000$

$0.12x-60,000 \ge 30,000$

$0.12x \ge 90,000$

$x \ge 750,000$

The company should produce and sell at least 750,000 compact discs.

### 4.4 Exercise Set

**1.** $x+y \ge 3$
First, graph the equation $x+y=3$.
Rewrite the equation in slope-intercept form by solving for $y$.

$x+y=3$

$y=-x+3$

$y$-intercept = 3
slope = –1
Next, use the origin as a test point.

$x+y \ge 3$

$0+0 \ge 3$

$0 \ge 3$

This is a false statement. This means that the point $(0,0)$ will not fall in the shaded half-plane.

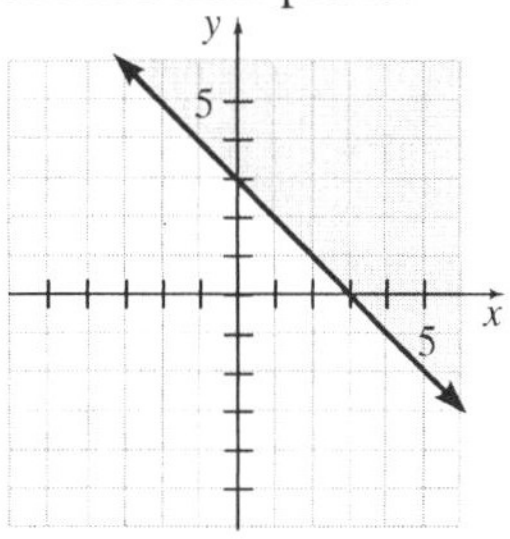


**3.** $x-y<5$
First, graph the equation $x-y=5$.
Rewrite the equation in slope-intercept form by solving for $y$.

$x-y=5$

$-y=-x+5$

$y=x-5$

$y$-intercept = –5
slope = 1
Next, use the origin as a test point.

$x-y<5$

$0-0<5$

$0<5$

This is a true statement. This means that the point $(0,0)$ will fall in the shaded half-plane.

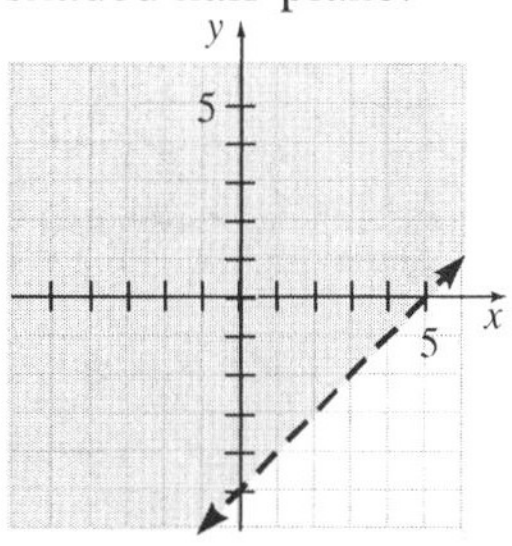

**5.** $x+2y>4$

First, graph the equation $x+2y=4$.

Rewrite the equation in slope-intercept form by solving for *y*.

$x+2y=4$

$2y=-x+4$

$y=-\frac{1}{2}x+2$

*y*-intercept = 2

slope = $-\frac{1}{2}$

Next, use the origin as a test point.

$0+2(0)>4$

$0+0>4$

$0>4$

This is a false statement. This means that the point $(0,0)$ will not fall in the shaded half-plane.

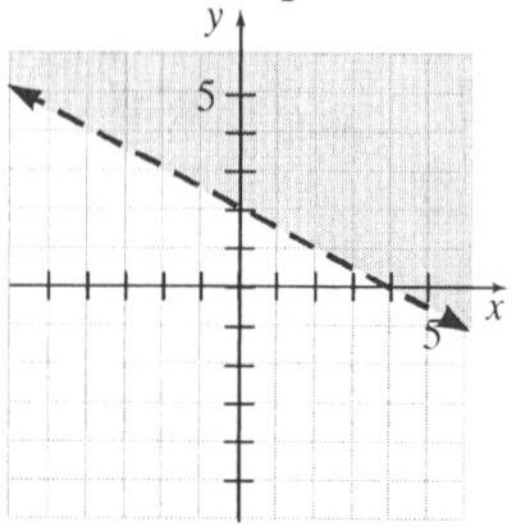


**7.** $3x-y\le 6$

First, graph the equation $3x-y=6$.

Rewrite the equation in slope-intercept form by solving for *y*.

$3x-y=6$

$-y=-3x+6$

$y=3x-6$

*y*-intercept = −6 slope = 3

Next, use the origin as a test point.

$3(0)-0\le 6$

$0-0\le 6$

$0\le 6$

This is a true statement. This means that the point $(0,0)$ will fall in the shaded half-plane.

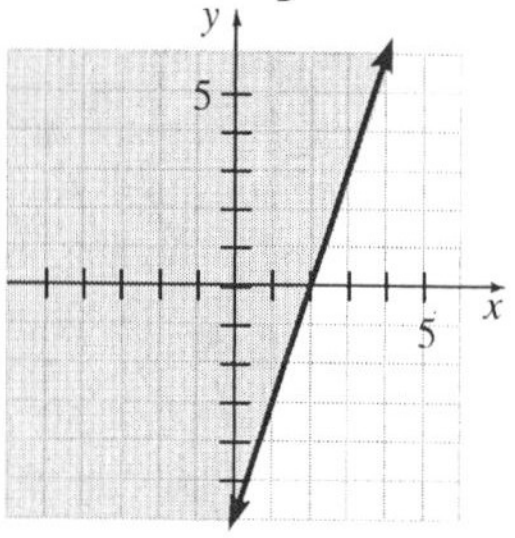


**9.** $\frac{x}{2}+\frac{y}{3}<1$

First, graph the equation $\frac{x}{2}+\frac{y}{3}=1$.

Rewrite the equation in slope-intercept form by solving for *y*.

$\frac{x}{2}+\frac{y}{3}=1$

$6\left(\frac{x}{2}\right)+6\left(\frac{y}{3}\right)=6(1)$

$3x+2y=6$

$2y=-3x+6$

$y=-\frac{3}{2}x+3$

*y*-intercept = 3 slope = $-\frac{3}{2}$

Next, use the origin as a test point.

$\frac{0}{2}+\frac{0}{3}<1$

$0+0<1$

$0<1$

This is a true statement. This means that the point $(0,0)$ will fall in the shaded half-plane.

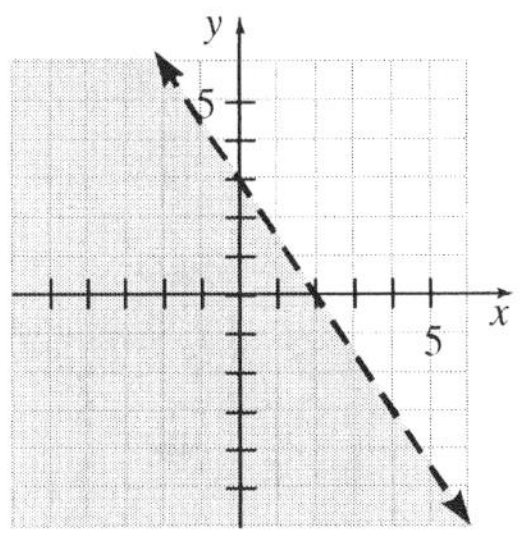


**11.** $y > \frac{1}{3}x$

Replacing the inequality symbol with an equal sign, we have $y > \frac{1}{3}x$. Since the equation is in slope-intercept form, use the slope and the intercept to graph the equation. The *y*–intercept is 0 and the slope is $\frac{1}{3}$.

Next, we need to find a test point. We cannot use the origin this time, because it lies on the line. Use $(1,1)$ as a test point.

$1 > \frac{1}{3}(1)$

$1 > \frac{1}{3}$

This is a true statement, so we know the point $(1,1)$ lies in the shaded half-plane.

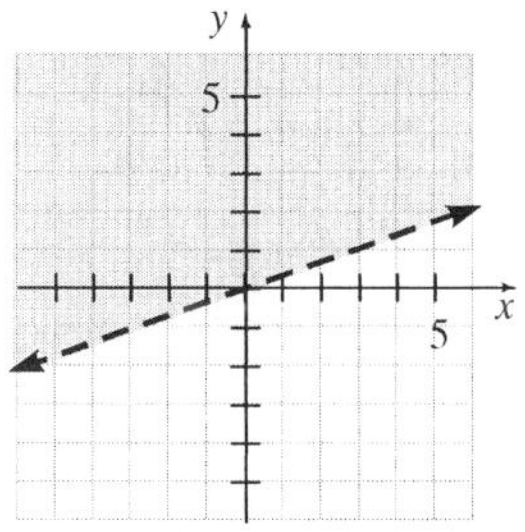


**13.** $y \le 3x + 2$

First, graph the equation $y = 3x + 2$.
Since the equation is in slope-intercept form, use the slope and the intercept to graph the equation. The *y*–intercept is 2 and the slope is 3.
Next, use the origin as a test point.

$0 \le 3(0) + 2$

$0 \le 2$

This is a true statement. This means that the point $(0,0)$ will fall in the shaded half-plane.

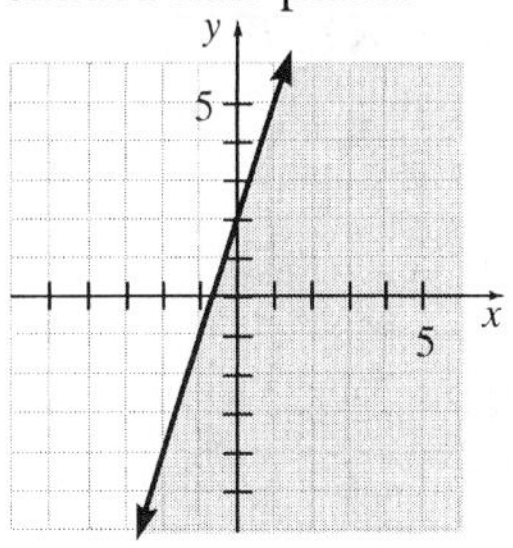


**15.** $y < -\frac{1}{4}x$

Replacing the inequality symbol with an equal sign, we have $y = -\frac{1}{4}x$. Since the equation is in slope-intercept form, use the slope and the intercept to graph the equation. The *y*–intercept is 0 and the slope is $-\frac{1}{4}$.

Next, we need to find a test point. We cannot use the origin this time, because it lies on the line. Use $(1,1)$ as a test point.

$1 < -\frac{1}{4}(1)$

$1 < -\frac{1}{4}$

This is a false statement, so we know

the point $(1,1)$ does not lie in the shaded half-plane.

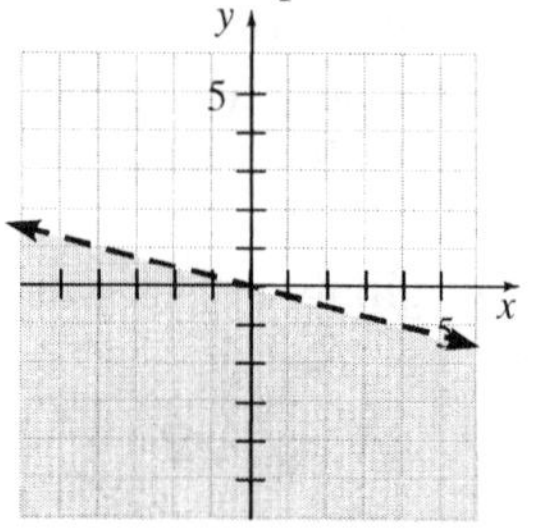


**17.** $x \leq 2$
Replacing the inequality symbol with an equal sign, we have $x = 2$. We know that equations of the form $x = a$ are vertical lines with $x$-intercept = $a$.
Next, use the origin as a test point.
$x \leq 2$
$0 \leq 2$
This is a true statement, so we know the point $(0,0)$ lies in the shaded half-plane.

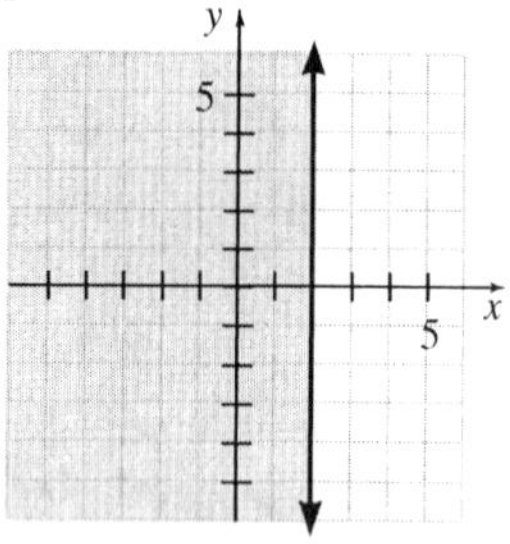


**19.** $y > -4$
Replacing the inequality symbol with an equal sign, we have $y = -4$. We know that equations of the form $y = b$ are horizontal lines with $y$-intercept = $b$.

Next, use the origin as a test point.
$y > -4$
$0 > -4$
This is a true statement, so we know the point $(0,0)$ lies in the shaded half-plane.

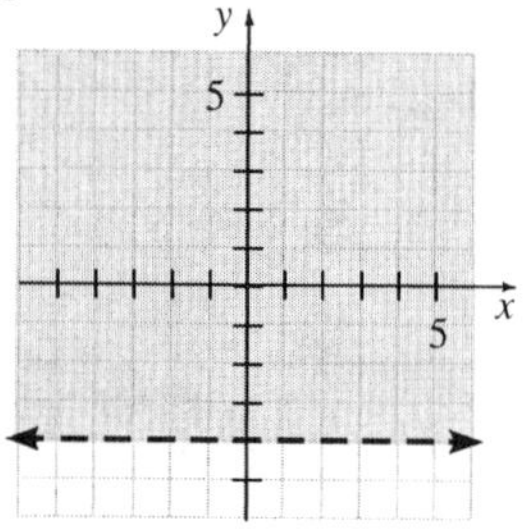


**21.** $y \geq 0$
Replacing the inequality symbol with an equal sign, we have $y = 0$. We know that equations of the form $y = b$ are vertical lines with $y$-intercept = $b$. In this case, we have $y = 0$, the equation of the $x$-axis.
Next, we need to find a test point. We cannot use the origin, because it lies on the line. Use $(1,1)$ as a test point.
$y \geq 0$
$1 \geq 0$
This is a true statement, so we know the point $(1,1)$ lies in the shaded half-plane.

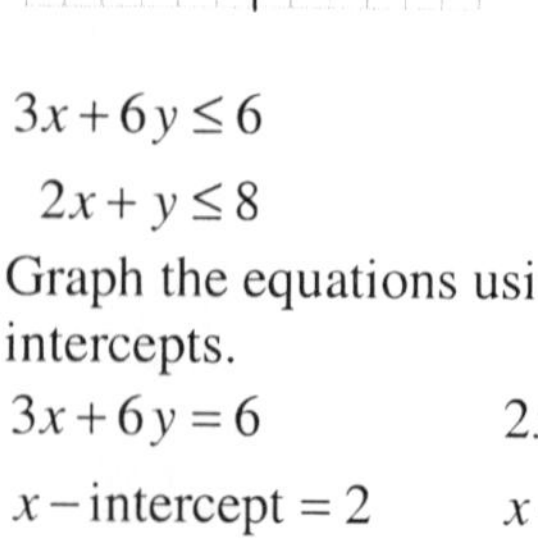


**23.** $3x + 6y \leq 6$
$2x + y \leq 8$
Graph the equations using the intercepts.

| | |
|---|---|
| $3x + 6y = 6$ | $2x + y = 8$ |
| $x$ − intercept = 2 | $x$ − intercept = 4 U |
| $y$ − intercept = 1 | $y$ − intercept = 8 |

se the origin as a test point to determine shading.

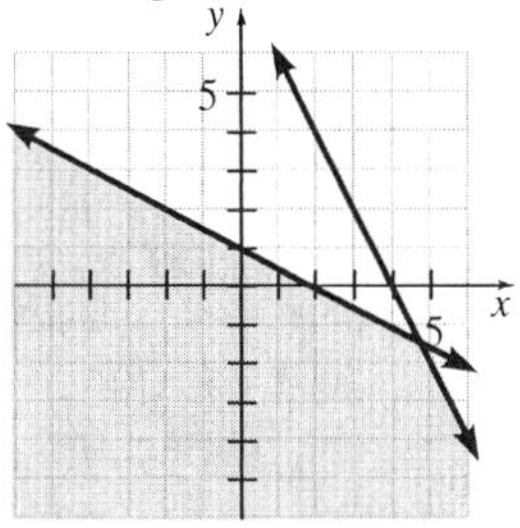

The solution set is the intersection of the shaded half-planes.

**25.** $2x-5y\le 10$

$3x-2y>6$

Graph the equations using the intercepts.

$2x-5y=10$ $\quad$ $3x-2y=6$

$x-\text{intercept}=5$ $\quad$ $x-\text{intercept}=2$ U

$y-\text{intercept}=2$ $\quad$ $y-\text{intercept}=-3$

se the origin as a test point to determine shading.

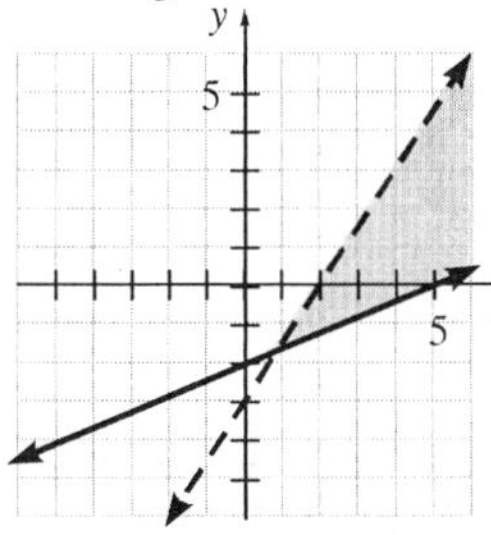

The solution set is the intersection of the shaded half-planes.

**27.** $y>2x-3$

$y<-x+6$

Graph the equations using the intercepts.

$y=2x-3$ $\quad$ $y=-x+6$

$x-\text{intercept}=\frac{3}{2}$ $\quad$ $x-\text{intercept}=6$ U

$y-\text{intercept}=6$

$y-\text{intercept}=-3$

se the origin as a test point to determine shading.

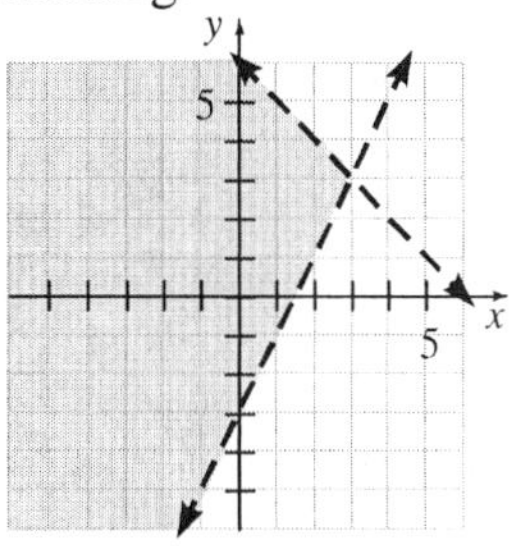

The solution set is the intersection of the shaded half-planes.

**29.** $x+2y\le 4$

$y\ge x-3$

Graph the equations using the intercepts.

$x+2y=4$ $\quad$ $y=x-3$

$x-\text{intercept}=4$ $\quad$ $x-\text{intercept}=3$ U

$y-\text{intercept}=2$ $\quad$ $y-\text{intercept}=-3$

se the origin as a test point to determine shading.

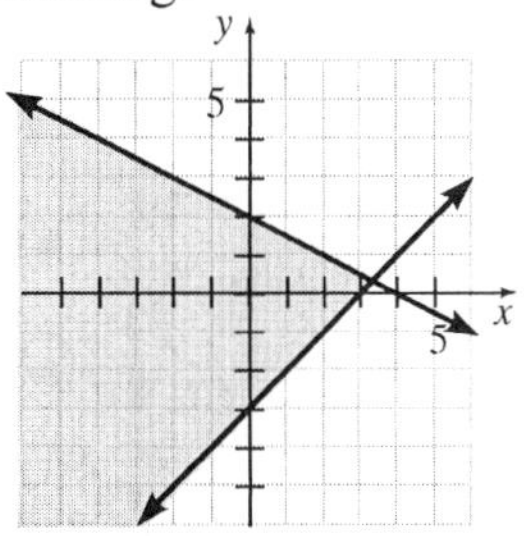

The solution set is the intersection of the shaded half-planes.

**31.** $x \le 2$

$y \ge -1$

Graph the vertical line, $x = 2$, and the horizontal line, $y = -1$. Use the origin as a test point to determine shading.

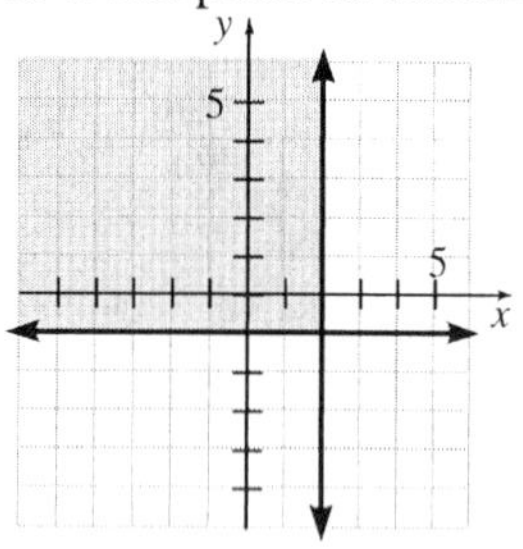


The solution set is the intersection of the shaded half-planes.

**33.** $-2 \le x < 5$

Since $x$ lies between –2 and 5, graph the two vertical lines, $x = -2$ and $x = 5$. Since $x$ lies between –2 and 5, shade between the two vertical lines.

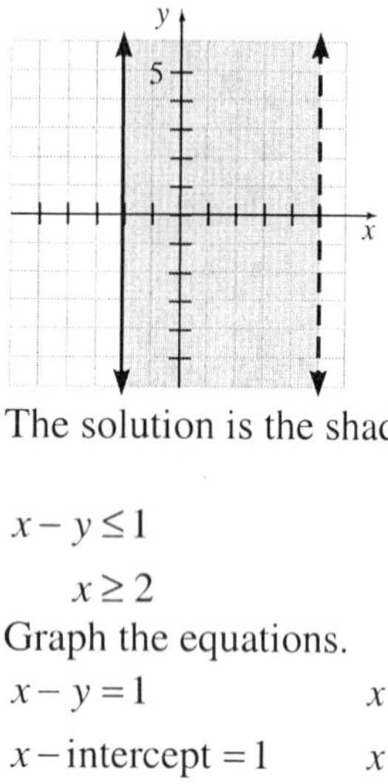


The solution is the shaded region.

**35.** $x - y \le 1$

$x \ge 2$

Graph the equations.

| | |
|---|---|
| $x - y = 1$ | $x = 2$ |
| $x-$intercept $= 1$ | $x-$intercept $= 2$U |
| $y-$intercept $= -1$ | vertical line |

se the origin as a test point to determine shading.

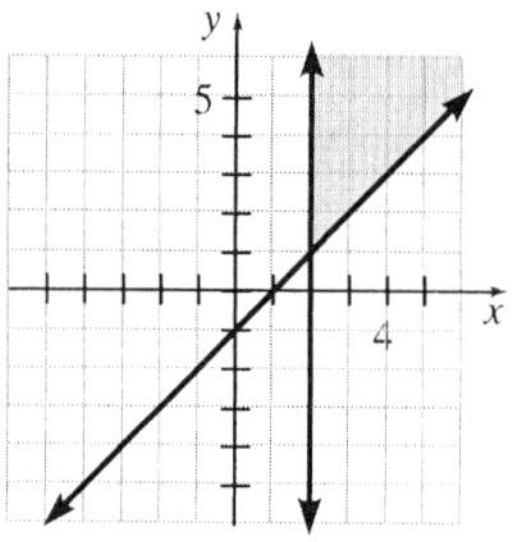


The solution set is the intersection of the shaded half-planes.

**37.** $x + y > 4$

$x + y < -1$

Graph the equations using the intercepts.

| | |
|---|---|
| $x + y = 4$ | $x + y = -1$ |
| $x-$intercept $= 4$ | $x-$intercept $= -1$ |
| $y-$intercept $= 4$ | $y-$intercept $= -1$ |

Use the origin as a test point to determine shading.

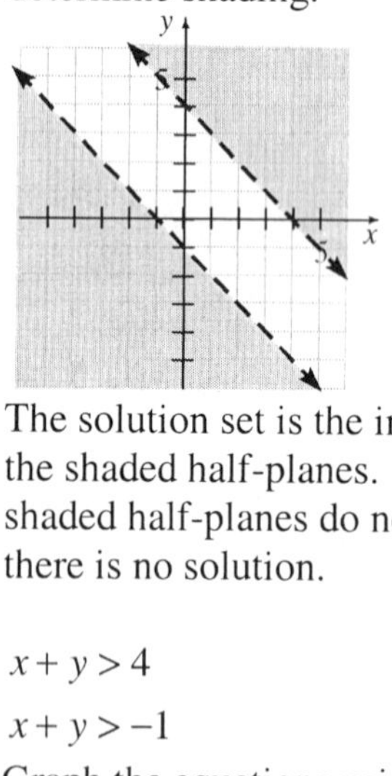


The solution set is the intersection of the shaded half-planes. Since the shaded half-planes do not intersect, there is no solution.

**39.** $x + y > 4$

$x + y > -1$

Graph the equations using the intercepts.

| | |
|---|---|
| $x + y = 4$ | $x + y = -1$ |
| $x-$intercept $= 4$ | $x-$intercept $= -1$ U |
| $y-$intercept $= 4$ | $y-$intercept $= -1$ |

se the origin as a test point to determine shading.

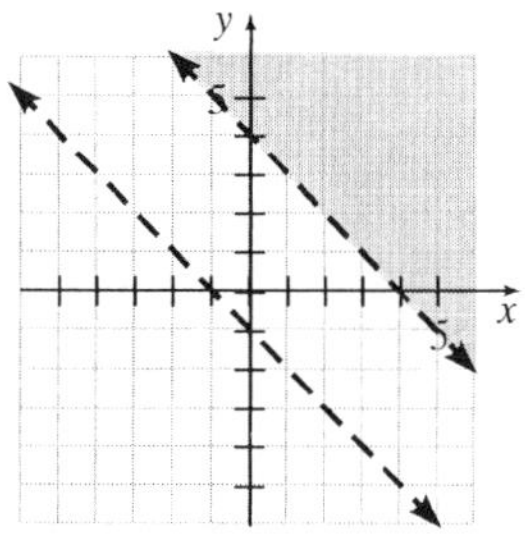

The solution set is the intersection of the shaded half-planes.

**41.** $x - y \leq 2$

$x \geq -2$

$y \leq 3$

Graph the equations using the intercepts.

| | |
|---|---|
| $x - y = 2$ | $y = 3$ |
| $x-$intercept $= 2$ | $y-$intercept $= 3$ |
| $y-$intercept $= -2$ | horizontal line |

$x = -2$

$x-$intercept $= -2$

vertical line

Use the origin as a test point to determine shading.

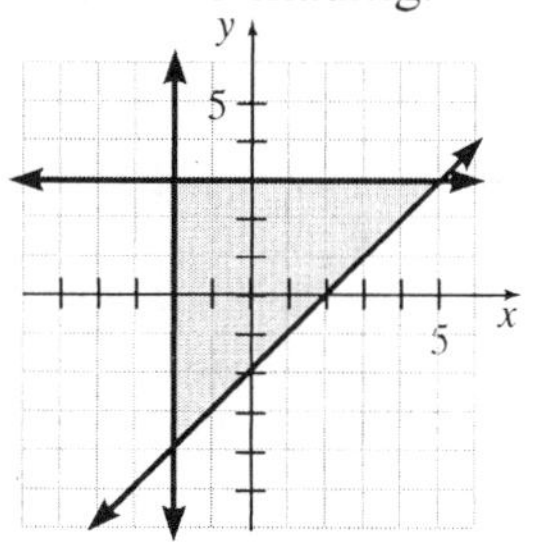

The solution set is the intersection of the shaded half-planes.

**43.** $x \geq 0$

$y \geq 0$

$2x + 5y \leq 10$

$3x + 4y \leq 12$

Since both $x$ and $y$ are greater than 0, we are concerned only with the first quadrant. Graph the other equations using the intercepts.

| | |
|---|---|
| $2x + 5y = 10$ | $3x + 4y = 12$ |
| $x-$intercept $= 5$ | $x-$intercept $= 4$ |
| $y-$intercept $= 2$ | $y-$intercept $= 3$ |

Use the origin as a test point to determine shading.

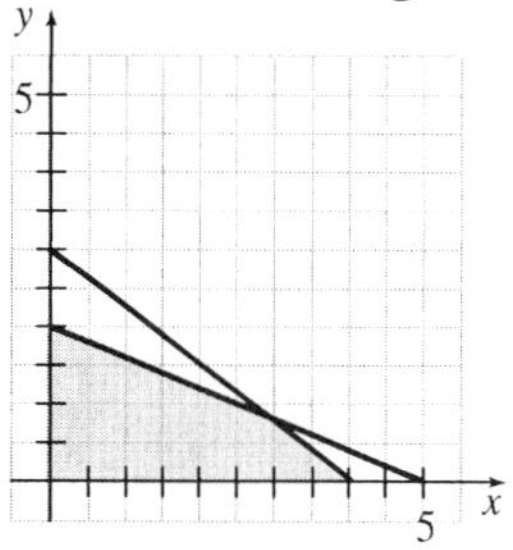

The solution set is the intersection of the shaded half-planes.

**45.** $3x + y \leq 6$

$2x - y \leq -1$

$x \geq -2$

$y \leq 4$

Graph the equations using the intercepts.

| | |
|---|---|
| $3x + y = 6$ | $x = -2$ |
| $x-$intercept $= 2$ | $x-$intercept $= -2$ |
| $y-$intercept $= 6$ | vertical line |

$2x - y = -1$ $\quad$ $y = 4$

$x - \text{intercept} = -\frac{1}{2}$ $\quad$ $y - \text{intercept} = 4$

$y - \text{intercept} = 1$ $\quad$ horizontal line

Use the origin as a test point to determine shading.

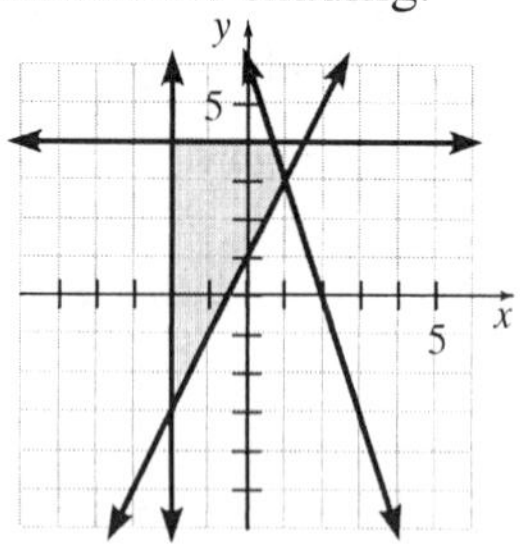

The solution set is the intersection of the shaded half-planes. Because all inequalities are greater than or equal to or less than or equal to, the boundaries of the shaded half-planes are also included in the solution set.

**47.** $y \geq -2x + 4$

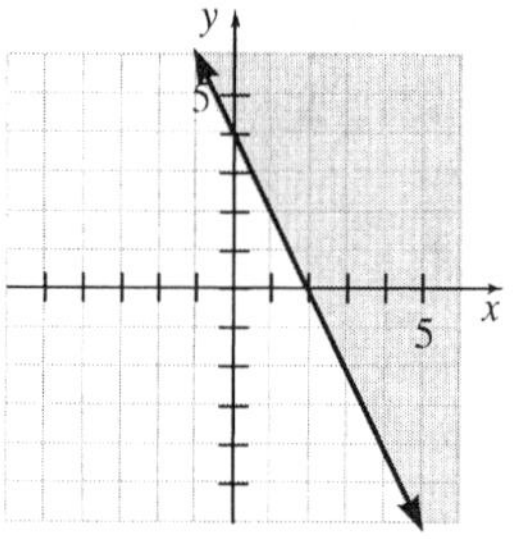

**48.** $y \geq -3x + 2$

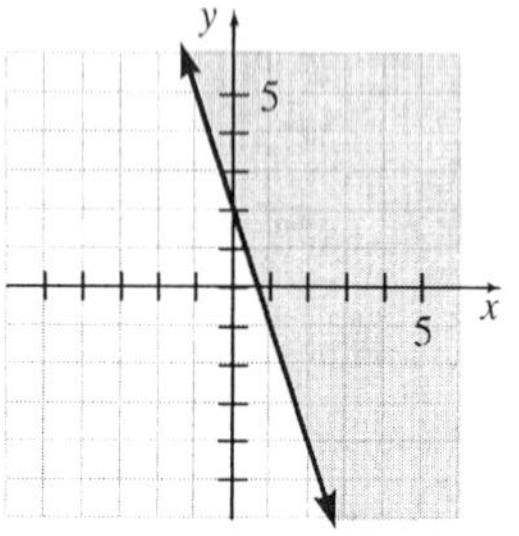

**49.** $x + y \leq 4$ and $3x + y \leq 6$

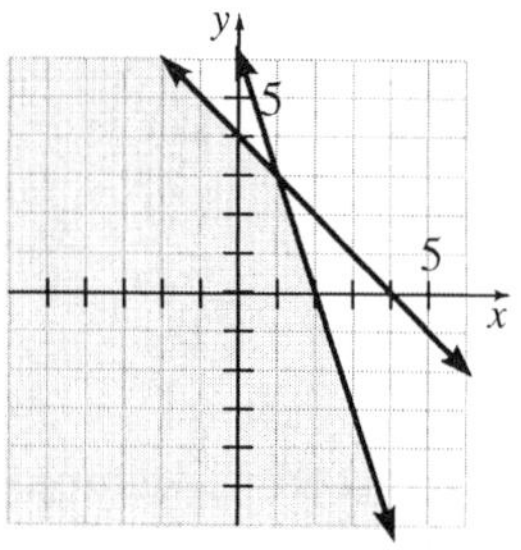

**50.** $x + y \leq 3$ and $4x + y \leq 6$

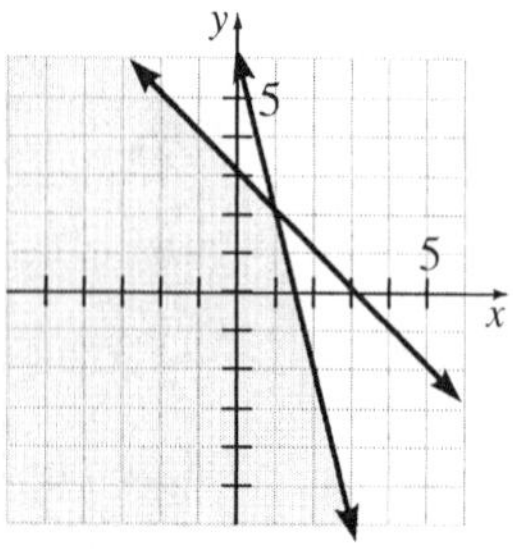

**51.** $-2 \leq x \leq 2$ and $-3 \leq y \leq 3$

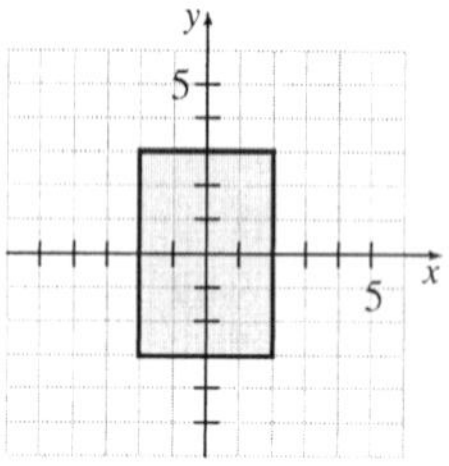

**52.** $-1 \leq x \leq 1$ and $-2 \leq y \leq 2$

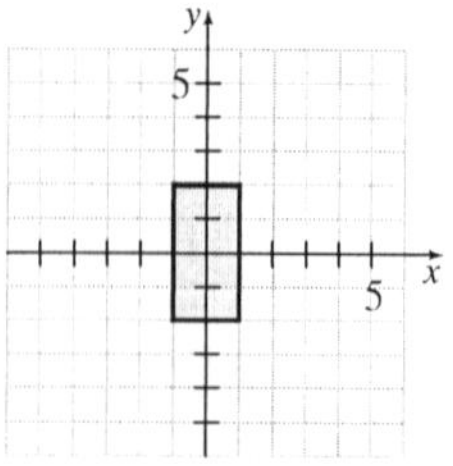

**53.** Find the union of solutions of $y > \frac{3}{2}x - 2$ and $y < 4$.

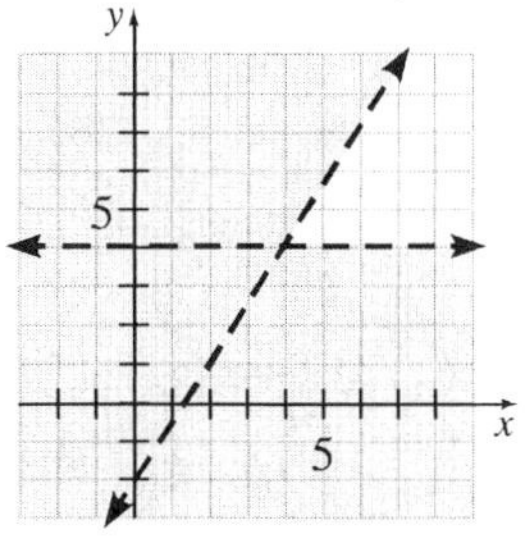

**54.** Find the union of solutions of $x - y \geq -1$ and $5x - 2y \leq 10$.

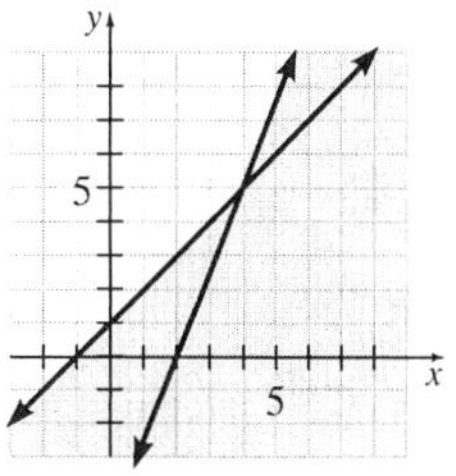

**55.** The system $\begin{array}{l} 3x + 3y < 9 \\ 3x + 3y > 9 \end{array}$ has no solution.

The number $3x + 3y$ cannot both be less than 9 and greater than 9 at the same time.

**56.** The system $\begin{array}{l} 6x - y \leq 24 \\ 6x - y > 24 \end{array}$ has no solution.

The number $6x - y$ cannot both be less than or equal to 24 and greater than 24 at the same time.

**57.** The system $\begin{array}{l} 3x + y \leq 9 \\ 3x + y \geq 9 \end{array}$ has infinitely many solutions. The solutions are all points on the line $3x + y = 9$.

**58.** The system $\begin{array}{l} 6x - y \leq 24 \\ 6x - y \geq 24 \end{array}$ has infinitely many solutions. The solutions are all points on the line $6x - y = 24$.

**59.** **a.** The coordinates of point A are $(20, 150)$. This means that a 20 year-old person with a pulse rate of 150 beats per minute falls within the target zone.

**b.** $10 \leq a \leq 70$

$10 \leq 20 \leq 70$

True

$150 \geq 0.7(220 - 20)$

True

$150 \leq 0.8(220 - 20)$

True

Since point *A* makes all three inequalities true, it is a solution of the system.

**61.** $10 \leq a \leq 70$

$H \geq 0.6(220 - a)$

$H \leq 0.7(220 - a)$

**63.** **a.** $y \geq 0$

$x + y \geq 5$

$x \geq 1$

$200x + 100y \leq 700$

**b.**

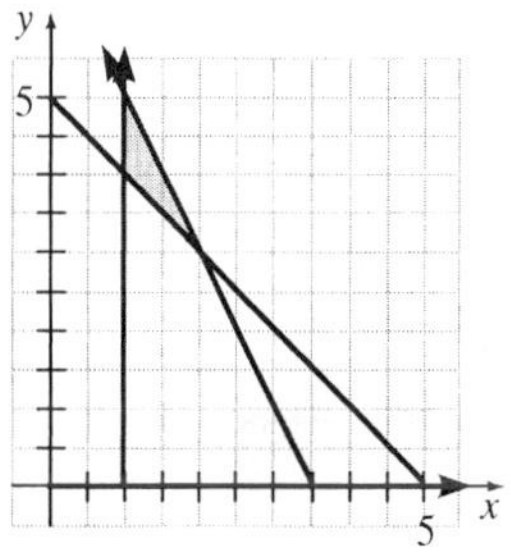

**c.** 2 nights

**65.-74. Answers will vary.**

**75.** $y \le 4x + 4$

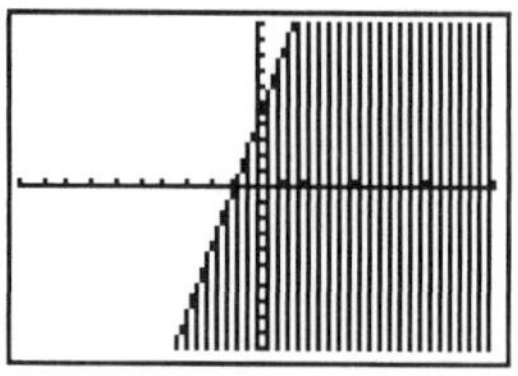

**77.** $2x + y \le 6$

$y \le -2x + 6$

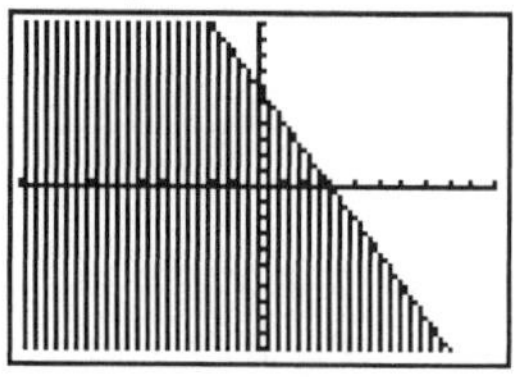

**79.** Answers will vary.

**81.** Answers will vary. For example, verify Exercise 23.

$3x + 6y \le 6$

$2x + y \le 8$

First solve both inequalities for $y$.

$3x + 6y \le 6$ $\quad$ $2x + y \le 8$

$6y \le -3x + 6$ $\quad$ $y \le -2x + 8$

$y \le -\frac{1}{2}x + 1$

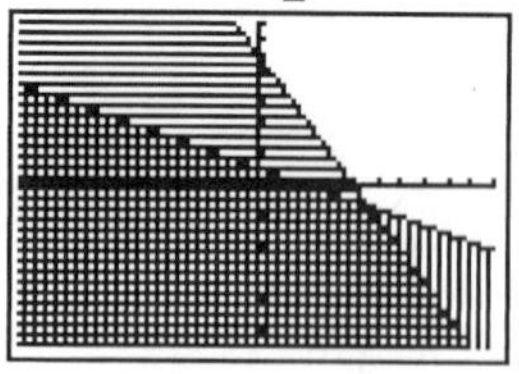

**83.** $x \ge -2, y > -1$

**85.** Answers will vary.
One simple example is

$x \ge 0$.

$x < 0$

**87.** $3x - y = 8$

$x - 5y = -2$

$$\left[\begin{array}{cc|c} 3 & -1 & 8 \\ 1 & -5 & -2 \end{array}\right] \quad R_1 \leftrightarrow R_2$$

$$= \left[\begin{array}{cc|c} 1 & -5 & -2 \\ 3 & -1 & 8 \end{array}\right] \quad -3R_1 + R_2$$

$$= \left[\begin{array}{cc|c} 1 & -5 & -2 \\ 0 & 14 & 14 \end{array}\right] \quad \frac{1}{14}R_2$$

$$= \left[\begin{array}{cc|c} 1 & -5 & -2 \\ 0 & 1 & 1 \end{array}\right]$$

The resulting system is:

$x - 5y = -2$

$y = 1.$

Since we know $y = 1$, we can use back-substitution to find $x$.

$x - 5(1) = -2$

$x - 5 = -2$

$x = 3$

The solution is $(3,1)$ and the solution set is $\{(3,1)\}$.

**88.** $y = 3x - 2$

$y = -2x + 8$

Both equations are in slope-intercept form, so use the slopes and $y$-intercepts to graph the lines.

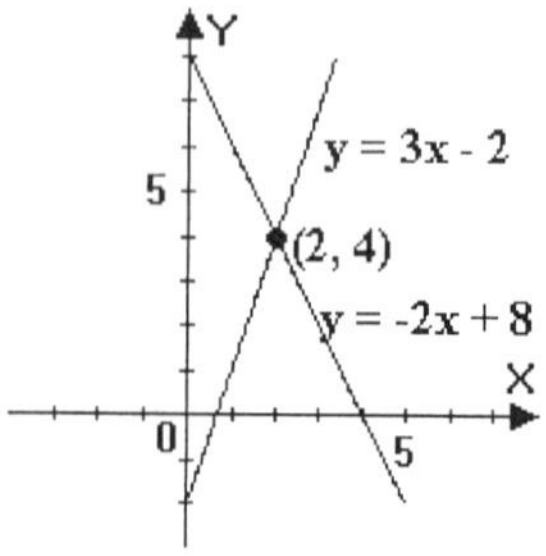


The solution is the intersection point $(2,4)$ and the solution set is $\{(2,4)\}$.

**89.** $\begin{vmatrix} 8 & 2 & -1 \\ 3 & 0 & 5 \\ 6 & -3 & 4 \end{vmatrix}$

$=8\begin{vmatrix} 0 & 5 \\ -3 & 4 \end{vmatrix} - 2\begin{vmatrix} 3 & 5 \\ 6 & 4 \end{vmatrix} - 1\begin{vmatrix} 3 & 0 \\ 6 & -3 \end{vmatrix}$

$=8(0+15)-2(12-30)-1(-9-0)$

$=8(15)-2(-18)-1(-9)$

$=120+36+9$

$=165$

## 4.5 Exercise Set

**1.**

| Corner $(x, y)$ | Objective Function $z=5x+6y$ |
|---|---|
| (1, 2) | $z=5x+6y$ <br> $=5(1)+6(2)$ <br> $=5+12=17$ |
| (8, 3) | $z=5x+6y$ <br> $=5(8)+6(3)$ <br> $=40+18=58$ |
| (7, 5) | $z=5x+6y$ <br> $=5(7)+6(5)$ <br> $=35+30=65$ |
| (2, 10) | $z=5x+6y$ <br> $=5(2)+6(10)$ <br> $=10+60$ <br> $=70$ |

The maximum value is 70 and the minimum is 17.

**3.**

| Corner $(x, y)$ | Objective Function $z=40x+50y$ |
|---|---|
| (0, 0) | $z=40x+50y$ <br> $=40(0)+50(0)$ <br> $=0+0=0$ |
| (8, 0) | $z=40x+50y$ <br> $=40(8)+\cancel{50(0)}=320$ |
| (4, 9) | $z=40x+50y$ <br> $=40(4)+50(9)$ <br> $=160+450=610$ |
| (0, 8) | $z=40x+50y$ <br> $=\cancel{40(0)}+50(8)=400$ |

The maximum value is 610 and the minimum value is 0.

**5.** Objective Function: $z=3x+2y$

Constraints: $x\ge 0,\ y\ge 0$

$2x+y\le 8$

$x+y\ge 4$

**a.**

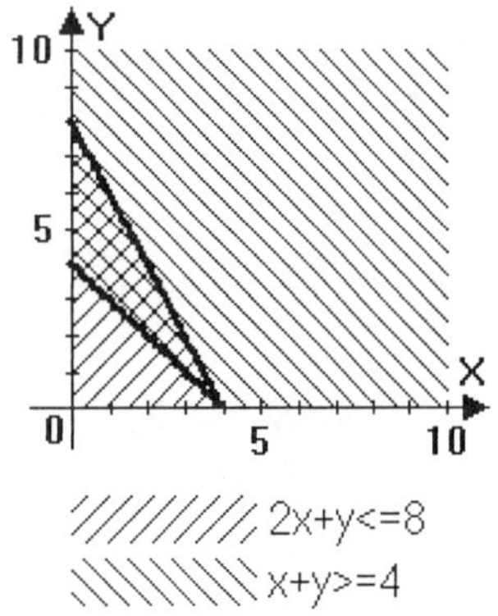

**b.**

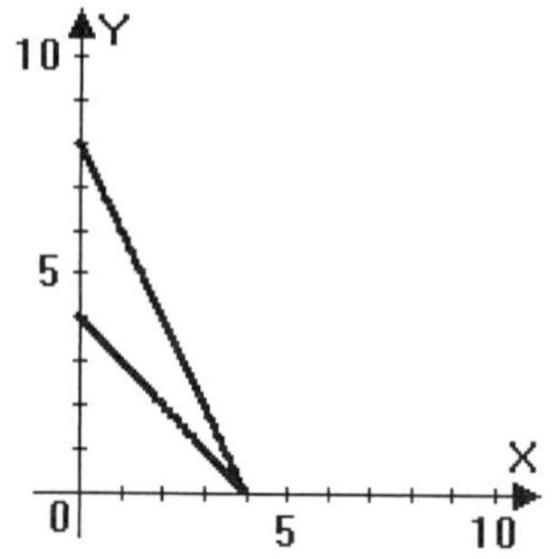


| Corner $(x, y)$ | Objective Function $z = 3x + 2y$ |
|---|---|
| (4, 0) | $z = 3x + 2y$ <br> $= 3(4) + 2(0) = 12$ |
| (0, 8) | $z = 3x + 2y$ <br> $= 3(0) + 2(8) = 16$ |
| (0, 4) | $z = 3x + 2y$ <br> $= 3(0) + 2(4) = 8$ |

**c.** The maximum value is 16. It occurs at the point (0, 8).

**7.** Objective Function: $z = 4x + y$

Constraints: $x \geq 0,\ y \geq 0$

$2x + 3y \leq 12$

$x + y \geq 3$

**a.**

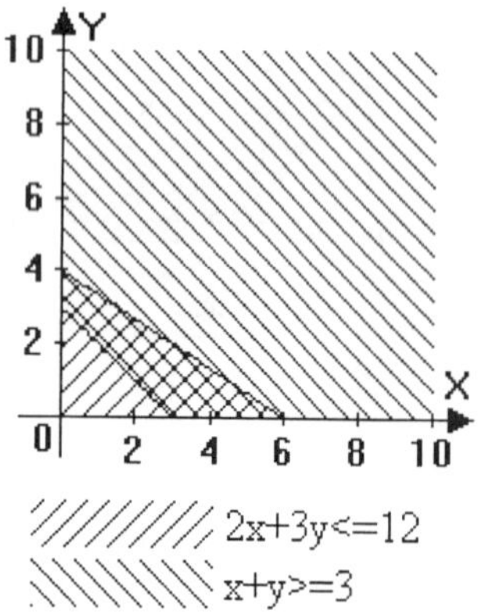


**b.**

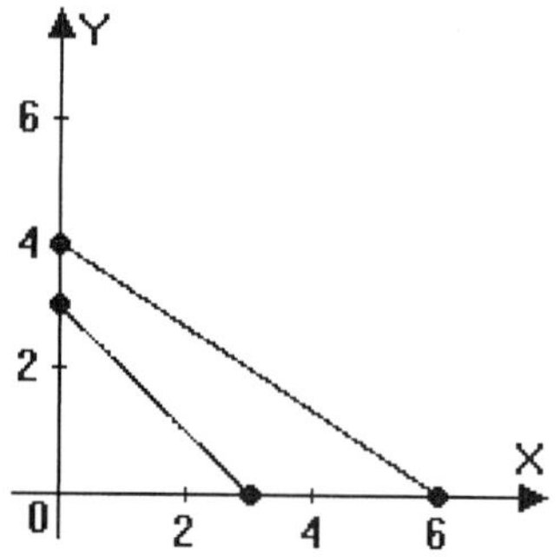


| Corner $(x, y)$ | Objective Function $z = 4x + y$ |
|---|---|
| (3, 0) | $z = 4x + y$ <br> $= 4(0) + 0 = 0$ |
| (6, 0) | $z = 4x + y$ <br> $= 4(6) + 0 = 24$ |
| (0, 3) | $z = 4x + y$ <br> $= 4(0) + 3 = 3$ |
| (0, 4) | $z = 4x + y$ <br> $= 4(0) + 4 = 4$ |

**c.** The maximum value is 24. It occurs at the point (6, 0).

**9.** Objective Function: $z = 3x - 2y$

Constraints: $1 \le x \le 5$

$y \ge 2$

$x - y \ge -3$

**a.**

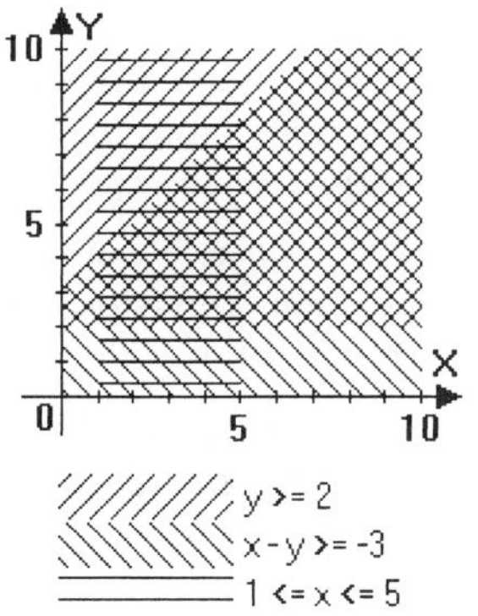


**b.**

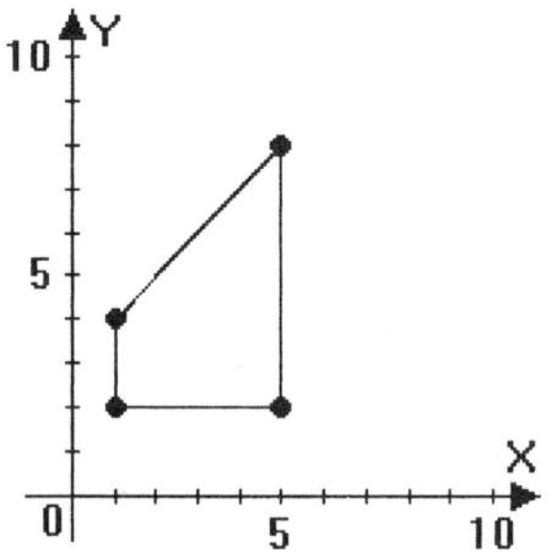

| Corner $(x, y)$ | Objective Function $z = 3x - 2y$ |
|---|---|
| (1, 2) | $z = 3x - 2y$ $= 3(1) - 2(2) = -1$ |
| (5, 2) | $z = 3x - 2y$ $= 3(5) - 2(2) = 11$ |
| (5, 8) | $z = 3x - 2y$ $= 3(5) - 2(8) = -1$ |
| (1, 4) | $z = 3x - 2y$ $= 3(1) - 2(4) = -5$ |

**c.** The maximum value is 11. It occurs at the point (5, 2).

**11.** Objective Function: $z = 4x + 2y$

Constraints: $x \ge 0,\ y \ge 0$

$2x + 3y \le 12$

$3x + 2y \le 12$

$x + y \ge 2$

**a.**

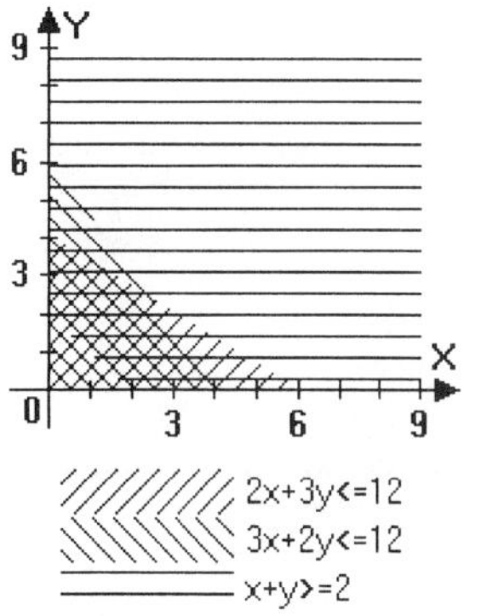


**b.**

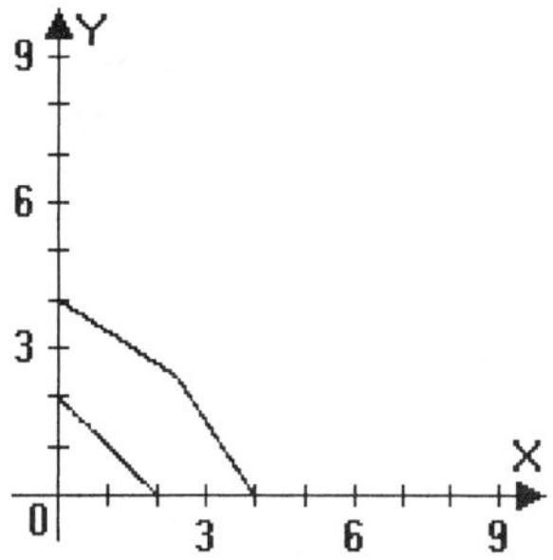

| Corner $(x, y)$ | Objective Function $z = 4x + 2y$ |
|---|---|
| (2, 0) | $z = 4x + 2y$ $= 4(2) + 2(0) = 8$ |
| (4, 0) | $z = 4x + 2y$ $= 4(4) + 2(0) = 16$ |
| (2.4,2.4) | $z = 4x + 2y$ $= 4(2.4) + 2(2.4)$ $= 9.6 + 4.8 = 14.4$ |
| (0, 4) | $z = 4x + 2y$ $= 4(0) + 2(4) = 8$ |
| (0, 2) | $z = 4x + 2y$ $= 4(0) + 2(2) = 4$ |

**c.** The maximum value is 16. It occurs at the point (4, 0).

**13.** Objective Function: $z = 10x + 12y$

Constraints: $x \geq 0, \quad y \geq 0$

$x + y \leq 7$

$2x + y \leq 10$

$2x + 3y \leq 18$

**a.**

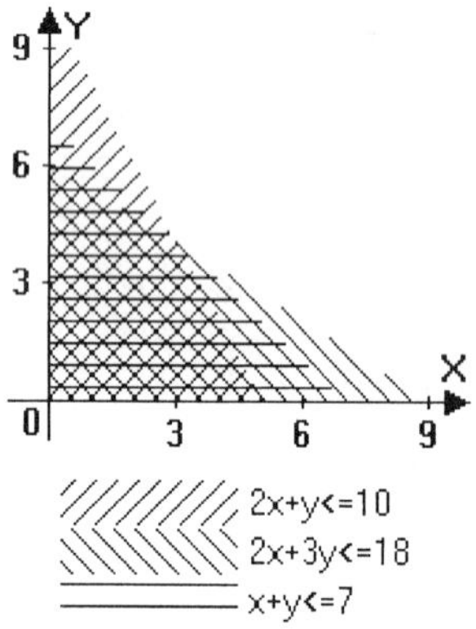


**b.**

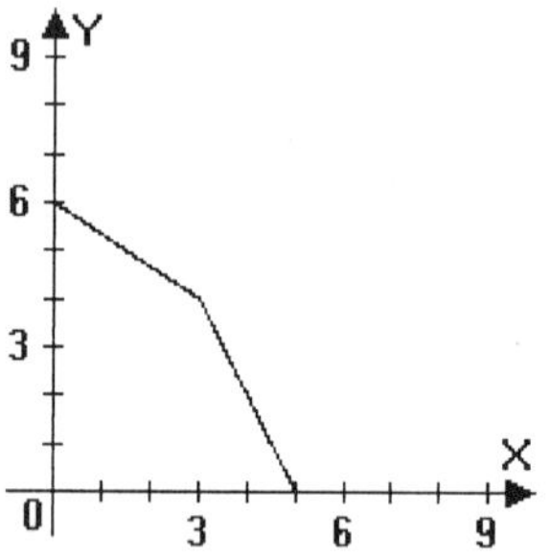


| Corner $(x, y)$ | Objective Function $z = 10x + 12y$ |
|---|---|
| (0, 0) | $10x + 12y$<br>$= 10(0) + 12(0) = 0$ |
| (5, 0) | $10x + 12y$<br>$= 10(5) + 12(0) = 50$ |
| (3, 4) | $10x + 12y$<br>$= 10(3) + 12(4) = 78$ |
| (0, 6) | $10x + 12y$<br>$= 10(0) + 12(6) = 72$ |

**c.** The maximum value is 78 and it occurs at the point (3, 4)

**15. a.** The objective function is $z = 125x + 200y$.

**b.** Since we can make at most 450 console televisions, we have $x \leq 450$.
Since we can make at most 200 wide screen televisions, we have $y \leq 200$.
Since we can spend at most \$360,000 per month, we have $600x + 900y \leq 360{,}000$.

**c.**

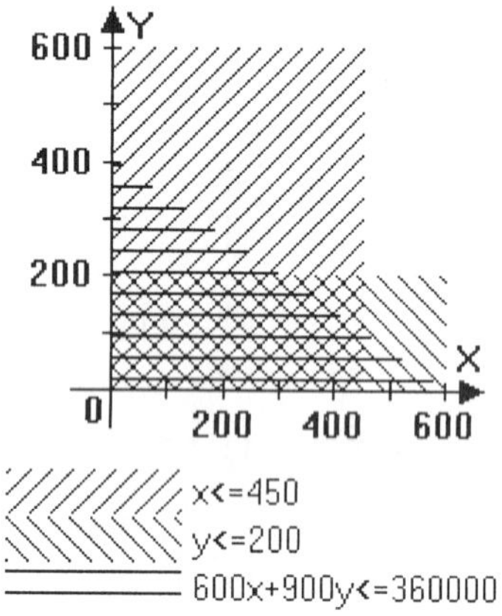


**d.**

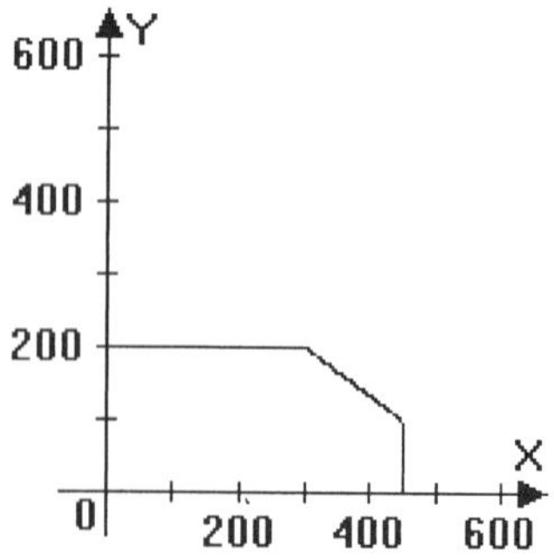

| Corner $(x, y)$ | Objective Function $z = 125x + 200y$ |
|---|---|
| (0, 0) | $125x + 200y$ <br> $= 125(0) + 200(0) = 0$ |
| (0, 200) | $125x + 200y$ <br> $= 125(0) + 200(200)$ <br> $= 40,000$ |
| (300, 200) | $125x + 200y$ <br> $= 125(300) + 200(200)$ <br> $= 37,500 + 40,000$ <br> $= 77,500$ |
| (450, 100) | $125x + 200y$ <br> $= 125(450) + 200(100)$ <br> $= 56,250 + 20,000$ <br> $= 76,250$ |
| (450, 0) | $125x + 200y$ <br> $= 125(450) + 200(0)$ <br> $= 56,250$ |

**e.** The television manufacturer will make the greatest profit by manufacturing 300 console televisions each month and 200 wide screen televisions each month. The maximum monthly profit is $77,500.

**17.** Let $x$ = the number of model $A$ bicycles produced
Let $y$ = the number of model $B$ bicycles produced
The objective function is
$z = 25x + 15y$.
The assembling constraint is
$5x + 4y \le 200$.
The painting constraint is
$2x + 3y \le 108$.

We also know that $x$ and $y$ must either be zero or a positive number. We cannot make a negative number of bicycles.
Next, graph the constraints.

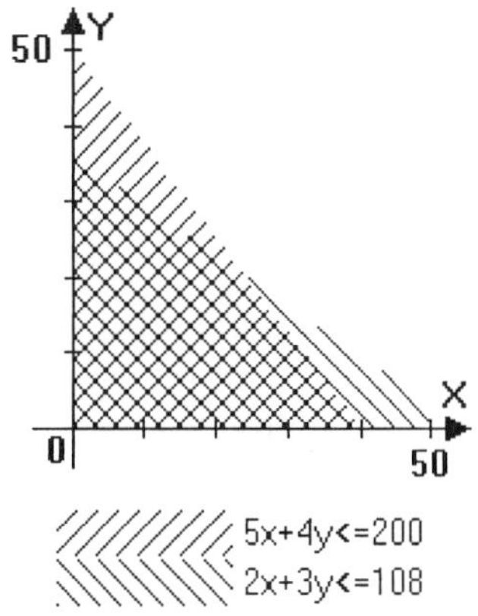


Using the graph, find the value of the objective function at each of the corner points.

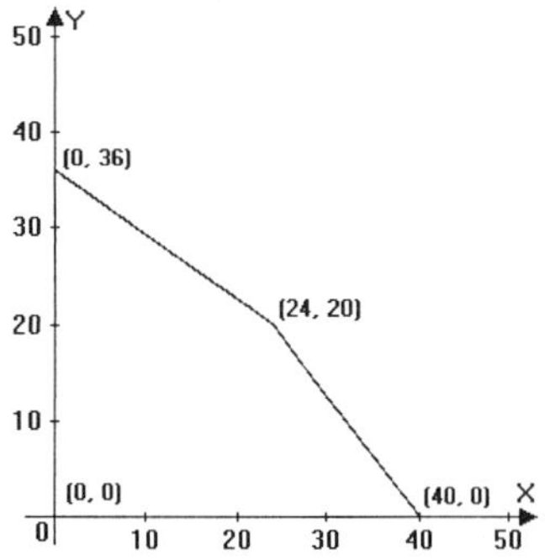


| Corner $(x, y)$ | Objective Function $z = 25x + 15y$ |
|---|---|
| (0, 0) | $25x + 15y$ <br> $= 25(0) + 15(0)$ <br> $= 0$ |
| (40, 0) | $25x + 15y$ <br> $= 25(40) + 15(0)$ <br> $= 1000$ |

| | |
|---|---|
| (24, 20) | $25x+15y$ <br> $=25(24)+15(20)$ <br> $=600+300$ <br> $=900$ |
| (0, 36) | $25x+15y$ <br> $=25(0)+15(36)$ <br> $=540$ |

The maximum of 1000 occurs at the point (40, 0). This means that the company should produce 40 of model *A* and none of model *B* each week for a profit of $1000.

**19.** Let $x$ = the number of cartons of food
Let $y$ = the number of cartons of clothing
The objective function is $z=12x+5y$.
The weight constraint is
$50x+20y\le 19{,}000$.
The volume constraint is
$20x+10y\le 8000$.
We also know that $x$ and $y$ must either be zero or a positive number. We cannot have a negative number of cartons of food or clothing.
Next, graph the constraints.

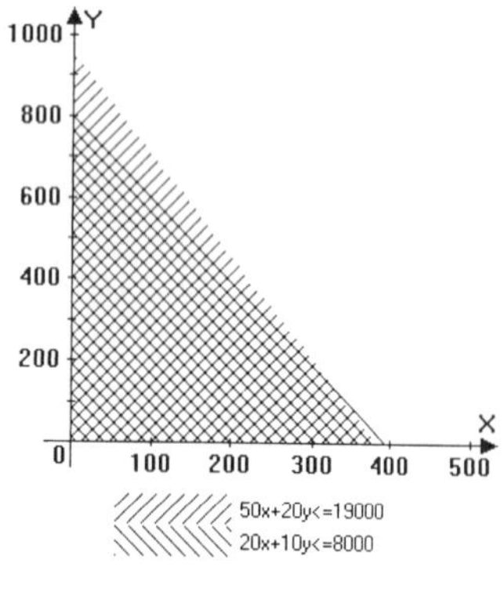


Using the graph, find the value of the objective function at each of the corner points.

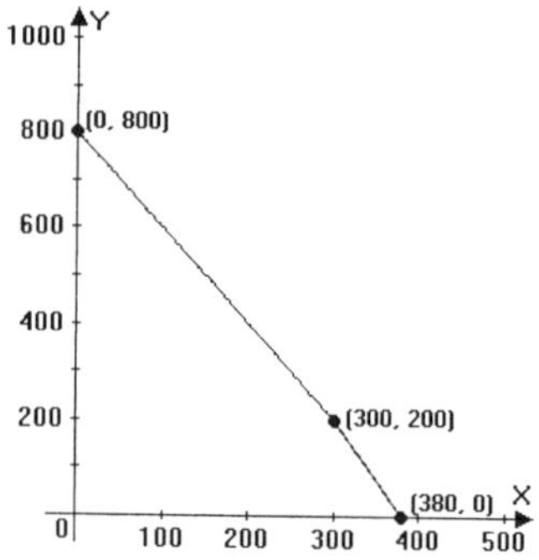


| Corner $(x, y)$ | Objective Function $z=12x+5y$ |
|---|---|
| (0, 0) | $12x+5y$ <br> $=12(0)+5(0)=0$ |
| (380, 0) | $12x+5y$ <br> $=12(380)+5(0)$ <br> $=4560$ |
| (300, 200) | $12x+5y$ <br> $=12(300)+5(200)$ <br> $=3600+1000=4600$ |
| (0, 600) | $12x+5y$ <br> $=12(0)+5(600)$ <br> $=3000$ |

The maximum of 4600 occurs at the point (300, 200). This means that to maximize the number of people who are helped, 300 boxes of food and 200 boxes of clothing should be sent.

**21.** Let $x$ = the number of parents
Let $y$ = the number of students
The objective function is $z=2x+y$.
The seating constraint is $x+y\le 150$.
The two parents per student constraint is $2x\ge y$.

We also know that $x$ and $y$ must either be zero or a positive number. We cannot have a negative number of parents or students.
Next, graph the constraints.

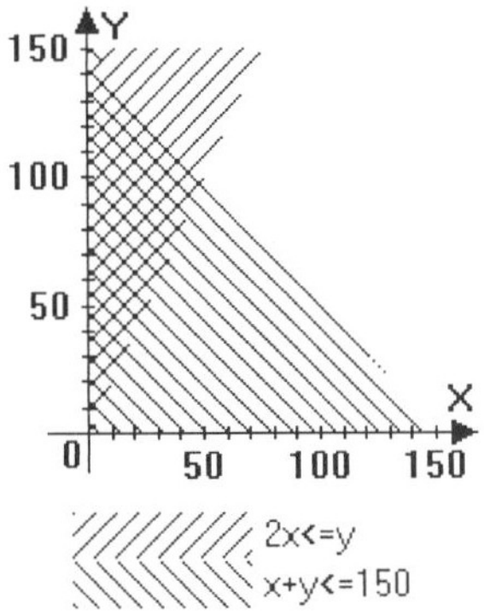


Using the graph, find the value of the objective function at each of the corner points.

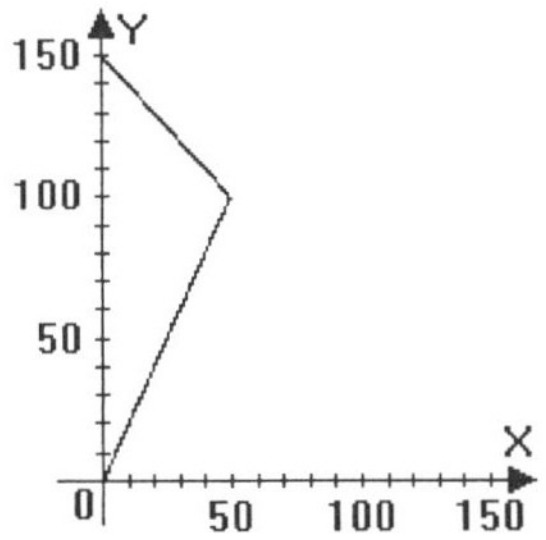


| Corner $(x, y)$ | Objective Function $z = 2x + y$ |
|---|---|
| (0, 0) | $2x+y$ $=2(0)+0=0$ |
| (50, 100) | $2x+y$ $=2(50)+100$ $=100+100=200$ |
| (0, 150) | $2x+y$ $=2(0)+100=100$ |

The maximum of 200 occurs at the point (50, 100). This means that to maximize the amount of money raised, 50 parents and 100 students should attend.

**23.** Let $x$ = the number of Boeing 727s
Let $y$ = the number of Falcon 20s
The objective function is $z = x + y$.
The hourly operating cost constraint is $1400x + 500y \leq 35000$.
The total payload constraint is $42000x + 6000y \geq 672,000$.
The 727 constraint is $x \leq 20$.
We also know that $x$ and $y$ must either be zero or a positive number. We cannot have a negative number of aircraft.
Next, graph the constraints.

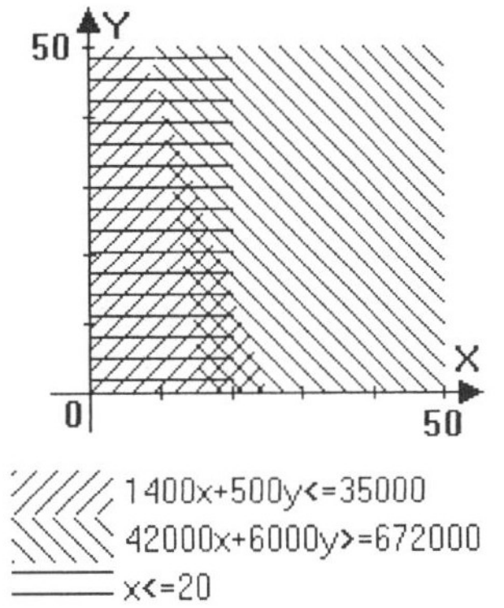


Using the graph, find the value of the objective function at each of the corner points.

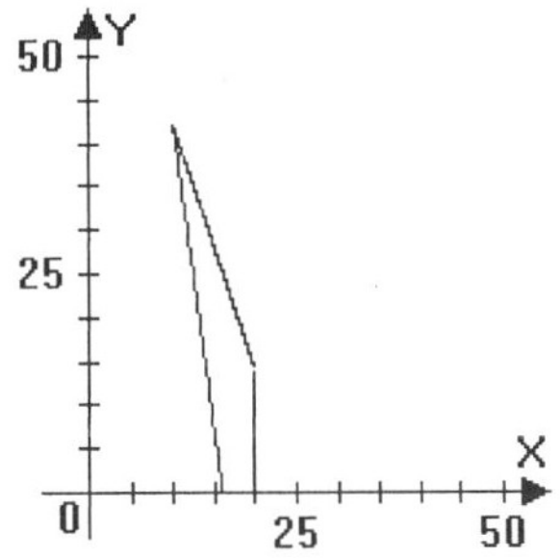

| Corner $(x, y)$ | Objective Function $z = x + y$ |
|---|---|
| (16, 0) | $z = x + y$ <br> $= 16 + 0 = 16$ |
| (20, 0) | $z = x + y$ <br> $= 20 + 0 = 20$ |
| (20, 14) | $z = x + y$ <br> $= 20 + 14 = 34$ |
| (10, 42) | $z = x + y$ <br> $= 10 + 42 = 52$ |

The maximum of 52 occurs at the point (10, 42). This means that to maximize the number of aircraft, 10 Boeing 727s and 42 Falcon 20s should be purchased.

**24.-28.** Answers will vary.

**29.** Let $x$ = the amount invested in stocks
Let $y$ = the amount invested in bonds
The objective function is
$z = 0.12x + 0.08y$.
The total money constraint is
$x + y \leq 10000$.
The minimum bond investment constraint is $y \geq 3000$.
The minimum stock investment constraint is $x \geq 2000$.
The stock versus bond constraint is
$y \geq x$.
We also know that $x$ and $y$ must either be zero or a positive number. We cannot invest a negative amount of money.

Next, graph the constraints.

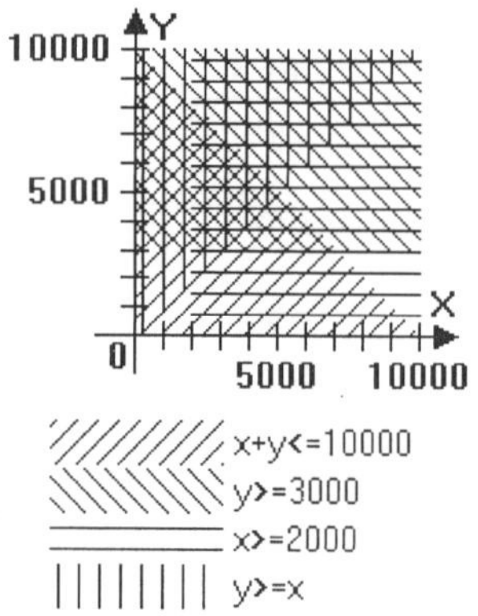


Using the graph, find the value of the objective function at each of the corner points.

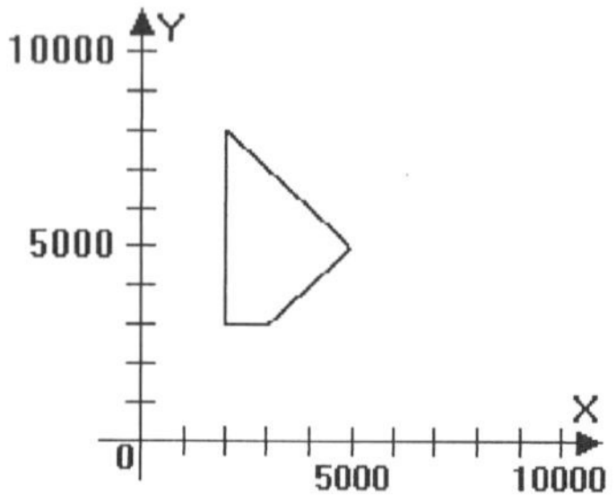


| Corner $(x, y)$ | Objective Function $z = 0.12x + 0.08y$ |
|---|---|
| (2000,3000) | $0.12x + 0.08y$ <br> $= 0.12(2000) + 0.08(3000)$ <br> $= 240 + 240 = 480$ |
| (3000,3000) | $0.12x + 0.08y$ <br> $= 0.12(3000) + 0.08(3000)$ <br> $= 360 + 240 = 600$ |
| (5000,5000) | $0.12x + 0.08y$ <br> $= 0.12(5000) + 0.08(5000)$ <br> $= 600 + 400 = 1000$ |
| (2000,8000) | $0.12x + 0.08y$ <br> $= 0.12(2000) + 0.08(8000)$ <br> $= 240 + 640 = 880$ |

The maximum of 1000 occurs at the point (5000, 5000). This means that to maximize the return on the investment, \$5000 should be invested in stocks and \$5000 should be invested in bonds.

**31.** $\left(2x^4y^3\right)\left(3xy^4\right)^3 = 2x^4y^3 3^3 x^3 y^{12}$
$= 2\cdot 3^3 x^4 x^3 y^3 y^{12}$
$= 2\cdot 27x^7 y^{15}$
$= 54x^7 y^{15}$

**32.**
$$3P = \frac{2L-W}{4}$$
$$12P = 2L - W$$
$$12P + W = 2L$$
$$L = \frac{12P+W}{2}$$
$$L = \frac{12P}{2} + \frac{W}{2} \text{ or } 6P + \frac{W}{2}$$

**33.**
$$f(x) = x^3 + 2x^2 - 5x + 4$$
$$f(-1) = (-1)^3 + 2(-1)^2 - 5(-1) + 4$$
$$= -1 + 2(1) + 5 + 4$$
$$= -1 + 2 + 5 + 4 = 10$$

## Chapter 4 Review

**1.** $x > 5$

**2.** $x \le 1$

**3.** $-3 \le x < 0$

**4.** $\{x \mid -2 < x \le 3\}$

**5.** $\{x \mid -1.5 \le x \le 2\}$

**6.** $\{x \mid x > -1\}$

**7.**
$$-6x + 3 \le 15$$
$$-6x \le 12$$
$$\frac{-6x}{-6} \ge \frac{12}{-6}$$
$$x \ge -2$$

The solution set is $\{x \mid x \ge -2\}$ or $[2, \infty)$.

**8.**
$$6x - 9 \ge -4x - 3$$
$$10x - 9 \ge -3$$
$$10x \ge 6$$
$$x \ge \frac{6}{10}$$
$$x \ge \frac{3}{5}$$

The solution set is $\left\{x \middle| x \ge \frac{3}{5}\right\}$ or $\left[\frac{3}{5}, \infty\right)$.

**9.**

$$\frac{x}{3}-\frac{3}{4}-1>\frac{x}{2}$$

$$12\left(\frac{x}{3}\right)-12\left(\frac{3}{4}\right)-12(1)>12\left(\frac{x}{2}\right)$$

$$4x-3(3)-12>6x$$

$$4x-9-12>6x$$

$$4x-21>6x$$

$$-2x-21>0$$

$$-2x>21$$

$$x<-\frac{21}{2}$$

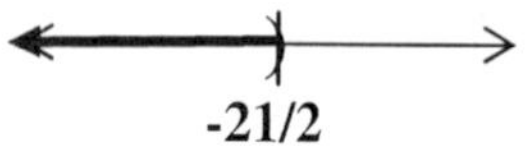


The solution set is $\left\{x \middle| x<-\frac{21}{2}\right\}$ or $\left(-\infty,-\frac{21}{2}\right)$.

**10.**

$$6x+5>-2(x-3)-25$$

$$6x+5>-2x+6-25$$

$$6x+5>-2x-19$$

$$8x+5>-19$$

$$8x>-24$$

$$x>-3$$

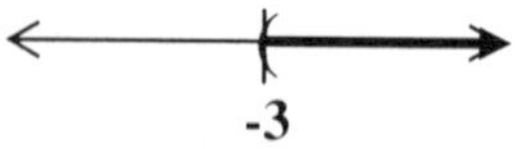


The solution set is $\{x|x>-3\}$ or $(-3,\infty)$.

**11.**

$$3(2x-1)-2(x-4)\geq 7+2(3+4x)$$

$$6x-3-2x+8\geq 7+6+8x$$

$$4x+5\geq 13+8x$$

$$-4x+5\geq 13$$

$$-4x\geq 8$$

$$x\leq -2$$

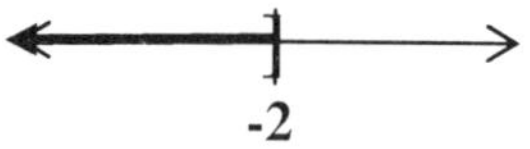


The solution set is $\{x|x\leq -2\}$ or $(-\infty,-2)$.

**12.**

$$2x+7\leq 5x-6-3x$$

$$2x+7\leq 2x-6$$

$$7\leq -6$$

This is a contradiction. Seven is not less than or equal to –6. There are no values of $x$ for which $7\leq -6$. The solution set is $\varnothing$ or $\{\ \}$.

**13.** Let $x$ = the number of checks written per month
The cost using the first method is $c_1=11+0.06x$.
The cost using the second method is $c_2=4+0.20x$.
The first method is a better deal if it costs less than the second method.

$$c_1<c_2$$

$$11+0.06x<4+0.20x$$

$$11-0.14x<4$$

$$-0.14x<-7$$

$$\frac{-0.14x}{-0.14}>\frac{-7}{-0.14}$$

$$x>50$$

The first method is a better deal when more than 50 checks per month are written.

**14.** Let $x$ = the amount of sales per month in dollars
The salesperson's commission is $c = 500 + 0.20x$.
We are looking for the amount of sales, $x$, the salesman must make to receive more than \$3200 in income.

$$c > 3200$$
$$500 + 0.20x > 3200$$
$$0.20x > 2700$$
$$x > 13500$$

The salesman must sell at least \$13,500 to receive a total income that exceeds \$3200 per month.

**15.** $A \cap B = \{a, c\}$

**16.** $A \cap C = \{a\}$

**17.** $A \cup B = \{a, b, c, d, e\}$

**18.** $A \cup C = \{a, b, c, d, f, g\}$

**19.** $x \le 3$

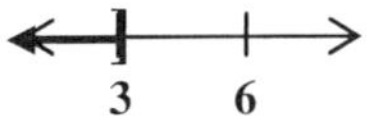

$x < 6$

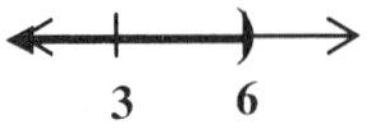

$x \le 3$ and $x < 6$

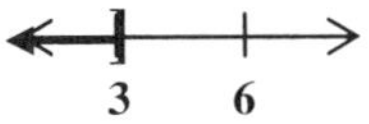

The solution set is $\{x | x \le 3\}$ or $(-\infty, 3]$.

**20.** $x \le 3$

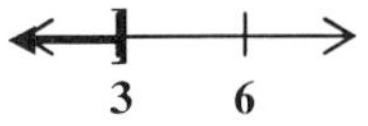

$x < 6$

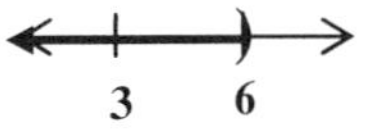

$x \le 3$ or $x < 6$

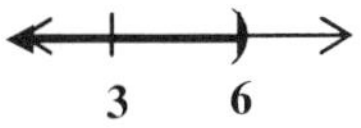

The solution set is $\{x | x < 6\}$ or $(-\infty, 6)$.

**21.** $-2x < -12$ and $x - 3 < 5$

$$\frac{-2x}{-2} > \frac{-12}{-2} \qquad x < 8$$
$$x > 6$$

$x < 8$

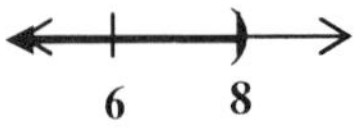

$x > 6$

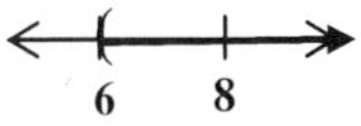

$x < 8$ and $x > 6$

The solution set is $\{x | 6 < x < 8\}$ or $(6, 8)$.

**22.** $5x + 3 \le 18$ and $2x - 7 \le -5$

$$5x \le 15 \qquad 2x \le 2$$
$$x \le 3 \qquad x \le 1$$

$x \le 3$

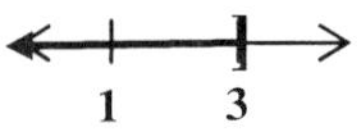

$x \le 1$

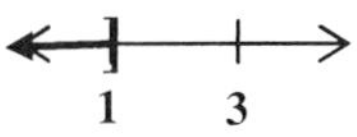

$x \le 3$ and $x \le 1$

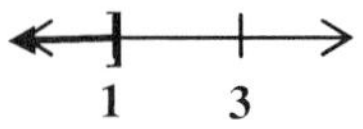

The solution set is $\{x | x \le 1\}$ or $(-\infty, 1]$.

**23.** $2x-5>-1$ and $3x<3$

$2x>4$ $\quad$ $x<1$

$x>2$

$x>2$ 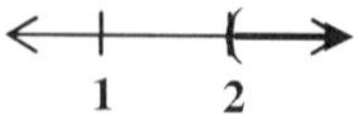

$x<1$ 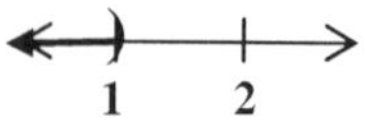

$x>2$ and $x<1$ 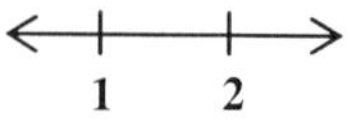

Since the two sets do not intersect, the solution set is $\varnothing$ or $\{\ \}$.

**24.** $2x-5>-1$ or $3x<3$

$2x>4$ $\quad$ $x<1$

$x>2$

$x>2$ 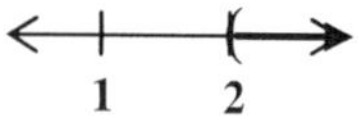

$x<1$ 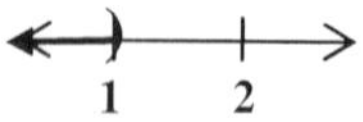

$x>2$ or $x<1$ 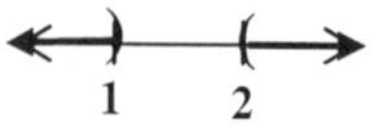

The solution set is $\{x|x<1 \text{ or } x>2\}$ or $(-\infty,1)\cup(2,\infty)$.

**25.** $x+1\le-3$ or $-4x+3<-5$

$x\le-4$ $\quad$ $-4x<-8$

$x>2$

$x\le-4$ 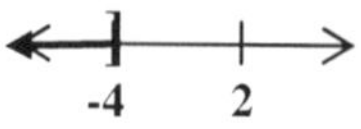

$x>2$ 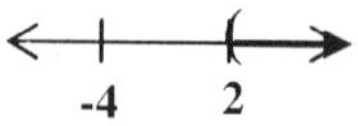

$x\le-4$ or $x>2$ 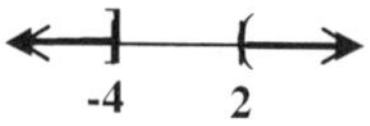

The solution set is $\{x|x\le-4$ and $x>2\}$ or $(-\infty,-4]\cup(2,\infty)$.

**26.** $5x-2\le-22$ or $-3x-2>4$

$5x\le-20$ $\quad$ $-3x>6$

$x\le-4$ $\quad$ $x<-2$

$x\le-4$ 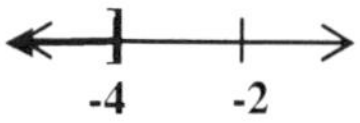

$x<-2$ 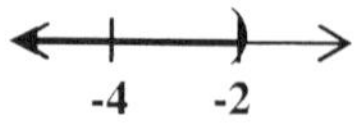

$x\le-4$ or $x<-2$ 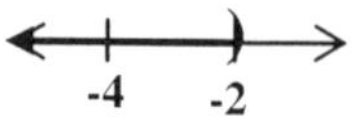

The solution set is $\{x|x<-2\}$ or $(-\infty,-2)$.

**27.** $5x+4\ge-11$ or $1-4x\ge9$

$5x\ge-15$ $\quad$ $-4x\ge8$

$x\ge-3$ $\quad$ $x\le-2$

$x\ge-3$ 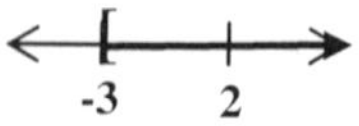

$x\le-2$ 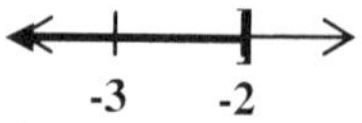

$x\ge-3$ or $x\le-2$ 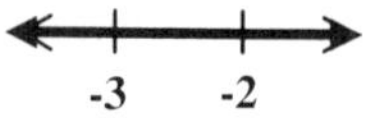

The solution set is $\mathbb{R}$, $(-\infty,\infty)$ or $\{x|x \text{ is a real number}\}$.

**28.**

$$-3 < x+2 \le 4$$
$$-3-2 < x+2-2 \le 4-2$$
$$-5 < x \le 2$$

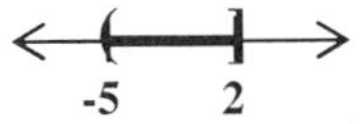

The solution set is $\{x|-5 < x \le 2\}$ or $(-5, 2]$.

**29.**

$$-1 \le 4x+2 \le 6$$
$$-1-2 \le 4x+2-2 \le 6-2$$
$$-3 \le 4x \le 4$$
$$-\frac{3}{4} \le \frac{4x}{4} \le \frac{4}{4}$$
$$-\frac{3}{4} \le x \le 1$$

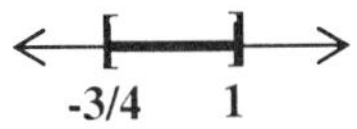

The solution set is $\left\{x \middle| -\frac{3}{4} \le x \le 1\right\}$ or $\left[-\frac{3}{4}, 1\right]$.

**30.** Let $x$ = the grade on the fifth exam

$$80 \le \frac{72+73+94+80+x}{5} < 90$$
$$80 \le \frac{319+x}{5} < 90$$
$$5(80) \le 5\left(\frac{319+x}{5}\right) < 5(90)$$
$$400 \le 319+x < 450$$
$$400-319 \le 319-319+x < 450-319$$
$$81 \le x < 131$$

You need to score at least 81 and less than 131 on the exam to receive a B. In interval notation, the range is $[81,131)$. If the highest score is 100, the range is $[81,100]$.

**31.** $|2x+1| = 7$

$$2x+1=7 \quad \text{or} \quad 2x+1=-7$$
$$2x=6 \qquad 2x=-8$$
$$x=3 \qquad x=-4$$

The solutions are –4 and 3 and the solution set is $\{-4,3\}$.

**32.** $|3x+2| = -5$

There are no values of $x$ for which the absolute value of $3x + 2$ is a negative number. By definition, absolute values are always positive. The solution set is $\varnothing$ or $\{\ \}$.

**33.**

$$2|x-3|-7=10$$
$$2|x-3|=17$$
$$|x-3|=8.5$$
$$x-3=8.5 \quad \text{or} \quad x-3=-8.5$$
$$x=11.5 \qquad x=-5.5$$

The solutions are –5.5 and 11.5 and the solution set is $\{-5.5, 11.5\}$.

**34.** $|4x-3| = |7x+9|$

$$4x-3=7x+9$$
$$-3x-3=9$$
$$-3x=12$$
$$x=-4$$

or

$$4x-3=-7x-9$$
$$11x-3=-9$$
$$11x=-6$$
$$x=-\frac{6}{11}$$

The solutions are $-4$ and $-\frac{6}{11}$ and the solution set is $\left\{-4, -\frac{6}{11}\right\}$.

**35.**

$$|2x+3| \le 15$$
$$-15 \le 2x+3 \le 15$$
$$-15-3 \le 2x+3-3 \le 15-3$$
$$-18 \le 2x \le 12$$
$$-\frac{18}{2} \le \frac{2x}{2} \le \frac{12}{2}$$
$$-9 \le x \le 6$$

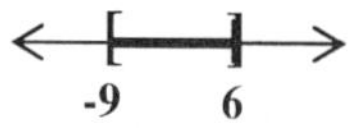


The solution set is $\{x|-9 \le x \le 6\}$ or $[-9,6]$.

**36.**

$$\left|\frac{2x+6}{3}\right| > 2$$

$$\frac{2x+6}{3} < -2 \quad \text{or} \quad \frac{2x+6}{3} > 2$$
$$2x+6 < -6 \qquad 2x+6 > 6$$
$$2x < -12 \qquad 2x > 0$$
$$x < -6 \qquad x > 0$$

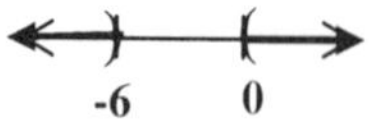


The solution set is $\{x|x < -6$ or $x > 0\}$ or $(-\infty,-6) \cup (0,\infty)$.

**37.**

$$|2x+5|-7 < -6$$
$$|2x+5| < 1$$
$$-1 < 2x+5 < 1$$
$$-1-5 < 2x+5-5 < 1-5$$
$$-6 < 2x < -4$$
$$-3 < x < -2$$

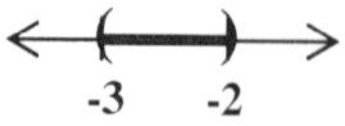


The solution set is $\{x|-3 < x < -2\}$ or $(-3,-2)$.

**38.**

$$|2x-3|+4 \le -10$$
$$|2x-3| \le -14$$

There are no values of $x$ for which the absolute value of $2x-3$ is a negative number. By definition, absolute values are always positive. The solution set is $\varnothing$ or $\{\ \}$.

**39.**

$$|h-6.5| \le 1$$
$$-1 \le h-6.5 \le 1$$
$$5.5 \le h \le 7.5$$

Approximately 90% of the population sleeps between 5.5 hours and 7.5 hours daily, inclusive.

**40.** $3x-4y > 12$

First, find the intercepts to the equation $3x-4y=12$.

Find the $x$–intercept by setting $y$ =0.

$$3x-4y=12$$
$$3x-4(0)=12$$
$$3x=12$$
$$x=4$$

Find the $y$–intercept by setting $x$ =0.

$$3x-4y=12$$
$$3(0)-4y=12$$
$$-4y=12$$
$$y=-3$$

Next, use the origin as a test point.

$$3x-4y>12$$
$$3(0)-4(0)>12$$
$$0>12$$

This is a false statement. This means that the point, $(0,0)$, will not fall in the shaded half-plane.

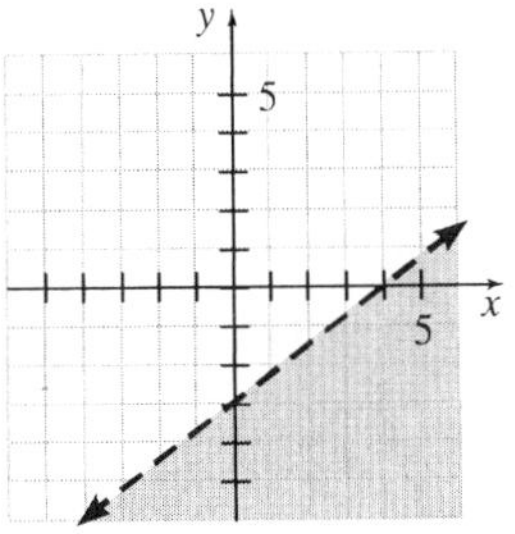


**41.** $x-3y\le 6$

First, find the intercepts to the equation $x-3y=6$.

Find the $x$–intercept by setting $y=0$, find the $y$–intercept by setting $x=0$.

$$x-3y=6 \qquad x-3y=6$$
$$x-3(0)=6 \qquad 0-3y=6$$
$$x=6 \qquad -3y=6$$
$$y=-2$$

Next, use the origin as a test point.

$$0-3(0)\le 6$$
$$0\le 6$$

This is a true statement. This means that the point, $(0,0)$, will fall in the shaded half-plane.

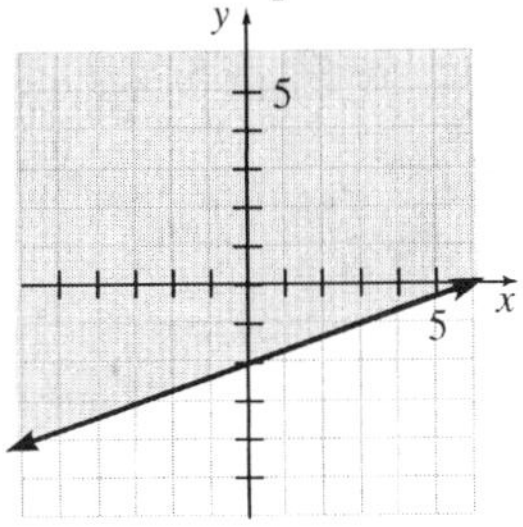


**42.** $y\le -\frac{1}{2}x+2$

Replacing the inequality symbol with an equal sign, we have $y=-\frac{1}{2}x+2$.

Since the equation is in slope-intercept form, use the slope and the intercept to graph the equation. The $y$–intercept is 2 and the slope is $-\frac{1}{2}$.

Next, use the origin as a test point.

$$y\le -\frac{1}{2}x+2$$
$$0\le -\frac{1}{2}(0)+2$$
$$0\le 2$$

This is a true statement. This means that the point $(0,0)$ will fall in the shaded half-plane.

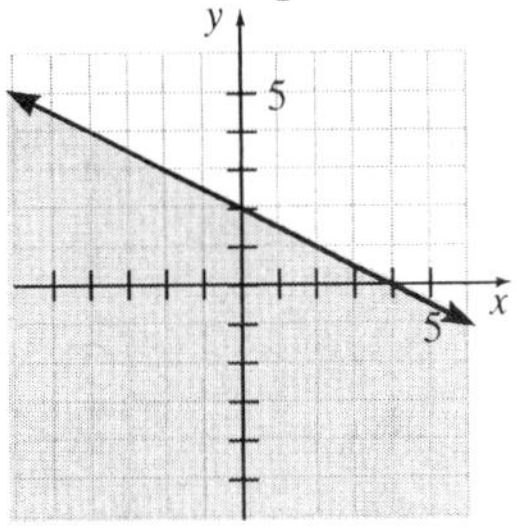


**43.** $y>\frac{3}{5}x$

Replacing the inequality symbol with an equal sign, we have $y=\frac{3}{5}x$.

Since the equation is in slope-intercept form, use the slope and the intercept to graph the equation. The $y$–intercept is 0 and the slope is $\frac{3}{5}$.

Next, we need to find a test point. We cannot use the origin this time, because it lies on the line.

Use $(1,1)$ as a test point.

$$1>\frac{3}{5}(1)$$
$$1>\frac{3}{5}$$

This is a true statement, so we know the point $(1,1)$ lies in the shaded half-plane.

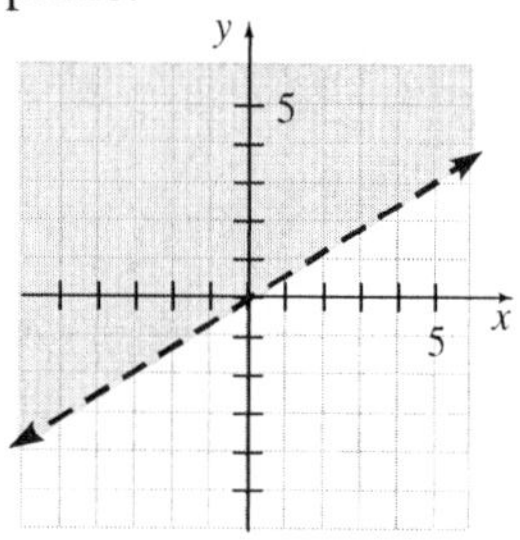

**44.** $x \le 2$

Replacing the inequality symbol with an equal sign, we have $x = 2$. We know that equations of the form $x = a$ are vertical lines with $x$-intercept = $a$.
Next, use the origin as a test point.

$x \le 2$

$0 \le 2$

This is a true statement, so we know the point $(0,0)$ lies in the shaded half-plane.

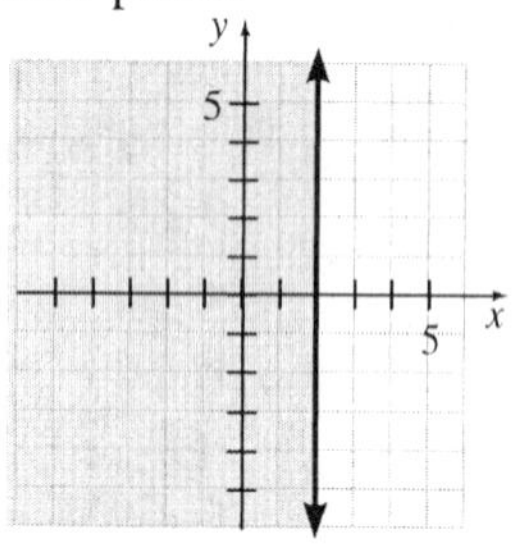

**45.** $y > -3$

Replacing the inequality symbol with an equal sign, we have $y = -3$. We

know that equations of the form $y = b$ are horizontal lines with $y$-intercept = $b$.
Next, use the origin as a test point.

$y > -3$

$0 > -3$

This is a true statement, so we know the point $(0,0)$ lies in the shaded half-plane.

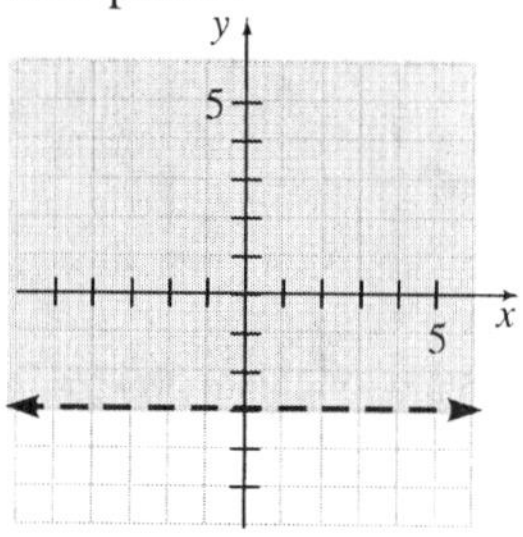

**46.** $3x - y \le 6$

$x + y \ge 2$

First consider $3x - y \le 6$. If we solve for $y$ in $3x - y = 6$, we can graph the line using the slope and the $y$-intercept.

$3x - y = 6$

$-y = -3x + 6$

$y = 3x - 6$

$y$-intercept = $-6$
slope = 3
Now, use the origin as a test point.

$3(0) - 0 \le 6$

$0 \le 6$

This is a true statement. This means that the point $(0,0)$ will fall in the shaded half-plane.

Next consider $x + y \ge 2$. If we solve for $y$ in $x + y = 2$, we can graph using the slope and the $y$-intercept.

$x + y = 2$

$y = -x + 2$

$y$-intercept = 2
slope = $-1$
Now, use the origin as a test point.

$0 + 0 \ge 2$

$0 \ge 2$

This is a false statement. This means

that the point $(0,0)$ will not fall in the shaded half-plane.
Next, graph each of the inequalities. The solution to the system is the intersection of the shaded half-planes.

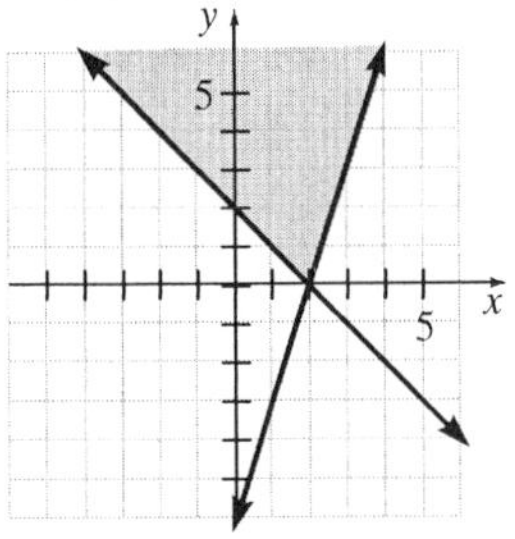


**47.** $y < -x + 4$

$y > x - 4$

First consider $y < -x + 4$. Change the inequality symbol to an equal sign. The line $y = -x + 4$ is in slope-intercept form and can be graphed using the slope and the $y$-intercept.
$y$-intercept = 4
slope = −1
Now, use the origin as a test point.

$0 < -0 + 4$

$0 < 4$

This is a true statement. This means that the point $(0,0)$ will fall in the shaded half-plane.
Next consider $y > x - 4$. Change the inequality symbol to an equal sign. The line $y = x - 4$ is in slope-intercept form and can be graphed using the slope and the $y$-intercept.
$y$-intercept = −4
slope = 1
Now, use the origin as a test point.

$0 > 0 - 4$

$0 > 4$

This is a false statement. This means that the point $(0,0)$ will not fall in the shaded half-plane.
Next, graph each of the inequalities. The solution to the system is the intersection of the shaded half-planes.

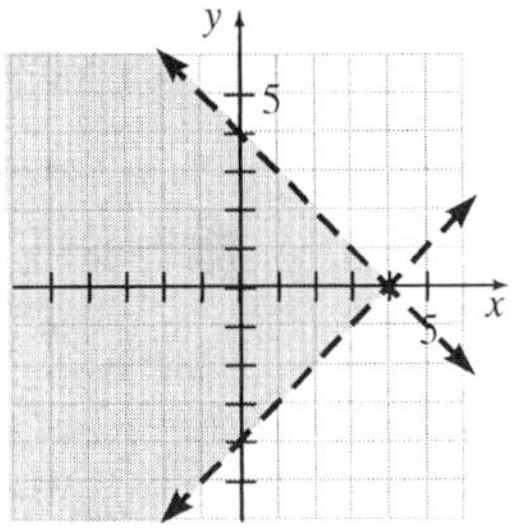


**48.** $-3 \le x < 5$

Rewrite the three part inequality as two separate inequalities. We have $-3 \le x$ and $x < 5$. We replace the inequality symbols with equal signs and obtain $-3 = x$ and $x = 5$. Equations of the form $x = a$ are vertical lines with $x$-intercept = $a$. We know the shading in the graph will be between $x = -3$ and $x = 5$ because in the original inequality we see that $x$ lies between −3 and 5.

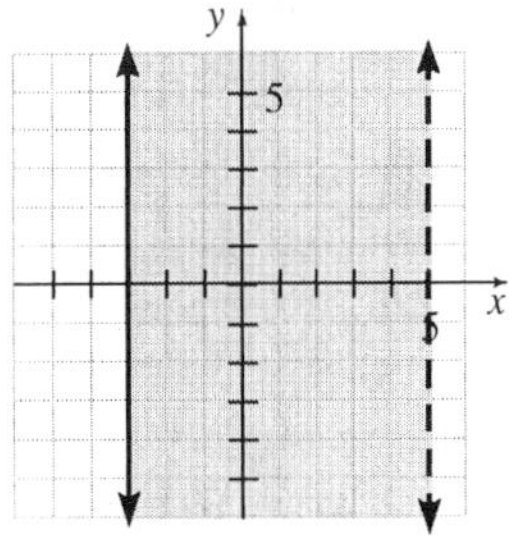


**49.** $-2 < y \le 6$

Rewrite the three part inequality as two separate inequalities. We have $-2 < y$ and $y \le 6$. We replace the inequality symbols with equal signs and obtain $-2 = y$ and $y = 6$. Equations of the form $y = b$ are vertical lines with $y$-intercept = $b$. We know the shading in the graph will be between $y = -2$ and $y = 6$

because in the original inequality we see that $y$ lies between –2 and 6.

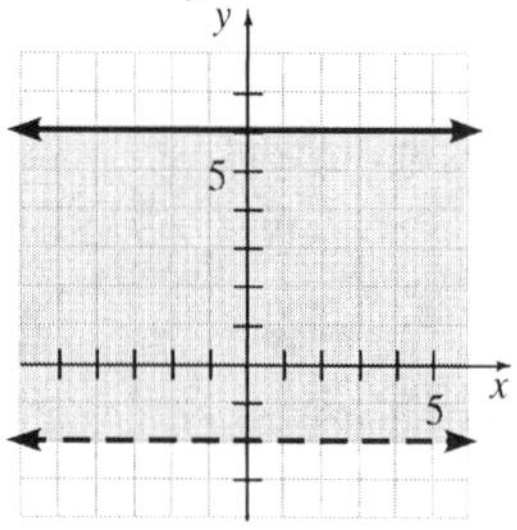

**50.** $x \geq 3$

$y \leq 0$

First consider $x \geq 3$. Change the inequality symbol to an equal sign and

we obtain the vertical line $x = 3$. Because we have $x \geq 3$, we know the shading is to the right of the line $x = 3$.

Next consider $y \leq 0$. Change the inequality symbol to an equal sign and we obtain the horizontal line $y = 0$. (Recall that this is the equation of the $x$–axis.) Because we have $y \leq 0$, we know that the shading will be below the $x$–axis.

Next, graph each of the inequalities. The solution to the system is the intersection of the shaded half-planes.

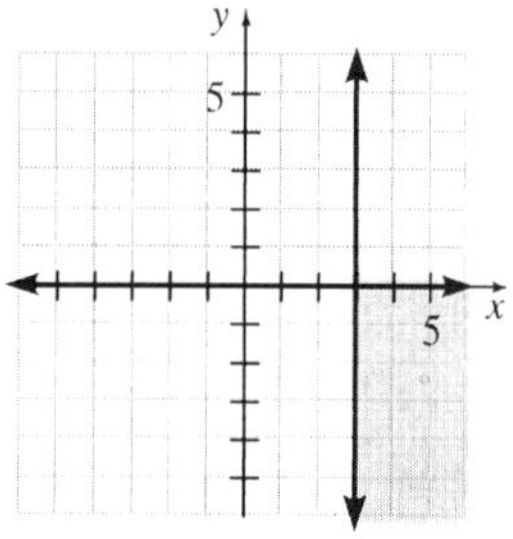

**51.** $2x - y > -4$

$x \geq 0$

First consider $2x - y > -4$. Replace the inequality symbol with an equal sign and we have $2x - y = -4$. Solve for $y$ to obtain slope-intercept form.

$$2x - y = -4$$
$$-y = -2x - 4$$
$$y = 2x + 4$$

$y$-intercept = 4
slope = 2

Now, use the origin as a test point.

$$2x - y > -4$$
$$2(0) - 0 > -4$$
$$0 > -4$$

This is a false statement. This means that the point $(0,0)$ will not fall in the shaded half-plane.

Next consider $x \geq 0$. Change the inequality symbol to an equal sign and we obtain the horizontal line $x = 0$. (Recall that this is the equation of the $y$–axis.) Because we have $x \geq 0$, we know that the shading will be above the $y$–axis.

Next, graph each of the inequalities. The solution to the system is the intersection of the shaded half-planes.

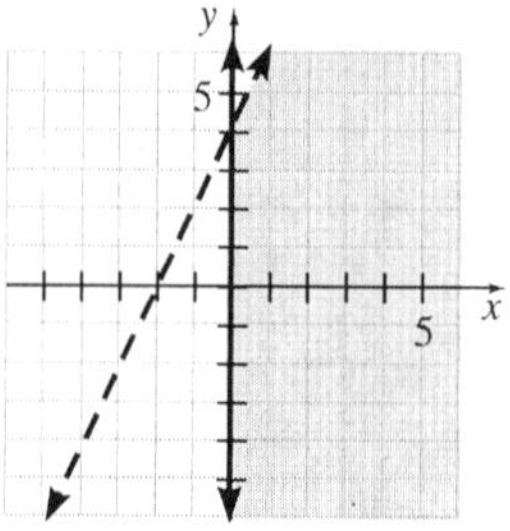

**52.** $x + y \le 6$

$y \ge 2x - 3$

First consider $x + y \le 6$. Replace the inequality symbol with an equal sign and we have $x + y = 6$. Solve for $y$ to obtain slope-intercept form.

$x + y = 6$

$y = -x + 6$

$y$-intercept = 6
slope = −1

Now, use the origin as a test point.

$0 + 0 \le 6$

$0 \le 6$

This is a true statement. This means that the point $(0,0)$ will fall in the shaded half-plane.

Next consider $y \ge 2x - 3$. Replace the inequality symbol with an equal sign and we have $y = 2x - 3$. The equation is in slope-intercept form, so we can use the slope and the $y$-intercept to graph the line.

$y$-intercept = −3
slope = 2

Now, use the origin as a test point.

$y \ge 2x - 3$

$0 \ge 2(0) - 3$

$0 \ge -3$

This is a true statement. This means that the point $(0,0)$ will fall in the shaded half-plane.

Next, graph each of the inequalities. The solution to the system is the intersection of the shaded half-planes.

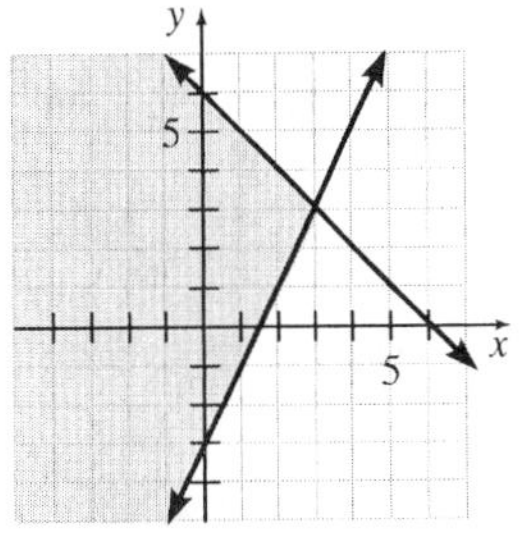


**53.** $3x + 2y \ge 4$

$x - y \le 3$

$x \ge 0, \quad y \ge 0$

First consider $3x + 2y \ge 4$. Replace the inequality symbol with an equal sign and we have $3x + 2y = 4$. Solve for $y$ to obtain slope-intercept form.

$3x + 2y = 4$

$2y = -3x + 4$

$y = -\frac{3}{2}x + 2$

$y$-intercept = 2 slope = $-\frac{3}{2}$

Now, use the origin as a test point.

$3x + 2y \ge 4$

$3(0) + 2(0) \ge 4$

$0 \ge 4$

This is a false statement. This means that the point $(0,0)$ will not fall in the shaded half-plane.

Now consider $x - y \le 3$. Replace the inequality symbol with an equal sign and we have $x - y = 3$. Solve for $y$ to obtain slope-intercept form.

$x - y = 3$

$-y = -x + 3$

$y = x - 3$

$y$-intercept = −3 slope = 1

Now, use the origin as a test point.

$$x - y \le 3$$
$$0 - 0 \le 3$$
$$0 \le 3$$

This is a true statement. This means that the point $(0,0)$ will fall in the shaded half-plane.

Now consider the inequalities $x \ge 0$ and $y \ge 0$. The inequalities mean that both $x$ and $y$ will be positive. This means that we only need to consider quadrant I.

Next, graph each of the inequalities. The solution to the system is the intersection of the shaded half-planes.

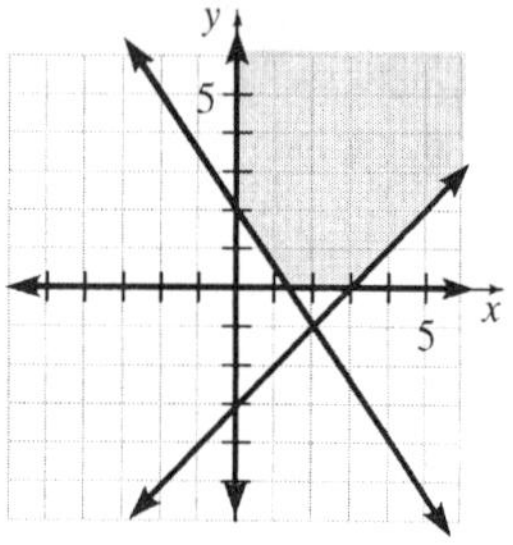


**54.** $2x - y > 2$

$2x - y < -2$

First consider $2x - y > 2$. Replace the inequality symbol with an equal sign and we have $2x - y = 2$. Solve for $y$ to obtain slope-intercept form.

$$2x - y = 2$$
$$-y = -2x + 2$$
$$y = 2x - 2$$

$y$-intercept = −2   slope = 2

Now, use the origin as a test point.

$$2x - y > 2$$
$$2(0) - 0 > 2$$
$$0 > 2$$

This is a false statement. This means that the point $(0,0)$ will not fall in the shaded half-plane.

Now consider $2x - y < -2$. Replace the inequality symbol with an equal sign and we have $2x - y = -2$. Solve for $y$ to obtain slope-intercept form.

$$2x - y = -2$$
$$-y = -2x - 2$$
$$y = 2x + 2$$

$y$-intercept = 2   slope = 2

Now, use the origin as a test point.

$$2x - y < -2$$
$$2(0) - 0 < -2$$
$$0 < -2$$

This is a false statement. This means that the point $(0,0)$ will not fall in the shaded half-plane.

Next, graph each of the inequalities. The solution to the system is the intersection of the shaded half-planes.

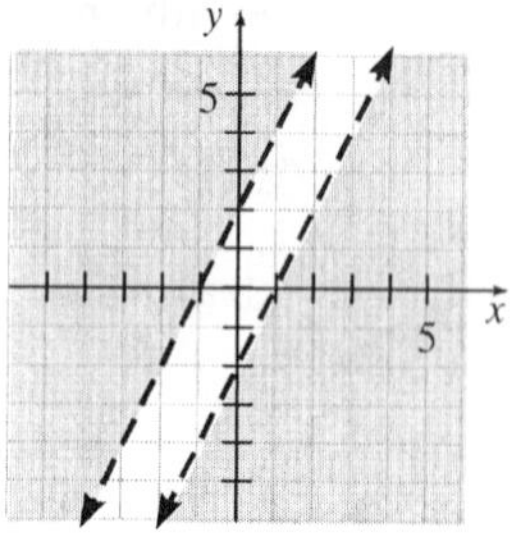


The graphs of the inequalities do not intersect, so there is no solution. The solution set is $\varnothing$ or $\{ \ \}$.

**55.**

| Corner $(x, y)$ | Objective Function $z = 2x + 3y$ |
|---|---|
| $(1, 0)$ | $z = 2x + 3y$<br>$= 2(1) + \cancel{3(0)} = 2$ |
| $(4, 0)$ | $z = 2x + 3y$<br>$= 2(4) + \cancel{3(0)} = 8$ |
| $(2, 2)$ | $z = 2x + 3y$<br>$= 2(2) + 3(2)$<br>$= 4 + 6 = 10$ |
| $\left(\frac{1}{2}, \frac{1}{2}\right)$ | $z = 2x + 3y$<br>$= 2\left(\frac{1}{2}\right) + 3\left(\frac{1}{2}\right)$<br>$= \frac{2}{2} + \frac{3}{2} = \frac{5}{2}$ |

The maximum value is 10 and the minimum is 2.

**56.** Objective Function: $z = 2x + 3y$

Constraints: $x \geq 0, \quad y \geq 0$

$x + y \leq 8$

$3x + 2y \geq 6$

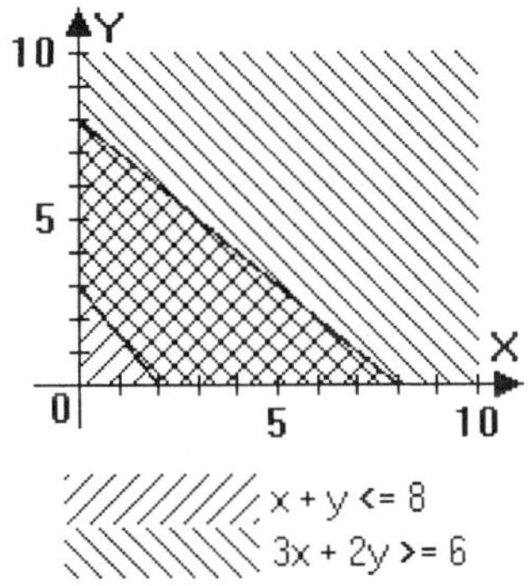


Using the graph, find the value of the objective function at each of the corner points.

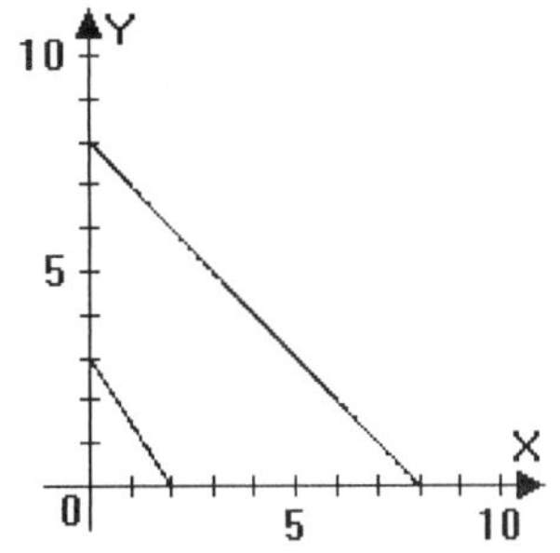


| Corner $(x, y)$ | Objective Function $z = 2x + 3y$ |
|---|---|
| $(2, 0)$ | $z = 2x + 3y$<br>$= 2(2) + \cancel{3(0)} = 4$ |
| $(8, 0)$ | $z = 2x + 3y$<br>$= 2(8) + \cancel{3(0)} = 16$ |
| $(0, 8)$ | $z = 2x + 3y$<br>$= \cancel{2(0)} + 3(8) = 24$ |
| $(0, 3)$ | $z = 2x + 3y$<br>$= \cancel{2(0)} + 3(3) = 9$ |

The maximum of 24 occurs at the point (0, 8).

**57.** Objective Function: $z = x + 4y$

Constraints: $0 \leq x \leq 5$

$0 \leq y \leq 7$

$x + y \geq 3$

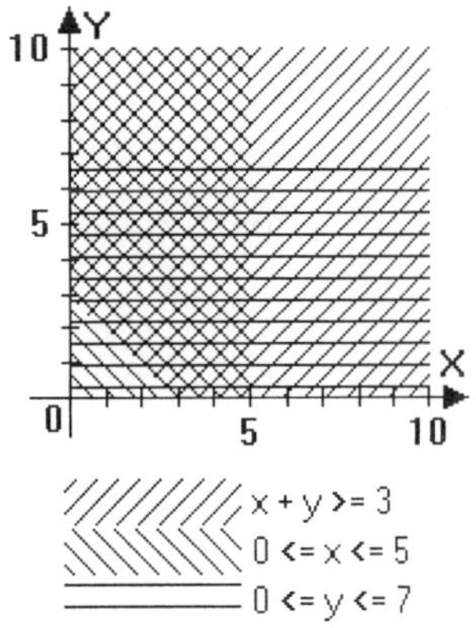

Using the graph, find the value of the objective function at each of the corner points.

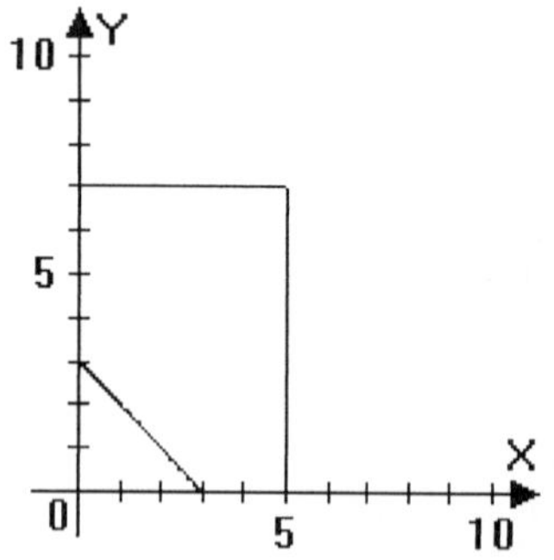


| Corner $(x, y)$ | Objective Function $z = x + 4y$ |
|---|---|
| (3, 0) | $z = x + 4y$ <br> $= 3 + 4(0) = 3$ |
| (5, 0) | $z = x + 4y$ <br> $= 5 + 4(0) = 5$ |
| (5, 7) | $z = x + 4y$ <br> $= 5 + 4(7)$ <br> $= 5 + 28 = 33$ |
| (0, 7) | $z = x + 4y$ <br> $= 0 + 4(7) = 28$ |
| (0, 3) | $z = x + 4y$ <br> $= 0 + 4(3) = 12$ |

The maximum of 33 occurs at the point (5, 7).

**58.** Objective Function: $z = 5x + 6y$

Constraints: $x \geq 0,\ y \geq 0$

$y \leq x$

$2x + y \leq 12$

$2x + 3y \geq 6$

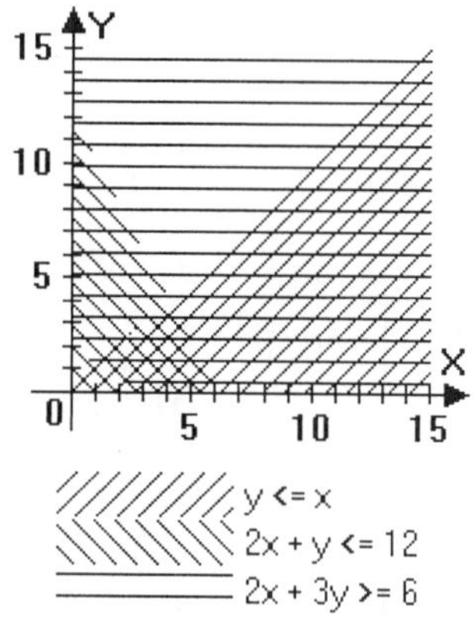


Using the graph, find the value of the objective function at each of the corner points.

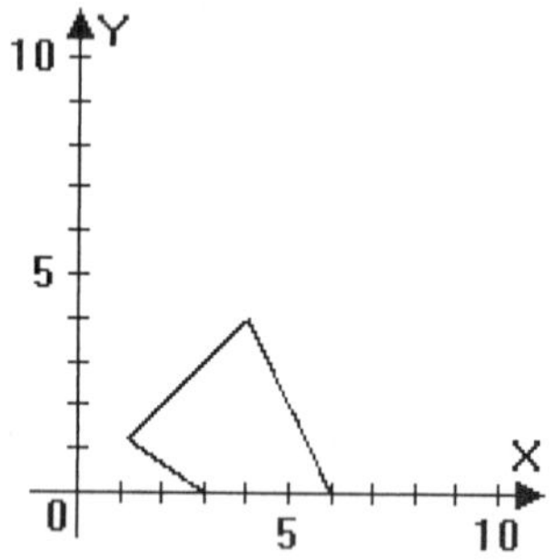


| Corner $(x, y)$ | Objective Function $z = 5x + 6y$ |
|---|---|
| (3, 0) | $z = 5x + 6y$ <br> $= 5(3) + 6(0) = 15$ |
| (6, 0) | $z = 5x + 6y$ <br> $= 5(6) + 6(0) = 30$ |
| (4, 4) | $z = 5x + 6y$ <br> $= 5(4) + 6(4)$ <br> $= 20 + 24 = 44$ |
| (1.2, 1.2) | $z = 5x + 6y$ <br> $= 5(1.2) + 6(1.2)$ <br> $= 6 + 7.2 = 13.2$ |

The maximum of 44 occurs at the point (4, 4).

**59.** **a.** The objective function is $z = 500x + 350y$.

**b.** The paper constraint is $x + y \le 200$.

The minimum writing paper constraint is $x \ge 10$.

The minimum newsprint constraint is $y \ge 80$.

**c.**

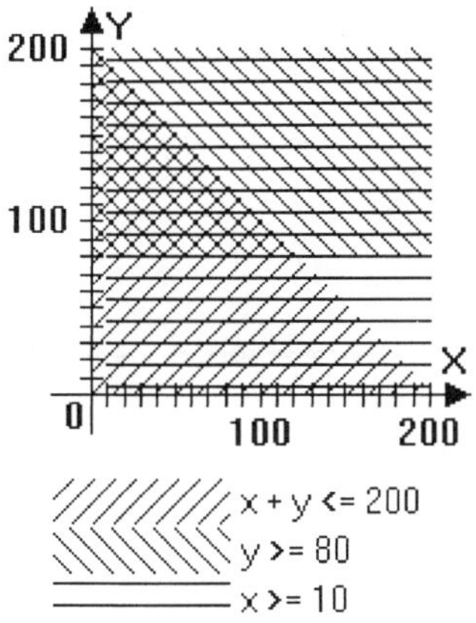


**d.**

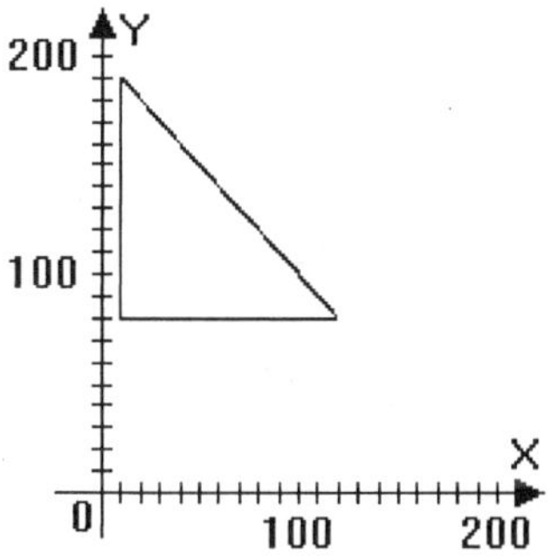


| Corner $(x, y)$ | Objective Function $z = 500x + 350y$ |
|---|---|
| (10, 80) | $z = 500x + 350y$ <br> $= 500(10) + 350(80)$ <br> $= 5000 + 28000$ <br> $= 33000$ |
| (120, 80) | $z = 500x + 350y$ <br> $= 500(120) + 350(80)$ <br> $= 60000 + 28000$ <br> $= 88000$ |
| (10, 190) | $z = 500x + 350y$ <br> $= 500(10) + 350(190)$ <br> $= 5000 + 66500$ <br> $= 71500$ |

**e.** The company will make the greatest profit by producing 120 units of writing paper and 80 units of newsprint each day. The maximum daily profit is \$88,000.

**60.** Let $x$ = the number of model $A$ produced
Let $y$ = the number of model $B$ produced
The objective function is $z = 25x + 40y$.
The cutting department labor constraint is $0.9x + 1.8y \le 864$.
The assembly department labor constraint is $0.8x + 1.2y \le 672$.
We also know that $x$ and $y$ are either zero or a positive number. We cannot have a negative number of units produced.
Next, graph the constraints.

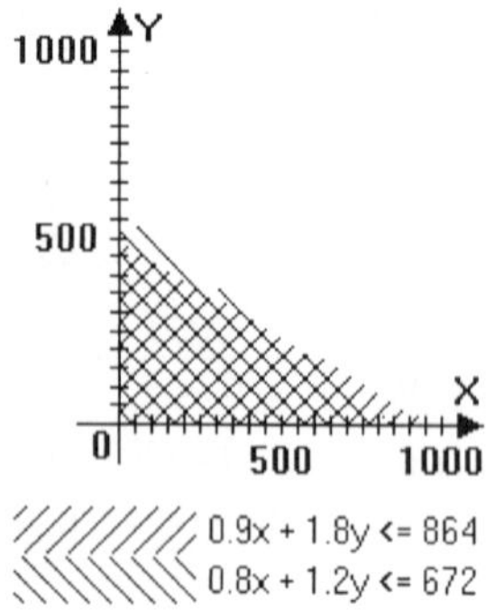


Using the graph, find the value of the objective function at each of the corner points.

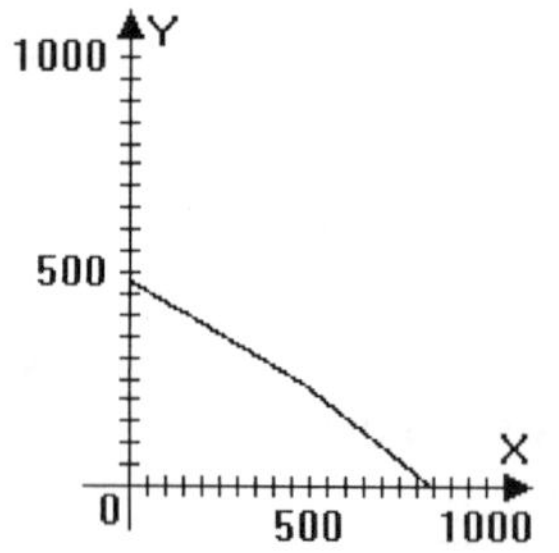


| Corner $(x, y)$ | Objective Function $z = 25x + 40y$ |
|---|---|
| (0, 0) | $25x+40y$ <br> $= 25(0) + 40(0)$ <br> $=0$ |
| (840, 0) | $25x+40y$ <br> $= 25(840) + 40(0)$ <br> $= 21000$ |
| (480, 240) | $25x+40y$ <br> $= 25(480)+40(240)$ <br> $= 12000+9600$ <br> $= 21600$ |
| (0, 480) | $25x+40y$ <br> $= 25(0) + 40(480)$ <br> $= 19200$ |

The maximum of 21,600 occurs at the point (480, 240). This means that to maximize the profit, 480 of model *A* and 240 of model *B* should be manufactured monthly. This would result in a profit of $21,600.

## Chapter 4 Test

**1.** $[-3,2)$

$\{x|-3 \le x < 2\}$

**2.** $(-\infty,-1]$

$\{x|x \le -1\}$

**3.**
$$3(x+4) \ge 5x-12$$
$$3x+12 \ge 5x-12$$
$$-2x+12 \ge -12$$
$$-2x \ge -24$$
$$\frac{-2x}{-2} \le \frac{-24}{-2}$$
$$x \le 12$$

The solution set is $\{x|x \le 12\}$ or $(-\infty,12]$.

**4.**

$$\frac{x}{6}+\frac{1}{8}\le\frac{x}{2}-\frac{3}{4}$$

$$24\left(\frac{x}{6}\right)+24\left(\frac{1}{8}\right)\le 24\left(\frac{x}{2}\right)-24\left(\frac{3}{4}\right)$$

$$4x+3\le 12x-6(3)$$

$$4x+3\le 12x-18$$

$$-8x+3\le -18$$

$$-8x\le -21$$

$$\frac{-8x}{-8}\ge\frac{-21}{-8}$$

$$x\ge\frac{21}{8}$$

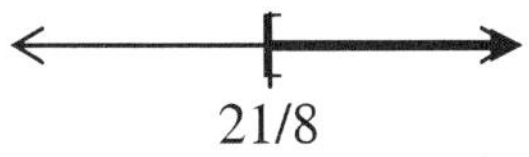


The solution set is $\left\{x \middle| x\ge\frac{21}{8}\right\}$ or $\left[\frac{21}{8},\infty\right)$.

**5.** Let $x$ = the number of local calls
The monthly cost using Plan $A$ is $C_A = 25$.
The monthly cost using Plan $B$ is $C_B = 13+0.06x$.
For Plan $A$ to be better deal, it must cost less than Plan $B$.

$$C_A < C_B$$

$$25 < 13+0.06x$$

$$12 < 0.06x$$

$$200 < x$$

$$x > 200$$

Plan $A$ is a better deal when more than 200 local calls are made per month.

**6.** $\{2,4,6,8,10\}\cap\{4,6,12,14\}=\{4,6\}$

**7.** $\{2,4,6,8,10\}\cup\{4,6,12,14\}$
$=\{2,4,6,8,10,12,14\}$

**8.** $2x+4<2$ and $x-3>-5$

$2x<-2$ $\quad$ $x>-2$

$x<-1$

$x<-1$

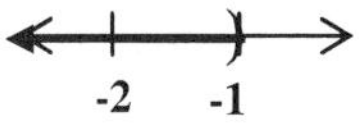


$x>-2$

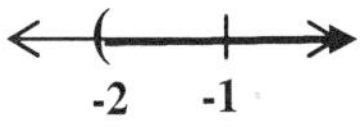


$x<-1$ and $x>-2$

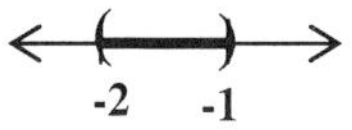


The solution set is $\{x|-2<x<-1\}$ or $(-2,-1)$.

**9.** $x+6\ge 4$ and $2x+3\ge -2$

$x\ge -2$ $\quad$ $2x\ge -5$

$x\ge -\frac{5}{2}$

$x\ge -2$

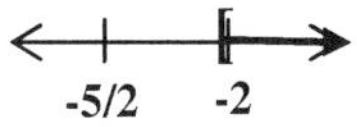


$x\ge -\frac{5}{2}$

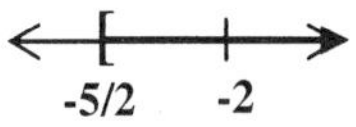


$x\ge -2$ and $x\ge -\frac{5}{2}$

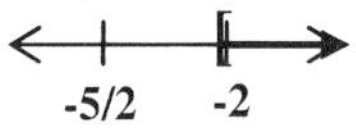


The solution set is $\{x|x\ge -2\}$ or $[-2,\infty)$.

**10.** $2x-3<5$ or $3x-6\le 4$

$2x<8 \qquad 3x\le 10$

$x<4 \qquad x\le \dfrac{10}{3}$

$x<4$

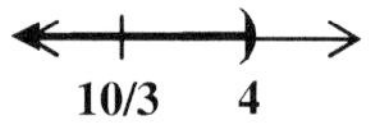


$x\le \dfrac{10}{3}$

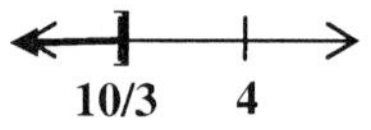


$x<4$ or $x\le \dfrac{10}{3}$

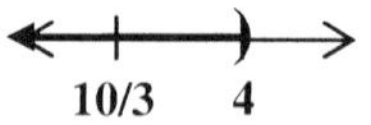


The solution set is $\{x|x<4\}$ or $(-\infty,4)$.

**11.** $x+3\le -1$ or $-4x+3<-5$

$x\le -4 \qquad -4x<-8$

$x>2$

$x\le -4$

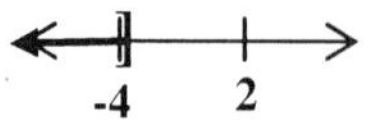


$x>8$

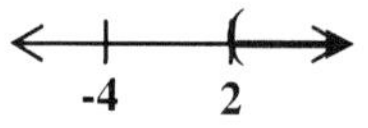


$x\le -4$ or $x>2$

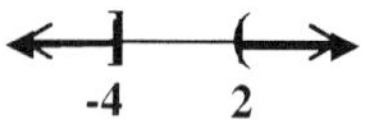


The solution set is $\{x|x\le -4$ or $x>2\}$ or $(-\infty,-4]\cup(2,\infty)$.

**12.**

$$-3\le \frac{2x+5}{3}<6$$

$$3(-3)\le 3\left(\frac{2x+5}{3}\right)<3(6)$$

$$-9\le 2x+5<18$$

$$-9-5\le 2x+5-5<18-5$$

$$-14\le 2x<13$$

$$-7\le x<\frac{13}{2}$$

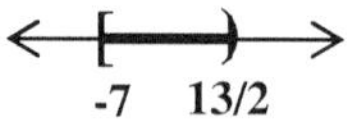


The solution set is $\left\{x\middle|-7\le x<\dfrac{13}{2}\right\}$ or $\left[-7,\dfrac{13}{2}\right)$.

**13.** $|5x+3|=7$

$5x+3=7$ or $5x+3=-7$

$5x=4 \qquad 5x=-10$

$x=\dfrac{4}{5} \qquad x=-2$

The solutions are $-2$ and $\dfrac{4}{5}$ and the solution set is $\left\{-2,\dfrac{4}{5}\right\}$.

**14.** $|6x+1|=|4x+15|$

$$6x+1=4x+15$$

$$2x+1=15$$

$$2x=14$$

$$x=7$$

or

$$6x+1=-(4x+15)$$
$$6x+1=-4x-15$$
$$10x+1=-15$$
$$10x=-16$$
$$x=-\frac{16}{10}=-\frac{8}{5}$$

The solutions are $-\frac{8}{5}$ and 7 and the solution set is $\left\{-\frac{8}{5},7\right\}$.

**15.** $$|2x-1|<7$$
$$-7<2x-1<7$$
$$-7+1<2x-1+1<7+1$$
$$-6<2x<8$$
$$-3<x<4$$

The solution set is $\{x|-3<x<4\}$ or $(-3,4)$.

**16.** $$|2x-3|\geq 5$$
$$2x-3\leq -5 \quad \text{or} \quad 2x-3\geq 5$$
$$2x\leq -2 \qquad 2x\geq 8$$
$$x\leq -1 \qquad x\geq 4$$

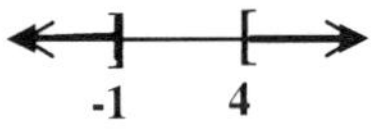


The solution set is $\{x|x\leq -1$ or $x\geq 4\}$ or $(-\infty,-1]\cup[4,\infty)$.

**17.** $$|T-74|\leq 8$$
$$-8\leq T-74\leq 8$$
$$-8+74\leq T-74+74\leq 8+74$$
$$66\leq T\leq 82$$

The monthly average temperature for Miami, Florida is between 66°F and 82°F, inclusive.

**18.** $3x-2y<6$

First, find the intercepts to the equation $3x-2y=6$.

Find the $x$–intercept by setting $y=0$.
$$3x-2y=6$$
$$3x-2(0)=6$$
$$3x=6$$
$$x=2$$

Find the $y$–intercept by setting $x=0$.
$$3x-2y=6$$
$$3(0)-2y=6$$
$$-2y=6$$
$$y=-3$$

Next, use the origin as a test point.
$$3x-2y<6$$
$$3(0)-2(0)<6$$
$$0<6$$

This is a true statement. This means that the point will fall in the shaded half-plane.

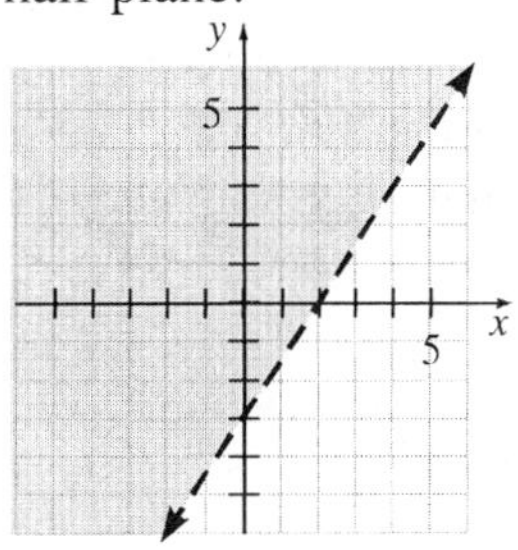

**19.** $y \geq \frac{1}{2}x - 1$

Replacing the inequality symbol with an equal sign, we have $y = \frac{1}{2}x - 1$. The equation is in slope-intercept form, so graph the line using the slope and the $y$-intercept.

$y$-intercept = −1  slope = $\frac{1}{2}$

Now, use the origin, $(0,0)$, as a test point.

$y \geq \frac{1}{2}x - 1$

$0 \geq \frac{1}{2}(0) - 1$

$0 \geq -1$

This is a true statement. This means that the point will fall in the shaded half-plane.

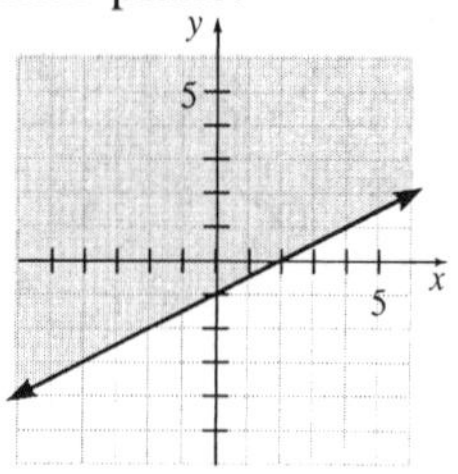

**20.** $y \leq -1$

Replacing the inequality symbol with an equal sign, we have $y = -1$. Equations of the form $y = b$ are horizontal lines with $y$-intercept = $b$, so this is a horizontal line at $y = -1$.

Next, use the origin as a test point.

$y \leq -1$

$0 \leq -1$

This is a false statement, so we know the point $(0,0)$ does not lie in the shaded half-plane.

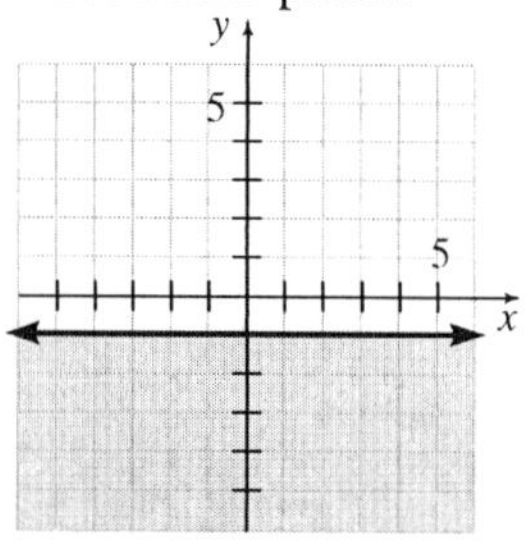

**21.** $x + y \geq 2$

$x - y \geq 4$

First consider $x + y \geq 2$. If we solve for $y$ in $x + y = 2$, we can graph the line using the slope and the $y$-intercept.

$x + y = 2$

$y = -x + 2$

$y$-intercept = 2  slope = −1

Now, use the origin as a test point.

$x + y \geq 2$

$0 + 0 \geq 2$

$0 \geq 2$

This is a false statement. This means that the point will not fall in the shaded half-plane.

Next consider $x - y \geq 4$. If we solve for $y$ in $x - y = 4$, we can graph using the slope and the $y$-intercept.

$x - y = 4$

$-y = -x + 4$

$y = x - 4$

$y$-intercept = −4  slope = 1

Now, use the origin as a test point.

$x - y \geq 4$

$0 - 0 \geq 4$

$0 \geq 4$

This is a false statement. This means that the point will not fall in the shaded half-plane.

Next, graph each of the inequalities.

The solution to the system is the intersection of the shaded half-planes.

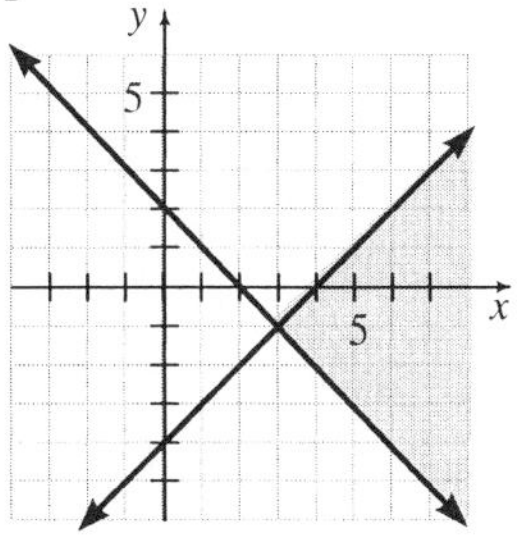

**22.** $3x + y \le 9$

$2x + 3y \ge 6$

$x \ge 0, \quad y \ge 0$

First consider $3x + y \le 9$. If we solve for $y$ in $3x + y = 9$, we can graph the line using the slope and the $y$-intercept.

$3x + y = 9$

$y = -3x + 9$

$y$-intercept = 9 slope = −3

Now, use the origin as a test point.

$3x + y \le 9$

$3(0) + 0 \le 9$

$0 \le 9$

This is a true statement. This means that the point will fall in the shaded half-plane.

Next consider $2x + 3y \ge 6$. If we solve for $y$ in $2x + 3y = 6$, we can graph using the slope and the $y$-intercept.

$2x + 3y = 6$

$3y = -2x + 6$

$y = -\frac{2}{3}x + 2$

$y$-intercept = 2 slope = $-\frac{2}{3}$

Now, use the origin as a test point.

$2x + 3y \ge 6$

$2(0) + 3(0) \ge 6$

$0 \ge 6$

This is a false statement. This means that the point will not fall in the shaded half-plane.

Next consider the inequalities $x \ge 0$ and $y \ge 0$. When $x$ and $y$ are both positive, we are only concerned with the first quadrant of the coordinate system.

Graph each of the inequalities. The solution to the system is the intersection of the shaded half-planes.

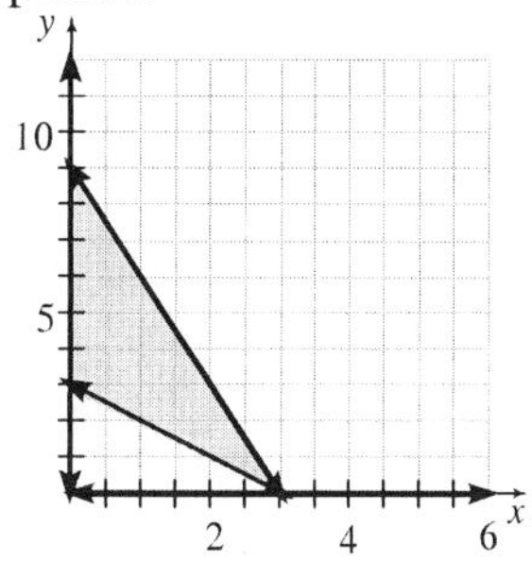

**23.** $-2 < x \le 4$

Rewrite the three part inequality as two separate inequalities. We have $-2 < x$ and $x \le 4$. We replace the inequality symbols with equal signs and obtain $-2 = x$ and $x = 4$. Equations of the form $x = a$ are vertical lines with $x$-intercept = $a$. We know the shading will be between $x = -2$ and $x = 4$ because in the original inequality we see that $x$ lies between −2 and 4.

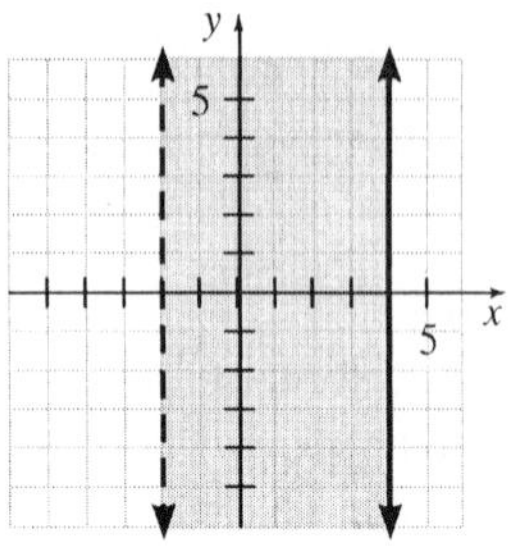


**24.** Objective Function: $z = 3x + 5y$

Constraints: $x \geq 0,\ y \geq 0$

$x + y \leq 6$

$x \geq 2$

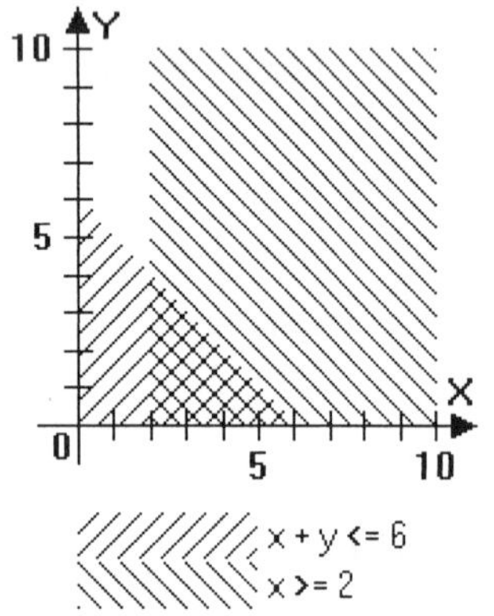


Using the graph, find the value of the objective function at each of the corner points.

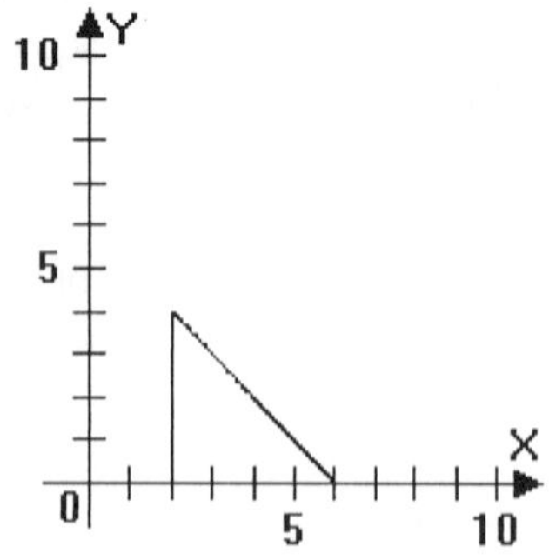


| Corner $(x, y)$ | Objective Function $z = 3x + 5y$ |
|---|---|
| (2, 0) | $z = 3x + 5y$ <br> $= 3(2) + 5(0) = 6$ |
| (6, 0) | $z = 3x + 5y$ <br> $= 3(6) + 5(0) = 18$ |
| (2, 4) | $z = 3x + 5y$ <br> $= 3(2) + 5(4)$ <br> $= 6 + 20 = 26$ |

The maximum of 26 occurs at the point (2, 4).

**25.** Let $x$ = the number of regular jet skis produced
Let $y$ = the number of deluxe jet skis produced
The objective function is
$z = 200x + 250y.$
The regular jet ski demand constraint is $x \geq 50.$
The deluxe jet ski demand constraint is $y \geq 75.$
The quality constraint is $x + y \leq 150.$

We also know that $x$ and $y$ are either zero or a positive number. We cannot have a negative number of units produced.
Next, graph the constraints.

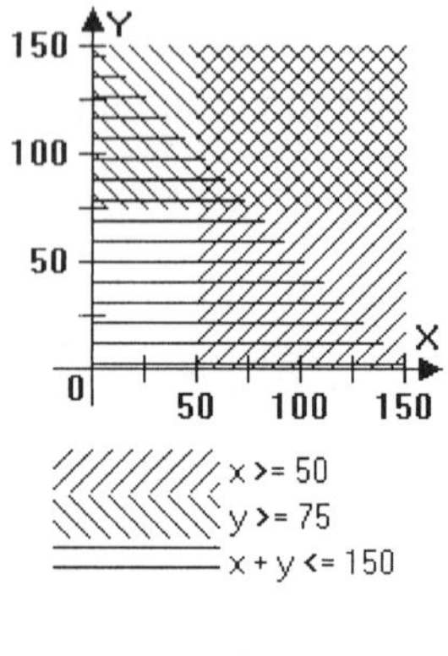

Using the graph, find the value of the objective function at each of the corner points.

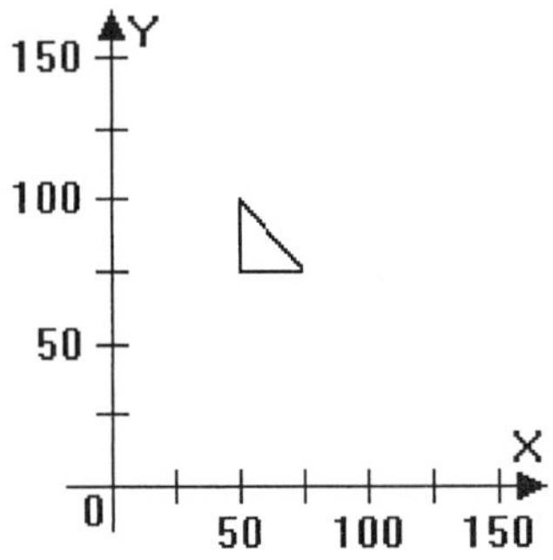


| Corner $(x, y)$ | Objective Function $z = 200x + 250y$ |
|---|---|
| (50, 75) | $z = 200x + 250y$<br>$= 200(50) + 250(75)$<br>$= 10000 + 18750$<br>$= 28750$ |
| (75, 75) | $z = 200x + 250y$<br>$= 200(75) + 250(75)$<br>$= 15000 + 18750$<br>$= 33750$ |
| (50, 100) | $z = 200x + 250y$<br>$= 200(50) + 250(100)$<br>$= 10000 + 25000$<br>$= 35000$ |

The maximum of 35,000 occurs at the point (50, 100). This means that to maximize the profit, 50 regular jet skis and 100 deluxe jet skis should be manufactured weekly. This would result in a profit of $35,000.

## Chapters 1-4
## Cumulative Review Exercises

**1.**
$$5(x+1)+2 = x-3(2x+1)$$
$$5x+5+2 = x-6x-3$$
$$5x+7 = -5x-3$$
$$10x+7 = -3$$
$$10x = -10$$
$$x = -1$$
The solution is $-1$ and the solution set is $\{-1\}$.

**2.**
$$\frac{2(x+6)}{3} = 1+\frac{4x-7}{3}$$
$$3\left(\frac{2(x+6)}{3}\right) = 3(1)+3\left(\frac{4x-7}{3}\right)$$
$$2(x+6) = 3+4x-7$$
$$2x+12 = 4x-4$$
$$-2x+12 = -4$$
$$-2x = -16$$
$$x = 8$$
The solution is 8 and the solution set is $\{8\}$.

**3.**
$$\frac{-10x^2y^4}{15x^7y^{-3}} = \frac{-10}{15}x^{2-7}y^{4-(-3)}$$
$$= -\frac{2}{3}x^{-5}y^7 = -\frac{2y^7}{3x^5}$$

**4.**
$$f(x) = x^2-3x+4$$
$$f(-3) = (-3)^2-3(-3)+4$$
$$= 9+9+4 = 22$$
$$f(2a) = (2a)^2-3(2a)+4$$
$$= 4a^2-6a+4$$

**5.** $f(x) = 3x^2 - 4x + 1$

$g(x) = x^2 - 5x - 1$

$(f-g)(x) = f(x) - g(x)$

$= (3x^2 - 4x + 1) - (x^2 - 5x - 1)$

$= 3x^2 - 4x + 1 - x^2 + 5x + 1$

$= 2x^2 + x + 2$

$(f-g)(2) = 2(2)^2 + 2 + 2$

$= 2(4) + 2 + 2$

$= 8 + 2 + 2 = 12$

**6.** Since the line we are concerned with is perpendicular to the line, $y = 2x - 3$, we know the slopes are negative reciprocals. The slope of the line will be the negative reciprocal of 2 which is $-\frac{1}{2}$. Using the slope and the point, $(2,3)$, write the equation of the line in point-slope form.

$y - y_1 = m(x - x_1)$

$y - 3 = -\frac{1}{2}(x - 2)$

Solve for $y$ to write the equation in function notation.

$y - 3 = -\frac{1}{2}(x - 2)$

$y - 3 = -\frac{1}{2}x + 1$

$y = -\frac{1}{2}x + 4$

$f(x) = -\frac{1}{2}x + 4$

**7.** $f(x) = 2x + 1$

$y = 2x + 1$

Find the $x$–intercept by setting $y = 0$, and the $y$–intercept by setting $x = 0$.

| | |
|---|---|
| $y = 2x + 1$ | $y = 2x + 1$ |
| $0 = 2x + 1$ | $y = 2(0) + 1$ |
| $-1 = 2x$ | $y = 1$ |
| $-\frac{1}{2} = x$ | |

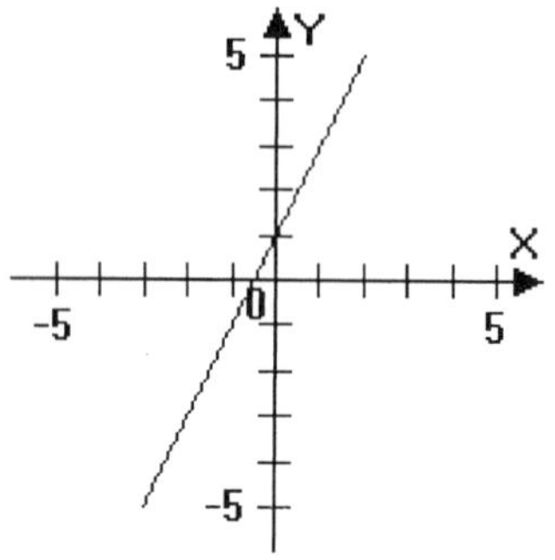


**8.** $y > 2x$

Consider the line $y = 2x$. Since the line is in slope-intercept form, we know that the slope is 2 and the $y$–intercept is 0. Use this information to graph the line.

Since the origin, (0, 0), lies on the line, we cannot use it as a test point. Instead, use the point (1, 1).

$y > 2x$

$1 > 2(1)$

$1 > 2$

This is a false statement. This means that the point (1, 1) does not lie in the shaded region.

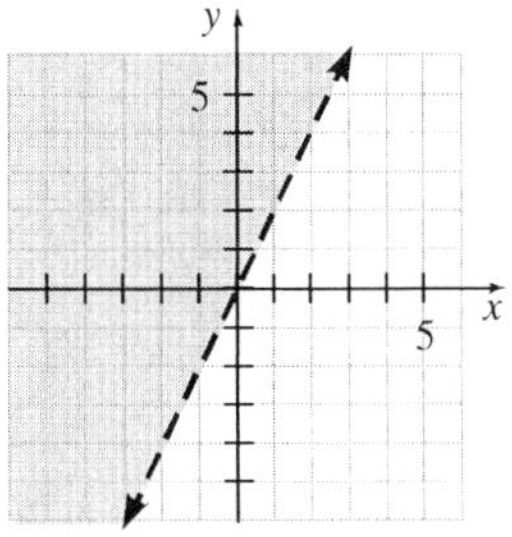


**9.** $2x - y \geq 6$

Graph the equation using the intercepts.

$2x - y = 6$

$x - \text{intercept} = 3$

$y - \text{intercept} = -6$

Use the origin as a test point to determine shading.

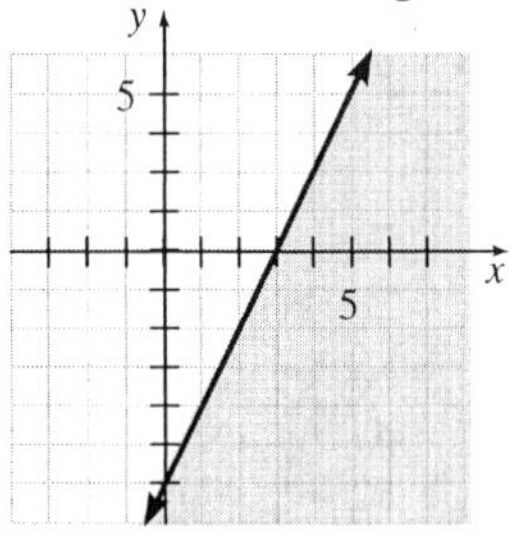


**10.** $f(x) = -1$

$y = -1$

Equations of the form $y = b$ are horizontal lines with $y$-intercept $= b$. This is the horizontal line at $y = -1$.

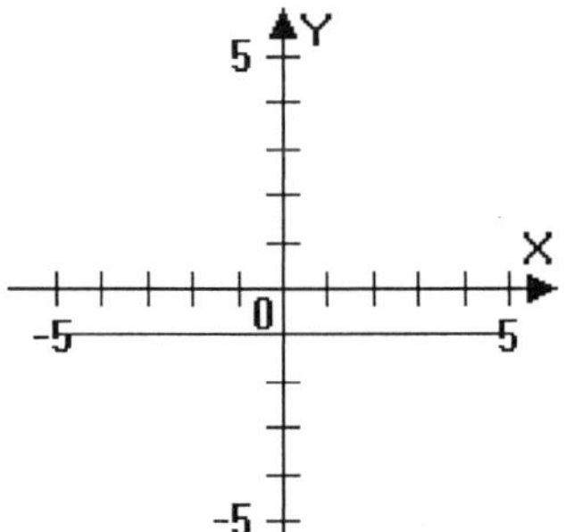


**11.** $3x - y + z = -15$

$x + 2y - z = 1$

$2x + 3y - 2z = 0$

Add the first two equations to eliminate $z$.

$3x - y + z = -15$

$x + 2y - z = 1$

$4x + y = -14$

Multiply the first equation by 2 and add to the third equation.

$6x - 2y + 2z = -30$

$2x + 3y - 2z = 0$

$8x + y = -30$

The system of two equations in two variables becomes is as follows.

$4x + y = -14$

$8x + y = -30$

Multiply the first equation by –1 and add to the second equation.

$-4x - y = 14$

$8x + y = -30$

$4x = -16$

$x = -4$

Back-substitute –4 for $x$ to find $y$.

$4(-4) + y = -14$

$-16 + y = -14$

$y = 2$

Back-substitute 2 for $y$ and –4 for $x$ to find $z$.

$3x - y + z = -15$

$3(-4) - 2 + z = -15$

$-12 - 2 + z = -15$

$-14 + z = -15$

$z = -1$

The solution is $(-4, 2, -1)$ and the solution set is $\{(-4, 2, -1)\}$.

**12.** $2x - y = -4$

$x + 3y = 5$

$$\left[\begin{array}{cc|c} 2 & -1 & -4 \\ 1 & 3 & 5 \end{array}\right] \quad R_1 \leftrightarrow R_2$$

$$= \left[\begin{array}{cc|c} 1 & 3 & 5 \\ 2 & -1 & -4 \end{array}\right] \quad -2R_1 + R_2$$

$$= \left[\begin{array}{cc|c} 1 & 3 & 5 \\ 0 & -7 & -14 \end{array}\right] \quad -\frac{1}{7}R_2$$

$$= \left[\begin{array}{cc|c} 1 & 3 & 5 \\ 0 & 1 & 2 \end{array}\right]$$

The resulting system is:

$x + 3y = 5$

$y = 2.$

Back-substitute 2 for $y$ to find $x$.

$x + 3(2) = 5$

$x + 6 = 5$

$x = -1$

The solution is $(-1, 2)$ and the solution set is $\{(-1, 2)\}$.

**13.** $\begin{vmatrix} 4 & 3 \\ -1 & -5 \end{vmatrix} = 4(-5) - (-1)3$

$= -20 + 3 = -17$

**14.** Let $x$ = the number of rooms with a kitchen

Let $y$ = the number of rooms without a kitchen

$x + y = 60$

$90x + 80y = 5260$

Solve the first equation for $y$.

$x + y = 60$

$y = 60 - x$

Substitute $60 - x$ for $y$ to find $x$.

$90x + 80y = 5260$

$90x + 80(60 - x) = 5260$

$90x + 4800 - 80x = 5260$

$10x + 4800 = 5260$

$10x = 460$

$x = 46$

Back-substitute 46 for $x$ to find $y$.

$y = 60 - x = 60 - 46 = 14$

There are 46 rooms with kitchens and 14 rooms without kitchens.

**15.** Using the vertical line test, we see that graphs a. and b. are functions.

**16.**

$$\frac{x}{4} - \frac{3}{4} - 1 \le \frac{x}{2}$$

$$4\left(\frac{x}{4}\right) - 4\left(\frac{3}{4}\right) - 4(1) \le 4\left(\frac{x}{2}\right)$$

$$x - 3 - 4 \le 2x$$

$$x - 7 \le 2x$$

$$x \le 2x + 7$$

$$-x \le 7$$

$$x \ge -7$$

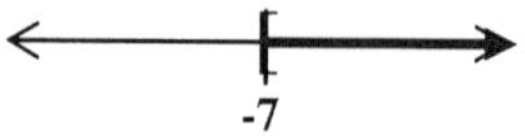


The solution set is $\{x | x \ge -7\}$ or $(-7, \infty)$.

**17.** $2x+5\le 11$ and $-3x>18$

$2x\le 6 \qquad x<-6$

$x\le 3$

$x\le 3$

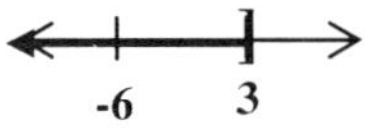


$x<-6$

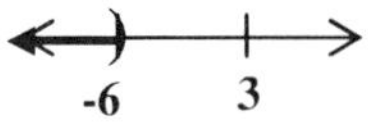


$x\le 3$ and $x<-6$

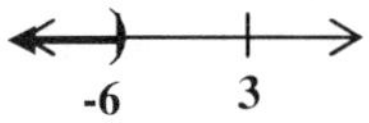


The solution set is $\{x|x<-6\}$ or $(-\infty,-6)$.

**18.** $x-4\ge 1$ or $-3x+1\ge -5-x$

$x\ge 5 \qquad -2x+1\ge -5$

$-2x\ge -6$

$x\le 3$

$x\ge 5$

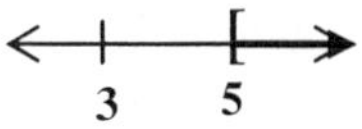


$x\le 3$

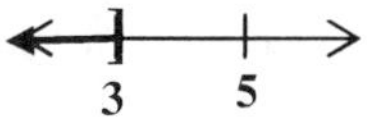


$x\ge 5$ or $x\le 3$

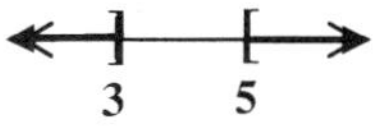


The solution set is $\{x|x\le 3 \text{ or } x\ge 5\}$ or $(-\infty,3]\cup[5,\infty)$.

**19.** $|2x+3|\le 17$

$-17\le 2x+3\le 17$

$-20\le 2x\le 14$

$-10\le x\le 7$

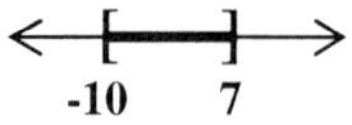


The solution set is $\{x|-10\le x\le 7\}$ or $[-10,7]$.

**20.** $|3x-8|>7$

$3x-8<-7$ or $3x-8>7$

$3x<1 \qquad 3x>15$

$x<\frac{1}{3} \qquad x>5$

$x<\frac{1}{3}$

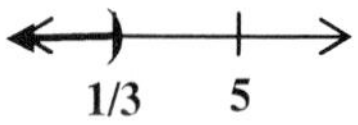


$x>5$

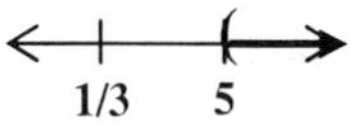


$x<\frac{1}{3}$ or $x>5$

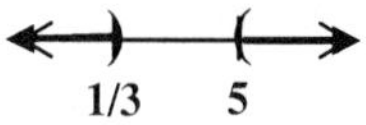


The solution set is $\left\{x\middle|x<\frac{1}{3} \text{ or } x>5\right\}$ or $\left(-\infty,\frac{1}{3}\right)\cup(5,\infty)$.

# Chapter 5
# Polynomials, Polynomial Functions, and Factoring

## 5.1 Exercise Set

**1.** The coefficient of $-x^4$ is –1 and the degree is 4.
The coefficient of $x^2$ is 1 and the degree is 2.
The degree of the polynomial is 4.
The leading term is $-x^4$ and the leading coefficient is –1.

**3.** The coefficient of $5x^3$ is 5 and the degree is 3.
The coefficient of $7x^2$ is 7 and the degree is 2.
The coefficient of $-x$ is –1 and the degree is 1.
The coefficient of 9 is 9 and the degree is 0.
The degree of the polynomial is 3.
The leading term is $5x^3$ and the leading coefficient is 5.

**5.** The coefficient of $3x^2$ is 3 and the degree is 2.
The coefficient of $-7x^4$ is –7 and the degree is 4.
The coefficient of $-x$ is –1 and the degree is 1.
The coefficient of 6 is 6 and the degree is 0.
The degree of the polynomial is 4.
The leading term is $-7x^4$ and the leading coefficient is –7.

**7.** The coefficient of $x^3y^2$ is 1 and the degree is 5.
The coefficient of $-5x^2y^7$ is –5 and the degree is 9.
The coefficient of $6y^2$ is 6 and the degree is 2.
The coefficient of –3 is –3 and the degree is 0.
The degree of the polynomial is 9.
The leading term is $-5x^2y^7$ and the leading coefficient is –5.

**9.** The coefficient of $x^5$ is 1 and the degree is 5.
The coefficient of $3x^2y^4$ is 3 and the degree is 6.
The coefficient of $7xy$ is 7 and the degree is 2.
The coefficient of $9x$ is 9 and the degree is 1.
The coefficient of –2 is –2 and the degree is 0.
The degree of the polynomial is 6.
The leading term is $3x^2y^4$ and the leading coefficient is 3.

**11.** $f(x) = x^2 - 5x + 6$

$$f(3) = (3)^2 - 5(3) + 6$$
$$= 9 - 15 + 6 = 0$$

**13.** $f(x) = x^2 - 5x + 6$

$$f(-1) = (-1)^2 - 5(-1) + 6$$
$$= 1 + 5 + 6 = 12$$

**15.** $g(x) = 2x^3 - x^2 + 4x - 1$

$$\begin{aligned} g(3) &= 2(3)^3 - (3)^2 + 4(3) - 1 \\ &= 2(27) - 9 + 12 - 1 \\ &= 54 - 9 + 12 - 1 = 56 \end{aligned}$$

**17.** $g(x) = 2x^3 - x^2 + 4x - 1$

$$\begin{aligned} g(-2) &= 2(-2)^3 - (-2)^2 + 4(-2) - 1 \\ &= 2(-8) - 4 - 8 - 1 \\ &= -16 - 4 - 8 - 1 \\ &= -29 \end{aligned}$$

**19.** $g(x) = 2x^3 - x^2 + 4x - 1$

$$\begin{aligned} g(0) &= 2(0)^3 - (0)^2 + 4(0) - 1 \\ &= 2(0) - 0 - 0 - 1 \\ &= 0 - 0 - 0 - 1 \\ &= -1 \end{aligned}$$

**21.** Polynomial function.

**23.** Not a polynomial function.

**25.** Since the degree of the polynomial is 4, an even number, and the leading coefficient is –1, the graph will fall to the left and to the right. The graph of the polynomial is graph (b).

**27.** Since the degree of the polynomial is 2, an even number, and the leading coefficient is 1, the graph will rise to the left and to the right. The graph of the polynomial is graph (a).

**29.**
$$\begin{aligned} (-6x^3 + 5x^2 - 8x + 9) + (17x^3 + 2x^2 - 4x - 13) &= -6x^3 + 5x^2 - 8x + 9 + 17x^3 + 2x^2 - 4x - 13 \\ &= -6x^3 + 17x^3 + 5x^2 + 2x^2 - 8x - 4x + 9 - 13 \\ &= 11x^3 + 7x^2 - 12x - 4 \end{aligned}$$

**31.**
$$\begin{aligned} \left(\frac{2}{5}x^4 + \frac{2}{3}x^3 + \frac{5}{8}x^2 + 7\right) + \left(-\frac{4}{5}x^4 + \frac{1}{3}x^3 - \frac{1}{4}x^2 - 7\right) &= \frac{2}{5}x^4 + \frac{2}{3}x^3 + \frac{5}{8}x^2 + 7 - \frac{4}{5}x^4 + \frac{1}{3}x^3 - \frac{1}{4}x^2 - 7 \\ &= \frac{2}{5}x^4 - \frac{4}{5}x^4 + \frac{2}{3}x^3 + \frac{1}{3}x^3 + \frac{5}{8}x^2 - \frac{1}{4}x^2 + \cancel{7 - 7} \\ &= -\frac{2}{5}x^4 + \frac{3}{3}x^3 + \left(\frac{5}{8} - \frac{2}{8}\right)x^2 \\ &= -\frac{2}{5}x^4 + \frac{3}{3}x^3 + \frac{3}{8}x^2 = -\frac{2}{5}x^4 + x^3 + \frac{3}{8}x^2 \end{aligned}$$

**33.**
$$\begin{aligned} (7x^2y - 5xy) + (2x^2y - xy) &= 7x^2y - 5xy + 2x^2y - xy \\ &= 7x^2y + 2x^2y - 5xy - xy \\ &= 9x^2y - 6xy \end{aligned}$$

**35.** $\left(5x^2y+9xy+12\right)+\left(-3x^2y+6xy+3\right)=5x^2y+9xy+12-3x^2y+6xy+3$
$=5x^2y-3x^2y+9xy+6xy+12+3$
$=2x^2y+15xy+15$

**37.** $\left(9x^4y^2-6x^2y^2+3xy\right)+\left(-18x^4y^2-5x^2y-xy\right)=9x^4y^2-6x^2y^2+3xy-18x^4y^2-5x^2y-xy$
$=9x^4y^2-18x^4y^2-6x^2y^2-5x^2y+3xy-xy$
$=-9x^4y^2-6x^2y^2-5x^2y+2xy$

**39.** $\left(x^{2n}+5x^n-8\right)+\left(4x^{2n}-7x^n+2\right)=x^{2n}+5x^n-8+4x^{2n}-7x^n+2$
$=x^{2n}+4x^{2n}+5x^n-7x^n-8+2$
$=5x^{2n}-2x^n-6$

**41.** $\left(17x^3-5x^2+4x-3\right)-\left(5x^3-9x^2-8x+11\right)=17x^3-5x^2+4x-3-5x^3+9x^2+8x-11$
$=17x^3-5x^3-5x^2+9x^2+4x+8x-3-11$
$=12x^3+4x^2+12x-14$

**43.** $\left(13y^5+9y^4-5y^2+3y+6\right)-\left(-9y^5-7y^3+8y^2+11\right)$
$=13y^5+9y^4-5y^2+3y+6+9y^5+7y^3-8y^2-11$
$=13y^5+9y^5+9y^4+7y^3-5y^2-8y^2+3y+6-11$
$=22y^5+9y^4+7y^3-13y^2+3y-5$

**45.** $\left(x^3+7xy-5y^2\right)-\left(6x^3-xy+4y^2\right)=x^3+7xy-5y^2-6x^3+xy-4y^2$
$=x^3-6x^3+7xy+xy-5y^2-4y^2$
$=-5x^3+8xy-9y^2$

**47.** $\left(3x^4y^2+5x^3y-3y\right)-\left(2x^4y^2-3x^3y-4y+6x\right)=3x^4y^2+5x^3y-3y-2x^4y^2+3x^3y+4y-6x$
$=3x^4y^2-2x^4y^2+5x^3y+3x^3y-3y+4y-6x$
$=x^4y^2+8x^3y+y-6x$

**49.** $\left(7y^{2n}+y^n-4\right)-\left(6y^{2n}-y^n-1\right)=7y^{2n}+y^n-4-6y^{2n}+y^n+1$
$=7y^{2n}-6y^{2n}+y^n+y^n-4+1$
$=y^{2n}+2y^n-3$

**51.** $\left(3a^2b^4-5ab^2+7ab\right)-\left(-5a^2b^4-8ab^2-ab\right)=3a^2b^4-5ab^2+7ab+5a^2b^4+8ab^2+ab$
$=3a^2b^4+5a^2b^4-5ab^2+8ab^2+7ab+ab$
$=8a^2b^4+3ab^2+8ab$

**53.** $\left(x^3+2x^2y-y^3\right)-\left(-4x^3-x^2y+xy^2+3y^3\right)=x^3+2x^2y-y^3+4x^3+x^2y-xy^2-3y^3$
$=x^3+4x^3+2x^2y+x^2y-xy^2-y^3-3y^3$
$=5x^3+3x^2y-xy^2-4y^3$

**55.** $\left(6x^4-5x^3+2x\right)+\left[\left(4x^3+3x^2-1\right)-\left(x^4-2x^2+7x-3\right)\right]$
$=6x^4-5x^3+2x+\left[4x^3+3x^2-1-x^4+2x^2-7x+3\right]$
$=6x^4-5x^3+2x+\left[-x^4+4x^3+3x^2+2x^2-7x-1+3\right]$
$=6x^4-5x^3+2x+\left[-x^4+4x^3+5x^2-7x+2\right]$
$=6x^4-5x^3+2x-x^4+4x^3+5x^2-7x+2$
$=6x^4-x^4-5x^3+4x^3+5x^2+2x-7x+2$
$=5x^4-x^3+5x^2-5x+2$

**56.** $\left(5x^4-2x^3+7x\right)+\left[\left(2x^3+5x^2-3\right)-\left(-x^4-x^2-x-1\right)\right]$
$=5x^4-2x^3+7x+\left[2x^3+5x^2-3+x^4+x^2+x+1\right]$
$=5x^4-2x^3+7x+\left[x^4+2x^3+5x^2+x^2+x-3+1\right]$
$=5x^4-2x^3+7x+\left[x^4+2x^3+6x^2+x-2\right]$
$=5x^4-2x^3+7x+x^4+2x^3+6x^2+x-2$
$=5x^4+x^4-2x^3+2x^3+6x^2+7x+x-2$
$=6x^4+6x^2+8x-2$

**57.** $\left[\left(-6x^2y^2-x^2-1\right)+\left(5x^2y^2+2x^2-1\right)\right]-\left(9x^2y^2-3x^2-5\right)$
$=\left[-6x^2y^2-x^2-1+5x^2y^2+2x^2-1\right]-9x^2y^2+3x^2+5$
$=\left[-6x^2y^2+5x^2y^2-x^2+2x^2-1-1\right]-9x^2y^2+3x^2+5$
$=\left[-x^2y^2+x^2-2\right]-9x^2y^2+3x^2+5$
$=-x^2y^2+x^2-2-9x^2y^2+3x^2+5$
$=-x^2y^2-9x^2y^2+x^2+3x^2-2+5$
$=-10x^2y^2+4x^2+3$

**58.** $\left[\left(-5x^2y^3+3x^2-4\right)+\left(4x^2y^3-2x^2-6\right)\right]-\left(6x^2y^3-2x^2-7\right)$

$=\left[-5x^2y^3+3x^2-4+4x^2y^3-2x^2-6\right]-6x^2y^3+2x^2+7$

$=\left[-5x^2y^3+4x^2y^3+3x^2-2x^2-4-6\right]-6x^2y^3+2x^2+7$

$=\left[-x^2y^3+x^2-10\right]-6x^2y^3+2x^2+7$

$=-x^2y^3+x^2-10-6x^2y^3+2x^2+7$

$=-x^2y^3-6x^2y^3+x^2+2x^2-10+7$

$=-7x^2y^3+3x^2-3$

**For Exercises 59-64,** $f(x)=-3x^3-2x^2-x+4$, $g(x)=x^3-x^2-5x-4$; $h(x)=-2x^3+5x^2-4x+1$

**59.** $(f-g)(x)=\left(-3x^3-2x^2-x+4\right)-\left(x^3-x^2-5x-4\right)$

$=-3x^3-2x^2-x+4-x^3+x^2+5x+4$

$=-3x^3-x^3-2x^2+x^2-x+5x+4+4$

$=-4x^3-x^2+4x+8$

$(f-g)(-1)=-4(-1)^3-(-1)^2+4(-1)+8$

$=4-1-4+8$

$=7$

**60.** $(g-h)(x)=\left(x^3-x^2-5x-4\right)-\left(-2x^3+5x^2-4x+1\right)$

$=x^3-x^2-5x-4+2x^3-5x^2+4x-1$

$=x^3+2x^3-x^2-5x^2-5x+4x-4-1$

$=3x^3-6x^2-x-5$

$(g-h)(-1)=3(-1)^3-6(-1)^2-(-1)-5$

$=-3-6+1-5$

$=-13$

**61.** $(f+g-h)(x)=\left(-3x^3-2x^2-x+4\right)+\left(x^3-x^2-5x-4\right)-\left(-2x^3+5x^2-4x+1\right)$

$=-3x^3-2x^2-x+4+x^3-x^2-5x-4+2x^3-5x^2+4x-1$

$=-3x^3+x^3+2x^3-2x^2-x^2-5x^2-x-5x+4x+4-4-1$

$=-8x^2-2x-1$

$(f+g-h)(-2)=-8(-2)^2-2(-2)-1=-8(4)+4-1=-32+3=-29$

**62.** $(g+h-f)(x)=(x^3-x^2-5x-4)+(-2x^3+5x^2-4x+1)-(-3x^3-2x^2-x+4)$

$=x^3-x^2-5x-4-2x^3+5x^2-4x+1+3x^3+2x^2+x-4$

$=x^3-2x^3+3x^3-x^2+5x^2+2x^2-5x-4x+x-4+1-4$

$=2x^3+6x^2-8x-7$

$(g+h-f)(-2)=2(-2)^3+6(-2)^2-8(-2)-7=2(-8)+6(4)+16-7$

$=-16+24+9=17$

**63.** $2f(x)-3g(x)=2(-3x^3-2x^2-x+4)-3(x^3-x^2-5x-4)$

$=-6x^3-4x^2-2x+8-3x^3+3x^2+15x+12$

$=-6x^3-3x^3-4x^2+3x^2-2x+15x+8+12$

$=-9x^3-x^2+13x+20$

**64.** $-2g(x)-3h(x)=-2(x^3-x^2-5x-4)-3(-2x^3+5x^2-4x+1)$

$=-2x^3+2x^2+10x+8+6x^3-15x^2+12x-3$

$=-2x^3+6x^3+2x^2-15x^2+10x+12x+8-3$

$=4x^3-13x^2+22x+5$

**65.** $f(x)=-2212x^2+57{,}575x+107{,}896$

$f(10)=-2212(10)^2+57{,}575(10)+107{,}896$

$=-2212(100)+575{,}750+107{,}896$

$=-221{,}200+575{,}750+107{,}896$

$=462{,}446$

In 2000 (10 years after 1990), the cumulative number of AIDS deaths in the U.S. was 464,446.

**67.** The result from Exercise 65 corresponds to the point $(10,\ 462{,}446)$ on the graph.

**69.** $f(x) = -2212x^2 + 57{,}575x + 107{,}896$

$f(10) = 462{,}446$ (from Exercise 65)

$g(x) = -84x^3 - 702x^2 + 50{,}609x + 113{,}435$

$$\begin{aligned} g(10) &= -84(10)^3 - 702(10)^2 + 50{,}609(10) + 113{,}435 \\ &= -84(1000) - 702(100) + 506{,}090 + 113{,}435 \\ &= -84{,}000 - 70{,}200 + 506{,}090 + 113{,}435 \\ &= 465{,}325 \end{aligned}$$

The graph indicates that *f* provides a better description because $f(10)$ is closer to the actual amount of 458,551.

**71.** The leading coefficient is negative and the degree is even. This means that the graph will fall to the right. This function will not be useful in modeling the number of cumulative AIDS deaths over an extended period of time because the cumulative number of deaths cannot decrease.

**73.** The leading coefficient is –0.75 and the degree is 4. This means that the graph will fall to the right. Over time, the number of viral particles in the body will go to zero.

**75.** In the polynomial, $f(x) = -x^4 + 21x^2 + 100$, the leading coefficient is –1 and the degree is 4. Applying the Leading Coefficient Test, we know that even-degree polynomials with negative leading coefficient will fall to the left and to the right. Since the graph falls to the right, we know that the elk population will die out over time.

**77.** Answers vary.

**79.** Answers vary.

**81.** Answers vary.

**83.** Answers vary.

**85.** Answers vary.

**87.** Answers vary.

**89.** Answers will vary. The graph falls to the right and left, so the polynomial must be of even degree with a negative leading coefficient. One example is $-x^4 - 2x^3$.

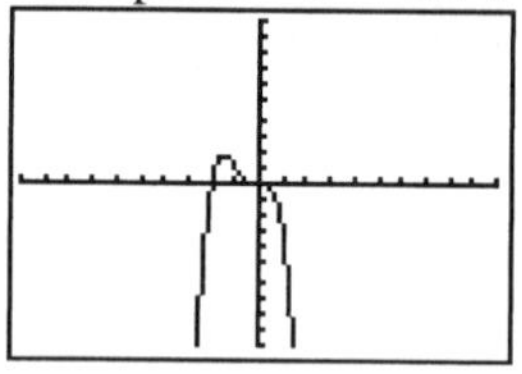

**91.** Answers will vary. The graph falls to the left and rises to the right, so the polynomial must be of odd degree with a positive leading coefficient. One example is $x^3 + 3x^2$.

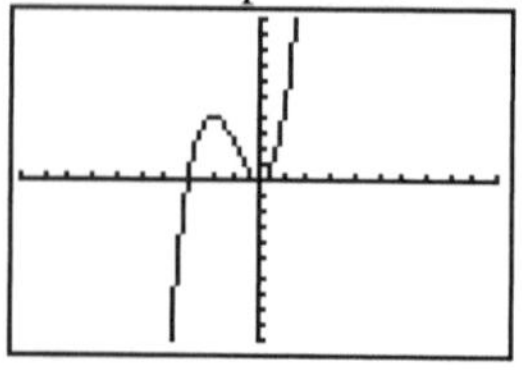

**93.** $f(x) = -2x^3 + 6x^2 + 3x - 1$

Since we have an odd-degree polynomial with a negative coefficient, the Leading Coefficient Test predicts that the graph will rise to the left and fall to the right.

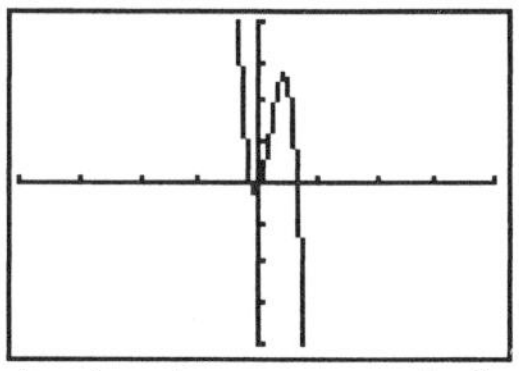

A viewing rectangle large enough to show the end behavior is shown.

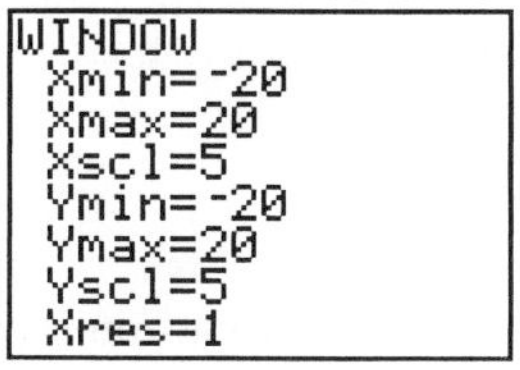


**95.** $f(x) = -x^5 + 5x^4 - 6x^3 + 2x + 20$

Since we have an odd-degree polynomial with a negative coefficient, the Leading Coefficient Test predicts that the graph will rise to the left and fall to the right.

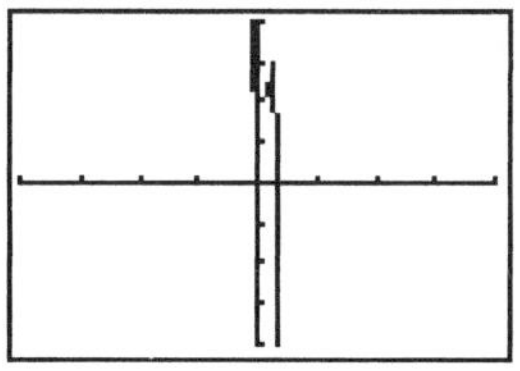

A viewing rectangle large enough to show the end behavior is shown.

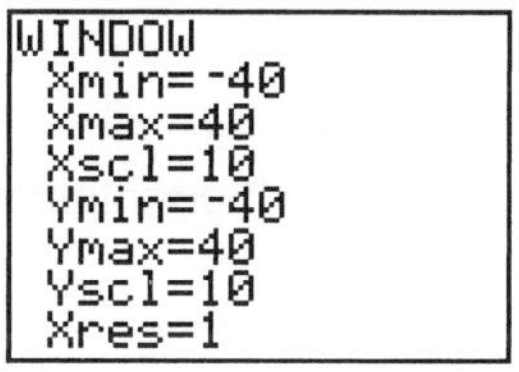


**97.** $f(x) = -x^4 + 2x^3 - 6x$

$g(x) = -x^4$

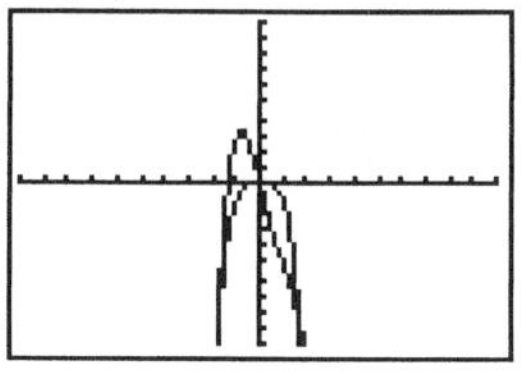

**99.** $\left(x^{2n} - 3x^n + 5\right) + \left(4x^{2n} - 3x^n - 4\right) - \left(2x^{2n} - 5x^n - 3\right)$

$= x^{2n} - 3x^n + 5 + 4x^{2n} - 3x^n - 4 - 2x^{2n} + 5x^n + 3$

$= x^{2n} + 4x^{2n} - 2x^{2n} - 3x^n - 3x^n + 5x^n + 5 - 4 + 3$

$= 3x^{2n} - x^n - 4$

**101.** $\left(4x^2 + 2x - 3\right) + \left(5x^2 - 5x + 8\right)$

$= 4x^2 + 2x - 3 + 5x^2 - 5x + 8$

$= 9x^2 - 3x + 5$

**102.** $9(x-1)=1+3(x-2)$

$9x-9=1+3x-6$

$9x-9=3x-5$

$6x-9=-5$

$6x=4$

$x=\frac{4}{6}=\frac{2}{3}$

The solution is $\frac{2}{3}$ and the solution set is $\left\{\frac{2}{3}\right\}$.

**103.** $2x-3y<-6$

Replacing the inequality symbol with an equal sign, we have $2x-3y=-6$.
Solve for $y$ to obtain slope-intercept form.

$2x-3y=-6$

$-3y=-2x-6$

$y=\frac{2}{3}x+2$

The slope is $\frac{2}{3}$ and the $y$–intercept is 2. Next, use the origin as a test point.

$2(0)-3(0)<-6$

$0<-6$

This is a false statement. This means that the point, $(0,0)$, will not fall in the shaded half-plane.

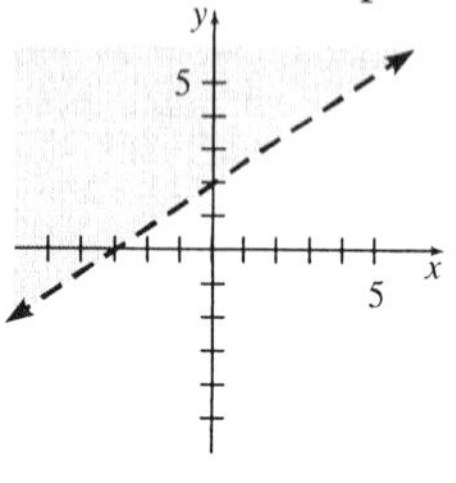


**104.** Since the line is parallel to $3x-y=9$, it will have the same slope. To find the slope, put the equation is slope intercept form.

$3x-y=9$

$-y=-3x+9$

$y=3x-9$

The slope is $m=3$.
Use the slope and the point, $(-2,5)$, to write the equation of the line in point-slope form.

$y-y_1=m(x-x_1)$

$y-5=3(x-(-2))$

$y-5=3(x+2)$

Solve for $y$ to obtain slope-intercept form.

$y-5=3x+6$

$y=3x+11$

## 5.2 Exercise Set

**1.** $(3x^2)(5x^4)=3(5)x^2\cdot x^4$

$=15x^{2+4}$

$=15x^6$

**3.** $(3x^2y^4)(5xy^7)=3(5)x^2\cdot x\cdot y^4\cdot y^7$

$=15x^{2+1}y^{4+7}$

$=15x^3y^{11}$

**5.** $(-3xy^2z^5)(2xy^7z^4)$

$=-3(2)x\cdot x\cdot y^2\cdot y^7\cdot z^5\cdot z^4$

$=-6x^{1+1}y^{2+7}z^{5+4}$

$=-6x^2y^9z^9$

**7.** $\left(-8x^{2n}y^{n-5}\right)\left(-\frac{1}{4}x^{n}y^{3}\right)$

$=(-8)\left(-\frac{1}{4}\right)x^{2n}\cdot x^{n}\cdot y^{n-5}\cdot y^{3}$

$=2x^{2n+n}y^{n-5+3}$

$=2x^{3n}y^{n-2}$

**9..** $4x^2(3x+2)=4x^2\cdot 3x+4x^2\cdot 2$

$=12x^3+8x^2$

**11.** $2y\left(y^2-5y\right)=2y\cdot y^2-2y\cdot 5y$

$=2y^3-10y^2$

**13.** $5x^3\left(2x^5-4x^2+9\right)$

$=5x^3\cdot 2x^5-5x^3\cdot 4x^2+5x^3\cdot 9$

$=10x^8-20x^5+45x^3$

**15.** $4xy(7x+3y)=4xy\cdot 7x+4xy\cdot 3y$

$=28x^2y+12xy^2$

**17.** $3ab^2\left(6a^2b^3+5ab\right)$

$=3ab^2\cdot 6a^2b^3+3ab^2\cdot 5ab$

$=18a^3b^5+15a^2b^3$

**19.** $-4x^2y\left(3x^4y^2-7xy^3+6\right)$

$=-4x^2y\cdot 3x^4y^2+4x^2y\cdot 7xy^3-4x^2y\cdot 6$

$=-12x^6y^3+28x^3y^4-24x^2y$

**21.** $-4x^n\left(3x^{2n}-5x^n+\frac{1}{2}x\right)$

$=-4x^n\cdot 3x^{2n}+4x^n\cdot 5x^n-4x^n\cdot\frac{1}{2}x$

$=-12x^{3n}+20x^{2n}-2x^{n+1}$

**23.**
$$\begin{array}{r} x^2+2x+5 \\ \underline{x-3} \\ x^3+2x^2+5x \\ \underline{-3x^2-6x-15} \\ x^3-x^2-x-15 \end{array}$$

**25.**
$$\begin{array}{r} x^2+x+1 \\ \underline{x-1} \\ x^3+x^2+x \\ \underline{-x^2-x-1} \\ x^3 \qquad\quad -1 \end{array}$$

**27.**
$$\begin{array}{r} a^2+ab+b^2 \\ \underline{a-b} \\ a^3+a^2b+ab^2 \\ \underline{-a^2b-ab^2-b^3} \\ a^3 \qquad\qquad -b^3 \end{array}$$

**29.**
$$\begin{array}{r} x^2+2x-1 \\ \underline{x^2+3x-4} \\ x^4+2x^3-x^2 \\ 3x^3+6x^2-3x \\ \underline{-4x^2-8x+4} \\ x^4+5x^3+x^2-11x+4 \end{array}$$

**31.**
$$\begin{array}{r} x^2-3xy+y^2 \\ \underline{x-y} \\ x^3-3x^2y+xy^2 \\ \underline{-x^2y+3xy^2-y^3} \\ x^3-4x^2y+4xy^2-y^3 \end{array}$$

**33.**
$$\begin{array}{r} x^2y^2-2xy+4 \\ xy+2 \\ \hline x^3y^3-2x^2y^2+4xy \\ 2x^2y^2-4xy+8 \\ \hline x^3y^3 \qquad\qquad\qquad +8 \end{array}$$

**35.** $(x+4)(x+7)=x^2+7x+4x+28$
$=x^2+11x+28$

**37.** $(y+5)(y-6)=y^2-6y+5y-30$
$=y^2-y-30$

**39.** $(5x+3)(2x+1)=10x^2+5x+6x+3$
$=10x^2+11x+3$

**41.** $(3y-4)(2y-1)=6y^2-3y-8y+4$
$=6y^2-11y+4$

**43.** $(3x-2)(5x-4)=15x^2-12x-10x+8$
$=15x^2-22x+8$

**45.** $(x-3y)(2x+7y)$
$=2x^2+7xy-6xy-21y^2$
$=2x^2+xy-21y^2$

**47.** $(7xy+1)(2xy-3)$
$=14x^2y^2-21xy+2xy-3$
$=14x^2y^2-19xy-3$

**49.** $(x-4)(x^2-5)=x^3-5x-4x^2+20$
$=x^3-4x^2-5x+20$

**51.** $(8x^3+3)(x^2-5)$
$=8x^3\cdot x^2-8x^3\cdot 5+3\cdot x^2-3\cdot 5$
$=8x^5-40x^3+3x^2-15$

**53.** $(3x^n-y^n)(x^n+2y^n)$
$=3x^n\cdot x^n+3x^n\cdot 2y^n-y^n\cdot x^n-y^n\cdot 2y^n$
$=3x^{2n}+6x^ny^n-x^ny^n-2y^{2n}$
$=3x^{2n}+5x^ny^n-2y^{2n}$

**55.** $(x+3)^2=x^2+2(3x)+9$
$=x^2+6x+9$

**57.** $(y-5)^2=y^2+2(-5y)+25$
$=y^2-10y+25$

**59.** $(2x+y)^2=4x^2+2(2xy)+y^2$
$=4x^2+4xy+y^2$

**61.** $(5x-3y)^2=25x^2+2(-15xy)+9y^2$
$=25x^2-30xy+9y^2$

**63.** $(2x^2+3y)^2=4x^4+2(6x^2y)+9y^2$
$=4x^4+12x^2y+9y^2$

**65.** $(4xy^2-xy)^2$
$=16x^2y^4+2(4x^2y^3)+x^2y^2$
$=16x^2y^4+8x^2y^3+x^2y^2$

**67.** $(a^n+4b^n)^2$
$=(a^n)^2+2(a^n)(4b^n)+(4b^n)^2$
$=a^{2n}+8a^nb^n+16b^{2n}$

**69.** $(x+4)(x-4)=(x)^2-(4)^2$
$=x^2-16$

**71.** $(5x+3)(5x-3)=(5x)^2-(3)^2$
$=25x^2-9$

**73.** $(4x+7y)(4x-7y)=(4x)^2-(7y)^2$
$=16x^2-49y^2$

**75.** $(y^3+2)(y^3-2)=(y^3)^2-(2)^2$
$=y^6-4$

**77.** $(1-y^5)(1+y^5)=(1)^2-(y^5)^2$
$=1-y^{10}$

**79.** $(7xy^2-10y)(7xy^2+10y)$
$=(7xy^2)^2-(10y)^2$
$=49x^2y^4-100y^2$

**81.** $(5a^n-7)(5a^n+7)=(5a^n)^2-(7)^2$
$=25a^{2n}-49$

**83.** $[(2x+3)+4y][(2x+3)-4y]$
$=(2x+3)^2-16y^2$
$=4x^2+2(6x)+9-16y^2$
$=4x^2+12x+9-16y^2$

**85.** $(x+y+3)(x+y-3)$
$=((x+y)+3)((x+y)-3)$
$=(x+y)^2-9$
$=x^2+2xy+y^2-9$

**87.** $(5x+7y-2)(5x+7y+2)$
$=((5x+7y)-2)((5x+7y)+2)$
$=(5x+7y)^2-4$
$=25x^2+2(35xy)+49y^2-4$
$=25x^2+70xy+49y^2-4$

**89.** $[5y+(2x+3)][5y-(2x+3)]$
$=25y^2-(2x+3)^2$
$=25y^2-(4x^2+2(6x)+9)$
$=25y^2-(4x^2+12x+9)$
$=25y^2-4x^2-12x-9$

**91.**

$$\begin{array}{lllll} x+y+1 \\ \underline{x+y+1} \\ x^2+xy & +x \\ \quad xy & & +y^2+y \\ \underline{\quad\quad x} & & +y & +1 \\ x^2+2xy+2x+y^2+2y+1 \end{array}$$

or

$x^2+2xy+y^2+2x+2y+1$

**93.** $(x+1)(x-1)(x^2+1)=(x^2-1)(x^2+1)$
$=(x^2)^2-(1)^2$
$=x^4-1$

**95.** **a.** $(fg)(x)=f(x)\cdot g(x)$
$=(x-2)(x+6)$
$=x^2+6x-2x-12$
$=x^2+4x-12$

**b.** $(fg)(-1)=(-1)^2+4(-1)-12$
$=1-4-12=-15$

**c.** $(fg)(0)=(0)^2+4(0)-12$
$=0+0-12=-12$

**97.** **a.**
$$\begin{aligned}&(fg)(x)\\&=f(x)\cdot g(x)\\&=(x-3)(x^2+3x+9)\\&=x(x^2+3x+9)-3(x^2+3x+9)\\&=x^3+3x^2+9x-3x^2-9x-27\\&=x^3-27\end{aligned}$$

**b.** $(fg)(-2)=(-2)^3-27$
$$=-8-27=-35$$

**c.** $(fg)(0)=(0)^3-27=-27$

**99.** **a.** $f(a+2)=(a+2)^2-3(a+2)+7$
$$=a^2+4a+4-3a-6+7=a^2+a+5$$

**b.** $f(a+h)-f(a)=(a+h)^2-3(a+h)+7-(a^2-3a+7)$
$$\begin{aligned}&=\cancel{a^2}+2ah+h^2-\cancel{3a}-3h+\cancel{7}-\cancel{a^2}+\cancel{3a}-\cancel{7}\\&=2ah+h^2-3h\end{aligned}$$

**101.** **a.** $f(a+2)=3(a+2)^2+2(a+2)-1$
$$\begin{aligned}&=3(a^2+4a+4)+2a+4-1\\&=3a^2+12a+12+2a+4-1\\&=3a^2+14a+15\end{aligned}$$

**b.** $f(a+h)-f(a)=3(a+h)^2+2(a+h)-1-(3a^2+2a-1)$
$$\begin{aligned}&=3(a^2+2ah+h^2)+2a+2h-1-3a^2-2a+1\\&=\cancel{3a^2}+6ah+3h^2+\cancel{2a}+2h-\cancel{1}-\cancel{3a^2}-\cancel{2a}+\cancel{1}\\&=6ah+3h^2+2h\end{aligned}$$

**103.** $(3x+4y)^2-(3x-4y)^2=\left[(3x)^2+2(3x)(4y)+(4y)^2\right]-\left[(3x)^2-2(3x)(4y)+(4y)^2\right]$
$$\begin{aligned}&=(9x^2+24xy+16y^2)-(9x^2-24xy+16y^2)\\&=9x^2+24xy+16y^2-9x^2+24xy-16y^2\\&=48xy\end{aligned}$$

**104.** $(5x+2y)^2-(5x-2y)^2=\left[(5x)^2+2(5x)(2y)+(2y)^2\right]-\left[(5x)^2-2(5x)(2y)+(2y)^2\right]$

$$=\left(25x^2+20xy+4y^2\right)-\left(25x^2-20xy+4y^2\right)$$
$$=25x^2+20xy+4y^2-25x^2+20xy-4y^2$$
$$=40xy$$

**105.** $(5x-7)(3x-2)-(4x-5)(6x-1)$

$$=\left[15x^2-10x-21x+14\right]-\left[24x^2-4x-30x+5\right]$$
$$=\left(15x^2-31x+14\right)-\left(24x^2-34x+5\right)$$
$$=15x^2-31x+14-24x^2+34x-5$$
$$=-9x^2+3x+9$$

**106.** $(3x+5)(2x-9)-(7x-2)(x-1)$

$$=\left(6x^2-27x+10x-45\right)-\left(7x^2-7x-2x+2\right)$$
$$=\left(6x^2-17x-45\right)-\left(7x^2-9x+2\right)$$
$$=6x^2-17x-45-7x^2+9x-2$$
$$=-x^2-8x-47$$

**107.** $(2x+5)(2x-5)\left(4x^2+25\right)$

$$=\left[(2x)^2-5^2\right]\left(4x^2+25\right)$$
$$=\left(4x^2-25\right)\left(4x^2+25\right)$$
$$=\left(4x^2\right)^2-(25)^2$$
$$=16x^4-625$$

**108.** $(3x+4)(3x-4)\left(9x^2+16\right)$

$$=\left[(3x)^2-4^2\right]\left(9x^2+16\right)$$
$$=\left(9x^2-16\right)\left(9x^2+16\right)$$
$$=\left(9x^2\right)^2-(16)^2$$
$$=81x^4-256$$

**109.** $(x-1)^3=(x-1)(x-1)^2$

$$=(x-1)\left(x^2-2x+1\right)$$
$$=x\left(x^2-2x+1\right)-1\left(x^2-2x+1\right)$$
$$=x^3-2x^2+x-x^2+2x-1$$
$$=x^3-3x^2+3x-1$$

**110.** $(x-2)^3=(x-2)(x-2)^2$

$$=(x-2)\left(x^2-4x+4\right)$$
$$=x\left(x^2-4x+4\right)-2\left(x^2-4x+4\right)$$
$$=x^3-4x^2+4x-2x^2+8x-8$$
$$=x^3-6x^2+12x-8$$

**111.** $\dfrac{(2x-7)^5}{(2x-7)^3}=(2x-7)^{5-3}$
$=(2x-7)^2$
$=(2x)^2-2(2x)(7)+(7)^2$
$=4x^2-28x+49$

**112.** $\dfrac{(5x-3)^6}{(5x-3)^4}=(5x-3)^{6-4}$
$=(5x-3)^2$
$=(5x)^2-2(5x)(3)+(3)^2$
$=25x^2-30x+9$

**113.** **a.** $x^2+6x+4x+24=x^2+10x+24$

**b.** $(x+6)(x+4)=x^2+4x+6x+24$
$=x^2+10x+24$

**115.** **a.** $(x+9)(x+3)=x^2+3x+9x+27$
$=x^2+12x+27$

**b.** $(x+5)(x+1)=x^2+x+5x+5$
$=x^2+6x+5$

**c.** $(x^2+12x+27)-(x^2+6x+5)$
$=x^2+12x+27-x^2-6x-5$
$=6x+22$

**117.** **a.** $(8-2x)(10-2x)=80-16x-20x+4x^2$
$=80-36x+4x^2$
$=4x^2-36x+80$

**b.** $x(80-36x+2x^2)$
$=80x-36x^2+4x^3$
$=4x^3-36x^2+80x$

**119.** Answers will vary.

**121.** Answers will vary.

**123.** Answers will vary.

**125.** Answers will vary.

**127.** Answers will vary.

**129.**

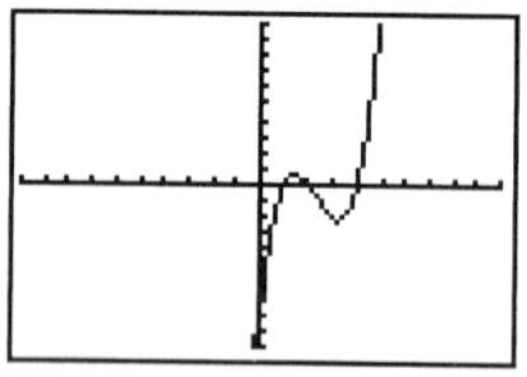

The functions $y_1$ and $y_2$ are the same. Verify by multiplying.

$$\begin{array}{r} x^2-3x+2 \\ \underline{x-4} \\ x^3-3x^2+2x \\ \underline{-4x^2+12x-8} \\ x^3-7x^2+14x-8 \end{array}$$

**131.**

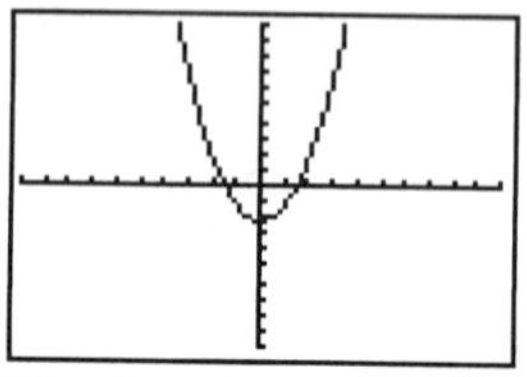

The functions $y_1$ and $y_2$ are the same. Verify by multiplying.
$(x+1.5)(x-1.5)=x^2-2.25$

**133.** Statement **d.** is true.
$(3x+7)(3x-2)$
$=9x^2-6x+21x-14$
$=9x^2+15x-14$

Statement **a.** is false. Do not confuse $(f+g)(x)$ with $f(a+h)$.

Statement **b.** is false.

$$(x-5)^2 = x^2 - 2\cdot x\cdot 5 + (-5)^2$$
$$= x^2 - 10x + 25$$

Statement **c.** is false.

$$(x+1)^2 = x^2 + 2\cdot x\cdot 1 + 1^2$$
$$= x^2 + 2x + 1$$

**135.** First divide the figure into two rectangular solids, one tall and one short. To find the total volume, we use $V_{\text{total}} = V_{\text{tall}} + V_{\text{short}}$. The only measurement that is missing is the width of the taller rectangular solid. Since the distance across the entire figure is $x+3$, and the distance across the shorter rectangular solid is $x$, the distance across the taller rectangular solid is $x+3-x=3$.

$$V_{\text{total}} = V_{\text{tall}} + V_{\text{short}}$$
$$= l_t w_t h_t + l_s w_s h_s$$
$$= 3\cdot x\cdot(2x-1) + x\cdot x\cdot(x+1)$$
$$= 3x(2x-1) + x^2(x+1)$$
$$= 6x^2 - 3x + x^3 + x^2$$
$$= x^3 + 7x^2 - 3x$$

**137.**

$$(y^n+2)(y^n-2)-(y^n-3)^2$$
$$= (y^n)^2 - 2^2 - (y^{2n} - 2(3y^n) + 9)$$
$$= y^{2n} - 4 - (y^{2n} - 6y^n + 9)$$
$$= y^{2n} - 4 - y^{2n} + 6y^n - 9$$
$$= 6y^n - 13$$

**139.** $|3x+4| \ge 10$

$$3x+4 \le -10 \qquad\qquad 3x+4 \ge 10$$
$$3x \le -14 \quad \text{or} \quad 3x \ge 6$$
$$x \le -\frac{14}{3} \qquad\qquad x \ge 2$$

The solution set is $\left\{x \middle| x \le -\frac{14}{3} \text{ or } x \ge 2\right\}$ or $\left(-\infty, -\frac{14}{3}\right] \cup [2, \infty)$.

**140.** $2 - 6x \le 20$

$$-6x \le 18$$
$$x \ge -3$$

The solution set is $\{x \mid x \ge -3\}$ or $[-3, \infty)$.

**141.** $8{,}034{,}000{,}000 = 8.034 \times 10^9$

## 5.3 Exercise Set

**1.** $10x^2 + 4x = 2x\cdot 5x + 2x\cdot 2$

$$= 2x(5x+2)$$

**3.** $y^2 - 4y = y\cdot y - y\cdot 4 = y(y-4)$

**5.** $x^3 + 5x^2 = x^2\cdot x + x^2\cdot 5 = x^2(x+5)$

**7.** $12x^4 - 8x^2 = 4x^2\cdot 3x^2 - 4x^2\cdot 2$

$$= 4x^2(3x^2-2)$$

**9.** $32x^4 + 2x^3 + 8x^2$

$$= 2x^2\cdot 16x^2 + 2x^2\cdot x + 2x^2\cdot 4$$
$$= 2x^2(16x^2 + x + 4)$$

**11.** $4x^2y^3+6xy=2xy\cdot 2xy^2+2xy\cdot 3$
$=2xy\left(2xy^2+3\right)$

**13.** $30x^2y^3-10xy^2$
$=10xy^2\cdot 3xy-10xy^2\cdot 1$
$=10xy^2\left(3xy-1\right)$

**15.** $12xy-6xz+4xw$
$=2x\cdot 6y-2x\cdot 3z+2x\cdot 2w$
$=2x\left(6y-3z+2w\right)$

**17.** $15x^3y^6-9x^4y^4+12x^2y^5$
$=3x^2y^4\cdot 5xy^2-3x^2y^4\cdot 3x^2+3x^2y^4\cdot 4y$
$=3x^2y^4\left(5xy^2-3x^2+4y\right)$

**19.** $25x^3y^6z^2-15x^4y^4z^4+25x^2y^5z^3$
$=5x^2y^4z^2\cdot 5xy^2-5x^2y^4z^2\cdot 3x^2z^2$
$+5x^2y^4z^2\cdot 5yz$
$=5x^2y^4z^2\left(5xy^2-3x^2z^2+5yz\right)$

**21.** $15x^{2n}-25x^n=5x^n\cdot 3x^n-5x^n\cdot 5$
$=5x^n\left(3x^n-5\right)$

**23.** $-4x+12=-4\cdot x+(-4)(-3)$
$=-4(x-3)$

**25.** $-8x-48=-8\cdot x+(-8)6$
$=-8(x+6)$

**27.** $-2x^2+6x-14$
$=-2\cdot x^2+(-2)(-3x)+(-2)(7)$
$=-2\left(x^2+(-3x)+7\right)$
$=-2\left(x^2-3x+7\right)$

**29.** $-5y^2+40x=-5\cdot y^2+(-5)(-8x)$
$=-5\left(y^2+(-8x)\right)$
$=-5\left(y^2-8x\right)$

**31.** $-4x^3+32x^2-20x$
$=-4x\cdot x^2+(-4x)(-8x)+(-4x)(5)$
$=-4x\left(x^2+(-8x)+5\right)$
$=-4x\left(x^2-8x+5\right)$

**33.** $-x^2-7x+5$
$=-1\cdot x^2+(-1)(7x)+(-1)(-5)$
$=-1\left(x^2+7x+(-5)\right)$
$=-1\left(x^2+7x-5\right)$

**35.** $4(x+3)+a(x+3)$
$=(x+3)(4+a)$

**37.** $x(y-6)-7(y-6)$
$=(y-6)(x-7)$

**39.** $3x(x+y)-(x+y)$
$=3x(x+y)-1(x+y)$
$=(x+y)(3x-1)$

**41.** $4x^2(3x-1)+3x-1$
$=4x^2(3x-1)+1(3x-1)$
$=(3x-1)\left(4x^2+1\right)$

**43.** $(x+2)(x+3)+(x-1)(x+3)$
$=(x+3)(x+2+x-1)$
$=(x+3)(2x+1)$

**45.** $x^2+3x+5x+15$
$=x(x+3)+5(x+3)$
$=(x+3)(x+5)$

**47.** $x^2+7x-4x-28$
$=x(x+7)-4(x+7)$
$=(x+7)(x-4)$

**49.** $x^3-3x^2+4x-12$
$=x^2(x-3)+4(x-3)$
$=(x-3)(x^2+4)$

**51.** $xy-6x+2y-12$
$=x(y-6)+2(y-6)$
$=(y-6)(x+2)$

**53.** $xy+x-7y-7$
$=x(y+1)-7(y+1)$
$=(y+1)(x-7)$

**55.** $10x^2-12xy+35xy-42y^2$
$=2x(5x-6y)+7y(5x-6y)$
$=(5x-6y)(2x+7y)$

**57.** $4x^3-x^2-12x+3$
$=x^2(4x-1)-3(4x-1)$
$=(4x-1)(x^2-3)$

**59.** $x^2-ax-bx+ab$
$=x(x-a)-b(x-a)$
$=(x-a)(x-b)$

**61.** $x^3-12-3x^2+4x$
$=x^3-3x^2+4x-12$
$=x^2(x-3)+4(x-3)$
$=(x-3)(x^2+4)$

**63.** $ay-by+bx-ax$
$=y(a-b)+x(b-a)$
$=y(a-b)+x(-1)(a-b)$
$=y(a-b)-x(a-b)$
$=(a-b)(y-x)$

**65.** $ay^2+2by^2-3ax-6bx$
$=y^2\cdot a+y^2\cdot 2b+(-3x)\cdot a+(-3x)\cdot 2b$
$=y^2(a+2b)-3x(a+2b)$
$=(a+2b)(y^2-3x)$

**67.** $x^ny^n+3x^n+y^n+3$
$=x^n\cdot y^n+x^n\cdot 3+1\cdot y^n+1\cdot 3$
$=x^n(y^n+3)+1(y^n+3)$
$=(y^n+3)(x^n+1)$

**69.** $ab-c-ac+b$
$=ab+b-ac-c$
$=a\cdot b+1\cdot b+(-c)\cdot a+(-c)\cdot 1$
$=b(a+1)+(-c)(a+1)$
$=(a+1)(b-c)$

**70.** $ab-3c-ac+3b=ab-ac+3b-3c$
$=a\cdot b-a\cdot c+3\cdot b-3\cdot c$
$=a(b-c)+3(b-c)$
$=(b-c)(a+3)$

**71.** $x^3-5+4x^3y-20y$
$=1\cdot x^3-1\cdot 5+4y\cdot x^3-4y\cdot 5$
$=1\left(x^3-5\right)+4y\left(x^3-5\right)$
$=\left(x^3-5\right)(1+4y)$

**72.** $x^3-2+3x^3y-6y$
$=1\cdot x^3-1\cdot 2+3y\cdot x^3-3y\cdot 2$
$=1\left(x^3-2\right)+3y\left(x^3-2\right)$
$=\left(x^3-2\right)(1+3y)$

**73.** $2y^7(3x-1)^5-7y^6(3x-1)^4$
$=y^6(3x-1)^4\cdot 2y(3x-1)-y^6(3x-1)^4\cdot 7$
$=y^6(3x-1)^4\left(2y(3x-1)-7\right)$
$=y^6(3x-1)^4(6xy-2y-7)$

**74.** $3y^9(3x-2)^7-5y^8(3x-2)^6$
$=y^8(3x-2)^6\cdot 3y(3x-2)-y^8(3x-2)^6\cdot 5$
$=y^8(3x-2)^6\left(3y(3x-2)-5\right)$
$=y^8(3x-2)^6(9xy-6y-5)$

**75.** $ax^2+5ax-2a+bx^2+5bx-2b$
$=a\cdot x^2+a\cdot 5x-a\cdot 2+b\cdot x^2+b\cdot 5x-b\cdot 2$
$=a\left(x^2+5x-2\right)+b\left(x^2+5x-2\right)$
$=\left(x^2+5x-2\right)(a+b)$

**76.** $ax^2+3ax-11a+bx^2+3bx-11b$
$=a\cdot x^2+a\cdot 3x-a\cdot 11+b\cdot x^2+b\cdot 3x-b\cdot 11$
$=a\left(x^2+3x-11\right)+b\left(x^2+2x-11\right)$
$=\left(x^2+3x-11\right)(a+b)$

**77.** $ax+ay+az-bx-by-bz+cx+cy+cz$
$=a\cdot x+a\cdot y+a\cdot z-b\cdot x-b\cdot y-b\cdot z$
$+c\cdot x+c\cdot y+c\cdot z$
$=a(x+y+z)-b(x+y+z)+c(x+y+z)$
$=(x+y+z)(a-b+c)$

**78.** $ax^2+ay^2-az^2+bx^2+by^2-bz^2$
$+cx^2+cy^2-cz^2$
$=a\cdot x^2+a\cdot y^2-a\cdot z^2+b\cdot x^2+b\cdot y^2$
$-b\cdot z^2+c\cdot x^{\cdot}+c\cdot y^2-c\cdot z^2$
$=a\left(x^2+y^2-z^2\right)+b\left(x^2+y^2-z^2\right)$
$+c\left(x^2+y^2-z^2\right)$
$=\left(x^2+y^2-z^2\right)(a+b+c)$

**79.** $f(t)=-16t^2+40t$

**a.** $f(2)=-16(2)^2+40(2)$
$=-16(4)+80$
$=-64+80$
$=16$

After 2 seconds, the ball will be at a height of 16 feet.

**b.** $f(2.5)=-16(2.5)^2+40(2.5)$
$=-16(6.25)+100$
$=-100+100=0$

After 2.5 seconds, the ball will hit the ground.

**c.** $-16t^2+40t=-4t\cdot 4t-4t(-10)$
$=-8t(2t-5)$

**d.** $f(t) = -4t(4t-10)$

$f(2) = -4(2)(4(2)-10)$
$= -8(8-10)$
$= -8(-2) = 16$

$f(2.5) = -4(2.5)(4(2.5)-10)$
$= -10(10-10)$
$= -10(0) = 0$

These are the same answers as parts **a.** and **b.** This shows that the factorization is equivalent to the original polynomial and that the factorization is correct.

**81.** **a.** $(x-0.4x)-0.4(x-0.4x)$
$=(x-0.4x)(1-0.4)$
$=(0.6x)(0.6)$
$=0.36x$

**b.** The computer is selling at 36% of its original price.

**83.** $A = P + Pr + (P+Pr)r$
$= (P+Pr)1 + (P+Pr)r$
$= (P+Pr)(1+r)$
$= (P\cdot 1 + P\cdot r)(1+r)$
$= P(1+r)(1+r)$
$= P(1+r)^2$

**85.** $A = \pi r^2 + 2rl$
$= r\cdot \pi r + r\cdot 2l$
$= r(\pi r + 2l)$

**87.** Answers will vary.

**89.** Answers will vary.

**91.** Answers will vary.

**93.** $x^2 - 4x = x(x-4)$

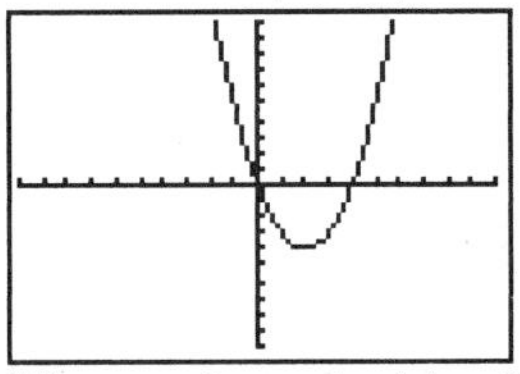

The graphs coincide. The polynomial is factored correctly.

**95.** $x^2 + 2x + x + 2 = x(x+2)+1$

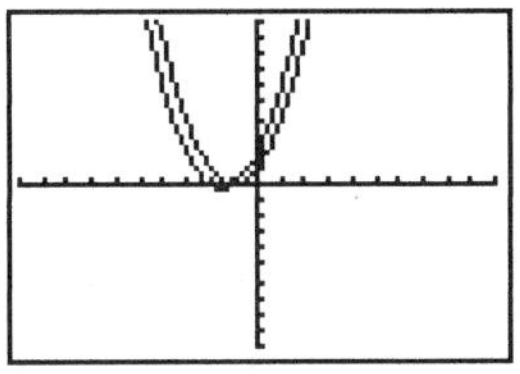

The graphs do not coincide. Factor the polynomial correctly.

$x^2 + 2x + x + 2 = x(x+2)+1(x+2)$
$= (x+2)(x+1)$

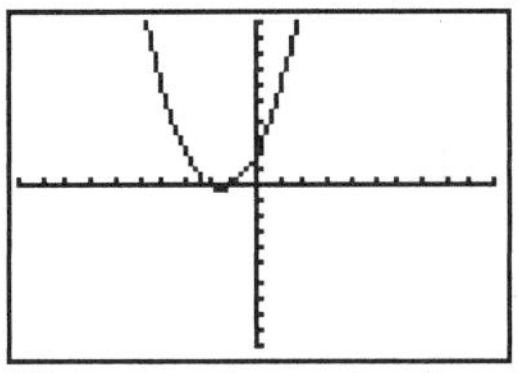

The new factorization is correct. The graphs coincide.

**97.** Statement **c.** is true. Some polynomials with four terms cannot be factored by grouping.

Statement **a.** is false. If all terms of a polynomial contain the same variable raised to different powers, the exponent on the variable of the GCF is the lowest power that appears in all the terms.

Statement **b.** is false. It is necessary to write the 1 when the expression is written is factored form.

**99.** $3x^{3m}y^m - 6x^{2m}y^{2m}$
$= 3x^{2m}y^m \cdot x^m - 3x^{2m}y^m \cdot 2y^m$
$= 3x^{2m}y^m\left(x^m - 2y^m\right)$

**101.** Answers vary. One example is:
$-6a^2b + 2a^2b^2 - 8a^2b^3$.

**103.** $3x - 2y = 8$
$2x - 5y = 10$

Because $x = \dfrac{D_x}{D}$ and $y = \dfrac{D_y}{D}$, we need to set up and evaluate $D$, $D_x$, and $D_y$.

$D = \begin{vmatrix} 3 & -2 \\ 2 & -5 \end{vmatrix} = (3)(-5) - (2)(-2)$
$= -15 + 4 = -11$

$D_x = \begin{vmatrix} 8 & -2 \\ 10 & -5 \end{vmatrix} = (8)(-5) - (10)(-2)$
$= -40 + 20 = -20$

$D_y = \begin{vmatrix} 3 & 8 \\ 2 & 10 \end{vmatrix} = (3)(10) - (2)(8)$
$= 30 - 16 = 14$

$x = \dfrac{D_x}{D} = \dfrac{-20}{-11} = \dfrac{20}{11}$

$y = \dfrac{D_y}{D} = \dfrac{14}{-11} = -\dfrac{14}{11}$

The solution is $\left(\dfrac{20}{11}, -\dfrac{14}{11}\right)$ and the solution set is $\left\{\left(\dfrac{20}{11}, -\dfrac{14}{11}\right)\right\}$.

**104.** **a.** The relation is a function.

**b.** The relation is not a function.

**105.** Let $w$ = the width of the rectangle
Let $2w + 2$ = the length of the rectangle
$P = 2l + 2w$
$22 = 2(2w+2) + 2w$
$22 = 4w + 4 + 2w$
$22 = 6w + 4$
$18 = 6w$
$3 = w$
To find the length:
$2w + 2 = 2(3) + 2 = 6 + 2 = 8$
The length is 8 feet and the width is 3 feet.

## 5.4 Exercise Set

**1.** $x^2 + 5x + 6 = (x+3)(x+2)$
product: 3(2) = 6
sum: 3 + 2 = 5

**3.** $x^2 + 8x + 12 = (x+6)(x+2)$
product: 6(2) = 12
sum: 6 + 2 = 8

**5.** $x^2 + 9x + 20 = (x+5)(x+4)$
product: 5(4) = 20
sum: 5 + 4 = 9

**7.** $y^2 + 10y + 16 = (y+8)(y+2)$
product: 8(2) = 16
sum: 8 + 2 = 10

**9.** $x^2 - 8x + 15 = (x-5)(x-3)$
product: −5(−3)=15
sum: −5 + (−3) = −8

**11.** $y^2 - 12y + 20 = (y-10)(y-2)$
product: −10(−2) = 20
sum: −10 + (−2) = −12

**13.** $a^2+5a-14=(a+7)(a-2)$
product: 7(−2) = −14
sum: 7 + (−2) = 5

**15.** $x^2+x-30=(x+6)(x-5)$
product: 6(−5)= −30
sum: 6 + (−5) = 1

**17.** $x^2-3x-28=(x-7)(x+4)$
product: −7(4)= −28
sum: −7 + 4 = −3

**19.** $y^2-5y-36=(y-9)(y+4)$
product: −9(4) = −36
sum: −9 + 4 = −5

**21.** $x^2-x+7$
The trinomial is not factorable. There are no factors of 7 that add up to −1. The polynomial is prime.

**23.** $x^2-9xy+14y^2=(x-7y)(x-2y)$
product: $-7y(-2y)=14y^2$
sum: $-7y+(-2y)=-9y$

**25.** $x^2-xy-30y^2=(x-6y)(x+5y)$
product: $-6(5)=-30$
sum: $-6+5=-1$

**27.** $x^2+xy+y^2$
The trinomial is not factorable. There are no factors of 1 that add up to 1. The polynomial is prime.

**29.** $a^2-18ab+80b^2=(a-10b)(a-8b)$
Product: $-10(-8)=80$
Sum: $-10+(-8)=-18$

**31.** $3x^2+3x-18=3(x^2+x-6)$
$=3(x+3)(x-2)$

**33.** $2x^3-14x^2+24x=2x(x^2-7x+12)$
$=2x(x-4)(x-3)$

**35.** $3y^3-15y^2+18y$
$=3y(y^2-5y+6)$
$=3y(y-3)(y-2)$

**37.** $2x^4-26x^3-96x^2$
$=2x^2(x^2-13x-48)$
$=2x^2(x-16)(x+3)$

**39.** Let $t=x^3$
$x^6-x^3-6=(x^3)^2-x^3-6$
$=t^2-t-6$
$=(t-3)(t+2)$
Substitute $x^3$ for $t$.
$=(x^3-3)(x^3+2)$

**41.** Let $t=x^2$
$x^4-5x^2-6=(x^2)^2-5x^2-6$
$=t^2-5t-6$
$=(t-6)(t+1)$
Substitute $x^2$ for $t$.
$=(x^2-6)(x^2+1)$

**43.** Let $t = x+1$

$(x+1)^2 + 6(x+1) + 5$

$= t^2 + 6t + 5$

$= (t+5)(t+1)$

Substitute $x+1$ for $t$.

$= ((x+1)+5)((x+1)+1)$

$= (x+1+5)(x+1+1)$

$= (x+6)(x+2)$

**45.** $3x^2 + 8x + 5 = (3x+5)(x+1)$

**47.** $5x^2 + 56x + 11 = (5x+1)(x+11)$

**49.** $3y^2 + 22y - 16 = (3y-2)(y+8)$

**51.** $4y^2 + 9y + 2 = (y+2)(4y+1)$

**53.** $10x^2 + 19x + 6 = (5x+2)(2x+3)$

**55.** $8x^2 - 18x + 9 = (4x-3)(2x-3)$

**57.** $6y^2 - 23y + 15 = (6y-5)(y-3)$

**59.** $6y^2 + 14y + 3$

The trinomial is not factorable. The polynomial is prime.

**61.** $3x^2 + 4xy + y^2 = (3x+y)(x+y)$

**63.** $6x^2 - 7xy - 5y^2 = (2x+y)(3x-5y)$

**65.** $15x^2 - 31xy + 10y^2$

$= (3x-5y)(5x-2y)$

**67.** $3a^2 - ab - 14b^2 = (3a-7b)(a+2b)$

**69.** $15x^3 - 25x^2 + 10x$

$= 5x(3x^2 - 5x + 2)$

$= 5x(3x-2)(x-1)$

**71.** $24x^4 + 10x^3 - 4x^2$

$= 2x^2(12x^2 + 5x - 2)$

$= 2x^2(3x+2)(4x-1)$

**73.** $15y^5 - 2y^4 - y^3 = y^3(15y^2 - 2y - 1)$

$= y^3(3y-1)(5y+1)$

**75.** $24x^2 + 3xy - 27y^2$

$= 3(8x^2 + xy - 9y^2)$

$= 3(8x+9y)(x-y)$

**77.** $6a^2b - 2ab - 60b$

$= 2b(3a^2 - a - 30)$

$= 2b(3a-10)(a+3)$

**79.** $12x^2y - 34xy^2 + 14y^3$

$= 2y(6x^2 - 17xy + 7y^2)$

$= 2y(3x-7y)(2x-y)$

**81.** $13x^3y^3 + 39x^3y^2 - 52x^3y$

$= 13x^3y(y^2 + 3y - 4)$

$= 13x^3y(y+4)(y-1)$

**83.** Let $t = x^2$

$2x^4 - x^2 - 3$

$= 2(x^2)^2 - x^2 - 3$

$= 2t^2 - t - 3$

$= (2t-3)(t+1)$

Substitute $x^2$ for $t$.

$= (2x^2-3)(x^2+1)$

**85.** Let $t = x^3$

$2x^6 + 11x^3 + 15$

$= 2(x^3)^2 + 11x^3 + 15$

$= 2t^2 + 11t + 15$

$= (2t+5)(t+3)$

Substitute $x^3$ for $t$.

$= (2x^3+5)(x^3+3)$

**87.** Let $t = y^5$

$2y^{10} + 7y^5 + 3$

$= 2(y^5)^2 + 7y^5 + 3$

$= 2t^2 + 7t + 3$

$= (2t+1)(t+3)$

Substitute $y^5$ for $t$.

$= (2y^5+1)(y^5+3)$

**89.** Let $t = x+1$

$5(x+1)^2 + 12(x+1) + 7$

$= 5t^2 + 12t + 7$

$= (5t+7)(t+1)$

Substitute $x+1$ for $t$.

$= (5(x+1)+7)((x+1)+1)$

$= (5x+5+7)(x+1+1)$

$= (5x+12)(x+2)$

**91.** Let $t = x-3$

$2(x-3)^2 - 5(x-3) - 7$

$= 2t^2 - 5t - 7$

$= (2t-7)(t+1)$

Substitute $x-3$ for $t$.

$= (2(x-3)-7)((x-3)+1)$

$= (2x-6-7)(x-3+1)$

$= (2x-13)(x-2)$

**93.** $x^2 - 0.5x + 0.06$

Since $(0.3)\cdot(0.2) = 0.06$ and $0.3 + 0.2 = 0.5$, we get

$x^2 - 0.5x + 0.06 = (x-0.3)(x-0.2)$

**94.** $x^2 + 0.3x - 0.04$

Since $(0.4)(-0.1) = -0.04$ and $0.4 + (-0.1) = 0.3$, we get

$x^2 + 0.3x - 0.04 = (x+0.4)(x-0.1)$

**95.** $x^2 - \frac{3}{49} + \frac{2}{7}x = x^2 + \frac{2}{7}x - \frac{3}{49}$

Since $\left(\frac{3}{7}\right)\left(-\frac{1}{7}\right) = -\frac{3}{49}$ and $\frac{3}{7} + \left(-\frac{1}{7}\right) = \frac{2}{7}$, we get

$x^2 - \frac{3}{49} + \frac{2}{7}x = \left(x + \frac{3}{7}\right)\left(x - \frac{1}{7}\right)$

**96.** $x^2 - \frac{6}{25} + \frac{1}{5}x = x^2 + \frac{1}{5}x - \frac{6}{25}$

Since $\left(\frac{3}{5}\right)\left(-\frac{2}{5}\right) = -\frac{6}{25}$ and $\frac{3}{5} + \left(-\frac{2}{5}\right) = \frac{1}{5}$, we get

$x^2 + \frac{1}{5}x - \frac{6}{25} = \left(x + \frac{3}{5}\right)\left(x - \frac{2}{5}\right)$

**97.** $acx^2 - bcx + adx - bd$
$= cx(ax-b) + d(ax-b)$
$= (ax-b)(cx+d)$

**98.** $acx^2 - bcx - adx + bd$
$= cx(ax-b) - d(ax-b)$
$= (ax-b)(cx-d)$

**99.** $-4x^5y^2 + 7x^4y^3 - 3x^3y^4$
$= (-x^3y^2)\cdot 4x^2 - (-x^3y^2)\cdot 7xy + (-x^3y^2)\cdot 3y^2$
$= -x^3y^2(4x^2 - 7xy + 3y^2)$
$= -x^3y^2(4x-3y)(x-y)$

**100.** $-5x^4y^3 + 7x^3y^4 - 2x^2y^5$
$= (-x^2y^3)\cdot 5x^2 - (-x^2y^3)\cdot 7xy + (-x^2y^3)\cdot 2y^2$
$= -x^2y^3(5x^2 - 7xy + 2y^2)$
$= -x^2y^3(5x-2y)(x-y)$

**101.** $(fg)(x) = 3x^2 - 22x + 39$
$= (3x-13)(x-3)$
$f(x) = 3x - 13$ and $g(x) = x - 3$, or vice versa.

**102.** $(fg)(x) = 4x^2 - x - 5$
$= (4x-5)(x+1)$
$f(x) = 4x - 5$ and $g(x) = x + 1$, or vice versa.

**103.** $x^2 + x^2 + x + x + x + x + x + x + x + 1 + 1 + 1$
$= 2x^2 + 7x + 3$
$2x^2 + 7x + 3 = (2x+1)(x+3)$
The dimensions of the rectangle are $2x+1$ by $x+3$.

**104.** $x^2 + x + x + x + x + x + 1 + 1 + 1 + 1$
$= x^2 + 5x + 4$
$x^2 + 5x + 4 = (x+4)(x+1)$
The dimensions of the rectangle are $x+4$ by $x+1$.

**105. a.** $f(1) = -16(1)^2 + 16(1) + 32$
$= -16(1) + 16 + 32$
$= -16 + 16 + 32$
$= 32$
After 1 second, the diver will be 32 feet above the water.

**b.** $f(2) = -16(2)^2 + 16(2) + 32$
$= -16(4) + 32 + 32$
$= -64 + 32 + 32 = 0$
After 2 seconds, the diver will hit the water.

**c.** $-16t^2 + 16t + 32$
$= -16(t^2 - t - 2)$
$= -16(t-2)(t+1)$

**d.** $f(t) = -16(t-2)(t+1)$
$f(1) = -16(1-2)(1+1)$
$= -16(-1)(2) = 32$
$f(2) = -16(2-2)(2+1)$
$= -16(0)(3) = 0$

**107.** **a.** $x \cdot x + 1 \cdot x + 1 \cdot x + 1 \cdot 1 + 1 \cdot x + 1 \cdot 1$
$= x^2 + x + x + 1 + x + 1$
$= x^2 + 3x + 2$

**b.** $(x+1)(x+2) = x^2 + 2x + x + 2$
$= x^2 + 3x + 2$

**c.** Answers will vary.

**109.** Answers will vary.

**111.** Answers will vary.

**113.** Answers will vary.

**115.** Answers will vary.

**117.** $x^2 + 7x + 12 = (x+4)(x+3)$

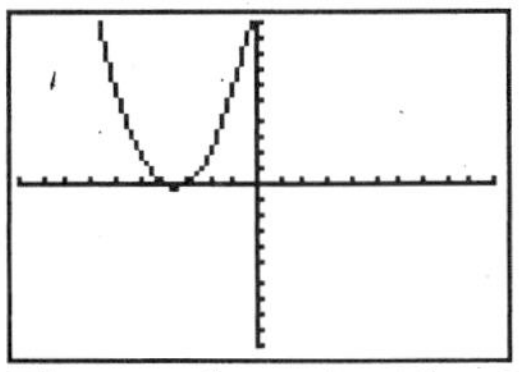

The graphs coincide. The polynomial is factored correctly.

**119.** $6x^3 + 5x^2 - 4x = x(3x+4)(2x-1)$

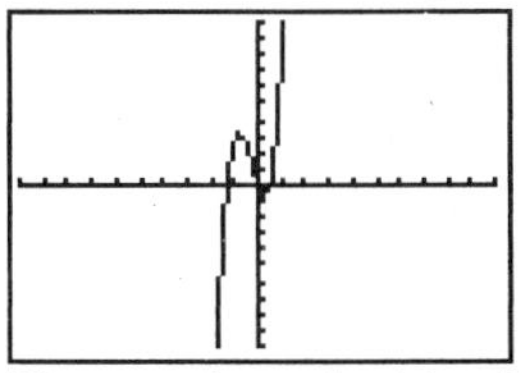

The graphs coincide. The polynomial is factored correctly.

**121.** Answers will vary.

**123.** Use the factors of 4 and 1 to determine find $b$. The factors of 4 are 4 and 1 or 2 and 2. The factors of $-1$ are 1 and $-1$. Multiply the combinations of the factors as follows.

$(4x-1)(x+1) = 4x^2 + 4x - x - 1$
$= 4x^2 + 3x - 1$

$(4x+1)(x-1) = 4x^2 - 4x + x - 1$
$= 4x^2 - 3x - 1$

$(2x+1)(2x-1) = 4x^2 - 2x + 2x - 1$
$= 4x^2 - 1$

The integers are $-3$, 0, and 3.

**125.** Let $t = x^n$

$9x^{2n} + x^n - 8 = 9(x^n)^2 + x^n - 8$
$= 9t^2 + t - 8$
$= (9t-8)(t+1)$
$= (9x^n - 8)(x^n + 1)$

**127.** $a^{2n+2} - a^{n+2} - 6a^2 = a^2(a^{2n} - a^n - 6)$
$= a^2(a-3)(a-2)$

**129.** $3c^{n+2} - 10c^{n+1} + 3c^n = c^n(3c^2 - 10c + 3)$
$= c^n(3c-1)(c-3)$

**131.** $-2x \le 6$ and $-2x + 3 < -7$
$x \ge -3$ $\quad -2x < -10$
$x > 5$

$x \ge -3$

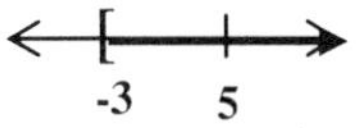


$x > 5$

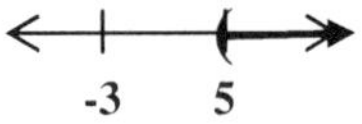


$x \ge -3$ and $x > 5$

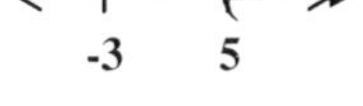


The solution set is $\{x | x > 5\}$ or $(5, \infty)$.

**132.** $2x - y - 2z = -1$
$x - 2y - z = 1$
$x + y + z = 4$

Multiply the second equation by –2 and add to the first equation.

$2x - y - 2z = -1$
$\underline{-2x + 4y + 2z = -2}$
$3y = -3$
$y = -1$

Add the second and third equations to eliminate $z$.

$x - 2y - z = 1$
$\underline{x + y + z = 4}$
$2x - y = 5$

Back-substitute –1 for $y$ to find $x$.

$2x - (-1) = 5$
$2x + 1 = 5$
$2x = 4$
$x = 2$

Back-substitute 2 for $x$ and –1 for $y$ to find $z$.

$x + y + z = 4$
$2 + (-1) + z = 4$
$1 + z = 4$
$z = 3$

The solution is $(2, -1, 3)$ or $\{(2, -1, 3)\}$.

**133.** $x - y \geq -4$
$x + 2y \geq 2$

First consider $x - y \geq -4$. Solve for $y$ in $x - y = -4$, and graph using the slope and $y$-intercept.

$x - y = -4$
$-y = -x - 4$
$y = x + 4$

$y$-intercept = 4 slope = 1

Now, use the origin, $(0,0)$, as a test point.

$0 - 0 \geq -4$
$0 \geq -4$

This is a true statement and the point $(0,0)$ will fall in the shaded half-plane.

Next consider $x + 2y \geq 2$. Solve for $y$ in $x + 2y = 2$, and graph using the slope and the $y$-intercept.

$x + 2y = 2$
$2y = -x + 2$
$y = -\frac{1}{2}x + 1$

$y$-intercept = 1 slope = $-\frac{1}{2}$

Now, use the origin, $(0,0)$, as a test point.

$0 + \cancel{2(0)} \geq 2$
$0 \geq 2$

This is a false statement, so the point $(0,0)$ will not fall in the shaded half-plane.

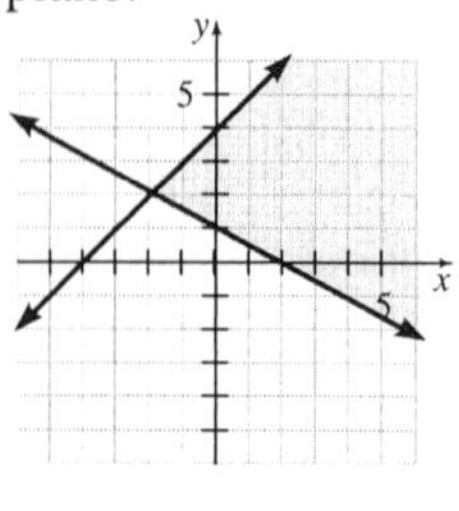

## Mid-Chapter 5 Check Point

**1.** $\left(-8x^3+6x^2-x+5\right)-\left(-7x^3+2x^2-7x-12\right)$

$=-8x^3+6x^2-x+5+7x^3-2x^2+7x+12$

$=-8x^3+7x^3+6x^2-2x^2-x+7x+5+12$

$=-x^3+4x^2+6x+17$

**2.** $\left(6x^2yz^4\right)\left(-\frac{1}{3}x^5y^2z\right)=(6)\left(-\frac{1}{3}\right)x^2\cdot x^5\cdot y\cdot y^2\cdot z^4\cdot z=-2x^7y^3z^5$

**3.** $5x^2y\left(6x^3y^2-7xy-\frac{2}{5}\right)=5x^2y\cdot 6x^3y^2-5x^2y\cdot 7xy-5x^2y\cdot\frac{2}{5}$

$=30x^5y^3-35x^3y^2-2x^2y$

**4.** $(3x-5)\left(x^2+3x-8\right)=3x\left(x^2+3x-8\right)-5\left(x^2+3x-8\right)$

$=3x^3+9x^2-24x-5x^2-15x+40$

$=3x^3+4x^2-39x+40$

**5.** $\left(x^2-2x+1\right)\left(2x^2+3x-4\right)=x^2\left(2x^2+3x-4\right)-2x\left(2x^2+3x-4\right)+1\left(2x^2+3x-4\right)$

$=2x^4+3x^3-4x^2-4x^3-6x^2+8x+2x^2+3x-4$

$=2x^4-x^3-8x^2+11x-4$

**6.** $\left(x^2-2x+1\right)-\left(2x^2+3x-4\right)=x^2-2x+1-2x^2-3x+4$

$=-x^2-5x+5$

**7.** $\left(6x^3y-11x^2y-4y\right)+\left(-11x^3y+5x^2y-y-6\right)-\left(-x^3y+2y-1\right)$

$=6x^3y-11x^2y-4y-11x^3y+5x^2y-y-6+x^3y-2y+1$

$=6x^3y-11x^3y+x^3y-11x^2y+5x^2y-4y-y-2y-6+1$

$=-4x^3y-6x^2y-7y-5$

**8.** $(2x+5)(4x-1)=2x\cdot 4x-2x\cdot 1+5\cdot 4x-5\cdot 1$

$=8x^2-2x+20x-5$

$=8x^2+18x-5$

**9.** $(2xy-3)(5xy+2) = 2xy\cdot 5xy + 2xy\cdot 2 - 3\cdot 5xy - 3\cdot 2 = 10x^2y^2 + 4xy - 15xy - 6$
$= 10x^2y^2 - 11xy - 6$

**10.** $(3x-2y)(3x+2y) = (3x)^2 - (2y)^2 = 9x^2 - 4y^2$

**11.** $(3xy+1)(2x^2-3y) = 3xy\cdot 2x^2 - 3xy\cdot 3y + 1\cdot 2x^2 - 1\cdot 3y = 6x^3y - 9xy^2 + 2x^2 - 3y$

**12.** $(7x^3y+5x)(7x^3y-5x) = (7x^3y)^2 - (5x)^2 = 49x^6y^2 - 25x^2$

**13.** $3(x+h)^2 - 2(x+h) + 5 - (3x^2 - 2x + 5) = 3(x^2 + 2xh + h^2) - 2x - 2h + 5 - 3x^2 + 2x - 5$
$= 3x^2 + 6xh + 3h^2 - 2x - 2h + 5 - 3x^2 + 2x - 5$
$= 3x^2 - 3x^2 - 2x + 2x + 6xh + 3h^2 - 2h + 5 - 5$
$= 6xh + 3h^2 - 2h$

**14.** $(x^2-3)^2 = (x^2)^2 - 2(x^2)(3) + 3^2$
$= x^4 - 6x^2 + 9$

**15.** $(x^2-3)(x^3+5x+2)$
$= x^2(x^3+5x+2) - 3(x^3+5x+2)$
$= x^5 + 5x^3 + 2x^2 - 3x^3 - 15x - 6$
$= x^5 + 2x^3 + 2x^2 - 15x - 6$

**16.** $(2x+5y)^2 = (2x)^2 + 2(2x)(5y) + (5y)^2$
$= 4x^2 + 20xy + 25y^2$

**17.** $(x+6+3y)(x+6-3y)$
$= ((x+6)+3y)((x+6)-3y)$
$= (x+6)^2 - (3y)^2$
$= x^2 + 12x + 36 - 9y^2$

**18.** $(x+y+5)^2$
$= ((x+y)+5)^2$
$= (x+y)^2 + 2(x+y)(5) + 5^2$
$= x^2 + 2xy + y^2 + 10x + 10y + 25$

**19.** $x^2 - 5x - 24$
We need two factors of $-24$ whose sum is $-5$. Since the product is negative, the factors will have opposite signs. Since the sum is negative, the factor with the larger absolute value will be negative.
$(-8)(3) = -24$ and $(-8)+(3) = -5$
$x^2 - 5x - 24 = (x-8)(x+3)$

**20.** $15xy + 5x + 6y + 2$
$= 5x(3y+1) + 2(3y+1)$
$= (3y+1)(5x+2)$

**21.** $5x^2+8x-4$

$a\cdot c=5\cdot(-4)=-20$

We need two factors of $-20$ whose sum is 8. Since the product is negative, the factors will have opposite signs. Since the sum is positive, the factor with the larger absolute value will be positive.

$(10)(-2)=-20$ and $(10)+(-2)=8$

$$\begin{aligned}5x^2+8x-4&=5x^2+10x-2x-4\\&=5x(x+2)-2(x+2)\\&=(x+2)(5x-2)\end{aligned}$$

**22.** $35x^2+10x-50=5(7x^2+2x-10)$

The reduced polynomial (in parentheses) cannot be factored further over the set of integers.

**23.** $$\begin{aligned}9x^2-9x-18&=9(x^2-x-2)\\&=9(x-2)(x+1)\end{aligned}$$

**24.** $10x^3y^2-20x^2y^2+35x^2y$

$=5x^2y(2xy-4y+7)$

**25.** $18x^2+21x+5$

$a\cdot c=18\cdot 5=90$

We need two factors of 90 whose sum is 21. Since both the product and sum are positive, the two factors will be positive.

$(15)(6)=90$ and $(15)+(6)=21$

$$\begin{aligned}18x^2+21x+5&=18x^2+15x+6x+5\\&=3x(6x+5)+1(6x+5)\\&=(6x+5)(3x+1)\end{aligned}$$

**26.** $12x^2-9xy-16x+12y$

$=3x(4x-3y)-4(4x-3y)$

$=(4x-3y)(3x-4)$

**27.** $9x^2-15x+4=(3x-1)(3x-4)$

$a\cdot c=9\cdot 4=36$

We need two factors of 36 whose sum is $-15$. Since the product is positive and the sum is negative, the two factors will be negative.

$(-12)(-3)=36$ and $(-12)+(-3)=-15$

$$\begin{aligned}9x^2-15x+4&=9x^2-3x-12x+4\\&=3x(3x-1)-4(3x-1)\\&=(3x-1)(3x-4)\end{aligned}$$

**28.** $3x^6+11x^3+10$

Let $t=x^3$. Then $t^2=\left(x^3\right)^2=x^6$.

$$\begin{aligned}3t^2+11t+10&=(3t+5)(t+2)\\&=\left(3x^3+5\right)\left(x^3+2\right)\end{aligned}$$

**29.** $$\begin{aligned}25x^3+25x^2-14x&=x(25x^2+25x-14)\\&=x(5x-2)(5x+7)\end{aligned}$$

**30.** $2x^4-6x-x^3y+3y$

$=2x(x^3-3)-y(x^3-3)$

$=(x^3-3)(2x-y)$

## 5.5 Exercise Set

**1.** $x^2 - 4 = x^2 - 2^2$
$= (x+2)(x-2)$

**3.** $9x^2 - 25 = (3x)^2 - 5^2$
$= (3x+5)(3x-5)$

**5.** $9 - 25y^2 = 3^2 - (5y)^2$
$= (3+5y)(3-5y)$

**7.** $36x^2 - 49y^2 = (6x)^2 - (7y)^2$
$= (6x+7y)(6x-7y)$

**9.** $x^2y^2 - 1 = (xy)^2 - 1^2$
$= (xy+1)(xy-1)$

**11.** $9x^4 - 25y^6 = (3x^2)^2 - (5y^3)^2$
$= (3x^2+5y^3)(3x^2-5y^3)$

**13.** $x^{14} - y^4 = (x^7)^2 - (y^2)^2$
$= (x^7+y^2)(x^7-y^2)$

**15.** $(x-3)^2 - y^2$
$= ((x-3)+y)((x-3)-y)$
$= (x-3+y)(x-3-y)$

**17.** $a^2 - (b-2)^2$
$= (a+(b-2))(a-(b-2))$
$= (a+b-2)(a-b+2)$

**19.** $x^{2n} - 25 = (x^n)^2 - 5^2$
$= (x^n+5)(x^n-5)$

**21.** $1 - a^{2n} = 1^2 - (a^n)^2$
$= (1+a^n)(1-a^n)$

**23.** $2x^3 - 8x = 2x(x^2-4)$
$= 2x(x^2-2^2)$
$= 2x(x+2)(x-2)$

**25.** $50 - 2y^2 = 2(25-y^2)$
$= 2(5^2-y^2)$
$= 2(5+y)(5-y)$

**27.** $8x^2 - 8y^2 = 8(x^2-y^2)$
$= 8(x+y)(x-y)$

**29.** $2x^3y - 18xy = 2xy(x^2-9)$
$= 2xy(x^2-3^2)$
$= 2xy(x+3)(x-3)$

**31.** $a^3b^2 - 49ac^2 = a(a^2b^2-49c^2)$
$= a((ab)^2-(7c)^2)$
$= a(ab+7c)(ab-7c)$

**33.** $5y - 5x^2y^7 = 5y(1-x^2y^6)$
$= 5y(1^2-(xy^3)^2)$
$= 5y(1+xy^3)(1-xy^3)$

**35.** $8x^2 + 8y^2 = 8(x^2+y^2)$

**37.** $x^2+25y^2$

Prime

The sum of two squares with no common factor other than 1 is a prime polynomial.

**39.** $$\begin{aligned} x^4-16 &= (x^2)^2-4^2 \\ &= (x^2+4)(x^2-4) \\ &= (x^2+4)(x^2-2^2) \\ &= (x^2+4)(x+2)(x-2) \end{aligned}$$

**41.** $$\begin{aligned} 81x^4-1 &= (9x^2)^2-1^2 \\ &= (9x^2+1)(9x^2-1) \\ &= (9x^2+1)((3x)^2-1^2) \\ &= (9x^2+1)(3x+1)(3x-1) \end{aligned}$$

**43.** $$\begin{aligned} &2x^5-2xy^4 \\ &= 2x(x^4-y^4) \\ &= 2x((x^2)^2-(y^2)^2) \\ &= 2x(x^2+y^2)(x^2-y^2) \\ &= 2x(x^2+y^2)(x+y)(x-y) \end{aligned}$$

**45.** $$\begin{aligned} &x^3+3x^2-4x-12 \\ &= (x^3+3x^2)+(-4x-12) \\ &= x^2(x+3)+(-4)(x+3) \\ &= (x+3)(x^2-4) \\ &= (x+3)(x^2-2^2) \\ &= (x+3)(x+2)(x-2) \end{aligned}$$

**47.** $$\begin{aligned} &x^3-7x^2-x+7 \\ &= (x^3-7x^2)+(-x+7) \\ &= x^2(x-7)+(-1)(x-7) \\ &= (x-7)(x^2-1) \\ &= (x-7)(x^2-1^2) \\ &= (x-7)(x+1)(x-1) \end{aligned}$$

**49.** $$\begin{aligned} x^2+4x+4 &= x^2+2\cdot x\cdot 2+2^2 \\ &= (x+2)^2 \end{aligned}$$

**51.** $$\begin{aligned} x^2-10x+25 &= x^2-2\cdot x\cdot 5+5^2 \\ &= (x-5)^2 \end{aligned}$$

**53.** $$\begin{aligned} x^4-4x^2+4 &= (x^2)^2-2\cdot x^2\cdot 2+2^2 \\ &= (x^2-2)^2 \end{aligned}$$

**55.** $$\begin{aligned} 9y^2+6y+1 &= (3y)^2+2\cdot y\cdot 3+1^2 \\ &= (3y+1)^2 \end{aligned}$$

**57.** $$\begin{aligned} 64y^2-16y+1 &= (8y)^2-2\cdot y\cdot 8+1^2 \\ &= (8y-1)^2 \end{aligned}$$

**59.** $$\begin{aligned} &x^2-12xy+36y^2 \\ &= x^2-2\cdot x\cdot 6y+(6y)^2 \\ &= (x-6y)^2 \end{aligned}$$

**61.** $x^2-8xy+64y^2$

Prime

Because the first and third terms of the polynomial are perfect squares, check to see if the trinomial is a perfect square. The middle term would have to be $-16xy$ for the polynomial to be a perfect square. Since this is not the case and the polynomial cannot be factored in any other way, we conclude that the polynomial is prime.

**63.** $9x^2+48xy+64y^2$
$=(3x)^2+2\cdot 3x\cdot 8y+(8y)^2$
$=(3x+8y)^2$

**65.** $x^2-6x+9-y^2$
$=(x^2-6x+9)-y^2$
$=(x^2-2\cdot x\cdot 3+3^2)-y^2$
$=(x-3)^2-y^2$
$=((x-3)+y)((x-3)-y)$
$=(x-3+y)(x-3-y)$

**67.** $x^2+20x+100-x^4$
$=(x^2+20x+100)-x^4$
$=(x^2+2\cdot x\cdot 10+10^2)-x^4$
$=(x+10)^2-(x^2)^2$
$=((x+10)+x^2)((x+10)-x^2)$
$=(x+10+x^2)(x+10-x^2)$

**69.** $9x^2-30x+25-36y^2$
$=(9x^2-30x+25)-36y^2$
$=((3x)^2-2\cdot 3x\cdot 5+5^2)-36y^2$
$=(3x-5)^2-(6y)^2$
$=((3x-5)+6y)((3x-5)-6y)$
$=(3x-5+6y)(3x-5-6y)$

**71.** $x^4-x^2-2x-1$
$=x^4-(x^2+2x+1)$
$=x^4-(x^2+2\cdot x\cdot 1+1^2)$
$=(x^2)^2-(x+1)^2$
$=(x^2+(x+1))(x^2-(x+1))$
$=(x^2+x+1)(x^2-x-1)$

**73.** $z^2-x^2+4xy-4y^2$
$=z^2-(x^2-4xy+4y^2)$
$=z^2-(x^2-2\cdot x\cdot 2y+(2y)^2)$
$=z^2-(x-2y)^2$
$=(z+(x-2y))(z-(x-2y))$
$=(z+x-2y)(z-x+2y)$

**75.** $x^3+64=x^3+4^3$
$=(x+4)(x^2-x\cdot 4+4^2)$
$=(x+4)(x^2-4x+16)$

**77.** $x^3-27=x^3-3^3$
$=(x-3)(x^2+3\cdot x+3^2)$
$=(x-3)(x^2+3x+9)$

**79.** $8y^3+1$
$=(2y)^3+1^3$
$=(2y+1)((2y)^2-2y\cdot 1+1^2)$
$=(2y+1)(4y^2-2y+1)$

**81.** $125x^3 - 8$
$= (5x)^3 - 2^3$
$= (5x-2)(25x^2 + 5x \cdot 2 + 2^2)$
$= (5x-2)(25x^2 + 10x + 4)$

**83.** $x^3y^3 + 27$
$= (xy)^3 + 3^3$
$= (xy+3)((xy)^2 - xy \cdot 3 + 3^2)$
$= (xy+3)(x^2y^2 - 3xy + 9)$

**85.** $64x - x^4$
$= x(64 - x^3)$
$= x(4^3 - x^3)$
$= x(4-x)(4^2 + 4 \cdot x + x^2)$
$= x(4-x)(16 + 4x + x^2)$

**87.** $x^6 + 27y^3$
$= (x^2)^3 + (3y)^3$
$= (x^2 + 3y)((x^2)^2 - x^2 \cdot 3y + (3y)^2)$
$= (x^2 + 3y)(x^4 - 3x^2y + 9y^2)$

**89.** $125x^6 - 64y^6$
$= (5x^2)^3 - (4y^2)^3$
$= (5x^2 - 4y^2)((5x^2)^2 + 5x^2 4y^2 + (4y^2)^2)$
$= (5x^2 - 4y^2)(25x^4 + 20x^2y^2 + 16y^4)$

**91.** $x^9 + 1$
$= (x^3)^3 + 1^3$
$= (x^3 + 1)((x^3)^2 - x^3 \cdot 1 + 1^2)$
$= (x^3 + 1)(x^6 - x^3 + 1)$
$= (x^3 + 1^3)(x^6 - x^3 + 1)$
$= (x+1)(x^2 - x \cdot 1 + 1^2)(x^6 - x^3 + 1)$
$= (x+1)(x^2 - x + 1)(x^6 - x^3 + 1)$

**93.** $(x-y)^3 - y^3$
$= ((x-y) - y)((x-y)^2 + y(x-y) + y^2)$
$= (x-2y)(x^2 - 2xy + y^2 + xy - y^2 + y^2)$
$= (x-2y)(x^2 - xy + y^2)$

**95.** $0.04x^2 + 0.12x + 0.09$
$= (0.2x + 0.03)^2$ or $\frac{1}{100}(2x+3)^2$

**96.** $0.09x^2 - 0.12x + 0.04$
$= (0.3x - 0.2)^2$ or $\frac{1}{100}(3x-2)^2$

**97.** $8x^4 - \frac{x}{8}$
$= x\left(8x^3 - \frac{1}{8}\right)$
$= x\left(2x - \frac{1}{2}\right)\left((2x)^2 + 2x \cdot \frac{1}{2} + \left(\frac{1}{2}\right)^2\right)$
$= x\left(2x - \frac{1}{2}\right)\left(4x^2 + x + \frac{1}{4}\right)$

**98.** $27x^4 + \dfrac{x}{27}$

$= x\left(27x^3 + \dfrac{1}{27}\right)$

$= x\left(3x + \dfrac{1}{3}\right)\left((3x)^2 - 3x \cdot \dfrac{1}{3} + \left(\dfrac{1}{3}\right)^2\right)$

$= x\left(3x + \dfrac{1}{3}\right)\left(9x^2 - x + \dfrac{1}{9}\right)$

**99.** $x^6 - 9x^3 + 8$

$= (x^3 - 1)(x^3 - 8)$

$= (x-1)(x^2 + x + 1)(x-2)(x^2 + 2x + 4)$

**100.** $x^6 + 9x^3 + 8$

$= (x^3 + 1)(x^3 + 8)$

$= (x+1)(x^2 - x + 1)(x+2)(x^2 - 2x + 4)$

**101.** $x^8 - 15x^4 - 16$

$= (x^4 + 1)(x^4 - 16)$

$= (x^4 + 1)(x^2 + 4)(x^2 - 4)$

$= (x^4 + 1)(x^2 + 4)(x+2)(x-2)$

**102.** $x^8 + 15x^4 - 16$

$= (x^4 + 16)(x^4 - 1)$

$= (x^4 + 16)(x^2 + 1)(x^2 - 1)$

$= (x^4 + 16)(x^2 + 1)(x+1)(x-1)$

**103.** $x^5 - x^3 - 8x^2 + 8$

$= x^3(x^2 - 1) - 8(x^2 - 1)$

$= (x^2 - 1)(x^3 - 8)$

$= (x+1)(x-1)(x-2)(x^2 + 2x + 4)$

**104.** $x^5 - x^3 + 27x^2 - 27$

$= x^3(x^2 - 1) + 27(x^2 - 1)$

$= (x^2 - 1)(x^3 + 27)$

$= (x+1)(x-1)(x+3)(x^2 - 3x + 9)$

**105.** $x^2 - 4^2 = (x+4)(x-4)$

**107.** $(3x)^2 - 4 \cdot 2^2 = (3x)^2 - 4 \cdot 4$

$= (3x)^2 - 4^2$

$= (3x+4)(3x-4)$

**109.** $V_{\text{shaded}} = V_{\text{outside}} - V_{\text{inside}}$

$= a \cdot a \cdot 4a - b \cdot b \cdot 4a$

$= 4a^3 - 4ab^2$

$= 4a(a^2 - b^2)$

$= 4a(a+b)(a-b)$

**In Exercises 111-113, answers will vary.**

**115.** $9x^2 - 4 = (3x+2)(3x-2)$

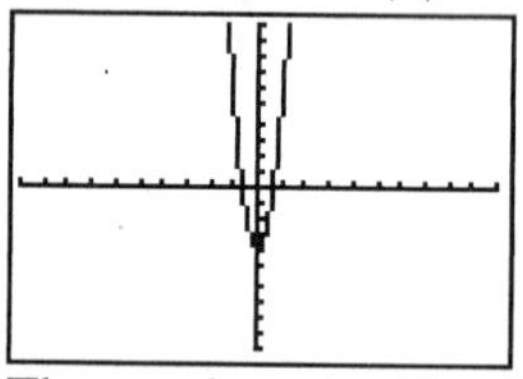

The graphs coincide. The polynomial is factored correctly.

**117.** $9x^2 + 12x + 4 = (3x+2)^2$

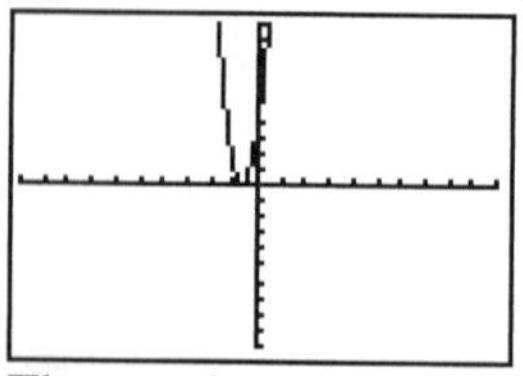

The graphs coincide. The polynomial is factored correctly.

**119.** $(2x+3)^2 - 9 = 4x(x+3)$

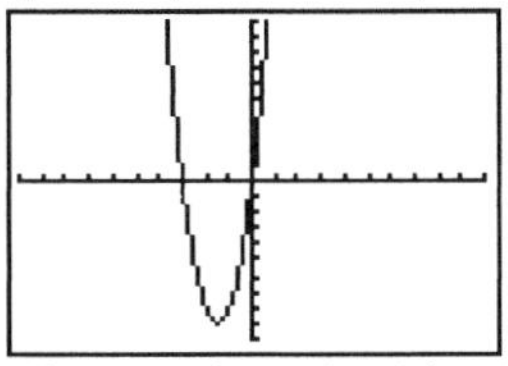

The graphs coincide. The polynomial is factored correctly.

**121.** $x^3 - 1 = (x-1)(x^2 - x + 1)$

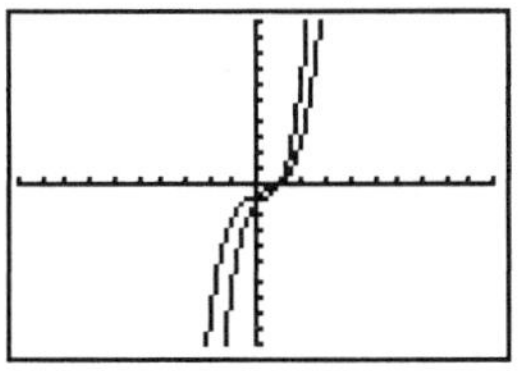

The graphs do not coincide. The factorization is not correct.

$x^3 - 1 = (x-1)(x^2 + x + 1)$

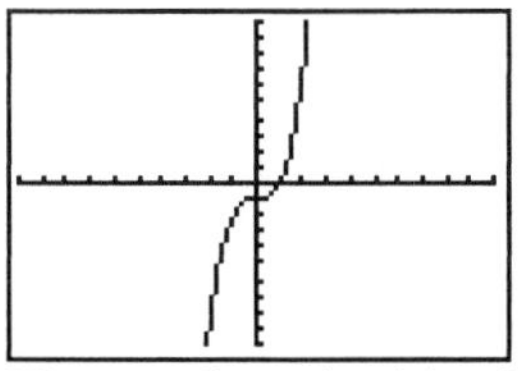

The graphs coincide. The new factorization is

**123.** Answers will vary.

**125.**

$$y^3 + x + x^3 + y$$
$$= y^3 + x^3 + y + x$$
$$= (y+x)(y^2 - xy + x^2) + (y+x)$$
$$= (y+x)((y^2 - xy + x^2 + 1)$$

**127.**

$$x^{3n} + y^{12n}$$
$$= (x^n)^3 + (y^{4n})^3$$
$$= (x^n + y^{4n})((x^n)^2 - x^n y^{4n} + (y^{4n})^2)$$
$$= (x^n + y^{4n})(x^{2n} - x^n y^{4n} + y^{8n})$$

**129.** Factoring as a difference of squares:

$$x^6 - y^6$$
$$= (x^3)^2 - (y^3)^2$$
$$= (x^3 + y^3)(x^3 - y^3)$$
$$= (x+y)(x^2 - xy + y^2)(x-y)(x^2 + xy + y^2)$$

Factoring as a difference of cubes:

$x^6 - y^6$

$(x^2)^3 - (y^2)^3$

$= (x^2 - y^2)(x^4 + x^2y^2 + y^4)$

$= (x + y)(x - y)(x^4 + x^2y^2 + y^4)$

Set the factorizations equal.

$$(x + y)(x - y)(x^4 + x^2y^2 + y^4) = (x + y)(x^2 - xy + y^2)(x - y)(x^2 + xy + y^2)$$

$$x^4 + x^2y^2 + y^4 = (x^2 - xy + y^2)(x^2 + xy + y^2)$$

**131.** In a perfect square trinomial, the middle term is $2AB$ and the first term is $A^2$. Rewrite the middle term in $2AB$ form and the first term in $A^2$ form..

$64x^2 - 16x + k$

$= (8x)^2 - 2 \cdot 8x \cdot 1 + k$

This means that $B$ must be 1. The last term is $B^2 = (1)^2 = 1$. From this, we see that $k = 1$.

**132.**

$2x + 2 \geq 12$ and $\dfrac{2x-1}{3} \leq 7$

$2x \geq 10$ $\qquad 2x - 1 \leq 21$

$x \geq 5$ $\qquad 2x \leq 22$

$x \leq 11$

$x \geq 5$

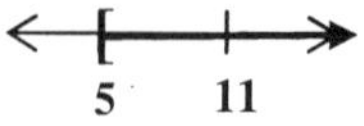

$x \leq 11$

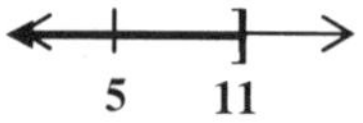

$x \geq 5$ and $x \leq 11$

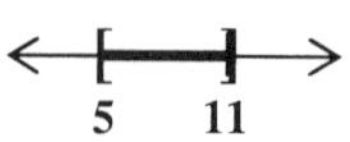

The solution set is $\{x | 5 \leq x \leq 11\}$ or $[5,11]$.

**133.** $3x - 2y = -8$

$x + 6y = 4$

Write the augmented matrix and solve by Gaussian elimination.

$$\left[\begin{array}{cc|c} 3 & -2 & -8 \\ 1 & 6 & 4 \end{array}\right] \quad R_1 \leftrightarrow R_2$$

$$= \left[\begin{array}{cc|c} 1 & 6 & 4 \\ 3 & -2 & -8 \end{array}\right] \quad -3R_1 + R_2$$

$$= \left[\begin{array}{cc|c} 1 & 6 & 4 \\ 0 & -20 & -20 \end{array}\right] \quad -\frac{1}{20}R_2$$

$$= \left[\begin{array}{cc|c} 1 & 6 & 4 \\ 0 & 1 & 1 \end{array}\right]$$

The system becomes

$x + 6y = 4$

$y = 1.$

Back-substitute 1 for $y$, to find $x$.

$x + 6(1) = 4$

$x + 6 = 4$

$x = -2$

The solution is set is $(-2,1)$ or $\{(-2,1)\}$.

**134.** $\dfrac{4^{-2}x^{-3}y^1}{x^2y^{-5}} = \dfrac{y^5y^1}{4^2x^3x^2} = \dfrac{y^6}{16x^5}$

## 5.6 Exercise Set

**1.** $x^3-16x=x(x^2-16)$
$=x(x+4)(x-4)$

**3.** $3x^2+18x+27=3(x^2+6x+9)$
$=3(x+3)^2$

**5.** $81x^3-3=3(27x^3-1)$
$=3(3x-1)(9x^2+3x+1)$

**7.** $x^2y-16y+32-2x^2$
$=(x^2y-16y)+(-2x^2+32)$
$=y(x^2-16)+(-2)(x^2-16)$
$=(x^2-16)(y+(-2))$
$=(x+4)(x-4)(y-2)$

**9.** $4a^2b-2ab-30b$
$=2b(2a^2-a-15)$
$=2b(2a+5)(a-3)$

**11.** $ay^2-4a-4y^2+16$
$=a(y^2-4)+(-4)(y^2-4)$
$=(y^2-4)(a+(-4))$
$=(y+2)(y-2)(a-4)$

**13.** $11x^5-11xy^2=11x(x^4-y^2)$
$=11x(x^2+y)(x^2-y)$

**15.** $4x^5-64x$
$=4x(x^4-16)$
$=4x(x^2+4)(x^2-4)$
$=4x(x^2+4)(x+2)(x-2)$

**17.** $x^3-4x^2-9x+36$
$=x^2(x-4)+(-9)(x-4)$
$=(x-4)(x^2+(-9))$
$=(x-4)(x^2-9)$
$=(x-4)(x+3)(x-3)$

**19.** $2x^5+54x^2=2x^2(x^3+27)$
$=2x^2(x+3)(x^2-3x+9)$

**21.** $3x^4y-48y^5$
$=3y(x^4-16y^4)$
$=3y(x^2+4y^2)(x^2-4y^2)$
$=3y(x^2+4y^2)(x+2y)(x-2y)$

**23.** $12x^3+36x^2y+27xy^2$
$=3x(4x^2+12xy+9y^2)$
$=3x(2x+3y)^2$

**25.** $x^2-12x+36-49y^2$
$=(x^2-12x+36)-49y^2$
$=(x-6)^2-(7y)^2$
$=((x-6)+7y)((x-6)-7y)$
$=(x-6+7y)(x-6-7y)$

**27.** $4x^2+25y^2$
Prime
The sum of two squares with no common factor other than 1 is a prime polynomial.

**29.** $12x^3y-12xy^3=12xy\left(x^2-y^2\right)$
$=12xy(x+y)(x-y)$

**31.** $6bx^2+6by^2=6b\left(x^2+y^2\right)$

**33.** $x^4-xy^3+x^3y-y^4$
$=x\left(x^3-y^3\right)+y\left(x^3-y^3\right)$
$=\left(x^3-y^3\right)(x+y)$
$=(x-y)\left(x^2+xy+y^2\right)(x+y)$

**35.** $x^2-4a^2+12x+36$
$=x^2+12x+36-4a^2$
$=\left(x^2+12x+36\right)-4a^2$
$=(x+6)^2-4a^2$
$=((x+6)+2a)((x+6)-2a)$
$=(x+6+2a)(x+6-2a)$

**37.** $5x^3+x^6-14$
$=x^6+5x^3-14$
$=\left(x^3+7\right)\left(x^3-2\right)$

**39.** $4x-14+2x^3-7x^2$
$=2(2x-7)+x^2(2x-7)$
$=(2x-7)\left(2+x^2\right)$

**41.** $54x^3-16y^3$
$=2\left(27x^3-8y^3\right)$
$=2(3x-2y)\left(9x^2+6xy+4y^2\right)$

**43.** $x^2+10x-y^2+25$
$=x^2+10x+25-y^2$
$=(x+5)^2-y^2$
$=((x+5)+y)((x+5)-y)$
$=(x+5+y)(x+5-y)$

**45.** $x^8-y^8$
$=\left(x^4+y^4\right)\left(x^4-y^4\right)$
$=\left(x^4+y^4\right)\left(x^2+y^2\right)\left(x^2-y^2\right)$
$=\left(x^4+y^4\right)\left(x^2+y^2\right)(x+y)(x-y)$

**47.** $x^3y-16xy^3$
$=xy\left(x^2-16y^2\right)$
$=xy(x+4y)(x-4y)$

**49.** $x+8x^4$
$=x\left(1+8x^3\right)$
$=x(1+2x)\left(1-2x+4x^2\right)$

**51.** $16y^2-4y-2$
$=2\left(8y^2-2y-1\right)$
$=2(4y+1)(2y-1)$

**53.** $14y^3+7y^2-10y$
$=y\left(14y^2+7y-10\right)$

**55.** $27x^2+36xy+12y^2$
$=3\left(9x^2+12xy+4y^2\right)$
$=3\left(3x+2y\right)^2$

**57.** $12x^3+3xy^2=3x\left(4x^2+y^2\right)$

**59.** $x^6y^6-x^3y^3$
$=x^3y^3\left(x^3y^3-1\right)$
$=x^3y^3\left(xy-1\right)\left(x^2y^2+xy+1\right)$

**61.** $(x+5)(x-3)+(x+5)(x-7)$
$=(x+5)\left((x-3)+(x-7)\right)$
$=(x+5)(2x-10)$
$=(x+5)\cdot 2(x-5)$
$=2(x+5)(x-5)$

**63.** $a^2(x-y)+4(y-x)$
$=a^2(x-y)+4(-1)(x-y)$
$=(x-y)\left(a^2-4\right)$
$=(x-y)(a+2)(a-2)$

**65.** $(c+d)^4-1$
$=\left((c+d)^2+1\right)\left((c+d)^2-1\right)$
$=\left((c+d)^2+1\right)\left((c+d)-1\right)\left((c+d)+1\right)$
$=\left((c+d)^2+1\right)\left(c+d-1\right)\left(c+d+1\right)$ or
$=\left(c^2+2cd+d^2+1\right)\left(c+d-1\right)\left(c+d+1\right)$

**67.** $p^3-pq^2+p^2q-q^3$
$=p^3+p^2q-pq^2-q^3$
$=p^2(p+q)-q^2(p+q)$
$=(p+q)\left(p^2-q^2\right)$
$=(p+q)(p+q)(p-q)$
$=\left(p+q\right)^2(p-q)$

**69.** $x^4-5x^2y^2+4y^4$
$=\left(x^2-4y^2\right)\left(x^2-y^2\right)$
$=\left(x-2y\right)\left(x+2y\right)\left(x+y\right)\left(x-y\right)$

**70.** $x^4-10x^2y^2+9y^4$
$=\left(x^2-9y^2\right)\left(x^2-y^2\right)$
$=(x-3y)(x+3y)(x-y)(x+y)$

**71.** $(x+y)^2+6(x+y)+9$
$=\left((x+y)+3\right)\left((x+y)+3\right)$
$=\left(x+y+3\right)^2$

**72.** $(x-y)^2-8(x-y)+16$
$=\left((x-y)-4\right)\left((x-y)-4\right)$
$=\left(x-y-4\right)^2$

**73.** $\left(x-y\right)^4-4\left(x-y\right)^2$
$=\left(x-y\right)^2\left((x-y)^2-4\right)$
$=\left(x-y\right)^2\left((x-y)+2\right)\left((x-y)-2\right)$
$=\left(x-y\right)^2\left(x-y+2\right)\left(x-y-2\right)$

**74.** $(x+y)^4-100(x+y)^2$
$=\left(x+y\right)^2\left((x+y)^2-100\right)$
$=(x+y)^2\left(x+y-10\right)\left(x+y+10\right)$

**75.** $2x^2 - 7xy^2 + 3y^4$
$= (2x - y^2)(x - 3y^2)$

**76.** $3x^2 + 5xy^2 + 2y^4$
$= (3x + 2y^2)(x + y^2)$

**77.** $x^3 - y^3 - x + y$
$= (x - y)(x^2 + xy + y^2) - (x - y)$
$= (x - y)(x^2 + xy + y^2 - 1)$

**78.** $x^3 + y^3 + x^2 - y^2$
$= (x + y)(x^2 - xy + y^2) + (x + y)(x - y)$
$= (x + y)(x^2 - xy + y^2 + x - y)$

**79.** $x^6y^3 + x^3 - 8x^3y^3 - 8$
$= x^3(x^3y^3 + 1) - 8(x^3y^3 + 1)$
$= (x^3y^3 + 1)(x^3 - 8) =$
$(xy + 1)(x^2y^2 - xy + 1)(x - 2)(x^2 + 2x + 4)$

**80.** $x^6y^3 - x^3 + x^3y^3 - 1$
$= x^3(x^3y^3 - 1) + (x^3y^3 - 1)$
$= (x^3y^3 - 1)(x^3 + 1) =$
$(xy - 1)(x^2y^2 + xy + 1)(x + 1)(x^2 - x + 1)$

**81.** **a.** $x(x + y) - y(x + y)$

**b.** $x(x + y) - y(x + y)$
$= (x + y)(x - y)$

**83.** **a.** $xy + xy + xy + 3x(x) = 3xy + 3x^2$

**b.** $3xy + 3x^2 = 3x(y + x)$

**85.** **a.** $4x(2x) - 2(\pi x^2) = 8x^2 - 2\pi x^2$

**b.** $8x^2 - 2\pi x^2 = 2x^2(4 - 2\pi)$

**87.** Answers will vary.

**89.** $4x^2 - 12x + 9 = (4x - 3)^2$

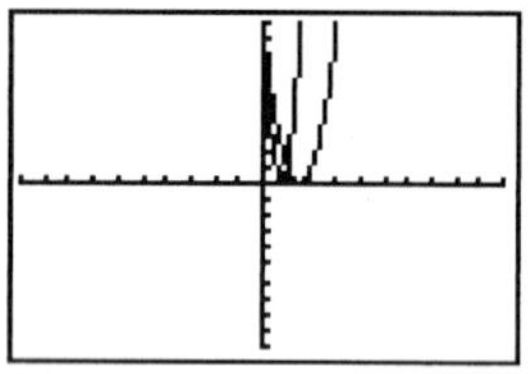

The graphs do not coincide. The factorization is not correct.

$4x^2 - 12x + 9 = (2x - 3)^2$

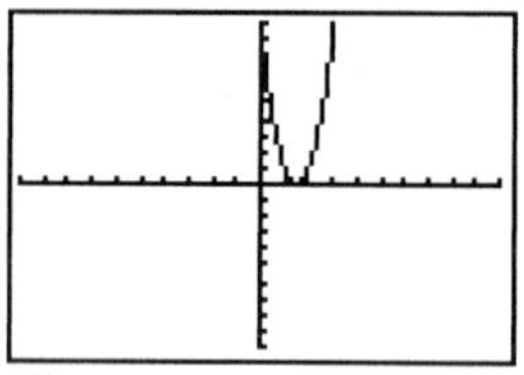

The graphs coincide. The new factorization is correct.

**91.** $x^4 - 16 = (x^2 + 4)(x + 2)(x - 2)$

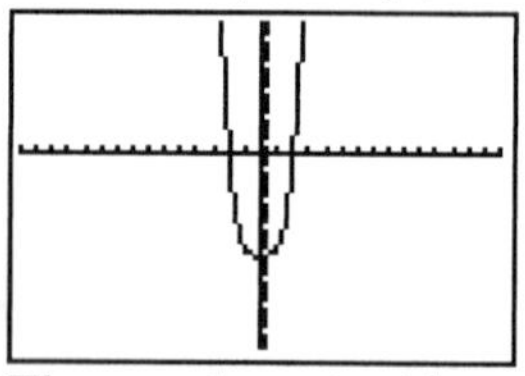

The graphs coincide so the factorization is correct.

**93.** Answers will vary.

**95.** $x^{2n+3} - 10x^{n+3} + 25x^3$
$= x^3(x^{2n} - 10x^n + 25)$
$= x^3(x^n - 5)^2$

**97.** $x^{4n+1} - xy^{4n}$

$= x^{4n}x - xy^{4n}$

$= x\left(x^{4n} - y^{4n}\right)$

$= x\left(x^{2n} + y^{2n}\right)\left(x^{2n} - y^{2n}\right)$

$= x\left(x^{2n} + y^{2n}\right)\left(x^{n} + y^{n}\right)\left(x^{n} - y^{n}\right)$

**99.** $x^3 + x + 2x^4 + 4x^2 + 2$

$= x(x^2 + 1) + 2\left(x^4 + 2x^2 + 1\right)$

$= x\left(x^2 + 1\right) + 2\left(x^2 + 1\right)^2$

$= \left(x^2 + 1\right)\left(x + 2(x^2 + 2)\right)$

$= \left(x^2 + 1\right)\left(2x^2 + x + 2\right)$

**100.** $\frac{3x-1}{5} + \frac{x+2}{2} = -\frac{3}{10}$

$10\left(\frac{3x-1}{5}\right) + 10\left(\frac{x+2}{2}\right) = 10\left(-\frac{3}{10}\right)$

$2(3x-1) + 5(x+2) = -3$

$6x - 2 + 5x + 10 = -3$

$11x + 8 = -3$

$11x = -11$

$x = -1$

The solution is –1 and the solution set is $\{-1\}$.

**101.** $\left(4x^3y^{-1}\right)^2\left(2x^{-3}y\right)^{-1}$

$= 4^2x^6y^{-2}2^{-1}x^3y^{-1}$

$= 4^2 2^{-1}x^6x^3y^{-2}y^{-1}$

$= 16\cdot\frac{1}{2}x^9y^{-3}$

$= \frac{8x^9}{y^3}$

**102.** $\begin{vmatrix} 0 & -3 & 2 \\ 1 & 5 & 3 \\ -2 & 1 & 4 \end{vmatrix}$

$= 0\begin{vmatrix} 5 & 3 \\ 1 & 4 \end{vmatrix} - 1\begin{vmatrix} -3 & 2 \\ 1 & 4 \end{vmatrix} - 2\begin{vmatrix} -3 & 2 \\ 5 & 3 \end{vmatrix}$

$= -1(-3\cdot4 - 1\cdot2) - 2(-3\cdot3 - 5\cdot2)$

$= -1(-12-2) - 2(-9-10)$

$= -1(-14) - 2(-19)$

$= 14 + 38 = 52$

### 5.7 Exercise Set

**1.** $x^2 + x - 12 = 0$

$(x+4)(x-3) = 0$

Apply the zero product principle.

$x + 4 = 0 \qquad x - 3 = 0$

$x = -4 \qquad x = 3$

The solutions are –4 and 3 and the solution set is $\{-4, 3\}$.

**3.** $x^2 + 6x = 7$

$x^2 + 6x - 7 = 0$

$(x+7)(x-1) = 0$

Apply the zero product principle.

$x + 7 = 0 \qquad x - 1 = 0$

$x = -7 \qquad x = 1$

The solutions are –7 and 1 and the solution set is $\{-7, 1\}$.

**5.** $3x^2+10x-8=0$

$(3x-2)(x+4)=0$

Apply the zero product principle.

$3x-2=0 \qquad x+4=0$

$3x=2 \qquad x=-4$

$x=\frac{2}{3}$

The solutions are $-4$ and $\frac{2}{3}$ and the solution set is $\left\{-4,\frac{2}{3}\right\}$.

**7.** $5x^2=8x-3$

$5x^2-8x+3=0$

$(5x-3)(x-1)=0$

Apply the zero product principle.

$5x-3=0 \qquad x-1=0$

$5x=3 \qquad x=1$

$x=\frac{3}{5}$

The solutions are $\frac{3}{5}$ and 1 and the solution set is $\left\{\frac{3}{5},1\right\}$.

**9.** $3x^2=2-5x$

$3x^2+5x-2=0$

$(3x-1)(x+2)=0$

Apply the zero product principle.

$3x-1=0 \qquad x+2=0$

$3x=1 \qquad x=-2$

$x=\frac{1}{3}$

The solutions are $-2$ and $\frac{1}{3}$ and the solution set is $\left\{-2,\frac{1}{3}\right\}$.

**11.** $x^2=8x$

$x^2-8x=0$

$x(x-8)=0$

Apply the zero product principle.

$x=0 \qquad x-8=0$

$x=8$

The solutions are 0 and 8 and the solution set is $\{0,8\}$.

**13.** $3x^2=5x$

$3x^2-5x=0$

$x(3x-5)=0$

Apply the zero product principle.

$x=0 \qquad 3x-5=0$

$3x=5$

$x=\frac{5}{3}$

The solutions are 0 and $\frac{5}{3}$ and the solution set is $\left\{0,\frac{5}{3}\right\}$.

**15.** $x^2+4x+4=0$

$(x+2)^2=0$

There are two identical factors. Apply the zero product principle and solve the equation one time to obtain the double-root.

$x+2=0$

$x=-2$

The solution is –2 and the solution set is $\{-2\}$.

**17.** $x^2 = 14x - 49$

$x^2 - 14x + 49 = 0$

$(x-7)^2 = 0$

Apply the zero product principle to obtain the double-root.

$x - 7 = 0$

$x = 7$

The solution is 7 and the solution set is $\{7\}$.

**19.** $9x^2 = 30x - 25$

$9x^2 - 30x + 25 = 0$

$(3x-5)^2 = 0$

Apply the zero product principle to obtain the double-root.

$3x - 5 = 0$

$3x = 5$

$x = \frac{5}{3}$

The solution is $\frac{5}{3}$ and the solution set is $\left\{\frac{5}{3}\right\}$.

**21.** $x^2 - 25 = 0$

$(x+5)(x-5) = 0$

Apply the zero product principle.

$x + 5 = 0 \qquad x - 5 = 0$

$x = -5 \qquad x = 5$

The solutions are –5 and 5 and the solution set is $\{-5, 5\}$.

**23.** $9x^2 = 100$

$9x^2 - 100 = 0$

$(3x+10)(3x-10) = 0$

Apply the zero product principle.

$3x + 10 = 0 \qquad 3x - 10 = 0$

$3x = -10 \qquad 3x = 10$

$x = -\frac{10}{3} \qquad x = \frac{10}{3}$

The solutions are $-\frac{10}{3}$ and $\frac{10}{3}$ and the solution set is $\left\{-\frac{10}{3}, \frac{10}{3}\right\}$.

**25.** $x(x-3) = 18$

$x^2 - 3x = 18$

$x^2 - 3x - 18 = 0$

$(x-6)(x+3) = 0$

Apply the zero product principle.

$x - 6 = 0 \qquad x + 3 = 0$

$x = 6 \qquad x = -3$

The solutions are –3 and 6 and the solution set is $\{-3, 6\}$.

**27.** $(x-3)(x+8) = -30$

$x^2 + 8x - 3x - 24 = -30$

$x^2 + 5x - 24 = -30$

$x^2 + 5x + 6 = 0$

$(x+3)(x+2) = 0$

Apply the zero product principle.

$x + 3 = 0 \qquad x + 2 = 0$

$x = -3 \qquad x = -2$

The solutions are –3 and –2 and the solution set is $\{-3, -2\}$.

**29.**
$$x(x+8)=16(x-1)$$
$$x^2+8x=16x-16$$
$$x^2-8x+16=0$$
$$(x-4)^2=0$$
Apply the zero product principle to obtain the double-root.
$$x-4=0$$
$$x=4$$
The solution is 4 and the solution set is $\{4\}$.

**31.**
$$(x+1)^2-5(x+2)=3x+7$$
$$(x^2+2x+1)-5x-10=3x+7$$
$$x^2+2x+1-5x-10=3x+7$$
$$x^2-3x-9=3x+7$$
$$x^2-6x-16=0$$
$$(x-8)(x+2)=0$$
Apply the zero product principle.
$x-8=0 \qquad x+2=0$
$x=8 \qquad x=-2$
The solutions are –2 and 8 and the solution set is $\{-2,8\}$.

**33.**
$$x(8x+1)=3x^2-2x+2$$
$$8x^2+x=3x^2-2x+2$$
$$5x^2+3x-2=0$$
$$(5x-2)(x+1)=0$$
Apply the zero product principle.
$5x-2=0 \qquad x+1=0$
$5x=2 \qquad x=-1$
$x=\frac{2}{5}$
The solutions are $-1$ and $\frac{2}{5}$ and the solution set is $\left\{-1,\frac{2}{5}\right\}$.

**35.**
$$\frac{x^2}{18}+\frac{x}{2}+1=0$$
$$18\left(\frac{x^2}{18}\right)+18\left(\frac{x}{2}\right)+18(1)=18(0)$$
$$x^2+9x+18=0$$
$$(x+3)(x+6)=0$$
Apply the zero product principle.
$x+3=0 \qquad x+6=0$
$x=-3 \qquad x=-6$
The solutions are –6 and –3 and the solution set is $\{-6,-3\}$.

**37.**
$$x^3+4x^2-25x-100=0$$
$$x^2(x+4)-25(x+4)=0$$
$$(x+4)(x^2-25)=0$$
$$(x+4)(x+5)(x-5)=0$$
Apply the zero product principle.
$x+4=0 \qquad x+5=0 \qquad x-5=0$
$x=-4 \qquad x=-5 \qquad x=5$
The solutions are –5, –4, and 5 and the solution set is $\{-5,-4,5\}$.

**39.**
$$x^3-x^2=25x-25$$
$$x^3-x^2-25x+25=0$$
$$x^2(x-1)-25(x-1)=0$$
$$(x-1)(x^2-25)=0$$
$$(x-1)(x+5)(x-5)=0$$
Apply the zero product principle.
$x-1=0 \qquad x+5=0 \qquad x-5=0$
$x=1 \qquad x=-5 \qquad x=5$
The solutions are –5, 1 and 5 and the solution set is $\{-5,1,5\}$.

**41.** $3x^4 - 48x^2 = 0$

$3x^2\left(x^2 - 16\right) = 0$

$3x^2\left(x+4\right)\left(x-4\right) = 0$

Apply the zero product principle.

$3x^2 = 0 \quad x+4=0 \quad x-4=0$

$x = 0 \quad x = -4 \quad x = 4$

The solutions are –4, 0 and 4 and the solution set is $\{-4, 0, 4\}$.

**43.** $2x^4 = 16x$

$2x^4 - 16x = 0$

$2x\left(x^3 - 8\right) = 0$

$2x\left(x-2\right)\left(x^2 + 2x + 4\right) = 0$

Apply the zero product principle.

$2x = 0 \quad x-2=0$

$x = 0 \quad x = 2$

$x^2 + 2x + 4 = 0$

Since $x^2 + 2x + 4$ will always be greater than zero, there is no solution.

The solutions are 0 and 2 and the solution set is $\{0, 2\}$.

**45.** $2x^3 + 16x^2 + 30x = 0$

$2x\left(x^2 + 8x + 15\right) = 0$

$2x\left(x+5\right)\left(x+3\right) = 0$

Apply the zero product principle.

$2x = 0 \quad x+5=0 \quad x+3=0$

$x = 0 \quad x = -5 \quad x = -3$

The solutions are –5, –3 and 0 and the solution set is $\{-5, -3, 0\}$.

**47.** $x^2 - 6x + 8 = 0$

$\left(x-4\right)\left(x-2\right) = 0$

Apply the zero product principle.

$x-4=0 \quad x-2=0$

$x = 4 \quad x = 2$

The $x$-intercepts are 2 and 4. This corresponds to graph (d).

**49.** $x^2 + 6x + 8 = 0$

$\left(x+4\right)\left(x+2\right) = 0$

Apply the zero product principle.

$x+4=0 \quad x+2=0$

$x = -4 \quad x = -2$

The $x$-intercepts are –4 and –2. This corresponds to graph (c).

**51.** $x(x+1)^3 - 42(x+1)^2 = 0$

$(x+1)^2\left(x(x+1) - 42\right) = 0$

$(x+1)^2\left(x^2 + x - 42\right) = 0$

Apply the zero product principle.

$(x+1)^2 = 0 \quad x^2 + x - 42 = 0$

$x = -1 \quad (x+7)(x-6) = 0$

$x = -7, x = 6$

The solutions are $-7, -1$, and 6, or $\{-7, -1, 6\}$.

**52.** $x(x-2)^3 - 35(x-2)^2 = 0$

$(x-2)^2\left(x(x-2) - 35\right) = 0$

$(x-2)^2\left(x^2 - 2x - 35\right) = 0$

$(x-2)^2(x-7)(x+5) = 0$

Apply the zero product principle.

$(x-2)^2 = 0 \quad x-7=0 \quad x+5=0$

$x = 2 \quad x = 7 \quad x = -5$

The solutions are $-5$, 2, and 7 or $\{-5, 2, 7\}$.

**53.** $-4x[x(3x-2)-8](25x^2-40x+16)=0$

$-4x[3x^2-2x-8](5x-4)(5x-4)=0$

$-4x(3x+4)(x-2)(5x-4)^2=0$

Apply the zero product principle.

$-7x=0 \quad 2x+3=0 \quad x-4=0 \quad 3x+5=0$

$x=0 \quad 2x=-3 \quad x=4 \quad 3x=-5$

$x=-\frac{3}{2} \qquad x=-\frac{5}{3}$

The solutions are

$-\frac{4}{3}, 0, \frac{4}{5},$ and 2, or $\left\{-\frac{4}{3}, 0, \frac{4}{5}, 2\right\}$.

**54.** $-7x[x(2x-5)-12](9x^2+30x+25)=0$

$-7x[2x^2-5x-12](3x+5)^2=0$

$-7x(2x+3)(x-4)(3x+5)^2=0$

Apply the zero product principle.

$-4x=0 \quad 3x+4=0 \quad x-2=0 \quad 5x-4=0$

$x=0 \quad 3x=-4 \quad x=2 \quad 5x=4$

$x=-\frac{4}{3} \quad x=2 \quad x=\frac{4}{5}$

The solutions are $\left\{-\frac{5}{3}, -\frac{3}{2}, 0, 4\right\}$.

**55.** $f(c)=c^2-4c-27$

$5=c^2-4c-27$

$0=c^2-4c-32$

$0=(c-8)(c+4)$

Apply the zero product principle.

$c-8=0 \quad c+4=0$

$c=8 \quad c=-4$

The solutions are $-4$ and $8$.

**56.** $f(c)=5c^2-11c+6$

$4=5c^2-11c+6$

$0=5c^2-11c+2$

$0=(5c-1)(c-2)$

Apply the zero product principle.

$5c-1=0 \quad c-2=0$

$5c=1 \quad c=2$

$c=\frac{1}{5}$

The solutions are $\frac{1}{5}$ and $2$.

**57.** $f(c)=2c^3+c^2-8c+2$

$6=2c^3+c^2-8c+2$

$0=2c^3+c^2-8c-4$

$0=c^2(2c+1)-4(2c+1)$

$0=(2c+1)(c^2-4)$

$0=(2c+1)(c-2)(c+2)$

Apply the zero product principle.

$2c+1=0 \quad c-2=0 \quad c+2=0$

$2c=-1 \quad c=2 \quad c=-2$

$c=-\frac{1}{2}$

The solutions are $-2, -\frac{1}{2},$ and $2$.

**58.** $f(c)=c^3+4c^2-c+6$

$10=c^3+4c^2-c+6$

$0=c^3+4c^2-c-4$

$0=c^2(c+4)-(c+4)$

$0=(c+4)(c^2-1)$

$0=(c+4)(c-1)(c+1)$

Apply the zero product principle.

$c+4=0 \quad c-1=0 \quad c+1=0$

$c=-4 \quad c=1 \quad c=-1$

The solutions are $-4, -1,$ and $1$.

**59.** Let $x$ be the number.

$(x-1)(x+4)=24$

$x^2+3x-4=24$

$x^2+3x-28=0$

$(x+7)(x-4)=0$

Apply the zero product principle.

$x+7=0 \quad x-4=0$

$x=-7 \qquad x=4$

**60.** Let $x$ be the number.

$(x-6)(x+2)=20$

$x^2-4x-12=20$

$x^2-4x-32=0$

$(x-8)(x+4)=0$

Apply the zero product principle.

$x-8=0 \quad x+4=0$

$x=8 \qquad x=-4$

**61.** Let $x$ be the number.

$3x-5=(x-1)^2$

$3x-5=(x-1)(x-1)$

$3x-5=x^2-2x+1$

$0=x^2-5x+6$

$0=(x-2)(x-3)$

Apply the zero product principle.

$x-2=0 \quad x-3=0$

$x=2 \qquad x=3$

**62.** $61-x^2=(x+1)^2$

$61-x^2=(x+1)(x+1)$

$61-x^2=x^2+2x+1$

$0=2x^2+2x-60$

$0=2(x^2+x-30)$

$0=2(x+6)(x-5)$

Apply the zero product principle.

$2=0 \quad x+6=0 \quad x-5=0$

No sol. $\quad x=-6 \qquad x=5$

**63.** $f(x)=\dfrac{3}{x^2+4x-45}$

The numbers that make the denominator 0 must be excluded from the domain. So, set the denominator equal to 0 and solve.

$x^2+4x-45=0$

$(x+9)(x-5)=0$

$x+9=0 \quad x-5=0$

$x=-9 \qquad x=5$

So, $-9$ and 5 must be excluded from the domain.

**64.** $f(x)=\dfrac{7}{x^2-3x-28}$

The numbers that make the denominator 0 must be excluded from the domain. So, set the denominator equal to 0 and solve.

$x^2-3x-28=0$

$(x-7)(x+4)=0$

$x-7=0 \quad x+4=0$

$x=7 \qquad x=-4$

So, $-4$ and 7 must be excluded from the domain.

**65.** $-16t^2 + 8t + 8 = 0$

$-8\left(2t^2 - t - 1\right) = 0$

$-8(2t+1)(t-1) = 0$

Apply the zero product principle.

$2t + 1 = 0 \qquad t - 1 = 0$

$2t + 1 = 0 \qquad t = 1$

$t = -\frac{1}{2}$

The solution set is $\left\{-\frac{1}{2}, 1\right\}$.

Disregard $-\frac{1}{2}$ because we can't have a negative time measurement. The gymnast will reach the ground at $t = 1$ second. The tick marks are then 0.25, 0.5, 0.75, and 1.

**67.** $-\frac{1}{4}x^2 + 3x + 17 = 26$

$-\frac{1}{4}x^2 + 3x - 9 = 0$

$-4\left(-\frac{1}{4}x^2 + 3x - 9\right) = -4(0)$

$x^2 - 12x + 36 = 0$

$(x-6)^2 = 0$

Apply the zero product principle by solving the equation one time to obtain the double-root.

$x - 6 = 0$

$x = 6$

The solution set is $\{6\}$. This means that 26 million people will receive food stamps in 1996.

**69.** In Exercise 67, we found that 6 years after 1990, 26 million people will receive food stamps. This corresponds to the point $(6, 26)$ on the graph. This is the graph's highest point; During this time period, the greatest number of recipients was 26 million in 1996.

**71.** Let $w$ = the width
Let $w + 3$ = the length

Area $= lw$

$54 = (w+3)w$

$54 = w^2 + 3w$

$0 = w^2 + 3w - 54$

$0 = (w+9)(w-6)$

Apply the zero product principle.

$w + 9 = 0 \qquad w - 6 = 0$

$w = -9 \qquad w = 6$

The solution set is $\{-9, 6\}$. Disregard $-9$ because we can't have a negative length measurement. The width is 6 feet and the length is $6 + 3 = 9$ feet.

**73.** Let $x$ = the length of the side of the original square
Let $x + 3$ = the length of the side of the new, larger square

$(x+3)^2 = 64$

$x^2 + 6x + 9 = 64$

$x^2 + 6x - 55 = 0$

$(x+11)(x-5) = 0$

Apply the zero product principle.

$x + 11 = 0 \qquad x - 5 = 0$

$x = -11 \qquad x = 5$

The solution set is $\{-11, 5\}$.

Disregard $-11$ because we can't have a negative length measurement. This means that $x$, the length of the side of the original square, is 5 inches.

**75.** Let $x$ = the width of the path

$$(20+2x)(10+2x)=600$$
$$200+40x+20x+4x^2=600$$
$$200+60x+4x^2=600$$
$$4x^2+60x+200=600$$
$$4x^2+60x-400=0$$
$$4(x^2+15x-100)=0$$
$$4(x+20)(x-5)=0$$

Apply the zero product principle.

$$4(x+20)=0 \qquad x-5=0$$
$$x+20=0 \qquad x=5$$
$$x=-20$$

The solution set is $\{-20,5\}$.

Disregard –20 because we can't have a negative width measurement. The width of the path is 5 meters.

**77.** **a.** $(2x+12)(2x+10)-10(12)$

$$=4x^2+20x+24x+\cancel{120}-\cancel{120}$$
$$=4x^2+44x$$

**b.**

$$4x^2+44x=168$$
$$4x^2+44x-168=0$$
$$4(x^2+11x-42)=0$$
$$4(x+14)(x-3)=0$$

Apply the zero product principle.

$$4(x+14)=0 \qquad x-3=0$$
$$x+14=0 \qquad x=3$$
$$x=-14$$

The solution set is $\{-14,3\}$. In this application, we disregard –14 because we can't have a negative width measurement. The width of the border is 3 feet.

**79.** Volume = $lwh$

$$200=x\cdot x\cdot 2$$
$$200=2x^2$$
$$0=2x^2-200$$
$$0=2(x^2-100)$$
$$0=2(x+10)(x-10)$$

Apply the zero product principle.

$$2(x+10)=0 \qquad x-10=0$$
$$x+10=0 \qquad x=10$$
$$x=-10$$

The solution set is $\{-10,10\}$.

Disregard –10 because we can't have a negative width measurement. The length and width of the open box are both 10 inches.

**81.**

$$\text{leg}^2+\text{leg}^2=\text{hypotenuse}^2$$
$$(2x+2)^2+x^2=13^2$$
$$4x^2+8x+4+x^2=169$$
$$5x^2+8x+4=169$$
$$5x^2+8x-165=0$$
$$(5x+33)(x-5)=0$$

Apply the zero product principle.

$$5x+33=0 \qquad x-5=0$$
$$5x=-33 \qquad x=5$$
$$x=-\frac{33}{5}$$

The solution set is $\left\{-\frac{33}{5},5\right\}$. In this application, we disregard $-\frac{33}{5}$ because we can't have a negative width measurement. The width of the closet is 5 feet. The length is $2x+2=2(5)+2=10+2=12$ feet.

**83.** Let $x$ = the length of the wire

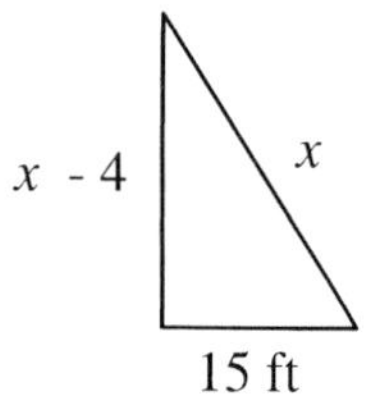


$$\text{leg}^2 + \text{leg}^2 = \text{hypotenuse}^2$$
$$(x-4)^2 + 15^2 = x^2$$
$$\cancel{x^2} - 8x + 16 + 225 = \cancel{x^2}$$
$$-8x + 241 = 0$$
$$-8x = -241$$
$$-8x = -241$$
$$x = \frac{241}{8} = 30\frac{1}{8}$$

The length of the wire is $30\frac{1}{8}$ or 30.125 feet.

**In Exercises 85-95, answers will vary.**

**97.** $y = x^3 + 3x^2 - x - 3$

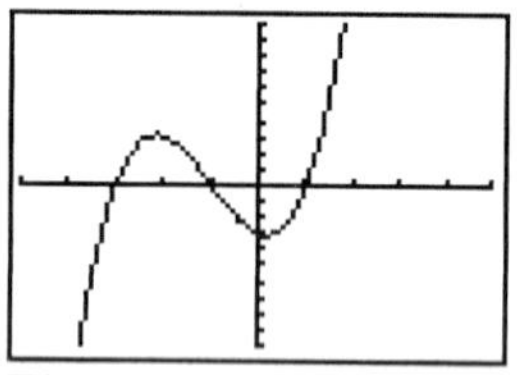

The $x$-intercepts are $-3$, $-1$, and 1. Use the intercepts to solve the equation.

$$x^3 + 3x^2 - x - 3 = 0$$
$$x^2(x+3) - 1(x+3) = 0$$
$$(x+3)(x^2-1) = 0$$
$$(x+3)(x+1)(x-1) = 0$$

Apply the zero product principle.

$$x+3=0 \qquad x+1=0 \qquad x-1=0$$
$$x=-3 \qquad x=-1 \qquad x=1$$

The solutions are –3, –1 and 1 and the solution set is $\{-3,-1,1\}$.

**99.** $y = -x^4 + 4x^3 - 4x^2$

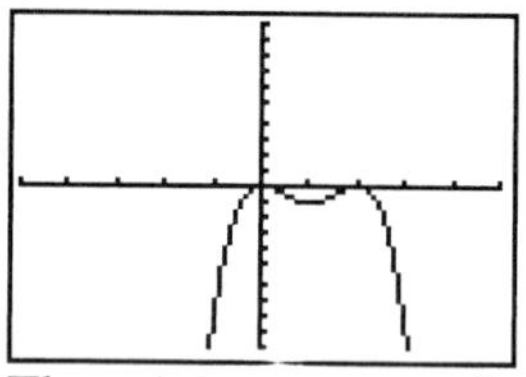

The $x$-intercepts are 0 and 2. Use the intercepts to solve the equation.

$$-x^4 + 4x^3 - 4x^2 = 0$$
$$-x^2(x^2 - 4x + 4) = 0$$
$$-x^2(x-2)^2 = 0$$

Apply the zero product principle.

$$-x^2 = 0 \qquad (x-2)^2 = 0$$
$$x^2 = 0 \qquad x-2 = 0$$
$$x = 0 \qquad x = 2$$

The solutions are 0 and 2 and the solution set is $\{0,2\}$.

**101.** Statement **c** is true. If –4 is a solution then

$$7y^2 + (2k-5)y - 20 = 0$$
$$7(-4)^2 + (2k-5)(-4) - 20 = 0$$
$$7(16) - 8k + 20 - 20 = 0$$
$$112 - 8k = 0$$
$$8k = 112$$
$$k = 14$$

Statement **a** is false. Not all quadratic equations solved by factoring will have two different solutions. For example, the quadratic equation $(x+2)^2 = 0$ has only one solution of $x = -2$.

Statement **b** is false.
$x = 7$ or $x = -7$ are not solutions to
$x^2 + 49 = 0$.

Statement **d** is false. Quadratic equations cannot have more than two solutions.

**103.** $|x^2 + 2x - 36| = 12$

$x^2 + 2x - 36 = 12$

$x^2 + 2x - 48 = 0$

$(x+8)(x-6) = 0$

Apply the zero product principle.

$x + 8 = 0 \quad x - 6 = 0$

$x = -8 \qquad x = 6$

or

$x^2 + 2x - 36 = -12$

$x^2 + 2x - 24 = 0$

$(x+6)(x-4) = 0$

Apply the zero product principle.

$x + 6 = 0 \quad x - 4 = 0$

$x = -6 \qquad x = 4$

The solutions are $\{-8, -6, 4, 6\}$.

**104.** $|3x - 2| = 8$

$3x - 2 = -8$ or $3x - 2 = 8$

$3x = -6 \qquad 3x = 10$

$x = -2 \qquad x = \frac{10}{3}$

The solutions are $-2$ and $\frac{10}{3}$ and the solution set is $\left\{-2, \frac{10}{3}\right\}$.

**105.** $3(5-7)^2 + \sqrt{16} + 12 \div (-3)$

$= 3(-2)^2 + 4 + (-4)$

$= 3(4) + 4 + (-4)$

$= 12 + 4 + (-4)$

$= 12$

**106.** Let $x$ = the amount invested at 5%
Let $y$ = the amount invested at 8%

$x + \quad y = 3000$

$0.05x + 0.08y = 189$

Multiply the first equation by –0.05 and add to the second equation.

$-0.05x - 0.05y = -0.05(3000)$

$-0.05x - 0.05y = -150$

$-0.05x - 0.05y = -150$

$0.05x + 0.08y = \ \ 189$

$0.03y = \quad 39$

$y = 1300$

Back-substitute 1300 for $y$ in one of the original equations.

$x + y = 3000$

$x + 1300 = 3000$

$x = 1700$

\$1700 was invested at 5% and \$1300 was invested at 8%.

## Chapter 5 Review

**1.** The coefficient of $-5x^3$ is –5 and the degree is 3.
The coefficient of $7x^2$ is 7 and the degree is 2.
The coefficient of $-x$ is –1 and the degree is 1.
The coefficient of 2 is 2 and the degree is 0.
The degree of the polynomial is 3.
The leading term is $-5x^3$ and the leading coefficient is –5.

**2.** The coefficient of $8x^4y^2$ is 8 and the degree is 6.
The coefficient of $-7xy^6$ is –7 and the degree is 7.
The coefficient of $-x^3y$ is –1 and the degree is 4.
The degree of the polynomial is 7.
The leading term is $-7xy^6$ and the leading coefficient is –7.

**3.**
$$f(x) = x^3 - 4x^2 + 3x - 1$$
$$\begin{aligned} f(-2) &= (-2)^3 - 4(-2)^2 + 3(-2) - 1 \\ &= -8 - 4(4) - 6 - 1 \\ &= -8 - 16 - 6 - 1 = -31 \end{aligned}$$

**4.**
$$f(x) = 0.26x^2 + 2.09x + 16.4$$
$$\begin{aligned} f(10) &= 0.26(10)^2 + 2.09(10) + 16.4 \\ &= 0.26(100) + 20.9 + 16.4 \\ &= 26 + 20.9 + 16.4 = 63.3 \end{aligned}$$
There were 63,300 people waiting for organ transplants in the United States in the year 1990 +10 = 2000.

**5.** Since the degree of the polynomial is 3, an odd number, and the leading coefficient is –1, the graph will rise to the left and fall to the right. The graph of the polynomial is graph **(c)**.

**6.** Since the degree of the polynomial is 6, an even number, and the leading coefficient is 1, the graph will rise to the left and rise to the right. The graph of the polynomial is graph **(b)**.

**7.** Since the degree of the polynomial is 5, an odd number, and the leading coefficient is 1, the graph will fall to the left and rise to the right. The graph of the polynomial is graph **(a)**.

**8.** Since the degree of the polynomial is 4, an even number, and the leading coefficient is –1, the graph will fall to the left and fall to the right. The graph of the polynomial is graph **(d)**.

**9.** Since the degree of the polynomial is 3, an odd number, and the leading coefficient is negative, the graph will fall to the right. This means that over time, the model predicts the number of families living below the poverty level will go to zero. This is probably not realistic and the model is valid only for a limited period of time.

**10.**
$$\begin{aligned}\left(-8x^3+5x^2-7x+4\right)+\left(9x^3-11x^2+6x-13\right)&=-8x^3+5x^2-7x+4+9x^3-11x^2+6x-13\\&=-8x^3+9x^3+5x^2-11x^2-7x+6x+4-13\\&=x^3-6x^2-x-9\end{aligned}$$

**11.**
$$\begin{aligned}\left(7x^3y-13x^2y-6y\right)+\left(5x^3y+11x^2y-8y-17\right)&=7x^3y-13x^2y-6y+5x^3y+11x^2y-8y-17\\&=7x^3y+5x^3y-13x^2y+11x^2y-6y-8y-17\\&=12x^3y-2x^2y-14y-17\end{aligned}$$

**12.**
$$\begin{aligned}\left(7x^3-6x^2+5x-11\right)-\left(-8x^3+4x^2-6x-3\right)&=7x^3-6x^2+5x-11+8x^3-4x^2+6x+3\\&=7x^3+8x^3-6x^2-4x^2+5x+6x-11+3\\&=15x^3-10x^2+11x-8\end{aligned}$$

**13.**
$$\begin{aligned}\left(4x^3y^2-7x^3y-4\right)-\left(6x^3y^2-3x^3y+4\right)&=4x^3y^2-7x^3y-4-6x^3y^2+3x^3y-4\\&=4x^3y^2-6x^3y^2-7x^3y+3x^3y-4-4\\&=-2x^3y^2-4x^3y-8\end{aligned}$$

**14.**
$$\begin{aligned}\left(x^3+4x^2y-y^3\right)-\left(-2x^3-x^2y+xy^2+7y^3\right)&=x^3+4x^2y-y^3+2x^3+x^2y-xy^2-7y^3\\&=x^3+2x^3+4x^2y+x^2y-xy^2-y^3-7y^3\\&=3x^3+5x^2y-xy^2-8y^3\end{aligned}$$

**15.**
$$\begin{aligned}&\left(4x^2yz^5\right)\left(-3x^4yz^2\right)\\&=4x^2yz^5(-3)x^4yz^2\\&=4(-3)x^2x^4yyz^5z^2\\&=-12x^6y^2z^7\end{aligned}$$

**16.**
$$\begin{aligned}&6x^3\left(\frac{1}{3}x^5-4x^2-2\right)\\&=6x^3\left(\frac{1}{3}x^5\right)-6x^3\left(4x^2\right)-6x^3(2)\\&=6\left(\frac{1}{3}\right)x^3x^5-6(4)x^3x^2-6(2)x^3\\&=2x^8-24x^5-12x^3\end{aligned}$$

**17.** $7xy^2(3x^4y^2 - 5xy - 1)$
$= 7xy^2(3x^4y^2) - 7xy^2(5xy) - 7xy^2(1)$
$= 7(3)xx^4y^2y^2 - 7(5)xxy^2y - 7xy^2$
$= 21x^5y^4 - 35x^2y^3 - 7xy^2$

**18.**
$$\begin{array}{r} 3x^2 + 7x - 4 \\ 2x + 5 \\ \hline 6x^3 + 14x^2 - 8x \\ 15x^2 + 35x - 20 \\ \hline 6x^3 + 29x^2 + 27x - 20 \end{array}$$

**19.**
$$\begin{array}{r} x^2 + x - 1 \\ x^2 + 3x + 2 \\ \hline x^4 + x^3 - x^2 \\ 3x^3 + 3x^2 - 3x \\ 2x^2 + 2x - 2 \\ \hline x^4 + 4x^3 + 4x^2 - x - 2 \end{array}$$

**20.** $(4x-1)(3x-5)$
$= 12x^2 - 20x - 3x + 5$
$= 12x^2 - 23x + 5$

**21.** $(3xy-2)(5xy+4)$
$= 15x^2y^2 + 12xy - 10xy - 8$
$= 15x^2y^2 + 2xy - 8$

**22.** Two methods can be used to multiply the binomials.
Using the FOIL Method:
$(3x+7y)^2$
$= (3x+7y)(3x+7y)$
$= 9x^2 + 21xy + 21xy + 49y^2$
$= 9x^2 + 42xy + 49y^2$
Recognizing a Perfect Square Trinomial:
$(3x+7y)^2$
$= (3x)^2 + (7y)^2$
$= 9x^2 + 2 \cdot 3x \cdot 7y + 49y^2$
$= 9x^2 + 42xy + 49y^2$

**23.** Two methods can be used to multiply the binomials.
Using the FOIL Method:
$(x^2 - 5y)^2$
$= (x^2 - 5y)(x^2 - 5y)$
$= x^4 - 5x^2y - 5x^2y + 25y^2$
$= x^4 - 10x^2y + 25y^2$
Recognizing a Perfect Square Trinomial:
$(x^2 - 5y)^2$
$= (x^2)^2 + (-5y)^2$
$= x^4 + 2 \cdot x^2(-5y) + 25y^2$
$= x^4 - 10x^2y + 25y^2$

**24.** Two methods can be used to multiply the binomials.
Using the FOIL Method:
$(2x+7y)(2x-7y)$
$= 4x^2 - 14xy + 14xy - 49y^2$
$= 4x^2 - 49y^2$
Recognizing the Difference of Two Squares:
$(2x+7y)(2x-7y)$
$= (2x)^2 - (7y)^2$
$= 4x^2 - 49y^2$

**25.** Two methods can be used to multiply the binomials.
Using the FOIL Method:

$$(3xy^2-4x)(3xy^2+4x)$$
$$=9x^2y^4+12x^2y^2-12x^2y^2-16x^2$$
$$=9x^2y^4-16x^2$$

Recognizing the Difference of Two Squares:

$$(3xy^2-4x)(3xy^2+4x)$$
$$=(3xy^2)^2-(4x)^2$$
$$=9x^2y^4-16x^2$$

**26.** Two methods can be used to multiply the binomials.
Using the FOIL Method:

$$[(x+3)+5y][(x+3)-5y]$$
$$=(x+3)^2-5(x+3)y+5(x+3)y-25y^2$$
$$=(x+3)^2-25y^2$$
$$=x^2+6x+9-25y^2$$

Recognizing the Difference of Two Squares:

$$[(x+3)+5y][(x+3)-5y]$$
$$=(x+3)^2-(5y)^2=x^2+6x+9-25y^2$$

**27.**
$$\begin{array}{l} x+y+4 \\ \underline{x+y+4} \\ x^2+xy+4x \\ \quad\;\; xy \quad\; +y^2+4y \\ \underline{\qquad\quad 4x \qquad +4y+16} \\ x^2+2xy+8x+y^2+8y+16 \end{array}$$
or
$$x^2+2xy+y^2+8x+8y+16$$

**28.** $f(x)=x-3$ and $g(x)=2x+5$

$$(fg)(x)=f(x)\cdot g(x)$$
$$=(x-3)(2x+5)$$
$$=2x^2+5x-6x-15$$
$$=2x^2-x-15$$
$$(fg)(-4)=2(-4)^2-(-4)-15$$
$$=2(16)+4-15$$
$$=32+4-15$$
$$=21$$

**29.** $f(x)=x^2-7x+2$

**a.** $f(a-1)$
$$=(a-1)^2-7(a-1)+2$$
$$=a^2-2a+1-7a+7+2$$
$$=a^2-9a+10$$

**b.** $f(a+h)-f(a)$
$$=(a+h)^2-7(a+h)+2-(a^2-7a+2)$$
$$=\cancel{a^2}+2ah+h^2\cancel{-7a}-7h\cancel{+2}\cancel{-a^2}\cancel{+7a}\cancel{-2}$$
$$=2ah+h^2-7h$$

**30.** $16x^3+24x^2=8x^2(2x+3)$

**31.** $2x-36x^2=2x(1-18x)$

**32.** $21x^2y^2-14xy^2+7xy$
$$=7xy(3xy-2y+1)$$

**33.** $18x^3y^2-27x^2y=9x^2y(2xy-3)$

**34.** $-12x^2+8x-48=-4\left(3x^2-2x+12\right)$

**35.** $-x^2-11x+14=-1\left(x^2+11x-14\right)$
$=-\left(x^2+11x-14\right)$

**36.** $x^3-x^2-2x+2=x^2(x-1)-2(x-1)$
$=(x-1)\left(x^2-2\right)$

**37.** $xy-3x-5y+15=x(y-3)-5(y-3)$
$=(y-3)(x-5)$

**38.** $5ax-15ay+2bx-6by$
$=5a(x-3y)+2b(x-3y)$
$=(x-3y)(5a+2b)$

**39.** $x^2+8x+15=(x+5)(x+3)$

**40.** $x^2+16x-80=(x+20)(x-4)$

**41.** $x^2+16xy-17y^2=(x+17y)(x-y)$

**42.** $3x^3-36x^2+33x=3x\left(x^2-12x+11\right)$
$=3x(x-11)(x-1)$

**43.** $3x^2+22x+7=(3x+1)(x+7)$

**44.** $6x^2-13x+6=(2x-3)(3x-2)$

**45.** $5x^2-6xy-8y^2=(5x+4y)(x-2y)$

**46.** $6x^3+5x^2-4x=x\left(6x^2+5x-4\right)$
$=x(2x-1)(3x+4)$

**47.** $2x^2+11x+15=(2x+5)(x+3)$

**48.** Let $t=x^3$
$x^6+x^3-30=\left(x^3\right)^2+x^3-30$
$=t^2+t-30$
$=(t+6)(t-5)$
$=\left(x^3+6\right)\left(x^3-5\right)$

**49.** Let $t=x^2$
$x^4-10x^2-39=\left(x^2\right)^2-10x^2-39$
$=t^2-10t-39$
$=(t-13)(t+3)$
$=\left(x^2-13\right)\left(x^2+3\right)$

**50.** Let $t=x+5$
$(x+5)^2+10(x+5)+24$
$=t^2+10t+24$
$=(t+6)(t+4)$
$=((x+5)+6)((x+5)+4)$
$=(x+5+6)(x+5+4)$
$=(x+11)(x+9)$

**51.** $5x^6+17x^3+6=5\left(x^3\right)^2+17x^3+6$
$=5t^2+17t+6$
$=(5t+2)(t+3)$
$=\left(5x^3+2\right)\left(x^3+3\right)$

**52.** $4x^2-25=(2x+5)(2x-5)$

**53.** $1-81x^2y^2=(1+9xy)(1-9xy)$

**54.** $x^8-y^6=\left(x^4+y^3\right)\left(x^4-y^3\right)$

**55.** $(x-1)^2 - y^2$
$= ((x-1)+y)((x-1)-y)$
$= (x-1+y)(x-1-y)$

**56.** $x^2 + 16x + 64$
$= x^2 + 2\cdot x\cdot 8 + 8^2$
$= (x+8)^2$

**57.** $9x^2 - 6x + 1 = (3x)^2 - 2\cdot 3x\cdot 1 + 1^2$
$= (3x-1)^2$

**58.** $25x^2 + 20xy + 4y^2$
$= (5x)^2 + 2\cdot 5x\cdot 2y + (2y)^2$
$= (5x+2y)^2$

**59.** $49x^2 + 7x + 1$
Prime

**60.** $25x^2 - 40xy + 16y^2$
$= (5x)^2 - 2\cdot 5x\cdot 4y + (4y)^2$
$= (5x-4y)^2$

**61.** $x^2 + 18x + 81 - y^2$
$= (x^2 + 18x + 81) - y^2$
$= (x+9)^2 - y^2$
$= ((x+9)+y)((x+9)-y)$
$= (x+9+y)(x+9-y)$

**62.** $z^2 - 25x^2 + 10x - 1$
$= z^2 - (25x^2 - 10x + 1)$
$= z^2 - ((5x)^2 - 2\cdot 5x\cdot 1 + 1^2)$
$= z^2 - (5x-1)^2$
$= (z + (5x-1))(z - (5x-1))$
$= (z + 5x - 1)(z - 5x + 1)$

**63.** $64x^3 + 27$
$= (4x)^3 + (3)^3$
$= (4x+3)((4x)^2 - 4x\cdot 3 + 3^2)$
$= (4x+3)(16x^2 - 12x + 9)$

**64.** $125x^3 - 8$
$= (5x)^3 - (2)^3$
$= (5x-2)((5x)^2 + 5x\cdot 2 + 2^2)$
$= (5x-2)(25x^2 + 10x + 4)$

**65.** $x^3y^3 + 1$
$= (xy)^3 + 1^3$
$= (xy+1)((xy)^2 - xy\cdot 1 + 1^2)$
$= (xy+1)(x^2y^2 - xy + 1)$

**66.** $15x^2 + 3x = 3x(5x+1)$

**67.** $12x^4 - 3x^2$
$= 3x^2(4x^2 - 1)$
$= 3x^2((2x)^2 - 1^2)$
$= 3x^2(2x+1)(2x-1)$

**68.** $20x^4 - 24x^3 + 28x^2 - 12x$
$= 4x(5x^3 - 6x^2 + 7x - 3)$

**69.** $x^3 - 15x^2 + 26x$
$= x(x^2 - 15x + 26)$
$= x(x-2)(x-13)$

**70.** $-2y^4 + 24y^3 - 54y^2$
$= -2y^2(y^2 - 12y + 27)$
$= -2y^2(y-9)(y-3)$

**71.** $9x^2 - 30x + 25$
$= (3x)^2 - 2\cdot 3x\cdot 5 + 5^2$
$= (3x-5)^2$

**72.** $5x^2 - 45 = 5(x^2 - 9)$
$= 5(x+3)(x-3)$

**73.** $2x^3 - x^2 - 18x + 9$
$= x^2(2x-1) - 9(2x-1)$
$= (2x-1)(x^2-9)$
$= (2x-1)(x+3)(x-3)$

**74.** $6x^2 - 23xy + 7y^2$
$= (3x-y)(2x-7y)$

**75.** $2y^3 + 12y^2 + 18y$
$= 2y(y^2 + 6y + 9)$
$= 2y(y^2 + 2\cdot y\cdot 3 + 3^2)$
$= 2y(y+3)^2$

**76.** $x^2 + 6x + 9 - 4a^2$
$= (x^2 + 6x + 9) - 4a^2$
$= (x^2 + 2\cdot x\cdot 3 + 3^2) - (2a)^2$
$= (x+3)^2 - (2a)^2$
$= ((x+3) + 2a)((x+3) - 2a)$
$= (x+3+2a)(x+3-2a)$

**77.** $8x^3 - 27$
$= (2x)^3 - 3^3$
$= (2x-3)((2x)^2 + 2x\cdot 3 + 3^2)$
$= (2x-3)(4x^2 + 6x + 9)$

**78.** $x^5 - x$
$= x(x^4 - 1)$
$= x((x^2)^2 - 1^2)$
$= x(x^2+1)(x^2-1)$
$= x(x^2+1)(x^2-1^2)$
$= x(x^2+1)(x+1)(x-1)$

**79.** Let $t = x^2$
$x^4 - 6x^2 + 9 = (x^2)^2 - 6x^2 + 9$
$= t^2 - 6t + 9$
$= t^2 - 2\cdot t\cdot 3 + 3^2$
$= (t-3)^2$
$= (x^2-3)^2$

**80.** $x^2 + xy + y^2$
Prime

**81.** $4a^3+32$
$=4(a^3+8)$
$=4(a^3+2^3)$
$=4(a+2)(a^2-a\cdot 2+2^2)$
$=4(a+2)(a^2-2a+4)$

**82.** $x^4-81$
$=(x^2)^2-9^2$
$=(x^2+9)(x^2-9)$
$=(x^2+9)(x^2-3^2)$
$=(x^2+9)(x+3)(x-3)$

**83.** $ax+3bx-ay-3by$
$=x(a+3b)-y(a+3b)$
$=(a+3b)(x-y)$

**84.** $27x^3-125y^3$
$=(3x)^3-(5y)^3$
$=(3x-5y)((3x)^2+3x\cdot 5y+(5y)^2)$
$=(3x-5y)(9x^2+15xy+25y^2)$

**85.** $10x^3y+22x^2y-24xy$
$=2xy(5x^2+11x-12)$
$=2xy(5x-4)(x+3)$

**86.** Let $t=x^3$
$6x^6+13x^3-5$
$=6(x^3)^2+13x^3-5$
$=6t^2+13t-5$
$=(2t+5)(3t-1)$
$=(2x^3+5)(3x^3-1)$

**87.** $2x+10+x^2y+5xy$
$=2(x+5)+xy(x+5)$
$=(x+5)(2+xy)$

**88.** $y^3+2y^2-25y-50$
$=y^2(y+2)-25(y+2)$
$=(y+2)(y^2-25)$
$=(y+2)(y+5)(y-5)$

**89.** Let $t=a^4$
$a^8-1$
$=(a^4)^2-1$
$=t^2-1$
$=(t+1)(t-1)$
$=(a^4+1)(a^4-1)$
$=(a^4+1)(a^2+1)(a^2-1)$
$=(a^4+1)(a^2+1)(a+1)(a-1)$

**90.** $9(x-4)+y^2(4-x)$
$=9(x-4)+(-1)y^2(x-4)$
$=(x-4)(9-y^2)$
$=(x-4)(3-y)(3+y)$

**91.** **a.** $2xy+2y^2$
**b.** $2xy+2y^2=2y(x+y)$

**92.** **a.** $x^2 - 4y^2$

**b.** $x - 4y^2 = x^2 - (2y)^2$
$= (x+2y)(x-2y)$

**93.** $x^2 + 6x + 5 = 0$
$(x+5)(x+1) = 0$
Apply the zero product principle.
$x+5=0 \qquad x+1=0$
$x=-5 \qquad x=-1$
The solutions are –5 and –1 and the solution set is $\{-5,-1\}$.

**94.** $3x^2 = 22x - 7$
$3x^2 - 22x + 7 = 0$
$(3x-1)(x-7) = 0$
Apply the zero product principle.
$3x-1=0 \qquad x-7=0$
$3x=1 \qquad x=7$
$x=\frac{1}{3}$
The solutions are $\frac{1}{3}$ and 7 and the solution set is $\left\{\frac{1}{3},7\right\}$.

**95.** $(x+3)(x-2) = 50$
$x^2 - 2x + 3x - 6 = 50$
$x^2 + x - 6 = 50$
$x^2 + x - 56 = 0$
$(x+8)(x-7) = 0$
Apply the zero product principle.
$x+8=0 \qquad x-7=0$
$x=-8 \qquad x=7$
The solutions are –8 and 7 and the solution set is $\{-8,7\}$.

**96.** $3x^2 = 12x$
$3x^2 - 12x = 0$
$3x(x-4) = 0$
Apply the zero product principle.
$3x=0 \qquad x-4=0$
$x=0 \qquad x=4$
The solutions are 0 and 4 and the solution set is $\{0,4\}$.

**97.** $x^3 + 5x^2 = 9x + 45$
$x^3 + 5x^2 - 9x - 45 = 0$
$(x^3 + 5x^2) + (-9x - 45) = 0$
$x^2(x+5) + (-9)(x+5) = 0$
$(x+5)(x^2-9) = 0$
$(x+5)(x+3)(x-3) = 0$
Apply the zero product principle.
$x+5=0 \qquad x+3=0 \qquad x-3=0$
$x=-5 \qquad x=-3 \qquad x=3$
The solutions are –5, –3 and 3 and the solution set is $\{-5,-3,3\}$.

**98.** $-16t^2 + 128t + 144 = 0$
$-16(t^2 - 8t - 9) = 0$
$-16(t-9)(t+1) = 0$
Apply the zero product principle.
$-16(t-9)=0 \qquad t+1=0$
$t-9=0 \qquad t=-1$
$t=9$
The solution set is $\{-1,9\}$. Disregard –1 because we can't have a negative time measurement. The rocket will hit the water at $t = 9$ seconds.

**99.** $3400 = 40x^2 - 120x + 600$

$0 = 40x^2 - 120x - 2800$

$0 = 40(x^2 - 3x - 70)$

$0 = 40(x+7)(x-10)$

Apply the zero product principle.

$x+7=0$ or $x-10=0$

$x=-7$ $\quad x=10$

The solution set is $\{-7, 10\}$.

Disregard –7 because we can't have a negative number of years. Since $x$ is the number of years after 1996, then the year that this amount will reach $3,400,000 is $1996+10=2006$.

**100.** Let $w$ = the width of the sign
Let $w$ + 3 = the length of the sign
Area = $lw$

$54 = (w+3)w$

$54 = w^2 + 3w$

$0 = w^2 + 3w - 54$

$0 = (w+9)(w-6)$

Apply the zero product principle.

$w+9=0$ $\quad w-6=0$

$w=-9$ $\quad w=6$

The solution set is $\{-9, 6\}$.

Disregard –9 because we can't have a negative length measurement. The width is 6 feet and the length is $6+3=9$ feet.

**101.** Let $x$ = the width of the border

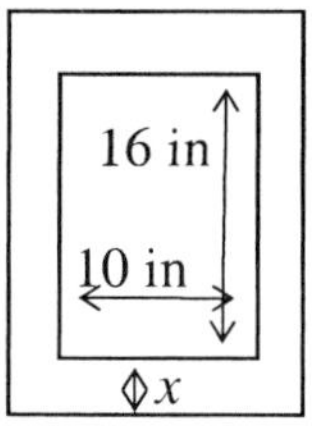


Area = $lw$

$280 = (2x+16)(2x+10)$

$280 = 4x^2 + 20x + 32x + 160$

$0 = 4x^2 + 52x - 120$

$0 = 4(x^2 + 13x - 30)$

$0 = 4(x+15)(x-2)$

Apply the zero product principle.

$4(x+15)=0$ $\quad x-2=0$

$x+15=0$ $\quad x=2$

$x=-15$

The solution set is $\{-15, 2\}$.

Disregard –15 because we can't have a negative length measurement. The width of the frame is 2 inches.

**102.** $\text{leg}^2 + \text{leg}^2 = \text{hypotenuse}^2$

$x^2 + (2x+20)^2 = (2x+30)^2$

$x^2 + 4x^2 + 80x + 400 = 4x^2 + 120x + 900$

$x^2 - 40x - 500 = 0$

$(x-50)(x+10) = 0$

Apply the zero product principle.

$x-50=0$ $\quad x+10=0$

$x=50$ $\quad x=-10$

The solution set is $\{-10, 50\}$.

Disregard –10 because we can't have a negative length measurement. We have $x$ = 50 yards. Find the lengths of the other sides.

| | |
|---|---|
| $2x+20$ | $2x+30$ |
| $=2(50)+20$ | $=2(50)+30$ |
| $=100+20$ | $=100+30$ |
| $=120$ | $=130$ |

The three sides are 50 yards, 120 yards, and 130 yards.

## Chapter 5 Test

**1.** The degree of the polynomial is 3 and the leading coefficient is –6.

**2.** The degree of the polynomial is 9 and the leading coefficient is 7.

**3.** $f(x) = 3x^3 + 5x^2 - x + 6$

$$\begin{aligned} f(0) &= 3(0)^3 + 5(0)^2 - 0 + 6 \\ &= 3(0) + 5(0) + 6 \\ &= 0 + 0 + 6 \\ &= 6 \end{aligned}$$

$$\begin{aligned} f(-2) &= 3(-2)^3 + 5(-2)^2 - (-2) + 6 \\ &= 3(-8) + 5(4) + 2 + 6 \\ &= -24 + 20 + 2 + 6 \\ &= 4 \end{aligned}$$

**4.** Since the degree of the polynomial is 2, an even number, and the leading coefficient is negative, the graph will fall to the left and fall to the right.

**5.** Since the degree of the polynomial is 3, an odd number, and the leading coefficient is positive, the graph will fall to the left and rise to the right.

**6.** $(4x^3y - 19x^2y - 7y) + (3x^3y + x^2y + 6y - 9)$

$$\begin{aligned} &= 4x^3y - 19x^2y - 7y + 3x^3y + x^2y + 6y - 9 \\ &= 4x^3y + 3x^3y - 19x^2y + x^2y - 7y + 6y - 9 \\ &= 7x^3y - 18x^2y - y - 9 \end{aligned}$$

**7.** $(6x^2 - 7x - 9) - (-5x^2 + 6x - 3)$

$$\begin{aligned} &= 6x^2 - 7x - 9 + 5x^2 - 6x + 3 \\ &= 6x^2 + 5x^2 - 7x - 6x - 9 + 3 \\ &= 11x^2 - 13x - 6 \end{aligned}$$

**8.** $(-7x^3y)(-5x^4y^2)$

$$\begin{aligned} &= -7(-5)x^3x^4yy^2 \\ &= 35x^7y^3 \end{aligned}$$

**9.**

$$\begin{array}{r} x^2 - 3xy - y^2 \\ \underline{x - y} \\ x^3 - 3x^2y - xy^2 \\ \underline{-x^2y + 3xy^2 + y^3} \\ x^3 - 4x^2y + 2xy^2 + y^3 \end{array}$$

**10.** $(7x - 9y)(3x + y)$

$$\begin{aligned} &= 21x^2 + 7xy - 27xy - 9y^2 \\ &= 21x^2 - 20xy - 9y^2 \end{aligned}$$

**11.** $(2x - 5y)(2x + 5y)$

$$\begin{aligned} &= 4x^2 + \cancel{10xy} - \cancel{10xy} - 25y^2 \\ &= 4x^2 - 25y^2 \end{aligned}$$

**12.** $(4y - 7)^2$

$$\begin{aligned} &= (4y)^2 + 2 \cdot 4y \cdot (-7) + (-7)^2 \\ &= 16y^2 - 56y + 49 \end{aligned}$$

**13.** $[(x + 2) + 3y][(x + 2) - 3y]$

$$\begin{aligned} &= (x + 2)^2 - 3(x + 2)y + 3(x + 2)y - 9y^2 \\ &= (x + 2)^2 - 9y^2 \\ &= x^2 + 4x + 4 - 9y^2 \end{aligned}$$

**14.** $f(x)=x+2$ and $g(x)=3x-5$

$$\begin{aligned}(fg)(x)&=f(x)\cdot g(x)\\&=(x+2)(3x-5)\\&=3x^2-5x+6x-10\\&=3x^2+x-10\end{aligned}$$

$$\begin{aligned}(fg)(-5)&=3(-5)^2+(-5)-10\\&=3(25)-5-10\\&=75-5-10\\&=60\end{aligned}$$

**15.** $f(x)=x^2-5x+3$

$$\begin{aligned}&f(a+h)-f(a)\\&=(a+h)^2-5(a+h)+3-(a^2-5a+3)\\&=a^2+2ah+h^2-5a-5h+3-a^2+5a-3\\&=2ah+h^2-5h\end{aligned}$$

**16.** $14x^3-15x^2=x^2(14x-15)$

**17.** $$\begin{aligned}81y^2-25&=(9y)^2-5^2\\&=(9y+5)(9y-5)\end{aligned}$$

**18.** $$\begin{aligned}&x^3+3x^2-25x-75\\&=x^2(x+3)-25(x+3)\\&=(x+3)(x^2-25)\\&=(x+3)(x^2-5^2)\\&=(x+3)(x+5)(x-5)\end{aligned}$$

**19.** $$\begin{aligned}25x^2-30x+9&=(5x)^2-2\cdot5x\cdot3+3^2\\&=(5x-3)^2\end{aligned}$$

**20.** $$\begin{aligned}&x^2+10x+25-9y^2\\&=(x^2+10x+25)-9y^2\\&=(x^2+2\cdot x\cdot5+5^2)-9y^2\\&=(x+5)^2-(3y)^2\\&=((x+5)+3y)((x+5)-3y)\\&=(x+5+3y)(x+5-3y)\end{aligned}$$

**21.** $x^4+1$
Prime

**22.** $$\begin{aligned}&y^2-16y-36\\&=(y-18)(y+2)\end{aligned}$$

**23.** $$\begin{aligned}&14x^2+41x+15\\&=(2x+5)(7x+3)\end{aligned}$$

**24.** $$\begin{aligned}5x^3-5&=5(x^3-1)\\&=5(x^3-1^3)\\&=5(x-1)(x^2+x\cdot1+1^2)\\&=5(x-1)(x^2+x+1)\end{aligned}$$

**25.** $$\begin{aligned}12x^2-3y^2&=3(4x^2-y^2)\\&=3((2x)^2-y^2)\\&=3(2x+y)(2x-y)\end{aligned}$$

**26.** $$\begin{aligned}&12x^2-34x+10\\&=2(6x^2-17x+5)\\&=2(3x-1)(2x-5)\end{aligned}$$

**27.** $3x^4 - 3$

$= 3(x^4 - 1)$

$= 3((x^2)^2 - 1^2)$

$= 3(x^2 + 1)(x^2 - 1)$

$= 3(x^2 + 1)(x^2 - 1^2)$

$= 3(x^2 + 1)(x + 1)(x - 1)$

**28.** $x^8 - y^8$

$= (x^4)^2 - (y^4)^2$

$= (x^4 + y^4)(x^4 - y^4)$

$= (x^4 + y^4)((x^2)^2 - (y^2)^2)$

$= (x^4 + y^4)(x^2 + y^2)(x^2 - y^2)$

$= (x^4 + y^4)(x^2 + y^2)(x + y)(x - y)$

**29.** $12x^2y^4 + 8x^3y^2 - 36x^2y$

$= 4x^2y(3y^3 + 2xy - 9)$

**30.** Let $t = x^3$

$x^6 - 12x^3 - 28$

$= (x^3)^2 - 12x^3 - 28$

$= t^2 - 12t - 28$

$= (t - 14)(t + 2)$ Substitute $x^3$ for $t$.

$= (x^3 - 14)(x^3 + 2)$

**31.** Let $t = x^2$

$x^4 - 2x^2 - 24$

$= (x^2)^2 - 2x^2 - 24$

$= t^2 - 2t - 24$

$= (t - 6)(t + 4)$ Substitute $x^2$ for $t$.

$= (x^2 - 6)(x^2 + 4)$

**32.** $12x^2y - 27xy + 6y$

$= 3y(4x^2 - 9x + 2)$

$= 3y(x - 2)(4x - 1)$

**33.** $y^4 - 3y^3 + 2y^2 - 6y$

$= y(y^3 - 3y^2 + 2y - 6)$

$= y(y^2(y - 3) + 2(y - 3))$

$= y(y - 3)(y^2 + 2)$

**34.** $3x^2 = 5x + 2$

$3x^2 - 5x - 2 = 0$

$(3x + 1)(x - 2) = 0$

Apply the zero product principle.

$3x + 1 = 0 \qquad x - 2 = 0$

$3x = -1 \qquad x = 2$

$x = -\dfrac{1}{3}$

The solutions are $-\dfrac{1}{3}$ and 2 and the solution set is $\left\{-\dfrac{1}{3}, 2\right\}$.

**35.** $(5x+4)(x-1)=2$

$5x^2-5x+4x-4=2$

$5x^2-x-4=2$

$5x^2-x-6=0$

$(5x-6)(x+1)=0$

Apply the zero product principle.

$5x-6=0 \qquad x+1=0$

$5x=6 \qquad x=-1$

$x=\frac{6}{5}$

The solutions are $-1$ and $\frac{6}{5}$ and the solution set is $\left\{-1,\frac{6}{5}\right\}$.

**36.** $15x^2-5x=0$

$5x(3x-1)=0$

Apply the zero product principle.

$5x=0 \qquad 3x-1=0$

$x=0 \qquad 3x=1$

$x=\frac{1}{3}$

The solutions are $0$ and $\frac{1}{3}$ and the solution set is $\left\{0,\frac{1}{3}\right\}$.

**37.** $x^3-4x^2-x+4=0$

$x^2(x-4)-1(x-4)=0$

$(x-4)(x^2-1)=0$

$(x-4)(x+1)(x-1)=0$

Apply the zero product principle.

$x-4=0 \qquad x+1=0 \qquad x-1=0$

$x=4 \qquad x=-1 \qquad x=1$

The solutions are $-1$, $1$ and $4$ and the solution set is $\{-1,1,4\}$.

**38.** $-16t^2+48t+448=0$

$-16(t^2-3t-28)=0$

$-16(t-7)(t+4)=0$

Apply the zero product principle.

$-16(t-7)=0 \qquad t+4=0$

$t-7=0 \qquad t=-4$

$t=7$

The solution set is $\{-4,7\}$. Disregard $-4$ because we can't have a negative time measurement. The baseball will hit the water at $t=7$ seconds.

**39.** Let $l$ be the length of the room
Let $w=2l-7$ be the width of the room

Area $=lw$

$15=l(2l-7)$

$15=2l^2-7l$

$0=2l^2-7l-15$

$0=(2l+3)(l-5)$

Apply the zero product principle.

$2l+3=0 \qquad l-5=0$

$2l=-3 \qquad l=5$

$l=-\frac{3}{2}$

The solution set is $\left\{-\frac{3}{2},5\right\}$.

Disregard $-\frac{3}{2}$ because we can't have a negative length measurement. The length is 5 yards and the width is $2l-7=2(5)-7=10-7=3$ yards.

**40.** $\text{leg}^2 + \text{leg}^2 = \text{hypotenuse}^2$

$$x^2 + 12^2 = (2x-3)^2$$
$$x^2 + 144 = 4x^2 - 12x + 9$$
$$0 = 3x^2 - 12x - 135$$
$$0 = 3(x^2 - 4x - 45)$$
$$0 = 3(x-9)(x+5)$$

Apply the zero product principle.

$3(x-9) = 0 \qquad x+5 = 0$

$x - 9 = 0 \qquad x = -5$

$x = 9$

The solution set is $\{-5, 9\}$. Disregard –5 because we can't have a negative length measurement. The lengths of the sides are 9 and 12 units and the length of the hypotenuse is $2x - 3 = 2(9) - 3 = 18 - 3 = 15$ units.

**Cumulative Review Exercises Chapters 1-5**

**1.** $8(x+2) - 3(2-x) = 4(2x+6) - 2$

$$8x + 16 - 6 + 3x = 8x + 24 - 2$$
$$11x + 10 = 8x + 22$$
$$3x = 12$$
$$x = 4$$

The solution is 4 and the solution set is $\{4\}$.

**2.** $2x + 4y = -6$

$x = 2y - 5$

Rewrite the system in standard form.

$2x + 4y = -6$

$x - 2y = -5$

Multiply the second equation by 2 and add to the first equation.

$$2x + 4y = -6$$
$$\underline{2x - 4y = -10}$$
$$4x = -16$$
$$x = -4$$

Back-substitute –4 for $x$ in one of the original equations.

$$2x + 4y = -6$$
$$2(-4) + 4y = -6$$
$$-8 + 4y = -6$$
$$4y = 2$$
$$y = \frac{2}{4} = \frac{1}{2}$$

The solution is $\left(-4, \frac{1}{2}\right)$ and the solution set is $\left\{\left(-4, \frac{1}{2}\right)\right\}$.

**3.** $2x - y + 3z = 0$

$2y + z = 1$

$x + 2y - z = 5$

Multiply the third equation by –2 and add to the first equation.

$$2x - y + 3z = 0$$
$$\underline{-2x - 4y + 2z = -10}$$
$$-5y + 5z = -10$$

We now have 2 equations in 2 variables.

$2y + z = 1$

$-5y + 5z = -10$

Multiply the first equation by –5 and add to the second equation.

$$-10y - 5z = -5$$
$$\underline{-5y + 5z = -10}$$
$$-15y = -15$$
$$y = 1$$

Back-substitute 1 for $y$ in one of the equations in 2 variables.

$$2y + z = 1$$
$$2(1) + z = 1$$
$$2 + z = 1$$
$$z = -1$$

Back-substitute 1 for $y$ and $-1$ for $z$ in one of the original equations in 3 variables.

$$2x - y + 3z = 0$$
$$2x - 1 + 3(-1) = 0$$
$$2x - 1 - 3 = 0$$
$$2x - 4 = 0$$
$$2x = 4$$
$$x = 2$$

The solution is $(2,1,-1)$ and the solution set is $\{(2,1,-1)\}$.

**4.** $2x + 4 < 10$ and $3x - 1 > 5$

$2x < 6 \qquad 3x > 6$

$x < 3 \qquad x > 2$

$x < 3$

$x > 2$

$x < 3$ and $x > 2$

The solution set is $\{x|\ 2 < x < 3\}$ or $(2,3)$.

**5.** $|2x - 5| \geq 9$

$2x - 5 \leq -9$ or $2x - 5 \geq 9$

$2x \leq -4 \qquad 2x \geq 14$

$x \leq -2 \qquad x \geq 7$

The solution set is $\{x|x \leq -2 \text{ or } x \geq 7\}$ or $(-\infty, -2] \cup [7, \infty)$.

**6.**
$$2x^2 = 7x - 5$$
$$2x^2 - 7x + 5 = 0$$
$$(2x - 5)(x - 1) = 0$$

Apply the zero product principle.

$2x - 5 = 0 \qquad x - 1 = 0$

$2x = 5 \qquad x = 1$

$x = \frac{5}{2}$

The solutions are 1 and $\frac{5}{2}$ or the solution set is $\left\{1, \frac{5}{2}\right\}$.

**7.**
$$2x^3 + 6x^2 = 20x$$
$$2x^3 + 6x^2 - 20x = 0$$
$$2x(x^2 + 3x - 10) = 0$$
$$2x(x + 5)(x - 2) = 0$$

Apply the zero product principle.

$2x = 0 \qquad x + 5 = 0 \qquad x - 2 = 0$

$x = 0 \qquad x = -5 \qquad x = 2$

The solutions are –5, 0, and 2 and the solution set is $\{-5, 0, 2\}$.

**8.** $$x = \frac{ax+b}{c}$$
$$cx = ax + b$$
$$cx - ax = b$$
$$x(c-a) = b$$
$$x = \frac{b}{c-a}$$

**9.** First, find the slope.
$$m = \frac{5-(-3)}{2-(-2)} = \frac{5+3}{2+2} = \frac{8}{4} = 2$$

Write the equation in point-slope form.
$$y - y_1 = m(x - x_1)$$
$$y - 5 = 2(x-2)$$
$$y - 5 = 2x - 4$$
$$y = 2x + 1$$
Write the equation of the line using function notation.
$$f(x) = 2x + 1$$

**10.** Let $x$ = the number of votes for the loser
Let $y$ = the number of votes for the winner
$$x + y = 2800$$
$$x + 160 = y$$
Rewrite the system in standard form.
$$x + y = 2800$$
$$x - y = -160$$
Add the equations to eliminate $y$.
$$x + y = 2800$$
$$\underline{x - y = -160}$$
$$2x = 2640$$
$$x = 1320$$
Back-substitute 1320 for $x$ to find $y$.
$$x + 160 = y$$
$$1320 + 160 = y$$
$$1480 = y$$
The solution set is $\{(1320, 1480)\}$.
The loser received 1320 votes and the winner received 1480 votes.

**11.** $$f(x) = -\frac{1}{3}x + 1$$
First, find the intercepts. To find the $y$–intercept, set $x = 0$.
$$y = -\frac{1}{3}(0) + 1$$
$$y = 0 + 1$$
$$y = 1$$
To find the $x$–intercept, set $y = 0$.
$$0 = -\frac{1}{3}x + 1$$
$$-3(-1) = -3\left(-\frac{1}{3}x\right)$$
$$3 = x$$
Graph the line using the intercepts.

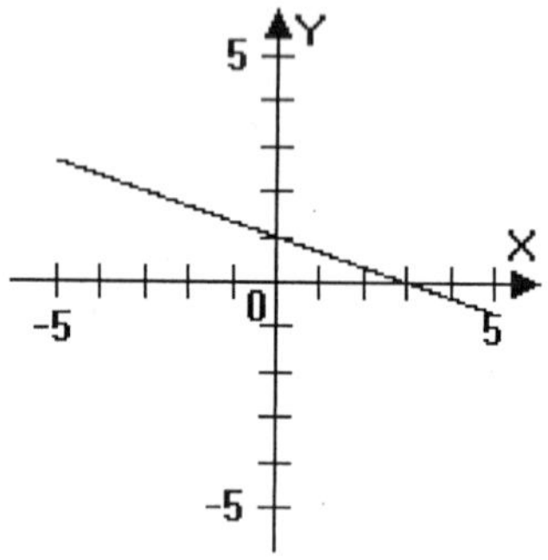


**12.** $4x - 5y < 20$
First, find the intercepts. To find the $y$–intercept, set $x = 0$.
$$4(0) - 5y = 20$$
$$0 - 5y = 20$$
$$-5y = 20$$
$$y = -4$$

To find the $x$–intercept, set $y = 0$.

$$4x - 5(0) = 20$$
$$4x - 0 = 20$$
$$4x = 20$$
$$x = 5$$

Graph the inequality using the intercepts.

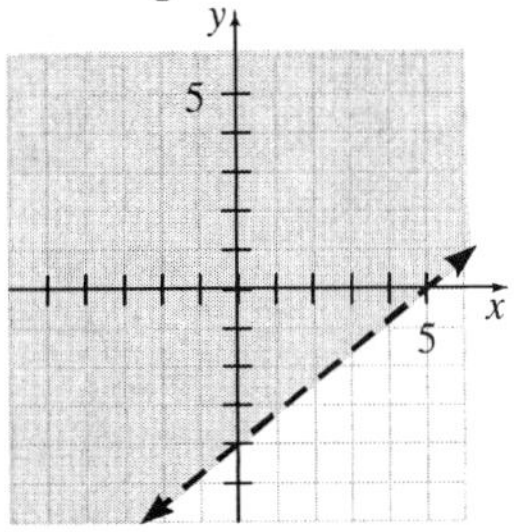


**13.** $y \leq -1$

We know that $y = -1$ is a horizontal line and the $y$–intercept is –1. This is sufficient information to graph the inequality.

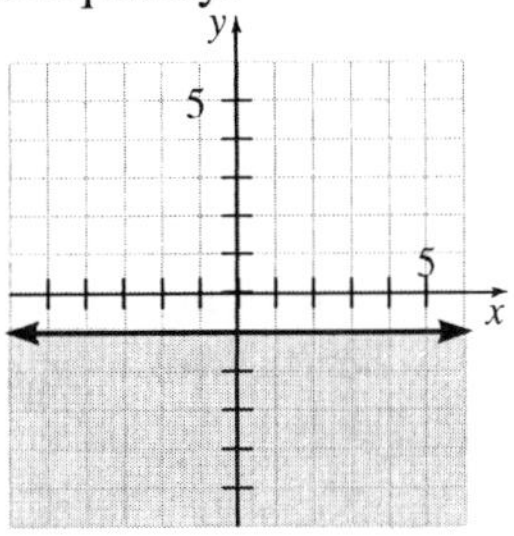


**14.**
$$\frac{-8x^3y^6}{16x^9y^{-4}} = \frac{-8}{16}x^{3-9}y^{6-(-4)}$$
$$= \frac{-8}{16}x^{3-9}y^{6+4}$$
$$= \frac{-1}{2}x^{-6}y^{10}$$
$$= -\frac{y^{10}}{2x^6}$$

**15.** $0.0000706 = 7.06 \times 10^{-5}$

**16.**
$$\left(3x^2 - y\right)^2 = \left(3x^2\right)^2 - 2 \cdot 3x^2 \cdot y + y^2$$
$$= 9x^4 - 6x^2y + y^2$$

**17.**
$$\left(3x^2 - y\right)\left(3x^2 + y\right)$$
$$= \left(3x^2\right)^2 + \cancel{3x^2y} - \cancel{3x^2y} - y^2$$
$$= 9x^4 - y^2$$

**18.**
$$x^3 - 3x^2 - 9x + 27$$
$$= x^2(x-3) - 9(x-3)$$
$$= (x-3)\left(x^2 - 9\right)$$
$$= (x-3)\left(x^2 - 3^2\right)$$
$$= (x-3)(x+3)(x-3)$$
$$= (x-3)^2(x+3)$$

**19.**
$$x^6 - x^2$$
$$= x^2\left(x^4 - 1\right)$$
$$= x^2\left(\left(x^2\right)^2 - 1^2\right)$$
$$= x^2\left(x^2+1\right)\left(x^2-1\right)$$
$$= x^2\left(x^2+1\right)\left(x^2-1^2\right)$$
$$= x^2\left(x^2+1\right)(x+1)(x-1)$$

**20.**
$$14x^3y^2 - 28x^4y^2$$
$$= 14x^3y^2(1-2x)$$

# Chapter 6
# Rational Expressions, Functions, and Equations

## 6.1 Exercise Set

**1.** $f(-2)=\frac{(-2)^2-9}{-2+3}=\frac{4-9}{1}=\frac{-5}{1}=-5$

$f(0)=\frac{0^2-9}{0+3}=-3$

$f(5)=\frac{5^2-9}{5+3}=\frac{25-9}{8}=\frac{16}{8}=2$

**3.** $f(x)=\frac{x^2-2x-3}{4-x}$

$f(-1)=\frac{(-1)^2-2(-1)-3}{4-(-1)}$

$=\frac{1+2-3}{4+1}=\frac{0}{5}=0$

$f(4)$ does not exist because we get division by 0.

$f(6)=\frac{6^2-2(6)-3}{4-6}=\frac{36-12-3}{-2}=-\frac{21}{2}$

**5.** $g(t)=\frac{2t^3-5}{t^2+1}$

$g(-1)=\frac{2(-1)^3-5}{(-1)^2+1}=\frac{2(-1)-5}{1+1}$

$=\frac{-2-5}{2}=-\frac{7}{2}$

$g(0)=\frac{2(0)^3-5}{0^2+1}=\frac{-5}{1}=-5$

$g(2)=\frac{2(2)^3-5}{(2)^2+1}=\frac{2(8)-5}{4+1}=\frac{16-5}{5}=\frac{11}{5}$

**7.** The domain is all real numbers except those which make the denominator zero. Set the denominator equal to zero and solve.

$x-5=0$

$x=5$

Domain of $f=\{x|x$ is a real number and $x\neq 5\}$ or $(-\infty,5)\cup(5,\infty)$.

**9.** The domain is all real numbers except those which make the denominator equal to zero. Set the denominator equal to zero and solve.

$x-1=0 \qquad x+3=0$

$x=1 \qquad x=-3$

Domain of $f=\{x|x$ is a real number and $x\neq -3$ and $x\neq 1\}$ or $(-\infty,-3)\cup(-3,1)\cup(1,\infty)$.

**11.** The domain is all real numbers except those which make the denominator equal to zero. Set the denominator equal to zero and solve. Since $x + 5$ is a double root, we only need solve the equation once.

$x+5=0$

$x=-5$

Domain of $f=\{x|x$ is a real number and $x\neq -5\}$ or $(-\infty,-5)\cup(-5,\infty)$.

**13.** The domain is all real numbers except those which make the denominator equal to zero. Set the denominator equal to zero and solve.

$x^2-8x+15=0$

$(x-5)(x-3)=0$

Apply the zero product principle.

$x-5=0 \qquad x-3=0$

$x=5 \qquad x=3$

Domain of $f=\{x|x$ is a real number and $x\neq 3$ and $x\neq 5\}$ or $(-\infty,3)\cup(3,5)\cup(5,\infty)$.

**15.** The domain is all real numbers except those which make the denominator equal to zero. Set the denominator equal to zero and solve.

$3x^2-2x-8=0$

$(3x+4)(x-2)=0$

Apply the zero product principle.

$3x+4=0 \qquad x-2=0$

$3x=-4 \qquad x=2$

$x=-\frac{4}{3}$

Domain of $f=\left\{x|x \text{ is a real number and } x\neq -\frac{4}{3} \text{ and } x\neq 2\right\}$ or $\left(-\infty,-\frac{4}{3}\right)\cup\left(-\frac{4}{3},2\right)\cup(2,\infty)$.

**17.** $f(4)=4$

**19.** Domain of $f=\{x|x$ is a real number and $x\neq -2$ and $x\neq 2\}$ or $(-\infty,-2)\cup(-2,2)\cup(2,\infty)$.

Range of $f=\{y\mid y\leq 0 \text{ or } y>3\}$, or $(-\infty,0]\cup(3,\infty)$

**21.** As $x$ decreases, the value of the function approaches 3. The equation of the horizontal asymptote is $y=3$.

**23.** There is no point on the graph with an $x$–coordinate of $-2$.

**25.** The graph is not continuous. Furthermore, it neither rises nor falls to the left or the right.

**27.**
$$\frac{x^2-4}{x-2}=\frac{(x+2)\cancel{(x-2)}}{1\cancel{(x-2)}}=x+2$$

**29.**
$$\frac{x+2}{x^2-x-6}=\frac{1\cancel{(x+2)}}{(x-3)\cancel{(x+2)}}=\frac{1}{x-3}$$

**31.**
$$\frac{4x+20}{x^2+5x}=\frac{4\cancel{(x+5)}}{x\cancel{(x+5)}}=\frac{4}{x}$$

**33.**
$$\frac{4y-20}{y^2-25}=\frac{4\cancel{(y-5)}}{(y+5)\cancel{(y-5)}}=\frac{4}{y+5}$$

**35.**
$$\frac{3x-5}{25-9x^2}=\frac{1\overset{-1}{\cancel{(3x-5)}}}{(5+3x)\cancel{(5-3x)}}$$
$$=\frac{-1}{5+3x} \text{ or } -\frac{1}{5+3x}$$
Or, by the Commutative Property of Addition, we have $\frac{-1}{3x+5}$ or $-\frac{1}{3x+5}$.

**37.**
$$\frac{y^2-49}{y^2-14y+49}=\frac{(y+7)\cancel{(y-7)}}{(y-7)\cancel{(y-7)}}$$
$$=\frac{y+7}{y-7}$$

**39.**
$$\frac{x^2+7x-18}{x^2-3x+2}=\frac{(x+9)\cancel{(x-2)}}{\cancel{(x-2)}(x-1)}$$
$$=\frac{x+9}{x-1}$$

**41.**
$$\frac{3x+7}{3x+10}$$
The rational expression cannot be simplified.

**43.**
$$\frac{x^2-x-12}{16-x^2}=\frac{\overset{-1}{\cancel{(x-4)}}(x+3)}{\cancel{(4-x)}(4+x)}$$
$$=-\frac{x+3}{4+x} \text{ or } -\frac{x+3}{x+4}$$

**45.**
$$\frac{x^2+3xy-10y^2}{3x^2-7xy+2y^2}=\frac{(x+5y)\cancel{(x-2y)}}{(3x-y)\cancel{(x-2y)}}$$
$$=\frac{x+5y}{3x-y}$$

**47.**
$$\frac{x^3-8}{x^2-4}=\frac{\cancel{(x-2)}(x^2+2x+4)}{(x+2)\cancel{(x-2)}}$$
$$=\frac{x^2+2x+4}{x+2}$$

**49.**
$$\frac{x^3+4x^2-3x-12}{x+4}$$
$$=\frac{x^2(x+4)-3(x+4)}{x+4}$$
$$=\frac{\cancel{(x+4)}(x^2-3)}{1\cancel{(x+4)}}$$
$$=x^2-3$$

**51.**
$$\frac{x-3}{x+7}\cdot\frac{3x+21}{2x-6}=\frac{1\cancel{(x-3)}}{1\cancel{(x+7)}}\cdot\frac{3\cancel{(x+7)}}{2\cancel{(x-3)}}$$
$$=\frac{3}{2}$$

**53.**
$$\frac{x^2-49}{x^2-4x-21}\cdot\frac{x+3}{x}$$
$$=\frac{(x+7)\cancel{(x-7)}}{\cancel{(x-7)}\cancel{(x+3)}}\cdot\frac{1\cancel{(x+3)}}{x}$$
$$=\frac{x+7}{x}$$

**55.**
$$\frac{x^2-9}{x^2-x-6}\cdot\frac{x^2+5x+6}{x^2+x-6}$$
$$=\frac{(x+3)\cancel{(x-3)}}{\cancel{(x-3)}\cancel{(x+2)}}\cdot\frac{\cancel{(x+3)}\cancel{(x+2)}}{\cancel{(x+3)}(x-2)}$$
$$=\frac{x+3}{x-2}$$

**57.**

$$\frac{x^2+4x+4}{x^2+8x+16}\cdot\frac{(x+4)^3}{(x+2)^3}$$

$$=\frac{(x+2)^2}{(x+4)^2}\cdot\frac{(x+4)^3}{(x+2)^3}$$

$$=\frac{\cancel{(x+2)^2}}{\cancel{(x+4)^2}}\cdot\frac{\overset{x+4}{\cancel{(x+4)^3}}}{\underset{x+2}{\cancel{(x+2)^3}}}$$

$$=\frac{x+4}{x+2}$$

**59.**

$$\frac{8y+2}{y^2-9}\cdot\frac{3-y}{4y^2+y}$$

$$=\frac{2\cancel{(4y+1)}}{(y+3)\cancel{(y-3)}}\cdot\frac{1\overset{-1}{\cancel{(3-y)}}}{y\cancel{(4y+1)}}$$

$$=\frac{-2}{y(y+3)}\text{ or }-\frac{2}{y(y+3)}$$

**61.**

$$\frac{y^3-8}{y^2-4}\cdot\frac{y+2}{2y}$$

$$=\frac{\cancel{(y-2)}(y^2+2y+4)}{\cancel{(y+2)}\cancel{(y-2)}}\cdot\frac{1\cancel{(y+2)}}{2y}$$

$$=\frac{y^2+2y+4}{2y}$$

**63.**

$$(x-3)\cdot\frac{x^2+x+1}{x^2-5x+6}$$

$$=\frac{1\cancel{(x-3)}}{1}\cdot\frac{x^2+x+1}{\cancel{(x-3)}(x-2)}$$

$$=\frac{x^2+x+1}{x-2}$$

**65.**

$$\frac{x^2+xy}{x^2-y^2}\cdot\frac{4x-4y}{x}$$

$$=\frac{x(x+y)}{(x+y)(x-y)}\cdot\frac{4(x-y)}{x}$$

$$=\frac{\cancel{x}\cancel{(x+y)}}{\cancel{(x+y)}\cancel{(x-y)}}\cdot\frac{4\cancel{(x-y)}}{\cancel{x}}=4$$

**67.**

$$\frac{x^2+2xy+y^2}{x^2-2xy+y^2}\cdot\frac{4x-4y}{3x+3y}$$

$$=\frac{\overset{x+y}{\cancel{(x+y)^2}}}{\underset{x-y}{\cancel{(x-y)^2}}}\cdot\frac{4\cancel{(x-y)}}{3\cancel{(x+y)}}$$

$$=\frac{4(x+y)}{3(x-y)}$$

**69.**

$$\frac{4a^2+2ab+b^2}{2a+b}\cdot\frac{4a^2-b^2}{8a^3-b^3}$$

$$=\frac{\cancel{4a^2+2ab+b^2}}{\cancel{2a+b}}\cdot\frac{\cancel{(2a-b)}\cancel{(2a+b)}}{\cancel{(2a-b)}\cancel{(4a^2+2ab+b^2)}}$$

$$=1$$

**71.** $\frac{10z^2+13z-3}{3z^2-8z+5}\cdot\frac{2z^2-3z-2z+3}{25z^2-10z+1}\cdot\frac{15z^2-28z+5}{4z^2-9}$

$=\frac{(5z-1)(2z+3)}{(3z-5)(z-1)}\cdot\frac{z(2z-3)-1(2z-3)}{(5z-1)(5z-1)}\cdot\frac{(5z-1)(3z-5)}{(2z-3)(2z+3)}$

$=\frac{\cancel{(5z-1)}\cancel{(2z+3)}}{\cancel{(3z-5)}\cancel{(z-1)}}\cdot\frac{\cancel{(z-1)}\cancel{(2z-3)}}{\cancel{(5z-1)}\cancel{(5z-1)}}\cdot\frac{\cancel{(5z-1)}\cancel{(3z-5)}}{\cancel{(2z-3)}\cancel{(2z+3)}}$

$=1$

**73.** $\frac{x+5}{7}\div\frac{4x+20}{9}$

$=\frac{x+5}{7}\cdot\frac{9}{4x+20}$

$=\frac{1\cancel{(x+5)}}{7}\cdot\frac{9}{4\cancel{(x+5)}}=\frac{9}{28}$

**75.** $\frac{4}{y-6}\div\frac{40}{7y-42}$

$=\frac{4}{y-6}\cdot\frac{7y-42}{40}$

$=\frac{\cancel{4}}{1\cancel{(y-6)}}\cdot\frac{7\cancel{(y-6)}}{\cancel{4}\cdot 10}=\frac{7}{10}$

**77.** $\frac{x^2-2x}{15}\div\frac{x-2}{5}$

$=\frac{x^2-2x}{15}\cdot\frac{5}{x-2}$

$=\frac{x\cancel{(x-2)}}{3\cdot\cancel{5}}\cdot\frac{\cancel{5}}{1\cancel{(x-2)}}=\frac{x}{3}$ or $\frac{1}{3}x$

**79.** $\frac{y^2-25}{2y-2}\div\frac{y^2+10y+25}{y^2+4y-5}$

$=\frac{y^2-25}{2y-2}\cdot\frac{y^2+4y-5}{y^2+10y+25}$

$=\frac{\cancel{(y+5)}(y-5)}{2\cancel{(y-1)}}\cdot\frac{\cancel{(y+5)}\cancel{(y-1)}}{\cancel{(y+5)^2}}$

$=\frac{y-5}{2}$

**81.** $(x^2-16)\div\frac{x^2+3x-4}{x^2+4}$

$=\frac{x^2-16}{1}\div\frac{x^2+3x-4}{x^2+4}$

$=\frac{x^2-16}{1}\cdot\frac{x^2+4}{x^2+3x-4}$

$=\frac{\cancel{(x+4)}(x-4)}{1}\cdot\frac{x^2+4}{\cancel{(x+4)}(x-1)}$

$=\frac{(x-4)(x^2+4)}{x-1}$

**83.** $\dfrac{y^2-4y-21}{y^2-10y+25}\div\dfrac{y^2+2y-3}{y^2-6y+5}$

$=\dfrac{y^2-4y-21}{y^2-10y+25}\cdot\dfrac{y^2-6y+5}{y^2+2y-3}$

$=\dfrac{(y-7)(y+3)}{(y-5)^2}\cdot\dfrac{(y-5)(y-1)}{(y+3)(y-1)}$

$=\dfrac{y-7}{y-5}$

**85.** $\dfrac{8x^3-1}{4x^2+2x+1}\div\dfrac{x-1}{(x-1)^2}$

$=\dfrac{8x^3-1}{4x^2+2x+1}\cdot\dfrac{(x-1)^2}{x-1}$

$=\dfrac{(2x-1)(4x^2+2x+1)}{1(4x^2+2x+1)}\cdot\dfrac{(x-1)^2}{1(x-1)}$

$=(2x-1)(x-1)$

**87.** $\dfrac{x^2-4y^2}{x^2+3xy+2y^2}\div\dfrac{x^2-4xy+4y^2}{x+y}$

$=\dfrac{x^2-4y^2}{x^2+3xy+2y^2}\cdot\dfrac{x+y}{x^2-4xy+4y^2}$

$=\dfrac{(x+2y)(x-2y)}{(x+2y)(x+y)}\cdot\dfrac{1(x+y)}{(x-2y)^2}$

$=\dfrac{1}{x-2y}$

**89.** $\dfrac{x^4-y^8}{x^2+y^4}\div\dfrac{x^2-y^4}{3x^2}$

$=\dfrac{x^4-y^8}{x^2+y^4}\cdot\dfrac{3x^2}{x^2-y^4}$

$=\dfrac{(x^2+y^4)(x^2-y^4)}{1(x^2+y^4)}\cdot\dfrac{3x^2}{(x+y^2)(x-y^2)}$

$=\dfrac{3x^2(x^2-y^4)}{(x+y^2)(x-y^2)}$

$=\dfrac{3x^2(x+y^2)(x-y^2)}{1(x+y^2)(x-y^2)}=3x^2$

**91.** $$\frac{x^3-4x^2+x-4}{2x^3-8x^2+x-4}\cdot\frac{2x^3+2x^2+x+1}{x^4-x^3+x^2-x}$$
$$=\frac{x^2(x-4)+1(x-4)}{2x^2(x-4)+1(x-4)}\cdot\frac{2x^2(x+1)+1(x+1)}{x^3(x-1)+x(x-1)}$$
$$=\frac{(x-4)(x^2+1)}{(x-4)(2x^2+1)}\cdot\frac{(x+1)(2x^2+1)}{(x-1)(x^3+x)}$$
$$=\frac{\cancel{(x-4)}(x^2+1)}{\cancel{(x-4)}\cancel{(2x^2+1)}}\cdot\frac{(x+1)\cancel{(2x^2+1)}}{(x-1)(x^3+x)}$$
$$=\frac{(x^2+1)(x+1)}{(x-1)x(x^2+1)}=\frac{\cancel{(x^2+1)}(x+1)}{(x-1)x\cancel{(x^2+1)}}$$
$$=\frac{x+1}{x(x-1)}$$

**92.** $$\frac{y^3+y^2+yz^2+z^2}{y^3+y+y^2+1}\cdot\frac{y^3+y+y^2z+z}{2y^2+2yz-yz^2-z^3}$$
$$=\frac{y^2(y+1)+z^2(y+1)}{y(y^2+1)+1(y^2+1)}\cdot\frac{y(y^2+1)+z(y^2+1)}{2y(y+z)-z^2(y+z)}$$
$$=\frac{(y+1)(y^2+z^2)(y^2+1)(y+z)}{(y^2+1)(y+1)(y+z)(2y-z^2)}$$
$$=\frac{\cancel{(y+1)}(y^2+z^2)\cancel{(y^2+1)}\cancel{(y+z)}}{\cancel{(y^2+1)}\cancel{(y+1)}\cancel{(y+z)}(2y-z^2)}$$
$$=\frac{y^2+z^2}{2y-z^2}$$

**93.** $$\frac{ax-ay+3x-3y}{x^3+y^3}\div\frac{ab+3b+ac+3c}{xy-x^2-y^2}$$

$$=\frac{ax-ay+3x-3y}{x^3+y^3}\cdot\frac{xy-x^2-y^2}{ab+3b+ac+3c}$$

$$=\frac{a(x-y)+3(x-y)}{(x+y)(x^2-xy+y^2)}\cdot\frac{(-1)(x^2-xy+y^2)}{b(a+3)+c(a+3)}$$

$$=\frac{(-1)(x-y)(a+3)(x^2-xy+y^2)}{(x+y)(x^2-xy+y^2)(a+3)(b+c)}$$

$$=\frac{(-1)(x-y)\cancel{(a+3)}\cancel{(x^2-xy+y^2)}}{(x+y)\cancel{(x^2-xy+y^2)}\cancel{(a+3)}(b+c)}$$

$$=\frac{-(x-y)}{(x+y)(b+c)}$$

**94.** $$\frac{a^3+b^3}{ac-ad-bc+bd}\div\frac{ab-a^2-b^2}{ac-ad+bc-bd}$$

$$=\frac{a^3+b^3}{ac-ad-bc+bd}\cdot\frac{ac-ad+bc-bd}{ab-a^2-b^2}$$

$$=\frac{(a+b)(a^2-ab+b^2)}{a(c-d)-b(c-d)}\cdot\frac{a(c-d)+b(c-d)}{(-1)(a^2-ab+b^2)}$$

$$=\frac{(a+b)\cancel{(a^2-ab+b^2)}\cancel{(c-d)}(a+b)}{(-1)\cancel{(c-d)}(a-b)\cancel{(a^2-ab+b^2)}}$$

$$=\frac{(a+b)(a+b)}{(-1)(a-b)}$$

$$=\frac{-(a+b)^2}{a-b}\quad\text{or}\quad\frac{(a+b)^2}{b-a}$$

**95.**

$$\frac{a^2b+b}{3a^2-4a-20}\cdot\frac{a^2+5a}{2a^2+11a+5}\div\frac{ab^2}{6a^2-17a-10}$$

$$=\frac{a^2b+b}{3a^2-4a-20}\cdot\frac{a^2+5a}{2a^2+11a+5}\cdot\frac{6a^2-17a-10}{ab^2}$$

$$=\frac{b\left(a^2+1\right)}{(3a-10)(a+2)}\cdot\frac{a(a+5)}{(2a+1)(a+5)}\cdot\frac{(3a-10)(2a+1)}{ab^2}$$

$$=\frac{\cancel{b}\left(a^2+1\right)}{\cancel{(3a-10)}(a+2)}\cdot\frac{\cancel{a}\cancel{(a+5)}}{\cancel{(2a+1)}\cancel{(a+5)}}\cdot\frac{\cancel{(3a-10)}\cancel{(2a+1)}}{\cancel{a}\cancel{b^2}_{b}}$$

$$=\frac{a^2+1}{b(a+2)}$$

**96.**

$$\frac{a^2-8a+15}{2a^3-10a^2}\cdot\frac{2a^2+3a}{3a^3-27a}\div\frac{14a+21}{a^2-6a-27}$$

$$=\frac{a^2-8a+15}{2a^3-10a^2}\cdot\frac{2a^2+3a}{3a^3-27a}\cdot\frac{a^2-6a-27}{14a+21}$$

$$=\frac{(a-5)(a-3)}{2a^2(a-5)}\cdot\frac{a(2a+3)}{3a\left(a^2-9\right)}\cdot\frac{(a-9)(a+3)}{7(2a+3)}$$

$$=\frac{(a-5)(a-3)}{2a^2(a-5)}\cdot\frac{a(2a+3)}{3a(a+3)(a-3)}\cdot\frac{(a-9)(a+3)}{7(2a+3)}$$

$$=\frac{\cancel{(a-5)}\cancel{(a-3)}}{2a^2\cancel{(a-5)}}\cdot\frac{\cancel{a}\cancel{(2a+3)}}{3\cancel{a}\cancel{(a+3)}\cancel{(a-3)}}\cdot\frac{(a-9)\cancel{(a+3)}}{7\cancel{(2a+3)}}$$

$$=\frac{a-9}{42a^2}$$

**97.**
$$\frac{a-b}{4c}\div\left(\frac{b-a}{c}\div\frac{a-b}{c^2}\right)=\frac{a-b}{4c}\div\left(\frac{b-a}{c}\cdot\frac{c^2}{a-b}\right)$$
$$=\frac{a-b}{4c}\div\left(\frac{-\cancel{(a-b)}}{\cancel{c}}\cdot\frac{\cancel{c}\cdot c}{\cancel{(a-b)}}\right)$$
$$=\frac{a-b}{4c}\div\frac{-c}{1}$$
$$=\frac{a-b}{4c}\cdot\frac{-1}{c}$$
$$=-\frac{a-b}{4c^2}\text{ or }\frac{b-a}{4c^2}$$

**98.**
$$\left(\frac{a-b}{4c}\div\frac{b-a}{c}\right)\div\frac{a-b}{c^2}=\left(\frac{a-b}{4c}\cdot\frac{c}{b-a}\right)\div\frac{a-b}{c^2}$$
$$=\left(\frac{\cancel{(a-b)}}{4\cancel{c}}\cdot\frac{\cancel{c}}{-\cancel{(a-b)}}\right)\div\frac{a-b}{c^2}$$
$$=\frac{1}{-4}\div\frac{a-b}{c^2}$$
$$=-\frac{1}{4}\cdot\frac{c^2}{(a-b)}$$
$$=-\frac{c^2}{4(a-b)}\text{ or }\frac{c^2}{4(b-a)}$$

**99.**
$$\frac{f(a+h)-f(a)}{h}=\frac{[7(a+h)-4]-[7a-4]}{h}$$
$$=\frac{\cancel{7a}+7h-\cancel{4}-\cancel{7a}+\cancel{4}}{h}$$
$$=\frac{7\cancel{h}}{\cancel{h}}$$
$$=7$$

**100.** $$\frac{f(a+h)-f(a)}{h}=\frac{[-3(a+h)+5]-[-3a+5]}{h}$$
$$=\frac{-3a-3h+5+3a-5}{h}$$
$$=\frac{-3h}{h}$$
$$=-3$$

**101.** $$\frac{f(a+h)-f(a)}{h}=\frac{[(a+h)^2-5(a+h)+3]-[a^2-5a+3]}{h}$$
$$=\frac{a^2+2ah+h^2-5a-5h+3-a^2+5a-3}{h}$$
$$=\frac{2ah+h^2-5h}{h}$$
$$=\frac{h(2a+h-5)}{h}$$
$$=2a+h-5$$

**102.** $$\frac{f(a+h)-f(a)}{h}=\frac{[3(a+h)^2-4(a+h)+7]-[3a^2-4a+7]}{h}$$
$$=\frac{[3(a^2+2ah+h^2-4a-4h+7)]-3a^2+4a-7}{h}$$
$$=\frac{3a^2+6ah+3h^2-4a-4h+7-3a^2+4a-7}{h}$$
$$=\frac{6ah+3h^2-4h}{h}$$
$$=\frac{h(6a+3h-4)}{h}$$
$$=6a+3h-4$$

**103.** $f(x)=\dfrac{(x+2)^2}{1-2x}$ and $g(x)=\dfrac{x+2}{2x-1}$

$$\left(\frac{f}{g}\right)(x)=\frac{f(x)}{g(x)}=\frac{\frac{(x+2)^2}{1-2x}}{\frac{x+2}{2x-1}}$$

$$=\frac{(x+2)^2}{1-2x}\cdot\frac{2x-1}{x+2}$$

$$=\frac{(x+2)^2(2x-1)}{(-1)(2x-1)(x+2)}$$

$$=\frac{(x+2)^{\not{2}}\,\cancel{(2x-1)}}{(-1)\cancel{(2x-1)}\cancel{(x+2)}}$$

$$=-(x+2)$$

To determine the domain, we need to exclude any values for $x$ that make either $f(x)$ or $g(x)$ have division by 0. In addition, we need to exclude all values for $x$ such that $g(x)=0$.

$1-2x=0$

$-2x=-1 \qquad 2x-1=0 \qquad x+2=0$

$x=\dfrac{1}{2} \qquad 2x=1 \qquad x=-2$

$x=\dfrac{1}{2}$

We need to exclude the values $x=\dfrac{1}{2}$ and $x=-2$.

Domain: $\left\{x \mid x \text{ is a real number and } x\neq -2 \text{ and } x\neq\dfrac{1}{2}\right\}$ or $(-\infty,-2)\cup\left(-2,\dfrac{1}{2}\right)\cup\left(\dfrac{1}{2},\infty\right)$.

**104.** $f(x)=\dfrac{(x+2)^2}{1-2x}$ and $g(x)=\dfrac{x+2}{2x-1}$

$$\left(\frac{g}{f}\right)(x)=\frac{g(x)}{f(x)}=\frac{\dfrac{x+2}{2x-1}}{\dfrac{(x+2)^2}{1-2x}}$$

$$=\frac{x+2}{2x-1}\cdot\frac{1-2x}{(x+2)^2}$$

$$=\frac{(x+2)(-1)(2x-1)}{(2x-1)(x+2)^2}$$

$$=\frac{\cancel{(x+2)}(-1)\cancel{(2x-1)}}{\cancel{(2x-1)}(x+2)^{\cancel{2}}}$$

$$=-\frac{1}{x+2} \text{ or } \frac{1}{-x-2}$$

To determine the domain, we need to exclude any values for $x$ that make either $f(x)$ or $g(x)$ have division by 0. In addition, we need to exclude all values for $x$ such that $f(x)=0$.

$$1-2x=0 \qquad 2x-1=0 \qquad (x+2)^2=0$$
$$-2x=-1 \qquad 2x=1 \qquad x=-2$$
$$x=\frac{1}{2} \qquad x=\frac{1}{2}$$

We need to exclude the values $x=\dfrac{1}{2}$ and $x=-2$.

Domain: $\left\{x \mid x \text{ is a real number and } x\neq -2 \text{ and } x\neq \dfrac{1}{2}\right\}$ or $(-\infty,-2)\cup\left(-2,\dfrac{1}{2}\right)\cup\left(\dfrac{1}{2},\infty\right)$.

**105.** From the graph, we see that $f(60)=195$. This corresponds to the point $(60,195)$. The cost to inoculate 60% of the population against a particular strain of flu is \$195,000,000.

**107.** The value 100 must be excluded from the domain. We cannot inoculate 100% of the population.

**109.** From the graph, we see that $P(10) = 7$. This is represented by the point $(10, 7)$. 7% of the people with 10 years of education are unemployed.

**111.** The function's value is approaching 0 to the far right of the graph. There is no education level that leads to guaranteed employment. This is indicated by the horizontal asymptote $y = 0$, showing that unemployment will never be 0%.

**113.** From the graph, we see that $P(10) = 90$. This is represented by the point $(10, 90)$ and refers to an incidence ratio of 10. From the chart, we know that smokers between the ages of 55 and 64 are 10 times more likely than nonsmokers to die from lung cancer. Also, 90% of the deaths from lung cancer in this group are smoking-related.

**115.** The horizontal asymptote of the graph is $y = 100$. This means that as incidence ratio increases the percentage of smoking-related deaths increases towards 100%, although the percentage will never actually reach 100%.

**117.** Answers will vary.

**119.** Answers will vary.

**121.** Answers will vary.

**123.** Answers will vary.

**125.** Answers will vary.

**127.** Answers will vary.

**129.** $\dfrac{x^2+x}{3x} \cdot \dfrac{6x}{x+1} = 2x$

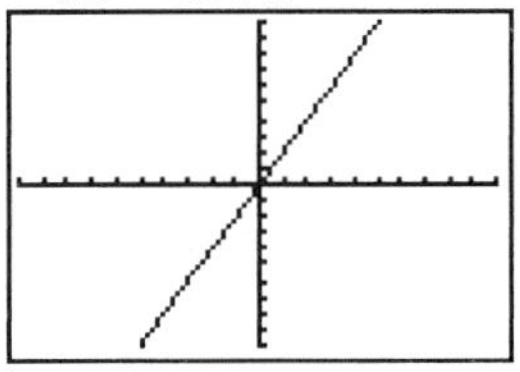

The graphs coincide. The multiplication is correct.

**131.** $\dfrac{x^2-9}{x+4} \div \dfrac{x-3}{x+4} = x-3$

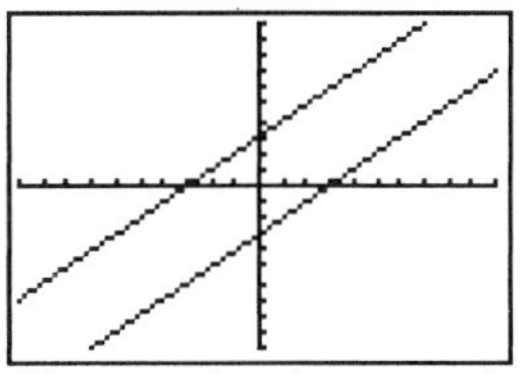

The graphs do not coincide. The division is incorrect.

$$\frac{x^2-9}{x+4} \div \frac{x-3}{x+4}$$
$$= \frac{x^2-9}{x+4} \cdot \frac{x+4}{x-3}$$
$$= \frac{(x+3)\cancel{(x-3)}}{\cancel{x+4}} \cdot \frac{\cancel{x+4}}{\cancel{x-3}}$$
$$= x+3$$

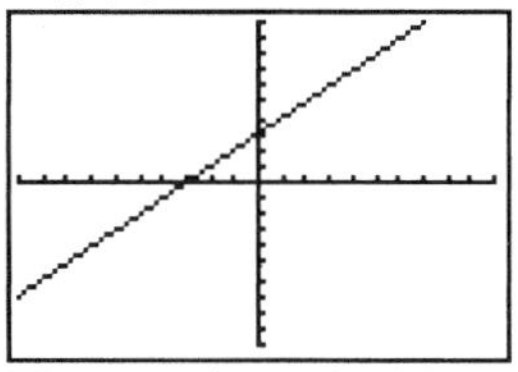

The graphs coincide. The division is now correct.

**133. a.** Answers will vary.

$$f(x)=\frac{x^2-x-2}{x-2}$$

$$g(x)=x+1$$

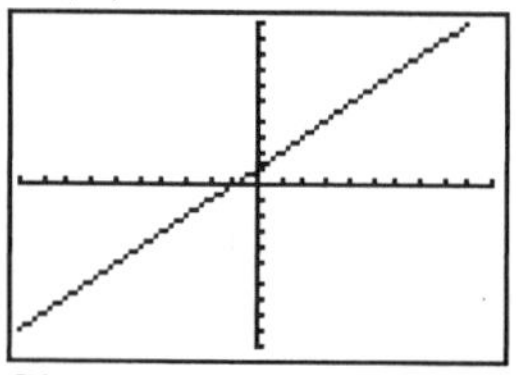

Since the graphs coincide, we know that $f(x)=g(x)$.

**b.** Answers will vary.

$$f(x)=\frac{x^2-x-2}{x-2}$$

$$=\frac{\cancel{(x-2)}(x+1)}{\cancel{x-2}}$$

$$=x+1$$

Although when simplified the expressions are the same, $f$ and $g$ do not represent the same function. The domains of the two functions are not the same since the domain of $f$ excludes $x=2$. Specifically, $g(2)=3$, while $f(2)$ is undefined.

**c.**

$$f(x)=\frac{x^2-x-2}{x-2}$$

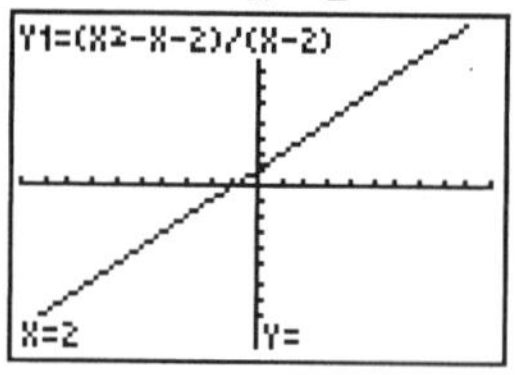


$$g(x)=x+1$$

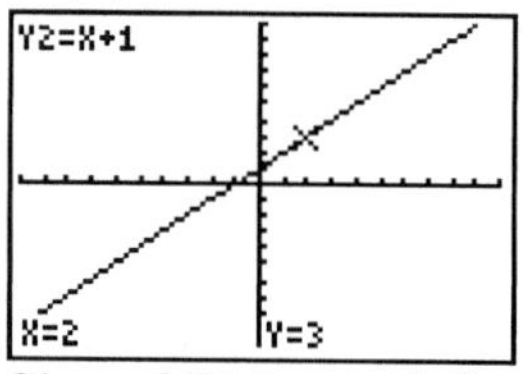


Since $f(2)$ is not defined, we know that 2 is not in the domain of $f$. Since $g(2) = 3$, we know that 2 is in the domain of $g$.

**135. a.**

$$f(x)=\frac{27{,}725(x-14)}{x^2+9}-5x$$

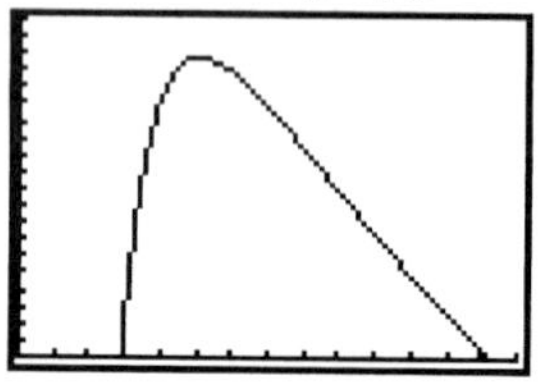

**b.** As $x$ increases, $y$ first increases quickly then decreases but at a slightly slower rate.

**c.**

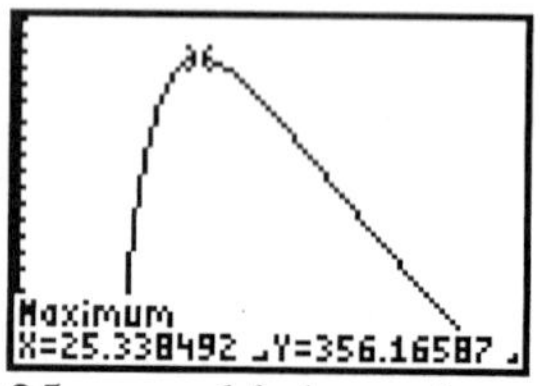


25-year olds have the greatest number of arrests; about 356 arrests per 100,000 drivers.

**137.**

| $x$ | $y = f(x) = \dfrac{x^2 - x - 2}{x - 2}$ | $(x, y)$ |
|---|---|---|
| $-2$ | $y = \dfrac{(-2)^2 - (-2) - 2}{-2 - 2} = -1$ | $(-2, -1)$ |
| $0$ | $y = \dfrac{0^2 - 0 - 2}{0 - 2} = 1$ | $(0, 1)$ |
| $2$ | $y = \dfrac{2^2 - 2 - 2}{2 - 2} = \dfrac{0}{0}$ indeterminate | hole in graph at $(2, 3)$ |
| $4$ | $y = \dfrac{4^2 - 4 - 2}{4 - 2} = 5$ | $(4, 5)$ |

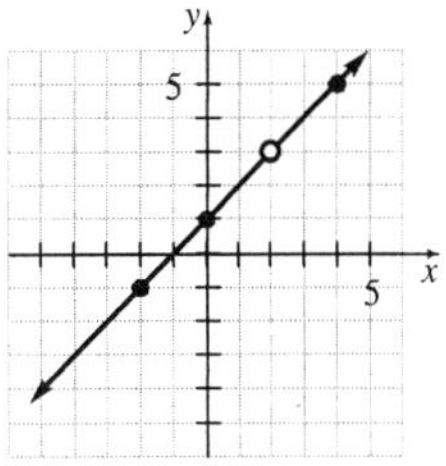


**139.** $$\frac{y^{2n} - 1}{y^{2n} + 3y^n + 2} \div \frac{y^{2n} + y^n - 12}{y^{2n} - y^n - 6} = \frac{y^{2n} - 1}{y^{2n} + 3y^n + 2} \cdot \frac{y^{2n} - y^n - 6}{y^{2n} + y^n - 12}$$

$$= \frac{(y^n + 1)(y^n - 1)}{(y^n + 2)(y^n + 1)} \cdot \frac{(y^n - 3)(y^n + 2)}{(y^n + 4)(y^n - 3)} = \frac{y^n - 1}{y^n + 4}$$

**141.** $4x - 5y \geq 20$

First, find the intercepts to the equation $4x - 5y = 20$.

Find the $x$–intercept by setting $y = 0$.

$$4x - 5(0) = 20$$
$$4x = 20$$
$$x = 5$$

Find the $y$–intercept by setting $x = 0$.

$$4(0) - 5y = 20$$
$$-5y = 20$$
$$y = -4$$

Next, use the origin, $(0, 0)$, as a test point.

$$4(0) - 5(0) \geq 20$$
$$0 \geq 20$$

This is a false statement. This means that the point, $(0, 0)$, will not fall in the shaded half-plane.

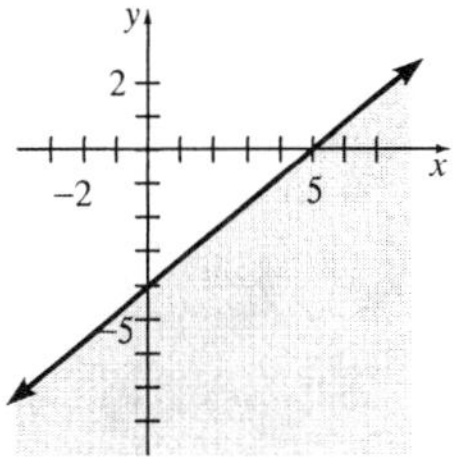


**142.** $(2x - 5)(x^2 - 3x - 6)$

$$= 2x(x^2 - 3x - 6) - 5(x^2 - 3x - 6)$$
$$= 2x^3 - 6x^2 - 12x - 5x^2 + 15x + 30$$
$$= 2x^3 - 11x^2 + 3x + 30$$

**143.** $\left(\dfrac{ab^{-3}c^{-4}}{4a^5b^{10}c^{-3}}\right)^{-2}$

$$= \left(4^{-1}a^{1-5}b^{-3-10}c^{-4-(-3)}\right)^{-2}$$
$$= \left(4^{-1}a^{-4}b^{-13}c^{-1}\right)^{-2}$$
$$= 4^{-1(-2)}a^{-4(-2)}b^{-13(-2)}c^{-1(-2)}$$
$$= 4^2a^8b^{26}c^2 = 16a^8b^{26}c^2$$

## 6.2 Exercise Set

**1.** $\frac{2}{9x}+\frac{4}{9x}=\frac{2+4}{9x}=\frac{6}{9x}=\frac{2\cdot\cancel{3}}{3\cdot\cancel{3}x}=\frac{2}{3x}$

**3.** $\frac{x}{x-5}+\frac{9x+3}{x-5}=\frac{x+9x+3}{x-5}=\frac{10x+3}{x-5}$

**5.** $\frac{x^2-2x}{x^2+3x}+\frac{x^2+x}{x^2+3x}=\frac{x^2-2x+x^2+x}{x^2+3x}$

$=\frac{2x^2-x}{x^2+3x}$

$=\frac{\cancel{x}(2x-1)}{\cancel{x}(x+3)}$

$=\frac{2x-1}{x+3}$

**7.** $\frac{y^2}{y^2-9}+\frac{9-6y}{y^2-9}=\frac{y^2+9-6y}{y^2-9}$

$=\frac{\overset{y-3}{\cancel{(y-3)^2}}}{(y+3)\cancel{(y-3)}}$

$=\frac{y-3}{y+3}$

**9.** $\frac{3x}{4x-3}-\frac{2x-1}{4x-3}=\frac{3x-(2x-1)}{4x-3}$

$=\frac{3x-2x+1}{4x-3}$

$=\frac{x+1}{4x-3}$

**11.** $\frac{x^2-2}{x^2+6x-7}-\frac{19-4x}{x^2+6x-7}$

$=\frac{x^2-2-(19-4x)}{x^2+6x-7}=\frac{x^2-2-19+4x}{x^2+6x-7}$

$=\frac{x^2+4x-21}{x^2+6x-7}=\frac{\cancel{(x+7)}(x-3)}{\cancel{(x+7)}(x-1)}$

$=\frac{x-3}{x-1}$

**13.** $\frac{20y^2+5y+1}{6y^2+y-2}-\frac{8y^2-12y-5}{6y^2+y-2}$

$=\frac{20y^2+5y+1-(8y^2-12y-5)}{6y^2+y-2}$

$=\frac{20y^2+5y+1-8y^2+12y+5}{6y^2+y-2}$

$=\frac{12y^2+17y+6}{6y^2+y-2}$

$=\frac{(4y+3)\cancel{(3y+2)}}{\cancel{(3y+2)}(2y-1)}$

$=\frac{4y+3}{2y-1}$

**15.** $\frac{2x^3-3y^3}{x^2-y^2}-\frac{x^3-2y^3}{x^2-y^2}$

$=\frac{2x^3-3y^3-(x^3-2y^3)}{x^2-y^2}$

$=\frac{2x^3-3y^3-x^3+2y^3}{x^2-y^2}$

$=\frac{x^3-y^3}{x^2-y^2}$

$=\frac{\cancel{(x-y)}(x^2+xy+y^2)}{(x+y)\cancel{(x-y)}}$

$=\frac{x^2+xy+y^2}{x+y}$

**17.** $25x^2=5^2\cdot x^2$

$35x=5\cdot 7x$

$\text{LCD}=5^2\cdot 7x^2=175x^2$

**19.** $x-5 = x-5$

$x^2-25=(x+5)(x-5)$

$\text{LCD}=(x+5)(x-5)$

**21.** $y^2-100=(y+10)(y-10)$

$y(y-10)=y(y-10)$

$\text{LCD}=y(y+10)(y-10)$

**23.** $x^2-16=(x+4)(x-4)$

$x^2-8x+16=(x-4)^2$

$\text{LCD}=(x+4)(x-4)^2$

**25.** $y^2-5y-6=(y-6)(y+1)$

$y^2-4y-5=(y-5)(y+1)$

$\text{LCD}=(y-6)(y-5)(y+1)$

**27.** $2y^2+7y+6=(2y+3)(y+2)$

$y^2-4=(y+2)(y-2)$

$2y^2-3y-2=(y-2)(2y+1)$

$\text{LCD}=(2y+3)(y+2)(y-2)(2y+1)$

**29.** $\frac{3}{5x^2}+\frac{10}{x}$

The LCD is $5x^2$

$=\frac{3}{5x^2}+\frac{10\cdot 5x}{x\cdot 5x}$

$=\frac{3}{5x^2}+\frac{50x}{5x^2}=\frac{3+50x}{5x^2}$

**31.** $\frac{4}{x-2}+\frac{3}{x+1}$

The LCD is $(x-2)(x+1)$.

$=\frac{4}{(x-2)}\cdot\frac{(x+1)}{(x+1)}+\frac{3}{(x+1)}\cdot\frac{(x-2)}{(x-2)}$

$=\frac{4(x+1)+3(x-2)}{(x-2)(x+1)}$

$=\frac{4x+4+3x-6}{(x-2)(x+1)}=\frac{7x-2}{(x-2)(x+1)}$

**33.** $\dfrac{3x}{x^2+x-2}+\dfrac{2}{x^2-4x+3}=\dfrac{3x}{(x+2)(x-1)}+\dfrac{2}{(x-1)(x-3)}$

The LCD is $(x+2)(x-1)(x-3)$.

$=\dfrac{3x(x-3)}{(x+2)(x-1)(x-3)}+\dfrac{2(x+2)}{(x+2)(x-1)(x-3)}=\dfrac{3x(x-3)+2(x+2)}{(x+2)(x-1)(x-3)}$

$=\dfrac{3x^2-9x+2x+4}{(x+2)(x-1)(x-3)}=\dfrac{3x^2-7x+4}{(x+2)(x-1)(x-3)}=\dfrac{(3x-4)\cancel{(x-1)}}{(x+2)\cancel{(x-1)}(x-3)}=\dfrac{3x-4}{(x+2)(x-3)}$

**35.** $\dfrac{x-6}{x+5}+\dfrac{x+5}{x-6}$

Since the denominators have no common factors, the LCD is $(x+5)(x-6)$.

$=\dfrac{(x-6)(x-6)}{(x+5)(x-6)}+\dfrac{(x+5)(x+5)}{(x+5)(x-6)}$

$=\dfrac{(x-6)(x-6)+(x+5)(x+5)}{(x+5)(x-6)}$

$=\dfrac{x^2-12x+36+x^2+10x+25}{(x+5)(x-6)}$

$=\dfrac{2x^2-2x+61}{(x+5)(x-6)}$

**37.** $\dfrac{3x}{x^2-25}-\dfrac{4}{x+5}$

$=\dfrac{3x}{(x+5)(x-5)}-\dfrac{4}{x+5}$

The LCD is $(x+5)(x-5)$.

$=\dfrac{3x}{(x+5)(x-5)}-\dfrac{4(x-5)}{(x+5)(x-5)}$

$=\dfrac{3x-4(x-5)}{(x+5)(x-5)}=\dfrac{3x-4x+20}{(x+5)(x-5)}$

$=\dfrac{-x+20}{(x+5)(x-5)}=\dfrac{20-x}{x^2-25}$

**39.** $\dfrac{3y+7}{y^2-5y+6}-\dfrac{3}{y-3}$

$=\dfrac{3y+7}{(y-3)(y-2)}-\dfrac{3}{y-3}$

The LCD is $(y-3)(y-2)$.

$=\dfrac{3y+7}{(y-3)(y-2)}-\dfrac{3(y-2)}{(y-3)(y-2)}$

$=\dfrac{3y+7-3(y-2)}{(y-3)(y-2)}$

$=\dfrac{3y+7-3y+6}{(y-3)(y-2)}=\dfrac{13}{(y-3)(y-2)}$

**41.** $\dfrac{x^2-6}{x^2+9x+18}-\dfrac{x-4}{x+6}$

$=\dfrac{x^2-6}{(x+3)(x+6)}-\dfrac{x-4}{x+6}$

The LCD is $(x+3)(x+6)$.

$=\dfrac{x^2-6}{(x+3)(x+6)}-\dfrac{(x-4)(x+3)}{(x+3)(x+6)}$

$=\dfrac{x^2-6-(x-4)(x+3)}{(x+3)(x+6)}$

$=\dfrac{x^2-6-(x^2-x-12)}{(x+3)(x+6)}$

$=\dfrac{x^2-6-x^2+x+12}{(x+3)(x+6)}=\dfrac{x+6}{(x+3)(x+6)}$

$=\dfrac{1\cancel{(x+6)}}{(x+3)\cancel{(x+6)}}=\dfrac{1}{x+3}$

**43.** $\frac{4x+1}{x^2+7x+12}+\frac{2x+3}{x^2+5x+4}=\frac{4x+1}{(x+3)(x+4)}+\frac{2x+3}{(x+4)(x+1)}$

The LCD is $(x+3)(x+4)(x+1)$.

$=\frac{(4x+1)(x+1)}{(x+3)(x+4)(x+1)}+\frac{(2x+3)(x+3)}{(x+3)(x+4)(x+1)}=\frac{(4x+1)(x+1)+(2x+3)(x+3)}{(x+3)(x+4)(x+1)}$

$=\frac{4x^2+5x+1+2x^2+9x+9}{(x+3)(x+4)(x+1)}=\frac{6x^2+14x+10}{(x+3)(x+4)(x+1)}=\frac{2(3x^2+7x+5)}{(x+3)(x+4)(x+1)}$

**45.** $\frac{x+4}{x^2-x-2}-\frac{2x+3}{x^2+2x-8}=\frac{x+4}{(x-2)(x+1)}-\frac{2x+3}{(x+4)(x-2)}$

The LCD is $(x-2)(x+1)(x+4)$.

$=\frac{(x+4)(x+4)}{(x-2)(x+1)(x+4)}-\frac{(2x+3)(x+1)}{(x+4)(x+1)(x-2)}=\frac{(x+4)(x+4)-(2x+3)(x+1)}{(x-2)(x+1)(x+4)}$

$=\frac{x^2+8x+16-(2x^2+5x+3)}{(x-2)(x+1)(x+4)}=\frac{x^2+8x+16-2x^2-5x-3}{(x-2)(x+1)(x+4)}=\frac{-x^2+3x+13}{(x-2)(x+1)(x+4)}$

$=\frac{-(x^2-3x-13)}{(x-2)(x+1)(x+4)}=-\frac{x^2-3x-13}{(x-2)(x+1)(x+4)}$

**47.** $4+\frac{1}{x-3}=\frac{4}{1}+\frac{1}{x-3}$

The LCD is $(x-3)$.

$=\frac{4(x-3)}{x-3}+\frac{1}{x-3}=\frac{4(x-3)+1}{x-3}$

$=\frac{4x-12+1}{x-3}$

$=\frac{4x-11}{x-3}$

**49.** $\frac{y-7}{y^2-16}+\frac{7-y}{16-y^2}$

The LCD is $y^2-16$.

$=\frac{y-7}{y^2-16}+\frac{(-1)}{(-1)}\cdot\frac{7-y}{16-y^2}$

$=\frac{y-7}{y^2-16}+\frac{-7+y}{-16+y^2}$

$=\frac{y-7}{y^2-16}+\frac{y-7}{y^2-16}$

$=\frac{y-7+y-7}{y^2-16}=\frac{2y-14}{(y+4)(y-4)}$

**51.** $\frac{x+7}{3x+6}+\frac{x}{4-x^2}=\frac{x+7}{3(x+2)}+\frac{x}{(2+x)(2-x)}=\frac{x+7}{3(x+2)}+\frac{x}{(x+2)(2-x)}$

The LCD is $3(x+2)(2-x)$.

$=\frac{(x+7)(2-x)}{3(x+2)(2-x)}+\frac{3x}{3(x+2)(2-x)}=\frac{(x+7)(2-x)+3x}{3(x+2)(2-x)}=\frac{2x-x^2+14-7x+3x}{3(x+2)(2-x)}$

$=\frac{-x^2-2x+14}{3(x+2)(2-x)}=\frac{-1}{-1}\cdot\frac{-x^2-2x+14}{3(x+2)(2-x)}=\frac{x^2+2x-14}{3(x+2)(x-2)}$

**53.** $\frac{2x}{x-4}+\frac{64}{x^2-16}-\frac{2x}{x+4}=\frac{2x}{x-4}+\frac{64}{(x+4)(x-4)}-\frac{2x}{x+4}$

The LCD is $(x+4)(x-4)$.

$=\frac{2x(x+4)}{(x+4)(x-4)}+\frac{64}{(x+4)(x-4)}-\frac{2x(x-4)}{(x+4)(x-4)}=\frac{2x(x+4)+64-2x(x-4)}{(x+4)(x-4)}$

$=\frac{2x^2+8x+64-2x^2+8x}{(x+4)(x-4)}=\frac{16x+64}{(x+4)(x-4)}=\frac{16\cancel{(x+4)}}{\cancel{(x+4)}(x-4)}=\frac{16}{x-4}$

**55.** $\frac{5x}{x^2-y^2}-\frac{7}{y-x}=\frac{5x}{(x+y)(x-y)}-\frac{(-1)7}{(-1)(y-x)}$

$=\frac{5x}{(x+y)(x-y)}-\frac{-7}{(-y+x)}=\frac{5x}{(x+y)(x-y)}-\frac{-7}{(x-y)}$

The LCD is $(x+y)(x-y)$.

$=\frac{5x}{(x+y)(x-y)}-\frac{-7(x+y)}{(x+y)(x-y)}=\frac{5x-(-7)(x+y)}{(x+y)(x-y)}=\frac{5x+7(x+y)}{(x+y)(x-y)}$

$=\frac{5x+7x+7y}{(x+y)(x-y)}=\frac{12x+7y}{(x+y)(x-y)}=\frac{12x+7y}{x^2-y^2}$

**57.** $\frac{3}{5x+6}-\frac{4}{x-2}+\frac{x^2-x}{5x^2-4x-12}=\frac{3}{5x+6}-\frac{4}{x-2}+\frac{x^2-x}{(5x+6)(x-2)}$

The LCD is $(5x+6)(x-2)$.

$=\frac{3(x-2)}{(5x+6)(x-2)}-\frac{4(5x+6)}{(5x+6)(x-2)}+\frac{x^2-x}{(5x+6)(x-2)}=\frac{3(x-2)-4(5x+6)+x^2-x}{(5x+6)(x-2)}$

$=\frac{3x-6-20x-24+x^2-x}{(5x+6)(x-2)}=\frac{x^2-18x-30}{(5x+6)(x-2)}$

**59.** $\dfrac{3x-y}{x^2-9xy+20y^2}+\dfrac{2y}{x^2-25y^2}=\dfrac{3x-y}{(x-5y)(x-4y)}+\dfrac{2y}{(x+5y)(x-5y)}$

The LCD is $(x+5y)(x-5y)(x-4y)$.

$=\dfrac{(3x-y)(x+5y)}{(x+5y)(x-5y)(x-4y)}+\dfrac{2y(x-4y)}{(x+5y)(x-5y)(x-4y)}=\dfrac{(3x-y)(x+5y)+2y(x-4y)}{(x+5y)(x-5y)(x-4y)}$

$=\dfrac{3x^2+14xy-5y^2+2xy-8y^2}{(x+5y)(x-5y)(x-4y)}=\dfrac{3x^2+16xy-13y^2}{(x+5y)(x-5y)(x-4y)}$

**61.** $\dfrac{3x}{x^2-4}+\dfrac{5x}{x^2+x-2}-\dfrac{3}{x^2-4x+4}=\dfrac{3x}{(x+2)(x-2)}+\dfrac{5x}{(x+2)(x-1)}-\dfrac{3}{(x-2)^2}$

The LCD is $(x+2)(x-2)^2(x-1)$.

$=\dfrac{3x(x-2)(x-1)}{(x+2)(x-2)^2(x-1)}+\dfrac{5x(x-2)^2}{(x+2)(x-2)^2(x-1)}-\dfrac{3(x+2)(x-1)}{(x+2)(x-2)^2(x-1)}$

$=\dfrac{3x(x-2)(x-1)+5x(x-2)^2-3(x+2)(x-1)}{(x+2)(x-2)^2(x-1)}$

$=\dfrac{3x(x^2-3x+2)+5x(x^2-4x+4)-3(x^2+x-2)}{(x+2)(x-2)^2(x-1)}$

$=\dfrac{3x^3-9x^2+6x+5x^3-20x^2+20x-3x^2-3x+6}{(x+2)(x-2)^2(x-1)}=\dfrac{8x^3-32x^2+23x+6}{(x+2)(x-2)^2(x-1)}$

**63.** $\dfrac{6a+5b}{6a^2+5ab-4b^2}-\dfrac{a+2b}{9a^2-16b^2}=\dfrac{6a+5b}{(3a+4b)(2a-b)}-\dfrac{a+2b}{(3a+4b)(3a-4b)}$

The LCD is $(3a+4b)(2a-b)(3a-4b)$.

$=\dfrac{(6a+5b)(3a-4b)}{(3a+4b)(2a-b)(3a-4b)}-\dfrac{(a+2b)(2a-b)}{(3a+4b)(2a-b)(3a-4b)}$

$=\dfrac{(6a+5b)(3a-4b)-(a+2b)(2a-b)}{(3a+4b)(2a-b)(3a-4b)}=\dfrac{18a^2-9ab-20b^2-(2a^2+3ab-2b^2)}{(3a+4b)(2a-b)(3a-4b)}$

$=\dfrac{18a^2-9ab-20b^2-2a^2-3ab+2b^2}{(3a+4b)(2a-b)(3a-4b)}=\dfrac{16a^2-12ab-18b^2}{(3a+4b)(2a-b)(3a-4b)}$

$=\dfrac{2(8a^2-6ab-9b^2)}{(3a+4b)(2a-b)(3a-4b)}=\dfrac{2(4a+3b)(2a-3b)}{(3a+4b)(2a-b)(3a-4b)}$

**65.**

$$\frac{1}{m^2+m-2}-\frac{3}{2m^2+3m-2}+\frac{2}{2m^2-3m+1}$$

$$=\frac{1}{(m+2)(m-1)}-\frac{3}{(2m-1)(m+2)}+\frac{2}{(2m-1)(m-1)}$$

The LCD is $(m+2)(m-1)(2m-1)$.

$$=\frac{1(2m-1)}{(m+2)(m-1)(2m-1)}-\frac{3(m-1)}{(2m-1)(m+2)}+\frac{2(m+2)}{(2m-1)(m-1)}$$

$$=\frac{2m-1-3m+3+2m+4}{(m+2)(m-1)(2m-1)}$$

$$=\frac{m+6}{(m+2)(m-1)(2m-1)}$$

**67.**

$$\left(\frac{2x+3}{x+1}\cdot\frac{x^2+4x-5}{2x^2+x-3}\right)-\frac{2}{x+2}=\left(\frac{\cancel{(2x+3)}}{x+1}\cdot\frac{(x+5)\cancel{(x-1)}}{\cancel{(2x+3)}\cancel{(x-1)}}\right)-\frac{2}{x+2}$$

$$=\frac{x+5}{x+1}-\frac{2}{x+2}$$

$$=\frac{(x+5)(x+2)}{(x+1)(x+2)}-\frac{2(x+1)}{(x+1)(x+2)}$$

$$=\frac{(x+5)(x+2)-2(x+1)}{(x+1)(x+2)}$$

$$=\frac{x^2+2x+5x+10-2x-2}{(x+1)(x+2)}$$

$$=\frac{x^2+5x+8}{(x+1)(x+2)}$$

**68.**

$$\frac{1}{x^2-2x-8}\cdot\left(\frac{1}{x-4}-\frac{1}{x+2}\right)=\frac{1}{(x-4)(x+2)}\div\left(\frac{(x+2)}{(x-4)(x+2)}-\frac{(x-4)}{(x-4)(x+2)}\right)$$

$$=\frac{1}{(x-4)(x+2)}\div\left(\frac{x+2-x+4}{(x-4)(x+2)}\right)$$

$$=\frac{1}{(x-4)(x+2)}\div\left(\frac{6}{(x-4)(x+2)}\right)$$

$$=\frac{1}{(x-4)(x+2)}\cdot\frac{(x-4)(x+2)}{6}=\frac{1}{6}$$

**69.**
$$\left(2-\frac{6}{x+1}\right)\left(1+\frac{3}{x-2}\right)=\left(\frac{2(x+1)}{(x+1)}-\frac{6}{(x+1)}\right)\left(\frac{(x-2)}{(x-2)}+\frac{3}{(x-2)}\right)$$
$$=\left(\frac{2x+2-6}{x+1}\right)\left(\frac{x-2+3}{x-2}\right)$$
$$=\left(\frac{2x-4}{x+1}\right)\left(\frac{x+1}{x-2}\right)$$
$$=\frac{2\cancel{(x-2)}\cancel{(x+1)}}{\cancel{(x+1)}\cancel{(x-2)}}$$
$$=2$$

**70.**
$$\left(4-\frac{3}{x+2}\right)\left(1+\frac{5}{x-1}\right)=\left(\frac{4(x+2)}{x+2}-\frac{3}{x+2}\right)\left(\frac{(x-1)}{x-1}+\frac{5}{x-1}\right)$$
$$=\left(\frac{4x+8-3}{x+2}\right)\left(\frac{x-1+5}{x-1}\right)$$
$$=\frac{4x+5}{x+2}\cdot\frac{x+4}{x-1}$$
$$=\frac{(4x+5)(x+4)}{(x+2)(x-1)}$$

**71.**
$$\left(\frac{1}{x+h}-\frac{1}{x}\right)\div h=\left(\frac{x}{x(x+h)}-\frac{(x+h)}{x(x+h)}\right)\div h$$
$$=\left(\frac{x-x-h}{x(x+h)}\right)\div\frac{h}{1}$$
$$=\frac{-h}{x(x+h)}\cdot\frac{1}{h}$$
$$=-\frac{1}{x(x+h)}$$

**72.** $\left(\frac{5}{x-5}-\frac{2}{x+3}\right)\div(3x+25)=\left(\frac{5(x+3)}{(x-5)(x+3)}-\frac{2(x-5)}{(x-5)(x+3)}\right)\div(3x+25)$

$$=\left(\frac{5x+15-2x+10}{(x-5)(x+3)}\right)\div(3x+25)$$

$$=\left(\frac{\cancel{3x+25}}{(x-5)(x+3)}\right)\cdot\frac{1}{\cancel{(3x+25)}}$$

$$=\frac{1}{(x-5)(x+3)}$$

**73.** $\left(\frac{1}{a^3-b^3}\cdot\frac{ac+ad-bc-bd}{1}\right)-\frac{c-d}{a^2+ab+b^2}$

$$=\left(\frac{1}{(a-b)(a^2+ab+b^2)}\cdot\frac{a(c+d)-b(c+d)}{1}\right)-\frac{c-d}{a^2+ab+b^2}$$

$$=\left(\frac{1}{\cancel{(a-b)}(a^2+ab+b^2)}\cdot\frac{(c+d)\cancel{(a-b)}}{1}\right)-\frac{c-d}{a^2+ab+b^2}$$

$$=\frac{c+d}{a^2+ab+b^2}-\frac{c-d}{a^2+bd+b^2}$$

$$=\frac{c+d-c+d}{a^2+ab+b^2}$$

$$=\frac{2d}{a^2+ab+b^2}$$

**74.**

$$\frac{ab}{a^2+ab+b^2}+\left(\frac{ac-ad-bc+bd}{ac-ad+bc-bd}\div\frac{a^3-b^3}{a^3+b^3}\right)$$

$$=\frac{ab}{a^2+ab+b^2}+\left(\frac{a(c-d)-b(c-d)}{a(c-d)+b(c-d)}\cdot\frac{a^3+b^3}{a^3-b^3}\right)$$

$$=\frac{ab}{a^2+ab+b^2}+\left(\frac{\cancel{(c-d)}\cancel{(a-b)}}{\cancel{(c-d)}\cancel{(a+b)}}\cdot\frac{\cancel{(a+b)}(a^2-ab+b^2)}{\cancel{(a-b)}(a^2+ab+b^2)}\right)$$

$$=\frac{ab}{a^2+ab+b^2}+\frac{a^2-ab+b^2}{a^2+ab+b^2}$$

$$=\frac{ab+a^2-ab+b^2}{a^2+ab+b^2}$$

$$=\frac{a^2+b^2}{a^2+ab+b^2}$$

**75.**

$$f(x)=\frac{2x-3}{x+5} \text{ and } g(x)=\frac{x^2-4x-19}{x^2+8x+15}$$

$$(f-g)(x)=f(x)-g(x)$$

$$=\frac{2x-3}{x+5}-\frac{x^2-4x-19}{x^2+8x+15}=\frac{2x-3}{x+5}-\frac{x^2-4x-19}{(x+5)(x+3)}$$

$$=\frac{(2x-3)(x+3)}{(x+5)(x+3)}-\frac{x^2-4x-19}{(x+5)(x+3)}=\frac{2x^2+6x-3x-9-x^2+4x+19}{(x+5)(x+3)}$$

$$=\frac{x^2+7x+10}{(x+5)(x+3)}=\frac{\cancel{(x+5)}(x+2)}{\cancel{(x+5)}(x+3)}$$

$$=\frac{x+2}{x+3}$$

To find the domain of $(f-g)(x)$ we need to find the intersection of the domains for the individual functions. The domain of $f$ is $\{x \mid x \text{ is a real number and } x \neq -5\}$. The domain of $g$ is $\{x \mid x \text{ is a real number and } x \neq -5, x \neq -3\}$. Therefore, the domain of $(f-g)(x)$ is $\{x \mid x \text{ is a real number and } x \neq -5, x \neq -3\}$ or $(-\infty,-5)\cup(-5,-3)\cup(-3,\infty)$.

**76.** $f(x)=\dfrac{2x-1}{x^2+x-6}$ and $g(x)=\dfrac{x+2}{x^2+5x+6}$

$(f-g)(x)=f(x)-g(x)$

$$=\frac{2x-1}{x^2+x-6}-\frac{x+2}{x^2+5x+6}=\frac{2x-1}{(x+3)(x-2)}-\frac{x+2}{(x+3)(x+2)}$$

$$=\frac{(2x-1)(x+2)}{(x+3)(x+2)(x-2)}-\frac{(x+2)(x-2)}{(x+3)(x+2)(x-2)}$$

$$=\frac{2x^2+4x-x-2}{(x+3)(x+2)(x-2)}-\frac{x^2-4}{(x+3)(x+2)(x-2)}$$

$$=\frac{2x^2+3x-2-x^2+4}{(x+3)(x+2)(x-2)}=\frac{x^2+3x+2}{(x+3)(x+2)(x-2)}$$

$$=\frac{\cancel{(x+2)}(x+1)}{(x+3)\cancel{(x+2)}(x-2)}=\frac{x+1}{(x+3)(x-2)}$$

To find the domain of $(f-g)(x)$ we need to find the intersection of the domains for the individual functions. The domain of *f* is $\{x \mid x \text{ is a real number and } x\neq -3, x\neq 2\}$. The domain of *g* is $\{x \mid x \text{ is a real number and } x\neq -3, x\neq -2\}$. Therefore, the domain of $(f-g)(x)$ is $\{x \mid x \text{ is a real number and } x\neq -3, x\neq -2, x\neq 2\}$ or $(-\infty,-3)\cup(-3,-2)\cup(-2,2)\cup(2,\infty)$.

**77.** $T(0)=\dfrac{470}{0+70}+\dfrac{250}{0+65}$

$=\dfrac{470}{70}+\dfrac{250}{65}$

$=6.7+3.8=10.5\approx 11$

This corresponds to the point $(0,11)$ on the graph. If you drive zero miles per hour over the speed limit, total driving time is approximately 11 hours.

**79.** $T(x)=\dfrac{470}{x+70}+\dfrac{250}{x+65}$

$$=\frac{470(x+65)}{(x+70)(x+65)}+\frac{250(x+70)}{(x+70)(x+65)}$$

$$=\frac{470(x+65)+250(x+70)}{(x+70)(x+65)}$$

$$=\frac{470x+30550+250x+17500}{(x+70)(x+65)}$$

$$=\frac{720x+48050}{(x+70)(x+65)}$$

$$T(0)=\frac{\cancel{720(0)}+48050}{(0+70)(0+65)}=\frac{48050}{(70)(65)}$$

$\approx 11$

**81.** Answers will vary. In order to make the trip in 9 hours, you need to drive approximately 12 miles per hour over the speed limit.

**83.**
$$P = 2\left(\frac{x}{x+4}\right) + 2\left(\frac{x}{x+5}\right)$$
$$= \frac{2x}{x+4} + \frac{2x}{x+5}$$
$$= \frac{2x(x+5)}{(x+4)(x+5)} + \frac{2x(x+4)}{(x+4)(x+5)}$$
$$= \frac{2x(x+5) + 2x(x+4)}{(x+4)(x+5)}$$
$$= \frac{2x^2 + 10x + 2x^2 + 8x}{(x+4)(x+5)}$$
$$= \frac{4x^2 + 18x}{(x+4)(x+5)}$$
$$= \frac{2x(2x+9)}{(x+4)(x+5)}$$

**85.**
$$f(x) = \frac{27{,}725(x-14)}{x^2+9} - 5x$$
$$= \frac{27{,}725(x-14)}{x^2+9} - \frac{5x}{1}$$
$$= \frac{27{,}725(x-14)}{x^2+9} - \frac{5x(x^2+9)}{(x^2+9)}$$
$$= \frac{27{,}725(x-14) - 5x(x^2+9)}{x^2+9}$$
$$= \frac{27{,}725x - 388{,}150 - 5x^3 - 45x}{x^2+9}$$
$$= \frac{-5x^3 + 27{,}680x - 388{,}150}{x^2+9}$$

**87.** Answers will vary.

**89.** Answers will vary.

**91.** Answers will vary; $\frac{b+a}{ab}$

**93.**
$$f(x) = \frac{27{,}725(x-14)}{x^2+9} - 5x$$

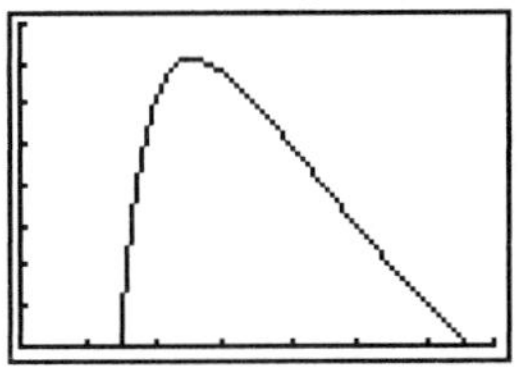

$$f(x) = \frac{-5x^3 + 27{,}680x - 388{,}150}{x^2+9}$$

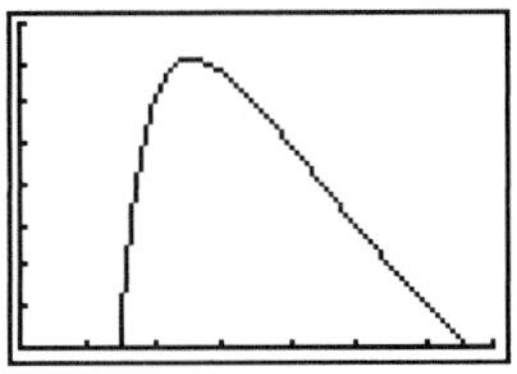

The graphs are identical.

As age increases, the number of arrests per 100,000 drivers increases until about age 25 and then decreases.

**95.**
$$\frac{1}{x^n-1} - \frac{1}{x^n+1} - \frac{1}{x^{2n}-1}$$
$$= \frac{x^n+1}{x^{2n}-1} - \frac{x^n-1}{x^{2n}-1} - \frac{1}{x^{2n}-1}$$
$$= \frac{x^n+1-x^n+1-1}{x^{2n}-1}$$
$$= \frac{1}{x^{2n}-1}$$

**97.** $(x-y)^{-1}+(x-y)^{-2}$

$$=\frac{1}{(x-y)}+\frac{1}{(x-y)^2}$$

$$=\frac{(x-y)}{(x-y)(x-y)}+\frac{1}{(x-y)^2}$$

$$=\frac{x-y+1}{(x-y)^2}$$

**98.** $\left(\frac{3x^2y^{-2}}{y^3}\right)^{-2}=\left(\frac{3x^2}{y^2y^3}\right)^{-2}=\left(\frac{3x^2}{y^5}\right)^{-2}$

$$=\left(\frac{y^5}{3x^2}\right)^2=\frac{y^{10}}{9x^4}$$

**99.**

$$|3x-1|\le 14$$
$$-14\le 3x-1\le 14$$
$$-14+1\le 3x-1+1\le 14+1$$
$$-13\le 3x\le 15$$
$$-\frac{13}{3}\le x\le 5$$

The solution set is $\left\{x\middle|-\frac{13}{3}\le x\le 5\right\}$ or $\left[-\frac{13}{3},5\right]$.

**100.** $50x^3-18x=2x\left(25x^2-9\right)$

$$=2x(5x+3)(5x-3)$$

## 6.3 Exercise Set

**1.** $\dfrac{4+\frac{2}{x}}{1-\frac{3}{x}}=\dfrac{x}{x}\cdot\dfrac{4+\frac{2}{x}}{1-\frac{3}{x}}=\dfrac{x\cdot 4+\cancel{x}\cdot\frac{2}{\cancel{x}}}{x\cdot 1-\cancel{x}\cdot\frac{3}{\cancel{x}}}=\dfrac{4x+2}{x-3}$

**3.** $\dfrac{\frac{3}{x}+\frac{x}{3}}{\frac{x}{3}-\frac{3}{x}}=\dfrac{3x}{3x}\cdot\dfrac{\frac{3}{x}+\frac{x}{3}}{\frac{x}{3}-\frac{3}{x}}=\dfrac{3\cancel{x}\cdot\frac{3}{\cancel{x}}+\cancel{3}x\cdot\frac{x}{\cancel{3}}}{\cancel{3}x\cdot\frac{x}{\cancel{3}}-3\cancel{x}\cdot\frac{3}{\cancel{x}}}=\dfrac{3\cdot 3+x\cdot x}{x\cdot x-3\cdot 3}=\dfrac{9+x^2}{x^2-9}=\dfrac{x^2+9}{(x+3)(x-3)}$

**5.** $\dfrac{\frac{1}{x}+\frac{1}{y}}{\frac{1}{x}-\frac{1}{y}}=\dfrac{xy}{xy}\cdot\dfrac{\frac{1}{x}+\frac{1}{y}}{\frac{1}{x}-\frac{1}{y}}=\dfrac{\cancel{x}y\cdot\frac{1}{\cancel{x}}+x\cancel{y}\cdot\frac{1}{\cancel{y}}}{\cancel{x}y\cdot\frac{1}{\cancel{x}}-x\cancel{y}\cdot\frac{1}{\cancel{y}}}=\dfrac{y+x}{y-x}$

**7.**

$$\frac{8x^{-2}-2x^{-1}}{10x^{-1}-6x^{-2}}=\frac{\frac{8}{x^2}-\frac{2}{x}}{\frac{10}{x}-\frac{6}{x^2}}=\frac{x^2}{x^2}\cdot\frac{\frac{8}{x^2}-\frac{2}{x}}{\frac{10}{x}-\frac{6}{x^2}}=\frac{x^2\cdot\frac{8}{x^2}-x^2\cdot\frac{2}{x}}{x^2\cdot\frac{10}{x}-x^2\cdot\frac{6}{x^2}}=\frac{8-x\cdot 2}{x\cdot 10-6}$$

$$=\frac{8-2x}{10x-6}=\frac{2(4-x)}{2(5x-3)}=\frac{4-x}{5x-3}$$

**9.**

$$\frac{\frac{1}{x-2}}{1-\frac{1}{x-2}}=\frac{x-2}{x-2}\cdot\frac{\frac{1}{x-2}}{1-\frac{1}{x-2}}=\frac{(x-2)\cdot\frac{1}{x-2}}{(x-2)\cdot 1-(x-2)\cdot\frac{1}{x-2}}=\frac{1}{x-2-1}=\frac{1}{x-3}$$

**11.**

$$\frac{\frac{1}{x+5}-\frac{1}{x}}{5}=\frac{x(x+5)}{x(x+5)}\cdot\frac{\frac{1}{x+5}-\frac{1}{x}}{5}=\frac{x(x+5)\cdot\frac{1}{x+5}-x(x+5)\cdot\frac{1}{x}}{5x(x+5)}=\frac{x-(x+5)}{5x(x+5)}$$

$$=\frac{x-x-5}{5x(x+5)}=-\frac{5}{5x(x+5)}=-\frac{1}{x(x+5)}$$

**13.**

$$\frac{\frac{4}{x+4}}{\frac{1}{x+4}-\frac{1}{x}}=\frac{x(x+4)}{x(x+4)}\cdot\frac{\frac{4}{x+4}}{\frac{1}{x+4}-\frac{1}{x}}=\frac{x(x+4)\cdot\frac{4}{x+4}}{x(x+4)\cdot\frac{1}{x+4}-x(x+4)\cdot\frac{1}{x}}$$

$$=\frac{x\cdot 4}{x-(x+4)}=\frac{4x}{x-x-4}=-\frac{4x}{4}=-x$$

**15.**

$$\frac{\frac{1}{x-1}+1}{\frac{1}{x+1}-1}=\frac{(x+1)(x-1)}{(x+1)(x-1)}\cdot\frac{\frac{1}{x-1}+1}{\frac{1}{x+1}-1}=\frac{(x+1)(x-1)\cdot\frac{1}{x-1}+(x+1)(x-1)\cdot 1}{(x+1)(x-1)\cdot\frac{1}{x+1}-(x+1)(x-1)\cdot 1}=\frac{(x+1)+(x^2-1)}{(x-1)-(x^2-1)}$$

$$=\frac{x+1+x^2-1}{x-1-x^2+1}=\frac{x^2+x}{x-x^2}=\frac{x(x+1)}{x(1-x)}=\frac{x+1}{-1(x-1)}=-\frac{x+1}{x-1}$$

**17.**

$$\frac{x^{-1}+y^{-1}}{(x+y)^{-1}}=\frac{\frac{1}{x}+\frac{1}{y}}{\frac{1}{(x+y)}}=\frac{xy(x+y)}{xy(x+y)}\cdot\frac{\frac{1}{x}+\frac{1}{y}}{\frac{1}{(x+y)}}=\frac{\cancel{x}y(x+y)\cdot\frac{1}{\cancel{x}}+x\cancel{y}(x+y)\cdot\frac{1}{\cancel{y}}}{xy\cancel{(x+y)}\cdot\frac{1}{\cancel{(x+y)}}}$$

$$=\frac{y(x+y)+x(x+y)}{xy}=\frac{(x+y)(y+x)}{xy}=\frac{(x+y)(x+y)}{xy}$$

**19.**

$$\frac{\frac{x+2}{x-2}-\frac{x-2}{x+2}}{\frac{x-2}{x+2}+\frac{x+2}{x-2}}=\frac{(x+2)(x-2)}{(x+2)(x-2)}\cdot\frac{\frac{x+2}{x-2}-\frac{x-2}{x+2}}{\frac{x-2}{x+2}+\frac{x+2}{x-2}}$$

$$=\frac{(x+2)\cancel{(x-2)}\cdot\frac{x+2}{\cancel{x-2}}-\cancel{(x+2)}(x-2)\cdot\frac{x-2}{\cancel{x+2}}}{\cancel{(x+2)}(x-2)\cdot\frac{x-2}{\cancel{x+2}}+(x+2)\cancel{(x-2)}\cdot\frac{x+2}{\cancel{x-2}}}$$

$$=\frac{(x+2)(x+2)-(x-2)(x-2)}{(x-2)(x-2)+(x+2)(x+2)}=\frac{x^2+4x+4-(x^2-4x+4)}{x^2-\cancel{4x}+4+x^2+\cancel{4x}+4}$$

$$=\frac{\cancel{x^2}+4x+\cancel{4}-\cancel{x^2}+4x-\cancel{4}}{2x^2+8}=\frac{\overset{4}{\cancel{8}}x}{\cancel{2}(x^2+4)}=\frac{4x}{x^2+4}$$

**21.**

$$\frac{\frac{2}{x^3y}+\frac{5}{xy^4}}{\frac{5}{x^3y}-\frac{3}{xy}}=\frac{x^3y^4}{x^3y^4}\cdot\frac{\frac{2}{x^3y}+\frac{5}{xy^4}}{\frac{5}{x^3y}-\frac{3}{xy}}=\frac{\cancel{x^3}\overset{y^3}{\cancel{y^4}}\cdot\frac{2}{\cancel{x^3}\cancel{y}}+\overset{x^2}{\cancel{x^3}}\cancel{y^4}\cdot\frac{5}{\cancel{x}\cancel{y^4}}}{\cancel{x^3}\underset{y^3}{\cancel{y^4}}\cdot\frac{5}{\cancel{x^3}\cancel{y}}-\underset{x^2}{\cancel{x^3}}\underset{y^3}{\cancel{y^4}}\cdot\frac{3}{\cancel{x}\cancel{y}}}$$

$$=\frac{y^3\cdot2+x^2\cdot5}{y^3\cdot5-x^2y^3\cdot3}=\frac{2y^3+5x^2}{5y^3-3x^2y^3}=\frac{2y^3+5x^2}{y^3(5-3x^2)}$$

**23.**
$$\frac{\frac{3}{x+2}-\frac{3}{x-2}}{\frac{5}{x^2-4}}=\frac{\frac{3}{x+2}-\frac{3}{x-2}}{\frac{5}{(x+2)(x-2)}}=\frac{(x+2)(x-2)}{(x+2)(x-2)}\cdot\frac{\frac{3}{x+2}-\frac{3}{x-2}}{\frac{5}{(x+2)(x-2)}}$$
$$=\frac{(x+2)(x-2)\cdot\frac{3}{x+2}-(x+2)(x-2)\cdot\frac{3}{x-2}}{(x+2)(x-2)\cdot\frac{5}{(x+2)(x-2)}}$$
$$=\frac{(x-2)\cdot 3-(x+2)\cdot 3}{5}=\frac{3x-6-(3x+6)}{5}=\frac{3x-6-3x-6}{5}=-\frac{12}{5}$$

**25.**
$$\frac{3a^{-1}+3b^{-1}}{4a^{-2}-9b^{-2}}=\frac{\frac{3}{a}+\frac{3}{b}}{\frac{4}{a^2}-\frac{9}{b^2}}=\frac{a^2b^2}{a^2b^2}\cdot\frac{\frac{3}{a}+\frac{3}{b}}{\frac{4}{a^2}-\frac{9}{b^2}}=\frac{a^2b^2\cdot\frac{3}{a}+a^2b^2\cdot\frac{3}{b}}{a^2b^2\cdot\frac{4}{a^2}-a^2b^2\cdot\frac{9}{b^2}}$$
$$=\frac{ab^2\cdot 3+a^2b\cdot 3}{b^2\cdot 4-a^2\cdot 9}=\frac{3ab^2+3a^2b}{4b^2-9a^2}=\frac{3ab(b+a)}{(2b+3a)(2b-3a)}$$

**27.**
$$\frac{\frac{4x}{x^2-4}-\frac{5}{x-2}}{\frac{2}{x-2}+\frac{3}{x+2}}=\frac{\frac{4x}{(x+2)(x-2)}-\frac{5}{x-2}}{\frac{2}{x-2}+\frac{3}{x+2}}=\frac{(x+2)(x-2)}{(x+2)(x-2)}\cdot\frac{\frac{4x}{(x+2)(x-2)}-\frac{5}{x-2}}{\frac{2}{x-2}+\frac{3}{x+2}}$$
$$=\frac{(x+2)(x-2)\cdot\frac{4x}{(x+2)(x-2)}-(x+2)(x-2)\cdot\frac{5}{x-2}}{(x+2)(x-2)\cdot\frac{2}{x-2}+(x+2)(x-2)\cdot\frac{3}{x+2}}$$
$$=\frac{4x-(x+2)\cdot 5}{(x+2)\cdot 2+(x-2)\cdot 3}=\frac{4x-5x-10}{2x+4+3x-6}=\frac{-x-10}{5x-2}=-\frac{x+10}{5x-2}$$

**29.** $$\frac{\frac{2y}{y^2+4y+3}}{\frac{1}{y+3}+\frac{2}{y+1}}=\frac{\frac{2y}{(y+3)(y+1)}}{\frac{1}{y+3}+\frac{2}{y+1}}=\frac{(y+3)(y+1)}{(y+3)(y+1)}\cdot\frac{\frac{2y}{(y+3)(y+1)}}{\frac{1}{y+3}+\frac{2}{y+1}}$$

$$=\frac{(y+3)(y+1)\cdot\frac{2y}{(y+3)(y+1)}}{(y+3)(y+1)\cdot\frac{1}{y+3}+(y+3)(y+1)\cdot\frac{2}{y+1}}$$

$$=\frac{2y}{(y+1)+(y+3)\cdot 2}=\frac{2y}{y+1+2y+6}=\frac{2y}{3y+7}$$

**31.** $$\frac{\frac{2}{a^2}-\frac{1}{ab}-\frac{1}{b^2}}{\frac{1}{a^2}-\frac{3}{ab}+\frac{2}{b^2}}=\frac{a^2b^2}{a^2b^2}\cdot\frac{\frac{2}{a^2}-\frac{1}{ab}-\frac{1}{b^2}}{\frac{1}{a^2}-\frac{3}{ab}+\frac{2}{b^2}}=\frac{a^2b^2\cdot\frac{2}{a^2}-a^2b^2\cdot\frac{1}{ab}-a^2b^2\cdot\frac{1}{b^2}}{a^2b^2\cdot\frac{1}{a^2}-a^2b^2\cdot\frac{3}{ab}+a^2b^2\cdot\frac{2}{b^2}}$$

$$=\frac{b^2\cdot 2-ab-a^2}{b^2-ab\cdot 3+a^2\cdot 2}=\frac{2b^2-ab-a^2}{b^2-3ab+2a^2}=\frac{(2b+a)(b-a)}{(b-2a)(b-a)}=\frac{2b+a}{b-2a}$$

**33.** $$\frac{\frac{2x}{x^2-25}+\frac{1}{3x-15}}{\frac{5}{x-5}+\frac{3}{4x-20}}=\frac{\frac{2x}{(x+5)(x-5)}+\frac{1}{3(x-5)}}{\frac{5}{x-5}+\frac{3}{4(x-5)}}=\frac{12(x+5)(x-5)}{12(x+5)(x-5)}\cdot\frac{\frac{2x}{(x+5)(x-5)}+\frac{1}{3(x-5)}}{\frac{5}{x-5}+\frac{3}{4(x-5)}}$$

$$=\frac{12(x+5)(x-5)\cdot\frac{2x}{(x+5)(x-5)}+12(x+5)(x-5)\cdot\frac{1}{3(x-5)}}{12(x+5)(x-5)\cdot\frac{5}{x-5}+12(x+5)(x-5)\cdot\frac{3}{4(x-5)}}$$

$$=\frac{12\cdot 2x+4(x+5)}{12(x+5)\cdot 5+3(x+5)\cdot 3}=\frac{24x+4x+20}{60(x+5)+9(x+5)}$$

$$=\frac{28x+20}{(x+5)(60+9)}=\frac{4(7x+5)}{69(x+5)}$$

**35.**

$$\frac{\dfrac{3}{x+2y}-\dfrac{2y}{x^2+2xy}}{\dfrac{3y}{x^2+2xy}+\dfrac{5}{x}}=\frac{\dfrac{3}{x+2y}-\dfrac{2y}{x(x+2y)}}{\dfrac{3y}{x(x+2y)}+\dfrac{5}{x}}=\frac{x(x+2y)}{x(x+2y)}\cdot\frac{\dfrac{3}{x+2y}-\dfrac{2y}{x(x+2y)}}{\dfrac{3y}{x(x+2y)}+\dfrac{5}{x}}$$

$$=\frac{x(x+2y)\cdot\dfrac{3}{x+2y}-x(x+2y)\cdot\dfrac{2y}{x(x+2y)}}{x(x+2y)\cdot\dfrac{3y}{x(x+2y)}+x(x+2y)\cdot\dfrac{5}{x}}$$

$$=\frac{x\cdot 3-2y}{3y+(x+2y)\cdot 5}=\frac{3x-2y}{3y+5x+10y}=\frac{3x-2y}{5x+13y}$$

**37.**

$$\frac{\dfrac{2}{m^2-3m+2}+\dfrac{2}{m^2-m-2}}{\dfrac{2}{m^2-1}+\dfrac{2}{m^2+4m+3}}=\frac{\dfrac{2}{(m-2)(m-1)}+\dfrac{2}{(m-2)(m+1)}}{\dfrac{2}{(m-1)(m+1)}+\dfrac{2}{(m+3)(m+1)}}$$

$$=\frac{\dfrac{2(m+1)}{(m-2)(m-1)(m+1)}+\dfrac{2(m-1)}{(m-2)(m-1)(m+1)}}{\dfrac{2(m+3)}{(m+3)(m-1)(m+1)}+\dfrac{2(m-1)}{(m+3)(m-1)(m+1)}}$$

$$=\frac{\dfrac{2m+2+2m-2}{(m-2)(m-1)(m+1)}}{\dfrac{2m+6+2m-2}{(m+3)(m-1)(m+1)}}=\frac{\dfrac{4m}{(m-2)(m-1)(m+1)}}{\dfrac{4m+4}{(m+3)(m-1)(m+1)}}$$

$$=\frac{4m}{(m-2)(m-1)(m+1)}\cdot\frac{(m+3)(m-1)(m+1)}{4(m+1)}$$

$$=\frac{m(m+3)}{(m-2)(m+1)}$$

**39.** $$\frac{\frac{2}{a^2+2a-8}+\frac{1}{a^2+5a+4}}{\frac{1}{a^2-5a+6}+\frac{2}{a^2-a-2}}=\frac{\frac{2}{(a+4)(a-2)}+\frac{1}{(a+4)(a+1)}}{\frac{1}{(a-3)(a-2)}+\frac{2}{(a+1)(a-2)}}$$

$$=\frac{\frac{2(a+1)}{(a+4)(a+1)(a-2)}+\frac{1(a-2)}{(a+4)(a+1)(a-2)}}{\frac{1(a+1)}{(a+1)(a-3)(a-2)}+\frac{2(a-3)}{(a+1)(a-3)(a-2)}}$$

$$=\frac{\frac{2a+2+a-2}{(a+4)(a+1)(a-2)}}{\frac{a+1+2a-6}{(a+1)(a-3)(a-2)}}=\frac{3a}{(a+4)\cancel{(a+1)}\cancel{(a-2)}}\cdot\frac{\cancel{(a+1)}(a-3)\cancel{(a-2)}}{(3a-5)}$$

$$=\frac{3a(a-3)}{(a+4)(3a-5)}$$

**41.** $$\frac{x}{1-\frac{1}{1+\frac{1}{x}}}=\frac{x}{1-\frac{x}{x}\cdot\frac{1}{1+\frac{1}{x}}}=\frac{x}{1-\frac{x\cdot 1}{x\cdot 1+\cancel{x}\cdot\frac{1}{\cancel{x}}}}=\frac{x}{1-\frac{x}{x+1}}=\frac{x+1}{x+1}\cdot\frac{x}{1-\frac{x}{x+1}}$$

$$=\frac{(x+1)x}{(x+1)1-(x+1)\frac{x}{x+1}}=\frac{x(x+1)}{(x+1)-\cancel{(x+1)}\frac{x}{\cancel{x+1}}}=\frac{x(x+1)}{x+1-x}=\frac{x(x+1)}{1}=x(x+1)$$

**43.** $$\frac{\frac{x-1}{x^2-4}}{1+\frac{1}{x-2}}-\frac{1}{x-2}=\frac{\frac{x-1}{(x-2)(x+2)}}{\frac{x-2}{x-2}+\frac{1}{x-2}}-\frac{1}{x-2}=\frac{\frac{x-1}{(x-2)(x+2)}}{\frac{x-1}{x-2}}-\frac{1}{x-2}$$

$$=\frac{\cancel{(x-1)}}{\cancel{(x-2)}(x+2)}\cdot\frac{\cancel{(x-2)}}{\cancel{(x-1)}}-\frac{1}{x-2}$$

$$=\frac{1}{x+2}-\frac{1}{x-2}=\frac{x-2}{(x+2)(x-2)}-\frac{x+2}{(x+2)(x-2)}$$

$$=\frac{x-2-x-2}{(x+2)(x-2)}=\frac{-4}{(x+2)(x-2)}$$

**44.** $$\frac{\frac{x-3}{x^2-16}}{1+\frac{1}{x-4}}-\frac{1}{x-4}=\frac{\frac{x-3}{(x-4)(x+4)}}{\frac{x-4}{x-4}+\frac{1}{x-4}}-\frac{1}{x-4}=\frac{\frac{x-3}{(x-4)(x+4)}}{\frac{x-3}{x-4}}-\frac{1}{x-4}$$

$$=\frac{\cancel{(x-3)}}{\cancel{(x-4)}(x+4)}\cdot\frac{\cancel{(x-4)}}{\cancel{(x-3)}}-\frac{1}{x-4}$$

$$=\frac{1}{x+4}-\frac{1}{x-4}=\frac{x-4}{(x+4)(x-4)}-\frac{x+4}{(x+4)(x-4)}$$

$$=\frac{x-4-x-4}{(x+4)(x-4)}=\frac{-8}{(x+4)(x-4)}$$

**45.** $$\frac{3}{1-\frac{3}{3+x}}-\frac{3}{\frac{3}{3-x}-1}=\frac{3}{\frac{3+x}{3+x}-\frac{3}{3+x}}-\frac{3}{\frac{3}{3-x}-\frac{3-x}{3-x}}$$

$$=\frac{3}{\frac{3+x-3}{3+x}}-\frac{3}{\frac{3-3+x}{3-x}}=\frac{3}{\frac{x}{3+x}}-\frac{3}{\frac{x}{3-x}}$$

$$=3\left(\frac{3+x}{x}\right)-3\left(\frac{3-x}{x}\right)=\frac{9+3x}{x}-\frac{9-3x}{x}$$

$$=\frac{9+3x-9+3x}{x}=\frac{6x}{x}=6$$

**46.** $$\frac{5}{1-\frac{5}{5+x}}-\frac{5}{\frac{5}{5-x}-1}=\frac{5}{\frac{5+x}{5+x}-\frac{5}{5+x}}-\frac{5}{\frac{5}{5-x}-\frac{5-x}{5-x}}$$

$$=\frac{5}{\frac{5+x-5}{5+x}}-\frac{5}{\frac{5-5+x}{5-x}}=\frac{5}{\frac{x}{5+x}}-\frac{5}{\frac{x}{5-x}}$$

$$=5\left(\frac{5+x}{x}\right)-5\left(\frac{5-x}{x}\right)=\frac{25+5x}{x}-\frac{25-5x}{x}$$

$$=\frac{25+5x-25+5x}{x}=\frac{10x}{x}=10$$

**47.** $f(x)=\dfrac{1+x}{1-x}$

$$f\left(\frac{1}{x+3}\right)=\frac{1+\dfrac{1}{x+3}}{1-\dfrac{1}{x+3}}=\frac{x+3}{x+3}\cdot\frac{1+\dfrac{1}{x+3}}{1-\dfrac{1}{x+3}}$$

$$=\frac{(x+3)+1}{(x+3)-1}=\frac{x+3+1}{x+3-1}$$

$$=\frac{x+4}{x+2}$$

**48.** $f(x)=\dfrac{1+x}{1-x}$

$$f\left(\frac{1}{x-6}\right)=\frac{1+\dfrac{1}{x-6}}{1-\dfrac{1}{x-6}}=\frac{x-6}{x-6}\cdot\frac{1+\dfrac{1}{x-6}}{1-\dfrac{1}{x-6}}$$

$$=\frac{(x-6)+1}{(x-6)-1}=\frac{x-6+1}{x-6-1}$$

$$=\frac{x-5}{x-7}$$

**49.** $f(x)=\dfrac{3}{x}$

$$\frac{f(a+h)-f(a)}{h}=\frac{\dfrac{3}{a+h}-\dfrac{3}{a}}{h}=\frac{\dfrac{3a}{a(a+h)}-\dfrac{3(a+h)}{a(a+h)}}{h}$$

$$=\frac{\dfrac{3a-3a-3h}{a(a+h)}}{h}=\frac{3a-3a-3h}{a(a+h)}\cdot\frac{1}{h}$$

$$=\frac{-3\cancel{h}}{a(a+h)}\cdot\frac{1}{\cancel{h}}$$

$$=-\frac{3}{a(a+h)}$$

**50.** $f(x)=\dfrac{1}{x^2}$

$$\frac{f(a+h)-f(a)}{h}=\frac{\dfrac{1}{(a+h)^2}-\dfrac{1}{a^2}}{h}=\frac{\dfrac{a^2}{a^2(a+h)^2}-\dfrac{(a+h)^2}{a^2(a+h)^2}}{h}$$

$$=\frac{\dfrac{a^2-\left(a^2+2ah+h^2\right)}{a^2(a+h)^2}}{h}=\frac{a^2-a^2-2ah-h^2}{a^2(a+h)^2}\cdot\frac{1}{h}$$

$$=\frac{-2ah-h^2}{a^2(a+h)^2}\cdot\frac{1}{h}=\frac{\cancel{h}(-2a-h)}{a^2(a+h)^2\cancel{h}}$$

$$=\frac{-2a-h}{a^2(a+h)^2}=-\frac{2a+h}{a^2(a+h)^2}$$

**51. a.**

$$A=\frac{Pi}{1-\dfrac{1}{(1+i)^n}}=\frac{(1+i)^n}{(1+i)^n}\cdot\frac{Pi}{1-\dfrac{1}{(1+i)^n}}=\frac{Pi(1+i)^n}{(1+i)^n\cdot 1-\cancel{(1+i)^n}\cdot\dfrac{1}{\cancel{(1+i)^n}}}=\frac{Pi(1+i)^n}{(1+i)^n-1}$$

**b.**

$$A=\frac{Pi(1+i)^n}{(1+i)^n-1}=\frac{(20000)(0.01)(1+0.01)^{48}}{(1+0.01)^{48}-1}=\frac{(20000)(0.01)(1.01)^{48}}{(1.01)^{48}-1}$$

$$=\frac{(20000)(0.01)(1.612)}{(1.612)-1}=\frac{322.4}{0.612}=526.80$$

You will pay approximately \$527 each month.

**53.**

$$R=\frac{1}{\dfrac{1}{R_1}+\dfrac{1}{R_2}+\dfrac{1}{R_3}}=\frac{R_1R_2R_3}{R_1R_2R_3}\cdot\frac{1}{\dfrac{1}{R_1}+\dfrac{1}{R_2}+\dfrac{1}{R_3}}$$

$$=\frac{R_1R_2R_3\cdot 1}{\cancel{R_1}R_2R_3\cdot\dfrac{1}{\cancel{R_1}}+R_1\cancel{R_2}R_3\cdot\dfrac{1}{\cancel{R_2}}+R_1R_2\cancel{R_3}\cdot\dfrac{1}{\cancel{R_3}}}=\frac{R_1R_2R_3}{R_2R_3+R_1R_3+R_1R_2}$$

$$R=\frac{R_1R_2R_3}{R_2R_3+R_1R_3+R_1R_2}=\frac{4\cdot 8\cdot 12}{8\cdot 12+4\cdot 12+4\cdot 8}=\frac{384}{96+48+32}=\frac{384}{176}\approx 2.18$$

The combined resistance is approximately 2.18 ohms.

**55.** Answers will vary.

**57.** Answers will vary.

**59.** $$\frac{x-\frac{1}{2x+1}}{1-\frac{x}{2x+1}}=2x-1$$

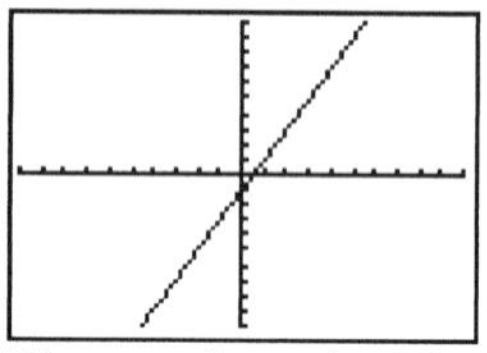

The graphs coincide. The simplification is correct.

**61.** $$\frac{\frac{1}{x}+\frac{1}{3}}{\frac{1}{3x}}=x+\frac{1}{3}$$

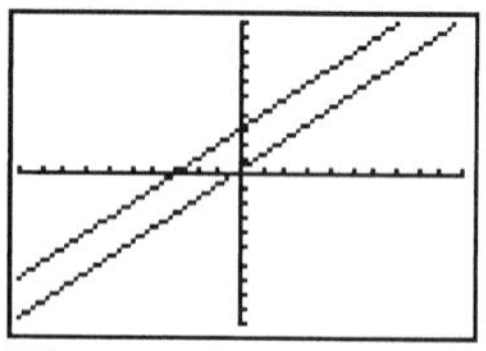

The graphs do not coincide. Simplify the complex fraction.

$$\frac{\frac{1}{x}+\frac{1}{3}}{\frac{1}{3x}}=\frac{3x}{3x}\cdot\frac{\frac{1}{x}+\frac{1}{3}}{\frac{1}{3x}}$$

$$=\frac{3\cancel{x}\cdot\frac{1}{\cancel{x}}+\cancel{3}x\cdot\frac{1}{\cancel{3}}}{\cancel{3x}\cdot\frac{1}{\cancel{3x}}}$$

$$=\frac{3+x}{1}=x+3$$

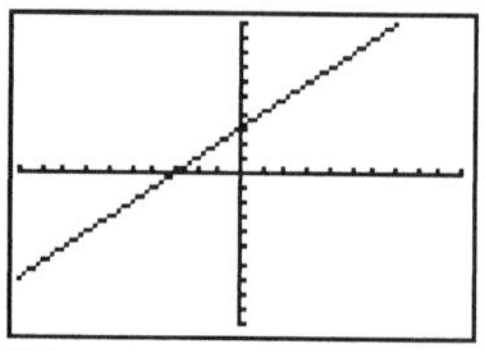

The graphs coincide. The revised simplification is correct.

**63.** $$\frac{\frac{x+h}{x+h+1}-\frac{x}{x+1}}{h}$$

$$=\frac{\frac{(x+h)(x+1)}{(x+h+1)(x+1)}-\frac{x(x+h+1)}{(x+h+1)(x+1)}}{h}$$

$$=\frac{\frac{(x+h)(x+1)-x(x+h+1)}{(x+h+1)(x+1)}}{h}$$

$$=\frac{\frac{\cancel{x^2}+\cancel{x}+\cancel{hx}+h-\cancel{x^2}-\cancel{hx}-\cancel{x}}{(x+h+1)(x+1)}}{h}$$

$$=\frac{\frac{h}{(x+h+1)(x+1)}}{h}$$

$$=\frac{\frac{h}{(x+h+1)(x+1)}}{\frac{h}{1}}$$

$$=\frac{\cancel{h}}{(x+h+1)(x+1)}\cdot\frac{1}{\cancel{h}}$$

$$=\frac{1}{(x+h+1)(x+1)}$$

**65.** $$f(x)=\frac{1}{x+1};\ f(a)=\frac{1}{a+1}$$

$$f(f(a))=\frac{1}{\frac{1}{a+1}+1}$$

$$=\frac{1}{\frac{1}{a+1}+\frac{a+1}{a+1}}$$

$$=\frac{1}{\frac{a+2}{a+1}}=\frac{a+1}{a+2}$$

Therefore, $f(f(a))=\dfrac{a+1}{a+2}$.

**67.** $x^2+27=12x$

$x^2-12x+27=0$

$(x-9)(x-3)=0$

Apply the zero product principle.

$x-9=0 \qquad x-3=0$

$x=9 \qquad x=3$

The solutions are 3 and 9 and the solution set is $\{3,9\}$.

**68.** $(4x^2-y)^2=(4x^2)^2+2\cdot 4x^2(-y)+(-y)^2$

$=16x^4-8x^2y+y^2$

**69.** $-4<3x-7<8$

$-4+7<3x-7+7<8+7$

$3<3x<15$

$1<x<5$

The solution set is $\{x|1<x<5\}$ or $(1,5)$.

### 6.4 Exercise Set

**1.** $\dfrac{25x^7-15x^5+10x^3}{5x^3}$

$=\dfrac{25x^7}{5x^3}-\dfrac{15x^5}{5x^3}+\dfrac{10x^3}{5x^3}$

$=5x^4-3x^2+2$

**3.** $\dfrac{18x^3+6x^2-9x-6}{3x}$

$=\dfrac{18x^3}{3x}+\dfrac{6x^2}{3x}-\dfrac{9x}{3x}-\dfrac{6}{3x}$

$=6x^2+2x-3-\dfrac{2}{x}$

**5.** $\dfrac{28x^3-7x^2-16x}{4x^2}=\dfrac{28x^3}{4x^2}-\dfrac{7x^2}{4x^2}-\dfrac{16x}{4x^2}$

$=7x-\dfrac{7}{4}-\dfrac{4}{x}$

**7.** $\dfrac{25x^8-50x^7+3x^6-40x^5}{-5x^5}$

$=\dfrac{25x^8}{-5x^5}-\dfrac{50x^7}{-5x^5}+\dfrac{3x^6}{-5x^5}-\dfrac{40x^5}{-5x^5}$

$=-5x^3+10x^2-\dfrac{3}{5}x+8$

**9.** $\dfrac{18a^3b^2-9a^2b-27ab^2}{9ab}$

$=\dfrac{18a^3b^2}{9ab}-\dfrac{9a^2b}{9ab}-\dfrac{27ab^2}{9ab}$

$=2a^2b-a-3b$

**11.** $\dfrac{36x^4y^3-18x^3y^2-12x^2y}{6x^3y^3}$

$=\dfrac{36x^4y^3}{6x^3y^3}-\dfrac{18x^3y^2}{6x^3y^3}-\dfrac{12x^2y}{6x^3y^3}$

$=6x-\dfrac{3}{y}-\dfrac{2}{xy^2}$

**13.**

$$\begin{array}{r} x+3 \\ x+5\overline{)x^2+8x+15} \\ \underline{x^2+5x} \\ 3x+15 \\ \underline{3x+15} \\ 0 \end{array}$$

$\dfrac{x^2+8x+15}{x+5}=x+3$

**15.**

$$\begin{array}{r} x^2+x-2 \\ x-3\overline{)x^3-2x^2-5x+6} \\ \underline{x^3-3x^2} \\ x^2-5x \\ \underline{x^2-3x} \\ -2x+6 \\ \underline{-2x+6} \\ 0 \end{array}$$

$$\frac{x^3-2x^2-5x+6}{x-3}=x^2+x-2$$

**17.**

$$\begin{array}{r} x-2 \\ x-5\overline{)x^2-7x+12} \\ \underline{x^2-5x} \\ -2x+12 \\ \underline{-2x+10} \\ 2 \end{array}$$

$$\frac{x^2-7x+12}{x-5}=x-2+\frac{2}{x-5}$$

**19.**

$$\begin{array}{r} x+5 \\ 2x+3\overline{)2x^2+13x+5} \\ \underline{2x^2+3x} \\ 10x+5 \\ \underline{10x+15} \\ -10 \end{array}$$

$$\frac{2x^2+13x+5}{2x+3}=x+5-\frac{10}{2x+3}$$

**21.**

$$\begin{array}{r} x^2+2x+3 \\ x+1\overline{)x^3+3x^2+5x+4} \\ \underline{x^3+x^2} \\ 2x^2+5x \\ \underline{2x^2+2x} \\ 3x+4 \\ \underline{3x+3} \\ 1 \end{array}$$

$$\frac{x^3+3x^2+5x+4}{x+1}$$

$$=x^2+2x+3+\frac{1}{x+1}$$

**23.**

$$\begin{array}{r} 2y^2+3y-1 \\ 2y+3\overline{)4y^3+12y^2+7y-3} \\ \underline{4y^3+6y^2} \\ 6y^2+7y \\ \underline{6y^2+9y} \\ -2y-3 \\ \underline{-2y-3} \\ 0 \end{array}$$

$$\frac{4y^3+12y^2+7y-3}{2y+3}=2y^2+3y-1$$

**25.**

$$\begin{array}{r} 3x^2-3x+1 \\ 3x+2\overline{)9x^3-3x^2-3x+4} \\ \underline{9x^3+6x^2} \\ -9x^2-3x \\ \underline{-9x^2-6x} \\ 3x+4 \\ \underline{3x+2} \\ 2 \end{array}$$

$$\frac{9x^3-3x^2-3x+4}{3x+2}=3x^2-3x+1+\frac{2}{3x+2}$$

**27.** $(4x^3-6x-11)\div(2x-4)$

Rewrite the dividend with the missing power of $x$ and divide.

$$
\begin{array}{r l}
 & \phantom{)}2x^2+4x+5 \\
2x-4 & \overline{\big)\,4x^3+0x^2-6x-11} \\
 & \underline{4x^3-8x^2} \\
 & \phantom{4x^3-}8x^2-6x \\
 & \phantom{4x^3-}\underline{8x^2-16x} \\
 & \phantom{4x^3-8x^2-}10x-11 \\
 & \phantom{4x^3-8x^2-}\underline{10x-20} \\
 & \phantom{4x^3-8x^2-10x-1}9
\end{array}
$$

$$\frac{4x^3-6x-11}{2x-4}=2x^2+4x+5+\frac{9}{2x-4}$$

**29.** $(4y^3-5y)\div(2y-1)$

Rewrite the dividend with the missing powers of $y$ and divide. Notice: If there is no constant term in the polynomial, it should also be added for the purposes of long division.

$$
\begin{array}{r l}
 & \phantom{)}2y^2+y-2 \\
2y-1 & \overline{\big)\,4y^3+0y^2-5y+0} \\
 & \underline{4y^3-2y^2} \\
 & \phantom{4y^3-}2y^2-5y \\
 & \phantom{4y^3-}\underline{2y^2-y} \\
 & \phantom{4y^3-2y^2-}-4y+0 \\
 & \phantom{4y^3-2y^2-}\underline{-4y+2} \\
 & \phantom{4y^3-2y^2-4y+}-2
\end{array}
$$

$$\frac{4y^3-5y}{2y-1}=2y^2+y-2-\frac{2}{2y-1}$$

**31.** $(4y^4-17y^2+14y-3)\div(2y-3)$

Rewrite the dividend with the missing power of $y$ and divide.

$$
\begin{array}{r l}
 & \phantom{)}2y^3+3y^2-4y+1 \\
2y-3 & \overline{\big)\,4y^4+0y^3-17y^2+14y-3} \\
 & \underline{4y^3-6y^3} \\
 & \phantom{4y^4-}6y^3-17y^2 \\
 & \phantom{4y^4-}\underline{6y^3-9y^2} \\
 & \phantom{4y^4-6y^3}-8y^2+14y \\
 & \phantom{4y^4-6y^3}\underline{-8y^2+12y} \\
 & \phantom{4y^4-6y^3-8y^2+}2y-3 \\
 & \phantom{4y^4-6y^3-8y^2+}\underline{2y-3} \\
 & \phantom{4y^4-6y^3-8y^2+2y-}0
\end{array}
$$

$$\frac{4y^4-17y^2+14y-3}{2y-3}=2y^3+3y^2-4y+1$$

**33.** $(4x^4+3x^3+4x^2+9x-6)\div(x^2+3)$

Rewrite the divisor with the missing power of $x$ and divide.

$$
\begin{array}{r l}
 & \phantom{)}4x^2+3x-8 \\
x^2+0x+3 & \overline{\big)\,4x^4+3x^3+4x^2+9x-6} \\
 & \underline{4x^4+0x^3+12x^2} \\
 & \phantom{4x^4+}3x^3-8x^2+9x \\
 & \phantom{4x^4+}\underline{3x^3+0x^2+9x} \\
 & \phantom{4x^4+3x^3}-8x^2+0x-6 \\
 & \phantom{4x^4+3x^3}\underline{-8x^2+0x-24} \\
 & \phantom{4x^4+3x^3-8x^2+0x-}18
\end{array}
$$

$$\frac{4x^4+3x^3+4x^2+9x-6}{x^2+3}$$
$$=4x^2+3x-8+\frac{18}{x^2+3}$$

**35.** $\left(15x^4+3x^3+4x^2+4\right)\div\left(3x^2-1\right)$

Rewrite the dividend and the divisor with the missing powers of $x$ and divide.

$$\begin{array}{r} 5x^2+\ x+3 \\ 3x^2+0x-1\overline{)15x^4+3x^3+4x^2+0x+4} \\ \underline{15x^4+0x^3-5x^2} \\ 3x^3+9x^2+0x \\ \underline{3x^3+0x^2-\ x} \\ 9x^2+\ x+4 \\ \underline{9x^2+0x-3} \\ x+7 \end{array}$$

$$\frac{15x^4+3x^3+4x^2+4}{3x^2-1}$$
$$=5x^2+x+3+\frac{7+x}{3x^2-1}$$
$$\text{or } 5x^2+x+3+\frac{x+7}{3x^2-1}$$

**37.** $f(x)=8x^3-38x^2+49x-10$

$g(x)=4x-1$

$$\left(\frac{f}{g}\right)(x)=\frac{f(x)}{g(x)}=\frac{8x^3-38x^2+49x-10}{4x-1}$$

$$\begin{array}{r} 2x^2-9x+10 \\ 4x-1\overline{)8x^3-38x^2+49x-10} \\ \underline{8x^3-2x^2} \\ -36x^2+49x \\ \underline{-36x^2+\ 9x} \\ 40x-10 \\ \underline{40x-10} \\ 0 \end{array}$$

$$\left(\frac{f}{g}\right)(x)=2x^2-9x+10$$

**39.** $f(x)=2x^4-7x^3+7x^2-9x+10$

$g(x)=2x-5$

$$\left(\frac{f}{g}\right)(x)=\frac{f(x)}{g(x)}$$
$$=\frac{2x^4-7x^3+7x^2-9x+10}{2x-5}$$

$$\begin{array}{r} x^3-x^2+x-2 \\ 2x-5\overline{)2x^4-7x^3+7x^2-9x+10} \\ \underline{2x^4-5x^3} \\ -2x^3+7x^2 \\ \underline{-2x^3+5x^2} \\ 2x^2-9x \\ \underline{2x^2-5x} \\ -4x+10 \\ \underline{-4x+10} \\ 0 \end{array}$$

$$\left(\frac{f}{g}\right)(x)=\frac{f(x)}{g(x)}=x^3-x^2+x-2$$

**41.**

$$\begin{array}{r} x^3-x^2y+xy^2-y^3 \\ x+y\overline{)x^4+0x^3+0x^2+0x+y^4} \\ \underline{x^4+x^3y} \\ -x^3y+0x^2 \\ \underline{-x^3y-x^2y^2} \\ x^2y^2+0x \\ \underline{x^2y^2+xy^3} \\ -xy^3+y^4 \\ \underline{-xy^3-y^4} \\ 2y^4 \end{array}$$

$$\frac{x^4+y^4}{x+y}=x^3-x^2y+xy^2-y^3+\frac{2y^4}{x+y}$$

**42.**

$$\begin{array}{r}
x^4 - x^3y + x^2y^2 - xy^3 + y^4 \\
x+y\overline{\big)x^5 + 0x^4 + 0x^3 + 0x^2 + 0x + y^4} \\
\underline{x^5 + x^4y} \\
-x^4y + 0x^3 \\
\underline{-x^4y - x^3y^2} \\
x^3y^2 + 0x^2 \\
\underline{x^3y^2 + x^2y^3} \\
-x^2y^3 + 0x \\
\underline{-x^2y^3 - xy^4} \\
xy^4 + y^5 \\
\underline{xy^4 + y^5} \\
0
\end{array}$$

$$\frac{x^5 + y^5}{x+y} = x^4 - x^3y + x^2y^2 - xy^3 + y^4$$

**43.**

$$\begin{array}{r}
3x^2 + 2x - 1 \\
x^2 + x + 2\overline{\big)3x^4 + 5x^3 + 7x^2 + 3x - 2} \\
\underline{3x^4 + 3x^3 + 6x^2} \\
2x^3 + x^2 + 3x \\
\underline{2x^3 + 2x^2 + 4x} \\
-x^2 - x - 2 \\
\underline{-x^2 - x - 2} \\
0
\end{array}$$

$$\frac{3x^4 + 5x^3 + 7x^2 + 3x - 2}{x^2 + x + 2} = 3x^2 + 2x - 1$$

**44.**

$$\begin{array}{r}
x^2 + 2x + 1 \\
x^2 - 3x - 2\overline{\big)x^4 - x^3 - 7x^2 - 7x - 2} \\
\underline{x^4 - 3x^3 - 2x^2} \\
2x^3 - 5x^2 - 7x \\
\underline{2x^3 - 6x^2 - 4x} \\
x^2 - 3x - 2 \\
\underline{x^2 - 3x - 2} \\
0
\end{array}$$

$$\frac{x^4 - x^3 - 7x^2 - 7x - 2}{x^2 - 3x - 2} = x^2 + 2x + 1$$

**45.**

$$\begin{array}{r}
4x - 7 \\
x^2 + x + 1\overline{\big)4x^3 - 3x^2 + x + 1} \\
\underline{4x^3 + 4x^2 + 4x} \\
-7x^2 - 3x + 1 \\
\underline{-7x^2 - 7x - 7} \\
4x + 8
\end{array}$$

$$\frac{4x^3 - 3x^2 + x + 1}{x^2 + x + 1} = 4x - 7 + \frac{4x + 8}{x^2 + x + 1}$$

**46.**

$$\begin{array}{r}
x^2 - x - 1 \\
x^2 + x + 1\overline{\big)x^4 + 0x^3 - x^2 + 0x + 1} \\
\underline{x^4 + x^3 + x^2} \\
-x^3 - 2x^2 + 0x \\
\underline{-x^3 - x^2 - x} \\
-x^2 + x + 1 \\
\underline{-x^2 - x - 1} \\
2x + 2
\end{array}$$

$$\frac{x^4 - x^2 + 1}{x^2 + x + 1} = x^2 - x - 1 + \frac{2x + 2}{x^2 + x + 1}$$

**47.**

$$x^2-x+2\overline{\smash{)}x^5+0x^4+0x^3+0x^2+0x-1}$$

Quotient: $x^3+x^2-x-3$

$$\underline{x^5-x^4+2x^3}$$
$$x^4-2x^3+0x^2$$
$$\underline{x^4-x^3+2x^2}$$
$$-x^3-2x^2+0x$$
$$\underline{-x^3+x^2-2x}$$
$$-3x^2+2x-1$$
$$\underline{-3x^2+3x-6}$$
$$-x+5$$

$$\frac{x^5-1}{x^2-x+2}=x^3+x^2-x-3+\frac{-x+5}{x^2-x+2}$$

**48.**

$$x^3-4\overline{\smash{)}5x^5-7x^4+3x^3-20x^2+28x-12}$$

Quotient: $5x^2-7x+3$

$$\underline{5x^5 \qquad -20x^2}$$
$$-7x^4+3x^3 \qquad +28x$$
$$\underline{-7x^4 \qquad +28x}$$
$$3x^3 \qquad -12$$
$$\underline{3x^3 \qquad -12}$$
$$0$$

$$\frac{5x^5-7x^4+3x^3-20x^2+28x-12}{x^3-4}$$
$$=5x^2-7x+3$$

**49.**

$$x-3y\overline{\smash{)}4x^3-7x^2y-16xy^2+3y^3}$$

Quotient: $4x^2+5xy-y^2$

$$\underline{4x^3-12x^2y}$$
$$5x^2y-16xy^2$$
$$\underline{5x^2y-15xy^2}$$
$$-xy^2+3y^3$$
$$\underline{-xy^2+3y^3}$$
$$0$$

$$\frac{4x^3-7x^2y\ \ 16xy^2+3y^3}{x-3y}$$
$$=4x^2+5xy-y^2$$

**50.**

$$4x-5y\overline{\smash{)}12x^3-19x^2y+13xy^2-10y^3}$$

Quotient: $3x^2-xy+2y^2$

$$\underline{12x^3-15x^2y}$$
$$-4x^2y+13xy^2$$
$$\underline{-4x^2y+5xy^2}$$
$$8xy^2-10y^3$$
$$\underline{8xy^2-10y^3}$$
$$0$$

$$\frac{12x^3-19x^2y+13xy^2-10y^3}{4x-5y}$$
$$=3x^2-xy+2y^2$$

**51.** $f(x)=3x^3+4x^2-x-4$

$g(x)=-5x^3+22x^2-28x-12$

$h(x)=4x+1$

$$\left(\frac{f-g}{h}\right)(x)=\frac{f(x)-g(x)}{h(x)}$$

$$=\frac{(3x^3+4x^2-x-4)-(-5x^3+22x^2-28x-12)}{4x+1}$$

$$=\frac{3x^3+4x^2-x-4+5x^3-22x^2+28x+12}{4x+1}$$

$$=\frac{8x^3-18x^2+27x+8}{4x+1}$$

$$\begin{array}{r} 2x^2-5x+8 \\ 4x+1\overline{)8x^3-18x^2+27x+8} \\ \underline{8x^3+2x^2} \\ -20x^2+27x \\ \underline{-20x^2-5x} \\ 32x+8 \\ \underline{32x+8} \\ 0 \end{array}$$

Therefore, $\left(\frac{f-g}{h}\right)(x)=2x^2-5x+8$.

Domain: $\left\{x \mid x \text{ is a real number } x \neq -\frac{1}{4}\right\}$ or $\left(-\infty,-\frac{1}{4}\right)\cup\left(-\frac{1}{4},\infty\right)$.

**52.** $f(x)=x^3+9x^2-6x+25$

$g(x)=-3x^3+2x^2-14x+5$

$h(x)=2x+4$

$$\left(\frac{f-g}{h}\right)(x)=\frac{f(x)-g(x)}{h(x)}$$

$$=\frac{(x^3+9x^2-6x+25)-(-3x^3+2x^2-14x+5)}{2x+4}$$

$$=\frac{x^3+9x^2-6x+25+3x^3-2x^2+14x-5}{2x+4}$$

$$=\frac{4x^3+7x^2+8x+20}{2x+4}$$

$$\begin{array}{r} 2x^2-\frac{1}{2}x+5 \\ 2x+4\overline{)4x^3+7x^2+8x+20} \\ \underline{4x^3+8x^2} \\ -x^2+8x \\ \underline{-x^2-2x} \\ 10x+20 \\ \underline{10x+20} \\ 0 \end{array}$$

Therefore, $\left(\frac{f-g}{h}\right)(x)=2x^2-\frac{1}{2}x+5$.

Domain: $\{x \mid x \text{ is a real number } x \neq -2\}$ or $(-\infty,-2)\cup(-2,\infty)$

**53.**
$$\begin{aligned} ax+2x+4&=3a^3+10a^2+6a\\ x(a+2)+4&=3a^3+10a^2+6a\\ x(a+2)&=3a^3+10a^2+6a-4\\ x&=\frac{3a^3+10a^2+6a-4}{a+2}\end{aligned}$$

$$\begin{array}{r} 3a^2+4a-2\\ a+2\overline{)3a^3+10a^2+6a-4}\\ \underline{3a^3+6a^2}\phantom{+6a-4}\\ 4a^2+6a\phantom{-4}\\ \underline{4a^2+8a}\phantom{-4}\\ -2a-4\\ \underline{-2a-4}\\ 0\end{array}$$

Therefore, $x=3a^2+4a-2$.

**54.**
$$\begin{aligned} ax-3x+6&=a^3-6a^2+11a\\ x(a-3)+6&=a^3-6a^2+11a\\ x(a-3)&=a^3-6a^2+11a-6\\ x&=\frac{a^3-6a^2+11a-6}{a-3}\end{aligned}$$

$$\begin{array}{r} a^2-3a+2\\ a-3\overline{)a^3-6a^2+11a-6}\\ \underline{a^3-3a^2}\phantom{+11a-6}\\ -3a^2+11a\phantom{-6}\\ \underline{-3a^2+9a}\phantom{-6}\\ 2a-6\\ \underline{2a-6}\\ 0\end{array}$$

Therefore, $x=a^2-3a+2$.

**55.**
$$\begin{aligned} f(30)&=\frac{80(30)-8000}{30-110}\\ &=\frac{2400-8000}{-80}=\frac{-5600}{-80}=70\end{aligned}$$

With a tax rate percentage of 30, the government tax revenue will be \$70 ten billions or \$700 billion. This is shown on the graph as the point $(30,70)$.

**57.** $(80x-8000)\div(x-110)$

$$\begin{array}{r} 80\\ x-110\overline{)80x-8000}\\ \underline{80x-8800}\\ 800\end{array}$$

$$\frac{80x-8000}{x-110}=80+\frac{800}{x-110}$$

$$\begin{aligned} f(x)&=80+\frac{800}{x-110}\\ f(30)&=80+\frac{800}{30-110}\\ &=80+\frac{800}{-80}\\ &=80-10\\ &=70\end{aligned}$$

This is the same answer obtained in Exercise 37.

**59.** Answers will vary.

**61.** Answers will vary.

**63.** Answers will vary.

**65.** $(6x^2+16x+8)\div(3x+2)=2x+4$

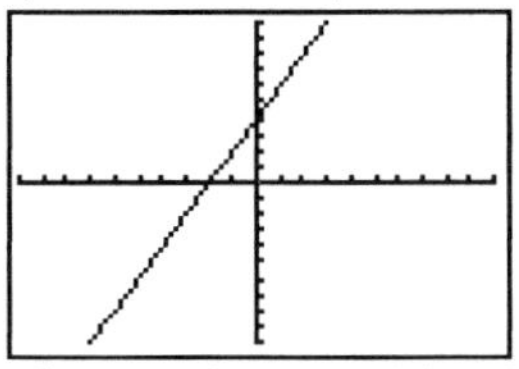

The graphs coincide. The division is correct.

**67.** $(3x^4+4x^3-32x^2-5x-20)\div(x+4)$

$=3x^3-8x^2+5$

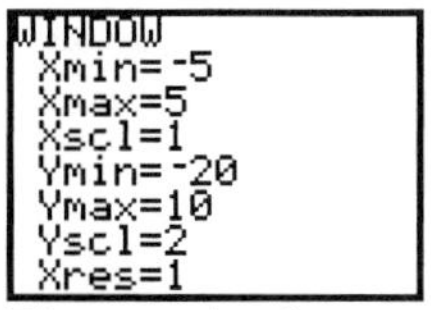


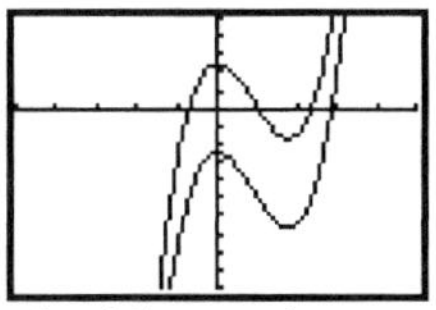

The graphs do not coincide, so the division has not been performed correctly.

$$
\begin{array}{r}
3x^3-8x^2-5 \\
x+4\overline{)3x^4+4x^3-32x^2-5x-20} \\
\underline{3x^4+12x^3} \qquad\qquad\qquad \\
-8x^3-32x^2 \qquad\quad \\
\underline{-8x^3-32x^2} \qquad\quad \\
0-5x-20 \\
\underline{-5x-20} \\
0
\end{array}
$$

$(3x^4+4x^3-32x^2-5x-20)\div(x+4)$

$=3x^3-8x^2-5$

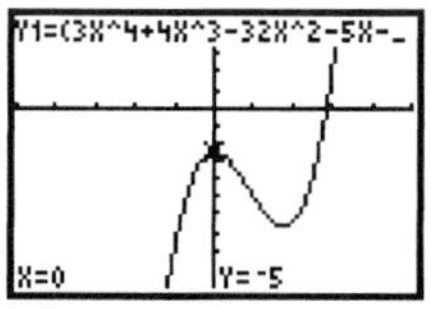


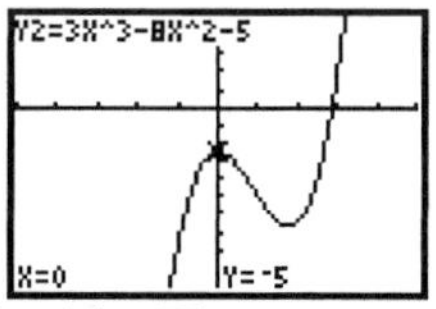


The graphs now coincide. The new result is correct.

**69.** Statement **d.** is true. If long division results in a zero remainder, the divisor is a factor of the dividend.

Statement **a.** is false. Since some polynomials cannot be factored this method cannot always be used.

Statement **b.** is false. In many cases, long-division results in a remainder with a variable term in the denominator. Polynomials do not have variable terms in any denominator.

Statement **c.** is false. There is no such restriction in long division.

**71.** $(x^{3n}+1)\div(x^n+1)$

Rewrite the dividend with the missing power of $x$ and divide.

$$
\begin{array}{r}
x^{2n}-x^n+1 \\
x^n+1\overline{)x^{3n}+0x^{2n}+0x^n+1} \\
\underline{x^{3n}+x^{2n}} \qquad\qquad\quad \\
-x^{2n}+0x^n \qquad \\
\underline{-x^{2n}-x^n} \qquad \\
x^n+1 \\
\underline{x^n+1} \\
0
\end{array}
$$

$$\frac{x^{3n}+1}{x^n+1}=x^{2n}-x^n+1$$

**73.** $\left(20x^3+23x^2-10x+k\right)\div(4x+3)$

$$\begin{array}{r} 5x^2+2x-4 \\ 4x+3\overline{)20x^3+23x^2-10x+k} \\ \underline{20x^3+15x^2} \\ 8x^2-10x \\ \underline{8x^2+6x} \\ -16x+k \\ \underline{-16x-12} \\ k-(-12) \end{array}$$

Set the remainder equal to zero to find $k$.

$k-(-12)=0$

$k+12=0$

$k=-12$

When $k=-12$, the remainder will be zero.

**74.** $|2x-3|>4$

$2x-3<-4 \qquad 2x-3>4$

$x<-\frac{1}{2} \qquad 2x>7$

$x>\frac{7}{2}$

The solution set is $\left\{x \middle| x<-\frac{1}{2} \text{ or } x>\frac{7}{2}\right\}$ or $\left(-\infty,-\frac{1}{2}\right)\cup\left(\frac{7}{2},\infty\right)$.

**75.** $40{,}610{,}000=4.061\times10^7$

**76.** $2x-4[x-3(2x+1)]$

$=2x-4[x-6x-3]$

$=2x-4[-5x-3]$

$=2x+20x+12$

$=22x+12$

## Mid-Chapter 6 Check Point

**1.** $\dfrac{x^2-x-6}{x^2+3x-18}=\dfrac{(x-3)(x-2)}{(x+6)(x-3)}$

$=\dfrac{x-2}{x+6}$

**2.** $\dfrac{2x^2-8x-11}{x^2+3x-4}+\dfrac{x^2+14x-13}{x^2+3x-4}$

$=\dfrac{2x^2-8x-11+x^2+14x-13}{x^2+3x-4}$

$=\dfrac{2x^2+x^2-8x+14x-11-13}{x^2+3x-4}$

$=\dfrac{3x^2+6x-24}{x^2+3x-4}$

$=\dfrac{3(x^2+2x-8)}{x^2+3x-4}$

$=\dfrac{3(x+4)(x-2)}{(x+4)(x-1)}$

$=\dfrac{3(x-2)}{x-1}$

**3.** $\dfrac{x^3-27}{4x^2-4x}\cdot\dfrac{4x}{x-3}$

$=\dfrac{(x-3)(x^2+3x+9)}{4x(x-1)}\cdot\dfrac{4x}{(x-3)}$

$=\dfrac{x^2+3x+9}{x-1}$

**4.** $5+\dfrac{7}{x-2}=\dfrac{5(x-2)}{x-2}+\dfrac{7}{x-2}$

$=\dfrac{5x-10+7}{x-2}$

$=\dfrac{5x-3}{x-2}$

**5.** $$\frac{x-\frac{4}{x+6}}{\frac{1}{x+6}+x}=\frac{\frac{x(x+6)}{x+6}-\frac{4}{x+6}}{\frac{1}{x+6}+\frac{x(x+6)}{x+6}}=\frac{\frac{x^2+6x-4}{x+6}}{\frac{x^2+6x+1}{x+6}}=\frac{x^2+6x-4}{\cancel{(x+6)}}\cdot\frac{\cancel{(x+6)}}{x^2+6x+1}=\frac{x^2+6x-4}{x^2+6x+1}$$

**6.**
$$\begin{array}{r} 2x^3-5x^2-3x+6 \\ x-4\overline{)2x^4-13x^3+17x^2+18x-24} \\ \underline{2x^4-8x^3} \\ -5x^3+17x^2 \\ \underline{-5x^3+20x^2} \\ -3x^2+18x \\ \underline{-3x^2+12x} \\ 6x-24 \\ \underline{6x-24} \\ 0 \end{array}$$

$$\left(2x^4-13x^3+17x^2+18x-24\right)\div(x-4)=2x^3-5x^2-3x+6$$

**7.** $$\frac{x^3y-y^3x}{x^2y-xy^2}=\frac{xy\left(x^2-y^2\right)}{xy(x-y)}=\frac{\cancel{xy}\,\cancel{(x-y)}\,(x+y)}{\cancel{xy}\,\cancel{(x-y)}}=x+y$$

**8.** $$\frac{28x^8y^3-14x^6y^2+3x^2y^2}{7x^2y}=\frac{28x^8y^3}{7x^2y}-\frac{14x^6y^2}{7x^2y}+\frac{3x^2y^2}{7x^2y}$$
$$=4x^6y^2-2x^4y+\frac{3}{7}y$$

**9.** $$\frac{2x-1}{x+6}-\frac{x+3}{x-2}=\frac{(2x-1)(x-2)}{(x+6)(x-2)}-\frac{(x+6)(x+3)}{(x+6)(x-2)}=\frac{2x^2-4x-x+2}{(x+6)(x-2)}-\frac{x^2+3x+6x+18}{(x+6)(x-2)}$$
$$=\frac{2x^2-5x+2}{(x+6)(x-2)}-\frac{x^2+9x+18}{(x+6)(x-2)}=\frac{2x^2-5x+2-x^2-9x-18}{(x+6)(x-2)}$$
$$=\frac{2x^2-x^2-5x-9x+2-18}{(x+6)(x-2)}$$
$$=\frac{x^2-14x-16}{(x+6)(x-2)}$$

**10.**
$$
\begin{aligned}
\frac{3}{x-2}-\frac{2}{x+2}-\frac{x}{x^2-4} &= \frac{3}{x-2}-\frac{2}{x+2}-\frac{x}{(x-2)(x+2)} \\
&= \frac{3(x+2)}{(x-2)(x+2)}-\frac{2(x-2)}{(x-2)(x+2)}-\frac{x}{(x-2)(x+2)} \\
&= \frac{3x+6}{(x-2)(x+2)}-\frac{2x-4}{(x-2)(x+2)}-\frac{x}{(x-2)(x+2)} \\
&= \frac{3x+6-2x+4-x}{(x-2)(x+2)}=\frac{3x-2x-x+6+4}{(x-2)(x+2)} \\
&= \frac{10}{(x-2)(x+2)}
\end{aligned}
$$

**11.**
$$
\begin{aligned}
\frac{3x^2-7x-6}{3x^2-12x-10}\div\frac{2x^2-x-1}{4x^2-18x-10} &= \frac{3x^2-7x-6}{3x^2-12x-10}\cdot\frac{4x^2-18x-10}{2x^2-x-1} \\
&= \frac{(3x+2)(x-3)}{(3x+2)(x-5)}\cdot\frac{2(2x^2-9x-5)}{(2x+1)(x-1)} \\
&= \frac{2(x-3)(2x+1)(x-5)}{(x-5)(2x+1)(x-1)} \\
&= \frac{2(x-3)}{x-1}
\end{aligned}
$$

**12.**
$$
\frac{3}{7-x}+\frac{x-2}{x-7}=\frac{-3}{-(7-x)}+\frac{x-2}{x-7}=\frac{-3}{x-7}+\frac{x-2}{x-7}=\frac{-3+x-2}{x-7}=\frac{x-5}{x-7}
$$

**13.**
$$
\begin{array}{r}
2x^2-x-3 \\
3x^2-1\overline{\smash{)}\,6x^4-3x^3-11x^2+2x+4} \\
\underline{6x^4\qquad\quad-2x^2} \\
-3x^3-9x^2 \\
\underline{-3x^3\qquad\quad+x} \\
-9x^2\;+x \\
\underline{-9x^2\qquad+3} \\
x+1
\end{array}
$$

$$
\left(6x^4-3x^3-11x^2+2x+4\right)\div\left(3x^2-1\right)=2x^2-x-3+\frac{x+1}{3x^2-1}
$$

**14.** $$\frac{5+\frac{2}{x}}{3-\frac{1}{x}}=\frac{\frac{5x}{x}+\frac{2}{x}}{\frac{3x}{x}-\frac{1}{x}}=\frac{\frac{5x+2}{x}}{\frac{3x-1}{x}}=\frac{5x+2}{\cancel{x}}\cdot\frac{\cancel{x}}{3x-1}=\frac{5x+2}{3x-1}$$

**15.** $$\begin{aligned}\frac{x}{x^2-7x+6}-\frac{x}{x^2-2x-24}&=\frac{x}{(x-6)(x-1)}-\frac{x}{(x-6)(x+4)}\\&=\frac{x(x+4)}{(x-6)(x-1)(x+4)}-\frac{x(x-1)}{(x-6)(x-1)(x+4)}\\&=\frac{x(x+4)-x(x-1)}{(x-6)(x-1)(x+4)}=\frac{x^2+4x-x^2+x}{(x-6)(x-1)(x+4)}\\&=\frac{x^2-x^2+4x+x}{(x-6)(x-1)(x+4)}=\frac{5x}{(x-6)(x-1)(x+4)}\end{aligned}$$

**16.** $$\begin{aligned}\frac{\frac{3}{x+1}+\frac{4}{x}}{\frac{4}{x}}&=\frac{x(x+1)}{x(x+1)}\cdot\frac{\frac{3}{x+1}+\frac{4}{x}}{\frac{4}{x}}\\&=\frac{3x+4(x+1)}{4(x+1)}=\frac{3x+4x+4}{4(x+1)}=\frac{7x+4}{4(x+1)}\end{aligned}$$

**17.** $$\begin{aligned}\frac{x^2-x-6}{x+1}\div\left(\frac{x^2-9}{x^2-1}\cdot\frac{x-1}{x+3}\right)&=\frac{(x-3)(x+2)}{(x+1)}\div\left(\frac{(x-3)\cancel{(x+3)}\cancel{(x-1)}}{\cancel{(x-1)}(x+1)\cancel{(x+3)}}\right)\\&=\frac{(x-3)(x+2)}{(x+1)}\div\left(\frac{x-3}{x+1}\right)\\&=\frac{\cancel{(x-3)}(x+2)}{\cancel{(x+1)}}\cdot\frac{\cancel{(x+1)}}{\cancel{(x-3)}}\\&=x+2\end{aligned}$$

**18.**

$$\begin{array}{r} 16x^2-8x+4 \\ 4x+2\overline{)64x^3+0x^2+\ 0x+4} \\ \underline{64x^3+32x^2} \qquad\qquad \\ -32x^2+0x \qquad \\ \underline{-32x-16x} \qquad \\ 16x+4 \\ \underline{16x+8} \\ -4 \end{array}$$

$$\begin{aligned}\left(64x^3+4\right)\div(4x+2) &= 16x^2-8x+4-\frac{4}{4x+2} \\ &= 16x^2-8x+4-\frac{2(2)}{2(2x+1)} \\ &= 16x^2-8x+4-\frac{2}{2x+1}\end{aligned}$$

**19.**

$$\frac{x+1}{x^2+x-2}-\frac{1}{x^2-3x+2}+\frac{2x}{x^2-4}$$

$$=\frac{x+1}{(x+2)(x-1)}-\frac{1}{(x-2)(x-1)}+\frac{2x}{(x-2)(x+2)}$$

$$=\frac{(x+1)(x-2)}{(x-2)(x+2)(x-1)}-\frac{1(x+2)}{(x-2)(x+2)(x-1)}+\frac{2x(x-1)}{(x-2)(x+2)(x-1)}$$

$$=\frac{x^2-2x+x-2}{(x-2)(x+2)(x-1)}-\frac{x+2}{(x-2)(x+2)(x-1)}+\frac{2x^2-2x}{(x-2)(x+2)(x-1)}$$

$$=\frac{x^2-x-2-x-2+2x^2-2x}{(x-2)(x+2)(x-1)}$$

$$=\frac{x^2+2x^2-x-x-2x-2-2}{(x-2)(x+2)(x-1)}$$

$$=\frac{3x^2-4x-4}{(x-2)(x+2)(x-1)}$$

$$=\frac{(3x+2)\cancel{(x-2)}}{\cancel{(x-2)}(x+2)(x-1)}$$

$$=\frac{3x+2}{(x+2)(x-1)}$$

**20.** $f(x)=\dfrac{5x-10}{x^2+5x-14}$

To find the domain, we first set the denominator equal to 0 and solve the resulting equation.

$$x^2+5x-14=0$$
$$(x+7)(x-2)=0$$
$$x+7=0 \quad \text{or} \quad x-2=0$$
$$x=-7 \qquad\qquad x=2$$

Since the values $-7$ and 2 make the denominator 0, we must exclude these values from the domain. Therefore, we have

Domain: $\{x\,|\,x \text{ is a real number and } x\neq -7, x\neq 2\}$

$$f(x)=\frac{5x-10}{x^2+5x-14}=\frac{5\cancel{(x-2)}}{(x+7)\cancel{(x-2)}}=\frac{5}{x+7} \quad \text{where } x\neq -7.x\neq 2.$$

## 6.5 Exercise Set

**1.** $(2x^2+x-10)\div(x-2)$

$$\begin{array}{r|rrr} 2 & 2 & 1 & -10 \\ & & 4 & 10 \\ \hline & 2 & 5 & 0 \end{array}$$

$(2x^2+x-10)\div(x-2)=2x+5$

**3.** $(3x^2+7x-20)\div(x+5)$

$$\begin{array}{r|rrr} -5 & 3 & 7 & -20 \\ & & -15 & 40 \\ \hline & 3 & -8 & 20 \end{array}$$

$(3x^2+7x-20)\div(x+5)=3x-8+\dfrac{20}{x+5}$

**5.** $(4x^3-3x^2+3x-1)\div(x-1)$

$$\begin{array}{r|rrrr} 1 & 4 & -3 & 3 & -1 \\ & & 4 & 1 & 4 \\ \hline & 4 & 1 & 4 & 3 \end{array}$$

$(4x^3-3x^2+3x-1)\div(x-1)$

$=4x^2+x+4+\dfrac{3}{x-1}$

**7.** $(6x^5-2x^3+4x^2-3x+1)\div(x-2)$

$$\begin{array}{r|rrrrrr} 2 & 6 & 0 & -2 & 4 & -3 & 1 \\ & & 12 & 24 & 44 & 96 & 186 \\ \hline & 6 & 12 & 22 & 48 & 93 & 187 \end{array}$$

$(6x^5-2x^3+4x^3-3x+1)\div(x-2)$

$=6x^4+12x^3+22x^2+48x+93+\dfrac{187}{x-2}$

**9.** $\left(x^2-5x-5x^3+x^4\right)\div(5+x)$

Rewrite the polynomials in descending order.

$\left(x^4-5x^3+x^2-5x\right)\div(x+5)$

$$\begin{array}{r|rrrrr} -5 & 1 & -5 & 1 & -5 & 0 \\ & & -5 & 50 & -255 & 1300 \\ \hline & 1 & -10 & 51 & -260 & 1300 \end{array}$$

$\left(x^2-5x-5x^3+x^4\right)\div(5+x)$

$=x^3-10x^2+51x-260+\dfrac{1300}{5+x}$

**11.** $\left(3x^3+2x^2-4x+1\right)\div\left(x-\dfrac{1}{3}\right)$

$$\begin{array}{r|rrrr} \frac{1}{3} & 3 & 2 & -4 & 1 \\ & & 1 & 1 & -1 \\ \hline & 3 & 3 & -3 & 0 \end{array}$$

$\left(3x^3+2x^2-4x-1\right)\div\left(x-\dfrac{1}{3}\right)=3x^2+3x-3$

**13.** $\dfrac{x^5+x^3-2}{x-1}$

$$\begin{array}{r|rrrrrr} 1 & 1 & 0 & 1 & 0 & 0 & -2 \\ & & 1 & 1 & 2 & 2 & 2 \\ \hline & 1 & 1 & 2 & 2 & 2 & 0 \end{array}$$

$\dfrac{x^5+x^3-2}{x-1}=x^4+x^3+2x^2+2x+2$

**15.** $\dfrac{x^4-256}{x-4}$

$$\begin{array}{r|rrrrr} 4 & 1 & 0 & 0 & 0 & -256 \\ & & 4 & 16 & 64 & 256 \\ \hline & 1 & 4 & 16 & 64 & 0 \end{array}$$

$\dfrac{x^4-256}{x-4}=x^3+4x^2+16x+64$

**17.** $\dfrac{2x^5-3x^4+x^3-x^2+2x-1}{x+2}$

$$\begin{array}{r|rrrrrr} -2 & 2 & -3 & 1 & -1 & 2 & -1 \\ & & -4 & 14 & -30 & 62 & -128 \\ \hline & 2 & -7 & 15 & -31 & 64 & -129 \end{array}$$

$\dfrac{2x^5-3x^4+x^3-x^2+2x-1}{x+2}$

$=2x^4-7x^3+15x^2-31x+64-\dfrac{129}{x+2}$

**19.** $f(x)=2x^3-11x^2+7x-5$

$$\begin{array}{r|rrrr} 4 & 2 & -11 & 7 & -5 \\ & & 8 & -12 & -20 \\ \hline & 2 & -3 & -5 & -25 \end{array}$$

$f(4)=-25$

**21.** $f(x)=3x^3-7x^2-2x+5$

$$\begin{array}{r|rrrr} -3 & 3 & -7 & -2 & 5 \\ & & -9 & 48 & -138 \\ \hline & 3 & -16 & 46 & -133 \end{array}$$

$f(-3)=-133$

**23.** $f(x)=x^4+5x^3+5x^2-5x-6$

$$\begin{array}{r|rrrrr} 3 & 1 & 5 & 5 & -5 & -6 \\ & & 3 & 24 & 87 & 246 \\ \hline & 1 & 8 & 29 & 82 & 240 \end{array}$$

$f(3)=240$

**25.** $f(x)=2x^4-5x^3-x^2+3x+2$

$$\begin{array}{r|rrrrr} -\frac{1}{2} & 2 & -5 & -1 & 3 & 2 \\ & & -1 & 3 & -1 & -1 \\ \hline & 2 & -6 & 2 & 2 & 1 \end{array}$$

$f\left(-\frac{1}{2}\right)=1$

**27.** $x^3 - 4x^2 + x + 6 = 0$

$$\begin{array}{r|rrrr} -1 & 1 & -4 & 1 & 6 \\ & & -1 & 5 & -6 \\ \hline & 1 & -5 & 6 & 0 \end{array}$$

The remainder is zero and –1 is a solution to the equation.

$x^3 - 4x^2 + x + 6 = (x+1)(x^2 - 5x + 6)$

To solve the equation, we set it equal to zero and factor.

$$x^3 - 4x^2 + x + 6 = 0$$
$$(x+1)(x^2 - 5x + 6) = 0$$
$$(x+1)(x-3)(x-2) = 0$$

Apply the zero product principle.

$x+1=0 \quad x-3=0 \quad x-2=0$

$x=-1 \quad x=3 \quad x=2$

The solutions are –1, 2 and 3 and the solution set is $\{-1, 2, 3\}$.

**29.** $2x^3 - 5x^2 + x + 2 = 0$

$$\begin{array}{r|rrrr} 2 & 2 & -5 & 1 & 2 \\ & & 4 & -2 & -2 \\ \hline & 2 & -1 & -1 & 0 \end{array}$$

The remainder is zero and 2 is a solution to the equation.

$2x^3 - 5x^2 + x + 2 = (x-2)(2x^2 - x - 1)$ To solve the equation, we set it equal to zero and factor.

$$2x^3 - 5x^2 + x + 2 = 0$$
$$(x-2)(2x^2 - x - 1) = 0$$
$$(x-2)(2x+1)(x-1) = 0$$

Apply the zero product principle.

$x-2=0 \quad 2x+1=0 \quad x-1=0$

$x=2 \quad 2x=-1 \quad x=1$

$x=-\frac{1}{2}$

The solutions are $-\frac{1}{2}$, 1 and 2 and the solution set is $\left\{-\frac{1}{2}, 1, 2\right\}$.

**31.** $6x^3 + 25x^2 - 24x + 5 = 0$

$$\begin{array}{r|rrrr} -5 & 6 & 25 & -24 & 5 \\ & & -30 & 25 & -5 \\ \hline & 6 & -5 & 1 & 0 \end{array}$$

The remainder is zero and 3 is a solution to the equation.

$6x^3 + 25x^2 - 24x + 5$

$= (x+5)(6x^2 - 5x + 1)$

To solve the equation, we set it equal to zero and factor.

$$6x^3 + 25x^2 - 24x + 5 = 0$$
$$(x+5)(6x^2 - 5x + 1) = 0$$
$$(x+5)(3x-1)(2x-1) = 0$$

Apply the zero product principle.

$x+5=0 \quad 3x-1=0 \quad 2x-1=0$

$x=-5 \quad 3x=1 \quad 2x=1$

$x=\frac{1}{3} \quad x=\frac{1}{2}$

The solutions are –5, $\frac{1}{3}$ and $\frac{1}{2}$ and the solution set is $\left\{-5, \frac{1}{3}, \frac{1}{2}\right\}$.

**33.** The graph indicates that 2 is a solution to the equation.

$$\begin{array}{r|rrrr} 2 & 1 & 2 & -5 & -6 \\ & & 2 & 8 & 6 \\ \hline & 1 & 4 & 3 & 0 \end{array}$$

The remainder is 0, so 2 is a solution.

$$x^3+2x^2-5x-6=0$$
$$(x-2)(x^2+4x+3)=0$$
$$(x-2)(x+3)(x+1)=0$$

The solutions are 2, $-3$, and $-1$, or $\{-3,-1,2\}$.

**34.** The graph indicates that $-3$ is a solution to the equation.

$$\begin{array}{r|rrrr} -3 & 2 & 1 & -13 & 6 \\ & & -6 & 15 & -6 \\ \hline & 2 & -5 & 2 & 0 \end{array}$$

The remainder is 0, so $-3$ is a solution.

$$2x^3+x^2-13x+6=0$$
$$(x+3)(2x^2-5x+2)=0$$
$$(x+3)(2x-1)(x-2)=0$$

The solutions are $-3$, $\frac{1}{2}$, and 2, or $\{-3,\frac{1}{2},2\}$.

**35.** The table indicates that 1 is a solution to the equation.

$$\begin{array}{r|rrrr} 1 & 6 & -11 & 6 & -1 \\ & & 6 & -5 & 1 \\ \hline & 6 & -5 & 1 & 0 \end{array}$$

The remainder is 0, so 1 is a solution.

$$6x^3-11x^2+6x-1=0$$
$$(x-1)(6x^2-5x+1)=0$$
$$(x-1)(3x-1)(2x-1)=0$$

The solutions are 1, $\frac{1}{3}$, and $\frac{1}{2}$, or $\left\{\frac{1}{3},\frac{1}{2},1\right\}$.

**36.** The table indicates that 1 is a solution to the equation.

$$\begin{array}{r|rrrr} 1 & 2 & 11 & -7 & -6 \\ & & 2 & 13 & -6 \\ \hline & 2 & 13 & 6 & 0 \end{array}$$

The remainder is 0, so 1 is a solution.

$$2x^3+11x^2-7x-6=0$$
$$(x-1)(2x^2+13x+6)=0$$
$$(x-1)(2x+1)(x+6)=0$$

The solutions are 1, $-\frac{1}{2}$, and $-6$, or $\left\{-6,-\frac{1}{2},1\right\}$.

**37.** $$(22x-24+7x^3-29x^2+4x^4)(x+4)^{-1}$$
$$=\frac{22x-24+7x^3-29x^2+4x^4}{x+4}$$

$$\begin{array}{r|rrrrr} -4 & 4 & 7 & -29 & 22 & -24 \\ & & -16 & 36 & -28 & 24 \\ \hline & 4 & -9 & 7 & -6 & 0 \end{array}$$

Therefore,
$$(22x-24+7x^3-29x^2+4x^4)(x+4)^{-1}$$
$$=4x^3-9x^2+7x-6$$

**38.** $\left(9-x^2+6x+2x^3\right)\left(x+1\right)^{-1}$

$=\dfrac{2x^3-x^2+6x+9}{x+1}$

$$\begin{array}{r|rrrr} -1 & 2 & -1 & 6 & 9 \\ & & -2 & 3 & -9 \\ \hline & 2 & -3 & 9 & 0 \end{array}$$

Therefore,

$\left(9-x^2+6x+2x^3\right)\left(x+1\right)^{-1}$

$=2x^2-3x+9$

**39.** $A=l\cdot w$ so

$l=\dfrac{A}{w}=\dfrac{0.5x^3-0.3x^2+0.22x+0.06}{x+0.2}$

$$\begin{array}{r|rrrr} -0.2 & 0.5 & -0.3 & 0.22 & 0.06 \\ & & -0.1 & 0.08 & -0.06 \\ \hline & 0.5 & -0.4 & 0.3 & 0 \end{array}$$

Therefore, the length of the rectangle is $0.5x^2-0.4x+0.3$ units.

**40.** $A=l\cdot w$ so,

$l=\dfrac{A}{w}=\dfrac{8x^3-6x^2-5x+3}{x+\frac{3}{4}}$

$$\begin{array}{r|rrrr} -\frac{3}{4} & 8 & -6 & -5 & 3 \\ & & -6 & 9 & -3 \\ \hline & 8 & -12 & 4 & 0 \end{array}$$

Therefore, the length of the rectangle is $8x^2-12x+4$ units.

**41.** **a.** $14x^3-17x^2-16x-177=0$

$$\begin{array}{r|rrrr} 3 & 14 & -17 & -16 & -177 \\ & & 42 & 75 & 177 \\ \hline & 14 & 25 & 59 & 0 \end{array}$$

The remainder is 0 so 3 is a solution.

$14x^3-17x^2-16x-177$

$=\left(x-3\right)\left(14x^2+25x+59\right)$

**b.** $f\left(x\right)=14x^3-17x^2-16x+34$

We need to find $x$ when $f\left(x\right)=211$.

$f\left(x\right)=14x^3-17x^2-16x+34$

$211=14x^3-17x^2-16x+34$

$0=14x^3-17x^2-16x-177$

This is the equation obtained in part **a.** One solution is 3. It can be used to find other solutions (if they exist).

$14x^3-17x^2-16x-177=0$

$\left(x-3\right)\left(14x^2+25x+59\right)=0$

The polynomial $14x^2+25x+59$ cannot be factored, so the only solution is $x=3$. The female moth's abdominal width is 3 millimeters.

**43.** Answers will vary.

**45.** Answers will vary.

**47.** Exercise 27:

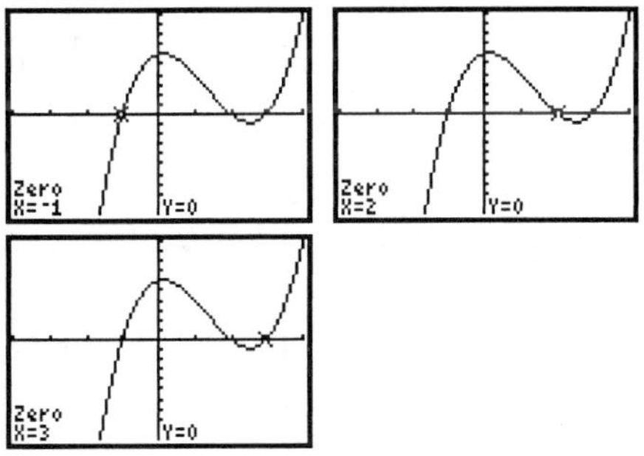

Exercise 28:

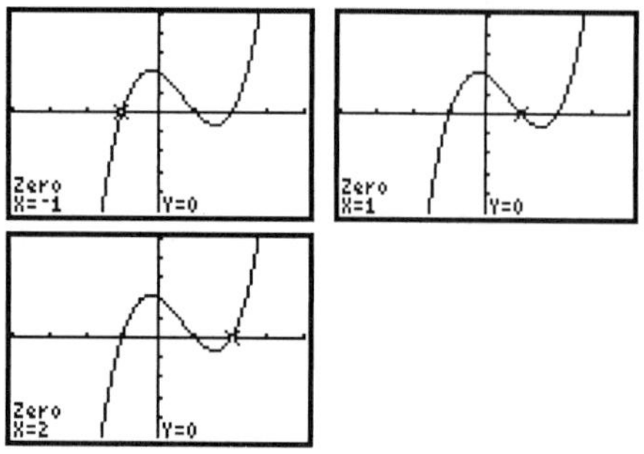


Exercise 29:

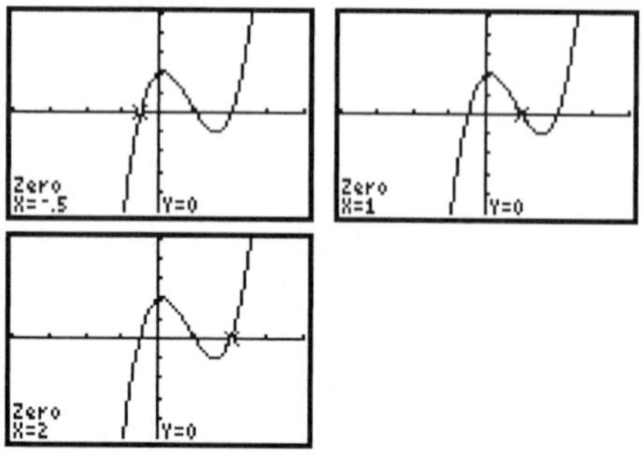


Exercise 30:

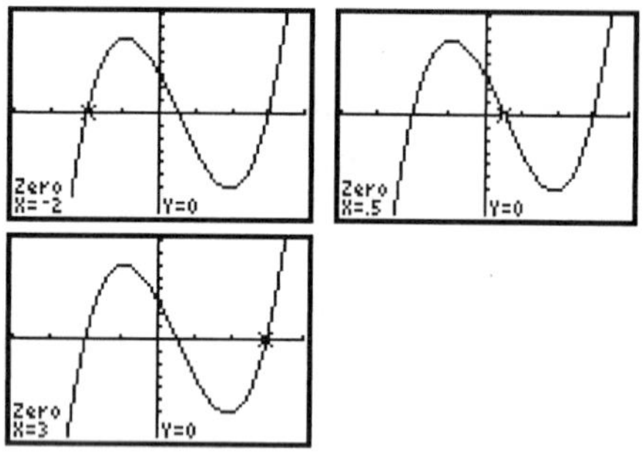


Exercise 31:

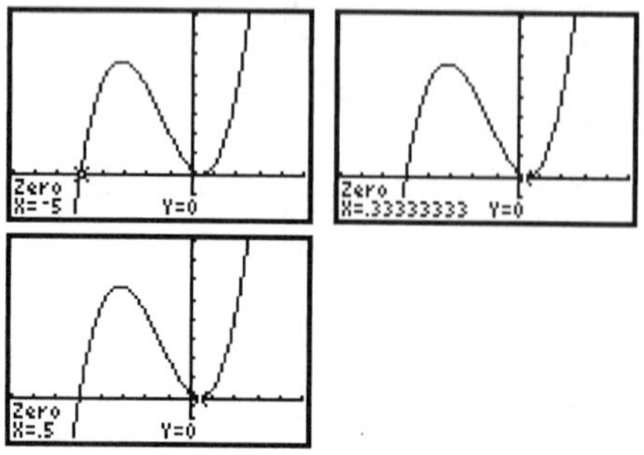


Exercise 32:

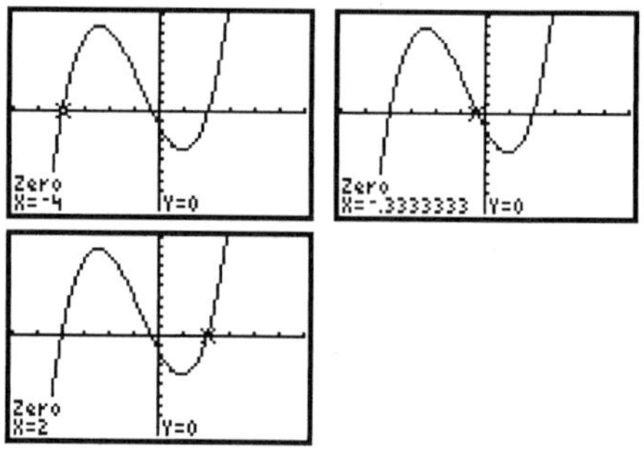


**49.** $x^4 - 4x^3 - 9x^2 + 16x + 20 = 0$

| | | | | | |
|---|---|---|---|---|---|
| 5 | 1 | −4 | −9 | 16 | 20 |
| | | 5 | 5 | −20 | −20 |
| | 1 | 1 | −4 | −4 | 0 |

The remainder is zero and 5 is a solution to the equation.

$x^4 - 4x^3 - 9x^2 + 16x + 20$

$= (x-5)(x^3 + x^2 - 4x - 4)$

To solve the equation, we set it equal to zero and factor.

$$(x-5)(x^3 + x^2 - 4x - 4) = 0$$

$$(x-5)(x^2(x+1) - 4(x+1)) = 0$$

$$(x-5)(x+1)(x^2 - 4) = 0$$

$$(x-5)(x+1)(x+2)(x-2) = 0$$

Apply the zero product principle.

$x - 5 = 0 \qquad x + 1 = 0$

$x = 5 \qquad x = -1$

$x + 2 = 0 \qquad x - 2 = 0$

$x = -2 \qquad x = 2$

The solutions are –2, –1, 2 and 5 and the solution set is $\{-2, -1, 2, 5\}$.

**50.** $4x+3-13x-7<2(3-4x)$

$$-9x-4<6-8x$$
$$-x-4<6$$
$$-x<10$$
$$x>-10$$

The solution set is $\{x|x>-10\}$ or $(-10,\infty)$.

**51.** $2x(x+3)+6(x-3)=-28$

$$2x^2+6x+6x-18=-28$$
$$2x^2+12x-18=-28$$
$$2x^2+12x+10=0$$
$$2(x^2+6x+5)=0$$
$$2(x+5)(x+1)=0$$

Apply the zero product principle

$2(x+5)=0 \qquad x+1=0$

$x+5=0 \qquad x=-1$

$x=-5$

The solutions are –5 and –1 and the solution set is $\{-5,-1\}$.

**52.** $7x-6y=17$

$3x+\ \ y=18$

$$D=\begin{vmatrix}7 & -6\\ 3 & 1\end{vmatrix}=7(1)-3(-6)$$
$$=7+18=25$$
$$D_x=\begin{vmatrix}17 & -6\\ 18 & 1\end{vmatrix}=17(1)-18(-6)$$
$$=17+108=125$$
$$D_y=\begin{vmatrix}7 & 17\\ 3 & 18\end{vmatrix}=7(18)-3(17)$$
$$=126-51=75$$
$$x=\frac{D_x}{D}=\frac{125}{25}=5$$
$$y=\frac{D_y}{D}=\frac{75}{25}=3$$

The solution is $(5,3)$ and the solution set is $\{(5,3)\}$.

## 6.6 Exercise Set

**1.** $\frac{1}{x}+2=\frac{3}{x}$

So that the denominator will not equal zero, $x$ cannot be zero. To eliminate fractions, multiply by the LCD, $x$.

$$x\left(\frac{1}{x}+2\right)=x\left(\frac{3}{x}\right)$$

$$x\cdot\frac{1}{x}+x\cdot 2=3$$

$$1+2x=3$$

$$2x=2$$

$$x=1$$

The solution is 1 and the solution set is $\{1\}$.

**3.** $\frac{5}{x}+\frac{1}{3}=\frac{6}{x}$

So that the denominator will not equal zero, $x$ cannot equal 0. To eliminate fractions, multiply by the LCD, $3x$.

$$3x\left(\frac{5}{x}+\frac{1}{3}\right)=3x\left(\frac{6}{x}\right)$$

$$3x\cdot\frac{5}{x}+3x\cdot\frac{1}{3}=3\cdot 6$$

$$3\cdot 5+x\cdot 1=18$$

$$15+x=18$$

$$x=3$$

The solution is 3 and the solution set is $\{3\}$.

**5.** $\frac{x-2}{2x}+1=\frac{x+1}{x}$

So that the denominator will not equal zero, $x$ cannot equal 0. To eliminate fractions, multiply by the LCD, $2x$.

$$2x\left(\frac{x-2}{2x}+1\right)=2x\left(\frac{x+1}{x}\right)$$

$$2x\cdot\frac{x-2}{2x}+2x\cdot 1=2(x+1)$$

$$x-2+2x=2x+2$$

$$-2+3x=2x+2$$

$$-2+x=2$$

$$x=4$$

The solution is 4 and the solution set is $\{4\}$.

**7.** $\frac{3}{x+1}=\frac{5}{x-1}$

So that the denominator will not equal zero, $x$ cannot equal 1 or –1. To eliminate fractions, multiply by the LCD, $(x+1)(x-1)$.

$$(x+1)(x-1)\frac{3}{x+1}=(x+1)(x-1)\frac{5}{x-1}$$

$$(x-1)\cdot 3=(x+1)\cdot 5$$

$$3x-3=5x+5$$

$$-3=2x+5$$

$$-8=2x$$

$$-4=x$$

The solution is –4 and the solution set is $\{-4\}$.

**9.** $\frac{x-6}{x+5}=\frac{x-3}{x+1}$

So that the denominator will not equal zero, $x$ cannot equal –5 or –1. To eliminate fractions, multiply by the LCD, $(x+5)(x+1)$.

$$(x+5)(x+1)\frac{x-6}{x+5}=(x+5)(x+1)\frac{x-3}{x+1}$$
$$(x+1)(x-6)=(x+5)(x-3)$$
$$\cancel{x^2}-5x-6=\cancel{x^2}+2x-15$$
$$-5x-6=2x-15$$
$$-7x-6=-15$$
$$-7x=-9$$
$$x=\frac{9}{7}$$

The solution is $\frac{9}{7}$ and the solution set is $\left\{\frac{9}{7}\right\}$.

**11.** $\frac{x+6}{x+3}=\frac{3}{x+3}+2$

So that the denominator will not equal zero, $x$ cannot equal –3. To eliminate fractions, multiply by the LCD, $x+3$.

$$(x+3)\left(\frac{x+6}{x+3}\right)=(x+3)\left(\frac{3}{x+3}+2\right)$$
$$x+6=3+2x+6$$
$$x+6=2x+9$$
$$6=x+9$$
$$-3=x$$

Since –3 would result in a zero denominator, we disregard it and conclude that the solution set is $\varnothing$ or $\{\ \}$.

**13.** $1-\frac{4}{x+7}=\frac{5}{x+7}$

So that the denominator will not equal zero, $x$ cannot equal –7. To eliminate fractions, multiply by the LCD, $x+7$.

$$(x+7)\left(1-\frac{4}{x+7}\right)=(x+7)\left(\frac{5}{x+7}\right)$$
$$(x+7)\cdot 1-(x+7)\cdot\frac{4}{x+7}=5$$
$$x+7-4=5$$
$$x+3=5$$
$$x=2$$

The solution is 2 and the solution set is $\{2\}$.

**15.** $\frac{4x}{x+2}+\frac{2}{x-1}=4$

So that the denominator will not equal zero, $x$ cannot equal –2 or 1. To eliminate fractions, multiply by the LCD, $(x+2)(x-1)$.

$$(x+2)(x-1)\cdot\left(\frac{4x}{x+2}+\frac{2}{x-1}\right)=(x+2)(x-1)\cdot 4$$
$$(x-1)4x+(x+2)2=(x^2+x-2)4$$
$$4x^2-4x+2x+4=4x^2+4x-8$$
$$\cancel{4x^2}-2x+4=\cancel{4x^2}+4x-8$$
$$-2x+4=4x-8$$
$$-6x+4=-8$$
$$-6x=-12$$
$$x=2$$

The solution is 2 and the solution set is $\{2\}$.

**17.** $\frac{8}{x^2-9}+\frac{4}{x+3}=\frac{2}{x-3}$

$\frac{8}{(x+3)(x-3)}+\frac{4}{x+3}=\frac{2}{x-3}$

So that the denominator will not equal zero, $x$ cannot equal –3 or 3. To eliminate fractions, multiply by the LCD, $(x+3)(x-3)$.

$$(x+3)(x-3)\cdot\left(\frac{8}{(x+3)(x-3)}+\frac{4}{x+3}\right)=(x+3)(x-3)\cdot\frac{2}{x-3}$$
$$8+4(x-3)=2(x+3)$$
$$8+4x-12=2x+6$$
$$4x-4=2x+6$$
$$2x-4=6$$
$$2x=10$$
$$x=5$$

The solution is 5 and the solution set is $\{5\}$.

**19.** $x+\frac{7}{x}=-8$

So that the denominator will not equal zero, $x$ cannot equal 0. To eliminate fractions, multiply by the LCD, $x$.

$$x\left(x+\frac{7}{x}\right)=x(-8)$$

$$x^2+7=-8x$$

$$x^2+8x+7=0$$

$$(x+7)(x+1)=0$$

Apply the zero product principle.

$x+7=0 \qquad x+1=0$

$x=-7 \qquad x=-1$

The solutions are –7 and –1 and the solution set is $\{-7,-1\}$.

**21.** $\frac{6}{x}-\frac{x}{3}=1$

So that the denominator will not equal zero, $x$ cannot equal 0. To eliminate fractions, multiply by the LCD, $3x$.

$$3x\left(\frac{6}{x}-\frac{x}{3}\right)=3x(1)$$

$$3x\left(\frac{6}{x}\right)-3x\left(\frac{x}{3}\right)=3x$$

$$3(6)-x(x)=3x$$

$$18-x^2=3x$$

$$0=x^2+3x-18$$

$$0=(x+6)(x-3)$$

Apply the zero product principle.

$x+6=0 \qquad x-3=0$

$x=-6 \qquad x=3$

The solutions are –6 and 3 and the solution set is $\{-6,3\}$.

**23.** $\frac{x+6}{3x-12}=\frac{5}{x-4}+\frac{2}{3}$

$\frac{x+6}{3(x-4)}=\frac{5}{x-4}+\frac{2}{3}$

So that the denominator will not equal zero, $x$ cannot equal 4. To eliminate fractions, multiply by the LCD, $3(x-4)$.

$$3(x-4)\left(\frac{x+6}{3(x-4)}\right)=3(x-4)\left(\frac{5}{x-4}+\frac{2}{3}\right)$$

$$x+6=3(x-4)\cdot\frac{5}{x-4}+3(x-4)\cdot\frac{2}{3}$$

$$x+6=3\cdot5+2(x-4)$$

$$x+6=15+2x-8$$

$$x+6=7+2x$$

$$6=7+x$$

$$-1=x$$

The solution is $-1$ and the solution set is $\{-1\}$.

**25.** $\frac{1}{x-1}+\frac{1}{x+1}=\frac{2}{x^2-1}$

$\frac{1}{x-1}+\frac{1}{x+1}=\frac{2}{(x+1)(x-1)}$

So that the denominator will not equal zero, $x$ cannot equal $-1$ or 1. To eliminate fractions, multiply by the LCD, $(x+1)(x-1)$.

$$(x+1)(x-1)\left(\frac{1}{x-1}+\frac{1}{x+1}\right)=(x+1)(x-1)\cdot\frac{2}{(x+1)(x-1)}$$

$$(x+1)+(x-1)=2$$

$$x+1+x-1=2$$

$$x+x=2$$

$$2x=2$$

$$x=1$$

Since 1 would result in a zero denominator, we disregard it and conclude that the solution set is $\varnothing$ or $\{\ \}$.

**27.** $\frac{5}{x+4}+\frac{3}{x+3}=\frac{12x+19}{x^2+7x+12}$

$\frac{5}{x+4}+\frac{3}{x+3}=\frac{12x+19}{(x+4)(x+3)}$

So that the denominator will not equal zero, $x$ cannot equal –4 or –3. To eliminate fractions, multiply by the LCD, $(x+4)(x+3)$.

$$(x+4)(x+3)\left(\frac{5}{x+4}+\frac{3}{x+3}\right)=(x+4)(x+3)\left(\frac{12x+19}{(x+4)(x+3)}\right)$$

$$(x+4)(x+3)\cdot\left(\frac{5}{x+4}\right)+(x+4)(x+3)\cdot\left(\frac{3}{x+3}\right)=12x+19$$

$$(x+3)\cdot 5+(x+4)\cdot 3=12x+19$$

$$5x+15+3x+12=12x+19$$

$$8x+27=12x+19$$

$$-4x+27=19$$

$$-4x=-8$$

$$x=2$$

The solution is 2 and the solution set is $\{2\}$.

**29.** $\frac{4x}{x+3}-\frac{12}{x-3}=\frac{4x^2+36}{x^2-9}$

$\frac{4x}{x+3}-\frac{12}{x-3}=\frac{4x^2+36}{(x+3)(x-3)}$

So that the denominator will not equal zero, $x$ cannot equal –3 or 3. To eliminate fractions, multiply by the LCD, $(x+3)(x-3)$.

$$(x+3)(x-3)\left(\frac{4x}{x+3}-\frac{12}{x-3}\right)=(x+3)(x-3)\left(\frac{4x^2+36}{(x+3)(x-3)}\right)$$

$$(x+3)(x-3)\left(\frac{4x}{x+3}\right)-(x+3)(x-3)\left(\frac{12}{x-3}\right)=4x^2+36$$

$$(x-3)(4x)-(x+3)(12)=4x^2+36$$

$$4x^2-12x-(12x+36)=4x^2+36$$

$$\cancel{4x^2}-12x-12x-36=\cancel{4x^2}+36$$

$$-24x-36=36$$

$$-24x=72$$

$$x=-3$$

Since –3 would result in a zero denominator, we disregard it and conclude that the solution set is $\varnothing$ or $\{\ \}$.

**31.**

$$\frac{4}{x^2+3x-10}+\frac{1}{x^2+9x+20}=\frac{2}{x^2+2x-8}$$

$$\frac{4}{(x+5)(x-2)}+\frac{1}{(x+5)(x+4)}=\frac{2}{(x+4)(x-2)}$$

So that the denominator will not equal zero, $x$ cannot equal –5, –4, or 2. To eliminate fractions, multiply by the LCD, $(x+5)(x+4)(x-2)$.

$$(x+5)(x+4)(x-2)\left(\frac{4}{(x+5)(x-2)}+\frac{1}{(x+5)(x+4)}\right)=(x+5)(x+4)(x-2)\left(\frac{2}{(x+4)(x-2)}\right)$$

$$(x+4)\cdot 4+(x-2)\cdot 1=(x+5)\cdot 2$$

$$4x+16+x-2=2x+10$$

$$5x+14=2x+10$$

$$3x+14=10$$

$$3x=-4$$

$$x=-\frac{4}{3}$$

The solution is $-\frac{4}{3}$ and the solution set is $\left\{-\frac{4}{3}\right\}$.

**33.**

$$\frac{3y}{y^2+5y+6}+\frac{2}{y^2+y-2}=\frac{5y}{y^2+2y-3}$$

$$\frac{3y}{(y+2)(y+3)}+\frac{2}{(y+2)(y-1)}=\frac{5y}{(y+3)(y-1)}$$

So that the denominator will not equal zero, $x$ cannot equal $-3, -2,$ or $1$. To eliminate fractions, multiply by the LCD $(y+2)(y+3)(y-1)$.

$$(y+2)(y+3)(y-1)\left(\frac{3y}{(y+2)(y+3)}+\frac{2}{(y+2)(y-1)}\right)+=(y+2)(y+3)(y-1)\frac{5y}{(y+3)(y-1)}$$

$$3y(y-1)+2(y+3)=5y(y+2)$$

$$3y^2-3y+2y+6=5y^2+10y+10y$$

$$0=2y^2+11y-6$$

$$0=(2y-1)(y+6)$$

$$y=\frac{1}{2} \text{ or } y=-6$$

The solution set is $\left\{-6, \frac{1}{2}\right\}$.

**35.** $g(x)=\frac{x}{2}+\frac{20}{x}$; Since $g(a)=7$, then

$$7=\frac{a}{2}+\frac{20}{a}$$
$$(2a)7=(2a)\left(\frac{a}{2}+\frac{20}{a}\right)$$
$$14a=a^2+40$$
$$a^2-14a+40=0$$
$$(a-10)(a-4)=0$$
$$a=10 \text{ or } a=4$$

**37.** $g(x)=\frac{5}{x+2}+\frac{25}{x^2+4x+4}$;

$$=\frac{5}{x+2}+\frac{25}{(x+2)^2}$$

Since $g(a)=20$ then

$$20=\frac{5}{a+2}+\frac{25}{(a+2)^2}$$
$$20(a+2)^2=(a+2)^2\left(\frac{5}{a+2}+\frac{25}{(a+2)^2}\right)$$
$$20(a+2)^2=5(a+2)+25$$
$$20(a^2+4a+4)=5a+10+25$$
$$20a^2+80a+80=5a+35$$
$$20a^2+75a+45=0$$
$$5(4a^2+15a+9)=0$$
$$5(4a+3)(a+3)=0$$
$$a=-\frac{3}{4} \text{ or } a=-3$$

**39.**

$$\frac{x+2}{x^2-x}-\frac{6}{x^2-1}$$
$$=\frac{x+2}{x(x-1)}-\frac{6}{(x-1)(x+1)}$$
$$=\frac{(x+1)}{(x+1)}\frac{x+2}{x(x-1)}-\frac{x}{x}\cdot\frac{6}{(x-1)(x+1)}$$
$$=\frac{(x+1)(x+2)}{x(x-1)(x+1)}-\frac{6x}{x(x-1)(x+1)}$$
$$=\frac{(x+1)(x+2)-6x}{x(x-1)(x+1)}$$
$$=\frac{x^2+3x+2-6x}{x(x-1)(x+1)}$$
$$=\frac{x^2-3x+2}{x(x-1)(x+1)}$$
$$=\frac{(x-2)(x-1)}{x(x-1)(x+1)}$$
$$=\frac{x-2}{x(x+1)}$$

**40.**

$$\frac{x+3}{x^2-x}-\frac{8}{x^2-1}$$
$$=\frac{x+3}{x(x-1)}-\frac{8}{(x-1)(x+1)}$$
$$=\frac{(x+1)}{(x+1)}\frac{x+3}{x(x-1)}-\frac{x}{x}\frac{8}{(x-1)(x+1)}$$
$$=\frac{(x+1)(x+3)}{x(x-1)(x+1)}-\frac{8x}{x(x-1)(x+1)}$$
$$=\frac{(x+1)(x+3)-8x}{x(x-1)(x+1)}$$
$$=\frac{x^2+4x+3-8x}{x(x-1)(x+1)}$$
$$=\frac{x^2-4x+3}{x(x-1)(x+1)}$$
$$=\frac{(x-3)(x-1)}{x(x-1)(x+1)}$$
$$=\frac{x-3}{x(x+1)}$$

**41.** In Exercise 39, the left side of this equation was simplified;

$$\frac{x+2}{x^2-x}-\frac{6}{x^2-1}=0$$

$$\frac{x-2}{x(x+1)}=0$$

$$x(x+1)\frac{x-2}{x(x+1)}=0\cdot x(x+1)$$

$$x-2=0$$

$$x=2$$

The solution is 2 or $\{2\}$.

**42.** In Exercise 40, the left side of this equation was simplified;

$$\frac{x-3}{x(x+1)}=0$$

$$x(x+1)\frac{x-3}{x(x+1)}=0$$

$$x-3=0$$

$$x=3$$

The solution is 3 or $\{3\}$.

**43.**

$$\frac{1}{x^3-8}+\frac{3}{(x-2)(x^2+2x+4)}=\frac{2}{x^2+2x+4}$$

$$\frac{1}{x^3-8}+\frac{3}{(x-2)(x^2+2x+4)}-\frac{2}{x^2+2x+4}=0$$

$$\frac{1}{(x-2)(x^2+2x+4)}+\frac{3}{(x-2)(x^2+2x+4)}-\frac{2}{x^2+2x+4}=0$$

$$\frac{4}{(x-2)(x^2+2x+4)}-\frac{(x-2)}{(x-2)}\frac{2}{x^2+2x+4}=0$$

$$\frac{4-2(x-2)}{(x-2)\left(x^2+2x+4\right)}=0$$

$$\frac{4-2x+4}{(x-2)\left(x^2+2x+4\right)}=0$$

$$\frac{-2x+8}{(x-2)(x^2+2x+4)}=0$$

$$(x-2)\left(x^2+2x+4\right)\frac{-2x+8}{(x-2)\left(x^2+2x+4\right)}=0(x-2)\left(x^2+2x+4\right)$$

$$-2x+8=0$$

$$-2x=-8$$

$$x=4$$

The solution is 4 or $\{4\}$.

**44.**

$$\frac{2}{x^3-1}+\frac{4}{(x-1)(x^2+x+1)}=-\frac{1}{x^2+x+1}$$
$$\frac{2}{x^3-1}+\frac{4}{(x-1)(x^2+x+1)}+\frac{1}{x^2+x+1}=0$$
$$\frac{2}{(x-1)(x^2+x+1)}+\frac{4}{(x-1)(x^2+x+1)}+\frac{1}{x^2+x+1}=0$$
$$\frac{6}{(x-1)(x^2+x+1)}+\frac{(x-1)}{(x-1)}\frac{1}{x^2+x+1}=0$$
$$\frac{6+(x-1)}{(x-1)\left(x^2+x+1\right)}=0$$
$$\frac{x+5}{(x-1)\left(x^2+x+1\right)}=0$$
$$\frac{x+5}{(x-1)(x^2+x+1)}=0$$
$$(x-1)\left(x^2+x+1\right)\frac{x+5}{(x-1)\left(x^2+x+1\right)}=0(x-1)\left(x^2+x+1\right)$$
$$x+5=0$$
$$x=-5$$

The solution is $-5$ or $\{-5\}$.

**45.**

$$\frac{1}{x^3-8}+\frac{3}{(x-2)(x^2+2x+4)}-\frac{2}{x^2+2x+4}$$
$$=\frac{1}{(x-2)(x^2+2x+4)}+\frac{3}{(x-2)(x^2+2x+4)}-\frac{2}{x^2+2x+4}$$
$$=\frac{4}{(x-2)(x^2+2x+4)}-\frac{(x-2)}{(x-2)}\frac{2}{x^2+2x+4}$$
$$=\frac{4-2(x-2)}{(x-2)\left(x^2+2x+4\right)}=\frac{4-2x+4}{(x-2)\left(x^2+2x+4\right)}$$
$$=\frac{-2x+8}{(x-2)(x^2+2x+4)}=(x-2)\left(x^2+2x+4\right)\frac{-2x+8}{(x-2)\left(x^2+2x+4\right)}$$
$$=\frac{-2x+8}{(x-2)\left(x^2+2x+4\right)}=\frac{-2(x-4)}{(x-2)\left(x^2+2x+4\right)}$$

**46.**

$$\frac{2}{x^3-1}+\frac{4}{(x-1)(x^2+x+1)}+\frac{1}{x^2+x+1}=\frac{2}{(x-1)(x^2+x+1)}+\frac{4}{(x-1)(x^2+x+1)}+\frac{1}{x^2+x+1}$$

$$=\frac{6}{(x-1)(x^2+x+1)}+\frac{(x-1)}{(x-1)}\frac{1}{x^2+x+1}$$

$$=\frac{6+(x-1)}{(x-1)\left(x^2+x+1\right)}$$

$$\frac{x+5}{(x-1)\left(x^2+x+1\right)}$$

**47.** First find $f(a)$ and $g(a)$;

$$f(a)=\frac{a+2}{a+3};\quad g(a)=\frac{a+1}{a^2+2a-3}=\frac{a+1}{(a+3)(a-1)}$$

Then

$$f(a)=g(a)+1$$

$$\frac{a+2}{a+3}=\frac{a+1}{(a+3)(a-1)}+1$$

$$\frac{a+2}{a+3}-\frac{a+1}{(a+3)(a-1)}-1=0$$

$$(a+3)(a-1)\left(\frac{a+2}{a+3}-\frac{a+1}{(a+3)(a-1)}-1\right)=0\left(a+3\right)\left(a-1\right)$$

$$(a-1)(a+2)-(a+1)-(a+3)(a-1)=0$$

$$(a^2-a-2)-a-1-\left(a^2+2a-3\right)=0$$

$$a^2-2a-3-a^2-2a+3=0$$

$$-4a=0$$

$$a=0$$

**48.** First find $f(a)$ and $g(a)$;

$$f(a)=\frac{4}{a-3};\ g(a)=\frac{10}{a^2+a-12}=\frac{10}{(a+4)(a-3)}$$

Then

$$f(a)=g(a)+1$$

$$\frac{4}{a-3}=\frac{10}{(a+4)(a-3)}+1$$

$$\frac{4}{a-3}-\frac{10}{(a+4)(a-3)}-1=0$$

$$(a-3)(a+4)\left(\frac{4}{a-3}-\frac{10}{(a+4)(a-3)}-1\right)=0(a-3)(a+4)$$

$$4(a+4)-10-1(a-3)(a+4)=0$$

$$4a+16-10-\left(a^2+a-12\right)=0$$

$$4a+6-a^2-a+12=0$$

$$-a^2+3a+18=0$$

$$-(a^2-3a-18)=0$$

$$-(a-6)(a+3)=0$$

$$a=-3 \text{ or } a=6$$

**49.** First find $(f+g)(a)$ and $h(a)$;

$$\begin{aligned}(f+g)(a)&=f(a)+g(a)\\ &=\frac{5}{a-4}+\frac{3}{a-3}\\ &=\frac{(a-3)}{(a-3)}\cdot\frac{5}{a-4}+\frac{(a-4)}{(a-4)}\cdot\frac{3}{a-3}\\ &=\frac{5(a-3)+3(a-4)}{(a-3)(a-4)}\\ &=\frac{5a-15+3a-12}{(a-3)(a-4)}\\ &=\frac{8a-27}{(a-3)(a-4)}\end{aligned}$$

$$h(a)=\frac{a^2-20}{a^2-7x+12}=\frac{a^2-20}{(a-4)(a-3)}$$

$$\frac{8a-27}{(a-3)(a-4)}=\frac{a^2-20}{(a-4)(a-3)}$$

$$(a-3)(a-4)\frac{8a-27}{(a-3)(a-4)}=(a-3)(a-4)\frac{a^2-20}{(a-4)(a-3)}$$

$$8a-27=a^2-20$$

$$a^2-8a+7=0$$

$$(a-1)(a-7)=0$$

$$a=1 \text{ or } a=7$$

**50.** First find $(f+g)(a)$ and $h(a)$;

$$\begin{aligned}(f+g)(a)&=f(a)+g(a)\\&=\frac{6}{a+3}+\frac{2a}{a-3}\\&=\frac{(a-3)}{(a-3)}\cdot\frac{6}{a+3}+\frac{(a+3)}{(a+3)}\cdot\frac{2a}{a-3}\\&=\frac{6(a-3)+2a(a+3)}{(a-3)(a+3)}\\&=\frac{6a-18+2a^2+6a}{(a-3)(a+3)}\\&=\frac{2a^2+12a-18}{(a-3)(a+3)}\\&=\frac{2(a^2+6a-9)}{(a-3)(a+3)}\end{aligned}$$

$$h(a)=-\frac{28}{a^2-9}=-\frac{28}{(a-3)(a+3)}$$

$$\frac{2(a^2+6a-9)}{(a-3)(a+3)} = -\frac{28}{(a-3)(a+3)}$$

$$(a-3)(a+3)\frac{2(a^2+6a-9)}{(a-3)(a+3)} = (a-3)(a+3)\cdot -\frac{28}{(a-3)(a+3)}$$

$$2(a^2+6a-9) = -28$$

$$2a^2+12a-18 = -28$$

$$2a^2+12a+10 = 0$$

$$2(a^2+6a+5) = 0$$

$$2(a+1)(a+5) = 0$$

$$a = -1 \text{ or } a = -5$$

**51.**

$$f(x) = \frac{250x}{100-x}$$

$$375 = \frac{250x}{100-x}$$

$$(100-x)(375) = (100-x)\left(\frac{250x}{100-x}\right)$$

$$37500 - 375x = 250x$$

$$37500 = 625x$$

$$60 = x$$

If the government commits \$375 million for the project, 60% of pollutants can be removed.

**53.**

$$f(x) = \frac{5x+30}{x}$$

$$8 = \frac{5x+30}{x}$$

$$x(8) = x\left(\frac{5x+30}{x}\right)$$

$$8x = 5x+30$$

$$3x = 30$$

$$x = 10$$

The students will remember an average of 8 words after 10 days. This is shown on the graph as the point (10, 8).

**55.** The horizontal asymptote is $y = 5$. The means that on average, the students will remember more than 5 words over an extended period of time.

**57.**

$$f(x) = \frac{0.9x-0.4}{0.9x+0.1}$$

$$0.95 = \frac{0.9x-0.4}{0.9x+0.1}$$

$$\frac{0.95}{1} = \frac{0.9x-0.4}{0.9x+0.1}$$

$$0.95(0.9x+0.1) = 0.9x-0.4$$

$$0.855x+0.095 = 0.9x-0.4$$

$$0.095 = 0.045x-0.4$$

$$0.495 = 0.045x$$

$$11 = x$$

It will take 11 learning trials for 0.95 or 95% of the responses to be correct. This is shown on the graph as the point (11, 0.95).

**59.** As the number of learning trials increases, the proportion of correct responses increases. Initially the proportion of correct responses increases rapidly, but slows down over time.

**61.**
$$C(x) = \frac{x + 0.1(500)}{x + 500}$$
$$0.28 = \frac{x + 0.1(500)}{x + 500}$$
$$0.28 = \frac{x + 50}{x + 500}$$
$$0.28(x + 500) = (x + 500)\frac{x + 50}{x + 500}$$
$$0.28x + 140 = x + 50$$
$$0.72x = 90$$
$$x = 125$$
125 liters of pure peroxide should be added to produce a new product that is 28% peroxide.

**In Exercises 63-69, answers will vary.**

**71.** $\frac{50}{x} = 2x$

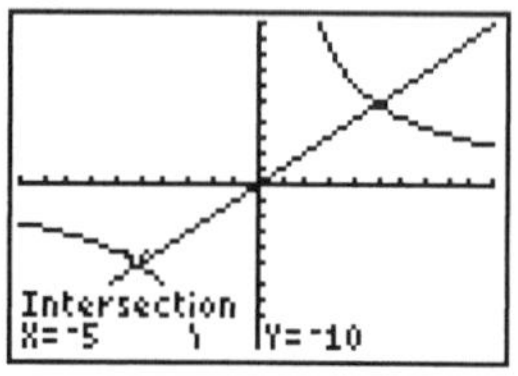


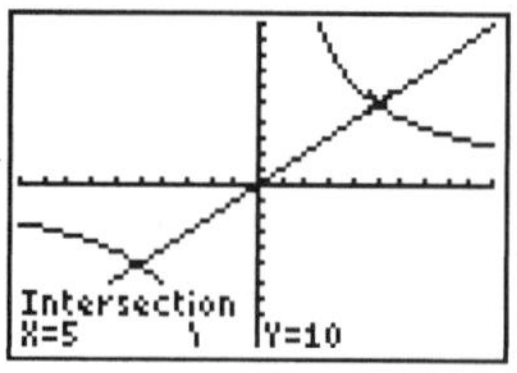


Check the solutions by substituting –5 and 5 in the original equation.

$x = 5$      $x = -5$

$\frac{50}{5} = 2(5)$      $\frac{50}{-5} = 2(-5)$

$10 = 10$      $-10 = -10$

The solutions check. The solutions are –5 and 5 and the solution set is $\{-5, 5\}$.

**73.** $\frac{2}{x} = x + 1$

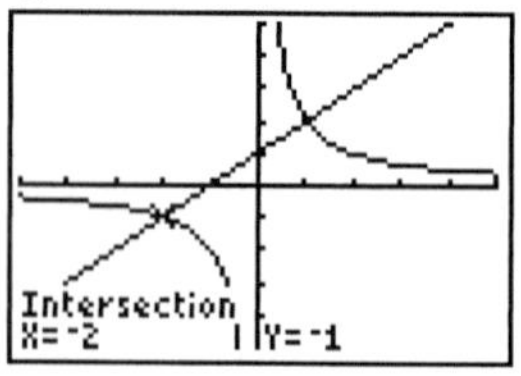


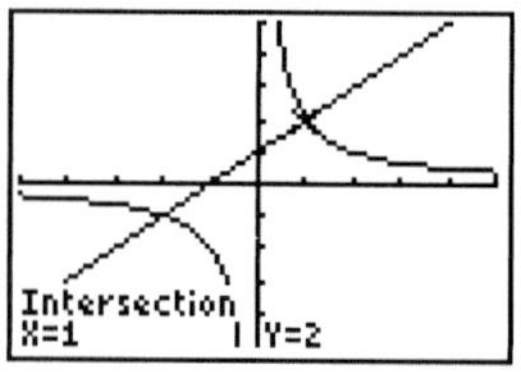


Check the solutions by substituting –2 and 1 in the original equation.

$x = -2$      $x = 1$

$\frac{2}{-2} = -2 + 1$      $\frac{2}{1} = 1 + 1$

$-1 = -1$      $2 = 2$

The solutions check. The solutions are –2 and 1 and the solution set is $\{-2, 1\}$.

**75.** Statement **d.** is true.

Statement **a.** is false. Not all of the solutions of the resulting equation are solutions of the rational equation. Solutions which will result in a denominator of zero must be discarded.

Statement **b.** is false. The only time you can multiply by the LCD to clear fractions is in an equation. The expression given here does not have an equal sign and is not an equation.

Statement **c.** is false. Zero does not satisfy the equation. Zero results in a denominator of zero.

**77.** $\left|\frac{x+1}{x+8}\right| = \frac{2}{3}$

The solutions are found by solving the equations $\frac{x+1}{x+8} = \frac{2}{3}$ and $\frac{x+1}{x+8} = -\frac{2}{3}$:

$$\frac{x+1}{x+8} = \frac{2}{3}$$
$$(x+8)\frac{x+1}{x+8} = \frac{2}{3}(x+8)$$
$$x+1 = \frac{2}{3}x + \frac{16}{3}$$
$$\frac{1}{3}x = \frac{13}{3}$$
$$(3)\frac{1}{3}x = (3)\frac{13}{3}$$
$$x = 13$$

$$\frac{x+1}{x+8} = -\frac{2}{3}$$
$$(x+8)\frac{x+1}{x+8} = -\frac{2}{3}(x+8)$$
$$x+1 = -\frac{2}{3}x - \frac{16}{3}$$
$$\frac{5}{3}x = -\frac{19}{3}$$
$$\left(\frac{3}{5}\right)\frac{5}{3}x = \left(\frac{3}{5}\right)\left(-\frac{19}{5}\right)$$
$$x = -\frac{19}{5}$$

The solution is $-\frac{19}{5}$ and 13 or $\left\{-\frac{19}{5}, 13\right\}$.

**79.**
$$\left(\frac{4}{x-1}\right)^2 + 2\left(\frac{4}{x-1}\right) + 1 = 0$$
$$\frac{4^2}{(x-1)^2} + \frac{8}{x-1} + 1 = 0$$
$$\frac{16}{(x-1)^2} + \frac{8}{x-1} = -1$$
$$(x-1)^2\left(\frac{16}{(x-1)^2} + \frac{8}{x-1}\right) = -1(x-1)^2$$
$$16 + 8(x-1) = -(x^2 - 2x + 1)$$
$$16 + 8x - 8 = -x^2 + 2x - 1$$
$$x^2 + 6x + 9 = 0$$
$$(x+3)(x+3) = 0$$
$$x = -3$$

The solution is $-3$ or $\{-3\}$.

**80.** $x+2y\geq 2$

$x-y\geq -4$

First consider $x+2y\geq 2$. If we solve for $y$ in $x+2y=2$, we can graph using the slope and the $y$-intercept.

$$x+2y=2$$
$$2y=-x+2$$
$$y=-\frac{1}{2}x+1$$

$y$-intercept = 1    slope = $-\frac{1}{2}$

Now, use the origin, $(0,0)$, as a test point.

$$x+2y\geq 2$$
$$0+2(0)\geq 2$$
$$0\geq 2$$

This is a false statement. This means that the point $(0,0)$ will not fall in the shaded half-plane.

Next consider $x-y\geq -4$. If we solve for $y$ in $x-y=-4$, we can graph using the slope and the $y$-intercept.

$$x-y=-4$$
$$-y=-x-4$$
$$y=x+4$$

$y$-intercept = 4    slope = 1

Now, use the origin, $(0,0)$, as a test point.

$$x-y\geq -4$$
$$0-0\geq -4$$
$$0\geq -4$$

This is a true statement. This means that the point $(0,0)$ will fall in the shaded half-plane.

Next, graph each of the inequalities. The solution to the system is the intersection of the shaded half-planes.

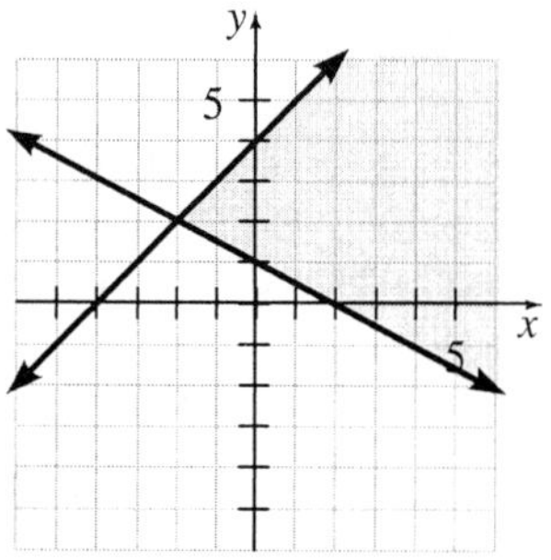


**81.** $$\frac{x-4}{2}-\frac{1}{5}=\frac{7x+1}{20}$$
$$20\left(\frac{x-4}{2}-\frac{1}{5}\right)=20\left(\frac{7x+1}{20}\right)$$
$$20\left(\frac{x-4}{2}\right)-20\left(\frac{1}{5}\right)=7x+1$$
$$10(x-4)-4(1)=7x+1$$
$$10x-40-4=7x+1$$
$$10x-44=7x+1$$
$$3x-44=1$$
$$3x=45$$
$$x=15$$

The solution is 15 and the solution set is $\{15\}$.

**82.** $$C=\frac{5F-160}{9}$$
$$9C=(9)\frac{5F-160}{9}$$
$$9C=5F-160$$
$$5F=9C+160$$
$$\frac{5F}{5}=\frac{9C+160}{5}$$
$$F=\frac{9C+160}{5}$$

## 6.7 Exercise Set

**1.**

$$\frac{V_1}{V_2} = \frac{P_2}{P_1}$$

$$P_1V_2\left(\frac{V_1}{V_2}\right) = P_1V_2\left(\frac{P_2}{P_1}\right)$$

$$P_1(V_1) = V_2(P_2)$$

$$P_1V_1 = P_2V_2$$

$$\frac{P_1V_1}{V_1} = \frac{P_2V_2}{V_1}$$

$$P_1 = \frac{P_2V_2}{V_1}$$

**3.**

$$\frac{1}{p} + \frac{1}{q} = \frac{1}{f}$$

$$fpq\left(\frac{1}{p} + \frac{1}{q}\right) = fpq\left(\frac{1}{f}\right)$$

$$fpq\left(\frac{1}{p}\right) + fpq\left(\frac{1}{q}\right) = pq$$

$$fq + fp = pq$$

$$f(q+p) = pq$$

$$\frac{f(q+p)}{q+p} = \frac{pq}{q+p}$$

$$f = \frac{pq}{q+p}$$

**5.**

$$P = \frac{A}{1+r}$$

$$(1+r)(P) = (1+r)\left(\frac{A}{1+r}\right)$$

$$P + Pr = A$$

$$Pr = A - P$$

$$\frac{Pr}{P} = \frac{A-P}{P}$$

$$r = \frac{A-P}{P}$$

**7.**

$$F = \frac{Gm_1m_2}{d^2}$$

$$d^2(F) = d^2\left(\frac{Gm_1m_2}{d^2}\right)$$

$$d^2F = Gm_1m_2$$

$$\frac{d^2F}{Gm_2} = \frac{Gm_1m_2}{Gm_2}$$

$$m_1 = \frac{d^2F}{Gm_2}$$

**9.**

$$z = \frac{x - \overline{x}}{s}$$

$$s(z) = s\left(\frac{x - \overline{x}}{s}\right)$$

$$zs = x - \overline{x}$$

$$x = \overline{x} + zs$$

**11.**

$$I = \frac{E}{R+r}$$

$$(R+r)(I) = (R+r)\left(\frac{E}{R+r}\right)$$

$$IR + Ir = E$$

$$IR = E - Ir$$

$$\frac{IR}{I} = \frac{E - Ir}{I}$$

$$R = \frac{E - Ir}{I}$$

**13.**

$$f = \frac{f_1 f_2}{f_1 + f_2}$$

$$(f_1 + f_2)(f) = (f_1 + f_2)\left(\frac{f_1 f_2}{f_1 + f_2}\right)$$

$$ff_1 + ff_2 = f_1 f_2$$

$$ff_1 = f_1 f_2 - ff_2$$

$$ff_1 + ff_2 = f_1 f_2$$

$$ff_2 = f_1 f_2 - ff_1$$

$$ff_2 = f_1(f_2 - f)$$

$$\frac{ff_2}{f_2 - f} = \frac{f_1(f_2 - f)}{f_2 - f}$$

$$f_1 = \frac{ff_2}{f_2 - f}$$

**15.** Approximately 50,000 wheelchairs must be produced each month for the average cost to be $410 per chair.

**17.** The horizontal asymptote is $y = 400$. This means that despite the number of wheelchairs produced, the cost will approach but never reach $400.

**19.** **a.** $C(x) = 100{,}000 + 100x$

**b.** $\overline{C}(x) = \dfrac{100{,}000 + 100x}{x}$

**c.**

$$300 = \frac{100{,}000 + 100x}{x}$$

$$x(300) = x\left(\frac{100{,}000 + 100x}{x}\right)$$

$$300x = 100{,}000 + 100x$$

$$200x = 100{,}000$$

$$x = 500$$

500 mountain bikes must be produced each month for the average cost to be $300.

**21.** The running rate is 5 miles per hour.

**23.** The time increases as the running rate is close to zero miles per hour.

**25.**

| | $d$ | $r$ | $t = \frac{d}{r}$ |
|---|---|---|---|
| Car | 300 | $x$ | $\frac{300}{x}$ |
| Bus | 180 | $x - 20$ | $\frac{180}{x-20}$ |

$$\frac{300}{x} = \frac{180}{x - 20}$$

$$x(x-20)\left(\frac{300}{x}\right) = x(x-20)\left(\frac{180}{x-20}\right)$$

$$(x-20)(300) = 180x$$

$$300x - 6000 = 180x$$

$$120x = 6000$$

$$x = 50$$

The average rate for the car is 50 miles per hour and the average rate for the bus is $x - 20 = 50 - 20 = 30$ miles per hour.

**27.**

| | $d$ | $r$ | $t = \frac{d}{r}$ |
|---|---|---|---|
| To Campus | 5 | $x + 9$ | $\frac{5}{x+9}$ |
| From Campus | 5 | $x$ | $\frac{5}{x}$ |

$$\frac{5}{x+9} + \frac{5}{x} = \frac{7}{6}$$

$$6x(x+9)\left(\frac{5}{x+9} + \frac{5}{x}\right) = 6x(x+9)\left(\frac{7}{6}\right)$$

$$6x(x+9)\left(\frac{5}{x+9}\right) + 6x(x+9)\left(\frac{5}{x}\right) = x(x+9)(7)$$

$$6x(5) + 6(x+9)(5) = 7x(x+9)$$

$$30x + 30(x+9) = 7x^2 + 63x$$

$$30x + 30x + 270 = 7x^2 + 63x$$

$$60x + 270 = 7x^2 + 63x$$

$$0 = 7x^2 + 3x - 270$$

$$0 = (7x+45)(x-6)$$

Apply the zero product principle.

$$7x + 45 = 0 \qquad x - 6 = 0$$

$$7x = -45 \qquad x = 6$$

$$x = -\frac{45}{7}$$

The solution set is $\left\{-\frac{45}{7}, 6\right\}$. We disregard $-\frac{45}{7}$ because we can't have a negative time measurement. The average rate on the trip home is 6 miles per hour.

**29.** The time with the current is $\frac{20}{7+x}$. The time against the current is $\frac{8}{7-x}$.

$$\frac{20}{7+x} = \frac{8}{7-x}$$

$$(7+x)(7-x)\left(\frac{20}{7+x}\right) = (7+x)(7-x)\left(\frac{8}{7-x}\right)$$

$$(7-x)(20) = (7+x)(8)$$

$$140 - 20x = 56 + 8x$$

$$84 = 28x$$

$$3 = x$$

The rate of the current is 3 miles per hour.

**31.**

| | $d$ | $r$ | $t=\frac{d}{r}$ |
|---|---|---|---|
| With Stream | 2400 | $x+100$ | $\frac{2400}{x+100}$ |
| Against Stream | 1600 | $x-100$ | $\frac{1600}{x-100}$ |

$$\frac{2400}{x+100}=\frac{1600}{x-100}$$
$$(x+100)(x-100)\left(\frac{2400}{x+100}\right)=(x+100)(x-100)\left(\frac{1600}{x-100}\right)$$
$$(x-100)2400=1600(x+100)$$
$$2400x-240000=1600x+160000$$
$$800x=4000000$$
$$x=500$$

The airplane's average rate in calm air is 500 miles per hour.

**33.** Think of the speed of the sidewalk as a "current." Walking with or against the movement of the sidewalk is the same as paddling with or against a current.

| | $d$ | $r$ | $t=\frac{d}{r}$ |
|---|---|---|---|
| With Sidewalk | 100 | $x+1.8$ | $\frac{100}{x+1.8}$ |
| Against Sidewalk | 40 | $x-1.8$ | $\frac{40}{x-1.8}$ |

$$\frac{100}{x+1.8}=\frac{40}{x-1.8}$$
$$(x+1.8)(x-1.8)\left(\frac{100}{x+1.8}\right)=(x+1.8)(x-1.8)\left(\frac{40}{x-1.8}\right)$$
$$(x-1.8)(100)=(x+1.8)(40)$$
$$100x-180=40x+72$$
$$60x=252$$
$$x=4.2$$

The walking speed on a nonmoving sidewalk is 4.2 miles per hour.

**35.**

| | $d$ | $r$ | $t=\frac{d}{r}$ |
|---|---|---|---|
| Fast Runner | $x$ | 8 | $\frac{x}{8}$ |
| Slow Runner | $x$ | 6 | $\frac{x}{6}$ |

$$\frac{x}{6}-\frac{x}{8}=\frac{1}{2}$$
$$24\left(\frac{x}{6}-\frac{x}{8}\right)=24\left(\frac{1}{2}\right)$$
$$24\left(\frac{x}{6}\right)-24\left(\frac{x}{8}\right)=12$$
$$4x-3x=12$$
$$x=12$$

Each person ran 12 miles.

**37.**

| | Part Done in 1 Minute | Time Working Together | Part Done in $x$ Minutes |
|---|---|---|---|
| You | $\frac{1}{45}$ | $x$ | $\frac{x}{45}$ |
| Your Sister | $\frac{1}{30}$ | $x$ | $\frac{x}{30}$ |

$$\frac{x}{45}+\frac{x}{30}=1$$
$$120\left(\frac{x}{45}+\frac{x}{30}\right)=120(1)$$
$$120\left(\frac{x}{45}\right)+120\left(\frac{x}{30}\right)=120$$
$$\frac{8}{3}x+4x=120$$
$$\frac{20}{3}x=120$$
$$(3)\frac{20}{3}x=120(3)$$
$$20x=360$$
$$x=18$$

If they work together, it will take about 18 minutes to wash the car. There is not enough time to finish the job before the parents return home.

**39.**

| | Part Done in 1hr. | Time Working Together | Part Done in $x$ Hours |
|---|---|---|---|
| First Pipe | $\frac{1}{6}$ | $x$ | $\frac{x}{6}$ |
| Second Pipe | $\frac{1}{12}$ | $x$ | $\frac{x}{12}$ |

$$\frac{x}{6}+\frac{x}{12}=1$$
$$12\left(\frac{x}{6}+\frac{x}{12}\right)=12(1)$$
$$12\left(\frac{x}{6}\right)+12\left(\frac{x}{12}\right)=12$$
$$2x+x=12$$
$$3x=12$$
$$x=4$$

If both pipes are used, the pool will be filled in 4 hours.

**41.**

| | Part Done in 1hr. | Time Working Together | Part Done in 3 Hours |
|---|---|---|---|
| You | $\frac{1}{x}$ | 3 | $\frac{3}{x}$ |
| Your Cousin | $\frac{1}{4}$ | 3 | $\frac{3}{4}$ |

$$\frac{3}{x}+\frac{3}{4}=1$$
$$4x\left(\frac{3}{x}\right)+4x\left(\frac{3}{4}\right)=4x(1)$$
$$4(3)+x(3)=4x$$
$$12+3x=4x$$
$$12=x$$

Working alone, it would take you 12 hours to finish the job.

**43.** $60\left(\frac{x}{20}\right)+60\left(\frac{x}{30}\right)+60\left(\frac{x}{60}\right)=60$

$$3x+2x+x=60$$
$$6x=60$$
$$x=10$$

If the three crews work together, it will take 10 hours to dispense the food and water.

**45.**

| | Part done in 1 hour | Time together | Part done in 6 hrs. |
|---|---|---|---|
| Old | $\frac{1}{x+5}$ | 6 | $\frac{6}{x+5}$ |
| New | $\frac{1}{x}$ | 6 | $\frac{6}{x}$ |

$$\frac{6}{x+5}+\frac{6}{x}=1$$
$$x(x+5)\left(\frac{6}{x+5}+\frac{6}{x}\right)=x(x+5)(1)$$
$$x(x+5)\left(\frac{6}{x+5}\right)+x(x+5)\left(\frac{6}{x}\right)=x(x+5)$$
$$x(6)+(x+5)(6)=x^2+5x$$
$$6x+6x+30=x^2+5x$$
$$12x+30=x^2+5x$$
$$0=x^2-7x-30$$
$$0=(x-10)(x+3)$$

Apply the zero product principle.

$x-10=0 \qquad x+3=0$

$x=10 \qquad x=-3$

We disregard –3 because we can't have a negative time measurement. Working alone, it would take the new copying machine 10 hours to make all the copies.

**47.**

| | Part Done in 1 Minute | Time Working Together | Part Done in $x$ Minutes |
|---|---|---|---|
| Faucet | $\frac{1}{5}$ | $x$ | $\frac{x}{5}$ |
| Drain | $\frac{1}{10}$ | $x$ | $\frac{x}{10}$ |

$$\frac{x}{5}-\frac{x}{10}=1$$
$$10\left(\frac{x}{5}-\frac{x}{10}\right)=10(1)$$
$$10\left(\frac{x}{5}\right)-10\left(\frac{x}{10}\right)=10$$
$$2x-x=10$$
$$x=10$$

It will take 10 minutes for the sink to fill if the drain is left open.

**49.** Let $x$ be the number.

$$\frac{4x}{x+5} = \frac{3}{2}$$
$$2(x+5)\frac{4x}{x+5} = 2(x+5)\frac{3}{2}$$
$$8x = 3(x+5)$$
$$8x = 3x+15$$
$$5x = 15$$
$$x = 3$$

**51.** Let $x$ be the number.

$$2x + 2\cdot\frac{1}{x} = \frac{20}{3}$$
$$(3x)2x + (3x)2\cdot\frac{1}{x} = (3x)\frac{20}{3}$$
$$6x^2 + 6 = 20x$$
$$6x^2 - 20x + 6 = 0$$
$$2(3x^2 - 10x + 3) = 0$$
$$2(3x-1)(x-3) = 0$$

Using the zero product principle, $x = \frac{1}{3}$ and $x = 3$.

**53.** Let $x$ be the number of consecutive hits. You already have 35 hits in 140 times at bat.

$$\frac{x+35}{x+140} = 0.30$$
$$(x+140)\frac{x+35}{x+140} = 0.30(x+140)$$
$$x+35 = 0.30x + 42$$
$$0.70x = 7$$
$$x = 10$$

You must have 10 consecutive hits to increase your batting average to 10.

**55.**

| | Part Done in 1 Hour | Time Working Together | Part Done in $x$ Hours |
|---|---|---|---|
| First Pipe | $\frac{1}{a}$ | $x$ | $\frac{x}{a}$ |
| Second Pipe | $\frac{1}{b}$ | $x$ | $\frac{x}{b}$ |

$$\frac{x}{a} + \frac{x}{b} = 1$$
$$ab\left(\frac{x}{a} + \frac{x}{b}\right) = ab(1)$$
$$ab\left(\frac{x}{a}\right) + ab\left(\frac{x}{b}\right) = ab$$
$$bx + ax = ab$$
$$(a+b)x = ab$$
$$x = \frac{ab}{a+b}$$

If both pipes are used, the pool will be filled in $\frac{ab}{a+b}$ hours

**In Exercises 57-63, answers will vary.**

**65.** Exercise 45.

$$\frac{6}{x+5} + \frac{6}{x} = 1$$

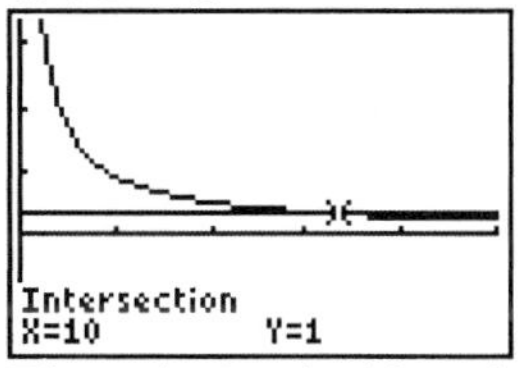


The solution is 10 and the solution set is $\{10\}$.

Exercise 46.

$$\frac{12}{x+10}+\frac{12}{x}=1$$

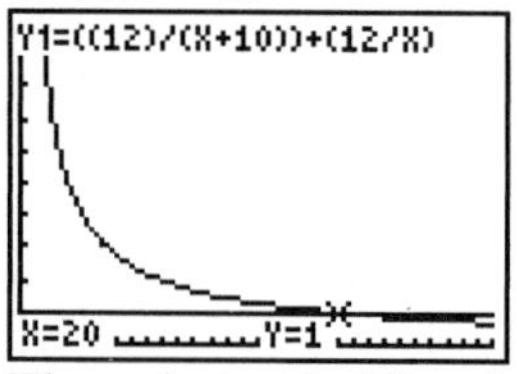


The solution is 20 and the solution set is $\{20\}$.

**67.** Statement **c.** is true. You can account for anything that "takes away" from the job getting done by subtracting it instead of adding it to the right hand side of the equation.

Statement **a.** is false. As production level increases, the average cost for a company to produce each unit of its product decreases.

Statement **b.** is false. The equation contains two terms involving $p$. If you subtract $qf$ from both sides and then divide by $f$, there will still be a $p$ term on the other side of the equation.

Statement **d.** is false. Statement **c.** is true.

**69.**

| | $d$ | $r$ | $t=\frac{d}{r}$ |
|---|---|---|---|
| Old schedule | 351 | $x-2$ | $\frac{351}{x-2}$ |
| New schedule | 351 | $x$ | $\frac{351}{x}$ |

$$\frac{351}{x-2}-\frac{351}{x}=\frac{1}{4}$$

$$4x(x-2)\left(\frac{351}{x-2}-\frac{351}{x}\right)=4x(x-2)\left(\frac{1}{4}\right)$$

$$4x(x-2)\left(\frac{351}{x-2}\right)-4x(x-2)\left(\frac{351}{x}\right)=x(x-2)$$

$$4x(351)-4(x-2)(351)=x^2-2x$$

$$1404x-1404(x-2)=x^2-2x$$

$$\cancel{1404x}-\cancel{1404x}+2808=x^2-2x$$

$$2808=x^2-2x$$

$$0=x^2-2x-2808$$

$$0=(x-54)(x+52)$$

Apply the zero product principle.

$x-54=0$ $\quad$ $x+52=0$

$x=54$ $\quad$ $x=-52$

We disregard –52 because we can't have a negative average rate. The average rate of the train on the new schedule is 54 miles per hour.

**71.** $x^2+4x+4-9y^2$

$=(x^2+4x+4)-9y^2$

$=(x+2)^2-(3y)^2$

$=((x+2)+3y)((x+2)-3y)$

$=(x+2+3y)(x+2-3y)$

**72.** $2x+5y=-5$

$x+2y=-1$

$$\left[\begin{array}{cc|c} 2 & 5 & -5 \\ 1 & 2 & -1 \end{array}\right] \quad R_1 \leftrightarrow R_2$$

$$=\left[\begin{array}{cc|c} 1 & 2 & -1 \\ 2 & 5 & -5 \end{array}\right] \quad -2R_1+R_2$$

$$=\left[\begin{array}{cc|c} 1 & 2 & -1 \\ 0 & 1 & -3 \end{array}\right]$$

The resulting system is:

$x+2y=-1$

$y=-3.$

Since we know $y = -3$, we can use back-substitution to find $x$.

$$x + 2y = -1$$
$$x + 2(-3) = -1$$
$$x - 6 = -1$$
$$x = 5$$

The solution is (5, −3) and the solution set is $\{(5, -3)\}$.

**73.** $$\begin{aligned} x + y + z &= 4 \\ 2x + 5y \quad &= 1 \\ x - y - 2z &= 0 \end{aligned}$$

Multiply the first equation by 2 and add to the third equation.

$$\begin{aligned} 2x + 2y + 2z &= 8 \\ x - y - 2z &= 0 \\ \hline 3x + y \quad &= 8 \end{aligned}$$

We can use this equation and the second equation in the original system to form a system of two equations in two variables.

$$2x + 5y = 1$$
$$3x + y = 8$$

Multiply the second equation by −5 and solve by addition.

$$\begin{aligned} 2x + 5y &= 1 \\ -15x - 5y &= -40 \\ \hline -13x \quad &= -39 \\ x \quad &= 3 \end{aligned}$$

Back-substitute 3 for $x$ to find $y$.

$$2x + 5y = 1$$
$$2(3) + 5y = 1$$
$$6 + 5y = 1$$
$$5y = -5$$
$$y = -1$$

Back-substitute −1 for $y$ and 3 for $x$ to find $z$.

$$x + y + z = 4$$
$$3 + (-1) + z = 4$$
$$2 + z = 4$$
$$z = 2$$

The solution is $(3, -1, 2)$ and the solution set is $\{(3, -1, 2)\}$.

### 6.8 Exercise Set

**1.** Since $y$ varies directly with $x$, we have $y = kx$.

Use the given values to find $k$.

$$y = kx$$
$$65 = k \cdot 5$$
$$\frac{65}{5} = \frac{k \cdot 5}{5}$$
$$13 = k$$

The equation becomes $y = 13x$.

When $x = 12$, $y = 13x = 13 \cdot 12 = 156$.

**3.** Since $y$ varies inversely with $x$, we have $y = \frac{k}{x}$.

Use the given values to find $k$.

$$y = \frac{k}{x}$$
$$12 = \frac{k}{5}$$
$$5 \cdot 12 = 5 \cdot \frac{k}{5}$$
$$60 = k$$

The equation becomes $y = \frac{60}{x}$.

When $x = 2$, $y = \frac{60}{2} = 30$.

**5.** Since $y$ varies inversely as $x$ and inversely as the square of $z$, we have $y=\frac{kx}{z^2}$.

Use the given values to find $k$.

$$y=\frac{kx}{z^2}$$

$$20=\frac{k(50)}{5^2}$$

$$20=\frac{k(50)}{25}$$

$$20=2k$$

$$10=k$$

The equation becomes $y=\frac{10x}{z^2}$.

When $x=3$ and $z=6$,

$$y=\frac{10x}{z^2}=\frac{10(3)}{6^2}=\frac{10(3)}{36}=\frac{30}{36}=\frac{5}{6}.$$

**7.** Since $y$ varies jointly as $x$ and $y$, we have $y=kxy$.

Use the given values to find $k$.

$$y=kxy$$

$$25=k(2)(5)$$

$$25=k(10)$$

$$\frac{25}{10}=\frac{k(10)}{10}$$

$$\frac{5}{2}=k$$

The equation becomes $y=\frac{5}{2}xy$.

When $x=8$ and $z=12$,

$$y=\frac{5}{2}(8)(12)=\frac{5}{\cancel{2}}(\overset{4}{\cancel{8}})(12)=240.$$

**9.** Since $y$ varies jointly as $a$ and $b$ and inversely as the square root of $c$, we have $y=\frac{kab}{\sqrt{c}}$.

Use the given values to find $k$.

$$y=\frac{kab}{\sqrt{c}}$$

$$12=\frac{k(3)(2)}{\sqrt{25}}$$

$$12=\frac{k(6)}{5}$$

$$12(5)=\frac{k(6)}{5}(5)$$

$$60=6k$$

$$\frac{60}{6}=\frac{6k}{6}$$

$$10=k$$

The equation becomes $y=\frac{10ab}{\sqrt{c}}$.

When $a=5$, $b=3$, $c=9$,

$$y=\frac{10ab}{\sqrt{c}}=\frac{10(5)(3)}{\sqrt{9}}=\frac{150}{3}=50.$$

**11.** $x=kyz$;

Solving for $y$:

$$x=kyz$$

$$\frac{x}{kz}=\frac{kyz}{yz}.$$

$$y=\frac{x}{kz}$$

**12.** $x=kyz^2$;

Solving for $y$:

$$x=kyz^2$$

$$\frac{x}{kz^2}=\frac{kyz^2}{kz^2}$$

$$y=\frac{x}{kz^2}$$

**13.** $x = \dfrac{kz^3}{y}$;

Solving for $y$

$$x = \frac{kz^3}{y}$$
$$xy = y \cdot \frac{kz^3}{y}$$
$$xy = kz^3$$
$$\frac{xy}{x} = \frac{kz^3}{x}$$
$$y = \frac{kz^3}{x}$$

**14.**
$$x = \frac{k\sqrt[3]{z}}{y}$$
$$yx = y \cdot \frac{k\sqrt[3]{z}}{y}$$
$$yx = k\sqrt[3]{z}$$
$$\frac{yx}{x} = \frac{k\sqrt[3]{z}}{x}$$
$$y = \frac{k\sqrt[3]{z}}{x}$$

**15.** $x = \dfrac{kyz}{\sqrt{w}}$;

Solving for $y$:

$$x = \frac{kyz}{\sqrt{w}}$$
$$x\left(\sqrt{w}\right) = \left(\sqrt{w}\right)\frac{kyz}{\sqrt{w}}$$
$$x\sqrt{w} = kyz$$
$$\frac{x\sqrt{w}}{kz} = \frac{kyz}{kz}$$
$$y = \frac{x\sqrt{w}}{kz}$$

**16.**
$$x = \frac{kyz}{w^2}$$
$$\left(\frac{w^2}{kz}\right)x = \frac{w^2}{kz}\frac{kyz}{w^2}$$
$$y = \frac{xw^2}{kz}$$

**17.** $x = kz(y + w)$;

Solving for $y$:

$$x = kz(y + w)$$
$$x = kzy + kzw$$
$$x - kzw = kzy$$
$$\frac{x - kzw}{kz} = \frac{kzy}{kz}$$
$$y = \frac{x - kzw}{kz}$$

**18.**
$$x = kz(y - w)$$
$$x = kzy - kzw$$
$$x + kzw = kzy$$
$$\frac{x + kzw}{kz} = \frac{kzy}{kz}$$
$$y = \frac{x + kzw}{kz}$$

**19.** $x = \dfrac{kz}{y-w}$;

Solving for *y*:

$$x = \frac{kz}{y-w}$$

$$(y-w)x = (y-w)\frac{kz}{y-w}$$

$$xy - wx = kz$$

$$xy = kz + wx$$

$$\frac{xy}{x} = \frac{kz+wx}{x}$$

$$y = \frac{kz+wx}{x}$$

**20.**

$$x = \frac{kz}{y+w}$$

$$(y+w)x = (y+w)\frac{kz}{y+w}$$

$$yx + xw = kz$$

$$yz = kz - xw$$

$$\frac{yz}{z} = \frac{kz-xw}{z}$$

$$y = \frac{kz-xw}{z}$$

**21.** Since *T* varies directly as *B*, we have $T = kB$.

Use the given values to find *k*.

$$T = kB$$

$$3.6 = k(4)$$

$$\frac{3.6}{4} = \frac{k(4)}{4}$$

$$0.9 = k$$

The equation becomes $T = 0.9B$. When $B = 6$, $T = 0.9(6) = 5.4$. The tail length is 5.4 feet.

**23.** Since *B* varies directly as *D*, we have $B = kD$.

Use the given values to find *k*.

$$B = kD$$

$$8.4 = k(12)$$

$$\frac{8.4}{12} = \frac{k(12)}{12}$$

$$k = \frac{8.4}{12} = 0.7$$

The equation becomes $B = 0.7D$.

When $B = 56$,

$$56 = 0.7D$$

$$\frac{56}{0.7} = \frac{0.7D}{0.7}$$

$$D = \frac{56}{0.7} = 80$$

It was dropped from 80 inches.

**25.** Since a man's weight varies directly as the cube of his height, we have $w = kh^3$.

Use the given values to find *k*.

$$w = kh^3$$

$$170 = k(70)^3$$

$$170 = k(343{,}000)$$

$$\frac{170}{343{,}000} = \frac{k(343{,}000)}{343{,}000}$$

$$0.000496 = k$$

The equation becomes $w = 0.000496h^3$.

When $h = 107$,

$$w = 0.000496(107)^3$$

$$= 0.000496(1{,}225{,}043) \approx 607.$$

Robert Wadlow's weight was approximately 607 pounds.

**27.** Since the banking angle varies inversely as the turning radius, we have $B = \frac{k}{r}$.

Use the given values to find *k*.

$$B = \frac{k}{r}$$
$$28 = \frac{k}{4}$$
$$28(4) = 28\left(\frac{k}{4}\right)$$
$$112 = k$$

The equation becomes $B = \frac{112}{r}$.

When $r = 3.5$, $B = \frac{112}{r} = \frac{112}{3.5} = 32$.

The banking angle is $32^\circ$ when the turning radius is 3.5 feet.

**29.** Since intensity varies inversely as the square of the distance, we have pressure, we have $I = \frac{k}{d}$.

Use the given values to find *k*.

$$I = \frac{k}{d^2}.$$
$$62.5 = \frac{k}{3^2}$$
$$62.5 = \frac{k}{9}$$
$$9(62.5) = 9\left(\frac{k}{9}\right)$$
$$562.5 = k$$

The equation becomes $I = \frac{562.5}{d^2}$.

When $d = 2.5$,

$$I = \frac{562.5}{2.5^2} = \frac{562.5}{6.25} = 90$$

The intensity is 90 milliroentgens per hour.

**31.** Since index varies directly as weight and inversely as the square of one's height, we have $I = \frac{kw}{h^2}$.

Use the given values to find *k*.

$$I = \frac{kw}{h^2}$$
$$35.15 = \frac{k(180)}{60^2}$$
$$35.15 = \frac{k(180)}{3600}$$
$$(3600)35.15 = \frac{k(180)}{3600}$$
$$126540 = k(180)$$
$$k = \frac{126540}{180} = 703$$

The equation becomes $I = \frac{703w}{h^2}$.

When $w = 170$ and $h = 70$,

$$I = \frac{703(170)}{(70)^2} \approx 24.4.$$

This person has a BMI of 24.4 and is not overweight.

**33.** Since heat loss varies jointly as the area and temperature difference, we have $L = kAD$.

Use the given values to find *k*.

$$L = kAD$$
$$1200 = k(3 \cdot 6)(20)$$
$$1200 = 360k$$
$$\frac{1200}{360} = \frac{360k}{360}$$
$$k = \frac{10}{3}$$

The equation becomes $L = \frac{10}{3}AD$

When $A = 6 \cdot 9 = 54$, $D = 10$,

$$L = \frac{10}{3}(9 \cdot 6)(10) = 1800.$$

The heat loss is 1800 Btu.

**35.** Since intensity varies inversely as the square of the distance from the sound source, we have $I = \frac{k}{d^2}$. If you move to a seat twice as far, then $d = 2d$ .So we have $I = \frac{k}{(2d)^2} = \frac{k}{4d^2} = \frac{1}{4} \cdot \frac{k}{d^2}$. The intensity will be multiplied by a factor of $\frac{1}{4}$. So the sound intensity is $\frac{1}{4}$ of what it was originally.

**37.** **a.** Since the average number of phone calls varies jointly as the product of the populations and inversely as the square of the distance, we have $C = \frac{kP_1P_2}{d^2}$.

**b.** Use the given values to find *k*.

$$C = \frac{kP_1P_2}{d^2}$$

$$326{,}000 = \frac{k(777{,}000)(3{,}695{,}000)}{(420)^2}$$

$$326{,}000 = \frac{k(2.87 \times 10^{12})}{176{,}400}$$

$$326{,}000 = 16269841.27k$$

$$0.02 \approx k$$

The equation becomes

$$C = \frac{0.02P_1P_2}{d^2}.$$

**c.**

$$C = \frac{0.02(650{,}000)(490{,}000)}{(400)^2}$$

$$\approx 39813$$

The average number of calls is approximately 39,813 daily phone calls.

**39.** **a.**

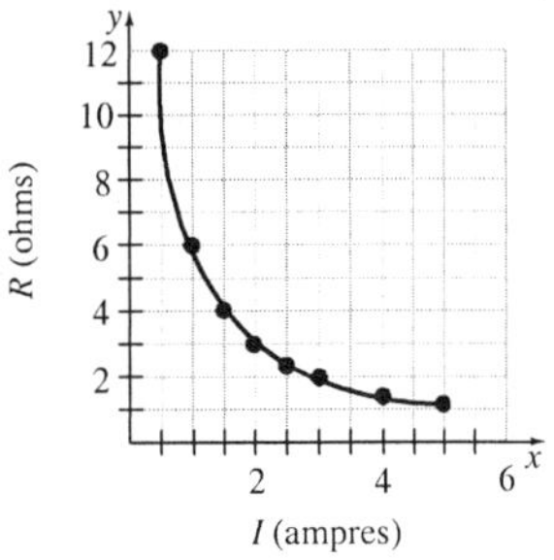


**b.** Current varies inversely as resitance. Answers will vary.

**c.** Since the current varies inversely as resistance we have $R = \frac{k}{I}$. Using one of the given ordered pairs to find $k$.

$$12 = \frac{k}{0.5}$$

$$12(0.5) = \frac{k}{0.5}(0.5)$$

$$k = 6$$

The equation becomes $R = \frac{6}{I}$.

**41.** Answers will vary.

**43.** Answers will vary.

**45.** *z* varies directly as the square root of *x* and inversely as the square root of *y*.

**47.** Answers will vary.

**49.** Since wind pressure varies directly as the square of the wind velocity, we have $P = kv^2$. If the wind speed doubles then the value of *v* has been multiplied by two. In the formula, $P = k(2v)^2 = k(4v^2) = 4kv^2$. Then the wind pressure will be multiplied by a factor of 4. So if the wind speed doubles, the wind pressure is 4 times more destructive.

**51.** Since the brightness of a source point varies inversely as the square of its distance from an observer, we have $B=\frac{k}{d^2}$. We can now see things that are only $\frac{1}{50}$ as bright.

$$B=\frac{1}{50}\cdot\frac{k}{d^2}=\frac{k}{50d^2}$$
$$=\frac{k}{(7.07)^2 d^2}=\frac{k}{(7.07d)^2}$$

The distance that can be seen is about 7.07 times farther with the space telescope.

**52.** $\begin{vmatrix} -1 & 2 \\ 3 & -4 \end{vmatrix}=-1(-4)-3(2)=4-6=-2$

**53.** $x^2y-9y-3x^2+27$
$$=y(x^2-9)-3(x^2-9)$$
$$=(x^2-9)(y-3)$$
$$=(x+3)(x-3)(y-3)$$

**54.** $7xy+x^2y^2-5x^3-7$

The degree of the polynomial is 4.

## Chapter 6 Review

**1.** $f(x)=\frac{x^2+2x-3}{x^2-4}$

**a.** $f(4)=\frac{(4)^2+2(4)-3}{(4)^2-4}$
$$=\frac{16+8-3}{16-4}=\frac{21}{12}=\frac{7}{4}$$

**b.** $f(0)=\frac{0^2+2(0)-3}{0^2-4}$
$$=\frac{0+0-3}{0-4}=\frac{-3}{-4}=\frac{3}{4}$$

**c.** $f(-2)=\frac{(-2)^2+2(-2)-3}{(-2)^2-4}$
$$=\frac{4-4-3}{4-4}=\frac{-3}{0}$$

Division by zero is undefined. $f(-2)$ does not exist.

**d.** $f(-3)=\frac{(-3)^2+2(-3)-3}{(-3)^2-4}$
$$=\frac{9-6-3}{9-4}=\frac{0}{5}=0$$

**2.** The domain is all real numbers except those that make the denominator zero. To find these values, set the denominator equal to zero and solve.

$(x-3)(x+4)=0$

Apply the zero product principle.

$x-3=0 \qquad x+4=0$

$x=3 \qquad x=-4$

The domain of $f=\{x|x$ is a real number and $x\neq -4$ and $x\neq 3\}$ or $(-\infty,-4)\cup(-4,3)\cup(3,\infty)$.

**3.** The domain is all real numbers except those that make the denominator zero. To find these values, set the denominator equal to zero and solve.

$$x^2+x-2=0$$
$$(x+2)(x-1)=0$$

Apply the zero product principle.

$$x+2=0 \qquad x-1=0$$
$$x=-2 \qquad x=1$$

The domain of $f = \{x | x$ is a real number and $x \neq -2$ and $x \neq 1\}$ or $(-\infty,-2)\cup(-2,1)\cup(1,\infty)$.

**4.** $$\frac{5x^3-35x}{15x^2}=\frac{\cancel{5x}(x^2-7)}{3x\cdot\cancel{5x}}=\frac{x^2-7}{3x}$$

**5.** $$\frac{x^2+6x-7}{x^2-49}=\frac{(x+7)(x-1)}{(x+7)(x-7)}$$
$$=\frac{\cancel{(x+7)}(x-1)}{\cancel{(x+7)}(x-7)}$$
$$=\frac{x-1}{x-7}$$

**6.** $$\frac{6x^2+7x+2}{2x^2-9x-5}=\frac{(3x+2)(2x+1)}{(2x+1)(x-5)}$$
$$=\frac{(3x+2)\cancel{(2x+1)}}{\cancel{(2x+1)}(x-5)}$$
$$=\frac{3x+2}{x-5}$$

**7.** $\frac{x^2+4}{x^2-4}$; cannot be simplified

**8.** $$\frac{x^3-8}{x^2-4}=\frac{\cancel{(x-2)}(x^2+2x+4)}{(x+2)\cancel{(x-2)}}$$
$$=\frac{x^2+2x+4}{x+2}$$

**9.** $$\frac{5x^2-5}{3x+12}\cdot\frac{x+4}{x-1}=\frac{5(x^2-1)}{3\cancel{(x+4)}}\cdot\frac{\cancel{x+4}}{x-1}$$
$$=\frac{5(x^2-1)}{3(x-1)}$$
$$=\frac{5(x+1)\cancel{(x-1)}}{3\cancel{(x-1)}}$$
$$=\frac{5(x+1)}{3}$$

**10.** $$\frac{2x+5}{4x^2+8x-5}\cdot\frac{4x^2-4x+1}{x+1}$$
$$=\frac{\cancel{2x+5}}{\cancel{(2x+5)}\cancel{(2x-1)}}\cdot\frac{\cancel{(2x-1)}(2x-1)}{x+1}$$
$$=\frac{2x-1}{x+1}$$

**11.** $$\frac{x^2-9x+14}{x^3+2x^2}\cdot\frac{x^2-4}{x^2-4x+4}$$
$$=\frac{(x-7)\cancel{(x-2)}}{x^2\cancel{(x+2)}}\cdot\frac{\cancel{(x+2)}\cancel{(x-2)}}{\cancel{(x-2)}\cancel{(x-2)}}$$
$$=\frac{x-7}{x^2}$$

**12.**
$$\frac{1}{x^2+8x+15}\div\frac{3}{x+5}$$
$$=\frac{1}{x^2+8x+15}\cdot\frac{x+5}{3}$$
$$=\frac{1}{(x+5)(x+3)}\cdot\frac{x+5}{3}$$
$$=\frac{1}{3(x+3)}$$

**13.**
$$\frac{x^2+16x+64}{2x^2-128}\div\frac{x^2+10x+16}{x^2-6x-16}$$
$$=\frac{x^2+16x+64}{2x^2-128}\cdot\frac{x^2-6x-16}{x^2+10x+16}$$
$$=\frac{(x+8)(x+8)}{2(x^2-64)}\cdot\frac{(x-8)(x+2)}{(x+8)(x+2)}$$
$$=\frac{(x+8)(x-8)}{2(x^2-64)}$$
$$=\frac{(x+8)(x-8)}{2(x+8)(x-8)}=\frac{1}{2}$$

**14.**
$$\frac{y^2-16}{y^3-64}\div\frac{y^2-3y-18}{y^2+5y+6}$$
$$=\frac{y^2-16}{y^3-64}\cdot\frac{y^2+5y+6}{y^2-3y-18}$$
$$=\frac{(y+4)(y-4)}{(y-4)(y^2+4y+16)}\cdot\frac{(y+3)(y+2)}{(y-6)(y+3)}$$
$$=\frac{(y+4)(y+2)}{(y-6)(y^2+4y+16)}$$

**15.**
$$\frac{x^2-4x+4-y^2}{2x^2-11x+15}\cdot\frac{x^4y}{x-2+y}\div\frac{x^3y-2x^2y-x^2y^2}{3x-9}$$
$$=\frac{(x-2)(x-2)-y^2}{(2x-5)(x-3)}\cdot\frac{x^4y}{x-2+y}\div\frac{x^2y(x-2-y)}{3(x-3)}$$
$$=\frac{(x-2)^2-y^2}{(2x-5)(x-3)}\cdot\frac{x^4y}{(x-2+y)}\cdot\frac{3(x-3)}{x^2y(x-2-y)}$$
$$=\frac{((x-2)+y)((x-2)-y)}{2x-5}\cdot\frac{x^2}{(x-2+y)}\cdot\frac{3}{(x-2-y)}=\frac{3x^2}{2x-5}$$

**16.**
**a.** 50 deer were introduced into the habitat.
**b.** After 10 years, the population is 150 deer.
**c.** The equation of the horizontal asymptote is $y = 225$. This means that the deer population will increase over time to 225, but will never reach it.

**17.** $\dfrac{4x+1}{3x-1}+\dfrac{8x-5}{3x-1}=\dfrac{4x+1+8x-5}{3x-1}$

$=\dfrac{12x-4}{3x-1}=\dfrac{4(3x-1)}{3x-1}$

$=4$

**18.** $\dfrac{2x-7}{x^2-9}-\dfrac{x-4}{x^2-9}=\dfrac{2x-7-(x-4)}{x^2-9}$

$=\dfrac{2x-7-x+4}{x^2-9}$

$=\dfrac{\cancel{x-3}}{(x+3)(\cancel{x-3})}$

$=\dfrac{1}{x+3}$

**19.** $\dfrac{4x^2-11x+4}{x-3}-\dfrac{x^2-4x+10}{x-3}$

$=\dfrac{4x^2-11x+4-(x^2-4x+10)}{x-3}$

$=\dfrac{4x^2-11x+4-x^2+4x-10}{x-3}$

$=\dfrac{3x^2-7x-6}{x-3}$

$=\dfrac{(\cancel{x-3})(3x+2)}{\cancel{x-3}}$

$=3x+2$

**20.** $9x^3 = \quad 3^2x^2$

$12x = 2^2\cdot 3\,x$

$\text{LCD} = 2^2\cdot 3^2x^3 = 4\cdot 9x^3 = 36x^3$

**21.** $x^2+2x-35=(x+7)(x-5)$

$x^2+9x+14=(x+7)(x+2)$

$\text{LCD}=(x+7)(x-5)(x+2)$

**22.** $\dfrac{1}{x}+\dfrac{2}{x-5}=\dfrac{1(x-5)}{x(x-5)}+\dfrac{2x}{x(x-5)}$

$=\dfrac{1(x-5)+2x}{x(x-5)}$

$=\dfrac{x-5+2x}{x(x-5)}$

$=\dfrac{3x-5}{x(x-5)}$

**23.** $\dfrac{2}{x^2-5x+6}+\dfrac{3}{x^2-x-6}=\dfrac{2}{(x-3)(x-2)}+\dfrac{3}{(x-3)(x+2)}$

$=\dfrac{2(x+2)}{(x-3)(x-2)(x+2)}+\dfrac{3(x-2)}{(x-3)(x-2)(x+2)}$

$=\dfrac{2(x+2)+3(x-2)}{(x-3)(x-2)(x+2)}=\dfrac{2x+4+3x-6}{(x-3)(x-2)(x+2)}$

$=\dfrac{5x-2}{(x-3)(x-2)(x+2)}$

**24.** $$\frac{x-3}{x^2-8x+15}+\frac{x+2}{x^2-x-6}=\frac{x-3}{(x-3)(x-5)}+\frac{\cancel{x+2}}{(x-3)(\cancel{x+2})}=\frac{x-3}{(x-3)(x-5)}+\frac{1}{x-3}$$
$$=\frac{x-3}{(x-3)(x-5)}+\frac{1(x-5)}{(x-3)(x-5)}=\frac{x-3+1(x-5)}{(x-3)(x-5)}$$
$$=\frac{x-3+x-5}{(x-3)(x-5)}=\frac{2x-8}{(x-3)(x-5)}\text{ or }\frac{2(x-4)}{(x-3)(x-5)}$$

**25.** $$\frac{3x^2}{9x^2-16}-\frac{x}{3x+4}=\frac{3x^2}{(3x+4)(3x-4)}-\frac{x}{3x+4}=\frac{3x^2}{(3x+4)(3x-4)}-\frac{x(3x-4)}{(3x+4)(3x-4)}$$
$$=\frac{3x^2-x(3x-4)}{(3x+4)(3x-4)}=\frac{3x^2-3x^2+4x}{(3x+4)(3x-4)}=\frac{4x}{(3x+4)(3x-4)}$$

**26.** $$\frac{y}{y^2+5y+6}-\frac{2}{y^2+3y+2}=\frac{y}{(y+3)(y+2)}-\frac{2}{(y+2)(y+1)}$$
$$=\frac{y(y+1)}{(y+3)(y+2)(y+1)}-\frac{2(y+3)}{(y+3)(y+2)(y+1)}$$
$$=\frac{y(y+1)-2(y+3)}{(y+3)(y+2)(y+1)}=\frac{y^2+y-2y-6}{(y+3)(y+2)(y+1)}$$
$$=\frac{y^2-y-6}{(y+3)(y+2)(y+1)}=\frac{(y-3)\cancel{(y+2)}}{(y+3)\cancel{(y+2)}(y+1)}$$
$$=\frac{y-3}{(y+3)(y+1)}$$

**27.** $$\frac{x}{x+3}+\frac{x}{x-3}-\frac{9}{x^2-9}=\frac{x}{x+3}+\frac{x}{x-3}-\frac{9}{(x+3)(x-3)}$$
$$=\frac{x(x-3)}{(x+3)(x-3)}+\frac{x(x+3)}{(x-3)(x+3)}-\frac{9}{(x+3)(x-3)}$$
$$=\frac{x(x-3)+x(x+3)-9}{(x+3)(x-3)}=\frac{x^2-3x+x^2+3x-9}{(x+3)(x-3)}=\frac{2x^2-9}{(x+3)(x-3)}$$

**28.** $$\frac{3x^2}{x-y}+\frac{3y^2}{y-x}=\frac{3x^2}{x-y}+\frac{-1\left(3y^2\right)}{-1\left(y-x\right)}=\frac{3x^2}{x-y}+\frac{-3y^2}{x-y}=\frac{3x^2-3y^2}{x-y}=\frac{3\left(x^2-y^2\right)}{x-y}$$
$$=\frac{3\left(x+y\right)\cancel{\left(x-y\right)}}{\cancel{x-y}}=3\left(x+y\right)\text{ or }3x+3y$$

**29.** $$\frac{\frac{3}{x}-3}{\frac{8}{x}-8}=\frac{x}{x}\cdot\frac{\frac{3}{x}-3}{\frac{8}{x}-8}=\frac{x\cdot\frac{3}{x}-x\cdot 3}{x\cdot\frac{8}{x}-x\cdot 8}=\frac{3-3x}{8-8x}=\frac{3\cancel{\left(1-x\right)}}{8\cancel{\left(1-x\right)}}=\frac{3}{8}$$

**30.** $$\frac{\frac{5}{x}+1}{1-\frac{25}{x^2}}=\frac{x^2}{x^2}\cdot\frac{\frac{5}{x}+1}{1-\frac{25}{x^2}}=\frac{x^2\cdot\frac{5}{x}+x^2\cdot 1}{x^2\cdot 1-x^2\cdot\frac{25}{x^2}}=\frac{5x+x^2}{x^2-25}=\frac{x\cancel{\left(5+x\right)}}{\cancel{\left(x+5\right)}\left(x-5\right)}=\frac{x}{x-5}$$

**31.** $$\frac{3-\frac{1}{x+3}}{3+\frac{1}{x+3}}=\frac{x+3}{x+3}\cdot\frac{3-\frac{1}{x+3}}{3+\frac{1}{x+3}}=\frac{\left(x+3\right)\cdot 3-\left(x+3\right)\cdot\frac{1}{x+3}}{\left(x+3\right)\cdot 3+\left(x+3\right)\cdot\frac{1}{x+3}}=\frac{3x+9-1}{3x+9+1}=\frac{3x+8}{3x+10}$$

**32.** $$\frac{\frac{4}{x+3}}{\frac{2}{x-2}-\frac{1}{x^2+x-6}}=\frac{\frac{4}{x+3}}{\frac{2}{x-2}-\frac{1}{\left(x+3\right)\left(x-2\right)}}=\frac{\left(x+3\right)\left(x-2\right)}{\left(x+3\right)\left(x-2\right)}\cdot\frac{\frac{4}{x+3}}{\frac{2}{x-2}-\frac{1}{\left(x+3\right)\left(x-2\right)}}$$
$$=\frac{\left(x-2\right)4}{\left(x+3\right)2-1}=\frac{4x-8}{2x+6-1}=\frac{4x-8}{2x+5}=\frac{4\left(x-2\right)}{2x+5}$$

**33.** $$\frac{\frac{2}{x^2-x-6}+\frac{1}{x^2-4x+3}}{\frac{3}{x^2+x-2}-\frac{2}{x^2+5x+6}}=\frac{\frac{2}{(x-3)(x+2)}+\frac{1}{(x-3)(x-1)}}{\frac{3}{(x+2)(x-1)}-\frac{2}{(x+2)(x+3)}}$$

$$=\frac{\frac{2(x-1)}{(x-3)(x+2)(x-1)}+\frac{1(x+2)}{(x-3)(x+2)(x-1)}}{\frac{3(x+3)}{(x+2)(x-1)(x+3)}-\frac{2(x-1)}{(x+2)(x-1)(x+3)}}$$

$$=\frac{\frac{2(x-1)+(x+2)}{(x-3)(x+2)(x-1)}}{\frac{3(x+3)-2(x-1)}{(x+2)(x-1)(x+3)}}=\frac{\frac{2x-2+x+2}{(x-3)(x+2)(x-1)}}{\frac{3x+9-2x+2}{(x+2)(x-1)(x+3)}}$$

$$=\frac{\frac{3x}{(x-3)(x+2)(x-1)}}{\frac{x+11}{(x+2)(x-1)(x+3)}}$$

$$=\frac{3x}{(x-3)\cancel{(x+2)}\cancel{(x-1)}}\cdot\frac{\cancel{(x+2)}\cancel{(x-1)}(x+3)}{x+11}$$

$$=\frac{3x(x+3)}{(x-3)(x+11)}=\frac{3x^2+9x}{x^2+8x-33}$$

**34.** $$\frac{x^{-2}+x^{-1}}{x^{-2}-x^{-1}}=\frac{\frac{1}{x^2}+\frac{1}{x}}{\frac{1}{x^2}-\frac{1}{x}}=\frac{x^2}{x^2}\cdot\frac{\frac{1}{x^2}+\frac{1}{x}}{\frac{1}{x^2}-\frac{1}{x}}=\frac{x^2\cdot\frac{1}{x^2}+x^2\cdot\frac{1}{x}}{x^2\cdot\frac{1}{x^2}-x^2\cdot\frac{1}{x}}=\frac{1+x}{1-x}$$

**35.** $$\frac{15x^3-30x^2+10x-2}{5x^2}=\frac{15x^3}{5x^2}-\frac{30x^2}{5x^2}+\frac{10x}{5x^2}-\frac{2}{5x^2}=3x-6+\frac{2}{x}-\frac{2}{5x^2}$$

**36.** $$\frac{36x^4y^3+12x^2y^3-60x^2y^2}{6xy^2}=\frac{36x^4y^3}{6xy^2}+\frac{12x^2y^3}{6xy^2}-\frac{60x^2y^2}{6xy^2}=6x^3y+2xy-10x$$

**37.**

$$\begin{array}{r} 3x-7 \\ 2x+3\overline{)6x^2-5x+5} \\ \underline{6x^2+9x} \\ -14x+5 \\ \underline{-14x-21} \\ 26 \end{array}$$

$$\frac{6x^2-5x+5}{2x+3} = 3x-7+\frac{26}{2x+3}$$

**38.**

$$\begin{array}{r} 2x^2-4x+1 \\ 5x-3\overline{)10x^3-26x^2+17x-13} \\ \underline{10x^3-6x^2} \\ -20x^2+17x \\ \underline{-20x^2+12x} \\ 5x-13 \\ \underline{5x-3} \\ -10 \end{array}$$

$$\frac{10x^3-26x^2+17x-13}{5x-3} = 2x^2-4x+1-\frac{10}{5x-3}$$

**39.**

$$\begin{array}{r} x^5+5x^4+8x^3+16x^2+33x+63 \\ x-2\overline{)x^6+3x^5-2x^4+0x^3+x^2-3x+2} \\ \underline{x^6-2x^5} \\ 5x^5-2x^4 \\ \underline{5x^5-10x^4} \\ 8x^4+0x^3 \\ \underline{8x^4-16x^3} \\ 16x^3+x^2 \\ \underline{16x^3-32x^2} \\ 33x^2-3x \\ \underline{33x^2-66x} \\ 63x+2 \\ \underline{63x-126} \\ 128 \end{array}$$

$$\frac{x^6+3x^5-2x^4+x^2-3x+2}{x-2} = x^5+5x^4+8x^3+16x^2+33x+63+\frac{128}{x-2}$$

**40.**

$$\begin{array}{r} 2x^2+3x-1 \\ 2x^2+0x+1\overline{\big)4x^4+6x^3+0x^2+3x-1} \\ \underline{4x^4+0x^3+2x^2\phantom{+3x-1}} \\ 6x^3-2x^2+3x\phantom{-1} \\ \underline{6x^3+0x^2+3x}\phantom{-1} \\ -2x^2+0x-1 \\ \underline{-2x^2+0x-1} \\ 0 \end{array}$$

$$\frac{4x^4+6x^3+3x-1}{2x^2+1}=2x^2+3x-1$$

**41.** $(4x^3-3x^2-2x+1)\div(x+1)$

$$\begin{array}{r|rrrr} \underline{-1} & 4 & -3 & -2 & 1 \\ & & -4 & 7 & -5 \\ \hline & 4 & -7 & 5 & -4 \end{array}$$

$(4x^3-3x^2-2x+1)\div(x+1)$

$=4x^2-7x+5-\dfrac{4}{x+1}$

**42.** $(3x^4-2x^2-10x-20)\div(x-2)$

$$\begin{array}{r|rrrrr} \underline{2} & 3 & 0 & -2 & -10 & -20 \\ & & 6 & 12 & 20 & 20 \\ \hline & 3 & 6 & 10 & 10 & 0 \end{array}$$

$(3x^4-2x^2-10x-20)\div(x-2)$

$=3x^3+6x^2+10x+10$

**43.** $(x^4+16)\div(x+4)$

$$\begin{array}{r|rrrrr} \underline{-4} & 1 & 0 & 0 & 0 & 16 \\ & & -4 & 16 & -64 & 256 \\ \hline & 1 & -4 & 16 & -64 & 272 \end{array}$$

$(x^4+16)\div(x+4)$

$=x^3-4x^2+16x-64+\dfrac{272}{x+4}$

**44.** $f(x)=2x^3-5x^2+4x-1$

Divide $f(x)$ by $x-2$.

$$\begin{array}{r|rrrr} \underline{2} & 2 & -5 & 4 & -1 \\ & & 4 & -2 & 4 \\ \hline & 2 & -1 & 2 & 3 \end{array}$$

$f(2)=3$

**45.** $f(x)=3x^4+7x^3+8x^2+2x+4$

$$\begin{array}{r|rrrrr} \underline{-\frac{1}{3}} & 3 & 7 & 8 & 2 & 4 \\ & & -1 & -2 & -2 & 0 \\ \hline & 3 & 6 & 6 & 0 & 4 \end{array}$$

$f\left(-\dfrac{1}{3}\right)=4$

**46.** To show that –2 is a solution to the equation, show that when the polynomial is divided by $x+2$ the remainder is zero.

$$\begin{array}{r|rrrr} \underline{-2} & 2 & -1 & -8 & 4 \\ & & -4 & 10 & -4 \\ \hline & 2 & -5 & 2 & 0 \end{array}$$

Since the remainder is zero, $-2$ is a solution to the equation.

**47.** $x^4 - x^3 - 7x^2 + x + 6 = 0$

$$\begin{array}{r|rrrrr} 4 & 1 & -1 & -7 & 1 & 6 \\ & & 4 & 12 & 20 & 84 \\ \hline & 1 & 3 & 5 & 21 & 90 \end{array}$$

Since the remainder is not zero, 4 is not a solution to the equation.

**48.** To show that $\frac{1}{2}$ is a solution to the equation, show that when the polynomial is divided by $x - \frac{1}{2}$ the remainder is zero.

$$\begin{array}{r|rrrr} \frac{1}{2} & 6 & 1 & -4 & 1 \\ & & 3 & 2 & -1 \\ \hline & 6 & 4 & -2 & 0 \end{array}$$

$6x^3 + x^2 - 4x + 1$

$= \left(x - \frac{1}{2}\right)\left(6x^2 + 4x - 2\right)$

To solve the equation, we set it equal to zero and factor.

$$\left(x - \frac{1}{2}\right)\left(6x^2 + 4x - 2\right) = 0$$

$$\left(x - \frac{1}{2}\right)\left(2\left(3x^2 + 2x - 1\right)\right) = 0$$

$$\left(x - \frac{1}{2}\right)\left(2(3x - 1)(x + 1)\right) = 0$$

$$2\left(x - \frac{1}{2}\right)(3x - 1)(x + 1) = 0$$

$x = -1, \frac{1}{3}, \frac{1}{2}$ or $\left\{-1, \frac{1}{3}, \frac{1}{2}\right\}$

**49.** So that denominators will not equal zero, $x$ cannot equal zero. To eliminate fractions, multiply by the LCD, $3x$.

$$\frac{3}{x} + \frac{1}{3} = \frac{5}{x}$$

$$3x\left(\frac{3}{x} + \frac{1}{3}\right) = 3x\left(\frac{5}{x}\right)$$

$$3x\left(\frac{3}{x}\right) + 3x\left(\frac{1}{3}\right) = 3(5)$$

$$3(3) + x(1) = 15$$

$$9 + x = 15$$

$$x = 6$$

The solution set is $\{6\}$.

**50.** $\frac{5}{3x+4}=\frac{3}{2x-8}$

To find the restrictions on $x$, set the denominators equal to zero and solve.

$$\begin{aligned} 3x+4&=0 & 2x-8&=0 \\ 3x&=-4 & 2x&=8 \\ x&=-\tfrac{4}{3} & x&=4 \end{aligned}$$

To eliminate fractions, multiply by the LCD, $3x$.

$$\begin{aligned} \frac{5}{3x+4}&=\frac{3}{2x-8} \\ (3x+4)(2x-8)\left(\frac{5}{3x+4}\right)&=(3x+4)(2x-8)\left(\frac{3}{2x-8}\right) \\ (2x-8)(5)&=(3x+4)(3) \\ 10x-40&=9x+12 \\ x-40&=12 \\ x&=52 \end{aligned}$$

The solution set is $\{52\}$.

**51.** $\frac{1}{x-5}-\frac{3}{x+5}=\frac{6}{x^2-25}$

$\frac{1}{x-5}-\frac{3}{x+5}=\frac{6}{(x+5)(x-5)}$

So that denominators will not equal zero, $x$ cannot equal 5 or –5. To eliminate fractions, multiply by the LCD, $(x+5)(x-5)$.

$$\begin{aligned} \frac{1}{x-5}-\frac{3}{x+5}&=\frac{6}{(x+5)(x-5)} \\ (x+5)(x-5)\left(\frac{1}{x-5}-\frac{3}{x+5}\right)&=(x+5)(x-5)\left(\frac{6}{(x+5)(x-5)}\right) \\ (x+5)(x-5)\left(\frac{1}{x-5}\right)-(x+5)(x-5)\left(\frac{3}{x+5}\right)&=6 \\ (x+5)(1)-(x-5)(3)&=6 \\ x+5-(3x-15)&=6 \\ x+5-3x+15&=6 \\ -2x+20&=6 \\ -2x&=-14 \\ x&=7 \end{aligned}$$

The solution set is $\{7\}$.

**52.** $\dfrac{x+5}{x+1}-\dfrac{x}{x+2}=\dfrac{4x+1}{x^2+3x+2}$

$\dfrac{x+5}{x+1}-\dfrac{x}{x+2}=\dfrac{4x+1}{(x+2)(x+1)}$

So that denominators will not equal zero, $x$ cannot equal –1 or –2. To eliminate fractions, multiply by the LCD, $(x+2)(x+1)$.

$$\frac{x+5}{x+1}-\frac{x}{x+2}=\frac{4x+1}{(x+2)(x+1)}$$

$$(x+2)(x+1)\left(\frac{x+5}{x+1}-\frac{x}{x+2}\right)=(x+2)(x+1)\left(\frac{4x+1}{(x+2)(x+1)}\right)$$

$$(x+2)(x+1)\left(\frac{x+5}{x+1}\right)-(x+2)(x+1)\left(\frac{x}{x+2}\right)=4x+1$$

$$(x+2)(x+5)-(x+1)(x)=4x+1$$

$$x^2+7x+10-(x^2+x)=4x+1$$

$$x^2+7x+10-x^2-x=4x+1$$

$$6x+10=4x+1$$

$$2x+10=1$$

$$2x=-9$$

$$x=-\frac{9}{2}$$

The solution set is $\left\{-\dfrac{9}{2}\right\}$.

**53.** $\dfrac{2}{3}-\dfrac{5}{3x}=\dfrac{1}{x^2}$

So that denominators will not equal zero, $x$ cannot equal zero. To eliminate fractions, multiply by the LCD, $3x^2$.

$$3x^2\left(\frac{2}{3}-\frac{5}{3x}\right)=3x^2\left(\frac{1}{x^2}\right)$$

$$3x^2\left(\frac{2}{3}\right)-3x^2\left(\frac{5}{3x}\right)=3(1)$$

$$x^2(2)-x(5)=3$$

$$2x^2-5x=3$$

$$2x^2-5x-3=0$$

$$(2x+1)(x-3)=0$$

Apply the zero product principle.

$2x+1=0 \qquad x-3=0$

$2x=-1 \qquad x=3$

$x=-\frac{1}{2}$

The solution set is $\left\{-\frac{1}{2},3\right\}$.

**54.** $\frac{2}{x-1}=\frac{1}{4}+\frac{7}{x+2}$

So that denominators will not equal zero, $x$ cannot equal 1 or –2. To eliminate fractions, multiply by the LCD, $4(x-1)(x+2)$.

$$4(x-1)(x+2)\left(\frac{2}{x-1}\right)=4(x-1)(x+2)\left(\frac{1}{4}+\frac{7}{x+2}\right)$$

$$4(x+2)(2)=4(x-1)(x+2)\left(\frac{1}{4}\right)+4(x-1)(x+2)\left(\frac{7}{x+2}\right)$$

$$8x+16=(x-1)(x+2)+4(x-1)(7)$$

$$8x+16=x^2+x-2+28x-28$$

$$8x+16=x^2+29x-30$$

$$0=x^2+21x-46$$

$$0=(x+23)(x-2)$$

Apply the zero product principle.

$x+23=0 \qquad x-2=0$

$x=-23 \qquad x=2$

The solution set is $\{-23,2\}$.

**55.** $\frac{2x+7}{x+5}-\frac{x-8}{x-4}=\frac{x+18}{x^2+x-20}$

$\frac{2x+7}{x+5}-\frac{x-8}{x-4}=\frac{x+18}{(x+5)(x-4)}$

So that denominators will not equal zero, $x$ cannot equal –5 or 4. To eliminate fractions, multiply by the LCD, $(x+5)(x-4)$.

$$\frac{2x+7}{x+5}-\frac{x-8}{x-4}=\frac{x+18}{(x+5)(x-4)}$$

$$(x+5)(x-4)\left(\frac{2x+7}{x+5}-\frac{x-8}{x-4}\right)=(x+5)(x-4)\left(\frac{x+18}{(x+5)(x-4)}\right)$$

$$(x+5)(x-4)\left(\frac{2x+7}{x+5}\right)-(x+5)(x-4)\left(\frac{x-8}{x-4}\right)=x+18$$

$$(x-4)(2x+7)-(x+5)(x-8)=x+18$$

$$2x^2-x-28-(x^2-3x-40)=x+18$$

$$2x^2-x-28-x^2+3x+40=x+18$$

$$x^2+2x+12=x+18$$

$$x^2+x-6=0$$

$$(x+3)(x-2)=0$$

Apply the zero product principle: $x+3=0$ $\quad$ $x-2=0$

$x=-3$ $\quad$ $x=2$

The solution set is $\{-3,2\}$.

**56.**

$$f(x)=\frac{4x}{100-x}$$

$$16=\frac{4x}{100-x}$$

$$(100-x)(16)=(100-x)\left(\frac{4x}{100-x}\right)$$

$$1600-16x=4x$$

$$1600=20x$$

$$80=x$$

80% of the pollutants can be removed for \$16 million.

**57.**

$$P=\frac{R-C}{n}$$

$$n(P)=n\left(\frac{R-C}{n}\right)$$

$$nP=R-C$$

$$nP+C=R$$

$$C=R-nP$$

**58.**

$$\frac{P_1V_1}{T_1}=\frac{P_2V_2}{T_2}$$

$$T_1T_2\left(\frac{P_1V_1}{T_1}\right)=T_1T_2\left(\frac{P_2V_2}{T_2}\right)$$

$$T_2(P_1V_1)=T_1(P_2V_2)$$

$$P_1T_2V_1=P_2T_1V_2$$

$$\frac{P_1T_2V_1}{P_2V_2}=\frac{P_2T_1V_2}{P_2V_2}$$

$$T_1=\frac{P_1T_2V_1}{P_2V_2}$$

**59.**

$$T = \frac{A-P}{Pr}$$
$$Pr(T) = Pr\left(\frac{A-P}{Pr}\right)$$
$$PrT = A - P$$
$$PrT + P = A$$
$$P(rT+1) = A$$
$$\frac{P(rT+1)}{rT+1} = \frac{A}{rT+1}$$
$$P = \frac{A}{rT+1}$$

**60.**

$$\frac{1}{R} = \frac{1}{R_1} + \frac{1}{R_2}$$
$$RR_1R_2\left(\frac{1}{R}\right) = RR_1R_2\left(\frac{1}{R_1} + \frac{1}{R_2}\right)$$
$$R_1R_2 = RR_1R_2\left(\frac{1}{R_1}\right) + RR_1R_2\left(\frac{1}{R_2}\right)$$
$$R_1R_2 = RR_2 + RR_1$$
$$R_1R_2 = R(R_2 + R_1)$$
$$\frac{R_1R_2}{R_2 + R_1} = \frac{R(R_2 + R_1)}{R_2 + R_1}$$
$$R = \frac{R_1R_2}{R_2 + R_1}$$

**61.**

$$I = \frac{nE}{R+nr}$$
$$(R+nr)(I) = (R+nr)\left(\frac{nE}{R+nr}\right)$$
$$IR + Inr = nE$$
$$IR = nE - Inr$$
$$IR = n(E - Ir)$$
$$\frac{IR}{E-Ir} = \frac{n(E-Ir)}{E-Ir}$$
$$n = \frac{IR}{E-Ir}$$

**62.** **a.** $C(x) = 50{,}000 + 25x$

**b.** $\overline{C}(x) = \dfrac{50{,}000 + 25x}{x}$

**c.**

$$35 = \frac{50{,}000 + 25x}{x}$$
$$x(35) = x\left(\frac{50{,}000 + 25x}{x}\right)$$
$$35x = 50{,}000 + 25x$$
$$10x = 50{,}000$$
$$x = 5000$$

5000 graphing calculators must be produced each month to have an average cost of \$35.

**63.**

| | $d$ | $r$ | $t = \frac{d}{r}$ |
|---|---|---|---|
| Riding | 60 | $3x$ | $\frac{60}{3x}$ |
| Walking | 8 | $x$ | $\frac{8}{x}$ |

$$\frac{60}{3x} + \frac{8}{x} = 7$$
$$3x\left(\frac{60}{3x} + \frac{8}{x}\right) = 3x(7)$$
$$3x\left(\frac{60}{3x}\right) + 3x\left(\frac{8}{x}\right) = 21x$$
$$60 + 3(8) = 21x$$
$$60 + 24 = 21x$$
$$84 = 21x$$
$$4 = x$$

The cyclist was riding at a rate of $3x = 3(4) = 12$ miles per hour.

**64.**

| | $d$ | $r$ | $t = \frac{d}{r}$ |
|---|---|---|---|
| Down Stream | 12 | $x+3$ | $\frac{12}{x+3}$ |
| Up Stream | 12 | $x-3$ | $\frac{12}{x-3}$ |

$$\frac{12}{x+3}+\frac{12}{x-3}=3$$
$$(x+3)(x-3)\left(\frac{12}{x+3}+\frac{12}{x-3}\right)=(x+3)(x-3)(3)$$
$$(x+3)(x-3)\left(\frac{12}{x+3}\right)+(x+3)(x-3)\left(\frac{12}{x-3}\right)=3(x^2-9)$$
$$(x-3)(12)+(x+3)(12)=3x^2-27$$
$$12x-36+12x+36=3x^2-27$$
$$24x=3x^2-27$$
$$0=3x^2-24x-27$$
$$0=3(x^2-8x-9)$$
$$0=3(x-9)(x+1)$$

Apply the zero product principle.

$3(x-9)=0 \qquad x+1=0$

$x=9 \qquad x=-1$

The solutions are –1 and 9. We disregard –1 because we cannot have a negative rate. The boat's rate in still water is 9 miles per hour.

**65.**

| | Part done in 1 hour | Time working together | Part done in $x$ hours |
|---|---|---|---|
| First Person | $\frac{1}{3}$ | $x$ | $\frac{x}{3}$ |
| Second Person | $\frac{1}{6}$ | $x$ | $\frac{x}{6}$ |

$$\frac{x}{3}+\frac{x}{6}=1$$
$$6\left(\frac{x}{3}+\frac{x}{6}\right)=6(1)$$
$$6\left(\frac{x}{3}\right)+6\left(\frac{x}{6}\right)=6$$
$$2x+x=6$$
$$3x=6$$
$$x=2$$

If they work together, it will take 2 hours to clean the house. There is not enough time to finish the job before the TV program starts.

**66.**

| | Part Done in 1 Hour | Time Working Together | Part Done in 20 Hours |
|---|---|---|---|
| Fast Crew | $\frac{1}{x-9}$ | 20 | $\frac{20}{x-9}$ |
| Slow Crew | $\frac{1}{x}$ | 20 | $\frac{20}{x}$ |

$$\frac{20}{x-9}+\frac{20}{x}=1$$

$$x(x-9)\left(\frac{20}{x-9}+\frac{20}{x}\right)=x(x-9)(1)$$

$$x(x-9)\left(\frac{20}{x-9}\right)+x(x-9)\left(\frac{20}{x}\right)=x^2-9x$$

$$x(20)+(x-9)(20)=x^2-9x$$

$$20x+20x-180=x^2-9x$$

$$40x-180=x^2-9x$$

$$0=x^2-49x+180$$

$$0=x^2-49x+180$$

$$0=(x-45)(x-4)$$

Apply the zero product principle.

$x-45=0 \qquad x-4=0$

$x=45 \qquad x=4$

The solutions are 45 and 4. We disregard 4, because the fast crew's rate would be 4 – 9 = – 5. No crew can do the job in a negative number of hours. It would take the slow crew 45 hours and the fast crew 45 – 9 = 36 hours to complete the job working alone.

**67.**

| | Part done in 1 minute | Time working together | Part done in $x$ minutes |
|---|---|---|---|
| Faucet | $\frac{1}{60}$ | $x$ | $\frac{x}{60}$ |
| Drain | $\frac{1}{80}$ | $x$ | $\frac{x}{80}$ |

$$\frac{x}{60}-\frac{x}{80}=1$$

$$240\left(\frac{x}{60}-\frac{x}{80}\right)=240(1)$$

$$240\left(\frac{x}{60}\right)-240\left(\frac{x}{80}\right)=240$$

$$4x-3x=240$$

$$x=240$$

It will take 240 minutes or 4 hours to fill the pond.

**68.** Since the profit varies directly as the number of products sold, we have
$p = kn.$
Use the given values to find $k$.

$$p = kn.$$
$$1175 = k(25)$$
$$\frac{1175}{25} = \frac{k(25)}{25}$$
$$47 = k$$

The equation becomes $p = 47n$
When $n = 105$ products,
$p = 47(105) = 4935$.
If 105 products are sold, the company's profit is $4935.

**69.** Since distance varies directly as the square of the time, we have $d = kt^2$.
Use the given values to find $k$.

$$d = kt^2$$
$$144 = k(3)^2$$
$$144 = k(9)$$
$$\frac{144}{9} = \frac{k(9)}{9}$$
$$16 = k$$

The equation becomes $d = 16t^2$.
The equation becomes $d = 16t^2$.
When $t = 10$,
$d = 16(10)^2 = 16(100) = 1600.$
A skydiver will fall 1600 feet in 10 seconds.

**70.** Since the pitch of a musical tone varies inversely as its wavelength, we have $p = \frac{k}{w}$.
Use the given values to find $k$.

$$p = \frac{k}{w}$$
$$660 = \frac{k}{1.6}$$
$$660(1.6) = 1.6\left(\frac{k}{1.6}\right)$$
$$1056 = k$$

The equation becomes $p = \frac{1056}{w}$.
When $w = 2.4$, $p = \frac{1056}{2.4} = 440.$
The tone's pitch is 440 vibrations per second.

**71.** Since loudness varies inversely as the square of the distance, we have
$l = \frac{k}{d^2}$.
Use the given values to find $k$.

$$l = \frac{k}{d^2}$$
$$28 = \frac{k}{8^2}$$
$$28 = \frac{k}{64}$$
$$64(28) = 64\left(\frac{k}{64}\right)$$
$$1792 = k$$

The equation becomes $l = \frac{1792}{d^2}$.
When $d = 4$, $l = \frac{1792}{(4)^2} = \frac{1792}{16} = 112.$
At a distance of 4 feet, the loudness of the stereo is 112 decibels.

**72.** Since time varies directly as the number of computers and inversely as the number of workers, we have $t = \frac{kn}{w}$.

Use the given values to find *k*.

$$t = \frac{kn}{w}$$
$$10 = \frac{k(30)}{6}$$
$$10 = 5k$$
$$\frac{10}{5} = \frac{5k}{5}$$
$$2 = k$$

The equation becomes $t = \frac{2n}{w}$. When $n = 40$ and $w = 5$,

$$t = \frac{2(40)}{5} = \frac{80}{5} = 16.$$

It will take 16 hours for 5 workers to assemble 40 computers.

**73.** Since the volume varies jointly as height and the area of the base, we have $v = kha$.

Use the given values to find *k*.

$$175 = k(15)(35)$$
$$175 = k(525)$$
$$\frac{175}{525} = \frac{k(525)}{525}$$
$$\frac{1}{3} = k$$

The equation becomes $v = \frac{1}{3}ha$.

When $h$ = 20 feet and $a$ = 120 square feet, $v = \frac{1}{3}(20)(120) = 800.$

If the height is 20 feet and the area is 120 square feet, the volume will be 800 cubic feet.

## Chapter 6 Test

**1.** The domain is all real numbers except those that make the denominator zero. To find these values, set the denominator equal to zero and solve.

$$x^2 - 7x + 10 = 0$$
$$(x-5)(x-2) = 0$$

Apply the zero product principle.

$$x - 5 = 0 \qquad x - 2 = 0$$
$$x = 5 \qquad x = 2$$

The domain of $f = \{x \mid x$ is a real number and $x \neq 2$ and $x \neq 5\}$ or $(-\infty, 2) \cup (2, 5) \cup (5, \infty)$.

$$f(x) = \frac{x^2 - 2x}{x^2 - 7x + 10}$$
$$= \frac{x(\cancel{x-2})}{(x-5)(\cancel{x-2})} = \frac{x}{x-5}$$

**2.**
$$\frac{x^2}{x^2-16} \cdot \frac{x^2+7x+12}{x^2+3x}$$
$$= \frac{x^2}{(x+4)(x-4)} \cdot \frac{(x+4)(x+3)}{x(x+3)}$$
$$= \frac{x^{\cancel{2}}}{(\cancel{x+4})(x-4)} \cdot \frac{(\cancel{x+4})(\cancel{x+3})}{x(\cancel{x+3})}$$
$$= \frac{x}{x-4}$$

**3.**

$$\frac{x^3+27}{x^2-1}\div\frac{x^2-3x+9}{x^2-2x+1}$$

$$=\frac{x^3+27}{x^2-1}\cdot\frac{x^2-2x+1}{x^2-3x+9}$$

$$=\frac{(x+3)\cancel{(x^2-3x+9)}}{(x+1)\cancel{(x-1)}}\cdot\frac{(x-1)^{\cancel{2}}}{\cancel{x^2-3x+9}}$$

$$=\frac{(x+3)(x-1)}{x+1}$$

**4.**

$$\frac{x^2+3x-10}{x^2+4x+3}\cdot\frac{x^2+x-6}{x^2+10x+25}\cdot\frac{x+1}{x-2}$$

$$=\frac{\cancel{(x+5)}\cancel{(x-2)}}{\cancel{(x+3)}\cancel{(x+1)}}\cdot\frac{\cancel{(x+3)}(x-2)}{\cancel{(x+5)}(x+5)}\cdot\frac{\cancel{x+1}}{\cancel{x-2}}$$

$$=\frac{x-2}{x+5}$$

**5.**

$$\frac{x^2-6x-16}{x^3+3x^2+2x}\cdot\left(x^2-3x-4\right)\div\frac{x^2-7x+12}{3x}$$

$$=\frac{(x-8)(x+2)}{x(x^2+3x+2)}\cdot(x-4)(x+1)\div\frac{(x-4)(x-3)}{3x}$$

$$=\frac{(x-8)\cancel{(x+2)}}{\cancel{x}\cancel{(x+1)}\cancel{(x+2)}}\cdot\cancel{(x-4)}\cancel{(x+1)}\cdot\frac{3\cancel{x}}{\cancel{(x-4)}(x-3)}$$

$$=\frac{3(x-8)}{x-3}=\frac{3x-24}{x-3}$$

**6.**

$$\frac{x^2-5x-2}{6x^2-11x-35}-\frac{x^2-7x+5}{6x^2-11x-35}$$

$$=\frac{x^2-5x-2-\left(x^2-7x+5\right)}{6x^2-11x-35}$$

$$=\frac{x^2-5x-2-x^2+7x-5}{6x^2-11x-35}$$

$$=\frac{2x-7}{6x^2-11x-35}$$

$$=\frac{\cancel{2x-7}}{\cancel{(2x-7)}(3x+5)}=\frac{1}{3x+5}$$

**7.**

$$\frac{x}{x+3}+\frac{5}{x-3}$$

$$=\frac{x(x-3)}{(x+3)(x-3)}+\frac{5(x+3)}{(x+3)(x-3)}$$

$$=\frac{x(x-3)+5(x+3)}{(x+3)(x-3)}$$

$$=\frac{x^2-3x+5x+15}{(x+3)(x-3)}$$

$$=\frac{x^2+2x+15}{(x+3)(x-3)}$$

**8.** 
$$\frac{2}{x^2-4x+3}+\frac{3x}{x^2+x-2}=\frac{2}{(x-3)(x-1)}+\frac{3x}{(x-1)(x+2)}$$
$$=\frac{2(x+2)}{(x-3)(x-1)(x+2)}+\frac{3x(x-3)}{(x-3)(x-1)(x+2)}$$
$$=\frac{2(x+2)+3x(x-3)}{(x-3)(x-1)(x+2)}=\frac{2x+4+3x^2-9x}{(x-3)(x-1)(x+2)}$$
$$=\frac{3x^2-7x+4}{(x-3)(x-1)(x+2)}=\frac{(3x-4)\cancel{(x-1)}}{(x-3)\cancel{(x-1)}(x+2)}=\frac{3x-4}{(x+2)(x-3)}$$

**9.** 
$$\frac{5x}{x^2-4}-\frac{2}{x^2+x-2}=\frac{5x}{(x+2)(x-2)}-\frac{2}{(x+2)(x-1)}=$$
$$\frac{5x(x-1)}{(x+2)(x-2)(x-1)}-\frac{2(x-2)}{(x+2)(x-2)(x-1)}$$
$$=\frac{5x(x-1)-2(x-2)}{(x+2)(x-2)(x-1)}=\frac{5x^2-5x-(2x-4)}{(x+2)(x-2)(x-1)}$$
$$=\frac{5x^2-5x-2x+4}{(x+2)(x-2)(x-1)}=\frac{5x^2-7x+4}{(x+2)(x-2)(x-1)}$$

**10.** 
$$\frac{x-4}{x-5}-\frac{3}{x+5}-\frac{10}{x^2-25}=\frac{(x-4)(x+5)}{(x+5)(x-5)}-\frac{3(x-5)}{(x+5)(x-5)}-\frac{10}{(x+5)(x-5)}$$
$$=\frac{(x-4)(x+5)-3(x-5)-10}{(x+5)(x-5)}=\frac{x^2+x-20-3x+15-10}{(x+5)(x-5)}$$
$$=\frac{x^2-2x-15}{(x+5)(x-5)}=\frac{\cancel{(x-5)}(x+3)}{(x+5)\cancel{(x-5)}}=\frac{x+3}{x+5}$$

**11.** 
$$\frac{1}{10-x}+\frac{x-1}{x-10}=\frac{-1(1)}{-1(10-x)}+\frac{x-1}{x-10}=\frac{-1}{x-10}+\frac{x-1}{x-10}=\frac{-1+x-1}{x-10}=\frac{x-2}{x-10}$$

**12.** 
$$\frac{\frac{x}{4}-\frac{1}{x}}{1+\frac{x+4}{x}}=\frac{4x}{4x}\cdot\frac{\frac{x}{4}-\frac{1}{x}}{1+\frac{x+4}{x}}=\frac{\cancel{4}x\cdot\frac{x}{\cancel{4}}-4\cancel{x}\cdot\frac{1}{\cancel{x}}}{4x\cdot 1+4\cancel{x}\cdot\frac{x+4}{\cancel{x}}}=\frac{x^2-4}{4x+4(x+4)}=\frac{(x+2)(x-2)}{4x+4x+16}$$
$$=\frac{(x+2)(x-2)}{8x+16}=\frac{\cancel{(x+2)}(x-2)}{8\cancel{(x+2)}}=\frac{x-2}{8}$$

**13.** $$\frac{\frac{1}{x}-\frac{3}{x+2}}{\frac{2}{x^2+2x}}=\frac{\frac{1}{x}-\frac{3}{x+2}}{\frac{2}{x(x+2)}}$$

$$=\frac{x(x+2)}{x(x+2)}\cdot\frac{\frac{1}{x}-\frac{3}{x+2}}{\frac{2}{x(x+2)}}$$

$$=\frac{x(x+2)\cdot\frac{1}{x}-x(x+2)\cdot\frac{3}{x+2}}{x(x+2)\cdot\frac{2}{x(x+2)}}$$

$$=\frac{x+2-3x}{2}=\frac{-2x+2}{2}$$

$$=\frac{-\cancel{2}(x-1)}{\cancel{2}}=-(x-1)=1-x$$

**14.** $$\frac{12x^4y^3+16x^2y^3-10x^2y^2}{4x^2y}$$

$$=\frac{12x^4y^3}{4x^2y}+\frac{16x^2y^3}{4x^2y}-\frac{10x^2y^2}{4x^2y}$$

$$=3x^2y^2+4y^2-\frac{5y}{2}$$

**15.**

$$\begin{array}{r} 3x^2-3x+1 \\ 3x+2\,\overline{)\,9x^3-3x^2-3x+4} \\ \underline{9x^3+6x^2\phantom{-3x+4}} \\ -9x^2-3x\phantom{+4} \\ \underline{-9x^2-6x\phantom{+4}} \\ 3x+4 \\ \underline{3x+2} \\ 2 \end{array}$$

$$\frac{9x^3-3x^2-3x+4}{3x+2}$$

$$=3x^2-3x+1+\frac{2}{3x+2}$$

**16.**

$$\begin{array}{r} 3x^2+2x+3 \\ x^2+0x-1\,\overline{)\,3x^4+2x^3+0x^2-8x+6} \\ \underline{3x^4+0x^3-3x^2\phantom{-8x+6}} \\ 2x^3+3x^2-8x\phantom{+6} \\ \underline{2x^3+0x^2-2x\phantom{+6}} \\ 3x^2-6x+6 \\ \underline{3x^2+0x-3} \\ -6x+9 \end{array}$$

$$\frac{3x^4+2x^3-8x+6}{x^2-1}$$

$$=3x^2+2x+3+\frac{-6x+9}{x^2-1}$$

$$=3x^2+2x+3-\frac{6x-9}{x^2-1}$$

**17.** $(3x^4+11x^3-20x^2+7x+35)\div(x+5)$

$$\begin{array}{r|rrrrr} \underline{-5} & 3 & 11 & -20 & 7 & 35 \\ & & -15 & 20 & 0 & -35 \\ \hline & 3 & -4 & 0 & 7 & 0 \end{array}$$

$$(3x^4+11x^3-20x^2+7x+35)\div(x+5)$$

$$=3x^3-4x^2+7$$

**18.** Divide $f(x)$ by $x-(-2)=x+2$.

$$\begin{array}{r|rrrrr} \underline{-2} & 1 & -2 & -11 & 5 & 34 \\ & & -2 & 8 & 6 & -22 \\ \hline & 1 & -4 & -3 & 11 & 12 \end{array}$$

$f(-2)=12$

**19.** $2x^3-3x^2-11x+6$

$$\begin{array}{r|rrrr} \underline{-2} & 2 & -3 & -11 & 6 \\ & & -4 & 14 & -6 \\ \hline & 2 & -7 & 3 & 0 \end{array}$$

Since the remainder is 0, $-2$ is a solution.

**20.** $\frac{x}{x+4}=\frac{11}{x^2-16}+2$

$\frac{x}{x+4}=\frac{11}{(x-4)(x+4)}+\frac{2(x-4)(x+4)}{(x-4)(x+4)}$

$\frac{x}{x+4}=\frac{11+2(x-4)(x+4)}{(x-4)(x+4)}$

So that denominators will not equal zero, $x$ cannot equal 4 and $-4$. To eliminate fractions, multiply by the LCD, $(x-4)(x+4)$.

$(x-4)(x+4)\frac{x}{x+4}=(x-4)(x+4)\frac{11+2(x-4)(x+4)}{(x-4)(x+4)}$

$x(x-4)=11+2(x-4)(x+4)$

$x^2-4x=11+2(x^2-16)$

$x^2-4x=11+2x^2-32$

$0=x^2+4x-21$

$0=(x+7)(x-3)$

$x=-7$ or $x=3$

The solution set is $\{-7, 3\}$.

**21.** $\frac{x+1}{x^2+2x-3}-\frac{1}{x+3}=\frac{1}{x-1}$

$\frac{x+1}{(x+3)(x-1)}-\frac{1}{x+3}=\frac{1}{x-1}$

So that denominators will not equal zero, $x$ cannot equal 1 or –3. To eliminate fractions, multiply by the LCD, $(x+3)(x-1)$.

$$(x+3)(x-1)\left(\frac{x+1}{(x+3)(x-1)}-\frac{1}{x+3}\right)=(x+3)(x-1)\left(\frac{1}{x-1}\right)$$

$$(x+3)(x-1)\left(\frac{x+1}{(x+3)(x-1)}\right)-(x+3)(x-1)\left(\frac{1}{x+3}\right)=(x+3)(1)$$

$$x+1-(x-1)=x+3$$

$$x+1-x+1=x+3$$

$$2=x+3$$

$$-1=x$$

The solution set is $\{-1\}$.

**22.**

$$f(t)=\frac{250(3t+5)}{t+25}$$

$$125=\frac{250(3t+5)}{t+25}$$

$$(t+25)(125)=(t+25)\left(\frac{250(3t+5)}{t+25}\right)$$

$$125t+3125=250(3t+5)$$

$$125t+3125=750t+1250$$

$$-625t+3125=1250$$

$$-625t=-1875$$

$$t=3$$

It will take 3 years for the elk population to reach 125.

**23.**

$$R=\frac{as}{a+s}$$

$$(a+s)R=(a+s)\left(\frac{as}{a+s}\right)$$

$$aR+Rs=as$$

$$aR=as-Rs$$

$$aR-as=-Rs$$

$$a(R-s)=-Rs$$

$$\frac{a(R-s)}{R-s}=-\frac{Rs}{R-s}$$

$$a=-\frac{Rs}{R-s} \text{ or } \frac{Rs}{s-R}$$

**24.** **a.** $C(x)=300{,}000+10x$

**b.** $\overline{C}(x)=\dfrac{300{,}000+10x}{x}$

**c.**

$$25=\frac{300{,}000+10x}{x}$$

$$x(25)=x\left(\frac{300{,}000+10x}{x}\right)$$

$$25x=300{,}000+10x$$

$$15x=300{,}000$$

$$x=20{,}000$$

20,000 televisions must be produced for the average cost to be \$25.

**25.**

| | Part Done in 1 Hour | Time Working Together | Part Done in $x$ Hours |
|---|---|---|---|
| Fill Pipe | $\frac{1}{3}$ | $x$ | $\frac{x}{3}$ |
| Drain Pipe | $\frac{1}{4}$ | $x$ | $\frac{x}{4}$ |

$$\frac{x}{3}-\frac{x}{4}=1$$

$$12\left(\frac{x}{3}-\frac{x}{4}\right)=12(1)$$

$$12\left(\frac{x}{3}\right)-12\left(\frac{x}{4}\right)=12$$

$$4x-3x=12$$

$$x=12$$

It will take 12 hours to fill the pool.

**26.**

| | $d$ | $r$ | $t = \dfrac{d}{r}$ |
|---|---|---|---|
| Down Stream | 3 | $20 + x$ | $\dfrac{3}{20 + x}$ |
| Up Stream | 2 | $20 - x$ | $\dfrac{2}{20 - x}$ |

$$\frac{3}{20+x} = \frac{2}{20-x}$$
$$(20+x)(20-x)\left(\frac{3}{20+x}\right) = (20+x)(20-x)\left(\frac{2}{20-x}\right)$$
$$(20-x)(3) = (20+x)(2)$$
$$60 - 3x = 40 + 2x$$
$$60 - 5x = 40$$
$$-5x = -20$$
$$x = 4$$

The current's rate in still water is 4 miles per hour.

**27.** Since intensity varies inversely as the square of the distance, we have $I = \dfrac{k}{d^2}$.

Use the given values to find $k$.

$$I = \frac{k}{d^2}$$
$$20 = \frac{k}{15^2}$$
$$20 = \frac{k}{225}$$
$$225(20) = 225\left(\frac{k}{225}\right)$$
$$4500 = k$$

The equation becomes $I = \dfrac{4500}{d^2}$. When $d = 10$, $I = \dfrac{4500}{10^2} = \dfrac{4500}{100} = 45$.

At a distance of 10 feet, the light's intensity if 45 foot-candles.

## Cumulative Review Exercises (Chapters 1-6)

**1.** $2x+5\le 11$ and $-3x>18$

$2x\le 6$ $\qquad \dfrac{-3x}{-3}<\dfrac{18}{-3}$

$x\le 3$ $\qquad x<-6$

$x\le 3$

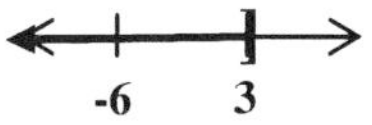


$x<-6$

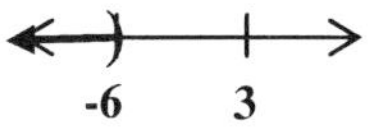


$x\le 3$ and $x<-6$

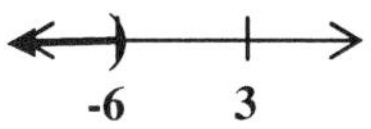


The solution set is $\{x|x<-6\}$ or $(-\infty,-6)$.

**2.**

$$2x^2=7x+4$$
$$2x^2-7x-4=0$$
$$(2x+1)(x-4)=0$$

Apply the zero product principle.

$2x+1=0 \qquad x-4=0$

$2x=-1 \qquad x=4$

$x=-\dfrac{1}{2}$

The solutions are $-\dfrac{1}{2}$ and 4 and the solution set is $\left\{-\dfrac{1}{2},4\right\}$.

**3.**

$$4x+3y+3z=4$$
$$3x+2z=2$$
$$2x-5y=-4$$

Multiply the first equation by 5 and the third equation by 3.

$$20x+15y+15z=20$$
$$6x-15y=-12$$
$$26x+15z=8$$

We now have a system of two equations in two variables.

$$3x+2z=2$$
$$26x+15z=8$$

Multiply the first equation by –15 and the second equation by 2.

$$-45x-30z=-30$$
$$52x+30z=16$$
$$7x=-14$$
$$x=-2$$

Back-substitute –2 for $x$ to find $z$.

$$3x+2z=2$$
$$3(-2)+2z=2$$
$$-6+2z=2$$
$$2z=8$$
$$z=4$$

Back-substitute–2 for $x$ to find $y$.

$$2x-5y=-4$$
$$2(-2)-5y=-4$$
$$-4-5y=-4$$
$$-5y=0$$
$$y=0$$

The solution is $(-2,0,4)$ and the solution set is $\{(-2,0,4)\}$.

**4.**

$$|3x-4| \le 10$$
$$-10 \le 3x-4 \le 10$$
$$-10+4 \le 3x-4+4 \le 10+4$$
$$-6 \le 3x \le 14$$
$$\frac{-6}{3} \le \frac{3x}{3} \le \frac{14}{3}$$
$$-2 \le x \le \frac{14}{3}$$

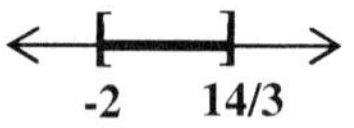


The solution set is $\left\{ x \middle| -2 \le x \le \frac{14}{3} \right\}$ or $\left[-2, \frac{14}{3}\right]$.

**5.**

$$\frac{x}{x-8} + \frac{6}{x-2} = \frac{x^2}{x^2-10x+16}$$
$$\frac{x}{x-8} + \frac{6}{x-2} = \frac{x^2}{(x-8)(x-2)}$$

So that denominators will not equal zero, $x$ cannot equal 2 or 8. To eliminate fractions, multiply by the LCD, $(x-8)(x-2)$.

$$(x-8)(x-2)\left(\frac{x}{x-8} + \frac{6}{x-2}\right) = (x-8)(x-2)\left(\frac{x^2}{(x-8)(x-2)}\right)$$
$$(x-8)(x-2)\left(\frac{x}{x-8}\right) + (x-8)(x-2)\left(\frac{6}{x-2}\right) = x^2$$
$$(x-2)(x) + (x-8)(6) = x^2$$
$$\cancel{x^2} - 2x + 6x - 48 = \cancel{x^2}$$
$$4x - 48 = 0$$
$$4x = 48$$
$$x = 12$$

The solution set is $\{12\}$.

**6.**

$$I = \frac{2R}{w+2s}$$

$$(w+2s)(I) = (w+2s)\left(\frac{2R}{w+2s}\right)$$

$$Iw + 2Is = 2R$$

$$2Is = 2R - Iw$$

$$\frac{2Is}{2I} = \frac{2R - Iw}{2I}$$

$$s = \frac{2R - Iw}{2I}$$

**7.** $2x - y = 4$
$x + y = 5$

$$2x - y = 4$$
$$-y = -2x + 4$$
$$y = 2x - 4$$

m = 2

$y -$ intercept $= -4$

$$x + y = 5$$
$$y = -x + 5$$

m = −1

$y -$ intercept $= 5$

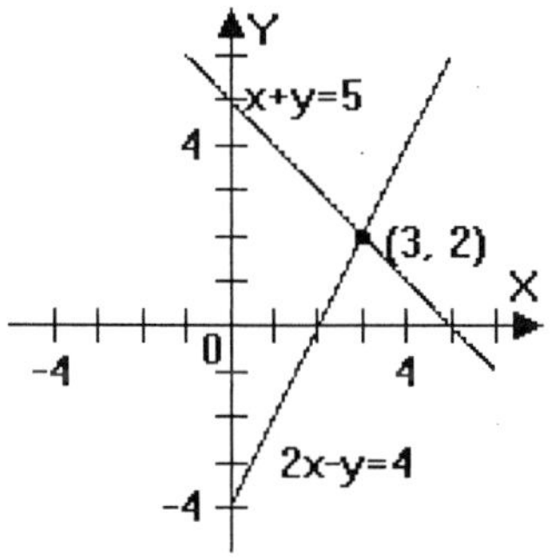


The solution set is $\{(3,2)\}$.

**8.** Slope $= -3$, passing through $(1, -5)$

Point-Slope Form

$$y - y_1 = m(x - x_1)$$
$$y - (-5) = -3(x - 1)$$
$$y + 5 = -3(x - 1)$$

Slope-Intercept Form

$$y + 5 = -3(x - 1)$$
$$y + 5 = -3x + 3$$
$$y = -3x - 2$$

In function notation, the equation of the line is $f(x) = -3x - 2$.

**9.** $y = |x| + 2$

| $x$ | $(x, y)$ |
|---|---|
| $-3$ | $(-3, 5)$ |
| $-2$ | $(-2, 4)$ |
| $-1$ | $(-1, 3)$ |
| $0$ | $(0, 2)$ |
| $1$ | $(1, 3)$ |
| $2$ | $(2, 4)$ |
| $3$ | $(3, 5)$ |

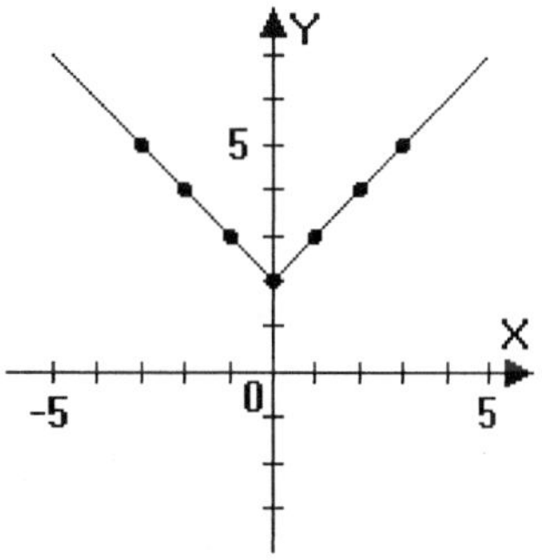

**10.** $y \geq 2x - 1$
$x \geq 1$
First consider $y \geq 2x - 1$. Replace the inequality symbol with an equal sign and we have $y = 2x - 1$. Since the equation is in slope-intercept form, we know that the $y$-intercept is –1 and the slope is 2.
Now, use the origin, $(0,0)$, as a test point.

$y \geq 2x - 1$

$0 \geq 2(0) - 1$

$0 \geq -1$

This is a true statement. This means that the point $(0,0)$ will fall in the shaded half-plane.
Next consider $x \geq 1$. Replace the inequality symbol with an equal sign and we have $x = 1$. We know that equations of the form $x = b$ are vertical lines through the point $(b, 0)$. Since the inequality is greater than or equal to, we know that the shading will extend from $x = 1$ toward $\infty$.
Next, graph each of the inequalities. The solution to the system is the intersection of the shaded half-planes.

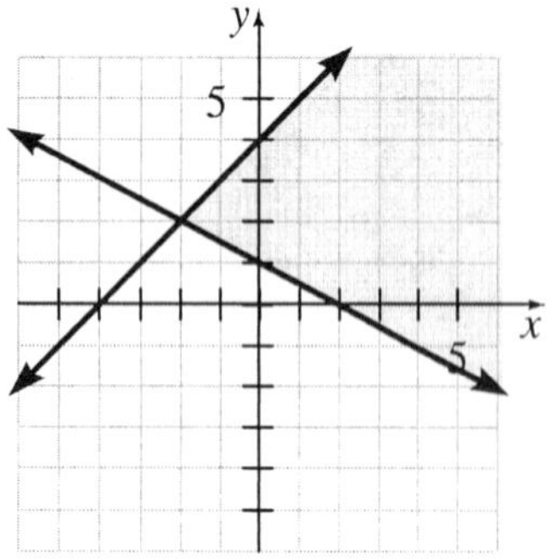


**11.** $2x - y < 4$
First graph the line, $2x - y = 4$. Solve for $y$ to obtain slope-intercept form.

$2x - y = 4$

$-y = -2x + 4$

$y = 2x - 4$

slope = 2
$y$–intercept = –4
Now, use the origin, $(0,0)$, as a test point.

$2x - y < 4$

$2(0) - 0 < 4$

$0 - 0 < 4$

$0 < 4$

This is a true statement. This means that the point $(0,0)$ will fall in the shaded half-plane.
Next, graph the inequality.

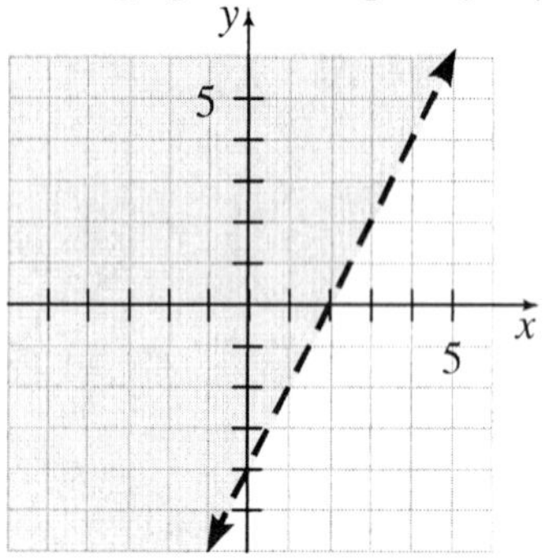


**12.** $[(x+2)+3y][(x+2)-3y]$

$= \left[(x+2)^2 - (3y)^2\right]$

$= \left[(x^2 + 4x + 4) - 9y^2\right]$

$= x^2 + 4x + 4 - 9y^2$

**13.** $\dfrac{2x^2+x-1}{2x^2-9x+4}\div\dfrac{6x+15}{3x^2-12x}$

$=\dfrac{2x^2+x-1}{2x^2-9x+4}\cdot\dfrac{3x^2-12x}{6x+15}$

$=\dfrac{\cancel{(2x-1)}(x+1)}{\cancel{(2x-1)}\cancel{(x-4)}}\cdot\dfrac{\cancel{3}x\cancel{(x-4)}}{\cancel{3}(2x+5)}$

$=\dfrac{x(x+1)}{2x+5}$

**14.** $\dfrac{3x}{x^2-9x+20}-\dfrac{5}{2x-8}$

$=\dfrac{3x}{(x-4)(x-5)}-\dfrac{5}{2(x-4)}$

The LCD is $2(x-4)(x-5)$.

$\dfrac{3x}{(x-4)(x-5)}-\dfrac{5}{2(x-4)}$

$=\dfrac{2\cdot 3x}{2(x-4)(x-5)}-\dfrac{5(x-5)}{2(x-4)(x-5)}$

$=\dfrac{6x-5(x-5)}{2(x-4)(x-5)}$

$=\dfrac{6x-5x+25}{2(x-4)(x-5)}$

$=\dfrac{x+25}{2(x-4)(x-5)}$

**15.**

$$\begin{array}{r}
3x+4 \\
x+2\overline{\smash{)}3x^2+10x+10} \\
\underline{3x^2+\ 6x\phantom{+10}} \\
4x+10 \\
\underline{4x+\ 8} \\
2
\end{array}$$

$\dfrac{3x^2+10x+10}{x+2}=3x+4+\dfrac{2}{x+2}$

**16.** $xy-6x+2y-12$

$=x(y-6)+2(y-6)$

$=(y-6)(x+2)$

**17.** $24x^3y+16x^2y-30xy$

$=2xy(12x^2+8x-15)$

$=2xy(2x+3)(6x-5)$

**18.** $s(t)=-16t^2+48t+64$

$0=-16t^2+48t+64$

$0=-16(t^2-3t-4)$

$0=-16(t-4)(t+1)$

Apply the zero product principle.

$-16(t-4)=0 \qquad t+1=0$

$t-4=0 \qquad t=-1$

$t=4$

The solutions are –1 and 4. Disregard –1 because we can't have a negative time measurement. The ball will hit the ground in 4 seconds.

**19.** Let $x$ = the cost for the basic cable service

Let $y$ = the cost for a movie channel

$x+\ \ y=35$

$x+2y=45$

Solve the first equation for $x$.

$x+y=35$

$x=35-y$

Substitute $35-y$ for $x$ in the second equation to find $y$.

$x+2y=45$

$(35-y)+2y=45$

$35-y+2y=45$

$35+y=45$

$y=10$

Back-substitute 10 for $y$ to find $x$.

$x = 35 - 10$

$x = 25$

The cost of basic cable is \$25 and the cost for each movie channel is \$10.

**20.**

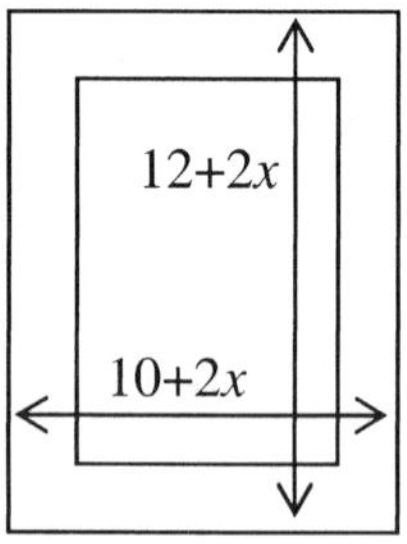


$$A = lw$$
$$168 = (12 + 2x)(10 + 2x)$$
$$168 = 120 + 44x + 4x^2$$
$$0 = -48 + 44x + 4x^2$$
$$0 = 4x^2 + 44x - 48$$
$$0 = 4(x^2 + 11x - 12)$$
$$0 = 4(x + 12)(x - 1)$$

Apply the zero product principle.

$4(x+12) = 0$ $\qquad$ $x - 1 = 0$

$x + 12 = 0$ $\qquad$ $x = 1$

$x = -12$

The solutions are –12 and 1. We disregard –12 because we cannot have a negative length measurement. The width of the rock border is 1 foot.

# Chapter 7
# Radicals, Radical Functions, and Rational Exponents

## 7.1 Exercise Set

**1.** $\sqrt{36} = 6$ because $6^2 = 36$

**3.** $-\sqrt{36} = -6$ because $6^2 = 36$

**5.** $\sqrt{-36}$
Not a real number

**7.** $\sqrt{\frac{1}{25}} = \frac{1}{5}$ because $\left(\frac{1}{5}\right)^2 = \frac{1}{25}$

**9.** $-\sqrt{\frac{9}{16}} = -\frac{3}{4}$ because $\left(\frac{3}{4}\right)^2 = \frac{9}{16}$

**11.** $\sqrt{0.81} = 0.9$ because $(0.9)^2 = 0.81$

**13.** $-\sqrt{0.04} = -0.2$ because $(0.2)^2 = 0.04$

**15.** $\sqrt{25-16} = \sqrt{9} = 3$

**17.** $\sqrt{25} - \sqrt{16} = 5 - 4 = 1$

**19.** $\sqrt{16-25} = \sqrt{-9}$
Not a real number

**21.** $f(x) = \sqrt{x-2}$
$f(18) = \sqrt{18-2} = \sqrt{16} = 4$
$f(3) = \sqrt{3-2} = \sqrt{1} = 1$
$f(2) = \sqrt{2-2} = \sqrt{0} = 0$
$f(-2) = \sqrt{-2-2} = \sqrt{-4}$
Not a real number

**23.** $g(x) = -\sqrt{2x+3}$
$g(11) = -\sqrt{2(11)+3}$
$= -\sqrt{22+3}$
$= -\sqrt{25} = -5$
$g(1) = -\sqrt{2(1)+3}$
$= -\sqrt{2+3}$
$= -\sqrt{5} \approx -2.24$
$g(-1) = -\sqrt{2(-1)+3}$
$= -\sqrt{-2+3}$
$= -\sqrt{1} = -1$
$g(-2) = -\sqrt{2(-2)+3}$
$= -\sqrt{-4+3} = -\sqrt{-1}$
Not a real number

**25.** $h(x) = \sqrt{(x-1)^2}$
$h(5) = \sqrt{(5-1)^2} = \sqrt{(4)^2} = |4| = 4$
$h(3) = \sqrt{(3-1)^2} = \sqrt{(2)^2} = |2| = 2$
$h(0) = \sqrt{(0-1)^2} = \sqrt{(-1)^2} = |-1| = 1$
$h(-5) = \sqrt{(-5-1)^2} = \sqrt{(-6)^2}$
$= |-6| = 6$

**27.** To find the domain, set the radicand greater than or equal to zero and solve.
$x - 3 \ge 0$
$x \ge 3$
The domain of $f$ is $\{x \mid x \ge 3\}$ or $[3, \infty)$. This corresponds to graph (c).

**29.** To find the domain, set the radicand greater than or equal to zero and solve.
$3x+15 \geq 0$
$3x \geq -15$
$x \geq -5$
The domain of $f$ is $\{x|x \geq -5\}$ or $[-5,\infty)$. This corresponds to graph (d).

**31.** To find the domain, set the radicand greater than or equal to zero and solve.
$6-2x \geq 0$
$-2x \geq -6$
$x \leq 3$
The domain of $f$ is $\{x|x \leq 3\}$ or $(-\infty,3]$. This corresponds to graph (e).

**33.** $\sqrt{5^2} = |5| = 5$

**35.** $\sqrt{(-4)^2} = |-4| = 4$

**37.** $\sqrt{(x-1)^2} = |x-1|$

**39.** $\sqrt{36x^4} = \sqrt{(6x^2)^2} = |6x^2| = 6x^2$

**41.** $-\sqrt{100x^6} = -\sqrt{(10x^3)^2}$
$= -|10x^3| = -10|x^3|$

**43.** $\sqrt{x^2+12x+36} = \sqrt{(x+6)^2} = |x+6|$

**45.** $-\sqrt{x^2-8x+16} = -\sqrt{(x-4)^2}$
$= -|x-4|$

**47.** $\sqrt[3]{27} = 3$ because $3^3 = 27$

**49.** $\sqrt[3]{-27} = -3$ because $(-3)^3 = -27$

**51.** $\sqrt[3]{\frac{1}{125}} = \frac{1}{5}$ because $\left(\frac{1}{5}\right)^3 = \frac{1}{125}$

**53.** $f(x) = \sqrt[3]{x-1}$
$f(28) = \sqrt[3]{28-1} = \sqrt[3]{27} = 3$
$f(9) = \sqrt[3]{9-1} = \sqrt[3]{8} = 2$
$f(0) = \sqrt[3]{0-1} = \sqrt[3]{-1} = -1$
$f(-63) = \sqrt[3]{-63-1} = \sqrt[3]{-64} = -4$

**55.** $g(x) = -\sqrt[3]{8x-8}$
$g(2) = -\sqrt[3]{8(2)-8} = -\sqrt[3]{16-8}$
$= -\sqrt[3]{8} = -2$
$g(1) = -\sqrt[3]{8(1)-8} = -\sqrt[3]{8-8}$
$= -\sqrt[3]{0} = -0 = 0$
$g(0) = -\sqrt[3]{8(0)-8} = -\sqrt[3]{-8}$
$= -(-2) = 2$

**57.** $\sqrt[4]{1} = 1$ because $1^4 = 1$

**59.** $\sqrt[4]{16} = 2$ because $2^4 = 16$

**61.** $-\sqrt[4]{16} = -2$ because $2^4 = 16$

**63.** $\sqrt[4]{-16}$
Not a real number

**65.** $\sqrt[5]{-1} = -1$ because $(-1)^5 = -1$

**67.** $\sqrt[6]{-1}$
Not a real number

**69.** $-\sqrt[4]{256} = -4$ because $4^4 = 256$

**71.** $\sqrt[6]{64} = 2$ because $2^6 = 64$

**73.** $-\sqrt[5]{32} = -2$ because $2^5 = 32$

**75.** $\sqrt[3]{x^3} = x$

**77.** $\sqrt[4]{y^4} = |y|$

**79.** $\sqrt[3]{-8x^3} = -2x$

**81.** $\sqrt[3]{(-5)^3} = -5$

**83.** $\sqrt[4]{(-5)^4} = |-5| = 5$

**85.** $\sqrt[4]{(x+3)^4} = |x+3|$

**87.** $\sqrt[5]{-32(x-1)^5} = -2(x-1)$

**89.**

| $x$ | $f(x) = \sqrt{x} + 3$ |
|---|---|
| 0 | $f(0) = \sqrt{0} + 3 = 0 + 3 = 3$ |
| 1 | $f(1) = \sqrt{1} + 3 = 1 + 3 = 4$ |
| 4 | $f(4) = \sqrt{4} + 3 = 2 + 3 = 5$ |
| 9 | $f(9) = \sqrt{9} + 3 = 3 + 3 = 6$ |

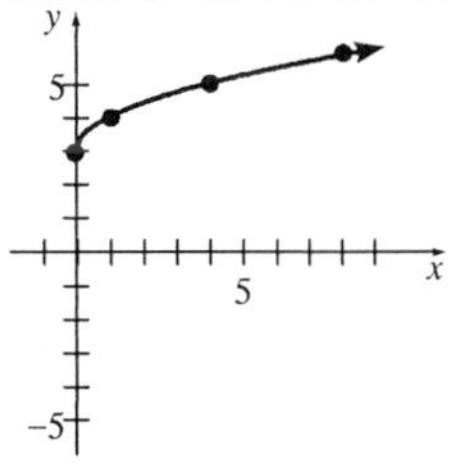


Domain: $\{x \mid x \geq 0\}$ or $[0, \infty)$

Range: $\{y \mid y \geq 3\}$ or $[3, \infty)$

**90.**

| $x$ | $f(x) = \sqrt{x} - 2$ |
|---|---|
| 0 | $f(0) = \sqrt{0} - 2 = 0 - 2 = -2$ |
| 1 | $f(1) = \sqrt{1} - 2 = 1 - 2 = -1$ |
| 4 | $f(4) = \sqrt{4} - 2 = 2 - 2 = 0$ |
| 9 | $f(9) = \sqrt{9} - 2 = 3 - 2 = 1$ |

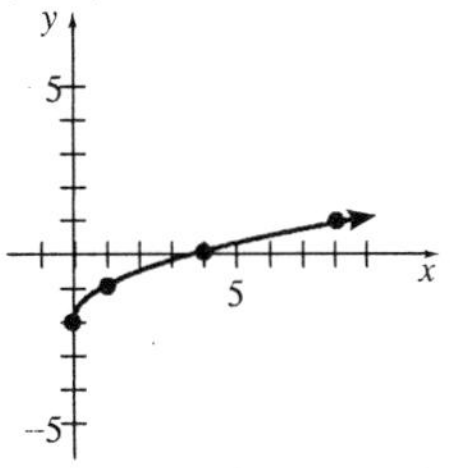


Domain: $\{x \mid x \geq 0\}$ or $[0, \infty)$

Range: $\{y \mid y \geq -2\}$ or $[-2, \infty)$

**91.**

| $x$ | $f(x) = \sqrt{x - 3}$ |
|---|---|
| 3 | $f(3) = \sqrt{3 - 3} = \sqrt{0} = 0$ |
| 4 | $f(4) = \sqrt{4 - 3} = \sqrt{1} = 1$ |
| 7 | $f(7) = \sqrt{7 - 3} = \sqrt{4} = 2$ |
| 12 | $f(12) = \sqrt{12 - 3} = \sqrt{9} = 3$ |

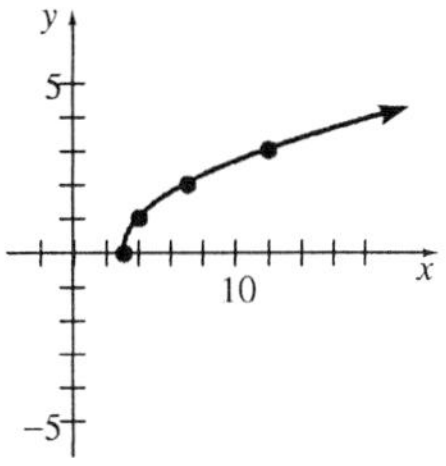


Domain: $\{x \mid x \geq 3\}$ or $[3, \infty)$

Range: $\{y \mid y \geq 0\}$ or $[0, \infty)$

**92.**

| $x$ | $f(x)=\sqrt{4-x}$ |
|---|---|
| $-5$ | $f(-5)=\sqrt{4-(-5)}=\sqrt{9}=3$ |
| $0$ | $f(0)=\sqrt{4-0}=\sqrt{4}=2$ |
| $3$ | $f(3)=\sqrt{4-3}=\sqrt{1}=1$ |
| $4$ | $f(4)=\sqrt{4-4}=\sqrt{0}=0$ |

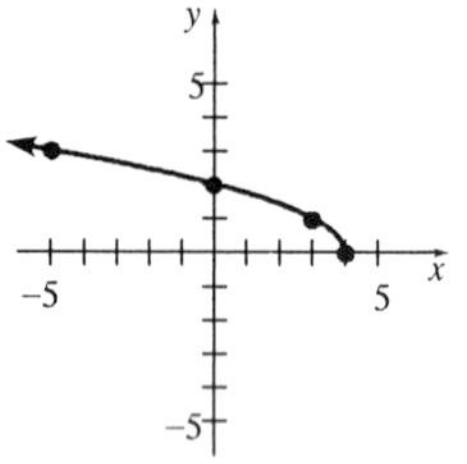


Domain: $\{x \mid x \le 4\}$ or $(-\infty, 4]$

Range: $\{y \mid y \ge 0\}$ or $[0, \infty)$

**93.** The domain of the cube root function is all real numbers, so we only need to worry about the square root in the denominator. We need the radicand of the square root to be $\ge 0$, but we also cannot divide by 0. Therefore, we have

$30-2x>0$

$-2x>-30$

$x<15$

The domain of $f$ is $\{x \mid x<15\}$ or $(-\infty, 15)$.

**94.** The domain of the cube root function is all real numbers, so we only need to worry about the square root in the denominator. We need the radicand of the square root to be $\ge 0$, but we also cannot divide by 0. Therefore, we have

$80-5x>0$

$-5x>-80$

$x<16$

The domain of $f$ is $\{x \mid x<16\}$ or $(-\infty, 16)$.

**95.** From the numerator, we need $x-1 \ge 0$. From the denominator, we need $3-x>0$. We need to solve the two inequalities. The domain of the function is the overlap of the two solution sets.

$x-1 \ge 0$ and $3-x>0$

$x \ge 1$ $\quad -x>-3$

$x<3$

We need $x \ge 1$ and $x<3$. Therefore, the domain of $f$ is $\{x \mid 1 \le x < 3\}$ or $[1, 3)$.

**96.** From the numerator, we need $x-2 \ge 0$. From the denominator, we need $7-x>0$. We need to solve the two inequalities. The domain of the function is the overlap of the two solution sets.

$x-2 \ge 0$ and $7-x>0$

$x \ge 2$ $\quad -x>-7$

$x<7$

We need $x \ge 2$ and $x<7$. Therefore, the domain of $f$ is $\{x \mid 2 \le x < 7\}$ or $[2, 7)$.

**97.** $\sqrt[3]{\sqrt[4]{16}+\sqrt{625}}=\sqrt[3]{2+25}=\sqrt[3]{27}=3$

**98.**

$$\sqrt[3]{\sqrt{\sqrt{169}+\sqrt{9}}+\sqrt{\sqrt[3]{1000}+\sqrt[3]{216}}}$$

$$=\sqrt[3]{\sqrt{13+3}+\sqrt{10+6}}$$

$$=\sqrt[3]{\sqrt{16}+\sqrt{16}}$$

$$=\sqrt[3]{4+4}=\sqrt[3]{8}$$

$$=2$$

**99.** $f(48) = 2.9\sqrt{48} + 20.1$
$= 2.9(6.9) + 20.1$
$= 20.1 + 20.1 \approx 40.2$

The model predicts the median height of boys who are 48 months old to be 40.2 inches. The model predicts the median height very well. According to the table, the median height is 40.8.

**101.** $f(245) = \sqrt{20(245)} = \sqrt{4900} = 70$

The officer should not believe the motorist. The model predicts that the motorist's speed was 70 miles per hour. This is well above the 50 miles per hour speed limit.

**103.** Answers will vary.

**105.** Answers will vary.

**107.** Answers will vary.

**109.** Answers will vary.

**111.** Answers will vary.

**113.** Answers will vary.

**115.** $y = \sqrt{x}$
$y = \sqrt{x} + 4$
$y = \sqrt{x} - 3$

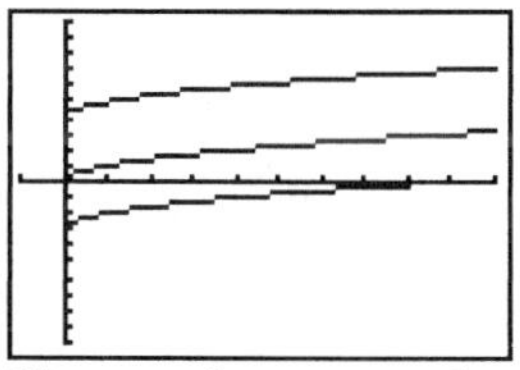

The graphs start at the y-axis, are increasing, and have the same shape. They differ in where they begin on the y-axis (i.e. different y-intercepts).

**117.** $y_1 = \sqrt{x^2}$
$y_2 = -x$

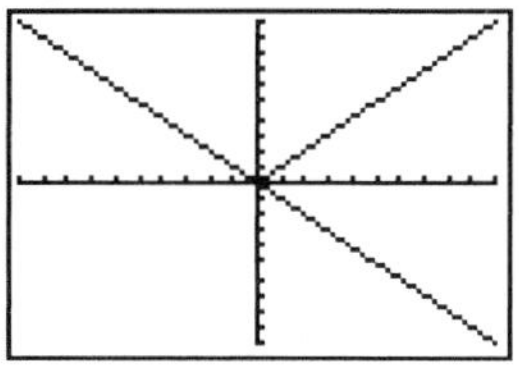

**a.** $\sqrt{x^2} = -x$ for $\{x|x \le 0\}$.

**b.** $\sqrt{x^2} \ne -x$ for $\{x|x > 0\}$.

**119.** Answers will vary. One example is $f(x) = \sqrt{5-x}$.

**121.** $\sqrt{(2x+3)^{10}} = \sqrt{\left((2x+3)^5\right)^2}$
$= \left|(2x+3)^5\right|$

**123.** $h(x) = \sqrt{x+3}$

| $x$ | $h(x) = \sqrt{x+3}$ |
|---|---|
| $-3$ | $h(-3) = \sqrt{-3+3} = \sqrt{0} = 0$ |
| $-2$ | $h(-2) = \sqrt{-2+3} = \sqrt{1} = 1$ |
| $1$ | $h(1) = \sqrt{1+3} = \sqrt{4} = 2$ |
| $6$ | $h(6) = \sqrt{6+3} = \sqrt{9} = 3$ |

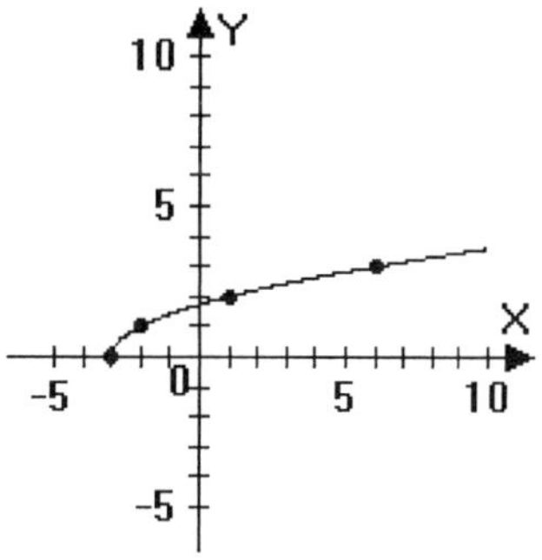


The graph of $h$ is the graph of $f$ shifted three units to the left.

**124.** $3x-2[x-3(x+5)]$
$=3x-2[x-3x-15]$
$=3x-2[-2x-15]$
$=3x+4x+30$
$=7x+30$

**125.** $(-3x^{-4}y^3)^{-2}=(-3)^{-2}(x^{-4})^{-2}(y^3)^{-2}$
$=\dfrac{1}{(-3)^2}x^8y^{-6}$
$=\dfrac{x^8}{(-3)^2y^6}=\dfrac{x^8}{9y^6}$

**126.** $|3x-4|>11$
$3x-4<-11$ or $3x-4>11$
$3x<-7$ $\quad$ $3x>15$
$x<-\dfrac{7}{3}$ $\quad$ $x>5$
The solution set is
$\left\{x \mid x<-\frac{7}{3} \text{ or } x>5\right\}$
or $\left(-\infty,-\dfrac{7}{3}\right)\cup(5,\infty)$.

## 7.2 Exercise Set

**1.** $49^{1/2}=\sqrt{49}=7$

**3.** $(-27)^{1/3}=\sqrt[3]{-27}=-3$

**5.** $-16^{1/4}=-\sqrt[4]{16}=-2$

**7.** $(xy)^{1/3}=\sqrt[3]{xy}$

**9.** $(2xy^3)^{1/5}=\sqrt[5]{2xy^3}$

**11.** $81^{3/2}=\left(\sqrt{81}\right)^3=9^3=729$

**13.** $125^{2/3}=\left(\sqrt[3]{125}\right)^2=5^2=25$

**15.** $(-32)^{3/5}=\left(\sqrt[5]{-32}\right)^3=(-2)^3=-8$

**17.** $27^{2/3}+16^{3/4}=\left(\sqrt[3]{27}\right)^2+\left(\sqrt[4]{16}\right)^3$
$=3^2+2^3$
$=9+8=17$

**19.** $(xy)^{4/7}=\left(\sqrt[7]{xy}\right)^4$ or $\sqrt[7]{(xy)^4}$

**21.** $\sqrt{7}=7^{1/2}$

**23.** $\sqrt[3]{5}=5^{1/3}$

**25.** $\sqrt[5]{11x}=(11x)^{1/5}$

**27.** $\sqrt{x^3}=x^{3/2}$

**29.** $\sqrt[5]{x^3}=x^{3/5}$

**31.** $\sqrt[5]{x^2y}=(x^2y)^{1/5}$

**33.** $\left(\sqrt{19xy}\right)^3=(19xy)^{3/2}$

**35.** $\left(\sqrt[6]{7xy^2}\right)^5=(7xy^2)^{5/6}$

**37.** $2x\sqrt[3]{y^2}=2xy^{2/3}$

**39.** $49^{-1/2}=\dfrac{1}{49^{1/2}}=\dfrac{1}{\sqrt{49}}=\dfrac{1}{7}$

**41.** $27^{-1/3} = \dfrac{1}{27^{1/3}} = \dfrac{1}{\sqrt[3]{27}} = \dfrac{1}{3}$

**43.** $16^{-3/4} = \dfrac{1}{16^{3/4}} = \dfrac{1}{\left(\sqrt[4]{16}\right)^3} = \dfrac{1}{2^3} = \dfrac{1}{8}$

**45.** $8^{-2/3} = \dfrac{1}{8^{2/3}} = \dfrac{1}{\left(\sqrt[3]{8}\right)^2} = \dfrac{1}{2^2} = \dfrac{1}{4}$

**47.** $\left(\dfrac{8}{27}\right)^{-1/3} = \left(\dfrac{27}{8}\right)^{1/3} = \sqrt[3]{\dfrac{27}{8}} = \dfrac{3}{2}$

**49.** $(-64)^{-2/3} = \dfrac{1}{(-64)^{2/3}} = \dfrac{1}{\left(\sqrt[3]{-64}\right)^2}$

$= \dfrac{1}{(-4)^2} = \dfrac{1}{16}$

**51.** $(2xy)^{-7/10} = \dfrac{1}{(2xy)^{7/10}}$

$= \dfrac{1}{\sqrt[10]{(2xy)^7}}$ or $\dfrac{1}{\left(\sqrt[10]{2xy}\right)^7}$

**53.** $5xz^{-1/3} = \dfrac{5xz^{-1/3}}{1} = \dfrac{5x}{z^{1/3}}$

**55.** $3^{3/4} \cdot 3^{1/4} = 3^{(3/4)+(1/4)}$

$= 3^{4/4} = 3^1 = 3$

**57.** $\dfrac{16^{3/4}}{16^{1/4}} = 16^{(3/4)-(1/4)} = 16^{2/4}$

$= 16^{1/2} = \sqrt{16} = 4$

**59.** $x^{1/2} \cdot x^{1/3} = x^{(1/2)+(1/3)}$

$= x^{(3/6)+(2/6)} = x^{5/6}$

**61.** $\dfrac{x^{4/5}}{x^{1/5}} = x^{(4/5)-(1/5)} = x^{3/5}$

**63.** $\dfrac{x^{1/3}}{x^{3/4}} = x^{(1/3)-(3/4)} = x^{(4/12)-(9/12)}$

$= x^{-5/12} = \dfrac{1}{x^{5/12}}$

**65.** $\left(5^{\frac{2}{3}}\right)^3 = 5^{\frac{2}{3}\cdot 3} = 5^2 = 25$

**67.** $\left(y^{-2/3}\right)^{1/4} = y^{(-2/3)\cdot(1/4)} = y^{-2/12}$

$= y^{-1/6} = \dfrac{1}{y^{1/6}}$

**69.** $\left(2x^{1/5}\right)^5 = 2^5 x^{(1/5)\cdot 5} = 32x^1 = 32x$

**71.** $\left(25x^4y^6\right)^{1/2} = 25^{1/2}\left(x^4\right)^{1/2}\left(y^6\right)^{1/2}$

$= \sqrt{25}x^{4(1/2)}y^{6(1/2)}$

$= 5x^2y^3$

**73.** $\left(x^{1/2}y^{-3/5}\right)^{1/2} = \left(\dfrac{x^{1/2}y^{-3/5}}{1}\right)^{1/2}$

$= \left(\dfrac{x^{1/2}}{y^{3/5}}\right)^{1/2}$

$= \dfrac{x^{(1/2)\cdot(1/2)}}{y^{(3/5)\cdot(1/2)}} = \dfrac{x^{1/4}}{y^{3/10}}$

**75.** $\dfrac{3^{1/2} \cdot 3^{3/4}}{3^{1/4}} = 3^{(1/2)+(3/4)-(1/4)}$

$= 3^{(2/4)+(3/4)-(1/4)}$

$= 3^{4/4} = 3^1 = 3$

**77.** $\dfrac{\left(3y^{1/4}\right)^3}{y^{1/12}} = \dfrac{3^3 y^{(1/4)\cdot 3}}{y^{1/12}} = \dfrac{27y^{3/4}}{y^{1/12}}$
$= 27y^{(3/4)-(1/12)}$
$= 27y^{(9/12)-(1/12)}$
$= 27y^{8/12} = 27y^{2/3}$

**79.** $\sqrt[8]{x^2} = x^{2/8} = x^{1/4} = \sqrt[4]{x}$

**81.** $\sqrt[3]{8a^6} = 8^{1/3}a^{6/3} = 2a^2$

**83.** $\sqrt[5]{x^{10}y^{15}} = x^{10/5}y^{15/5} = x^2y^3$

**85.** $\left(\sqrt[3]{xy}\right)^{18} = (xy)^{18/3} = (xy)^6 = x^6y^6$

**87.** $\sqrt[10]{(3y)^2} = (3y)^{2/10} = (3y)^{1/5} = \sqrt[5]{3y}$

**89.** $\left(\sqrt[6]{2a}\right)^4 = (2a)^{4/6} = (2a)^{2/3}$
$= \left(4a^2\right)^{1/3} = \sqrt[3]{4a^2}$

**91.** $\sqrt[9]{x^6y^3} = x^{6/9}y^{3/9}$
$= x^{2/3}y^{1/3} = \sqrt[3]{x^2y}$

**93.** $\sqrt{2}\cdot\sqrt[3]{2} = 2^{1/2}\cdot 2^{1/3} = 2^{(1/2)+(1/3)}$
$= 2^{(3/6)+(2/6)} = 2^{5/6}$
$= \sqrt[6]{2^5}$ or $\sqrt[6]{32}$

**95.** $\sqrt[5]{x^2}\cdot\sqrt{x} = \left(x^2\right)^{1/5}\cdot x^{1/2}$
$= x^{2/5}\cdot x^{1/2} = x^{(2/5)+(1/2)}$
$= x^{(4/10)+(5/10)} = x^{9/10}$
$= \sqrt[10]{x^9}$

**97.** $\sqrt[4]{a^2b}\cdot\sqrt[3]{ab} = \left(a^2b\right)^{1/4}\cdot(ab)^{1/3}$
$= a^{1/2}b^{1/4}\cdot a^{1/3}b^{1/3}$
$= a^{(1/2)+(1/3)}b^{(1/4)+(1/3)}$
$= a^{(6/12)+(4/12)}b^{(3/12)+(4/12)}$
$= a^{10/12}b^{7/12}$
$= \sqrt[12]{a^{10}b^7}$

**99.** $\dfrac{\sqrt[4]{x}}{\sqrt[5]{x}} = \dfrac{x^{1/4}}{x^{1/5}} = x^{(1/4)-(1/5)}$
$= x^{(5/20)-(4/20)}$
$= x^{1/20} = \sqrt[20]{x}$

**101.** $\dfrac{\sqrt[3]{y^2}}{\sqrt[6]{y}} = \dfrac{y^{2/3}}{y^{1/6}} = y^{(2/3)-(1/6)}$
$= y^{(4/6)-(1/6)} = y^{3/6}$
$= y^{1/2} = \sqrt{y}$

**103.** $\sqrt[4]{\sqrt{x}} = \sqrt[4]{x^{1/2}} = \left(x^{1/2}\right)^{1/4}$
$= x^{(1/2)\cdot(1/4)} = x^{1/8}$
$= \sqrt[8]{x}$

**105.** $\sqrt{\sqrt{x^2y}} = \sqrt{\left(x^2y\right)^{1/2}} = \left(\left(x^2y\right)^{1/2}\right)^{1/2}$
$= \left(x^2y\right)^{(1/2)\cdot(1/2)} = \left(x^2y\right)^{1/4}$
$= \sqrt[4]{x^2y}$

**107.** $\sqrt[4]{\sqrt[3]{2x}} = \sqrt[4]{(2x)^{1/3}} = \left((2x)^{1/3}\right)^{1/4}$
$= (2x)^{(1/3)\cdot(1/4)} = (2x)^{1/12}$
$= \sqrt[12]{2x}$

**109.** $\left(\sqrt[4]{x^3y^5}\right)^{12} = \left(\left(x^3y^5\right)^{1/4}\right)^{12}$

$= \left(x^3y^5\right)^{(1/4)\cdot 12}$

$= \left(x^3y^5\right)^{12/4} = \left(x^3y^5\right)^3$

$= x^{3\cdot 3}y^{5\cdot 3} = x^9y^{15}$

**111.** $\dfrac{\sqrt[4]{a^5b^5}}{\sqrt{ab}} = \dfrac{a^{5/4}b^{5/4}}{a^{1/2}b^{1/2}}$

$= a^{(5/4)-(1/2)}b^{(5/4)-(1/2)}$

$= a^{(5/4)-(2/4)}b^{(5/4)-(2/4)}$

$= a^{3/4}b^{3/4} = \left(a^3b^3\right)^{1/4}$

$= \sqrt[4]{a^3b^3}$

**113.** $x^{1/3}\left(x^{1/3} - x^{2/3}\right)$

$= x^{1/3}\cdot x^{1/3} - x^{1/3}\cdot x^{2/3}$

$= x^{(1/3)+(1/3)} - x^{(1/3)+(2/3)}$

$= x^{2/3} - x^{3/3}$

$= x^{2/3} - x$

**114.** $x^{-1/4}\left(x^{9/4} - x^{1/4}\right)$

$= x^{-1/4}\cdot x^{9/4} - x^{-1/4}\cdot x^{1/4}$

$= x^{(-1/4)+(9/4)} - x^{(-1/4)+(1/4)}$

$= x^{8/4} - x^0$

$= x^2 - 1$

**115.** $\left(x^{1/2} - 3\right)\left(x^{1/2} + 5\right)$

$= x^{1/2}\cdot x^{1/2} + x^{1/2}\cdot 5 - 3\cdot x^{1/2} - 3\cdot 5$

$= x^{(1/2)+(1/2)} + 5x^{1/2} - 3x^{1/2} - 15$

$= x^{2/2} + 2x^{1/2} - 15$

$= x + 2x^{1/2} - 15$

**116.** $\left(x^{1/3} - 2\right)\left(x^{1/3} + 6\right)$

$= x^{1/3}\cdot x^{1/3} + x^{1/3}\cdot 6 - 2\cdot x^{1/3} - 2\cdot 6$

$= x^{(1/3)+(1/3)} + 6x^{1/3} - 2x^{1/3} - 12$

$= x^{2/3} + 4x^{1/3} - 12$

**117.** $6x^{1/2} + 2x^{3/2}$

$= 3\cdot 2x^{1/2} + 2x^{(1/2)+(2/2)}$

$= 3\cdot 2x^{1/2} + 2x^{1/2}\cdot x^{2/2}$

$= 3\cdot 2x^{1/2} + 2x^{1/2}\cdot x$

$= 2x^{1/2}(3 + x)$

**118.** $8x^{1/4} + 4x^{5/4}$

$= 2\cdot 4x^{1/4} + 4x^{(1/4)+(4/4)}$

$= 2\cdot 4x^{1/4} + 4x^{1/4}\cdot x^{4/4}$

$= 2\cdot 4x^{1/4} + 4x^{1/4}\cdot x^1$

$= 4x^{1/4}(2 + x)$

**119.** $15x^{1/3} - 60x = 15x^{1/3} - 60x^{3/3}$

$= 15x^{1/3} - 60x^{(1/3)+(2/3)}$

$= 15x^{1/3} - 60x^{1/3}x^{2/3}$

$= 15x^{1/3}\cdot 1 - 15x^{1/3}\cdot 4x^{2/3}$

$= 15x^{1/3}\left(1 - 4x^{2/3}\right)$

**120.** $7x^{1/3} - 70x = 7x^{1/3} - 70x^{3/3}$

$= 7x^{1/3} - 70x^{(1/3)+(2/3)}$

$= 7x^{1/3} - 70x^{1/3}\cdot x^{2/3}$

$= 7x^{1/3}\cdot 1 - 7x^{1/3}\cdot 10x^{2/3}$

$= 7x^{1/3}\left(1 - 10x^{2/3}\right)$

**121.** $\left(49x^{-2}y^4\right)^{-1/2}\left(xy^{1/2}\right)$

$=\left(49\right)^{-1/2}\left(x^{-2}\right)^{-1/2}\left(y^4\right)^{-1/2}\left(xy^{1/2}\right)$

$=\dfrac{1}{49^{1/2}}x^{(-2)(-1/2)}y^{(4)(-1/2)}\left(xy^{1/2}\right)$

$=\dfrac{1}{7}x^1y^{-2}\cdot xy^{1/2}$

$=\dfrac{1}{7}x^{1+1}y^{-2+(1/2)}$

$=\dfrac{1}{7}x^2y^{-3/2}$

$=\dfrac{x^2}{7y^{3/2}}$

**122.** $\left(8x^{-6}y^3\right)^{1/3}\left(x^{5/6}y^{-1/3}\right)^6$

$=8^{1/3}x^{(-6)(1/3)}y^{(3)(1/3)}x^{(5/6)(6)}y^{(-1/3)(6)}$

$=2x^{-2}y^1x^5y^{-2}$

$=2x^{-2+5}y^{1+(-2)}$

$=2x^3y^{-1}$

$=\dfrac{2x^3}{y}$

**123.** $\left(\dfrac{x^{-5/4}y^{1/3}}{x^{-3/4}}\right)^{-6}$

$=\left(x^{(-5/4)-(-3/4)}y^{1/3}\right)^{-6}$

$=\left(x^{-2/4}y^{1/3}\right)^{-6}$

$=x^{(-2/4)(-6)}y^{(1/3)(-6)}$

$=x^3y^{-2}$

$=\dfrac{x^3}{y^2}$

**124.** $\left(\dfrac{x^{1/2}y^{-7/4}}{y^{-5/4}}\right)^{-4}$

$=\left(x^{1/2}y^{(-7/4)-(-5/4)}\right)^{-4}$

$=\left(x^{1/2}y^{-2/4}\right)^{-4}$

$=x^{(1/2)(-4)}y^{(-2/4)(-4)}$

$=x^{-2}y^2$

$=\dfrac{y^2}{x^2}$

**125.** $f(8)=29(8)^{1/3}$

$=29\sqrt[3]{8}$

$=29(2)=58$

There are 58 plant species on an 8 square mile island.

**127.** $f(x)=70x^{3/4}$

$f(80)=70(80)^{3/4}\approx 1872$

A person who weighs 80 kilograms needs about 1872 calories per day to maintain life.

**129.** $C = 35.74 + 0.6215t - 35.74v^{4/25} + 0.4275t \cdot v^{4/25}$

**a.** For $t = 0$, we get
$C(v) = 35.74 - 35.74v^{4/25}$

**b.** $C(25) = 35.74 - 35.74(25)^{4/25} \approx -24$
When the air temperature is $0°\text{F}$ and the wind speed is 25 miles per hour, the windchill temperature is $-24°\text{F}$.

**c.** The solution to part (b) is represented by the point $(25, -24)$ on the graph.

**131.** $L + 1.25\sqrt{S} - 9.8\sqrt[3]{D} \le 16.296$

**a.** $L + 1.25S^{1/2} - 9.8D^{1/3} \le 16.296$

**b.**
$$\begin{aligned} L + 1.25S^{1/2} - 9.8D^{1/3} &\le 16.296 \\ 20.85 + 1.25(276.4)^{1/2} - 9.8(18.55)^{1/3} &\le 16.296 \\ 20.85 + 1.25\sqrt{276.4} - 9.8\sqrt[3]{18.55} &\le 16.296 \\ 20.85 + 1.25(16.625) - 9.8(2.647) &\le 16.296 \\ 20.85 + 20.781 - 25.941 &\le 16.296 \\ 15.69 &\le 16.296 \end{aligned}$$
The yacht is eligible to enter the America's Cup.

**133.** Answers will vary.

**135.** Answers will vary.

**137.** Answers will vary.

**139.** Answers will vary.

**141.** Answers will vary depending on exercises selected. For example, consider Exercise 45.

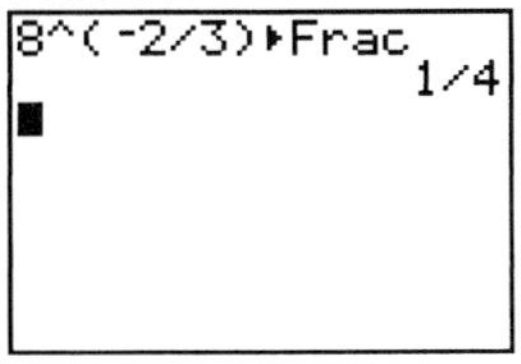


**143.**

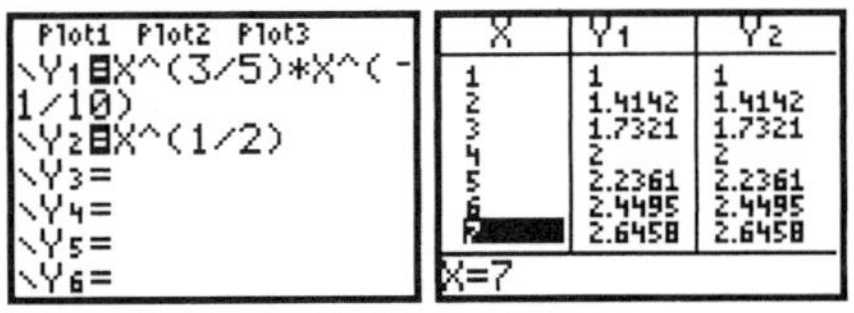


The simplification is correct.

**145.**

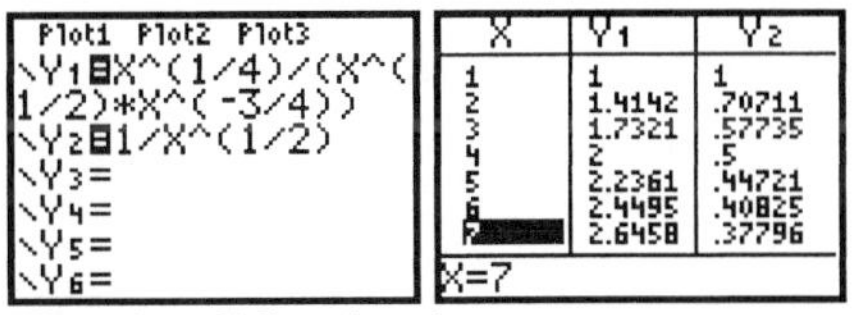


The simplification is not correct.

$$\frac{x^{1/4}}{x^{1/2}\cdot x^{-3/4}} = x^{(1/4)-[(1/2)+(-3/4)]}$$
$$= x^{(1/4)-(2/4)+(3/4)}$$
$$= x^{2/4}$$
$$= x^{1/2}$$

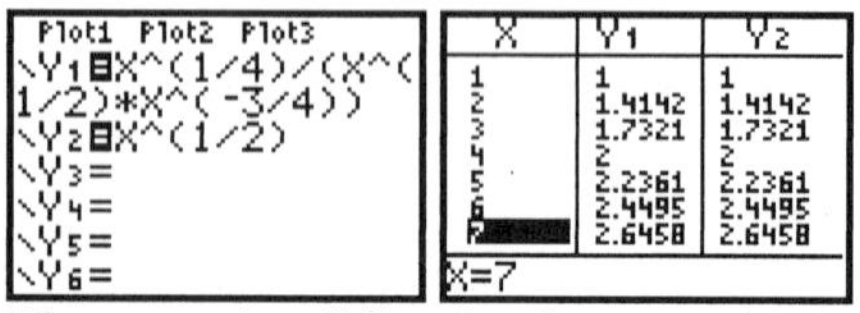


The new simplification is correct.

**147.** $2^{5/2}\cdot 2^{3/4} \div 2^{1/4} = 2^{(5/2)+(3/4)-(1/4)}$
$$= 2^{(10/4)+(3/4)-(1/4)}$$
$$= 2^{12/4} = 2^3 = 8$$

The son is 8 years old.

**149.**
$$\left[3+\left(27^{2/3}+32^{2/5}\right)\right]^{3/2}-9^{1/2}$$
$$=\left[3+\left(\left(\sqrt[3]{27}\right)^2+\left(\sqrt[5]{32}\right)^2\right)\right]^{3/2}-\sqrt{9}$$
$$=\left[3+\left(3^2+2^2\right)\right]^{3/2}-3$$
$$=\left[3+(9+4)\right]^{3/2}-3$$
$$=\left[3+(13)\right]^{3/2}-3=[16]^{3/2}-3$$
$$=\left(\sqrt{16}\right)^3-3=(4)^3-3$$
$$=64-3=61$$

**151.** First, find the slope.
$$m=\frac{y_2-y_1}{x_2-x_1}=\frac{3-1}{4-5}=\frac{2}{-1}=-2$$
Use the slope and one of the points to write the equation of the line in point-slope form.
$$y-y_1=m(x-x_1)$$
$$y-3=-2(x-4)$$
Solve for $y$ to write the equation in slope–intercept form.
$$y-3=-2(x-4)$$
$$y-3=-2x+8$$
$$y=-2x+11 \text{ or } f(x)=-2x+11$$

**152.** $y \le -\frac{3}{2}x+3$

First graph the line, $y=-\frac{3}{2}x+3$.

Since the line is in slope–intercept form we can identify the slope and *y*–intercept.

slope $= -\frac{3}{2}$ *y*–intercept = 3

Now, use the origin, $(0,0)$, as a test point.
$$0 \le -\frac{3}{2}(0)+3$$
$$0 \le 3$$
This is a true statement. This means that the point $(0,0)$ will fall in the shaded half-plane.

Next, graph the inequality.

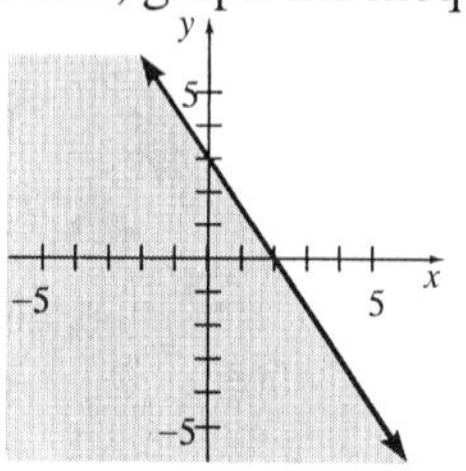

**153.** $5x - 3y = 3$
$7x + y = 25$

$D = \begin{vmatrix} 5 & -3 \\ 7 & 1 \end{vmatrix} = 5(1) - 7(-3)$
$= 5 + 21 = 26$

$D_x = \begin{vmatrix} 3 & -3 \\ 25 & 1 \end{vmatrix} = 3(1) - 25(-3)$
$= 3 + 75 = 78$

$D_y = \begin{vmatrix} 5 & 3 \\ 7 & 25 \end{vmatrix} = 5(25) - 7(3)$
$= 125 - 21 = 104$

$x = \frac{D_x}{D} = \frac{78}{26} = 3$

$y = \frac{D_y}{D} = \frac{104}{26} = 4$

The solution is $(3, 4)$.

### 7.3 Exercise Set

**1.** $\sqrt{3} \cdot \sqrt{5} = \sqrt{3 \cdot 5} = \sqrt{15}$

**3.** $\sqrt[3]{2} \cdot \sqrt[3]{9} = \sqrt[3]{2 \cdot 9} = \sqrt[3]{18}$

**5.** $\sqrt[4]{11} \cdot \sqrt[4]{3} = \sqrt[4]{11 \cdot 3} = \sqrt[4]{33}$

**7.** $\sqrt{3x} \cdot \sqrt{11y} = \sqrt{3x \cdot 11y} = \sqrt{33xy}$

**9.** $\sqrt[5]{6x^3} \cdot \sqrt[5]{4x} = \sqrt[5]{6x^3 \cdot 4x} = \sqrt[5]{24x^4}$

**11.** $\sqrt{x+3} \cdot \sqrt{x-3} = \sqrt{(x+3)(x-3)}$
$= \sqrt{x^2 - 9}$

**13.** $\sqrt[6]{x-4} \cdot \sqrt[6]{(x-4)^4}$
$= \sqrt[6]{(x-4)(x-4)^4} = \sqrt[6]{(x-4)^5}$

**15.** $\sqrt{\frac{2x}{3}} \cdot \sqrt{\frac{3}{2}} = \sqrt{\frac{2x}{3} \cdot \frac{3}{2}}$
$= \sqrt{\frac{{}^{1}\cancel{2}x}{{}_{1}\cancel{3}} \cdot \frac{{}^{1}\cancel{3}}{{}_{1}\cancel{2}}}$
$= \sqrt{x}$

**17.** $\sqrt[4]{\frac{x}{7}} \cdot \sqrt[4]{\frac{3}{y}} = \sqrt[4]{\frac{x}{7} \cdot \frac{3}{y}} = \sqrt[4]{\frac{3x}{7y}}$

**19.** $\sqrt[7]{7x^2y} \cdot \sqrt[7]{11x^3y^2} = \sqrt[7]{7x^2y \cdot 11x^3y^2}$
$= \sqrt[7]{7 \cdot 11x^2x^3yy^2}$
$= \sqrt[7]{77x^5y^3}$

**21.** $\sqrt{50} = \sqrt{25 \cdot 2} = \sqrt{25} \cdot \sqrt{2} = 5\sqrt{2}$

**23.** $\sqrt{45} = \sqrt{9 \cdot 5} = \sqrt{9} \cdot \sqrt{5} = 3\sqrt{5}$

**25.** $\sqrt{75x} = \sqrt{25 \cdot 3x}$
$= \sqrt{25} \cdot \sqrt{3x} = 5\sqrt{3x}$

**27.** $\sqrt[3]{16} = \sqrt[3]{8 \cdot 2} = \sqrt[3]{8} \cdot \sqrt[3]{2} = 2\sqrt[3]{2}$

**29.** $\sqrt[3]{27x^3} = \sqrt[3]{27 \cdot x^3} = \sqrt[3]{27} \cdot \sqrt[3]{x^3} = 3x$

**31.** $\sqrt[3]{-16x^2y^3} = \sqrt[3]{-8 \cdot 2x^2y^3}$
$= \sqrt[3]{-8y^3} \cdot \sqrt[3]{2x^2}$
$= -2y\sqrt[3]{2x^2}$

**33.** $f(x) = \sqrt{36(x+2)^2} = 6|x+2|$

**35.** $f(x)=\sqrt[3]{32(x+2)^3}$
$=\sqrt[3]{8\cdot 4(x+2)^3}$
$=\sqrt[3]{8(x+2)^3}\cdot\sqrt[3]{4}$
$=2(x+2)\sqrt[3]{4}$

**37.** $f(x)=\sqrt{3x^2-6x+3}$
$=\sqrt{3(x^2-2x+1)}$
$=\sqrt{3(x-1)^2}$
$=|x-1|\sqrt{3}$

**39.** $\sqrt{x^7}=\sqrt{x^6\cdot x}=\sqrt{x^6}\cdot\sqrt{x}=x^3\sqrt{x}$

**41.** $\sqrt{x^8y^9}=\sqrt{x^8y^8y}=\sqrt{x^8y^8}\sqrt{y}$
$=x^4y^4\sqrt{y}$

**43.** $\sqrt{48x^3}=\sqrt{16\cdot 3x^2x}=\sqrt{16x^2}\cdot\sqrt{3x}$
$=4x\sqrt{3x}$

**45.** $\sqrt[3]{y^8}=\sqrt[3]{y^6\cdot y^2}=\sqrt[3]{y^6}\cdot\sqrt[3]{y^2}$
$=y^2\sqrt[3]{y^2}$

**47.** $\sqrt[3]{x^{14}y^3z}=\sqrt[3]{x^{12}x^2y^3z}$
$=\sqrt[3]{x^{12}y^3}\cdot\sqrt[3]{x^2z}$
$=x^4y\sqrt[3]{x^2z}$

**49.** $\sqrt[3]{81x^8y^6}=\sqrt[3]{27\cdot 3x^6x^2y^6}$
$=\sqrt[3]{27x^6y^6}\cdot\sqrt[3]{3x^2}$
$=3x^2y^2\sqrt[3]{3x^2}$

**51.** $\sqrt[3]{(x+y)^5}=\sqrt[3]{(x+y)^3\cdot(x+y)^2}$
$=\sqrt[3]{(x+y)^3}\cdot\sqrt[3]{(x+y)^2}$
$=(x+y)\sqrt[3]{(x+y)^2}$

**53.** $\sqrt[5]{y^{17}}=\sqrt[5]{y^{15}\cdot y^2}=\sqrt[5]{y^{15}}\cdot\sqrt[5]{y^2}$
$=y^3\sqrt[5]{y^2}$

**55.** $\sqrt[5]{64x^6y^{17}}=\sqrt[5]{32\cdot 2x^5xy^{15}y^2}$
$=\sqrt[5]{32x^5y^{15}}\cdot\sqrt[5]{2xy^2}$
$=2xy^3\sqrt[5]{2xy^2}$

**57.** $\sqrt[4]{80x^{10}}=\sqrt[4]{16\cdot 5x^8x^2}$
$=\sqrt[4]{16x^8}\cdot\sqrt[4]{5x^2}$
$=2x^2\sqrt[4]{5x^2}$

**59.** $\sqrt[4]{(x-3)^{10}}=\sqrt[4]{(x-3)^8(x-3)^2}$
$=\sqrt[4]{(x-3)^8}\cdot\sqrt[4]{(x-3)^2}$
$=(x-3)^2\sqrt[4]{(x-3)^2}$
or
$=(x-3)^2\sqrt{x-3}$

**61.** $\sqrt{12}\cdot\sqrt{2}=\sqrt{12\cdot 2}=\sqrt{24}$
$=\sqrt{4\cdot 6}=\sqrt{4}\cdot\sqrt{6}$
$=2\sqrt{6}$

**63.** $\sqrt{5x}\cdot\sqrt{10y}=\sqrt{5x\cdot 10y}=\sqrt{50xy}$
$=\sqrt{25\cdot 2xy}=5\sqrt{2xy}$

**65.** $\sqrt{12x}\cdot\sqrt{3x}=\sqrt{12x\cdot 3x}$
$=\sqrt{36x^2}=6x$

**67.**
$$\sqrt{50xy}\cdot\sqrt{4xy^2}=\sqrt{50xy\cdot 4xy^2}$$
$$=\sqrt{200x^2y^3}$$
$$=\sqrt{100\cdot 2x^2y^2y}$$
$$=\sqrt{100x^2y^2}\cdot\sqrt{2y}$$
$$=10xy\sqrt{2y}$$

**69.**
$$2\sqrt{5}\cdot 3\sqrt{40}=2\cdot 3\sqrt{5\cdot 40}=6\sqrt{200}$$
$$=6\sqrt{100\cdot 2}=6\sqrt{100}\cdot\sqrt{2}$$
$$=6\cdot 10\sqrt{2}=60\sqrt{2}$$

**71.**
$$\sqrt[3]{12}\cdot\sqrt[3]{4}=\sqrt[3]{12\cdot 4}=\sqrt[3]{48}=\sqrt[3]{8\cdot 6}$$
$$=\sqrt[3]{8}\cdot\sqrt[3]{6}=2\sqrt[3]{6}$$

**73.**
$$\sqrt{5x^3}\cdot\sqrt{8x^2}=\sqrt{5x^3\cdot 8x^2}=\sqrt{40x^5}$$
$$=\sqrt{4\cdot 10x^4x}$$
$$=\sqrt{4x^4}\cdot\sqrt{10x}$$
$$=2x^2\sqrt{10x}$$

**75.**
$$\sqrt[3]{25x^4y^2}\cdot\sqrt[3]{5xy^{12}}$$
$$=\sqrt[3]{25x^4y^2\cdot 5xy^{12}}$$
$$=\sqrt[3]{125x^5y^{14}}$$
$$=\sqrt[3]{125x^3x^2y^{12}y^2}$$
$$=\sqrt[3]{125x^3y^{12}}\cdot\sqrt[3]{x^2y^2}$$
$$=5xy^4\sqrt[3]{x^2y^2}$$

**77.**
$$\sqrt[4]{8x^2y^3z^6}\cdot\sqrt[4]{2x^4yz}$$
$$=\sqrt[4]{8x^2y^3z^6\cdot 2x^4yz}$$
$$=\sqrt[4]{16x^6y^4z^7}$$
$$=\sqrt[4]{16x^4x^2y^4z^4z^3}$$
$$=\sqrt[4]{16x^4y^4z^4}\cdot\sqrt[4]{x^2z^3}$$
$$=2xyz\sqrt[4]{x^2z^3}$$

**79.**
$$\sqrt[5]{8x^4y^6z^2}\cdot\sqrt[5]{8xy^7z^4}$$
$$=\sqrt[5]{8x^4y^6z^2\cdot 8xy^7z^4}$$
$$=\sqrt[5]{64x^5y^{13}z^6}$$
$$=\sqrt[5]{32\cdot 2x^5y^{10}z^5z}$$
$$=\sqrt[5]{32x^5y^{10}z^5}\cdot\sqrt[5]{2y^3z}$$
$$=2xy^2z\sqrt[5]{2y^3z}$$

**81.**
$$\sqrt[3]{x-y}\cdot\sqrt[3]{(x-y)^7}$$
$$=\sqrt[3]{(x-y)\cdot(x-y)^7}$$
$$=\sqrt[3]{(x-y)^8}$$
$$=\sqrt[3]{(x-y)^6(x-y)^2}$$
$$=\sqrt[3]{(x-y)^6}\cdot\sqrt[3]{(x-y)^2}$$
$$=(x-y)^2\sqrt[3]{(x-y)^2}$$

**83.**
$$-2x^2y\left(\sqrt[3]{54x^3y^7z^2}\right)$$
$$=-2x^2y\sqrt[3]{27\cdot 2x^3y^6yz^2}$$
$$=-2x^2y\sqrt[3]{27x^3y^6}\cdot\sqrt[3]{2yz^2}$$
$$=-2x^2y\cdot 3xy^2\cdot\sqrt[3]{2yz^2}$$
$$=-6x^3y^3\sqrt[3]{2yz^2}$$

**84.** $\dfrac{-x^2y^7}{2}\left(\sqrt[3]{-32x^4y^9z^7}\right)$

$=\dfrac{-x^2y^7}{2}\sqrt[3]{-8\cdot 4x^3xy^9z^6z}$

$=\dfrac{-x^2y^7}{2}\sqrt[3]{-8x^3y^9z^6}\cdot\sqrt[3]{4xz}$

$=\dfrac{-x^2y^7}{2}\cdot\left(-2xy^3z^2\right)\cdot\sqrt[3]{4xz}$

$=x^3y^{10}z^2\sqrt[3]{4xz}$

**85.** $-3y\left(\sqrt[5]{64x^3y^6}\right)$

$=-3y\sqrt[5]{32\cdot 2x^3y^5y}$

$=-3y\sqrt[5]{32y^5}\cdot\sqrt[5]{2x^3y}$

$=-3y\cdot 2y\sqrt[5]{2x^3y}$

$=-6y^2\sqrt[5]{2x^3y}$

**86.** $-4x^2y^7\left(\sqrt[5]{-32x^{11}y^{17}}\right)$

$=-4x^2y^7\sqrt[5]{-32x^{10}xy^{15}y^2}$

$=-4x^2y^7\sqrt[5]{-32x^{10}y^{15}}\cdot\sqrt[5]{xy^2}$

$=-4x^2y^7\left(-2x^2y^3\right)\sqrt[5]{xy^2}$

$=8x^4y^{10}\sqrt[5]{xy^2}$

**87.** $\left(-2xy^2\sqrt{3x}\right)\left(xy\sqrt{6x}\right)$

$=-2x^2y^3\sqrt{3x\cdot 6x}$

$=-2x^2y^3\sqrt{18x^2}$

$=-2x^2y^3\sqrt{9x^2\cdot 2}$

$=-2x^2y^3\sqrt{9x^2}\cdot\sqrt{2}$

$=-2x^2y^3(3x)\sqrt{2}$

$=-6x^3y^3\sqrt{2}$

**88.** $\left(-5x^2y^3z\sqrt{2xyz}\right)\left(-x^4z\sqrt{10xz}\right)$

$=5x^6y^3z^2\sqrt{2xyz\cdot 10xz}$

$=5x^6y^3z^2\sqrt{20x^2yz^2}$

$=5x^6y^3z^2\sqrt{4x^2z^2\cdot 5y}$

$=5x^6y^3z^2\sqrt{4x^2z^2}\cdot\sqrt{5y}$

$=5x^6y^3z^2(2xz)\sqrt{5y}$

$=10x^7y^3z^3\sqrt{5y}$

**89.** $\left(2x^2y\sqrt[4]{8xy}\right)\left(-3xy^2\sqrt[4]{2x^2y^3}\right)$

$=-6x^3y^3\sqrt[4]{8xy\cdot 2x^2y^3}$

$=-6x^3y^3\sqrt[4]{16x^3y^4}$

$=-6x^3y^3\sqrt[4]{16y^4\cdot x^3}$

$=-6x^3y^3(2y)\sqrt[4]{x^3}$

$=-12x^3y^4\sqrt[4]{x^3}$

**90.** $\left(5a^2b\sqrt[4]{8a^2b}\right)\left(4ab\sqrt[4]{4a^3b^2}\right)$

$=20a^3b^2\sqrt[4]{8a^2b\cdot 4a^3b^2}$

$=20a^3b^2\sqrt[4]{32a^5b^3}$

$=20a^3b^2\sqrt[4]{16a^4\cdot 2ab^3}$

$=20a^3b^2(2a)\sqrt[4]{2ab^3}$

$=40a^4b^2\sqrt[4]{2ab^3}$

**91.** $\sqrt[5]{8x^4y^6}\left(\sqrt[5]{2xy^7}+\sqrt[5]{4x^6y^9}\right)$

$=\sqrt[5]{8x^4y^6\cdot 2xy^7}+\sqrt[5]{8x^4y^6\cdot 4x^6y^9}$

$=\sqrt[5]{16x^5y^{13}}+\sqrt[5]{32x^{10}y^{15}}$

$=\sqrt[5]{x^5y^{10}\cdot 16y^3}+\sqrt[5]{32x^{10}y^{15}}$

$=xy^2\sqrt[5]{16y^3}+2x^2y^3$

**92.** $\sqrt[4]{2x^3y^2}\left(\sqrt[4]{6x^5y^6}+\sqrt[4]{8x^5y^7}\right)$

$=\sqrt[4]{2x^3y^2\cdot 6x^5y^6}+\sqrt[4]{2x^3y^2\cdot 8x^5y^7}$

$=\sqrt[4]{12x^8y^8}+\sqrt[4]{16x^8y^9}$

$=\sqrt[4]{x^8y^8\cdot 12}+\sqrt[4]{16x^8y^8\cdot y}$

$=x^2y^2\sqrt[4]{12}+2x^2y^2\sqrt[4]{y}$

**93.** $d(x)=\sqrt{\frac{3x}{2}}$

$d(72)=\sqrt{\frac{3(72)}{2}}$

$=\sqrt{3(36)}$

$=\sqrt{3}\cdot\sqrt{36}$

$=6\sqrt{3}\approx 10.4$ miles

A passenger on the pool deck can see roughly 10.4 miles.

**95.** $W(x)=4\sqrt{2x}$

$W(6)=4\sqrt{2(6)}=4\sqrt{12}$

$=4\sqrt{4\cdot 3}=4\sqrt{4}\cdot\sqrt{3}$

$=4\cdot 2\sqrt{3}$

$=8\sqrt{3}\approx 14$ feet per second

A dinosaur with a leg length of 6 feet has a walking speed of about 14 feet per second.

**97.** **a.** $C(32)=\frac{7.644}{\sqrt[4]{32}}=\frac{7.644}{\sqrt[4]{16\cdot 2}}$

$=\frac{7.644}{2\sqrt[4]{2}}=\frac{3.822}{\sqrt[4]{2}}$

The cardiac index of a 32-year-old is $\frac{3.822}{\sqrt[4]{2}}$.

**b.** $\frac{3.822}{\sqrt[4]{2}}=\frac{3.822}{1.189}=\frac{3.822}{1.189}\approx 3.21$

The cardiac index of a 32-year-old is 3.21 liters per minute per square meter. This is shown on the graph as the point (32, 3.21).

**99.** Answers will vary.

**101.** Answers will vary.

**103.** Answers will vary.

**105.** $\sqrt{x^4}=x^2$

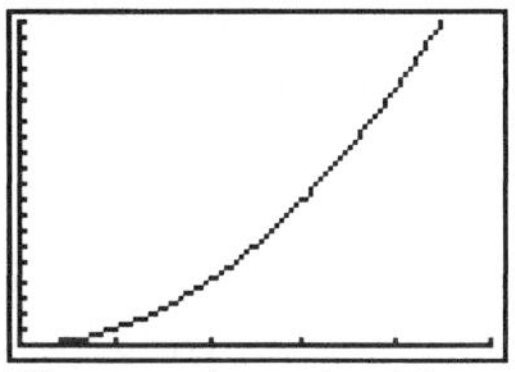

The graphs coincide, so the simplification is correct.

**107.** $\sqrt{3x^2-6x+3}=(x-1)\sqrt{3}$

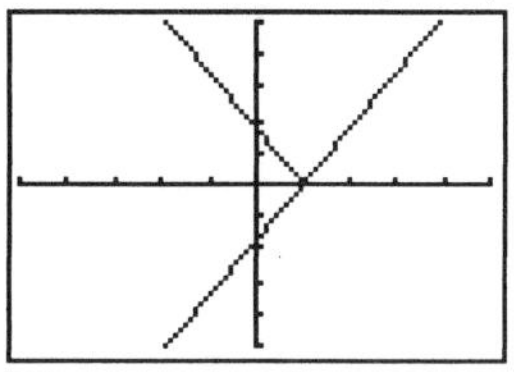

The graphs do not coincide.

To correct the simplification:

$\sqrt{3x^2-6x+3}=\sqrt{3(x^2-2x+1)}$

$=\sqrt{3(x-1)^2}=|x-1|\sqrt{3}$

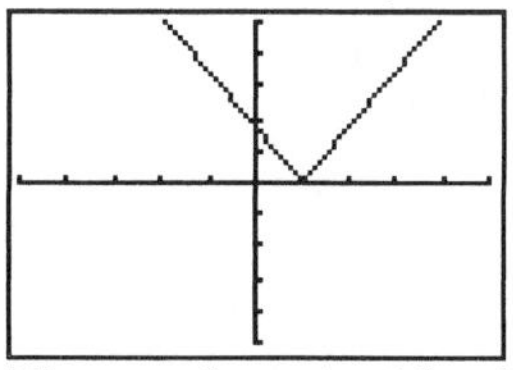

The graphs coincide. The simplification is correct.

**109.** Statement **d.** is true.

$\sqrt[5]{3^{25}} = 3^5 = 243$

Statement **a.** is false.

$2\sqrt{5} \cdot 6\sqrt{5} = 12\sqrt{5 \cdot 5} = 12 \cdot 5 = 60$

Statement **b.** is false.

$\sqrt[3]{4} \cdot \sqrt[3]{4} = \sqrt[3]{4 \cdot 4} = \sqrt[3]{16}$

Statement **c.** is false.

$\sqrt{12} = \sqrt{4 \cdot 3} = 2\sqrt{3}$

**111.** Assume the cube root of a number is $x$. Triple the cube root and work in reverse to find the number.

$3x = \sqrt[3]{(3x)^3} = \sqrt[3]{27x^3}$

The number must be multiplied by 27 for the cube root to be tripled.

**113.** $f(x) = \sqrt{(x-1)^2}$

| $x$ | $f(x) = \sqrt{(x-1)^2}$ |
|---|---|
| $-3$ | $f(-3) = \sqrt{(-3-1)^2} = \sqrt{16} = 4$ |
| $-2$ | $f(-2) = \sqrt{(-2-1)^2} = \sqrt{9} = 3$ |
| $-1$ | $f(-1) = \sqrt{(-1-1)^2} = \sqrt{4} = 2$ |
| 0 | $f(0) = \sqrt{(0-1)^2} = \sqrt{1} = 1$ |
| 1 | $f(1) = \sqrt{(1-1)^2} = \sqrt{0} = 0$ |
| 2 | $f(2) = \sqrt{(2-1)^2} = \sqrt{1} = 1$ |
| 3 | $f(3) = \sqrt{(3-1)^2} = \sqrt{4} = 2$ |

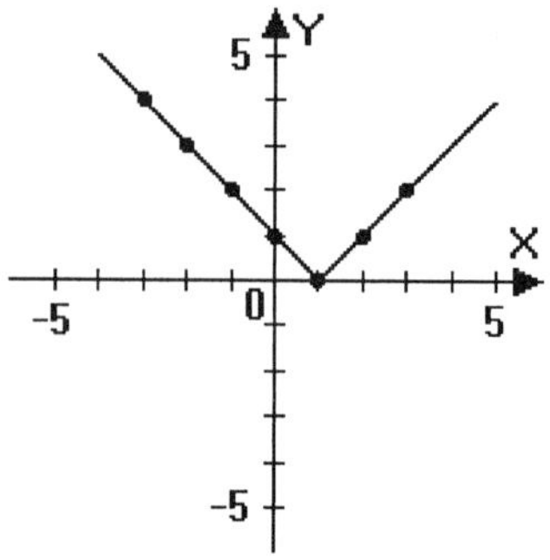


**114.** $2x - 1 \le 21$ and $2x + 2 \ge 12$

$2x \le 22 \qquad 2x \ge 10$

$x \le 11 \qquad x \ge 5$

The solution set is $\{x | 5 \le x \le 11\}$ or $[5, 11]$.

**115.** $5x + 2y = 2$

$4x + 3y = -4$

Multiply the first equation by –3, the second equation by 2 and solve by addition.

$$\begin{aligned} -15x - 6y &= -6 \\ 8x + 6y &= -8 \\ \hline -7x &= -14 \\ x &= 2 \end{aligned}$$

Back-substitute 2 for $x$ to find $y$.

$$\begin{aligned} 5x + 2y &= 2 \\ 5(2) + 2y &= 2 \\ 10 + 2y &= 2 \\ 2y &= -8 \\ y &= -4 \end{aligned}$$

The solution is $(2, -4)$.

**116.** $64x^3 - 27$
$= (4x)^3 - (3)^3$
$= (4x-3)\left((4x)^2 + 4x\cdot 3 + 3^2\right)$
$= (4x-3)\left(16x^2 + 12x + 9\right)$

## 7.4 Exercise Set

**1.** $8\sqrt{5} + 3\sqrt{5} = (8+3)\sqrt{5} = 11\sqrt{5}$

**3.** $9\sqrt[3]{6} - 2\sqrt[3]{6} = (9-2)\sqrt[3]{6} = 7\sqrt[3]{6}$

**5.** $4\sqrt[5]{2} + 3\sqrt[5]{2} - 5\sqrt[5]{2} = (4+3-5)\sqrt[5]{2}$
$= 2\sqrt[5]{2}$

**7.** $3\sqrt{13} - 2\sqrt{5} - 2\sqrt{13} + 4\sqrt{5}$
$= 3\sqrt{13} - 2\sqrt{13} - 2\sqrt{5} + 4\sqrt{5}$
$= (3-2)\sqrt{13} + (-2+4)\sqrt{5}$
$= \sqrt{13} + 2\sqrt{5}$

**9.** $3\sqrt{5} - \sqrt[3]{x} + 4\sqrt{5} + 3\sqrt[3]{x}$
$= 3\sqrt{5} + 4\sqrt{5} - \sqrt[3]{x} + 3\sqrt[3]{x}$
$= (3+4)\sqrt{5} + (-1+3)\sqrt[3]{x}$
$= 7\sqrt{5} + 2\sqrt[3]{x}$

**11.** $\sqrt{3} + \sqrt{27} = \sqrt{3} + \sqrt{9\cdot 3} = \sqrt{3} + 3\sqrt{3}$
$= (1+3)\sqrt{3} = 4\sqrt{3}$

**13.** $7\sqrt{12} + \sqrt{75} = 7\sqrt{4\cdot 3} + \sqrt{25\cdot 3}$
$= 7\cdot 2\sqrt{3} + 5\sqrt{3}$
$= 14\sqrt{3} + 5\sqrt{3}$
$= (14+5)\sqrt{3} = 19\sqrt{3}$

**15.** $3\sqrt{32x} - 2\sqrt{18x}$
$= 3\sqrt{16\cdot 2x} - 2\sqrt{9\cdot 2x}$
$= 3\cdot 4\sqrt{2x} - 2\cdot 3\sqrt{2x}$
$= 12\sqrt{2x} - 6\sqrt{2x} = 6\sqrt{2x}$

**17.** $5\sqrt[3]{16} + \sqrt[3]{54} = 5\sqrt[3]{8\cdot 2} + \sqrt[3]{27\cdot 2}$
$= 5\cdot 2\sqrt[3]{2} + 3\sqrt[3]{2}$
$= 10\sqrt[3]{2} + 3\sqrt[3]{2}$
$= (10+3)\sqrt[3]{2} = 13\sqrt[3]{2}$

**19.** $3\sqrt{45x^3} + \sqrt{5x} = 3\sqrt{9\cdot 5x^2x} + \sqrt{5x}$
$= 3\cdot 3x\sqrt{5x} + \sqrt{5x}$
$= 9x\sqrt{5x} + \sqrt{5x}$
$= (9x+1)\sqrt{5x}$

**21.** $\sqrt[3]{54xy^3} + y\sqrt[3]{128x}$
$= \sqrt[3]{27\cdot 2xy^3} + y\sqrt[3]{64\cdot 2x}$
$= 3y\sqrt[3]{2x} + 4y\sqrt[3]{2x}$
$= (3y+4y)\sqrt[3]{2x} = 7y\sqrt[3]{2x}$

**23.** $\sqrt[3]{54x^4} - \sqrt[3]{16x} = \sqrt[3]{27\cdot 2x^3x} - \sqrt[3]{8\cdot 2x}$
$= 3x\sqrt[3]{2x} - 2\sqrt[3]{2x}$
$= (3x-2)\sqrt[3]{2x}$

**25.** $\sqrt{9x-18} + \sqrt{x-2}$
$= \sqrt{9(x-2)} + \sqrt{x-2}$
$= 3\sqrt{x-2} + \sqrt{x-2}$
$= (3+1)\sqrt{x-2}$
$= 4\sqrt{x-2}$

**27.** $2\sqrt[3]{x^4y^2}+3x\sqrt[3]{xy^2}$
$=2\sqrt[3]{x^3xy^2}+3x\sqrt[3]{xy^2}$
$=2x\sqrt[3]{xy^2}+3x\sqrt[3]{xy^2}$
$=(2x+3x)\sqrt[3]{xy^2}$
$=5x\sqrt[3]{xy^2}$

**29.** $\sqrt{\frac{11}{4}}=\frac{\sqrt{11}}{\sqrt{4}}=\frac{\sqrt{11}}{2}$

**31.** $\sqrt[3]{\frac{19}{27}}=\frac{\sqrt[3]{19}}{\sqrt[3]{27}}=\frac{\sqrt[3]{19}}{3}$

**33.** $\sqrt{\frac{x^2}{36y^8}}=\frac{\sqrt{x^2}}{\sqrt{36y^8}}=\frac{x}{6y^4}$

**35.** $\sqrt{\frac{8x^3}{25y^6}}=\frac{\sqrt{8x^3}}{\sqrt{25y^6}}=\frac{\sqrt{4\cdot 2x^2x}}{5y^3}$
$=\frac{2x\sqrt{2x}}{5y^3}$

**37.** $\sqrt[3]{\frac{x^4}{8y^3}}=\frac{\sqrt[3]{x^4}}{\sqrt[3]{8y^3}}=\frac{\sqrt[3]{x^3x}}{2y}=\frac{x\sqrt[3]{x}}{2y}$

**39.** $\sqrt[3]{\frac{50x^8}{27y^{12}}}=\frac{\sqrt[3]{50x^8}}{\sqrt[3]{27y^{12}}}=\frac{\sqrt[3]{50x^6x^2}}{3y^4}$
$=\frac{x^2\sqrt[3]{50x^2}}{3y^4}$

**41.** $\sqrt[4]{\frac{9y^6}{x^8}}=\frac{\sqrt[4]{9y^6}}{\sqrt[4]{x^8}}=\frac{\sqrt[4]{9y^4y^2}}{x^2}=\frac{y\sqrt[4]{9y^2}}{x^2}$

**43.** $\sqrt[5]{\frac{64x^{13}}{y^{20}}}=\frac{\sqrt[5]{64x^{13}}}{\sqrt[5]{y^{20}}}=\frac{\sqrt[5]{32\cdot 2x^{10}x^3}}{y^4}$
$=\frac{2x^2\sqrt[5]{2x^3}}{y^4}$

**45.** $\frac{\sqrt{40}}{\sqrt{5}}=\sqrt{\frac{40}{5}}=\sqrt{8}=\sqrt{4\cdot 2}=2\sqrt{2}$

**47.** $\frac{\sqrt[3]{48}}{\sqrt[3]{6}}=\sqrt[3]{\frac{48}{6}}=\sqrt[3]{8}=2$

**49.** $\frac{\sqrt{54x^3}}{\sqrt{6x}}=\sqrt{\frac{54x^3}{6x}}=\sqrt{9x^2}=3x$

**51.** $\frac{\sqrt{x^5y^3}}{\sqrt{xy}}=\sqrt{\frac{x^5y^3}{xy}}=\sqrt{x^4y^2}=x^2y$

**53.** $\frac{\sqrt{200x^3}}{\sqrt{10x^{-1}}}=\sqrt{\frac{200x^3}{10x^{-1}}}=\sqrt{20x^{3-(-1)}}$
$=\sqrt{20x^4}=\sqrt{4\cdot 5x^4}=2x^2\sqrt{5}$

**55.** $\frac{\sqrt{48a^8b^7}}{\sqrt{3a^{-2}b^{-3}}}=\sqrt{\frac{48a^8b^7}{3a^{-2}b^{-3}}}$
$=\sqrt{16a^{10}b^{10}}$
$=4a^5b^5$

**57.** $\frac{\sqrt{72xy}}{2\sqrt{2}}=\frac{1}{2}\sqrt{\frac{72xy}{2}}=\frac{1}{2}\sqrt{36xy}$
$=\frac{1}{2}\cdot 6\sqrt{xy}=3\sqrt{xy}$

**59.** $\frac{\sqrt[3]{24x^3y^5}}{\sqrt[3]{3y^2}}=\sqrt[3]{\frac{24x^3y^5}{3y^2}}=\sqrt[3]{8x^3y^3}=2xy$

**61.** $\dfrac{\sqrt[4]{32x^{10}y^8}}{\sqrt[4]{2x^2y^{-2}}}=\sqrt[4]{\dfrac{32x^{10}y^8}{2x^2y^{-2}}}=\sqrt[4]{16x^8y^{8-(-2)}}$

$=\sqrt[4]{16x^8y^{10}}=\sqrt[4]{16x^8y^8y^2}$

$=2x^2y^2\sqrt[4]{y^2}$ or $2x^2y^2\sqrt{y}$

**63.** $\dfrac{\sqrt[3]{x^2+5x+6}}{\sqrt[3]{x+2}}=\sqrt[3]{\dfrac{x^2+5x+6}{x+2}}$

$=\sqrt[3]{\dfrac{\cancel{(x+2)}(x+3)}{\cancel{x+2}}}$

$=\sqrt[3]{x+3}$

**65.** $\dfrac{\sqrt[3]{a^3+b^3}}{\sqrt[3]{a+b}}=\sqrt[3]{\dfrac{a^3+b^3}{a+b}}$

$=\sqrt[3]{\dfrac{(a+b)\left(a^2-ab+b^2\right)}{a+b}}$

$=\sqrt[3]{a^2-ab+b^2}$

**67.** $\dfrac{\sqrt{32}}{5}+\dfrac{\sqrt{18}}{7}=\dfrac{\sqrt{16\cdot 2}}{5}+\dfrac{\sqrt{9\cdot 2}}{7}$

$=\dfrac{\sqrt{16}\cdot\sqrt{2}}{5}+\dfrac{\sqrt{9}\cdot\sqrt{2}}{7}$

$=\dfrac{4\sqrt{2}}{5}+\dfrac{3\sqrt{2}}{7}$

$=\dfrac{28\sqrt{2}}{35}+\dfrac{15\sqrt{2}}{35}$

$=\dfrac{28\sqrt{2}+15\sqrt{2}}{35}$

$=\dfrac{(28+15)\sqrt{2}}{35}$

$=\dfrac{43\sqrt{2}}{35}$

**68.** $\dfrac{\sqrt{27}}{2}+\dfrac{\sqrt{75}}{7}=\dfrac{\sqrt{9\cdot 3}}{2}+\dfrac{\sqrt{25\cdot 3}}{7}$

$=\dfrac{\sqrt{9}\cdot\sqrt{3}}{2}+\dfrac{\sqrt{25}\cdot\sqrt{3}}{7}$

$=\dfrac{3\sqrt{3}}{2}+\dfrac{5\sqrt{3}}{7}$

$=\dfrac{21\sqrt{3}}{14}+\dfrac{10\sqrt{3}}{14}$

$=\dfrac{(21+10)\sqrt{3}}{14}$

$=\dfrac{31\sqrt{3}}{14}$

**69.** $3x\sqrt{8xy^2}-5y\sqrt{32x^3}+\sqrt{18x^3y^2}$

$=3x\sqrt{4y^2\cdot 2x}-5y\sqrt{16x^2\cdot 2x}$

$+\sqrt{9x^2y^2\cdot 2x}$

$=3x(2y)\sqrt{2x}-5y(4x)\sqrt{2x}+3xy\sqrt{2x}$

$=6xy\sqrt{2x}-20xy\sqrt{2x}+3xy\sqrt{2x}$

$=(6-20+3)xy\sqrt{2x}$

$=-11xy\sqrt{2x}$

**70.** $6x\sqrt{3xy^2}-4x^2\sqrt{27xy}-5\sqrt{75x^5y}$

$=6x\sqrt{y^2\cdot 3x}-4x^2\sqrt{9\cdot 3xy}$

$-5\sqrt{25x^4\cdot 3xy}$

$=6x(y)\sqrt{3x}-4x^2(3)\sqrt{3xy}$

$-5\left(5x^2\right)\sqrt{3xy}$

$=6xy\sqrt{3x}-12x^2\sqrt{3xy}-25x^2\sqrt{3xy}$

$=6xy\sqrt{3x}-(12+25)x^2\sqrt{3xy}$

$=6xy\sqrt{3x}-37x^2\sqrt{3xy}$

**71.**
$$5\sqrt{2x^3}+\frac{30x^3\sqrt{24x^2}}{3x^2\sqrt{3x}}$$
$$=5\sqrt{2x^3}+10x\sqrt{\frac{24x^2}{3x}}$$
$$=5\sqrt{2x^3}+10x\sqrt{8x}$$
$$=5\sqrt{x^2\cdot 2x}+10x\sqrt{4\cdot 2x}$$
$$=5x\sqrt{2x}+10x(2)\sqrt{2x}$$
$$=5x\sqrt{2x}+20x\sqrt{2x}$$
$$=(5+20)x\sqrt{2x}=25x\sqrt{2x}$$

**72.**
$$7\sqrt{2x^3}+\frac{40x^3\sqrt{150x^2}}{5x^2\sqrt{3x}}$$
$$=7\sqrt{2x^3}+8x\sqrt{\frac{150x^2}{3x}}$$
$$=7\sqrt{2x^3}+8x\sqrt{50x}$$
$$=7\sqrt{x^2\cdot 2x}+8x\sqrt{25\cdot 2x}$$
$$=7x\sqrt{2x}+8x(5)\sqrt{2x}$$
$$=7x\sqrt{2x}+40x\sqrt{2x}$$
$$=(7+40)x\sqrt{2x}=47x\sqrt{2x}$$

**73.**
$$2x\sqrt{75xy}-\frac{\sqrt{81xy^2}}{\sqrt{3x^{-2}y}}$$
$$=2x\sqrt{75xy}-\sqrt{\frac{81xy^2}{3x^{-2}y}}$$
$$=2x\sqrt{75xy}-\sqrt{27x^3y}$$
$$=2x\sqrt{25\cdot 3xy}-\sqrt{9x^2\cdot 3xy}$$
$$=2x(5)\sqrt{3xy}-3x\sqrt{3xy}$$
$$=10x\sqrt{3xy}-3x\sqrt{3xy}$$
$$=(10-3)x\sqrt{3xy}$$
$$=7x\sqrt{3xy}$$

**74.**
$$5\sqrt{8x^2y^3}-\frac{9x^2\sqrt{64y}}{3x\sqrt{2y^{-2}}}$$
$$=5\sqrt{8x^2y^3}-3x\sqrt{\frac{64y}{2y^{-2}}}$$
$$=5\sqrt{8x^2y^3}-3x\sqrt{32y^3}$$
$$=5\sqrt{4x^2y^2\cdot 2y}-3x\sqrt{16y^2\cdot 2y}$$
$$=5(2xy)\sqrt{2y}-3x(4y)\sqrt{2y}$$
$$=10xy\sqrt{2y}-12xy\sqrt{2y}$$
$$=(10-12)xy\sqrt{2y}=-2xy\sqrt{2y}$$

**75.**
$$\frac{15x^4\sqrt[3]{80x^3y^2}}{5x^3\sqrt[3]{2x^2y}}-\frac{75\sqrt[3]{5x^3y}}{25\sqrt[3]{x^{-1}}}$$
$$=3x\sqrt[3]{\frac{80x^3y^2}{2x^2y}}-3\sqrt[3]{\frac{5x^3y}{x^{-1}}}$$
$$=3x\sqrt[3]{40xy}-3\sqrt[3]{5x^4y}$$
$$=3x\sqrt[3]{8\cdot 5xy}-3\sqrt[3]{x^3\cdot 5xy}$$
$$=3x(2)\sqrt[3]{5xy}-3x\sqrt[3]{5xy}$$
$$=6x\sqrt[3]{5xy}-3x\sqrt[3]{5xy}$$
$$=(6-3)x\sqrt[3]{5xy}=3x\sqrt[3]{5xy}$$

**76.**
$$\frac{16x^4\sqrt[3]{48x^3y^2}}{8x^3\sqrt[3]{3x^2y}}-\frac{20\sqrt[3]{2x^3y}}{4\sqrt[3]{x^{-1}}}$$
$$=2x\sqrt[3]{\frac{48x^3y^2}{3x^2y}}-5\sqrt[3]{\frac{2x^3y}{x^{-1}}}$$
$$=2x\sqrt[3]{16xy}-5\sqrt[3]{2x^4y}$$
$$=2x\sqrt[3]{8\cdot 2xy}-5\sqrt[3]{x^3\cdot 2xy}$$
$$=2x(2)\sqrt[3]{2xy}-5x\sqrt[3]{2xy}$$
$$=4x\sqrt[3]{2xy}-5x\sqrt[3]{2xy}$$
$$=(4-5)x\sqrt[3]{2xy}=-x\sqrt[3]{2xy}$$

**77.** $\left(\frac{f}{g}\right)(x) = \frac{\sqrt{48x^5}}{\sqrt{3x^2}} = \sqrt{\frac{48x^5}{3x^2}}$

$= \sqrt{16x^3} = \sqrt{16x^2 \cdot x}$

$= 4x\sqrt{x}$

To get the domain, we need $x \geq 0$ and $3x^2 > 0$. Combining these restrictions gives us $x > 0$.

Domain: $\{x \mid x > 0\}$ or $(0, \infty)$

**78.** $\left(\frac{f}{g}\right)(x) = \frac{\sqrt{x^2 - 25}}{\sqrt{x+5}}$

$= \sqrt{\frac{x^2 - 25}{x+5}} = \sqrt{\frac{(x-5)(x+5)}{x+5}}$

$= \sqrt{x-5}$

To get the domain, we need

$x + 5 > 0$ and $x - 5 \geq 0$

$x > -5$ $\quad$ $x \geq 5$

Therefore, we need $x \geq 5$.

Domain: $\{x \mid x \geq 5\}$ or $[5, \infty)$

**79.** $\left(\frac{f}{g}\right)(x) = \frac{\sqrt[3]{32x^6}}{\sqrt[3]{2x^2}} = \sqrt[3]{\frac{32x^6}{2x^2}}$

$= \sqrt[3]{16x^4} = \sqrt[3]{8x^3 \cdot 2x}$

$= 2x\sqrt[3]{2x}$

Our only restriction here is that we cannot divide by 0. Thus, we need $x \neq 0$.

Domain: $\{x \mid x \neq 0\}$ or $(-\infty, 0) \cup (0, \infty)$

**80.** $\left(\frac{f}{g}\right)(x) = \frac{\sqrt[3]{2x^6}}{\sqrt[3]{16x}} = \sqrt[3]{\frac{2x^6}{16x}}$

$= \sqrt[3]{\frac{x^5}{8}} = \frac{\sqrt[3]{x^3 \cdot x^2}}{\sqrt[3]{8}}$

$= \frac{x\sqrt[3]{x^2}}{2}$

Our only restriction here is that we cannot divide by 0. Thus, we need $x \neq 0$.

Domain: $\{x \mid x \neq 0\}$ or $(-\infty, 0) \cup (0, \infty)$

**81.** **a.** $R_f \frac{\sqrt{c^2 - v^2}}{\sqrt{c^2}} = R_f\sqrt{\frac{c^2 - v^2}{c^2}}$

$= R_f\sqrt{\frac{c^2}{c^2} - \frac{v^2}{c^2}}$

$= R_f\sqrt{1 - \left(\frac{v}{c}\right)^2}$

**b.** $R_f\sqrt{1 - \left(\frac{v}{c}\right)^2} = R_f\sqrt{1 - \left(\frac{c}{c}\right)^2}$

$= R_f\sqrt{1 - (1)^2}$

$= R_f\sqrt{1-1}$

$= R_f\sqrt{0}$

$= R_f \cdot 0 = 0$

Your aging rate is zero. This means that a person moving close to the speed of light does not age relative to a friend on Earth.

**83.** $P = 2l + 2w$

$$= 2\left(2\sqrt{20}\right) + 2\left(\sqrt{125}\right)$$
$$= 4\sqrt{20} + 2\sqrt{125}$$
$$= 4\sqrt{4\cdot 5} + 2\sqrt{25\cdot 5}$$
$$= 4\cdot 2\sqrt{5} + 2\cdot 5\sqrt{5}$$
$$= 8\sqrt{5} + 10\sqrt{5}$$
$$= (8+10)\sqrt{5} = 18\sqrt{5}$$

The perimeter is $18\sqrt{5}$ feet.

$$A = lw = 2\sqrt{20}\cdot\sqrt{125}$$
$$= 2\sqrt{20\cdot 125} = 2\sqrt{2500}$$
$$= 2\cdot 50 = 100$$

The area is 100 square feet.

**85.** Answers will vary.

**87.** We can add $\sqrt{2}$ and $\sqrt{8}$ because we can simplify $\sqrt{8} = \sqrt{4\cdot 2} = 2\sqrt{2}$.

**89.** Answers will vary.

**91.** Answers will vary. For example, consider Exercise 11.

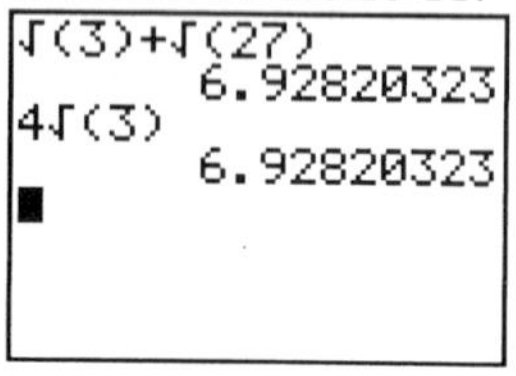


**93.** $\sqrt{16x} - \sqrt{9x} = \sqrt{7x}$

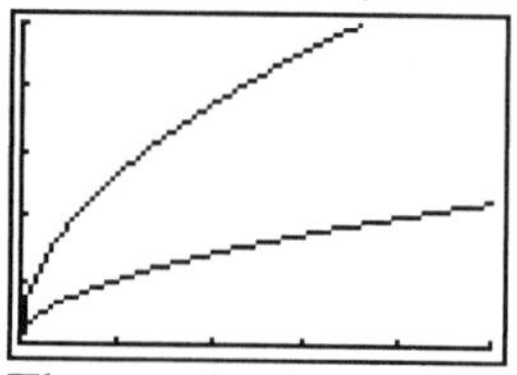

The graphs do not coincide. Correct the simplification.

$$\sqrt{16x} - \sqrt{9x} = 4\sqrt{x} - 3\sqrt{x}$$
$$= (4-3)\sqrt{x} = \sqrt{x}$$

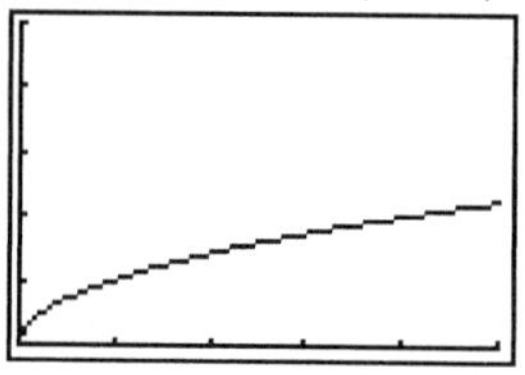

**95.** Statement **d.** is true.

Statement **a.** is false.
$\sqrt{5} + \sqrt{5} = 2\sqrt{5}$

Statement **b.** is false.
$4\sqrt{3} + 5\sqrt{3} = 9\sqrt{3}$

Statement **c.** is false. In order for two radical expressions to be combined, both the index and the radicand must be the same. Just because two radical expressions are completely simplified does guarantee that the index and radicands match.

**97.**
$$\frac{\sqrt{20}}{3} + \frac{\sqrt{45}}{4} - \sqrt{80}$$
$$= \frac{\sqrt{4\cdot 5}}{3} + \frac{\sqrt{9\cdot 5}}{4} - \sqrt{16\cdot 5}$$
$$= \frac{2\sqrt{5}}{3} + \frac{3\sqrt{5}}{4} - 4\sqrt{5}$$
$$= \frac{2}{3}\sqrt{5} + \frac{3}{4}\sqrt{5} - 4\sqrt{5}$$
$$= \left(\frac{2}{3} + \frac{3}{4} - 4\right)\sqrt{5}$$
$$= \left(\frac{2}{3}\cdot\frac{4}{4} + \frac{3}{4}\cdot\frac{3}{3} - \frac{4}{1}\cdot\frac{12}{12}\right)\sqrt{5}$$
$$= \left(\frac{8}{12} + \frac{9}{12} - \frac{48}{12}\right)\sqrt{5}$$
$$= -\frac{31}{12}\sqrt{5} \text{ or } -\frac{31\sqrt{5}}{12}$$

**99.** $$\begin{aligned} 2(3x-1)-4 &= 2x-(6-x) \\ 6x-2-4 &= 2x-6+x \\ 6x-6 &= x-6 \\ 5x &= 0 \\ x &= 0 \end{aligned}$$

The solution is 0.

**100.** $x^2-8xy+12y^2=(x-6y)(x-2y)$

**101.** $$\begin{aligned} &\frac{2}{x^2+5x+6}+\frac{3x}{x^2+6x+9} \\ &=\frac{2}{(x+3)(x+2)}+\frac{3x}{(x+3)^2} \\ &=\frac{2(x+3)}{(x+3)^2(x+2)}+\frac{3x(x+2)}{(x+3)^2(x+2)} \\ &=\frac{2(x+3)+3x(x+2)}{(x+3)^2(x+2)} \\ &=\frac{2x+6+3x^2+6x}{(x+3)^2(x+2)} \\ &=\frac{3x^2+8x+6}{(x+3)^2(x+2)} \end{aligned}$$

## Mid-Chapter 7 Check Point

**1.** $$\begin{aligned} \sqrt{100}-\sqrt[3]{-27} &= 10-(-3) \\ &= 10+3 \\ &= 13 \end{aligned}$$

**2.** $$\begin{aligned} \sqrt{8x^5y^7} &= \sqrt{4x^4y^6\cdot 2xy} \\ &= 2x^2y^3\sqrt{2xy} \end{aligned}$$

**3.** $$\begin{aligned} 3\sqrt[3]{4x^2}+2\sqrt[3]{4x^2} &= (3+2)\sqrt[3]{4x^2} \\ &= 5\sqrt[3]{4x^2} \end{aligned}$$

**4.** $$\begin{aligned} \left(3\sqrt[3]{4x^2}\right)\left(2\sqrt[3]{4x^2}\right) &= 6\sqrt[3]{4x^2\cdot 4x^2} \\ &= 6\sqrt[3]{16x^4} \\ &= 6\sqrt[3]{8x^3\cdot 2x} \\ &= 6(2x)\sqrt[3]{2x} \\ &= 12x\sqrt[3]{2x} \end{aligned}$$

**5.** $$\begin{aligned} 27^{2/3}+(-32)^{3/5} &= \left(\sqrt[3]{27}\right)^2+\left(\sqrt[5]{-32}\right)^3 \\ &= (3)^2+(-2)^3 \\ &= 9+(-8) \\ &= 1 \end{aligned}$$

**6.** $$\begin{aligned} \left(64x^3y^{1/4}\right)^{1/3} &= (64)^{1/3}\left(x^3\right)^{1/3}\left(y^{1/4}\right)^{1/3} \\ &= \sqrt[3]{64}\cdot x^{3(1/3)}\cdot y^{(1/4)(1/3)} \\ &= 4xy^{1/12} \end{aligned}$$

**7.** $$\begin{aligned} 5\sqrt{27}-4\sqrt{48} &= 5\sqrt{9\cdot 3}-4\sqrt{16\cdot 3} \\ &= 5(3)\sqrt{3}-4(4)\sqrt{3} \\ &= 15\sqrt{3}-16\sqrt{3} \\ &= (15-16)\sqrt{3} \\ &= -\sqrt{3} \end{aligned}$$

**8.** $$\begin{aligned} \sqrt{\frac{500x^3}{4y^4}} &= \frac{\sqrt{500x^3}}{\sqrt{4y^4}}=\frac{\sqrt{100x^2\cdot 5x}}{\sqrt{4y^4}} \\ &= \frac{10x\sqrt{5x}}{2y^2}=\frac{5x\sqrt{5x}}{y^2} \end{aligned}$$

**9.** $$\begin{aligned} \frac{x}{\sqrt[4]{x}} &= \frac{x}{x^{1/4}}=x^{1-(1/4)} \\ &= x^{(4/4)-(1/4)}=x^{3/4} \\ &= \sqrt[4]{x^3} \end{aligned}$$

**10.** $$\begin{aligned} \sqrt[3]{54x^5} &= \sqrt[3]{27x^3\cdot 2x^2} \\ &= 3x\sqrt[3]{2x^2} \end{aligned}$$

**11.** $\dfrac{\sqrt[3]{160}}{\sqrt[3]{2}} = \sqrt[3]{\dfrac{160}{2}} = \sqrt[3]{80}$

$= \sqrt[3]{8 \cdot 10} = 2\sqrt[3]{10}$

**12.** $\sqrt[5]{\dfrac{x^{10}}{y^{20}}} = \left(\dfrac{x^{10}}{y^{20}}\right)^{1/5} = \dfrac{\left(x^{10}\right)^{1/5}}{\left(y^{20}\right)^{1/5}}$

$= \dfrac{x^{10(1/5)}}{y^{20(1/5)}} = \dfrac{x^2}{y^4}$

**13.** $\dfrac{\left(x^{2/3}\right)^2}{\left(x^{1/4}\right)^3} = \dfrac{x^{(2/3)\cdot 2}}{x^{(1/4)\cdot 3}} = \dfrac{x^{4/3}}{x^{3/4}}$

$= x^{(4/3)-(3/4)} = x^{(16/12)-(9/12)}$

$= x^{7/12}$

**14.** $\sqrt[6]{x^6 y^4} = \left(x^6 y^4\right)^{1/6} = \left(x^6\right)^{1/6}\left(y^4\right)^{1/6}$

$= x^{6(1/6)} y^{4(1/6)} = x^1 y^{2/3}$

$= x\sqrt[3]{y^2}$

**15.** $\sqrt[7]{(x-2)^3} \cdot \sqrt[7]{(x-2)^6}$

$= \sqrt[7]{(x-2)^3 \cdot (x-2)^6}$

$= \sqrt[7]{(x-2)^9}$

$= \sqrt[7]{(x-2)^7 \cdot (x-2)^2}$

$= (x-2)\sqrt[7]{(x-2)^2}$

**16.** $\sqrt[4]{32x^{11}y^{17}} = \sqrt[4]{16x^8 y^{16} \cdot 2x^3 y}$

$= \sqrt[4]{16x^8 y^{16}} \cdot \sqrt[4]{2x^3 y}$

$= 2x^2 y^4 \sqrt[4]{2x^3 y}$

**17.** $4\sqrt[3]{16} + 2\sqrt[3]{54} = 4\sqrt[3]{8 \cdot 2} + 2\sqrt[3]{27 \cdot 2}$

$= 4\sqrt[3]{8} \cdot \sqrt[3]{2} + 2\sqrt[3]{27} \cdot \sqrt[3]{2}$

$= 4(2)\sqrt[3]{2} + 2(3)\sqrt[3]{2}$

$= 8\sqrt[3]{2} + 6\sqrt[3]{2}$

$= (8+6)\sqrt[3]{2}$

$= 14\sqrt[3]{2}$

**18.** $\dfrac{\sqrt[7]{x^4 y^9}}{\sqrt[7]{x^{-5} y^7}} = \sqrt[7]{\dfrac{x^4 y^9}{x^{-5} y^7}} = \sqrt[7]{x^9 y^2}$

$= \sqrt[7]{x^7 \cdot x^2 y^2}$

$= x\sqrt[7]{x^2 y^2}$

**19.** $(-125)^{-2/3} = \dfrac{1}{(-125)^{2/3}} = \dfrac{1}{\left(\sqrt[3]{-125}\right)^2}$

$= \dfrac{1}{(-5)^2} = \dfrac{1}{25}$

**20.** $\sqrt{2} \cdot \sqrt[3]{2} = 2^{1/2} \cdot 2^{1/3} = 2^{(1/2)+(1/3)}$

$= 2^{(3/6)+(2/6)} = 2^{5/6}$

$= \sqrt[6]{2^5} = \sqrt[6]{32}$

**21.** $\sqrt[3]{\dfrac{32x}{y^4}} \cdot \sqrt[3]{\dfrac{2x^2}{y^2}} = \sqrt[3]{\dfrac{32x}{y^4} \cdot \dfrac{2x^2}{y^2}}$

$= \sqrt[3]{\dfrac{64x^3}{y^6}} = \dfrac{\sqrt[3]{64x^3}}{\sqrt[3]{y^6}}$

$= \dfrac{4x}{y^2}$

**22.** $\sqrt{32xy^2} \cdot \sqrt{2x^3 y^5} = \sqrt{32xy^2 \cdot 2x^3 y^5}$

$= \sqrt{64x^4 y^7}$

$= \sqrt{64x^4 y^6 \cdot y}$

$= 8x^2 y^3 \sqrt{y}$

**23.** $4x\sqrt{6x^4y^3}-7y\sqrt{24x^6y}$
$=4x\sqrt{x^4y^2\cdot 6y}-7y\sqrt{4x^6\cdot 6y}$
$=4x\left(x^2y\right)\sqrt{6y}-7y\left(2x^3\right)\sqrt{6y}$
$=4x^3y\sqrt{6y}-14x^3y\sqrt{6y}$
$=(4-14)x^3y\sqrt{6y}$
$=-10x^3y\sqrt{6y}$

**24.** $f(x)=\sqrt{30-5x}$
Do find the domain, we set the integrand greater than or equal to 0.
$30-5x\ge 0$
$-5x\ge -30$
$x\le 6$
Domain: $\{x\,|\,x\le 6\}$ or $(-\infty,6]$

**25.** $g(x)=\sqrt[3]{3x-15}$
The domain of a cube root is all real numbers. Since there are no other restrictions, we have
Domain: $\{x\,|\,x \text{ is a real number}\}$ or $(-\infty,\infty)$

**7.5 Exercise Set**

**1.** $\sqrt{2}\left(x+\sqrt{7}\right)=\sqrt{2}\cdot x+\sqrt{2}\sqrt{7}$
$=x\sqrt{2}+\sqrt{14}$

**3.** $\sqrt{6}\left(7-\sqrt{6}\right)=\sqrt{6}\cdot 7-\sqrt{6}\sqrt{6}$
$=7\sqrt{6}-\sqrt{36}=7\sqrt{6}-6$

**5.** $\sqrt{3}\left(4\sqrt{6}-2\sqrt{3}\right)$
$=\sqrt{3}\cdot 4\sqrt{6}-\sqrt{3}\cdot 2\sqrt{3}$
$=4\sqrt{18}-2\sqrt{9}$
$=4\sqrt{9\cdot 2}-2\cdot 3$
$=4\cdot 3\sqrt{2}-6=12\sqrt{2}-6$

**7.** $\sqrt[3]{2}\left(\sqrt[3]{6}+4\sqrt[3]{5}\right)=\sqrt[3]{2}\cdot\sqrt[3]{6}+\sqrt[3]{2}\cdot 4\sqrt[3]{5}$
$=\sqrt[3]{12}+4\sqrt[3]{10}$

**9.** $\sqrt[3]{x}\left(\sqrt[3]{16x^2}-\sqrt[3]{x}\right)$
$=\sqrt[3]{x}\cdot\sqrt[3]{16x^2}-\sqrt[3]{x}\cdot\sqrt[3]{x}$
$=\sqrt[3]{x}\cdot\sqrt[3]{8\cdot 2x^2}-\sqrt[3]{x^2}$
$=\sqrt[3]{8\cdot 2x^3}-\sqrt[3]{x^2}$
$=2x\sqrt[3]{2}-\sqrt[3]{x^2}$

**11.** $\left(5+\sqrt{2}\right)\left(6+\sqrt{2}\right)$
$=5\cdot 6+5\sqrt{2}+6\sqrt{2}+\sqrt{2}\sqrt{2}$
$=30+(5+6)\sqrt{2}+2$
$=32+11\sqrt{2}$

**13.** $\left(6+\sqrt{5}\right)\left(9-4\sqrt{5}\right)$
$=6\cdot 9-6\cdot 4\sqrt{5}+9\sqrt{5}-4\sqrt{5}\sqrt{5}$
$=54-24\sqrt{5}+9\sqrt{5}-4\cdot 5$
$=54+(-24+9)\sqrt{5}-20$
$=34+(-15)\sqrt{5}$
$=34-15\sqrt{5}$

**15.** $\left(6-3\sqrt{7}\right)\left(2-5\sqrt{7}\right)$
$=6\cdot 2-6\cdot 5\sqrt{7}-2\cdot 3\sqrt{7}+3\sqrt{7}\cdot 5\sqrt{7}$
$=12-30\sqrt{7}-6\sqrt{7}+15\cdot 7$
$=12+(-30-6)\sqrt{7}+105$
$=117+(-36)\sqrt{7}$
$=117-36\sqrt{7}$

**17.** $\left(\sqrt{2}+\sqrt{7}\right)\left(\sqrt{3}+\sqrt{5}\right)$
$=\sqrt{2}\sqrt{3}+\sqrt{2}\sqrt{5}+\sqrt{7}\sqrt{3}+\sqrt{7}\sqrt{5}$
$=\sqrt{6}+\sqrt{10}+\sqrt{21}+\sqrt{35}$

**19.** $\left(\sqrt{2}-\sqrt{7}\right)\left(\sqrt{3}-\sqrt{5}\right)$
$=\sqrt{2}\sqrt{3}-\sqrt{2}\sqrt{5}-\sqrt{7}\sqrt{3}+\sqrt{7}\sqrt{5}$
$=\sqrt{6}-\sqrt{10}-\sqrt{21}+\sqrt{35}$

**21.** $\left(3\sqrt{2}-4\sqrt{3}\right)\left(2\sqrt{2}+5\sqrt{3}\right)$
$=3\sqrt{2}\left(2\sqrt{2}\right)+3\sqrt{2}\left(5\sqrt{3}\right)$
$\quad-4\sqrt{3}\left(2\sqrt{2}\right)-4\sqrt{3}\left(5\sqrt{3}\right)$
$=6\cdot 2+15\sqrt{6}-8\sqrt{6}-20\cdot 3$
$=12+7\sqrt{6}-60$
$=7\sqrt{6}-48$ or $-48+7\sqrt{6}$

**23.** $\left(\sqrt{3}+\sqrt{5}\right)^2=\left(\sqrt{3}\right)^2+2\sqrt{3}\sqrt{5}+\left(\sqrt{5}\right)^2$
$=3+2\sqrt{15}+5$
$=8+2\sqrt{15}$

**25.** $\left(\sqrt{3x}-\sqrt{y}\right)^2$
$=\left(\sqrt{3x}\right)^2-2\sqrt{3x}\sqrt{y}+\left(\sqrt{y}\right)^2$
$=3x-2\sqrt{3xy}+y$

**27.** $\left(\sqrt{5}+7\right)\left(\sqrt{5}-7\right)$
$=\sqrt{5}\sqrt{5}-7\sqrt{5}+7\sqrt{5}-7\cdot 7$
$=5-\cancel{7\sqrt{5}}+\cancel{7\sqrt{5}}-49$
$=5-49$
$=-44$

**29.** $\left(2-5\sqrt{3}\right)\left(2+5\sqrt{3}\right)$
$=2\cdot 2+2\cdot 5\sqrt{3}-2\cdot 5\sqrt{3}-5\sqrt{3}\cdot 5\sqrt{3}$
$=4+\cancel{10\sqrt{3}}-\cancel{10\sqrt{3}}-25\cdot 3$
$=4-75$
$=-71$

**31.** $\left(3\sqrt{2}+2\sqrt{3}\right)\left(3\sqrt{2}-2\sqrt{3}\right)$
$=3\sqrt{2}\cdot 3\sqrt{2}-3\sqrt{2}\cdot 2\sqrt{3}$
$\quad+3\sqrt{2}\cdot 2\sqrt{3}-2\sqrt{3}\cdot 2\sqrt{3}$
$=9\cdot 2-\cancel{6\sqrt{6}}+\cancel{6\sqrt{6}}-4\cdot 3$
$=18-12$
$=6$

**33.** $\left(3-\sqrt{x}\right)\left(2-\sqrt{x}\right)$
$=3\cdot 2-3\sqrt{x}-2\sqrt{x}+\sqrt{x}\sqrt{x}$
$=6+(-3-2)\sqrt{x}+x$
$=6+(-5)\sqrt{x}+x$
$=6-5\sqrt{x}+x$

**35.** $\left(\sqrt[3]{x}-4\right)\left(\sqrt[3]{x}+5\right)$
$=\sqrt[3]{x}\sqrt[3]{x}+5\sqrt[3]{x}-4\sqrt[3]{x}-4\cdot 5$
$=\sqrt[3]{x^2}+(5-4)\sqrt[3]{x}-20$
$=\sqrt[3]{x^2}+\sqrt[3]{x}-20$

**37.** $\left(x+\sqrt[3]{y^2}\right)\left(2x-\sqrt[3]{y^2}\right)$
$=x\cdot 2x-x\sqrt[3]{y^2}+2x\sqrt[3]{y^2}-\sqrt[3]{y^2}\sqrt[3]{y^2}$
$=2x^2+(-x+2x)\sqrt[3]{y^2}-\sqrt[3]{y^4}$
$=2x^2+x\sqrt[3]{y^2}-\sqrt[3]{y^3y}$
$=2x^2+x\sqrt[3]{y^2}-y\sqrt[3]{y}$

**39.** $\frac{\sqrt{2}}{\sqrt{5}} = \frac{\sqrt{2}}{\sqrt{5}} \cdot \frac{\sqrt{5}}{\sqrt{5}} = \frac{\sqrt{2 \cdot 5}}{\sqrt{5 \cdot 5}} = \frac{\sqrt{10}}{5}$

**41.** $\sqrt{\frac{11}{x}} = \frac{\sqrt{11}}{\sqrt{x}} = \frac{\sqrt{11}}{\sqrt{x}} \cdot \frac{\sqrt{x}}{\sqrt{x}}$

$= \frac{\sqrt{11x}}{\sqrt{x^2}} = \frac{\sqrt{11x}}{x}$

**43.** $\frac{9}{\sqrt{3y}} = \frac{9}{\sqrt{3y}} \cdot \frac{\sqrt{3y}}{\sqrt{3y}} = \frac{9\sqrt{3y}}{\sqrt{3y \cdot 3y}}$

$= \frac{\overset{3}{\cancel{9}}\sqrt{3y}}{\underset{1}{\cancel{3}}y} = \frac{3\sqrt{3y}}{y}$

**45.** $\frac{1}{\sqrt[3]{2}} = \frac{1}{\sqrt[3]{2}} \cdot \frac{\sqrt[3]{2^2}}{\sqrt[3]{2^2}} = \frac{\sqrt[3]{2^2}}{\sqrt[3]{2^3}} = \frac{\sqrt[3]{4}}{2}$

**47.** $\frac{6}{\sqrt[3]{4}} = \frac{6}{\sqrt[3]{4}} \cdot \frac{\sqrt[3]{4^2}}{\sqrt[3]{4^2}} = \frac{6\sqrt[3]{4^2}}{\sqrt[3]{4}\sqrt[3]{4^2}}$

$= \frac{6\sqrt[3]{16}}{\sqrt[3]{4^3}} = \frac{6\sqrt[3]{8 \cdot 2}}{4} = \frac{6 \cdot 2\sqrt[3]{2}}{4}$

$= \frac{\overset{3}{\cancel{12}}\sqrt[3]{2}}{\underset{1}{\cancel{4}}} = 3\sqrt[3]{2}$

**49.** $\sqrt[3]{\frac{2}{3}} = \frac{\sqrt[3]{2}}{\sqrt[3]{3}} = \frac{\sqrt[3]{2}}{\sqrt[3]{3}} \cdot \frac{\sqrt[3]{3^2}}{\sqrt[3]{3^2}} = \frac{\sqrt[3]{2 \cdot 3^2}}{\sqrt[3]{3^3}}$

$= \frac{\sqrt[3]{2 \cdot 9}}{3} = \frac{\sqrt[3]{18}}{3}$

**51.** $\frac{4}{\sqrt[3]{x}} = \frac{4}{\sqrt[3]{x}} \cdot \frac{\sqrt[3]{x^2}}{\sqrt[3]{x^2}} = \frac{4\sqrt[3]{x^2}}{\sqrt[3]{x}\sqrt[3]{x^2}}$

$= \frac{4\sqrt[3]{x^2}}{\sqrt[3]{x^3}} = \frac{4\sqrt[3]{x^2}}{x}$

**53.** $\sqrt[3]{\frac{2}{y^2}} = \frac{\sqrt[3]{2}}{\sqrt[3]{y^2}} = \frac{\sqrt[3]{2}}{\sqrt[3]{y^2}} \cdot \frac{\sqrt[3]{y}}{\sqrt[3]{y}}$

$= \frac{\sqrt[3]{2y}}{\sqrt[3]{y^3}} = \frac{\sqrt[3]{2y}}{y}$

**55.** $\frac{7}{\sqrt[3]{2x^2}} = \frac{7}{\sqrt[3]{2x^2}} \cdot \frac{\sqrt[3]{2^2x}}{\sqrt[3]{2^2x}} = \frac{7\sqrt[3]{2^2x}}{\sqrt[3]{2x^2}\sqrt[3]{2^2x}}$

$= \frac{7\sqrt[3]{4x}}{\sqrt[3]{2^3x^3}} = \frac{7\sqrt[3]{4x}}{2x}$

**57.** $\sqrt[3]{\frac{2}{xy^2}} = \frac{\sqrt[3]{2}}{\sqrt[3]{xy^2}} = \frac{\sqrt[3]{2}}{\sqrt[3]{xy^2}} \cdot \frac{\sqrt[3]{x^2y}}{\sqrt[3]{x^2y}}$

$= \frac{\sqrt[3]{2}\sqrt[3]{x^2y}}{\sqrt[3]{xy^2}\sqrt[3]{x^2y}} = \frac{\sqrt[3]{2x^2y}}{\sqrt[3]{x^3y^3}}$

$= \frac{\sqrt[3]{2x^2y}}{xy}$

**59.** $\frac{3}{\sqrt[4]{x}} = \frac{3}{\sqrt[4]{x}} \cdot \frac{\sqrt[4]{x^3}}{\sqrt[4]{x^3}} = \frac{3\sqrt[4]{x^3}}{\sqrt[4]{xx^3}}$

$= \frac{3\sqrt[4]{x^3}}{\sqrt[4]{x^4}} = \frac{3\sqrt[4]{x^3}}{x}$

**61.** $\frac{6}{\sqrt[5]{8x^3}} = \frac{6}{\sqrt[5]{2^3x^3}} \cdot \frac{\sqrt[5]{2^2x^2}}{\sqrt[5]{2^2x^2}} = \frac{6\sqrt[5]{4x^2}}{\sqrt[5]{2^5x^5}}$

$= \frac{6\sqrt[5]{4x^2}}{2x} = \frac{3\sqrt[5]{4x^2}}{x}$

**63.** $\frac{2x^2y}{\sqrt[5]{4x^2y^4}}=\frac{2x^2y}{\sqrt[5]{2^2x^2y^4}}\cdot\frac{\sqrt[5]{2^3x^3y}}{\sqrt[5]{2^3x^3y}}$

$=\frac{2x^2y\sqrt[5]{8x^3y}}{\sqrt[5]{2^5x^5y^5}}$

$=\frac{\not{2}x^{\not{2}}\not{y}\sqrt[5]{8x^3y}}{\not{2}\not{x}\not{y}}$

$=x\sqrt[5]{8x^3y}$

**65.** $\frac{9}{\sqrt{3x^2y}}=\frac{9}{\sqrt{x^2\cdot 3y}}=\frac{9}{x\sqrt{3y}}$

$=\frac{9}{x\sqrt{3y}}\cdot\frac{\sqrt{3y}}{\sqrt{3y}}$

$=\frac{9\sqrt{3y}}{x\sqrt{(3y)^2}}=\frac{9\sqrt{3y}}{x(3y)}$

$=\frac{\not{3}\cdot 3\sqrt{3y}}{\not{3}xy}=\frac{3\sqrt{3y}}{xy}$

**67.** $-\sqrt{\frac{75a^5}{b^3}}=-\frac{\sqrt{75a^5}}{\sqrt{b^3}}=-\frac{\sqrt{25a^4\cdot 3a}}{\sqrt{b^2\cdot b}}$

$=-\frac{5a^2\sqrt{3a}}{b\sqrt{b}}=-\frac{5a^2\sqrt{3a}}{b\sqrt{b}}\cdot\frac{\sqrt{b}}{\sqrt{b}}$

$=-\frac{5a^2\sqrt{3ab}}{b\sqrt{b^2}}=-\frac{5a^2\sqrt{3ab}}{b(b)}$

$=-\frac{5a^2\sqrt{3ab}}{b^2}$

**69.** $\sqrt{\frac{7m^2n^3}{14m^3n^2}}=\sqrt{\frac{n}{2m}}=\frac{\sqrt{n}}{\sqrt{2m}}$

$=\frac{\sqrt{n}}{\sqrt{2m}}\cdot\frac{\sqrt{2m}}{\sqrt{2m}}=\frac{\sqrt{2mn}}{\sqrt{(2m)^2}}$

$=\frac{\sqrt{2mn}}{2m}$

**71.** $\frac{3}{\sqrt[4]{x^5y^3}}=\frac{3}{\sqrt[4]{x^4\cdot xy^3}}=\frac{3}{x\sqrt[4]{xy^3}}$

$=\frac{3}{x\sqrt[4]{xy^3}}\cdot\frac{\sqrt[4]{x^3y}}{\sqrt[4]{x^3y}}$

$=\frac{3\sqrt[4]{x^3y}}{x\sqrt[4]{x^4y^4}}=\frac{3\sqrt[4]{x^3y}}{x(xy)}$

$=\frac{3\sqrt[4]{x^3y}}{x^2y}$

**73.** $\frac{12}{\sqrt[3]{-8x^5y^8}}=\frac{12}{\sqrt[3]{-8x^3y^6\cdot x^2y^2}}$

$=\frac{12}{-2xy^2\sqrt[3]{x^2y^2}}$

$=\frac{12}{-2xy^2\sqrt[3]{x^2y^2}}\cdot\frac{\sqrt[3]{xy}}{\sqrt[3]{xy}}$

$=\frac{12\sqrt[3]{xy}}{-2xy^2\sqrt[3]{x^3y^3}}=\frac{12\sqrt[3]{xy}}{-2xy^2(xy)}$

$=\frac{12\sqrt[3]{xy}}{-2x^2y^3}$

$=-\frac{6\sqrt[3]{xy}}{x^2y^3}$

**75.**
$$\frac{8}{\sqrt{5}+2}=\frac{8}{\sqrt{5}+2}\cdot\frac{\sqrt{5}-2}{\sqrt{5}-2}$$
$$=\frac{8\sqrt{5}-8\cdot 2}{\sqrt{5}\sqrt{5}-2\sqrt{5}+2\sqrt{5}-2\cdot 2}$$
$$=\frac{8\sqrt{5}-16}{5-\cancel{2\sqrt{5}}+\cancel{2\sqrt{5}}-4}$$
$$=\frac{8\sqrt{5}-16}{5-4}=\frac{8\sqrt{5}-16}{1}$$
$$=8\sqrt{5}-16$$

**77.**
$$\frac{13}{\sqrt{11}-3}=\frac{13}{\sqrt{11}-3}\cdot\frac{\sqrt{11}+3}{\sqrt{11}+3}$$
$$=\frac{13\left(\sqrt{11}+3\right)}{\left(\sqrt{11}-3\right)\left(\sqrt{11}+3\right)}$$
$$=\frac{13\sqrt{11}+13\cdot 3}{\sqrt{11}\cdot\sqrt{11}+3\sqrt{11}-3\sqrt{11}-3\cdot 3}$$
$$=\frac{13\sqrt{11}+39}{11+\cancel{3\sqrt{11}}-\cancel{3\sqrt{11}}-9}$$
$$=\frac{13\sqrt{11}+39}{11-9}$$
$$=\frac{13\sqrt{11}+39}{2}$$

**79.**
$$\frac{6}{\sqrt{5}+\sqrt{3}}$$
$$=\frac{6}{\sqrt{5}+\sqrt{3}}\cdot\frac{\sqrt{5}-\sqrt{3}}{\sqrt{5}-\sqrt{3}}$$
$$=\frac{6\left(\sqrt{5}-\sqrt{3}\right)}{\left(\sqrt{5}+\sqrt{3}\right)\left(\sqrt{5}-\sqrt{3}\right)}$$
$$=\frac{6\sqrt{5}-6\sqrt{3}}{\sqrt{5}\sqrt{5}-\sqrt{3}\sqrt{5}+\sqrt{3}\sqrt{5}-\sqrt{3}\sqrt{3}}$$
$$=\frac{6\sqrt{5}-6\sqrt{3}}{5-\cancel{\sqrt{15}}+\cancel{\sqrt{15}}-3}$$
$$=\frac{6\sqrt{5}-6\sqrt{3}}{5-3}$$
$$=\frac{6\sqrt{5}-6\sqrt{3}}{2}$$
$$=\frac{\cancel{2}\left(3\sqrt{5}-3\sqrt{3}\right)}{\cancel{2}}$$
$$=3\sqrt{5}-3\sqrt{3}$$

**81.**
$$\frac{\sqrt{a}}{\sqrt{a}-\sqrt{b}}$$
$$=\frac{\sqrt{a}}{\sqrt{a}-\sqrt{b}}\cdot\frac{\sqrt{a}+\sqrt{b}}{\sqrt{a}+\sqrt{b}}$$
$$=\frac{\sqrt{a}\left(\sqrt{a}+\sqrt{b}\right)}{\left(\sqrt{a}-\sqrt{b}\right)\left(\sqrt{a}+\sqrt{b}\right)}$$
$$=\frac{\sqrt{a}\sqrt{a}+\sqrt{a}\sqrt{b}}{\sqrt{a}\sqrt{a}+\sqrt{a}\sqrt{b}-\sqrt{a}\sqrt{b}-\sqrt{b}\sqrt{b}}$$
$$=\frac{a+\sqrt{ab}}{a-b}$$

**83.**

$$\frac{25}{5\sqrt{2}-3\sqrt{5}}=\frac{25}{5\sqrt{2}-3\sqrt{5}}\cdot\frac{5\sqrt{2}+3\sqrt{5}}{5\sqrt{2}+3\sqrt{5}}=\frac{25\left(5\sqrt{2}+3\sqrt{5}\right)}{\left(5\sqrt{2}-3\sqrt{5}\right)\left(5\sqrt{2}+3\sqrt{5}\right)}$$

$$=\frac{125\sqrt{2}+75\sqrt{5}}{5\cdot5\sqrt{2\cdot2}+\cancel{5\cdot3\sqrt{2\cdot5}}-\cancel{5\cdot3\sqrt{2\cdot5}}-3\cdot3\sqrt{5\cdot5}}=\frac{125\sqrt{2}+75\sqrt{5}}{25\cdot2-9\cdot5}$$

$$=\frac{125\sqrt{2}+75\sqrt{5}}{50-45}=\frac{125\sqrt{2}+75\sqrt{5}}{5}=\frac{\cancel{5}\left(25\sqrt{2}+15\sqrt{5}\right)}{\cancel{5}}=25\sqrt{2}+15\sqrt{5}$$

**85.**

$$\frac{\sqrt{5}+\sqrt{3}}{\sqrt{5}-\sqrt{3}}=\frac{\sqrt{5}+\sqrt{3}}{\sqrt{5}-\sqrt{3}}\cdot\frac{\sqrt{5}+\sqrt{3}}{\sqrt{5}+\sqrt{3}}=\frac{\left(\sqrt{5}+\sqrt{3}\right)^2}{\left(\sqrt{5}-\sqrt{3}\right)\left(\sqrt{5}+\sqrt{3}\right)}$$

$$=\frac{\left(\sqrt{5}\right)^2+2\sqrt{5}\sqrt{3}+\left(\sqrt{3}\right)^2}{\sqrt{5}\cdot\sqrt{5}+\sqrt{5}\cdot\sqrt{3}-\sqrt{5}\cdot\sqrt{3}-\sqrt{3}\sqrt{3}}=\frac{5+2\sqrt{15}+3}{5+\cancel{\sqrt{15}}-\cancel{\sqrt{15}}-3}$$

$$=\frac{8+2\sqrt{15}}{5-3}=\frac{\cancel{2}\left(4+\sqrt{15}\right)}{\cancel{2}}=4+\sqrt{15}$$

**87.**

$$\frac{\sqrt{x}+1}{\sqrt{x}+3}=\frac{\sqrt{x}+1}{\sqrt{x}+3}\cdot\frac{\sqrt{x}-3}{\sqrt{x}-3}=\frac{\sqrt{x}\cdot\sqrt{x}-3\sqrt{x}+1\sqrt{x}-3\cdot1}{\sqrt{x}\cdot\sqrt{x}-\cancel{3\sqrt{x}}+\cancel{3\sqrt{x}}-3\cdot3}$$

$$=\frac{\sqrt{x^2}+(-3+1)\sqrt{x}-3}{\sqrt{x^2}-9}=\frac{x+(-2)\sqrt{x}-3}{x-9}=\frac{x-2\sqrt{x}-3}{x-9}$$

**89.**

$$\frac{5\sqrt{3}-3\sqrt{2}}{3\sqrt{2}-2\sqrt{3}}=\frac{5\sqrt{3}-3\sqrt{2}}{3\sqrt{2}-2\sqrt{3}}\cdot\frac{3\sqrt{2}+2\sqrt{3}}{3\sqrt{2}+2\sqrt{3}}=\frac{5\sqrt{3}\cdot3\sqrt{2}+5\sqrt{3}\cdot2\sqrt{3}-3\sqrt{2}\cdot3\sqrt{2}-3\sqrt{2}\cdot2\sqrt{3}}{3\sqrt{2}\cdot3\sqrt{2}+3\sqrt{2}\cdot2\sqrt{3}-3\sqrt{2}\cdot2\sqrt{3}-2\sqrt{3}\cdot2\sqrt{3}}$$

$$=\frac{15\sqrt{6}+10\cdot3-9\cdot2-6\sqrt{6}}{9\cdot2+\cancel{6\sqrt{6}}-\cancel{6\sqrt{6}}-4\cdot3}=\frac{15\sqrt{6}+30-18-6\sqrt{6}}{18-12}$$

$$=\frac{9\sqrt{6}+12}{6}=\frac{\cancel{3}\left(3\sqrt{6}+4\right)}{\cancel{3}\cdot2}=\frac{3\sqrt{6}+4}{2}$$

**91.**

$$\frac{2\sqrt{x}+\sqrt{y}}{\sqrt{y}-2\sqrt{x}}=\frac{2\sqrt{x}+\sqrt{y}}{\sqrt{y}-2\sqrt{x}}\cdot\frac{\sqrt{y}+2\sqrt{x}}{\sqrt{y}+2\sqrt{x}}=\frac{2\sqrt{x}\sqrt{y}+2\sqrt{x}\cdot2\sqrt{x}+\sqrt{y}\sqrt{y}+2\sqrt{x}\sqrt{y}}{\sqrt{y}\sqrt{y}+2\sqrt{x}\sqrt{y}-2\sqrt{x}\sqrt{y}-2\sqrt{x}\cdot2\sqrt{x}}$$

$$=\frac{2\sqrt{xy}+4\sqrt{x^2}+\sqrt{y^2}+2\sqrt{xy}}{\sqrt{y^2}+\cancel{2\sqrt{xy}}-\cancel{2\sqrt{xy}}-4\sqrt{x^2}}=\frac{2\sqrt{xy}+4x+y+2\sqrt{xy}}{y-4x}=\frac{4\sqrt{xy}+4x+y}{y-4x}$$

**93.** $$\sqrt{\frac{3}{2}} = \frac{\sqrt{3}}{\sqrt{2}} \cdot \frac{\sqrt{3}}{\sqrt{3}} = \frac{\sqrt{3}\sqrt{3}}{\sqrt{2}\sqrt{3}} = \frac{3}{\sqrt{6}}$$

**95.** $$\frac{\sqrt[3]{4x}}{\sqrt[3]{y}} = \frac{\sqrt[3]{4x}}{\sqrt[3]{y}} \cdot \frac{\sqrt[3]{4^2 x^2}}{\sqrt[3]{4^2 x^2}} = \frac{\sqrt[3]{4^3 x^3}}{\sqrt[3]{4^2 x^2 y}} = \frac{4x}{\sqrt[3]{16x^2 y}} = \frac{4x}{\sqrt[3]{8 \cdot 2x^2 y}} = \frac{4x}{2\sqrt[3]{2x^2 y}} = \frac{2x}{\sqrt[3]{2x^2 y}}$$

**97.** $$\frac{\sqrt{x}+3}{\sqrt{x}} = \frac{\sqrt{x}+3}{\sqrt{x}} \cdot \frac{\sqrt{x}-3}{\sqrt{x}-3} = \frac{\sqrt{x}\cdot\sqrt{x} - \cancel{3\sqrt{x}} + \cancel{3\sqrt{x}} - 3\cdot 3}{\sqrt{x}\cdot\sqrt{x} - 3\sqrt{x}} = \frac{\sqrt{x^2} - 9}{\sqrt{x^2} - 3\sqrt{x}} = \frac{x-9}{x-3\sqrt{x}}$$

**99.** $$\frac{\sqrt{a}+\sqrt{b}}{\sqrt{a}-\sqrt{b}} = \frac{\sqrt{a}+\sqrt{b}}{\sqrt{a}-\sqrt{b}} \cdot \frac{\sqrt{a}-\sqrt{b}}{\sqrt{a}-\sqrt{b}} = \frac{\sqrt{a}\cdot\sqrt{a} - \sqrt{a}\sqrt{b} + \sqrt{a}\sqrt{b} - \sqrt{b}\sqrt{b}}{\sqrt{a}\cdot\sqrt{a} - \sqrt{a}\sqrt{b} - \sqrt{a}\sqrt{b} + \sqrt{b}\sqrt{b}}$$
$$= \frac{\sqrt{a^2} - \cancel{\sqrt{ab}} + \cancel{\sqrt{ab}} - \sqrt{b^2}}{\sqrt{a^2} - \sqrt{ab} - \sqrt{ab} + \sqrt{b^2}} = \frac{a-b}{a - 2\sqrt{ab} + b}$$

**101.** $$\frac{\sqrt{x+5}-\sqrt{x}}{5} = \frac{\sqrt{x+5}-\sqrt{x}}{5} \cdot \frac{\sqrt{x+5}+\sqrt{x}}{\sqrt{x+5}+\sqrt{x}} = \frac{\left(\sqrt{x+5}\right)^2 + \sqrt{x+5}\cdot\sqrt{x} - \sqrt{x+5}\cdot\sqrt{x} - \left(\sqrt{x}\right)^2}{5\left(\sqrt{x+5}+\sqrt{x}\right)}$$
$$= \frac{x+5+\sqrt{x(x+5)} - \sqrt{x(x+5)} - x}{5\left(\sqrt{x+5}+\sqrt{x}\right)} = \frac{5}{5\left(\sqrt{x+5}+\sqrt{x}\right)} = \frac{1}{\sqrt{x+5}+\sqrt{x}}$$

**103.** $$\frac{\sqrt{x}+\sqrt{y}}{x^2-y^2} = \frac{\sqrt{x}+\sqrt{y}}{x^2-y^2} \cdot \frac{\sqrt{x}-\sqrt{y}}{\sqrt{x}-\sqrt{y}} = \frac{\left(\sqrt{x}\right)^2 - \cancel{\sqrt{xy}} + \cancel{\sqrt{xy}} - \left(\sqrt{y}\right)^2}{x^2\sqrt{x} - x^2\sqrt{y} - y^2\sqrt{x} + y^2\sqrt{y}}$$
$$= \frac{x-y}{x^2\left(\sqrt{x}-\sqrt{y}\right) - y^2\left(\sqrt{x}-\sqrt{y}\right)} = \frac{x-y}{\left(\sqrt{x}-\sqrt{y}\right)\left(x^2-y^2\right)}$$
$$= \frac{\cancel{x-y}}{\left(\sqrt{x}-\sqrt{y}\right)(x+y)\cancel{(x-y)}} = \frac{1}{\left(\sqrt{x}-\sqrt{y}\right)(x+y)}$$

**105.**
$$\sqrt{2}+\frac{1}{\sqrt{2}}=\sqrt{2}+\frac{1}{\sqrt{2}}\cdot\frac{\sqrt{2}}{\sqrt{2}}$$
$$=\sqrt{2}+\frac{\sqrt{2}}{2}$$
$$=\frac{2\sqrt{2}}{2}+\frac{\sqrt{2}}{2}$$
$$=\frac{2\sqrt{2}+\sqrt{2}}{2}$$
$$=\frac{3\sqrt{2}}{2}$$

**106.**
$$\sqrt{5}+\frac{1}{\sqrt{5}}=\sqrt{5}+\frac{1}{\sqrt{5}}\cdot\frac{\sqrt{5}}{\sqrt{5}}$$
$$=\sqrt{5}+\frac{\sqrt{5}}{5}$$
$$=\frac{5\sqrt{5}}{5}+\frac{\sqrt{5}}{5}$$
$$=\frac{5\sqrt{5}+\sqrt{5}}{5}$$
$$=\frac{6\sqrt{5}}{5}$$

**107.**
$$\sqrt[3]{25}-\frac{15}{\sqrt[3]{5}}=\sqrt[3]{25}-\frac{15}{\sqrt[3]{5}}\cdot\frac{\sqrt[3]{5^2}}{\sqrt[3]{5^2}}$$
$$=\sqrt[3]{25}-\frac{15\sqrt[3]{25}}{5}$$
$$=\sqrt[3]{25}-3\sqrt[3]{25}$$
$$=-2\sqrt[3]{25}$$

**108.**
$$\sqrt[4]{8}-\frac{20}{\sqrt[4]{2}}=\sqrt[4]{8}-\frac{20}{\sqrt[4]{2}}\cdot\frac{\sqrt[4]{2^3}}{\sqrt[4]{2^3}}$$
$$=\sqrt[4]{8}-\frac{20\sqrt[4]{8}}{2}$$
$$=\sqrt[4]{8}-10\sqrt[4]{8}$$
$$=-9\sqrt[4]{8}$$

**109.**
$$\sqrt{6}-\sqrt{\frac{1}{6}}+\sqrt{\frac{2}{3}}$$
$$=\sqrt{6}-\frac{\sqrt{1}}{\sqrt{6}}+\frac{\sqrt{2}}{\sqrt{3}}$$
$$=\sqrt{6}-\frac{1}{\sqrt{6}}\cdot\frac{\sqrt{6}}{\sqrt{6}}+\frac{\sqrt{2}}{\sqrt{3}}\cdot\frac{\sqrt{3}}{\sqrt{3}}$$
$$=\sqrt{6}-\frac{\sqrt{6}}{6}+\frac{\sqrt{6}}{3}$$
$$=\frac{6\sqrt{6}}{6}-\frac{\sqrt{6}}{6}+\frac{2\sqrt{6}}{6}$$
$$=\frac{6\sqrt{6}-\sqrt{6}+2\sqrt{6}}{6}=\frac{7\sqrt{6}}{6}$$

**110.**
$$\sqrt{15}-\sqrt{\frac{5}{3}}+\sqrt{\frac{3}{5}}$$
$$=\sqrt{15}-\frac{\sqrt{5}}{\sqrt{3}}+\frac{\sqrt{3}}{\sqrt{5}}$$
$$=\sqrt{15}-\frac{\sqrt{5}}{\sqrt{3}}\cdot\frac{\sqrt{3}}{\sqrt{3}}+\frac{\sqrt{3}}{\sqrt{5}}\cdot\frac{\sqrt{5}}{\sqrt{5}}$$
$$=\sqrt{15}-\frac{\sqrt{15}}{3}+\frac{\sqrt{15}}{5}$$
$$=\frac{15\sqrt{15}}{15}-\frac{5\sqrt{15}}{15}+\frac{3\sqrt{15}}{15}$$
$$=\frac{15\sqrt{15}-5\sqrt{15}+3\sqrt{15}}{15}$$
$$=\frac{13\sqrt{15}}{15}$$

**111.** $\dfrac{2}{\sqrt{2}+\sqrt{3}}+\sqrt{75}-\sqrt{50}$

$=\dfrac{2}{\sqrt{2}+\sqrt{3}}\cdot\dfrac{\sqrt{2}-\sqrt{3}}{\sqrt{2}-\sqrt{3}}+\sqrt{25\cdot 3}-\sqrt{25\cdot 2}$

$=\dfrac{2\sqrt{2}-2\sqrt{3}}{\left(\sqrt{2}\right)^2-\left(\sqrt{3}\right)^2}+5\sqrt{3}-5\sqrt{2}$

$=\dfrac{2\sqrt{2}-2\sqrt{3}}{2-3}+5\sqrt{3}-5\sqrt{2}$

$=\dfrac{2\sqrt{2}-2\sqrt{3}}{-1}+5\sqrt{3}-5\sqrt{2}$

$=2\sqrt{3}-2\sqrt{2}+5\sqrt{3}-5\sqrt{2}$

$=7\sqrt{3}-7\sqrt{2}$

**112.** $\dfrac{5}{\sqrt{2}+\sqrt{7}}-2\sqrt{32}+\sqrt{28}$

$=\dfrac{5}{\sqrt{2}+\sqrt{7}}\cdot\dfrac{\sqrt{2}-\sqrt{7}}{\sqrt{2}-\sqrt{7}}-2\sqrt{16\cdot 2}+\sqrt{4\cdot 7}$

$=\dfrac{5\sqrt{2}-5\sqrt{7}}{\left(\sqrt{2}\right)^2-\left(\sqrt{7}\right)^2}-2(4)\sqrt{2}+2\sqrt{7}$

$=\dfrac{5\sqrt{2}-5\sqrt{7}}{2-7}-8\sqrt{2}+2\sqrt{7}$

$=\dfrac{5\sqrt{2}-5\sqrt{7}}{-5}-8\sqrt{2}+2\sqrt{7}$

$=\sqrt{7}-\sqrt{2}-8\sqrt{2}+2\sqrt{7}$

$=3\sqrt{7}-9\sqrt{2}$

**113.** $f(x)=x^2-6x-4$

$f\left(3-\sqrt{13}\right)=\left(3-\sqrt{13}\right)^2-6\left(3-\sqrt{13}\right)-4$

$=9-6\sqrt{13}+13-18+6\sqrt{13}-4$

$=0$

**114.** $f(x)=x^2+4x-2$

$f\left(-2+\sqrt{6}\right)$

$=\left(-2+\sqrt{6}\right)^2+4\left(-2+\sqrt{6}\right)-2$

$=4-4\sqrt{6}+6-8+4\sqrt{6}-2$

$=0$

**115.** $f(x)=\sqrt{9+x}$

$f\left(3\sqrt{5}\right)\cdot f\left(-3\sqrt{5}\right)$

$=\sqrt{9+3\sqrt{5}}\cdot\sqrt{9-3\sqrt{5}}$

$=\sqrt{\left(9+3\sqrt{5}\right)\left(9-3\sqrt{5}\right)}$

$=\sqrt{9^2-\left(3\sqrt{5}\right)^2}$

$=\sqrt{81-9\cdot 5}$

$=\sqrt{81-45}$

$=\sqrt{36}$

$=6$

**116.** $f(x)=x^2$

$f\left(\sqrt{a+1}-\sqrt{a-1}\right)$

$=\left(\sqrt{a+1}-\sqrt{a-1}\right)^2$

$=\left(\sqrt{a+1}\right)^2-2\sqrt{a+1}\cdot\sqrt{a-1}+\left(\sqrt{a-1}\right)^2$

$=a+1-2\sqrt{(a+1)(a-1)}+a-1$

$=2a-2\sqrt{a^2-1}$

**117.** $P(t)=15.92\sqrt{t}+19$

$P(4)=15.92\sqrt{4}+19$

$=15.92(2)+19$

$=50.84\approx 51\%$

In 2001 (4 years after 1997), the percentage of U.S. households online was roughly 51%. This answer models the data well. According to the graph, 51% of U.S. households were online in 2001.

**119.** $\text{a.r.c.}=\dfrac{\text{change in percent}}{\text{change in time}}$

$=\dfrac{\left[15.92\sqrt{6}+19\right]-\left[15.92\sqrt{0}+19\right]}{2003-1997}$

$=\dfrac{15.92\sqrt{6}+19-19}{6}$

$=\dfrac{15.92\sqrt{6}}{6}\approx 6.5\%$

The average yearly increase in the percentage of online households from 1997 to 2003 was about 6.5%.

**121.** $15.92\left(\dfrac{\sqrt{0+6}-\sqrt{0}}{6}\right)=\dfrac{15.92\sqrt{6}}{6}$

This is exactly the same result as that obtained in Exercise 119. This answer models the actual yearly percentage increase very well.

**123. a.** $15.92\left(\dfrac{\sqrt{t+h}-\sqrt{t}}{h}\right)$

$=15.92\left(\dfrac{\sqrt{t+h}-\sqrt{t}}{h}\cdot\dfrac{\sqrt{t+h}+\sqrt{t}}{\sqrt{t+h}+\sqrt{t}}\right)$

$=15.92\left(\dfrac{\left(\sqrt{t+h}\right)^2-\left(\sqrt{t}\right)^2}{h\left(\sqrt{t+h}+\sqrt{t}\right)}\right)$

$=15.92\left(\dfrac{t+h-t}{h\left(\sqrt{t+h}+\sqrt{t}\right)}\right)$

$=\dfrac{15.92h}{h\left(\sqrt{t+h}+\sqrt{t}\right)}$

$=\dfrac{15.92}{\sqrt{t+h}+\sqrt{t}}$

**b.** $\dfrac{15.92}{\sqrt{t+0}+\sqrt{t}}=\dfrac{15.92}{\sqrt{t}+\sqrt{t}}$

$=\dfrac{15.92}{2\sqrt{t}}$

$=\dfrac{7.96}{\sqrt{t}}$

**c.** In 2003, we have $t=6$.

$\dfrac{7.96}{\sqrt{6}}\approx 3.2497\%$

The rate of change in the percentage of households online in 2003 was roughly 3.2%.

**125.** Perimeter $= 2l + 2w$

$$= 2\left(\sqrt{8}+1\right)+2\left(\sqrt{8}-1\right)$$
$$= 2\sqrt{8}+\not{2}+2\sqrt{8}-\not{2}$$
$$= (2+2)\sqrt{8} = 4\sqrt{8}$$
$$= 4\sqrt{4\cdot 2} = 4\cdot 2\sqrt{2} = 8\sqrt{2}$$

The perimeter is $8\sqrt{2}$ inches.

Area $= lw = \left(\sqrt{8}+1\right)\left(\sqrt{8}-1\right)$

$$= \left(\sqrt{8}\right)^2 - \not{\sqrt{8}} + \not{\sqrt{8}} - 1$$
$$= 8 - 1 = 7$$

The area is 7 square inches.

**127.** $\dfrac{7\sqrt{2\cdot 2\cdot 3}}{6} = \dfrac{7\cdot 2\sqrt{3}}{6}$

$$= \frac{7\cdot \not{2}\sqrt{3}}{\not{2}\cdot 3} = \frac{7}{3}\sqrt{3}$$

**129.** Answers will vary. You will need to distribute the $\sqrt{2}$ across the sum.

**131.** Answers will vary. Use the form $(a+b)^2 = a^2 + 2ab + b^2$.

**133.** Answers will vary. One approach is to FOIL like you would when multiplying two binomials.

**135.** Answers will vary. The value remains the same because we are just multiplying by 1. However, we do so in a way that changes the *form*, not the value.

**137.** Answers will vary. The percentage has been increasing each year, but the rate of increase began to slow.

**139.** $\left(\sqrt{x}+2\right)\left(\sqrt{x}-2\right) = x^2 - 4 \quad$ for $x \geq 0$

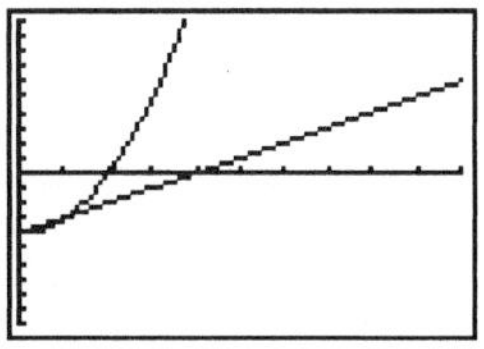

The graphs do not coincide. Correct the simplification.

$\left(\sqrt{x}+2\right)\left(\sqrt{x}-2\right) = x - 4 \quad$ for $x \geq 0$

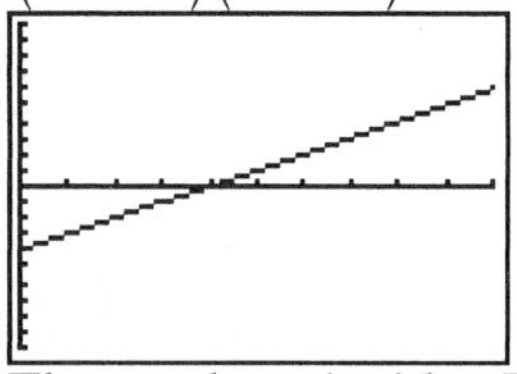

The graphs coincide. The new simplification is correct.

**141.** $\dfrac{3}{\sqrt{x+3}-\sqrt{x}} = \sqrt{x+3}+\sqrt{x}$

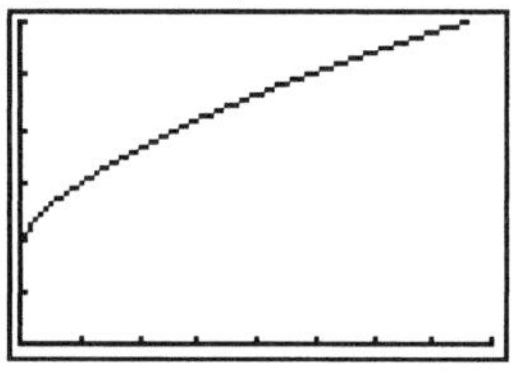

The graphs coincide, so the simplification is correct.

**143.** $7\left[(2x-5)-(x+1)\right] = \left(\sqrt{7}+2\right)\left(\sqrt{7}-2\right)$

$$7[2x-5-x-1] = \left(\sqrt{7}\right)^2 - (2)^2$$
$$7(x-6) = 7-4$$
$$7x - 42 = 3$$
$$7x = 45$$
$$x = \frac{45}{7}$$

The solution is $\frac{45}{7}$ and the solution set is $\left\{\frac{45}{7}\right\}$.

**145.** $\dfrac{1}{\sqrt{2}+\sqrt{3}+\sqrt{4}}$

$= \dfrac{1}{\left(\sqrt{2}+\sqrt{3}\right)+2} \cdot \dfrac{\left(\sqrt{2}+\sqrt{3}\right)-2}{\left(\sqrt{2}+\sqrt{3}\right)-2}$

$= \dfrac{\sqrt{2}+\sqrt{3}-2}{\left(\sqrt{2}+\sqrt{3}\right)^2-2^2} = \dfrac{\sqrt{2}+\sqrt{3}-2}{2+2\sqrt{6}+3-4}$

$= \dfrac{\sqrt{2}+\sqrt{3}-2}{2\sqrt{6}+1} \cdot \dfrac{2\sqrt{6}-1}{2\sqrt{6}-1}$

$= \dfrac{2\sqrt{12}+2\sqrt{18}-4\sqrt{6}-\sqrt{2}-\sqrt{3}+2}{\left(2\sqrt{6}\right)^2-1^2}$

$= \dfrac{4\sqrt{3}+6\sqrt{2}-4\sqrt{6}-\sqrt{2}-\sqrt{3}+2}{4\cdot 6-1}$

$= \dfrac{3\sqrt{3}+5\sqrt{2}-4\sqrt{6}+2}{24+1}$

$= \dfrac{3\sqrt{3}+5\sqrt{2}-4\sqrt{6}+2}{24-1}$

$= \dfrac{3\sqrt{3}+5\sqrt{2}-4\sqrt{6}+2}{23}$

**146.** $\dfrac{2}{x-2}+\dfrac{3}{x^2-4}$

$= \dfrac{2}{x-2}+\dfrac{3}{(x+2)(x-2)}$

$= \dfrac{2}{x-2}\cdot\dfrac{(x+2)}{(x+2)}+\dfrac{3}{(x+2)(x-2)}$

$= \dfrac{2(x+2)}{(x-2)(x+2)}+\dfrac{3}{(x+2)(x-2)}$

$= \dfrac{2(x+2)+3}{(x-2)(x+2)}$

$= \dfrac{2x+4+3}{(x-2)(x+2)}$

$= \dfrac{2x+7}{(x-2)(x+2)}$ or $\dfrac{2x+7}{x^2-4}$

**147.** Using the results from Exercise 146, we know that the left side of the equation simplifies to $\dfrac{2x+7}{(x-2)(x+2)}$.

Setting this equal to zero, we have

$\dfrac{2x+7}{(x-2)(x+2)}=0$.

To solve, set each factor equal to zero.

$2x+7=0 \qquad x-2=0$

$2x=-7 \qquad x=2$

$x=-\dfrac{7}{2}$

$x+2=0$

$x=-2$

The solutions are $-\dfrac{7}{2}$, $-2$ and 2. We disregard $-2$ and 2 because these values result in a zero denominator.

The solution set is $\left\{-\dfrac{7}{2}\right\}$.

**148.** $f(x)=x^4-3x^2-2x+5$

To find $f(-2)$, we use synthetic division to divide by $-(-2)=2$.

| | | | | | |
|---|---|---|---|---|---|
| −2 | 1 | 0 | −3 | −2 | 5 |
| | | −2 | 4 | −2 | 8 |
| | 1 | −2 | 1 | −4 | 13 |

The remainder is 13. This means that $f(-2)=13$.

## 7.6 Exercise Set

**1.** $\sqrt{3x-2}=4$

$\left(\sqrt{3x-2}\right)^2=4^2$

$3x-2=16$

$3x=18$

$x=6$

Check:

$\sqrt{3(6)-2}=4$

$\sqrt{18-2}=4$

$\sqrt{16}=4$

$4=4$

The solution is 6 and the solution set is $\{6\}$.

**3.** $\sqrt{5x-4}-9=0$

$\sqrt{5x-4}=9$

$\left(\sqrt{5x-4}\right)^2=9^2$

$5x-4=81$

$5x=85$

$x=17$

Check:

$\sqrt{5(17)-4}-9=0$

$\sqrt{85-4}-9=0$

$\sqrt{81}-9=0$

$9-9=0$

$0=0$

The solution is 17 and the solution set is $\{17\}$.

**5.** $\sqrt{3x+7}+10=4$

$\sqrt{3x+7}=-6$

Since the square root of a number is always positive, the solution set is $\{\ \}$ or $\varnothing$.

**7.** $x=\sqrt{7x+8}$

$x^2=\left(\sqrt{7x+8}\right)^2$

$x^2=7x+8$

$x^2-7x-8=0$

$(x-8)(x+1)=0$

Apply the zero product principle.

$x-8=0 \qquad x+1=0$

$x=8 \qquad x=-1$

Check:

$8=\sqrt{7(8)+8} \qquad -1=\sqrt{7(-1)+8}$

$8=\sqrt{56+8}$

$8=\sqrt{64}$

$8=8$

We disregard –1 because square roots are always positive. The solution is 8 and the solution set is $\{8\}$.

**9.** $\sqrt{5x+1}=x+1$

$\left(\sqrt{5x+1}\right)^2=(x+1)^2$

$5x+\cancel{1}=x^2+2x+\cancel{1}$

$0=x^2-3x$

$0=x(x-3)$

Apply the zero product principle.

$x=0 \qquad x-3=0$

$x=3$

Both solutions check. The solutions are 0 and 3 and the solution set is $\{0,3\}$.

**11.**
$$x = \sqrt{2x-2}+1$$
$$x-1 = \sqrt{2x-2}$$
$$(x-1)^2 = \left(\sqrt{2x-2}\right)^2$$
$$x^2-2x+1 = 2x-2$$
$$x^2-4x+3 = 0$$
$$(x-3)(x-1) = 0$$
Apply the zero product principle.
$x-3=0 \qquad x-1=0$
$x=3 \qquad x=1$
Both solutions check. The solutions are 1 and 3 and the solution set is $\{1,3\}$.

**13.**
$$x-2\sqrt{x-3} = 3$$
$$x-3 = 2\sqrt{x-3}$$
$$(x-3)^2 = \left(2\sqrt{x-3}\right)^2$$
$$x^2-6x+9 = 4(x-3)$$
$$x^2-6x+9 = 4x-12$$
$$x^2-10x+21 = 0$$
$$(x-7)(x-3) = 0$$
Apply the zero product principle.
$x-7=0 \qquad x-3=0$
$x=7 \qquad x=3$
Both solutions check. The solutions are 3 and 7 and the solution set is $\{3,7\}$.

**15.**
$$\sqrt{2x-5} = \sqrt{x+4}$$
$$\left(\sqrt{2x-5}\right)^2 = \left(\sqrt{x+4}\right)^2$$
$$2x-5 = x+4$$
$$x-5 = 4$$
$$x = 9$$
The solution checks. The solution is 9 and the solution set is $\{9\}$.

**17.**
$$\sqrt[3]{2x+11} = 3$$
$$\left(\sqrt[3]{2x+11}\right)^3 = 3^3$$
$$2x+11 = 27$$
$$2x = 16$$
$$x = 8$$
The solution checks. The solution is 8 and the solution set is $\{8\}$.

**19.**
$$\sqrt[3]{2x-6}-4 = 0$$
$$\sqrt[3]{2x-6} = 4$$
$$\left(\sqrt[3]{2x-6}\right)^3 = 4^3$$
$$2x-6 = 64$$
$$2x = 70$$
$$x = 35$$
The solution checks. The solution is 35 and the solution set is $\{35\}$.

**21.**
$$\sqrt{x-7} = 7-\sqrt{x}$$
$$\left(\sqrt{x-7}\right)^2 = \left(7-\sqrt{x}\right)^2$$
$$\cancel{x}-7 = 49-14\sqrt{x}+\cancel{x}$$
$$-7 = 49-14\sqrt{x}$$
$$-56 = -14\sqrt{x}$$
$$\frac{-56}{-14} = \frac{-14\sqrt{x}}{-14}$$
$$4 = \sqrt{x}$$
$$4^2 = \left(\sqrt{x}\right)^2$$
$$16 = x$$
The solution checks. The solution is 16 and the solution set is $\{16\}$.

**23.** $\sqrt{x+2}+\sqrt{x-1}=3$

$$\sqrt{x+2}=3-\sqrt{x-1}$$
$$\left(\sqrt{x+2}\right)^2=\left(3-\sqrt{x-1}\right)^2$$
$$\cancel{x}+2=9-6\sqrt{x-1}+\cancel{x}-1$$
$$2=8-6\sqrt{x-1}$$
$$-6=-6\sqrt{x-1}$$
$$\frac{-6}{-6}=\frac{-6\sqrt{x-1}}{-6}$$
$$1=\sqrt{x-1}$$
$$1^2=\left(\sqrt{x-1}\right)^2$$
$$1=x-1$$
$$2=x$$

The solution checks. The solution is 2 and the solution set is $\{2\}$.

**25.** $2\sqrt{4x+1}-9=x-5$

$$2\sqrt{4x+1}=x+4$$
$$\left(2\sqrt{4x+1}\right)^2=(x+4)^2$$
$$2^2\left(\sqrt{4x+1}\right)^2=x^2+8x+16$$
$$4(4x+1)=x^2+8x+16$$
$$16x+4=x^2+8x+16$$
$$0=x^2-8x+12$$
$$0=(x-6)(x-2)$$
$$x-6=0 \quad \text{or} \quad x-2=0$$
$$x=6 \qquad\qquad x=2$$

Check $x=6$:
$$2\sqrt{4(6)+1}-9=6-5$$
$$2\sqrt{25}-9=1$$
$$2(5)-9=1$$
$$10-9=1$$
$$1=1$$

Check $x=2$:
$$2\sqrt{4(2)+1}-9=2-5$$
$$2\sqrt{8+1}-9=-3$$
$$2\sqrt{9}-9=-3$$
$$2(3)-9=-3$$
$$6-9=-3$$
$$-3=-3$$

Both solutions check. The solutions are 2 and 6, and the solution set is $\{2,6\}$.

**27.** $(2x+3)^{1/3}+4=6$

$$(2x+3)^{1/3}=2$$
$$\left((2x+3)^{1/3}\right)^3=2^3$$
$$2x+3=8$$
$$2x=5$$
$$x=\frac{5}{2}$$

The solution checks. The solution is $\frac{5}{2}$ and the solution set is $\left\{\frac{5}{2}\right\}$.

**29.** $(3x+1)^{1/4}+7=9$

$$(3x+1)^{1/4}=2$$

$$\left((3x+1)^{1/4}\right)^4=2^4$$

$$3x+1=16$$

$$3x=15$$

$$x=5$$

The solution checks. The solution is 5 and the solution set is $\{5\}$.

**31.** $(x+2)^{1/2}+8=4$

$$(x+2)^{1/2}=-4$$

$$\sqrt{x+2}=-4$$

The square root of a number must be positive. The solution set is $\varnothing$.

**33.** $\sqrt{2x-3}-\sqrt{x-2}=1$

$$\sqrt{2x-3}=\sqrt{x-2}+1$$

$$\left(\sqrt{2x-3}\right)^2=\left(\sqrt{x-2}+1\right)^2$$

$$2x-3=x-2+2\sqrt{x-2}+1$$

$$2x-3=x-1+2\sqrt{x-2}$$

$$x-2=2\sqrt{x-2}$$

$$(x-2)^2=\left(2\sqrt{x-2}\right)^2$$

$$x^2-4x+4=4(x-2)$$

$$x^2-4x+4=4x-8$$

$$x^2-8x+12=0$$

$$(x-6)(x-2)=0$$

Apply the zero product principle.

$x-6=0 \qquad x-2=0$

$x=6 \qquad\quad x=2$

Both solutions check. The solutions are 2 and 6 and the solution set is $\{2,6\}$.

**35.** $3x^{1/3}=\left(x^2+17x\right)^{1/3}$

$$\left(3x^{1/3}\right)^3=\left(\left(x^2+17x\right)^{1/3}\right)^3$$

$$3^3x=x^2+17x$$

$$27x=x^2+17x$$

$$0=x^2-10x$$

$$0=x(x-10)$$

$x=0 \qquad x-10=0$

$\qquad\qquad x=10$

Both solutions check. The solutions are 0 and 10 and the solution set is $\{0,10\}$.

**37.** $(x+8)^{1/4}=(2x)^{1/4}$

$$\left((x+8)^{1/4}\right)^4=\left((2x)^{1/4}\right)^4$$

$$x+8=2x$$

$$8=x$$

The solution checks. The solution is 8 and the solution set is $\{8\}$.

**39.** $f(x)=x+\sqrt{x+5}$

$$7=x+\sqrt{x+5}$$

$$7-x=\sqrt{x+5}$$

$$(7-x)^2=\left(\sqrt{x+5}\right)^2$$

$$49-14x+x^2=x+5$$

$$x^2-15x+44=0$$

$$(x-11)(x-4)=0$$

$x-11=0 \quad \text{or} \quad x-4=0$

$x=11 \qquad\qquad x=4$

Check $x=11$: $11+\sqrt{11+5}=11+\sqrt{16}$
$=15\neq 7$

Check $x=4$: $4+\sqrt{4+5}=4+\sqrt{9}$
$=7$

Discard 11. The solution is 4.

**40.** $f(x)=x-\sqrt{x-2}$

$$4=x-\sqrt{x-2}$$
$$\sqrt{x-2}=x-4$$
$$\left(\sqrt{x-2}\right)^2=(x-4)^2$$
$$x-2=x^2-8x+16$$
$$0=x^2-9x+18$$
$$0=(x-6)(x-3)$$
$$x-6=0 \quad \text{or} \quad x-3=0$$
$$x=6 \qquad\qquad x=3$$

Check $x=6$: $6-\sqrt{6-2}=6-\sqrt{4}$
$=4$

Check $x=3$: $3-\sqrt{3-2}=3-\sqrt{1}$
$=2\neq 4$

Discard 3. The solution is 6.

**41.** $f(x)=(5x+16)^{1/3}$; $g(x)=(x-12)^{1/3}$

$$(5x+16)^{1/3}=(x-12)^{1/3}$$
$$\left[(5x+16)^{1/3}\right]^3=\left[(x-12)^{1/3}\right]^3$$
$$5x+16=x-12$$
$$4x=-28$$
$$x=-7$$

The solution is $-7$.

**42.** $f(x)=(9x+2)^{1/4}$; $g(x)=(5x+18)^{1/4}$

$$(9x+2)^{1/4}=(5x+18)^{1/4}$$
$$\left[(9x+2)^{1/4}\right]^4=\left[(5x+18)^{1/4}\right]^4$$
$$9x+2=5x+18$$
$$4x=16$$
$$x=4$$

Check:
$$(9(4)+2)^{1/4}=(5(4)+18)^{1/4}$$
$$(36+2)^{1/4}=(20+18)^{1/4}$$
$$38^{1/4}=38^{1/4}$$

The solution is 4.

**43.**
$$r=\sqrt{\frac{3V}{\pi h}}$$
$$r^2=\left(\sqrt{\frac{3V}{\pi h}}\right)^2$$
$$r^2=\frac{3V}{\pi h}$$
$$\pi r^2 h=3V$$
$$\frac{\pi r^2 h}{3}=V \quad \text{or} \quad V=\frac{\pi r^2 h}{3}$$

**44.**
$$r=\sqrt{\frac{A}{4\pi}}$$
$$r^2=\left(\sqrt{\frac{A}{4\pi}}\right)^2$$
$$r^2=\frac{A}{4\pi}$$
$$4\pi r^2=A \quad \text{or} \quad A=4\pi r^2$$

**45.**

$$t = 2\pi\sqrt{\frac{l}{32}}$$

$$\frac{t}{2\pi} = \sqrt{\frac{l}{32}}$$

$$\left(\frac{t}{2\pi}\right)^2 = \left(\sqrt{\frac{l}{32}}\right)^2$$

$$\frac{t^2}{4\pi^2} = \frac{l}{32}$$

$$\frac{32t^2}{4\pi^2} = l$$

$$\frac{8t^2}{\pi^2} = l \quad \text{or} \quad l = \frac{8t^2}{\pi^2}$$

**46.**

$$v = \sqrt{\frac{FR}{m}}$$

$$v^2 = \left(\sqrt{\frac{FR}{m}}\right)^2$$

$$v^2 = \frac{FR}{m}$$

$$mv^2 = FR$$

$$m = \frac{FR}{v^2}$$

**47.** Let $x$ = the number.

$$\sqrt{5x-4} = x-2$$

$$\left(\sqrt{5x-4}\right)^2 = (x-2)^2$$

$$5x-4 = x^2-4x+4$$

$$0 = x^2-9x+8$$

$$0 = (x-8)(x-1)$$

$$x-8=0 \quad \text{or} \quad x-1=0$$

$$x=8 \qquad\qquad x=1$$

Check $x=8$: $\sqrt{5(8)-4} = 8-2$

$$\sqrt{40-4} = 6$$

$$\sqrt{36} = 6$$

$$6 = 6$$

Check $x=1$: $\sqrt{5(1)-4} = 1-2$

$$\sqrt{5-4} = -1$$

$$\sqrt{-1} \neq -1$$

Discard $x=1$. The number is 8.

**48.** Let $x$ = the number.

$$\sqrt{x-3} = x-5$$

$$\left(\sqrt{x-3}\right)^2 = (x-5)^2$$

$$x-3 = x^2-10x+25$$

$$0 = x^2-11x+28$$

$$0 = (x-7)(x-4)$$

$$x-7=0 \quad \text{or} \quad x-4=0$$

$$x=7 \qquad\qquad x=4$$

Check $x=7$: $\sqrt{7-3} = 7-5$

$$\sqrt{4} = 2$$

$$2 = 2$$

Check $x=4$: $\sqrt{4-3} = 4-5$

$$\sqrt{1} = -1$$

$$1 \neq -1$$

Discard 4. The number is 7.

**49.** $f(x)=\sqrt{x+16}-\sqrt{x}-2$

To find the x-intercepts, set the function equal to 0 and solve for $x$.

$$0=\sqrt{x+16}-\sqrt{x}-2$$
$$\sqrt{x}+2=\sqrt{x+16}$$
$$\left(\sqrt{x}+2\right)^2=\left(\sqrt{x+16}\right)^2$$
$$x+4\sqrt{x}+4=x+16$$
$$4\sqrt{x}=12$$
$$\sqrt{x}=3$$
$$\left(\sqrt{x}\right)^2=3^2$$
$$x=9$$

Check $x=9$:

$$\sqrt{9+16}-\sqrt{9}-2$$
$$=\sqrt{25}-\sqrt{9}-2$$
$$=5-3-2$$
$$=0$$

The only x-intercept is 9.

**50.** $f(x)=\sqrt{2x-3}-\sqrt{2x}+1$

To find the x-intercepts, set the function equal to 0 and solve for $x$.

$$0=\sqrt{2x-3}-\sqrt{2x}+1$$
$$\sqrt{2x}-1=\sqrt{2x-3}$$
$$\left(\sqrt{2x}-1\right)^2=\left(\sqrt{2x-3}\right)^2$$
$$2x-2\sqrt{2x}+1=2x-3$$
$$-2\sqrt{2x}=-4$$
$$\sqrt{2x}=2$$
$$\left(\sqrt{2x}\right)^2=2^2$$
$$2x=4$$
$$x=2$$

Check $x=2$:

$$\sqrt{2(2)-3}-\sqrt{2(2)}+1$$
$$=\sqrt{4-3}-\sqrt{4}+1$$
$$=\sqrt{1}-\sqrt{4}+1$$
$$=1-2+1$$
$$=0$$

The only x-intercept is 2.

**51.** For the year 2100, we use $x=98$.

$$f(98)=0.083(98)+57.9$$
$$=66.034$$
$$g(98)=0.36\sqrt{98}+57.9$$
$$\approx 61.464$$

In the year 2100, the projected high end temperature is about $66°$ and the projected low end temperature is about $61.5°$.

**53.** Using *f*:

$$0.083x=1$$
$$x=\frac{1}{0.083}\approx 12.05$$

The projected global temperature will exceed the 2002 average by 1 degree in 2014 (12 years after 2002).

Using *g*:

$$0.36\sqrt{x}=1$$
$$\sqrt{x}=\frac{1}{0.36}$$
$$\left(\sqrt{x}\right)^2=\left(\frac{1}{0.36}\right)^2$$
$$x\approx 7.716 \text{ (roughly 8)}$$

The projected global temperature will exceed the 2002 average by 1 degree in 2010 (8 years after 2002).

**55.**

$$40000 = 5000\sqrt{100-x}$$
$$\frac{40000}{5000} = \frac{5000\sqrt{100-x}}{5000}$$
$$8 = \sqrt{100-x}$$
$$8^2 = \left(\sqrt{100-x}\right)^2$$
$$64 = 100 - x$$
$$-36 = -x$$
$$36 = x$$

40,000 people in the group will survive to age 36. This is shown on the graph as the point $(36,\ 40{,}000)$.

**57.**

$$87 = 29x^{1/3}$$
$$\frac{87}{29} = \frac{29x^{1/3}}{29}$$
$$3 = x^{1/3}$$
$$3^3 = \left(x^{1/3}\right)^3$$
$$27 = x$$

A Galápagos island with an area of 27 square miles will have 87 plant species.

**59.**

$$365 = 0.2x^{3/2}$$
$$\frac{365}{0.2} = \frac{0.2x^{3/2}}{0.2}$$
$$1825 = x^{3/2}$$
$$1825^2 = \left(x^{3/2}\right)^2$$
$$3{,}330{,}625 = x^3$$
$$\sqrt[3]{3{,}330{,}625} = \sqrt[3]{x^3}$$
$$149.34 \approx x$$

The average distance of the Earth from the sun is approximately 149 million kilometers.

**61.** Answers will vary.

**63.** Answers will vary. An extraneous solution is a solution to an equation that does not satisfy the equation in its original form. Typically these arise due to some manipulation of the original equation.

**65.** Answers will vary.

**67.** Answers will vary. The graph indicates that the number of survivors at a given age decreases at a faster rate as the age increases. The decrease is most rapid just before the last of the survivors die.

**69.** $\sqrt{x} + 3 = 5$

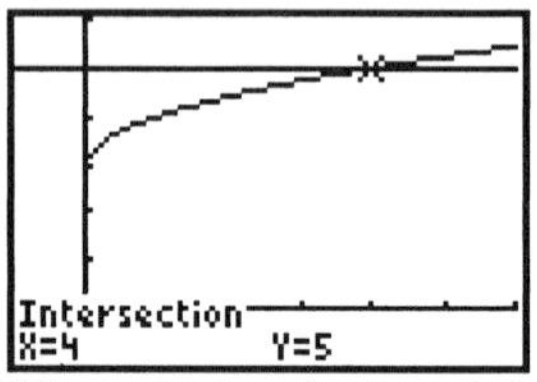


The solution is 4 and the solution set is $\{4\}$.

**71.** $4\sqrt{x} = x + 3$

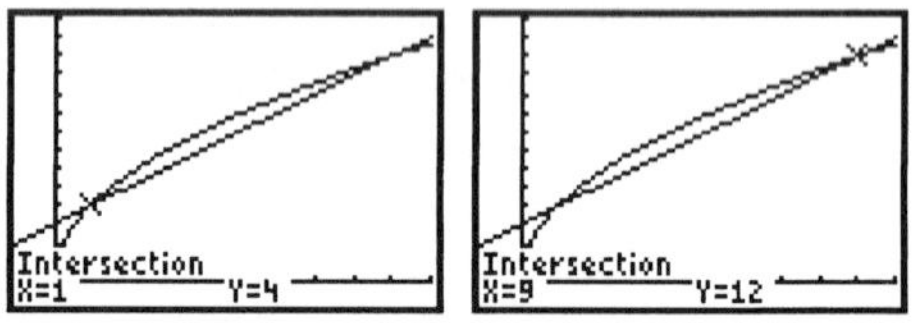


The solutions are 1 and 9. The solution set is $\{1, 9\}$.

**73.** Statement **c.** is true. To show this, substitute for $T$ in the equation for $L$ and simplify.

$$L = \frac{8T^2}{\pi^2} = \frac{8\left(2\pi\sqrt{\frac{L}{32}}\right)^2}{\pi^2} = \frac{8\left(4\pi^2 \frac{L}{32}\right)}{\pi^2}$$

$$= \frac{32\pi^2 \frac{L}{32}}{\pi^2} = \frac{\pi^2 L}{\pi^2} = L$$

Statement **a.** is false. The first step is to square both sides, obtaining $x+6 = x^2 + 4x + 4$.

Statement **b.** is false. The equation $\sqrt{x+4} = -5$ has no solution. By definition, absolute values are positive.

Statement **d.** is false. We know that an equation with an absolute value equal to a negative number has no solution. In this case, however, we do not know that $-x$ represents a negative number. If $x$ is negative, then $-x$ is positive.

**75.**

$$\sqrt[3]{x\sqrt{x}} = 9$$
$$\left(\sqrt[3]{x\sqrt{x}}\right)^3 = 9^3$$
$$x\sqrt{x} = 729$$
$$\left(x\sqrt{x}\right)^2 = 729^2$$
$$x^2x = 531441$$
$$x^3 = 531441$$
$$x = 81$$

Check:

$$\sqrt[3]{81\sqrt{81}} = 9$$
$$9 = 9$$

The solution checks. The solution is 81 and the solution set is $\{81\}$.

**77.**

$$(x-4)^{2/3} = 25$$
$$\left((x-4)^{2/3}\right)^{3/2} = 25^{3/2}$$
$$x - 4 = \left(\sqrt{25}\right)^3$$
$$x - 4 = 5^3$$
$$x - 4 = 125$$
$$x = 129$$

Check:

$$(129-4)^{2/3} = 25$$
$$(125)^{2/3} = 25$$
$$\left(\sqrt[3]{125}\right)^2 = 25$$
$$5^2 = 25$$
$$25 = 25$$

The solution checks. The solution is 129 and the solution set is $\{129\}$.

**78.**

$$\frac{4x^4 - 3x^3 + 2x^2 - x - 1}{x+3} = \frac{4x^4 - 3x^3 + 2x^2 - x - 1}{x-(-3)}$$

| | | | | | |
|---|---|---|---|---|---|
| $-3$ | 4 | $-3$ | 2 | $-1$ | $-1$ |
| | | $-12$ | 45 | $-141$ | 426 |
| | 4 | $-15$ | 47 | $-142$ | 425 |

$$\frac{4x^4 - 3x^3 + 2x^2 - x - 1}{x+3} = 4x^3 - 15x^2 + 47x - 142 + \frac{425}{x+3}$$

**79.** $\dfrac{3x^2-12}{x^2+2x-8} \div \dfrac{6x+18}{x+4}$

$= \dfrac{3x^2-12}{x^2+2x-8} \cdot \dfrac{x+4}{6x+18}$

$= \dfrac{3(x^2-4)}{(x+4)(x-2)} \cdot \dfrac{x+4}{6(x+3)}$

$= \dfrac{3(x+2)(x-2)}{(x-2)} \cdot \dfrac{1}{6(x+3)}$

$= \dfrac{3(x+2)}{1} \cdot \dfrac{1}{6(x+3)}$

$= \dfrac{3(x+2)}{6(x+3)} = \dfrac{x+2}{2(x+3)}$

**80.** $y^2-6y+9-25x^2$

$= (y^2-6y+9)-25x^2$

$= (y-3)^2-(5x)^2$

$= ((y-3)+5x)((y-3)-5x)$

$= (y-3+5x)(y-3-5x)$

## 7.7 Exercise Set

**1.** $\sqrt{-100} = \sqrt{100\cdot -1} = \sqrt{100}\cdot\sqrt{-1} = 10i$

**3.** $\sqrt{-23} = \sqrt{23\cdot -1} = \sqrt{23}\cdot\sqrt{-1} = \sqrt{23}\,i$

**5.** $\sqrt{-18} = \sqrt{9\cdot 2\cdot -1}$

$= \sqrt{9}\cdot\sqrt{2}\cdot\sqrt{-1}$

$= 3\sqrt{2}\,i$

**7.** $\sqrt{-63} = \sqrt{9\cdot 7\cdot -1}$

$= \sqrt{9}\cdot\sqrt{7}\cdot\sqrt{-1}$

$= 3\sqrt{7}\,i$

**9.** $-\sqrt{-108} = -\sqrt{36\cdot 3\cdot -1}$

$= -\sqrt{36}\cdot\sqrt{3}\cdot\sqrt{-1}$

$= -6\sqrt{3}\,i$

**11.** $5+\sqrt{-36} = 5+\sqrt{36\cdot -1}$

$= 5+\sqrt{36}\cdot\sqrt{-1}$

$= 5+6i$

**13.** $15+\sqrt{-3} = 15+\sqrt{3\cdot -1}$

$= 15+\sqrt{3}\cdot\sqrt{-1}$

$= 15+\sqrt{3}\,i$

**15.** $-2-\sqrt{-18} = -2-\sqrt{9\cdot 2\cdot -1}$

$= -2-\sqrt{9}\cdot\sqrt{2}\cdot\sqrt{-1}$

$= -2-3\sqrt{2}\,i$

**17.** $(3+2i)+(5+i)$

$= 3+2i+5+i = 3+5+2i+i$

$= (3+5)+(2+1)i = 8+3i$

**19.** $(7+2i)+(1-4i)$

$= 7+2i+1-4i = 7+1+2i-4i$

$= (7+1)+(2-4)i = 8-2i$

**21.** $(10+7i)-(5+4i)$

$= 10+7i-5-4i = 10-5+7i-4i$

$= (10-5)+(7-4)i = 5+3i$

**23.** $(9-4i)-(10+3i)$

$= 9-4i-10-3i = 9-10-4i-3i$

$= (9-10)+(-4-3)i$

$= -1+(-7)i = -1-7i$

**25.** $(3+2i)-(5-7i)$
$=3+2i-5+7i=3-5+2i+7i$
$=(3-5)+(2+7)i=-2+9i$

**27.** $(-5+4i)-(-13-11i)$
$=-5+4i+13+11i$
$=-5+13+4i+11i$
$=(-5+13)+(4+11)i=8+15i$

**29.** $8i-(14-9i)$
$=8i-14+9i=-14+8i+9i$
$=-14+(8+9)i=-14+17i$

**31.** $\left(2+\sqrt{3}\,i\right)+\left(7+4\sqrt{3}\,i\right)$
$=2+\sqrt{3}\,i+7+4\sqrt{3}\,i$
$=2+7+\sqrt{3}\,i+4\sqrt{3}\,i$
$=(2+7)+\left(\sqrt{3}+4\sqrt{3}\right)i=9+5\sqrt{3}\,i$

**33.** $2i(5+3i)$
$=2i\cdot 5+2i\cdot 3i=10i+6i^2$
$=10i+6(-1)=-6+10i$

**35.** $3i(7i-5)$
$=3i\cdot 7i-3i\cdot 5=21i^2-15i$
$=21(-1)-15i=-21-15i$

**37.** $-7i(2-5i)$
$=-7i\cdot 2-(-7i)5i=-14i+35i^2$
$=-14i+35(-1)=-35-14i$

**39.** $(3+i)(4+5i)=12+15i+4i+5i^2$
$=12+15i+4i+5(-1)$
$=12-5+15i+4i$
$=7+19i$

**41.** $(7-5i)(2-3i)$
$=14-21i-10i+15i^2$
$=14-21i-10i+15(-1)$
$=14-15-21i-10i=-1-31i$

**43.** $(6-3i)(-2+5i)$
$=-12+30i+6i-15i^2$
$=-12+30i+6i-15(-1)$
$=-12+15+30i+6i=3+36i$

**45.** $(3+5i)(3-5i)$
$=9-\cancel{15i}+\cancel{15i}-25i^2$
$=9-25(-1)=9+25$
$=34=34+0i$

**47.** $(-5+3i)(-5-3i)$
$=25+\cancel{15i}-\cancel{15i}-9i^2$
$=25-9(-1)=25+9$
$=34=34+0i$

**49.** $\left(3-\sqrt{2}\,i\right)\left(3+\sqrt{2}\,i\right)$
$=9+\cancel{3\sqrt{2}\,i}-\cancel{3\sqrt{2}\,i}-2i^2$
$=9-2(-1)=9+2$
$=11=11+0i$

**51.** $(2+3i)^2$
$=4+2\cdot 6i+9i^2=4+12i+9(-1)$
$=4-9+12i=-5+12i$

**53.** $(5-2i)^2=25-2\cdot 10i+4i^2$
$=25-20i+4(-1)$
$=25-4-20i=21-20i$

**55.** $\sqrt{-7}\cdot\sqrt{-2}=\sqrt{7}\sqrt{-1}\cdot\sqrt{2}\sqrt{-1}$
$=\sqrt{7}\,i\cdot\sqrt{2}\,i=\sqrt{14}\,i^2$
$=\sqrt{14}(-1)=-\sqrt{14}$
$=-\sqrt{14}+0i$

**57.** $\sqrt{-9}\cdot\sqrt{-4}$
$=\sqrt{9}\sqrt{-1}\cdot\sqrt{4}\sqrt{-1}=3i\cdot 2i=6i^2$
$=6(-1)=-6=-6+0i$

**59.** $\sqrt{-7}\cdot\sqrt{-25}=\sqrt{7}\sqrt{-1}\cdot\sqrt{25}\sqrt{-1}$
$=\sqrt{7}\,i\cdot 5i=5\sqrt{7}\,i^2$
$=5\sqrt{7}(-1)=-5\sqrt{7}$
$=-5\sqrt{7}+0i$

**61.** $\sqrt{-8}\cdot\sqrt{-3}=\sqrt{4\cdot 2}\sqrt{-1}\cdot\sqrt{3}\sqrt{-1}$
$=2\sqrt{2}\,i\cdot\sqrt{3}\,i=2\sqrt{6}\,i^2$
$=2\sqrt{6}(-1)=-2\sqrt{6}$
$=-2\sqrt{6}+0i$

**63.** $\frac{2}{3+i}=\frac{2}{3+i}\cdot\frac{3-i}{3-i}=\frac{6-2i}{3^2-i^2}$
$=\frac{6-2i}{9-(-1)}=\frac{6-2i}{9+1}$
$=\frac{6-2i}{10}=\frac{6}{10}-\frac{2i}{10}$
$=\frac{3}{5}-\frac{1}{5}i$

**65.** $\frac{2i}{1+i}=\frac{2i}{1+i}\cdot\frac{1-i}{1-i}=\frac{2i-2i^2}{1^2-i^2}$
$=\frac{2i-2(-1)}{1-(-1)}=\frac{2+2i}{1+1}$
$=\frac{2+2i}{2}=\frac{2}{2}+\frac{2i}{2}=1+i$

**67.** $\frac{7}{4-3i}=\frac{7}{4-3i}\cdot\frac{4+3i}{4+3i}=\frac{28+21i}{4^2-(3i)^2}$
$=\frac{28+21i}{16-9i^2}=\frac{28+21i}{16-9(-1)}$
$=\frac{28+21i}{16+9}=\frac{28+21i}{25}$
$=\frac{28}{25}+\frac{21}{25}i$

**69.** $\frac{6i}{3-2i}=\frac{6i}{3-2i}\cdot\frac{3+2i}{3+2i}=\frac{18i+12i^2}{3^2-(2i)^2}$
$=\frac{18i+12(-1)}{9-4i^2}=\frac{-12+18i}{9-4(-1)}$
$=\frac{-12+18i}{9+4}=\frac{-12+18i}{13}$
$=-\frac{12}{13}+\frac{18}{13}i$

**71.** $\frac{1+i}{1-i}=\frac{1+i}{1-i}\cdot\frac{1+i}{1+i}=\frac{1+2i+i^2}{1^2-i^2}$
$=\frac{1+2i+(-1)}{1-(-1)}=\frac{2i}{2}$
$=i$ or $0+i$

**73.** $\frac{2-3i}{3+i}=\frac{2-3i}{3+i}\cdot\frac{3-i}{3-i}$
$=\frac{6-2i-9i+3i^2}{3^2-i^2}$
$=\frac{6-11i+3(-1)}{9-(-1)}$
$=\frac{6-3-11i}{9+1}$
$=\frac{3-11i}{10}=\frac{3}{10}-\frac{11}{10}i$

**75.** $\dfrac{5-2i}{3+2i}=\dfrac{5-2i}{3+2i}\cdot\dfrac{3-2i}{3-2i}$

$=\dfrac{15-10i-6i+4i^2}{3^2-(2i)^2}$

$=\dfrac{15-10i-6i+4i^2}{3^2-(2i)^2}$

$=\dfrac{15-16i+4(-1)}{9-4i^2}$

$=\dfrac{15-4-16i}{9-4(-1)}$

$=\dfrac{11-16i}{9+4}$

$=\dfrac{11-16i}{13}=\dfrac{11}{13}-\dfrac{16}{13}i$

**77.** $\dfrac{4+5i}{3-7i}=\dfrac{4+5i}{3-7i}\cdot\dfrac{3+7i}{3+7i}$

$=\dfrac{12+28i+15i+35i^2}{3^2-(7i)^2}$

$=\dfrac{12+43i+35(-1)}{9-49i^2}$

$=\dfrac{12-35+43i}{9-49(-1)}$

$=\dfrac{-23+43i}{9+49}=\dfrac{-23+43i}{58}$

$=-\dfrac{23}{58}+\dfrac{43}{58}i$

**79.** $\dfrac{7}{3i}=\dfrac{7}{3i}\cdot\dfrac{-3i}{-3i}=\dfrac{-21i}{-9i^2}=\dfrac{-21i}{-9(-1)}$

$=\dfrac{-21i}{9}=-\dfrac{7}{3}i$ or $0-\dfrac{7}{3}i$

**81.** $\dfrac{8-5i}{2i}=\dfrac{8-5i}{2i}\cdot\dfrac{-2i}{-2i}=\dfrac{-16i+10i^2}{-4i^2}$

$=\dfrac{-16i+10(-1)}{-4(-1)}=\dfrac{-10-16i}{4}$

$=-\dfrac{10}{4}-\dfrac{16}{4}i=-\dfrac{5}{2}-4i$

**83.** $\dfrac{4+7i}{-3i}=\dfrac{4+7i}{-3i}\cdot\dfrac{3i}{3i}=\dfrac{12i+21i^2}{-9i^2}$

$=\dfrac{12i+21(-1)}{-9(-1)}=\dfrac{-21+12i}{9}$

$=-\dfrac{21}{9}+\dfrac{12}{9}i=-\dfrac{7}{3}+\dfrac{4}{3}i$

**85.** $i^{10}=(i^2)^5=(-1)^5=-1$

**87.** $i^{11}=(i^2)^5 i=(-1)^5 i=-i$

**89.** $i^{22}=(i^2)^{11}=(-1)^{11}=-1$

**91.** $i^{200}=(i^2)^{100}=(-1)^{100}=1$

**93.** $i^{17}=(i^2)^8 i=(-1)^8 i=i$

**95.** $(-i)^4=(-1)^4 i^4=i^4=(i^2)^2$

$=(-1)^2=1$

**97.** $(-i)^9=(-1)^9 i^9=(-1)(i^2)^4 i$

$=(-1)(-1)^4 i=(-1)i$

$=-i$

**99.** $i^{24}+i^2=\left(i^2\right)^{12}+(-1)$
$=(-1)^{12}+(-1)$
$=1+(-1)=0$

**101.** $(2-3i)(1-i)-(3-i)(3+i)$
$=\left(2-2i-3i+3i^2\right)-\left(3^2-i^2\right)$
$=2-5i+3i^2-9+i^2$
$=-7-5i+4i^2$
$=-7-5i+4(-1)$
$=-11-5i$

**102.** $(8+9i)(2-i)-(1-i)(1+i)$
$=\left(16-8i+18i-9i^2\right)-\left(1^2-i^2\right)$
$=16+10i-9i^2-1+i^2$
$=15+10i-8i^2$
$=15+10i-8(-1)$
$=23+10i$

**103.** $(2+i)^2-(3-i)^2$
$=\left(4+4i+i^2\right)-\left(9-6i+i^2\right)$
$=4+4i+i^2-9+6i-i^2$
$=-5+10i$

**104.** $(4-i)^2-(1+2i)^2$
$=\left(16-8i+i^2\right)-\left(1+4i+4i^2\right)$
$=16-8i+i^2-1-4i-4i^2$
$=15-12i-3i^2$
$=15-12i-3(-1)$
$=18-12i$

**105.** $5\sqrt{-16}+3\sqrt{-81}$
$=5\sqrt{16}\sqrt{-1}+3\sqrt{81}\sqrt{-1}$
$=5\cdot 4i+3\cdot 9i$
$=20i+27i$
$=47i$ or $0+47i$

**106.** $5\sqrt{-8}+3\sqrt{-18}$
$=5\sqrt{4}\sqrt{2}\sqrt{-1}+3\sqrt{9}\sqrt{2}\sqrt{-1}$
$=5\cdot 2\sqrt{2}\,i+3\cdot 3\sqrt{2}\,i$
$=10\sqrt{2}\,i+9\sqrt{2}\,i$
$=(10+9)\sqrt{2}\,i$
$=19\sqrt{2}\,i$ or $0+19\sqrt{2}\,i$

**107.** $\dfrac{i^4+i^{12}}{i^8-i^7}=\dfrac{i^4+\left(i^4\right)^3}{\left(i^4\right)^2-\left(i^2\right)^3 i}$
$=\dfrac{1+1^3}{1^2-(-1)^3 i}=\dfrac{1+1}{1+i}$
$=\dfrac{2}{1+i}=\dfrac{2}{1+i}\cdot\dfrac{1-i}{1-i}$
$=\dfrac{2-2i}{1^2-i^2}=\dfrac{2-2i}{1+1}$
$=\dfrac{2-2i}{2}=1-i$

**108.** $\dfrac{i^8+i^{40}}{i^4+i^3}=\dfrac{\left(i^4\right)^2+\left(i^4\right)^{10}}{i^4+i^2\cdot i}$
$=\dfrac{1^2+1^{10}}{1+(-1)i}=\dfrac{1+1}{1-i}$
$=\dfrac{2}{1-i}=\dfrac{2}{1-i}\cdot\dfrac{1+i}{1+i}$
$=\dfrac{2+2i}{1^2-i^2}=\dfrac{2+2i}{1+1}$
$=\dfrac{2+2i}{2}=1+i$

**109.** $f(x) = x^2 - 2x + 2$

$$f(1+i) = (1+i)^2 - 2(1+i) + 2$$
$$= 1 + 2i + i^2 - 2 - 2i + 2$$
$$= 1 + i^2$$
$$= 1 - 1$$
$$= 0$$

**110.** $f(x) = x^2 - 2x + 5$

$$f(1-2i) = (1-2i)^2 - 2(1-2i) + 5$$
$$= 1 - 4i + 4i^2 - 2 + 4i + 5$$
$$= 4 + 4i^2$$
$$= 4 - 4$$
$$= 0$$

**111.** $f(x) = x - 3i$; $g(x) = 4x + 2i$

$$f(-1) = -1 - 3i$$
$$g(-1) = -4 + 2i$$
$$(fg)(-1) = (-1-3i)(-4+2i)$$
$$= 4 - 2i + 12i - 6i^2$$
$$= 4 + 10i - 6(-1)$$
$$= 10 + 10i$$

**112.** $f(x) = 12x - i$; $g(x) = 6x + 3i$

$$f\left(-\frac{1}{3}\right) = 12\left(-\frac{1}{3}\right) - i = -4 - i$$
$$g\left(-\frac{1}{3}\right) = 6\left(-\frac{1}{3}\right) + 3i = -2 + 3i$$
$$(fg)\left(-\tfrac{1}{3}\right) = (-4-i)(-2+3i)$$
$$= 8 - 12i + 2i - 3i^2$$
$$= 8 - 10i - 3(-1)$$
$$= 11 - 10i$$

**113.** $f(x) = \dfrac{x^2 + 19}{2 - x}$

$$f(3i) = \frac{(3i)^2 + 19}{2 - 3i} = \frac{9i^2 + 19}{2 - 3i}$$
$$= \frac{9(-1) + 19}{2 - 3i} = \frac{10}{2 - 3i}$$
$$= \frac{10}{2 - 3i} \cdot \frac{2 + 3i}{2 + 3i}$$
$$= \frac{20 + 30i}{2^2 - (3i)^2} = \frac{20 + 30i}{4 - 9i^2}$$
$$= \frac{20 + 30i}{4 - 9(-1)} = \frac{20 + 30i}{13}$$
$$= \frac{20}{13} + \frac{30}{13}i$$

**114.** $f(x) = \dfrac{x^2 + 11}{3 - x}$

$$f(4i) = \frac{(4i)^2 + 11}{3 - 4i} = \frac{16i^2 + 11}{3 - 4i}$$
$$= \frac{16(-1) + 11}{3 - 4i} = \frac{-5}{3 - 4i}$$
$$= \frac{-5}{3 - 4i} \cdot \frac{3 + 4i}{3 + 4i}$$
$$= \frac{-15 - 20i}{3^2 - (4i)^2} = \frac{-15 - 20i}{9 - 16i^2}$$
$$= \frac{-15 - 20i}{9 - 16(-1)} = \frac{-15 - 20i}{25}$$
$$= \frac{-15}{25} - \frac{20}{25}i$$
$$= -\frac{3}{5} - \frac{4}{5}i$$

**115.** $E = IR = (4-5i)(3+7i)$
$= 12 + 28i - 15i - 35i^2$
$= 12 + 13i - 35(-1)$
$= 12 + 35 + 13i = 47 + 13i$
The voltage of the circuit is $(47 + 13i)$ volts.

**117.** Sum:
$(5+\sqrt{15}\,i)+(5-\sqrt{15}\,i)$
$= 5 + \cancel{\sqrt{15}\,i} + 5 - \cancel{\sqrt{15}\,i}$
$= 5 + 5 = 10$
Product:
$(5+\sqrt{15}\,i)(5-\sqrt{15}\,i)$
$= 25 - \cancel{5\sqrt{15}\,i} + \cancel{5\sqrt{15}\,i} - 15i^2$
$= 25 - 15(-1) = 25 + 15 = 40$

**119.** Answers will vary.
Write $-64$ as the product of 64 and $-1$. Then split up the radical as
$\sqrt{-64} = \sqrt{64 \cdot -1} = \sqrt{64}\sqrt{-1} = 8i$

**121.** Answers will vary. Add corresponding real parts and corresponding imaginary parts.

**123.** Answers will vary. Use the distributive property to distribute $2i$. Then simplify by converting powers of $i$ as appropriate and collecting like terms.

**125.** Answers will vary. The product is in the form $(a+b)(a-b)$ so the result is the difference of two squares. Simplify by converting powers of $i$ as appropriate and collecting like terms.

**127.** The conjugate of $2+3i$ is $2-3i$. Answers may vary. When you multiply conjugates, the result is a pure real number.

**129.** Answers will vary.

**131.** Answers may vary. The two radicands were incorrectly added together. First write the numbers in terms of $i$ and then see if they can be combined.
$\sqrt{-9} + \sqrt{-16} = \sqrt{9}\sqrt{-1} + \sqrt{16}\sqrt{-1}$
$= 3i + 4i$
$= 7i$

**133.** Statement **d.** is true.
$(x+yi)(x-yi)$
$= x^2 - \cancel{xyi} + \cancel{xyi} - y^2i^2$
$= x^2 - y^2(-1)$
$= x^2 + y^2$

Statement **a.** is false. All irrational numbers are complex numbers.

Statement **b.** is false.
$(3+7i)(3-7i)$
$= 3^2 - (7i)^2 = 9 - 49i^2$
$= 9 - 49(-1)$
$= 9 + 49 = 58$

Statement **c.** is false.
$$\frac{7+3i}{5+3i} = \frac{7+3i}{5+3i} \cdot \frac{5-3i}{5-3i}$$
$$= \frac{35 - 21i + 15i - 9i^2}{5^2 - (3i)^2} = \frac{35 - 6i - 9(-1)}{25 - 9i^2}$$
$$= \frac{35 - 6i + 9}{25 - 9(-1)} = \frac{44 - 6i}{25 + 9} = \frac{44 - 6i}{34}$$
$$= \frac{44}{34} - \frac{6}{34}i = \frac{22}{17} - \frac{3}{17}i$$

**135.** $\dfrac{1+i}{1+2i}+\dfrac{1-i}{1-2i}$

$=\dfrac{(1+i)(1-2i)}{(1+2i)(1-2i)}+\dfrac{(1-i)(1+2i)}{(1+2i)(1-2i)}$

$=\dfrac{(1+i)(1-2i)+(1-i)(1+2i)}{(1+2i)(1-2i)}$

$=\dfrac{1-2i+i-2i^2+1+2i-i-2i^2}{1^2-(2i)^2}$

$=\dfrac{2-4i^2}{1-4i^2}=\dfrac{2-4(-1)}{1-4(-1)}$

$=\dfrac{2+4}{1+4}=\dfrac{6}{5}=\dfrac{6}{5}+0i$

**137.** $\dfrac{\frac{x}{y^2}+\frac{1}{y}}{\frac{y}{x^2}+\frac{1}{x}}=\dfrac{\frac{x}{y^2}+\frac{1}{y}}{\frac{y}{x^2}+\frac{1}{x}}\cdot\dfrac{x^2y^2}{x^2y^2}$

$=\dfrac{\frac{x}{\cancel{y^2}}\cdot x^2\cancel{y^2}+\frac{1}{\cancel{y}}\cdot x^2y^{\cancel{2}}}{\frac{y}{\cancel{x^2}}\cdot\cancel{x^2}y^2+\frac{1}{\cancel{x}}\cdot x^{\cancel{2}}y^2}$

$=\dfrac{x^3+x^2y}{y^3+xy^2}=\dfrac{x^2\cancel{(x+y)}}{y^2\cancel{(y+x)}}$

$=\dfrac{x^2}{y^2}$

**138.** $\dfrac{1}{x}+\dfrac{1}{y}=\dfrac{1}{z}$

$\dfrac{1}{\cancel{x}}\cdot\cancel{x}yz+\dfrac{1}{\cancel{y}}\cdot x\cancel{y}z=\dfrac{1}{\cancel{z}}\cdot xy\cancel{z}$

$yz+xz=xy$

$yz=xy-xz$

$yz=x(y-z)$

$x=\dfrac{yz}{y-z}$

**139.** $2x-\dfrac{x-3}{8}=\dfrac{1}{2}+\dfrac{x+5}{2}$

$2x\cdot 8-\dfrac{x-3}{\cancel{8}}\cdot\cancel{8}=\dfrac{1}{2}\cdot 8+\dfrac{x+5}{2}\cdot 8$

$16x-x+3=4+4(x+5)$

$16x-x+3=4+4x+20$

$15x+3=4x+24$

$11x+3=24$

$11x=21$

$x=\dfrac{21}{11}$

The solution is $\dfrac{21}{11}$ and the solution set is $\left\{\dfrac{21}{11}\right\}$.

## Chapter 7 Review Exercises

**1.** $\sqrt{81}=9$ because $9^2=81$

**2.** $-\sqrt{\dfrac{1}{100}}=-\dfrac{1}{10}$ because $\left(\dfrac{1}{10}\right)^2=\dfrac{1}{100}$

**3.** $\sqrt[3]{-27}=-3$ because $(-3)^3=-27$

**4.** $\sqrt[4]{-16}$
Not a real number
The index is even and the radicand is negative.

**5.** $\sqrt[5]{-32}=-2$ because $(-2)^5=-32$

**6.** $f(15)=\sqrt{2(15)-5}=\sqrt{30-5}$
$=\sqrt{25}=5$
$f(4)=\sqrt{2(4)-5}=\sqrt{8-5}=\sqrt{3}$
$f\left(\frac{5}{2}\right)=\sqrt{2\left(\frac{5}{2}\right)-5}=\sqrt{5-5}$
$=\sqrt{0}=0$
$f(1)=\sqrt{2(1)-5}=\sqrt{2-5}=\sqrt{-3}$
Not a real number

**7.** $g(4)=\sqrt[3]{4(4)-8}=\sqrt[3]{16-8}=\sqrt[3]{8}=2$
$g(0)=\sqrt[3]{4(0)-8}=\sqrt[3]{-8}=-2$
$g(-14)=\sqrt[3]{4(-14)-8}=\sqrt[3]{-56-8}$
$=\sqrt[3]{-64}=-4$

**8.** To find the domain, set the radicand greater than or equal to zero and solve the resulting inequality.
$x-2\geq 0$
$x\geq 2$
The domain of $f$ is $\{x|x\geq 2\}$ or $[2,\infty)$.

**9.** To find the domain, set the radicand greater than or equal to zero and solve the resulting inequality.
$100-4x\geq 0$
$-4x\geq -100$
$\frac{-4x}{-4}\leq\frac{-100}{-4}$
$x\leq 25$
The domain of $g$ is $\{x|x\leq 25\}$ or $(-\infty,25]$.

**10.** $\sqrt{25x^2}=5|x|$

**11.** $\sqrt{(x+14)^2}=|x+14|$

**12.** $\sqrt{x^2-8x+16}=\sqrt{(x-4)^2}=|x-4|$

**13.** $\sqrt[3]{64x^3}=4x$

**14.** $\sqrt[4]{16x^4}=2|x|$

**15.** $\sqrt[5]{-32(x+7)^5}=-2(x+7)$

**16.** $(5xy)^{\frac{1}{3}}=\sqrt[3]{5xy}$

**17.** $16^{\frac{3}{2}}=\left(\sqrt{16}\right)^3=(4)^3=64$

**18.** $32^{\frac{4}{5}}=\left(\sqrt[5]{32}\right)^4=(2)^4=16$

**19.** $\sqrt{7x}=(7x)^{\frac{1}{2}}$

**20.** $\left(\sqrt[3]{19xy}\right)^5=(19xy)^{\frac{5}{3}}$

**21.** $8^{-\frac{2}{3}}=\frac{1}{8^{\frac{2}{3}}}=\frac{1}{\left(\sqrt[3]{8}\right)^2}=\frac{1}{(2)^2}=\frac{1}{4}$

**22.** $3x(ab)^{-\frac{4}{5}}=\frac{3x}{(ab)^{\frac{4}{5}}}$
$=\frac{3x}{\left(\sqrt[5]{ab}\right)^4}$ or $\frac{3x}{\sqrt[5]{(ab)^4}}$

**23.** $x^{\frac{1}{3}}\cdot x^{\frac{1}{4}}=x^{\frac{1}{3}+\frac{1}{4}}=x^{\frac{4}{12}+\frac{3}{12}}=x^{\frac{7}{12}}$

**24.** $\dfrac{5^{\frac{1}{2}}}{5^{\frac{1}{3}}} = 5^{\frac{1}{2}-\frac{1}{3}} = 5^{\frac{3}{6}-\frac{2}{6}} = 5^{\frac{1}{6}}$

**25.** $\left(8x^6y^3\right)^{\frac{1}{3}} = 8^{\frac{1}{3}}x^{6\cdot\frac{1}{3}}y^{3\cdot\frac{1}{3}} = 2x^2y$

**26.** $\left(x^{-\frac{2}{3}}y^{\frac{1}{4}}\right)^{\frac{1}{2}} = x^{-\frac{2}{3}\cdot\frac{1}{2}}y^{\frac{1}{4}\cdot\frac{1}{2}}$

$= x^{-\frac{1}{3}}y^{\frac{1}{8}} = \dfrac{y^{\frac{1}{8}}}{x^{\frac{1}{3}}}$

**27.** $\sqrt[3]{x^9y^{12}} = \left(x^9y^{12}\right)^{\frac{1}{3}}$

$= x^{9\cdot\frac{1}{3}}y^{12\cdot\frac{1}{3}} = x^3y^4$

**28.** $\sqrt[9]{x^3y^9} = \left(x^3y^9\right)^{\frac{1}{9}} = x^{3\cdot\frac{1}{9}}y^{9\cdot\frac{1}{9}}$

$= x^{\frac{1}{3}}y = y\sqrt[3]{x}$

**29.** $\sqrt{x}\cdot\sqrt[3]{x} = x^{\frac{1}{2}}x^{\frac{1}{3}} = x^{\frac{1}{2}+\frac{1}{3}} = x^{\frac{3}{6}+\frac{2}{6}}$

$= x^{\frac{5}{6}} = \sqrt[6]{x^5}$

**30.** $\dfrac{\sqrt[3]{x^2}}{\sqrt[4]{x^2}} = \dfrac{x^{\frac{2}{3}}}{x^{\frac{2}{4}}} = x^{\frac{2}{3}-\frac{1}{2}}$

$= x^{\frac{4}{6}-\frac{3}{6}} = x^{\frac{1}{6}} = \sqrt[6]{x}$

**31.** $\sqrt[5]{\sqrt[3]{x}} = \sqrt[5]{x^{\frac{1}{3}}} = \left(x^{\frac{1}{3}}\right)^{\frac{1}{5}} = x^{\frac{1}{3}\cdot\frac{1}{5}}$

$= x^{\frac{1}{15}} = \sqrt[15]{x}$

**32.** Since 2012 is 27 years after 1985, find $f(27)$.

$f(27) = 350(27)^{\frac{2}{3}} = 350\left(\sqrt[3]{27}\right)^2$

$= 350(3)^2 = 350(9) = 3150$

Expenditures will be \$3150 million or \$3,150,000,000 in the year 2012.

**33.** $\sqrt{3x}\cdot\sqrt{7y} = \sqrt{21xy}$

**34.** $\sqrt[5]{7x^2}\cdot\sqrt[5]{11x} = \sqrt[5]{77x^3}$

**35.** $\sqrt[6]{x-5}\cdot\sqrt[6]{(x-5)^4} = \sqrt[6]{(x-5)^5}$

**36.** $f(x) = \sqrt{7x^2-14x+7}$

$= \sqrt{7\left(x^2-2x+1\right)}$

$= \sqrt{7(x-1)^2} = \sqrt{7}\,|x-1|$

**37.** $\sqrt{20x^3} = \sqrt{4\cdot 5\cdot x^2\cdot x} = \sqrt{4x^2\cdot 5x}$

$= 2x\sqrt{5x}$

**38.** $\sqrt[3]{54x^8y^6} = \sqrt[3]{27\cdot 2\cdot x^6\cdot x^2y^6}$

$= \sqrt[3]{27x^6y^6\cdot 2x^2}$

$= 3x^2y^2\sqrt[3]{2x^2}$

**39.** $\sqrt[4]{32x^3y^{11}z^5} = \sqrt[4]{16\cdot 2\cdot x^3y^8\cdot y^3\cdot z^5\cdot z}$

$= \sqrt[4]{16y^8z^4\cdot 2x^3y^3z}$

$= 2y^2z\sqrt[4]{2x^3y^3z}$

**40.** $\sqrt{6x^3}\cdot\sqrt{4x^2} = \sqrt{24x^5} = \sqrt{4\cdot 6\cdot x^4\cdot x}$

$= \sqrt{4x^4\cdot 6x} = 2x^2\sqrt{6x}$

**41.** $\sqrt[3]{4x^2y}\cdot\sqrt[3]{4xy^4} = \sqrt[3]{16x^3y^5}$
$= \sqrt[3]{8\cdot 2\cdot x^3\cdot y^3\cdot y^2}$
$= \sqrt[3]{8x^3y^3\cdot 2y^2}$
$= 2xy\sqrt[3]{2y^2}$

**42.** $\sqrt[5]{2x^4y^3z^4}\cdot\sqrt[5]{8xy^6z^7}$
$= \sqrt[5]{16x^5y^9z^{11}}$
$= \sqrt[5]{16\cdot x^5\cdot y^5\cdot y^4\cdot z^{10}\cdot z}$
$= \sqrt[5]{x^5y^5z^{10}\cdot 16y^4z}$
$= xyz^2\sqrt[5]{16y^4z}$

**43.** $\sqrt{x+1}\cdot\sqrt{x-1} = \sqrt{(x+1)(x-1)}$
$= \sqrt{x^2-1}$

**44.** $6\sqrt[3]{3}+2\sqrt[3]{3} = (6+2)\sqrt[3]{3} = 8\sqrt[3]{3}$

**45.** $5\sqrt{18}-3\sqrt{8} = 5\sqrt{9\cdot 2}-3\sqrt{4\cdot 2}$
$= 5\cdot 3\sqrt{2}-3\cdot 2\sqrt{2}$
$= 15\sqrt{2}-6\sqrt{2}$
$= (15-6)\sqrt{2} = 9\sqrt{2}$

**46.** $\sqrt[3]{27x^4}+\sqrt[3]{xy^6}$
$= \sqrt[3]{27x^3x}+\sqrt[3]{xy^6}$
$= 3x\sqrt[3]{x}+y^2\sqrt[3]{x}$
$= (3x+y^2)\sqrt[3]{x}$

**47.** $2\sqrt[3]{6}-5\sqrt[3]{48} = 2\sqrt[3]{6}-5\sqrt[3]{8\cdot 6}$
$= 2\sqrt[3]{6}-5\cdot 2\sqrt[3]{6}$
$= 2\sqrt[3]{6}-10\sqrt[3]{6}$
$= (2-10)\sqrt[3]{6} = -8\sqrt[3]{6}$

**48.** $\sqrt[3]{\dfrac{16}{125}} = \sqrt[3]{\dfrac{8\cdot 2}{125}} = \dfrac{2}{5}\sqrt[3]{2}$

**49.** $\sqrt{\dfrac{x^3}{100y^4}} = \sqrt{\dfrac{x^2\cdot x}{100y^4}}$
$= \dfrac{x}{10y^2}\sqrt{x}$ or $\dfrac{x\sqrt{x}}{10y^2}$

**50.** $\sqrt[4]{\dfrac{3y^5}{16x^{20}}} = \sqrt[4]{\dfrac{y^4\cdot 3y}{16x^{20}}}$
$= \dfrac{y}{2x^5}\sqrt[4]{3y}$ or $\dfrac{y\sqrt[4]{3y}}{2x^5}$

**51.** $\dfrac{\sqrt{48}}{\sqrt{2}} = \sqrt{\dfrac{48}{2}} = \sqrt{24} = \sqrt{4\cdot 6} = 2\sqrt{6}$

**52.** $\dfrac{\sqrt[3]{32}}{\sqrt[3]{2}} = \sqrt[3]{\dfrac{32}{2}} = \sqrt[3]{16} = \sqrt[3]{8\cdot 2} = 2\sqrt[3]{2}$

**53.** $\dfrac{\sqrt[4]{64x^7}}{\sqrt[4]{2x^2}} = \sqrt[4]{\dfrac{64x^7}{2x^2}} = \sqrt[4]{32x^5}$
$= \sqrt[4]{16\cdot 2\cdot x^4\cdot x}$
$= \sqrt[4]{16x^4\cdot 2x} = 2x\sqrt[4]{2x}$

**54.** $\dfrac{\sqrt{200x^3y^2}}{\sqrt{2x^{-2}y}} = \sqrt{\dfrac{200x^3y^2}{2x^{-2}y}} = \sqrt{100x^5y}$
$= \sqrt{100x^4xy} = 10x^2\sqrt{xy}$

**55.** $\sqrt{3}\left(2\sqrt{6}+4\sqrt{15}\right) = 2\sqrt{18}+4\sqrt{45}$
$= 2\sqrt{9\cdot 2}+4\sqrt{9\cdot 5}$
$= 2\cdot 3\sqrt{2}+4\cdot 3\sqrt{5}$
$= 6\sqrt{2}+12\sqrt{5}$

**56.** $\sqrt[3]{5}\left(\sqrt[3]{50}-\sqrt[3]{2}\right)=\sqrt[3]{250}-\sqrt[3]{10}$
$=\sqrt[3]{125\cdot 2}-\sqrt[3]{10}$
$=5\sqrt[3]{2}-\sqrt[3]{10}$

**57.** $\left(\sqrt{7}-3\sqrt{5}\right)\left(\sqrt{7}+6\sqrt{5}\right)$
$=7+6\sqrt{35}-3\sqrt{35}-18\cdot 5$
$=7+3\sqrt{35}-90$
$=3\sqrt{35}-83$ or $-83+3\sqrt{35}$

**58.** $\left(\sqrt{x}-\sqrt{11}\right)\left(\sqrt{y}-\sqrt{11}\right)$
$=\sqrt{xy}-\sqrt{11x}-\sqrt{11y}+11$

**59.** $\left(\sqrt{5}+\sqrt{8}\right)^2=5+2\cdot\sqrt{5}\cdot\sqrt{8}+8$
$=13+2\sqrt{40}$
$=13+2\sqrt{4\cdot 10}$
$=13+2\cdot 2\sqrt{10}$
$=13+4\sqrt{10}$

**60.** $\left(2\sqrt{3}-\sqrt{10}\right)^2$
$=4\cdot 3-2\cdot 2\sqrt{3}\cdot\sqrt{10}+10$
$=12-4\sqrt{30}+10=22-4\sqrt{30}$

**61.** $\left(\sqrt{7}+\sqrt{13}\right)\left(\sqrt{7}-\sqrt{13}\right)$
$=\left(\sqrt{7}\right)^2-\left(\sqrt{13}\right)^2=7-13=-6$

**62.** $\left(7-3\sqrt{5}\right)\left(7+3\sqrt{5}\right)=7^2-\left(3\sqrt{5}\right)^2$
$=49-9\cdot 5$
$=49-45=4$

**63.** $\frac{4}{\sqrt{6}}=\frac{4}{\sqrt{6}}\cdot\frac{\sqrt{6}}{\sqrt{6}}=\frac{4\sqrt{6}}{6}=\frac{2\sqrt{6}}{3}$

**64.** $\sqrt{\frac{2}{7}}=\frac{\sqrt{2}}{\sqrt{7}}=\frac{\sqrt{2}}{\sqrt{7}}\cdot\frac{\sqrt{7}}{\sqrt{7}}=\frac{\sqrt{14}}{7}$

**65.** $\frac{12}{\sqrt[3]{9}}=\frac{12}{\sqrt[3]{3^2}}\cdot\frac{\sqrt[3]{3}}{\sqrt[3]{3}}=\frac{12\sqrt[3]{3}}{\sqrt[3]{3^3}}$
$=\frac{12\sqrt[3]{3}}{3}=4\sqrt[3]{3}$

**66.** $\sqrt{\frac{2x}{5y}}=\frac{\sqrt{2x}}{\sqrt{5y}}\cdot\frac{\sqrt{5y}}{\sqrt{5y}}=\frac{\sqrt{10xy}}{\sqrt{5^2y^2}}=\frac{\sqrt{10xy}}{5y}$

**67.** $\frac{14}{\sqrt[3]{2x^2}}=\frac{14}{\sqrt[3]{2x^2}}\cdot\frac{\sqrt[3]{2^2x}}{\sqrt[3]{2^2x}}=\frac{14\sqrt[3]{2^2x}}{\sqrt[3]{2^3x^3}}$
$=\frac{14\sqrt[3]{4x}}{2x}=\frac{7\sqrt[3]{4x}}{x}$

**68.** $\sqrt[4]{\frac{7}{3x}}=\frac{\sqrt[4]{7}}{\sqrt[4]{3x}}=\frac{\sqrt[4]{7}}{\sqrt[4]{3x}}\cdot\frac{\sqrt[4]{3^3x^3}}{\sqrt[4]{3^3x^3}}$
$=\frac{\sqrt[4]{7\cdot 3^3x^3}}{\sqrt[4]{3^4x^4}}=\frac{\sqrt[4]{7\cdot 27x^3}}{3x}$
$=\frac{\sqrt[4]{189x^3}}{3x}$

**69.** $\frac{5}{\sqrt[5]{32x^4y}}=\frac{5}{\sqrt[5]{2^5x^4y}}\cdot\frac{\sqrt[5]{xy^4}}{\sqrt[5]{xy^4}}$
$=\frac{5\sqrt[5]{xy^4}}{\sqrt[5]{2^5x^5y^5}}=\frac{5\sqrt[5]{xy^4}}{2xy}$

**70.** $$\frac{6}{\sqrt{3}-1}=\frac{6}{\sqrt{3}-1}\cdot\frac{\sqrt{3}+1}{\sqrt{3}+1}$$
$$=\frac{6\left(\sqrt{3}+1\right)}{\left(\sqrt{3}\right)^2-1^2}=\frac{6\left(\sqrt{3}+1\right)}{3-1}$$
$$=\frac{6\left(\sqrt{3}+1\right)}{2}=3\left(\sqrt{3}+1\right)$$
$$=3\sqrt{3}+3$$

**71.** $$\frac{\sqrt{7}}{\sqrt{5}+\sqrt{3}}=\frac{\sqrt{7}}{\sqrt{5}+\sqrt{3}}\cdot\frac{\sqrt{5}-\sqrt{3}}{\sqrt{5}-\sqrt{3}}$$
$$=\frac{\sqrt{35}-\sqrt{21}}{\left(\sqrt{5}\right)^2-\left(\sqrt{3}\right)^2}$$
$$=\frac{\sqrt{35}-\sqrt{21}}{5-3}=\frac{\sqrt{35}-\sqrt{21}}{2}$$

**72.** $$\frac{10}{2\sqrt{5}-3\sqrt{2}}$$
$$=\frac{10}{2\sqrt{5}-3\sqrt{2}}\cdot\frac{2\sqrt{5}+3\sqrt{2}}{2\sqrt{5}+3\sqrt{2}}$$
$$=\frac{10\left(2\sqrt{5}+3\sqrt{2}\right)}{\left(2\sqrt{5}\right)^2-\left(3\sqrt{2}\right)^2}=\frac{10\left(2\sqrt{5}+3\sqrt{2}\right)}{4\cdot5-9\cdot2}$$
$$=\frac{10\left(2\sqrt{5}+3\sqrt{2}\right)}{20-18}=\frac{10\left(2\sqrt{5}+3\sqrt{2}\right)}{2}$$
$$=5\left(2\sqrt{5}+3\sqrt{2}\right)=10\sqrt{5}+15\sqrt{2}$$

**73.** $$\frac{\sqrt{x}+5}{\sqrt{x}-3}=\frac{\sqrt{x}+5}{\sqrt{x}-3}\cdot\frac{\sqrt{x}+3}{\sqrt{x}+3}$$
$$=\frac{x+3\sqrt{x}+5\sqrt{x}+15}{\left(\sqrt{x}\right)^2-3^2}$$
$$=\frac{x+8\sqrt{x}+15}{x-9}$$

**74.** $$\frac{\sqrt{7}+\sqrt{3}}{\sqrt{7}-\sqrt{3}}=\frac{\sqrt{7}+\sqrt{3}}{\sqrt{7}-\sqrt{3}}\cdot\frac{\sqrt{7}+\sqrt{3}}{\sqrt{7}+\sqrt{3}}$$
$$=\frac{7+2\cdot\sqrt{7}\cdot\sqrt{3}+3}{\left(\sqrt{7}\right)^2-\left(\sqrt{3}\right)^2}$$
$$=\frac{10+2\sqrt{21}}{7-3}=\frac{10+2\sqrt{21}}{4}$$
$$=\frac{2\left(5+\sqrt{21}\right)}{4}=\frac{5+\sqrt{21}}{2}$$

**75.** $$\frac{2\sqrt{3}+\sqrt{6}}{2\sqrt{6}+\sqrt{3}}$$
$$=\frac{2\sqrt{3}+\sqrt{6}}{2\sqrt{6}+\sqrt{3}}\cdot\frac{2\sqrt{6}-\sqrt{3}}{2\sqrt{6}-\sqrt{3}}$$
$$=\frac{4\sqrt{18}-2\cdot3+2\cdot6-\sqrt{18}}{\left(2\sqrt{6}\right)^2-\left(\sqrt{3}\right)^2}$$
$$=\frac{3\sqrt{18}-6+12}{4\cdot6-3}=\frac{3\sqrt{9\cdot2}+6}{24-3}$$
$$=\frac{3\cdot3\sqrt{2}+6}{21}=\frac{9\sqrt{2}+6}{21}$$
$$=\frac{3\left(3\sqrt{2}+2\right)}{21}=\frac{3\sqrt{2}+2}{7}$$

**76.** $$\sqrt{\frac{2}{7}}=\frac{\sqrt{2}}{\sqrt{7}}=\frac{\sqrt{2}}{\sqrt{7}}\cdot\frac{\sqrt{2}}{\sqrt{2}}=\frac{2}{\sqrt{14}}$$

**77.** $$\frac{\sqrt[3]{3x}}{\sqrt[3]{y}}=\frac{\sqrt[3]{3x}}{\sqrt[3]{y}}\cdot\frac{\sqrt[3]{3^2x^2}}{\sqrt[3]{3^2x^2}}$$
$$=\frac{\sqrt[3]{3^3x^3}}{\sqrt[3]{3^2x^2y}}=\frac{3x}{\sqrt[3]{9x^2y}}$$

**78.** $$\frac{\sqrt{7}}{\sqrt{5}+\sqrt{3}} = \frac{\sqrt{7}}{\sqrt{5}+\sqrt{3}} \cdot \frac{\sqrt{7}}{\sqrt{7}} = \frac{7}{\sqrt{35}+\sqrt{21}}$$

**79.** $$\frac{\sqrt{7}+\sqrt{3}}{\sqrt{7}-\sqrt{3}}$$
$$= \frac{\sqrt{7}+\sqrt{3}}{\sqrt{7}-\sqrt{3}} \cdot \frac{\sqrt{7}-\sqrt{3}}{\sqrt{7}-\sqrt{3}}$$
$$= \frac{\left(\sqrt{7}\right)^2-\left(\sqrt{3}\right)^2}{7-2\sqrt{7}\sqrt{3}+3} = \frac{7-3}{10-2\sqrt{21}}$$
$$= \frac{4}{10-2\sqrt{21}} = \frac{4}{2\left(5-\sqrt{21}\right)}$$
$$= \frac{2}{5-\sqrt{21}}$$

**80.** $$\sqrt{2x+4} = 6$$
$$\left(\sqrt{2x+4}\right)^2 = 6^2$$
$$2x+4 = 36$$
$$2x = 32$$
$$x = 16$$
The solution checks. The solution is 16 and the solution set is $\{16\}$.

**81.** $$\sqrt{x-5}+9 = 4$$
$$\sqrt{x-5} = -5$$
The square root of a number is always positive. The solution set is $\varnothing$ or $\{\ \ \}$.

**82.** $$\sqrt{2x-3}+x = 3$$
$$\sqrt{2x-3} = 3-x$$
$$\left(\sqrt{2x-3}\right)^2 = (3-x)^2$$
$$2x-3 = 9-6x+x^2$$
$$0 = 12-8x+x^2$$
$$0 = x^2-8x+12$$
$$0 = (x-6)(x-2)$$
Apply the zero product principle.
$x-6=0 \qquad x-2=0$
$x=6 \qquad x=2$
6 is an extraneous solution. The solution is 2 and the solution set is $\{2\}$.

**83.** $$\sqrt{x-4}+\sqrt{x+1} = 5$$
$$\sqrt{x-4} = 5-\sqrt{x+1}$$
$$\left(\sqrt{x-4}\right)^2 = \left(5-\sqrt{x+1}\right)^2$$
$$x-4 = 25-10\sqrt{x+1}+x+1$$
$$-30 = -10\sqrt{x+1}$$
$$\frac{-30}{-10} = \frac{-10\sqrt{x+1}}{-10}$$
$$3 = \sqrt{x+1}$$
$$3^2 = \left(\sqrt{x+1}\right)^2$$
$$9 = x+1$$
$$8 = x$$
The solution checks. The solution is 8 and the solution set is $\{8\}$.

**84.** $\left(x^2+6x\right)^{\frac{1}{3}}+2=0$

$\left(x^2+6x\right)^{\frac{1}{3}}=-2$

$\sqrt[3]{x^2+6x}=-2$

$\left(\sqrt[3]{x^2+6x}\right)^3=(-2)^3$

$x^2+6x=-8$

$x^2+6x+8=0$

$(x+4)(x+2)=0$

Apply the zero product principle.

$x+4=0 \qquad x+2=0$

$x=-4 \qquad x=-2$

Both solutions check. The solutions are –4 and –2, and the solution set is $\{-4,-2\}$.

**85.** $4=\sqrt{\dfrac{x}{16}}$

$4^2=\left(\sqrt{\dfrac{x}{16}}\right)^2$

$16=\dfrac{x}{16}$

$256=x$

The hammer was dropped from a height of 256 feet.

**86.** $20{,}000=5000\sqrt{100-x}$

$\dfrac{20{,}000}{5000}=\dfrac{5000\sqrt{100-x}}{5000}$

$4=\sqrt{100-x}$

$4^2=\left(\sqrt{100-x}\right)^2$

$16=100-x$

$-84=-x$

$84=x$

20,000 people in the group will survive to 84 years old.

**87.** $\sqrt{-81}=\sqrt{81\cdot -1}=\sqrt{81}\sqrt{-1}=9i$

**88.** $\sqrt{-63}=\sqrt{9\cdot 7\cdot -1}$

$=\sqrt{9}\sqrt{7}\sqrt{-1}=3\sqrt{7}i$

**89.** $-\sqrt{-8}=-\sqrt{4\cdot 2\cdot -1}$

$=-\sqrt{4}\sqrt{2}\sqrt{-1}=-2\sqrt{2}i$

**90.** $(7+12i)+(5-10i)$

$=7+12i+5-10i=12+2i$

**91.** $(8-3i)-(17-7i)=8-3i-17+7i$

$=-9+4i$

**92.** $4i(3i-2)=4i\cdot 3i-4i\cdot 2$

$=12i^2-8i$

$=12(-1)-8i$

$=-12-8i$

**93.** $(7-5i)(2+3i)=14+21i-10i-15i^2$

$=14+11i-15(-1)$

$=14+11i+15$

$=29+11i$

**94.** $(3-4i)^2=3^2-2\cdot 3\cdot 4i+(4i)^2$

$=9-24i+16i^2$

$=9-24i+16(-1)$

$=9-24i-16$

$=-7-24i$

**95.** $(7+8i)(7-8i)$

$=7^2-(8i)^2=49-64i^2$

$=49-64(-1)=49+64$

$=113=113+0i$

**96.** $\sqrt{-8}\cdot\sqrt{-3}=\sqrt{4\cdot2\cdot-1}\cdot\sqrt{3\cdot-1}$
$=2\sqrt{2}i\cdot\sqrt{3}i=2\sqrt{6}i^2$
$=2\sqrt{6}(-1)=-2\sqrt{6}$
$=-2\sqrt{6}+0i$

**97.** $\frac{6}{5+i}=\frac{6}{5+i}\cdot\frac{5-i}{5-i}=\frac{30-6i}{25-i^2}$
$=\frac{30-6i}{25-(-1)}=\frac{30-6i}{25+1}$
$=\frac{30-6i}{26}=\frac{30}{26}-\frac{6}{26}i$
$=\frac{15}{13}-\frac{3}{13}i$

**98.** $\frac{3+4i}{4-2i}=\frac{3+4i}{4-2i}\cdot\frac{4+2i}{4+2i}$
$=\frac{12+6i+16i+8i^2}{16-4i^2}$
$=\frac{12+22i+8(-1)}{16-4(-1)}$
$=\frac{12+22i-8}{16+4}=\frac{4+22i}{20}$
$=\frac{4}{20}+\frac{22}{20}i=\frac{1}{5}+\frac{11}{10}i$

**99.** $\frac{5+i}{3i}=\frac{5+i}{3i}\cdot\frac{i}{i}=\frac{5i+i^2}{3i^2}$
$=\frac{5i+(-1)}{3(-1)}=\frac{5i-1}{-3}$
$=\frac{-1}{-3}+\frac{5}{-3}i=\frac{1}{3}-\frac{5}{3}i$

**100.** $i^{16}=\left(i^2\right)^8=(-1)^8=1$

**101.** $i^{23}=i^{22}\cdot i=\left(i^2\right)^{11}i=(-1)^{11}i$
$=(-1)i=-i$

## Chapter 7 Test

**1. a.** $f(-14)=\sqrt{8-2(-14)}$
$=\sqrt{8+28}=\sqrt{36}=6$

**b.** To find the domain, set the radicand greater than or equal to zero and solve the resulting inequality.
$8-2x\geq0$
$-2x\geq-8$
$x\leq4$
The domain of $f$ is $\{x|x\leq4\}$ or $(-\infty,4]$.

**2.** $27^{-\frac{4}{3}}=\frac{1}{27^{\frac{4}{3}}}=\frac{1}{\left(\sqrt[3]{27}\right)^4}=\frac{1}{(3)^4}=\frac{1}{81}$

**3.** $\left(25x^{-\frac{1}{2}}y^{\frac{1}{4}}\right)^{\frac{1}{2}}=25^{\frac{1}{2}}x^{-\frac{1}{4}}y^{\frac{1}{8}}=5x^{-\frac{1}{4}}y^{\frac{1}{8}}$
$=\frac{5y^{\frac{1}{8}}}{x^{\frac{1}{4}}}=\frac{5\sqrt[8]{y}}{\sqrt[4]{x}}$

**4.** $\sqrt[8]{x^4}=\left(x^4\right)^{\frac{1}{8}}=x^{4\cdot\frac{1}{8}}=x^{\frac{1}{2}}=\sqrt{x}$

**5.** $\sqrt[4]{x}\cdot\sqrt[5]{x}=x^{\frac{1}{4}}\cdot x^{\frac{1}{5}}=x^{\frac{1}{4}+\frac{1}{5}}=x^{\frac{5}{20}+\frac{4}{20}}$
$=x^{\frac{9}{20}}=\sqrt[20]{x^9}$

**6.** $\sqrt{75x^2}=\sqrt{25\cdot3x^2}=5|x|\sqrt{3}$

**7.** $\sqrt{x^2-10x+25}=\sqrt{(x-5)^2}$
$=|x-5|$

**8.** $\sqrt[3]{16x^4y^8} = \sqrt[3]{8\cdot 2\cdot x^3\cdot x\cdot y^6\cdot y^2}$
$= \sqrt[3]{8x^3y^6\cdot 2xy^2}$
$= 2xy^2\sqrt[3]{2xy^2}$

**9.** $\sqrt[5]{-\frac{32}{x^{10}}} = \sqrt[5]{-\frac{2^5}{(x^2)^5}} = -\frac{2}{x^2}$

**10.** $\sqrt[3]{5x^2}\cdot\sqrt[3]{10y} = \sqrt[3]{50x^2y}$

**11.** $\sqrt[4]{8x^3y}\cdot\sqrt[4]{4xy^2} = \sqrt[4]{32x^4y^3}$
$= \sqrt[4]{16\cdot 2\cdot x^4\cdot y^3}$
$= \sqrt[4]{16x^4\cdot 2y^3}$
$= 2x\sqrt[4]{2y^3}$

**12.** $3\sqrt{18} - 4\sqrt{32} = 3\sqrt{9\cdot 2} - 4\sqrt{16\cdot 2}$
$= 3\cdot 3\sqrt{2} - 4\cdot 4\sqrt{2}$
$= 9\sqrt{2} - 16\sqrt{2} = -7\sqrt{2}$

**13.** $\sqrt[3]{8x^4} + \sqrt[3]{xy^6} = \sqrt[3]{8x^3\cdot x} + \sqrt[3]{xy^6}$
$= 2x\sqrt[3]{x} + y^2\sqrt[3]{x}$
$= (2x + y^2)\sqrt[3]{x}$

**14.** $\frac{\sqrt[3]{16x^8}}{\sqrt[3]{2x^4}} = \sqrt[3]{\frac{16x^8}{2x^4}} = \sqrt[3]{8x^4}$
$= \sqrt[3]{8x^3\cdot x} = 2x\sqrt[3]{x}$

**15.** $\sqrt{3}\left(4\sqrt{6} - \sqrt{5}\right) = \sqrt{3}\cdot 4\sqrt{6} - \sqrt{3}\cdot\sqrt{5}$
$= 4\sqrt{18} - \sqrt{15}$
$= 4\sqrt{9\cdot 2} - \sqrt{15}$
$= 4\cdot 3\sqrt{2} - \sqrt{15}$
$= 12\sqrt{2} - \sqrt{15}$

**16.** $\left(5\sqrt{6} - 2\sqrt{2}\right)\left(\sqrt{6} + \sqrt{2}\right)$
$= 5\cdot 6 + 5\sqrt{12} - 2\sqrt{12} - 2\cdot 2$
$= 30 + 3\sqrt{12} - 4 = 26 + 3\sqrt{4\cdot 3}$
$= 26 + 3\cdot 2\sqrt{3} = 26 + 6\sqrt{3}$

**17.** $\left(7 - \sqrt{3}\right)^2 = 49 - 2\cdot 7\cdot\sqrt{3} + 3$
$= 52 - 14\sqrt{3}$

**18.** $\sqrt{\frac{5}{x}} = \frac{\sqrt{5}}{\sqrt{x}}\cdot\frac{\sqrt{x}}{\sqrt{x}} = \frac{\sqrt{5x}}{x}$

**19.** $\frac{5}{\sqrt[3]{5x^2}} = \frac{5}{\sqrt[3]{5x^2}}\cdot\frac{\sqrt[3]{5^2x}}{\sqrt[3]{5^2x}} = \frac{5\sqrt[3]{5^2x}}{\sqrt[3]{5^3x^3}}$
$= \frac{5\sqrt[3]{25x}}{5x} = \frac{\sqrt[3]{25x}}{x}$

**20.** $\frac{\sqrt{2} - \sqrt{3}}{\sqrt{2} + \sqrt{3}} = \frac{\sqrt{2} - \sqrt{3}}{\sqrt{2} + \sqrt{3}}\cdot\frac{\sqrt{2} - \sqrt{3}}{\sqrt{2} - \sqrt{3}}$
$= \frac{2 - 2\sqrt{2}\sqrt{3} + 3}{2 - 3}$
$= \frac{5 - 2\sqrt{6}}{-1} = -5 + 2\sqrt{6}$

**21.** $3 + \sqrt{2x - 3} = x$
$\sqrt{2x - 3} = x - 3$
$\left(\sqrt{2x - 3}\right)^2 = (x - 3)^2$
$2x - 3 = x^2 - 6x + 9$
$0 = x^2 - 8x + 12$
$0 = (x - 6)(x - 2)$

Apply the zero product rule.

$x - 6 = 0 \qquad x - 2 = 0$
$x = 6 \qquad x = 2$

2 is an extraneous solution. The solution is 6 and the solution set is $\{6\}$.

**22.** $\sqrt{x+9}-\sqrt{x-7}=2$

$$\sqrt{x+9}=2+\sqrt{x-7}$$
$$\left(\sqrt{x+9}\right)^2=\left(2+\sqrt{x-7}\right)^2$$
$$x+9=4+2\cdot 2\cdot\sqrt{x-7}+x-7$$
$$x+9=4\sqrt{x-7}+x-3$$
$$12=4\sqrt{x-7}$$
$$3=\sqrt{x-7}$$
$$3^2=\left(\sqrt{x-7}\right)^2$$
$$9=x-7$$
$$16=x$$

The solution is 16 and the solution set is $\{16\}$.

**23.** $(11x+6)^{\frac{1}{3}}+3=0$

$$(11x+6)^{\frac{1}{3}}=-3$$
$$\sqrt[3]{11x+6}=-3$$
$$\left(\sqrt[3]{11x+6}\right)^3=(-3)^3$$
$$11x+6=-27$$
$$11x=-33$$
$$x=-3$$

The solution is –3 and the solution set is $\{-3\}$.

**24.** $40.4=2.9\sqrt{x}+20.1$

$$20.3=2.9\sqrt{x}$$
$$7=\sqrt{x}$$
$$7^2=\left(\sqrt{x}\right)^2$$
$$49=x$$

Boys who are 49 months of age have an average height of 40.4 inches.

**25.** $\sqrt{-75}=\sqrt{25\cdot 3\cdot -1}$

$$=\sqrt{25}\cdot\sqrt{3}\cdot\sqrt{-1}=5\sqrt{3}i$$

**26.** $(5-3i)-(6-9i)=5-3i-6+9i$

$$=5-6-3i+9i$$
$$=-1+6i$$

**27.** $(3-4i)(2+5i)=6+15i-8i-20i^2$

$$=6+7i-20(-1)$$
$$=6+7i+20$$
$$=26+7i$$

**28.** $\sqrt{-9}\cdot\sqrt{-4}=\sqrt{9\cdot -1}\cdot\sqrt{4\cdot -1}$

$$=\sqrt{9}\cdot\sqrt{-1}\cdot\sqrt{4}\cdot\sqrt{-1}$$
$$=3\cdot i\cdot 2\cdot i=6i^2=6(-1)$$
$$=-6 \text{ or } -6+0i$$

**29.** $\dfrac{3+i}{1-2i}=\dfrac{3+i}{1-2i}\cdot\dfrac{1+2i}{1+2i}=\dfrac{3+6i+i+2i^2}{1-4i^2}$

$$=\frac{3+7i+2(-1)}{1-4(-1)}=\frac{3+7i-2}{1+4}$$
$$=\frac{1+7i}{5}=\frac{1}{5}+\frac{7}{5}i$$

**30.** $i^{35}=i^{34}\cdot i=\left(i^2\right)^{17}\cdot i=(-1)^{17}\cdot i$

$$=(-1)i=-i$$

**Cumulative Review Exercises (Chapters 1-7)**

**1.** $2x - y + z = -5$

$x - 2y - 3z = 6$

$x + y - 2z = 1$

Add the first and third equations to eliminate $y$.

$2x - y + z = -5$

$\underline{x + y - 2z = 1}$

$3x - z = -4$

Multiply the third equation by 2 and add to the second equation.

$x - 2y - 3z = 6$

$\underline{2x + 2y - 4z = 2}$

$3x - 7z = 8$

We now have a system of two equations in two variables.

$3x - z = -4$

$3x - 7z = 8$

Multiply the first equation by –1 and add to the second equation.

$-3x + z = 4$

$\underline{3x - 7z = 8}$

$-6z = 12$

$z = -2$

Back-substitute –2 for $z$ to find $x$.

$-3x + z = 4$

$-3x - 2 = 4$

$-3x = 6$

$x = -2$

Back-substitute –2 for $x$ and $z$ in one of the original equations to find $y$.

$2x - y + z = -5$

$2(-2) - y - 2 = -5$

$-4 - y - 2 = -5$

$-y - 6 = -5$

$-y = 1$

$y = -1$

The solution is $(-2,-1,-2)$ or the solution set is $\{(-2,-1,-2)\}$.

**2.** $3x^2 - 11x = 4$

$3x^2 - 11x - 4 = 0$

$(3x+1)(x-4) = 0$

Apply the zero product principle.

$3x + 1 = 0 \qquad x - 4 = 0$

$3x = -1 \qquad x = 4$

$x = -\frac{1}{3}$

The solutions are $-\frac{1}{3}$ and 4 and the solution set is $\left\{-\frac{1}{3}, 4\right\}$.

**3.** $2(x+4) < 5x + 3(x+2)$

$2x + 8 < 5x + 3x + 6$

$2x + 8 < 8x + 6$

$-6x + 8 < 6$

$-6x < -2$

$\frac{-6x}{-6} > \frac{-2}{-6}$

$x > \frac{1}{3}$

The solution set is $\left\{x \middle| x > \frac{1}{3}\right\}$ or $\left(\frac{1}{3}, \infty\right)$.

**4.**
$$\frac{1}{x+2}+\frac{15}{x^2-4}=\frac{5}{x-2}$$
$$\frac{1}{x+2}+\frac{15}{(x+2)(x-2)}=\frac{5}{x-2}$$

So that denominators will not equal zero, $x$ cannot equal 2 or –2. To eliminate fractions, multiply by the LCD, $(x+2)(x-2)$.

$$(x+2)(x-2)\left(\frac{1}{x+2}+\frac{15}{(x+2)(x-2)}\right)=(x+2)(x-2)\left(\frac{5}{x-2}\right)$$
$$(x+2)(x-2)\left(\frac{1}{x+2}\right)+(x+2)(x-2)\left(\frac{15}{(x+2)(x-2)}\right)=(x+2)(5)$$
$$x-2+15=5x+10$$
$$x+13=5x+10$$
$$-4x+13=10$$
$$-4x=-3$$
$$x=\frac{3}{4}$$

The solution is $\frac{3}{4}$ and the solution set is $\left\{\frac{3}{4}\right\}$.

**5.**
$$\sqrt{x+2}-\sqrt{x+1}=1$$
$$\sqrt{x+2}=1+\sqrt{x+1}$$
$$\left(\sqrt{x+2}\right)^2=\left(1+\sqrt{x+1}\right)^2$$
$$x+2=1+2\sqrt{x+1}+x+1$$
$$x+2=2+2\sqrt{x+1}+x$$
$$0^2=\left(2\sqrt{x+1}\right)^2$$
$$0=4(x+1)$$
$$0=4x+4$$
$$-4=4x$$
$$-1=x$$

The solution checks. The solution is –1 and the solution set is $\{-1\}$.

**6.**
$$x+2y<2$$
$$2y-x>4$$

First consider $x+2y<2$. Replace the inequality symbol with an equal sign and we have $x+2y=2$. Solve for $y$ to put the equation in slope-intercept form.

$$x+2y=2$$
$$2y=-x+2$$
$$y=-\frac{1}{2}x+1$$

slope = $-\frac{1}{2}$ $\qquad$ $y$–intercept = 1

Now, use the origin as a test point.

$$0+2(0)<2$$
$$0<2$$

This is a true statement. This means that the point $(0,0)$ will fall in the shaded half-plane.

Next consider $2y - x > 4$. Replace the inequality symbol with an equal sign and we have $2y - x = 4$. Solve for $y$ to put the equation in slope-intercept form.

$$2y - x = 4$$
$$2y = x + 4$$
$$y = \frac{1}{2}x + 2$$

slope = $\frac{1}{2}$   $y$–intercept = 2

Now, use the origin as a test point.

$$2(0) - 0 > 4$$
$$0 > 4$$

This is a false statement. This means that the point $(0,0)$ will not fall in the shaded half-plane.

Next, graph each of the inequalities. The solution to the system is the intersection of the shaded half-planes.

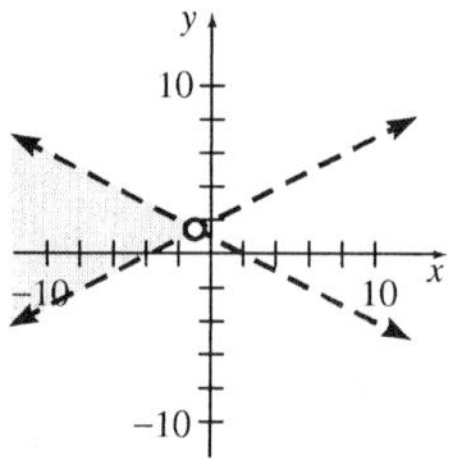


**7.** $$\frac{8x^2}{3x^2-12} \div \frac{40}{x-2}$$
$$= \frac{8x^2}{3x^2-12} \cdot \frac{x-2}{40}$$
$$= \frac{8x^2}{3(x^2-4)} \cdot \frac{x-2}{40}$$
$$= \frac{\overset{1}{\cancel{8}}x^2}{3(x+2)\cancel{(x-2)}} \cdot \frac{\cancel{x-2}}{\underset{5}{\cancel{40}}}$$
$$= \frac{x^2}{3 \cdot 5(x+2)} = \frac{x^2}{15(x+2)}$$

**8.** $$\frac{x+\frac{1}{y}}{y+\frac{1}{x}} = \frac{xy}{xy} \cdot \frac{x+\frac{1}{y}}{y+\frac{1}{x}} = \frac{xy \cdot x + xy \cdot \frac{1}{y}}{xy \cdot y + xy \cdot \frac{1}{x}}$$
$$= \frac{x^2y + x}{xy^2 + y} = \frac{x(xy+1)}{y(xy+1)} = \frac{x}{y}$$

**9.** $$(2x-3)(4x^2-5x-2)$$
$$= 2x \cdot 4x^2 - 2x \cdot 5x - 2x \cdot 2 - 3 \cdot 4x^2 + 3 \cdot 5x + 3 \cdot 2$$
$$= 8x^3 - 10x^2 - 4x - 12x^2 + 15x + 6$$
$$= 8x^3 - 22x^2 + 11x + 6$$

**10.** $$\frac{7x}{x^2-2x-15} - \frac{2}{x-5}$$
$$= \frac{7x}{(x-5)(x+3)} - \frac{2}{x-5}$$
$$= \frac{7x}{(x-5)(x+3)} - \frac{2(x+3)}{(x-5)(x+3)}$$
$$= \frac{7x - 2(x+3)}{(x-5)(x+3)} = \frac{7x - 2x - 6}{(x-5)(x+3)}$$
$$= \frac{5x-6}{(x-5)(x+3)}$$

**11.** $$7(8-10)^3 - 7 + 3 \div (-3)$$
$$= 7(-2)^3 - 7 + 3 \div (-3)$$
$$= 7(-8) - 7 + 3 \div (-3)$$
$$= -56 - 7 + (-1) = -64$$

**12.** $$\sqrt{80x} - 5\sqrt{20x} + 2\sqrt{45x}$$
$$= \sqrt{16 \cdot 5x} - 5\sqrt{4 \cdot 5x} + 2\sqrt{9 \cdot 5x}$$
$$= 4\sqrt{5x} - 5 \cdot 2\sqrt{5x} + 2 \cdot 3\sqrt{5x}$$
$$= 4\sqrt{5x} - 10\sqrt{5x} + 6\sqrt{5x} = 0$$

**13.** 
$$\frac{\sqrt{3}-2}{2\sqrt{3}+5}=\frac{\sqrt{3}-2}{2\sqrt{3}+5}\cdot\frac{2\sqrt{3}-5}{2\sqrt{3}-5}$$
$$=\frac{2\cdot 3-5\sqrt{3}-4\sqrt{3}+10}{4\cdot 3-25}$$
$$=\frac{6-9\sqrt{3}+10}{12-25}$$
$$=\frac{16-9\sqrt{3}}{-13}$$
$$=-\frac{16-9\sqrt{3}}{13}$$

**14.** 
$$\begin{array}{r} 2x^2+\ x+5 \\ x-2\overline{)2x^3-3x^2+3x-4} \\ \underline{2x^3-4x^2}\qquad\qquad \\ x^2+3x\qquad \\ \underline{x^2-2x}\qquad \\ 5x-\ 4 \\ \underline{5x-10} \\ 6 \end{array}$$

$$\frac{2x^3-3x^2+3x-4}{x-2}=2x^2+x+5+\frac{6}{x-2}$$

**15.** 
$$\left(2\sqrt{3}+5\sqrt{2}\right)\left(\sqrt{3}-4\sqrt{2}\right)$$
$$=2\cdot 3-8\sqrt{6}+5\sqrt{6}-20\cdot 2$$
$$=6-3\sqrt{6}-40=-34-3\sqrt{6}$$

**16.** 
$$24x^2+10x-4=2\left(12x^2+5x-2\right)$$
$$=2(3x+2)(4x-1)$$

**17.** 
$$16x^4-1=\left(4x^2+1\right)\left(4x^2-1\right)$$
$$=\left(4x^2+1\right)(2x+1)(2x-1)$$

**18.** Since light varies inversely as the square of the distance, we have $l=\frac{k}{d^2}$.

Use the given values to find $k$.
$$l=\frac{k}{d^2}$$
$$120=\frac{k}{10^2}$$
$$120=\frac{k}{100}$$
$$12{,}000=k$$

The equation becomes $l=\frac{12{,}000}{d^2}$.

When $d = 15$, $l=\frac{12{,}000}{15^2}\approx 53.3$.

At a distance of 15 feet, approximately 53 lumens are provided.

**19.** Let $x$ = the amount invested at 7%
Let $y$ = the amount invested at 9%
$$x+\qquad y=6000$$
$$0.07x+0.09y=\ 510$$

Solve the first equation for $y$.
$$x+y=6000$$
$$y=6000-x$$

Substitute and solve.
$$0.07x+0.09(6000-x)=510$$
$$0.07x+540-0.09x=510$$
$$540-0.02x=510$$
$$-0.02x=-30$$
$$x=1500$$

Back-substitute 1500 for $x$ to find $y$.
$$y=6000-x$$
$$y=6000-1500=4500$$

\$1500 was invested at 7% and \$4500 was invested at 9%.

**20.** Let $x$ = the number of students enrolled last year
$$x-0.12x=2332$$
$$0.88x=2332$$
$$x=2650$$

2650 students were enrolled last year.

# Chapter 8
# Quadratic Equations and Functions

## 8.1 Exercise Set

**1.** $3x^2 = 75$

$x^2 = 25$

Apply the square root property.

$x = \pm\sqrt{25}$

$x = \pm 5$

The solutions are $\pm 5$ and the solution st is $\{\pm 5\}$.

**3.** $7x^2 = 42$

$x^2 = 6$

Apply the square root property.

$x = \pm\sqrt{6}$

The solutions are $\pm\sqrt{6}$ and the solution set is $\{\pm\sqrt{6}\}$.

**5.** $16x^2 = 25$

$x^2 = \frac{25}{16}$

Apply the square root property.

$x = \pm\sqrt{\frac{25}{16}}$

$x = \pm\frac{5}{4}$

The solutions are $\pm\frac{5}{4}$ and the solution set is $\left\{\pm\frac{5}{4}\right\}$.

**7.** $3x^2 - 2 = 0$

$3x^2 = 2$

$x^2 = \frac{2}{3}$

Apply the square root property.

$x = \pm\sqrt{\frac{2}{3}}$

Because the proposed solutions are opposites, rationalize both denominators at once.

$x = \pm\sqrt{\frac{2}{3}} = \pm\frac{\sqrt{2}}{\sqrt{3}} \cdot \frac{\sqrt{3}}{\sqrt{3}} = \pm\frac{\sqrt{6}}{3}$

The solutions are $\pm\frac{\sqrt{6}}{3}$ and the solution set is $\left\{\pm\frac{\sqrt{6}}{3}\right\}$.

**9.** $25x^2 + 16 = 0$

$25x^2 = -16$

$x^2 = -\frac{16}{25}$

Apply the square root property.

$x = \pm\sqrt{-\frac{16}{25}}$

$x = \pm\sqrt{\frac{16}{25}}\sqrt{-1}$

$x = \pm\frac{4}{5}i$

$x = 0 \pm \frac{4}{5}i = \pm\frac{4}{5}i$

The solutions are $\pm\frac{4}{5}i$ and the solution set is $\left\{\pm\frac{4}{5}i\right\}$.

**11.** $(x+7)^2 = 9$

Apply the square root property.

$x+7=\sqrt{9}$ or $x+7=-\sqrt{9}$

$x+7=3$ $\quad$ $x+7=-3$

$x=-4$ $\quad$ $x=-10$

The solutions are $-4$ and $-10$ and the solution set is $\{-10,-4\}$.

**13.** $(x-3)^2 = 5$

Apply the square root property.

$x-3=\pm\sqrt{5}$

$x=3\pm\sqrt{5}$

The solutions are $3\pm\sqrt{5}$ and the solution set is $\{3\pm\sqrt{5}\}$.

**15.** $(x+2)^2 = 8$

Apply the square root property.

$x+2=\pm\sqrt{8}$

$x+2=\pm\sqrt{4\cdot 2}$

$x+2=\pm 2\sqrt{2}$

$x=-2\pm 2\sqrt{2}$

The solutions are $-2\pm 2\sqrt{2}$ and the solution set is $\{-2\pm 2\sqrt{2}\}$.

**17.** $(x-5)^2 = -9$

Apply the square root property.

$x-5=\pm\sqrt{-9}$

$x-5=\pm 3i$

$x=5\pm 3i$

The solutions are $5\pm 3i$ and the solution set is $\{5\pm 3i\}$.

**19.** $\left(x+\frac{3}{4}\right)^2 = \frac{11}{16}$

Apply the square root property.

$x+\frac{3}{4}=\pm\sqrt{\frac{11}{16}}$

$x+\frac{3}{4}=\pm\frac{\sqrt{11}}{4}$

$x=-\frac{3}{4}\pm\frac{\sqrt{11}}{4}=\frac{-3\pm\sqrt{11}}{4}$

The solutions are $\frac{-3\pm\sqrt{11}}{4}$ and the solution set is $\left\{\frac{-3\pm\sqrt{11}}{4}\right\}$.

**21.** $x^2-6x+9=36$

$(x-3)^2=36$

Apply the square root property.

$x-3=\sqrt{36}$ or $x-3=-\sqrt{36}$

$x-3=6$ $\quad$ $x-3=-6$

$x=9$ $\quad$ $x=-3$

The solutions are 9 and $-3$ and the solution set is $\{-3,9\}$.

**23.** $x^2+2x+$____

Since $b = 2$, we add

$\left(\frac{b}{2}\right)^2=\left(\frac{2}{2}\right)^2=(1)^2=1$.

$x^2+2x+1=(x+1)^2$

**25.** $x^2-14x+$____

Since $b=-14$, we add

$\left(\frac{b}{2}\right)^2=\left(\frac{-14}{2}\right)^2=(-7)^2=49$.

$x^2-14x+49=(x-7)^2$

**27.** $x^2+7x+\_\_\_\_$

Since $b = 7$, we add

$\left(\frac{b}{2}\right)^2=\left(\frac{7}{2}\right)^2=\frac{49}{4}$.

$x^2+7x+\frac{49}{4}=\left(x+\frac{7}{2}\right)^2$

**29.** $x^2-\frac{1}{2}x+\_\_\_\_$

Since $b=-\frac{1}{2}$, we add

$\left(\frac{b}{2}\right)^2=\left(\frac{-1}{2}\div 2\right)^2$

$=\left(\frac{-1}{2}\cdot\frac{1}{2}\right)^2=\left(\frac{-1}{4}\right)^2=\frac{1}{16}$.

$x^2-\frac{1}{2}x+\frac{1}{16}=\left(x-\frac{1}{4}\right)^2$

**31.** $x^2+\frac{4}{3}x+\_\_\_\_$

Since $b=\frac{4}{3}$, we add

$\left(\frac{b}{2}\right)^2=\left(\frac{4}{3}\div 2\right)^2=\left(\frac{4}{3}\cdot\frac{1}{2}\right)^2=\left(\frac{2}{3}\right)^2=\frac{4}{9}$.

$x^2+\frac{4}{3}x+\frac{4}{9}=\left(x+\frac{2}{3}\right)^2$

**33.** $x^2-\frac{9}{4}x+\_\_\_\_$

Since $b=-\frac{9}{4}$, we add

$\left(\frac{b}{2}\right)^2=\left(-\frac{9}{4}\div 2\right)^2$

$=\left(-\frac{9}{4}\cdot\frac{1}{2}\right)^2=\left(-\frac{9}{8}\right)^2=\frac{81}{64}$.

$x^2-\frac{9}{4}x+\frac{81}{64}=\left(x-\frac{9}{8}\right)^2$

**35.** $x^2+4x=32$

$x^2+4x \quad =32$

Since $b = 4$, we add

$\left(\frac{b}{2}\right)^2=\left(\frac{4}{2}\right)^2=(2)^2=4$.

$x^2+4x+4=32+4$

$(x+2)^2=36$

Apply the square root property.

$x+2=\sqrt{36}$ or $x+2=-\sqrt{36}$

$x+2=6 \qquad x+2=-6$

$x=4 \qquad x=-8$

The solutions are 4 and $-8$ and the solution set is $\{-8, 4\}$.

**37.** $x^2+6x=-2$

$x^2+6x \quad =-2$

Since $b = 6$, we add

$\left(\frac{b}{2}\right)^2=\left(\frac{6}{2}\right)^2=(3)^2=9$.

$x^2+6x+9=-2+9$

$(x+3)^2=7$

Apply the square root property.

$x+3=\pm\sqrt{7}$

$x=-3\pm\sqrt{7}$

The solutions are $-3\pm\sqrt{7}$ and the solution set is $\left\{-3\pm\sqrt{7}\right\}$.

**39.** $x^2-8x+1=0$

$x^2-8x \quad =-1$

Since $b=-8$, we add

$\left(\frac{b}{2}\right)^2=\left(\frac{-8}{2}\right)^2=(-4)^2=16$.

$x^2-8x+16=-1+16$

$(x-4)^2=15$

Apply the square root property.

$x - 4 = \pm\sqrt{15}$

$x = 4 \pm \sqrt{15}$

The solutions are $4 \pm \sqrt{15}$ and the solution set is $\{4 \pm \sqrt{15}\}$.

**41.** $x^2 + 2x + 2 = 0$

$x^2 + 2x \quad = -2$

Since $b = 2$, we add

$\left(\frac{b}{2}\right)^2 = \left(\frac{2}{2}\right)^2 = (1)^2 = 1.$

$x^2 + 2x + 1 = -2 + 1$

$(x+1)^2 = -1$

Apply the square root property.

$x + 1 = \pm\sqrt{-1} = \pm i$

$x = -1 \pm i$

The solutions are $-1 \pm i$ and the solution set is $\{-1 \pm i\}$.

**43.** $x^2 + 3x - 1 = 0$

$x^2 + 3x \quad = 1$

Since $b = 3$, we add

$\left(\frac{b}{2}\right)^2 = \left(\frac{3}{2}\right)^2 = \frac{9}{4}.$

$x^2 + 3x + \frac{9}{4} = 1 + \frac{9}{4}$

$\left(x + \frac{3}{2}\right)^2 = \frac{13}{4}$

Apply the square root property.

$x + \frac{3}{2} = \pm\sqrt{\frac{13}{4}} = \pm\frac{\sqrt{13}}{2}$

$x = -\frac{3}{2} \pm \frac{\sqrt{13}}{2} = \frac{-3 \pm \sqrt{13}}{2}$

The solutions are $\frac{-3 \pm \sqrt{13}}{2}$ and the solution set is $\left\{\frac{-3 \pm \sqrt{13}}{2}\right\}$.

**45.** $x^2 + \frac{4}{7}x + \frac{3}{49} = 0$

$x^2 + \frac{4}{7}x \quad = -\frac{3}{49}$

Since $b = \frac{4}{7}$, we add

$\left(\frac{1}{2}b\right)^2 = \left(\frac{1}{2}\cdot\frac{4}{7}\right)^2 \left(\frac{2}{7}\right)^2 = \frac{4}{49}.$

$x^2 + \frac{4}{7}x + \frac{4}{49} = -\frac{3}{49} + \frac{4}{49}$

$\left(x + \frac{2}{7}\right)^2 = \frac{1}{49}$

Apply the square root property.

$x + \frac{2}{7} = \pm\sqrt{\frac{1}{49}}$

$x + \frac{2}{7} = \pm\frac{1}{7}$

$x = -\frac{2}{7} \pm \frac{1}{7}$

$x = -\frac{2}{7} + \frac{1}{7} = -\frac{1}{7}$ or $-\frac{2}{7} - \frac{1}{7} = -\frac{3}{7}$

The solutions are $-\frac{1}{7}$ and $-\frac{3}{7}$, and the solution set is $\left\{-\frac{3}{7}, -\frac{1}{7}\right\}$.

**47.** $x^2 + x - 1 = 0$

$x^2 + x \quad = 1$

Since $b = 1$, we add

$\left(\frac{b}{2}\right)^2 = \left(\frac{1}{2}\right)^2 = \frac{1}{4}.$

$x^2 + x + \frac{1}{4} = 1 + \frac{1}{4}$

$\left(x + \frac{1}{2}\right)^2 = \frac{5}{4}$

Apply the square root property.

$x+\frac{1}{2}=\pm\sqrt{\frac{5}{4}}$

$x+\frac{1}{2}=\pm\frac{\sqrt{5}}{2}$

$x=-\frac{1}{2}\pm\frac{\sqrt{5}}{2}=\frac{-1\pm\sqrt{5}}{2}$

The solutions are $\frac{-1\pm\sqrt{5}}{2}$ and the solution set is $\left\{\frac{-1\pm\sqrt{5}}{2}\right\}$.

**49.** $2x^2+3x-5=0$

$x^2+\frac{3}{2}x-\frac{5}{2}=0$

$x^2+\frac{3}{2}x \quad =\frac{5}{2}$

Since $b=\frac{3}{2}$, we add

$\left(\frac{1}{2}b\right)^2=\left(\frac{1}{2}\cdot\frac{3}{2}\right)^2=\left(\frac{3}{4}\right)^2=\frac{9}{16}.$

$x^2+\frac{3}{2}x+\frac{9}{16}=\frac{5}{2}+\frac{9}{16}$

$\left(x+\frac{3}{4}\right)^2=\frac{40}{16}+\frac{9}{16}=\frac{49}{16}$

Apply the square root property.

$x+\frac{3}{4}=\pm\sqrt{\frac{49}{16}}=\pm\frac{7}{4}$

$x=-\frac{3}{4}\pm\frac{7}{4}$

$x=-\frac{3}{4}+\frac{7}{4}$ or $x=-\frac{3}{4}-\frac{7}{4}$

$x=\frac{4}{4}=1$ or $x=-\frac{10}{4}=-\frac{5}{2}$

The solutions are $-\frac{5}{2}$ and $1$, and the solution set is $\left\{-\frac{5}{2},\ 1\right\}$.

**51.** $3x^2+6x+1=0$

$x^2+2x+\frac{1}{3}=0$

$x^2+2x \quad =-\frac{1}{3}$

Since $b = 2$, we add

$\left(\frac{b}{2}\right)^2=\left(\frac{2}{2}\right)^2=1^2=1.$

$x^2+2x+1=-\frac{1}{3}+1$

$(x+1)^2=-\frac{1}{3}+\frac{3}{3}=\frac{2}{3}$

Apply the square root property.

$x+1=\pm\sqrt{\frac{2}{3}}$

$x+1=\pm\frac{\sqrt{2}}{\sqrt{3}}\cdot\frac{\sqrt{3}}{\sqrt{3}}=\pm\frac{\sqrt{6}}{3}$

$x=-1\pm\frac{\sqrt{6}}{3}=\frac{-3\pm\sqrt{6}}{3}$

The solutions are $\frac{-3\pm\sqrt{6}}{3}$ and the solution set is $\left\{\frac{-3\pm\sqrt{6}}{3}\right\}$.

**53.** $3x^2-8x+1=0$

$x^2-\frac{8}{3}x+\frac{1}{3}=0$

$x^2-\frac{8}{3}x \quad =-\frac{1}{3}$

Since $b=-\frac{8}{3}$, we add

$\left(\frac{1}{2}b\right)^2=\left[\frac{1}{2}\left(-\frac{8}{3}\right)\right]^2=\left(-\frac{4}{3}\right)^2=\frac{16}{9}.$

$x^2-\frac{8}{3}x+\frac{16}{9}=-\frac{1}{3}+\frac{16}{9}$

$\left(x-\frac{4}{3}\right)^2=-\frac{3}{9}+\frac{16}{9}=\frac{13}{9}$

Apply the square root property.

$$x-\frac{4}{3}=\pm\sqrt{\frac{13}{9}}$$

$$x-\frac{4}{3}=\pm\frac{\sqrt{13}}{3}$$

$$x=\frac{4}{3}\pm\frac{\sqrt{13}}{3}=\frac{4\pm\sqrt{13}}{3}$$

The solutions are $\frac{4\pm\sqrt{13}}{3}$ and the solution set is $\left\{\frac{4\pm\sqrt{13}}{3}\right\}$.

**55.** $8x^2-4x+1=0$

$$x^2-\frac{1}{2}x+\frac{1}{8}=0$$

$$x^2-\frac{1}{2}x \quad =-\frac{1}{8}$$

Since $b=-\frac{1}{2}$, we add

$$\left(\frac{1}{2}b\right)^2=\left[\frac{1}{2}\left(-\frac{1}{2}\right)\right]^2=\left(-\frac{1}{4}\right)^2=\frac{1}{16}.$$

$$x^2-\frac{1}{2}x+\frac{1}{16}=-\frac{1}{8}+\frac{1}{16}$$

$$\left(x-\frac{1}{4}\right)^2=-\frac{2}{16}+\frac{1}{16}=-\frac{1}{16}$$

Apply the square root property.

$$x-\frac{1}{4}=\pm\sqrt{-\frac{1}{16}}$$

$$x-\frac{1}{4}=\pm\frac{1}{4}i$$

$$x=\frac{1}{4}\pm\frac{1}{4}i$$

The solutions are $\frac{1}{4}\pm\frac{1}{4}i$ and the solution set is $\left\{\frac{1}{4}\pm\frac{1}{4}i\right\}$.

**57.** $f(x)=36$

$$(x-1)^2=36$$

Apply the square root property.

$$x-1=\pm\sqrt{36}$$

$$x-1=\pm6$$

$$x=1\pm6$$

$$x=1+6=7 \text{ or } 1-6=-5$$

The values are $-5$ and 7.

**59.** $g(x)=\frac{9}{25}$

$$\left(x-\frac{2}{5}\right)^2=\frac{9}{25}$$

Apply the square root property.

$$x-\frac{2}{5}=\pm\sqrt{\frac{9}{25}}=\pm\frac{3}{5}$$

$$x=\frac{2}{5}\pm\frac{3}{5}$$

$$x=\frac{2}{5}+\frac{3}{5} \text{ or } x=\frac{2}{5}-\frac{3}{5}$$

$$x=\frac{5}{5}=1 \text{ or } x=-\frac{1}{5}$$

The values are $-\frac{1}{5}$ and 1.

**61.** $h(x)=-125$

$$5(x+2)^2=-125$$

$$(x+2)^2=-25$$

Apply the square root property.

$$x+2=\pm\sqrt{-25}$$

$$x+2=\pm5i$$

$$x=-2\pm5i$$

The values are $-2\pm5i$.

**63.** Let $x$ = the number.

$$3(x-2)^2 = -12$$
$$(x-2)^2 = -4$$

Apply the square root property.

$$x-2 = \pm\sqrt{-4} = \pm 2i$$
$$x = 2 \pm 2i$$

The values are $2+2i$ and $2-2i$.

**64.** Let $x$ = the number.

$$3(x-9)^2 = -27$$
$$(x-9)^2 = -9$$

Apply the square root property.

$$x-9 = \pm\sqrt{-9} = \pm 3i$$
$$x = 9 \pm 3i$$

The values are $9+3i$ and $9-3i$.

**65.**

$$h = \frac{v^2}{2g}$$
$$2gh = v^2$$

Apply the square root property and keep only the principal square root.

$$v = \sqrt{2gh}$$

**66.**

$$s = \frac{kwd^2}{l}$$
$$sl = kwd^2$$
$$\frac{sl}{kw} = d^2$$

Apply the square root property and keep only the principal square root.

$$d = \sqrt{\frac{sl}{kw}} = \frac{\sqrt{sl}}{\sqrt{kw}} \cdot \frac{\sqrt{kw}}{\sqrt{kw}} = \frac{\sqrt{slkw}}{kw}$$

**67.** $A = P(1+r)^2$

$$\frac{A}{P} = (1+r)^2$$

Apply the square root property and keep only the principal square root.

$$1+r = \sqrt{\frac{A}{P}}$$
$$1+r = \frac{\sqrt{A}}{\sqrt{P}} \cdot \frac{\sqrt{P}}{\sqrt{P}} = \frac{\sqrt{AP}}{P}$$
$$r = \frac{\sqrt{AP}}{P} - 1$$

**68.**

$$C = \frac{kP_1P_2}{d^2}$$
$$Cd^2 = kP_1P_2$$
$$d^2 = \frac{kP_1P_2}{C}$$

Apply the square root property and keep only the principal square root.

$$d = \sqrt{\frac{kP_1P_2}{C}}$$
$$d = \frac{\sqrt{kP_1P_2}}{\sqrt{C}} \cdot \frac{\sqrt{C}}{\sqrt{C}} = \frac{\sqrt{kP_1P_2C}}{C}$$

**69.**

$$\frac{x^2}{3} + \frac{x}{9} - \frac{1}{6} = 0$$
$$3\left(\frac{x^2}{3} + \frac{x}{9} - \frac{1}{6}\right) = 3(0)$$
$$x^2 + \frac{1}{3}x - \frac{1}{2} = 0$$
$$x^2 + \frac{1}{3}x = \frac{1}{2}$$

Since $b = \frac{1}{3}$, we add

$$\left(\frac{1}{2}b\right)^2 = \left(\frac{1}{2}\cdot\frac{1}{3}\right)^2 = \left(\frac{1}{6}\right)^2 = \frac{1}{36}.$$
$$x^2 + \frac{1}{3}x + \frac{1}{36} = \frac{1}{2} + \frac{1}{36}$$
$$\left(x + \frac{1}{6}\right)^2 = \frac{18}{36} + \frac{1}{36} = \frac{19}{36}$$

Apply the square root property.

$$x+\frac{1}{6}=\pm\sqrt{\frac{19}{36}}=\pm\frac{\sqrt{19}}{6}$$

$$x=-\frac{1}{6}\pm\frac{\sqrt{19}}{6}=\frac{-1\pm\sqrt{19}}{6}$$

The solutions are $\frac{-1\pm\sqrt{19}}{6}$ and the solution set is $\left\{\frac{-1\pm\sqrt{19}}{6}\right\}$.

**70.** $$\frac{x^2}{2}-\frac{x}{6}-\frac{3}{4}=0$$

$$2\left(\frac{x^2}{2}-\frac{x}{6}-\frac{3}{4}\right)=2(0)$$

$$x^2-\frac{1}{3}x-\frac{3}{2}=0$$

$$x^2-\frac{1}{3}x \quad =\frac{3}{2}$$

Since $b=-\frac{1}{3}$, we add

$$\left(\frac{1}{2}b\right)^2=\left[\frac{1}{2}\left(-\frac{1}{3}\right)\right]^2=\left(-\frac{1}{6}\right)^2=\frac{1}{36}.$$

$$x^2-\frac{1}{3}x+\frac{1}{36}=\frac{3}{2}+\frac{1}{36}$$

$$\left(x-\frac{1}{6}\right)^2=\frac{54}{36}+\frac{1}{36}=\frac{55}{36}$$

Apply the square root property.

$$x-\frac{1}{6}=\pm\sqrt{\frac{55}{36}}=\pm\frac{\sqrt{55}}{6}$$

$$x=\frac{1}{6}\pm\frac{\sqrt{55}}{6}=\frac{1\pm\sqrt{55}}{6}$$

The solutions are $\frac{1\pm\sqrt{55}}{6}$ and the solution set is $\left\{\frac{1\pm\sqrt{55}}{6}\right\}$.

**71.** $$x^2-bx=2b^2$$

$$x^2-bx \quad =2b^2$$

Since $-b$ is the linear coefficient, we add

$$\left(\frac{-b}{2}\right)^2=\frac{b^2}{4}.$$

$$x^2-bx+\frac{b^2}{4}=2b^2+\frac{b^2}{4}$$

$$\left(x-\frac{b}{2}\right)^2=\frac{8b^2}{4}+\frac{b^2}{4}=\frac{9b^2}{4}$$

Apply the square root property.

$$x-\frac{b}{2}=\pm\sqrt{\frac{9b^2}{4}}=\pm\frac{3b}{2}$$

$$x=\frac{b}{2}\pm\frac{3b}{2}$$

$$x=\frac{b}{2}+\frac{3b}{2} \quad \text{or} \quad x=\frac{b}{2}-\frac{3b}{2}$$

$$x=\frac{4b}{2}=2b \quad \text{or} \quad x=-\frac{2b}{2}=-b$$

The solutions are $2b$ and $-b$, and the solution set is $\{-b,\ 2b\}$.

**72.** $$x^2-bx=6b^2$$

$$x^2-bx \quad =6b^2$$

Since $-b$ is the linear coefficient, we add

$$\left(\frac{-b}{2}\right)^2=\frac{b^2}{4}.$$

$$x^2-bx+\frac{b^2}{4}=6b^2+\frac{b^2}{4}$$

$$\left(x-\frac{b}{2}\right)^2=\frac{24b^2}{4}+\frac{b^2}{4}=\frac{25b^2}{4}$$

Apply the square root property.

$$x-\frac{b}{2}=\pm\sqrt{\frac{25b^2}{4}}=\pm\frac{5b}{2}$$

$$x=\frac{b}{2}\pm\frac{5b}{2}$$

$x = \frac{b}{2} + \frac{5b}{2}$ or $x = \frac{b}{2} - \frac{5b}{2}$

$x = \frac{6b}{2} = 3b$ or $x = -\frac{4b}{2} = -2b$

The solutions are $3b$ and $-2b$, and the solution set is $\{-2b, 3b\}$.

**73.** $2880 = 2000(1+r)^2$

$\frac{2880}{2000} = (1+r)^2$

$1.44 = (1+r)^2$

Apply the square root property.

$1+r = \pm\sqrt{1.44}$

$1+r = \pm 1.2$

$r = -1 \pm 1.2$

$r = -1+1.2$ or $-1-1.2$

$r = 0.2$ or $-2.2$

We reject –2.2 because we cannot have a negative interest rate. The solution is 0.2 and we conclude that the annual interest rate is 20%.

**75.** $1445 = 1280(1+r)^2$

$\frac{1445}{1280} = (1+r)^2$

$1.12890625 = (1+r)^2$

Apply the square root property.

$1+r = \pm\sqrt{1.12890625}$

$1+r = \pm 1.0625$

$r = -1 \pm 1.0625$

$r = -1+1.0625$ or $-1-1.0625$

$r = 0.0625$ or $-2.0625$

We reject –2.0625 because we cannot have a negative interest rate. The solution is 0.0625 and we conclude that the annual interest rate is 6.25%.

**77.** $92,000 = 62.2x^2 + 7000$

$85,000 = 62.2x^2$

$\frac{85,000}{62.2} = x^2$

Apply the square root property.

$x = \pm\sqrt{\frac{85,000}{62.2}} \approx \pm 37$

We disregard –37 because we can't have a negative number of years. The solution is 37 and we conclude that there will be 92,000 multinational corporation in 1970 + 37 = 2007.

**79.** $4800 = 16t^2$

$\frac{4800}{16} = t^2$

$300 = t^2$

Apply the square root property.

$t = \pm\sqrt{300}$

$t = \pm 10\sqrt{3} \approx \pm 17.3$

We disregard –17.3 because we can't have a negative time measurement. The solution is 17.3 and we conclude that the sky diver was in a free fall for $10\sqrt{3}$ or approximately 17.3 seconds.

**81.**

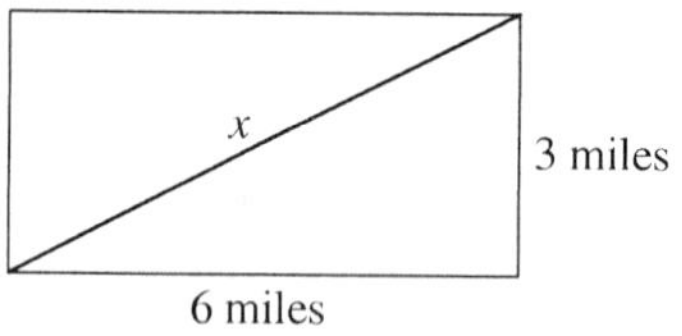


$x^2 = 6^2 + 3^2 = 36 + 9 = 45$

Apply the square root property.

$x = \pm\sqrt{45} = \pm\sqrt{9 \cdot 5} = \pm 3\sqrt{5}$

We disregard $-3\sqrt{5}$ because we can't have a negative length measurement. The solution is $3\sqrt{5}$ and we conclude that the pedestrian route is $3\sqrt{5}$ or approximately 6.7 miles long.

**83.** $x^2 + 10^2 = 30^2$

$x^2 + 100 = 900$

$x^2 = 800$

Apply the square root property.

$x = \pm\sqrt{800} = \pm\sqrt{400 \cdot 2} = \pm 20\sqrt{2}$

We disregard $-20\sqrt{2}$ because we can't have a negative length measurement. The solution is $20\sqrt{2}$. We conclude that the ladder reaches $20\sqrt{2}$ feet, or approximately 28.3 feet, up the house.

**85.**

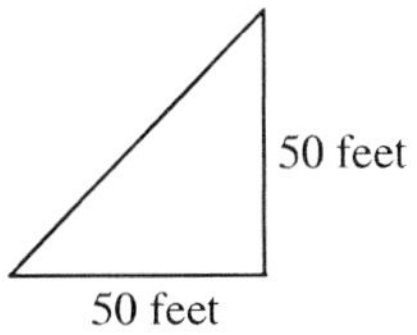


$50^2 + 50^2 = x^2$

$2500 + 2500 = x^2$

$5000 = x^2$

Apply the square root property.

$x = \pm\sqrt{5000}$

$x = \pm\sqrt{2500 \cdot 2} = \pm 50\sqrt{2} \approx \pm 70.7$

We disregard $-50\sqrt{2}$ because we cannot have a negative length measurement. The solution is $50\sqrt{2}$. We conclude that a supporting wire of $50\sqrt{2}$ feet, or approximately 70.7 feet, is required.

**87.** $A = lw$

$196 = (x+2+2)(x+2+2)$

$196 = (x+4)(x+4)$

$196 = x^2 + 8x + 16$

$180 = x^2 + 8x$

$x^2 + 8x \quad = 180$

Since $b = 8$, we add

$\left(\frac{b}{2}\right)^2 = \left(\frac{8}{2}\right)^2 = 4^2 = 16.$

$x^2 + 8x + 16 = 180 + 16$

$(x+4)^2 = 196$

Apply the square root property.

$x + 4 = \pm\sqrt{196} = \pm 14$

$x = -4 \pm 14$

$x = -4 + 14 = 10$ or $x = -4 - 14 = -18$

We disregard $-18$ because we can't have a negative length measurement. The solution is 10. We conclude that the length of the original square is 10 meters.

**89. – 95.** Answers will vary.

**97.** $4 - (x+1)^2 = 0$

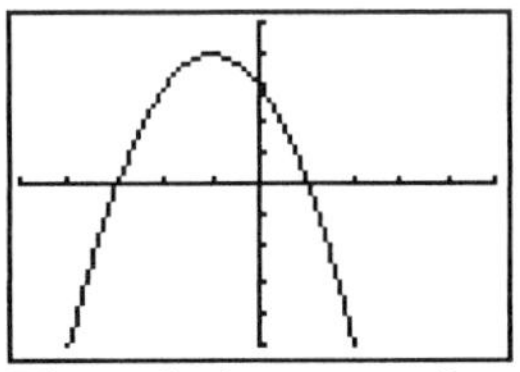

The solutions are –3 and 1, and the solution set is $\{-3, 1\}$.

Check:

| $x = -3$ | $x = 1$ |
|---|---|
| $4 - (-3+1)^2 = 0$ | $4 - (1+1)^2 = 0$ |
| $4 - (-2)^2 = 0$ | $4 - 2^2 = 0$ |
| $4 - 4 = 0$ | $4 - 4 = 0$ |
| $0 = 0$ | $0 = 0$ |
| True | True |

**99.** Answers will vary

**101.** $\frac{x^2}{a^2}+\frac{y^2}{b^2}=1$

$\frac{y^2}{b^2}=1-\frac{x^2}{a^2}$

$y^2=b^2\left(1-\frac{x^2}{a^2}\right)$

$\sqrt{y^2}=\pm\sqrt{b^2\left(1-\frac{x^2}{a^2}\right)}$

$y=\pm b\sqrt{1-\frac{x^2}{a^2}}$

$y=\pm b\sqrt{\frac{a^2}{a^2}-\frac{x^2}{a^2}}$

$y=\pm b\sqrt{\frac{a^2-x^2}{a^2}}=\pm\frac{b\sqrt{a^2-x^2}}{a}$

**103.** $x^2+bx+c=0$

$x^2+bx \quad =-c$

$x^2+bx+\frac{b^2}{4}=-c+\frac{b^2}{4}$

$\left(x+\frac{b}{2}\right)^2=-c+\frac{b^2}{4}$

$x+\frac{b}{2}=\pm\sqrt{-c+\frac{b^2}{4}}$

$x=-\frac{b}{2}\pm\sqrt{-\frac{4c}{4}+\frac{b^2}{4}}$

$x=-\frac{b}{2}\pm\sqrt{-\frac{4c+b^2}{4}}$

$x=-\frac{b}{2}\pm\frac{\sqrt{-4c+b^2}}{2}$

$x=\frac{-b\pm\sqrt{b^2-4c}}{2}$

**105.** $4x-2-3[4-2(3-x)]$

$=4x-2-3[4-6+2x]$

$=4x-2-3[-2+2x]$

$=4x-2+6-6x$

$=4-2x$

**106.** $1-8x^3=1^3-(2x)^3$

$=(1-2x)(1^2+1\cdot 2x+(2x)^2)$

$=(1-2x)(1+2x+4x^2)$

**107.** $(x^4-5x^3+2x^2-6)\div(x-3)$

| 3 | 1 | −5 | 2 | 0 | −6 |
|---|---|---|---|---|---|
| | | 3 | −6 | −12 | −36 |
| | 1 | −2 | −4 | −12 | −42 |

$(x^4-5x^3+2x^2-6)\div(x-3)$

$=x^3-2x^2-4x-12-\frac{42}{x-3}$

## 8.2 Exercise Set

**1.** $x^2+8x+12=0$

$a=1 \quad b=8 \quad c=12$

$x=\frac{-8\pm\sqrt{8^2-4(1)(12)}}{2(1)}$

$=\frac{-8\pm\sqrt{64-48}}{2}$

$=\frac{-8\pm\sqrt{16}}{2}=\frac{-8\pm 4}{2}$

Evaluate the expression to obtain two solutions.

$x=\frac{-8-4}{2}=-6$ or $x=\frac{-8+4}{2}=-2$

The solutions are –2 and –6 and the solution set is $\{-6,-2\}$.

**3.** $2x^2-7x=-5$

$2x^2-7x+5=0$

$a=2 \quad b=-7 \quad c=5$

$$x = \frac{-(-7) \pm \sqrt{(-7)^2 - 4(2)(5)}}{2(2)}$$

$$= \frac{7 \pm \sqrt{49-40}}{4} = \frac{7 \pm \sqrt{9}}{4} = \frac{7 \pm 3}{4}$$

Evaluate the expression to obtain two solutions.

$x = \frac{7-3}{4} = 1$ or $x = \frac{7+3}{4} = \frac{5}{2}$

The solutions are 1 and $\frac{5}{2}$ and the solution set is $\left\{1, \frac{5}{2}\right\}$.

**5.** $$x^2 + 3x - 20 = 0$$

$a = 1 \quad b = 3 \quad c = -20$

$$x = \frac{-3 \pm \sqrt{3^2 - 4(1)(-20)}}{2(1)}$$

$$= \frac{-3 \pm \sqrt{9-(-80)}}{2} = \frac{-3 \pm \sqrt{89}}{2}$$

The solutions are $\frac{-3 \pm \sqrt{89}}{2}$ and the solution set is $\left\{\frac{-3 \pm \sqrt{89}}{2}\right\}$.

**7.** $$3x^2 - 7x = 3$$

$$3x^2 - 7x - 3 = 0$$

$a = 3 \quad b = -7 \quad c = -3$

$$x = \frac{-(-7) \pm \sqrt{(-7)^2 - 4(3)(-3)}}{2(3)}$$

$$= \frac{7 \pm \sqrt{49-(-36)}}{6} = \frac{7 \pm \sqrt{85}}{6}$$

The solutions are $\frac{7 \pm \sqrt{85}}{6}$ and the solution set is $\left\{\frac{7 \pm \sqrt{85}}{6}\right\}$.

**9.** $$6x^2 = 2x + 1$$

$$6x^2 - 2x - 1 = 0$$

$a = 6 \quad b = -2 \quad c = -1$

$$x = \frac{-(-2) \pm \sqrt{(-2)^2 - 4(6)(-1)}}{2(6)}$$

$$= \frac{2 \pm \sqrt{4+24}}{12}$$

$$= \frac{2 \pm \sqrt{28}}{12}$$

$$= \frac{2 \pm 2\sqrt{7}}{12} = \frac{2\left(1 \pm \sqrt{7}\right)}{12} = \frac{1 \pm \sqrt{7}}{6}$$

The solutions are $\frac{1 \pm \sqrt{7}}{6}$ and the solution set is $\left\{\frac{1 \pm \sqrt{7}}{6}\right\}$.

**11.** $$4x^2 - 3x = -6$$

$$4x^2 - 3x + 6 = 0$$

$a = 4 \quad b = -3 \quad c = 6$

$$x = \frac{-(-3) \pm \sqrt{(-3)^2 - 4(4)(6)}}{2(4)}$$

$$= \frac{3 \pm \sqrt{9-96}}{8}$$

$$= \frac{3 \pm \sqrt{-87}}{8}$$

$$= \frac{3 \pm \sqrt{87(-1)}}{8}$$

$$= \frac{3 \pm \sqrt{87}i}{8} = \frac{3}{8} \pm \frac{\sqrt{87}}{8}i$$

The solutions are $\frac{3}{8} \pm \frac{\sqrt{87}}{8}i$ and the solution set is $\left\{\frac{3}{8} \pm \frac{\sqrt{87}}{8}i\right\}$.

**13.** $x^2-4x+8=0$

$a=1 \quad b=-4 \quad c=8$

$$x=\frac{-(-4)\pm\sqrt{(-4)^2-4(1)(8)}}{2(1)}$$

$$=\frac{4\pm\sqrt{16-32}}{2}$$

$$=\frac{4\pm\sqrt{-16}}{2}$$

$$=\frac{4\pm 4i}{2}=\frac{4}{2}\pm\frac{4}{2}i=2\pm 2i$$

The solutions are $2\pm 2i$ and the solution set is $\{2\pm 2i\}$.

**15.** $3x^2=8x-7$

$3x^2-8x+7=0$

$a=3 \quad b=-8 \quad c=7$

$$x=\frac{-(-8)\pm\sqrt{(-8)^2-4(3)(7)}}{2(3)}$$

$$=\frac{8\pm\sqrt{64-84}}{6}$$

$$=\frac{8\pm\sqrt{-20}}{6}$$

$$=\frac{8\pm\sqrt{4\cdot 5(-1)}}{6}$$

$$=\frac{8\pm 2\sqrt{5}i}{6}=\frac{8}{6}\pm\frac{2}{6}\sqrt{5}i=\frac{4}{3}\pm\frac{\sqrt{5}}{3}i$$

The solutions are $\frac{4}{3}\pm\frac{\sqrt{5}}{3}i$ and the solution set is $\left\{\frac{4}{3}\pm\frac{\sqrt{5}}{3}i\right\}$.

**17.** $2x(x-2)=x+12$

$2x^2-4x=x+12$

$2x^2-5x-12=0$

$a=2 \quad b=-5 \quad c=-12$

$$x=\frac{-(-5)\pm\sqrt{(-5)^2-4(2)(-12)}}{2(2)}$$

$$=\frac{5\pm\sqrt{25+96}}{4}=\frac{5\pm\sqrt{121}}{4}=\frac{5\pm 11}{4}$$

Evaluate the expression to obtain two solutions.

$$x=\frac{5-11}{4}=-\frac{3}{2} \quad \text{or} \quad x=\frac{5+11}{4}=4$$

The solutions are $-\frac{3}{2}$ and 4, and the solution set is $\left\{-\frac{3}{2},\ 4\right\}$.

**19.** $x^2+8x+3=0$

$a=1 \quad b=8 \quad c=3$

$b^2-4ac=8^2-4(1)(3)=64-12=52$

Since the discriminant is positive and not a perfect square, there are two irrational solutions.

**21.** $x^2+6x+8=0$

$a=1 \quad b=6 \quad c=8$

$b^2-4ac=(6)^2-4(1)(8)$

$=36-32=4$

Since the discriminant is greater than zero, there are two unequal real solutions. Also, since the discriminant is a perfect square, the solutions are rational.

**23.** $2x^2+x+3=0$

$a=2 \quad b=1 \quad c=3$

$b^2-4ac=1^2-4(2)(3)=1-24=-23$

Since the discriminant is negative, there are no real solutions. There are two imaginary solutions that are complex conjugates.

**25.** $2x^2 + 6x = 0$

$a = 2 \quad b = 6 \quad c = 0$

$b^2 - 4ac = (6)^2 - 4(1)(0)$

$= 36 - 0 = 36$

Since the discriminant is greater than zero, there are two unequal real solutions. Also, since the discriminant is a perfect square, the solutions are rational.

**27.** $5x^2 + 3 = 0$

$a = 5 \quad b = 0 \quad c = 3$

$b^2 - 4ac = 0^2 - 4(5)(3) = 0 - 60 = -60$

Since the discriminant is negative, there are no real solutions. There are two imaginary solutions that are complex conjugates.

**29.** $9x^2 = 12x - 4$

$9x^2 - 12x + 4 = 0$

$a = 9 \quad b = -12 \quad c = 4$

$b^2 - 4ac = (-12)^2 - 4(9)(4)$

$= 144 - 144 = 0$

Since the discriminant is zero, there is one repeated rational solution.

**31.** $3x^2 - 4x = 4$

$3x^2 - 4x - 4 = 0$

$(3x + 2)(x - 2) = 0$

Apply the zero product principle.

$3x + 2 = 0 \quad \text{or} \quad x - 2 = 0$

$3x = -2 \qquad x = 2$

$x = -\frac{2}{3}$

The solutions are $-\frac{2}{3}$ and $2$, and the solution set is $\left\{-\frac{2}{3}, 2\right\}$.

**33.** $x^2 - 2x = 1$

Since $b = -2$, we add

$\left(\frac{b}{2}\right)^2 = \left(\frac{-2}{2}\right)^2 = (-1)^2 = 1.$

$x^2 - 2x + 1 = 1 + 1$

$(x - 1)^2 = 2$

Apply the square root principle.

$x - 1 = \pm\sqrt{2}$

$x = 1 \pm \sqrt{2}$

The solutions are $1 \pm \sqrt{2}$, and the solution set is $\left\{1 \pm \sqrt{2}\right\}$.

**35.** $(2x - 5)(x + 1) = 2$

$2x^2 + 2x - 5x - 5 = 2$

$2x^2 - 3x - 7 = 0$

Apply the quadratic formula.

$a = 2 \quad b = -3 \quad c = -7$

$$x = \frac{-(-3) \pm \sqrt{(-3)^2 - 4(2)(-7)}}{2(2)}$$

$$= \frac{3 \pm \sqrt{9 - (-56)}}{4} = \frac{3 \pm \sqrt{65}}{4}$$

The solutions are $\frac{3 \pm \sqrt{65}}{4}$, and the solution set is $\left\{\frac{3 \pm \sqrt{65}}{4}\right\}$.

**37.** $(3x - 4)^2 = 16$

Apply the square root property.

$3x - 4 = \sqrt{16} \quad \text{or} \quad 3x - 4 = -\sqrt{16}$

$3x - 4 = 4 \qquad 3x - 4 = -4$

$3x = 8 \qquad 3x = 0$

$x = \frac{8}{3} \qquad x = 0$

The solutions are $\frac{8}{3}$ and 0, and the solution set is $\left\{0, \frac{8}{3}\right\}$.

**39.** $\frac{x^2}{2}+2x+\frac{2}{3}=0$

Multiply both sides of the equation by 6 to clear fractions.

$3x^2+12x+4=0$

Apply the quadratic formula.

$a=3 \quad b=12 \quad c=4$

$$x=\frac{-12\pm\sqrt{12^2-4(3)(4)}}{2(3)}$$
$$=\frac{-12\pm\sqrt{144-48}}{6}$$
$$=\frac{-12\pm\sqrt{96}}{6}$$
$$=\frac{-12\pm\sqrt{16\cdot 6}}{6}$$
$$=\frac{-12\pm 4\sqrt{6}}{6}$$
$$=\frac{2\left(-6\pm 2\sqrt{6}\right)}{6}=\frac{-6\pm 2\sqrt{6}}{3}$$

The solutions are $\frac{-6\pm 2\sqrt{6}}{3}$, and the solution set is $\left\{\frac{-6\pm 2\sqrt{6}}{3}\right\}$.

**41.** $(3x-2)^2=10$

Apply the square root property.

$$3x-2=\pm\sqrt{10}$$
$$3x=2\pm\sqrt{10}$$
$$x=\frac{2\pm\sqrt{10}}{3}$$

The solutions are $\frac{2\pm\sqrt{10}}{3}$, and the solution set is $\left\{\frac{2\pm\sqrt{10}}{3}\right\}$.

**43.** $\frac{1}{x}+\frac{1}{x+2}=\frac{1}{3}$

The LCD is $3x(x+2)$.

$$3x(x+2)\left(\frac{1}{x}+\frac{1}{x+2}\right)=3x(x+2)\left(\frac{1}{3}\right)$$
$$3(x+2)+3x=x(x+2)$$
$$3x+6+3x=x^2+2x$$
$$0=x^2-4x-6$$

Apply the quadratic formula.

$a=1 \quad b=-4 \quad c=-6$

$$x=\frac{-(-4)\pm\sqrt{(-4)^2-4(1)(-6)}}{2(1)}$$
$$=\frac{4\pm\sqrt{16-(-24)}}{2}$$
$$=\frac{4\pm\sqrt{40}}{2}=\frac{4\pm 2\sqrt{10}}{2}=2\pm\sqrt{10}$$

The solutions are $2\pm\sqrt{10}$, and the solution set is $\left\{2\pm\sqrt{10}\right\}$.

**45.**
$$(2x-6)(x+2)=5(x-1)-12$$
$$2x^2+4x-6x-12=5x-5-12$$
$$2x^2-2x-12=5x-17$$
$$2x^2-7x+5=0$$
$$(2x-5)(x-1)=0$$

Apply the zero product principle.

$2x-5=0 \quad \text{or} \quad x-1=0$

$2x=5 \qquad x=1$

$x=\frac{5}{2}$

The solutions are 1 and $\frac{5}{2}$, and the solution set is $\left\{1,\frac{5}{2}\right\}$.

**47.** Because the solution set is $\{-3,5\}$, we have

$x=-3$ or $x=5$

$x+3=0$ $\quad$ $x-5=0.$

Use the zero-product principle in reverse.

$$(x+3)(x-5)=0$$
$$x^2-5x+3x-15=0$$
$$x^2-2x-15=0$$

**49.** Because the solution set is $\left\{-\frac{2}{3},\frac{1}{4}\right\}$, we have

$x=-\frac{2}{3}$ or $x=\frac{1}{4}$

$3x=-2$ $\quad$ $4x=1$

$3x+2=0$ $\quad$ $4x-1=0.$

Use the zero-product principle in reverse.

$$(3x+2)(4x-1)=0$$
$$12x^2-3x+8x-2=0$$
$$12x^2+5x-2=0$$

**51.** Because the solution set is $\{-6i,6i\}$, we have

$x=6i$ or $x=-6i$

$x-6i=0$ $\quad$ $x+6i=0.$

Use the zero-product principle in reverse.

$$(x-6i)(x+6i)=0$$
$$x^2+\cancel{6i}-\cancel{6i}-36i^2=0$$
$$x^2-36(-1)=0$$
$$x^2+36=0$$

**53.** Because the solution set is $\{-\sqrt{2},\sqrt{2}\}$, we have

$x=\sqrt{2}$ or $x=-\sqrt{2}$

$x-\sqrt{2}=0$ $\quad$ $x+\sqrt{2}=0.$

Use the zero-product principle in reverse.

$$\left(x-\sqrt{2}\right)\left(x+\sqrt{2}\right)=0$$
$$x^2+\cancel{x\sqrt{2}}-\cancel{x\sqrt{2}}-2=0$$
$$x^2-2=0$$

**55.** Because the solution set is $\{-2\sqrt{5},\ 2\sqrt{5}\}$ we have

$x=2\sqrt{5}$ or $x=-2\sqrt{5}$

$x-2\sqrt{5}=0$ $\quad$ $x+2\sqrt{5}=0.$

Use the zero-product principle in reverse.

$$\left(x-2\sqrt{5}\right)\left(x+2\sqrt{5}\right)=0$$
$$x^2+\cancel{2x\sqrt{5}}-\cancel{2x\sqrt{5}}-4\cdot5=0$$
$$x^2-20=0$$

**57.** Because the solution set is $\{1+i,\ 1-i\}$, we have

$x=1+i$ or $x=1-i$

$x-(1+i)=0$ $\quad$ $x-(1-i)=0.$

Use the zero-product principle in reverse.

$$[x-(1+i)][x-(1-i)]=0$$
$$x^2-x(1-i)-x(1+i)+(1+i)(1-i)=0$$
$$x^2-x+\cancel{xi}-x-\cancel{xi}+1-i^2=0$$
$$x^2-x-x+1-(-1)=0$$
$$x^2-2x+2=0$$

**59.** Because the solution set is $\{1+\sqrt{2}, 1-\sqrt{2}\}$, we have

$$x = 1+\sqrt{2} \quad \text{or} \quad x = 1-\sqrt{2}$$

$$x-(1+\sqrt{2}) = 0 \qquad x-(1-\sqrt{2}) = 0.$$

Use the zero-product principle in reverse.

$$(x-(1+\sqrt{2}))(x-(1-\sqrt{2})) = 0$$

$$x^2 - x(1-\sqrt{2}) - x(1+\sqrt{2}) + (1+\sqrt{2})(1-\sqrt{2}) = 0$$

$$x^2 - x + \cancel{x\sqrt{2}} - x - \cancel{x\sqrt{2}} + 1 - 2 = 0$$

$$x^2 - 2x - 1 = 0$$

**61.** b. If the solutions are imaginary number, then the graph will not cross the $x$-axis.

**62.** c. If the discriminant is 0, then the equation has one real solution (a repeated solution). Thus, the graph only touches the $x$-axis at one point.

**63.** a. The equation has two non-integer solutions $3\pm\sqrt{2}$, so the graph crosses the $x$-axis at $3-\sqrt{2}$ and $3+\sqrt{2}$.

**64.** d. The equation has two integer solutions, so the graph has two integer $x$-intercepts.

**65.** Let $x$ = the number.

$$x^2 - (6+2x) = 0$$

$$x^2 - 2x - 6 = 0$$

Apply the quadratic formula.

$a = 1 \quad b = -2 \quad c = -6$

$$x = \frac{-(-2) \pm \sqrt{(-2)^2 - 4(1)(-6)}}{2(1)}$$

$$= \frac{2 \pm \sqrt{4-(-24)}}{2}$$

$$= \frac{2 \pm \sqrt{28}}{2}$$

$$= \frac{2 \pm \sqrt{4 \cdot 7}}{2} = \frac{2 \pm 2\sqrt{7}}{2} = 1 \pm \sqrt{7}$$

We disregard $1-\sqrt{7}$ because it is negative, and we are looking for a positive number. Thus, the number is $1+\sqrt{7}$.

**66.** Let $x$ = the number.

$$2x^2 - (1+2x) = 0$$

$$2x^2 - 2x - 1 = 0$$

Apply the quadratic formula.

$a = 2 \quad b = -2 \quad c = -1$

$$x = \frac{-(-2) \pm \sqrt{(-2)^2 - 4(2)(-1)}}{2(2)}$$

$$= \frac{2 \pm \sqrt{4-(-8)}}{4}$$

$$= \frac{2 \pm \sqrt{12}}{4} = \frac{2 \pm \sqrt{4 \cdot 3}}{4} = \frac{2 \pm 2\sqrt{3}}{4} = \frac{1 \pm \sqrt{3}}{2}$$

We disregard $\frac{1+\sqrt{3}}{2}$ because it is positive, and we are looking for a negative number. The number is $\frac{1-\sqrt{3}}{2}$.

**67.**
$$\frac{1}{x^2-3x+2}=\frac{1}{x+2}+\frac{5}{x^2-4}$$
$$\frac{1}{(x-1)(x-2)}=\frac{1}{x+2}+\frac{5}{(x+2)(x-2)}$$

Multiply both sides of the equation by the least common denominator $(x-1)(x-2)(x+2)$:

$$(x-1)(x-2)(x+2)\left[\frac{1}{(x-1)(x-2)}\right]=(x-1)(x-2)(x+2)\left[\frac{1}{x+2}+\frac{5}{(x-2)(x+2)}\right]$$
$$x+2=(x-1)(x-2)+5(x-1)$$
$$x+2=x^2-2x-x+2+5x-5$$
$$x+2=x^2+2x-3$$
$$0=x^2+x-5$$

Apply the quadratic formula: $a=1 \quad b=1 \quad c=-5$.

$$x=\frac{-1\pm\sqrt{1^2-4(1)(-5)}}{2(1)}=\frac{-1\pm\sqrt{1-(-20)}}{2}=\frac{-1\pm\sqrt{21}}{2}$$

The solutions are $\frac{-1\pm\sqrt{21}}{2}$, and the solution set is $\left\{\frac{-1\pm\sqrt{21}}{2}\right\}$.

**68.**
$$\frac{x-1}{x-2}+\frac{x}{x-3}=\frac{1}{x^2-5x+6}$$
$$\frac{x-1}{x-2}+\frac{x}{x-3}=\frac{1}{(x-2)(x-3)}$$

Multiply both sides of the equation by the least common denominator $(x-2)(x-3)$:

$$(x-2)(x-3)\left[\frac{x-1}{x-2}+\frac{x}{x-3}\right]=(x-2)(x-3)\left[\frac{1}{(x-2)(x-3)}\right]$$
$$(x-3)(x-1)+x(x-2)=1$$
$$x^2-x-3x+3+x^2-2x=1$$
$$2x^2-6x+3=1$$
$$2x^2-6x+2=0$$

Apply the quadratic formula: $a=2 \quad b=-6 \quad c=2$.

$$x=\frac{-(-6)\pm\sqrt{(-6)^2-4(2)(2)}}{2(2)}=\frac{6\pm\sqrt{36-16}}{4}=\frac{6\pm\sqrt{20}}{4}=\frac{6\pm\sqrt{4\cdot5}}{4}=\frac{6\pm2\sqrt{5}}{4}=\frac{3\pm\sqrt{5}}{2}$$

The solutions are $\frac{3\pm\sqrt{5}}{2}$, and the solution set is $\left\{\frac{3\pm\sqrt{5}}{2}\right\}$.

**69.** $\sqrt{2}x^2+3x-2\sqrt{2}=0$

Apply the quadratic formula:

$a=\sqrt{2}\quad b=3\quad c=-2\sqrt{2}$

$$x=\frac{-3\pm\sqrt{3^2-4\left(\sqrt{2}\right)\left(-2\sqrt{2}\right)}}{2\left(\sqrt{2}\right)}$$

$$=\frac{-3\pm\sqrt{9-(-16)}}{2\sqrt{2}}$$

$$=\frac{-3\pm\sqrt{25}}{2\sqrt{2}}=\frac{-3\pm 5}{2\sqrt{2}}$$

Evaluate the expression to obtain two solutions.

$$x=\frac{-3-5}{2\sqrt{2}} \quad\text{or}\quad x=\frac{-3+5}{2\sqrt{2}}$$

$$=\frac{-8}{2\sqrt{2}}\cdot\frac{\sqrt{2}}{\sqrt{2}} \qquad =\frac{2}{2\sqrt{2}}\cdot\frac{\sqrt{2}}{\sqrt{2}}$$

$$=\frac{-8\sqrt{2}}{4} \qquad =\frac{2\sqrt{2}}{4}$$

$$=-2\sqrt{2} \qquad =\frac{\sqrt{2}}{2}$$

The solutions are $-2\sqrt{2}$ and $\frac{\sqrt{2}}{2}$, and the solution set is $\left\{-2\sqrt{2},\ \frac{\sqrt{2}}{2}\right\}$.

**70.** $\sqrt{3}x^2+6x+7\sqrt{3}=0$

Apply the quadratic formula:

$a=\sqrt{3}\quad b=6\quad c=7\sqrt{3}$

$$x=\frac{-6\pm\sqrt{6^2-4\left(\sqrt{3}\right)\left(7\sqrt{3}\right)}}{2\left(\sqrt{3}\right)}$$

$$=\frac{-6\pm\sqrt{36-84}}{2\sqrt{3}}$$

$$=\frac{-6\pm\sqrt{-48}}{2\sqrt{3}}$$

$$=\frac{-6\pm\sqrt{16\cdot 3\cdot(-1)}}{2\sqrt{3}}$$

$$=\frac{-6\pm 4\sqrt{3}i}{2\sqrt{3}}$$

$$=\frac{-6}{2\sqrt{3}}\pm\frac{4\sqrt{3}i}{2\sqrt{3}}=-\sqrt{3}\pm 2i$$

The solutions are $-\sqrt{3}\pm 2i$, and the solution set is $\left\{-\sqrt{3}\pm 2i\right\}$.

**71.** $\left|x^2+2x\right|=3$

$$x^2+2x=-3 \quad\text{or}\quad x^2+2x=3$$

$$x^2+2x+3=0 \qquad x^2+2x-3=0$$

Apply the quadratic formula to solve $x^2+2x+3=0$:

$a=1\quad b=2\quad c=3$.

$$x=\frac{-2\pm\sqrt{2^2-4(1)(3)}}{2(1)}$$

$$=\frac{-2\pm\sqrt{4-12}}{2}$$

$$=\frac{-2\pm\sqrt{-8}}{2}$$

$$=\frac{-2\pm\sqrt{2\cdot 4\cdot(-1)}}{2}$$

$$=\frac{-2\pm 2\sqrt{2}i}{2}=-1\pm\sqrt{2}i$$

Apply the zero product principle to solve $x^2+2x-3=0$:

$(x+3)(x-1)=0$

$x+3=0 \quad\text{or}\quad x-1=0$

$x=-3 \quad\text{or}\quad x=1$

The solutions are $-1\pm\sqrt{2}i$, $-3$, and 1, and the solution set is $\left\{-3,\ 1,\ -1\pm\sqrt{2}i\right\}$.

**72.** $|x^2 + 3x| = 2$

$x^2 + 3x = -2$ or $x^2 + 3x = 2$

$x^2 + 3x + 2 = 0$ $\quad$ $x^2 + 3x - 2 = 0$

Apply the zero product principle to solve $x^2 + 3x + 2 = 0$:

$(x+2)(x+1) = 0$

$x + 2 = 0$ or $x + 1 = 0$

$x = -2$ or $x = -1$

Apply the quadratic formula to solve $x^2 + 3x - 2 = 0$:

$a = 1 \quad b = 3 \quad c = -2$.

$$x = \frac{-3 \pm \sqrt{3^2 - 4(1)(-2)}}{2(1)}$$

$$= \frac{-3 \pm \sqrt{9-(-8)}}{2} = \frac{-3 \pm \sqrt{17}}{2}$$

The solutions are $-2$, $-1$, and $\frac{-3 \pm \sqrt{17}}{2}$, and the solution set is $\left\{-2,\ -1,\ \frac{-3 \pm \sqrt{17}}{2}\right\}$.

**73.** $f(x) = 0.013x^2 - 1.19x + 28.24$

$3 = 0.013x^2 - 1.19x + 28.24$

$0 = 0.013x^2 - 1.19x + 25.24$

Apply the quadratic formula:

$a = 0.013 \quad b = -1.19 \quad c = 25.24$

$$x = \frac{-(-1.19) \pm \sqrt{(-1.19)^2 - 4(0.013)(25.24)}}{2(0.013)}$$

$$= \frac{1.19 \pm \sqrt{1.4161 - 1.31248}}{0.026}$$

$$= \frac{1.19 \pm \sqrt{0.10362}}{0.026}$$

$$\approx \frac{1.19 \pm 0.32190}{0.026}$$

$$\approx 58.15 \text{ or } 33.39$$

The solutions are approximately 33.39 and 58.15. Thus, 33 year olds and 58 year olds are expected to be in 3 fatal crashes per 100 million miles driven. The function models the actual data well.

**75.** Let $y_1 = -0.01x^2 + 0.7x + 6.1$

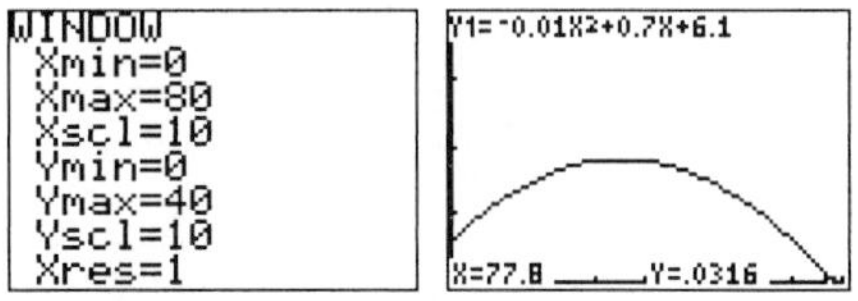


Using the TRACE feature, we find that the height of the shot put is approximately 0 feet when the distance is 77.8 feet. Graph (b) shows the shot' path.

**77.** Let $x$ = the width of the rectangle;
$x + 4$ = the length of the rectangle

$A = lw$

$8 = x(x+4)$

$0 = x^2 + 4x - 8$

Apply the quadratic formula:

$a = 1 \quad b = 4 \quad c = -8$

$$x = \frac{-4 \pm \sqrt{4^2 - 4(1)(-8)}}{2(1)}$$

$$= \frac{-4 \pm \sqrt{16-(-32)}}{2}$$

$$= \frac{-4 \pm \sqrt{48}}{2}$$

$$= \frac{-4 \pm 4\sqrt{3}}{2}$$

$$= -2 \pm 2\sqrt{3} \approx 1.5 \text{ or } -5.5$$

Disregard $-5.5$ because the width of a rectangle cannot be negative. Thus, the solution is 1.5, and we conclude that the rectangle's dimensions are 1.5 meters by $1.5 + 4 = 5.5$ meters.

**79.** Let $x$ = the length of the longer leg;
$x-1$ = the length of the shorter leg;
$x+7$ = the length of the hypotenuse.

$$x^2+(x-1)^2=(x+7)^2$$
$$x^2+x^2-2x+1=x^2+14x+49$$
$$2x^2-2x+1=x^2+14x+49$$
$$x^2-16x-48=0$$

Apply the quadratic formula:
$a=1 \quad b=-16 \quad c=-48$

$$x=\frac{-(-16)\pm\sqrt{(-16)^2-4(1)(-48)}}{2(1)}$$
$$=\frac{16\pm\sqrt{256-(-192)}}{2}$$
$$=\frac{16\pm\sqrt{448}}{2}$$
$$=\frac{16\pm 8\sqrt{7}}{2}$$
$$=8\pm 4\sqrt{7}\approx 18.6 \text{ or } -2.6$$

Disregard −2.6 because the length of a leg cannot be negative. The solution is 18.6, and we conclude that the lengths of the triangle's legs are approximately 18.6 inches and $18.6-1=17.6$ inches.

**81.** $x(20-2x)=13$

$$20x-2x^2=13$$
$$0=2x^2-20x+13$$

Apply the quadratic formula:
$a=2 \quad b=-20 \quad c=13$

$$x=\frac{-(-20)\pm\sqrt{(-20)^2-4(2)(13)}}{2(2)}$$
$$=\frac{20\pm\sqrt{400-104}}{4}$$
$$=\frac{20\pm\sqrt{296}}{4}$$
$$=\frac{20\pm 2\sqrt{74}}{4}=\frac{10\pm\sqrt{74}}{2}\approx 9.3 \text{ or } 0.7$$

A gutter with depth 9.3 or 0.7 inches will have a cross-sectional area of 13 square inches.

**83.** Let $x$ = the time for the first person to mow the yard alone;
$x+1$ = the time for the second person to mow the yard alone.

| | Fractional part of job completed in 1 hour | Time working together | Fractional part of job completed in 4 hour |
|---|---|---|---|
| 1st person | $\frac{1}{x}$ | 4 | $\frac{4}{x}$ |
| 2nd person | $\frac{1}{x+1}$ | 4 | $\frac{1}{x+1}$ |

$$\frac{4}{x}+\frac{4}{x+1}=1$$
$$x(x+1)\left(\frac{4}{x}+\frac{4}{x+1}\right)=x(x+1)1$$
$$4(x+1)+4x=x^2+x$$
$$4x+4+4x=x^2+x$$
$$0=x^2-7x-4$$

Apply the quadratic formula:
$a=1 \quad b=-7 \quad c=-4$

$$x=\frac{-(-7)\pm\sqrt{(-7)^2-4(1)(-4)}}{2(1)}$$
$$=\frac{7\pm\sqrt{49-(-16)}}{2}$$
$$=\frac{7\pm\sqrt{65}}{2}\approx 7.5 \text{ or } -0.5$$

Disregard −0.5 because time cannot be negative. Thus, the solution is 7.5, and we conclude that the first person can mow the lawn alone in 7.5 hours, and the second mow the yare alone in 7.5 + 1 = 8.5 hours.

**85. – 91.** Answers will vary.

**93.** Answers will vary. For example, consider Exercise 23.

$$2x^2 + x + 3 = 0$$

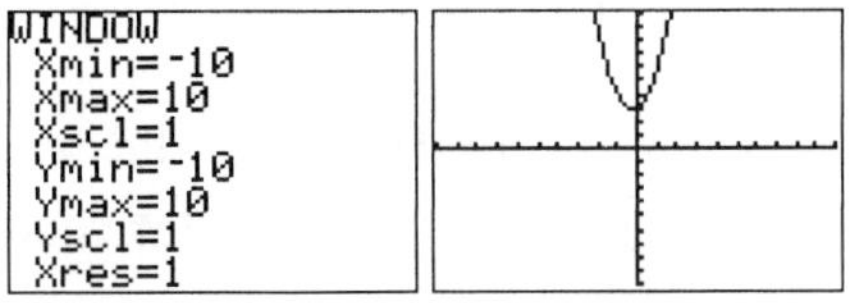


The graph does not cross the $x$–axis and we conclude that there are 2 imaginary solutions which are complex conjugates.

**95.** Statement **d** is true. Any quadratic equation that can be solved by completing the square can be solved by the quadratic formula.

Statement **a** is false. The quadratic equation is developed by completing the square and the zero product principle.

Statement **b** is false. Before using the quadratic equation to solve $5x^2 = 2x - 7$, the equation must be rewritten in standard form.

$$5x^2 = 2x - 7$$

$$5x^2 - 2x + 7 = 0$$

We now have, $a = 5,\ b = -2,$ and $c = 7$.

Statement **c** is false. The quadratic equation can be used to solve $x^2 - 9 = 0,$ with $a = 1,\ b = 0$ and $c = -9.$ (It would be easier, however, to factor or use the square root property.)

**97.** The dimensions of the pool are 12 meters by 8 meters. With the tile, the dimensions will be 12 + $2x$ meters by 8 + $2x$ meters. If we take the area of the pool with the tile and subtract the area of the pool without the tile, we are left with the area of the tile only.

$$(12 + 2x)(8 + 2x) - 12(8) = 120$$

$$\cancel{96} + 24x + 16x + 4x^2 - \cancel{96} = 120$$

$$4x^2 + 40x - 120 = 0$$

$$x^2 + 10x - 30 = 0$$

$$a = 1 \qquad b = 10 \qquad c = -30$$

$$x = \frac{-10 \pm \sqrt{10^2 - 4(1)(-30)}}{2(1)}$$

$$= \frac{-10 \pm \sqrt{100 + 120}}{2}$$

$$= \frac{-10 \pm \sqrt{220}}{2} \approx \frac{-10 \pm 14.8}{2}$$

Evaluate the expression to obtain two solutions.

$$x = \frac{-10 + 14.8}{2} \quad \text{or} \quad x = \frac{-10 - 14.8}{2}$$

$$x = \frac{4.8}{2} \qquad\qquad x = \frac{-24.8}{2}$$

$$x = 2.4 \qquad\qquad x = -12.4$$

We disregard –12.4 because we can't have a negative width measurement. The solution is 2.4 and we conclude that the width of the uniform tile border is 2.4 meters. This is more than the 2-meter requirement, so the tile meets the zoning laws.

**99.** $|5x+2|=|4-3x|$

$5x+2=4-3x$ or $5x+2=-(4-3x)$

$8x+2=4$ $\quad$ $5x+2=-4+3x$

$8x=2$ $\quad$ $2x+2=-4$

$x=\frac{1}{4}$ $\quad$ $2x=-6$

$x=-3$

The solutions are $\frac{1}{4}$ and $-3$, and the solution set is $\left\{-3,\frac{1}{4}\right\}$.

**100.** $\sqrt{2x-5}-\sqrt{x-3}=1$

$$\sqrt{2x-5}=\sqrt{x-3}+1$$
$$\left(\sqrt{2x-5}\right)^2=\left(\sqrt{x-3}+1\right)^2$$
$$2x-5=x-3+2\sqrt{x-3}+1$$
$$2x-5=x-2+2\sqrt{x-3}$$
$$x-3=2\sqrt{x-3}$$
$$(x-3)^2=\left(2\sqrt{x-3}\right)^2$$
$$x^2-6x+9=4(x-3)$$
$$x^2-6x+9=4x-12$$
$$x^2-10x+21=0$$
$$(x-7)(x-3)=0$$

Apply the zero product principle.

$x-7=0$ or $x-3=0$

$x=7$ $\quad$ $x=3$

Both check. The solutions are 3 and 7, and the solution set is $\{3,7\}$.

**101.** $$\frac{5}{\sqrt{3}+x}=\frac{5}{\sqrt{3}+x}\cdot\frac{\sqrt{3}-x}{\sqrt{3}-x}$$
$$=\frac{5\left(\sqrt{3}-x\right)}{3-x^2}=\frac{5\sqrt{3}-5x}{3-x^2}$$

## 8.3 Exercise Set

**1.** The vertex of the graph is the point $(1, 1)$. This means that the equation is $h(x)=(x-1)^2+1$.

**3.** The vertex of the graph is the point $(1,-1)$. This means that the equation is $j(x)=(x-1)^2-1$.

**5.** The vertex of the graph is the point $(0,-1)$. This means that the equation is $h(x)=(x-0)^2-1=x^2-1$.

**7.** The vertex of the graph is the point $(1, 0)$. This means that the equation is $f(x)=(x-1)^2+0$
$$=(x-1)^2=x^2-2x+1$$

**9.** $f(x)=2(x-3)^2+1$

The vertex is (3, 1).

**11.** $f(x)=-2(x+1)^2+5$

The vertex is $(-1,5)$.

**13.** $f(x)=2x^2-8x+3$

The $x$–coordinate of the vertex of the parabola is $-\frac{b}{2a}=-\frac{-8}{2(2)}=-\frac{-8}{4}=2$,

and the $y$–coordinate of the vertex of the parabola is

$$f\left(-\frac{b}{2a}\right)=f(2)=2(2)^2-8(2)+3$$
$$=2(4)-16+3$$
$$=8-16+3=-5.$$

The vertex is $(2,-5)$.

**15.** $f(x) = -x^2 - 2x + 8$

The $x$–coordinate of the vertex of the parabola is

$$-\frac{b}{2a} = -\frac{-2}{2(-1)} = -\frac{-2}{-2} = -1,$$

and the $y$–coordinate of the vertex of the parabola is

$$f\left(-\frac{b}{2a}\right) = f(-1)$$
$$= -(-1)^2 - 2(-1) + 8$$
$$= -1 + 2 + 8 = 9.$$

The vertex is $(-1, 9)$.

**17.** $f(x) = (x-4)^2 - 1$

Since $a = 1$ is positive, the parabola opens upward. The vertex of the parabola is $(h,k) = (4,-1)$. Replace $f(x)$ with 0 to find $x$–intercepts.

$0 = (x-4)^2 - 1$

$1 = (x-4)^2$

Apply the square root property.

$x - 4 = \pm\sqrt{1} = \pm 1$

$x = 4 \pm 1 = 5 \text{ or } 3$

The $x$–intercepts are 5 and 3.
Set $x = 0$ and solve for $y$ to obtain the $y$–intercept. $y = (0-4)^2 - 1 = 15$

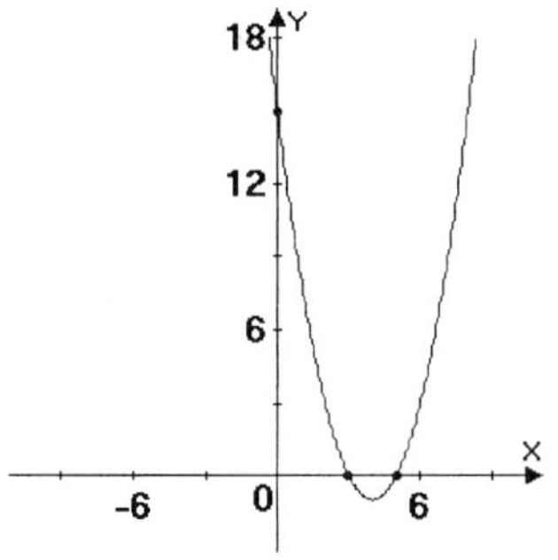


Axis of symmetry: $x = 4$.
Range: $\{y \mid y \geq -1\}$ or $[-1, \infty)$.

**19.** $f(x) = (x-1)^2 + 2$

Since $a = 1$ is positive, the parabola opens upward. The vertex of the parabola is $(h,k) = (1,2)$. Replace $f(x)$ with 0 to find $x$–intercepts.

$0 = (x-1)^2 + 2$

$-2 = (x-1)^2$

Because the solutions to the equation are imaginary, we know that there are no $x$–intercepts. Set $x = 0$ and solve for $y$ to obtain the $y$–intercept.

$y = (0-1)^2 + 2 = (-1)^2 + 2 = 1 + 2 = 3$

The $y$–intercept is 3.

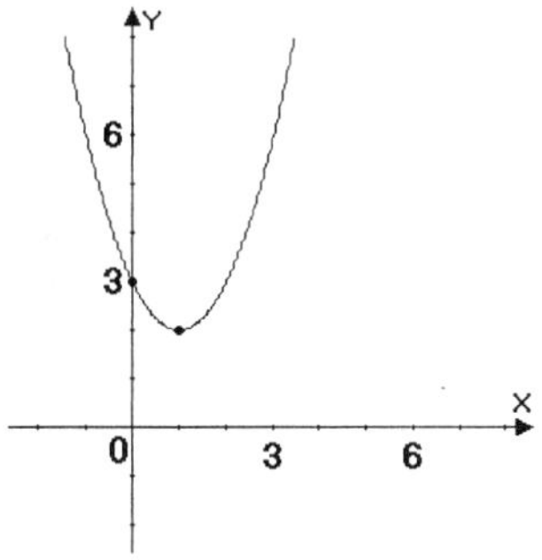


Axis of symmetry: $x = 1$.
Range: $\{y \mid y \geq 2\}$ or $[2, \infty)$.

**21.** $y - 1 = (x-3)^2$

$y = (x-3)^2 + 1$

$f(x) = (x-3)^2 + 1$

Since $a = 1$ is positive, the parabola opens upward. The vertex of the parabola is $(h,k) = (3,1)$. Replace $f(x)$ with 0 to find $x$–intercepts.

$0 = (x-3)^2 + 1$

$-1 = (x-3)^2$

Because the solutions to the equation are imaginary, we know that there are no $x$–intercepts. Set $x = 0$ and solve for $y$ to obtain the $y$–intercept.

$y = (0-3)^2 + 1 = (-3)^2 + 1 = 9 + 1 = 10$

The $y$–intercept is 10.

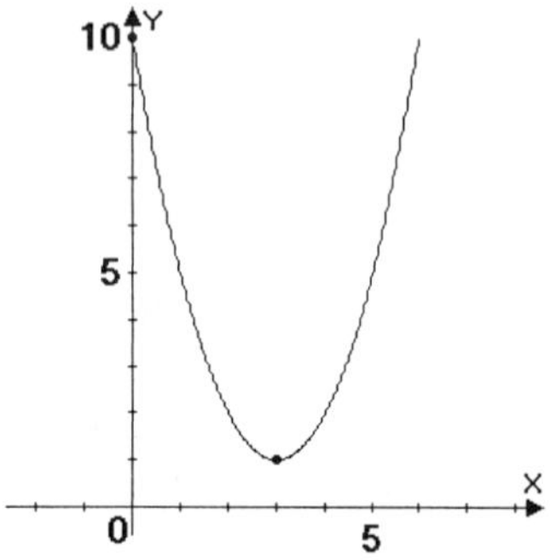


Axis of symmetry: $x = 3$.
Range: $\{y \mid y \geq 1\}$ or $[1, \infty)$.

**23.** $f(x) = 2(x+2)^2 - 1$

Since $a = 2$ is positive, the parabola opens upward. The vertex of the parabola is $(h,k) = (-2,-1)$. Replace $f(x)$ with 0 to find $x$–intercepts.

$0 = 2(x+2)^2 - 1$

$1 = 2(x+2)^2$

$\frac{1}{2} = (x+2)^2$

Apply the square root property.

$x + 2 = \pm\sqrt{\frac{1}{2}}$

$x = -2 \pm \sqrt{\frac{1}{2}}$

$x \approx -2 - \sqrt{\frac{1}{2}}$ or $-2 + \sqrt{\frac{1}{2}}$

$x \approx -2.7$ or $-1.3$

The $x$–intercepts are –1.3 and –2.7.
Set $x = 0$ and solve for $y$ to obtain the $y$–intercept.

$y = 2(0+2)^2 - 1$

$= 2(2)^2 - 1 = 2(4) - 1 = 8 - 1 = 7$

The $y$–intercept is 7.

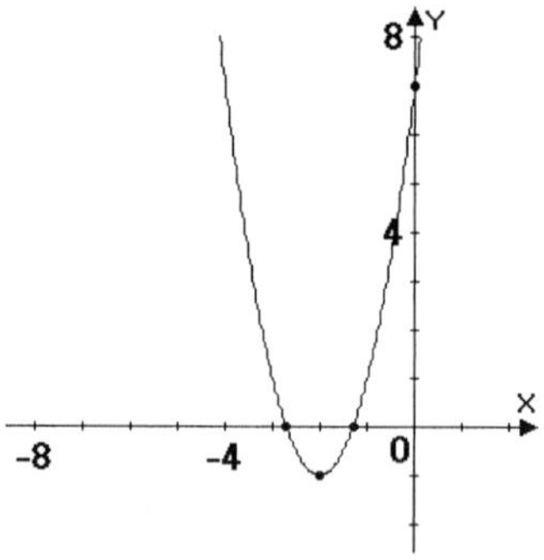


Axis of symmetry: $x = -2$.
Range: $\{y \mid y \geq -1\}$ or $[-1, \infty)$.

**25.** $f(x) = 4 - (x-1)^2$

$f(x) = -(x-1)^2 + 4$

Since $a = -1$ is negative, the parabola opens downward. The vertex of the parabola is $(h,k) = (1,4)$. Replace $f(x)$ with 0 to find $x$–intercepts.

$0 = -(x-1)^2 + 4$

$-4 = -(x-1)^2$

$4 = (x-1)^2$

Apply the square root property.

$\sqrt{4} = x - 1$ or $-\sqrt{4} = x - 1$

$2 = x - 1$ $\quad$ $-2 = x - 1$

$3 = x$ $\quad$ $-1 = x$

The $x$–intercepts are –1 and 3.
Set $x = 0$ and solve for $y$ to obtain the $y$–intercept.

$y = -(0-1)^2 + 4$

$= -(-1)^2 + 4 = -1 + 4 = 3$

The $y$–intercept is 3.

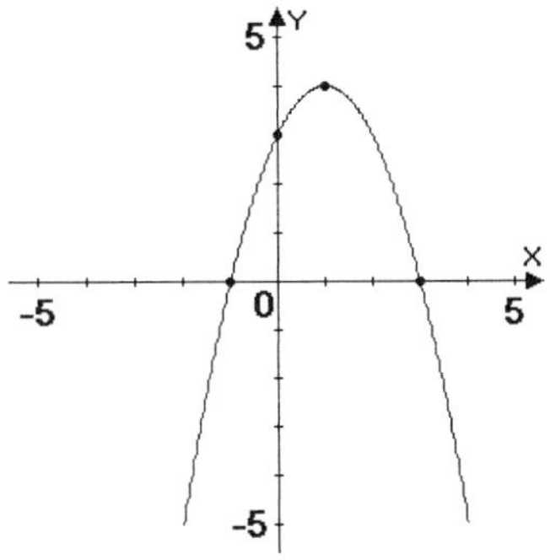


Axis of symmetry: $x = 1$.
Range: $\{y \mid y \le 4\}$ or $(-\infty, 4]$.

**27.** $f(x) = x^2 + 2x - 3$
Since $a = 1$ is positive, the parabola opens upward. The $x$–coordinate of the vertex of the parabola is

$$-\frac{b}{2a} = -\frac{2}{2(1)} = -\frac{2}{2} = -1 \text{ and the}$$

$y$–coordinate of the vertex of the parabola is

$$f\left(-\frac{b}{2a}\right) = f(-1)$$
$$= (-1)^2 + 2(-1) - 3$$
$$= 1 - 2 - 3 = -4.$$

The vertex is $(-1, -4)$. Replace $f(x)$ with 0 to find $x$–intercepts.

$0 = x^2 + 2x - 3$

$0 = (x+3)(x-1)$

Apply the zero product principle.

$x + 3 = 0$ or $x - 1 = 0$

$x = -3$ $\qquad x = 1$

The $x$–intercepts are $-3$ and 1. Set $x = 0$ and solve for $y$ to obtain the $y$–intercept.

$y = (0)^2 + 2(0) - 3 = -3$

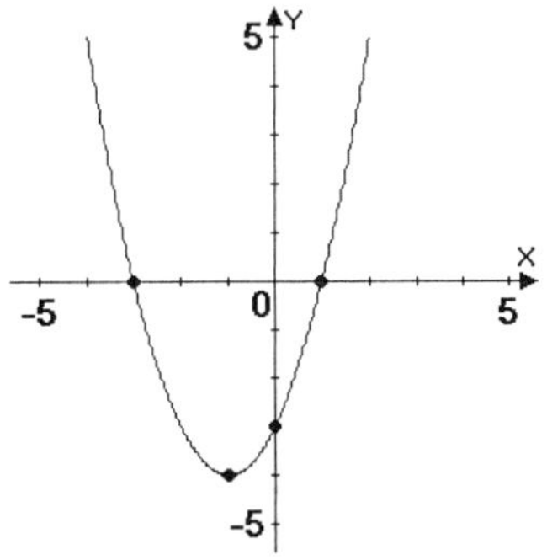


Axis of symmetry: $x = -1$.
Range: $\{y \mid y \ge -4\}$ or $[-4, \infty)$.

**29.** $f(x) = x^2 + 3x - 10$
Since $a = 1$ is positive, the parabola opens upward. The $x$–coordinate of the vertex of the parabola is

$$-\frac{b}{2a} = -\frac{3}{2(1)} = -\frac{3}{2} \text{ and the}$$

$y$–coordinate of the vertex of the parabola is

$$f\left(-\frac{b}{2a}\right) = f\left(-\frac{3}{2}\right)$$
$$= \left(-\frac{3}{2}\right)^2 + 3\left(-\frac{3}{2}\right) - 10$$
$$= \frac{9}{4} - \frac{9}{2} - 10 = -\frac{49}{4}.$$

The vertex is $\left(-\frac{3}{2}, -\frac{49}{4}\right)$.

Replace $f(x)$ with 0 to find the $x$–intercepts.

$0 = x^2 + 3x - 10$

$0 = (x+5)(x-2)$

Apply the zero product principle.

$x + 5 = 0$ or $x - 2 = 0$

$x = -5$ $\qquad x = 2$

The $x$–intercepts are –5 and 2.
Set $x = 0$ and solve for $y$ to obtain the $y$–intercept.

$y = 0^2 + 3(0) - 10 = -10$

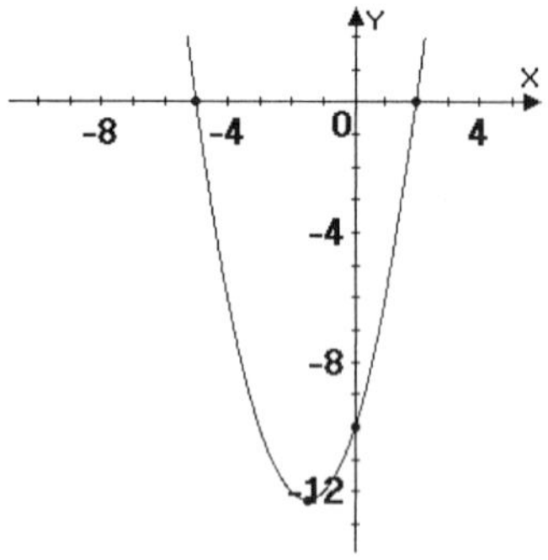


Axis of symmetry: $x = -\frac{3}{2}$.

Range: $\left\{y \mid y \geq -\frac{49}{4}\right\}$ or $\left[-\frac{49}{4}, \infty\right)$.

**31.** $f(x) = 2x - x^2 + 3$

$f(x) = -x^2 + 2x + 3$

Since $a = -1$ is negative, the parabola opens downward. The $x$–coordinate of the vertex of the parabola is $-\frac{b}{2a} = -\frac{2}{2(-1)} = -\frac{2}{-2} = 1$ and the $y$–coordinate of the vertex of the parabola is

$$f\left(-\frac{b}{2a}\right) = f(1) = -(1)^2 + 2(1) + 3$$
$$= -1 + 2 + 3 = 4.$$

The vertex is $(1, 4)$. Replace $f(x)$ with 0 to find $x$–intercepts.

$0 = -x^2 + 2x + 3$

$0 = x^2 - 2x - 3$

$0 = (x-3)(x+1)$

Apply the zero product principle.

$x - 3 = 0$ or $x + 1 = 0$

$x = 3$ $\quad$ $x = -1$

The $x$–intercepts are 3 and –1. Set $x = 0$ and solve for $y$ to obtain the $y$–intercept. $y = -(0)^2 + 2(0) + 3 = 3$

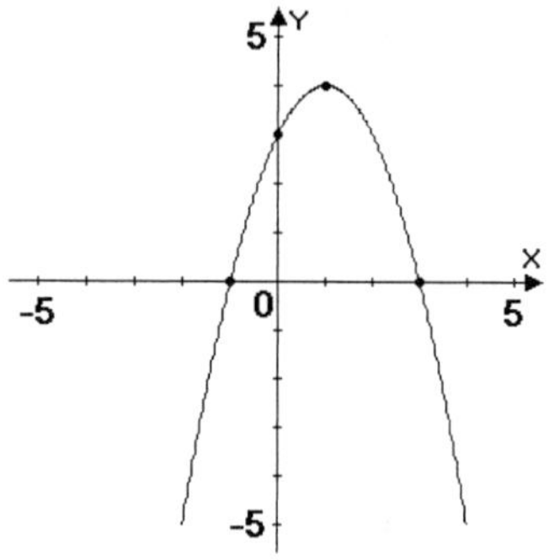


Axis of symmetry: $x = 1$.

Range: $\{y \mid y \leq 4\}$ or $(-\infty, 4]$.

**33.** $f(x) = 2x - x^2 - 2$

$f(x) = -x^2 + 2x - 2$

Since $a = -1$ is negative, the parabola opens downward. The $x$–coordinate of the vertex is

$$-\frac{b}{2a} = -\frac{2}{2(-1)} = -\frac{2}{-2} = 1$$

and the $y$–coordinate of the vertex is

$$f\left(-\frac{b}{2a}\right) = f(1)$$
$$= -(1)^2 + 2(1) - 2$$
$$= -1 + 2 - 2 = -1.$$

The vertex is $(1, -1)$. Replace $f(x)$ with 0 to find $x$–intercepts.

$0 = -x^2 + 2x - 2$

$x^2 - 2x = -2$

Since $b = -2$, we add

$$\left(\frac{b}{2}\right)^2 = \left(\frac{-2}{2}\right)^2 = (-1)^2 = 1$$

$x^2 - 2x + 1 = -2 + 1$

$(x-1)^2 = -1$

Because the solutions to the equation are imaginary, we know that there are no $x$–intercepts. Set $x = 0$ and solve for $y$ to obtain the $y$–intercept.

$y = 2(0) - 0^2 - 2 = -2$

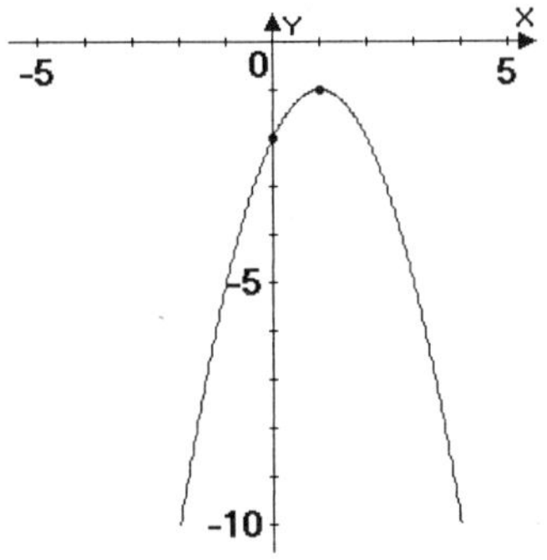


Axis of symmetry: $x = 1$.
Range: $\{y \mid y \leq -1\}$ or $(-\infty, -1]$.

**35.** $f(x) = 3x^2 - 12x - 1$
Since $a = 3$, the parabola opens upward and has a minimum. The $x$–coordinate of the minimum is
$-\frac{b}{2a} = -\frac{-12}{2(3)} = -\frac{-12}{6} = 2$ and the
$y$–coordinate of the minimum is

$$f\left(-\frac{b}{2a}\right) = f(2)$$
$$= 3(2)^2 - 12(2) - 1$$
$$= 12 - 24 - 1 = -13.$$

The minimum is $(2, -13)$.

**37.** $f(x) = -4x^2 + 8x - 3$
Since $a = -4$, the parabola opens downward and has a maximum. The $x$–coordinate of the maximum is
$-\frac{b}{2a} = -\frac{8}{2(-4)} = -\frac{8}{-8} = 1$ and the
$y$–coordinate of the maximum is

$$f\left(-\frac{b}{2a}\right) = f(1)$$
$$= -4(1)^2 + 8(1) - 3$$
$$= -4 + 8 - 3 = 1.$$

The maximum is (1, 1).

**39.** $f(x) = 5x^2 - 5x$
Since $a = 5$, the parabola opens upward and has a minimum. The $x$–coordinate of the minimum is
$-\frac{b}{2a} = -\frac{-5}{2(5)} = -\frac{-5}{10} = \frac{1}{2}$ and the
$y$–coordinate of the minimum is

$$f\left(-\frac{b}{2a}\right) = f\left(\frac{1}{2}\right)$$
$$= 5\left(\frac{1}{2}\right)^2 - 5\left(\frac{1}{2}\right)$$
$$= 5\left(\frac{1}{4}\right) - \frac{5}{2} = \frac{5}{4} - \frac{10}{4} = -\frac{5}{4}.$$

The minimum is $\left(\frac{1}{2}, -\frac{5}{4}\right)$.

**41.** Since the parabola opens up, the vertex $(-1, -2)$ is a minimum point. The domain is $\{x|x \text{ is a real number}\}$ or $(-\infty, \infty)$. The range is $\{y \mid y \geq -2\}$ or $[-2, \infty)$.

**42.** Since the parabola opens down, the vertex $(-3, -4)$ is a maximum point. The domain is $\{x|x \text{ is a real number}\}$ or $(-\infty, \infty)$. The range is $\{y \mid y \leq -4\}$ or $(-\infty, -4]$.

**43.** Since the parabola has a maximum point of (10, 3), it opens down. The domain is $\{x|x \text{ is a real number}\}$ or $(-\infty, \infty)$. The range is $\{y \mid y \leq 3\}$ or $(-\infty, 3]$.

**44.** Since the parabola has a minimum point of (5, 12), it opens up. The domain is $\{x|x$ is a real number$\}$ or $(-\infty, \infty)$. The range is $\{y \mid y \geq 12\}$ or $[12, \infty)$.

**45.** $(h,k)=(5,3)$

$f(x)=2(x-h)^2+k=2(x-5)^2+3$

**46.** $(h,k)=(7,4)$

$f(x)=2(x-h)^2+k=2(x-7)^2+4$

**47.** $(h,k)=(-10,-5)$

$$\begin{aligned} f(x) &= 2(x-h)^2+k \\ &= 2[x-(-10)]^2+(-5) \\ &= 2(x+10)^2-5 \end{aligned}$$

**48.** $(h,k)=(-8,-6)$

$$\begin{aligned} f(x) &= 2(x-h)^2+k \\ &= 2[x-(-8)]^2+(-6) \\ &= 2(x+8)^2-6 \end{aligned}$$

**49.** Since the vertex is a maximum, the parabola opens down and $a=-3$.

$(h,k)=(-2,4)$

$$\begin{aligned} f(x) &= -3(x-h)^2+k \\ &= -3[x-(-2)]^2+4 \\ &= -3(x+2)^2+4 \end{aligned}$$

**50.** Since the vertex is a maximum, the parabola opens down and $a=-3$.

$(h,k)=(5,-7)$

$$\begin{aligned} f(x) &= -3(x-h)^2+k \\ &= -3(x-5)^2+(-7) \\ &= -3(x-5)^2-7 \end{aligned}$$

**51.** Since the vertex is a minimum, the parabola opens up and $a=3$.

$(h,k)=(11,0)$

$$\begin{aligned} f(x) &= 3(x-h)^2+k \\ &= 3(x-11)^2+0 \\ &= 3(x-11)^2 \end{aligned}$$

**52.** Since the vertex is a minimum, the parabola opens up and $a=3$.

$(h,k)=(9,0)$

$$\begin{aligned} f(x) &= 3(x-h)^2+k \\ &= 3(x-9)^2+0 \\ &= 3(x-9)^2 \end{aligned}$$

**53.** From the graph, the vertex (maximum point) appears to be approximately $(14, 19.5)$. This means that in the year 2016, the projected mortality from *HCV* will reach a maximum of 19,500.

**55.** $s(t)=-16t^2+64t+160$

**a.** The $t$-coordinate of the minimum is

$$t=-\frac{b}{2a}=-\frac{64}{2(-16)}=-\frac{64}{-32}=2.$$

The $s$-coordinate of the minimum is

$$\begin{aligned} s(2) &= -16(2)^2+64(2)+160 \\ &= -16(4)+128+160 \\ &= -64+128+160=224 \end{aligned}$$

The ball reaches a maximum height of 244 feet 2 seconds after it is thrown.

**b.**

$$\begin{aligned} 0 &= -16t^2+64t+160 \\ 0 &= t^2-4t-10 \end{aligned}$$

$a=1 \quad b=-4 \quad c=-10$

$$t=\frac{-(-4)\pm\sqrt{(-4)^2-4(1)(-10)}}{2(1)}$$
$$=\frac{4\pm\sqrt{16+40}}{2}$$
$$=\frac{4\pm\sqrt{56}}{2}\approx\frac{4\pm7.48}{2}$$

Evaluate the expression to obtain two solutions.

$x=\frac{4+7.48}{2}$ or $x=\frac{4-7.48}{2}$

$x=\frac{11.48}{2}$ $\quad x=\frac{-3.48}{2}$

$x=5.74$ $\quad x=-1.74$

We disregard –1.74 because we can't have a negative time measurement. The solution is 5.74 and we conclude that the ball will hit the ground in approximately 5.7 seconds.

**c.** $s(0)=-16(0)^2+64(0)+160$
$$=-16(0)+0+160=160$$

At $t=0$, the ball has not yet been thrown and is at a height of 160 feet. This is the height of the building.

**d.**

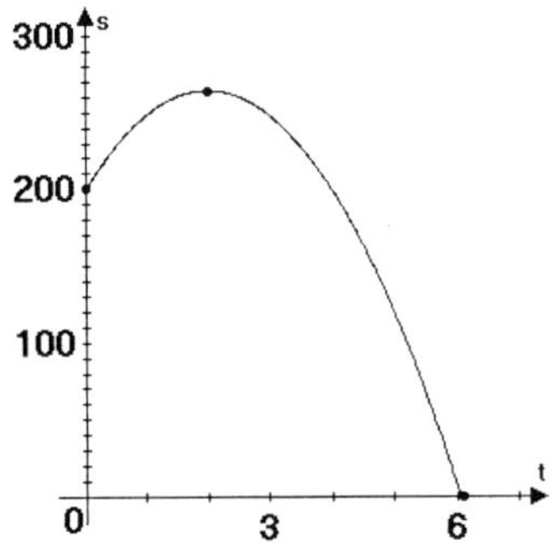


**57.** $f(x)=104.5x^2-1501.5x+6016$

The $x$-coordinate of the minimum is
$$x=-\frac{b}{2a}=-\frac{-1501.5}{2(104.5)}\approx7.2.$$
$f(7.2)$
$$=104.5(7.2)^2-1501.5(7.2)+6016$$
$$=104.5(51.84)-10810.8+6016$$
$$=622.48$$

The minimum death rate is about 622 per year per 100,000 males among U.S. men who average 7.2 hours of sleep per night.

**59.** Let $x$ = one of the numbers;
$16-x$ = the other number.
The product is
$$f(x)=x(16-x)$$
$$=16x-x^2=-x^2+16x$$
The $x$-coordinate of the maximum is
$$x=-\frac{b}{2a}=-\frac{16}{2(-1)}=-\frac{16}{-2}=8.$$
$$f(8)=-8^2+16(8)=-64+128=64$$
The vertex is (8, 64). The maximum product is 64. This occurs when the two number are 8 and $16-8=8$.

**61.** Let $x$ = one of the numbers;
$x-16$ = the other number.
The product is
$$f(x)=x(x-16)=x^2-16x$$
The $x$-coordinate of the minimum is
$$x=-\frac{b}{2a}=-\frac{16}{2(1)}=-\frac{16}{2}=-8.$$
$$f(-8)=(-8)^2+16(-8)$$
$$=64-128=-64$$
The vertex is $(-8,-64)$. The minimum product is $-64$. This occurs when the two number are $-8$ and $-8+16=8$.

**63.** Maximize the area of a rectangle constructed along a river with 600 feet of fencing.
Let $x$ = the width of the rectangle;
$600-2x$ = the length of the rectangle
We need to maximize.

$$A(x)=x(600-2x)$$
$$=600x-2x^2=-2x^2+600x$$

Since $a=-2$ is negative, we know the function opens downward and has a maximum at

$$x=-\frac{b}{2a}=-\frac{600}{2(-2)}=-\frac{600}{-4}=150.$$

When the width is $x=150$ feet, the length is

$$600-2(150)=600-300=300 \text{ feet.}$$

The dimensions of the rectangular plot with maximum area are 150 feet by 300 feet. This gives an area of $150\cdot300=45{,}000$ square feet.

**65.** Maximize the area of a rectangle constructed with 50 yards of fencing.
Let $x$ = the length of the rectangle.
Let $y$ = the width of the rectangle.
Since we need an equation in one variable, use the perimeter to express $y$ in terms of $x$.

$$2x+2y=50$$
$$2y=50-2x$$
$$y=\frac{50-2x}{2}=25-x$$

We need to maximize $A=xy=x(25-x)$. Rewrite $A$ as a function of $x$.

$$A(x)=x(25-x)=-x^2+25x$$

Since $a=-1$ is negative, we know the function opens downward and has a maximum at

$$x=-\frac{b}{2a}=-\frac{25}{2(-1)}=-\frac{25}{-2}=12.5.$$

When the length $x$ is 12.5, the width $y$ is $y=25-x=25-12.5=12.5$.
The dimensions of the rectangular region with maximum area are 12.5 yards by 12.5 yards. This gives an area of $12.5\cdot12.5=156.25$ square feet.

**67.** Maximize the cross-sectional area of the gutter:

$$A(x)=x(20-2x)$$
$$=20x-2x^2=-2x^2+20x.$$

Since $a=-2$ is negative, we know the function opens downward and has a maximum at

$$x=-\frac{b}{2a}=-\frac{20}{2(-2)}=-\frac{20}{-4}=5.$$

When the height $x$ is 5, the width is $20-2x=20-2(5)=20-10=10.$

$$A(5)=-2(5)^2+20(5)$$
$$=-2(25)+100=-50+100=50$$

The maximum cross-sectional area is 50 square inches. This occurs when the gutter is 5 inches deep and 10 inches wide.

**69.** **a.** $C(x)=525+0.55x$

**b.**
$$P(x)=R(x)-C(x)$$
$$=(-0.001x^2+3x)-(525+0.55x)$$
$$=-0.001x^2+3x-525-0.55x$$
$$=-0.001x^2+2.45x-525$$

**c.**
$$P(x)=R(x)-C(x)$$
$$=(-0.001x^2+3x)-(525+0.55x)$$
$$=-0.001x^2+3x-525-0.55x$$
$$=-0.001x^2+2.45x-525$$

Since $a=-0.001$ is negative, we know the function opens down and has a maximum at

$x = -\frac{b}{2a}$

$= -\frac{2.45}{2(-0.001)} = -\frac{2.45}{-0.002} = 1225.$

When the number of units $x$ is 1225, the profit is

$P(1225)$

$= -0.001(1225)^2 + 2.45(1225) - 525$

$= -0.001(1500625) + 3001.25 - 525$

$= -1500.625 + 3001.25 - 525$

$= 975.625$

The store maximizes its weekly profit when 1225 roast beef sandwiches are made and sold, resulting in a profit of \$975.63.

**71. – 75.** Answers will vary.

**77.** **a.** $y = 2x^2 - 82x + 720$

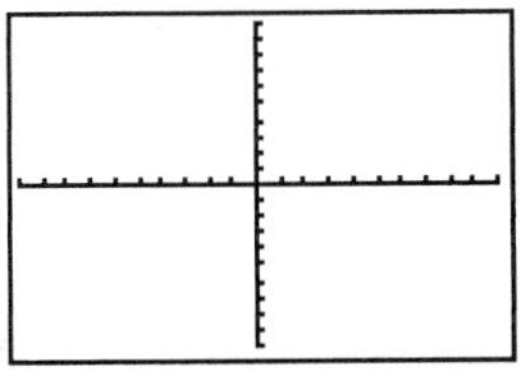

The function has no values that fall within the window.

**b.** $y = 2x^2 - 82x + 720$

The $x$–coordinate of the vertex of the parabola is

$-\frac{b}{2a} = -\frac{-82}{2(2)} = -\frac{-82}{4} = 20.5$

and the $y$–coordinate of the vertex of the parabola is

$f\left(-\frac{b}{2a}\right) = f(20.5)$

$= 2(20.5)^2 - 82(20.5) + 720$

$= 2(420.25) - 1681 + 720$

$= 840.5 - 1681 + 720 = -120.5.$

The vertex is $(20.5, -120.5)$.

**c.** Using the viewing window $[0, 30, 10]$ by $[-130, 10, 20]$, we have the following.

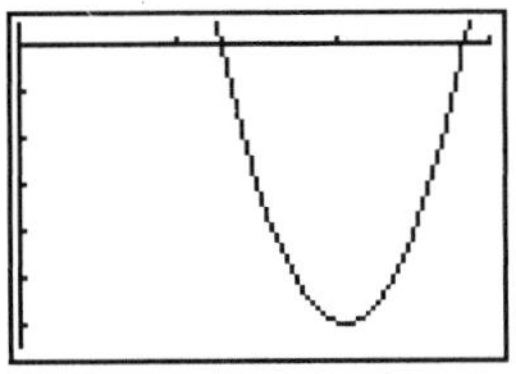

**d.** Answers will vary.

**79.** $y = -4x^2 + 20x + 160$

The $x$–coordinate of the vertex of the parabola is

$-\frac{b}{2a} = -\frac{20}{2(-4)} = -\frac{20}{-8} = 2.5$

and the $y$–coordinate of the vertex of the parabola is

$f\left(-\frac{b}{2a}\right) = f(2.5)$

$= -4(2.5)^2 + 20(2.5) + 160$

$= -4(6.25) + 50 + 160$

$= -25 + 50 + 160$

$= 185.$

The vertex is $(2.5, 185)$.

Using the viewing window $[-20, 20, 10]$ by $[0, 200, 20]$, we have the following.

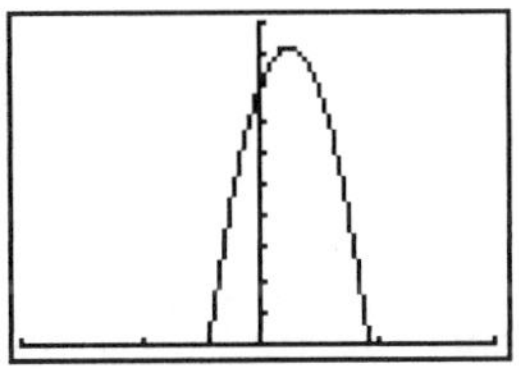

**81.** $y = 0.01x^2 + 0.6x + 100$

The $x$–coordinate of the vertex of the parabola is

$$-\frac{b}{2a} = -\frac{0.6}{2(0.01)} = -\frac{0.6}{0.02} = -30$$

and the $y$–coordinate of the vertex of the parabola is

$$f\left(-\frac{b}{2a}\right) = f(-30)$$
$$= 0.01(-30)^2 + 0.6(-30) + 100$$
$$= 0.01(900) - 18 + 100$$
$$= 9 - 18 + 100 = 91.$$

The vertex is $(-30, 91)$.

Using the viewing window $[-150, 150, 20]$ by $[0, 200, 20]$, we have the following.

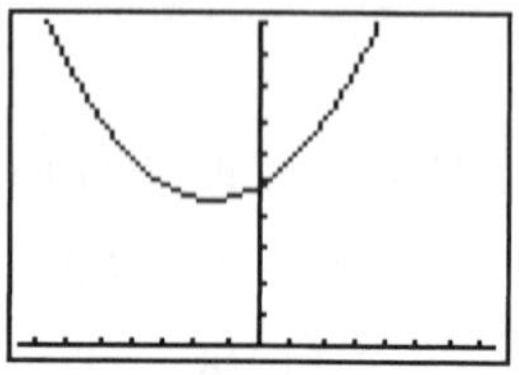

**83.** Statement **a** is true. Since quadratic functions represent parabolas, we know that the function has a maximum or a minimum. This means that the range cannot be $(-\infty, \infty)$.

Statement **b** is false. The vertex is $(5, -1)$.

Statement **c** is false. The graph has no $x$–intercepts. To find $x$–intercepts, set $y = 0$ and solve for $x$.

$$0 = -2(x+4)^2 - 8$$
$$2(x+4)^2 = -8$$
$$(x+4)^2 = -4$$

Because the solutions to the equation are imaginary, we know that there are no $x$–intercepts.

Statement **d** is false. The $x$-coordinate of the maximum is

$$-\frac{b}{2a} = -\frac{1}{2(-1)} = -\frac{1}{-2} = \frac{1}{2}$$ and the $y$–coordinate of the vertex of the parabola is

$$f\left(-\frac{b}{2a}\right) = f\left(\frac{1}{2}\right)$$
$$= -\left(\frac{1}{2}\right)^2 + \frac{1}{2} + 1$$
$$= -\frac{1}{4} + \frac{1}{2} + 1$$
$$= -\frac{1}{4} + \frac{2}{4} + \frac{4}{4} = \frac{5}{4}.$$

The maximum $y$–value is $\frac{5}{4}$.

**85.** $f(x) = (x-3)^2 + 2$

Since the vertex is $(3, 2)$, we know that the axis of symmetry is the line $x = 3$. The point (6, 11) is on the parabola and lies three units to the right of the axis of symmetry. This means that the point (0, 11) will also lie on the parabola since it lies 3 units to the right of the axis of symmetry.

**87.** We know $(h, k) = (-3, -4)$, so the equation is of the form

$$f(x) = a(x-h)^2 + k$$
$$= a[x - (-3)]^2 + (-1)$$
$$= a(x+3)^2 - 1$$

We use the point $(-2,-3)$ on the graph to determine the value of $a$:

$$f(x) = a(x+3)^2 - 1$$
$$-3 = a(-2+3)^2 - 1$$
$$-3 = a(1)^2 - 1$$
$$-3 = a - 1$$
$$-2 = a$$

Thus, the equation of the parabola is $f(x) = -2(x+3)^2 - 1$.

**89.** Let $x$ = the number of trees over 50 that will be planted.
The function describing the annual yield per lemon tree when $x + 50$ trees are planted per acre is

$$f(x) = (x+50)(320-4x)$$
$$= 320x - 4x^2 + 16000 - 200x$$
$$= -4x^2 + 120x + 16000.$$

This represents the number of lemon trees planted per acre multiplied by yield per tree.

The $x$–coordinate of the maximum is $-\frac{b}{2a} = -\frac{120}{2(-4)} = -\frac{120}{-8} = 15$ and the $y$–coordinate of the vertex of the parabola is

$$f\left(-\frac{b}{2a}\right) = f(15)$$
$$= -4(15)^2 + 120(15) + 16000$$
$$= -4(225) + 1800 + 16000$$
$$= -900 + 1800 + 16000$$
$$= 16900$$

The maximum lemon yield is 16,900 pounds when 50 + 15 = 65 lemon trees are planted per acre.

**90.** $$\frac{2}{x+5}+\frac{1}{x-5}=\frac{16}{x^2-25}$$
$$\frac{2}{x+5}+\frac{1}{x-5}=\frac{16}{(x+5)(x-5)}$$

The LCD is $(x+5)(x-5)$, so multiply each side of the equation by the LCD and clear fractions.

$$(x+5)(x-5)\left(\frac{2}{x+5}+\frac{1}{x-5}\right)=(x+5)(x-5)\left[\frac{16}{(x+5)(x-5)}\right]$$
$$(x-5)(2)+(x+5)(1)=16$$
$$2x-10+x+5=16$$
$$3x-5=16$$
$$3x=21$$
$$x=7$$

Since $x = 7$ will not make any of the denominators zero, the solution is 7, and the solution set is $\{7\}$.

**91.** $$\frac{1+\frac{2}{x}}{1-\frac{4}{x^2}}=\frac{x^2}{x^2}\cdot\frac{1+\frac{2}{x}}{1-\frac{4}{x^2}}$$
$$=\frac{x^2\cdot 1+x^2\cdot\frac{2}{x}}{x^2\cdot 1-x^2\cdot\frac{4}{x^2}}$$
$$=\frac{x^2+x\cdot 2}{x^2-4}$$
$$=\frac{x^2+2x}{(x+2)(x-2)}$$
$$=\frac{x\cancel{(x+2)}}{\cancel{(x+2)}(x-2)}=\frac{x}{x-2}$$

**92.** $$2x+3y=6$$
$$x-4y=14$$
$$D=\begin{vmatrix}2 & 3\\ 1 & -4\end{vmatrix}$$
$$=2(-4)-1(3)=-8-3=-11$$
$$D_x=\begin{vmatrix}6 & 3\\ 14 & -4\end{vmatrix}$$
$$=6(-4)-14(3)=-24-42=-66$$
$$D_y=\begin{vmatrix}2 & 6\\ 1 & 14\end{vmatrix}$$
$$=2(14)-1(6)=28-6=22$$
$$x=\frac{D_x}{D}=\frac{-66}{-11}=6;\ y=\frac{D_y}{D}=\frac{22}{-11}=-2$$

The solution is $(6,-2)$, and the solution set is $\{(6,-2)\}$.

## Mid-Chapter Check Point – Chapter 8

**1.** $(3x-5)^2 = 36$

Apply the square root principle:

$3x-5 = \pm\sqrt{36} = \pm 6$

$3x = 5 \pm 6$

$x = \frac{5 \pm 6}{3} = \frac{11}{3}$ or $-\frac{1}{3}$

The solutions are $\frac{11}{3}$ and $-\frac{1}{3}$, and the solution set is $\left\{\frac{11}{3}, -\frac{1}{3}\right\}$.

**2.** $5x^2 - 2x = 7$

$5x^2 - 2x - 7 = 0$

$(5x-7)(x+1) = 0$

Apply the zero-product principle:

$5x - 7 = 0$ or $x + 1 = 0$

$5x = 7$ $\quad$ $x = -1$

$x = \frac{7}{5}$

The solutions are $-1$ and $\frac{7}{5}$, and the solution set is $\left\{-1, \frac{7}{5}\right\}$.

**3.** $3x^2 - 6x - 2 = 0$

Apply the quadratic formula:

$a = 3 \quad b = -6 \quad c = -2$

$$x = \frac{-(-6) \pm \sqrt{(-6)^2 - 4(3)(-2)}}{2(3)}$$

$$= \frac{6 \pm \sqrt{36-(-24)}}{6}$$

$$= \frac{6 \pm \sqrt{60}}{6}$$

$$= \frac{6 \pm \sqrt{4 \cdot 15}}{6} = \frac{6 \pm 2\sqrt{15}}{6} = \frac{3 \pm \sqrt{15}}{3}$$

The solutions are $\frac{3 \pm \sqrt{15}}{3}$, and the solution set is $\left\{\frac{3 \pm \sqrt{15}}{3}\right\}$.

**4.** $x^2 + 6x = -2$

$x^2 + 6x + 2 = 0$

Apply the quadratic formula:

$a = 1 \quad b = 6 \quad c = 2$

$$x = \frac{-6 \pm \sqrt{6^2 - 4(1)(2)}}{2(1)}$$

$$= \frac{-6 \pm \sqrt{36-8}}{2}$$

$$= \frac{-6 \pm \sqrt{28}}{2}$$

$$= \frac{-6 \pm \sqrt{4 \cdot 7}}{2} = \frac{-6 \pm 2\sqrt{7}}{2} = -3 \pm \sqrt{7}$$

The solutions are $-3 \pm \sqrt{7}$, and the solution set is $\left\{-3 \pm \sqrt{7}\right\}$.

**5.** $5x^2 + 1 = 37$

$5x^2 = 36$

$x^2 = \frac{36}{5}$

Apply the square root principle:

$$x = \pm\sqrt{\frac{36}{5}} = \pm\frac{6}{\sqrt{5}} \cdot \frac{\sqrt{5}}{\sqrt{5}} = \pm\frac{6\sqrt{5}}{5}$$

The solutions are $\pm\frac{6\sqrt{5}}{5}$, and the solution set is $\left\{\pm\frac{6\sqrt{5}}{5}\right\}$.

**6.** $x^2-5x+8=0$

Apply the quadratic formula:

$a=1 \quad b=-5 \quad c=8$

$$x=\frac{-(-5)\pm\sqrt{(-5)^2-4(1)(8)}}{2(1)}$$
$$=\frac{5\pm\sqrt{25-32}}{2}$$
$$=\frac{5\pm\sqrt{-7}}{2}$$
$$=\frac{5\pm\sqrt{7\cdot(-1)}}{2}=\frac{5\pm\sqrt{7}i}{2}=\frac{5}{2}\pm\frac{\sqrt{7}}{2}i$$

The solutions are $\frac{5}{2}\pm\frac{\sqrt{7}}{2}i$, and the solution set is $\left\{\frac{5}{2}\pm\frac{\sqrt{7}}{2}i\right\}$.

**7.** $2x^2+26=0$
$$2x^2=-26$$
$$x^2=-13$$

Apply the square root principle:

$x=\pm\sqrt{-13}=\pm\sqrt{13(-1)}=\pm\sqrt{13}i$.

The solutions are $\pm\sqrt{13}i$, and the solution set is $\left\{\pm\sqrt{13}i\right\}$.

**8.** $(2x+3)(x+2)=10$
$$2x^2+4x+3x+6=10$$
$$2x^2+7x-4=0$$
$$(2x-1)(x+4)=0$$

Apply the zero-product principle:

$2x-1=0 \quad \text{or} \quad x+4=0$

$2x=1 \qquad x=-4$

$x=\frac{1}{2}$

The solutions are $-4$ and $\frac{1}{2}$, and the solution set is $\left\{-4,\ \frac{1}{2}\right\}$.

**9.** $(x+3)^2=24$

Apply the square root principle:

$x+3=\pm\sqrt{24}$

$x+3=\pm\sqrt{4\cdot 6}=\pm 2\sqrt{6}$

$x=-3\pm 2\sqrt{6}$

The solutions are $-3\pm 2\sqrt{6}$, and the solution set is $\left\{-3\pm 2\sqrt{6}\right\}$.

**10.** $\frac{1}{x^2}-\frac{4}{x}+1=0$

Multiply both sides of the equation by the least common denominator $x^2$.

$$x^2\left(\frac{1}{x^2}-\frac{4}{x}+1\right)=x^2(0)$$
$$1-4x+x^2=0$$
$$x^2-4x+1=0$$

Apply the quadratic formula:

$a=1 \quad b=-4 \quad c=1$

$$x=\frac{-(-4)\pm\sqrt{(-4)^2-4(1)(1)}}{2(1)}$$
$$=\frac{4\pm\sqrt{16-4}}{2}$$
$$=\frac{4\pm\sqrt{12}}{2}$$
$$=\frac{4\pm\sqrt{4\cdot 3}}{2}=\frac{4\pm 2\sqrt{3}}{2}=2\pm\sqrt{3}$$

The solutions are $2\pm\sqrt{3}$, and the solution set is $\left\{2\pm\sqrt{3}\right\}$.

**11.** $x(2x-3)=-4$

$2x^2-3x=-4$

$2x^2-3x+4=0$

Apply the quadratic formula:

$a=2 \quad b=-3 \quad c=4$

$$x=\frac{-(-3)\pm\sqrt{(-3)^2-4(2)(4)}}{2(2)}$$

$$=\frac{3\pm\sqrt{9-32}}{4}$$

$$=\frac{3\pm\sqrt{-23}}{4}$$

$$=\frac{3\pm\sqrt{23(-1)}}{4}=\frac{3\pm\sqrt{23}i}{4}=\frac{3}{4}\pm\frac{\sqrt{23}}{4}i$$

The solutions are $\frac{3}{4}\pm\frac{\sqrt{23}}{4}i$, and the solution set is $\left\{\frac{3}{4}\pm\frac{\sqrt{23}}{4}i\right\}$.

**12.** $\frac{x^2}{3}+\frac{x}{2}=\frac{2}{3}$

Multiply both sides of the equation by the least common denominator 6.

$$6\left(\frac{x^2}{3}+\frac{x}{2}\right)=6\left(\frac{2}{3}\right)$$

$2x^2+3x=4$

$2x^2+3x-4=0$

Apply the quadratic formula:

$a=3 \quad b=3 \quad c=-4$

$$x=\frac{-3\pm\sqrt{3^2-4(2)(-4)}}{2(2)}$$

$$=\frac{-3\pm\sqrt{9-(-32)}}{4}=\frac{-3\pm\sqrt{41}}{4}$$

The solutions are $\frac{-3\pm\sqrt{41}}{4}$, and the solution set is $\left\{\frac{-3\pm\sqrt{41}}{4}\right\}$

**13.**

$$\frac{2x}{x^2+6x+8}=\frac{x}{x+4}-\frac{2}{x+2}$$

$$\frac{2x}{(x+4)(x+2)}=\frac{x}{x+4}-\frac{2}{x+2}$$

Multiply both sides of the equation by the least common denominator $(x+4)(x+2)$.

$$(x+4)(x+2)\left[\frac{2x}{(x+4)(x+2)}\right]=(x+4)(x+2)\left[\frac{x}{x+4}-\frac{2}{x+2}\right]$$

$$2x=x(x+2)-2(x+4)$$

$$2x=x^2+2x-2x-8$$

$$0=x^2-2x-8$$

$$0=(x-4)(x+2)$$

Apply the zero-product principle:

$x-4=0$ or $x+2=0$

$x=4 \qquad x=-2$

The solutions are $-2$ and 4, and the solution set is $\{-2,\ 4\}$.

**14.** $x^2 + 10x - 3 = 0$

$x^2 + 10x \quad = 3$

Since $b = 10$, we add $\left(\frac{10}{2}\right)^2 = 5^2 = 25$.

$x^2 + 10x + 25 = 3 + 25$

$(x+5)^2 = 28$

Apply the square root principle:

$x + 5 = \pm\sqrt{28}$

$x + 5 = \pm\sqrt{4 \cdot 7} = \pm 2\sqrt{7}$

$x = -5 \pm 2\sqrt{7}$

The solutions are $-5 \pm 2\sqrt{7}$, and the solution set is $\left\{-5 \pm 2\sqrt{7}\right\}$.

**15.** $f(x) = (x-3)^2 - 4$

Since $a = 1$ is positive, the parabola opens upward. The vertex of the parabola is $(h,k) = (3,-4)$. Replace $f(x)$ with 0 to find $x$–intercepts.

$0 = (x-3)^2 - 4$

$4 = (x-3)^2$

Apply the square root property.

$x - 3 = \pm\sqrt{4} = \pm 2$

$x = 3 \pm 2 = 5$ or $1$

The $x$–intercepts are 5 and 1.

Set $x = 0$ and solve for $y$ to obtain the $y$–intercept. $y = (0-3)^2 - 4 = 5$

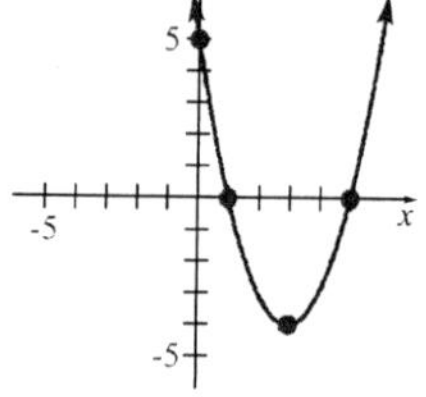


Domain: $\{x | x \text{ is a real number}\}$ or $(-\infty, \infty)$

Range: $\{y \mid y \ge -4\}$ or $[-4, \infty)$.

**16.** $g(x) = 5 - (x+2)^2$

$g(x) = -[x-(-2)]^2 + 5$

Since $a = -1$ is negative, the parabola opens downward. The vertex of the parabola is $(h,k) = (-2,5)$. Replace $g(x)$ with 0 to find $x$–intercepts.

$0 = -(x+2)^2 + 5$

$-5 = -(x+2)^2$

$5 = (x+2)^2$

Apply the square root property.

$\sqrt{5} = x + 2$ or $-\sqrt{5} = x + 2$

$-2 + \sqrt{5} = x \qquad -2 - \sqrt{5} = x$

$0.24 \approx x \qquad -4.24 \approx x$

The $x$–intercepts are $-2 + \sqrt{5} \approx 0.24$ and $-2 - \sqrt{5} \approx -4.24$.

Set $x = 0$ and solve for $y$ to obtain the $y$–intercept.

$y = 5 - (0+2)^2 = 5 - 4 = 1$

The $y$–intercept is 1.

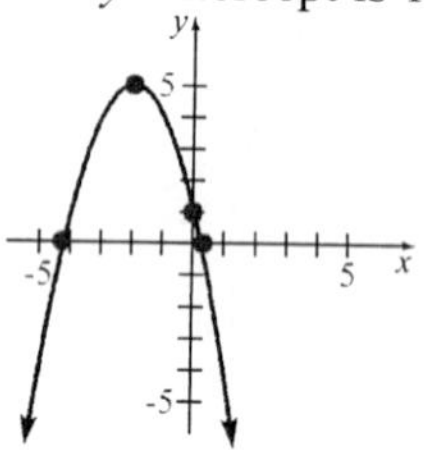


Domain: $\{x | x \text{ is a real number}\}$ or $(-\infty, \infty)$

Range: $\{y \mid y \le 5\}$ or $(-\infty, 5]$.

**17.** $h(x) = -x^2 - 4x + 5$

Since $a = -1$ is negative, the parabola opens downward. The $x$–coordinate of the vertex of the parabola is

$-\frac{b}{2a} = -\frac{-4}{2(-1)} = -\frac{-4}{-2} = -2$ and the

$y$–coordinate of the vertex of the parabola is

$$h\left(-\frac{b}{2a}\right) = h(-2)$$
$$= -(-2)^2 - 4(-2) + 5$$
$$= -4 + 8 + 5 = 9$$

The vertex is $(-2, 9)$. Replace $h(x)$ with 0 to find teh $x$–intercepts:

$0 = -x^2 - 4x + 5$

$0 = x^2 + 4x - 5$

$0 = (x+5)(x-1)$

Apply the zero product principle.

$x + 5 = 0$ or $x - 1 = 0$

$x = -5$ $\quad$ $x = 1$

The $x$–intercepts are $-5$ and 1. Set $x = 0$ and solve for $y$ to obtain the $y$–intercept. $y = -(0)^2 - 4(0) + 5 = 5$

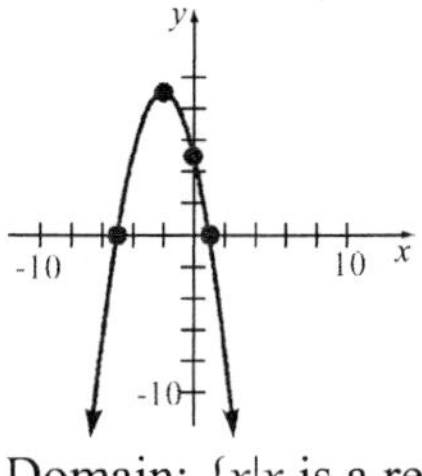


Domain: $\{x|x$ is a real number$\}$ or $(-\infty, \infty)$

Range: $\{y \mid y \le 9\}$ or $(-\infty, 9]$.

**18.** $f(x) = 3x^2 - 6x + 1$

Since $a = 3$, the parabola opens upward. The $x$–coordinate of the vertex of the parabola is

$-\frac{b}{2a} = -\frac{-6}{2(3)} = -\frac{-6}{6} = 1$ and the

$y$–coordinate of the minimum is

$$f\left(-\frac{b}{2a}\right) = f(1)$$
$$= 3(1)^2 - 6(1) + 1$$
$$= 3 - 6 + 1 = -2$$

The vertex $(1, 1)$. Replace $f(x)$ with 0 to find the $x$-intercepts:

$0 = 3x^2 - 6x + 1$

Apply the quadratic formula:

$a = 3 \quad b = -6 \quad c = 1$

$$x = \frac{-(-6) \pm \sqrt{(-6)^2 - 4(3)(1)}}{2(3)}$$
$$= \frac{6 \pm \sqrt{36 - 12}}{6}$$
$$= \frac{6 \pm \sqrt{24}}{6}$$
$$= \frac{6 \pm 2\sqrt{6}}{6} = \frac{3 \pm \sqrt{6}}{3} \approx 0.18 \text{ or } 1.82$$

Set $x = 0$ and solve for $y$ to obtain the $y$-intercept: $y = 3(0)^2 - 6(0) + 1 = 1$

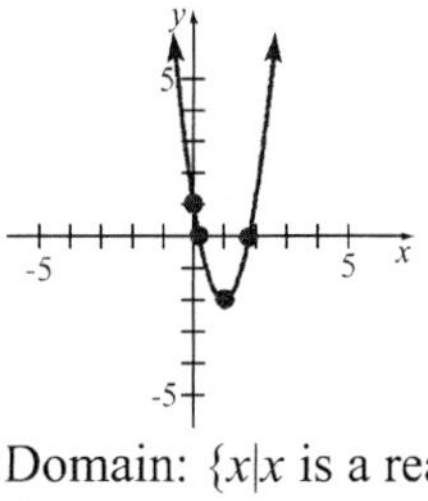


Domain: $\{x|x$ is a real number$\}$ or $(-\infty, \infty)$

Range: $\{y \mid y \ge -2\}$ or $[-2, \infty)$.

**19.** $2x^2 + 5x + 4 = 0$

$a = 2 \quad b = 5 \quad c = 4$

$b^2 - 4ac = 5^2 - 4(2)(4)$

$= 25 - 32 = -7$

Since the discriminant is negative, there are no real solutions. There are two imaginary solutions that are complex conjugates.

**20.** $10x(x+4) = 15x - 15$

$10x^2 + 40x = 15x - 15$

$10x^2 - 25x + 15 = 0$

$a = 10 \quad b = -25 \quad c = 15$

$b^2 - 4ac = (-25)^2 - 4(10)(15)$

$= 625 - 600 = 25$

Since the discriminant is positive and a perfect square, there are two rational solutions.

**21.** Because the solution set is $\left\{-\frac{1}{2}, \frac{3}{4}\right\}$, we have

$x = -\frac{1}{2}$ or $x = \frac{3}{4}$

$2x = -1 \qquad 4x = 3$

$2x + 1 = 0 \qquad 4x - 3 = 0$

Use the zero-product principle in reverse.

$(2x+1)(4x-3) = 0$

$8x^2 - 6x + 4x - 3 = 0$

$8x^2 - 2x - 3 = 0$

**22.** Because the solution set is $\{-2\sqrt{3}, 2\sqrt{3}\}$ we have

$x = -2\sqrt{3}$ or $x = 2\sqrt{3}$

$x + 2\sqrt{3} = 0 \qquad x - 2\sqrt{3} = 0$

Use the zero-product principle in reverse.

$(x + 2\sqrt{3})(x - 2\sqrt{3}) = 0$

$x^2 + 2x\sqrt{3} - 2x\sqrt{3} - 4 \cdot 3 = 0$

$x^2 - 12 = 0$

**23.** $P(x) = -x^2 + 150x - 4425$

Since $a = -1$ is negative, we know the function opens down and has a maximum at

$x = -\frac{b}{2a} = -\frac{150}{2(-1)} = -\frac{150}{-2} = 75$.

$P(75) = -75^2 + 150(75) - 4425$

$= -5625 + 11{,}250 - 4425 = 1200$

The company will maximize its profit by manufacturing and selling 75 cabinets per day. The maximum daily profit is \$1200.

**24.** Let $x$ = one of the numbers;

$-18 - x$ = the other number

The product is

$f(x) = x(-18 - x) = -x^2 - 18x$

The $x$-coordinate of the maximum is

$x = -\frac{b}{2a} = -\frac{-18}{2(-1)} = -\frac{-18}{-2} = -9.$

$f(-9) = -9[-18 - (-9)]$

$= -9(-18 + 9) = -9(-9) = 81$

The vertex is $(-9, 81)$. The maximum product is 81. This occurs when the two number are $-9$ and $-18 - (-9) = -9$.

**25.** Let $x$ = the measure of the height;
$40-2x$ = the measure of the base.

$$A=\frac{1}{2}bh$$

$$A(x)=\frac{1}{2}(40-2x)x$$

$$A(x)=-x^2+20x$$

Since $a=-1$ is negative, we know the function opens down and has a maximum at

$$x=-\frac{b}{2a}=-\frac{20}{2(-1)}=-\frac{20}{-2}=10.$$

$$A(10)=-10^2+20(10)$$
$$=-100+200=100$$

A height of 10 inches will maximize the area of the triangle. The maximum area will be 100 square inches.

## 8.4 Exercise Set

**1.** Let $t=x^2$.

$$x^4-5x^2+4=0$$
$$\left(x^2\right)^2-5x^2+4=0$$
$$t^2-5t+4=0$$
$$(t-4)(t-1)=0$$

Apply the zero product principle.

$t-4=0$ or $t-1=0$

$t=4$ $\quad t=1$

Replace $t$ by $x^2$.

$x^2=4$ or $x^2=1$

$x=\pm 2$ $\quad x=\pm 1$

The solutions are $\pm 1$ and $\pm 2$, and the solution set is $\{-2,-1,1,2\}$.

**3.** Let $t=x^2$.

$$x^4-11x^2+18=0$$
$$\left(x^2\right)^2-11x^2+18=0$$
$$t^2-11t+18=0$$
$$(t-9)(t-2)=0$$

Apply the zero product principle.

$t-9=0$ or $t-2=0$

$t=9$ $\quad t=2$

Replace $t$ by $x^2$.

$x^2=9$ or $x^2=2$

$x=\pm 3$ $\quad x=\pm\sqrt{2}$

The solutions are $\pm\sqrt{2}$ and $\pm 3$, and the solution set is $\left\{-3,-\sqrt{2},\sqrt{2},3\right\}$.

**5.** Let $t=x^2$.

$$x^4+2x^2=8$$
$$x^4+2x^2-8=0$$
$$\left(x^2\right)^2+2x^2-8=0$$
$$t^2+2t-8=0$$
$$(t+4)(t-2)=0$$

Apply the zero product principle.

$t+4=0$ or $t-2=0$

$t=-4$ $\quad t=2$

Replace $t$ by $x^2$.

$x^2=-4$ or $x^2=2$

$x=\pm\sqrt{-4}$ $\quad x=\pm\sqrt{2}$

$x=\pm 2i$

The solutions are $\pm 2i$ and $\pm\sqrt{2}$, and the solution set is $\left\{-\sqrt{2}i,-\sqrt{2},\sqrt{2},\sqrt{2}i\right\}$.

**7.** Let $t=\sqrt{x}$.

$$x+\sqrt{x}-2=0$$
$$\left(\sqrt{x}\right)^2+\sqrt{x}-2=0$$
$$t^2+t-2=0$$
$$(t+2)(t-1)=0$$

Apply the zero product principle.

$t+2=0$ or $t-1=0$

$t=-2$ $\quad$ $t=1$

Replace $t$ by $\sqrt{x}$.

~~$\sqrt{x}=-2$~~ or $\sqrt{x}=1$

$x=1$

We disregard –2 because the square root of $x$ cannot be a negative number. We need to check the solution, 1, because both sides of the equation were raised to an even power.

Check:

$$1+\sqrt{1}-2=0$$
$$1+1-2=0$$
$$2-2=0$$
$$0=0$$

The solution is 1, and the solution set is $\{1\}$.

**9.** Let $t=x^{\frac{1}{2}}$.

$$x-4x^{\frac{1}{2}}-21=0$$
$$\left(x^{\frac{1}{2}}\right)^2-4x^{\frac{1}{2}}-21=0$$
$$t^2-4t-21=0$$
$$(t-7)(t+3)=0$$

Apply the zero product principle.

$t-7=0$ or $t+3=0$

$t=7$ $\quad$ $t=-3$

Replace $t$ by $x^{\frac{1}{2}}$.

$x^{\frac{1}{2}}=7$ or $x^{\frac{1}{2}}=-3$

$\sqrt{x}=7$ $\quad$ ~~$\sqrt{x}=-3$~~

$x=49$

We disregard –2 because the square root of $x$ cannot be a negative number. We need to check the solution, 49, because both sides of the equation were raised to an even power.

Check:

$$49-4(49)^{\frac{1}{2}}-21=0$$
$$49-4(7)-21=0$$
$$49-28-21=0$$
$$49-49=0$$
$$0=0$$

The solution is 49, and the solution set is $\{49\}$.

**11.** Let $t=\sqrt{x}$.

$$x-13\sqrt{x}+40=0$$
$$\left(\sqrt{x}\right)^2-13\sqrt{x}+40=0$$
$$t^2-13t+40=0$$
$$(t-5)(t-8)=0$$

Apply the zero product principle.

$t-5=0$ or $t-8=0$

$t=5$ $\quad$ $t=8$

Replace $t$ by $\sqrt{x}$.

$\sqrt{x}=5$ or $\sqrt{x}=8$

$x=25$ $\quad$ $x=64$

Both solutions must be checked since both sides of the equation were raised to an even power.

$x = 25$

$$25 - 13\sqrt{25} + 40 = 0$$
$$25 - 13(5) + 40 = 0$$
$$25 - 65 + 40 = 0$$
$$65 - 65 = 0$$
$$0 = 0$$

$x = 64$

$$64 - 13\sqrt{64} + 40 = 0$$
$$64 - 13(8) + 40 = 0$$
$$64 - 104 + 40 = 0$$
$$104 - 104 = 0$$
$$0 = 0$$

Both solutions check. The solutions are 25 and 64, and the solution set is $\{25, 64\}$.

**13.** Let $t = x - 5$.

$$(x-5)^2 - 4(x-5) - 21 = 0$$
$$t^2 - 4t - 21 = 0$$
$$(t-7)(t+3) = 0$$

Apply the zero product principle.

$t - 7 = 0$ or $t + 3 = 0$

$t = 7$ $\qquad t = -3$

Replace $t$ by $x - 5$.

$x - 5 = 7$ or $x - 5 = -3$

$x = 12$ $\qquad x = 2$

The solutions are 2 and 12, and the solution set is $\{2, 12\}$.

**15.** Let $t = x^2 - 1$.

$$\left(x^2 - 1\right)^2 - \left(x^2 - 1\right) = 2$$
$$\left(x^2 - 1\right)^2 - \left(x^2 - 1\right) - 2 = 0$$
$$t^2 - t - 2 = 0$$
$$(t-2)(t+1) = 0$$

Apply the zero product principle.

$t - 2 = 0$ or $t + 1 = 0$

$t = 2$ $\qquad t = -1$

Replace $t$ by $x^2 - 1$.

$x^2 - 1 = 2$ or $x^2 - 1 = -1$

$x^2 = 3$ $\qquad x^2 = 0$

$x = \pm\sqrt{3}$ $\qquad x = 0$

The solutions are $\pm\sqrt{3}$ and $0$, and the solution set is $\left\{-\sqrt{3}, 0, \sqrt{3}\right\}$.

**17.** Let $t = x^2 + 3x$.

$$\left(x^2 + 3x\right)^2 - 8\left(x^2 + 3x\right) - 20 = 0$$
$$t^2 - 8t - 20 = 0$$
$$(t-10)(t+2) = 0$$

Apply the zero product principle.

$t - 10 = 0$ or $t + 2 = 0$

$t = 10$ $\qquad t = -2$

Replace $t$ by $x^2 + 3x$.

First, consider $t = 10$.

$$x^2 + 3x = 10$$
$$x^2 + 3x - 10 = 0$$
$$(x+5)(x-2) = 0$$

Apply the zero product principle.

$x + 5 = 0$ or $x - 2 = 0$

$x = -5$ $\qquad x = 2$

Next, consider $t = -2$.

$$x^2 + 3x = -2$$
$$x^2 + 3x + 2 = 0$$
$$(x+2)(x+1) = 0$$

Apply the zero product principle.

$x + 2 = 0$ or $x + 1 = 0$

$x = -2$ $\qquad x = -1$

The solutions are $-5, -2, -1$, and $2$, and the solution set is $\{-5, -2, -1, 2\}$.

**19.** Let $t = x^{-1}$.

$$x^{-2} - x^{-1} - 20 = 0$$
$$\left(x^{-1}\right)^2 - x^{-1} - 20 = 0$$
$$t^2 - t - 20 = 0$$
$$(t-5)(t+4) = 0$$

Apply the zero product principle.

$t - 5 = 0$ or $t + 4 = 0$

$t = 5$ $\quad t = -4$

Replace $t$ by $x^{-1}$.

$x^{-1} = 5$ or $x^{-1} = -4$

$\frac{1}{x} = 5$ $\quad \frac{1}{x} = -4$

$5x = 1$ $\quad -4x = 1$

$x = \frac{1}{5}$ $\quad x = -\frac{1}{4}$

The solutions are $-\frac{1}{4}$ and $\frac{1}{5}$, and the solution set is $\left\{-\frac{1}{4}, \frac{1}{5}\right\}$.

**21.** Let $t = x^{-1}$.

$$2x^{-2} - 7x^{-1} + 3 = 0$$
$$2\left(x^{-1}\right)^2 - 7x^{-1} + 3 = 0$$
$$2t^2 - 7t + 3 = 0$$
$$(2t-1)(t-3) = 0$$

Apply the zero product principle.

$2t - 1 = 0$ or $t - 3 = 0$

$2t = 1$ $\quad t = 3$

$t = \frac{1}{2}$

Replace $t$ by $x^{-1}$.

$x^{-1} = \frac{1}{2}$ or $x^{-1} = 3$

$\frac{1}{x} = \frac{1}{2}$ $\quad \frac{1}{x} = 3$

$x = 2$ $\quad 3x = 1$

$x = \frac{1}{3}$

The solutions are $\frac{1}{3}$ and $2$, and the solution set is $\left\{\frac{1}{3}, 2\right\}$.

**23.** Let $t = x^{-1}$.

$$x^{-2} - 4x^{-1} = 3$$
$$x^{-2} - 4x^{-1} - 3 = 0$$
$$\left(x^{-1}\right)^2 - 4x^{-1} - 3 = 0$$
$$t^2 - 4t - 3 = 0$$

$a = 1 \quad b = -4 \quad c = -3$

Use the quadratic formula.

$$t = \frac{-(-4) \pm \sqrt{(-4)^2 - 4(1)(-3)}}{2(1)}$$
$$= \frac{4 \pm \sqrt{16+12}}{2} = \frac{4 \pm \sqrt{28}}{2}$$
$$= \frac{4 \pm 2\sqrt{7}}{2} = \frac{2\left(2 \pm \sqrt{7}\right)}{2} = 2 \pm \sqrt{7}$$

Replace $t$ by $x^{-1}$.

$$x^{-1} = 2 \pm \sqrt{7}$$
$$\frac{1}{x} = 2 \pm \sqrt{7}$$
$$\left(2 \pm \sqrt{7}\right)x = 1$$
$$x = \frac{1}{2 \pm \sqrt{7}}$$

Rationalize the denominator.

$$x = \frac{1}{2 \pm \sqrt{7}} \cdot \frac{2 \mp \sqrt{7}}{2 \mp \sqrt{7}} = \frac{2 \mp \sqrt{7}}{2^2 - \left(\sqrt{7}\right)^2}$$
$$= \frac{2 \mp \sqrt{7}}{4-7} = \frac{2 \mp \sqrt{7}}{-3} = \frac{-2 \pm \sqrt{7}}{3}$$

The solutions are $\frac{-2 \pm \sqrt{7}}{3}$, and the solution set is $\left\{\frac{-2 \pm \sqrt{7}}{3}\right\}$.

**25.** Let $t = x^{\frac{1}{3}}$.

$$x^{\frac{2}{3}} - x^{\frac{1}{3}} - 6 = 0$$
$$\left(x^{\frac{1}{3}}\right)^2 - x^{\frac{1}{3}} - 6 = 0$$
$$t^2 - t - 6 = 0$$
$$(t-3)(t+2) = 0$$

Apply the zero product principle.

$t - 3 = 0$ or $t + 2 = 0$

$t = 3$ $\quad$ $t = -2$

Replace $t$ by $x^{\frac{1}{3}}$.

$x^{\frac{1}{3}} = 3$ or $x^{\frac{1}{3}} = -2$

$\left(x^{\frac{1}{3}}\right)^3 = 3^3$ $\quad$ $\left(x^{\frac{1}{3}}\right)^3 = (-2)^3$

$x = 27$ $\quad$ $x = -8$

The solutions are $-8$ and $27$, and the solution set is $\{-8, 27\}$.

**27.** Let $t = x^{\frac{1}{5}}$.

$$x^{\frac{2}{5}} + x^{\frac{1}{5}} - 6 = 0$$
$$\left(x^{\frac{1}{5}}\right)^2 + x^{\frac{1}{5}} - 6 = 0$$
$$t^2 + t - 6 = 0$$
$$(t+3)(t-2) = 0$$

Apply the zero product principle.

$t + 3 = 0$ or $t - 2 = 0$

$t = -3$ $\quad$ $t = 2$

Replace $t$ by $x^{\frac{1}{5}}$.

$x^{\frac{1}{5}} = -3$ or $x^{\frac{1}{5}} = 2$

$\left(x^{\frac{1}{5}}\right)^5 = (-3)^5$ $\quad$ $\left(x^{\frac{1}{5}}\right)^5 = (2)^5$

$x = -243$ $\quad$ $x = 32$

The solutions are $-243$ and $32$, and the solution set is $\{-243, 32\}$.

**29.** Let $t = x^{\frac{1}{4}}$.

$$2x^{\frac{1}{2}} - x^{\frac{1}{4}} = 1$$
$$2\left(x^{\frac{1}{4}}\right)^2 - x^{\frac{1}{4}} - 1 = 0$$
$$2t^2 - t - 1 = 0$$
$$(2t+1)(t-1) = 0$$

Apply the zero product principle.

$2t + 1 = 0$ or $t - 1 = 0$

$2t = -1$ $\quad$ $t = 1$

$t = -\frac{1}{2}$

Replace $t$ by $x^{\frac{1}{4}}$.

$x^{\frac{1}{4}} = -\frac{1}{2}$ or $x^{\frac{1}{4}} = 1$

$\left(x^{\frac{1}{4}}\right)^4 = \left(-\frac{1}{2}\right)^4$ $\quad$ $\left(x^{\frac{1}{4}}\right)^4 = 1^4$

$x = \frac{1}{16}$ $\quad$ $x = 1$

Since both sides of the equations were raised to an even power, the solutions must be checked.

First, check $x = \frac{1}{16}$.

$$2\left(\frac{1}{16}\right)^{\frac{1}{2}}-\left(\frac{1}{16}\right)^{\frac{1}{4}}=1$$
$$2\left(\frac{1}{4}\right)-\frac{1}{2}=1$$
$$\frac{1}{2}-\frac{1}{2}=1$$
$$0\neq 1$$

The solution does not check, so disregard $x=\frac{1}{16}$.

Next, check $x=1$.

$$2(1)^{\frac{1}{2}}-(1)^{\frac{1}{4}}=1$$
$$2(1)-1=1$$
$$1=1$$

The solution checks. The solution is 1, and the solution set is $\{1\}$.

**31.** Let $t=x-\frac{8}{x}$.

$$\left(x-\frac{8}{x}\right)^2+5\left(x-\frac{8}{x}\right)-14=0$$
$$t^2+5t-14=0$$
$$(t+7)(t-2)=0$$

Apply the zero product principle.

$t+7=0$ or $t-2=0$

$t=-7$ $\qquad t=2$

Replace $t$ by $x-\frac{8}{x}$.

First, consider $t=-7$.

$$x-\frac{8}{x}=-7$$
$$x\left(x-\frac{8}{x}\right)=x(-7)$$
$$x^2-8=-7x$$
$$x^2+7x-8=0$$
$$(x+8)(x-1)=0$$

Apply the zero product principle.

$x+8=0$ or $x-1=0$

$x=-8$ $\qquad x=1$

Next, consider $t=2$.

$$x-\frac{8}{x}=2$$
$$x\left(x-\frac{8}{x}\right)=x(2)$$
$$x^2-8=2x$$
$$x^2-2x-8=0$$
$$(x-4)(x+2)=0$$

Apply the zero product principle.

$x-4=0$ or $x+2=0$

$x=4$ $\qquad x=-2$

The solutions are $-8, -2, 1$, and 4, and the solution set is $\{-8,-2,1,4\}$.

**33.** $f(x)=x^4-5x^2+4$

$$y=x^4-5x^2+4$$

Set $y=0$ to find the $x$–intercept(s).

$0=x^4-5x^2+4$

Let $t=x^2$.

$$x^4-5x^2+4=0$$
$$\left(x^2\right)^2-5x^2+4=0$$
$$t^2-5t+4=0$$
$$(t-4)(t-1)=0$$

Apply the zero product principle.

$t-1=0$ or $t-4=0$

$t=1$ $\qquad t=4$

Substitute $x^2$ for $t$.

$x^2=1$ or $x^2=4$

$x=\pm1$ $\qquad x=\pm2$

The intercepts are $\pm1$ and $\pm2$. The corresponding graph is graph **c.**

**35.** $f(x) = x^{\frac{1}{3}} + 2x^{\frac{1}{6}} - 3$

$y = x^{\frac{1}{3}} + 2x^{\frac{1}{6}} - 3$

Set $y = 0$ to find the $x$–intercept(s).

$0 = x^{\frac{1}{3}} + 2x^{\frac{1}{6}} - 3$

Let $t = x^{\frac{1}{6}}$.

$x^{\frac{1}{3}} + 2x^{\frac{1}{6}} - 3 = 0$

$\left(x^{\frac{1}{6}}\right)^2 + 2x^{\frac{1}{6}} - 3 = 0$

$t^2 + 2t - 3 = 0$

$(t+3)(t-1) = 0$

Apply the zero product principle.

$t + 3 = 0$ or $t - 1 = 0$

$t = -3$ $\quad$ $t = 1$

Substitute $x^{\frac{1}{6}}$ for $t$.

$x^{\frac{1}{6}} = -3$ or $x^{\frac{1}{6}} = 1$

$\left(x^{\frac{1}{6}}\right)^6 = (-3)^6$ $\quad$ $\left(x^{\frac{1}{6}}\right)^6 = (1)^6$

$x = 729$ $\quad$ $x = 1$

Since both sides of the equations were raised to an even power, the solutions must be checked.

First check $x = 729$.

$(729)^{\frac{1}{3}} + 2(729)^{\frac{1}{6}} - 3 = 0$

$9 + 2(3) - 3 = 0$

$9 + 6 - 3 = 0$

$12 \neq 0$

Next check $x = 1$.

$(1)^{\frac{1}{3}} + 2(1)^{\frac{1}{6}} - 3 = 0$

$1 + 2(1) - 3 = 0$

$1 + 2 - 3 = 0$

$0 = 0$

Since 729 does not check, we disregard it. The intercept is 1. The corresponding graph is graph **e.**

**37.** $f(x) = (x+2)^2 - 9(x+2) + 20$

$y = (x+2)^2 - 9(x+2) + 20$

Set $y = 0$ to find the $x$–intercept(s).

$(x+2)^2 - 9(x+2) + 20 = 0$

Let $t = x + 2$.

$(x+2)^2 - 9(x+2) + 20 = 0$

$t^2 - 9t + 20 = 0$

$(t-5)(t-4) = 0$

Apply the zero product principle.

$t - 5 = 0$ or $t - 4 = 0$

$t = 5$ $\quad$ $t = 4$

Substitute $x + 2$ for $t$.

$x + 2 = 5$ or $x + 2 = 4$

$x = 3$ $\quad$ $x = 2$

The intercepts are 2 and 3. The corresponding graph is graph **f.**

**39.** Let $t = x^2 + 3x - 2$

$f(x) = -16$

$(x^2 + 3x - 2)^2 - 10(x^2 + 3x - 2) = -16$

$t^2 - 10t = -16$

$t^2 - 10t + 16 = 0$

$(t-8)(t-2) = 0$

Apply the zero product principle:

$t - 8 = 0$ or $t - 2 = 0$

$t = 8$ $\quad$ $t = 2$

Replace $t$ by $x^2 + 3x - 2$.

First, consider $t = 8$.

$x^2 + 3x - 2 = 8$

$x^2 + 3x - 10 = 0$

$(x+5)(x-2) = 0$

Apply the zero product principle.

$x+5=0$ or $x-2=0$

$x=-5$ $\quad x=2$

Next, consider $t=2$.

$$x^2+3x-2=2$$
$$x^2+3x-4=0$$
$$(x+4)(x-1)=0$$

Apply the zero product principle.

$x+4=0$ or $x-1=0$

$x=-4$ $\quad x=1$

The solutions are $-5, -4$, 1, and 2.

**40.** Let $t=x^2+2x-2$

$$f(x)=-6$$
$$\left(x^2+2x-2\right)^2-7\left(x^2+2x-2\right)=-6$$
$$t^2-7t=-6$$
$$t^2-7t+6=0$$
$$(t-1)(t-6)=0$$

Apply the zero product principle:

$t-1=0$ or $t-6=0$

$t=1$ $\quad t=6$

Replace $t$ by $x^2+2x-2$.

First, consider $t=1$.

$$x^2+2x-2=1$$
$$x^2+2x-3=0$$
$$(x+3)(x-1)=0$$

Apply the zero product principle.

$x+3=0$ or $x-1=0$

$x=-3$ $\quad x=1$

Next, consider $t=6$.

$$x^2+2x-2=6$$
$$x^2+2x-8=0$$
$$(x+4)(x-2)=0$$

Apply the zero product principle.

$x+4=0$ or $x-2=0$

$x=-4$ $\quad x=2$

The solutions are $-4, -3$, 1, and 2.

**41.** Let $t=\dfrac{1}{x}+1$.

$$f(x)=2$$
$$3\left(\frac{1}{x}+1\right)^2+5\left(\frac{1}{x}+1\right)=2$$
$$3t^2+5t=2$$
$$3t^2+5t-2=0$$
$$(3t-1)(t+2)=0$$

Apply the zero product principle.

$3t-1=0$ or $t+2=0$

$3t=1$ $\quad t=-2$

$t=\dfrac{1}{3}$

Replace $t$ by $\dfrac{1}{x}+1$.

First, consider $t=\dfrac{1}{3}$.

$$\frac{1}{x}+1=\frac{1}{3}$$
$$3x\left(\frac{1}{x}+1\right)=3x\left(\frac{1}{3}\right)$$
$$3+3x=x$$
$$2x=-3$$
$$x=-\frac{3}{2}$$

Next, consider $t=-2$.

$$\frac{1}{x}+1=-2$$
$$3x\left(\frac{1}{x}+1\right)=3x(-2)$$
$$3+3x=-6x$$
$$9x=-3$$
$$x=-\frac{3}{9}=-\frac{1}{3}$$

The solutions are $-\dfrac{3}{2}$ and $-\dfrac{1}{3}$.

**42.** Let $t = x^{\frac{1}{3}}$.

$$f(x) = 2$$
$$2x^{\frac{2}{3}} + 3x^{\frac{1}{3}} = 2$$
$$2\left(x^{\frac{1}{3}}\right)^2 + 3x^{\frac{1}{3}} = 2$$
$$2t^2 + 3t - 2 = 0$$
$$(2t-1)(t+2) = 0$$

Apply the zero product principle.

$$2t - 1 = 0 \quad \text{or} \quad t + 2 = 0$$
$$2t = 1 \qquad\qquad t = -2$$
$$t = \frac{1}{2}$$

Replace $t$ by $x^{\frac{1}{3}}$.

$$x^{\frac{1}{3}} = \frac{1}{2} \quad \text{or} \quad x^{\frac{1}{3}} = -2$$
$$\left(x^{\frac{1}{3}}\right)^3 = \left(\frac{1}{2}\right)^3 \qquad \left(x^{\frac{1}{3}}\right)^3 = (-2)^3$$
$$x = \frac{1}{8} \qquad\qquad x = -8$$

The solutions are $-8$ and $\frac{1}{8}$.

**43.** Let $t = \sqrt{\frac{x}{x-4}}$.

$$f(x) = g(x)$$
$$\frac{x}{x-4} = 13\sqrt{\frac{x}{x-4}} - 36$$
$$\left(\sqrt{\frac{x}{x-4}}\right)^2 = 13\sqrt{\frac{x}{x-4}} - 36$$
$$t^2 = 13t - 36$$
$$t^2 - 13t + 36 = 0$$
$$(t-9)(t-4) = 0$$

Apply the zero product principle:

$$t - 9 = 0 \quad \text{or} \quad t - 4 = 0$$
$$t = 9 \qquad\qquad t = 4$$

Replace $t$ by $\sqrt{\frac{x}{x-4}}$.

First, consider $t = 9$:

$$\sqrt{\frac{x}{x-4}} = 9$$
$$\left(\sqrt{\frac{x}{x-4}}\right)^2 = 9^2$$
$$\frac{x}{x-4} = 81$$
$$81(x-4) = x$$
$$81x - 324 = x$$
$$80x = 324$$
$$x = \frac{324}{80} = \frac{81}{20}$$

Next, consider $t = 4$:

$$\sqrt{\frac{x}{x-4}} = 4$$
$$\left(\sqrt{\frac{x}{x-4}}\right)^2 = 4^2$$
$$\frac{x}{x-4} = 16$$
$$16(x-4) = x$$
$$16x - 64 = x$$
$$15x = 64$$
$$x = \frac{64}{15}$$

Since both sides of the equations were raised to an even power, the solutions must be checked. In this case, both check, so the solutions are $\frac{81}{20}$ and $\frac{64}{15}$.

**44.** Let $t = \sqrt{\frac{x}{x-2}}$.

$$f(x) = g(x)$$
$$\frac{x}{x-2} + 10 = -11\sqrt{\frac{x}{x-2}}$$
$$\left(\sqrt{\frac{x}{x-2}}\right)^2 + 10 = -11\sqrt{\frac{x}{x-2}}$$
$$t^2 + 10 = -11t$$
$$t^2 + 11t + 10 = 0$$
$$(t+10)(t+1) = 0$$

Apply the zero product principle:

$t + 10 = 0$ or $t + 1 = 0$

$t = -10$ $\quad$ $t = -1$

Replace $t$ by $\sqrt{\frac{x}{x-2}}$.

First, consider $t = -10$:

$$\sqrt{\frac{x}{x-2}} = -10$$ (crossed out)

We disregard $-10$ because the square root cannot be a negative number.

Next, consider $t = -1$:

$$\sqrt{\frac{x}{x-2}} = -1$$ (crossed out)

We disregard $-1$ because the square root cannot be a negative number.
Thus, there are no values of $x$ that satisfy $f(x) = g(x)$.

**45.** Let $t = (x-4)^{-1}$

$$f(x) = g(x) + 12$$
$$3(x-4)^{-2} = 16(x-4)^{-1} + 12$$
$$3\left[(x-4)^{-1}\right]^2 = 16(x-4)^{-1} + 12$$
$$3t^2 = 16t + 12$$
$$3t^2 - 16t - 12 = 0$$
$$(3t+2)(t-6) = 0$$

Apply the zero product principle:

$3t + 2 = 0$ or $t - 6 = 0$

$3t = -2$ $\quad$ $t = 6$

$t = -\frac{2}{3}$

Replace $t$ by $(x-4)^{-1}$.

First, consider $t = -\frac{2}{3}$.

$$(x-4)^{-1} = -\frac{2}{3}$$
$$\frac{1}{x-4} = -\frac{2}{3}$$
$$-2(x-4) = 1(3)$$
$$-2x + 8 = 3$$
$$-2x = -5$$
$$x = \frac{-5}{-2} = \frac{5}{2}$$

Next, consider $t = 6$.

$$(x-4)^{-1} = 6$$
$$\frac{1}{x-4} = 6$$
$$6(x-4) = 1$$
$$6x - 24 = 1$$
$$6x = 25$$
$$x = \frac{25}{6}$$

The solutions are $\frac{5}{2}$ and $\frac{25}{6}$.

**46.** Let $t = \frac{2x}{x-3}$

$$f(x) = g(x) + 6$$
$$6\left(\frac{2x}{x-3}\right)^2 = 5\left(\frac{2x}{x-3}\right) + 6$$
$$6t^2 = 5t + 6$$
$$6t^2 - 5t - 6 = 0$$
$$(3t+2)(2t-3) = 0$$

Apply the zero product principle:

$3t + 2 = 0$ or $2t - 3 = 0$

$3t = -2$ $\quad$ $2t = 3$

$t = -\frac{2}{3}$ $\quad$ $t = \frac{3}{2}$

Replace $t$ by $\frac{2x}{x-3}$.

First, consider $t = -\frac{2}{3}$.

$$\frac{2x}{x-3} = -\frac{2}{3}$$
$$-2(x-3) = 2x(3)$$
$$-2x + 6 = 6x$$
$$-8x = -6$$
$$x = \frac{-6}{-8} = \frac{3}{4}$$

Next, consider $t = \frac{3}{2}$.

$$\frac{2x}{x-3} = \frac{3}{2}$$
$$3(x-3) = 2x(2)$$
$$3x - 9 = 4x$$
$$-9 = x$$

The solutions are $-9$ and $\frac{3}{4}$.

**47.** $P(x) = 0.04(x+40)^2 - 3(x+40) + 104$

$$60 = 0.04(x+40)^2 - 3(x+40) + 104$$
$$0 = 0.04(x+40)^2 - 3(x+40) + 44$$

Let $t = x + 40$.

$$0.04(x+40)^2 - 3(x+40) + 44 = 0$$
$$0.04t^2 - 3t + 44 = 0$$

Solve using the quadratic formula.

$a = 0.04 \quad b = -3 \quad c = 44$

$$t = \frac{-(-3) \pm \sqrt{(-3)^2 - 4(0.04)(44)}}{2(0.04)}$$
$$= \frac{3 \pm \sqrt{9 - 7.04}}{0.08}$$
$$= \frac{3 \pm \sqrt{1.96}}{0.08} = \frac{3 \pm 1.4}{0.08} = 55 \text{ or } 20$$

Since $x$ represents the number of years a person's age is above or below 40, $t = x + 40$ is the percentage we are looking for. The ages at which 60% of us feel that having a clean house is very important are 20 and 55. From the graph, we see that at 20, 58%, and at 55, 52% feel that a clean house if very important. The function models the data fairly well.

**49. – 51.** Answers will vary.

**53.** $x^6 - 7x^3 - 8 = 0$

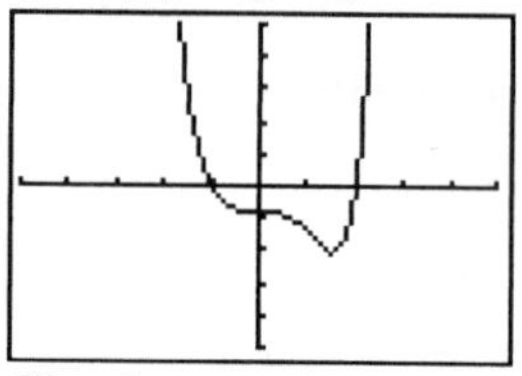

Check $x = -1$:

$$(-1)^6 - 7(-1)^3 - 8 = 0$$
$$1 - 7(-1) - 8 = 0$$
$$1 + 7 - 8 = 0$$
$$8 - 8 = 0$$
$$0 = 0$$

Check $x = 2$ using the same method. The solutions are –1 and 2, and the solution set is $\{-1, 2\}$.

**55.** $x^4 - 10x^2 + 9 = 0$

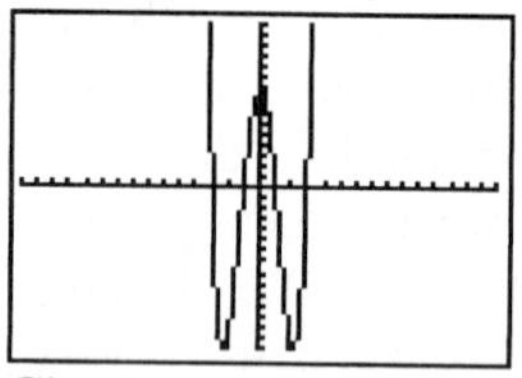

Check $x = -3$:

$$(-3)^4 - 10(-3)^2 + 9 = 0$$
$$81 - 10(9) + 9 = 0$$
$$81 - 90 + 9 = 0$$
$$90 - 90 = 0$$
$$0 = 0$$

Check $x =$ –1, 1, and 3 using the same method. The solutions are –3, –1, 1 and 3, and the solution set is $\{-3, -1, 1, 3\}$.

**57.**

$$2(x+1)^2 = 5(x+1) + 3$$
$$2(x+1)^2 - 5(x+1) - 3 = 0$$

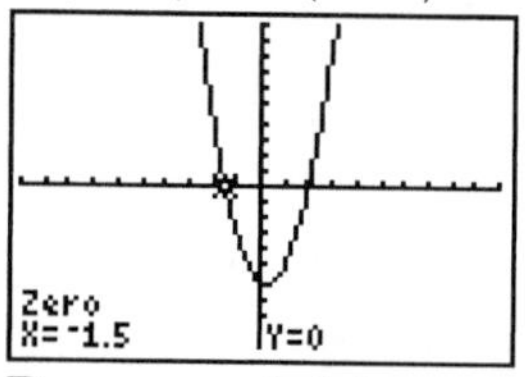


Because –1.5 is not an integer, the calculate zero feature was used to determine the intercept.

Check: $x =$ –1.5:

$$2(-1.5+1)^2 = 5(-1.5+1) + 3$$
$$2(-0.5)^2 = 5(-0.5) + 3$$
$$2(0.25) = -2.5 + 3$$
$$0.5 = 0.5$$

Check $x = 2$ using the same method.

The solutions are $-1.5 = -\frac{3}{2}$ and 2, and the solution set is $\left\{-\frac{3}{2}, 2\right\}$.

**59.**

$$x^{\frac{1}{2}} + 4x^{\frac{1}{4}} = 5$$
$$x^{\frac{1}{2}} + 4x^{\frac{1}{4}} - 5 = 0$$

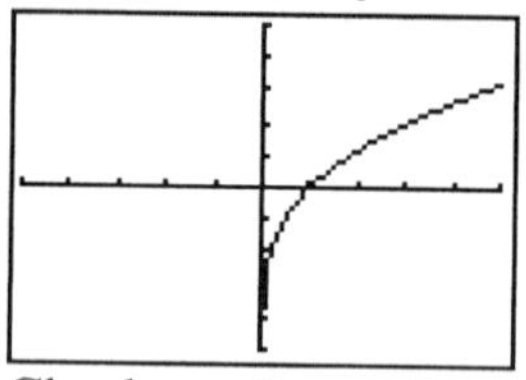

Check $x = 1$:

$$1^{\frac{1}{2}} + 4(1)^{\frac{1}{4}} = 5$$
$$1 + 4(1) = 5$$
$$1 + 4 = 5$$
$$5 = 5$$

The solution is 1, and the solution set is $\{1\}$.

**61.** Statement **b.** is true.

Statement **a.** is false. Any method that can be used to solve a quadratic equation can also be used to solve a quadratic-in-form equation.

Statement **c.** is false. The equation would be quadratic in form if the third term were a constant.

Statement **d.** is false. To solve the equation, let $t = \sqrt{x}$.

**63.**

$$5x^6 + x^3 = 18$$
$$5x^6 + x^3 - 18 = 0$$

Let $t = x^3$.

$$5x^6 + x^3 - 18 = 0$$
$$5\left(x^3\right)^2 + x^3 - 18 = 0$$
$$5t^2 + t - 18 = 0$$
$$(5t-9)(t+2) = 0$$

Apply the zero product principle.

$$5t - 9 = 0 \quad \text{or} \quad t + 2 = 0$$
$$5t = 9 \qquad t = -2$$
$$t = \frac{9}{5}$$

Substitute $x^3$ for $t$.

$$x^3 = \frac{9}{5} \quad \text{or} \quad x^3 = -2$$
$$x = \sqrt[3]{\frac{9}{5}} \qquad x = \sqrt[3]{-2}$$

Rationalize the denominator.

$$\sqrt[3]{\frac{9}{5}} = \frac{\sqrt[3]{9}}{\sqrt[3]{5}} \cdot \frac{\sqrt[3]{5^2}}{\sqrt[3]{5^2}} = \frac{\sqrt[3]{9 \cdot 5^2}}{\sqrt[3]{5^3}} = \frac{\sqrt[3]{225}}{5}$$

The solutions are $\sqrt[3]{-2}$ and $\frac{\sqrt[3]{225}}{5}$, and the solution set is $\left\{\sqrt[3]{-2}, \frac{\sqrt[3]{225}}{5}\right\}$.

**65.**

$$\frac{2x^2}{10x^3 - 2x^2} = \frac{2x^2}{2x^2(5x-1)} = \frac{1}{5x-1}$$

**66.**

$$\frac{2+i}{1-i} = \frac{2+i}{1-i} \cdot \frac{1+i}{1+i}$$
$$= \frac{2+2i+i+i^2}{1^2 - i^2}$$
$$= \frac{2+3i-1}{1-(-1)} = \frac{1+3i}{2} = \frac{1}{2} + \frac{3}{2}i$$

**67.**

$$2x + y = 6$$
$$x - 2y = 8$$

$$\left[\begin{array}{cc|c} 2 & 1 & 6 \\ 1 & -2 & 8 \end{array}\right] \quad R_1 \leftrightarrow R_2$$

$$= \left[\begin{array}{cc|c} 1 & -2 & 8 \\ 2 & 1 & 6 \end{array}\right] \quad -2R_1 + R_2$$

$$= \left[\begin{array}{cc|c} 1 & -2 & 8 \\ 0 & 5 & -10 \end{array}\right] \quad \frac{1}{5}R_2$$

$$= \left[\begin{array}{cc|c} 1 & -2 & 8 \\ 0 & 1 & -2 \end{array}\right]$$

The resulting system is:

$$x - 2y = 8$$
$$y = -2.$$

Back-substitute –2 for $y$ to find $x$.

$$x - 2(-2) = 8$$
$$x + 4 = 8$$
$$x = 4$$

The solution is $(4, -2)$, and the solution set is $\{(4, -2)\}$.

## 8.5 Exercise Set

**1.** $(x-4)(x+2)>0$

Solve the related quadratic equation.

$(x-4)(x+2)=0$

Apply the zero product principle.

$x-4=0 \quad \text{or} \quad x+2=0$

$x=4 \qquad\qquad x=-2$

The boundary points are –2 and 4.

| Test Interval | Test Number | Test | Conclusion |
|---|---|---|---|
| $(-\infty,-2)$ | -3 | $(-3-4)(-3+2)>0$<br>$7>0$, true | $(-\infty,-2)$ belongs to the solution set. |
| $(-2,4)$ | 0 | $(0-4)(0+2)>0$<br>$-8>0$, false | $(-2,4)$ does not belong to the solution set. |
| $(4,\infty)$ | 5 | $(5-4)(5+2)>0$<br>$7>0$, true | $(4,\infty)$ belongs to the solution set. |

The solution set is $(-\infty,-2)\cup(4,\infty)$ or $\{x\,|\,x<-2 \text{ or } x>4\}$.

**3.** $(x-7)(x+3)\le 0$

Solve the related quadratic equation.

$(x-7)(x+3)=0$

Apply the zero product principle.

$x-7=0 \quad \text{or} \quad x+3=0$

$x=7 \qquad\qquad x=-3$

The boundary points are –3 and 7.

| Test Interval | Test Number | Test | Conclusion |
|---|---|---|---|
| $(-\infty,-3)$ | -4 | $(-4-7)(-4+3)\le 0$<br>$11\le 0$, false | $(-\infty,-3)$ does not belong to the solution set. |
| $(-3,7)$ | 0 | $(0-7)(0+3)\le 0$<br>$-21\le 0$, true | $(-3,7)$ belongs to the solution set. |
| $(7,\infty)$ | 8 | $(8-7)(8+3)\le 0$<br>$11\le 0$, false | $(7,\infty)$ does not belong to the solution set. |

The solution set is $[-3,7]$ or $\{x\,|\,-3\le x\le 7\}$.

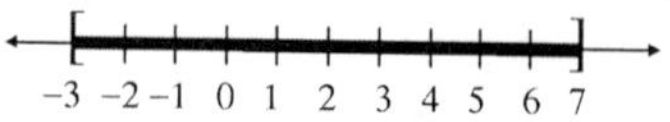

**5.** $x^2 - 5x + 4 > 0$

Solve the related quadratic equation.

$$x^2 - 5x + 4 = 0$$
$$(x-4)(x-1) = 0$$

Apply the zero product principle.

$x - 4 = 0$ or $x - 1 = 0$

$x = 4$ $\quad$ $x = 1$

The boundary points are 1 and 4.

| Test Interval | Test Number | Test | Conclusion |
|---|---|---|---|
| $(-\infty, 1)$ | 0 | $0^2 - 5(0) + 4 > 0$<br>$4 > 0$, true | $(-\infty, 1)$ belongs to the solution set. |
| $(1, 4)$ | 2 | $2^2 - 5(2) + 4 > 0$<br>$-2 > 0$, false | $(1, 4)$ does not belong to the solution set. |
| $(4, \infty)$ | 5 | $5^2 - 5(5) + 4 > 0$<br>$4 > 0$, true | $(4, \infty)$ belongs to the solution set. |

The solution set is $(-\infty, 1) \cup (4, \infty)$ or $\{x \mid x < 1 \text{ or } x > 4\}$.

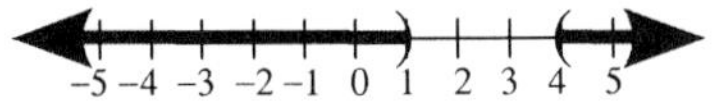


**7.** $x^2 + 5x + 4 > 0$

Solve the related quadratic equation.

$$x^2 + 5x + 4 = 0$$
$$(x+4)(x+1) = 0$$

Apply the zero product principle.

$x + 4 = 0$ or $x + 1 = 0$

$x = -4$ $\quad$ $x = -1$

The boundary points are –1 and –4.

| Test Interval | Test Number | Test | Conclusion |
|---|---|---|---|
| $(-\infty, -4)$ | -5 | $(-5)^2 + 5(-5) + 4 > 0$<br>$4 > 0$, true | $(-\infty, -4)$ belongs to the solution set. |
| $(-4, -1)$ | -2 | $(-2)^2 + 5(-2) + 4 > 0$<br>$-2 > 0$, false | $(-4, -1)$ does not belong to the solution set. |
| $(-1, \infty)$ | 0 | $0^2 + 5(0) + 4 > 0$<br>$4 > 0$, true | $(-1, \infty)$ belongs to the solution set. |

The solution set is $(-\infty, -4) \cup (-1, \infty)$ or $\{x \mid x < -4 \text{ or } x > -1\}$.

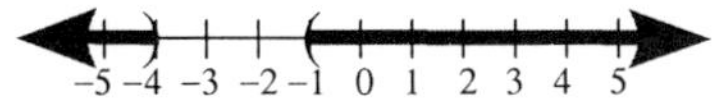

**9.** $x^2 - 6x + 8 \le 0$

Solve the related quadratic equation.

$$x^2 - 6x + 8 = 0$$

$$(x-4)(x-2) = 0$$

Apply the zero product principle.

$x - 4 = 0$ or $x - 2 = 0$

$x = 4$ $\qquad$ $x = 2$

The boundary points are 2 and 4.

| Test Interval | Test Number | Test | Conclusion |
|---|---|---|---|
| $(-\infty, 2)$ | 0 | $0^2 - 6(0) + 8 \le 0$ <br> $8 \le 0$, false | $(-\infty, 2)$ does not belong to the solution set. |
| $(2, 4)$ | 3 | $3^2 - 6(3) + 8 \le 0$ <br> $-1 \le 0$, true | $(2, 4)$ belongs to the solution set. |
| $(4, \infty)$ | 5 | $5^2 - 6(5) + 8 \le 0$ <br> $3 \le 0$, false | $(4, \infty)$ does not belong to the solution set. |

The solution set is $[2, 4]$ or $\{x \mid 2 \le x \le 4\}$.

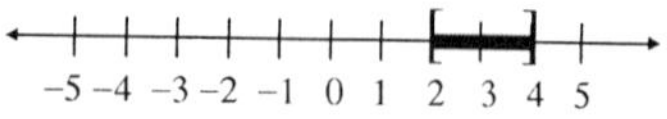


**11.** $3x^2 + 10x - 8 \le 0$

Solve the related quadratic equation.

$$3x^2 + 10x - 8 = 0$$

$$(3x-2)(x+4) = 0$$

Apply the zero product principle.

$3x - 2 = 0$ or $x + 4 = 0$

$3x = 2$ $\qquad$ $x = -4$

$x = \frac{2}{3}$

The boundary points are $-4$ and $\frac{2}{3}$.

| Test Interval | Test Number | Test | Conclusion |
|---|---|---|---|
| $(-\infty,-4)$ | -5 | $3(-5)^2+10(-5)-8\le 0$<br>$17\le 0$, false | $(-\infty,-4)$ does not belong to the solution set. |
| $\left(-4,\frac{2}{3}\right)$ | 0 | $3(0)^2+10(0)-8\le 0$<br>$-8\le 0$, true | $\left(-4,\frac{2}{3}\right)$ belongs to the solution set. |
| $\left(\frac{2}{3},\infty\right)$ | 1 | $3(1)^2+10(1)-8\le 0$<br>$5\le 0$, false | $\left(\frac{2}{3},\infty\right)$ does not belong to the solution set. |

The solution set is $\left[-4,\frac{2}{3}\right]$ or $\left\{x \middle| -4\le x\le\frac{2}{3}\right\}$.

**13.** $2x^2+x<15$

$2x^2+x-15<0$

Solve the related quadratic equation.

$2x^2+x-15=0$

$(2x-5)(x+3)=0$

Apply the zero product principle.

$2x-5=0$ or $x+3=0$

$2x=5$ $\quad x=-3$

$x=\frac{5}{2}$

The boundary points are $-3$ and $\frac{5}{2}$.

| Test Interval | Test Number | Test | Conclusion |
|---|---|---|---|
| $(-\infty,-3)$ | -4 | $2(-4)^2+(-4)<15$<br>$28<15$, false | $(-\infty,-3)$ does not belong to the solution set. |
| $\left(-3,\frac{5}{2}\right)$ | 0 | $2(0)^2+0<15$<br>$0<15$, true | $\left(-3,\frac{5}{2}\right)$ belongs to the solution set. |
| $\left(\frac{5}{2},\infty\right)$ | 3 | $2(3)^2+3<15$<br>$21<15$, false | $\left(\frac{5}{2},\infty\right)$ does not belong to the solution set. |

The solution set is $\left(-3,\frac{5}{2}\right)$ or $\left\{x \middle| -3<x<\frac{5}{2}\right\}$.

**15.** $4x^2 + 7x < -3$

$4x^2 + 7x + 3 < 0$

Solve the related quadratic equation.

$4x^2 + 7x + 3 = 0$

$(4x+3)(x+1) = 0$

Apply the zero product principle.

$4x + 3 = 0$ or $x + 1 = 0$

$4x = -3$ $\quad x = -1$

$x = -\frac{3}{4}$

The boundary points are $-1$ and $-\frac{3}{4}$.

| Test Interval | Test Number | Test | Conclusion |
|---|---|---|---|
| $(-\infty, -1)$ | -2 | $4(-2)^2 + 7(-2) < -3$<br>$2 < -3$, false | $(-\infty, -1)$ does not belong to the solution set. |
| $\left(-1, -\frac{3}{4}\right)$ | $-\frac{7}{8}$ | $4\left(-\frac{7}{8}\right)^2 + 7\left(-\frac{7}{8}\right) < -3$<br>$-3\frac{1}{16} < -3$, true | $\left(-1, -\frac{3}{4}\right)$ belongs to the solution set. |
| $\left(-\frac{3}{4}, \infty\right)$ | 0 | $4(0)^2 + 7(0) < -3$<br>$0 < -3$, false | $\left(-\frac{3}{4}, \infty\right)$ does not belong to the solution set. |

The solution set is $\left(-1, -\frac{3}{4}\right)$ or $\left\{x \middle| -1 < x < -\frac{3}{4}\right\}$.

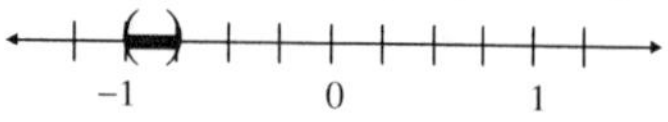


**17.** $x^2 - 4x \geq 0$

Solve the related quadratic equation.

$x^2 - 4x = 0$

$x(x-4) = 0$

Apply the zero product principle.

$x = 0$ or $x - 4 = 0$

$x = 4$

The boundary points are 0 and 4.

| Test Interval | Test Number | Test | Conclusion |
|---|---|---|---|
| $(-\infty,0)$ | -1 | $(-1)^2-4(-1)\geq 0$<br>$5\geq 0$, true | $(-\infty,0)$ belongs to the solution set. |
| $(0,4)$ | 1 | $(1)^2-4(1)\geq 0$<br>$-3\geq 0$, false | $(0,4)$ does not belong to the solution set. |
| $(4,\infty)$ | 5 | $(5)^2-4(5)\geq 0$<br>$5\geq 0$, true | $(4,\infty)$ belongs to the solution set. |

The solution set is $(-\infty,0]\cup[4,\infty)$ or $\{x|x\leq 0 \text{ or } x\geq 4\}$.

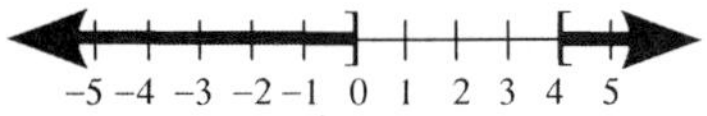

**19.** $2x^2+3x>0$

Solve the related quadratic equation.

$$2x^2+3x=0$$
$$x(2x+3)=0$$

Apply the zero product principle.

$$x=0 \quad \text{or} \quad 2x+3=0$$
$$2x=-3$$
$$x=-\frac{3}{2}$$

The boundary points are $-\frac{3}{2}$ and 0.

| Test Interval | Test Number | Test | Conclusion |
|---|---|---|---|
| $\left(-\infty,-\frac{3}{2}\right)$ | -2 | $2(-2)^2+3(-2)>0$<br>$2>0$, true | $\left(-\infty,-\frac{3}{2}\right)$ belongs to the solution set. |
| $\left(-\frac{3}{2},0\right)$ | -1 | $2(-1)^2+3(-1)>0$<br>$-1>0$, false | $\left(-\frac{3}{2},0\right)$ does not belong to the solution set. |
| $(0,\infty)$ | 1 | $2(1)^2+3(1)>0$<br>$5>0$, true | $(0,\infty)$ belongs to the solution set. |

The solution set is $\left(-\infty,-\frac{3}{2}\right)\cup(0,\infty)$ or $\left\{x\middle|x<-\frac{3}{2} \text{ or } x>0\right\}$.

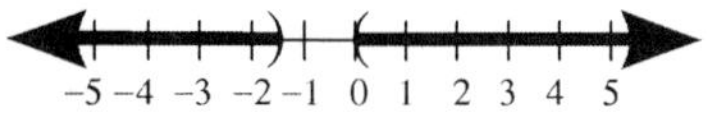

**21.** $-x^2 + x \geq 0$

Solve the related quadratic equation.

$$-x^2 + x = 0$$
$$-x(x-1) = 0$$

Apply the zero product principle.

$-x = 0$ or $x - 1 = 0$

$x = 0$ $\quad$ $x = 1$

The boundary points are 0 and 1.

| Test Interval | Test Number | Test | Conclusion |
|---|---|---|---|
| $(-\infty, 0)$ | -1 | $-(-1)^2 + (-1) \geq 0$ <br> $-2 \geq 0$, false | $(-\infty, 0)$ does not belong to the solution set. |
| $(0,1)$ | $\frac{1}{2}$ | $-\left(\frac{1}{2}\right)^2 + \frac{1}{2} \geq 0$ <br> $\frac{1}{4} \geq 0$, true | $(0,1)$ belongs to the solution set. |
| $(1, \infty)$ | 2 | $-(2)^2 + 2 \geq 0$ <br> $-2 \geq 0$, false | $(1, \infty)$ does not belong to the solution set. |

The solution set is $[0,1]$ or $\{x \mid 0 \leq x \leq 1\}$.

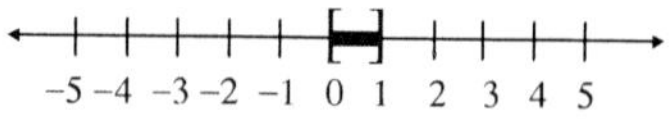


**23.** $x^2 \leq 4x - 2$

$x^2 - 4x + 2 \leq 0$

Solve the related quadratic equation, using the quadratic formula.

$x^2 - 4x + 2 = 0$

$a = 1 \quad b = -4 \quad c = 2$

$$x = \frac{-(-4) \pm \sqrt{(-4)^2 - 4(1)(2)}}{2(1)} = \frac{4 \pm \sqrt{16-8}}{2} = \frac{4 \pm \sqrt{8}}{2} = \frac{4 \pm \sqrt{4 \cdot 2}}{2}$$

$$= \frac{4 \pm 2\sqrt{2}}{2} = \frac{2\left(2 \pm \sqrt{2}\right)}{2} = 2 \pm \sqrt{2}$$

The boundary points are $2 - \sqrt{2}$ and $2 + \sqrt{2}$.

| Test Interval | Test Number | Test | Conclusion |
|---|---|---|---|
| $\left(-\infty, 2-\sqrt{2}\right)$ | 0 | $0^2 \le 4(0)-2$<br>$0 \le -2$, false | $\left(-\infty, 2-\sqrt{2}\right)$ does not belong to the solution set. |
| $\left(2-\sqrt{2}, 2+\sqrt{2}\right)$ | 2 | $2^2 \le 4(2)-2$<br>$4 \le 6$, true | $\left(2-\sqrt{2}, 2+\sqrt{2}\right)$ belongs to the solution set. |
| $\left(2+\sqrt{2}, \infty\right)$ | 4 | $4^2 \le 4(4)-2$<br>$16 \le 14$, false | $\left(2+\sqrt{2}, \infty\right)$ does not belong to the solution set. |

The solution set is $\left[2-\sqrt{2}, 2+\sqrt{2}\right]$ or $\left\{x \middle| 2-\sqrt{2} \le x \le 2+\sqrt{2}\right\}$.

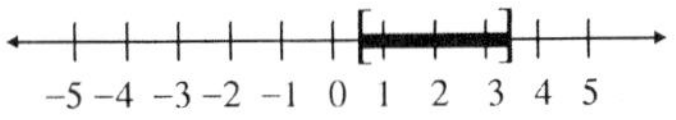


**25.** $x^2 - 6x + 9 < 0$
Solve the related quadratic equation.
$x^2 - 6x + 9 = 0$
$(x-3)^2 = 0$
Apply the zero product principle to obtain the double root.
$x - 3 = 0$
$x = 3$
The boundary point is 3.

| Test Interval | Test Number | Test | Conclusion |
|---|---|---|---|
| $(-\infty, 3)$ | 0 | $0^2 - 6(0) + 9 < 0$<br>$9 < 0$, False | $(-\infty, 3)$ does not belong to the solution set. |
| $(3, \infty)$ | 4 | $4^2 - 6(4) + 9 < 0$<br>$1 < 0$, false | $(3, \infty)$ does not belong to the solution set. |

There is no solution. The solution set is $\varnothing$ or $\{ \quad \}$.

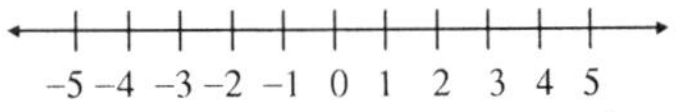


**27.** $(x-1)(x-2)(x-3) \ge 0$
Solve the related quadratic equation.
$(x-1)(x-2)(x-3) = 0$
Apply the zero product principle.
$x - 1 = 0$ or $x - 2 = 0$ or $x - 3 = 0$
$x = 1$ $\quad x = 2$ $\quad x = 3$
The boundary points are 1, 2, and 3.

| Test Interval | Test Number | Test | Conclusion |
|---|---|---|---|
| $(-\infty,1)$ | 0 | $(0-1)(0-2)(0-3)\ge 0$ <br> $-6\ge 0$, False | $(-\infty,1)$ does not belong to the solution set. |
| $(1,2)$ | 1.5 | $(1.5-1)(1.5-2)(1.5-3)\ge 0$ <br> $0.375\ge 0$, True | $(1,2)$ belongs to the solution set. |
| $(2,3)$ | 2.5 | $(2.5-1)(2.5-2)(2.5-3)\ge 0$ <br> $-0.375\ge 0$, False | $(2,3)$ does not belong to the solution set. |
| $(3,\infty)$ | 4 | $(4-1)(4-2)(4-3)\ge 0$ <br> $6\ge 0$, True | $(3,\infty)$ belongs to the solution set. |

The solution set is $[1,2]\cup[3,\infty)$ or $\{x|1\le x\le 2 \text{ or } x\ge 3\}$.

**29.** $x^3+2x^2-x-2\ge 0$

Solve the related quadratic equation.

$$x^3+2x^2-x-2=0$$
$$x^2(x+2)-1(x+2)=0$$
$$(x^2-1)(x+2)=0$$
$$(x-1)(x+1)(x+2)=0$$

Apply the zero product principle.

$x-1=0$ or $x+1=0$ or $x+2=0$

$x=1$ $\quad$ $x=-1$ $\quad$ $x=-2$

The boundary points are $-2,-1,$ and $1$.

| Test Interval | Test Number | Test | Conclusion |
|---|---|---|---|
| $(-\infty,-2)$ | $-3$ | $(-3)^3+2(-3)^2-(-3)-2\ge 0$ <br> $-8\ge 0$, False | $(-\infty,-2)$ does not belong to the solution set. |
| $(-2,-1)$ | $-1.5$ | $(-1.5)^3+2(-1.5)^2-(-1.5)-2\ge 0$ <br> $0.625\ge 0$, True | $(-2,-1)$ belongs to the solution set. |
| $(-1,1)$ | 0 | $0^3+2(0)^2-0-2\ge 0$ <br> $-2\ge 0$, False | $(-1,1)$ does not belong to the solution set. |
| $(1,\infty)$ | 2 | $2^3+2(2)^2-2-2\ge 0$ <br> $12\ge 0$, True | $(1,\infty)$ belongs to the solution set. |

The solution set is $[-2,-1]\cup[1,\infty)$ or $\{x|-2\le x\le -1 \text{ or } x\ge 1\}$.

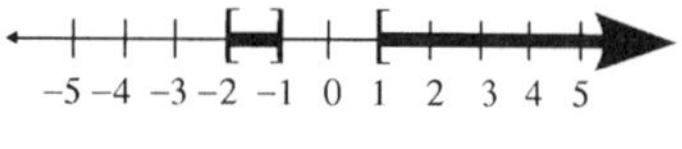

**31.** $x^3 - 3x^2 - 9x + 27 < 0$

Solve the related quadratic equation.

$$x^3 - 3x^2 - 9x + 27 = 0$$
$$x^2(x-3) - 9(x-3) = 0$$
$$(x^2 - 9)(x-3) = 0$$
$$(x-3)(x+3)(x-3) = 0$$
$$(x-3)^2(x+3) = 0$$

Apply the zero product principle.

$x - 3 = 0$ or $x + 3 = 0$

$x = 3$ $\quad\quad x = -3$

The boundary points are $-3$ and $3$.

| Test Interval | Test Number | Test | Conclusion |
|---|---|---|---|
| $(-\infty, -3)$ | $-4$ | $(-4)^3 - 3(-4)^2 - 9(-4) + 27 < 0$<br>$-49 < 0$, True | $(-\infty, -3)$ belongs to the solution set. |
| $(-3, 3)$ | $0$ | $0^3 - 3(0)^2 - 9(0) + 27 < 0$<br>$27 < 0$, False | $(-3, 3)$ does not belong to the solution set. |
| $(3, \infty)$ | $4$ | $4^3 - 3(4)^2 - 9(4) + 27 < 0$<br>$7 < 0$, False | $(3, \infty)$ does not belong to the solution set. |

The solution set is $(-\infty, -3)$ or $\{x | x < -3\}$.

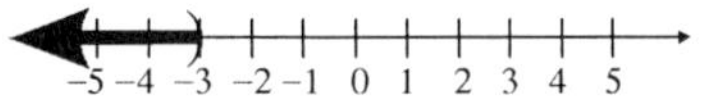


**33.** $x^3 + x^2 + 4x + 4 > 0$

Solve the related quadratic equation.

$$x^3 + x^2 + 4x + 4 = 0$$
$$x^2(x+1) + 4(x+1) = 0$$
$$(x^2 + 4)(x+1) = 0$$

Apply the zero product principle.

$x^2 + 4 = 0$ or $x + 1 = 0$

$x^2 = -4$ $\quad\quad x = -1$

$x = \pm\sqrt{-4}$

$= \pm 2i$

The imaginary solutions will not be boundary points, so the only boundary point is $-1$.

| Test Interval | Test Number | Test | Conclusion |
|---|---|---|---|
| $(-\infty,-1)$ | $-2$ | $(-2)^3+(-2)^2+4(-2)+4>0$<br>$-8>0$, False | $(-\infty,-1)$ does not belong to the solution set. |
| $(-1,\infty)$ | 0 | $0^3+0^2+4(0)+4>0$<br>$4>0$, True | $(-1,\infty)$ belongs to the solution set. |

The solution set is $(-1,\infty)$ or $\{x \mid x>-1\}$.

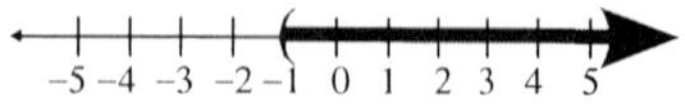

**35.** $x^3 \geq 9x^2$

$x^3-9x^2 \geq 0$

Solve the related quadratic equation.

$x^3-9x^2=0$

$x^2(x-9)=0$

Apply the zero product principle.

$x^2=0$ or $x-9=0$

$x=\pm\sqrt{0}=0$ $\quad x=9$

The boundary points are 0 and 9.

| Test Interval | Test Number | Test | Conclusion |
|---|---|---|---|
| $(-\infty,0)$ | $-1$ | $(-1)^3 \geq 9(-1)^2$<br>$-1 \geq 9$, False | $(-\infty,0)$ does not belong to the solution set. |
| $(0,9)$ | 1 | $1^3 \geq 9(1)^2$<br>$1 \geq 9$, False | $(0,9)$ does not belong to the solution set. |
| $(9,\infty)$ | 10 | $10^3 \geq 9(10)^2$<br>$1000 \geq 900$, True | $(9,\infty)$ belongs to the solution set. |

The solution set is $\{0\}\cup[9,\infty)$ or $\{x \mid x=0 \text{ or } x\geq 9\}$.

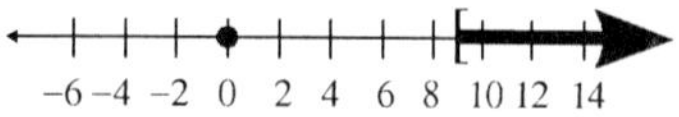

**37.** $\dfrac{x-4}{x+3}>0$

Find the values of $x$ that make the numerator and denominator zero.

$x-4=0$ $\quad x+3=0$

$x=4$ $\quad x=-3$

The boundary points are –3 and 4. We exclude –3 from the solution set, since this would make the denominator zero.

| Test Interval | Test Number | Test | Conclusion |
|---|---|---|---|
| $(-\infty,-3)$ | -4 | $\frac{-4-4}{-4+3}>0$ <br> $8>0$, true | $(-\infty,-3)$ belongs to the solution set. |
| $(-3,4)$ | 0 | $\frac{0-4}{0+3}>0$ <br> $\frac{-4}{3}>0$, false | $(-3,4)$ does not belong to the solution set. |
| $(4,\infty)$ | 5 | $\frac{5-4}{5+3}>0$ <br> $\frac{1}{8}>0$, true | $(4,\infty)$ belongs to the solution set. |

The solution set is $(-\infty,-3)\cup(4,\infty)$ or $\{x \mid x<-3 \text{ or } x>4\}$.

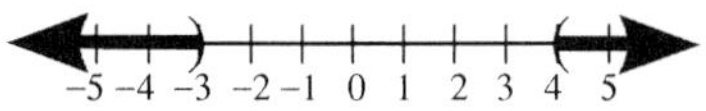


**39.** $\frac{x+3}{x+4}<0$

Find the values of $x$ that make the numerator and denominator zero.

$x+3=0 \qquad x+4=0$

$x=-3 \qquad x=-4$

The boundary points are –4 and –3.

| Test Interval | Test Number | Test | Conclusion |
|---|---|---|---|
| $(-\infty,-4)$ | -5 | $\frac{-5+3}{-5+4}<0$ <br> $2<0$, false | $(-\infty,-4)$ does not belong to the solution set. |
| $(-4,-3)$ | -3.5 | $\frac{-3.5+3}{-3.5+4}<0$ <br> $-1<0$, true | $(-4,-3)$ belongs to the solution set. |
| $(-3,\infty)$ | 0 | $\frac{0+3}{0+4}<0$ <br> $\frac{3}{4}<0$, false | $(-3,\infty)$ does not belong to the solution set. |

The solution set is $(-4,-3)$ or $\{x \mid -4<x<-3\}$.

**41.** $\dfrac{-x+2}{x-4} \geq 0$

Find the values of $x$ that make the numerator and denominator zero.

$-x+2=0$ and $x-4=0$

$-x=-2$ $\qquad x=4$

$x=2$

The boundary points are 2 and 4.

| Test Interval | Test Number | Test | Conclusion |
|---|---|---|---|
| $(-\infty, 2)$ | 0 | $\dfrac{-0+2}{0-4} \geq 0$<br>$-\dfrac{1}{2} \geq 0$, false | $(-\infty, 2)$ does not belong to the solution set. |
| $(2,4)$ | 3 | $\dfrac{-3+2}{3-4} \geq 0$<br>$1 \geq 0$, true | $(2,4)$ belongs to the solution set. |
| $(4, \infty)$ | 5 | $\dfrac{-5+2}{5-4} \geq 0$<br>$-3 \geq 0$, false | $(4, \infty)$ does not belong to the solution set. |

We exclude 4 from the solution set because 4 would make the denominator zero. The solution set is $[2,4)$ or $\{x \mid 2 \leq x < 4\}$.

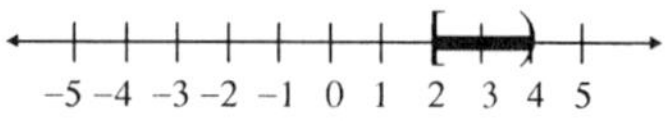


**33.** $\dfrac{4-2x}{3x+4} \leq 0$

Find the values of $x$ that make the numerator and denominator zero.

$4-2x=0$ and $3x+4=0$

$-2x=-4$ $\qquad 3x=-4$

$x=2$ $\qquad x=-\dfrac{4}{3}$

The boundary points are $-\dfrac{4}{3}$ and $2$.

| Test Interval | Test Number | Test | Conclusion |
|---|---|---|---|
| $\left(-\infty, -\frac{4}{3}\right)$ | -2 | $\frac{4-2(-2)}{3(-2)+4} \le 0$ <br> $-4 \le 0$, true | $\left(-\infty, -\frac{4}{3}\right)$ belongs to the solution set. |
| $\left(-\frac{4}{3}, 2\right)$ | 0 | $\frac{4-2(0)}{3(0)+4} \le 0$ <br> $1 \le 0$, false | $\left(-\frac{4}{3}, 2\right)$ does not belong to the solution set. |
| $[2, \infty)$ | 3 | $\frac{4-2(3)}{3(3)+4} \le 0$ <br> $-\frac{2}{13} \le 0$, true | $[2, \infty)$ belongs to the solution set. |

We exclude $-\frac{4}{3}$ from the solution set because $-\frac{4}{3}$ would make the denominator zero. The solution set is $\left(-\infty, -\frac{4}{3}\right) \cup [2, \infty)$ or $\left\{ x \middle| x < -\frac{4}{3} \text{ or } x \ge 2 \right\}$.

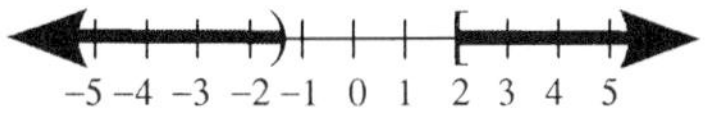


**45.** $\frac{x}{x-3} > 0$

Find the values of $x$ that make the numerator and denominator zero.

$x = 0$ and $x - 3 = 0$

$x = 3$

The boundary points are 0 and 3.

| Test Interval | Test Number | Test | Conclusion |
|---|---|---|---|
| $(-\infty, 0)$ | -1 | $\frac{-1}{-1-3} > 0$ <br> $\frac{1}{4} > 0$, true | $\left(-\infty, -\frac{4}{3}\right)$ belongs to the solution set. |
| $(0, 3)$ | 1 | $\frac{1}{1-3} > 0$ <br> $-\frac{1}{2} > 0$, false | $(0, 3)$ does not belong to the solution set. |
| $(3, \infty)$ | 4 | $\frac{4}{4-3} > 0$ <br> $4 > 0$, true | $(3, \infty)$ belongs to the solution set. |

The solution set is $(-\infty, 0) \cup (3, \infty)$ or $\{ x | x < 0 \text{ or } x > 3 \}$.

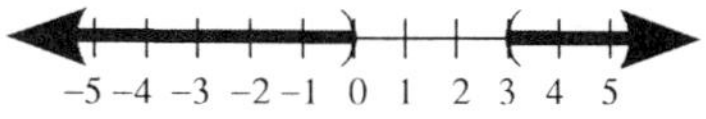

**47.** $\dfrac{x+1}{x+3} < 2$

Express the inequality so that one side is zero.

$$\frac{x+1}{x+3} - 2 < 0$$

$$\frac{x+1}{x+3} - \frac{2(x+3)}{x+3} < 0$$

$$\frac{x+1-2(x+3)}{x+3} < 0$$

$$\frac{x+1-2x-6}{x+3} < 0$$

$$\frac{-x-5}{x+3} < 0$$

Find the values of $x$ that make the numerator and denominator zero.

$-x-5=0 \qquad x+3=0$

$-x=5 \qquad x=-3$

$x=-5$

The boundary points are –5 and –3.

| Test Interval | Test Number | Test | Conclusion |
|---|---|---|---|
| $(-\infty,-5)$ | -6 | $\dfrac{-6+1}{-6+3} < 2$<br>$\dfrac{5}{3} < 2$, true | $(-\infty,-5)$ belongs to the solution set. |
| $(-5,-3)$ | -4 | $\dfrac{-4+1}{-4+3} < 2$<br>$3 < 2$, false | $(-5,-3)$ does not belong to the solution set. |
| $(-3,\infty)$ | 0 | $\dfrac{0+1}{0+3} < 2$<br>$\dfrac{1}{3} < 2$, true | $(-3,\infty)$ belongs to the solution set. |

The solution set is $(-\infty,-5)\cup(-3,\infty)$ or $\{x \mid x<-5 \text{ or } x>-3\}$.

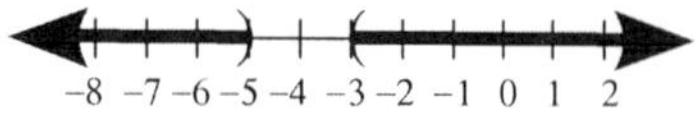

**49.** $\frac{x+4}{2x-1} \le 3$

Express the inequality so that one side is zero.

$$\frac{x+4}{2x-1} - 3 \le 0$$

$$\frac{x+4}{2x-1} - \frac{3(2x-1)}{2x-1} \le 0$$

$$\frac{x+4-3(2x-1)}{2x-1} \le 0$$

$$\frac{x+4-6x+3}{2x-1} \le 0$$

$$\frac{-5x+7}{2x-1} \le 0$$

Find the values of $x$ that make the numerator and denominator zero.

$$-5x+7=0 \qquad 2x-1=0$$

$$-5x=-7 \qquad 2x=1$$

$$x=\frac{7}{5} \qquad x=\frac{1}{2}$$

The boundary points are $\frac{1}{2}$ and $\frac{7}{5}$.

| Test Interval | Test Number | Test | Conclusion |
|---|---|---|---|
| $\left(-\infty, \frac{1}{2}\right)$ | 0 | $\frac{0+4}{2(0)-1} \le 3$ <br> $-4 \le 3$, true | $\left(-\infty, \frac{1}{2}\right)$ belongs to the solution set. |
| $\left(\frac{1}{2}, \frac{7}{5}\right)$ | 1 | $\frac{1+4}{2(1)-1} \le 3$ <br> $5 \le 3$, false | $\left(\frac{1}{2}, \frac{7}{5}\right)$ does not belong to the solution set. |
| $\left(\frac{7}{5}, \infty\right)$ | 2 | $\frac{2+4}{2(2)-1} \le 3$ <br> $2 \le 3$, true | $\left(\frac{7}{5}, \infty\right)$ belongs to the solution set. |

We exclude $\frac{1}{2}$ from the solution set because $\frac{1}{2}$ would make the denominator zero. The solution set is $\left(-\infty, \frac{1}{2}\right) \cup \left[\frac{7}{5}, \infty\right)$ or $\left\{x \,\middle|\, x < \frac{1}{2} \text{ or } x \ge \frac{7}{5}\right\}$.

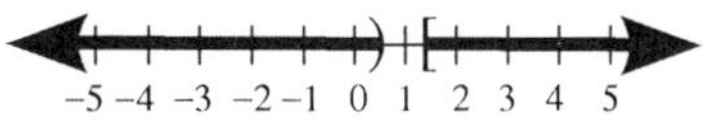

**51.** $\dfrac{x-2}{x+2} \le 2$

Express the inequality so that one side is zero.

$$\frac{x-2}{x+2} - 2 \le 0$$

$$\frac{x-2}{x+2} - \frac{2(x+2)}{x+2} \le 0$$

$$\frac{x-2-2(x+2)}{x+2} \le 0$$

$$\frac{x-2-2x-4}{x+2} \le 0$$

$$\frac{-x-6}{x+2} \le 0$$

Find the values of $x$ that make the numerator and denominator zero.

$-x-6=0 \qquad x+2=0$

$-x=6 \qquad x=-2$

$x=-6$

The boundary points are $-6$ and $-2$.

| Test Interval | Test Number | Test | Conclusion |
|---|---|---|---|
| $(-\infty,-6)$ | -7 | $\frac{-7-2}{-7+2} \le 2$ <br> $\frac{9}{5} \le 2$, true | $(-\infty,-6)$ belongs to the solution set. |
| $(-6,-2)$ | -3 | $\frac{-3-2}{-3+2} \le 2$ <br> $5 \le 2$, false | $(-6,-2)$ does not belong to the solution set. |
| $(-2,\infty)$ | 0 | $\frac{0-2}{0+2} \le 2$ <br> $-1 \le 2$, true | $(-2,\infty)$ belongs to the solution set. |

We exclude $-2$ from the solution set because $-2$ would make the denominator zero. The solution set is $(-\infty,-6] \cup (-2,\infty)$ or $\{x \mid x \le -6 \text{ or } x > -2\}$.

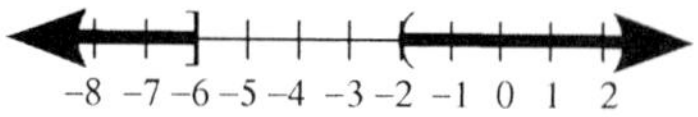

**53.** $f(x) \geq g(x)$

$2x^2 \geq 5x - 2$

$2x^2 - 5x + 2 \geq 0$

Solve the related quadratic equation.

$2x^2 - 5x + 2 = 0$

$(2x-1)(x-2) = 0$

Apply the zero product principle.

$2x - 1 = 0$ or $x - 2 = 0$

$2x = 1$ $\quad x = 2$

$x = \frac{1}{2}$

The boundary points are $\frac{1}{2}$ and 2.

| Test Interval | Test Number | Test | Conclusion |
|---|---|---|---|
| $\left(-\infty, \frac{1}{2}\right)$ | 0 | $2(0)^2 \geq 5(0) - 2$<br>$0 \geq -2$, True | $\left(-\infty, \frac{1}{2}\right)$ belongs to the solution set. |
| $\left(\frac{1}{2}, 2\right)$ | 1 | $2(1)^2 \geq 5(1) - 2$<br>$2 \geq 3$, False | $\left(\frac{1}{2}, 2\right)$ does not belong to the solution set. |
| $(2, \infty)$ | 3 | $2(3)^2 \geq 5(3) - 2$<br>$18 \geq 13$, True | $(2, \infty)$ does not belong to the solution set. |

The solution set is $\left(-\infty, \frac{1}{2}\right] \cup [2, \infty)$ or $\left\{x \middle| x \leq \frac{1}{2} \text{ or } x \geq 2\right\}$.

**55.** $f(x) \leq g(x)$

$\frac{2x}{x+1} \leq 1$

Express the inequality so that one side is zero.

$\frac{2x}{x+1} - 1 \leq 0$

$\frac{2x}{x+1} - \frac{x+1}{x+1} \leq 0$

$\frac{2x - x - 1}{x+1} \leq 0$

$\frac{x-1}{x+1} \leq 0$

Find the values of $x$ that make the numerator and denominator zero.

$x-1=0$ or $x+1=0$

$x=1$ $\quad x=-1$

The boundary points are $-1$ and $1$.

| Test Interval | Test Number | Test | Conclusion |
|---|---|---|---|
| $(-\infty,-1)$ | $-3$ | $\frac{2(-3)}{-3+1}\le 1$ <br> $3\le 1$, false | $(-\infty,-1)$ does not belong to the solution set. |
| $(-1,1)$ | 0 | $\frac{2(0)}{0+1}\le 1$ <br> $0\le 1$, true | $(-1,1)$ belongs to the solution set. |
| $(1,\infty)$ | 2 | $\frac{2(3)}{3+1}\le 1$ <br> $\frac{3}{2}\le 1$, false | $(1,\infty)$ does not belong to the solution set. |

The solution set is $(-1,1)$ or $\{x|-1<x<1\}$.

**57.** $|x^2+2x-36|>12$

Express the inequality without the absolute value symbol:

$x^2+2x-36<-12$ or $x^2+2x-36>12$

$x^2+2x-24<0$ $\quad x^2+2x-48>0$

Solve the related quadratic equations.

$x^2+2x-24=0$ or $x^2+2x-48=0$

$(x+6)(x-4)=0$ $\quad (x+8)(x-6)=0$

Apply the zero product principle.

$x+6=0$ or $x-4=0$ or $x+8=0$ or $x-6=0$

$x=-6$ $\quad x=4$ $\quad x=-8$ $\quad x=6$

The boundary points are $-8$, $-6$, $4$ and $6$.

| Test Interval | Test Number | Test | Conclusion |
|---|---|---|---|
| $(-\infty,-8)$ | $-9$ | $\|(-9)^2+2(-9)-36\|>12$ <br> $27>12$, True | $(-\infty,-8)$ belongs to the solution set. |
| $(-8,-6)$ | $-7$ | $\|(-7)^2+2(-7)-36\|>12$ <br> $1>12$, False | $(-8,-6)$ does not belong to the solution set. |
| $(-6,4)$ | 0 | $\|0^2+2(0)-36\|>12$ <br> $36>12$, True | $(-6,4)$ belongs to the solution set. |

| | | | |
|---|---|---|---|
| $(4,6)$ | 5 | $\lvert 5^2+2(5)-36\rvert >12$ <br> $1>12$, False | $(4,6)$ does not belong to the solution set. |
| $(6,\infty)$ | 7 | $\lvert 7^2+2(7)-36\rvert >12$ <br> $27>12$, True | $(6,\infty)$ belongs to the solution set. |

The solution set is $(-\infty,-8)\cup(-6,4)\cup(6,\infty)$ or $\{x \mid x<-8 \text{ or } -6<x<4 \text{ or } x>6\}$.

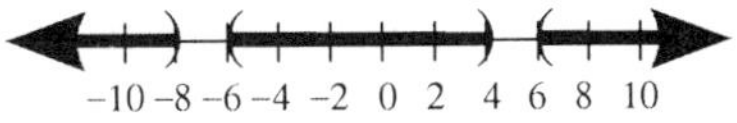


**58.** $\lvert x^2+6x+1\rvert >8$

Express the inequality without the absolute value symbol:

$x^2+6x+1<-8$ or $x^2+6x+1>8$

$x^2+6x+9<0$ $\quad$ $x^2+6x-7>0$

Solve the related quadratic equations.

$x^2+6x+9=0$ or $x^2+6x-7=0$

$(x+3)^2=0$ $\quad$ $(x+7)(x-1)=0$

$x+3=\pm\sqrt{0}$ or $x+7=0$ or $x-1=0$

$x+3=0$ $\quad$ $x=-7$ $\quad$ $x=1$

$x=-3$

The boundary points are $-7,-3,$ and $1$.

| Test Interval | Test Number | Test | Conclusion |
|---|---|---|---|
| $(-\infty,-7)$ | $-8$ | $\lvert(-8)^2+6(-8)+1\rvert >8$ <br> $17\geq 8$, True | $(-\infty,-7)$ belongs to the solution set. |
| $(-7,-3)$ | $-5$ | $\lvert(-5)^2+6(-5)+1\rvert >8$ <br> $4\geq 8$, False | $(-7,-3)$ does not belong to the solution set. |
| $(-3,1)$ | 0 | $\lvert 0^2+6(0)+1\rvert >8$ <br> $1\geq 8$, False | $(-3,1)$ does not belong to the solution set. |
| $(1,\infty)$ | 2 | $\lvert 2^2+6(2)+1\rvert >8$ <br> $17\geq 8$, True | $(1,\infty)$ belongs to the solution set. |

The solution set is $(-\infty,-7)\cup(1,\infty)$ or $\{x \mid x<-7 \text{ or } x>1\}$.

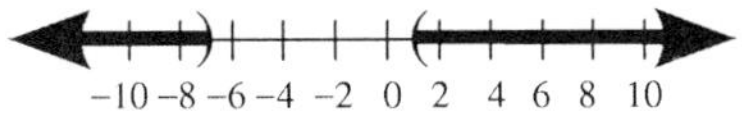

**59.** $\dfrac{3}{x+3} > \dfrac{3}{x-2}$

Express the inequality so that one side is zero.

$$\frac{3}{x+3} - \frac{3}{x-2} > 0$$

$$\frac{3(x-2)}{(x+3)(x-2)} - \frac{3(x+3)}{(x+3)(x-2)} > 0$$

$$\frac{3x-6-3x-9}{(x+3)(x-2)} < 0$$

$$\frac{-15}{(x+3)(x-2)} < 0$$

Find the values of $x$ that make the denominator zero.

$x+3=0 \qquad x-2=0$

$x=-3 \qquad x=2$

The boundary points are $-3$ and 2.

| Test Interval | Test Number | Test | Conclusion |
|---|---|---|---|
| $(-\infty,-3)$ | $-4$ | $\frac{3}{-4+3} > \frac{3}{-4-2}$ <br> $-3 > \frac{1}{2}$, False | $(-\infty,-3)$ does not belong to the solution set. |
| $(-3,2)$ | 0 | $\frac{3}{0+3} > \frac{3}{0-2}$ <br> $1 > -\frac{3}{2}$, True | $(-3,2)$ belongs to the solution set. |
| $(2,\infty)$ | 3 | $\frac{3}{3+3} > \frac{3}{3-2}$ <br> $\frac{1}{2} > 3$, False | $(2,\infty)$ does not belong to the solution set. |

The solution set is $(-3,2)$ or $\{x \mid -3 < x < 2\}$.

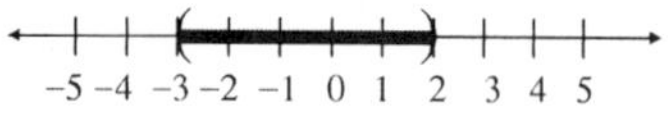

**60.** $\dfrac{1}{x+1} > \dfrac{2}{x-1}$

Express the inequality so that one side is zero.

$$\frac{1}{x+1} - \frac{2}{x-1} > 0$$

$$\frac{x-1}{(x+1)(x-1)} - \frac{2(x+1)}{(x+1)(x-1)} > 0$$

$$\frac{x-1-2x-2}{(x+1)(x-1)} < 0$$

$$\frac{-x-3}{(x+1)(x-1)} < 0$$

Find the values of $x$ that make the numerator and denominator zero.

$-x-3=0 \qquad x+1=0 \qquad x-1=0$

$-3=x \qquad x=-1 \qquad x=1$

The boundary points are $-3$, $-1$, and $1$.

| Test Interval | Test Number | Test | Conclusion |
|---|---|---|---|
| $(-\infty,-3)$ | $-4$ | $\frac{1}{-4+1} > \frac{2}{-3-1}$ <br> $-\frac{1}{3} > -\frac{1}{2}$, True | $(-\infty,-3)$ belongs to the solution set. |
| $(-3,-1)$ | $-2$ | $\frac{1}{-2+1} > \frac{2}{-2-1}$ <br> $-1 > -\frac{2}{3}$, False | $(-3,-1)$ does not belong to the solution set. |
| $(-1,1)$ | $0$ | $\frac{1}{0+1} > \frac{2}{0-1}$ <br> $1 > -2$, True | $(-3,1)$ belongs to the solution set. |
| $(1,\infty)$ | $2$ | $\frac{1}{2+1} > \frac{2}{2-1}$ <br> $\frac{1}{3} > 1$, False | $(1,\infty)$ does not belong to the solution set. |

The solution set is $(-\infty,-3)\cup(-1,1)$ or $\{x \mid x<-3 \text{ or } -1<x<1\}$.

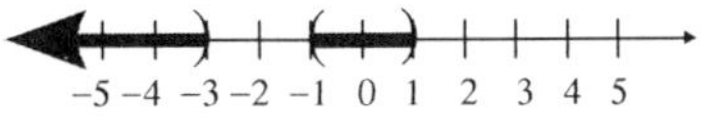

**61.** $\dfrac{x^2-x-2}{x^2-4x+3}>0$

Find the values of $x$ that make the numerator and denominator zero.

$x^2-x-2=0 \qquad x^2-4x+3=0$

$(x-2)(x+1)=0 \qquad (x-3)(x-1)=0$

Apply the zero product principle.

$x-2=0$ or $x+1=0 \qquad x-3=0$ or $x-1=0$

$x=2 \qquad x=-1 \qquad x=3 \qquad x=1$

The boundary points are $-1$, 1, 2 and 3.

| Test Interval | Test Number | Test | Conclusion |
|---|---|---|---|
| $(-\infty,-1)$ | $-2$ | $\dfrac{(-2)^2-(-2)-2}{(-2)^2-4(-2)+3}>0$ <br> $\dfrac{4}{15}>0$, True | $(-\infty,-1)$ belongs to the solution set. |
| $(-1,1)$ | 0 | $\dfrac{0^2-0-2}{0^2-4(0)+3}>0$ <br> $-\dfrac{2}{3}>0$, False | $(-1,1)$ does not belong to the solution set. |
| $(1,2)$ | 1.5 | $\dfrac{1.5^2-1.5-2}{1.5^2-4(1.5)+3}>0$ <br> $\dfrac{5}{3}>0$, True | $(1,2)$ belongs to the solution set. |
| $(2,3)$ | 2.5 | $\dfrac{2.5^2-2.5-2}{2.5^2-4(2.5)+3}>0$ <br> $-\dfrac{7}{3}>0$, False | $(2,3)$ does not belong to the solution set. |
| $(3,\infty)$ | 4 | $\dfrac{4^2-4-2}{4^2-4(4)+3}>0$ <br> $\dfrac{10}{3}>0$, True | $(3,\infty)$ belongs to the solution set. |

The solution set is $(-\infty,-1)\cup(1,2)\cup(3,\infty)$ or $\{x \mid x<-1 \text{ or } 1<x<2 \text{ or } x>3\}$.

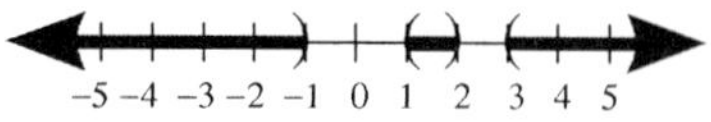

**62.** $\dfrac{x^2-3x+2}{x^2-2x-3}>0$

Find the values of $x$ that make the numerator and denominator zero.

$x^2-3x+2=0 \qquad x^2-2x-3=0$

$(x-2)(x-1)=0 \qquad (x-3)(x+1)=0$

Apply the zero product principle.

$x-2=0$ or $x-1=0 \qquad x-3=0$ or $x+1=0$

$x=2 \qquad x=1 \qquad x=3 \qquad x=-1$

The boundary points are $-1$, 1, 2 and 3.

| Test Interval | Test Number | Test | Conclusion |
|---|---|---|---|
| $(-\infty,-1)$ | $-2$ <br> $\dfrac{x^2-3x+2}{x^2-2x-3}>0$ | $\dfrac{(-2)^2-3(-2)+2}{(-2)^2-2(-2)-3}>0$ <br> $\dfrac{12}{5}>0$, True | $(-\infty,-1)$ belongs to the solution set. |
| $(-1,1)$ | 0 | $\dfrac{0^2-3(0)+2}{0^2-2(0)-3}>0$ <br> $-\dfrac{2}{3}>0$, False | $(-1,1)$ does not belong to the solution set. |
| $(1,2)$ | 1.5 | $\dfrac{1.5^2-3(1.5)+2}{1.5^2-2(1.5)-3}>0$ <br> $\dfrac{1}{15}>0$, True | $(1,2)$ belongs to the solution set. |
| $(2,3)$ | 2.5 | $\dfrac{2.5^2-3(2.5)+2}{2.5^2-2(2.5)-3}>0$ <br> $-\dfrac{3}{7}>0$, False | $(2,3)$ does not belong to the solution set. |
| $(3,\infty)$ | 4 | $\dfrac{4^2-3(4)+2}{4^2-2(4)-3}>0$ <br> $\dfrac{6}{5}>0$, True | $(3,\infty)$ belongs to the solution set. |

The solution set is $(-\infty,-1)\cup(1,2)\cup(3,\infty)$ or $\{x \mid x<-1 \text{ or } 1<x<2 \text{ or } x>3\}$.

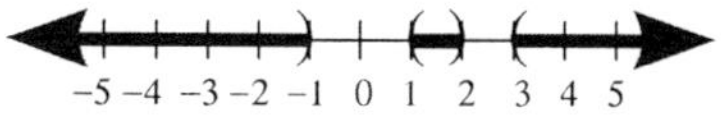

**63.** $$2x^3+11x^2\ge 7x+6$$
$$2x^3+11x^2-7x-6\ge 0$$

The graph of $f(x)=2x^3+11x^2-7x-6$ appears to cross the $x$-axis at $-6$, $-\frac{1}{2}$, and 1. We verify this numerically by substituting these values into the function:

$$f(-6)=2(-6)^3+11(-6)^2-7(-6)-6=2(-216)+11(36)-(-42)-6=-432+396+42-6=0$$
$$f\left(-\frac{1}{2}\right)=2\left(-\frac{1}{2}\right)^3+11\left(-\frac{1}{2}\right)^2-7\left(-\frac{1}{2}\right)-6=2\left(-\frac{1}{8}\right)+11\left(\frac{1}{4}\right)-\left(-\frac{7}{2}\right)-6=-\frac{1}{4}+\frac{11}{4}+\frac{7}{2}-6=0$$
$$f(1)=2(1)^3+11(1)^2-7(1)-6=2(1)+11(1)-7-6=2+11-7-6=0$$

Thus, the boundaries are $-6$, $-\frac{1}{2}$, and 1. We need to find the intervals on which $f(x)\ge 0$. These intervals are indicated on the graph where the curve is above the $x$-axis. Now, the curve is above the $x$-axis when $-6<x<-\frac{1}{2}$ and when $x>1$. Thus, the solution set is

$\left\{x \middle| -6\le x\le -\frac{1}{2} \text{ or } x\ge 1\right\}$ or $\left[-6,-\frac{1}{2}\right]\cup[1,\infty)$.

**64.** $$2x^3+11x^2<7x+6$$
$$2x^3+11x^2-7x-6<0$$

In Problem 63, we verified that the boundaries are $-6$, $-\frac{1}{2}$, and 1. We need to find the intervals on which $f(x)<0$. These intervals are indicated on the graph where the curve is below the $x$-axis. Now, the curve is below the $x$-axis when $x<-6$ and when $-\frac{1}{2}<x<1$.

Thus, the solution set is $\left\{x \middle| x<-6 \text{ or } -\frac{1}{2}<x<1\right\}$ or $(-\infty,-6)\cup\left(-\frac{1}{2},1\right)$.

**65.** $$\frac{1}{4(x+2)}\le -\frac{3}{4(x-2)}$$
$$\frac{1}{4(x+2)}+\frac{3}{4(x-2)}\le 0$$

Simplify the left side of the inequality:

$$\frac{x-2}{4(x+2)}+\frac{3(x+2)}{4(x-2)}=\frac{x-2+3x+6}{4(x+2)(x-2)}=\frac{4x+4}{4(x+2)(x-2)}=\frac{4(x+1)}{4(x+2)(x-2)}=\frac{x+1}{x^2-4}.$$

The graph of $f(x)=\frac{x+1}{x^2-4}$ crosses the $x$-axis at $-1$, and has vertical asymptotes at $x=-2$ and $x=2$. Thus, the boundaries are $-2$, $-1$, and 1. We need to find the intervals on which

$f(x) \le 0$. These intervals are indicated on the graph where the curve is below the $x$-axis. Now, the curve is below the $x$-axis when $x < -2$ and when $-1 < x < 2$. Thus, the solution set is $\{x | x < -2 \text{ or } -1 \le x < 2\}$ or $(-\infty, -2) \cup [-1, 2)$.

**66.**
$$\frac{1}{4(x+2)} > -\frac{3}{4(x-2)}$$
$$\frac{1}{4(x+2)} + \frac{3}{4(x-2)} > 0$$

In Problem 65, we found that the boundaries are $-2$, $-1$, and $1$. We need to find the intervals on which $f(x) > 0$. These intervals are indicated on the graph where the curve is above the $x$-axis. Now, the curve is above the $x$-axis when $-2 < x < -1$ and when $x > 2$. Thus, the solution set is $\{x | -2 < x \le -1 \text{ or } x > 2\}$ or $(-2, -1] \cup (2, \infty)$.

**67.** $s(t) = -16t^2 + 48t + 160$

To find when the height exceeds the height of the building, solve the inequality
$-16t^2 + 48t + 160 > 160$.
Solve the related quadratic equation.
$$-16t^2 + 48t + 160 = 160$$
$$-16t^2 + 48t = 0$$
$$t^2 - 3t = 0$$
$$t(t-3) = 0$$
Apply the zero product principle.
$t = 0$ or $t - 3 = 0$
$t = 3$

The boundary points are 0 and 3.

| Test Interval | Test Number | Test | Conclusion |
|---|---|---|---|
| $(0,3)$ | 1 | $-16(1)^2 + 48(1) + 160 > 160$<br>$192 > 160$, true | $(0,3)$ belongs to the solution set. |
| $(3,\infty)$ | 4 | $-16(4)^2 + 48(4) + 160 > 160$<br>$96 > 160$, true | $(3,\infty)$ does not belong to the solution set. |

The solution set is $(0,3)$. This means that the ball exceeds the height of the building between 0 and 3 seconds.

**69.** $f(8) = 27(8) + 163 = 216 + 163 = 379$

$g(8) = 1.2(8)^2 + 15.2(8) + 181.4 = 1.2(64) + 121.6 + 181.4$

$= 76.8 + 121.6 + 181.4 = 379.8$

Since the graph indicates that Medicare spending will reach $379 billion, we conclude that both functions model the data quite well.

**71.** $g(x) = 1.2x^2 + 15.2x + 181.4$

To find when spending exceeds $536.6 billion, solve the inequality

$1.2x^2 + 15.2x + 181.4 > 536.6.$

Solve the related quadratic equation using the quadratic formula.

$1.2x^2 + 15.2x + 181.4 = 536.6$

$1.2x^2 + 15.2x - 355.2 = 0$

$a = 1.2 \quad b = 15.2 \quad c = -355.2$

$$x = \frac{-15.2 \pm \sqrt{15.2^2 - 4(1.2)(-355.2)}}{2(1.2)} = \frac{-15.2 \pm \sqrt{231.04 + 1704.96}}{2.4}$$

$$= \frac{-15.2 \pm \sqrt{1936}}{2.4} = \frac{-15.2 \pm 44}{2.4}$$

$$= \frac{-15.2 - 44}{2.4} \text{ or } \frac{-15.2 + 44}{2.4} = -24\frac{2}{3} \text{ or } 12$$

We disregard $-24\frac{2}{3}$ since $x$ represents the number of years after 1995 and cannot be negative.

The boundary point is 12.

| Test Interval | Test Number | Test | Conclusion |
|---|---|---|---|
| $(0, 12)$ | 1 | $1.2(1)^2 + 15.2(1) + 181.4 > 536.6$<br>$197.8 > 536.6$, false | $(0, 12)$ does not belong to the solution set. |
| $(12, \infty)$ | 13 | $1.2(13)^2 + 15.2(13) + 181.4 > 536.6$<br>$581.8 > 536.6$, true | $(12, \infty)$ belongs to the solution set. |

The solution set is $(12, \infty)$. This means that spending will exceed $536.6 billion after 1995 + 12 = 2007.

**73.** $\overline{C}(x) = \dfrac{500{,}000 + 400x}{x}$

To find when the cost of producing each wheelchair does not exceed $425, solve the inequality

$\dfrac{500{,}000 + 400x}{x} \le 425.$

Express the inequality so that one side is zero.

$$\frac{500,000+400x}{x}-425\le 0$$
$$\frac{500,000+400x}{x}-\frac{425x}{x}\le 0$$
$$\frac{500,000+400x-425x}{x}\le 0$$
$$\frac{500,000-25x}{x}\le 0$$

Find the values of $x$ that make the numerator and denominator zero.

$$500,000-25x=0 \qquad x=0$$
$$500,000=25x$$
$$20,000=x$$

The boundary points are 0 and 20,000.

| Test Interval | Test Number | Test | Conclusion |
|---|---|---|---|
| $[0,20000]$ | 1 | $\frac{500,000+400(1)}{1}\le 425$ <br> $500,400\le 425$, false | $[0,20000]$ does not belong to the solution set. |
| $[20000,\infty)$ | 25,000 | $\frac{500,000+400(25,000)}{25,000}\le 425$ <br> $420\le 425$, true | $[20000,\infty)$ belongs to the solution set. |

The solution set is $[20000,\infty)$. This means that the company's production level will have to be at least 20,000 wheelchairs per week. The boundary corresponds to the point (20,000, 425) on the graph. When production is 20,000 or more per month, the average cost is \$425 or less.

**75.** Let $x$ = the length of the rectangle.
Since $\text{Perimeter}=2(\text{length})+2(\text{width})$, we know

$$50=2x+2(\text{width})$$
$$50-2x=2(\text{width})$$
$$\text{width}=\frac{50-2x}{2}=25-x$$

Now, $A=(\text{length})(\text{width})$, so we have that

$$A(x)\le 114$$
$$x(25-x)\le 114$$
$$25x-x^2\le 114$$

Solve the related equation

$$25x - x^2 = 114$$
$$0 = x^2 - 25x + 114$$
$$0 = (x-19)(x-6)$$

Apply the zero product principle:

$x - 19 = 0$ or $x - 6 = 0$

$x = 19$ $\quad$ $x = 6$

The boundary points are 6 and 19.

| Test Interval | Test Number | Test | Conclusion |
|---|---|---|---|
| $(-\infty, 6)$ | 0 | $25(0) - 0^2 \le 114$<br>$0 \le 114$, True | $(-\infty, 6)$ belongs to the solution set. |
| $(6, 19)$ | 10 | $25(10) - 10^2 \le 114$<br>$150 \le 114$, False | $(6, 19)$ does not belong to the solution set. |
| $(19, \infty)$ | 20 | $25(20) - 20^2 \le 114$<br>$100 \le 114$, True | $(19, \infty)$ belongs to the solution set. |

If the length is 6 feet, then the width is 19 feet. If the length is less than 6 feet, then the width is greater than 19 feet. Thus, if the area of the rectangle is not to exceed 114 square feet, the length of the shorter side must be 6 feet or less.

**77. – 79.** Answers will vary.

**81.** $2x^2 + 5x - 3 \le 0$

Let $y_1 = 2x^2 + 5x - 3$.

```
WINDOW
 Xmin=-10
 Xmax=10
 Xscl=1
 Ymin=-10
 Ymax=10
 Yscl=1
 Xres=1
```

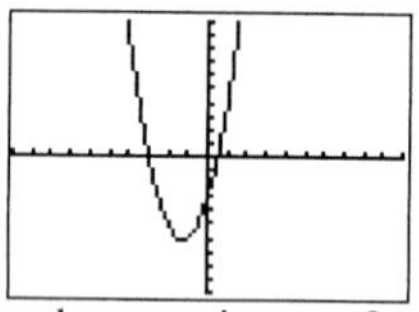

The graph is crosses the $x$-axis at $-3$ and $\frac{1}{2}$. The graph is below the $x$-axis when $-3 < x < \frac{1}{2}$. Thus, the solution set is $\left\{x \middle| -3 \le x \le \frac{1}{2}\right\}$ or $\left[-3, \frac{1}{2}\right]$.

**83.** $\frac{x+2}{x-3} \le 2$

$\frac{x+2}{x-3} - 2 \le 0$

Let $y_1 = \frac{x+2}{x-3} - 2$.

```
WINDOW
 Xmin=-10
 Xmax=10
 Xscl=1
 Ymin=-10
 Ymax=10
 Yscl=1
 Xres=1
```

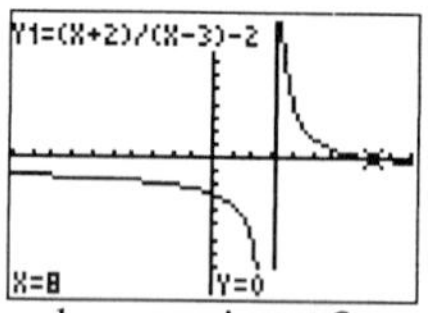


The graph is crosses the $x$-axis at 8. The function has a vertical asymptote at $x = 3$. The graph is below the $x$-axis when $x < 3$ and when $x > 8$. Thus, the solution set is $\{x \mid x < 3$ or $x \ge 8\}$ or $(-\infty, 3) \cup [8, \infty)$.

**85.** $x^3 + 2x^2 - 5x - 6 > 0$

Let $y_1 = x^3 + 2x^2 - 5x - 6$

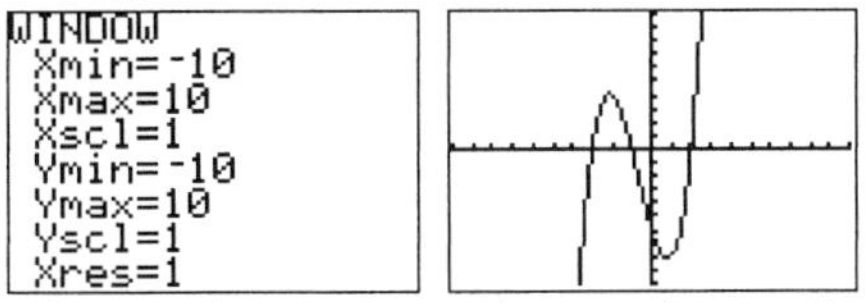


The graph is crosses the $x$-axis at $-3$, $-1$, and 2. The graph is above the $x$-axis when $-3 < x < -1$ and when $x > 2$. Thus, the solution set is $\{x \mid -3 < x < -1 \text{ or } x > 2\}$ or $(-3,-1) \cup (2,\infty)$.

**87.** Answers will vary. An example is as follows.

$$x = -3 \qquad x = 5$$
$$x + 3 = 0 \qquad x - 5 = 0$$
$$(x+3)(x-5) = 0$$
$$x^2 - 2x - 15 = 0.$$

Since $x$ falls between two numbers the inequality symbol should be less than or equal to, so the inequality is $x^2 - 2x - 15 \le 0$.

**89.** The left hand side of the inequality is zero when $x$ is 2, so the solution set is $\{x \mid x \text{ is a real number and } x \ne 2\}$ or $(-\infty, 2) \cup (2, \infty)$.

**91.** There is no value of $x$ for which the left hand side will be less than $-1$. The solution set is $\varnothing$ or $\{\ \}$.

**93.** **a.** The $x$–axis is the divider between $y$–values that are positive and $y$–values that are negative. Since the entire graph falls above the $x$–axis, we know that all of the corresponding $y$–values are positive. Since the inequality is greater than zero, the solution set is $\{x \mid x \text{ is a real number}\}$ or $(-\infty, \infty)$.

**b.** Since the entire graph falls above the $x$–axis, we know that all of the corresponding $y$–values are negative. Since the inequality is less than zero, the solution set is $\varnothing$ or $\{\ \}$.

**c.** First, consider $4x^2 - 8x + 7 > 0$ from part **a.**

$$4x^2 - 8x + 7 = 0$$
$$a = 4 \qquad b = -8 \qquad c = 7$$
$$x = \frac{-(-8) \pm \sqrt{(-8)^2 - 4(4)(7)}}{2(4)}$$
$$= \frac{8 \pm \sqrt{64 - 112}}{8}$$
$$= \frac{8 \pm \sqrt{-48}}{8}$$
$$= \frac{8 \pm \sqrt{-16 \cdot 3}}{8}$$
$$= \frac{8 \pm 4\sqrt{3}i}{8} = \frac{8}{8} \pm \frac{4\sqrt{3}i}{8} = 1 \pm \frac{\sqrt{3}}{2}i$$

The graph of $4x^2 - 8x + 7 = 0$ is a parabola. Since the values of $x$ are complex numbers, we know there are no $x$–intercepts. This means that the entire graph lies above the $x$–axis and the value of $4x^2 - 8x + 7$ will always be greater than zero. As a result, the solution set is $\{x \mid x \text{ is a real number}\}$ or $(-\infty, \infty)$.

Next, consider $4x^2 - 8x + 7 < 0$ from part **b.** From above, we know that $4x^2 - 8x + 7$ will always be greater than zero and never less than zero. As a result, the solution set is the solution set is $\varnothing$ or $\{\ \}$.

**95.** $\left|\frac{x-5}{3}\right| < 8$

$-8 < \frac{x-5}{3} < 8$

$-24 < x-5 < 24$

$-19 < x < 29$

The solution set is $\{x|-19 < x < 29\}$ or $(-19, 29)$.

**96.** $\frac{2x+6}{x^2+8x+16} \div \frac{x^2-9}{x^2+3x-4}$

$= \frac{2x+6}{x^2+8x+16} \cdot \frac{x^2+3x-4}{x^2-9}$

$= \frac{2\cancel{(x+3)}}{\cancel{(x+4)}(x+4)} \cdot \frac{\cancel{(x+4)}(x-1)}{\cancel{(x+3)}(x-3)}$

$= \frac{2(x-1)}{(x+4)(x-3)}$

**97.** $x^4 - 16y^4$

$= (x^2+4y^2)(x^2-4y^2)$

$= (x^2+4y^2)(x+2y)(x-2y)$

## Chapter 8 Review Exercises

**1.** $2x^2 - 3 = 125$

$2x^2 = 128$

$x^2 = 64$

$x = \pm 8$

The solutions are –8 and 8, and the solution set is $\{-8, 8\}$.

**2.** $3x^2 - 150 = 0$

$3x^2 = 150$

$x^2 = 50$

$x = \pm\sqrt{50}$

$x = \pm\sqrt{25 \cdot 2}$

$x = \pm 5\sqrt{2}$

The solutions are $-5\sqrt{2}$ and $5\sqrt{2}$, and the solution set is $\{-5\sqrt{2}, 5\sqrt{2}\}$.

**3.** $3x^2 - 2 = 0$

$3x^2 = 2$

$x^2 = \frac{2}{3}$

$x = \pm\sqrt{\frac{2}{3}}$

Rationalize the denominator.

$x = \pm\frac{\sqrt{2}}{\sqrt{3}} \cdot \frac{\sqrt{3}}{\sqrt{3}} = \pm\frac{\sqrt{6}}{3}$

The solutions are $-\frac{\sqrt{6}}{3}$ and $\frac{\sqrt{6}}{3}$, and the solution set is $\left\{-\frac{\sqrt{6}}{3}, \frac{\sqrt{6}}{3}\right\}$.

**4.** $(x-4)^2 = 18$

$x - 4 = \pm\sqrt{18}$

$x = 4 \pm \sqrt{9 \cdot 2}$

$x = 4 \pm 3\sqrt{2}$

The solutions are $4 - 3\sqrt{2}$ and $4 + 3\sqrt{2}$, and the solution set is $\{4 - 3\sqrt{2}, 4 + 3\sqrt{2}\}$.

**5.** $(x+7)^2 = -36$

$x+7 = \pm\sqrt{-36}$

$x = -7 \pm 6i$

The solutions are $-7-6i$ and $-7+6i$, and the solution set is $\{-7-6i, -7+6i\}$.

**6.** $x^2 + 20x +$ _____

Since $b = 20$, we add

$\left(\frac{b}{2}\right)^2 = \left(\frac{20}{2}\right)^2 = (10)^2 = 100.$

$x^2 + 20x + 100 = (x+10)^2$

**7.** $x^2 - 3x +$ _____

Since $b = 3$, we add

$\left(\frac{b}{2}\right)^2 = \left(\frac{3}{2}\right)^2 = \frac{9}{4}.$

$x^2 - 3x + \frac{9}{4} = \left(x - \frac{3}{2}\right)^2$

**8.** $x^2 - 12x + 27 = 0$

$x^2 - 12x \quad = -27$

Since $b = !12$, we add

$\left(\frac{b}{2}\right)^2 = \left(\frac{-12}{2}\right)^2 = (-6)^2 = 36.$

$x^2 - 12x + 27 = 0$

$x^2 - 12x + 36 = -27 + 36$

$(x-6)^2 = 9$

Apply the square root property.

$x - 6 = 3 \qquad x - 6 = -3$

$x = 9 \qquad x = 3$

The solutions are 3 and 9 and the solution set is $\{3, 9\}$.

**9.** $x^2 - 7x - 1 = 0$

$x^2 - 7x \quad = 1$

Since $b = !7$, we add

$\left(\frac{b}{2}\right)^2 = \left(\frac{-7}{2}\right)^2 = \frac{49}{4}.$

$x^2 - 7x + \frac{49}{4} = 1 + \frac{49}{4}$

$\left(x - \frac{7}{2}\right)^2 = \frac{4}{4} + \frac{49}{4}$

$\left(x - \frac{7}{2}\right)^2 = \frac{53}{4}$

Apply the square root property.

$x - \frac{7}{2} = \pm\sqrt{\frac{53}{4}}$

$x = \frac{7}{2} \pm \frac{\sqrt{53}}{2} = \frac{7 \pm \sqrt{53}}{2}$

The solutions are $\frac{7 \pm \sqrt{53}}{2}$ and the solution set is $\left\{\frac{7 \pm \sqrt{53}}{2}\right\}$.

**10.** $2x^2 + 3x - 4 = 0$

$x^2 + \frac{3}{2}x - 2 = 0$

$x^2 + \frac{3}{2}x \quad = 2$

Since $b = \frac{3}{2}$, we add

$\left(\frac{b}{2}\right)^2 = \left(\frac{\frac{3}{2}}{2}\right)^2 = \left(\frac{3}{2} \div 2\right)^2$

$= \left(\frac{3}{2} \cdot \frac{1}{2}\right)^2 = \left(\frac{3}{4}\right)^2 = \frac{9}{16}.$

$$x^2+\frac{3}{2}x+\frac{9}{16}=2+\frac{9}{16}$$

$$\left(x+\frac{3}{4}\right)^2=\frac{32}{16}+\frac{9}{16}$$

$$\left(x+\frac{3}{4}\right)^2=\frac{41}{16}$$

Apply the square root property.

$$x+\frac{3}{4}=\pm\sqrt{\frac{41}{16}}$$

$$x=-\frac{3}{4}\pm\frac{\sqrt{41}}{4}$$

$$x=\frac{-3\pm\sqrt{41}}{4}$$

The solutions are $\frac{-3\pm\sqrt{41}}{4}$ and the solution set is $\left\{\frac{-3\pm\sqrt{41}}{4}\right\}$.

**11.** $A=P(1+r)^t$

$$2916=2500(1+r)^2$$

$$\frac{2916}{2500}=(1+r)^2$$

Apply the square root property.

$$1+r=\pm\sqrt{\frac{2916}{2500}}$$

$$r=-1\pm\sqrt{1.1664}$$

$$r=-1\pm1.08$$

The solutions are –1 – 1.08 = –2.08 and –1 + 1.08 = 0.08. We disregard –2.08 since we cannot have a negative interest rate. The interest rate is 0.08 or 8%.

**12.** $W(t)=3t^2$

$$588=3t^2$$

$$196=t^2$$

Apply the square root property.

$$t^2=196$$

$$t=\pm\sqrt{196}$$

$$t=\pm14$$

The solutions are –14 and 14. We disregard –14, because we cannot have a negative time measurement. The fetus will weigh 588 grams after 14 weeks.

**13.**

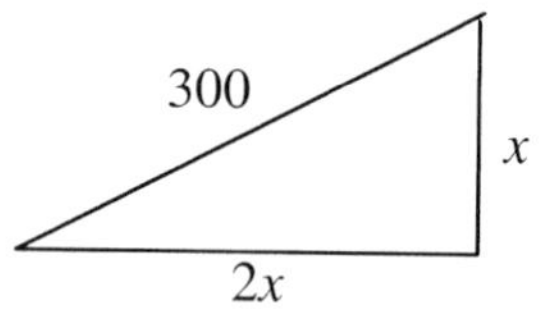


Use the Pythagorean Theorem.

$$(2x)^2+x^2=300^2$$

$$4x^2+x^2=90,000$$

$$5x^2=90,000$$

$$x^2=18,000$$

$$x=\pm\sqrt{18,000}$$

$$x=\pm\sqrt{3600\cdot5}$$

$$x=\pm60\sqrt{5}$$

The solutions are $\pm60\sqrt{5}$ meters. We disregard $-60\sqrt{5}$ meters, because we can't have a negative length measurement. Therefore, the building is $60\sqrt{5}$ meters, or approximately 134.2 meters high.

**14.**

$$x^2 = 2x + 4$$
$$x^2 - 2x - 4 = 0$$
$$a = 1 \quad b = -2 \quad c = -4$$
$$x = \frac{-(-2) \pm \sqrt{(-2)^2 - 4(1)(-4)}}{2(1)}$$
$$= \frac{2 \pm \sqrt{4 + 16}}{2}$$
$$= \frac{2 \pm \sqrt{20}}{2}$$
$$= \frac{2 \pm \sqrt{4 \cdot 5}}{2}$$
$$= \frac{2 \pm 2\sqrt{5}}{2} = \frac{2(1 \pm \sqrt{5})}{2} = 1 \pm \sqrt{5}$$

The solutions are $1 \pm \sqrt{5}$, and the solution set is $\{1 \pm \sqrt{5}\}$.

**15.**

$$x^2 - 2x + 19 = 0$$
$$a = 1 \quad b = -2 \quad c = 19$$
$$x = \frac{-(-2) \pm \sqrt{(-2)^2 - 4(1)(19)}}{2(1)}$$
$$= \frac{2 \pm \sqrt{4 - 76}}{2}$$
$$= \frac{2 \pm \sqrt{-72}}{2}$$
$$= \frac{2 \pm \sqrt{-36 \cdot 2}}{2}$$
$$= \frac{2 \pm 6\sqrt{2}i}{2} = \frac{2(1 \pm 3\sqrt{2}i)}{2} = 1 \pm 3\sqrt{2}i$$

The solutions are $1 \pm 3\sqrt{2}i$, and the solution set is $\{1 \pm 3\sqrt{2}i\}$.

**16.**

$$2x^2 = 3 - 4x$$
$$2x^2 + 4x - 3 = 0$$
$$a - 2 \quad b = 4 \quad c = -3$$
$$x = \frac{-4 \pm \sqrt{4^2 - 4(2)(-3)}}{2(2)}$$
$$= \frac{-4 \pm \sqrt{16 + 24}}{4}$$
$$= \frac{-4 \pm \sqrt{40}}{4}$$
$$= \frac{-4 \pm \sqrt{4 \cdot 10}}{4}$$
$$= \frac{-4 \pm 2\sqrt{10}}{4}$$
$$= \frac{2(-2 \pm \sqrt{10})}{4} = \frac{-2 \pm \sqrt{10}}{2}$$

The solutions are $\frac{-2 \pm \sqrt{10}}{2}$, and the solution set is $\left\{\frac{-2 \pm \sqrt{10}}{2}\right\}$.

**17.**

$$x^2 - 4x + 13 = 0$$
$$a = 1 \quad b = -4 \quad c = 13$$

Find the discriminant.

$$b^2 - 4ac = (-4)^2 - 4(1)(13)$$
$$= 16 - 52 = -36$$

Since the discriminant is negative, there are two imaginary solutions which are complex conjugates.

**18.**

$$9x^2 = 2 - 3x$$
$$9x^2 + 3x - 2 = 0$$
$$a = 9 \quad b = 3 \quad c = -2$$

Find the discriminant.

$$b^2 - 4ac = 3^2 - 4(9)(-2)$$
$$= 9 + 72 = 81$$

Since the discriminant is greater than zero, there are two unequal real solutions. Also, since the discriminant is a perfect square, the solutions are rational.

**19.** $2x^2+4x=3$

$2x^2+4x-3=0$

$a=2 \quad b=4 \quad c=-3$

Find the discriminant.

$b^2-4ac=4^2-4(2)(-3)$

$=16+24=40$

Since the discriminant is greater than zero, there are two unequal real solutions. Also, since the discriminant is not a perfect square, the solutions are irrational.

**20.** $3x^2-10x-8=0$

$(3x+2)(x-4)=0$

Apply the zero product principle.

$3x+2=0$ and $x-4=0$

$3x=-2 \qquad x=4$

$x=-\frac{2}{3}$

The solutions are $-\frac{2}{3}$ and 4, and the solution set is $\left\{-\frac{2}{3},4\right\}$.

**21.** $(2x-3)(x+2)=x^2-2x+4$

$2x^2+4x-3x-6=x^2-2x+4$

$x^2+3x-10=0$

Use the quadratic formula.

$a=1 \quad b=3 \quad c=-10$

$$x=\frac{-3\pm\sqrt{3^2-4(1)(-10)}}{2(1)}$$

$$=\frac{-3\pm\sqrt{9-(-40)}}{2}$$

$$=\frac{-3\pm\sqrt{49}}{2}=\frac{-3\pm7}{2}=-5 \text{ or } 2$$

The solutions are $-5$ and 2, and the solution set is $\{-5,\ 2\}$.

**22.** $5x^2-x-1=0$

Use the quadratic formula.

$a=5 \quad b=-1 \quad c=-1$

$$x=\frac{-(-1)\pm\sqrt{(-1)^2-4(5)(-1)}}{2(5)}$$

$$=\frac{1\pm\sqrt{1-(-20)}}{10}=\frac{1\pm\sqrt{21}}{10}$$

The solutions are $\frac{1\pm\sqrt{21}}{10}$, and the solution set is $\left\{\frac{1\pm\sqrt{21}}{10}\right\}$.

**23.** $x^2-16=0$

$x^2=16$

Apply the square root principle.

$x=\pm\sqrt{16}=\pm4$

The solutions are –4 and 4, and the solution set is $\{-4,4\}$.

**24.** $(x-3)^2-8=0$

$(x-3)^2=8$

Apply the square root principle.

$x-3=\pm\sqrt{8}$

$x=3\pm\sqrt{4\cdot2}$

$x=3\pm2\sqrt{2}$

The solutions are $3\pm2\sqrt{2}$, and the solution set is $\{3\pm2\sqrt{2}\}$.

**25.** $3x^2-x+2=0$

Use the quadratic formula.

$a=3 \quad b=-1 \quad c=2$

$$x=\frac{-(-1)\pm\sqrt{(-1)^2-4(3)(2)}}{2(3)}$$

$$=\frac{1\pm\sqrt{1-24}}{6}$$

$$=\frac{1\pm\sqrt{-23}}{6}=\frac{1}{6}\pm\frac{\sqrt{23}}{6}i$$

The solutions are $\frac{1}{6} \pm \frac{\sqrt{23}}{6}i$, and the solution set is $\left\{\frac{1}{6} \pm \frac{\sqrt{23}}{6}i\right\}$.

**26.**
$$\frac{5}{x+1} + \frac{x-1}{4} = 2$$
$$4(x+1)\left(\frac{5}{x+1} + \frac{x-1}{4}\right) = 4(x+1)(2)$$
$$20 + (x+1)(x-1) = 8x+8$$
$$20 + x^2 - 1 = 8x+8$$
$$x^2 - 8x + 11 = 0$$

Use the quadratic formula.
$a = 1 \quad b = -8 \quad c = 11$
$$x = \frac{-(-8) \pm \sqrt{(-8)^2 - 4(1)(11)}}{2(1)}$$
$$= \frac{8 \pm \sqrt{64-44}}{2}$$
$$= \frac{8 \pm \sqrt{20}}{2}$$
$$= \frac{8 \pm \sqrt{4 \cdot 5}}{2}$$
$$= \frac{8 \pm 2\sqrt{5}}{2} = \frac{2(4 \pm \sqrt{5})}{2} = 4 \pm \sqrt{5}$$

The solutions are $4 \pm \sqrt{5}$, and the solution set is $\{4 \pm \sqrt{5}\}$.

**27.** Because the solution set is $\left\{-\frac{1}{3}, \frac{3}{5}\right\}$, we have
$$x = -\frac{1}{3} \quad \text{or} \quad x = \frac{3}{5}$$
$$3x = -1 \qquad 5x = 3$$
$$3x + 1 = 0 \qquad 5x - 3 = 0.$$

Apply the zero-product principle in reverse.
$$(3x+1)(5x-3) = 0$$
$$15x^2 - 9x + 5x - 3 = 0$$
$$15x^2 - 4x - 3 = 0$$

**28.** Because the solution set is $\{-9i,\ 9i\}$, we have
$$x = -9i \quad \text{or} \quad x = 9i$$
$$x + 9i = 0 \qquad x - 9i = 0.$$

Apply the zero-product principle in reverse.
$$(x+9i)(x-9i) = 0$$
$$x^2 - 9ix + 9ix - 81i^2 = 0$$
$$x^2 - 81(-1) = 0$$
$$x^2 + 81 = 0$$

**29.** Because the solution set is $\{-4\sqrt{3}, 4\sqrt{3}\}$, we have
$$x = -4\sqrt{3} \quad \text{or} \quad x = 4\sqrt{3}$$
$$x + 4\sqrt{3} = 0 \qquad x - 4\sqrt{3} = 0.$$

Apply the zero product principle in reverse.
$$(x + 4\sqrt{3})(x - 4\sqrt{3}) = 0$$
$$x^2 - (4\sqrt{3})^2 = 0$$
$$x^2 - 16 \cdot 3 = 0$$
$$x^2 - 48 = 0$$

**30.** $1020 = 23x^2 - 259x + 816$

$0 = 23x^2 - 259x - 204$

Apply the Pythagorean Theorem.

$a = 23 \quad b = -259 \quad c = -204$

$$x = \frac{-(-259) \pm \sqrt{(-259)^2 - 4(23)(-204)}}{2(23)}$$

$$= \frac{259 \pm \sqrt{67{,}081 - (-18{,}768)}}{46}$$

$$= \frac{259 \pm \sqrt{85{,}849}}{46}$$

$$= \frac{259 \pm 293}{46} = 12 \text{ or } -\frac{17}{23}$$

We disregard $-\frac{17}{23}$ because we cannot have a negative number of years. The solution is 12. We conclude that 1020 police officers were convicted in the year 1990 + 12 = 2002.

**31.** $0 = -16t^2 + 140t + 3$

Apply the Pythagorean Theorem.

$a = -16 \quad b = 140 \quad c = 3$

$$= \frac{-140 \pm \sqrt{19{,}600 + 192}}{-32}$$

$$= \frac{-140 \pm \sqrt{19{,}792}}{-32}$$

$$\approx \frac{-140 \pm 140.7}{-32}$$

$$\approx \frac{-140 - 140.7}{-32} \text{ or } \frac{-140 + 140.7}{-32}$$

$$\approx \frac{-280.7}{-32} \text{ or } \frac{0.7}{-32}$$

$\approx 8.8$ or $-0.02$

We disregard –0.02 because we cannot have a negative time measurement. The solution is approximately 8.8. We conclude that the ball will hit the ground in about 8.8 seconds.

**32.** $f(x) = -(x+1)^2 + 4$

Since $a = -1$ is negative, the parabola opens downward. The vertex of the parabola is $(h,k) = (-1,4)$ and the axis of symmetry is $x = -1$. Replace $f(x)$ with 0 to find $x$–intercepts.

$$0 = -(x+1)^2 + 4$$

$(x+1)^2 = 4$

Apply the square root property.

$x + 1 = \sqrt{4}$ or $x + 1 = -\sqrt{4}$

$x + 1 = 2 \qquad x + 1 = -2$

$x = 1 \qquad x = -3$

The $x$–intercepts are 1 and –3. Set $x = 0$ and solve for $y$ to obtain the $y$–intercept.

$y = -(0+1)^2 + 4$

$y = -(1)^2 + 4$

$y = -1 + 4 = 3$

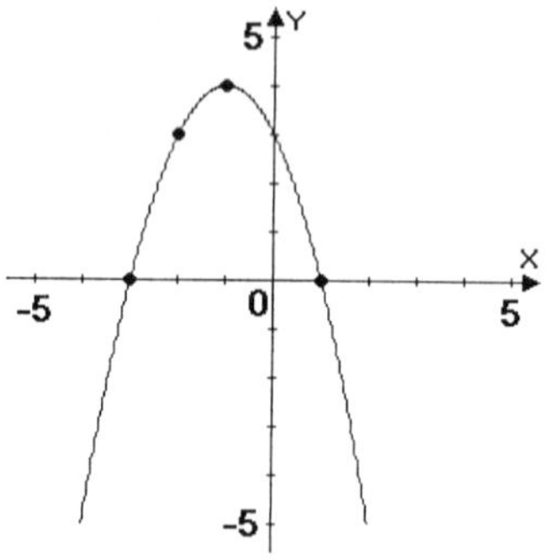


Axis of symmetry: $x = -1$.

**33.** $f(x) = (x+4)^2 - 2$

Since $a = 1$ is positive, the parabola opens upward. The vertex of the parabola is $(h,k) = (-4,-2)$ and the axis of symmetry is $x = -4$. Replace $f(x)$ with 0 to find $x$–intercepts.

$0=(x+4)^2-2$

$2=(x+4)^2$

Apply the square root property.

$x+4=\sqrt{2}$ or $x+4=-\sqrt{2}$

$x=-4+\sqrt{2}$ $\quad x=-4-\sqrt{2}$

The $x$–intercepts are $-4-\sqrt{2}$ and $-4+\sqrt{2}$. Set $x=0$ and solve for $y$ to obtain the $y$–intercept.

$y=(0+4)^2-2$

$y=4^2-2$

$y=16-2$

$y=14$

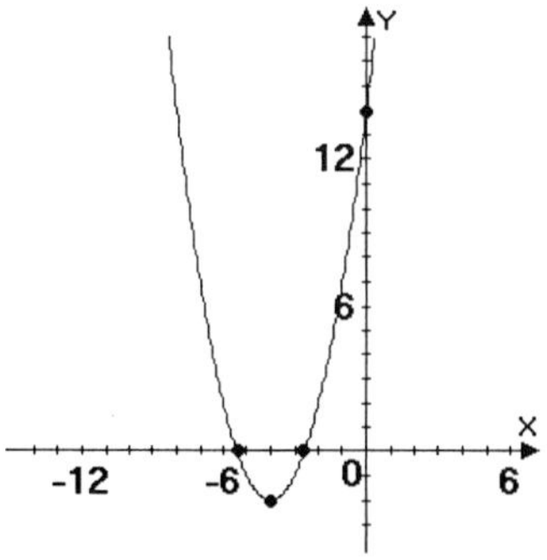


Axis of symmetry: $x=-4$.

**34.** $f(x)=-x^2+2x+3$

Since $a=-1$ is negative, the parabola opens downward. The $x$–coordinate of the vertex of the parabola is

$-\frac{b}{2a}=-\frac{2}{2(-1)}=-\frac{2}{-2}=1$ and the

$y$–coordinate of the vertex of the parabola is $f\left(-\frac{b}{2a}\right)=f(1)$

$=-1^2+2(1)+3$

$=-1+2+3=4.$

The vertex is (1, 4). Replace $f(x)$ with 0 to find $x$–intercepts.

$0=-x^2+2x+3$

$0=x^2-2x-3$

$0=(x-3)(x+1)$

Apply the zero product principle.

$x-3=0$ or $x+1=0$

$x=3$ $\quad x=-1$

The $x$–intercepts are –1 and 3. Set $x=0$ and solve for $y$ to obtain the $y$–intercept.

$y=-0^2+2(0)+3$

$y=0+0+3$

$y=3$

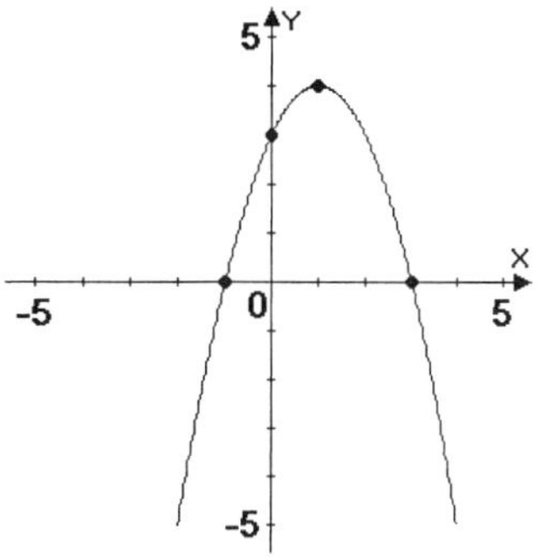


Axis of symmetry: $x=1$.

**35.** $f(x)=2x^2-4x-6$

Since $a=2$ is positive, the parabola opens upward. The $x$–coordinate of the vertex of the parabola is

$-\frac{b}{2a}=-\frac{-4}{2(2)}=-\frac{-4}{4}=1$ and the

$y$–coordinate of the vertex of the parabola is

$f\left(-\frac{b}{2a}\right)=f(1)$

$=2(1)^2-4(1)-6$

$=2(1)-4-6$

$=2-4-6=-8.$

The vertex is $(1,-8)$. Replace $f(x)$

with 0 to find $x$–intercepts.

$0 = 2x^2 - 4x - 6$

$0 = x^2 - 2x - 3$

$0 = (x-3)(x+1)$

Apply the zero product principle.

$x - 3 = 0$ or $x + 1 = 0$

$x = 3$ $\quad x = -1$

The $x$–intercepts are –1 and 3. Set $x = 0$ and solve for $y$ to obtain the $y$–intercept.

$y = 2(0)^2 - 4(0) - 6$

$y = 2(0) - 0 - 6$

$y = 0 - 0 - 6 = -6$

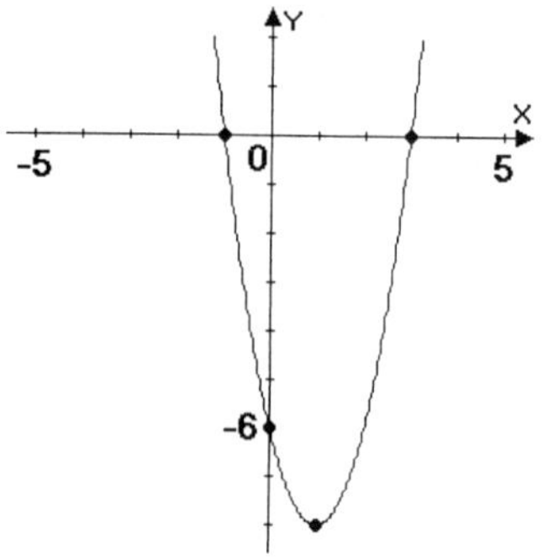


Axis of symmetry: $x = 1$.

**36.** $f(x) = -0.02x^2 + x + 1$

Since $a = -0.02$ is negative, we know the function opens downward and has a maximum at

$$x = -\frac{b}{2a} = -\frac{1}{2(-0.02)} = -\frac{1}{-0.04} = 25.$$

When 25 inches of rain falls, the maximum growth will occur. The maximum growth is

$$\begin{aligned} f(25) &= -0.02(25)^2 + 25 + 1 \\ &= -0.02(625) + 25 + 1 \\ &= -12.5 + 25 + 1 = 13.5. \end{aligned}$$

A maximum yearly growth of 13.5 inches occurs when 25 inches of rain falls per year.

**37.** $s(t) = -16t^2 + 400t + 40$

Since $a = -16$ is negative, we know the function opens downward and has a maximum at

$$x = -\frac{b}{2a} = -\frac{400}{2(-16)} = -\frac{400}{-32} = 12.5.$$

At 12.5 seconds, the rocket reaches its maximum height. The maximum height is

$$\begin{aligned} s(12.5) &= -16(12.5)^2 + 400(12.5) + 40 \\ &= -16(156.25) + 5000 + 40 \\ &= -2500 + 5000 + 40 = 2540. \end{aligned}$$

The rocket reaches a maximum height of 2540 feet in 12.5 seconds.

**38.** According to the graph, the vertex (maximum point) is (25, 5). This means that the maximum divorce rate of 5 divorces per 1000 people occurred in the year 1960 + 25 = 1985.

**39.** Maximize the area using $A = lw$.

$A(x) = x(1000 - 2x)$

$A(x) = -2x^2 + 1000x$

Since $a = -2$ is negative, we know the function opens downward and has a maximum at

$$x = -\frac{b}{2a} = -\frac{1000}{2(-2)} = -\frac{1000}{-4} = 250.$$

The maximum area is achieved when the width is 250 yards. The maximum area is

$$\begin{aligned} A(250) &= 250(1000 - 2(250)) \\ &= 250(1000 - 500) \\ &= 250(500) = 125{,}000. \end{aligned}$$

The area is maximized at 125,000 square yards when the width is 250 yards and the length is $1000 - 2 \cdot 250 = 500$ yards.

**40.** Let $x$ = one of the numbers
Let $14 + x$ = the other number

We need to minimize the function

$$P(x) = x(14+x) = 14x + x^2 = x^2 + 14x.$$

The minimum is at

$$x = -\frac{b}{2a} = -\frac{14}{2(1)} = -\frac{14}{2} = -7.$$

The other number is

$14 + x = 14 + (-7) = 7.$

The numbers which minimize the product are –7 and 7. The minimum product is $-7 \cdot 7 = -49$.

**41.** Let $t = x^2$.

$$x^4 - 6x^2 + 8 = 0$$
$$(x^2)^2 - 6x^2 + 8 = 0$$
$$t^2 - 6t + 8 = 0$$
$$(t-4)(t-2) = 0$$

Apply the zero product principle.

$t - 4 = 0$ or $t - 2 = 0$

$t = 4$ $\quad$ $t = 2$

Replace $t$ by $x^2$.

$x^2 = 4$ or $x^2 = 2$

$x = \pm 2$ $\quad$ $x = \pm\sqrt{2}$

The solutions are $\pm\sqrt{2}$ and $\pm 2$, and the solution set is $\{-2, -\sqrt{2}, \sqrt{2}, 2\}$.

**42.** Let $t = \sqrt{x}$.

$$x + 7\sqrt{x} - 8 = 0$$
$$(\sqrt{x})^2 + 7\sqrt{x} - 8 = 0$$
$$t^2 + 7t - 8 = 0$$
$$(t+8)(t-1) = 0$$

Apply the zero product principle.

$t + 8 = 0$ or $t - 1 = 0$

$t = -8$ $\quad$ $t = 1$

Replace $t$ by $\sqrt{x}$.

~~$\sqrt{x} = -8$~~ or $\sqrt{x} = 1$

$x = 1$

We disregard –8 because the square root of $x$ cannot be a negative number. We need to check the solution, 1, because both sides of the equation were raised to an even power.

Check:

$$1 + 7\sqrt{1} - 8 = 0$$
$$1 + 7(1) - 8 = 0$$
$$1 + 7 - 8 = 0$$
$$8 - 8 = 0$$
$$0 = 0$$

The solution is 1 and the solution set is $\{1\}$.

**43.** Let $t = x^2 + 2x$.

$$(x^2+2x)^2 - 14(x^2+2x) = 15$$
$$(x^2+2x)^2 - 14(x^2+2x) - 15 = 0$$
$$t^2 - 14t - 15 = 0$$
$$(t-15)(t+1) = 0$$

Apply the zero product principle.

$t - 15 = 0$ or $t + 1 = 0$

$t = 15$ $\quad$ $t = -1$

Replace $t$ by $x^2 + 2x$.

First, consider $t = 15$.

$$x^2 + 2x = 15$$
$$x^2 + 2x - 15 = 0$$
$$(x+5)(x-3) = 0$$

Apply the zero product principle.

$x + 5 = 0$ or $x - 3 = 0$

$x = -5$ $\quad$ $x = 3$

Next, consider $t = -1$.

$x^2 + 2x = -1$

$x^2 + 2x + 1 = 0$

$(x+1)^2 = 0$

Apply the zero product principle to find the double root.

$x + 1 = 0$

$x = -1$

The solutions are $-5, -1$, and 3, and the solution set is $\{-5, -1, 3\}$.

**44.** Let $t = x^{-1}$.

$x^{-2} + x^{-1} - 56 = 0$

$(x^{-1})^2 + x^{-1} - 56 = 0$

$t^2 + t - 56 = 0$

$(t+8)(t-7) = 0$

Apply the zero product principle.

$t + 8 = 0$ or $t - 7 = 0$

$t = -8$ $\quad$ $t = 7$

Replace $t$ by $x^{-1}$.

$x^{-1} = -8$ or $x^{-1} = 7$

$\frac{1}{x} = -8$ $\quad$ $\frac{1}{x} = 7$

$-8x = 1$ $\quad$ $7x = 1$

$x = -\frac{1}{8}$ $\quad$ $x = \frac{1}{7}$

The solutions are $-\frac{1}{8}$ and $\frac{1}{7}$, and the solution set is $\left\{-\frac{1}{8}, \frac{1}{7}\right\}$.

**45.** Let $t = x^{\frac{1}{3}}$.

$x^{\frac{2}{3}} - x^{\frac{1}{3}} - 12 = 0$

$\left(x^{\frac{1}{3}}\right)^2 - x^{\frac{1}{3}} - 12 = 0$

$t^2 - t - 12 = 0$

$(t-4)(t+3) = 0$

Apply the zero product principle.

$t - 4 = 0$ or $t + 3 = 0$

$t = 4$ $\quad$ $t = -3$

Replace $t$ by $x^{\frac{1}{3}}$.

$x^{\frac{1}{3}} = 4$ or $x^{\frac{1}{3}} = -3$

$\left(x^{\frac{1}{3}}\right)^3 = 4^3$ $\quad$ $\left(x^{\frac{1}{3}}\right)^3 = (-3)^3$

$x = 64$ $\quad$ $x = -27$

The solutions are $-27$ and 64, and the solution set is $\{-27, 64\}$.

**46.** Let $t = x^{\frac{1}{4}}$.

$x^{\frac{1}{2}} + 3x^{\frac{1}{4}} - 10 = 0$

$\left(x^{\frac{1}{4}}\right)^2 + 3x^{\frac{1}{4}} - 10 = 0$

$t^2 + 3t - 10 = 0$

$(t+5)(t-2) = 0$

Apply the zero product principle.

$t + 5 = 0$ or $t - 2 = 0$

$t = -5$ $\quad$ $t = 2$

Replace $t$ by $x^{\frac{1}{4}}$.

$x^{\frac{1}{4}} = -5$ or $x^{\frac{1}{4}} = 2$

~~$\sqrt[4]{x} = -5$~~ $\quad$ $\left(x^{\frac{1}{4}}\right)^4 = 2^4$

$x = 16$

We disregard –5 because the fourth root of $x$ cannot be a negative number. We need to check the solution, 16, because both sides of the equation were raised to an even power.

Check $x = 16$.

$$16^{\frac{1}{2}} + 3(16)^{\frac{1}{4}} - 10 = 0$$
$$4 + 3(2) - 10 = 0$$
$$4 + 6 - 10 = 0$$
$$10 - 10 = 0$$
$$0 = 0$$

The solution checks. The solution is 16 and the solution set is $\{16\}$.

**47.** $2x^2 + 5x - 3 < 0$

Solve the related quadratic equation.

$$2x^2 + 5x - 3 = 0$$
$$(2x-1)(x+3) = 0$$

Apply the zero product principle.

$2x - 1 = 0$ or $x + 3 = 0$

$2x = 1$ $\quad x = -3$

$x = \frac{1}{2}$

The boundary points are $-3$ and $\frac{1}{2}$.

| Test Interval | Test Number | Test | Conclusion |
|---|---|---|---|
| $(-\infty, -3)$ | -4 | $2(-4)^2 + 5(-4) - 3 < 0$ <br> $9 < 0$, false | $(-\infty, -3)$ does not belong to the solution set. |
| $\left(-3, \frac{1}{2}\right)$ | 0 | $2(0)^2 + 5(0) - 3 < 0$ <br> $-3 < 0$, true | $\left(-3, \frac{1}{2}\right)$ belongs to the solution set. |
| $\left(\frac{1}{2}, \infty\right)$ | 1 | $2(1)^2 + 5(1) - 3 < 0$ <br> $4 < 0$, false | $\left(\frac{1}{2}, \infty\right)$ does not belong to the solution set. |

The solution set is $\left(-3, \frac{1}{2}\right)$ or $\left\{x \middle| -3 < x < \frac{1}{2}\right\}$.

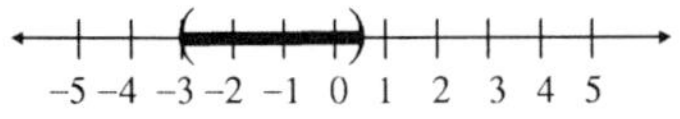

**48.** $2x^2+9x+4\ge 0$

Solve the related quadratic equation.

$$2x^2+9x+4=0$$
$$(2x+1)(x+4)=0$$

Apply the zero product principle.

$2x+1=0$ or $x+4=0$

$2x=-1$ $\quad x=-4$

$x=-\frac{1}{2}$

The boundary points are $-4$ and $-\frac{1}{2}$.

| Test Interval | Test Number | Test | Conclusion |
|---|---|---|---|
| $(-\infty,-4]$ | -5 | $2(-5)^2+9(-5)+4\ge 0$ <br> $9\ge 0$, true | $(-\infty,-4]$ belongs to the solution set. |
| $\left[-4,-\frac{1}{2}\right]$ | -1 | $2(-1)^2+9(-1)+4\ge 0$ <br> $-3\ge 0$, false | $\left[-4,-\frac{1}{2}\right]$ does not belong to the solution set. |
| $\left[-\frac{1}{2},\infty\right)$ | 0 | $2(0)^2+9(0)+4\ge 0$ <br> $4\ge 0$, true | $\left[-\frac{1}{2},\infty\right)$ belongs to the solution set. |

The solution set is $(-\infty,-4]\cup\left[-\frac{1}{2},\infty\right)$ or $\left\{x \middle| x\le -4 \text{ or } x\ge -\frac{1}{2}\right\}$.

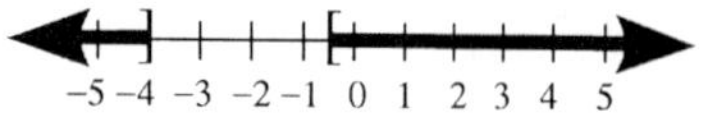


**49.** $x^3+2x^2>3x$

Solve the related quadratic equation.

$$x^3+2x^2=3x$$
$$x^3+2x^2-3x=0$$
$$x(x^2+2x-3)=0$$
$$x(x+3)(x-1)=0$$

Apply the zero product principle.

$x=0$ or $x+3=0$ or $x-1=0$

$x=-3$ $\quad x=1$

The boundary points are $-3$, 0, and 1.

| Test Interval | Test Number | Test | Conclusion |
|---|---|---|---|
| $(-\infty,-3)$ | $-4$ | $(-4)^3+2(-4)^2>3(-4)$ <br> $-32>-12$, False | $(-\infty,-3)$ does not belong to the solution set. |
| $(-3,0)$ | $-2$ | $(-2)^3+2(-2)^2>3(-2)$ <br> $0>-6$, True | $(-3,0)$ belongs to the solution set. |
| $(0,1)$ | 0.5 | $0.5^3+2(0.5)^2>3(0.5)$ <br> $0.625>1.5$, False | $(0,1)$ does not belong to the solution set. |
| $(1,\infty)$ | 2 | $2^3+2(2)^2>3(2)$ <br> $16>6$, True | $(1,\infty)$ belongs to the solution set. |

The solution set is $(-3,0)\cup(1,\infty)$ or $\{x \mid -3<x<0 \text{ or } x>1\}$.

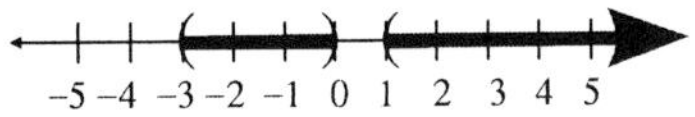

**50.** $\dfrac{x-6}{x+2}>0$

Find the values of $x$ that make the numerator and denominator zero.

$x-6=0 \qquad x+2=0$

$x=6 \qquad x=-2$

The boundary points are –2 and 6.

| Test Interval | Test Number | Test | Conclusion |
|---|---|---|---|
| $(-\infty,-2)$ | -3 | $\dfrac{-3-6}{-3+2}>0$ <br> $9>0$, true | $(-\infty,-2)$ belongs to the solution set. |
| $(-2,6)$ | 0 | $\dfrac{0-6}{0+2}>0$ <br> $-3>0$, false | $(-2,6)$ does not belong to the solution set. |
| $(6,\infty)$ | 7 | $\dfrac{7-6}{7+2}>0$ <br> $\dfrac{1}{9}>0$, true | $(6,\infty)$ belongs to the solution set. |

The solution set is $(-\infty,-2)\cup(6,\infty)$ or $\{x \mid x<-2 \text{ or } x>6\}$.

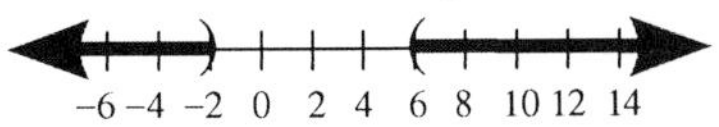

**51.** $\dfrac{x+3}{x-4} \le 5$

Express the inequality so that one side is zero.

$$\frac{x+3}{x-4} - 5 \le 0$$

$$\frac{x+3}{x-4} - \frac{5(x-4)}{x-4} \le 0$$

$$\frac{x+3-5(x-4)}{x-4} \le 0$$

$$\frac{x+3-5x+20}{x-4} \le 0$$

$$\frac{-4x+23}{x-4} \le 0$$

Find the values of $x$ that make the numerator and denominator zero.

$-4x+23=0$ and $x-4=0$

$-4x=-23$ $\quad x=4$

$x=\dfrac{23}{4}$

The boundary points are 4 and $\dfrac{23}{4}$. We exclude 4 from the solution set, since this would make the denominator zero.

| Test Interval | Test Number | Test | Conclusion |
|---|---|---|---|
| $(-\infty, 4)$ | 0 | $\dfrac{0+3}{0-4} \le 5$ <br> $\dfrac{3}{-4} \le 5$, true | $(-\infty, 4)$ belongs to the solution set. |
| $\left(4, \dfrac{23}{4}\right]$ | 5 | $\dfrac{5+3}{5-4} \le 5$ <br> $8 \le 5$, false | $\left(4, \dfrac{23}{4}\right]$ does not belong to the solution set. |
| $\left[\dfrac{23}{4}, \infty\right)$ | 6 | $\dfrac{6+3}{6-4} \le 5$ <br> $\dfrac{9}{2} \le 5$, true | $\left[\dfrac{23}{4}, \infty\right)$ belongs to the solution set. |

The solution set is $(-\infty, 4) \cup \left[\dfrac{23}{4}, \infty\right)$ or $\left\{x \,\middle|\, x < 4 \text{ or } x \ge \dfrac{23}{4}\right\}$.

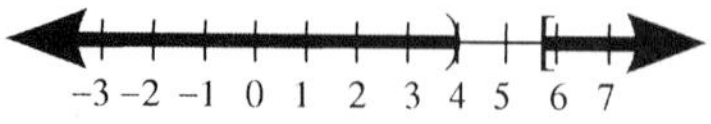

**52.** $s(t) = -16t^2 + 48t$

To find when the height is more than 32 feet above the ground, solve the inequality $-16t^2 + 48t > 32$.

Solve the related quadratic equation.

$$-16t^2 + 48t = 32$$
$$-16t^2 + 48t - 32 = 0$$
$$t^2 - 3t + 2 = 0$$
$$(t-2)(t-1) = 0$$

Apply the zero product principle.

$t - 2 = 0$ or $t - 1 = 0$

$t = 2$ $\qquad t = 1$

The boundary points are 1 and 2.

| Test Interval | Test Number | Test | Conclusion |
|---|---|---|---|
| $(0,1)$ | 0.5 | $-16(0.5)^2 + 48(0.5) > 32$<br>$20 > 32$, false | $(0,1)$ does not belong to the solution set. |
| $(1,2)$ | 1.5 | $-16(1.5)^2 + 48(1.5) > 32$<br>$36 > 32$, true | $(1,2)$ belongs to the solution set. |
| $(2,\infty)$ | 3 | $-16(3)^2 + 48(3) > 32$<br>$0 > 32$, false | $(2,\infty)$ does not belong to the solution set. |

The solution set is $(1,2)$. This means that the ball will be more than 32 feet above the graph between 1 and 2 seconds.

**53.** **a.** $H(0) = \frac{15}{8}(0)^2 - 30(0) + 200 = \frac{15}{8}(0) - 0 + 200 = 0 - 0 + 200 = 200$

The heart rate is 200 beats per minute immediately following the workout.

**b.**

$$\frac{15}{8}x^2 - 30x + 200 > 110$$
$$\frac{15}{8}x^2 - 30x + 90 > 0$$
$$\frac{8}{15}\left(\frac{15}{8}x^2 - 30x + 90\right) > \frac{8}{15}(0)$$
$$x^2 - \frac{8}{15}(30x) + \frac{8}{15}(90) > 0$$
$$x^2 - 16x + 48 > 0$$
$$(x-12)(x-4) > 0$$

Apply the zero product principle.

$x-12=0$ or $x-4=0$

$x=12$ $\quad$ $x=4$

The boundary points are 4 and 12.

| Test Interval | Test Number | Test | Conclusion |
|---|---|---|---|
| $(0,4)$ | 1 | $\frac{15}{8}(1)^2-30(1)+200>110$ <br> $171\frac{7}{8}>110$, true | $(0,4)$ belongs to the solution set. |
| $(4,12)$ | 5 | $\frac{15}{8}(5)^2-30(5)+200>110$ <br> $96\frac{7}{8}>110$, false | $(4,\infty)$ does not belong to the solution set. |
| $(12,\infty)$ | 13 | $\frac{15}{8}(13)^2-30(13)+200>110$ <br> $126\frac{7}{8}>110$, false | $(12,\infty)$ does not belong to the solution set. |

The solution set is $(0,4)\cup(12,\infty)$. This means that the heart rate exceeds 110 beats per minute between 0 and 4 minutes after the workout and more than 12 minutes after the workout. Between 0 and 4 minutes provides a more realistic answer since it is unlikely that the heart rate will begin to climb again without further exertion. Model breakdown occurs for the interval $(12,\infty)$.

## Chapter 8 Test

**1.** $2x^2-5=0$

$2x^2=5$

$x^2=\frac{5}{2}$

$x=\pm\sqrt{\frac{5}{2}}$

Rationalize the denominators.

$x=\pm\frac{\sqrt{5}}{\sqrt{2}}\cdot\frac{\sqrt{2}}{\sqrt{2}}=\pm\frac{\sqrt{10}}{2}$

The solutions are $\pm\frac{\sqrt{10}}{2}$, and the solution set is $\left\{\pm\frac{\sqrt{10}}{2}\right\}$.

**2.** $(x-3)^2=20$

$x-3=\pm\sqrt{20}$

$x=3\pm\sqrt{4\cdot 5}$

$x=3\pm 2\sqrt{5}$

The solutions are $3\pm 2\sqrt{5}$, and the solution set is $\{3\pm 2\sqrt{5}\}$.

**3.** $x^2-16x+$ _____

Since $b=-16$, we add

$\left(\frac{b}{2}\right)^2=\left(\frac{-16}{2}\right)^2=(-8)^2=64$.

$x^2-16x+64=(x-8)^2$

**4.** $x^2 + \frac{2}{5}x + \_\_\_\_\_$

Since $b = \frac{2}{5}$, we add

$\left(\frac{1}{2}b\right)^2 = \left(\frac{1}{2} \cdot \frac{2}{5}\right)^2 = \left(\frac{1}{5}\right)^2 = \frac{1}{25}$.

$x^2 + \frac{2}{5}x + \frac{1}{25} = \left(x + \frac{1}{5}\right)^2$

**5.** $x^2 - 6x + 7 = 0$

$x^2 - 6x \quad = -7$

Since $b = -6$, we add

$\left(\frac{b}{2}\right)^2 = \left(\frac{-6}{2}\right)^2 = (-3)^2 = 9$.

$x^2 - 6x + 9 = -7 + 9$

$(x-3)^2 = 2$

Apply the square root property.

$x - 3 = \pm\sqrt{2}$

$x = 3 \pm \sqrt{2}$

The solutions are $3 \pm \sqrt{2}$ and the solution set is $\left\{3 \pm \sqrt{2}\right\}$.

**6.** Use the Pythagorean Theorem.

$50^2 + 50^2 = x^2$

$2500 + 2500 = x^2$

$5000 = x^2$

$\pm\sqrt{5000} = x$

$\pm\sqrt{2500 \cdot 2} = x$

$\pm 50\sqrt{2} = x$

The solutions are $\pm 50\sqrt{2}$ feet. We disregard $-50\sqrt{2}$ feet because we can't have a negative length measurement. The width of the pond is $50\sqrt{2}$ feet.

**7.** $3x^2 + 4x - 2 = 0$

$a = 3 \quad b = 4 \quad c = -2$

Find the discriminant.

$b^2 - 4ac = 4^2 - 4(3)(-2)$

$= 16 + 24 = 40$

Since the discriminant is greater than zero, there are two unequal real solutions. Also, since the discriminant is not a perfect, the solutions are irrational.

**8.** $x^2 = 4x - 8$

$x^2 - 4x + 8 = 0$

$a = 1 \quad b = -4 \quad c = 8$

Find the discriminant.

$b^2 - 4ac = (-4)^2 - 4(1)(8)$

$= 16 - 32 = -16$

Since the discriminant is negative, there are two imaginary solutions which are complex conjugates.

**9.** $2x^2 + 9x = 5$

$2x^2 + 9x - 5 = 0$

$(2x-1)(x+5) = 0$

Apply the zero product principle.

$2x - 1 = 0$ and $x + 5 = 0$

$2x = 1 \qquad x = -5$

$x = \frac{1}{2}$

The solutions are $\frac{1}{2}$ and –5, and the solution set is $\left\{-5, \frac{1}{2}\right\}$.

**10.** $x^2 + 8x + 5 = 0$

Solve using the quadratic formula.

$a = 1 \quad b = 8 \quad c = 5$

$$x = \frac{-8 \pm \sqrt{8^2 - 4(1)(5)}}{2(1)}$$

$$= \frac{-8 \pm \sqrt{64-20}}{2}$$
$$= \frac{-8 \pm \sqrt{44}}{2}$$
$$= \frac{-8 \pm \sqrt{4 \cdot 11}}{2}$$
$$= \frac{-8 \pm 2\sqrt{11}}{2}$$
$$= \frac{2\left(-4 \pm \sqrt{11}\right)}{2} = -4 \pm \sqrt{11}$$

The solutions are $-4 \pm \sqrt{11}$, and the solution set is $\{-4 \pm \sqrt{11}\}$.

**11.** $(x+2)^2 + 25 = 0$

$$(x+2)^2 = -25$$

Apply the square root principle.

$$x+2 = \pm\sqrt{-25}$$
$$x = -2 \pm 5i$$

The solutions are $-2 \pm 5i$, and the solution set is $\{-2 \pm 5i\}$.

**12.** $2x^2 - 6x + 5 = 0$

$a = 2 \quad b = -6 \quad c = 5$

$$x = \frac{-(-6) \pm \sqrt{(-6)^2 - 4(2)(5)}}{2(2)}$$
$$= \frac{6 \pm \sqrt{36-40}}{4}$$
$$= \frac{6 \pm \sqrt{-4}}{4}$$
$$= \frac{6 \pm 2i}{4} = \frac{6}{4} \pm \frac{2}{4}i = \frac{3}{2} \pm \frac{1}{2}i$$

The solutions are $\frac{3}{2} \pm \frac{1}{2}i$, and the solution set is $\left\{\frac{3}{2} \pm \frac{1}{2}i\right\}$.

**13.** Because the solution set is $\{-3, 7\}$, we have

$x = -3 \quad$ or $\quad x = 7$

$x + 3 = 0 \qquad x - 7 = 0$

Apply the zero-product principle in reverse.

$$(x+3)(x-7) = 0$$
$$x^2 - 7x + 3x - 21 = 0$$
$$x^2 - 4x - 21 = 0$$

**14.** Because the solution set is $\{-10i, 10i\}$, we have

$x = -10i \quad$ or $\quad x = 10i$

$x + 10i = 0 \qquad x - 10i = 0$

Apply the zero-product principle in reverse.

$$(x+10i)(x-10i) = 0$$
$$x^2 - 100i^2 = 0$$
$$x^2 - 100(-1) = 0$$
$$x^2 + 100 = 0$$

**15.** $f(x) = -0.5x^2 + 4x + 19$

$$20 = -0.5x^2 + 4x + 19$$

$0.5x^2 - 4x + 1 = 0$

Solve using the quadratic formula.

$a = 0.5 \quad b = -4 \quad c = 1$

$$x = \frac{-(-4) \pm \sqrt{(-4)^2 - 4(0.5)(1)}}{2(0.5)}$$
$$= \frac{4 \pm \sqrt{16-2}}{1}$$
$$= 4 \pm \sqrt{14}$$
$$= 7.7 \text{ or } 0.3$$
$$\approx 8 \text{ or } 0$$

In the years 1900 and 1998, 20 million people were receiving food stamps.

**16.** $f(x)=(x+1)^2+4$

Since $a=1$ is negative, the parabola opens upward. The vertex of the parabola is $(h,k)=(-1,4)$ and the axis of symmetry is $x=-1$. Replace $f(x)$ with 0 to find $x$–intercepts.

$0=(x+1)^2+4$

$-4=(x+1)^2$

This will be result in complex solutions. As a result, there are no $x$–intercepts. Set $x = 0$ and solve for $y$ to obtain the $y$–intercept.

$y=(0+1)^2+4=1+4=5$

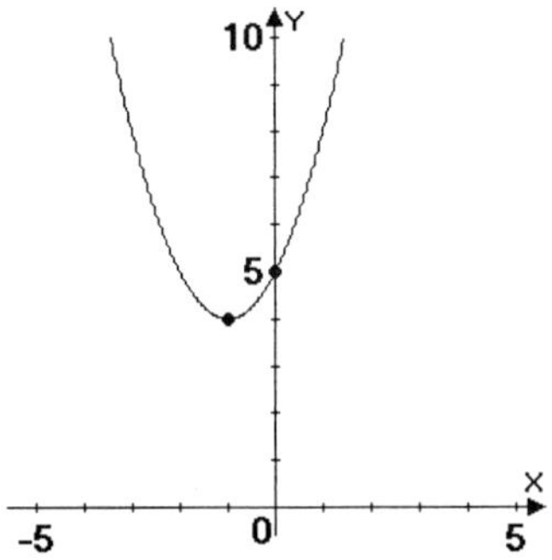


Axis of symmetry: $x=-1$.

**17.** $f(x)=x^2-2x-3$

Since $a=1$ is positive, the parabola opens upward. The $x$–coordinate of the vertex of the parabola is

$-\frac{b}{2a}=-\frac{-2}{2(1)}=-\frac{-2}{2}=1$ and the

$y$–coordinate of the vertex of the parabola is

$$f\left(-\frac{b}{2a}\right)=f(1)=1^2-2(1)-3$$
$$=1-2-3=-4.$$

The vertex is $(1,-4)$. Replace $f(x)$ with 0 to find $x$–intercepts.

$0=x^2-2x-3$

$0=(x-3)(x+1)$

Apply the zero product principle.

$x-3=0$ or $x+1=0$

$x=3$ $\qquad$ $x=-1$

The $x$–intercepts are –1 and 3. Set $x = 0$ and solve for $y$ to obtain the $y$–intercept.

$y=0^2-2(0)-3=-3$

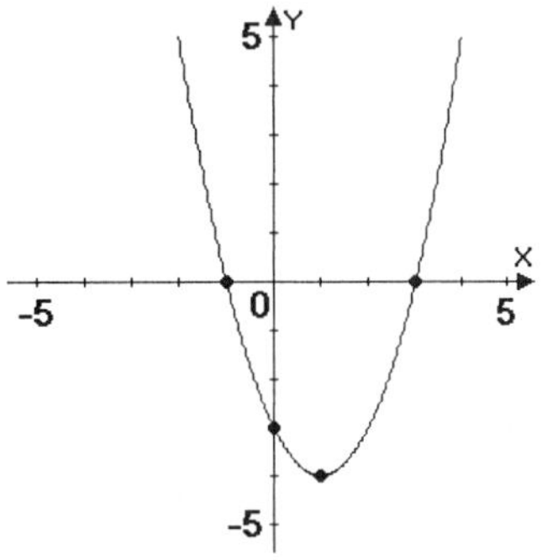


Axis of symmetry: $x=1$.

**18.** $s(t)=-16t^2+64t+5$

Since $a=-16$ is negative, we know the function opens downward and has a maximum at

$x=-\frac{b}{2a}=-\frac{64}{2(-16)}=-\frac{64}{-32}=2.$

The ball reaches its maximum height a two seconds. The maximum height is

$$s(2)=-16(2)^2+64(2)+5$$
$$=-16(4)+128+5$$
$$=-64+128+5=69.$$

The baseball reaches a maximum height of 69 feet after 2 seconds.

**19.** $0=-16t^2+64t+5$

Solve using the quadratic formula.

$a=-16 \quad b=64 \quad c=5$

$$x=\frac{-64\pm\sqrt{64^2-4(-16)(5)}}{2(-16)}$$
$$=\frac{-64\pm\sqrt{4096+320}}{-32}$$
$$=\frac{-64\pm\sqrt{4416}}{-32}$$
$$\approx\frac{-64-66.5}{-32} \text{ or } \frac{-64+66.5}{-32}$$
$$\approx\frac{-130.5}{-32} \text{ or } \frac{2.5}{-32}$$
$$\approx 4.1 \text{ or } -0.1$$

We disregard –0.1 since we cannot have a negative time measurement. The solution is 4.1 and we conclude that the baseball hits the ground in approximately 4.1 seconds.

**20.** $f(x)=-x^2+46x-360$

Since $a=-1$ is negative, we know the function opens downward and has a maximum at

$$x=-\frac{b}{2a}=-\frac{46}{2(-1)}=-\frac{46}{-2}=23.$$

$f(23)=-23^2+46(23)-360=169$

Profit is maximized when 23 computers are manufactured. This produces a profit of \$169 hundreds or \$16,900.

**21.** Let $t=2x-5$.

$$(2x-5)^2+4(2x-5)+3=0$$
$$t^2+4t+3=0$$
$$(t+3)(t+1)=0$$

Apply the zero product principle.

$t+3=0 \quad \text{or} \quad t+1=0$

$t=-3 \qquad\qquad t=-1$

Replace $t$ by $2x-5$.

First, consider $t=15$.

$2x-5=-3 \quad \text{or} \quad 2x-5=-1$

$2x=2 \qquad\qquad 2x=4$

$x=1 \qquad\qquad x=2$

The solutions are 1 and 2 and the solution set is $\{1,2\}$.

**22.** Let $t=x^2$.

$$x^4-13x^2+36=0$$
$$\left(x^2\right)^2-13x^2+36=0$$
$$t^2-13t+36=0$$
$$(t-9)(t-4)=0$$

Apply the zero product principle.

$t-9=0 \quad \text{or} \quad t-4=0$

$t=9 \qquad\qquad t=4$

Replace $t$ by $x^2$.

$x^2=9 \quad \text{or} \quad x^2=4$

$x=\pm3 \qquad\qquad x=\pm2$

The solutions are $\pm2$ and $\pm3$ and the solution set is $\{-3,-2,2,3\}$.

**23.** Let $t=x^{1/3}$.

$$x^{2/3}-9x^{1/3}+8=0$$
$$\left(x^{1/3}\right)^2-9x^{1/3}+8=0$$
$$t^2-9t+8=0$$
$$(t-8)(t-1)=0$$

Apply the zero product principle.

$t-8=0 \quad \text{or} \quad t-1=0$

$t=8 \qquad\qquad t=1$

Replace $t$ by $x^{1/3}$.

$x^{1/3}=8 \quad \text{or} \quad x^{1/3}=1$

$x=8^3=512 \qquad\qquad x=1^3=1$

The solutions are 1 and 512 and the solution set is $\{1,\ 512\}$.

**24.** $x^2 - x - 12 < 0$

Solve the related quadratic equation.

$$x^2 - x - 12 = 0$$

$$(x-4)(x+3) = 0$$

Apply the zero product principle.

$x - 4 = 0$ or $x + 3 = 0$

$x = 4$ $\qquad$ $x = -3$

The boundary points are $-3$ and 4.

| Test Interval | Test Number | Test | Conclusion |
|---|---|---|---|
| $(-\infty, -3)$ | -4 | $(-4)^2 - (-4) - 12 < 0$ <br> $8 < 0$, false | $(-\infty, -3)$ does not belong to the solution set. |
| $(-3, 4)$ | 0 | $0^2 - 0 - 12 < 0$ <br> $-12 < 0$, true | $(-3, 4)$ belongs to the solution set. |
| $(4, \infty)$ | 5 | $5^2 - 5 - 12 < 0$ <br> $8 < 0$, false | $(4, \infty)$ does not belong to the solution set. |

The solution set is $(-3, 4)$ or $\{x \mid -3 < x < 4\}$.

**25.** $\dfrac{2x+1}{x-3} \le 3$

Express the inequality so that one side is zero.

$$\frac{2x+1}{x-3} - 3 \le 0$$

$$\frac{2x+1}{x-3} - \frac{3(x-3)}{x-3} \le 0$$

$$\frac{2x+1-3(x-3)}{x-3} \le 0$$

$$\frac{2x+1-3x+9}{x-3} \le 0$$

$$\frac{-x+10}{x-3} \le 0$$

Find the values of $x$ that make the numerator and denominator zero.

$-x + 10 = 0$ and $x - 3 = 0$

$-x = -10$ $\qquad$ $x = 3$

$x = 10$

The boundary points are 3 and 10. We exclude 3 from the solution set(s), since this would make the denominator zero.

| Test Interval | Test Number | Test | Conclusion |
|---|---|---|---|
| $(-\infty, 3)$ | 0 | $\frac{2(0)+1}{0-3} \le 3$ <br> $-\frac{1}{3} \le 3$, true | $(-\infty, 3)$ belongs to the solution set. |
| $(3, 10]$ | 4 | $\frac{2(4)+1}{4-3} \le 3$ <br> $9 \le 3$, false | $(3, 10]$ does not belong to the solution set. |
| $[10, \infty)$ | 11 | $\frac{2(10)+1}{10-3} \le 3$ <br> $3 \le 3$, true | $[10, \infty)$ belongs to the solution set. |

The solution set is $(-\infty, 3) \cup [10, \infty)$ or $\{x \mid x < 3 \text{ or } x \ge 10\}$.

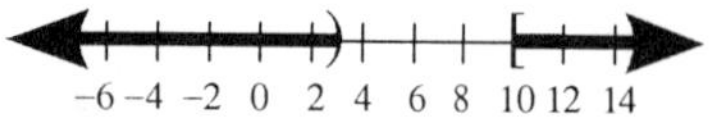

**Cumulative Review Exercises (Chapters 1 – 8)**

**1.**
$$8-(4x-5)=x-7$$
$$8-4x+5=x-7$$
$$13-4x=x-7$$
$$13=5x-7$$
$$20=5x$$
$$4=x$$
The solution is 4, and the solution set is $\{4\}$.

**2.**
$$5x+4y=22$$
$$3x-8y=-18$$
Multiply the first equation by 2 and solve by addition.
$$10x+8y=44$$
$$3x-8y=-18$$
$$13x=26$$
$$x=2$$
Back-substitute 2 for $x$ to find $y$.
$$5(2)+4y=22$$
$$10+4y=22$$
$$4y=12$$
$$y=3$$
The solution is $(2,3)$ and the solution set is $\{(2,3)\}$.

**3.** $-3x+2y+4z=6$

$7x-y+3z=23$

$2x+3y+z=7$

Multiply the second equation by 2 and add to the first equation to eliminate $y$.

$-3x+2y+4z=6$

$14x-2y+6z=46$

$11x+10z=52$

Multiply the second equation by 3 and add to the second equation to eliminate $y$.

$21x-3y+9z=69$

$2x+3y+z=7$

$23x+10z=76$

The system of two variables in two equations is:

$11x+10z=52$

$23x+10z=76$

Multiply the first equation by –1 and add to the second equation.

$-11x-10z=-52$

$23x+10z=76$

$12x=24$

$x=2$

Back-substitute 2 for $x$ to find $z$.

$11(2)+10z=52$

$22+10z=52$

$10z=30$

$z=3$

Back-substitute 2 for $x$ and 3 for $z$ to find $y$.

$-3(2)+2y+4(3)=6$

$-6+2y+12=6$

$2y=0$

$y=0$

The solution is $(2,0,3)$, and the solution set is $\{(2, 0, 3)\}$.

**4.** $|x-1|>3$

$x-1<-3$ or $x-1>3$

$x<-2$ $\quad$ $x>4$

The solution set is $\{x|x<-2$ and $x>4\}$ or $(-\infty,-2)\cup(4,\infty)$.

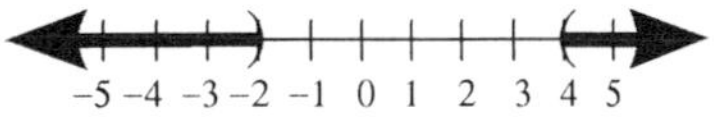


**5.** $\sqrt{x+4}-\sqrt{x-4}=2$

$\sqrt{x+4}=2+\sqrt{x-4}$

$\left(\sqrt{x+4}\right)^2=\left(2+\sqrt{x-4}\right)^2$

$x+4=4+4\sqrt{x-4}+x-4$

$\cancel{x}+4=4\sqrt{x-4}+\cancel{x}$

$4=4\sqrt{x-4}$

$1=\sqrt{x-4}$

$1^2=\left(\sqrt{x-4}\right)^2$

$1=x-4$

$5=x$

The solution is 5, and the solution set is $\{5\}$.

**6.** $x-4\ge 0$ and $-3x\le -6$

$x\ge 4$ $\quad$ $x\ge 2$

For a value to be in the solution set, it must satisfy both of the conditions $x\ge 4$ and $x\ge 2$. Now any value that is 4 or larger is also larger than 2. But values between 2 and 4 do not satisfy both conditions. Therefore, only values that are 4 or larger will be in the solution set. Thus, the solution set is $\{x|x\ge 4\}$ or $[4,\infty)$.

**7.** $$2x^2 = 3x - 2$$
$$2x^2 - 3x + 2 = 0$$

Solve using the quadratic formula.

$a = 2 \qquad b = -3 \qquad c = 2$

$$x = \frac{-(-3) \pm \sqrt{(-3)^2 - 4(2)(2)}}{2(2)}$$
$$= \frac{3 \pm \sqrt{9-16}}{4}$$
$$= \frac{3 \pm \sqrt{-7}}{4} = \frac{3 \pm \sqrt{7}i}{4} = \frac{3}{4} \pm \frac{\sqrt{7}}{4}i$$

The solutions are $\frac{3}{4} \pm \frac{\sqrt{7}}{4}i$, and the solution set is $\left\{\frac{3}{4} - \frac{\sqrt{7}}{4}i, \frac{3}{4} + \frac{\sqrt{7}}{4}i\right\}$.

**8.** $3x = 15 + 5y$

Find the $x$–intercept by setting $y = 0$ and solving.

$$3x = 15 + 5(0)$$
$$3x = 15$$
$$x = 5$$

Find the $y$–intercept by setting $x = 0$ and solving.

$$3(0) = 15 + 5y$$
$$0 = 15 + 5y$$
$$-15 = 5y$$
$$-3 = y$$

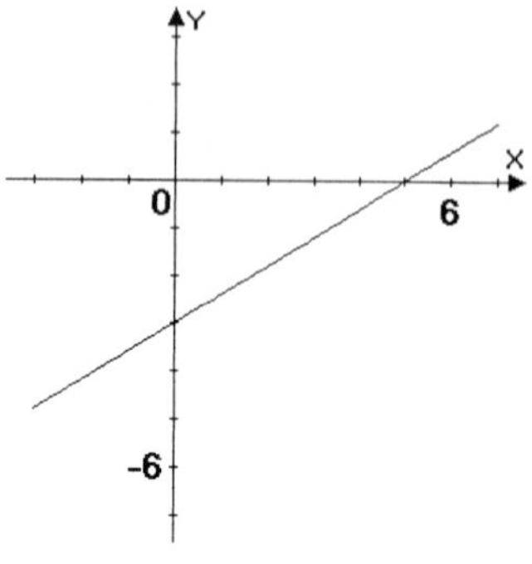


**9.** $2x - 3y > 6$

First, find the intercepts to the equation $2x - 3y = 6$.

Find the $x$–intercept by setting $y = 0$ and solving.

$$2x - 3(0) = 6$$
$$2x = 6$$
$$x = 3$$

Find the $y$–intercept by setting $x = 0$ and solving.

$$2(0) - 3y = 6$$
$$-3y = 6$$
$$y = -2$$

Next, use the origin as a test point.

$$2(0) - 3(0) > 6$$
$$0 > 6$$

This is a false statement. This means that the origin will not fall in the shaded half-plane.

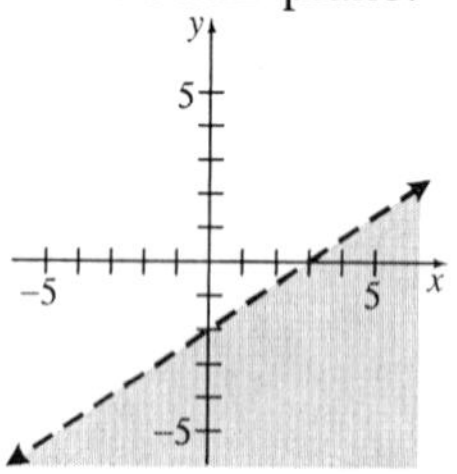


**10.** $$f(x) = -\frac{1}{2}x + 1$$

$$m = -\frac{1}{2};\ y\text{-intercept} = 1$$

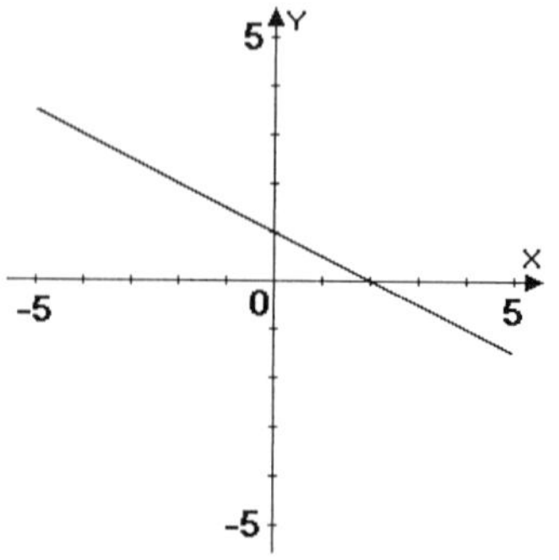

**11.** $f(x)=x^2+6x+8$

Since $a=1$ is positive, the parabola opens upward. The $x$–coordinate of the vertex of the parabola is

$-\frac{b}{2a}=-\frac{6}{2(1)}=-\frac{6}{2}=-3$ and the

$y$–coordinate of the vertex of the parabola is

$$f\left(-\frac{b}{2a}\right)=f(-3)=(-3)^2+6(-3)+8$$
$$=9-18+8=-1.$$

The vertex is $(-3,-1)$. Replace $f(x)$ with 0 to find $x$–intercepts.

$0=x^2+6x+8$

$0=(x+4)(x+2)$

Apply the zero product principle.

$x+4=0$ or $x+2=0$

$x=-4$ $\quad$ $x=-2$

The $x$–intercepts are –4 and –2. Set $x = 0$ and solve for $y$ to obtain the $y$–intercept.

$y=0^2+6(0)+8$

$y=0+0+8$

$y=8$

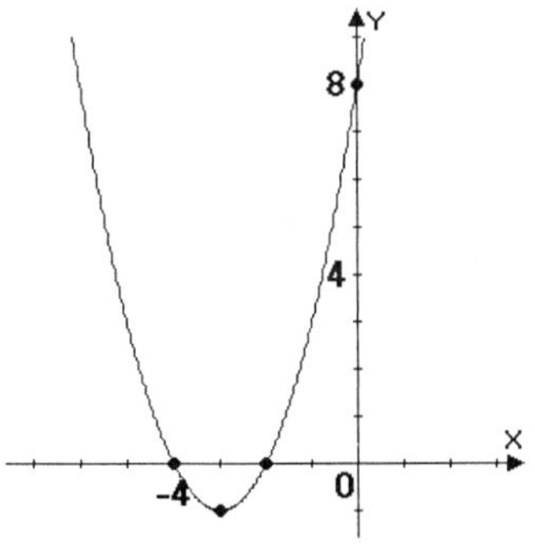


**12.** $f(x)=(x-3)^2-4$

Since $a=1$ is positive, the parabola opens upward. The vertex of the parabola is $(h,k)=(3,-4)$ and the axis of symmetry is $x=3$. Replace $f(x)$ with 0 to find $x$–intercepts.

$0=(x-3)^2-4$

$4=(x-3)^2$

Apply the square root property.

$x-3=-2$ and $x-3=2$

$x=1$ $\quad$ $x=5$

The $x$–intercepts are 1 and 5.

Set $x = 0$ and solve for $y$ to obtain the $y$–intercept.

$y=(0-3)^2-4$

$y=(-3)^2-4$

$y=9-4$

$y=5$

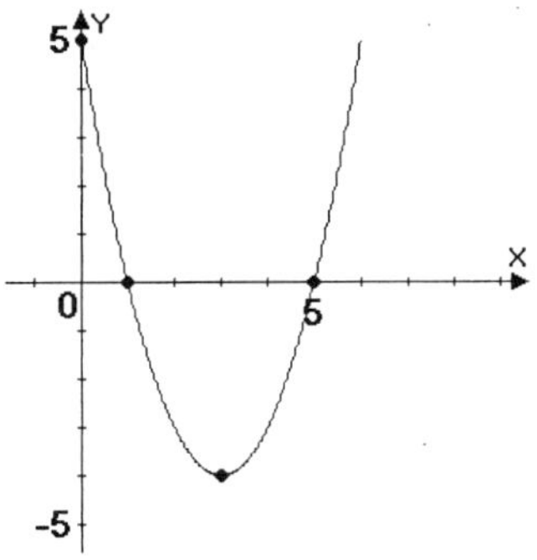


**13.** $\begin{vmatrix} 3 & 1 & 0 \\ 0 & 5 & -6 \\ -2 & -1 & 0 \end{vmatrix}$

$=3\begin{vmatrix} 5 & -6 \\ -1 & 0 \end{vmatrix}-0\begin{vmatrix} 1 & 0 \\ -1 & 0 \end{vmatrix}+(-2)\begin{vmatrix} 1 & 0 \\ 5 & -6 \end{vmatrix}$

$=3(5(0)-(-1)(-6))+(-2)(1(-6)-5(0))$

$=3(-6)+(-2)(-6)=-18+12=-6$

**14.**
$$A=\frac{cd}{c+d}$$
$$A(c+d)=cd$$
$$Ac+Ad=cd$$
$$Ac-cd=-Ad$$
$$c(A-d)=-Ad$$
$$c=-\frac{Ad}{A-d} \text{ or } \frac{Ad}{d-A}$$

**15.** First, solve for $y$ to obtain the slope of the line whose equation is $2x+y=10$.
$$2x+y=10$$
$$y=-2x+10$$
The slope is –2. The line we want to find is perpendicular to this line, so we know the slope will be $\frac{1}{2}$.
Using the point, $(-2,4)$, and the slope, $\frac{1}{2}$, we can write the equation in point-slope form.
$$y-y_1=m(x-x_1)$$
$$y-4=\frac{1}{2}(x-(-2))$$
$$y-4=\frac{1}{2}(x+2)$$
Solve for $y$ to obtain slope-intercept form.
$$y-4=\frac{1}{2}(x+2)$$
$$y-4=\frac{1}{2}x+1$$
$$y=\frac{1}{2}x+5$$
$$f(x)=\frac{1}{2}x+5$$

**16.** $\dfrac{-5x^3y^7}{15x^4y^{-2}}=\dfrac{-y^7y^2}{3x}=\dfrac{-y^9}{3x}=-\dfrac{y^9}{3x}$

**17.** $(4x^2-5y)^2$
$$=(4x^2)^2+2(4x^2)(-5y)+(-5y)^2$$
$$=16x^4-40x^2y+26y^2$$

**18.**
$$\begin{array}{r} x^2-5x+1 \\ 5x+1\overline{)5x^3-24x^2+0x+9} \\ \underline{5x^3+\phantom{0}x^2\phantom{+0x+9}} \\ -25x^2+0x\phantom{+9} \\ \underline{-25x^2-5x\phantom{+9}} \\ 5x+9 \\ \underline{5x+1} \\ 8 \end{array}$$
$$\frac{5x^3-24x^2+9}{5x+1}=x^2-5x+1+\frac{8}{5x+1}$$

**19.** $\dfrac{\sqrt[3]{32xy^{10}}}{\sqrt[3]{2xy^2}}=\sqrt[3]{\dfrac{32xy^{10}}{2xy^2}}$
$$=\sqrt[3]{16y^8}$$
$$=\sqrt[3]{8\cdot 2y^6y^2}=2y^2\sqrt[3]{2y^2}$$

**20.** $$\frac{x+2}{x^2-6x+8}+\frac{3x-8}{x^2-5x+6}=\frac{x+2}{(x-4)(x-2)}+\frac{3x-8}{(x-2)(x-3)}$$
$$=\frac{(x+2)(x-3)}{(x-4)(x-2)(x-3)}+\frac{(3x-8)(x-4)}{(x-4)(x-2)(x-3)}$$
$$=\frac{x^2-3x+2x-6+3x^2-12x-8x+32}{(x-4)(x-2)(x-3)}$$
$$=\frac{4x^2-21x+26}{(x-4)(x-2)(x-3)}$$
$$=\frac{(4x-13)\cancel{(x-2)}}{(x-4)\cancel{(x-2)}(x-3)}=\frac{4x-13}{(x-4)(x-3)}$$

**21.** $x^4-4x^3+8x-32$
$=x^3(x-4)+8(x-4)$
$=(x-4)(x^3+8)$
$=(x-4)(x+2)(x^2-2x+4)$

**22.** $2x^2+12xy+18y^2$
$=2(x^2+6xy+9y^2)=2(x+3y)^2$

**23.** Let $x$ = the width of the carpet
Let $2x+4$ = the length of the carpet
$x(2x+4)=48$
$2x^2+4x=48$
$2x^2+4x-48=0$
$x^2+2x-24=0$
$(x+6)(x-4)=0$
Apply the zero product principle.
$x+6=0$ and $x-4=0$
$x=-6$ $\quad$ $x=4$
We disregard –6 because we can't have a negative length measurement. The width of the carpet is 4 feet and the length of the carpet is
$2x+4=2(4)+4=8+4=12$ feet.

**24.**

| | Part Done in 1 Hour | Time Working Together | Part Done in $x$ Hours |
|---|---|---|---|
| You | $\frac{1}{2}$ | $x$ | $\frac{x}{2}$ |
| Your Sister | $\frac{1}{3}$ | $x$ | $\frac{x}{3}$ |

$$\frac{x}{2}+\frac{x}{3}=1$$
$$6\left(\frac{x}{2}+\frac{x}{3}\right)=6(1)$$
$$6\left(\frac{x}{2}\right)+6\left(\frac{x}{3}\right)=6$$
$$3x+2x=6$$
$$5x=6$$
$$x=\frac{6}{5}$$
If you and your sister work together, it will take $\frac{6}{5}$ hours, or 1 hour and 12 minutes, to clean the house.

**25.**

| | $d$ | $r$ | $t = \frac{d}{r}$ |
|---|---|---|---|
| Down Stream | 20 | $15 + x$ | $\frac{20}{15 + x}$ |
| Up Stream | 10 | $15 - x$ | $\frac{10}{15 - x}$ |

$$\frac{20}{15+x} = \frac{10}{15-x}$$

$$20(15-x) = 10(15+x)$$

$$300 - 20x = 150 + 10x$$

$$300 = 150 + 30x$$

$$150 = 30x$$

$$5 = x$$

The rate of the current is 5 miles per hour.

# Chapter 9

## 9.1 Exercise Set

**1.** $2^{3.4} \approx 10.556$

**3.** $3^{\sqrt{5}} \approx 11.665$

**5.** $4^{-1.5} = 0.125$

**7.** $e^{2.3} \approx 9.974$

**9.** $e^{-0.95} \approx 0.387$

**11.** $f(x) = 3^x$

| $x$ | $f(x)$ |
|---|---|
| -2 | $3^{-2} = \frac{1}{3^2} = \frac{1}{9}$ |
| -1 | $3^{-1} = \frac{1}{3^1} = \frac{1}{3}$ |
| 0 | $3^0 = 1$ |
| 1 | $3^1 = 3$ |
| 2 | $3^2 = 9$ |

This functions matches graph (**d**).

**13.** $f(x) = 3^x - 1$

| $x$ | $f(x)$ |
|---|---|
| -2 | $3^{-2} - 1 = \frac{1}{3^2} - 1 = \frac{1}{9} - 1 = -\frac{8}{9}$ |
| -1 | $3^{-1} - 1 = \frac{1}{3^1} - 1 = \frac{1}{3} - 1 = -\frac{2}{3}$ |
| 0 | $3^0 - 1 = 1 - 1 = 0$ |
| 1 | $3^1 - 1 = 3 - 1 = 2$ |
| 2 | $3^2 - 1 = 9 - 1 = 8$ |

This functions matches graph (**e**).

**15.** $f(x) = 3^{-x}$

| $x$ | $f(x)$ |
|---|---|
| -2 | $3^{-(-2)} = 3^2 = 9$ |
| -1 | $3^{-(-1)} = 3^1 = 3$ |
| 0 | $3^{-(0)} = 3^0 = 1$ |
| 1 | $3^{-(1)} = 3^{-1} = \frac{1}{3}$ |
| 2 | $3^{-(2)} = 3^{-2} = \frac{1}{3^2} = \frac{1}{9}$ |

This functions matches graph (**f**).

**17.** $f(x) = 4^x$

| $x$ | $f(x)$ |
|---|---|
| -2 | $4^{-2} = \frac{1}{4^2} = \frac{1}{16}$ |
| -1 | $4^{-1} = \frac{1}{4^1} = \frac{1}{4}$ |
| 0 | $4^0 = 1$ |
| 1 | $4^1 = 4$ |
| 2 | $4^2 = 16$ |

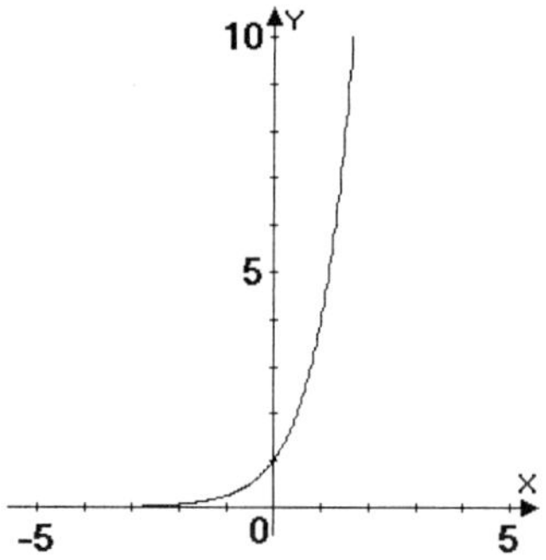

**19.** $g(x)=\left(\frac{3}{2}\right)^x$

| $x$ | $g(x)$ |
|---|---|
| -2 | $\left(\frac{3}{2}\right)^{-2}=\left(\frac{2}{3}\right)^2=\frac{4}{9}$ |
| -1 | $\left(\frac{3}{2}\right)^{-1}=\left(\frac{2}{3}\right)^1=\frac{2}{3}$ |
| 0 | $\left(\frac{3}{2}\right)^0=1$ |
| 1 | $\left(\frac{3}{2}\right)^1=\frac{3}{2}$ |
| 2 | $\left(\frac{3}{2}\right)^2=\frac{9}{4}$ |

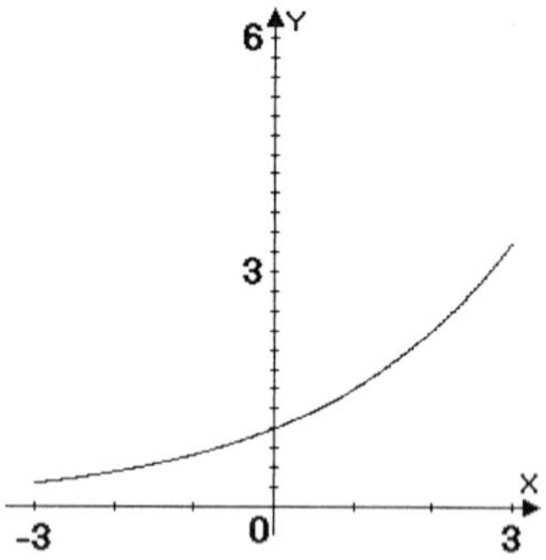


**21.** $h(x)=\left(\frac{1}{2}\right)^x$

| $x$ | $h(x)$ |
|---|---|
| -2 | $\left(\frac{1}{2}\right)^{-2}=\left(\frac{2}{1}\right)^2=\frac{4}{1}=4$ |
| -1 | $\left(\frac{1}{2}\right)^{-1}=\left(\frac{2}{1}\right)^1=\frac{2}{1}=2$ |
| 0 | $\left(\frac{1}{2}\right)^0=1$ |
| 1 | $\left(\frac{1}{2}\right)^1=\frac{1}{2}$ |
| 2 | $\left(\frac{1}{2}\right)^2=\frac{1}{4}$ |

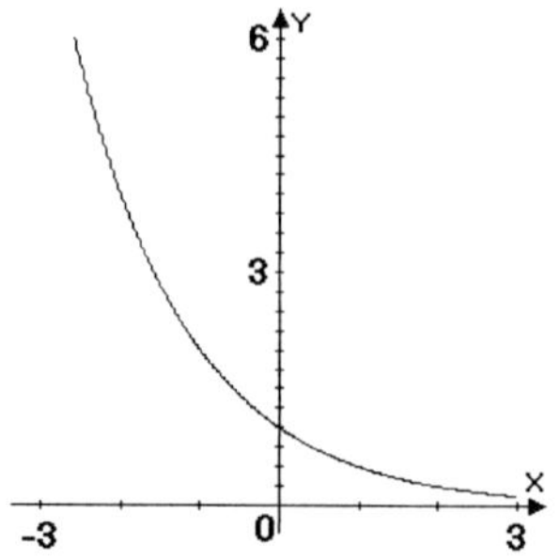


**23.** $f(x)=(0.6)^x=\left(\frac{6}{10}\right)^x=\left(\frac{3}{5}\right)^x$

| $x$ | $f(x)$ |
|---|---|
| -2 | $\left(\frac{3}{5}\right)^{-2}=\left(\frac{5}{3}\right)^2=\frac{25}{9}$ |
| -1 | $\left(\frac{3}{5}\right)^{-1}=\left(\frac{5}{3}\right)^1=\frac{5}{3}$ |
| 0 | $\left(\frac{3}{5}\right)^0=1$ |
| 1 | $\left(\frac{3}{5}\right)^1=\frac{3}{5}$ |
| 2 | $\left(\frac{3}{5}\right)^2=\frac{9}{25}$ |

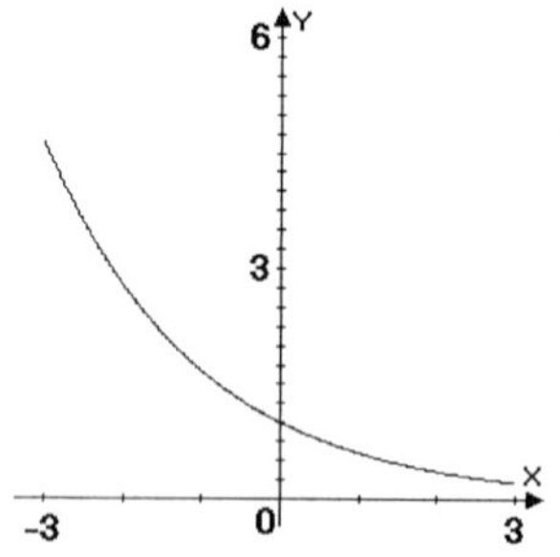

**25.**

| $x$ | $f(x)=2^x$ | $g(x)=2^{x+1}$ |
|---|---|---|
| $-2$ | $\frac{1}{4}$ | $\frac{1}{2}$ |
| $-1$ | $\frac{1}{2}$ | $1$ |
| $0$ | $1$ | $2$ |
| $1$ | $2$ | $4$ |
| $2$ | $4$ | $8$ |

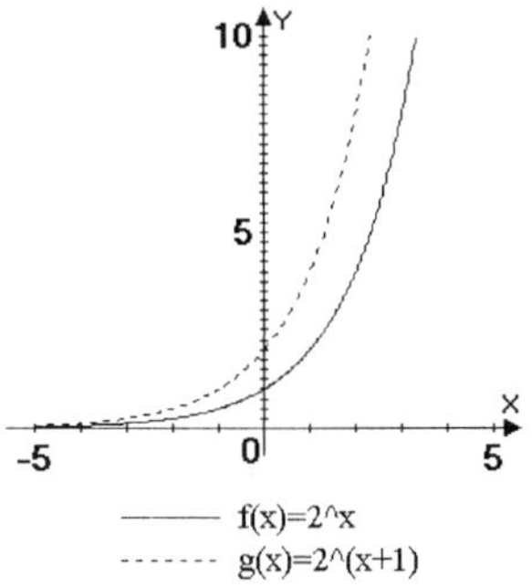


The graph of $g$ is the graph of $f$ shifted 1 unit to the left.

**27.**

| $x$ | $f(x)=2^x$ | $g(x)=2^{x-2}$ |
|---|---|---|
| $-2$ | $\frac{1}{4}$ | $\frac{1}{16}$ |
| $-1$ | $\frac{1}{2}$ | $\frac{1}{8}$ |
| $0$ | $1$ | $\frac{1}{2}$ |
| $1$ | $2$ | $1$ |
| $2$ | $4$ | $2$ |

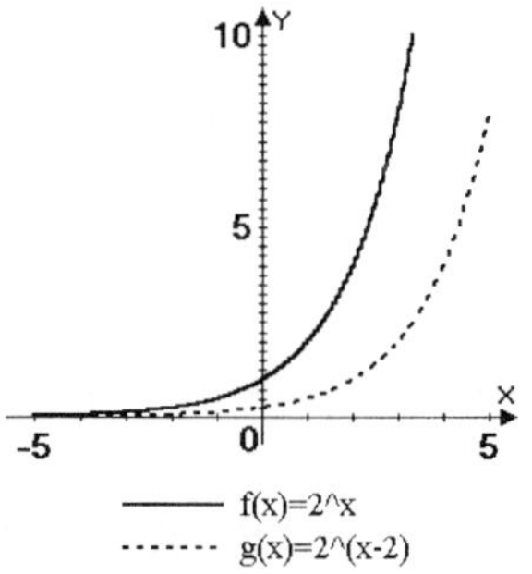


The graph of $g$ is the graph of $f$ shifted 2 units to the right.

**29.**

| $x$ | $f(x)=2^x$ | $g(x)=2^x+1$ |
|---|---|---|
| $-2$ | $\frac{1}{4}$ | $\frac{5}{4}$ |
| $-1$ | $\frac{1}{2}$ | $\frac{3}{2}$ |
| $0$ | $1$ | $2$ |
| $1$ | $2$ | $3$ |
| $2$ | $4$ | $5$ |

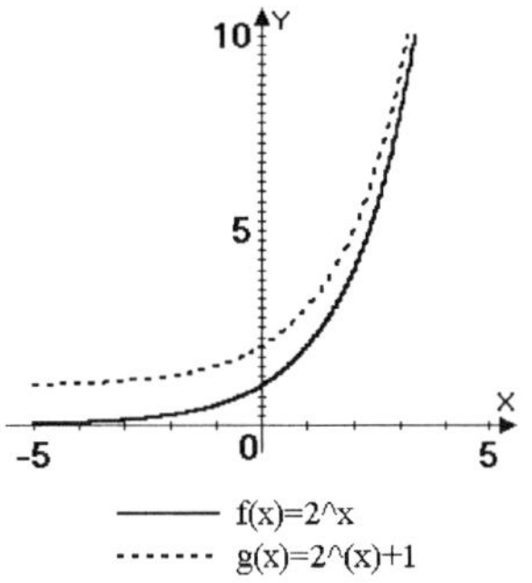


The graph of $g$ is the graph of $f$ shifted up 1 unit.

**31.**

| $x$ | $f(x)=2^x$ | $g(x)=2^x-2$ |
|---|---|---|
| $-2$ | $\frac{1}{4}$ | $-\frac{7}{4}$ |
| $-1$ | $\frac{1}{2}$ | $-\frac{3}{2}$ |
| $0$ | $1$ | $-1$ |
| $1$ | $2$ | $0$ |
| $2$ | $4$ | $2$ |

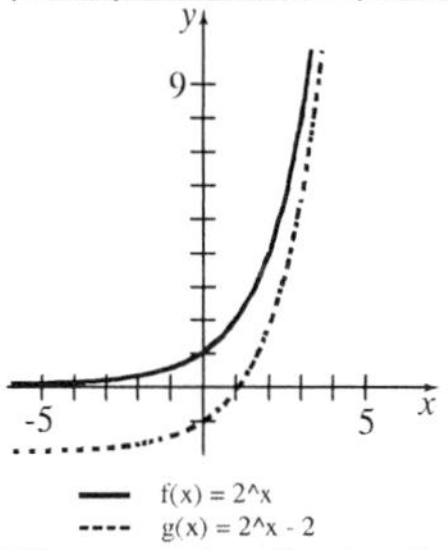


The graph of $g$ is the graph of $f$ shifted down 2 units.

**33.**

| $x$ | $f(x)=3^x$ | $g(x)=-3^x$ |
|---|---|---|
| $-2$ | $\frac{1}{9}$ | $-\frac{1}{9}$ |
| $-1$ | $\frac{1}{3}$ | $-\frac{1}{3}$ |
| 0 | 1 | $-1$ |
| 1 | 3 | $-3$ |
| 2 | 9 | $-9$ |

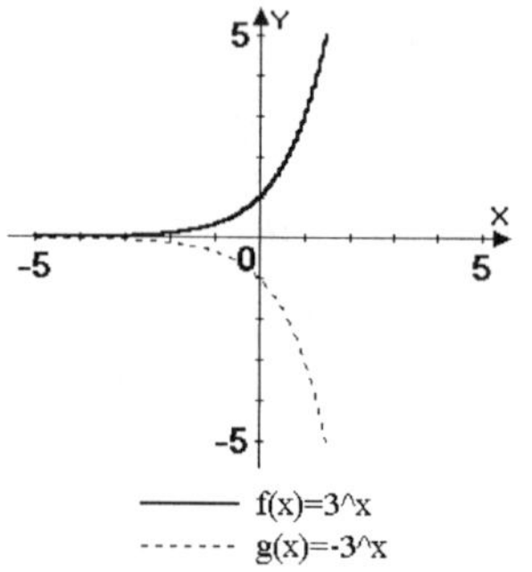


The graph of $g$ is the graph of $f$ reflected across the $x$–axis.

**35.**

| $x$ | $f(x)=2^x$ | $g(x)=2^{x+1}-1$ |
|---|---|---|
| $-2$ | $\frac{1}{4}$ | $-\frac{1}{2}$ |
| $-1$ | $\frac{1}{2}$ | 0 |
| 0 | 1 | 1 |
| 1 | 2 | 3 |
| 2 | 4 | 7 |

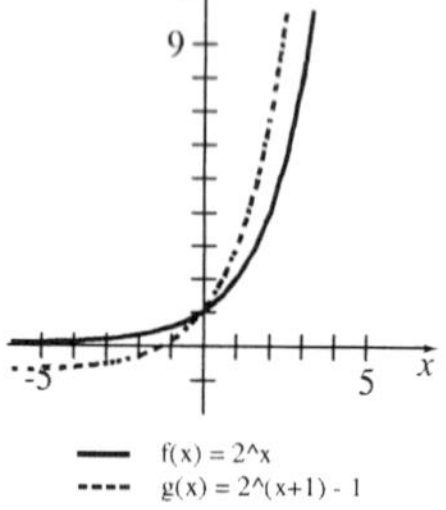


The graph of $g$ is the graph of $f$ shifted 1 unit down and 1 unit to the left.

**37.**

| $x$ | $f(x)=3^x$ | $g(x)=\frac{1}{3}\cdot 3^x$ |
|---|---|---|
| $-2$ | $\frac{1}{9}$ | $\frac{1}{27}$ |
| $-1$ | $\frac{1}{3}$ | $\frac{1}{9}$ |
| 0 | 1 | $\frac{1}{3}$ |
| 1 | 3 | 1 |
| 2 | 9 | 3 |

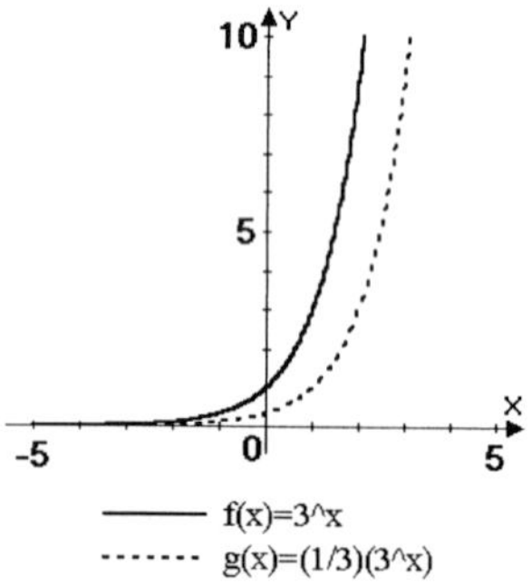


The graph of $g$ is the graph of $f$ compressed vertically by a factor of $\frac{1}{3}$.

**39.** **a.**

$$A=10{,}000\left(1+\frac{0.055}{2}\right)^{2(5)}$$
$$=10{,}000(1.0275)^{10}$$
$$=13116.51$$

The balance in the account is \$13,116.51 after 5 years of semiannual compounding.

**b.**

$$A=10{,}000\left(1+\frac{0.055}{12}\right)^{12(5)}$$
$$=10{,}000(1.0045833)^{60}$$
$$=13157.04$$

The balance in the account is \$13,157.04 after 5 years of monthly compounding.

**c.** $A = Pe^{rt} = 10{,}000e^{0.055(5)}$
$= 10{,}000e^{0.275} = 13165.31$
The balance in the account is \$13,165.31 after 5 years of continuous compounding.

**41.** Monthly Compounding

$$A = 12{,}000\left(1+\frac{0.07}{12}\right)^{12(3)}$$
$$= 10{,}000(1+0.0058333)^{36}$$
$$= 10{,}000(1.0058333)^{36}$$
$$= 12329.24$$

Continuous Compounding

$$A = 12{,}000e^{0.0685(3)}$$
$$= 10{,}000e^{0.2055} = 12281.39$$

Monthly compounding at 7% yields the greatest return.

**43.** Domain: $\{x \mid x \text{ is a real number}\}$ or $(-\infty,\infty)$
Range: $\{y \mid y > -2\}$ or $(-2,\infty)$

**44.** Domain: $\{x \mid x \text{ is a real number}\}$ or $(-\infty,\infty)$
Range: $\{y \mid y > -3\}$ or $(-3,\infty)$

**45.** Domain: $\{x \mid x \text{ is a real number}\}$ or $(-\infty,\infty)$
Range: $\{y \mid y > 1\}$ or $(1,\infty)$

**46.** Domain: $\{x \mid x \text{ is a real number}\}$ or $(-\infty,\infty)$
Range: $\{y \mid y > 2\}$ or $(2,\infty)$

**47.** Domain: $\{x \mid x \text{ is a real number}\}$ or $(-\infty,\infty)$
Range: $\{y \mid y > 0\}$ or $(0,\infty)$

**48.** Domain: $\{x \mid x \text{ is a real number}\}$ or $(-\infty,\infty)$
Range: $\{y \mid y > 0\}$ or $(0,\infty)$

**49.**

| $x$ | $f(x) = 2^x$ | $g(x) = 2^{-x}$ |
|---|---|---|
| $-2$ | $\frac{1}{4}$ | 4 |
| $-1$ | $\frac{1}{2}$ | 2 |
| 0 | 1 | 1 |
| 1 | 2 | $\frac{1}{2}$ |
| 2 | 4 | $\frac{1}{4}$ |

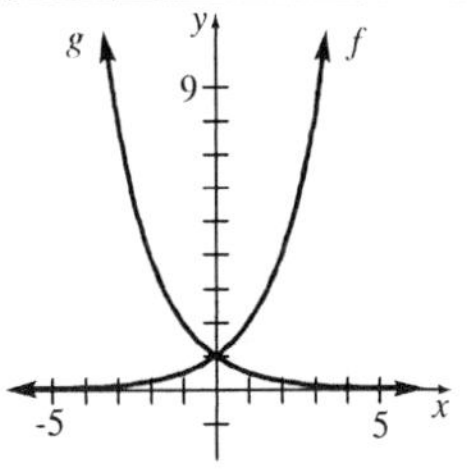


The point of intersection is $(0,1)$.

**50.**

| $x$ | $f(x) = 2^{x+1}$ | $g(x) = 2^{-x+1}$ |
|---|---|---|
| $-2$ | $\frac{1}{2}$ | 8 |
| $-1$ | 1 | 4 |
| 0 | 2 | 2 |
| 1 | 4 | 1 |
| 2 | 8 | $\frac{1}{2}$ |

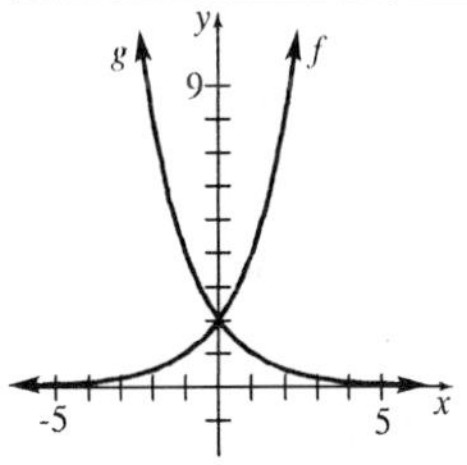


The point of intersection is $(0,1)$.

**51.**

| $x$ | $y = 2^x$ |
|---|---|
| $-2$ | $\frac{1}{4}$ |
| $-1$ | $\frac{1}{2}$ |
| 0 | 1 |
| 1 | 2 |
| 2 | 4 |

| $y$ | $x = 2^y$ |
|---|---|
| $-2$ | $\frac{1}{4}$ |
| $-1$ | $\frac{1}{2}$ |
| 0 | 1 |
| 1 | 2 |
| 2 | 4 |

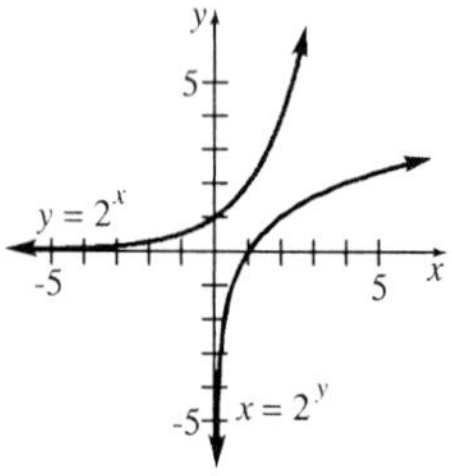


**52.**

| $x$ | $y = 3^x$ |
|---|---|
| $-2$ | $\frac{1}{9}$ |
| $-1$ | $\frac{1}{3}$ |
| 0 | 1 |
| 1 | 3 |
| 2 | 9 |

| $y$ | $x = 3^y$ |
|---|---|
| $-2$ | $\frac{1}{9}$ |
| $-1$ | $\frac{1}{3}$ |
| 0 | 1 |
| 1 | 3 |
| 2 | 9 |

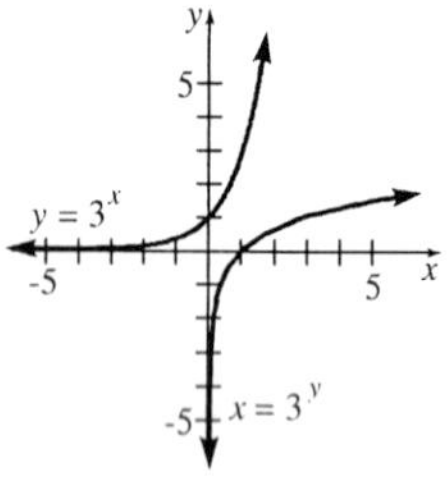


**53.** **a.** $f(0) = 574(1.026)^0$

$= 574(1) = 574$

India's population in 1974 was 574 million.

**b.** $f(27) = 574(1.026)^{27} \approx 1148$

India's population in 2001 will be 1148 million.

**c.** Since $2028 - 1974 = 54$, find

$f(54) = 574(1.026)^{54} \approx 2295$.

India's population in 2028 will be 2295 million.

**d.** $2055 - 1974 = 81$, find

$f(54) = 574(1.026)^{81} \approx 4590$.

India's population in 2055 will be 4590 million.

**e.** India's population appears to be doubling every 27 years.

**55.** $S = 65{,}000(1 + 0.06)^{10}$

$= 65{,}000(1.06)^{10} \approx 116{,}405$

In 10 years, the house will be worth \$116,405.

**57.** Since $2001 - 1995 = 6$, find

$f(6) = 35.86e^{0.207(6)}$

$= 35.86e^{1.242} \approx 124.2$

According to the model, there were 124.2 million cellular telephone subscribers in 2001. The actual number of cellular telephone subscribers was 128.4 million, so the model describes the actual data fairly well.

**59.** **a.** $f(0) = 80e^{-0.5(0)} + 20$

$= 80e^0 + 20$

$= 80(1) + 20$

$= 80 + 20 = 100$

100% of information is remembered at the moment it is first learned.

**b.** $f(1) = 80e^{-0.5(1)} + 20$

$= 80e^{-0.5} + 20 \approx 68.522$

About 68.5% of information is remembered after one week.

**c.** $f(4) = 80e^{-0.5(4)} + 20$
$= 80e^{-2} + 20$
$= 10.827 + 20 = 30.827$
Approximately 30.8% of information is remembered after four weeks.

**d.** $f(52) = 80e^{-0.5(52)} + 20$
$= 80e^{-26} + 20$
$= (4.087 \times 10^{-10}) + 20$
$\approx 20$
Approximately 20% of information is remembered after one year. ($4.087 \times 10^{-10}$ will be eliminated in rounding.)

**61.** $f(30) = \dfrac{90}{1 + 270e^{-0.122(30)}}$
$= \dfrac{90}{1 + 270e^{-3.66}}$
$= \dfrac{90}{1 + 6.948} = \dfrac{90}{7.948} \approx 11.3$
Approximately 11.3% of 30-year-olds have some coronary heart disease.

**63.** **a.** $N(0) = \dfrac{30,000}{1 + 20e^{-1.5(0)}}$
$= \dfrac{30,000}{1 + 20e^{0}}$
$= \dfrac{30,000}{1 + 20(1)}$
$= \dfrac{30,000}{1 + 20}$
$= \dfrac{30,000}{21} \approx 1428.6$
Approximately 1429 people became ill with the flu when the epidemic began.

**b.** $N(3) = \dfrac{30,000}{1 + 20e^{-1.5(3)}}$
$= \dfrac{30,000}{1 + 20e^{-4.5}} \approx 24,546$
Approximately 24,546 people became ill with the flu by the end of the third week.

**c.** The epidemic cannot grow indefinitely because there are a limited number of people that can become ill. Because there are 30,000 people in the town, the limit is 30,000.

**65. – 69.** Answers will vary

**71.** **a.** $Q(t) = 10000\left(1 + \dfrac{0.05}{4}\right)^{4t}$

$M(t) = 10000\left(1 + \dfrac{0.045}{12}\right)^{12t}$

**b.** 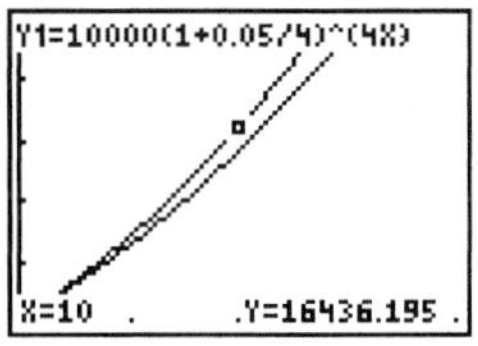


The bank paying 5% compounded quarterly offers a better return.

**73.** Statement **d** is true.

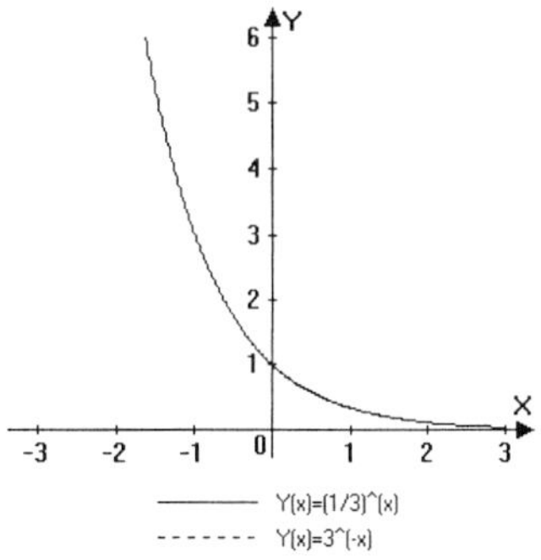

The graphs coincide, so the functions are equivalent.

Statement **a** is false. The amount of money will not increase without bound.

Statement **b** is false.

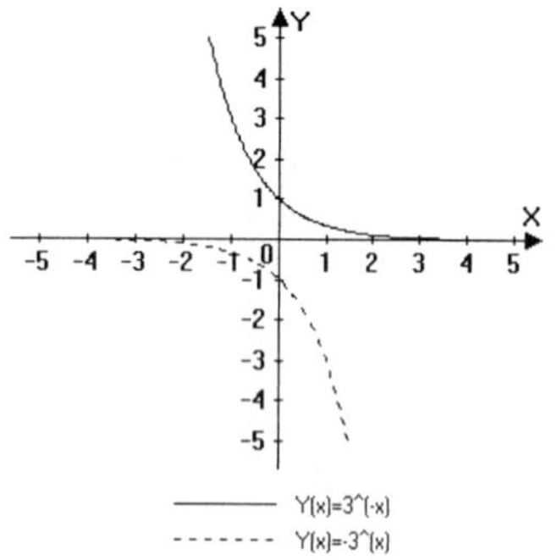


The graphs do not coincide.

Statement **c** is false. 2.718 is an approximation of *e*.

**75.** $(\cosh x)^2 - (\sinh x)^2$

$$= \left(\frac{e^x + e^{-x}}{2}\right)^2 - \left(\frac{e^x - e^{-x}}{2}\right)^2$$

$$= \frac{e^{2x} + 2e^x e^{-x} + e^{-2x}}{4} - \frac{e^{2x} - 2e^x e^{-x} + e^{-2x}}{4}$$

$$= \frac{\left(e^{2x} + 2e^x e^{-x} + e^{-2x}\right) - \left(e^{2x} - 2e^x e^{-x} + e^{-2x}\right)}{4}$$

$$= \frac{\cancel{e^{2x}} + 2e^x e^{-x} + \cancel{e^{-2x}} - \cancel{e^{2x}} + 2e^x e^{-x} - \cancel{e^{-2x}}}{4}$$

$$= \frac{2e^x e^{-x} + 2e^x e^{-x}}{4} = \frac{4e^x e^{-x}}{4} = e^x e^{-x} = \frac{e^x}{e^x} = 1$$

**76.**

$$D = \frac{ab}{a+b}$$

$$D(a+b) = ab$$

$$Da + Db = ab$$

$$Da = ab - Db$$

$$Da = (a - D)b$$

$$b = \frac{Da}{a - D}$$

**77.** $\begin{vmatrix} 3 & -2 \\ 7 & -5 \end{vmatrix} = 3(-5) - 7(-2)$

$$= -15 + 14 = -1$$

**78.**

$$x(x-3) = 10$$

$$x^2 - 3x = 10$$

$$x^2 - 3x - 10 = 0$$

$$(x-5)(x+2) = 0$$

Apply the zero product principle.

$x - 5 = 0$ or $x + 2 = 0$

$x = 5$ $\qquad x = -2$

The solutions are 5 and –2, and the solution set is $\{-2, 5\}$.

## 9.2 Exercise Set

**1.** **a.** $(f \circ g)(x) = f(g(x))$

$$= f(x+7)$$

$$= 2(x+7) = 2x + 14$$

**b.** $(g \circ f)(x) = g(f(x))$

$$= g(2x) = 2x + 7$$

**c.** $(f \circ g)(2) = 2(2) + 14$

$$= 4 + 14 = 18$$

**3.** **a.** $(f \circ g)(x) = f(g(x))$

$$= f(2x+1)$$

$$= (2x+1) + 4 = 2x + 5$$

**b.** $(g \circ f)(x) = g(f(x))$

$$= g(x+4)$$

$$= 2(x+4) + 1$$

$$= 2x + 8 + 1 = 2x + 9$$

**c.** $(f \circ g)(2) = 2(2) + 5 = 4 + 5 = 9$

**5.** **a.** $(f \circ g)(x) = f(g(x))$
$= f(5x^2 - 2)$
$= 4(5x^2 - 2) - 3$
$= 20x^2 - 8 - 3$
$= 20x^2 - 11$

**b.** $(g \circ f)(x) = g(f(x))$
$= g(4x - 3)$
$= 5(4x - 3)^2 - 2$
$= 5(16x^2 - 24x + 9) - 2$
$= 80x^2 - 120x + 45 - 2$
$= 80x^2 - 120x + 43$

**c.** $(f \circ g)(2) = 20(2)^2 - 11$
$= 20(4) - 11$
$= 80 - 11 = 69$

**7.** **a.** $(f \circ g)(x) = f(g(x))$
$= f(x^2 - 2)$
$= (x^2 - 2)^2 + 2$
$= x^4 - 4x^2 + 4 + 2$
$= x^4 - 4x^2 + 6$

**b.** $(g \circ f)(x) = g(f(x))$
$= g(x^2 + 2)$
$= (x^2 + 2)^2 - 2$
$= x^4 + 4x^2 + 4 - 2$
$= x^4 + 4x^2 + 2$

**c.** $(f \circ g)(2) = 2^4 - 4(2)^2 + 6$
$= 16 - 4(4) + 6$
$= 16 - 16 + 6 = 6$

**9.** **a.** $(f \circ g)(x) = f(g(x))$
$= f(x - 1) = \sqrt{x - 1}$

**b.** $(g \circ f)(x) = g(f(x))$
$= g(\sqrt{x}) = \sqrt{x} - 1$

**c.** $(f \circ g)(2) = \sqrt{2 - 1} = \sqrt{1} = 1$

**11.** **a.** $(f \circ g)(x) = f(g(x))$
$= f\left(\frac{x+3}{2}\right)$
$= 2\left(\frac{x+3}{2}\right) - 3$
$= x + 3 - 3 = x$

**b.** $(g \circ f)(x) = g(f(x))$
$= g(2x - 3)$
$= \frac{(2x - 3) + 3}{2}$
$= \frac{2x - 3 + 3}{2} = \frac{2x}{2} = x$

**c.** $(f \circ g)(2) = 2$

**13.** **a.** $(f \circ g)(x) = f(g(x))$
$= f\left(\frac{1}{x}\right)$
$= \frac{1}{\frac{1}{x}} = 1 \cdot \frac{x}{1} = x$

**b.** $(g \circ f)(x) = g(f(x))$
$= g\left(\frac{1}{x}\right)$
$= \frac{1}{\frac{1}{x}} = 1 \cdot \frac{x}{1} = x$

**c.** $(f \circ g)(2) = 2$

**15.** $f(g(x)) = f\left(\frac{x}{4}\right) = 4\left(\frac{x}{4}\right) = x$

$g(f(x)) = g(4x) = \frac{4x}{4} = x$

The functions are inverses.

**17.** $f(g(x)) = f\left(\frac{x-8}{3}\right)$

$= 3\left(\frac{x-8}{3}\right) + 8$

$= x - 8 + 8 = x$

$g(f(x)) = g(3x+8)$

$= \frac{(3x+8)-8}{3}$

$= \frac{3x+8-8}{3} = \frac{3x}{3} = x$

The functions are inverses.

**19.** $f(g(x)) = f\left(\frac{x+5}{9}\right)$

$= 5\left(\frac{x+5}{9}\right) - 9$

$= \frac{5x+25}{9} - \frac{81}{9}$

$= \frac{5x+25-81}{9} = \frac{5x-56}{9}$

$g(f(x)) = g(5x-9)$

$= \frac{(5x-9)+5}{9} = \frac{5x-4}{9}$

Since $f(g(x)) \neq g(f(x)) \neq x$, we conclude the functions are not inverses.

**21.** $f(g(x)) = f\left(\frac{3}{x}+4\right)$

$= \frac{3}{\left(\frac{3}{x}+4\right)-4}$

$= \frac{3}{\frac{3}{x}+4-4} = \frac{3}{\frac{3}{x}} = 3 \cdot \frac{x}{3} = x$

$g(f(x)) = g\left(\frac{3}{x-4}\right)$

$= \frac{3}{\frac{3}{x-4}} + 4$

$= 3 \cdot \frac{x-4}{3} + 4 = x - 4 + 4 = x$

The functions are inverses.

**23.** $f(g(x)) = f(-x) = -(-x) = x$

$g(f(x)) = g(-x) = -(-x) = x$

The functions are inverses.

**25.** **a.** $f(x) = x+3$

$y = x + 3$

Interchange $x$ and $y$ and solve for $y$.

$x = y + 3$

$x - 3 = y$

$f^{-1}(x) = x - 3$

**b.** $f(f^{-1}(x)) = f(x-3)$

$= (x-3)+3$

$= x - 3 + 3 = x$

$f^{-1}(f(x)) = f(x+3)$

$= (x+3) - 3$

$= x + 3 - 3 = x$

**27.** **a.** $f(x) = 2x$

$y = 2x$

Interchange $x$ and $y$ and solve for $y$.

$x = 2y$

$\frac{x}{2} = y$

$f^{-1}(x) = \frac{x}{2}$

**b.** $f(f^{-1}(x)) = f\left(\frac{x}{2}\right) = 2\left(\frac{x}{2}\right) = x$

$f^{-1}(f(x)) = f(2x) = \frac{2x}{2} = x$

**29.** **a.** $f(x) = 2x + 3$

$y = 2x + 3$

Interchange $x$ and $y$ and solve for $y$.

$x = 2y + 3$

$x - 3 = 2y$

$\frac{x-3}{2} = y$

$f^{-1}(x) = \frac{x-3}{2}$

**b.** 
$$\begin{aligned} f(f^{-1}(x)) &= f\left(\frac{x-3}{2}\right) \\ &= 2\left(\frac{x-3}{2}\right) + 3 \\ &= x - 3 + 3 = x \end{aligned}$$

$$\begin{aligned} f^{-1}(f(x)) &= f^{-1}(2x+3) \\ &= \frac{(2x+3)-3}{2} \\ &= \frac{2x+3-3}{2} = \frac{2x}{2} = x \end{aligned}$$

**31.** **a.** $f(x) = x^3 + 2$

$y = x^3 + 2$

Interchange $x$ and $y$ and solve for $y$.

$x = y^3 + 2$

$x - 2 = y^3$

$\sqrt[3]{x-2} = y$

$f^{-1}(x) = \sqrt[3]{x-2}$

**b.** 
$$\begin{aligned} f(f^{-1}(x)) &= f\left(\sqrt[3]{x-2}\right) \\ &= \left(\sqrt[3]{x-2}\right)^3 + 2 \\ &= x - 2 + 2 = x \end{aligned}$$

$$\begin{aligned} f^{-1}(f(x)) &= f^{-1}(x^3 + 2) \\ &= \sqrt[3]{(x^3+2) - 2} \\ &= \sqrt[3]{x^3 + 2 - 2} \\ &= \sqrt[3]{x^3} = x \end{aligned}$$

**33.** **a.** $f(x) = (x+2)^3$

$y = (x+2)^3$

Interchange $x$ and $y$ and solve for $y$.

$x = (y+2)^3$

$\sqrt[3]{x} = \sqrt[3]{(y+2)^3}$

$\sqrt[3]{x} = y + 2$

$\sqrt[3]{x} - 2 = y$

$f^{-1}(x) = \sqrt[3]{x} - 2$

**b.** $f\left(f^{-1}(x)\right) = f\left(\sqrt[3]{x}-2\right)$

$= \left(\left(\sqrt[3]{x}-2\right)+2\right)^3$

$= \left(\sqrt[3]{x}-2+2\right)^3$

$= \left(\sqrt[3]{x}\right)^3 = x$

$f^{-1}\left(f(x)\right) = f^{-1}\left((x+2)^3\right)$

$= \sqrt[3]{(x+2)^3} - 2$

$= x+2-2 = x$

**35.** **a.** $f(x) = \dfrac{1}{x}$

$y = \dfrac{1}{x}$

Interchange $x$ and $y$ and solve for $y$.

$x = \dfrac{1}{y}$

$xy = 1$

$y = \dfrac{1}{x}$

$f^{-1}(x) = \dfrac{1}{x}$

**b.** $f\left(f^{-1}(x)\right) = f\left(\dfrac{1}{x}\right)$

$= \dfrac{1}{\frac{1}{x}} = 1 \cdot \dfrac{x}{1} = x$

$f^{-1}\left(f(x)\right) = f^{-1}\left(\dfrac{1}{x}\right)$

$= \dfrac{1}{\frac{1}{x}} = 1 \cdot \dfrac{x}{1} = x$

**37.** **a.** $f(x) = \sqrt{x}$

$y = \sqrt{x}$

Interchange $x$ and $y$ and solve for $y$.

$x = \sqrt{y}$

$x^2 = y$

$f^{-1}(x) = x^2, \quad x \geq 0$

**b.** $f\left(f^{-1}(x)\right) = f\left(x^2\right) = \sqrt{x^2} = x$

$f^{-1}\left(f(x)\right) = f^{-1}\left(\sqrt{x}\right)$

$= \left(\sqrt{x}\right)^2 = x$

**39.** **a.** $f(x) = x^2 + 1$

$y = x^2 + 1$

Interchange $x$ and $y$ and solve for $y$.

$x = y^2 + 1$

$x - 1 = y^2$

$\sqrt{x-1} = y$

$f^{-1}(x) = \sqrt{x-1}$

**b.** $f\left(f^{-1}(x)\right) = f\left(\sqrt{x-1}\right)$

$= \left(\sqrt{x-1}\right)^2 + 1$

$= x - 1 + 1 = x$

$f^{-1}\left(f(x)\right) = f^{-1}\left(x^2+1\right)$

$= \sqrt{\left(x^2+1\right)-1}$

$= \sqrt{x^2+1-1}$

$= \sqrt{x^2} = x$

**41.** **a.**
$$f(x)=\frac{2x+1}{x-3}$$
$$y=\frac{2x+1}{x-3}$$
Interchange $x$ and $y$ and solve for $y$.
$$x=\frac{2y+1}{y-3}$$
$$x(y-3)=2y+1$$
$$xy-3x=2y+1$$
$$xy-2y=3x+1$$
$$(x-2)y=3x+1$$
$$y=\frac{3x+1}{x-2}$$
$$f^{-1}(x)=\frac{3x+1}{x-2}$$

**b.**
$$f\left(f^{-1}(x)\right)$$
$$=f\left(\frac{3x+1}{x-2}\right)$$
$$=\frac{2\left(\frac{3x+1}{x-2}\right)+1}{\left(\frac{3x+1}{x-2}\right)-3}$$
$$=\frac{x-2}{x-2}\cdot\frac{2\left(\frac{3x+1}{x-2}\right)+1}{\left(\frac{3x+1}{x-2}\right)-3}$$
$$=\frac{2(3x+1)+1(x-2)}{(3x+1)-3(x-2)}$$
$$=\frac{6x+2+x-2}{3x+1-3x+6}=\frac{7x}{7}=x$$

$$f^{-1}\left(f(x)\right)$$
$$=f^{-1}\left(\frac{2x+1}{x-3}\right)$$
$$=\frac{3\left(\frac{2x+1}{x-3}\right)+1}{\left(\frac{2x+1}{x-3}\right)-2}$$
$$=\frac{x-3}{x-3}\cdot\frac{3\left(\frac{2x+1}{x-3}\right)+1}{\left(\frac{2x+1}{x-3}\right)-2}$$
$$=\frac{3(2x+1)+1(x-3)}{(2x+1)-2(x-3)}$$
$$=\frac{6x+3+x-3}{2x+1-2x+6}=\frac{7x}{7}=x$$

**43.** **a.**
$$f(x)=\sqrt[3]{x-4}+3$$
$$y=\sqrt[3]{x-4}+3$$
Interchange $x$ and $y$ and solve for $y$.
$$x=\sqrt[3]{y-4}+3$$
$$x-3=\sqrt[3]{y-4}$$
$$(x-3)^3=y-4$$
$$(x-3)^3+4=y$$
$$f^{-1}(x)=(x-3)^3+4$$

**b.**
$$f\left(f^{-1}(x)\right)$$
$$=f\left((x-3)^3+4\right)$$
$$=\sqrt[3]{\left((x-3)^3+4\right)-4}+3$$
$$=\sqrt[3]{(x-3)^3+4-4}+3$$
$$=\sqrt[3]{(x-3)^3}+3=x-3+3=x$$

$$f^{-1}(f(x))$$
$$= f^{-1}\left(\sqrt[3]{x-4}+3\right)$$
$$= \left(\left(\sqrt[3]{x-4}+3\right)-3\right)^3+4$$
$$= \left(\sqrt[3]{x-4}+3-3\right)^3+4$$
$$= \left(\sqrt[3]{x-4}\right)^3+4 = x-4+4 = x$$

**45.** The graph does not satisfy the horizontal line test so the function does not have an inverse.

**47.** The graph does not satisfy the horizontal line test so the function does not have an inverse.

**49.** The graph satisfies the horizontal line test so the function has an inverse.

**51.**

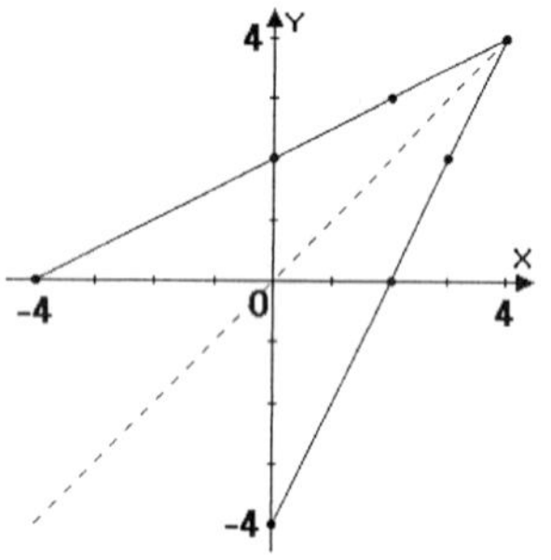


**53.**

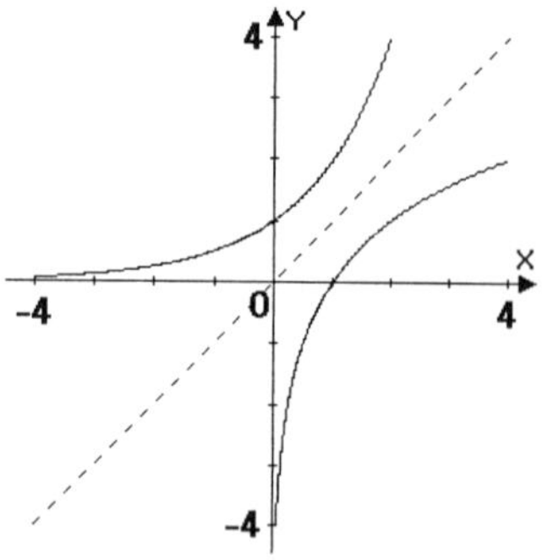


**55.** $f(g(1)) = f(1) = 5$

**56.** $f(g(4)) = f(2) = -1$

**57.** $(g \circ f)(-1) = g(f(-1)) = g(1) = 1$

**58.** $(g \circ f)(0) = g(f(0)) = g(4) = 2$

**59.** $f^{-1}(g(10)) = f^{-1}(-1) = 2$, since $f(2) = -1$.

**60.** $f^{-1}(g(1)) = f^{-1}(1) = -1$, since $f(-1) = 1$.

**61.** $(f \circ g)(-1) = f(g(-1)) = f(-3) = 1$

**62.** $(f \circ g)(1) = f(g(1)) = f(-5) = 3$

**63.** $(g \circ f)(0) = g(f(0)) = g(2) = -6$

**64.** $(g \circ f)(-1) = g(f(-1)) = g(1) = -5$

**65.** $(f \circ g)(0) = f(g(0))$
$$= f(4 \cdot 0 - 1)$$
$$= f(-1) = 2(-1) - 5 = -7$$

**66.** $(g \circ f)(0) = g(f(0))$
$$= g(2 \cdot 0 - 5)$$
$$= g(-5) = 4(-5) - 1 = -21$$

**67.** Let $f^{-1}(1) = x$. Then
$$f(x) = 1$$
$$2x - 5 = 1$$
$$2x = 6$$
$$x = 3$$
Thus, $f^{-1}(1) = 3$

**68.** Let $g^{-1}(7) = x$. Then

$$g(x) = 7$$
$$4x - 1 = 7$$
$$4x = 8$$
$$x = 2$$

Thus, $g^{-1}(7) = 2$

**69.**
$$\begin{aligned} g(f[h(1)]) &= g\left(f\left[1^2 + 1 + 2\right]\right) \\ &= g(f(4)) \\ &= g(2 \cdot 4 - 5) \\ &= g(3) \\ &= 4 \cdot 3 - 1 = 11 \end{aligned}$$

**70.**
$$\begin{aligned} f(g[h(1)]) &= f\left(g\left[1^2 + 1 + 2\right]\right) \\ &= f(g(4)) \\ &= f(4 \cdot 4 - 1) \\ &= f(15) \\ &= 2 \cdot 15 - 5 = 25 \end{aligned}$$

**71.** **a.** $f$ represents the price after a $400 discount; $g$ represents the price after a 25% discount (75% of the regular price).

**b.**
$$\begin{aligned} (f \circ g)(x) &= f(g(x)) \\ &= f(0.75x) \\ &= 0.75x - 400 \end{aligned}$$

$f \circ g$ represents and additional $400 discount on a price that has already been reduced by 25%.

**c.**
$$\begin{aligned} (g \circ f)(x) &= g(f(x)) \\ &= g(x - 400) \\ &= 0.75(x - 400) \\ &= 0.75x - 300 \end{aligned}$$

$g \circ f$ represents an additional 25% discount on a price that has already been reduced by $400.

**d.** $0.75x - 400 < 0.75x - 300$, so $f \circ g$ models the greater discount. It has a savings of $100 over $g \circ f$.

**e.** $f(x) = x - 400$

$$y = x - 400$$

Interchange $x$ and $y$ and solve for $y$.

$$x = y - 400$$
$$x + 400 = y$$
$$f^{-1}(x) = x + 400$$

$f^{-1}$ represents the regular price of the computer, since the value of $x$ here is the price after a $400 discount.

**73.** **a.** $f$: {(Zambia, 4.2), (Columbia, 4.5), (Poland, 3.3), (Italy, 3.3), (U.S., 2.5)}

**b.** Inverse: {(4.2, Zambia), (4.5, Columbia), (3.3, Poland), (3.3, Italy), (2.5, U.S.)}; The inverse is not a function because the input 3.3 is associated with two different outputs: Poland and Italy.

**75.** **a.** We know that $f$ has an inverse because no horizontal line intersects the graph of $f$ in more than one point.

**b.** $f^{-1}(0.25)$, or approximately 15, represents the number of people who must be in a room so that the probability of two sharing a birthday would be 0.25; $f^{-1}(0.5)$, or approximately 23, represents the number of people who must be in a room so that the probability of 2 sharing a birthday would be 0.5;

$f^{-1}(0.7)$, or approximately 30, represents the number of people who must be in a room so that the probability of two sharing a birthday would be 0.70.

**77. – 81.** Answers will vary.

**83.** $f(x) = x^2 - 1$

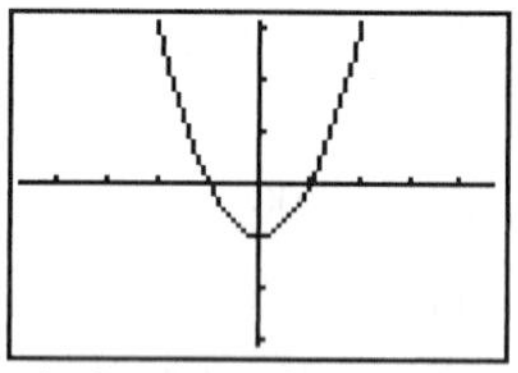

$f$ does not have an inverse function.

**85.** $f(x) = \dfrac{x^3}{2}$

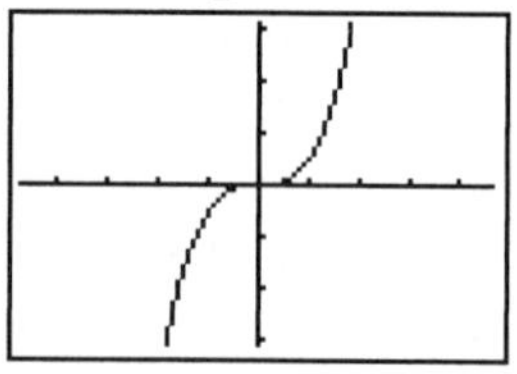

$f$ has an inverse function.

**87.** $f(x) = |x - 2|$

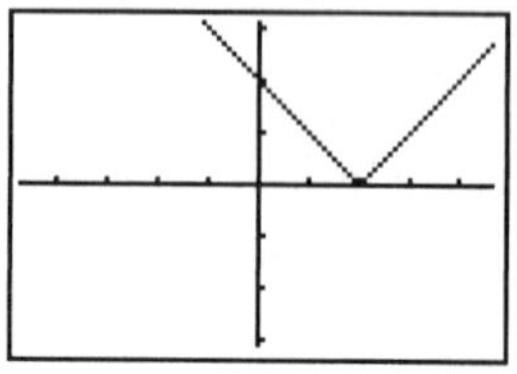

$f$ does not have an inverse function.

**89.** $f(x) = -\sqrt{16 - x^2}$

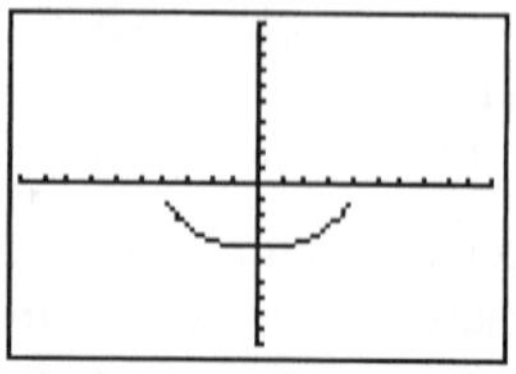

$f$ does not have an inverse function.

**91.** $f(x) = 4x + 4$

$g(x) = 0.25x - 1$

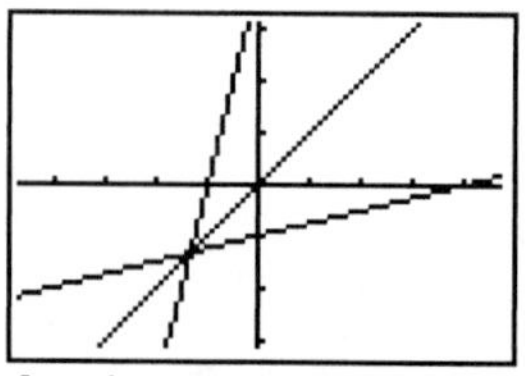

$f$ and $g$ are inverses.

**93.** $f(x) = \sqrt[3]{x} - 2$

$g(x) = (x + 2)^3$

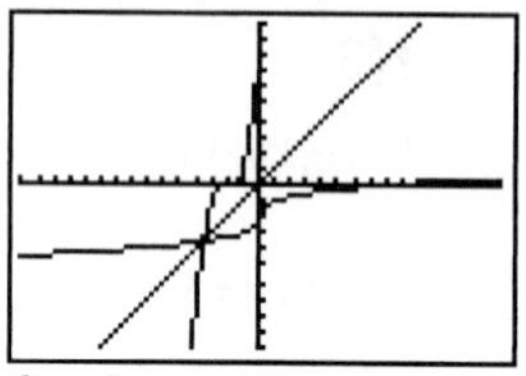

$f$ and $g$ are inverses.

**95.** Answers will vary. One example is $f(x) = \sqrt{x + 5}$ and $g(x) = 3x^2$.

**97.**
$$f(f(x)) = \frac{3(f(x)) - 2}{5(f(x)) - 3}$$
$$= \frac{3\left(\dfrac{3x-2}{5x-3}\right) - 2}{5\left(\dfrac{3x-2}{5x-3}\right) - 3}$$
$$= \frac{3\left(\dfrac{3x-2}{5x-3}\right) - 2}{5\left(\dfrac{3x-2}{5x-3}\right) - 3} \cdot \frac{5x-3}{5x-3}$$
$$= \frac{3(3x-2) - 2(5x-3)}{5(3x-2) - 3(5x-3)}$$
$$= \frac{9x - 6 - 10x + 6}{15x - 10 - 15x + 9}$$
$$= \frac{-x}{-1} = x$$

Since $f(f(x)) = x$, $f$ is it own inverse.

**99.** $\dfrac{4.3\times10^{5}}{8.6\times10^{-4}}=\dfrac{4.3}{8.6}\times\dfrac{10^{5}}{10^{-4}}$
$=0.5\times10^{9}$
$=5\times10^{-1}\times10^{9}=5\times10^{8}$

**100.** $f(x)=x^{2}-4x+3$
Since $a=1$ is positive, the parabola opens upward. The $x$–coordinate of the vertex of the parabola is
$-\dfrac{b}{2a}=-\dfrac{-4}{2(1)}=-\dfrac{-4}{2}=2$ and the $y$–coordinate of the vertex of the parabola is
$f\left(-\dfrac{b}{2a}\right)=f(2)=2^{2}-4(2)+3$
$=4-8+3=-1.$
The vertex is at (2, !1). Replace $f(x)$ with 0 to find $x$–intercepts.
$0=x^{2}-4x+3$
$0=(x-3)(x-1)$
Apply the zero product principle.
$x-3=0$ or $x-1=0$
$x=3$ $\quad x=1$
The $x$–intercepts are 1 and 3. Set $x = 0$ and solve for $y$ to obtain the $y$–intercept.
$y=0^{2}-4(0)+3$
$y=0-0+3=3$

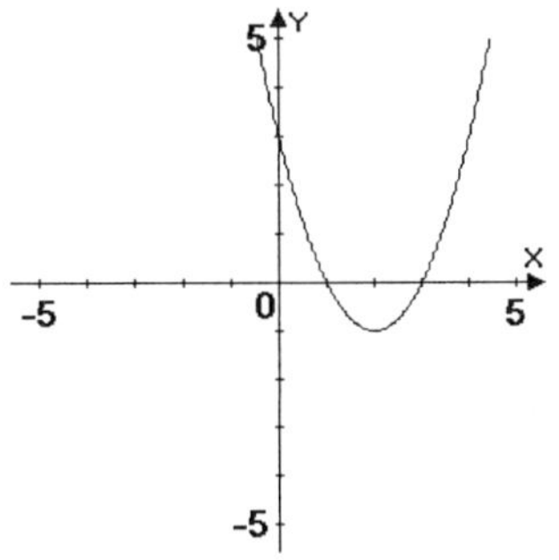


**101.** $\sqrt{x+4}-\sqrt{x-1}=1$
$\sqrt{x+4}=\sqrt{x-1}+1$
$\left(\sqrt{x+4}\right)^{2}=\left(\sqrt{x-1}+1\right)^{2}$
$x+4=x-1+2\sqrt{x-1}+1$
$x+4=x+2\sqrt{x-1}$
$4=2\sqrt{x-1}$
$2=\sqrt{x-1}$
$2^{2}=\left(\sqrt{x-1}\right)^{2}$
$4=x-1$
$5=x$
The solution is 5, and the solution set is $\{5\}$.

**9.3 Exercise Set**

**1.** $4=\log_{2}16$
$2^{4}=16$

**3.** $2=\log_{3}x$
$3^{2}=x$

**5.** $5=\log_{b}32$
$b^{5}=32$

**7.** $\log_{6}216=y$
$6^{y}=216$

**9.** $2^{3}=8$
$\log_{2}8=3$

**11.** $2^{-4}=\dfrac{1}{16}$
$\log_{2}\dfrac{1}{16}=-4$

**13.** $\sqrt[3]{8} = 2$

$8^{\frac{1}{3}} = 2$

$\log_8 2 = \frac{1}{3}$

**15.** $13^2 = x$

$\log_{13} x = 2$

**17.** $b^3 = 1000$

$\log_b 1000 = 3$

**19.** $7^y = 200$

$\log_7 200 = y$

**21.** $\log_4 16 = y$

$4^y = 16$

$4^y = 4^2$

$y = 2$

**23.** $\log_2 64 = y$

$2^y = 64$

$2^y = 2^6$

$y = 6$

**25.** $\log_5 \frac{1}{5} = y$

$5^y = \frac{1}{5}$

$5^y = 5^{-1}$

$y = -1$

**27.** $\log_2 \frac{1}{8} = y$

$2^y = \frac{1}{8}$

$2^y = \frac{1}{2^3}$

$2^y = 2^{-3}$

$y = -3$

**29.** $\log_7 \sqrt{7} = y$

$7^y = \sqrt{7}$

$7^y = 7^{\frac{1}{2}}$

$y = \frac{1}{2}$

**31.** $\log_2 \frac{1}{\sqrt{2}} = y$

$2^y = \frac{1}{\sqrt{2}}$

$2^y = \frac{1}{2^{\frac{1}{2}}}$

$2^y = 2^{-\frac{1}{2}}$

$y = -\frac{1}{2}$

**33.** $\log_{64} 8 = y$

$64^y = 8$

$64^y = 64^{\frac{1}{2}}$

$y = \frac{1}{2}$

**35.** $\log_5 5 = y$

$5^y = 5^1$

$y = 1$

**37.** $\log_4 1 = y$

$4^y = 1$

$4^y = 4^0$

$y = 0$

**39.** $\log_5 5^7 = y$

$5^y = 5^7$

$y = 7$

**41.** Since $b^{\log_b x} = x,\ 8^{\log_8 19} = 19.$

**43.** $f(x) = 4^x$

$g(x) = \log_4 x$

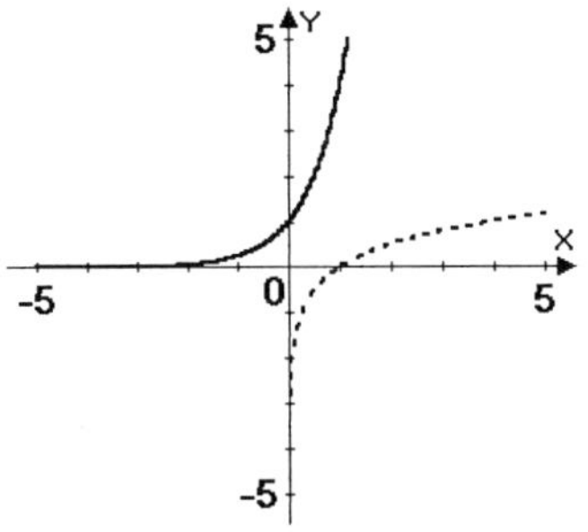


**45.** $f(x) = \left(\frac{1}{2}\right)^x$

$g(x) = \log_{\frac{1}{2}} x$

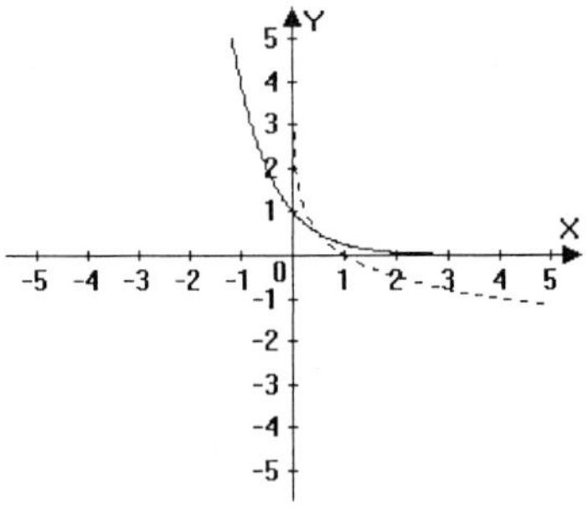


**47.** $f(x) = \log_5(x+4)$

$x + 4 > 0$

$x > -4$

The domain of *f* is $\{x | x > -4\}$ or $(-4, \infty)$.

**49.** $f(x) = \log_5(2-x)$

$2 - x > 0$

$-x > -2$

$x < 2$

The domain of *f* is $\{x | x < 2\}$ or $(-\infty, 2)$.

**51.** $f(x) = \ln(x-2)^2$

The domain of *g* is all real numbers for which $(x-2)^2 > 0$. The only number that must be excluded is 2. The domain of *f* is $\{x | x \neq 2\}$ or $(-\infty, 2) \cup (2, \infty)$.

**53.** $\log 100 = y$

$10^y = 100$

$10^y = 10^2$

$y = 2$

**55.** $\log 10^7 = y$

$10^y = 10^7$

$y = 7$

**57.** Since $10^{\log x} = x,\ 10^{\log 33} = 33.$

**59.** $\ln 1 = y$

$e^y = 1$

$e^y = e^0$

$y = 0$

**61.** Since $\ln e^x = x,\ \ln e^6 = 6$.

**63.** $\ln \frac{1}{e^6} = \ln e^{-6}$

Since $\ln e^x = x,\ \ln e^{-6} = -6$.

**65.** Since $e^{\ln x} = x,\ e^{\ln 125} = 125$.

**67.** Since $\ln e^x = x,\ \ln e^{9x} = 9x$.

**69.** Since $e^{\ln x} = x,\ e^{\ln 5x^2} = 5x^2$.

**71.** Since $10^{\log x} = x,\ 10^{\log \sqrt{x}} = \sqrt{x}$.

**73.** $\log_3(x-1) = 2$

$$3^2 = x-1$$
$$9 = x-1$$
$$10 = x$$

The solution is 10, and the solution set is $\{10\}$.

**74.** $\log_5(x+4) = 2$

$$5^2 = x+4$$
$$25 = x+4$$
$$21 = x$$

The solution is 21, and the solution set is $\{21\}$.

**75.** $\log_4 x = -3$

$$4^{-3} = x$$
$$x = \frac{1}{4^3} = \frac{1}{64}$$

The solution is $\frac{1}{64}$, and the solution set is $\left\{\frac{1}{64}\right\}$.

**76.** $\log_{64} x = \frac{2}{3}$

$$64^{\frac{2}{3}} = x$$
$$x = \left(\sqrt[3]{64}\right)^2 = 4^2 = 16$$

The solution is 16, and the solution set is $\{16\}$.

**77.** $\log_3(\log_7 7) = \log_3 1 = 0$

**78.** $\log_5(\log_2 32) = \log_5(\log_2 2^5)$
$= \log_5 5 = 1$

**79.** $\log_2(\log_3 81) = \log_2(\log_3 3^4)$
$= \log_2 4 = \log_2 2^2 = 2$

**80.** $\log(\ln e) = \log 1 = 0$

**81.** **(d)** The graph is similar to that of $y = \ln x$, but shifted left 2 units.

**82.** **(f)** The graph is similar to that of $y = \ln x$, but shifted right 2 units.

**83.** **(c)** The graph is similar to that of $y = \ln x$, but shifted up 2 units.

**84.** **(a)** The graph is similar to that of $y = \ln x$, but shifted down 2 units.

**85.** **(b)** The graph is similar to that of $y = \ln x$, but reflected across the $y$-axis and then shifted right 1 unit.

**86.** **(b)** The graph is similar to that of $y = \ln x$, but reflected across the $y$-axis and then shifted right 2 units.

**87.** $f(13) = 62 + 35\log(13-4)$
$= 62 + 35\log(9) \approx 95.4$

A 13-year-old girl is approximately 95.4% of her adult height.

**89.** Since 2003 – 1997 = 6, we find $f(6)$:

$f(6) = -4.9\ln 6 + 73.8 \approx 65$

In 2003, approximately 65% of U.S. companies performed drug tests. The function modeled the actual number very well. It gives the actual percent.

**91.** $D = 10\log\left(10^{12}\left(6.3\times 10^{6}\right)\right)$

$= 10\log\left(6.3\times 10^{18}\right) \approx 188.0$

The decibel level of a blue whale is approximately 188 decibels. At close range, the sound could rupture the human ear drum.

**93.** **a.** The original exam was at time, $t = 0$.

$f(0) = 88 - 15\ln(0+1)$

$= 88 - 15\ln(1) \approx 88$

The average score on the original exam was 88.

**b.** $f(2) = 88 - 15\ln(2+1)$

$= 88 - 15\ln(3) \approx 71.5$

$f(4) = 88 - 15\ln(4+1)$

$= 88 - 15\ln(5) \approx 63.9$

$f(6) = 88 - 15\ln(6+1)$

$= 88 - 15\ln(7) \approx 58.8$

$f(8) = 88 - 15\ln(8+1)$

$= 88 - 15\ln(9) \approx 55.0$

$f(10) = 88 - 15\ln(10+1)$

$= 88 - 15\ln(11) \approx 52.0$

$f(12) = 88 - 15\ln(12+1)$

$= 88 - 15\ln(13) \approx 49.5$

The average score for the tests is as follows:

2 months: 71.5
4 months: 63.9
6 months: 58.8
8 months: 55.0
10 months: 52.0
12 months: 49.5

**c.**

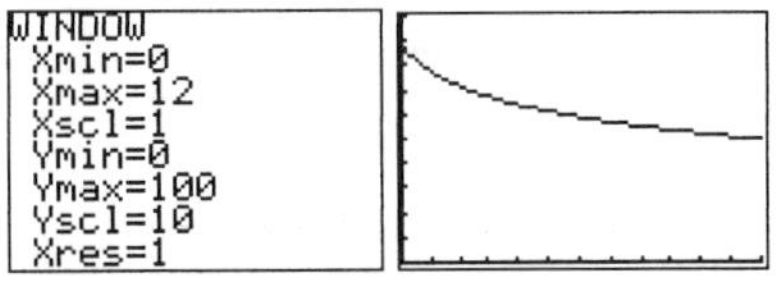


The students remembered less of the material over time.

**95. – 101.** Answers will vary.

**103.** $f(x) = \ln x \qquad g(x) = \ln x + 3$

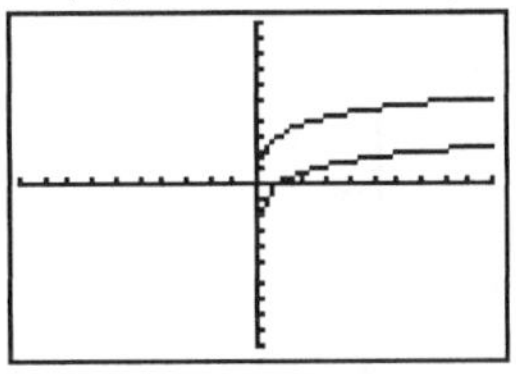

The graph of $g$ is the graph of $f$ shifted up 3 units.

**105.** $f(x) = \log x \qquad g(x) = \log(x-2) + 1$

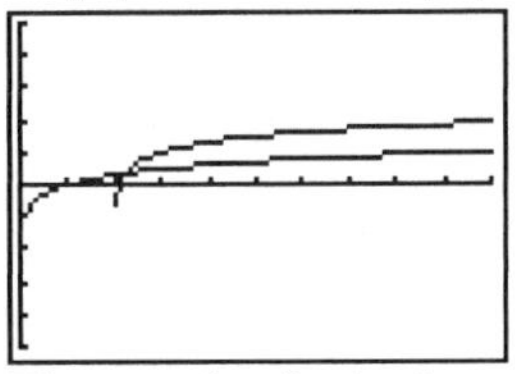

The graph of $g$ is the graph of $f$ shifted 2 units to the right and 1 unit up.

**107.** **a.** $f(x) = \ln(3x)$

$g(x) = \ln 3 + \ln x$

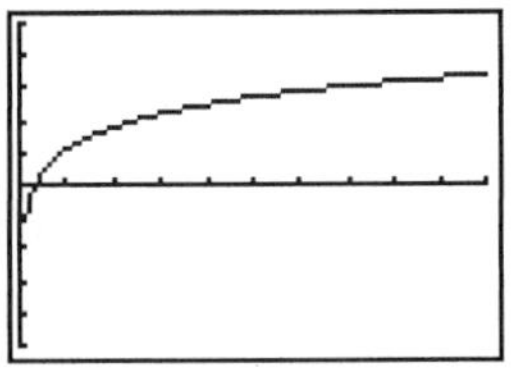

The graphs coincide.

**b.** $f(x)=\log(5x^2)$

$g(x)=\log 5+\log x^2$

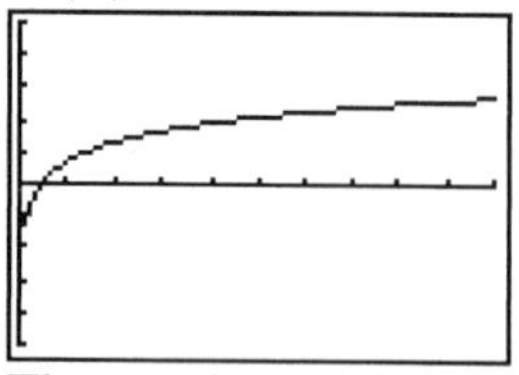

The graphs coincide.

**c.** $f(x)=\ln(2x^3)$

$g(x)=\ln 2+\ln x^3$

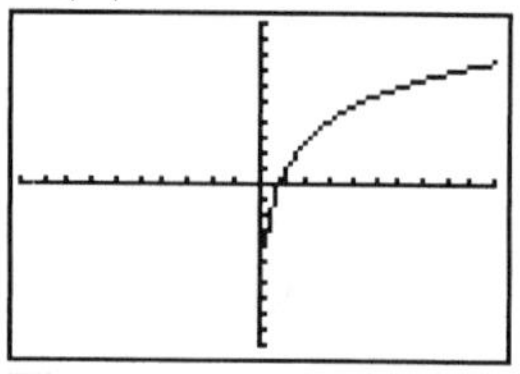

The graphs coincide.

**d.** In each case, the function, $f$, is equivalent to $g$. This means that $\log_b(MN)=\log_b M+\log_b N$.

**e.** The logarithm of a product is equal to the sum of the logarithms of the factors.

**109.** Statement **d.** is true. Recall: $b^{\log_b x}=x$.

Statement **a.** is false. To evaluate $\frac{\log_2 8}{\log_2 4}$, evaluate each term independently.

$\log_2 8=y \qquad \log_2 4=y$

$2^y=8 \qquad 2^y=4$

$2^y=2^3 \qquad 2^y=2^2$

$y=3 \qquad y=2$

Now substitute these values in the original expression.

$\frac{\log_2 8}{\log_2 4}=\frac{3}{2}$

Statement **b.** is false. We cannot take the log of a negative number.

Statement **c.** is false. The domain of $f(x)=\log_2 x$ is $(0,\infty)$.

**111.** $\log_4\left[\log_3(\log_2 8)\right]$

$=\log_4\left[\log_3(\log_2 2^3)\right]$

$=\log_4[\log_3 3]=\log_4 1=0$

**113.** $2x=11-5y$

$3x-2y=-12$

Rewrite the equations.

$2x+5y=\ 11$

$3x-2y=-12$

Multiply the first equation by 2 and the second equation by 5, and solve by addition.

$4x+10y=\ 22$

$15x-10y=-60$

$19x=-38$

$x=-2$

Back-substitute –2 for $x$ to find $y$.

$2(-2)+5y=11$

$-4+5y=11$

$5y=15$

$y=3$

The solution is $(-2,3)$, and the solution set is $\{(-2,3)\}$.

**114.** $6x^2-8xy+2y^2=2(3x^2-4xy+y^2)$

$=2(3x-y)(x-y)$

**115.** $x+3\le -4$ or $2-7x\le 16$

$x\le -7 \qquad -7x\le 14$

$x\ge -2$

The solution set is $\{x|x\le -7$ or $x\ge -2\}$ or $(-\infty,-7]\cup[-2,\infty)$.

## 9.4 Exercise Set

**1.** $\log_5(7\cdot 3)=\log_5 7+\log_5 3$

**3.** $\log_7(7x)=\log_7 7+\log_7 x=1+\log_7 x$

**5.** $\log(1000x)=\log 1000+\log x$
$=3+\log x$

**7.** $\log_7\left(\frac{7}{x}\right)=\log_7 7-\log_7 x=1-\log_7 x$

**9.** $\log\left(\frac{x}{100}\right)=\log x-\log 100=\log x-2$

**11.** $\log_4\left(\frac{64}{y}\right)=\log_4 64-\log_4 y$
$=3-\log_4 y$

**13.** $\ln\left(\frac{e^2}{5}\right)=\ln e^2-\ln 5=2-\ln 5$

**15.** $\log_b x^3=3\log_b x$

**17.** $\log N^{-6}=-6\log N$

**19.** $\ln\sqrt[5]{x}=\ln x^{\frac{1}{5}}=\frac{1}{5}\ln x$

**21.** $\log_b x^2y=\log_b x^2+\log_b y$
$=2\log_b x+\log_b y$

**23.** $\log_4\left(\frac{\sqrt{x}}{64}\right)=\log_4\sqrt{x}-\log_4 64$
$=\log_4 x^{\frac{1}{2}}-3$
$=\frac{1}{2}\log_4 x\ -3$

**25.** $\log_6\left(\frac{36}{\sqrt{x+1}}\right)=\log_6 36-\log_6\sqrt{x+1}$
$=2-\log_6(x+1)^{\frac{1}{2}}$
$=2-\frac{1}{2}\log_6(x+1)$

**27.** $\log_b\left(\frac{x^2y}{z^2}\right)=\log_b x^2y-\log_b z^2$
$=\log_b x^2+\log_b y-2\log_b z$
$=2\log_b x+\log_b y-2\log_b z$

**29.** $\log\sqrt{100x}=\log(100x)^{\frac{1}{2}}$
$=\frac{1}{2}\log(100x)$
$=\frac{1}{2}(\log 100+\log x)$
$=\frac{1}{2}(2+\log x)$
$=1+\frac{1}{2}\log x$

**31.** $\log\sqrt[3]{\frac{x}{y}}=\log\left(\frac{x}{y}\right)^{\frac{1}{3}}$
$=\frac{1}{3}\log\left(\frac{x}{y}\right)$
$=\frac{1}{3}(\log x-\log y)$
$=\frac{1}{3}\log x-\frac{1}{3}\log y$

**33.**
$$\log_b\left(\frac{\sqrt{x}y^3}{z^3}\right)$$
$$=\log_b\left(\frac{x^{\frac{1}{2}}y^3}{z^3}\right)$$
$$=\log_b\left(x^{\frac{1}{2}}y^3\right)-\log_b z^3$$
$$=\log_b x^{\frac{1}{2}}+\log_b y^3-\log_b z^3$$
$$=\frac{1}{2}\log_b x+3\log_b y-3\log_b z$$

**35.**
$$\log_5\sqrt[3]{\frac{x^2y}{25}}$$
$$=\log_5\left(\frac{x^2y}{25}\right)^{\frac{1}{3}}$$
$$=\frac{1}{3}\log_5\left(\frac{x^2y}{25}\right)$$
$$=\frac{1}{3}\left(\log_5\left[x^2y\right]-\log_5 25\right)$$
$$=\frac{1}{3}\left(\log_5 x^2+\log_5 y-\log_5 5^2\right)$$
$$=\frac{1}{3}(2\log_5 x+\log_5 y-2)$$
$$=\frac{2}{3}\log_5 x+\frac{1}{3}\log_5 y-\frac{2}{3}$$

**37.** $\log 5+\log 2=\log(5\cdot 2)=\log 10=1$

**39.** $\ln x+\ln 7=\ln(x\cdot 7)=\ln(7x)$

**41.** $\log_2 96-\log_2 3=\log_2\frac{96}{3}=\log_2 32=5$

**43.** $\log(2x+5)-\log x=\log\left(\frac{2x+5}{x}\right)$

**45.**
$$\log x+3\log y=\log x+\log y^3$$
$$=\log\left(xy^3\right)$$

**47.**
$$\frac{1}{2}\ln x+\ln y=\ln x^{\frac{1}{2}}+\ln y$$
$$=\ln\left(x^{\frac{1}{2}}y\right)=\ln\left(y\sqrt{x}\right)$$

**49.**
$$2\log_b x+3\log_b y=\log_b x^2+\log_b y^3$$
$$=\log_b\left(x^2y^3\right)$$

**51.**
$$5\ln x-2\ln y=\ln x^5-\ln y^2$$
$$=\ln\left(\frac{x^5}{y^2}\right)$$

**53.**
$$3\ln x-\frac{1}{3}\ln y=\ln x^3-\ln y^{\frac{1}{3}}$$
$$=\ln\left(\frac{x^3}{y^{\frac{1}{3}}}\right)=\ln\left(\frac{x^3}{\sqrt[3]{y}}\right)$$

**55.**
$$4\ln(x+6)-3\ln x=\ln(x+6)^4-\ln x^3$$
$$=\ln\left[\frac{(x+6)^4}{x^3}\right]$$

**57.**
$$3\ln x+5\ln y-6\ln z$$
$$=\ln x^3+\ln y^5-\ln z^6$$
$$=\ln\left(x^3y^5\right)-\ln z^6=\ln\left(\frac{x^3y^5}{z^6}\right)$$

**59.**
$$\frac{1}{2}(\log_5 x+\log_5 y)-2\log_5(x+1)$$
$$=\frac{1}{2}\log_5(xy)-\log_5(x+1)^2$$
$$=\log_5(xy)^{\frac{1}{2}}-\log_5(x+1)^2$$
$$=\log_5\sqrt{xy}-\log_5(x+1)^2$$
$$=\log_5\left[\frac{\sqrt{xy}}{(x+1)^2}\right]$$

**61.** $\log_5 13 = \dfrac{\log 13}{\log 5} \approx 1.5937$

**63.** $\log_{14} 87.5 = \dfrac{\log 87.5}{\log 14} \approx 1.6944$

**65.** $\log_{0.1} 17 = \dfrac{\log 17}{\log 0.1} \approx -1.2304$

**67.** $\log_\pi 63 = \dfrac{\log 63}{\log \pi} \approx 3.6193$

**69.** $\log_b \dfrac{3}{2} = \log_b 3 - \log_b 2 = C - A$

**70.** $\log_b 6 = \log_b (2 \cdot 3)$
$= \log_b 2 + \log_b 3 = A + C$

**71.** $\log_b 8 = \log_b 2^3 = 3\log_b 2 = 3A$

**72.** $\log_b 81 = \log_b 3^4 = 4\log_b 3 = 4C$

**73.**
$$\log_b \sqrt{\frac{2}{27}} = \log_b \left(\frac{2}{27}\right)^{\frac{1}{2}}$$
$$= \frac{1}{2}\log_b \left(\frac{2}{3^3}\right)$$
$$= \frac{1}{2}\left(\log_b 2 - \log_b 3^3\right)$$
$$= \frac{1}{2}\left(\log_b 2 - 3\log_b 3\right)$$
$$= \frac{1}{2}\log_b 2 - \frac{3}{2}\log_b 3$$
$$= \frac{1}{2}A - \frac{3}{2}C$$

**74.**
$$\log_b \sqrt{\frac{3}{16}} = \log_b \left(\frac{\sqrt{3}}{4}\right)$$
$$= \log_b \sqrt{3} - \log_b 4$$
$$= \log_b 3^{\frac{1}{2}} - \log 2^2$$
$$= \frac{1}{2}\log_b 3 - 2\log 2$$
$$= \frac{1}{2}C - 2A$$

**75.** $\log_3 x = \log_3 5 + \log_3 7$
$\log_3 x = \log_3 (5 \cdot 7)$
$\log_3 x = \log_3 35$
$x = 35$
The solution is 35, and the solution set is {35}.

**76.** $\log_7 x = \log_7 3 + \log_7 5$
$\log_7 x = \log_7 (3 \cdot 5)$
$\log_7 x = \log_7 15$
$x = 15$
The solution is 15, and the solution set is {15}.

**77.** $\log_5 x = 4\log_5 2 - \log_5 8$
$\log_5 x = \log_5 2^4 - \log_5 8$
$\log_5 x = \log_5 16 - \log_5 8$
$\log_5 x = \log_5 \dfrac{16}{8}$
$\log_5 x = \log_5 2$
$x = 2$
The solution is 2, and the solution set is {2}.

**78.** $\log_5 x = 5\log_5 2 - \log_5 4$

$\log_5 x = \log_5 2^5 - \log_5 4$

$\log_5 x = \log_5 32 - \log_5 4$

$\log_5 x = \log_5 \frac{32}{4}$

$\log_5 x = \log_5 8$

$x = 8$

The solution is 8, and the solution set is $\{8\}$.

**79.** $(f-g)(x)$

$= f(x) - g(x)$

$= \log x + \log 7 + \log(x^2-1) - \log(x+1)$

$= \log(7x) + \log(x^2-1) - \log(x+1)$

$= \log[7x(x^2-1)] - \log(x+1)$

$= \log\left[\frac{7x(x^2-1)}{x+1}\right]$

$= \log\left[\frac{7x(x-1)(x+1)}{x+1}\right]$

$= \log[7x(x-1)]$

**80.** $(f-g)(x)$

$= f(x) - g(x)$

$= \log x + \log 15 + \log(x^2-4) - \log(x+2)$

$= \log(15x) + \log(x^2-4) - \log(x+2)$

$= \log[15x(x^2-4)] - \log(x+2)$

$= \log\left[\frac{15x(x^2-4)}{x+2}\right]$

$= \log\left[\frac{15x(x-2)(x+2)}{x+2}\right]$

$= \log[15x(x-2)]$

**81.** **a.** $D = 10(\log I - \log I_0)$

$= 10\left(\log \frac{I}{I_0}\right)$

**b.** $D = 10\left(\log \frac{100}{1}\right) = 10(\log 100)$

$= 10(2) = 20$

The sound is 20 decibels louder on the decibel scale.

**83. – 89.** Answers will vary.

**91.** **a.** $y = \log_3 x = \frac{\log x}{\log 3}$

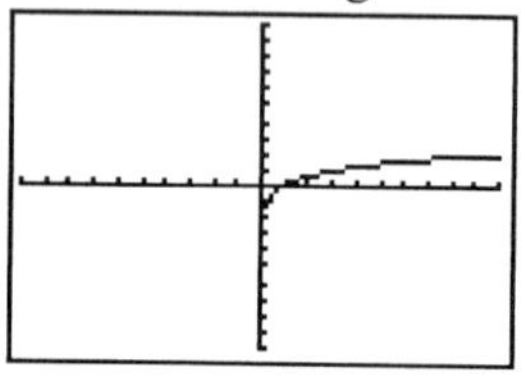

**b.** $y = 2 + \log_3 x$

$y = \log_3(x+2)$

$y = -\log_3 x$

$y = \log_3 x$

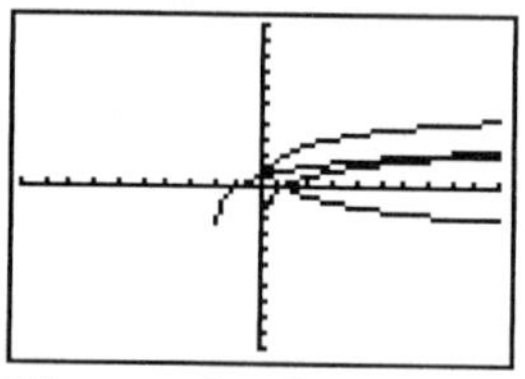

The graph of $y = 2 + \log_3 x$ is the graph of $y = \log_3 x$ shifted up two units.

The graph of $y = \log_3(x+2)$ is the graph of $y = \log_3 x$ shifted 2 units to the left.

The graph of $y = -\log_3 x$ is the graph of $y = \log_3 x$ reflected about the $x$–axis.

**93.** $y = \log_3 x$

$y = \log_{25} x$

$y = \log_{100} x$

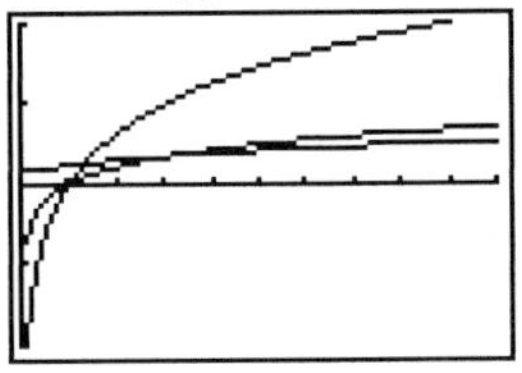

**a.** Change the window to focus on the (0, 1) interval.

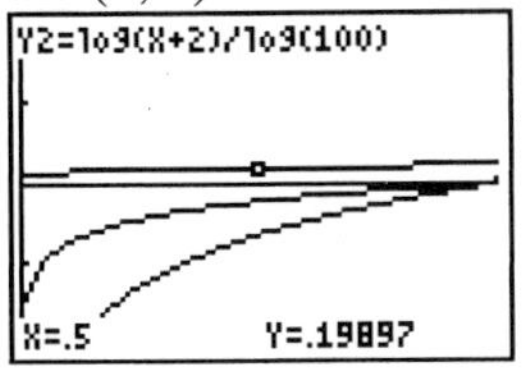


$y = \log_{100} x$ is on top.

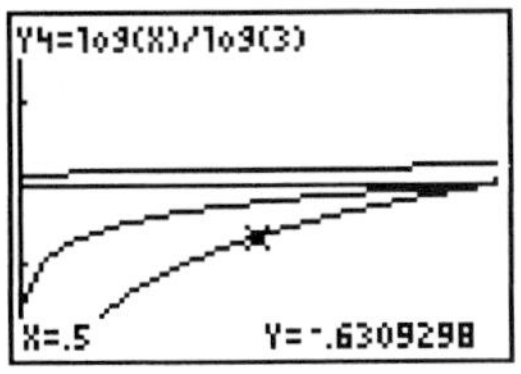


$y = \log_3 x$ is on the bottom.

**b.** Change the window to focus on the (1, 10) interval.

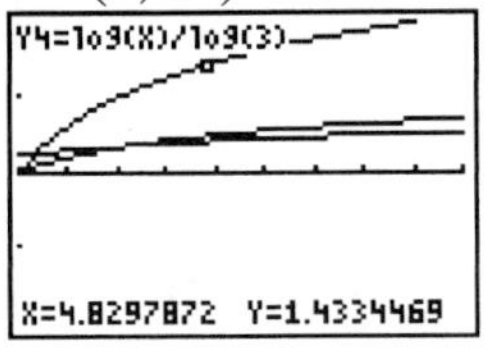


$y = \log_3 x$ is on top.

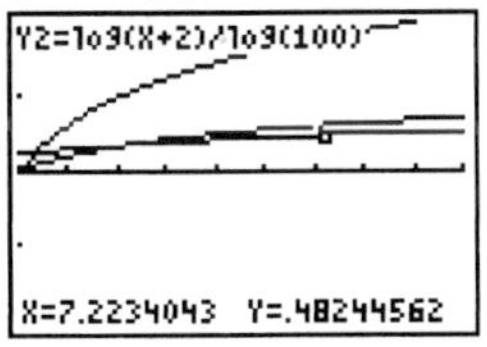


$y = \log_{100} x$ is on the bottom.

**c.** If $y = \log_b x$ is graphed for two different values of $b$, the graph of the one with the larger base will be on top in the interval (0, 1) and the one with the smaller base will be on top in the interval $(1, \infty)$. Likewise, if $y = \log_b x$ is graphed for two different values of $b$, the graph of the one with the smaller base will be on the bottom in the interval (0, 1) and the one with the larger base will be on the bottom in the interval $(1, \infty)$.

**95.** Answers will vary. One example follows. To disprove the statement $\log \frac{x}{y} = \frac{\log x}{\log y}$, let $y = 3$.

Graph $y = \log \frac{x}{3}$ and $y = \frac{\log x}{\log 3}$.

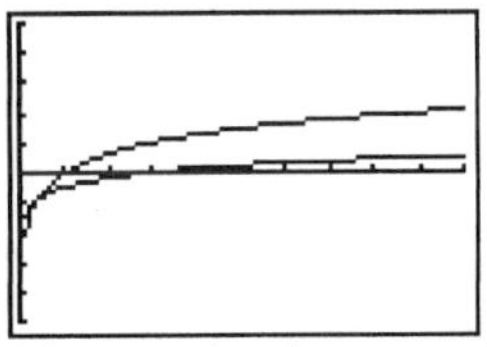

The graphs do not coincide, so the expressions are not equivalent.

**97.** Answers will vary. One example follows. To disprove the statement $\ln(xy) = (\ln x)(\ln y)$, let $y = 3$.

Graph $y = \ln(x \cdot 3)$ and $y = (\ln x)(\ln 3)$.

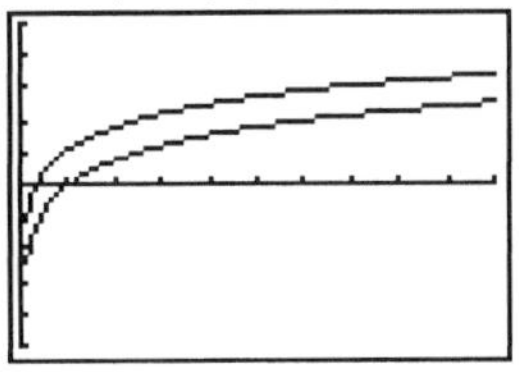

The graphs do not coincide, so the expressions are not equivalent.

**99.** Statement **d.** is true.

$$\ln\sqrt{2} = \ln 2^{\frac{1}{2}} = \frac{1}{2}\ln 2 = \frac{\ln 2}{2}$$

Statement **a.** is false.

$$\frac{\log_7 49}{\log_7 7} = \frac{\log_7 49}{1} = \log_7 49 = 2$$

Statement **b.** is false.

$\log_b\left(x^3 + y^3\right)$ cannot be simplified.

If we were taking the logarithm of a product and not a sum, we would have been able to simplify as follows.

$$\log_b\left(x^3y^3\right) = \log_b x^3 + \log_b y^3$$
$$= 3\log_b x + 3\log_b y$$

Statement **c.** is false.

$$\log_b(xy)^5 = 5\log_b(xy)$$
$$= 5(\log_b x + \log_b y)$$
$$= 5\log_b x + 5\log_b y$$

**101.**
$$\log_7 9 = \frac{\log 9}{\log 7} = \frac{\log 3^2}{\log 7} = \frac{2\log 3}{\log 7} = \frac{2A}{B}$$

**103.** $5x - 2y > 10$

First, find the intercepts to the equation $5x - 2y = 10$.

Find the $x$–intercept by setting y = 0.

$$5x - 2(0) = 10$$
$$5x = 10$$
$$x = 2$$

Find the $y$–intercept by setting $x = 0$.

$$5(0) - 2y = 10$$
$$-2y = 10$$
$$y = -5$$

Next, use the origin as a test point.

$$5(0) - 2(0) > 10$$
$$0 - 0 > 10$$
$$0 > 10$$

This is a false statement. This means that the origin will not fall in the shaded half-plane.

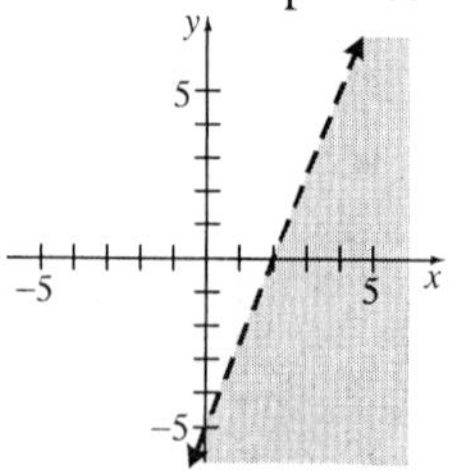


**104.**
$$x - 2(3x - 2) > 2x - 3$$
$$x - 6x + 4 > 2x - 3$$
$$-5x + 4 > 2x - 3$$
$$-7x + 4 > -3$$
$$-7x > -7$$
$$x < 1$$

The solution set is $\{x|x < 1\}$ or $(-\infty, 1)$.

**105.**
$$\frac{\sqrt[3]{40x^2y^6}}{\sqrt[3]{5xy}} = \sqrt[3]{\frac{40x^2y^6}{5xy}}$$
$$= \sqrt[3]{8xy^5}$$
$$= \sqrt[3]{8xy^3y^2} = 2y\sqrt[3]{xy^2}$$

## Mid-Chapter Check Points – Chapter 9

**1.**
$$(f \circ g)(x) = f(g(x)) = f(4x - 5)$$
$$= 3(4x - 5) + 2$$
$$= 12x - 15 + 2 = 12x - 13$$
$$(g \circ f)(x) = g(f(x)) = g(3x + 2)$$
$$= 4(3x + 2) - 5$$
$$= 12x + 8 - 5 = 12x + 3$$

Since $(f \circ g)(x) \neq (g \circ f)(x) \neq x$, we conclude the functions are not inverses.

**2.** $(f \circ g)(x) = f(g(x))$

$$= f\left(\frac{x^3-5}{7}\right)$$

$$= \sqrt[3]{7\left(\frac{x^3-5}{7}\right)+5}$$

$$= \sqrt[3]{(x^3-5)+5} = \sqrt[3]{x^3} = x$$

$(g \circ f)(x) = g(f(x))$

$$= g\left(\sqrt[3]{7x+5}\right)$$

$$= \frac{\left(\sqrt[3]{7x+5}\right)^3 - 5}{7}$$

$$= \frac{(7x+5)-5}{7} = \frac{7x}{7} = x$$

The functions are inverses.

**3.** $(f \circ g)(x) = f(g(x))$

$$= f(5^x) = \log_5 5^x = x$$

$(g \circ f)(x) = g(f(x))$

$$= g(\log_5 x) = 5^{\log_5 x} = x$$

The function are inverses.

**4.** **a.** $(f \circ g)(6) = f(g(6))$

$$= f\left(\sqrt{6+3}\right)$$

$$= f\left(\sqrt{9}\right)$$

$$= f(3) = \frac{3-1}{3} = \frac{2}{3}$$

**b.** $(g \circ f)(-1) = g(f(-1))$

$$= g\left(\frac{-1-1}{-1}\right)$$

$$= g(2)$$

$$= \sqrt{2+3} = \sqrt{5}$$

**c.** $(f \circ f)(5) = f(f(6))$

$$= f\left(\frac{5-1}{5}\right)$$

$$= f\left(\frac{4}{5}\right)$$

$$= \frac{\frac{4}{5}-1}{\frac{4}{5}} = \frac{-\frac{1}{5}}{\frac{4}{5}}$$

$$= -\frac{1}{5} \cdot \frac{5}{4} = -\frac{1}{4}$$

**d.** $(g \circ g)(-2) = g(g(-2))$

$$= g\left(\sqrt{-2+3}\right)$$

$$= g\left(\sqrt{1}\right) = g(1)$$

$$= \sqrt{1+3} = \sqrt{4} = 2$$

**5.**

$$f(x) = \frac{2x+5}{4}$$

$$y = \frac{2x+5}{4}$$

Interchange $x$ and $y$ and solve for $y$.

$$x = \frac{2y+5}{4}$$

$$4x = 2y+5$$

$$4x-5 = 2y$$

$$\frac{4x-5}{2} = y$$

$$f^{-1}(x) = \frac{4x-5}{2}$$

**6.** $f(x)=10x^3-7$

$y=10x^3-7$

Interchange $x$ and $y$ and solve for $y$.

$x=10y^3-7$

$x+7=10y^3$

$\frac{x+7}{10}=y^3$

$\sqrt[3]{\frac{x+7}{10}}=y$

$f^{-1}(x)=\sqrt[3]{\frac{x+7}{10}}$

**7.** Interchange the $x$ and $y$ coordinates of each ordered pair in the set: $f^{-1}=\{(5,2),(-7,10),(-10,11)\}$.

**8.** The graph passes the vertical line test but fails the horizontal line test. Thus, the graph represents a function, but its inverse is not a function.

**9.** The graph passes both the vertical line test and the horizontal line test. Thus, the graph represents a function, and its inverse is a function as well.

**10.** The graph fails the vertical line test, so it does not represent a function.

**11.**

| $x$ | $f(x)=2^x-3$ |
|---|---|
| $-2$ | $-\frac{11}{4}=-2.75$ |
| $-1$ | $-\frac{5}{2}=-2.5$ |
| 0 | $-3$ |
| 1 | $-1$ |
| 2 | 1 |
| 3 | 5 |

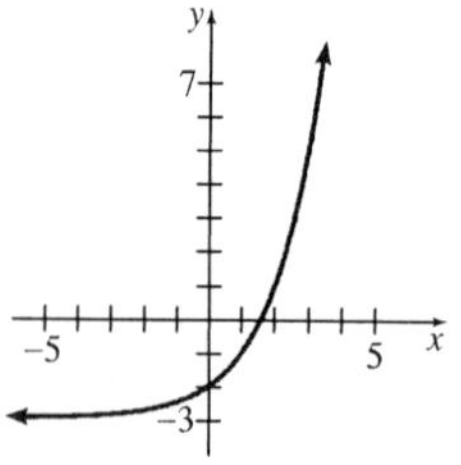


Domain: $\{x \mid x \text{ is a real number}\}$ or $(-\infty,\infty)$;

Range: $\{x \mid x>-3\}$ or $(-3,\infty)$.

**12.**

| $x$ | $f(x)=\left(\frac{1}{3}\right)^x$ |
|---|---|
| $-2$ | 9 |
| $-1$ | 3 |
| 0 | 1 |
| 1 | $\frac{1}{3}$ |
| 2 | $\frac{1}{9}$ |

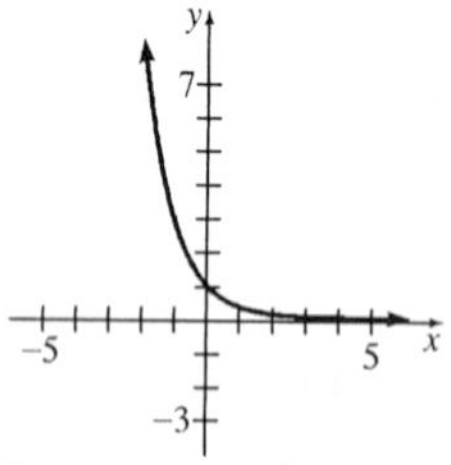


Domain: $\{x \mid x \text{ is a real number}\}$ or $(-\infty,\infty)$;

Range: $\{x \mid x>0\}$ or $(0,\infty)$.

**13.**

| $x$ | $f(x)=\log_2 x$ |
|---|---|
| $\frac{1}{4}$ | $-2$ |
| $\frac{1}{2}$ | $-1$ |
| 1 | 0 |
| 2 | 1 |
| 4 | 2 |

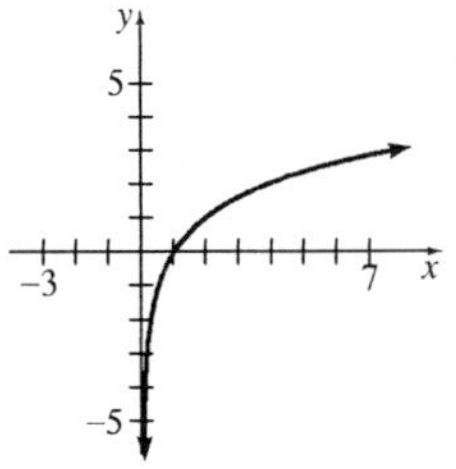


Domain: $\{x \mid x > 0\}$ or $(0, \infty)$;
Range: $\{x \mid x \text{ is a real number}\}$ or $(-\infty, \infty)$.

**14.**

| $x$ | $f(x) = \log_2 x + 1$ |
|---|---|
| $\frac{1}{4}$ | $-1$ |
| $\frac{1}{2}$ | 0 |
| 1 | 1 |
| 2 | 2 |
| 4 | 3 |

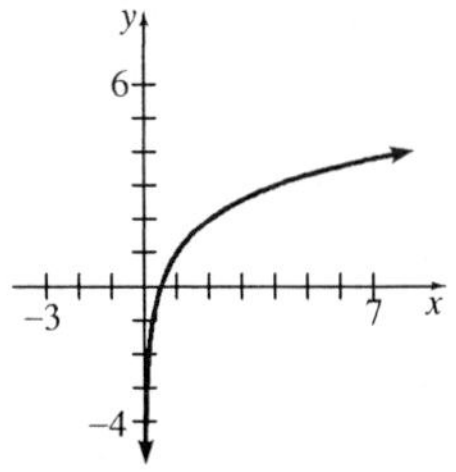


Domain: $\{x \mid x > 0\}$ or $(0, \infty)$;
Range: $\{x \mid x \text{ is a real number}\}$ or $(-\infty, \infty)$.

**15.** $f(x) = \log_3(x+6)$
The argument of the logarithm must be positive:
$x + 6 > 0$
$x > -6$
Domain: $\{x \mid x > -6\}$ or $(-6, \infty)$.

**16.** $f(x) = \log_3 x + 6$
The argument of the logarithm must be positive: $x > 0$
Domain: $\{x \mid x > 0\}$ or $(0, \infty)$.

**17.** $\log_3 (x+6)^2$
The argument of the logarithm must be positive. Now $(x+6)^2$ is always positive, except when $x = -6$
Domain: $\{x \mid x \neq 0\}$ or $(-\infty - 6) \cup (-6, \infty)$.

**18.** $f(x) = 3^{x+6}$
Domain: $\{x \mid x \text{ is a real number}\}$ or $(-\infty, \infty)$.

**19.** 
$$\begin{aligned}\log_2 8 + \log_5 25 &= \log_2 2^3 + \log_5 5^2 \\ &= 3 + 2 = 5\end{aligned}$$

**20.** $\log_3 \frac{1}{9} = \log_3 \frac{1}{3^2} = \log_3 3^{-2} = -2$

**21.** Let $\log_{100} 10 = y$
$$\begin{aligned}100^y &= 10 \\ (10^2)^y &= 10^1 \\ 10^{2y} &= 10^1 \\ 2y &= 1 \\ y &= \frac{1}{2}\end{aligned}$$

**22.** $\log \sqrt[3]{10} = \log 10^{\frac{1}{3}} = \frac{1}{3}$

**23.** 
$$\begin{aligned}\log_2(\log_3 81) &= \log_2(\log_3 3^4) \\ &= \log_2 4 = \log_2 2^2 = 2\end{aligned}$$

**24.** 
$$\begin{aligned}\log_3\left(\log_2 \frac{1}{8}\right) &= \log_3\left(\log_2 \frac{1}{2^3}\right) \\ &= \log_3(\log_2 2^{-3}) \\ &= \log_3(-3) \\ &= \text{not possible}\end{aligned}$$
This expression is impossible to evaluate.

**25.** $6^{\log_6 5} = 5$

**26.** $\ln e^{\sqrt{7}} = \sqrt{7}$

**27.** $10^{\log 13} = 13$

**28.** $\log_{100} 0.1 = y$

$$\begin{aligned} 100^y &= 0.1 \\ \left(10^2\right)^y &= \frac{1}{10} \\ 10^{2y} &= 10^{-1} \\ 2y &= -1 \\ y &= -\frac{1}{2} \end{aligned}$$

**29.** $\log_\pi \pi^{\sqrt{\pi}} = \sqrt{\pi}$

**30.**
$$\begin{aligned} \log\left(\frac{\sqrt{xy}}{1000}\right) &= \log\left(\sqrt{xy}\right) - \log 1000 \\ &= \log(xy)^{\frac{1}{2}} - \log 10^3 \\ &= \frac{1}{2}\log(xy) - 3 \\ &= \frac{1}{2}(\log x + \log y) - 3 \\ &= \frac{1}{2}\log x + \frac{1}{2}\log y - 3 \end{aligned}$$

**31.**
$$\begin{aligned} \ln\left(e^{19}x^{20}\right) &= \ln e^{19} + \ln x^{20} \\ &= 19 + 20\ln x \end{aligned}$$

**32.**
$$\begin{aligned} 8\log_7 x - \frac{1}{3}\log_7 y &= \log_7 x^8 - \log_7 y^{\frac{1}{3}} \\ &= \log_7\left(\frac{x^8}{y^{\frac{1}{3}}}\right) \\ &= \log_7\left(\frac{x^8}{\sqrt[3]{y}}\right) \end{aligned}$$

**33.**
$$\begin{aligned} 7\log_5 x + 2\log_5 x &= \log_5 x^7 + \log_5 x^2 \\ &= \log_5\left(x^7 \cdot x^2\right) \\ &= \log_5 x^9 \end{aligned}$$

**34.**
$$\begin{aligned} &\frac{1}{2}\ln x - 3\ln y - \ln(z-2) \\ &= \ln x^{\frac{1}{2}} - \ln y^2 - \ln(z-2) \\ &= \ln\sqrt{x} - \left[\ln y^2 + \ln(z-2)\right] \\ &= \ln\sqrt{x} - \ln\left[y(z-2)\right] \\ &= \ln\left[\frac{\sqrt{x}}{y(z-2)}\right] \end{aligned}$$

**35.** Continuously: $A = 8000e^{0.08(3)}$
$\approx 10{,}170$

Monthly: $A = 8000\left(1 + \frac{0.08}{12}\right)^{12 \cdot 3}$
$\approx 10{,}162$

$10{,}170 - 10{,}162 = 8$

Interest returned will be \$8 more if compounded continuously.

## 9.5 Exercise Set

**1.**
$$\begin{aligned} 2^x &= 64 \\ 2^x &= 2^6 \\ x &= 6 \end{aligned}$$

The solution is 6, and the solution set is $\{6\}$.

**3.**
$$\begin{aligned} 5^x &= 125 \\ 5^x &= 5^3 \\ x &= 3 \end{aligned}$$

The solution is 3, and the solution set is $\{3\}$.

**5.**
$$2^{2x-1}=32$$
$$2^{2x-1}=2^5$$
$$2x-1=5$$
$$2x=6$$
$$x=3$$
The solution is 3, and the solution set is $\{3\}$.

**7.**
$$4^{2x-1}=64$$
$$4^{2x-1}=4^3$$
$$2x-1=3$$
$$2x=4$$
$$x=2$$
The solution is 2, and the solution set is $\{2\}$.

**9.**
$$32^x=8$$
$$\left(2^5\right)^x=2^3$$
$$2^{5x}=2^3$$
$$5x=3$$
$$x=\frac{3}{5}$$
The solution is $\frac{3}{5}$, and the solution set is $\left\{\frac{3}{5}\right\}$.

**11.**
$$9^x=27$$
$$\left(3^2\right)^x=3^3$$
$$3^{2x}=3^3$$
$$2x=3$$
$$x=\frac{3}{2}$$
The solution is $\frac{3}{2}$, and the solution set is $\left\{\frac{3}{2}\right\}$.

**13.**
$$3^{1-x}=\frac{1}{27}$$
$$3^{1-x}=\frac{1}{3^3}$$
$$3^{1-x}=3^{-3}$$
$$1-x=-3$$
$$-x=-2$$
$$x=2$$
The solution is 2, and the solution set is $\{2\}$.

**15.**
$$6^{\frac{x-3}{4}}=\sqrt{6}$$
$$6^{\frac{x-3}{4}}=6^{\frac{1}{2}}$$
$$\frac{x-3}{4}=\frac{1}{2}$$
$$2(x-3)=4(1)$$
$$2x-6=4$$
$$2x=10$$
$$x=5$$
The solution is 5, and the solution set is $\{5\}$.

**17**
$$4^x=\frac{1}{\sqrt{2}}$$
$$\left(2^2\right)^x=\frac{1}{2^{\frac{1}{2}}}$$
$$2^{2x}=2^{-\frac{1}{2}}$$
$$2x=-\frac{1}{2}$$
$$x=\frac{1}{2}\left(-\frac{1}{2}\right)=-\frac{1}{4}$$
The solution is $-\frac{1}{4}$, and the solution set is $\left\{-\frac{1}{4}\right\}$.

**19.** $10^x = 3.91$

$\ln 10^x = \ln 3.91$

$x \ln 10 = \ln 3.91$

$x = \dfrac{\ln 3.91}{\ln 10} \approx 0.59$

The solution is $\dfrac{\ln 3.91}{\ln 10} \approx 0.59$, and the solution set is $\left\{\dfrac{\ln 3.91}{\ln 10} \approx 0.59\right\}$.

**21.** $e^x = 5.7$

$\ln e^x = \ln 5.7$

$x = \ln 5.7 \approx 1.74$

The solution is $\ln 5.7 \approx 1.74$, and the solution set is $\{\ln 5.7 \approx 1.74\}$.

**23.** $5^x = 17$

$\ln 5^x = \ln 17$

$x \ln 5 = \ln 17$

$x = \dfrac{\ln 17}{\ln 5} \approx 1.76$

The solution is $\dfrac{\ln 17}{\ln 5} \approx 1.76$, and the solution set is $\left\{\dfrac{\ln 17}{\ln 5} \approx 1.76\right\}$.

**25.** $5e^x = 25$

$e^x = 5$

$\ln e^x = \ln 5$

$x = \ln 5 \approx 1.61$

The solution is $\ln 5 \approx 1.61$, and the solution set is $\{\ln 5 \approx 1.61\}$.

**27.** $3e^{5x} = 1977$

$e^{5x} = 659$

$\ln e^{5x} = \ln 659$

$5x = \ln 659$

$x = \dfrac{\ln 659}{5} \approx 1.30$

The solution is $\dfrac{\ln 659}{5} \approx 1.30$, and the solution set is $\left\{\dfrac{\ln 659}{5} \approx 1.30\right\}$.

**29.** $e^{0.7x} = 13$

$\ln e^{0.7x} = \ln 13$

$0.7x = \ln 13$

$x = \dfrac{\ln 13}{0.7} \approx 3.66$

The solution is $\dfrac{\ln 13}{0.7} \approx 3.66$, and the solution set is $\left\{\dfrac{\ln 13}{0.7} \approx 3.66\right\}$.

**31.** $1250e^{0.055x} = 3750$

$e^{0.055x} = 3$

$\ln e^{0.055x} = \ln 3$

$0.055x = \ln 3$

$x = \dfrac{\ln 3}{0.055} \approx 19.97$

The solution is $\dfrac{\ln 3}{0.055} \approx 19.97$, and the solution set is $\left\{\dfrac{\ln 3}{0.055} \approx 19.97\right\}$.

**33.** $30 - (1.4)^x = 0$

$-1.4^x = -30$

$1.4^x = 30$

$\ln 1.4^x = \ln 30$

$x \ln 1.4 = \ln 30$

$x = \dfrac{\ln 30}{\ln 1.4} \approx 10.11$

The solution is $\dfrac{\ln 30}{\ln 1.4} \approx 10.11$, and the solution set is $\left\{\dfrac{\ln 30}{\ln 1.4} \approx 10.11\right\}$.

**35.**
$$e^{1-5x} = 793$$
$$\ln e^{1-5x} = \ln 793$$
$$1-5x = \ln 793$$
$$-5x = \ln 793 - 1$$
$$x = \frac{-(\ln 793 - 1)}{5}$$
$$x = \frac{1-\ln 793}{5} \approx -1.14$$
The solution is $\frac{1-\ln 793}{5} \approx -1.14$, and the solution set is $\left\{\frac{1-\ln 793}{5} \approx -1.14\right\}$.

**37.**
$$7^{x+2} = 410$$
$$\ln 7^{x+2} = \ln 410$$
$$(x+2)\ln 7 = \ln 410$$
$$x+2 = \frac{\ln 410}{\ln 7}$$
$$x = \frac{\ln 410}{\ln 7} - 2 \approx 1.09$$
The solution is $\frac{\ln 410}{\ln 7} - 2 \approx 1.09$, and the solution set is $\left\{\frac{\ln 410}{\ln 7} - 2 \approx 1.09\right\}$.

**39.**
$$2^{x+1} = 5^x$$
$$\ln 2^{x+1} = \ln 5^x$$
$$(x+1)\ln 2 = x\ln 5$$
$$x\ln 2 + \ln 2 = x\ln 5$$
$$x\ln 2 = x\ln 5 - \ln 2$$
$$x\ln 2 - x\ln 5 = -\ln 2$$
$$x(\ln 2 - \ln 5) = -\ln 2$$
$$x = \frac{-\ln 2}{\ln 2 - \ln 5}$$
$$x = \frac{\ln 2}{\ln 5 - \ln 2} \approx 0.76$$
The solution is $\frac{\ln 2}{\ln 5 - \ln 2} \approx 0.76$, and the solution set is $\left\{\frac{\ln 2}{\ln 5 - \ln 2} \approx 0.76\right\}$.

**41.**
$$\log_3 x = 4$$
$$x = 3^4$$
$$x = 81$$
The solution is 81, and the solution set is $\{81\}$.

**43.**
$$\log_2 x = -4$$
$$x = 2^{-4}$$
$$x = \frac{1}{2^4} = \frac{1}{16}$$
The solution is $\frac{1}{16}$, and the solution set is $\left\{\frac{1}{16}\right\}$.

**45.**
$$\log_9 x = \frac{1}{2}$$
$$x = 9^{\frac{1}{2}}$$
$$x = \sqrt{9} = 3$$
The solution is 3, and the solution set is $\{3\}$.

**47.**
$$\log x = 2$$
$$x = 10^2$$
$$x = 100$$
The solution is 100, and the solution set is $\{100\}$.

**49.**
$$\log_4(x+5) = 3$$
$$x+5 = 4^3$$
$$x+5 = 64$$
$$x = 59$$
The solution is 59, and the solution set is $\{59\}$.

**51.** $\log_3(x-4)=-3$

$$x-4=3^{-3}$$
$$x-4=\frac{1}{3^3}$$
$$x-4=\frac{1}{27}$$
$$x=\frac{1}{27}+4$$
$$x=\frac{1}{27}+\frac{108}{27}=\frac{109}{27}$$

The solution is $\frac{109}{27}$, and the solution set is $\left\{\frac{109}{27}\right\}$.

**53.** $\log_4(3x+2)=3$

$$3x+2=4^3$$
$$3x+2=64$$
$$3x=62$$
$$x=\frac{62}{3}$$

The solution is $\frac{62}{3}$, and the solution set is $\left\{\frac{62}{3}\right\}$.

**55.** $\log_5 x+\log_5(4x-1)=1$

$$\log_5(x(4x-1))=1$$
$$x(4x-1)=5^1$$
$$4x^2-x=5$$
$$4x^2-x-5=0$$
$$(4x-5)(x+1)=0$$

Apply the zero product principle.

$4x-5=0$ and $x+1=0$

$4x=5$ $\quad x=-1$

$x=\frac{5}{4}$

We disregard –1 because it would result in taking the logarithm of a negative number in the original equation. The solution is $\frac{5}{4}$, and the solution set is $\left\{\frac{5}{4}\right\}$.

**57.** $\log_3(x-5)+\log_3(x+3)=2$

$$\log_3((x-5)(x+3))=2$$
$$(x-5)(x+3)=3^2$$
$$x^2-2x-15=9$$
$$x^2-2x-24=0$$
$$(x-6)(x+2)=0$$

Apply the zero product principle.

$x-6=0$ and $x+2=0$

$x=6$ $\quad x=-2$

We disregard –2 because it would result in taking the logarithm of a negative number in the original equation. The solution is 6, and the solution set is $\{6\}$.

**59.** $\log_2(x+2)-\log_2(x-5)=3$

$$\log_2\frac{x+2}{x-5}=3$$
$$\frac{x+2}{x-5}=2^3$$
$$\frac{x+2}{x-5}=8$$
$$x+2=8(x-5)$$
$$x+2=8x-40$$
$$-7x+2=-40$$
$$-7x=-42$$
$$x=6$$

The solution is 6, and the solution set is $\{6\}$.

**61.** $\log(3x-5)-\log(5x)=2$

$$\log\frac{3x-5}{5x}=2$$
$$\frac{3x-5}{5x}=10^2$$
$$\frac{3x-5}{5x}=100$$
$$3x-5=500x$$
$$-5=497x$$
$$-\frac{5}{497}=x$$

We disregard $-\frac{5}{497}$ because it would result in taking the logarithm of a negative number in the original equation. Therefore, the equation has no solution. The solution set is $\varnothing$ or $\{\ \}$.

**63.** $\ln x=2$

$$e^{\ln x}=e^2$$
$$x=e^2\approx 7.39$$

The solution is $e^2\approx 7.39$, and the solution set is $\{e^2\approx 7.39\}$.

**65.** $\ln x=-3$

$$x=e^{-3}=\frac{1}{e^3}$$

The solution is $e^{-3}=\frac{1}{e^3}\approx 0.05$, and the solution set is $\left\{e^{-3}=\frac{1}{e^3}\approx 0.05\right\}$.

**67.** $5\ln(2x)=20$

$$\ln(2x)=4$$
$$e^{\ln(2x)}=e^4$$
$$2x=e^4$$
$$x=\frac{e^4}{2}\approx 27.30$$

The solution is $\frac{e^4}{2}\approx 27.30$, and the solution set is $\left\{\frac{e^4}{2}\approx 27.30\right\}$.

**69.** $6+2\ln x=5$

$$2\ln x=-1$$
$$e^{\ln x}=e^{-\frac{1}{2}}$$
$$x=e^{-\frac{1}{2}}\approx 0.61$$

The solution is $e^{-\frac{1}{2}}\approx 0.61$, and the solution set is $\left\{e^{-\frac{1}{2}}\approx 0.61\right\}$.

**71.** $\ln\sqrt{x+3}=1$

$$\ln(x+3)^{\frac{1}{2}}=1$$
$$\frac{1}{2}\ln(x+3)=1$$
$$\ln(x+3)=2$$
$$e^{\ln(x+3)}=e^2$$
$$x+3=e^2$$
$$x=e^2-3\approx 4.39$$

The solution is $e^2-3\approx 4.39$, and the solution set is $\{e^2-3\approx 4.39\}$.

**73.** $\ln(x+1)-\ln x=1$

$$\ln\left(\frac{x+1}{x}\right)=1$$

$$\frac{x+1}{x}=e^1$$

$$x+1=ex$$

$$1=ex-x$$

$$1=(e-1)x$$

$$x=\frac{1}{e-1}\approx 0.58$$

The solution is $\frac{1}{e-1}\approx 0.58$, and the solution set is $\left\{\frac{1}{e-1}\approx 0.58\right\}$.

**75.** $5^{2x}\cdot 5^{4x}=125$

$$5^{2x+4x}=5^3$$

$$5^{6x}=5^3$$

$$6x=3$$

$$x=\frac{1}{2}$$

The solution is $\frac{1}{2}$, and the solution set is $\left\{\frac{1}{2}\right\}$.

**76.** $3^{x+2}\cdot 3^x=81$

$$3^{(x+2)+x}=3^4$$

$$3^{2x+2}=3^4$$

$$2x+2=4$$

$$2x=2$$

$$x=1$$

The solution is 1, and the solution set is $\{1\}$.

**77.** $2\log_3(x+4)=\log_3 9+2$

$$2\log_3(x+4)=\log_3 3^2+2$$

$$2\log_3(x+4)=2+2$$

$$2\log_3(x+4)=4$$

$$\log_3(x+4)=2$$

$$x+4=3^2$$

$$x+4=9$$

$$x=5$$

The solution is 5, and the solution set is $\{5\}$.

**78.** $3\log_2(x-1)=5-\log_2 4$

$$3\log_2(x-1)=5-\log_2 2^2$$

$$3\log_2(x-1)=5-2$$

$$3\log_2(x-1)=3$$

$$\log_2(x-1)=1$$

$$x-1=2^1$$

$$x-1=2$$

$$x=3$$

The solution is 3, and the solution set is $\{3\}$.

**79.** $3^{x^2}=45$

$$\ln 3^{x^2}=\ln 45$$

$$x^2\ln 3=\ln 45$$

$$x^2=\frac{\ln 45}{\ln 3}$$

$$x=\pm\sqrt{\frac{\ln 45}{\ln 3}}\approx\pm 1.86$$

The solutions are $\pm\sqrt{\frac{\ln 45}{\ln 3}}\approx\pm 1.86$, and the solution set is $\left\{\pm\sqrt{\frac{\ln 45}{\ln 3}}\approx\pm 1.86\right\}$.

**80.**

$$5^{x^2} = 50$$
$$\ln 5^{x^2} = \ln 50$$
$$x^2 \ln 5 = \ln 50$$
$$x^2 = \frac{\ln 50}{\ln 5}$$
$$x = \pm\sqrt{\frac{\ln 50}{\ln 5}} \approx \pm 1.56$$

The solutions are $\pm\sqrt{\frac{\ln 50}{\ln 5}} \approx \pm 1.56$, and the solution set is $\left\{\pm\sqrt{\frac{\ln 50}{\ln 5}} \approx \pm 1.56\right\}$.

**81.**

$$\log_2(x-6)+\log_2(x-4)-\log_2 x = 2$$
$$\log_2\left[(x-6)(x-4)\right]-\log_2 x = 2$$
$$\log_2\left[\frac{(x-6)(x-4)}{x}\right] = 2$$
$$\log_2\left(\frac{x^2-10x+24}{x}\right) = 2$$
$$\frac{x^2-10x+24}{x} = 2^2$$
$$\frac{x^2-10x+24}{x} = 4$$
$$x^2-10x+24 = 4x$$
$$x^2-14x+24 = 0$$
$$(x-12)(x-2) = 0$$

Apply the zero product property:

$x-12=0$ or $x-2=0$

$x=12$ $\quad$ $x=2$

We disregard 2 because it would result in taking the logarithm of a negative number in the original equation. The solution is 12, and the solution set is $\{12\}$.

**82.**

$$\log_2(x-3)+\log_2 x-\log_2(x+2) = 2$$
$$\log_2\left[(x-3)x\right]-\log_2(x+2) = 2$$
$$\log_2\left[\frac{(x-3)x}{x+2}\right] = 2$$
$$\log_2\left(\frac{x^2-3x}{x+2}\right) = 2$$
$$\frac{x^2-3x}{x+2} = 2^2$$
$$\frac{x^2-3x}{x+2} = 4$$
$$x^2-3x = 4(x+2)$$
$$x^2-3x = 4x+8$$
$$x^2-7x-8 = 0$$
$$(x-8)(x+1) = 0$$

Apply the zero product property:

$x-8=0$ or $x+1=0$

$x=8$ $\quad$ $x=-1$

We disregard $-1$ because it would result in taking the logarithm of a negative number in the original equation. The solution is 8, and the solution set is $\{8\}$.

**83.**

$$5^{x^2-12} = 25^{2x}$$
$$5^{x^2-12} = \left(5^2\right)^{2x}$$
$$5^{x^2-12} = 5^{4x}$$
$$x^2-12 = 4x$$
$$x^2-4x-12 = 0$$
$$(x-6)(x+2) = 0$$

Apply the zero product property:

$x-6=0$ or $x+2=0$

$x=6$ $\quad$ $x=-2$

The solutions are $-2$ and 6, and the solution set is $\{-2,\ 6\}$.

**84.**

$$3^{x^2-12} = 9^{2x}$$
$$3^{x^2-12} = \left(3^2\right)^{2x}$$
$$3^{x^2-12} = 3^{4x}$$
$$x^2 - 12 = 4x$$
$$x^2 - 4x - 12 = 0$$
$$(x-6)(x+2) = 0$$

Apply the zero product property:

$x - 6 = 0$ or $x + 2 = 0$

$x = 6$ $\quad$ $x = -2$

The solutions are $-2$ and 6, and the solution set is $\{-2, 6\}$.

**85.**

$$R = 6e^{12.77x}$$
$$6e^{12.77x} = 100$$
$$e^{12.77x} = \frac{100}{6}$$
$$\ln e^{12.77x} = \ln\frac{100}{6}$$
$$12.77x = \ln\frac{100}{6}$$
$$x = \frac{\ln\frac{100}{6}}{12.77} \approx 0.22$$

A blood alcohol concentration of 0.22 corresponds to a 100% risk.

**87.** **a.** Since 2000 is 0 years after 2000, find $f(0)$.

$$f(0) = 18.9e^{0.0055(0)}$$
$$= 18.9e^0 = 18.9(1) = 18.9$$

The population of New York was 18.9 million in 2000.

**b.**

$$19.54 = 18.9e^{0.0055t}$$
$$e^{0.0055t} = \frac{19.54}{18.9}$$
$$\ln e^{0.0055t} = \ln\frac{19.54}{18.9}$$
$$0.0055t = \ln\frac{19.54}{18.9}$$
$$t = \frac{\ln\frac{19.54}{18.9}}{0.0055} \approx 6$$

$2000 + 6 = 2006$

The population of New York will reach 19.54 million in approximately the year 2006.

**89.**

$$20000 = 12500\left(1 + \frac{0.0575}{4}\right)^{4t}$$
$$20000 = 12500(1 + 0.014375)^{4t}$$
$$20000 = 12500(1.014375)^{4t}$$
$$\frac{20000}{12500} = (1.014375)^{4t}$$
$$1.6 = (1.014375)^{4t}$$
$$\ln 1.6 = \ln(1.014375)^{4t}$$
$$\ln 1.6 = 4t \ln 1.014375$$
$$\frac{4t \ln 1.014375}{4 \ln 1.014375} = \frac{\ln 1.6}{4 \ln 1.014375}$$
$$t = \frac{\ln 1.6}{4 \ln 1.014375} \approx 8.2$$

It will take approximately 8.2 years.

**91.**

$$1400 = 1000\left(1+\frac{r}{360}\right)^{360(2)}$$

$$\frac{1400}{1000} = \left(1+\frac{r}{360}\right)^{720}$$

$$1.4 = \left(1+\frac{r}{360}\right)^{720}$$

$$\ln 1.4 = \ln\left(1+\frac{r}{360}\right)^{720}$$

$$\ln 1.4 = 720\ln\left(1+\frac{r}{360}\right)$$

$$\frac{\ln 1.4}{720} = \ln\left(1+\frac{r}{360}\right)$$

$$e^{\frac{\ln 1.4}{720}} = e^{\ln\left(1+\frac{r}{360}\right)}$$

$$e^{\frac{\ln 1.4}{720}} = 1+\frac{r}{360}$$

$$1+\frac{r}{360} = e^{\frac{\ln 1.4}{720}}$$

$$\frac{r}{360} = e^{\frac{\ln 1.4}{720}} - 1$$

$$r = 360\left(e^{\frac{\ln 1.4}{720}} - 1\right) \approx 0.168$$

The annual interest rate is approximately 16.8%.

**93.**

$$16000 = 8000e^{0.08t}$$

$$\frac{16000}{8000} = e^{0.08t}$$

$$2 = e^{0.08t}$$

$$\ln 2 = \ln e^{0.08t}$$

$$\ln 2 = 0.08t$$

$$t = \frac{\ln 2}{0.08} \approx 8.7$$

It will take approximately 8.7 years to double the money.

**95.**

$$7050 = 2350e^{r7}$$

$$\frac{7050}{2350} = e^{7r}$$

$$3 = e^{7r}$$

$$\ln 3 = \ln e^{7r}$$

$$\ln 3 = 7r$$

$$r = \frac{\ln 3}{7} \approx 0.157$$

The annual interest rate would have to be 15.7% to triple the money.

**97.**

$$1270 = 2246 - 501.4\ln x$$

$$1270 + 501.4\ln x = 2246$$

$$501.4\ln x = 976$$

$$\ln x = \frac{976}{501.4}$$

$$e^{\ln x} = e^{\frac{976}{501.4}}$$

$$x = e^{\frac{976}{501.4}} \approx 7$$

$$1995 + 7 = 2002$$

The average price of a new computer was $1270 in approximately the year 2002

**99.**

$$50 = 95 - 30\log_2 x$$

$$-45 = -30\log_2 x$$

$$\frac{-45}{-30} = \log_2 x$$

$$\log_2 x = \frac{3}{2}$$

$$x = 2^{\frac{3}{2}} \approx 2.8$$

After approximately 2.8 days, only half the students recall the important features of the lecture. This is represented by the point (2.8, 50).

**101.** $\text{pH} = -\log x$

$2.4 = -\log x$

$-2.4 = \log x$

$x = 10^{-2.4} \approx .004$

The hydrogen ion concentration is $10^{-2.4}$, or approximately 0.004, moles per liter.

**103. – 107.** Answers will vary.

**109.** $2^{x+1} = 8$

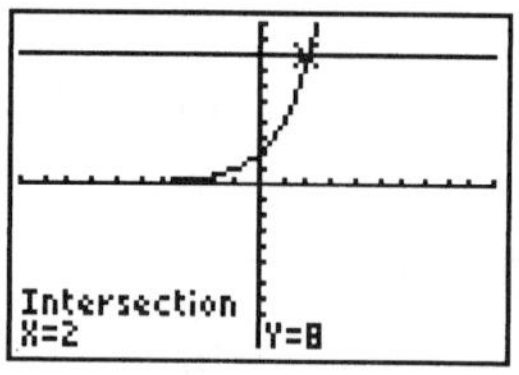


The solution is 2, and the solution set is $\{2\}$.
Verify the solution algebraically:

$2^{2+1} = 8$

$2^3 = 8$

$8 = 8$

**111.** $\log_3(4x-7) = 2$

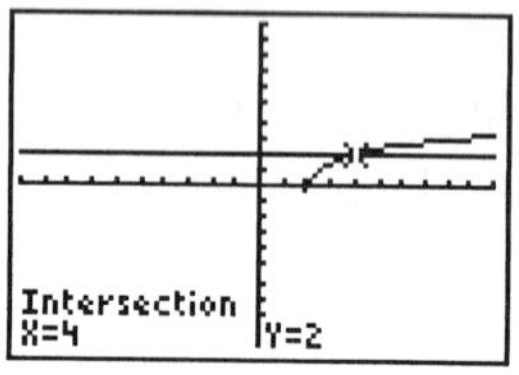


The solution is 4, and the solution set is $\{4\}$.
Verify the solution algebraically:

$\log_3(4\cdot4-7) = 2$

$\log_3(16-7) = 2$

$\log_3 9 = 2$

$\log_3 3^2 = 2$

$2 = 2$

**113.** $\log(x+3) + \log x = 1$

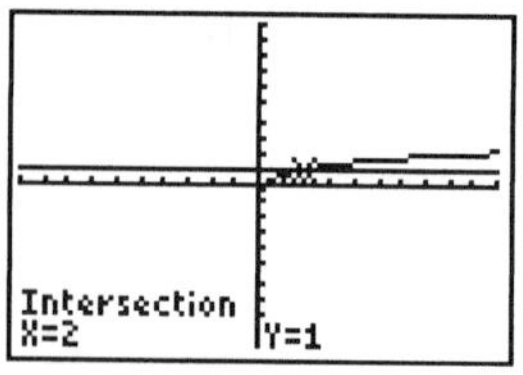


The solution is 2, and the solution set is $\{2\}$.
Verify the solution algebraically:

$\log(2+3) + \log 2 = 1$

$\log 5 + \log 2 = 1$

$\log(5\cdot2) = 1$

$\log 10 = 1$

$1 = 1$

**115.** $3^x = 2x + 3$

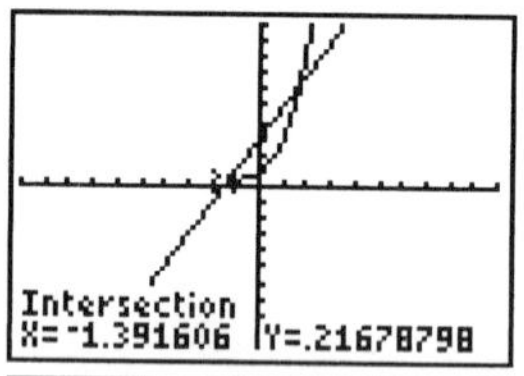


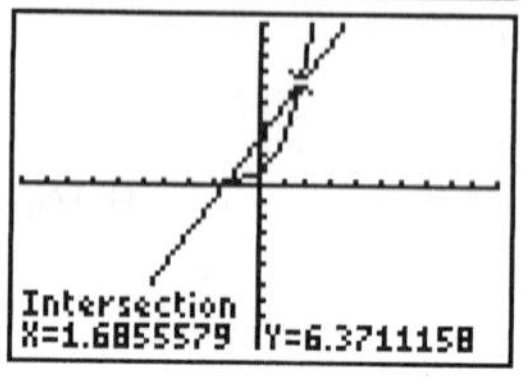


The solutions are $-1.39$ and $1.69$ and the solution set is $\{-1.39, 1.69\}$.
The solutions check algebraically.

**117.** $f(x) = 0.48\ln(x+1) + 27$

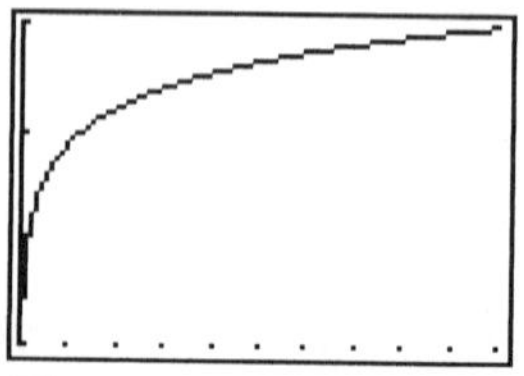

The barometric air pressure increases as the distance from the eye increases. It increases quickly at first, and the more slowly over time.

**119.** $P(t) = 145e^{-0.092t}$

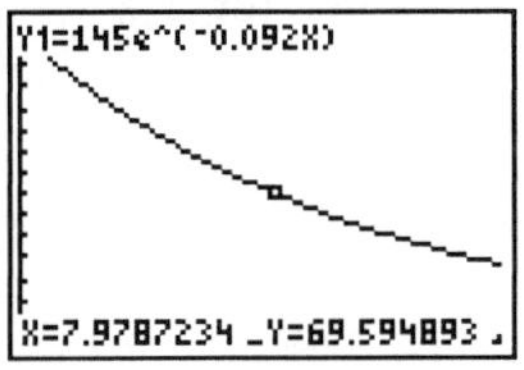


The runner's pulse will be 70 beats per minute after approximately 7.9 minutes.

Verifying algebraically:

$$P(7.9) = 145e^{-0.092(7.9)}$$
$$= 145e^{-07268} \approx 70$$

**121.** Statement **c.** is true.

$$x = \frac{1}{k}\ln y$$
$$kx = \ln y$$
$$e^{kx} = e^{\ln y}$$
$$y = e^{kx}$$

Statement **a.** is false. If $\log(x+3) = 2$, then $10^2 = x+3$.

Statement **b.** is false. If $\log(7x+3) - \log(2x+5) = 4$, then $\log\left(\frac{7x+3}{2x+5}\right) = 4$, and $10^4 = \frac{7x+3}{2x+5}$.

Statement **d.** is false. $x^{10} = 5.71$ is not an exponential equation, because there is not a variable in an exponent.

**123.** $(\ln x)^2 = \ln x^2$

$$(\ln x)^2 = 2\ln x$$

Let $t = \ln x$.

$$t^2 = 2t$$
$$t^2 - 2t = 0$$
$$t(t-2) = 0$$
$$t = 0 \quad \text{and} \quad t - 2 = 0$$
$$t = 2$$

Substitute $\ln x$ for $t$.

$$\ln x = 0 \quad \text{and} \quad \ln x = 2$$
$$x = e^0 \qquad x = e^2$$
$$x = 1$$

The solution are 1 and $e^2$, and the solution set is $\{1, e^2\}$.

**125.** $\ln(\ln x) = 0$

Let $t = \ln x$.

$$\ln(t) = 0$$
$$t = e^0$$
$$t = 1$$

Substitute $\ln x$ for $t$.

$$\ln(x) = 1$$
$$x = e^1$$
$$x = e$$

The solution is $e$, and the solution set is $\{e\}$.

**126.** $\sqrt{x+4} - \sqrt{x-1} = 1$

$$\sqrt{x+4} = 1 + \sqrt{x-1}$$
$$\left(\sqrt{x+4}\right)^2 = \left(1+\sqrt{x-1}\right)^2$$
$$x + 4 = 1 + 2\sqrt{x-1} + x - 1$$
$$4 = 2\sqrt{x-1}$$
$$2 = \sqrt{x-1}$$
$$2^2 = \left(\sqrt{x-1}\right)^2$$
$$4 = x - 1$$
$$5 = x$$

This value check, so the solution is 5, and the solution set is $\{5\}$.

**127.** $\frac{3}{x+1}-\frac{5}{x}=\frac{19}{x^2+x}$

$\frac{3}{x+1}-\frac{5}{x}=\frac{19}{x(x+1)}$

$x(x+1)\left(\frac{3}{x+1}-\frac{5}{x}\right)=x(x+1)\left(\frac{19}{x(x+1)}\right)$

$x(3)-5(x+1)=19$

$3x-5x-5=19$

$-2x-5=19$

$-2x=24$

$x=-12$

The solution is $-12$, and the solution set is $\{-12\}$.

**128.** $\left(-2x^3y^{-2}\right)^{-4}=\left(-\frac{2x^3}{y^2}\right)^{-4}$

$=\left(-\frac{y^2}{2x^3}\right)^4=\frac{y^8}{16x^{12}}$

## 9.6 Exercise Set

**1.** Since 2003 is 0 years after 2003, find $A$ when $t=0$:

$A=127.2e^{0.001t}$

$A=127.2e^{0.001(0)}$

$A=127.2e^0$

$A=127.2(1)$

$A=127.2$

In 2003, the population was 127.2 million.

**3.** Iraq has the greatest growth rate at 2.8% per year.

**5.** Substitute $A=1238$ into the model for India and solve for $t$:

$1238=1049.7e^{0.015t}$

$\frac{1238}{1049.7}=e^{0.015t}$

$\ln\frac{1238}{1049.7}=\ln e^{0.015t}$

$\ln\frac{1238}{1049.7}=0.015t$

$t=\frac{\ln\frac{1238}{1049.7}}{0.015}\approx 11$

Now, 2003 + 11 = 2014. The population of India will be 1238 million in approximately the year 2014.

**7.** **a.** $A_0=6.04$. Since 2050 is 50 years after 2000, when $t=50$, $A=10$.

$A=A_0e^{kt}$

$10=6.04e^{k(50)}$

$\frac{10}{6.04}=e^{50k}$

$\ln\left(\frac{10}{6.04}\right)=\ln e^{50k}$

$\ln\left(\frac{10}{6.04}\right)=50k$

$k=\frac{\ln\left(\frac{10}{6.04}\right)}{50}\approx 0.01$

Thus, the growth function is $A=6.04e^{0.01t}$.

**b.**

$$9 = 6.04e^{0.01t}$$

$$\frac{9}{6.04} = e^{0.01t}$$

$$\ln\left(\frac{9}{6.04}\right) = \ln e^{0.01t}$$

$$\ln\left(\frac{9}{6.04}\right) = 0.01t$$

$$t = \frac{\ln\left(\frac{9}{6.04}\right)}{0.01} \approx 40$$

Now, 2000 + 40 = 2040, so the population will be 9 million is approximately the year 2040.

**9.**

$$A = 16e^{-0.000121t}$$

$$A = 16e^{-0.000121(5715)}$$

$$A = 16e^{-0.691515}$$

$$A \approx 8.01$$

Approximately 8 grams of carbon-14 will be present in 5715 years.

**11.** After 10 seconds, there will be $16 \cdot \frac{1}{2} = 8$ grams present. After 20 seconds, there will be $8 \cdot \frac{1}{2} = 4$ grams present. After 30 seconds, there will be $4 \cdot \frac{1}{2} = 2$ grams present. After 40 seconds, there will be $2 \cdot \frac{1}{2} = 1$ grams present. After 50 seconds, there will be $1 \cdot \frac{1}{2} = \frac{1}{2}$ gram present.

**13.**

$$A = A_0 e^{-0.000121t}$$

$$15 = 100e^{-0.000121t}$$

$$\frac{15}{100} = e^{-0.000121t}$$

$$\ln 0.15 = \ln e^{-0.000121t}$$

$$\ln 0.15 = -0.000121t$$

$$t = \frac{\ln 0.15}{-0.000121} \approx 15{,}679$$

The paintings are approximately 15,679 years old.

**15.** **a.**

$$\frac{1}{2} = 1e^{k1.31}$$

$$\ln\frac{1}{2} = \ln e^{1.31k}$$

$$\ln\frac{1}{2} = 1.31k$$

$$k = \frac{\ln\frac{1}{2}}{1.31} \approx -0.52912$$

The exponential model is given by $A = A_0 e^{-0.52912t}$.

**b.**

$$A = A_0 e^{-0.52912t}$$

$$0.945A_0 = A_0 e^{-0.52912t}$$

$$0.945 = e^{-0.52912t}$$

$$\ln 0.945 = \ln e^{-0.52912t}$$

$$\ln 0.945 = -0.52912t$$

$$t = \frac{\ln 0.945}{-0.52912} \approx 0.1069$$

The age of the dinosaur ones is approximately 0.1069 billion or 106,900,000 years old.

**17.** $2A_0 = A_0e^{kt}$

$2 = e^{kt}$

$\ln 2 = \ln e^{kt}$

$\ln 2 = kt$

$t = \dfrac{\ln 2}{k}$

The population will double in $t = \dfrac{\ln 2}{k}$ years.

**19.** $A = e^{0.007t}$

**a.** $k = 0.007$, so New Zealand's growth rate is 0.7%.

**b.** $t = \dfrac{\ln 2}{k}$

$t = \dfrac{\ln 2}{0.007} \approx 99$

New Zealand's population will double in approximately 99 years.

**21. a.**

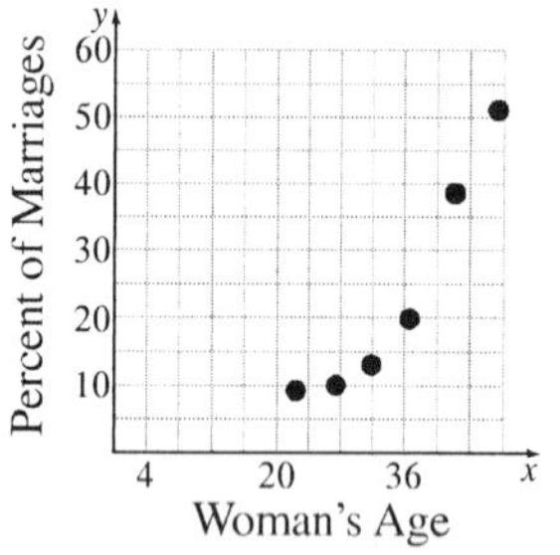


**b.** An exponential function appears to be the best choice for modeling the data.

**23. a.**

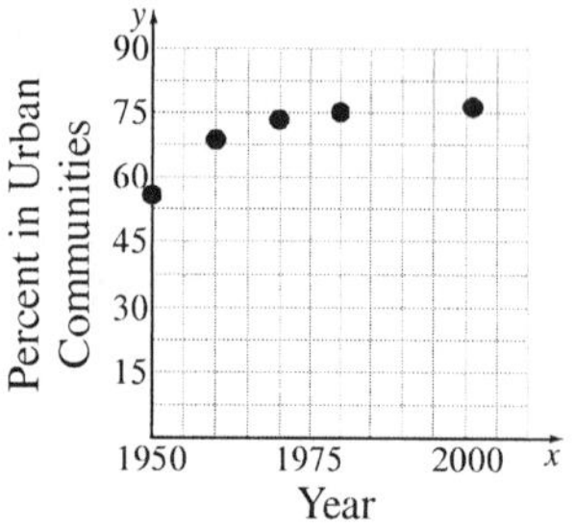


**b.** A logarithmic function appears to be the best choice for modeling the data.

**25. a.**

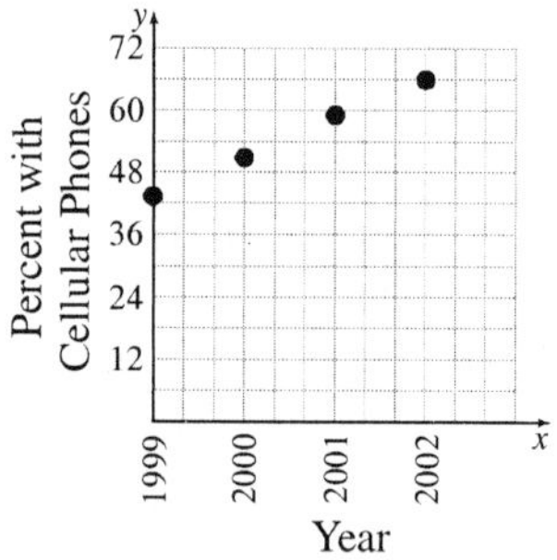


**b.** A linear function appears to be the best choice for modeling the data.

**27.** $y = 100(4.6)^x$

$y = 100e^{(\ln 4.6)x}$

$y = 100e^{1.526x}$

**29.** $y = 2.5(0.7)^x$

$y = 2.5e^{(\ln 0.7)x}$

$y = 100e^{-0.357x}$

**31. – 37.** Answers will vary.

**For Exercises 39. – 43, we enter the data in L1 and L2:**

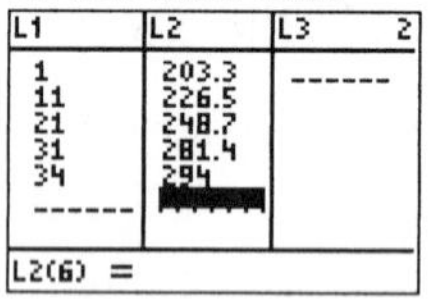
L1
L2
L3
2
1
11
21
31
34
203.3
226.5
248.7
281.4
294
------
------
L2(6) =

**39. a.**

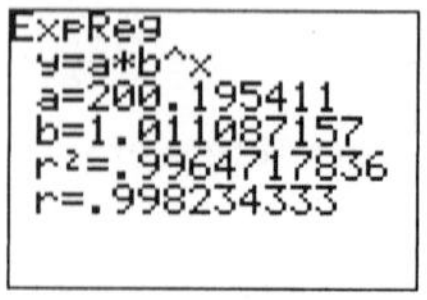
ExpReg
y=a*b^x
a=200.195411
b=1.011087157
r²=.9964717836
r=.998234333

The exponential model is $y = 200.2(1.011)^x$. Since $r \approx 0.998$ is very close to 1, the model fits the data very well.

**b.** $y = 200.2(1.011)^x$

$y = 200.2e^{(\ln 1.011)x}$

$y = 200.2e^{0.0109x}$

Since $k = .0109$, the population of the United States is increasing by about 1% each year.

**41.** 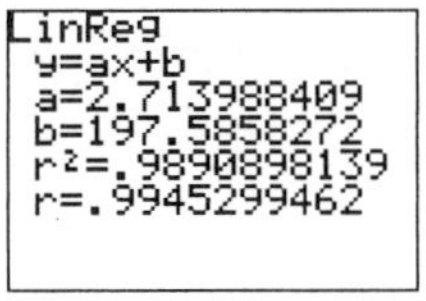


The linear model is $y = 2.714x + 197.586$. Since $r \approx 0.995$ is close to 1, the model fits the data very well.

**43.** Using *r*, the model of best fit is the exponential model $y = 200.2(1.011)^x$. The model of second best fit is the linear model $y = 2.714x + 197.586$.

Using the exponential model:

$$315 = 200.2(1.011)^x$$

$$\frac{315}{200.2} = (1.011)^x$$

$$\ln\left(\frac{315}{200.2}\right) = \ln(1.011)^x$$

$$\ln\left(\frac{315}{200.2}\right) = x\ln(1.011)$$

$$x = \frac{\ln\left(\frac{315}{200.2}\right)}{\ln(1.011)} \approx 41$$

$1969 + 41 = 2010$

Using the linear model:

$$315 = 2.714x + 197.586$$

$$117.414 = 2.714x$$

$$x = \frac{117.414}{2.714} \approx 43$$

$1969 + 43 = 2012$

According to the exponential model, the U.S. population will reach 315 million around the year 2010. According to the linear model, the U.S. population will reach 315 million around the year 2012. Both results are reasonably close to the result found in Example 1 (2010).

Explanations will vary.

**45.** For Exercise 21, we use an exponential model:

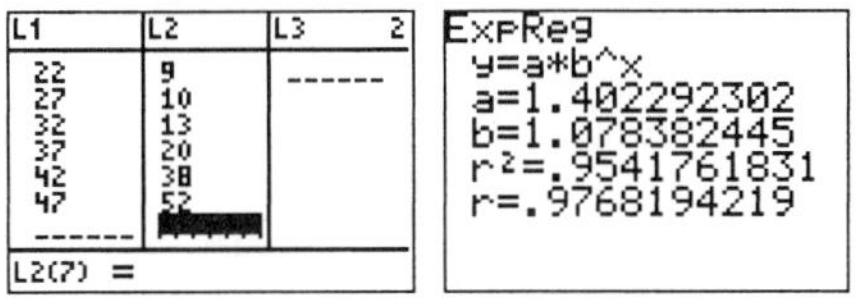


The model is $y = 1.402(1.078)^x$.

For Exercise 23, we use a logarithmic model with $x$ = the number of years after 1949:

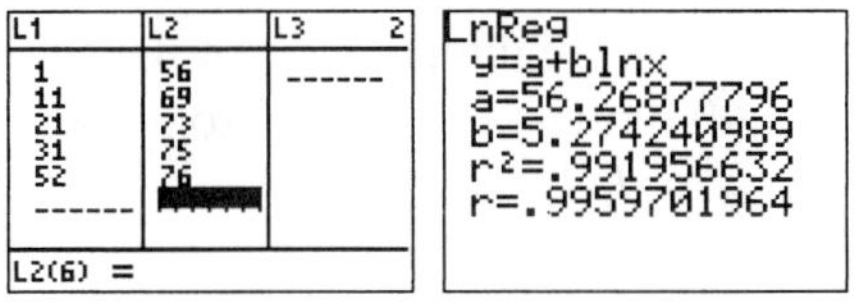


The model is $y = 56.269 + 5.274\ln x$.

For Exercise 25, we use a linear model with $x$ = the number of years after 1998:

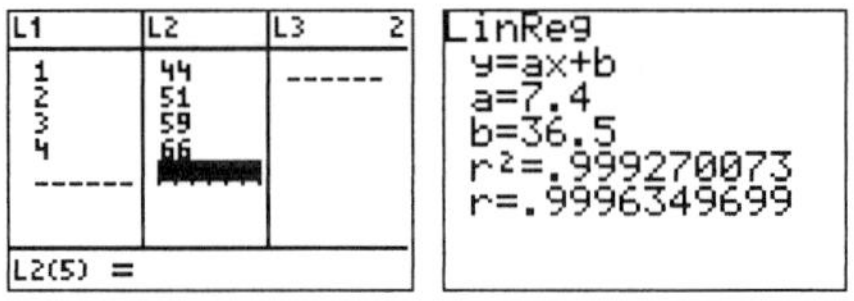


The model is $y = 7.4x + 36.5$.

Predictions will vary.

**47.** Use $T_0 = 210$, $C = 70$, $t = 30$, and $T = 140$ to determine the constant $k$:

$$T = C + (T_0 - C)e^{-kt}$$
$$140 = 70 + (210 - 70)e^{-k(30)}$$
$$140 = 70 + 140e^{-30k}$$
$$70 = 140e^{-30k}$$
$$\frac{70}{140} = \frac{140e^{-30k}}{140}$$
$$0.5 = e^{-30k}$$
$$\ln 0.5 = \ln e^{-30k}$$
$$\ln 0.5 = -30k$$
$$k = \frac{\ln 0.5}{-30} \approx 0.0231$$

Thus, the model for these conditions is $T = 70 + 140e^{-0.0231t}$.

Evaluate the model for $t = 40$:

$T = 70 + 140e^{-0.0231(40)} \approx 126$

Thus, the temperature of the cake after 40 minutes will be approximately $126°F$.

**48.**
$$\frac{x^2 - 9}{2x^2 + 7x + 3} \div \frac{x^2 - 3x}{2x^2 + 11x + 5}$$
$$= \frac{x^2 - 9}{2x^2 + 7x + 3} \cdot \frac{2x^2 + 11x + 5}{x^2 - 3x}$$
$$= \frac{\cancel{(x+3)}\cancel{(x-3)}}{\cancel{(2x+1)}\cancel{(x+3)}} \cdot \frac{\cancel{(2x+1)}(x+5)}{x\cancel{(x-3)}}$$
$$= \frac{x+5}{x}$$

**49.**
$$x^{\frac{2}{3}} + 2x^{\frac{1}{3}} - 3 = 0$$

Let $t = x^{\frac{1}{3}}$.

$$\left(x^{\frac{1}{3}}\right)^2 + 2x^{\frac{1}{3}} - 3 = 0$$
$$t^2 + 2t - 3 = 0$$
$$(t+3)(t-1) = 0$$

Apply the zero product principle.

$t + 3 = 0$ and $t - 1 = 0$

$t = -3$ $\quad$ $t = 1$

Substitute $x^{\frac{1}{3}}$ for $t$.

$x^{\frac{1}{3}} = -3$ and $x^{\frac{1}{3}} = 1$

$\left(x^{\frac{1}{3}}\right)^3 = (-3)^3$ $\quad$ $\left(x^{\frac{1}{3}}\right)^3 = (1)^3$

$x = -27$ $\quad$ $x = 1$

The solutions are –27 and 1, and the solution set is $\{-27, 1\}$.

**50.**
$$6\sqrt{2} - 2\sqrt{50} + 3\sqrt{98}$$
$$= 6\sqrt{2} - 2\sqrt{25 \cdot 2} + 3\sqrt{49 \cdot 2}$$
$$= 6\sqrt{2} - 2 \cdot 5\sqrt{2} + 3 \cdot 7\sqrt{2}$$
$$= 6\sqrt{2} - 10\sqrt{2} + 21\sqrt{2} = 17\sqrt{2}$$

## Chapter 9 Review Exercises

**1.** $f(x) = 4^x$

| $x$ | $f(x)$ |
|---|---|
| -2 | $4^{-2} = \frac{1}{4^2} = \frac{1}{16}$ |
| -1 | $4^{-1} = \frac{1}{4^1} = \frac{1}{4}$ |
| 0 | $4^0 = 1$ |
| 1 | $4^1 = 4$ |
| 2 | $4^2 = 16$ |

The coordinates match graph **d**.

**2.** $f(x)=4^{-x}$

| $x$ | $f(x)$ |
|---|---|
| -2 | $4^{-(-2)}=4^2=16$ |
| -1 | $4^{-(-1)}=4^1=4$ |
| 0 | $4^{-0}=4^0=1$ |
| 1 | $4^{-1}=\frac{1}{4^1}=\frac{1}{4}$ |
| 2 | $4^{-2}=\frac{1}{4^2}=\frac{1}{16}$ |

The coordinates match graph **a.**

**3.** $f(x)=-4^{-x}$

| $x$ | $f(x)$ |
|---|---|
| -2 | $-4^{-(-2)}=-4^2=-16$ |
| -1 | $-4^{-(-1)}=-4^1=-4$ |
| 0 | $-4^{-0}=-4^0=-1$ |
| 1 | $-4^{-1}=-\frac{1}{4^1}=-\frac{1}{4}$ |
| 2 | $-4^{-2}=-\frac{1}{4^2}=-\frac{1}{16}$ |

The coordinates match graph **b.**

**4.** $f(x)=-4^{-x}+3$

| $x$ | $f(x)$ |
|---|---|
| $-2$ | $-4^{-(-2)}+3=-4^2+3=-16+3=-13$ |
| $-1$ | $-4^{-(-1)}+3=-4^1+3=-4+3=-1$ |
| 0 | $-4^{-0}+3=-4^0+3=-1+3=2$ |
| 1 | $-4^{-1}+3=-\frac{1}{4^1}+3=-\frac{1}{4}+3=\frac{11}{4}$ |
| 2 | $-4^{-2}+3=-\frac{1}{4^2}+3=-\frac{1}{16}+3=\frac{47}{16}$ |

The coordinates match graph **c.**

**5.** $f(x)=2^x$ and $g(x)=2^{x-1}$

| $x$ | $f(x)$ | $g(x)$ |
|---|---|---|
| -2 | $\frac{1}{4}$ | $\frac{1}{8}$ |
| -1 | $\frac{1}{2}$ | $\frac{1}{4}$ |
| 0 | 1 | $\frac{1}{2}$ |
| 1 | 2 | 1 |
| 2 | 4 | 2 |

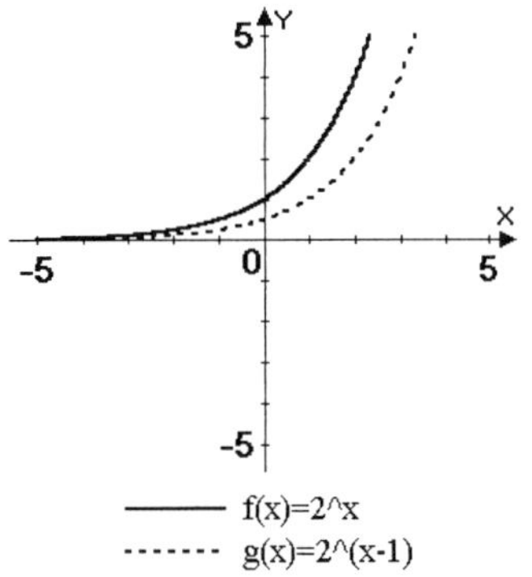


The graph of $g$ is the graph of $f$ shifted 1 unit to the right.

**6.** $f(x)=2^x$ and $g(x)=\left(\frac{1}{2}\right)^x$

| $x$ | $f(x)$ | $g(x)$ |
|---|---|---|
| -2 | $\frac{1}{4}$ | 4 |
| -1 | $\frac{1}{2}$ | 2 |
| 0 | 1 | 1 |
| 1 | 2 | $\frac{1}{2}$ |
| 2 | 4 | $\frac{1}{4}$ |

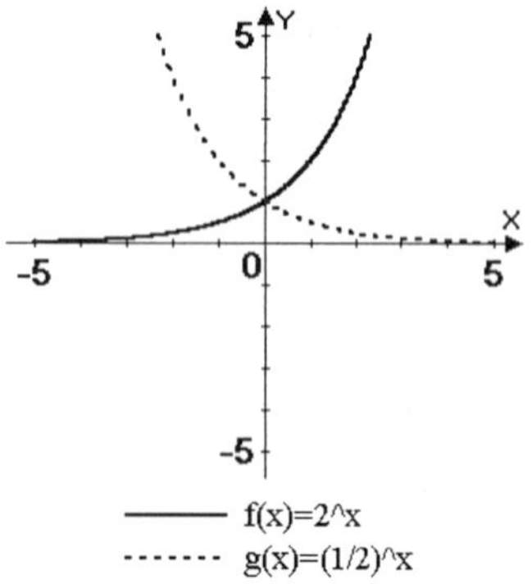


The graph of $g$ is the graph of $f$ reflected across the $y$–axis.

**7.** $f(x)=3^x$ and $g(x)=3^x-1$

| $x$ | $f(x)$ | $g(x)$ |
|---|---|---|
| -2 | $\frac{1}{9}$ | $-\frac{8}{9}$ |
| -1 | $\frac{1}{3}$ | $-\frac{2}{3}$ |
| 0 | 1 | 0 |
| 1 | 3 | 2 |
| 2 | 9 | 8 |

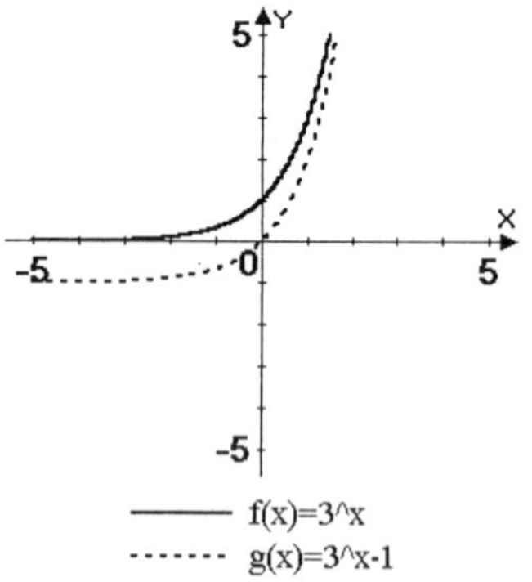


The graph of $g$ is the graph of $f$ shifted down 1 unit.

**8.** $f(x)=3^x$ and $g(x)=-3^x$

| $x$ | $f(x)$ | $g(x)$ |
|---|---|---|
| -2 | $\frac{1}{9}$ | $-\frac{1}{9}$ |
| -1 | $\frac{1}{3}$ | $-\frac{1}{3}$ |
| 0 | 1 | -1 |
| 1 | 3 | -3 |
| 2 | 9 | -9 |

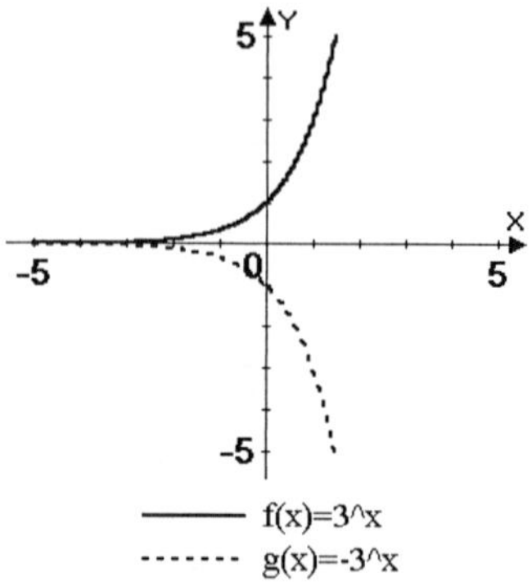


The graph of $g$ is the graph of $f$ reflected across the $x$–axis.

**9.** 5.5% Compounded Semiannually:

$$A=5000\left(1+\frac{0.055}{2}\right)^{2\cdot 5}$$
$$=5000(1+0.0275)^{10}$$
$$=5000(1.0275)^{10}\approx 6558.26$$

5.25% Compounded Monthly:

$$A=5000\left(1+\frac{0.0525}{12}\right)^{12\cdot 5}$$
$$=5000(1+0.004375)^{60}$$
$$=5000(1.004375)^{60}\approx 6497.16$$

5.5% compounded semiannually yields the greater return.

**10.** 7.0% Compounded Monthly:

$$A = 14000\left(1+\frac{0.07}{12}\right)^{12\cdot 10}$$

$$= 14000\left(1+\frac{7}{1200}\right)^{120}$$

$$= 14000\left(\frac{1207}{1200}\right)^{120} \approx 28135.26$$

6.85% Compounded Continuously:

$$A = 14000e^{0.0685\cdot 10}$$

$$= 14000e^{0.685} \approx 27772.81$$

7.0% compounded monthly yields the greater return.

**11.** **a.** The coffee was $200°F$ when it was first taken out of the microwave.

**b.** After 20 minutes, the temperature is approximately $119°F$.

**c.** The coffee will cool to a low of $70°F$. This means that the temperature of the room is $70°F$.

**12.** **a.**
$$(f \circ g)(x) = f(g(x))$$
$$= f(4x-1)$$
$$= (4x-1)^2 + 3$$
$$= 16x^2 - 8x + 1 + 3$$
$$= 16x^2 - 8x + 4$$

**b.**
$$(g \circ f)(x) = g(f(x))$$
$$= g(x^2+3)$$
$$= 4(x^2+3) - 1$$
$$= 4x^2 + 12 - 1$$
$$= 4x^2 + 11$$

**c.**
$$(f \circ g)(3) = 16(3)^2 - 8(3) + 4$$
$$= 16(9) - 24 + 4$$
$$= 144 - 24 + 4$$
$$= 124$$

**13.** **a.**
$$(f \circ g)(x) = f(g(x))$$
$$= f(x+1) = \sqrt{x+1}$$

**b.**
$$(g \circ f)(x) = g(f(x))$$
$$= g(\sqrt{x}) = \sqrt{x} + 1$$

**c.**
$$(f \circ g)(3) = \sqrt{3+1} = \sqrt{4} = 2$$

**14.**
$$f(x) = \frac{3}{5}x + \frac{1}{2} \text{ and } g(x) = \frac{5}{3}x - 2$$

$$f(g(x)) = f\left(\frac{5}{3}x - 2\right)$$
$$= \frac{3}{5}\left(\frac{5}{3}x - 2\right) + \frac{1}{2}$$
$$= \frac{3}{5}\left(\frac{5}{3}x\right) - \left(\frac{3}{5}\right)2 + \frac{1}{2}$$
$$= x - \frac{6}{5} + \frac{1}{2}$$
$$= x - \frac{7}{10}$$

$$g(f(x)) = g\left(\frac{3}{5}x + \frac{1}{2}\right)$$
$$= \frac{5}{3}\left(\frac{3}{5}x + \frac{1}{2}\right) - 2$$
$$= \frac{5}{3}\left(\frac{3}{5}x\right) + \left(\frac{5}{3}\right)\frac{1}{2} - 2$$
$$= x + \frac{5}{6} - 2$$
$$= x - \frac{7}{6}$$

The functions are not inverses.

**15.** $f(x) = 2 - 5x$ and $g(x) = \dfrac{2-x}{5}$

$$f(g(x)) = f\left(\frac{2-x}{5}\right)$$
$$= 2 - 5\left(\frac{2-x}{5}\right)$$
$$= 2 - (2-x) = 2 - 2 + x = x$$

$$g(f(x)) = g(2-5x)$$
$$= \frac{2-(2-5x)}{5}$$
$$= \frac{2-2+5x}{5} = \frac{5x}{5} = x$$

The functions are inverses.

**16. a.** $f(x) = 4x - 3$

$$y = 4x - 3$$

Interchange $x$ and $y$ and solve for $y$.

$$x = 4y - 3$$
$$x + 3 = 4y$$
$$\frac{x+3}{4} = y$$
$$f^{-1}(x) = \frac{x+3}{4}$$

**b.**

$$f(f^{-1}(x)) = f\left(\frac{x+3}{4}\right)$$
$$= 4\left(\frac{x+3}{4}\right) - 3$$
$$= x + 3 - 3 = x$$

$$f^{-1}(f(x)) = f(4x-3)$$
$$= \frac{(4x-3)+3}{4}$$
$$= \frac{4x-3+3}{4} = \frac{4x}{4} = x$$

**17. a.** $f(x) = \sqrt{x+2}$

$$y = \sqrt{x+2}$$

Interchange $x$ and $y$ and solve for $y$.

$$x = \sqrt{y+2}$$
$$x^2 = y + 2$$
$$x^2 - 2 = y$$
$$f^{-1}(x) = x^2 - 2 \text{ for } x \geq 0$$

**b.** $f(f^{-1}(x)) = f\left(\sqrt{x+2}\right)$

$$= \left(\sqrt{x+2}\right)^2 - 2$$
$$= x + 2 - 2 = x$$

$$f^{-1}(f(x)) = f(x^2 - 2)$$
$$= \sqrt{(x^2-2)+2}$$
$$= \sqrt{x^2 - 2 + 2}$$
$$= \sqrt{x^2} = x$$

**18. a.** $f(x) = 8x^3 + 1$

$$y = 8x^3 + 1$$

Interchange $x$ and $y$ and solve for $y$.

$$x = 8y^3 + 1$$
$$x - 1 = 8y^3$$
$$\frac{x-1}{8} = y^3$$
$$\sqrt[3]{\frac{x-1}{8}} = y$$
$$\frac{\sqrt[3]{x-1}}{2} = y$$
$$f^{-1}(x) = \frac{\sqrt[3]{x-1}}{2}$$

**b.**

$$f\left(f^{-1}(x)\right)=f\left(\frac{\sqrt[3]{x-1}}{2}\right)$$
$$=8\left(\frac{\sqrt[3]{x-1}}{2}\right)^3+1$$
$$=8\left(\frac{x-1}{8}\right)+1$$
$$=x-1+1=x$$

$$f^{-1}\left(f(x)\right)=f\left(8x^3+1\right)$$
$$=\frac{\sqrt[3]{\left(8x^3+1\right)-1}}{2}$$
$$=\frac{\sqrt[3]{8x^3+1-1}}{2}$$
$$=\frac{\sqrt[3]{8x^3}}{2}=\frac{2x}{2}=x$$

**19.** Since the graph satisfies the horizontal line test, its inverse is a function.

**20.** Since the graph does not satisfy the horizontal line test, its inverse in not a function.

**21.** Since the graph satisfies the horizontal line test, its inverse is a function.

**22.** Since the graph does not satisfy the horizontal line test, its inverse is not a function.

**23.** Since the points $(-3,-1),(0,0)$ and $(2,4)$ lie on the graph of the function, the points $(-1,-3)$, $(0,0)$ and $(4,2)$ lie on the inverse function.

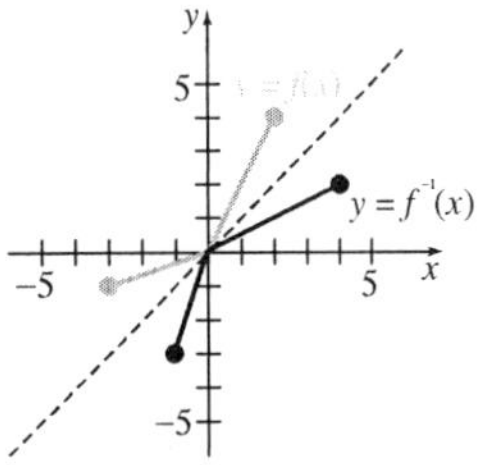


**24.** $\frac{1}{2}=\log_{49}7$

$49^{\frac{1}{2}}=7$

**25.** $3=\log_4 x$

$4^3=x$

**26.** $\log_3 81=y$

$3^y=81$

**27.** $6^3=216$

$\log_6 216=3$

**28.** $b^4=625$

$\log_b 625=4$

**29.** $13^y=874$

$\log_{13}874=y$

**30.** $\log_4 64=\log_4 4^3=3$ because $\log_b b^x=x$.

**31.** $\log_5\frac{1}{25}=\log_5\frac{1}{5^2}=\log_5 5^{-2}=-2$ because $\log_b b^x=x$.

**32.** $\log_3(-9)$

This logarithm cannot be evaluated because –9 is not in the domain of $y=\log_3(-9)$.

**33.** $\log_{16} 4 = y$

$16^y = 4$

$(4^2)^y = 4$

$4^{2y} = 4^1$

$2y = 1$

$y = \frac{1}{2}$

**34.** $\log_{17} 17 = 1$ because $17^1 = 17$.

**35.** $\log_3 3^8 = 8$ because $\log_b b^x = x$.

**36.** Because $\ln e^x = x$, we conclude that $\ln e^5 = 5$.

**37.** $\log_3 \frac{1}{\sqrt{3}} = \log_3 \frac{1}{3^{\frac{1}{2}}} = \log_3 3^{-\frac{1}{2}} = -\frac{1}{2}$

because $\log_b b^x = x$.

**38.** $\ln \frac{1}{e^2} = \ln e^{-2} = -2$ because $\log_b b^x = x$.

**39.** $\log \frac{1}{1000} = \log \frac{1}{10^3} = \log 10^{-3} = -3$

because $\log_b b^x = x$.

**40.** Recall that $\log_b b = 1$ and $\log_b 1 = 0$ for all $b > 0$, $b \neq 1$. Therefore, $\log_3(\log_8 8) = \log_3 1 = 0$.

**41.** $f(x) = 2^x$; $g(x) = \log_2 x$

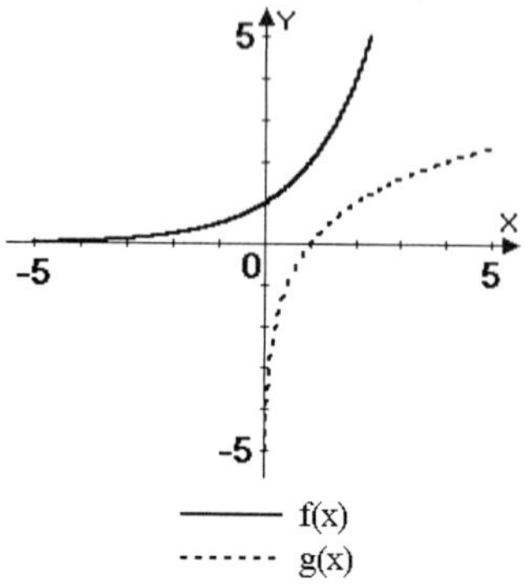


Domain of *f*: $\{x \mid x \text{ is a real number}\}$ or $(-\infty, \infty)$.

Range of *f*: $\{y \mid y > 0\}$ or $(0, \infty)$

Domain of *g*: $\{x \mid x > 0\}$ or $(0, \infty)$

Range of *g*: $\{y \mid y \text{ is a real number}\}$ or $(-\infty, \infty)$.

**42.** $f(x) = \left(\frac{1}{3}\right)^x$; $g(x) = \log_{\frac{1}{3}} x$

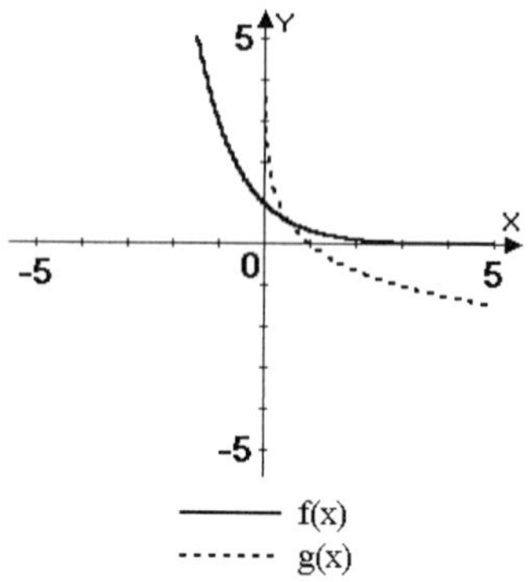


Domain of *f*: $\{x \mid x \text{ is a real number}\}$ or $(-\infty, \infty)$.

Range of *f*: $\{y \mid y > 0\}$ or $(0, \infty)$

Domain of *g*: $\{x \mid x > 0\}$ or $(0, \infty)$

Range of *g*: $\{y \mid y \text{ is a real number}\}$ or $(-\infty, \infty)$.

**43.** $f(x) = \log_8(x+5)$

$x + 5 > 0$

$x > -5$

The domain of *f* is $\{x \mid x > -5\}$ or $(-5, \infty)$.

**44.** $f(x) = \log(3 - x)$

$3 - x > 0$

$-x > -3$

$x < 3$

The domain of *f* is $\{x \mid x < 3\}$ or $(-\infty, 3)$.

**45.** $f(x)=\ln(x-1)^2$
The domain of $g$ is all real numbers for which $(x-1)^2>0$. The only number that must be excluded is 1. The domain of $f$ is $\{x|x\neq 1\}$ or $(-\infty,1)\cup(1,\infty)$.

**46.** Since $\ln e^x = x$, $\ln e^{6x}=6x$.

**47.** Since $e^{\ln x}=x$, $e^{\ln\sqrt{x}}=\sqrt{x}$.

**48.** Since $10^{\log x}=x$, $10^{\log 4x^2}=4x^2$.

**49.**
$$R=\log\frac{I}{I_0}$$
$$R=\log\frac{1000I_0}{I_0}$$
$$R=\log 1000$$
$$10^R=1000$$
$$10^R=10^3$$
$$R=3$$
The magnitude on the Richter scale is 3.

**50. a.** $f(0)=76-18\log(0+1)$
$=76-18\log(1)$
$=76-18(0)=76-0=76$
The average score when the exam was first given was 76.

$f(2)=76-18\log(2+1)$
$=76-18\log(3)\approx 67.4$
$f(4)=76-18\log(4+1)$
$=76-18\log(5)\approx 63.4$
$f(6)=76-18\log(6+1)$
$=76-18\log(7)\approx 60.8$
$f(8)=76-18\log(8+1)$
$=76-18\log(9)\approx 58.8$
$f(12)=76-18\log(12+1)$
$=76-18\log(13)\approx 55.9$
The average scores were as follows:

| | |
|---|---|
| 2 months | 67.4 |
| 4 months | 63.4 |
| 6 months | 60.8 |
| 8 months | 58.8 |
| 12 months | 55.9. |

**c.**
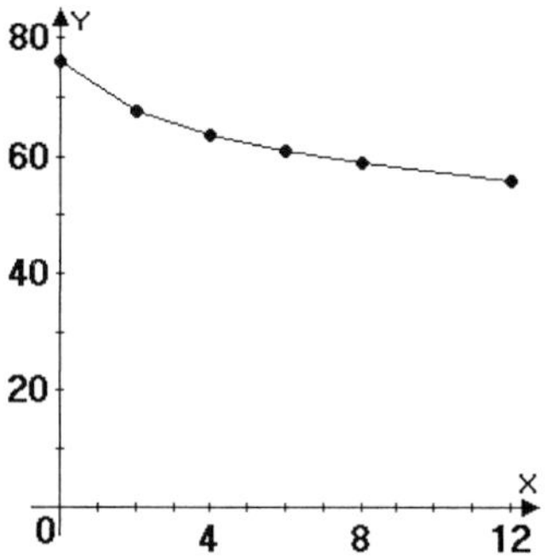


The students retain less material over time.

**51.**
$$t=\frac{1}{0.06}\ln\left(\frac{12}{12-5}\right)$$
$$=\frac{1}{0.06}\ln\left(\frac{12}{7}\right)\approx 9.0$$
It will take approximately 9 weeks for the man to run 5 miles per hour.

**52.** $\log_6(36x^3)=\log_6 36+\log_6 x^3$
$=2+3\log_6 x$

**53.**
$$\log_4\frac{\sqrt{x}}{64}=\log_4\sqrt{x}-\log_4 64$$
$$=\log_4 x^{\frac{1}{2}}-3=\frac{1}{2}\log_4 x-3$$

**54.** $\log_2\left(\frac{xy^2}{64}\right) = \log_2 xy^2 - \log_2 64$

$= \log_2 x + \log_2 y^2 - 6$

$= \log_2 x + 2\log_2 y - 6$

**55.** $\ln\sqrt[3]{\frac{x}{e}} = \ln\left(\frac{x}{e}\right)^{\frac{1}{3}} = \frac{1}{3}\ln\left(\frac{x}{e}\right)$

$= \frac{1}{3}(\ln x - \ln e)$

$= \frac{1}{3}(\ln x - 1) = \frac{1}{3}\ln x - \frac{1}{3}$

**56.** $\log_b 7 + \log_b 3 = \log_b(7 \cdot 3) = \log_b 21$

**57.** $\log 3 - 3\log x = \log 3 - \log x^3 = \log\frac{3}{x^3}$

**58.** $3\ln x + 4\ln y = \ln x^3 + \ln y^4 = \ln\left(x^3y^4\right)$

**59.** $\frac{1}{2}\ln x - \ln y = \ln x^{\frac{1}{2}} - \ln y$

$= \ln\sqrt{x} - \ln y = \ln\left(\frac{\sqrt{x}}{y}\right)$

**60.** $\log_6 72{,}348 = \frac{\log 72{,}348}{\log 6} \approx 6.2448$

**61.** $\log_4 0.863 = \frac{\log 0.863}{\log 4} \approx -0.1063$

**62.** $2^{4x-2} = 64$

$2^{4x-2} = 2^6$

$4x - 2 = 6$

$4x = 8$

$x = 2$

The solution is 2, and the solution set is $\{2\}$.

**63.** $125^x = 25$

$\left(5^3\right)^x = 5^2$

$5^{3x} = 5^2$

$3x = 2$

$x = \frac{2}{3}$

The solution is $\frac{2}{3}$, and the solution set is $\left\{\frac{2}{3}\right\}$.

**64.** $9^x = \frac{1}{27}$

$\left(3^2\right)^x = 3^{-3}$

$3^{2x} = 3^{-3}$

$2x = -3$

$x = -\frac{3}{2}$

The solution is $-\frac{3}{2}$, and the solution set is $\left\{-\frac{3}{2}\right\}$.

**65.** $8^x = 12{,}143$

$\ln 8^x = \ln 12{,}143$

$x\ln 8 = \ln 12{,}143$

$x = \frac{\ln 12{,}143}{\ln 8} \approx 4.52$

The solution is $\frac{\ln 12{,}143}{\ln 8} \approx 4.52$ and the solution set is $\left\{\frac{\ln 12{,}143}{\ln 8} \approx 4.52\right\}$.

**66.** $9e^{5x} = 1269$

$$e^{5x} = \frac{1269}{9}$$
$$\ln e^{5x} = \ln 141$$
$$5x = \ln 141$$
$$x = \frac{\ln 141}{5} \approx 0.99$$

The solution is $\frac{\ln 141}{5} \approx 0.99$, and the solution set is $\left\{\frac{\ln 141}{5} \approx 0.99\right\}$.

**67.** $30e^{0.045x} = 90$

$$e^{0.045x} = \frac{90}{30}$$
$$\ln e^{0.045x} = \ln 3$$
$$0.045x = \ln 3$$
$$x = \frac{\ln 3}{0.045} \approx 24.41$$

The solution is $\frac{\ln 3}{0.045} \approx 24.41$, and the solution set is $\left\{\frac{\ln 3}{0.045} \approx 24.41\right\}$.

**68.** $\log_5 x = -3$

$$x = 5^{-3}$$
$$x = \frac{1}{125}$$

The solution is $\frac{1}{125}$ ,and the solution set is $\left\{\frac{1}{125}\right\}$.

**66.** $\log x = 2$

$$x = 10^2$$
$$x = 100$$

The solution is $100$, and the solution set is $\{100\}$.

**70.** $\log_4(3x-5) = 3$

$$3x - 5 = 4^3$$
$$3x - 5 = 64$$
$$3x = 69$$
$$x = 23$$

The solution is $23$, and the solution set is $\{23\}$.

**71.** $\log_2(x+3) + \log_2(x-3) = 4$

$$\log_2((x+3)(x-3)) = 4$$
$$\log_2(x^2 - 9) = 4$$
$$x^2 - 9 = 2^4$$
$$x^2 - 9 = 16$$
$$x^2 = 25$$
$$x = \pm 5$$

We disregard –5 because it would result in taking the logarithm of a negative number in the original equation. The solution is 5 and the solution set is $\{5\}$.

**72.** $\log_3(x-1) - \log_3(x+2) = 2$

$$\log_3 \frac{x-1}{x+2} = 2$$
$$\frac{x-1}{x+2} = 3^2$$
$$\frac{x-1}{x+2} = 9$$
$$x - 1 = 9(x+2)$$
$$x - 1 = 9x + 18$$
$$-8x - 1 = 18$$
$$-8x = 19$$
$$x = -\frac{19}{8}$$

We disregard $-\frac{19}{8}$ because it would result in taking the logarithm of a negative number in the original equation. There is no solution. The solution set is $\varnothing$ or $\{\ \}$.

**73.** $\ln x = -1$

$$x = e^{-1}$$

$$x = \frac{1}{e}$$

The solutions are $\frac{1}{e}$, and the solution set is $\left\{\frac{1}{e}\right\}$.

**74.** $3 + 4\ln 2x = 15$

$$4\ln 2x = 12$$

$$\ln 2x = 3$$

$$2x = e^3$$

$$x = \frac{e^3}{2}$$

The solutions are $\frac{e^3}{2}$, and the solution set is $\left\{\frac{e^3}{2}\right\}$.

**75.** $P(x) = 14.7e^{-0.21x}$

$$4.6 = 14.7e^{-0.21x}$$

$$\frac{4.6}{14.7} = e^{-0.21x}$$

$$\ln\frac{4.6}{14.7} = \ln e^{-0.21x}$$

$$\ln\frac{4.6}{14.7} = -0.21x$$

$$t = \frac{\ln\frac{4.6}{14.7}}{-0.21} \approx 5.5$$

The peak of Mt. Everest is about 5.5 miles above sea level.

**76.** $f(t) = 364(1.005)^t$

$$560 = 364(1.005)^t$$

$$\frac{560}{364} = (1.005)^t$$

$$\ln\frac{560}{364} = \ln(1.005)^t$$

$$\ln\frac{560}{364} = t\ln 1.005$$

$$t = \frac{\ln\frac{560}{364}}{\ln 1.005} \approx 86.4$$

The carbon dioxide concentration will be double the pre-industrial level approximately 86 years after the year 2000 in the year 2086.

**77.** $W(x) = 0.37\ln x + 0.05$

$$3.38 = 0.37\ln x + 0.05$$

$$3.33 = 0.37\ln x$$

$$\frac{3.33}{0.37} = \ln x$$

$$9 = \ln x$$

$$e^9 = e^{\ln x}$$

$$x = e^9 \approx 8103$$

The population of New Your City is approximately 8103 thousand, or 8,103,000

**78.**

$$20{,}000 = 12{,}500\left(1+\frac{0.065}{4}\right)^{4t}$$

$$20{,}000 = 12{,}500(1+0.01625)^{4t}$$

$$20{,}000 = 12{,}500(1.01625)^{4t}$$

$$\frac{20{,}000}{12{,}500} = (1.01625)^{4t}$$

$$1.6 = (1.01625)^{4t}$$

$$\ln 1.6 = \ln(1.01625)^{4t}$$

$$\ln 1.6 = 4t \ln 1.01625$$

$$\frac{\ln 1.6}{4\ln 1.01625} = \frac{4t\ln 1.01625}{4\ln 1.01625}$$

$$t = \frac{\ln 1.6}{4\ln 1.01625} \approx 7.3$$

It will take approximately 7.3 years.

**79.** $3(50{,}000) = 50{,}000e^{0.075t}$

$$\frac{3(50{,}000)}{50{,}000} = e^{0.075t}$$

$$3 = e^{0.075t}$$

$$\ln 3 = \ln e^{0.075t}$$

$$\ln 3 = 0.075t$$

$$t = \frac{\ln 3}{0.075} \approx 14.6$$

The money will triple in approximately 14.6 years.

**80.** $3 = e^{r5}$

$$\ln 3 = \ln e^{5r}$$

$$\ln 3 = 5r$$

$$r = \frac{\ln 3}{5} \approx 0.220$$

The money will triple in 5 years if the interest rate is approximately 22%.

**81.** **a.** $t = 2000 - 1990 = 10$

$$A = 22.4e^{kt}$$

$$35.3 = 22.46e^{k(10)}$$

$$\frac{35.3}{22.46} = e^{10k}$$

$$\ln\frac{35.3}{22.46} = \ln e^{10k}$$

$$\ln\frac{35.3}{22.46} = 10k$$

$$k = \frac{\ln\frac{35.3}{22.46}}{10} \approx 0.045$$

**b.** Note that 2010 is 20 years after 1990, find $A$ for $t = 25$.

$A = 22.4e^{0.045(20)} = 22.4e^{0.9} \approx 55.1$

The population will be about 55.1 million the year 2010.

**c.** $60 = 22.4e^{0.045t}$

$$\frac{60}{22.4} = e^{0.045t}$$

$$\ln\frac{60}{22.4} = \ln e^{0.045t}$$

$$\ln\frac{60}{22.4} = 0.045t$$

$$t = \frac{\ln\frac{60}{22.4}}{0.045t} \approx 22$$

Now, $1990 + 22 = 2012$, so the Hispanic resident population will reach 60 million approximately 30 years after 1990, in the year 2012.

**82.** $A = A_0 e^{-0.000121t}$

$15 = 100e^{-0.000121t}$

$\frac{15}{100} = e^{-0.000121t}$

$\ln\frac{3}{20} = \ln e^{-0.000121t}$

$\ln\frac{3}{20} = -0.000121t$

$t = \frac{\ln\frac{3}{20}}{-0.000121} \approx 15{,}679$

The paintings are approximately 15,679 years old.

**83. a.**

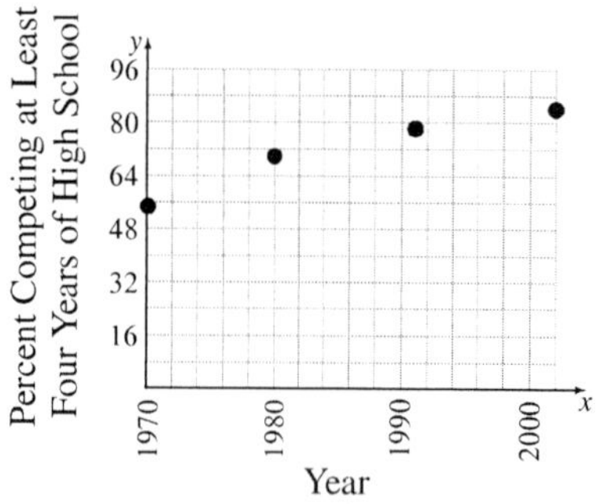


**b.** A logarithmic function appears to be the best choice for modeling the data.

**84. a.**

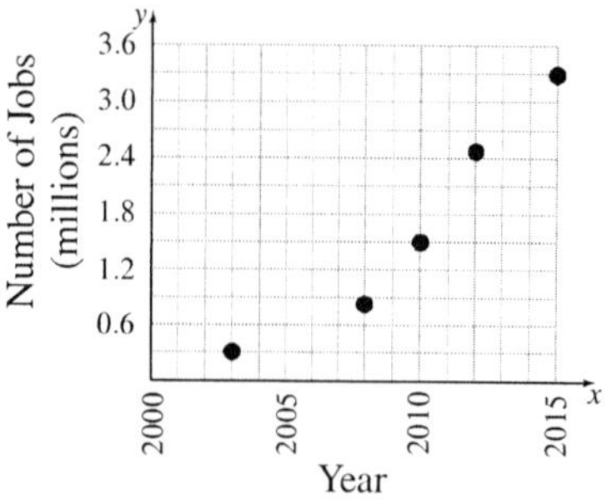


**b.** An exponential function appears to be the best choice for modeling the data.

**85.** $y = 73(2.6)^x$

$y = 73e^{(\ln 2.6)x}$

$y = 73e^{0.956x}$

**86.** $y = 6.5(0.43)^x$

$y = 6.5e^{(\ln 0.43)x}$

$y = 6.5e^{-0.844x}$

**87.** Answers will vary.

## Chapter 9 Test

**1.** $f(x) = 2^x$

$g(x) = 2^{x+1}$

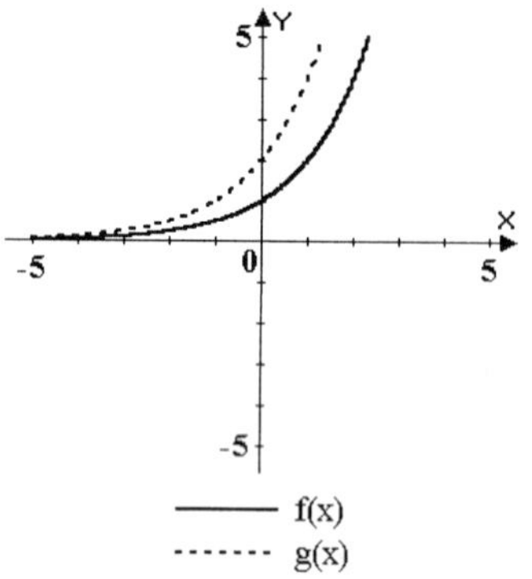


**2.** Semiannual Compounding:

$A = 3000\left(1 + \frac{0.065}{2}\right)^{2(10)}$

$= 3000(1.0325)^{20} \approx 5687.51$

Continuous Compounding:

$A = 3000e^{0.06(10)} = 3000e^{0.6} \approx 5466.36$

Semiannual compounding at 6.5% yields a greater return. The difference in the yields is \$221.

**3.** $f(x) = x^2 + x$ and $g(x) = 3x - 1$

$$(f \circ g)(x) = f(g(x)) = f(3x-1)$$
$$= (3x-1)^2 + (3x-1)$$
$$= 9x^2 - 6x + 1 + 3x - 1$$
$$= 9x^2 - 3x$$
$$(g \circ f)(x) = g(f(x))$$
$$= g(x^2 + x)$$
$$= 3(x^2 + x) - 1$$
$$= 3x^2 + 3x - 1$$

**4.** $f(x) = 5x - 7$

$$y = 5x - 7$$

Interchange $x$ and $y$ and solve for $y$.

$$x = 5y - 7$$
$$x + 7 = 5y$$
$$\frac{x+7}{5} = y$$
$$f^{-1}(x) = \frac{x+7}{5}$$

**5.** **a.** The function passes the horizontal line test (i.e., no horizontal line intersects the graph of *f* in more than one point), so we know its inverse is a function.

**b.** $f(80) = 2000$

**c.** $f^{-1}(2000)$ represents the income, $80 thousand, of a family that gives $2000 to charity.

**6.** $\log_5 125 = 3$

$$5^3 = 125$$

**7.** $\sqrt{36} = 6$

$$36^{\frac{1}{2}} = 6$$
$$\log_{36} 6 = \frac{1}{2}$$

**8.** $f(x) = 3^x$

$g(x) = \log_3 x$

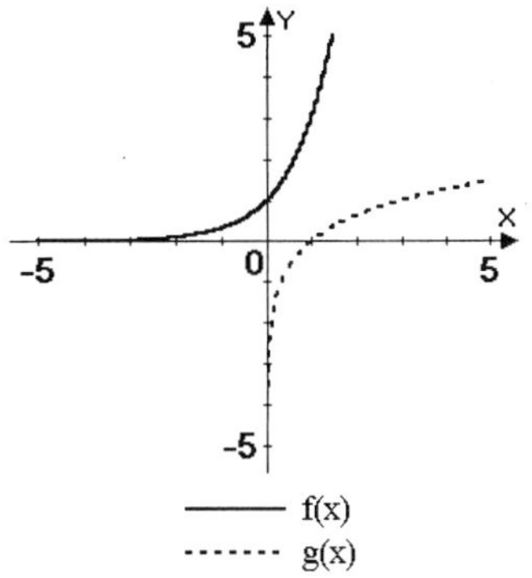


Domain of *f*: $\{x \mid x \text{ is a real number}\}$ or $(-\infty, \infty)$.

Range of *f*: $\{y \mid y > 0\}$ or $(0, \infty)$

Domain of *g*: $\{x \mid x > 0\}$ or $(0, \infty)$

Range of *g*: $\{y \mid y \text{ is a real number}\}$ or $(-\infty, \infty)$.

**9.** Since $\ln e^x = x$, $\ln e^{5x} = 5x$.

**10.** $\log_b b = 1$ because $b^1 = b$.

**11.** $\log_6 1 = 0$ because $6^0 = 1$.

**12.** $f(x) = \log_5(x - 7)$

$$x - 7 > 0$$
$$x > 7$$

The domain of *f* is $\{x \mid x > 7\}$ or $(7, \infty)$.

**13.** $D = 10\log\frac{I}{I_0}$

$D = 10\log\frac{10^{12}I_0}{I_0}$

$= 10\log 10^{12} = 10(12) = 120$

The sound has a loudness of 120 decibels.

**14.** $\log_4(64x^5) = \log_4 64 + \log_4 x^5$

$= 3 + 5\log_4 x$

**15.** $\log_3\frac{\sqrt[3]{x}}{81} = \log_3\sqrt[3]{x} - \log_3 81$

$= \log_3 x^{\frac{1}{3}} - 4 = \frac{1}{3}\log_3 x - 4$

**16.** $6\log x + 2\log y = \log x^6 + \log y^2$

$= \log x^6 y^2$

**17.** $\ln 7 - 3\ln x = \ln 7 - \ln x^3 = \ln\left(\frac{7}{x^3}\right)$

**18.** $\log_{15} 71 = \frac{\ln 71}{\ln 15} \approx 1.5741$

**19.** $3^{x-2} = 81$

$3^{x-2} = 3^4$

$x - 2 = 4$

$x = 6$

The solution is 6, and the solution set is $\{6\}$.

**20.** $5^x = 1.4$

$\ln 5^x = \ln 1.4$

$x\ln 5 = \ln 1.4$

$x = \frac{\ln 1.4}{\ln 5} \approx 0.21$

The solution is $\frac{\ln 1.4}{\ln 5} \approx 0.21$, and the solution set is $\left\{\frac{\ln 1.4}{\ln 5} \approx 0.21\right\}$.

**21.** $400e^{0.005x} = 1600$

$e^{0.005x} = \frac{1600}{400}$

$\ln e^{0.005x} = \ln 4$

$0.005x = \ln 4$

$x = \frac{\ln 4}{0.005} \approx 277.26$

The solution is $\frac{\ln 4}{0.005} \approx 277.26$, and the solution set is $\left\{\frac{\ln 4}{0.005} \approx 277.26\right\}$.

**22.** $\log_{25} x = \frac{1}{2}$

$x = 25^{\frac{1}{2}} = \sqrt{25} = 5$

The solution is 5, and the solution set is $\{5\}$.

**23.** $\log_6(4x - 1) = 3$

$4x - 1 = 6^3$

$4x - 1 = 216$

$4x = 217$

$x = \frac{217}{4}$

The solution is $\frac{217}{4}$, and the solution set is $\left\{\frac{217}{4}\right\}$.

**24.** $\log x + \log(x+15) = 2$

$$\log(x(x+15)) = 2$$
$$x(x+15) = 10^2$$
$$x^2 + 15 = 100$$
$$x^2 + 15 - 100 = 0$$
$$(x+20)(x-5) = 0$$

Apply the zero product principle.

$x + 20 = 0$ and $x - 5 = 0$

$x = -20$ $\quad$ $x = 5$

We disregard $-20$ because it would result in taking the logarithm of a negative number in the original equation. The solution is 5, and the solution set is $\{5\}$.

**25.** $2\ln 3x = 8$

$$\ln 3x = \frac{8}{2}$$
$$e^{\ln 3x} = e^4$$
$$3x = e^4$$
$$x = \frac{e^4}{3}$$

The solution is $\frac{e^4}{3}$, and the solution set is $\left\{\frac{e^4}{3}\right\}$.

**26. a.** $P(0) = 82.3e^{-0.002(0)}$

$$= 82.3e^0 = 82.3(1) = 82.3$$

In 2003, the population of Germany was 82.3 million.

**b.** The population of Germany is decreasing. We can tell the model has a negative, $k = -0.002$.

**c.** $81.5 = 82.3e^{-0.002t}$

$$\frac{81.5}{82.3} = e^{-0.002t}$$
$$\ln\frac{81.5}{82.3} = \ln e^{-0.002t}$$
$$\ln\frac{81.5}{82.3} = -0.002t$$
$$t = \frac{\ln\frac{81.5}{82.3}}{-0.002} \approx 5$$

The population of Germany will be 81.5 million approximately 5 years after 2003 in the year 2008.

**27.**

$$8000 = 4000\left(1 + \frac{0.05}{4}\right)^{4t}$$
$$\frac{8000}{4000} = (1 + 0.0125)^{4t}$$
$$2 = (1.0125)^{4t}$$
$$\ln 2 = \ln(1.0125)^{4t}$$
$$\ln 2 = 4t\ln(1.0125)$$
$$\frac{\ln 2}{4\ln(1.0125)} = \frac{4t\ln(1.0125)}{4\ln(1.0125)}$$
$$t = \frac{\ln 2}{4\ln(1.0125)} \approx 13.9$$

It will take approximately 13.9 years for the money to grow to $8000.

**28.** $2 = 1e^{r10}$

$$2 = e^{10r}$$
$$\ln 2 = \ln e^{10r}$$
$$\ln 2 = 10r$$
$$r = \frac{\ln 2}{10} \approx 0.069$$

The money will double in 10 years with an interest rate of approximately 6.9%.

**29.** Substitute $A_0 = 509$, $A = 729$, and $t = 2000 - 1990 = 10$ into the general growth function to determine the growth rate $k$:

$$A = A_0 e^{kt}$$

$$729 = 509e^{k(10)}$$

$$\frac{729}{509} = e^{10k}$$

$$\ln\frac{729}{509} = \ln e^{10k}$$

$$\ln\frac{729}{509} = 10k$$

$$k = \frac{\ln\frac{729}{509}}{10} \approx 0.036$$

The exponential growth function is $A = 484e^{0.005t}$

**30.**

$$A = A_0 e^{-0.000121t}$$

$$5 = 100e^{-0.000121t}$$

$$\frac{5}{100} = e^{-0.000121t}$$

$$\ln 0.05 = \ln e^{-0.000121t}$$

$$\ln 0.05 = -0.000121t$$

$$t = \frac{\ln 0.05}{-0.000121} \approx 24758$$

The man died approximately 24,758 years ago.

**31.** Plot the ordered pairs.

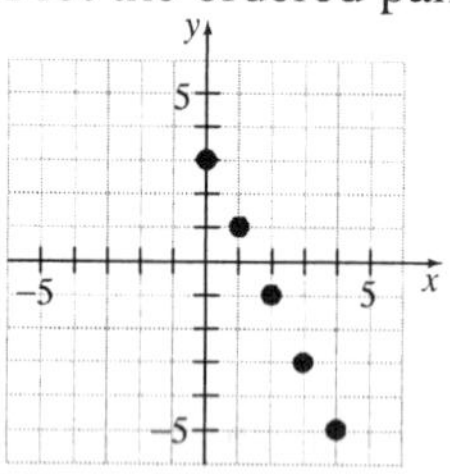


The values appear to belong to a linear function.

**32.** Plot the ordered pairs.

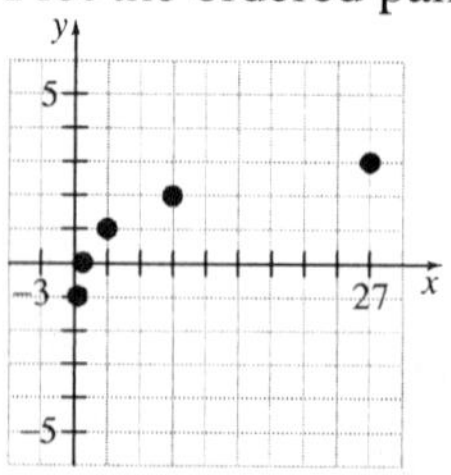


The values appear to belong to a logarithmic function.

**33.** Plot the ordered pairs.

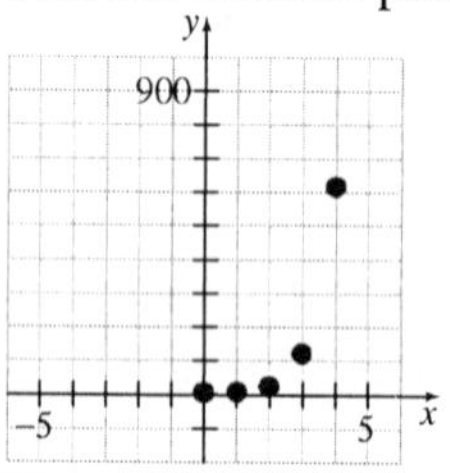


The values appear to belong to an exponential function.

**34.** Plot the ordered pairs.

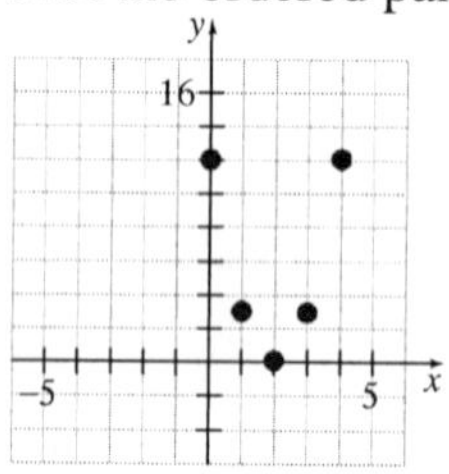


The values appear to belong to a quadratic function.

**35.**

$$y = 96(0.38)^x$$

$$y = 96e^{(\ln 0.38)x}$$

$$y = 96e^{-0.968x}$$

## Cumulative Review Exercises (Chapters 1 – 9)

**1.** $9(x-1)=1+3(x-2)$

$9x-9=1+3x-6$

$9x-9=3x-5$

$6x-9=-5$

$6x=4$

$x=\frac{4}{6}=\frac{2}{3}$

The solution is $\frac{2}{3}$, and the solution set is $\left\{\frac{2}{3}\right\}$.

**2.** $3x+4y=-7$

$x-2y=-9$

Multiply the second equation by 2 and add the result to the first equation:

$3x+4y=-7$

$2x-4y=-18$

$5x \quad =-25$

$x=-5$

Back substitute into the second equation:

$-5-2y=-9$

$-2y=-4$

$y=2$

The solution is $(-5,2)$, and the solution set is $\{(-5,2)\}$.

**3.** $x-y+3z=-9$

$2x+3y-z=16$

$5x+2y-z=15$

Multiply the second equation by 3 and add to the first equation to eliminate $z$.

$x-y+3z=-9$

$6x+9y-3z=48$

$7x+8y \quad =39$

Multiply the second equation by $-1$ and add to the third equation.

$-2x-3y+z=-16$

$5x+2y-z=15$

$3x-y \quad =-1$

We now have a system of two equations in two variables.

$7x+8y=39$

$3x-y=-1$

Multiply the second equation by 8 and add to the second equation.

$7x+8y=39$

$24x-8y=-8$

$31x \quad =31$

$x=1$

Back-substitute 1 for $x$ to find $y$.

$3x-y=-1$

$3(1)-y=-1$

$3-y=-1$

$-y=-4$

$y=4$

Back-substitute 1 for $x$ and 4 for $y$ to find $z$.

$x-y+3z=-9$

$1-4+3z=-9$

$-3+3z=-9$

$3z=-6$

$z=-2$

The solution is $(1,4,-2)$, and the solution set is $\{(1,4,-2)\}$.

**4.** $7x+18 \le 9x-2$

$-2x+18 \le -2$

$-2x \le -20$

$\dfrac{-2x}{-2} \ge \dfrac{-20}{-2}$

$x \ge 10$

The solution set is $\{x \mid x \ge 10\}$ or $[10,\infty)$.

**5.** $4x-3<13$ and $-3x-4 \ge 8$

$4x<16$ $\quad$ $-3x \ge 12$

$x<4$ $\quad$ $x \le -4$

For a value to be in the solution set, it must be both less than 4 and less than or equal to $-4$. Now only values that are less than or equal to $-4$ meet both conditions. Therefore, the solution set is $\{x \mid x \le -4\}$ or $(-\infty,-4]$.

**6.** $2x+4>8$ or $x-7 \ge 3$

$2x>4$ $\quad$ $x \ge 10$

$x>2$

For a value to be in the solution set, it must either one of the conditions. Now, all numbers that are greater than or equal to 10 are also greater than 2. Therefore, the solution set is $\{x \mid x>2\}$ or $(2,\infty)$.

**7.** $|2x-1|<5$

$-5<2x-1<5$

$-4<2x<6$

$-2<x<3$

The solution set is $\{x \mid -2<x<3\}$ or $(-2,3)$.

**8.** $\left|\dfrac{2}{3}x-4\right| = 2$

$\dfrac{2}{3}x-4=-2$ or $\dfrac{2}{3}x-4=2$

$3\left(\dfrac{2}{3}x-4\right)=3(-2)$ $\quad$ $3\left(\dfrac{2}{3}x-4\right)=3(2)$

$2x-12=-6$ $\quad$ $2x-12=6$

$2x=6$ $\quad$ $2x=18$

$x=3$ $\quad$ $x=9$

The solutions are 3 and 9, and the solution set is $\{3,9\}$.

**9.**

$$\frac{4}{x-3}-\frac{6}{x+3}=\frac{24}{x^2-9}$$

$$\frac{4}{x-3}-\frac{6}{x+3}=\frac{24}{(x-3)(x+3)}$$

$$(x-3)(x+3)\left[\frac{4}{x-3}-\frac{6}{x+3}\right]=(x-3)(x+3)\left[\frac{24}{(x-3)(x+3)}\right]$$

$$4(x+3)-6(x-3)=24$$

$$4x+12-6x+18=24$$

$$-2x+30=24$$

$$-2x=-6$$

$$x=3$$

We disregard 3 since it cause a result of 0 in the denominator of a fraction. Thus, the equation has no solution. The solution set is $\varnothing$ or $\{\ \}$.

**10.** $\sqrt{x+4}-\sqrt{x-3}=1$

$$\sqrt{x+4}=\sqrt{x-3}+1$$

$$\left(\sqrt{x+4}\right)^2=\left(\sqrt{x-3}+1\right)^2$$

$$x+4=(x-3)+2\sqrt{x-3}+1$$

$$x+4=x-2+2\sqrt{x-3}$$

$$6=2\sqrt{x-3}$$

$$3=\sqrt{x-3}$$

$$(3)^2=\left(\sqrt{x-3}\right)^2$$

$$9=x-3$$

$$12=x$$

Since we square both sides of the equation, we must check to make sure 12 is not extraneous:

$$\sqrt{12+4}-\sqrt{12-3}=1$$

$$\sqrt{16}-\sqrt{9}=1$$

$$4-3=1$$

$$1=1 \quad \text{True}$$

Thus, the solution is 12, and the solution set is $\{12\}$.

**11.**

$$2x^2=5-4x$$

$$2x^2+4x-5=0$$

Apply the quadratic formula:
$a=2 \quad b=4 \quad c=-5$

$$x=\frac{-4\pm\sqrt{4^2-4(2)(-5)}}{2(2)}$$

$$=\frac{-4\pm\sqrt{56}}{4}$$

$$=\frac{-4\pm\sqrt{4\cdot 14}}{4}$$

$$=\frac{-4\pm 2\sqrt{14}}{4}=\frac{-2\pm\sqrt{14}}{2}$$

The solutions are $\frac{-2\pm\sqrt{14}}{2}$, and the solution set is $\left\{\frac{-2\pm\sqrt{14}}{2}\right\}$.

**12.** $x^{\frac{2}{3}} - 5x^{\frac{1}{3}} + 6 = 0$

$\left(x^{\frac{1}{3}}\right)^2 - 5x^{\frac{1}{3}} + 6 = 0$

Let $t = x^{\frac{1}{3}}$.

$t^2 - 5t + 6 = 0$

$(t-3)(t-2) = 0$

Apply the zero-product principle:

$t-3=0$ or $t-2=0$

$t=3$ $\quad$ $t=2$

Substitute $x^{\frac{1}{3}}$ back in for $t$.

$x^{\frac{1}{3}} = 3$ or $x^{\frac{1}{3}} = 2$

$\left(x^{\frac{1}{3}}\right)^3 = 3^3$ $\quad$ $\left(x^{\frac{1}{3}}\right)^3 = 2^3$

$x = 27$ $\quad$ $x = 8$

The solutions are 8 and 27, and the solution set is $\{8, 27\}$.

**13.** $2x^2 + x - 6 \le 0$

Solve the related quadratic equation.

$2x^2 + x - 6 = 0$

$(2x-3)(x+2) = 0$

Apply the zero product principle.

$2x-3=0$ or $x+2=0$

$2x = 3$ $\quad$ $x=-2$

$x = \frac{3}{2}$

The boundary points are $-2$ and $\frac{3}{2}$.

| Test Interval | Test No. | Test | Conclusion |
|---|---|---|---|
| $(-\infty,-2)$ | $-3$ | $2(-3)^2 + (-3) - 6 \le 0$<br>$9 \le 0$,<br>False | $(-\infty,-2)$ does not belong to the solution set. |
| $\left(-2,\frac{3}{2}\right)$ | 0 | $2(0)^2 + 0 - 6 \le 0$<br>$-6 \le 0$,<br>True | $\left(-2,\frac{3}{2}\right)$ belongs to the solution set. |
| $\left(\frac{3}{2},\infty\right)$ | 2 | $2(2)^2 + 2 - 6 \le 0$<br>$4 \le 0$,<br>False | $\left(\frac{3}{2},\infty\right)$ does not belong to the solution set. |

The solution set is $\left[-2,\frac{3}{2}\right]$ or $\left\{x \middle| -2 \le x \le \frac{3}{2}\right\}$.

**14.** $\log_8 x + \log_8(x+2) = 1$

$\log_8\left[x(x+2)\right] = 1$

$\log_8\left[x^2 + 2x\right] = 1$

$x^2 + 2x = 8^1$

$x^2 + 2x - 8 = 0$

$(x+4)(x-2) = 0$

Apply the zero product principle.

$x+4=0$ and $x-2=0$

$x=-4$ $\quad$ $x=2$

We disregard $-4$ because it would result in taking the logarithm of a negative number in the original equation. The solution is 2, and the solution set is $\{2\}$.

**15.** $5^{2x+3} = 125$

$5^{2x+3} = 5^3$

$2x+3 = 3$

$2x = 0$

$x = 0$

The solution is 0, and the solution set is $\{0\}$.

**16.** $x - 3y = 6$

$$-3y = -x + 6$$

$$y = \frac{-x+6}{-3}$$

$$y = \frac{1}{3}x - 2$$

The slope is $m = \frac{1}{3}$ and the $y$-intercept is $b = -2$.

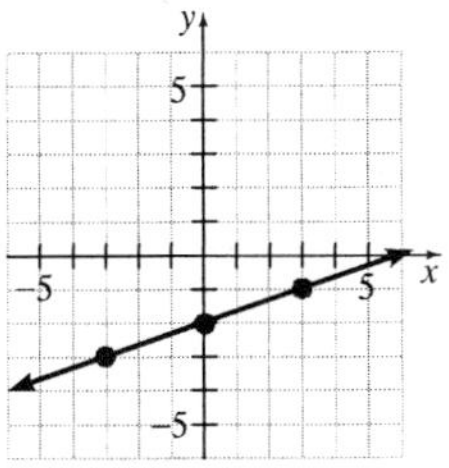

**17.** $f(x) = \frac{1}{2}x - 1$

This is a linear function with slope $m = \frac{1}{2}$ and $y$-intercept $b = -1$.

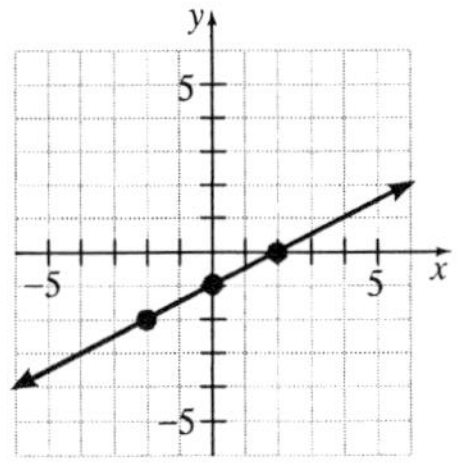

**18.** $3x - 2y > -6$

First, graph the equation $3x - 2y = -6$ as a dashed line.

$$3x - 2y = -6$$

$$-2y = -3x - 6$$

$$y = \frac{-3x-6}{-2}$$

$$y = \frac{3}{2}x + 3$$

The slope is $m = \frac{3}{2}$ and the $y$-intercept is $b = 3$. Next, use the origin as a test point.

$$3(0) - 2(0) > -6$$

$$0 > -6$$

This is a true statement. This means that the point, $(0,0)$, will fall in the shaded half-plane.

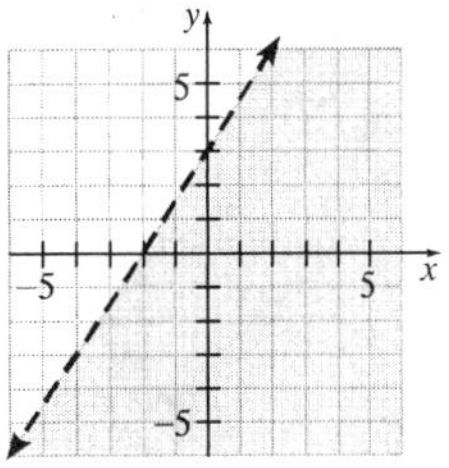

**19.** $f(x) = -2(x-3)^2 + 2$

Since $a = -2$ is negative, the parabola opens downward. The vertex of the parabola is $(h,k) = (3,2)$. Replace $f(x)$ with 0 to find $x$–intercepts.

$$0 = -2(x-3)^2 + 2$$

$$2(x-3)^2 = 2$$

$$(x-3)^2 = 1$$

$$x - 3 = \pm 1$$

$$x = 3 \pm 1$$

$$x = 3 - 1 \text{ or } 3 + 1$$

$$x = 2 \text{ or } 4$$

The $x$–intercepts are $-1$ and 4.

Set $x = 0$ to obtain the $y$–intercept.

$$f(0) = -2(0-3)^2 + 2$$

$$= -2(-3)^2 + 2$$

$$= -2(9) + 2 = -18 + 2 = -16$$

The $y$-intercept is $-16$.

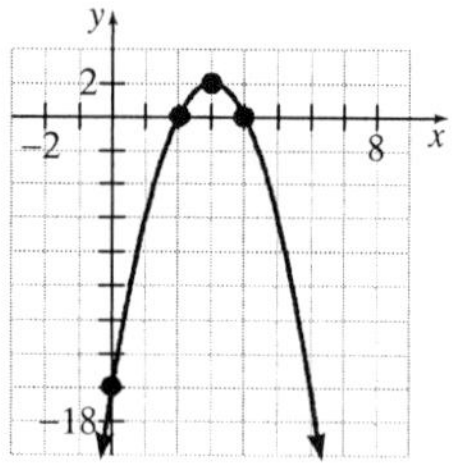


**20.** $y = \log_2 x$

| $x$ | $y = \log_2 x$ |
|---|---|
| $\frac{1}{4}$ | $-2$ |
| $\frac{1}{2}$ | $-1$ |
| 1 | 0 |
| 2 | 1 |
| 4 | 2 |

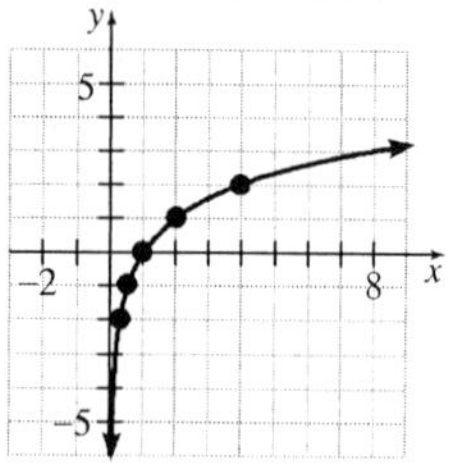


**21.** $$4[2x-6(x-y)] = 4(2x-6x+6y)$$
$$= 4(-4x+6y)$$
$$= -16x+24y$$

**22.** $$(-5x^3y^2)(4x^4y^{-6}) = -20x^{3+4}y^{2+(-6)}$$
$$= -20x^7y^{-4}$$
$$= -\frac{20x^7}{y^4}$$

**23.** $$(8x^2-9xy-11y^2)-(7x^2-4xy+5y^2)$$
$$= 8x^2-9xy-11y^2-7x^2+4xy-5y^2$$
$$= x^2-5xy-16y^2$$

**24.** $$(3x-1)(2x+5) = 6x^2+15x-2x-1$$
$$= 6x^2+13x-1$$

**25.** $$(3x^2-4y)^2$$
$$= (3x^2)^2-2(3x^2)(4y)+(4y)^2$$
$$= 9x^4-24x^2y+16y^2$$

**26.** $$\frac{3x}{x+5}-\frac{2}{x^2+7x+10}$$
$$= \frac{3x}{x+5}-\frac{2}{(x+5)(x+2)}$$
$$= \frac{3x}{x+5}\cdot\frac{(x+2)}{(x+2)}-\frac{2}{(x+5)(x+2)}$$
$$= \frac{3x^2+6x}{(x+5)(x+2)}-\frac{2}{(x+5)(x+2)}$$
$$= \frac{3x^2+6x-2}{(x-5)(x+2)}$$

**27.** $$\frac{1-\frac{9}{x^2}}{1+\frac{3}{x}} = \frac{1-\frac{9}{x^2}}{1+\frac{3}{x}}\cdot\frac{x^2}{x^2}$$
$$= \frac{x^2-9}{x^2+3x}$$
$$= \frac{(x-3)(x+3)}{x(x+3)} = \frac{x-3}{x}$$

**28.** $$\frac{x^2-6x+8}{3x+9} \div \frac{x^2-4}{x+3}$$
$$= \frac{x^2-6x+8}{3x+9}\cdot\frac{x+3}{x^2-4}$$
$$= \frac{(x-4)(x-2)}{3(x+3)}\cdot\frac{x+3}{(x-2)(x+2)}$$
$$= \frac{x-4}{3(x+2)} \text{ or } \frac{x-4}{3x+6}$$

**29.**
$$\begin{aligned}\sqrt{5xy}\cdot\sqrt{10x^2y} &= \sqrt{50x^3y^2}\\ &= \sqrt{25\cdot 2\cdot x^2\cdot x\cdot y^2}\\ &= 5xy\sqrt{2x}\end{aligned}$$

**30.**
$$\begin{aligned}4\sqrt{72}-3\sqrt{50} &= 4\sqrt{36\cdot 2}-3\sqrt{25\cdot 2}\\ &= 4\cdot 6\sqrt{2}-3\cdot 5\sqrt{2}\\ &= 45\sqrt{2}-15\sqrt{2}\\ &= 9\sqrt{2}\end{aligned}$$

**31.**
$$\begin{aligned}(5+3i)(7-3i) &= 35-15i+21i-9i^2\\ &= 35+6i-9(-1)\\ &= 35+6i+9\\ &= 44+6i\end{aligned}$$

**32.**
$$\begin{aligned}81x^4-1 &= (9x^2+1)(9x^2-1)\\ &= (9x^2+1)(3x+1)(3x-1)\end{aligned}$$

**33.**
$$\begin{aligned}24x^3-22x^2+4x &= 2x(12x^2-11x+2)\\ &= 2x(4x-1)(3x-2)\end{aligned}$$

**34.**
$$\begin{aligned}&x^3+27y^3 = x^3+(3y)^3\\ &= (x+3y)\left[x^2-x(3y)+(3y)^2\right]\\ &= (x+3y)(x^2-3xy+9y^2)\end{aligned}$$

**35.**
$$\begin{aligned}(f-g)(x) &= f(x)-g(x)\\ &= (x^2+3x-15)-(x-2)\\ &= x^2+3x-15-x+2\\ &= x^2+2x-13\end{aligned}$$

$$\begin{aligned}(f-g)(5) &= 5^2+2\cdot 5-13\\ &= 25+10-13 = 22\end{aligned}$$

**36.**
$$\left(\frac{f}{g}\right)(x) = \frac{f(x)}{g(x)} = \frac{x^2+3x-15}{x-2}$$

The domain of $\frac{f}{g}$ is

$\{x \mid x \text{ is a real number and } x \neq 2\}$ or $(-\infty, 2)\cup(2,\infty)$.

**37.**
$$\begin{aligned}f(g(x)) &= [g(x)]^2+3[g(x)]-15\\ &= (x-2)^2+3(x-2)-15\\ &= x^2-4x+4+3x-6-15\\ &= x^2-x-17\end{aligned}$$

**38.**
$$\begin{aligned}g(f(x)) &= f(x)-2\\ &= (x^2+3x-15)-2\\ &= x^2+3-17\end{aligned}$$

**39.**
$$\begin{aligned}f(x) &= 7x-3\\ y &= 7x-3\end{aligned}$$

Interchange $x$ and $y$, and solve for $y$.

$$\begin{aligned}x &= 7y-3\\ x+3 &= 7y\\ \frac{x+3}{7} &= y\end{aligned}$$

Thus, $f^{-1}(x) = \frac{x+3}{7}$.

**40.**

| | | | | |
|---|---|---|---|---|
| $\underline{-2}\rfloor$ | 3 | −1 | 4 | 8 |
| | | −6 | 14 | −36 |
| | 3 | −7 | 18 | −28 |

$$\begin{aligned}&(3x^3-x^2+4x+8)\div(x+2)\\ &= 3x^2-7x+18-\frac{28}{x+2}\end{aligned}$$

**41.**
$$I = \frac{R}{R+r}$$
$$I(R+r) = R$$
$$IR + Ir = R$$
$$Ir = R - IR$$
$$Ir = (1-I)R$$
$$\frac{Ir}{1-I} = R \quad \text{or} \quad R = -\frac{Ir}{I-1}$$

**42.** $3x + y = 9$
$$y = -3x + 9$$
The line whose equation we want to find has a slope of $m = -3$, the same as that of the line above. Using this slope with the point through with the line passes, $(-2, 5)$, we first find the point-slope equation and then put it in slope-intercept form.
$$y - y_1 = m(x - x_1)$$
$$y - 5 = -3(x - (-2))$$
$$y - 5 = -3(x + 2)$$
$$y - 5 = -3x - 6$$
$$y = -3x - 1$$
The slope-intercept equation of the line is through $(-2, 5)$ and parallel to $3x + y = 9$ is $y = -3x - 1$.

**43.** $\begin{vmatrix} -2 & -4 \\ 5 & 7 \end{vmatrix} = -2(7) - 5(-4)$
$$= -14 + 20 = 6$$

**44.**
$$2\ln x - \frac{1}{2}\ln y = \ln x^2 - \ln y^{\frac{1}{2}}$$
$$= \ln\left(\frac{x^2}{y^{\frac{1}{2}}}\right) = \ln\left(\frac{x^2}{\sqrt{y}}\right)$$

**45.** $f(x) = \dfrac{x-2}{x^2 - 3x + 2}$

Since a denominator cannot equal zero, exclude from the domain all values which make $x^2 - 3x + 2 = 0$.
$$x^2 - 3x + 2 = 0$$
$$(x-2)(x-1) = 0$$
Apply the zero-product principle.
$x - 2 = 0$ or $x - 1 = 0$

$x = 2$ $\qquad x = 1$

The domain of *f* is $\{x \mid x$ is a real number and $x \neq 1$ and $x \neq 2\}$ or $(-\infty, 1) \cup (1, 2) \cup (2, \infty)$.

**46.** $f(x) = \ln(2x - 8)$

To find the domain, find all values of $x$ for which $2x - 8$ is greater than zero.
$$2x - 8 > 0$$
$$2x > 8$$
$$x > 4$$
The domain of *f* is $\{x \mid x > 4\}$ or $(4, \infty)$.

**47.** Let $x$ = the computer's original price.
$$x - 0.30x = 434$$
$$0.70x = 434$$
$$x = \frac{434}{0.70} = 620$$
The original price of the computer was \$620.

**48.** Let $x$ = the width of the rectangle;
$3x + 1$ = the length of the rectangle.

$$x(3x+1)=52$$
$$3x^2+x-52=0$$
$$(3x+13)(x-4)=0$$

Apply the zero-product principle:

$3x+13=0$ or $x-4=0$

$x=-\frac{13}{3}$ $\quad x=4$

Disregard $-\frac{13}{3}$ because the width of a rectangle cannot be negative. If $x=4$, then $3x+1=3(4)+1=13$. Thus, the length of the rectangle is 13 yards and the width is 4 yards.

**49.** Let $x$ = the amount invested at 12%;
$4000-x$ = the amount invested at 14%.

$$0.12x+0.14(4000-x)=508$$
$$0.12x+560-0.14x=508$$
$$-0.02x+560=508$$
$$-0.02x=-52$$
$$x=2600$$

$4000-x=4000-2600=1400$

Thus, \$2600 was invested at 12% and \$1400 was invested at 14%.

**50.**

$$A=Pe^{rt}$$
$$18,000=6000e^{r(10)}$$
$$3=e^{10r}$$
$$\ln 3=\ln e^{10r}$$
$$\ln 3=10r$$
$$r=\frac{\ln 3}{10}\approx 0.11$$

An interest rate of approximately 11% compounded continuously would be required for \$6000 to grow to \$18,000 in 10 years.

**51.** Because $I$ varies inversely as $R$, we have the following for a constant $k$:

$$I=\frac{k}{R}$$

Use the fact that $I = 5$ when $R = 22$ to find $k$:

$$5=\frac{k}{22}$$
$$k=22\cdot 5=110$$

Thus, the equation relating $I$ and $R$ is $I=\frac{110}{R}$.

If $R = 10$, then $I=\frac{110}{10}=11$.

A current of 11 amperes is required when the resistance is 10 ohms.

# Chapter 10
# Conic Sections and Systems of Nonlinear Equations

## 10.1 Exercise Set

**1.**
$$d = \sqrt{(14-2)^2 + (8-3)^2}$$
$$= \sqrt{12^2 + 5^2} = \sqrt{144 + 25}$$
$$= \sqrt{169} = 13$$
The distance is 13 units.

**3.**
$$d = \sqrt{(6-4)^2 + (3-1)^2}$$
$$= \sqrt{2^2 + 2^2} = \sqrt{4+4}$$
$$= \sqrt{8} = \sqrt{4 \cdot 2} = 2\sqrt{2} \approx 2.83$$
The distance is $2\sqrt{2}$ or 2.83 units.

**5.**
$$d = \sqrt{(-3-0)^2 + (4-0)^2}$$
$$= \sqrt{(-3)^2 + 4^2} = \sqrt{9+16}$$
$$= \sqrt{25} = 5$$
The distance is 5 units.

**7.**
$$d = \sqrt{(3-(-2))^2 + (-4-(-6))^2}$$
$$= \sqrt{5^2 + 2^2} = \sqrt{25+4}$$
$$= \sqrt{29} \approx 5.39$$
The distance is $\sqrt{29}$ or 5.39 units.

**9.**
$$d = \sqrt{(4-0)^2 + (1-(-3))^2}$$
$$= \sqrt{4^2 + 4^2} = \sqrt{16+16}$$
$$= \sqrt{32} = \sqrt{16 \cdot 2} = 4\sqrt{2} \approx 5.66$$
The distance is $4\sqrt{2}$ or 5.66 units.

**11.**
$$d = \sqrt{(3.5-(-0.5))^2 + (8.2-6.2)^2}$$
$$= \sqrt{4^2 + 2^2} = \sqrt{16+4}$$
$$= \sqrt{20} = \sqrt{4 \cdot 5} = 2\sqrt{5} \approx 4.47$$
The distance is $2\sqrt{5}$ or 4.47 units.

**13.**
$$d = \sqrt{\left(\sqrt{5}-0\right)^2 + \left(0-\left(-\sqrt{3}\right)\right)^2}$$
$$= \sqrt{\left(\sqrt{5}\right)^2 + \left(\sqrt{3}\right)^2} = \sqrt{5+3}$$
$$= \sqrt{8} = \sqrt{4 \cdot 2} = 2\sqrt{2} \approx 2.83$$
The distance is $2\sqrt{2}$ or 2.83 units.

**15.**
$$d = \sqrt{\left(3\sqrt{3}-\left(-\sqrt{3}\right)\right)^2 + \left(\sqrt{5}-4\sqrt{5}\right)^2}$$
$$= \sqrt{\left(4\sqrt{3}\right)^2 + \left(-3\sqrt{5}\right)^2}$$
$$= \sqrt{16 \cdot 3 + 9 \cdot 5} = \sqrt{48+45}$$
$$= \sqrt{93} \approx 9.64$$
The distance is $\sqrt{93}$ or 9.64 units.

**17.**
$$d = \sqrt{\left(\frac{7}{3}-\frac{1}{3}\right)^2 + \left(\frac{1}{5}-\frac{6}{5}\right)^2}$$
$$= \sqrt{\left(\frac{6}{3}\right)^2 + \left(-\frac{5}{5}\right)^2}$$
$$= \sqrt{2^2 + (-1)^2} = \sqrt{4+1}$$
$$= \sqrt{5} \approx 2.24$$
The distance is $\sqrt{5}$ or 2.24 units.

**19.**
$$\text{Midpoint} = \left(\frac{6+2}{2}, \frac{8+4}{2}\right)$$
$$= \left(\frac{8}{2}, \frac{12}{2}\right) = (4,6)$$
The midpoint is $(4,6)$.

**21.**
$$\text{Midpoint} = \left(\frac{-2+(-6)}{2}, \frac{-8+(-2)}{2}\right)$$
$$= \left(\frac{-8}{2}, \frac{-10}{2}\right) = (-4,-5)$$
The midpoint is $(-4,-5)$.

**23.**
$$\text{Midpoint} = \left(\frac{-3+6}{2}, \frac{-4+(-8)}{2}\right)$$
$$= \left(\frac{3}{2}, \frac{-12}{2}\right) = \left(\frac{3}{2}, -6\right)$$
The midpoint is $\left(\frac{3}{2}, -6\right)$.

**25.**
$$\text{Midpoint} = \left(\frac{-\frac{7}{2}+\left(-\frac{5}{2}\right)}{2}, \frac{\frac{3}{2}+\left(-\frac{11}{2}\right)}{2}\right)$$
$$= \left(\frac{-\frac{12}{2}}{2}, \frac{-\frac{8}{2}}{2}\right)$$
$$= \left(-\frac{12}{2}\cdot\frac{1}{2}, -\frac{8}{2}\cdot\frac{1}{2}\right)$$
$$= \left(-\frac{12}{4}, -\frac{8}{4}\right) = (-3,-2)$$
The midpoint is $(-3,-2)$.

**27.**
$$\text{Midpoint} = \left(\frac{8+(-6)}{2}, \frac{3\sqrt{5}+7\sqrt{5}}{2}\right)$$
$$= \left(\frac{2}{2}, \frac{10\sqrt{5}}{2}\right) = \left(1, 5\sqrt{5}\right)$$
The midpoint is $\left(1, 5\sqrt{5}\right)$.

**29.**
$$\text{Midpoint} = \left(\frac{\sqrt{18}+\sqrt{2}}{2}, \frac{-4+4}{2}\right)$$
$$= \left(\frac{\sqrt{9\cdot 2}+\sqrt{2}}{2}, \frac{0}{2}\right)$$
$$= \left(\frac{3\sqrt{2}+\sqrt{2}}{2}, 0\right)$$
$$= \left(\frac{4\sqrt{2}}{2}, 0\right) = \left(2\sqrt{2}, 0\right)$$
The midpoint is $\left(2\sqrt{2}, 0\right)$.

**31.**
$$(x-h)^2+(y-k)^2=r^2$$
$$(x-0)^2+(y-0)^2=7^2$$
$$x^2+y^2=49$$

**33.**
$$(x-h)^2+(y-k)^2=r^2$$
$$(x-3)^2+(y-2)^2=5^2$$
$$(x-3)^2+(y-2)^2=25$$

**35.**
$$(x-h)^2+(y-k)^2=r^2$$
$$(x-(-1))^2+(y-4)^2=2^2$$
$$(x+1)^2+(y-4)^2=4$$

**37.**

$$(x-h)^2+(y-k)^2=r^2$$
$$(x-(-3))^2+(y-(-1))^2=(\sqrt{3})^2$$
$$(x+3)^2+(y+1)^2=3$$

**39.**

$$(x-h)^2+(y-k)^2=r^2$$
$$(x-(-4))^2+(y-0)^2=10^2$$
$$(x+4)^2+y^2=100$$

**41.**

$$x^2+y^2=16$$
$$(x-0)^2+(y-0)^2=4^2$$

The center is $(0,0)$ and the radius is 4 units.

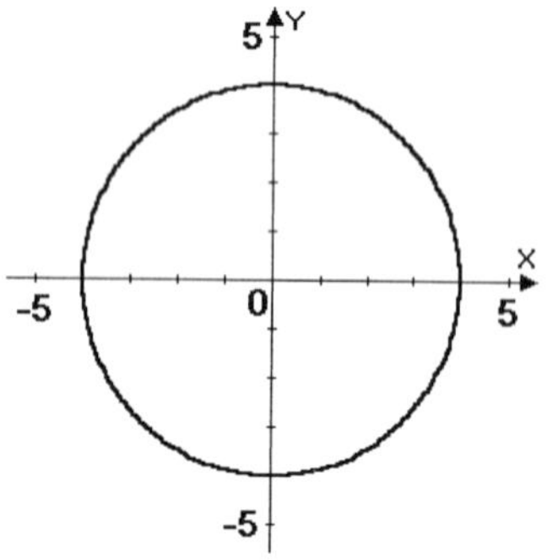


**43.**

$$(x-3)^2+(y-1)^2=36$$
$$(x-3)^2+(y-1)^2=6^2$$

The center is $(3,1)$ and the radius is 6 units.

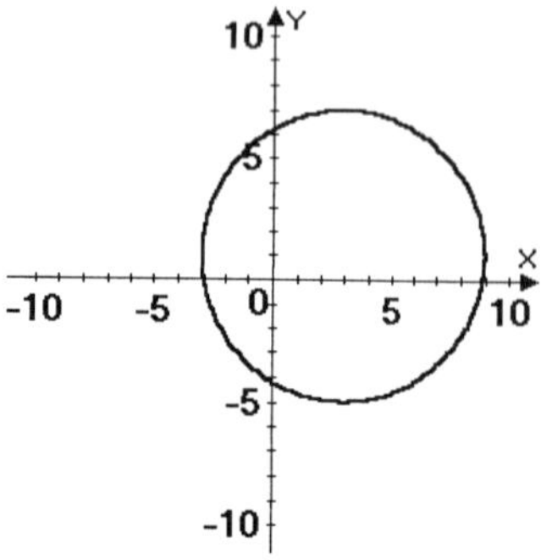


**45.**

$$(x+3)^2+(y-2)^2=4$$
$$(x-(-3))^2+(y-2)^2=2^2$$

The center is $(-3,2)$ and the radius is 2 units.

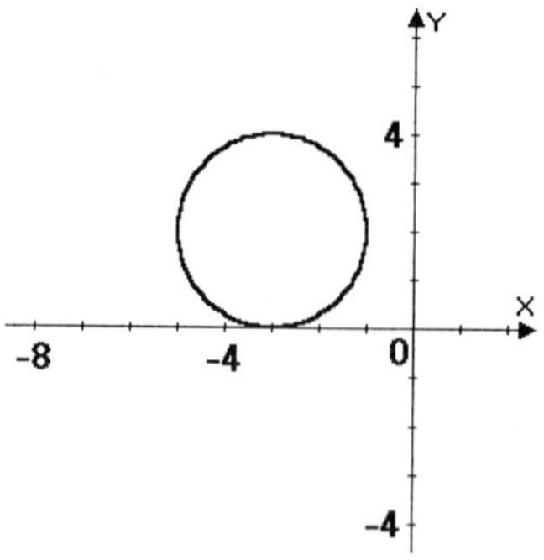


**47.**

$$(x+2)^2+(y+2)^2=4$$
$$(x-(-2))^2+(y-(-2))^2=2^2$$

The center is $(-2,-2)$ and the radius is 2 units.

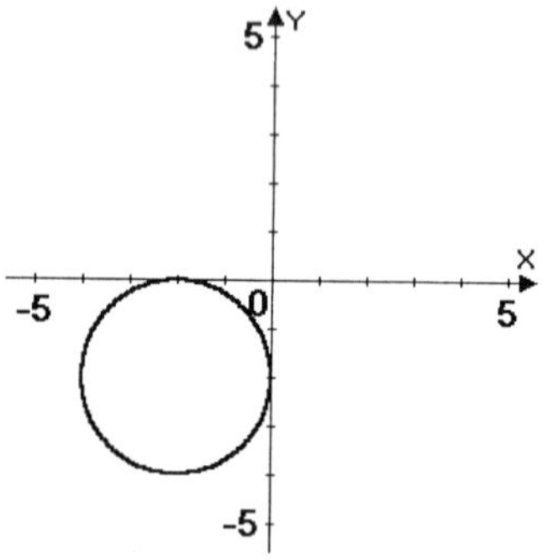


**49.**

$$x^2+y^2+6x+2y+6=0$$
$$(x^2+6x\quad)+(y^2+2y\quad)=-6$$

Complete the squares.

$$\left(\frac{b}{2}\right)^2=\left(\frac{6}{2}\right)^2=(3)^2=9$$
$$\left(\frac{b}{2}\right)^2=\left(\frac{2}{2}\right)^2=(1)^2=1$$

$$\left(x^2+6x+9\right)+\left(y^2+2y+1\right)=-6+9+1$$
T
$$(x+3)^2+(y+1)^2=4$$

he center is $(-3,-1)$ and the radius is 2 units.

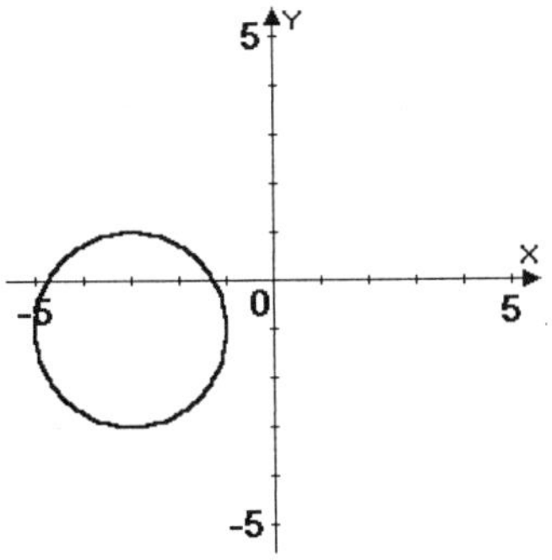


**51.** $$x^2+y^2-10x-6y-30=0$$

$$\left(x^2-10x\quad\right)+\left(y^2-6y\quad\right)=30$$

Complete the squares.

$$\left(\frac{b}{2}\right)^2=\left(\frac{-10}{2}\right)^2=(-5)^2=25$$

$$\left(\frac{b}{2}\right)^2=\left(\frac{-6}{2}\right)^2=(-3)^2=9$$

$$\left(x^2-10x+25\right)+\left(y^2-6y+9\right)=30+25+9$$

$$(x-5)^2+(y-3)^2=64$$

The center is $(5,3)$ and the radius is 8 units.

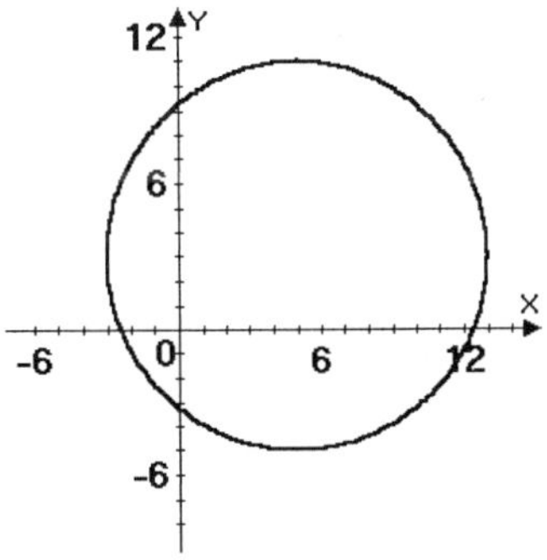


**53.** $$x^2+y^2+8x-2y-8=0$$

$$\left(x^2+8x\quad\right)+\left(y^2-2y\quad\right)=8$$

Complete the squares.

$$\left(\frac{b}{2}\right)^2=\left(\frac{8}{2}\right)^2=(4)^2=16$$

$$\left(\frac{b}{2}\right)^2=\left(\frac{-2}{2}\right)^2=(-1)^2=4$$

$$\left(x^2+8x+16\right)+\left(y^2-2y+1\right)=8+16+1$$

$$(x+4)^2+(y-1)^2=25$$

The center is $(-4,1)$ and the radius is 5 units.

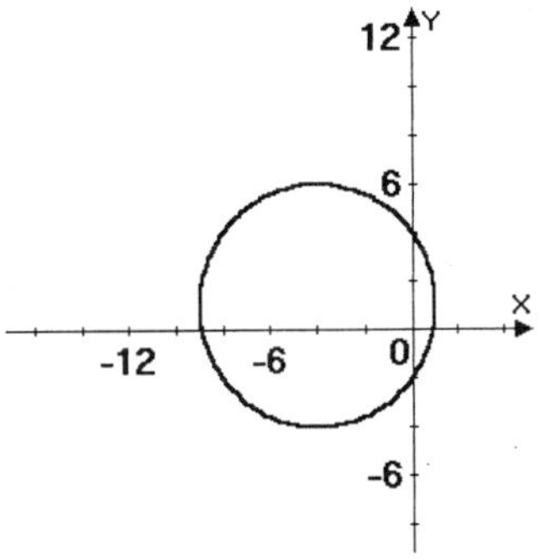


**55.** $$x^2-2x+y^2-15=0$$

$$\left(x^2-2x\quad\right)+y^2=15$$

Complete the squares.

$$\left(\frac{b}{2}\right)^2=\left(\frac{-2}{2}\right)^2=(-1)^2=1$$

$$\left(x^2-2x+1\right)+y^2=15+1$$

$$(x-1)^2+y^2=16$$

The center is $(1,0)$ and the radius is 4 units.

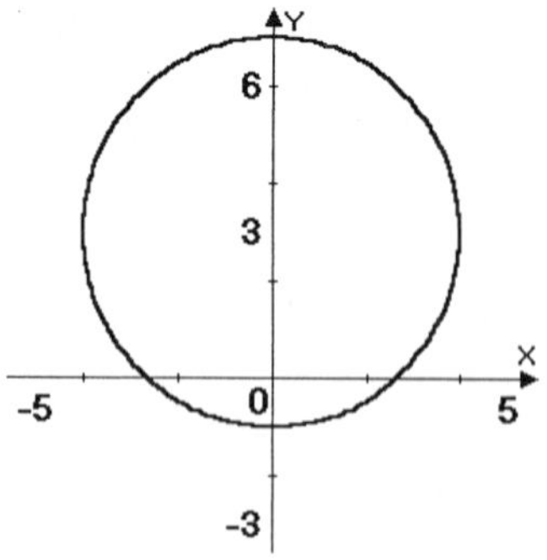


**57.**

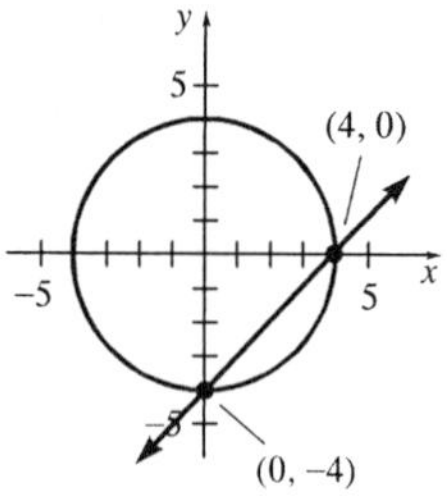


Intersection points: $(0,-4)$ and $(4,0)$

Check $(0,-4)$:

$0^2+(-4)^2=16 \qquad 0-(-4)=4$

$16=16$ true $\qquad 4=4$ true

Check $(4,0)$:

$4^2+0^2=16 \qquad 4-0=4$

$16=16$ true $\qquad 4=4$ true

The solution set is $\{(0,-4),(4,0)\}$.

**58.**

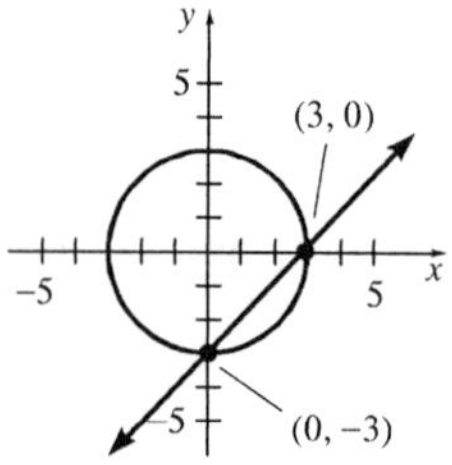


Intersection points: $(0,-3)$ and $(3,0)$

Check $(0,-3)$:

$0^2+(-3)^2=9 \qquad 0-(-3)=3$

$9=9$ true $\qquad 3=3$ true

Check $(3,0)$:

$3^2+0^2=9 \qquad 3-0=3$

$9=9$ true $\qquad 3=3$ true

The solution set is $\{(0,-3),(3,0)\}$.

**59.**

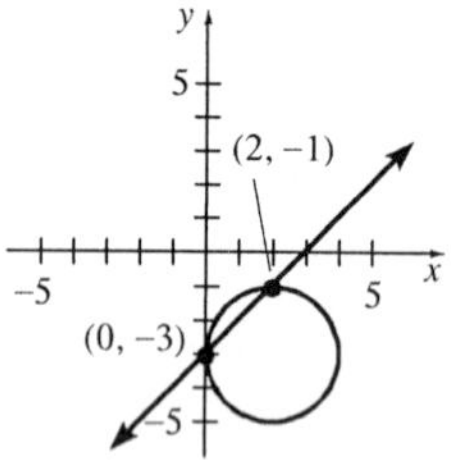


Intersection points: $(0,-3)$ and $(2,-1)$

Check $(0,-3)$:

$(0-2)^2+(-3+3)^2=9 \qquad -3=0-3$

$(-2)^2+0^2=4 \qquad -3=-3$ true

$4=4$

true

Check $(2,-1)$:

$(2-2)^2+(-1+3)^2=4 \qquad -1=2-3$

$0^2+2^2=4 \qquad -1=-1$ true

$4=4$

true

The solution set is $\{(0,-3),(2,-1)\}$.

**60.**

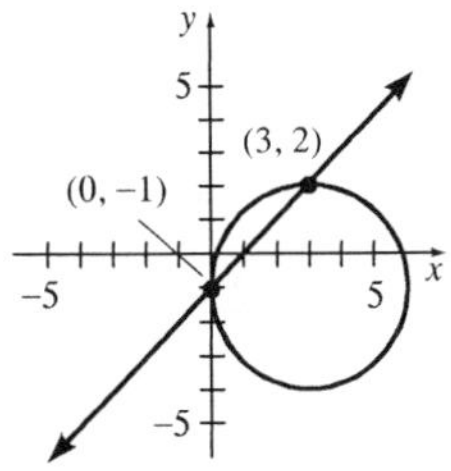


Intersection points: $(0,-1)$ and $(3,2)$

Check $(0,-1)$:

$$(0-3)^2+(-1+1)^2=9 \qquad -1=0-1$$
$$(-3)^2+0^2=9 \qquad -1=-1 \text{ true}$$
$$9=9$$
true

Check $(3,2)$:

$$(3-3)^2+(2+1)^2=9 \qquad 2=3-1$$
$$0^2+3^2=9 \qquad 2=2 \text{ true}$$
$$9=9$$
true

The solution set is $\{(0,-1),(3,2)\}$.

**61.** From the graph we can see that the center of the circle is at $(2,-1)$ and the radius is 2 units. Therefore, the equation is

$$(x-2)^2+(y-(-1))^2=2^2$$
$$(x-2)^2+(y+1)^2=4$$

**62.** From the graph we can see that the center of the circle is at $(3,-1)$ and the radius is 3 units. Therefore, the equation is

$$(x-3)^2+(y-(-1))^2=3^2$$
$$(x-3)^2+(y+1)^2=9$$

**63.** From the graph we can see that the center of the circle is at $(-3,-2)$ and the radius is 1 unit. Therefore, the equation is

$$(x-(-3))^2+(y-(-2))^2=1^2$$
$$(x+3)^2+(y+2)^2=1$$

**64.** From the graph we can see that the center of the circle is at $(-1,1)$ and the radius is 4 units. Therefore, the equation is

$$(x-(-1))^2+(y-1)^2=4^2$$
$$(x+1)^2+(y-1)^2=16$$

**65.** **a.** Since the line segment passes through the center, the center is the midpoint of the segment.

$$M=\left(\frac{x_1+x_2}{2},\frac{y_1+y_2}{2}\right)$$
$$=\left(\frac{3+7}{2},\frac{9+11}{2}\right)=\left(\frac{10}{2},\frac{20}{2}\right)$$
$$=(5,10)$$

The center is $(5,10)$.

**b.** The radius is the distance from the center to one of the points on the circle. Using the point $(3,9)$, we get:

$$d=\sqrt{(5-3)^2+(10-9)^2}$$
$$=\sqrt{2^2+1^2}=\sqrt{4+1}$$
$$=\sqrt{5}$$

The radius is $\sqrt{5}$ units.

**c.** $$(x-5)^2+(y-10)^2=\left(\sqrt{5}\right)^2$$
$$(x-5)^2+(y-10)^2=5$$

**66.** **a.** Since the line segment passes through the center, the center is the midpoint of the segment.

$$M=\left(\frac{x_1+x_2}{2},\frac{y_1+y_2}{2}\right)$$
$$=\left(\frac{3+5}{2},\frac{6+4}{2}\right)=\left(\frac{8}{2},\frac{10}{2}\right)$$
$$=(4,5)$$

The center is $(4,5)$.

**b.** The radius is the distance from the center to one of the points on the circle. Using the point $(3,6)$, we get:

$$d=\sqrt{(4-3)^2+(5-6)^2}$$
$$=\sqrt{1^2+(-1)^2}=\sqrt{1+1}$$
$$=\sqrt{2}$$

The radius is $\sqrt{2}$ units.

**c.**
$$(x-4)^2+(y-5)^2=\left(\sqrt{2}\right)^2$$
$$(x-4)^2+(y-5)^2=2$$

**67.** First find the distance from Bangkok to Phnom Penh.

$$d=\sqrt{(65-(-115))^2+(70-170)^2}$$
$$=\sqrt{180^2+(-100)^2}$$
$$=\sqrt{32400+10000}$$
$$=\sqrt{42400}\approx 205.9$$

The distance is approximately 205.9 miles.

$$t=\frac{d}{r}=\frac{205.9}{400}\approx 0.5$$

It will take approximately 0.5 hours or 30 minutes to make the flight.

**69.** If we place L.A. at the origin, then we want the equation of a circle with center at $(-2.4,-2.7)$ and radius 30.

$$(x-(-2.4))^2+(y-(-2.7))^2=30^2$$
$$(x+2.4)^2+(y+2.7)^2=900$$

**71.** Answers will vary.

**73.** Answers will vary.

**75.** Yes, according to the definition in the text, but the circle is reduced to the point $(3,5)$.

**77.**
$$x^2+y^2=25$$
$$y^2=25-x^2$$
$$y=\pm\sqrt{25-x^2}$$

Let $y_1=\sqrt{25-x^2}$ and $y_2=-\sqrt{25-x^2}$

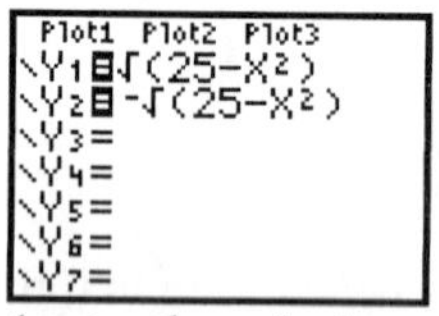


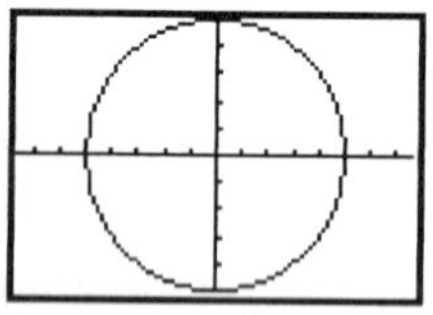

(note: the window was squared to avoid distorting the image)

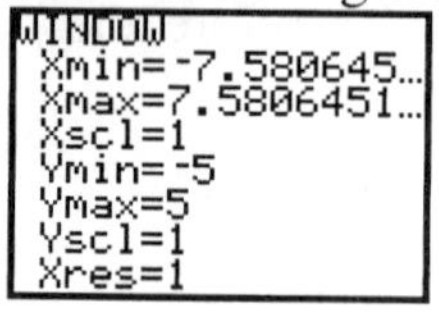


**79.**
$$x^2+10x+y^2-4y-20=0$$
$$(x^2+10x\quad)+(y^2-4y\quad)=20$$

Complete the squares.

$$\left(\frac{b}{2}\right)^2=\left(\frac{10}{2}\right)^2=(5)^2=25$$
$$\left(\frac{b}{2}\right)^2=\left(\frac{4}{2}\right)^2=(2)^2=4$$

$$\left(x^2+10x+25\right)+\left(y^2-4y+4\right)=20+25$$
$$(x+5)^2+(y-2)^2=49$$
$$(y-2)^2=49-(x+5)^2$$
$$y-2=\pm\sqrt{49-(x+5)^2}$$
$$y=2\pm\sqrt{49-(x+5)^2}$$

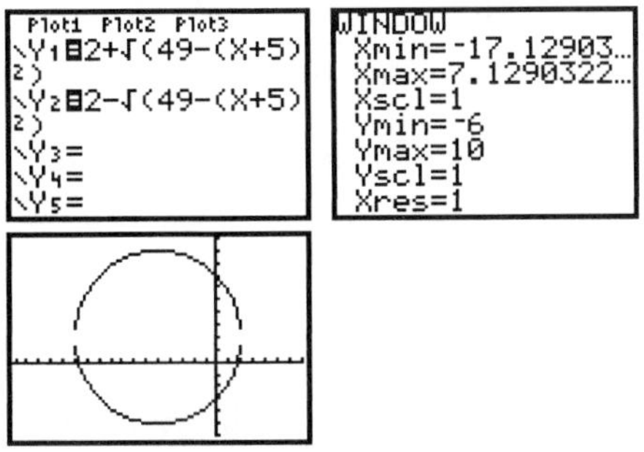


**81.** Distance from $A$ to $B$:

$$d=\sqrt{(3-1)^2+((3+d)-(1+d))^2}$$
$$=\sqrt{(2)^2+(3+d-1-d)^2}=\sqrt{4+(2)^2}$$
$$=\sqrt{4+4}=\sqrt{8}=\sqrt{4\cdot 2}=2\sqrt{2}$$

Distance from $B$ to $C$:

$$d=\sqrt{(6-3)^2+((6+d)-(3+d))^2}$$
$$=\sqrt{(3)^2+(6+d-3-d)^2}=\sqrt{9+(3)^2}$$
$$=\sqrt{9+9}=\sqrt{18}=\sqrt{9\cdot 2}=3\sqrt{2}$$

Distance from $A$ to $C$.

$$d=\sqrt{(6-1)^2+((6+d)-(1+d))^2}$$
$$=\sqrt{(5)^2+(6+d-1-d)^2}=\sqrt{25+(5)^2}$$
$$=\sqrt{25+25}=\sqrt{50}=\sqrt{25\cdot 2}=5\sqrt{2}$$

If the points are collinear,

$d_{AB}+d_{BC}=d_{AC}$.

$d_{AB}+d_{BC}=2\sqrt{2}+3\sqrt{2}=5\sqrt{2}$

Since this is the same as the distance from $A$ to $C$, we know that the points are collinear.

**83.**

$$d=\sqrt{(x_2-x_1)^2+(y_2-y_1)^2}$$
$$5=\sqrt{(x-2)^2+(2-(-1))^2}$$
$$5^2=(x-2)^2+3^2$$
$$25=(x-2)^2+9$$
$$16=(x-2)^2$$
$$\pm 4=x-2$$
$$2\pm 4=x$$

Therefore, $x=2-4=-2$ or $x=2+4=6$. There are two points with $y$-coordinate 2 whose distance is 5 units from the point $(2,-1)$. The two points are $(-2,2)$ and $(6,2)$.

**85.** The center of the circle with equation, $x^2+y^2=25$, is the point (0,0). First, find the slope of the line going through the center and the point, (3, −4).

$$m=\frac{y_2-y_1}{x_2-x_1}=\frac{-4-0}{3-0}=-\frac{4}{3}$$

Since the tangent line is perpendicular to the line going through the center and the point, (3, −4 ), we know that its slope will be $\frac{3}{4}$. We can now write the point-slope equation of the line.

$$y-(-4)=\frac{3}{4}(x-3)$$
$$y+4=\frac{3}{4}(x-3)$$

**86.**

$$f(g(x))=f(3x+4)=(3x+4)^2-2$$
$$=9x^2+24x+16-2$$
$$=9x^2+24x+14$$
$$g(f(x))=g\left(x^2-2\right)=3\left(x^2-2\right)+4$$
$$=3x^2-6+4=3x^2-2$$

**87.**

$$2x = \sqrt{7x-3} + 3$$
$$2x - 3 = \sqrt{7x-3}$$
$$(2x-3)^2 = 7x - 3$$
$$4x^2 - 12x + 9 = 7x - 3$$
$$4x^2 - 19x + 12 = 0$$
$$(4x-3)(x-4) = 0$$

Apply the zero product principle.

$4x - 3 = 0 \qquad x - 4 = 0$

$4x = 3 \qquad x = 4$

$x = \frac{3}{4}$

The solution $\frac{3}{4}$ does not check. The solution is 4 and the solution set is $\{4\}$.

**88.**

$$|2x-5| < 10$$
$$-10 < 2x - 5 < 10$$
$$-10 + 5 < 2x - 5 + 5 < 10 + 5$$
$$-5 < 2x < 15$$
$$-\frac{5}{2} < x < \frac{15}{2}$$

The solution set is $\left\{x \middle| -\frac{5}{2} < x < \frac{15}{2}\right\}$ or $\left(-\frac{5}{2}, \frac{15}{2}\right)$.

## 10.2 Exercise Set

**1.** $\frac{x^2}{16} + \frac{y^2}{4} = 1$

Because the denominator of the $x^2$ – term is greater than the denominator of the $y^2$ – term, the major axis is horizontal. Since $a^2 = 16$, $a = 4$ and the vertices are $(-4,0)$ and $(4,0)$. Since $b^2 = 4$, $b = 2$ and endpoints of the minor axis are $(0,-2)$ and $(0,2)$.

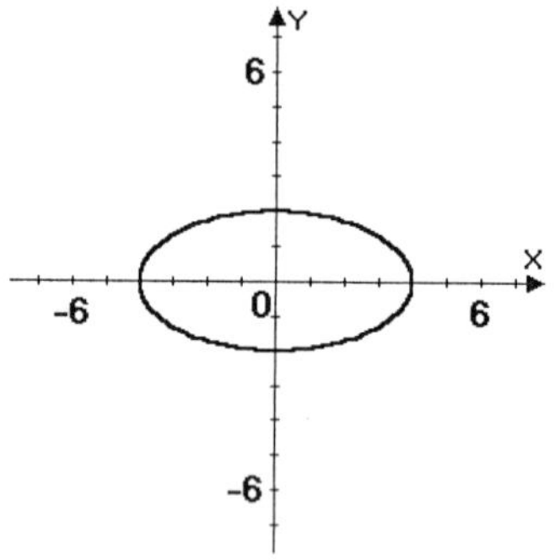


**3.** $\frac{x^2}{9} + \frac{y^2}{36} = 1$

Because the denominator of the $y^2$ – term is greater than the denominator of the $x^2$ – term, the major axis is vertical. Since $a^2 = 36$, $a = 6$ and the vertices are $(0,-6)$ and $(0,6)$. Since $b^2 = 9$, $b = 3$ and endpoints of the minor axis are $(-3,0)$ and $(3,0)$.

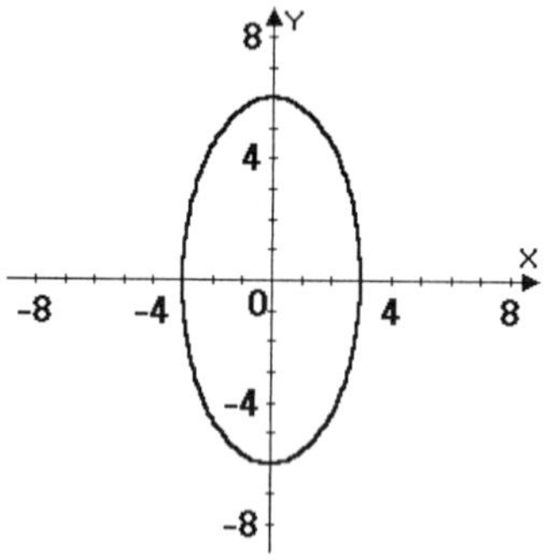

**5.** $\frac{x^2}{25}+\frac{y^2}{64}=1$

Because the denominator of the $y^2-$term is greater than the denominator of the $x^2-$term, the major axis is vertical. Since $a^2=64$, $a=8$ and the vertices are $(0,-8)$ and $(0,8)$. Since $b^2=25$, $b=5$ and endpoints of the minor axis are $(-5,0)$ and $(5,0)$.

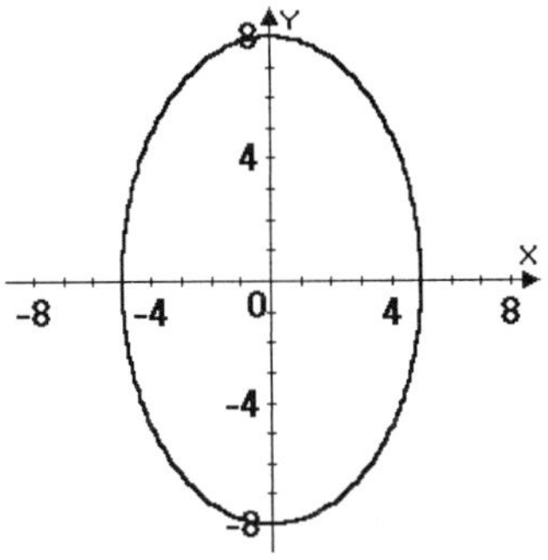


**7.** $\frac{x^2}{49}+\frac{y^2}{81}=1$

Because the denominator of the $y^2-$term is greater than the denominator of the $x^2-$term, the major axis is vertical. Since $a^2=81$, $a=9$ and the vertices are $(0,-9)$ and $(0,9)$. Since $b^2=49$, $b=7$ and endpoints of the minor axis are $(-7,0)$ and $(7,0)$.

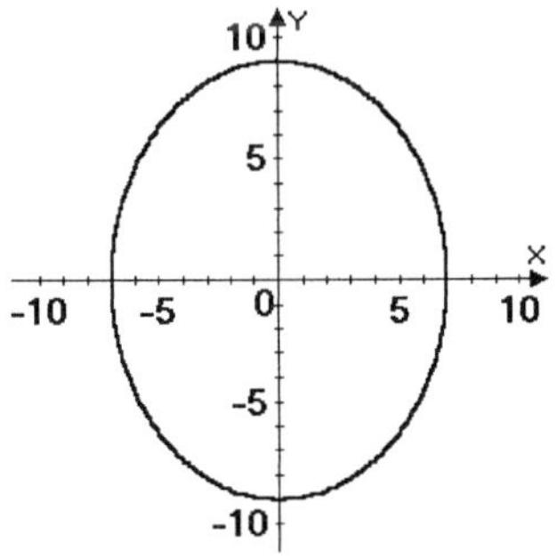


**9.** $25x^2+4y^2=100$

$$\frac{25x^2}{100}+\frac{4y^2}{100}=\frac{100}{100}$$

$$\frac{x^2}{4}+\frac{y^2}{25}=1$$

Because the denominator of the $y^2-$term is greater than the denominator of the $x^2-$term, the major axis is vertical. Since $a^2=25$, $a=5$ and the vertices are $(0,-5)$ and $(0,5)$. Since $b^2=4$, $b=2$ and endpoints of the minor axis are $(-2,0)$ and $(2,0)$.

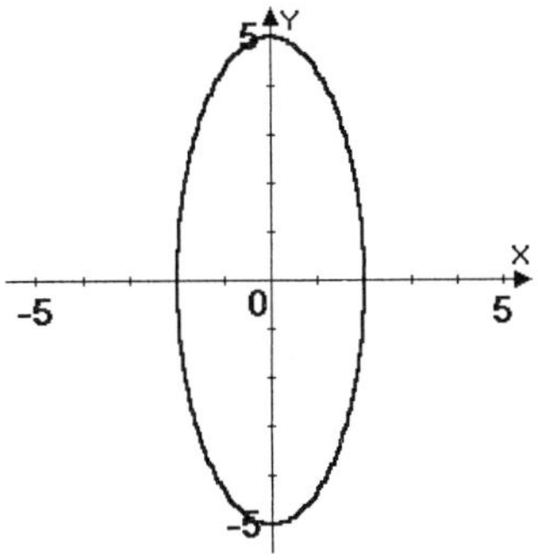


**11.** $4x^2+16y^2=64$

$$\frac{4x^2}{64}+\frac{16y^2}{64}=\frac{64}{64}$$

$$\frac{x^2}{16}+\frac{y^2}{4}=1$$

Because the denominator of the $x^2-$term is greater than the denominator of the $y^2-$term, the major axis is horizontal. Since $a^2=16$, $a=4$ and the vertices are $(-4,0)$ and $(4,0)$. Since $b^2=4$, $b=2$ and endpoints of the minor axis are $(0,-2)$ and $(0,2)$.

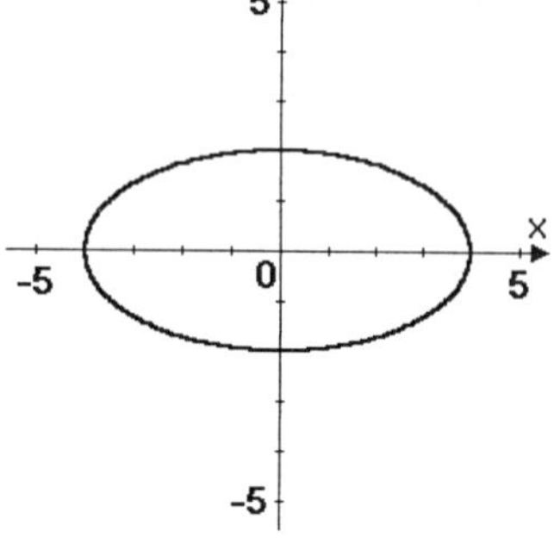


**13.** $25x^2 + 9y^2 = 225$

$$\frac{25x^2}{225} + \frac{9y^2}{225} = \frac{225}{225}$$

$$\frac{x^2}{9} + \frac{y^2}{25} = 1$$

Because the denominator of the $y^2$ – term is greater than the denominator of the $x^2$ – term, the major axis is vertical. Since $a^2 = 9$, $a = 3$ and the vertices are $(0,-3)$ and $(0,3)$. Since $b^2 = 25$, $b = 5$ and endpoints of the minor axis are $(-5,0)$ and $(5,0)$.

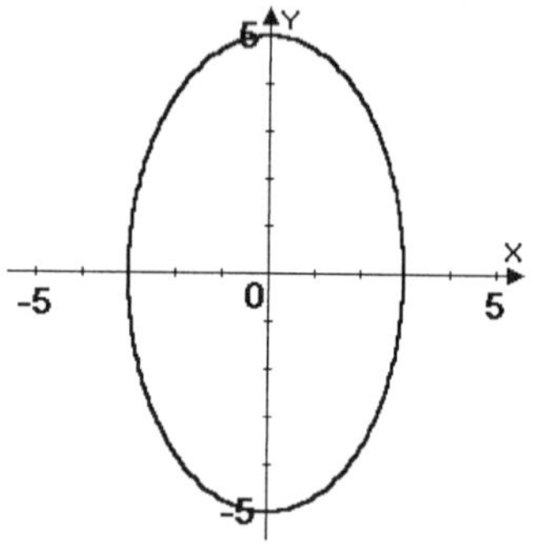


**15.** $x^2 + 2y^2 = 8$

$$\frac{x^2}{8} + \frac{2y^2}{8} = \frac{8}{8}$$

$$\frac{x^2}{8} + \frac{y^2}{4} = 1$$

Because the denominator of the $x^2$ – term is greater than the denominator of the $y^2$ – term, the major axis is horizontal. Since $a^2 = 8$, $a = \sqrt{8} = 2\sqrt{2}$ and the vertices are $(-2\sqrt{2},0)$ and $(2\sqrt{2},0)$. Since $b^2 = 4$, $b = 2$ and endpoints of the minor axis are $(0,-2)$ and $(0,2)$.

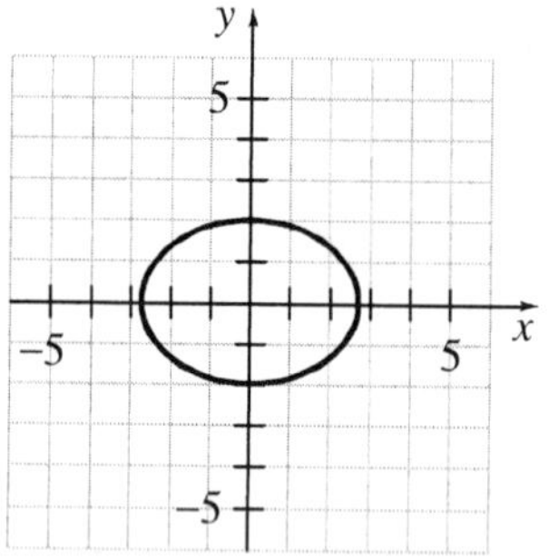


**17.** From the graph, we see that the center of the ellipse is the origin, the major axis is horizontal with $a = 2$, and $b = 1$.

$$\frac{x^2}{2^2} + \frac{y^2}{1^2} = 1$$

$$\frac{x^2}{4} + \frac{y^2}{1} = 1$$

**19.** From the graph, we see that the center of the ellipse is the origin, the major axis is vertical with $a = 2$, and $b = 1$.

$$\frac{x^2}{1^2} + \frac{y^2}{2^2} = 1$$

$$\frac{x^2}{1} + \frac{y^2}{4} = 1$$

**21.** $\frac{(x-2)^2}{9}+\frac{(y-1)^2}{4}=1$

The center of the ellipse is $(2,1)$. Because the denominator of the $x^2$ – term is greater than the denominator of the $y^2$ – term, the major axis is horizontal. Since $a^2=9$, $a=3$ and the vertices lie 3 units to the left and right of the center. Since $b^2=4$, $b=2$ and endpoints of the minor axis lie two units above and below the center.

| Center | Vertices | Endpoints of Minor Axis |
|---|---|---|
| $(2,1)$ | $(2-3,1)$ $=(-1,1)$ | $(2,1-2)$ $=(2,-1)$ |
| | $(2+3,1)$ $=(5,1)$ | $(2,1+2)$ $=(2,3)$ |

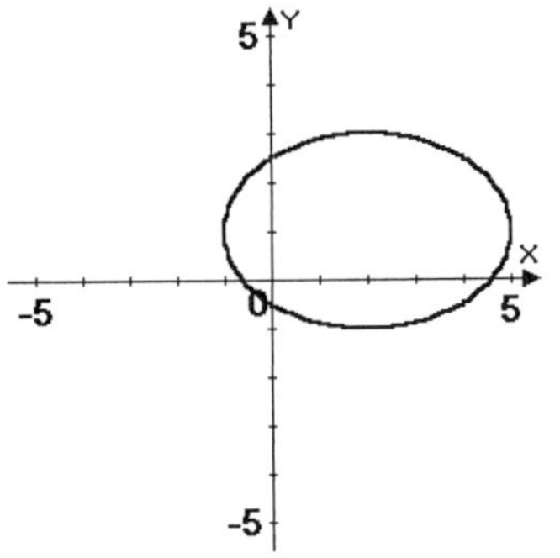


**23.** $(x+3)^2+4(y-2)^2=16$

$$\frac{(x+3)^2}{16}+\frac{4(y-2)^2}{16}=\frac{16}{16}$$

$$\frac{(x+3)^2}{16}+\frac{(y-2)^2}{4}=1$$

The center of the ellipse is $(-3,2)$. Because the denominator of the $x^2$ – term is greater than the denominator of the $y^2$ – term, the major axis is horizontal. Since $a^2=16$, $a=4$ and the vertices lie 4 units to the left and right of the center. Since $b^2=4$, $b=2$ and endpoints of the minor axis lie two units above and below the center.

| Center | Vertices | Endpoints of Minor Axis |
|---|---|---|
| $(-3,2)$ | $(-3-4,2)$ $=(-7,2)$ | $(-3,2-2)$ $=(-3,0)$ |
| | $(-3+4,2)$ $=(1,2)$ | $(-3,2+2)$ $=(-3,4)$ |

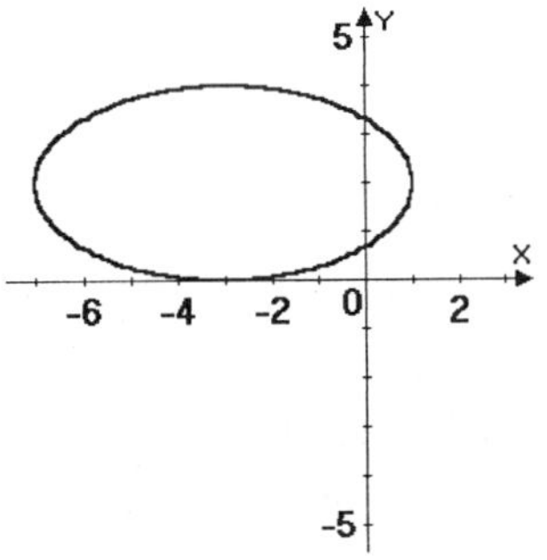


**25.** $\frac{(x-4)^2}{9}+\frac{(y+2)^2}{25}=1$

The center of the ellipse is $(4,-2)$. Because the denominator of the $y^2$ – term is greater than the denominator of the $x^2$ – term, the major axis is vertical. Since $a^2=25$, $a=5$ and the vertices lie 5 units to the above and below the center. Since $b^2=9$, $b=3$ and endpoints of the minor axis lie 3 units to the right and left of the center.

| Center | Vertices | Endpoints Minor Axis |
|---|---|---|
| $(4,-2)$ | $(4,-2-5)$ $=(4,-7)$ | $(4-3,-2)$ $=(1,-2)$ |
| | $(4,-2+5)$ $=(4,3)$ | $(4+3,-2)$ $=(7,-2)$ |

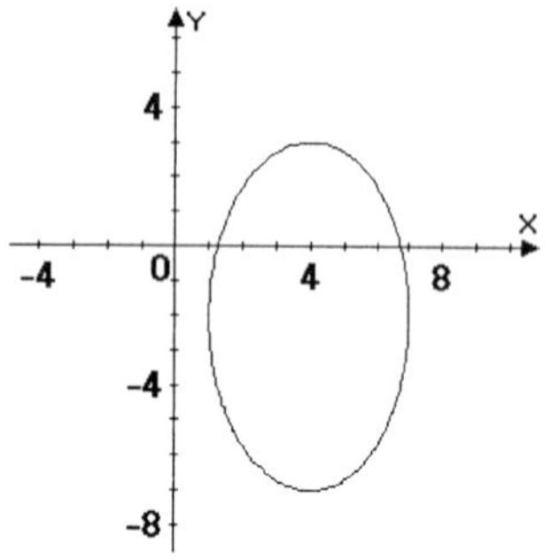


**27.** $\dfrac{x^2}{25}+\dfrac{(y-2)^2}{36}=1$

The center of the ellipse is $(0,2)$. Because the denominator of the $y^2-$term is greater than the denominator of the $x^2-$term, the major axis is vertical. Since $a^2=36$, $a=6$ and the vertices lie 6 units to the above and below the center. Since $b^2=25$, $b=5$ and endpoints of the minor axis lie 5 units to the left and right of the center.

| Center | Vertices | Endpoint Minor Axis |
|---|---|---|
| $(0,2)$ | $(0,2-6)$ $=(0,-4)$ | $(0-5,2)$ $=(-5,2)$ |
| | $(0,2+6)$ $=(0,8)$ | $(0+5,2)$ $=(5,2)$ |

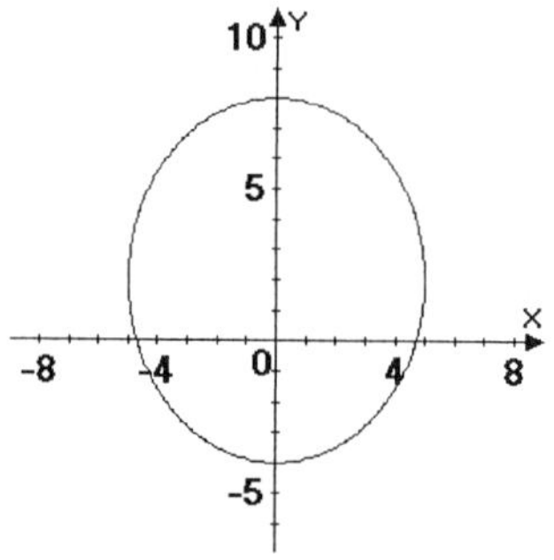


**29.** $\dfrac{(x+3)^2}{9}+(y-2)^2=1$

$\dfrac{(x+3)^2}{9}+\dfrac{(y-2)^2}{1}=1$

The center of the ellipse is $(-3,2)$. Because the denominator of the $x^2-$term is greater than the denominator of the $y^2-$term, the major axis is horizontal. Since $a^2=9$, $a=3$ and the vertices lie 3 units to the left and right of the center. Since $b^2=1$, $b=1$ and endpoints of the minor axis lie two units above and below the center.

| Center | Vertices | Endpoints of Minor Axis |
|---|---|---|
| $(-3,2)$ | $(-3+3,2)$ $=(0,2)$ | $(-3,2-1)$ $=(-3,1)$ |
| | $(-3-3,2)$ $=(-6,2)$ | $(-3,2+1)$ $=(-3,3)$ |

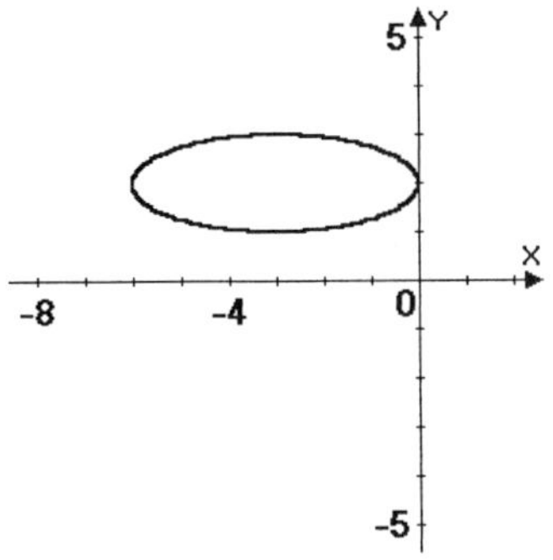


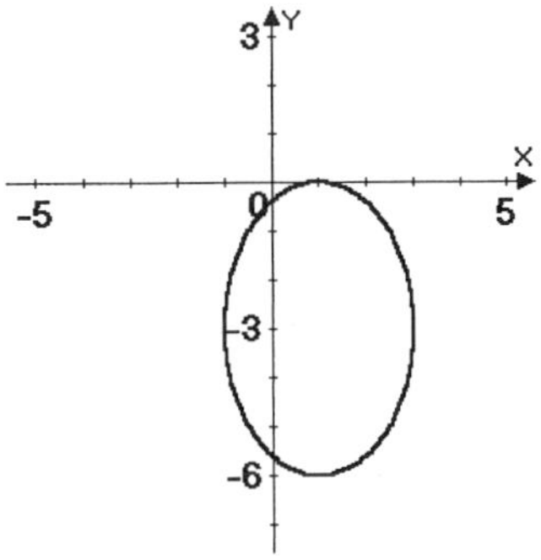


**31.** $9(x-1)^2+4(y+3)^2=36$

$$\frac{9(x-1)^2}{36}+\frac{4(y+3)^2}{36}=\frac{36}{36}$$

$$\frac{(x-1)^2}{4}+\frac{(y+3)^2}{9}=1$$

The center of the ellipse is $(1,-3)$. Because the denominator of the $y^2$ – term is greater than the denominator of the $x^2$ – term, the major axis is vertical. Since $a^2=9$, $a=3$ and the vertices lie 3 units to the above and below the center. Since $b^2=9$, $b=3$ and endpoints of the minor axis lie 3 units to the right and left of the center.

| Center | Vertices | Endpoints of Minor Axis |
|---|---|---|
| $(1,-3)$ | $(1,-3-3)$ $=(1,-6)$ | $(1-3,-3)$ $=(-2,-3)$ |
| | $(1,-3+3)$ $=(1,0)$ | $(1+3,-3)$ $=(4,-3)$ |

**33.** From the graph we see that the center of the ellipse is $(h,k)=(-1,1)$. We also see that the major axis is horizontal so we have $a>b$. The length of the major axis is 4 units, so $a=2$. The length of the minor axis is 2 units so $b=1$. Therefore, the equation of the ellipse is

$$\frac{(x-(-1))^2}{2^2}+\frac{(y-1)^2}{1^2}=1$$

$$\frac{(x+1)^2}{4}+\frac{(y-1)^2}{1}=1$$

**34.** From the graph we see that the center of the ellipse is $(h,k)=(-1,-1)$. We also see that the major axis is vertical so we have $b>a$. The length of the major axis is 4 units, so $b=2$. The length of the minor axis is 2 units so $a=1$. Therefore, the equation of the ellipse is

$$\frac{(x-(-1))^2}{1^2}+\frac{(y-(-1))^2}{2^2}=1$$

$$\frac{(x+1)^2}{1}+\frac{(y+1)^2}{4}=1$$

**35.** $x^2 + y^2 = 1$ $\quad x^2 + 9y^2 = 9$

$$\frac{x^2}{9} + \frac{9y^2}{9} = \frac{9}{9}$$

$$\frac{x^2}{9} + \frac{y^2}{1} = 1$$

The first equation is that of a circle with center at the origin and $r = 1$. The second equation is that of an ellipse with center at the origin, horizontal major axis of length 6 units $(a = 3)$, and vertical minor axis of length 2 units $(b = 1)$.

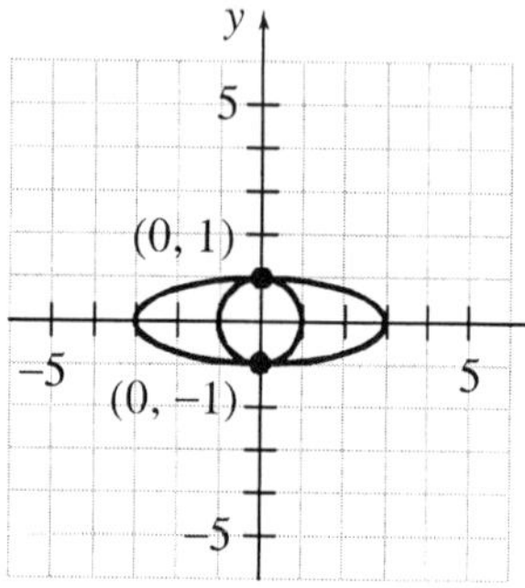


Check each intersection point.
The solutions set is $\{(0,-1),(0,1)\}$.

**36.** $x^2 + y^2 = 25$ $\quad 25x^2 + y^2 = 25$

$$\frac{25x^2}{25} + \frac{y^2}{25} = \frac{25}{25}$$

$$\frac{x^2}{1} + \frac{y^2}{25} = 1$$

The first equation is for a circle with center at the origin and $r = 5$. The second is for an ellipse with center at the origin, vertictal major axis of length 10 units $(b = 5)$, and horizontal minor axis of length 2 units $(a = 1)$.

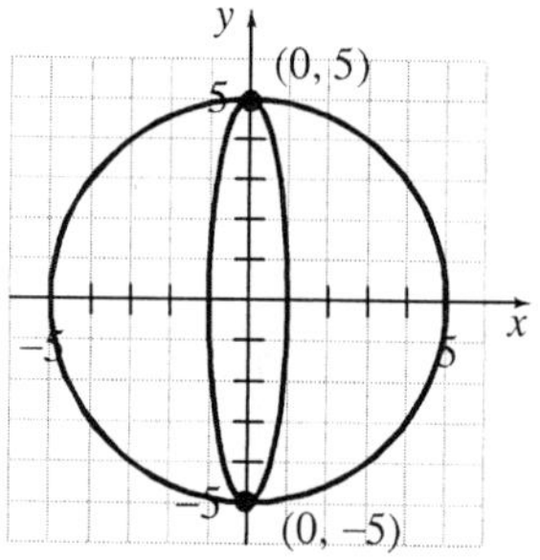


Check each intersection point.
The solutions set is $\{(0,-5),(0,5)\}$.

**37.** $\frac{x^2}{25} + \frac{y^2}{9} = 1$ $\quad y = 3$

The first equation is for an ellipse centered at the origin with horizontal major axis of length 10 units and vertical minor axis of length 6 units. The second equation is for a horizontal line with a y-intercept of 3.

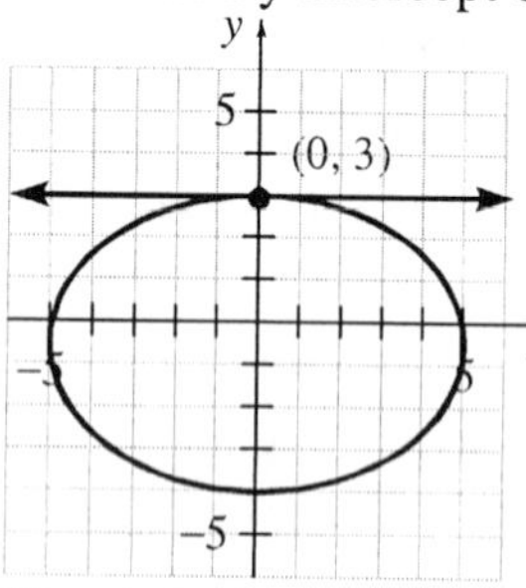


Check the intersection point.
The solution set is $\{(0,3)\}$.

**38.** $\frac{x^2}{4}+\frac{y^2}{36}=1 \qquad x=-2$

The first equation is for an ellipse centered at the origin with vertical major axis of length 12 units and horizontal minor axis of length 4 units. The second equation is for a horizontal line with an x-intercept of $-2$.

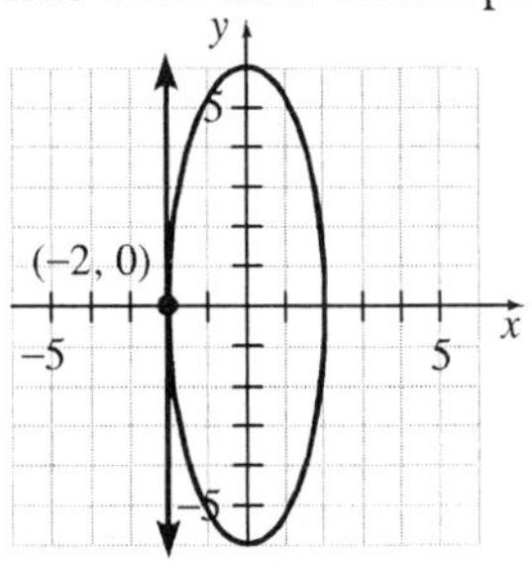


Check the intersection point.

The solution set is $\{(-2,0)\}$.

**39.** $4x^2+y^2=4 \qquad 2x-y=2$

$\frac{4x^2}{4}+\frac{y^2}{4}=\frac{4}{4} \qquad -y=-2x+2$

$\frac{x^2}{1}+\frac{y^2}{4}=1 \qquad y=2x-2$

The first equation is for an ellipse centered at the origin with vertical major axis of length 4 units ($b=2$) and horizontal minor axis of length 2 units $(a=1)$. The second equation is for a line with slope 2 and y-intercept $-2$.

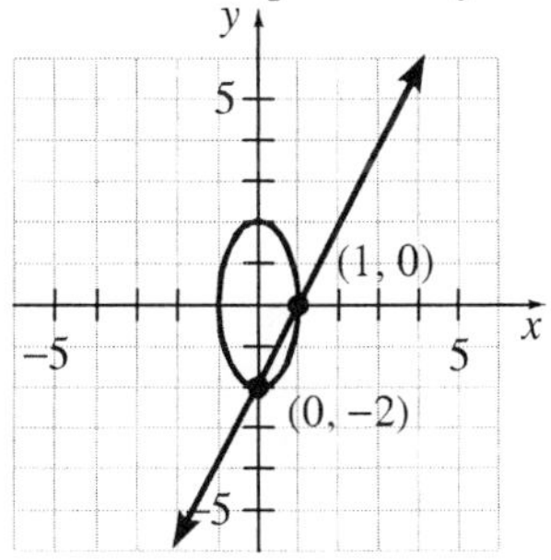


Check the intersection points.

The solution set is $\{(0,-2),(1,0)\}$.

**40.** $4x^2+y^2=4 \qquad x+y=3$

$\frac{4x^2}{4}+\frac{y^2}{4}=\frac{4}{4} \qquad y=-x+3$

$\frac{x^2}{1}+\frac{y^2}{4}=1$

The first equation is for an ellipse centered at the origin with vertical major axis of length 4 units ($b=2$) and horizontal minor axis of length 2 units $(a=1)$. The second equation is for a line with slope $-1$ and y-intercept 3.

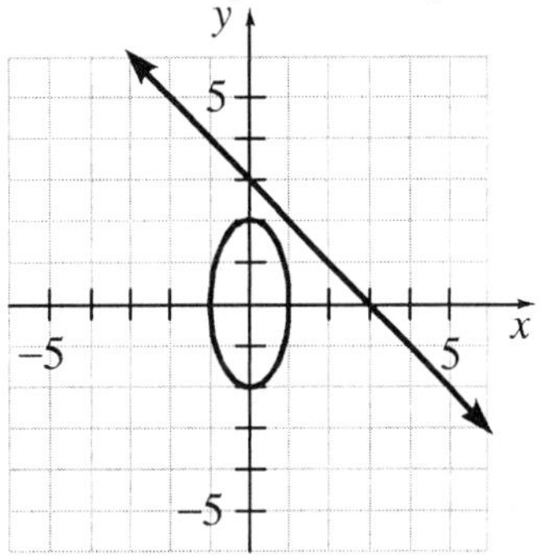


The two graphs never cross, so there are no intersection points.

The solution set is $\{\ \}$ or $\varnothing$.

**41.**

$$y=-\sqrt{16-4x^2}$$

$$y^2=\left(-\sqrt{16-4x^2}\right)^2$$

$$y^2=16-4x^2$$

$$4x^2+y^2=16$$

$$\frac{x^2}{4}+\frac{y^2}{16}=1$$

We want to graph the bottom half of an ellipse centered at the origin with a vertical major axis of length 8 units ($b=4$) and horizontal minor axis of length 4 units $(a=2)$.

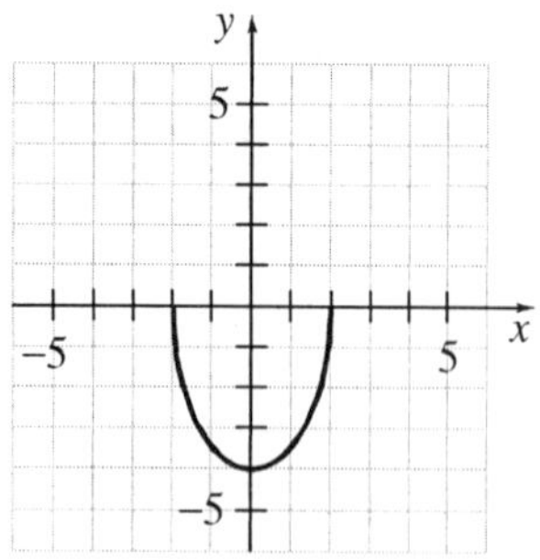


**42.** $y = -\sqrt{4-4x^2}$

$y^2 = \left(-\sqrt{4-4x^2}\right)^2$

$y^2 = 4-4x^2$

$4x^2 + y^2 = 4$

$\frac{x^2}{1} + \frac{y^2}{4} = 1$

We want to graph the bottom half of an ellipse centered at the origin with a vertical major axis of length 4 units ($b = 2$) and horizontal minor axis of length 2 units $(a = 1)$.

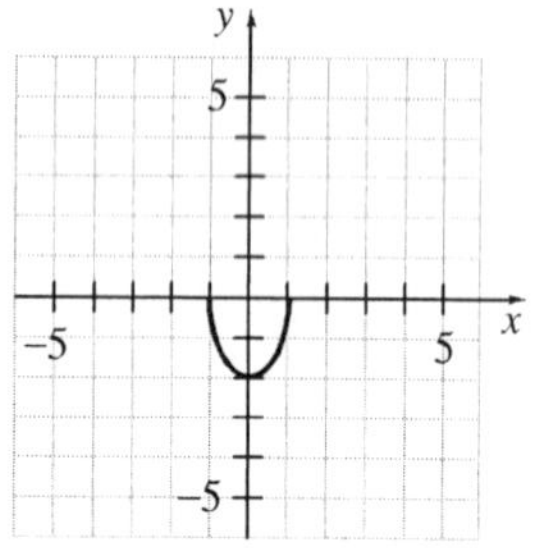


**43.** From the figure, we see that the major axis is horizontal with $a = 15$, and $b = 10$.

$\frac{x^2}{15^2} + \frac{y^2}{10^2} = 1$

$\frac{x^2}{225} + \frac{y^2}{100} = 1$

Since the truck is 8 feet wide, we need to determine the height of the archway at $\frac{8}{2} = 4$ feet from the center.

$\frac{4^2}{225} + \frac{y^2}{100} = 1$

$\frac{16}{225} + \frac{y^2}{100} = 1$

$900\left(\frac{16}{225} + \frac{y^2}{100}\right) = 900(1)$

$4(16) + 9y^2 = 900$

$64 + 9y^2 = 900$

$9y^2 = 836$

$y^2 = \frac{836}{9}$

$y = \sqrt{\frac{836}{9}} \approx 9.64$

The height of the archway 4 feet from the center is approximately 9.64 feet. Since the truck is 7 feet high, the truck will clear the archway.

**45.** **a.** $\frac{x^2}{48^2} + \frac{y^2}{23^2} = 1$

$\frac{x^2}{2304} + \frac{y^2}{529} = 1$

**b.** $c^2 = a^2 - b^2$

$c^2 = 48^2 - 23^2$

$c^2 = 2304 - 529$

$c^2 = 1775$

$c = \sqrt{1775} \approx 42.1$

The desk was situated approximately 42 feet from the center of the ellipse.

**47.** Answers will vary.

**49.** Answers will vary.

**51.** Answers will vary.

**53.** Answers will vary. For example, consider Exercise 21.

$$\frac{(x-2)^2}{9}+\frac{(y-1)^2}{4}=1$$

$$\frac{(y-1)^2}{4}=1-\frac{(x-2)^2}{9}$$

$$(y-1)^2=4\left(1-\frac{(x-2)^2}{9}\right)$$

$$(y-1)^2=4-\frac{4(x-2)^2}{9}$$

$$y-1=\pm\sqrt{4-\frac{4(x-2)^2}{9}}$$

$$y=1\pm\sqrt{4-\frac{4(x-2)^2}{9}}$$

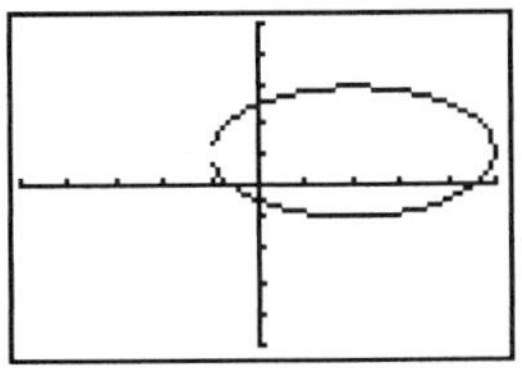

**55.** Graphing the points, we see that the center of the ellipse is at the origin and the major axis is vertical. We have $a=6$.

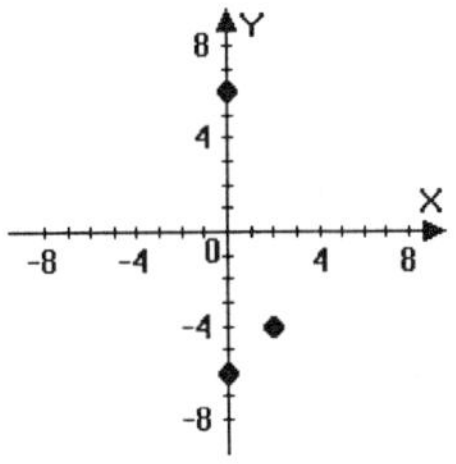


Using $a$ and the given point, we can solve for $b$.

$$\frac{x^2}{b^2}+\frac{y^2}{a^2}=1$$

$$\frac{2^2}{b^2}+\frac{(-4)^2}{6^2}=1$$

$$\frac{4}{b^2}+\frac{16}{36}=1$$

$$36b^2\left(\frac{4}{b^2}+\frac{16}{36}\right)=36b^2(1)$$

$$36(4)+16b^2=36b^2$$

$$144+16b^2=36b^2$$

$$144=20b^2$$

$$\frac{144}{20}=b^2$$

$$\frac{36}{5}=b^2$$

The equation ellipse in standard form is

$$\frac{x^2}{\frac{36}{5}}+\frac{y^2}{36}=1.$$

**57.**
$$4x^2+9y^2-32x+36y+64=0$$
$$\left(4x^2-32x\quad\right)+\left(9y^2+36y\quad\right)=-64$$
$$4\left(x^2-8x\quad\right)+9\left(y^2+4y\quad\right)=-64$$

Complete the squares.

$$\left(\frac{b}{2}\right)^2=\left(\frac{-8}{2}\right)^2=(-4)^2=16$$
$$\left(\frac{b}{2}\right)^2=\left(\frac{4}{2}\right)^2=(2)^2=4$$
$$4\left(x^2-8x+16\right)+9\left(y^2+4y+4\right)=-64+4(16)+9(4)$$
$$4(x-4)^2+9(y+2)^2=-64+64+36$$
$$4(x-4)^2+9(y+2)^2=36$$
$$\frac{4(x-4)^2}{36}+\frac{9(y+2)^2}{36}=\frac{36}{36}$$
$$\frac{(x-4)^2}{9}+\frac{(y+2)^2}{4}=1$$
$$\frac{(x-4)^2}{9}+\frac{(y+2)^2}{4}=1$$

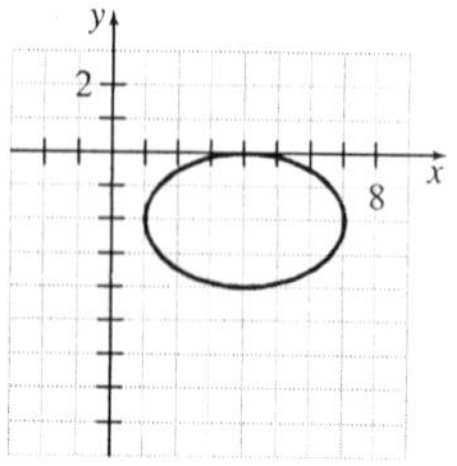


**59.** The ellipse's vertices lie on the larger circle. This means that $a$ is the radius of the circle. The equation of the larger circle is $x^2+y^2=25$. The endpoints of the ellipse's minor axis lie on the smaller circle. This means that $b$ is the radius of the smaller circle. The equation of the smaller circle is $x^2+y^2=9$.

**60.**
$$x^3+2x^2-4x-8=x^2(x+2)-4(x+2)=(x+2)\left(x^2-4\right)$$
$$=(x+2)(x+2)(x-2)=(x+2)^2(x-2)$$

**61.**
$$\sqrt[3]{40x^4y^7}=\sqrt[3]{8\cdot 5x^3xy^6y}=2xy^2\sqrt[3]{5xy}$$

**62.**

$$\frac{2}{x+2}+\frac{4}{x-2}=\frac{x-1}{x^2-4}$$

$$\frac{2}{x+2}+\frac{4}{x-2}=\frac{x-1}{(x+2)(x-2)}$$

$$(x+2)(x-2)\left(\frac{2}{x+2}+\frac{4}{x-2}\right)=(x+2)(x-2)\left(\frac{x-1}{(x+2)(x-2)}\right)$$

$$2(x-2)+4(x+2)=x-1$$

$$2x-4+4x+8=x-1$$

$$6x+4=x-1$$

$$5x=-5$$

$$x=-1$$

The solution is –1 and the solution set is $\{-1\}$.

## 10.3 Exercise Set

**1.** Since the $x^2$ – term is positive, the transverse axis lies along the $x$–axis. Also, since $a^2=4$ and $a=2,$ the vertices are $(-2,0)$ and $(2,0)$. This corresponds to graph (b).

**3.** Since the $y^2$ – term is positive, the transverse axis lies along the $y$–axis. Also, since $a^2=4$ and $a=2,$ the vertices are $(0,-2)$ and $(0,2)$. This corresponds to graph (a).

**5.** $\frac{x^2}{9}-\frac{y^2}{25}=1$

The equation is in the form $\frac{x^2}{a^2}-\frac{y^2}{b^2}=1$ with $a^2=9,$ and $b^2=25$. We know the transverse axis lies on the $x$-axis and the vertices are $(-3,0)$ and $(3,0)$.

Because $a^2=9$ and $b^2=25,$ $a=3$ and $b=5.$ Construct a rectangle using –3 and 3 on the $x$–axis, and –5 and 5 on the $y$–axis. Draw extended diagonals to obtain the asymptotes. Graph the hyperbola.

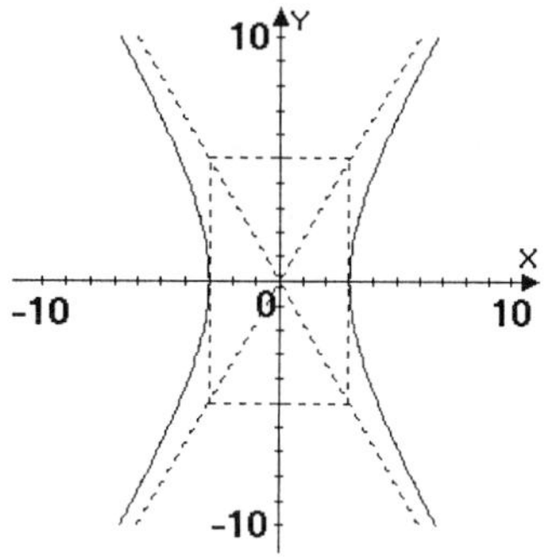

**7.** $\dfrac{x^2}{100} - \dfrac{y^2}{64} = 1$

The equation is in the form $\dfrac{x^2}{a^2} - \dfrac{y^2}{b^2} = 1$ with $a^2 = 100$, and $b^2 = 64$. We know the transverse axis lies on the $x$-axis and the vertices are $(-10,0)$ and $(10,0)$.

Because $a^2 = 100$ and $b^2 = 64$, $a = 10$ and $b = 8$. Construct a rectangle using –10 and 10 on the $x$–axis, and –8 and 8 on the $y$–axis. Draw extended diagonals to obtain the asymptotes. Graph the hyperbola.

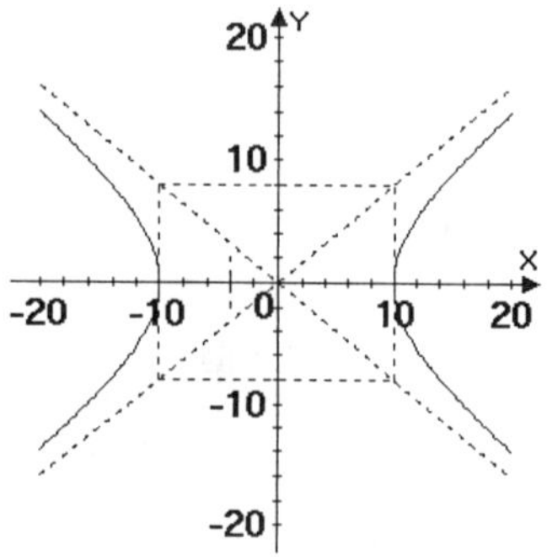


**9.** $\dfrac{y^2}{16} - \dfrac{x^2}{36} = 1$

The equation is in the form $\dfrac{y^2}{a^2} - \dfrac{x^2}{b^2} = 1$ with $a^2 = 16$, and $b^2 = 36$. We know the transverse axis lies on the $y$-axis and the vertices are $(0,-4)$ and $(0,4)$.

Because $a^2 = 16$ and $b^2 = 36$, $a = 4$ and $b = 6$. Construct a rectangle using –4 and 4 on the $x$–axis, and –6 and 6 on the $y$–axis. Draw extended diagonals to obtain the asymptotes. Graph the hyperbola.

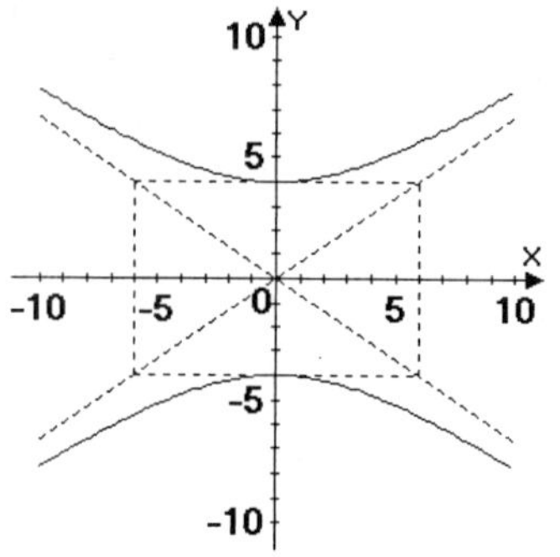


**11.** $\dfrac{y^2}{36} - \dfrac{x^2}{25} = 1$

The equation is in the form $\dfrac{y^2}{a^2} - \dfrac{x^2}{b^2} = 1$ with $a^2 = 36$, and $b^2 = 25$. We know the transverse axis lies on the $y$-axis and the vertices are $(0,-6)$ and $(0,6)$.

Because $a^2 = 36$ and $b^2 = 25$, $a = 6$ and $b = 5$. Construct a rectangle using –5 and 5 on the $x$–axis, and –6 and 6 on the $y$–axis. Draw extended diagonals to obtain the asymptotes. Graph the hyperbola.

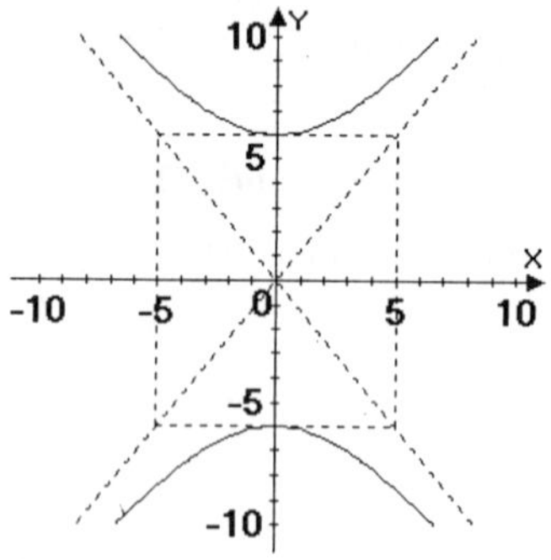

**13.** $9x^2 - 4y^2 = 36$

$$\frac{9x^2}{36} - \frac{4y^2}{36} = \frac{36}{36}$$

$$\frac{x^2}{4} - \frac{y^2}{9} = 1$$

The equation is in the form $\frac{x^2}{a^2} - \frac{y^2}{b^2} = 1$ with $a^2 = 4$ and $b^2 = 9$. We know the transverse axis lies on the $x$-axis and the vertices are $(-2,0)$ and $(2,0)$.

Because $a^2 = 4$ and $b^2 = 9$, $a = 2$ and $b = 3$. Construct a rectangle using –2 and 2 on the $x$–axis, and –3 and 3 on the $y$–axis. Draw extended diagonals to obtain the asymptotes. Graph the hyperbola.

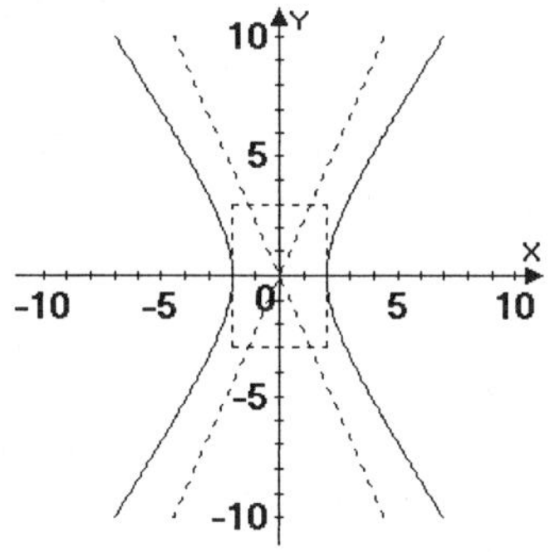


**15.** $9y^2 - 25x^2 = 225$

$$\frac{9y^2}{225} - \frac{25x^2}{225} = \frac{225}{225}$$

$$\frac{y^2}{25} - \frac{x^2}{9} = 1$$

The equation is in the form $\frac{y^2}{a^2} - \frac{x^2}{b^2} = 1$ with $a^2 = 25$ and $b^2 = 9$. We know the transverse axis lies on the $y$-axis and the vertices are $(0,-5)$ and $(0,5)$.

Because $a^2 = 25$ and $b^2 = 9$, $a = 5$ and $b = 3$. Construct a rectangle using –3 and 3 on the $x$–axis, and –5 and 5 on the $y$–axis. Draw extended diagonals to obtain the asymptotes. Graph the hyperbola.

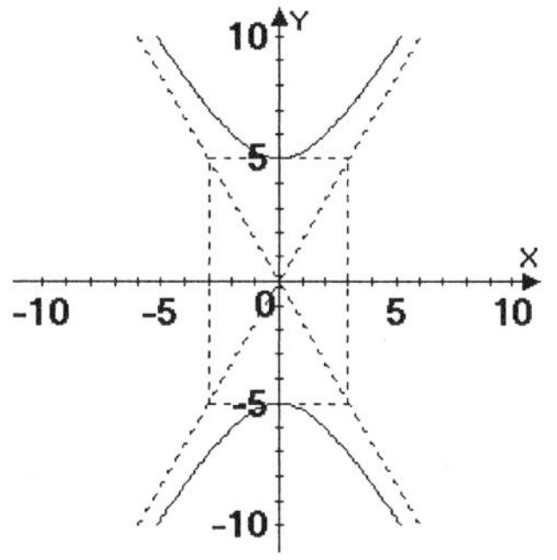


**17.** $4x^2 = 4 + y^2$

$$4x^2 - y^2 = 4$$

$$\frac{4x^2}{4} - \frac{y^2}{4} = \frac{4}{4}$$

$$\frac{x^2}{1} - \frac{y^2}{4} = 1$$

The equation is in the form $\frac{x^2}{a^2} - \frac{y^2}{b^2} = 1$ with $a^2 = 1$ and $b^2 = 4$. We know the transverse axis lies on the $x$-axis and the vertices are $(-1,0)$ and $(1,0)$. Because $a^2 = 1$ and $b^2 = 4$, $a = 1$ and $b = 2$. Construct a rectangle using –1 and 1 on the $x$–axis, and –2 and 2 on the $y$–axis. Draw extended diagonals to obtain the asymptotes. Graph the hyperbola.

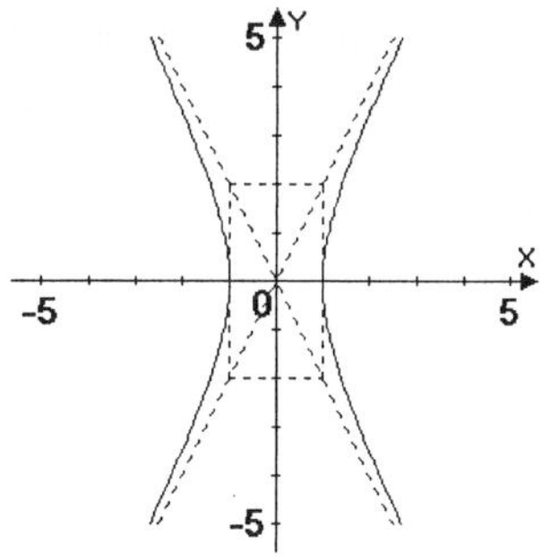

**19.** From the graph we see that the transverse axis lies along the $x$–axis and the vertices are $(-3,0)$ and $(3,0)$. This means that $a = 3$. We also see that $b = 5$.

$$\frac{x^2}{a^2} - \frac{y^2}{b^2} = 1$$
$$\frac{x^2}{3^2} - \frac{y^2}{5^2} = 1$$
$$\frac{x^2}{9} - \frac{y^2}{25} = 1$$

**21.** From the graph we see that the transverse axis lies along the $y$–axis and the vertices are $(0,-2)$ and $(0,2)$. This means that $a = 2$. We also see that $b = 3$.

$$\frac{y^2}{a^2} - \frac{x^2}{b^2} = 1$$
$$\frac{y^2}{2^2} - \frac{x^2}{3^2} = 1$$
$$\frac{y^2}{4} - \frac{x^2}{9} = 1$$

**23.** $\frac{x^2}{9} - \frac{y^2}{16} = 1$

The equation is for a hyperbola in standard form with the transverse axis on the $x$-axis. We have $a^2 = 9$ and $b^2 = 16$, so $a = 3$ and $b = 4$. Therefore, the vertices are at $(\pm a, 0)$ or $(\pm 3, 0)$. Using a dashed line, we construct a rectangle using the $\pm 3$ on the $x$-axis and $\pm 4$ on the $y$-axis. Then use dashed lines to draw extended diagonals for the rectangle. These represent the asymptotes of the graph.

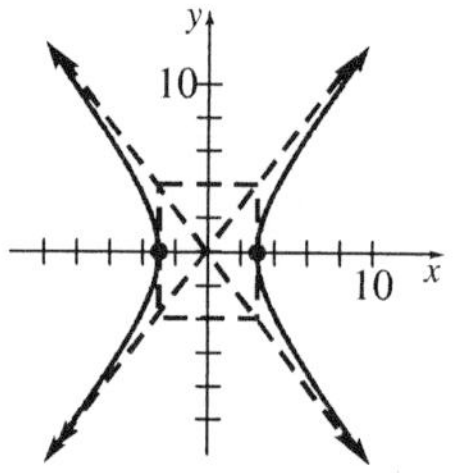


From the graph we determine the following:

Domain: $\{x \mid x \le -3 \text{ or } x \ge 3\}$ or $(-\infty, -3] \cup [3, \infty)$

Range: $\{y \mid y \text{ is a real number}\}$ or $(-\infty, \infty)$

**24.** $\frac{x^2}{25} - \frac{y^2}{4} = 1$

The equation is for a hyperbola in standard form with the transverse axis on the $x$-axis. We have $a^2 = 25$ and $b^2 = 4$, so $a = 5$ and $b = 2$. Therefore, the vertices are at $(\pm a, 0)$ or $(\pm 5, 0)$. Using a dashed line, we construct a rectangle using the $\pm 5$ on the $x$-axis and $\pm 2$ on the $y$-axis. Then use dashed lines to draw extended diagonals for the rectangle. These represent the asymptotes of the graph.

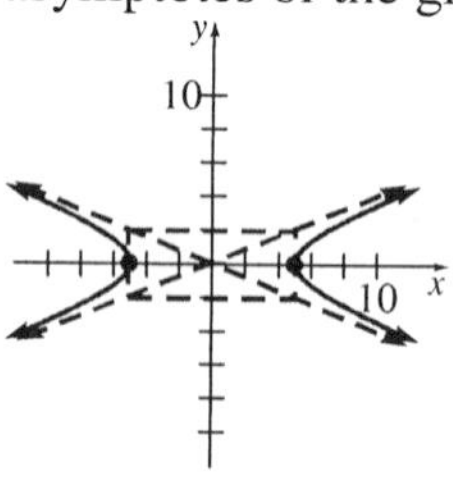


From the graph we determine the following:

Domain: $\{x \mid x \le -5 \text{ or } x \ge 5\}$ or $(-\infty, -5] \cup [5, \infty)$

Range: $\{y \mid y \text{ is a real number}\}$ or $(-\infty, \infty)$

**25.** $\frac{x^2}{9}+\frac{y^2}{16}=1$

The equation is for an ellipse in standard form with major axis along the y-axis. We have $a^2=16$ and $b^2=9$, so $a=4$ and $b=3$. Therefore, the vertices are $(0,\pm a)$ or $(0,\pm 4)$. The endpoints of the minor axis are $(\pm b,0)$ or $(\pm 3,0)$.

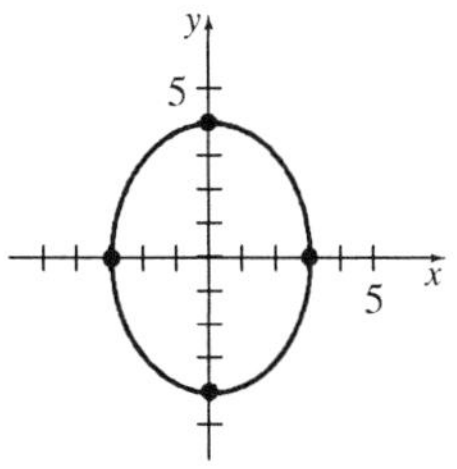


From the graph we determine the following:

Domain: $\{x\,|\,-3\le x\le 3\}$ or $[-3,3]$

Range: $\{y\,|\,-4\le y\le 4\}$ or $[-4,4]$.

**26.** $\frac{x^2}{25}+\frac{y^2}{4}=1$

The equation is for an ellipse in standard form with major axis along the y-axis. We have $a^2=25$ and $b^2=4$, so $a=5$ and $b=2$. Therefore, the vertices are $(\pm a,0)$ or $(\pm 5,0)$. The endpoints of the minor axis are $(0,\pm b)$ or $(0,\pm 2)$.

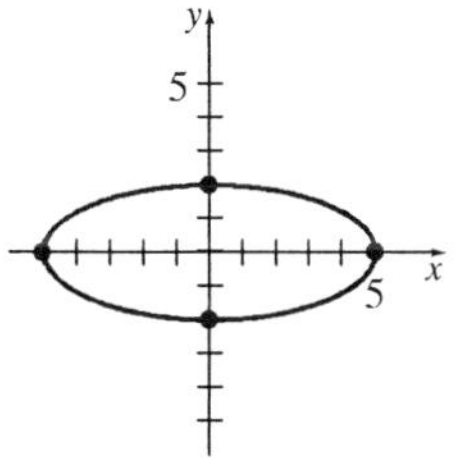


From the graph we determine the following:

Domain: $\{x\,|\,-5\le x\le 5\}$ or $[-5,5]$

Range: $\{y\,|\,-2\le y\le 2\}$ or $[-2,2]$.

**27.** $\frac{y^2}{16}-\frac{x^2}{9}=1$

The equation is in standard form with the transverse axis on the $y$-axis. We have $a^2=16$ and $b^2=9$, so $a=4$ and $b=3$. Therefore, the vertices are at $(0,\pm a)$ or $(0,\pm 4)$. Using a dashed line, we construct a rectangle using the $\pm 4$ on the $y$-axis and $\pm 3$ on the $x$-axis. Then use dashed lines to draw extended diagonals for the rectangle. These represent the asymptotes of the graph.

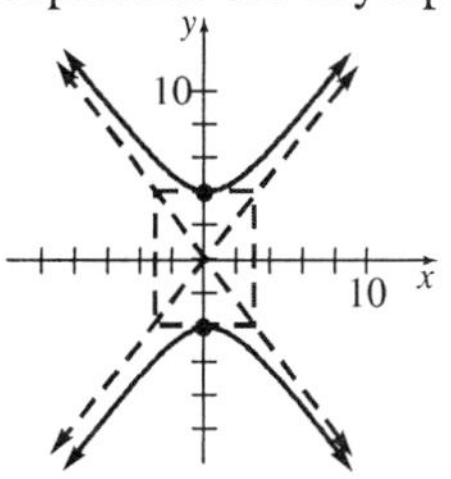


From the graph we determine the following:

Domain: $\{x\,|\,x \text{ is a real number}\}$ or $(-\infty,\infty)$

Range: $\{y\,|\,y\le -4 \text{ or } y\ge 4\}$ or $(-\infty,-4]\cup[4,\infty)$

**28.** $\frac{y^2}{4}-\frac{x^2}{25}=1$

The equation is in standard form with the transverse axis on the $y$-axis. We have $a^2=4$ and $b^2=25$, so $a=2$ and $b=5$. Therefore, the vertices are at $(0,\pm a)$ or $(0,\pm 2)$. Using a dashed line, we construct a rectangle using the $\pm 2$ on the $y$-axis and $\pm 5$ on the $x$-axis. Then use dashed lines to draw extended diagonals for the rectangle. These represent the asymptotes of the graph.

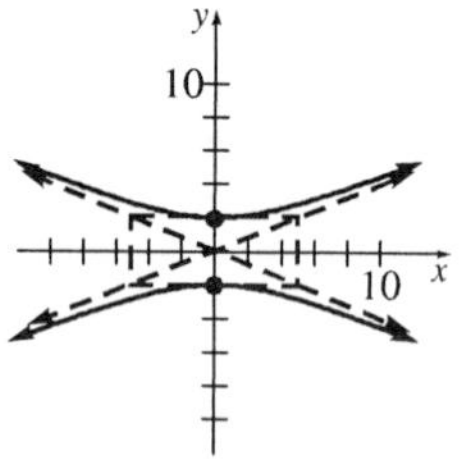


From the graph we determine the following:

Domain: $\{x \mid x \text{ is a real number}\}$ or $(-\infty, \infty)$

Range: $\{y \mid y \le -2 \text{ or } y \ge 2\}$ or $(-\infty, -2] \cup [2, \infty)$

**29.** $x^2 - y^2 = 4$

$x^2 + y^2 = 4$

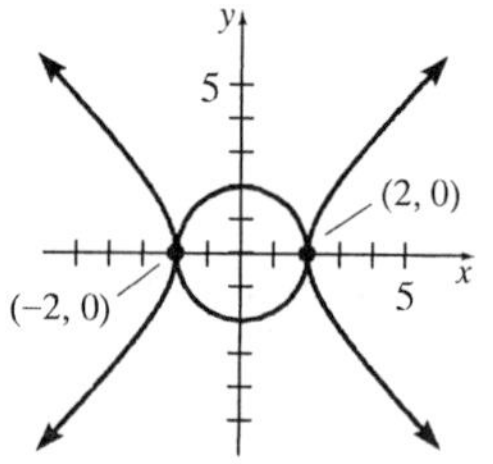


Check $(-2, 0)$:

$(-2)^2 - 0^2 = 4$ $\quad$ $(-2)^2 + 0^2 = 4$

$4 - 0 = 4$ $\quad$ $4 + 0 = 4$

$4 = 4$ true $\quad$ $4 = 4$ true

Check $(2, 0)$:

$(2)^2 - 0^2 = 4$ $\quad$ $(2)^2 + 0^2 = 4$

$4 - 0 = 4$ $\quad$ $4 + 0 = 4$

$4 = 4$ true $\quad$ $4 = 4$ true

The solution set is $\{(-2, 0), (2, 0)\}$.

**30.** $x^2 - y^2 = 9$

$x^2 + y^2 = 9$

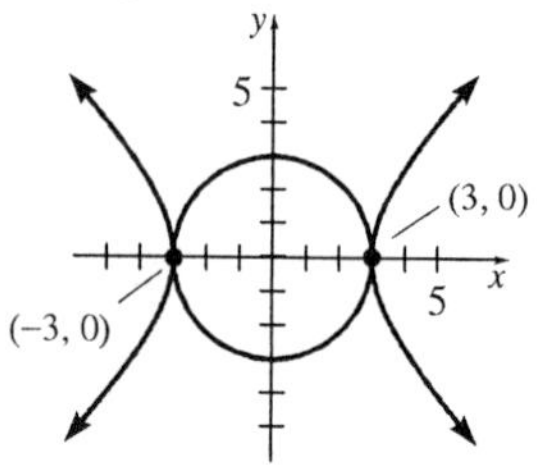


Check $(-3, 0)$:

$(-3)^2 - 0^2 = 9$ $\quad$ $(-3)^2 + 0^2 = 9$

$9 - 0 = 9$ $\quad$ $9 + 0 = 9$

$9 = 9$ true $\quad$ $9 = 9$ true

Check $(3, 0)$:

$(3)^2 - 0^2 = 9$ $\quad$ $(3)^2 + 0^2 = 9$

$9 - 0 = 9$ $\quad$ $9 + 0 = 9$

$9 = 9$ true $\quad$ $9 = 9$ true

The solution set is $\{(-3, 0), (3, 0)\}$.

**31.** $9x^2 + y^2 = 9$ or $\dfrac{x^2}{1} + \dfrac{y^2}{9} = 1$

$y^2 - 9x^2 = 9$ $\quad$ $\dfrac{y^2}{9} - \dfrac{x^2}{1} = 1$

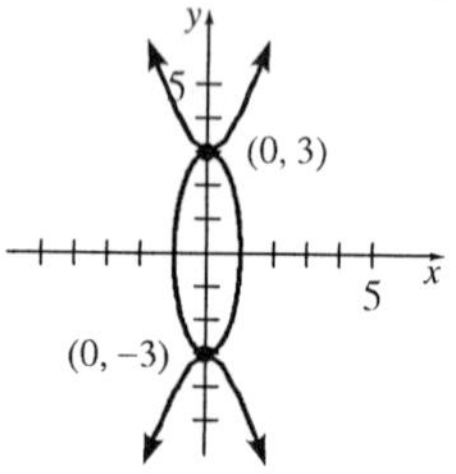


Check $(0, -3)$:

$9(0)^2 + (-3)^2 = 9$ $\quad$ $(-3)^2 - 9(0)^2 = 9$

$0 + 9 = 9$ $\quad$ $9 - 0 = 9$

$9 = 9$ $\quad$ $9 = 9$

true $\quad$ true

Check $(0,3)$:

$$9(0)^2+(3)^2=9 \qquad (3)^2-9(0)^2=9$$
$$0+9=9 \qquad 9-0=9$$
$$9=9 \qquad 9=9$$
true true

The solution set is $\{(0,-3),(0,3)\}$.

**32.** $4x^2+y^2=4$ or $\dfrac{x^2}{1}+\dfrac{y^2}{4}=1$

$y^2-4x^2=4$ $\quad \dfrac{y^2}{4}-\dfrac{x^2}{1}=1$

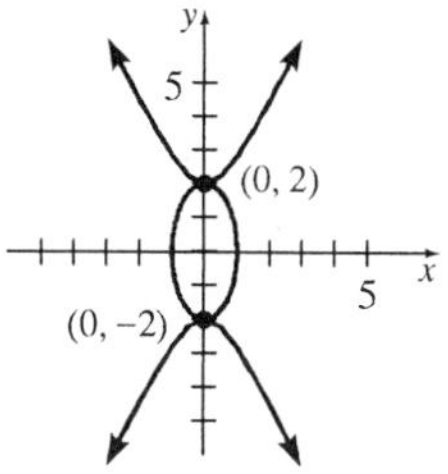


Check $(0,-2)$:

$$4(0)^2+(-2)^2=4 \qquad (-2)^2-4(0)^2=4$$
$$0+4=4 \qquad 4-0=4$$
$$4=4 \qquad 4=4$$
true true

Check $(0,2)$:

$$4(0)^2+(2)^2=4 \qquad (2)^2-4(0)^2=4$$
$$0+4=4 \qquad 4-0=4$$
$$4=4 \qquad 4=4$$
true true

The solution set is $\{(0,-2),(0,2)\}$.

**33.** $$625y^2-400x^2=250{,}000$$
$$\frac{625y^2}{250{,}000}-\frac{400x^2}{250{,}000}=\frac{250{,}000}{250{,}000}$$
$$\frac{y^2}{400}-\frac{x^2}{625}=1$$
Since the houses at the vertices of the hyperbola will be closest, find the distance between the vertices. Since $a^2=400$, $a=20$. The houses are $20+20=40$ yards apart.

**35.** Answers will vary.

**37.** Answers will vary.

**39.** Answers will vary.

**41.** $\dfrac{x^2}{4}-\dfrac{y^2}{9}=0$

Solve the equation for $y$.
$$\frac{x^2}{4}=\frac{y^2}{9}$$
$$9x^2=4y^2$$
$$\frac{9}{4}x^2=y^2$$
$$\pm\sqrt{\frac{9}{4}x^2}=y$$
$$\pm\frac{3}{2}x=y$$

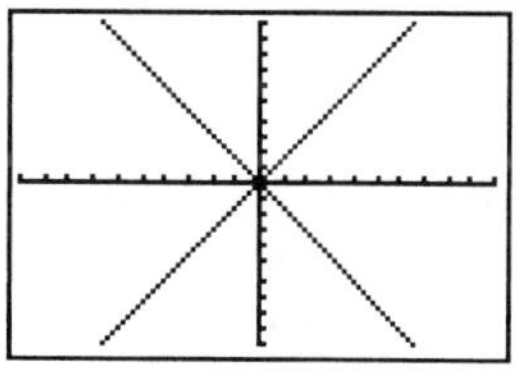

The graph is not a hyperbola. The graph is two lines.

**43.** Statement **c.** is true. Since $y = -\frac{2}{3}$ is an asymptote, the graph of the hyperbola does not intersect it.

Statement **a.** is false. If a hyperbola has a transverse axis along the *y*–axis and one of the branches is removed, the remaining branch does not define a function of *x*.
Statement **b**. is false. The points on the hyperbola's asymptotes do not satisfy the hyperbola's equation.

Statement **d.** is false. See Exercise 32 for two different hyperbolas that share the same asymptotes.

**45.** $\frac{(x+2)^2}{9} - \frac{(y-1)^2}{25} = 1$
This is the graph of a hyperbola with center $(-2,1)$. The equation is in the form $\frac{(x-h)^2}{a^2} - \frac{(y-k)^2}{b^2} = 1$ with $a^2 = 9$ and $b^2 = 25$. We know the transverse axis is horizontal and the vertices lie 3 units to the right and left of $(-2,1)$ at $(-2-3,1) = (-5,1)$ and $(-2+3,1) = (1,1)$. Because $a^2 = 9$ and $b^2 = 25$, $a = 3$ and $b = 5$.
Construct two sides of a rectangle using –5 and 1 (the *x*–coordinates of the vertices) on the *x*–axis. The remaining two sides of the rectangle are constructed 5 units above and 5 units below the center, $(-2,1)$, at $1-5 = -4$ and $1+5 = 6$. Draw extended diagonals to obtain the asymptotes. Graph the hyperbola.

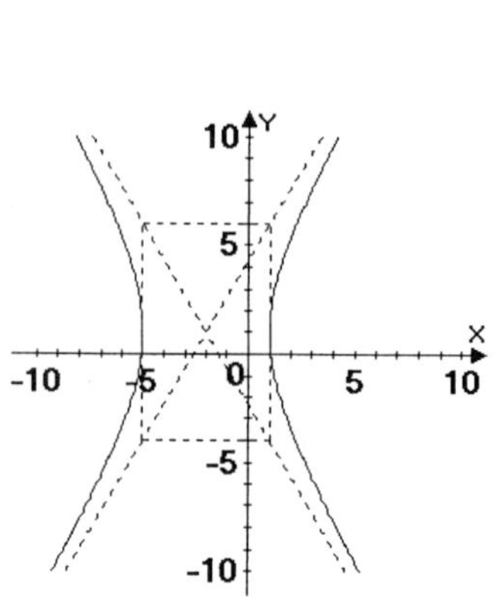


**47.** $x^2 - y^2 - 2x - 4y - 4 = 0$
Rearrange and complete the squares.
$$x^2 - 2x \quad - y^2 - 4y \quad = 4$$
$$(x^2 - 2x \quad) - (y^2 + 4y \quad) = 4$$
Complete the squares.
$$\left(\frac{b}{2}\right)^2 = \left(\frac{-2}{2}\right)^2 = (-1)^2 = 1$$
$$\left(\frac{b}{2}\right)^2 = \left(\frac{4}{2}\right)^2 = 2^2 = 4$$
$$(x^2 - 2x + 1) - (y^2 + 4y + 4) = 4 + 1 - 4$$
$$(x-1)^2 - (y+2)^2 = 1$$
$$\frac{(x-1)^2}{1} - \frac{(y+2)^2}{1} = 1$$
This is the graph of a hyperbola with center $(1,-2)$. The equation is in the form $\frac{(x-h)^2}{a^2} - \frac{(y-k)^2}{b^2} = 1$ with $a^2 = 1$ and $b^2 = 1$. We know the transverse axis is horizontal and the vertices lie 3 units to the right and left of $(1,-2)$ at $(1-1,-2) = (0,-2)$ and $(1+1,-2) = (2,-2)$. Because $a^2 = 1$ and $b^2 = 1$, $a = 1$ and $b = 1$.
Construct two sides of a rectangle using 0 and 2 (the *x*–coordinates of the vertices) on the *x*–axis. The remaining

two sides of the rectangle are constructed 1 unit above and 1 unit below the center, $(1,-2)$, at $-2-1=-3$ and $-2+1=-1$. Draw extended diagonals to obtain the asymptotes. Graph the hyperbola.

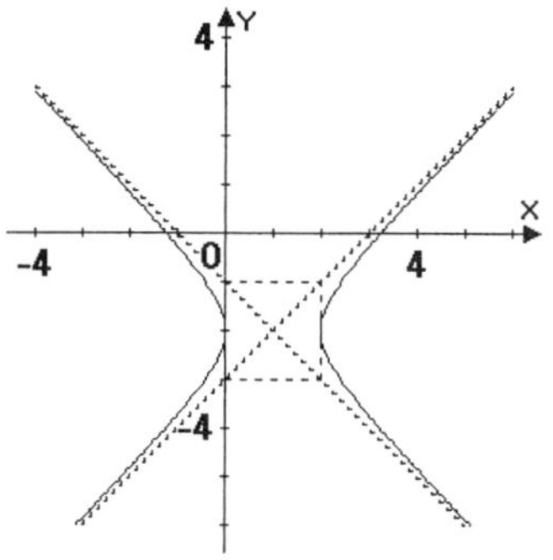


**49.** Since the vertices are $(0,7)$ and $(0,-7)$, we know that the transverse axis lies along the $y$–axis and $a = 7$. Use the equation of the asymptote, $y = 5x$, to find $b$. We need to find the $x$–coordinate that corresponds with $y = 7$.

$7 = 5x$

$\frac{7}{5} = x$

This means that $b = \pm\frac{7}{5}$. Using $a$ and $b$, write the equation of the hyperbola.

$$\frac{y^2}{7^2} - \frac{x^2}{\left(\frac{7}{5}\right)^2} = 1$$

$$\frac{y^2}{49} - \frac{x^2}{\frac{49}{25}} = 1$$

**50.** $y = -x^2 - 4x + 5$

Since $a = -1$ is negative, the parabola opens downward. The $x$–coordinate of the vertex of the parabola is $-\frac{b}{2a} = -\frac{-4}{2(-1)} = -2$ and the $y$–coordinate of the vertex of the parabola is

$$f\left(-\frac{b}{2a}\right) = f(-2)$$
$$= -(-2)^2 - 4(-2) + 5$$
$$= -4 + 8 + 5 = 9.$$

The vertex is at (−2, 9).

Replace $y$ with 0 to find $x$–intercepts.

$0 = -x^2 - 4x + 5$

$0 = x^2 + 4x - 5$

$0 = (x+5)(x-1)$

Apply the zero product principle.

$x+5=0 \quad$ or $\quad x-1=0$

$x=-5 \qquad\qquad x=1$

The $x$–intercepts are –5 and 1.
Set $x = 0$ and solve for $y$ to obtain the $y$–intercept.

$y = -0^2 - 4(0) + 5 = 5$

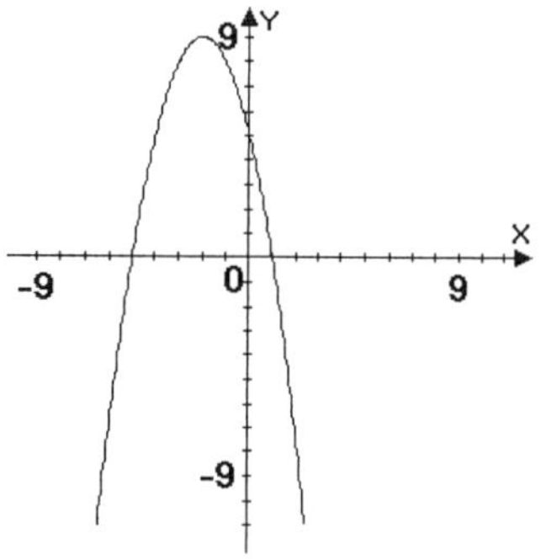

**51.** $3x^2 - 11x - 4 \geq 0$
Solve the related quadratic equation.
$3x^2 - 11x - 4 = 0$
$(3x+1)(x-4) = 0$
Use the zero product principle.
$3x + 1 = 0$ or $x - 4 = 0$
$3x = -1$ $\quad$ $x = 4$
$x = -\frac{1}{3}$

The boundary points are $-\frac{1}{3}$ and 4.

| Test Interval | Test Number | Substitution | Conclusion |
|---|---|---|---|
| $\left(-\infty, -\frac{1}{3}\right]$ | -1 | $3(-1)^2 - 11(-1) - 4 \geq 0$ <br> $10 \geq 0$, true | $\left(-\infty, -\frac{1}{3}\right]$ belongs in the solution set |
| $\left[-\frac{1}{3}, 4\right]$ | 0 | $3(0)^2 - 11(0) - 4 \geq 0$ <br> $-4 \geq 0$, false | $\left[-\frac{1}{3}, 4\right]$ does not belong in the solution set. |
| $[4, \infty)$ | 5 | $3(5)^2 - 11(5) - 4 \geq 0$ <br> $16 \geq 0$, true | $[4, \infty)$ belongs in the solution set. |

The solution set is $\left(-\infty, -\frac{1}{3}\right] \cup [4, \infty)$ or $\left\{x \middle| x \leq -\frac{1}{3} \text{ or } x \geq 4\right\}$.

**52.** $\log_4(3x+1) = 3$
$3x + 1 = 4^3$
$3x + 1 = 64$
$3x = 63$
$x = 21$
The solution is 21 and the solution set is $\{21\}$.

## Mid-Chapter 10 Check Point

**1.** $x^2 + y^2 = 9$

Center: $(0,0)$

Radius: $r = \sqrt{9} = 3$

We plot points that are 3 units to the left, right, above, and below the center. These points are $(-3,0)$, $(3,0)$, $(0,3)$ and $(0,-3)$.

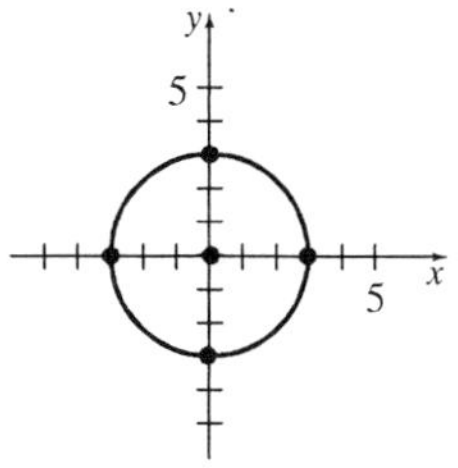


**2.** $(x-3)^2 + (y+2)^2 = 25$

Center: $(3,-2)$

Radius: $r = \sqrt{25} = 5$

We plot the points that are 5 units to the left, right, above and below the center.

These points are $(-2,-2)$, $(8,-2)$, $(3,3)$, and $(3,-7)$.

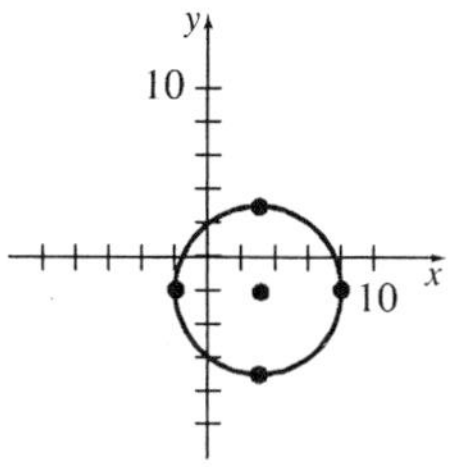


**3.** $x^2 + (y-1)^2 = 4$

Center: $(0,1)$

Radius: $r = \sqrt{4} = 2$

We plot the points that are 2 units to the left, right, above, and below the center. These points are $(-2,1)$, $(2,1)$, $(0,3)$, and $(0,-1)$.

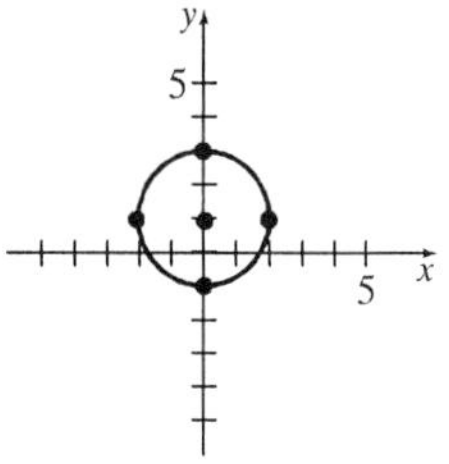


**4.** $x^2 + y^2 - 4x - 2y - 4 = 0$

Complete the square in both $x$ and $y$ to get the equation in standard form.

$$(x^2 - 4x) + (y^2 - 2y) = 4$$

$$(x^2 - 4x + 4) + (y^2 - 2y + 1) = 4 + 4 + 1$$

$$(x-2)^2 + (y-1)^2 = 9$$

Center: $(2,1)$

Radius: $r = \sqrt{9} = 3$

We plot the points that are 3 units to the left, right, above, and below the center. These points are $(-1,1)$, $(5,1)$, $(2,4)$, and $(2,-2)$.

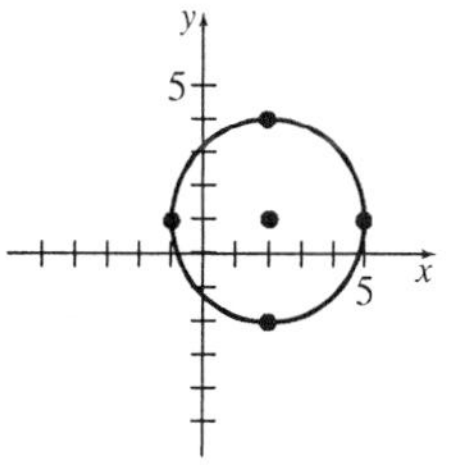

**5.** $\frac{x^2}{25}+\frac{y^2}{4}=1$

Center: $(0,0)$

Because the denominator of the $x^2$ – term is greater than the denominator of the $y^2$ – term, the major axis is horizontal. Since $a^2=25$, $a=5$ and the vertices are $(-5,0)$ and $(5,0)$. Since $b^2=4$, $b=2$ and endpoints of the minor axis are $(0,-2)$ and $(0,2)$.

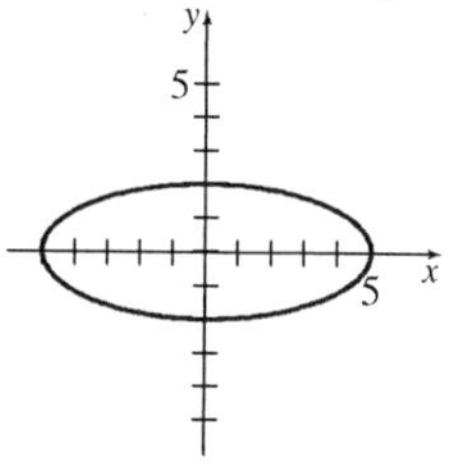


**6.** $9x^2+4y^2=36$

Divide both sides by 36 to get the standard form:

$\frac{x^2}{4}+\frac{y^2}{9}=1$

Center: $(0,0)$

Because the denominator of the $y^2$ – term is greater than the denominator of the $x^2$ – term, the major axis is vertical. Since $a^2=9$, $a=3$ and the vertices are $(0,-3)$ and $(0,3)$. Since $b^2=4$, $b=2$ and endpoints of the minor axis are $(-2,0)$ and $(2,0)$.

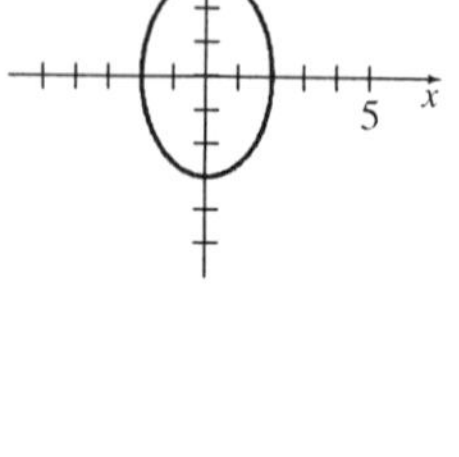


**7.** $\frac{(x-2)^2}{16}+\frac{(y+1)^2}{25}=1$

Center: $(2,-1)$

Because the denominator of the $y^2$ – term is greater than the denominator of the $x^2$ – term, the major axis is vertical. We have $a^2=25$ and $b^2=16$, so $a=5$ and $b=4$. The vertices lie 5 units above and below the center. The endpoints of the minor axis lie 4 units to the left and right of the center.

Vertices: $(2,4)$ and $(2,-6)$

Minor endpoints: $(-2,-1)$ and $(6,-1)$

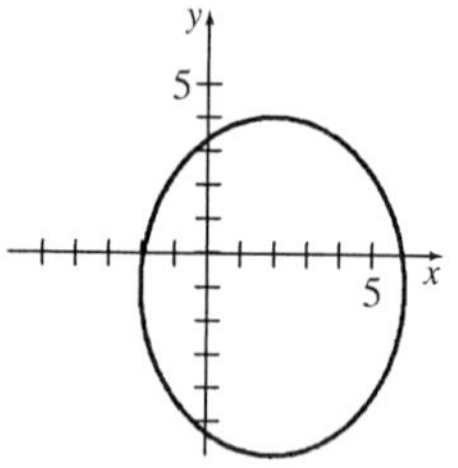


**8.** $\frac{(x+2)^2}{25}+\frac{(y-1)^2}{16}=1$

Center: $(-2,1)$

Because the denominator of the $x^2$ – term is greater than the denominator of the $y^2$ – term, the major axis is horizontal. We have $a^2=25$ and $b^2=16$, so $a=5$ and $b=4$. The vertices lie 5 units to the left and right of the center. The endpoints of the minor axis lie 4 units above and below the center.

Vertices: $(-7,1)$ and $(3,1)$

Minor endpoints: $(-2,5)$ and $(-2,-3)$

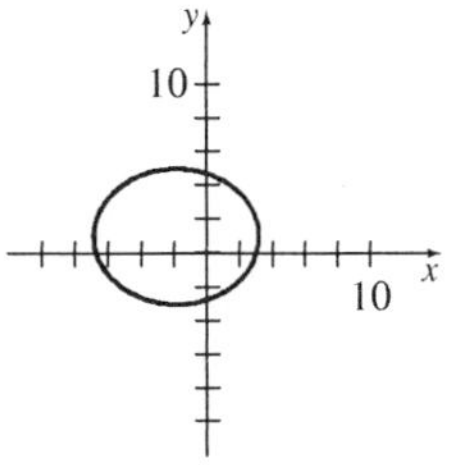


**9.** $\frac{x^2}{9} - y^2 = 1$

The equation is for a hyperbola in standard form with the transverse axis on the $x$-axis. We have $a^2 = 9$ and $b^2 = 1$, so $a = 3$ and $b = 1$. Therefore, the vertices are at $(\pm a, 0)$ or $(\pm 3, 0)$. Using a dashed line, we construct a rectangle using the $\pm 3$ on the $x$-axis and $\pm 1$ on the $y$-axis. Then use dashed lines to draw extended diagonals for the rectangle. These represent the asymptotes of the graph.
Graph the hyperbola.

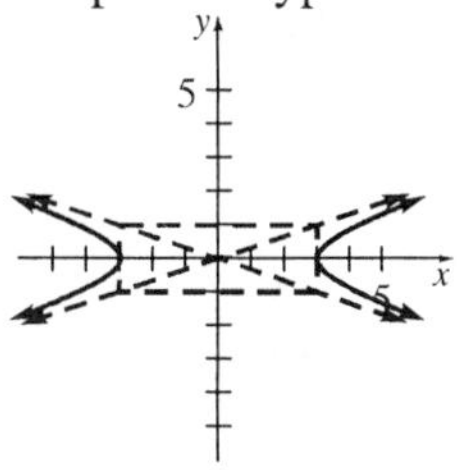


**10.** $\frac{y^2}{9} - x^2 = 1$

The equation is in the form $\frac{y^2}{a^2} - \frac{x^2}{b^2} = 1$ with $a^2 = 9$, and $b^2 = 1$. We know the transverse axis lies on the $y$-axis and the vertices are $(0, -3)$ and $(0, 3)$.
Because $a^2 = 9$ and $b^2 = 1$, $a = 3$ and $b = 1$. Construct a rectangle using –1 and 1 on the $x$–axis, and –3 and 3 on the $y$–axis. Draw extended diagonals to obtain the asymptotes.
Graph the hyperbola.

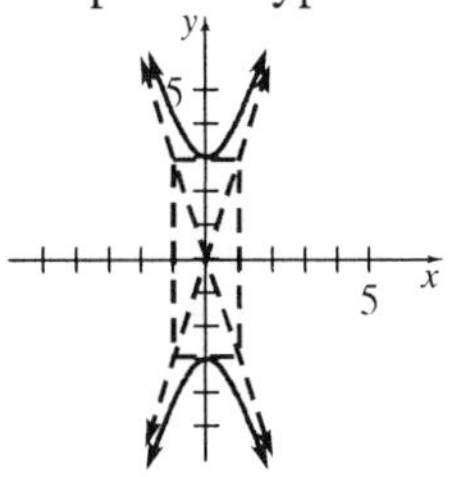


**11.** $y^2 - 4x^2 = 16$

$\frac{y^2}{16} - \frac{x^2}{4} = 1$

The equation is in the form $\frac{y^2}{a^2} - \frac{x^2}{b^2} = 1$ with $a^2 = 16$, and $b^2 = 4$. We know the transverse axis lies on the $y$-axis and the vertices are $(0, -4)$ and $(0, 4)$.
Because $a^2 = 16$ and $b^2 = 4$, $a = 4$ and $b = 2$. Construct a rectangle using –2 and 2 on the $x$–axis, and –4 and 4 on the $y$–axis. Draw extended diagonals to obtain the asymptotes.
Graph the hyperbola.

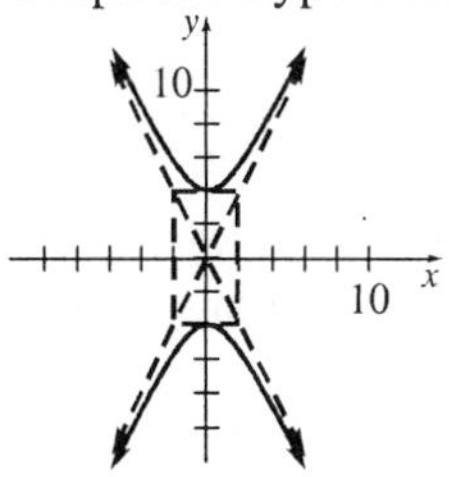

**12.** $4x^2 - 49y^2 = 196$

$$\frac{x^2}{49} - \frac{y^2}{4} = 1$$

The equation is for a hyperbola in standard form with the transverse axis on the $x$-axis. We have $a^2 = 49$ and $b^2 = 4$, so $a = 7$ and $b = 2$. Therefore, the vertices are at $(\pm a, 0)$ or $(\pm 7, 0)$. Using a dashed line, we construct a rectangle using the $\pm 7$ on the $x$-axis and $\pm 2$ on the $y$-axis. Then use dashed lines to draw extended diagonals for the rectangle. These represent the asymptotes of the graph.
Graph the hyperbola.

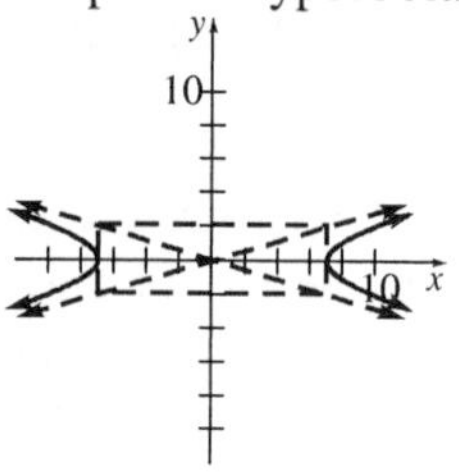


**13.** $x^2 + y^2 = 4$

This is the equation of a circle centered at the origin with radius $r = \sqrt{4} = 2$. We can plot points that are 2 units to the left, right, above, and below the origin and then graph the circle. The points are $(-2,0)$, $(2,0)$, $(0,2)$, and $(0,-2)$.

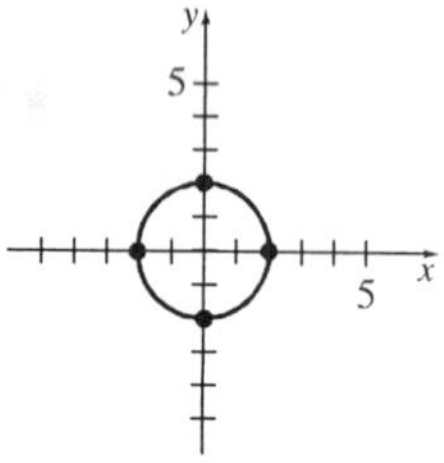


**14.** $x + y = 4$

$y = -x + 4$

This is the equation of a line with slope $m = -1$ and a $y$-intercept of 4. We can plot the point $(0,4)$, use the slope to get an additional point, connect the points with a straight line and then extend the line to represent the graph of the equation.

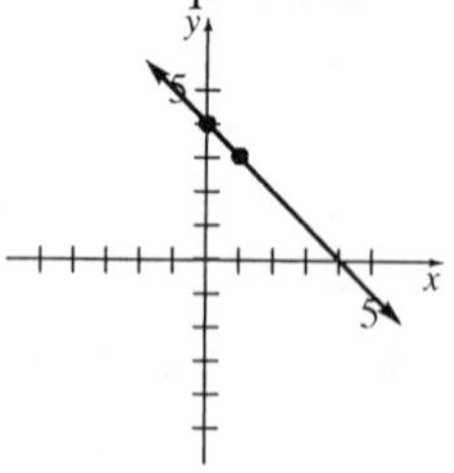


**15.** $x^2 - y^2 = 4$

$$\frac{x^2}{4} - \frac{y^2}{4} = 1$$

The equation is for a hyperbola in standard form with the transverse axis on the $x$-axis. We have $a^2 = 4$ and $b^2 = 4$, so $a = 2$ and $b = 2$. Therefore, the vertices are at $(\pm a, 0)$ or $(\pm 2, 0)$. Using a dashed line, we construct a rectangle using the $\pm 2$ on the $x$-axis and $\pm 2$ on the $y$-axis. Then use dashed lines to draw extended diagonals for the rectangle. These represent the asymptotes of the graph.
Graph the hyperbola.

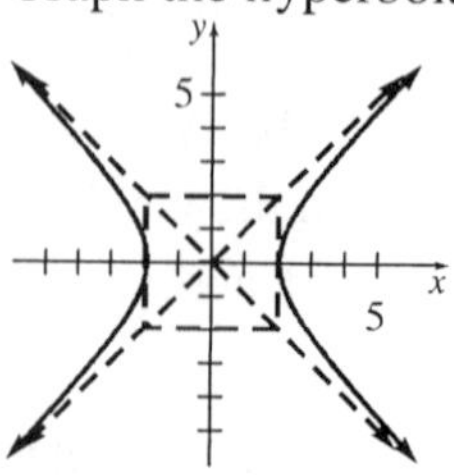

**16.** $x^2+4y^2=4$

$\frac{x^2}{4}+\frac{y^2}{1}=1$

Center: $(0,0)$

Because the denominator of the $x^2$ – term is greater than the denominator of the $y^2$ – term, the major axis is horizontal. We have $a^2=4$ and $b^2=1$, so $a=2$ and $b=1$. The vertices lie 2 units to the left and right of the center. The endpoints of the minor axis lie 1 unit above and below the center.

Vertices: $(-2,0)$ and $(2,0)$

Minor endpoints: $(0,-1)$ and $(0,1)$

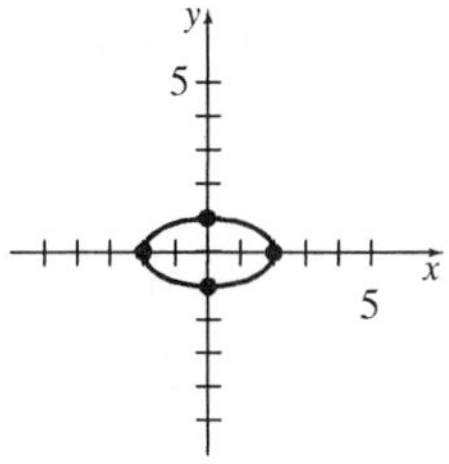


**17.** $(x+1)^2+(y-1)^2=4$

Center: $(-1,1)$

Radius: $r=\sqrt{4}=2$

We plot the points that are 2 units to the left, right, above and below the center.

These points are $(-3,1)$, $(1,1)$, $(-1,3)$, and $(-1.-1)$.

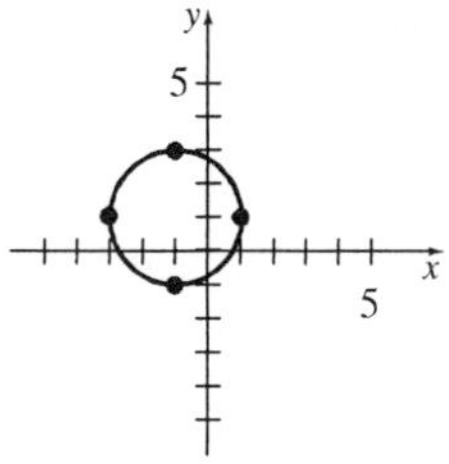


**18.** $x^2+4(y-1)^2=4$

$\frac{x^2}{4}+\frac{(y-1)^2}{1}=1$

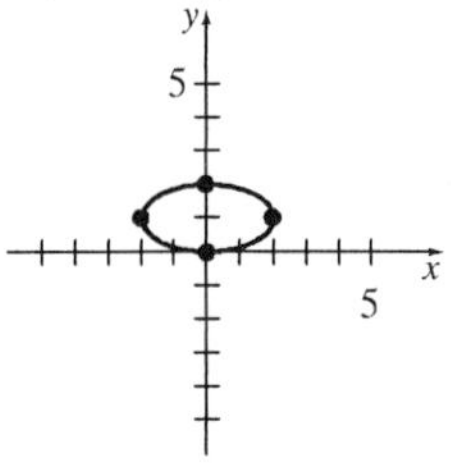


**19.** $d=\sqrt{(-2-2)^2+(2-(-2))^2}$

$=\sqrt{(-4)^2+(4)^2}$

$=\sqrt{16+16}=\sqrt{32}$

$=\sqrt{16\cdot 2}$

$=4\sqrt{2}\approx 5.66$ units

$\left(\frac{2+(-2)}{2},\frac{-2+2}{2}\right)=\left(\frac{0}{2},\frac{0}{2}\right)=(0,0)$

The length of the segment is $4\sqrt{2}\approx 5.66$ units and the midpoint is the origin, $(0,0)$.

**20.** $d=\sqrt{(-10-(-5))^2+(14-8)^2}$

$=\sqrt{(-5)^2+(6)^2}$

$=\sqrt{25+36}$

$=\sqrt{61}\approx 7.81$ units

$\left(\frac{-5+(-10)}{2},\frac{8+14}{2}\right)=\left(\frac{-15}{2},\frac{22}{2}\right)$

$=\left(-\frac{15}{2},11\right)$

The length of the segment is $\sqrt{61}\approx 7.81$ units and the midpoint is $\left(-\frac{15}{2},11\right)$.

## 10.4 Exercise Set

**1.** Since $a = 1$, the parabola opens to the right.
The vertex of the parabola is $(-1,2)$.
Graph **b.** is the equation's graph.

**3.** Since $a = 1$, the parabola opens to the right.
The vertex of the parabola is $(1,-2)$.
Graph **f** .is the equation's graph.

**5.** Since $a = -1$, the parabola opens to the left.
The vertex of the parabola is $(1,2)$.
Graph **a.** is the equation's graph. Either graph a or graph e will match this. One will be changed to open to the left.

**7.** $x = 2y^2$

$x = 2(y-0)^2 + 0$

The vertex is the point $(0,0)$.

**9.** $x = (y-2)^2 + 3$

The vertex is the point $(3,2)$.

**11.** $x = -4(y+2)^2 - 1$

The vertex is the point $(-1,-2)$.

**13.** $x = 2(y-6)^2$

$x = 2(y-6)^2 + 0$

The vertex is the point $(0,6)$.

**15.** $x = y^2 - 6y + 6$

The $y$–coordinate of the vertex is
$-\frac{b}{2a} = -\frac{-6}{2(1)} = -\frac{-6}{2} = 3.$

The $x$–coordinate of the vertex is
$f(3) = 3^2 - 6(3) + 6 = 9 - 18 + 6 = -3.$

The vertex is the point $(-3,3)$.

**17.** $x = 3y^2 + 6y + 7$

The $y$–coordinate of the vertex is
$-\frac{b}{2a} = -\frac{6}{2(3)} = -\frac{6}{6} = -1.$

The $x$–coordinate of the vertex is

$$f(-1) = 3(-1)^2 + 6(-1) + 7$$
$$= 3(1) - 6 + 7 = 3 - 6 + 7 = 4.$$

The vertex is the point $(4,-1)$.

**19.** $x = (y-2)^2 - 4$

This is a parabola of the form $x = a(y-k)^2 + h$. Since $a = 1$ is positive, the parabola opens to the right. The vertex of the parabola is $(-4,2)$. The axis of symmetry is $y = 2$. Replace $y$ with 0 to find the $x$–intercept.

$x = (0-2)^2 - 4 = 4 - 4 = 0$

The $x$–intercept is 0. Replace $x$ with 0 to find the $y$–intercepts.

$0=(y-2)^2-4$

$0=y^2-4y+4-4$

$0=y^2-4y$

$0=y(y-4)$

Apply the zero product principle.

$y=0$ and $y-4=0$

$y=4$

The $y$–intercepts are 0 and 4.

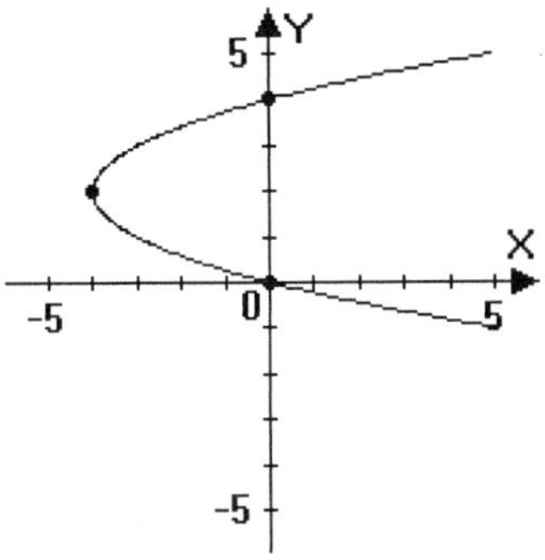


**21.** $x=(y-3)^2-5$

This is a parabola of the form $x=a(y-k)^2+h$. Since $a=1$ is positive, the parabola opens to the right. The vertex of the parabola is $(-5,3)$. The axis of symmetry is $y=3$. Replace $y$ with 0 to find the $x$–intercept.

$x=(0-3)^2-5=(-3)^2-5=9-5=4$

The $x$–intercept is 0. Replace $x$ with 0 to find the $y$–intercepts.

$0=(y-3)^2-5$

$0=y^2-6y+9-5$

$0=y^2-6y+4$

Solve using the quadratic formula.

$x=\dfrac{-b\pm\sqrt{b^2-4ac}}{2a}$

$=\dfrac{-(-6)\pm\sqrt{(-6)^2-4(1)4}}{2(1)}$

$=\dfrac{6\pm\sqrt{36-16}}{2}=\dfrac{6\pm\sqrt{20}}{2}$

$=\dfrac{6\pm2\sqrt{5}}{2}=3\pm\sqrt{5}$

The $y$–intercepts are $3-\sqrt{5}$ and $3+\sqrt{5}$.

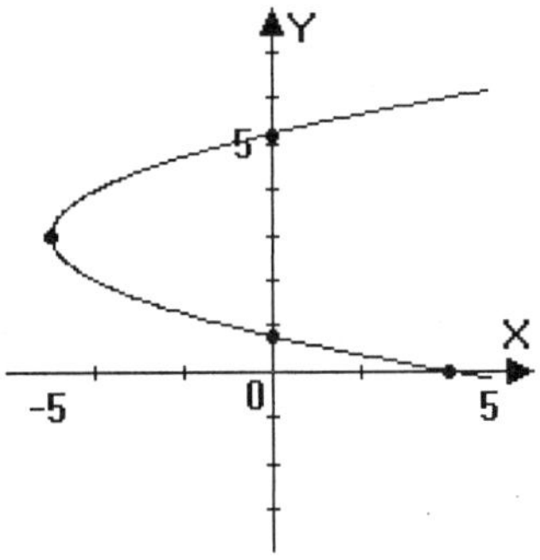


**23.** $x=-(y-5)^2+4$

This is a parabola of the form $x=a(y-k)^2+h$. Since $a=-1$ is negative, the parabola opens to the left. The vertex of the parabola is $(4,5)$.

The axis of symmetry is $y = 5$. Replace $y$ with 0 to find the $x$–intercept.

$x=-(0-5)^2+4=-(-5)^2+4$

$=-25+4=-21$

The $x$–intercept is 0. Replace $x$ with 0 to find the $y$–intercepts.

$0=-(y-5)^2+4$

$0=-(y^2-10y+25)+4$

$0=-y^2+10y-25+4$

$0=-y^2+10y-21$

$0=y^2-10y+21$

$0=(y-7)(y-3)$

Apply the zero product principle.

$y-7=0$ and $y-3=0$

$y=7$ $\quad\quad$ $y=3$

The $y$–intercepts are 3 and 7.

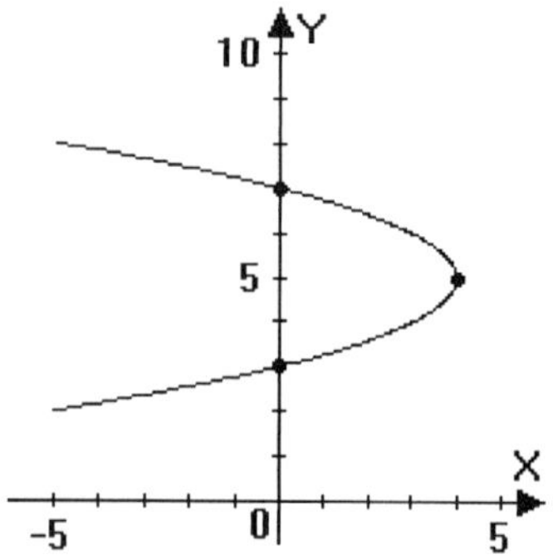


**25.** $x=(y-4)^2+1$

This is a parabola of the form $x=a(y-k)^2+h$. Since $a=1$ is positive, the parabola opens to the right. The vertex of the parabola is $(1,4)$. The axis of symmetry is $y=4$.

Replace $y$ with 0 to find the $x$–intercept.

$x=(0-4)^2+1=(-4)^2+1=16+1=17$

The $x$–intercept is 0. Replace $x$ with 0 to find the $y$–intercepts.

$0=(y-4)^2+1$

$0=y^2-8y+16+1$

$0=y^2-8y+17$

Solve using the quadratic formula.

$$y=\frac{-b\pm\sqrt{b^2-4ac}}{2a}$$

$$=\frac{-(-8)\pm\sqrt{(-8)^2-4(1)(17)}}{2(1)}$$

$$=\frac{8\pm\sqrt{64-68}}{2}=\frac{8\pm\sqrt{-4}}{2}$$

$$=\frac{8\pm 2i}{2}=4\pm i$$

The solutions are complex, so there are no $y$–intercepts.

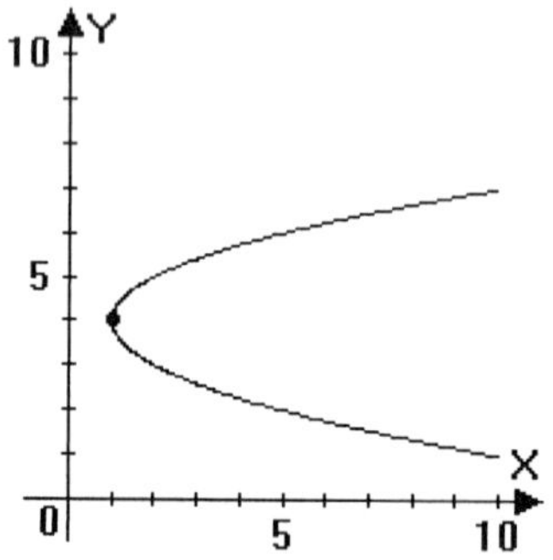


**27.** $x=-3(y-5)^2+3$

This is a parabola of the form $x=a(y-k)^2+h$. Since $a=-3$ is negative, the parabola opens to the left. The vertex of the parabola is $(3,5)$.

The axis of symmetry is $y=5$.

Replace $y$ with 0 to find the $x$–intercept.

$x=-3(0-5)^2+3=-3(-5)^2+3$

$=-3(25)+3=-75+3=-72$

The $x$–intercept is 0. Replace $x$ with 0 to find the $y$–intercepts.

$0 = -3(y-5)^2 + 3$

$0 = -3(y^2 - 10y + 25) + 3$

$0 = -3y^2 + 30y - 75 + 3$

$0 = -3y^2 + 30y - 72$

$0 = y^2 - 10y + 24$

$0 = (y-6)(y-4)$

Apply the zero product principle.

$y - 6 = 0$ and $y - 4 = 0$

$y = 6$ $\quad$ $y = 4$

The $y$–intercepts are 4 and 6.

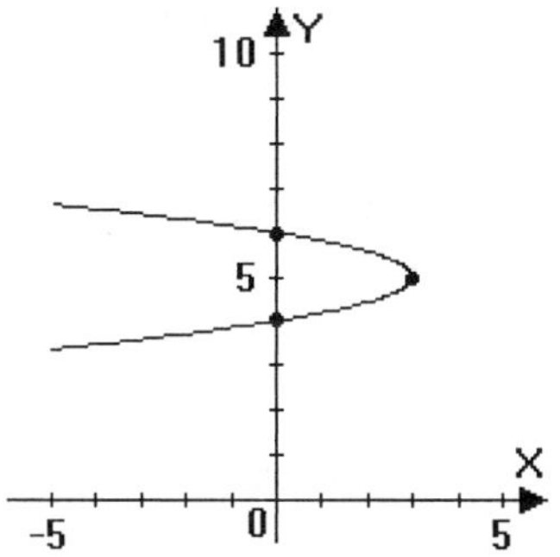


**29.** $x = -2(y+3)^2 - 1$

This is a parabola of the form $x = a(y-k)^2 + h$. Since $a = -2$ is negative, the parabola opens to the left. The vertex of the parabola is $(-1, -3)$. The axis of symmetry is $y = -3$. Replace $y$ with 0 to find the $x$–intercept.

$x = -2(0+3)^2 - 1 = -2(3)^2 - 1$

$= -2(9) - 1 = -18 - 1 = -19$

The $x$–intercept is 0. Replace $x$ with 0 to find the $y$–intercepts.

$0 = -2(y+3)^2 - 1$

$0 = -2(y^2 + 6x + 9) - 1$

$0 = -2y^2 - 12x - 18 - 1$

$0 = -2y^2 - 12x - 19$

$0 = 2y^2 + 12x + 19$

Solve using the quadratic formula.

$$y = \frac{-12 \pm \sqrt{12^2 - 4(2)(19)}}{2(2)}$$

$$= \frac{-12 \pm \sqrt{144 - 152}}{4}$$

$$= \frac{-12 \pm \sqrt{-8}}{4}$$

Since the solutions will be complex, there are no $y$–intercepts.

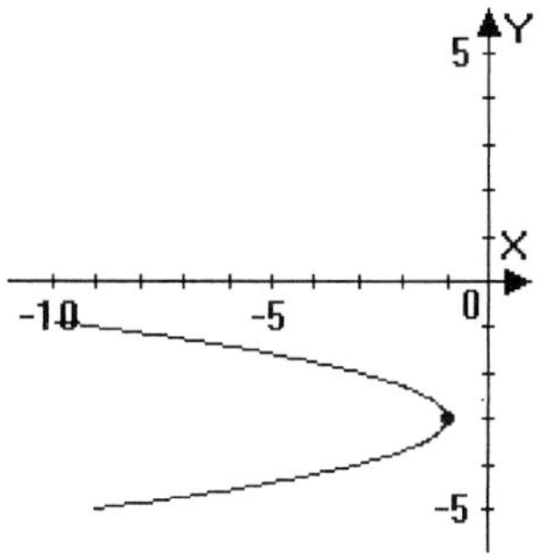


**31.** $x = \frac{1}{2}(y+2)^2 + 1$

This is a parabola of the form $x = a(y-k)^2 + h$. Since $a = \frac{1}{2}$ is positive, the parabola opens to the right. The vertex of the parabola is $(1, -2)$. The axis of symmetry is $y = -2$. Replace $y$ with 0 to find the $x$–intercept.

$x = \frac{1}{2}(0+2)^2 + 1$

$= \frac{1}{2}(4) + 1 = 2 + 1 = 3$

The $x$–intercept is 0. Replace $x$ with 0

to find the $y$–intercepts.

$$0=\frac{1}{2}(y+2)^2+1$$

$$0=\frac{1}{2}(y^2+2y+4)+1$$

$$0=\frac{1}{2}y^2+y+2+1$$

$$0=\frac{1}{2}y^2+y+3$$

$$0=y^2+2y+6$$

Solve using the quadratic formula.

$$y=\frac{-2\pm\sqrt{2^2-4(1)(6)}}{2(1)}$$

$$=\frac{-2\pm\sqrt{4-24}}{2}$$

$$=\frac{-2\pm\sqrt{-20}}{2}$$

Since the solutions will be complex, there are no $y$–intercepts.

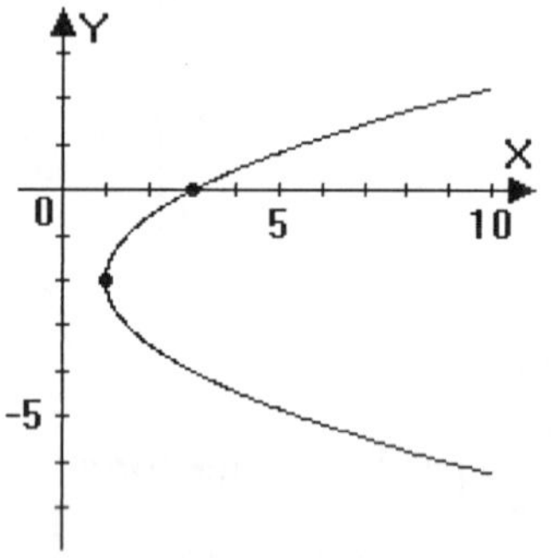


**33.** $x=y^2+2y-3$

This is a parabola of the form $x=ay^2+by+c$. Since $a=1$ is positive, the parabola opens to the right. The $y$–coordinate of the vertex is $-\frac{b}{2a}=-\frac{2}{2(1)}=-\frac{2}{2}=-1$. The $x$–coordinate of the vertex is $x=(-1)^2+2(-1)-3=1-2-3=-4$.

The vertex of the parabola is $(-4,-1)$.

The axis of symmetry is $y=-1$.

Replace $y$ with 0 to find the $x$–intercept.

$x=0^2+2(0)-3=0+0-3=-3$

The $x$–intercept is –3. Replace $x$ with 0 to find the $y$–intercepts.

$0=y^2+2y-3$

$0=(y+3)(y-1)$

Apply the zero product principle.

$y+3=0$ and $y-1=0$

$y=-3$ $\quad$ $y=1$

The $y$–intercepts are –3 and 1.

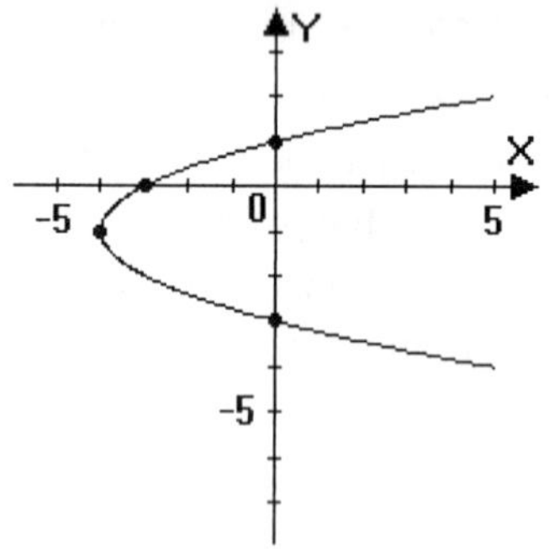

**35.** $x = -y^2 - 4y + 5$

This is a parabola of the form $x = ay^2 + by + c$. Since $a = -1$ is negative, the parabola opens to the left. The $y$–coordinate of the vertex is $-\frac{b}{2a} = -\frac{-4}{2(-1)} = -\frac{-4}{-2} = -2$. The $x$–coordinate of the vertex is

$x = -(-2)^2 - 4(-2) + 5 = -4 + 8 + 5 = 9.$

The vertex of the parabola is $(9, -2)$.

The axis of symmetry is $y = -2$.

Replace $y$ with 0 to find the $x$–intercept.

$x = -0^2 - 4(0) + 5 = 0 - 0 + 5 = 5$

The $x$–intercept is 5. Replace $x$ with 0 to find the $y$–intercepts.

$0 = -y^2 - 4y + 5$

$0 = y^2 + 4y - 5$

$0 = (y+5)(y-1)$

Apply the zero product principle.

$y + 5 = 0$ and $y - 1 = 0$

$y = -5$ $\quad$ $y = 1$

The $y$–intercepts are –5 and 1.

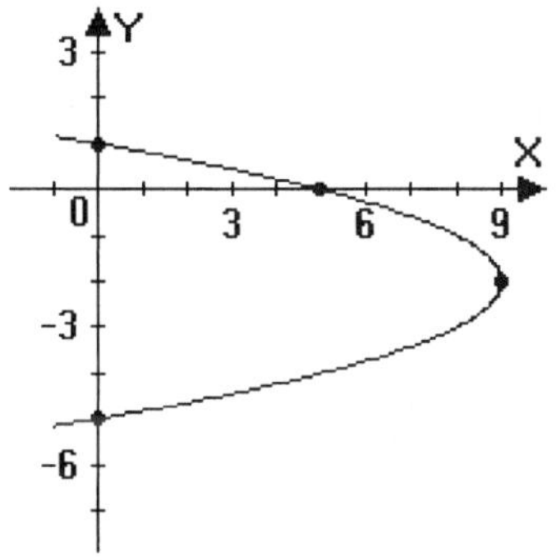


**37.** $x = y^2 + 6y$

This is a parabola of the form $x = ay^2 + by + c$. Since $a = 1$ is positive, the parabola opens to the right. The $y$–coordinate of the vertex is $-\frac{b}{2a} = -\frac{6}{2(1)} = -\frac{6}{2} = -3.$

The $x$–coordinate of the vertex is

$x = (-3)^2 + 6(-3) = 9 - 18 = -9.$

The vertex of the parabola is $(-9, -3)$.

The axis of symmetry is $y = -3$.

Replace $y$ with 0 to find the $x$–intercept.

$x = 0^2 + 6(0) = 0$

The $x$–intercept is 0. Replace $x$ with 0 to find the $y$–intercepts.

$0 = y^2 + 6y$

$0 = y(y+6)$

Apply the zero product principle.

$y = 0$ and $y + 6 = 0$

$y = -6$

The $y$–intercepts are –6 and 0.

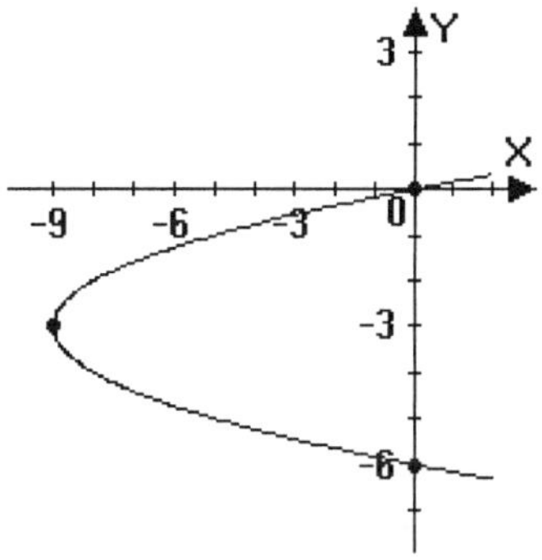

**39.** $x = -2y^2 - 4y$

This is a parabola of the form $x = ay^2 + by + c$. Since $a = -2$ is negative, the parabola opens to the left. The $y$–coordinate of the vertex is $-\frac{b}{2a} = -\frac{-4}{2(-2)} = -\frac{-4}{-4} = -1$. The $x$–coordinate of the vertex is

$$x = -2(-1)^2 - 4(-1) = -2(1) + 4$$
$$= -2 + 4 = 2$$

The vertex of the parabola is $(2, -1)$.

The axis of symmetry is $y = -1$.

Replace $y$ with 0 to find the $x$–intercept.

$$x = -2(0)^2 - 4(0) = -2(0) - 0 = 0$$

The $x$–intercept is 0. Replace $x$ with 0 to find the $y$–intercepts.

$$0 = -2y^2 - 4y$$
$$0 = y^2 + 2y$$
$$0 = y(y + 2)$$

Apply the zero product principle.

$y = 0$ and $y + 2 = 0$

$y = -2$

The $y$–intercepts are –2 and 0.

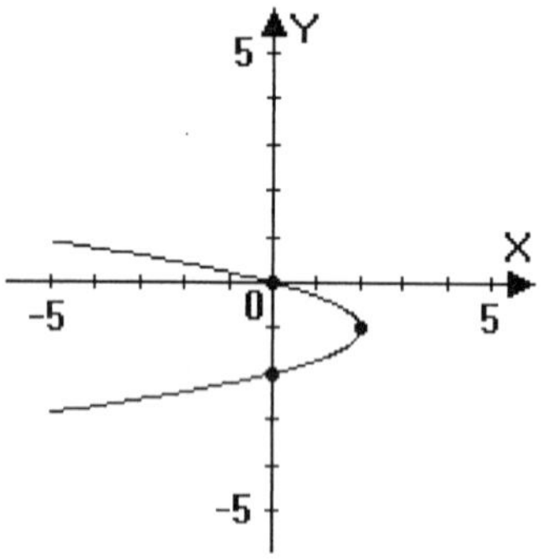


**41.** $x = -2y^2 - 4y + 1$

This is a parabola of the form $x = ay^2 + by + c$. Since $a = -2$ is negative, the parabola opens to the left. The $y$–coordinate of the vertex is $-\frac{b}{2a} = -\frac{-4}{2(-2)} = -\frac{-4}{-4} = -1$.

The $x$–coordinate of the vertex is

$$x = -2(-1)^2 - 4(-1) + 1 = -2(1) + 4 + 1$$
$$= -2 + 4 + 1 = 3$$

The vertex of the parabola is $(3, -1)$.

The axis of symmetry is $y = -1$.

Replace $y$ with 0 to find the $x$–intercept.

$$x = -2(0)^2 - 4(0) + 1 = -2(0) - 0 + 1$$
$$= 0 - 0 + 1 = 1$$

The $x$–intercept is 0. Replace $x$ with 0 to find the $y$–intercepts.

$$0 = -2y^2 - 4y + 1$$

Solve using the quadratic formula.

$$y = \frac{-b \pm \sqrt{b^2 - 4ac}}{2a}$$
$$= \frac{-(-4) \pm \sqrt{(-4)^2 - 4(-2)(1)}}{2(-2)}$$
$$= \frac{4 \pm \sqrt{16 + 8}}{-4} = \frac{4 \pm \sqrt{24}}{-4} = \frac{4 \pm 2\sqrt{6}}{-4}$$
$$= \frac{2(2 \pm \sqrt{6})}{-4} = \frac{2 \pm \sqrt{6}}{-2} = \frac{-(2 \pm \sqrt{6})}{2}$$
$$= \frac{-2 \pm \sqrt{6}}{2}$$

The $y$–intercepts are $\frac{-2 \pm \sqrt{6}}{2}$.

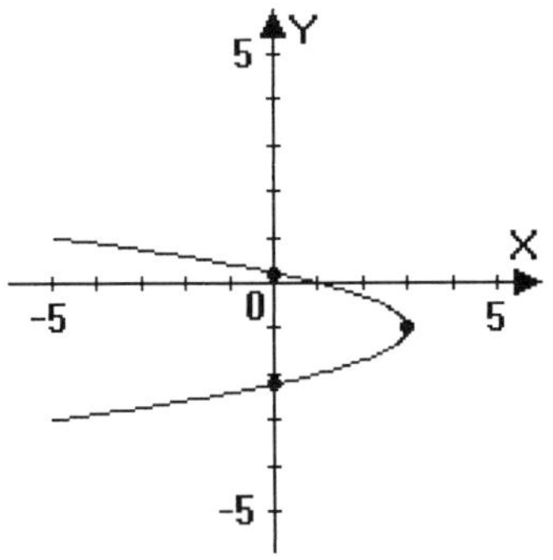


**43.** **a.** Since the squared term is $y$, the parabola is horizontal.

**b.** Since $a = 2$ is positive, the parabola opens to the right.

**c.** The vertex is the point $(2,1)$.

**45.** **a.** Since the squared term is $x$, the parabola is vertical.

**b.** Since $a = 2$ is positive, the parabola opens up.

**c.** The vertex is the point $(1,2)$.

**47.** **a.** Since the squared term is $x$, the parabola is vertical.

**b.** Since $a = -1$ is negative, the parabola opens down.

**c.** The vertex is the point $(-3,4)$.

**49.** **a.** Since the squared term is $y$, the parabola is horizontal.

**b.** Since $a = -1$ is negative, the parabola opens to the left.

**c.** The vertex is the point $(4,-3)$.

**51.** **a.** Since the squared term is $x$, the parabola is vertical.

**b.** Since $a = 1$ is positive, the parabola opens up.

**c.** The $x$–coordinate of the vertex is $-\frac{b}{2a} = -\frac{-4}{2(1)} = -\frac{-4}{2} = 2.$

The $y$–coordinate of the vertex is

$$f(2) = 2^2 - 4(2) - 1$$
$$= 4 - 8 - 1 = -5.$$

The vertex is the point $(2,-5)$.

**53.** **a.** Since the squared term is $y$, the parabola is horizontal.

**b.** Since $a = -1$ is negative, the parabola opens to the left.

**c.** The $y$–coordinate of the vertex is $-\frac{b}{2a} = -\frac{4}{2(-1)} = -\frac{4}{-2} = 2.$

The $x$–coordinate of the vertex is

$$f(2) = -(2)^2 + 4(2) + 1$$
$$= -4 + 8 + 1 = 5.$$

The vertex is the point $(5,2)$.

**55.** $x - 7 - 8y = y^2$

Since only one variable is squared, the graph of the equation is a parabola.

**57.** $4x^2 = 36 - y^2$

$4x^2 + y^2 = 36$

Because $x^2$ and $y^2$ have different positive coefficients, the equation's graph is an ellipse.

**59.** $x^2 = 36 + 4y^2$

$x^2 - 4y^2 = 36$

Because $x^2$ and $y^2$ have opposite signs, the equation's graph is a hyperbola.

**61.** $3x^2 = 12 - 3y^2$

$3x^2 + 3y^2 = 12$

Because $x^2$ and $y^2$ have the same positive coefficient, the equation's graph is a circle.

**63.** $3x^2 = 12 + 3y^2$

$3x^2 - 3y^2 = 12$

Because $x^2$ and $y^2$ have opposite signs, the equation's graph is a hyperbola.

**65.** $x^2 - 4y^2 = 16$

Because $x^2$ and $y^2$ have opposite signs, the equation's graph is a hyperbola.

$$\frac{x^2}{16} - \frac{4y^2}{16} = \frac{16}{16}$$

$$\frac{x^2}{16} - \frac{y^2}{4} = 1$$

The equation is in the form $\frac{x^2}{a^2} - \frac{y^2}{b^2} = 1$ with $a^2 = 16$, and $b^2 = 4$. We know the transverse axis lies on the $x$-axis and the vertices are $(-4,0)$ and $(4,0)$.

Because $a^2 = 16$ and $b^2 = 4$, $a = 4$ and $b = 2$. Construct a rectangle using –4 and 4 on the $x$–axis, and –2 and 2 on the $y$–axis. Draw extended diagonals to obtain the asymptotes. Graph the hyperbola.

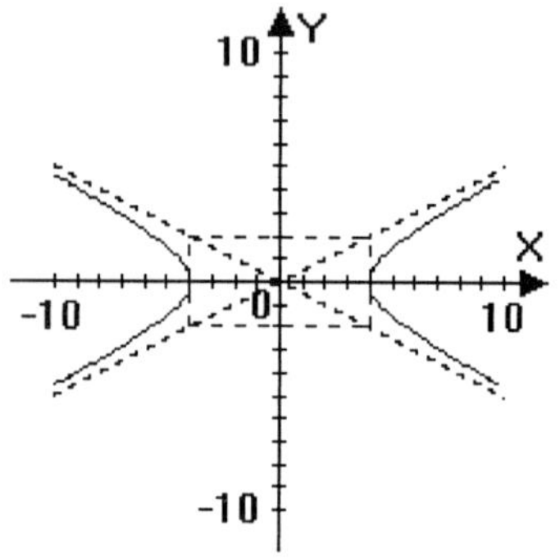


**67.** $4x^2 + 4y^2 = 16$

Because $x^2$ and $y^2$ have the same positive coefficient, the equation's graph is a circle.

$$\frac{4x^2}{4} + \frac{4y^2}{4} = \frac{16}{4}$$

$$x^2 + y^2 = 4$$

The center is $(0,0)$ and the radius is 2 units.

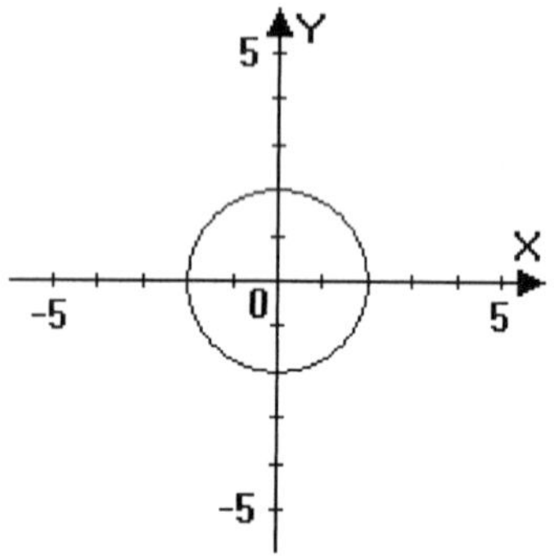


**69.** $x^2 + 4y^2 = 16$

Because $x^2$ and $y^2$ have different positive coefficients, the equation's graph is an ellipse.

$$\frac{x^2}{16} + \frac{4y^2}{16} = \frac{16}{16}$$

$$\frac{x^2}{16} + \frac{y^2}{4} = 1$$

Because the denominator of the $x^2$ – term is greater than the denominator of the $y^2$ – term, the major axis is horizontal. Since $a^2 = 16$, $a = 4$ and the vertices are $(-4,0)$ and $(4,0)$. Since $b^2 = 4$, $b = 2$ and endpoints of the minor axis are $(0,-2)$ and $(0,2)$.

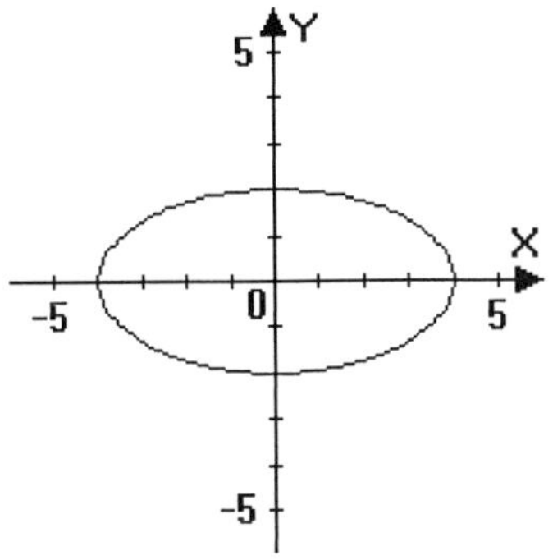


**71.** $x = (y-1)^2 - 4$

Since only one variable is squared, the graph of the equation is a parabola. This is a parabola of the form $x = a(y-k)^2 + h$. Since $a = 1$ is positive, the parabola opens to the right. The vertex of the parabola is $(-4,1)$. The axis of symmetry is $y = 1$. Replace $y$ with 0 to find the $x$–intercept.

$x = (0-1)^2 - 4 = (-1)^2 - 4 = 1 - 4 = -3$

The $x$–intercept is 0. Replace $x$ with 0 to find the $y$–intercepts.

$0 = (y-1)^2 - 4$

$0 = y^2 - 2y + 1 - 4$

$0 = y^2 - 2y - 3$

$0 = (y-3)(y+1)$

Apply the zero product principle.

$y - 3 = 0$ and $y + 1 = 0$

$y = 3$ $\qquad$ $y = -1$

The $y$–intercepts are –1 and 3.

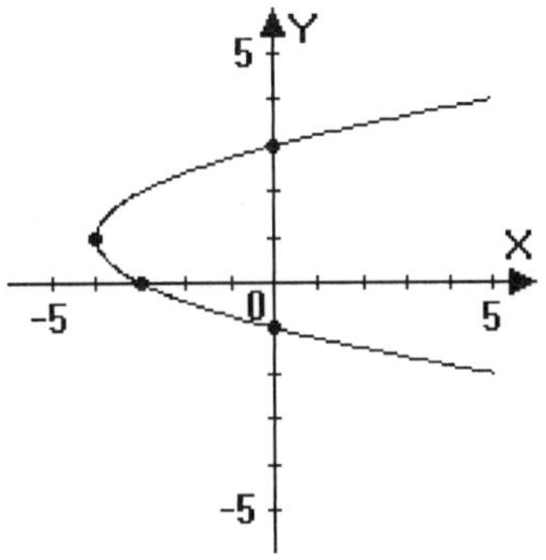


**73.** $(x-2)^2 + (y+1)^2 = 16$

Because $x^2$ and $y^2$ have the same positive coefficient, the equation's graph is a circle.

The center is $(2,-1)$ and the radius is 4 units.

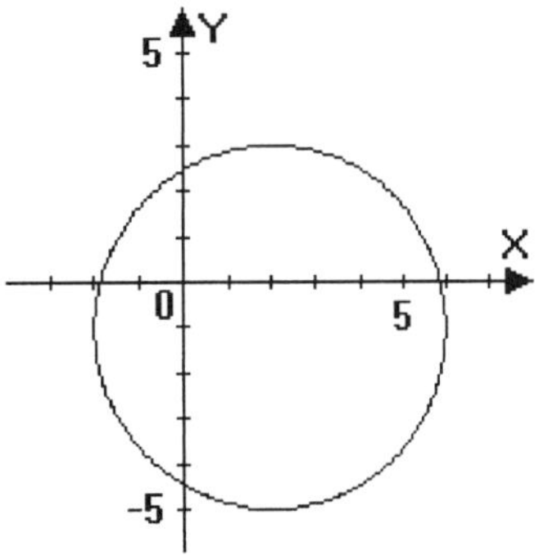


**75.** The y-coordinate of the vertex is

$y = -\frac{b}{2a} = -\frac{6}{2(1)} = -3$

The x-coordinate of the vertex is

$x = (-3)^2 + 6(-3) + 5$

$= 9 - 18 + 5$

$= -4$

The vertex is $(-4,-3)$.

Since the squared term is $y$ and $a > 0$, the graph opens to the right.
Domain: $\{x \mid x \geq -4\}$ or $[-4, \infty)$

Range: $\{y \mid y \text{ is a real number}\}$ or $(-\infty, \infty)$

The relation is not a function.

**76.** The y-coordinate of the vertex is

$$y = -\frac{b}{2a} = -\frac{(-2)}{2(1)} = 1$$

The x-coordinate of the vertex is

$$\begin{aligned} x &= (1)^2 - 2(1) - 5 \\ &= 1 - 2 - 5 \\ &= -6 \end{aligned}$$

The vertex is $(-6, 1)$.

Since the squared term is $y$ and $a > 0$, the graph opens to the right.
Domain: $\{x \mid x \geq -6\}$ or $[-6, \infty)$

Range: $\{y \mid y \text{ is a real number}\}$ or $(-\infty, \infty)$

The relation is not a function.

**77.** The x-coordinate of the vertex is

$$x = -\frac{b}{2a} = -\frac{(4)}{2(-1)} = 2$$

The y-coordinate of the vertex is

$$\begin{aligned} y &= -(2)^2 + 4(2) - 3 \\ &= -4 + 8 - 3 \\ &= 1 \end{aligned}$$

The vertex is $(2, 1)$.

Since the squared term is $x$ and $a < 0$, the graph opens down.
Domain: $\{x \mid x \text{ is a real number}\}$ or $(-\infty, \infty)$

Range: $\{y \mid y \leq 1\}$ or $(-\infty, 1]$

The relation is a function.

**78.** The x-coordinate of the vertex is

$$x = -\frac{b}{2a} = -\frac{(-4)}{2(-1)} = -2$$

The y-coordinate of the vertex is

$$\begin{aligned} y &= -(-2)^2 - 4(-2) + 4 \\ &= -4 + 8 + 4 \\ &= 8 \end{aligned}$$

The vertex is $(-2, 8)$.

Since the squared term is $x$ and $a < 0$, the graph opens down.
Domain: $\{x \mid x \text{ is a real number}\}$ or $(-\infty, \infty)$

Range: $\{y \mid y \leq 8\}$ or $(-\infty, 8]$

The relation is a function.

**79.** The equation is in the form

$x = a(y - k)^2 + h$

From the equation, we can see that the vertex is $(3, 1)$.

Since the squared term is $y$ and $a < 0$, the graph opens to the left.
Domain: $\{x \mid x \leq 3\}$ or $(-\infty, 3]$

Range: $\{y \mid y \text{ is a real number}\}$ or $(-\infty, \infty)$

The relation is not a function.

**80.** The equation is in the form

$x = a(y - k)^2 + h$

From the equation, we can see that the vertex is $(-2, 1)$.

Since the squared term is $y$ and $a < 0$, the graph opens to the left.
Domain: $\{x \mid x \leq -2\}$ or $(-\infty, -2]$

Range: $\{y \mid y \text{ is a real number}\}$ or $(-\infty, \infty)$

The relation is not a function.

**81.** $x = (y-2)^2 - 4$

$y = -\frac{1}{2}x$

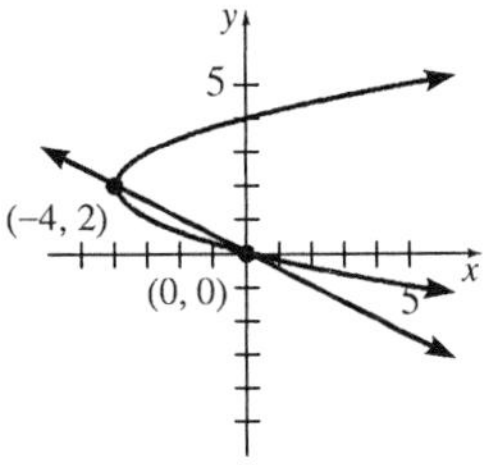


Check $(-4, 2)$:

$-4 = (2-2)^2 - 4$ $\quad$ $2 = -\frac{1}{2}(-4)$

$-4 = 0 - 4$ $\quad$ $2 = 2$

$-4 = -4$ $\quad$ true

true

Check $(0, 0)$:

$0 = (0-2)^2 - 4$ $\quad$ $0 = -\frac{1}{2}(0)$

$0 = 4 - 4$ $\quad$ $0 = 0$

$0 = 0$ $\quad$ true

true

The solution set is $\{(-4, 2), (0, 0)\}$.

**82.** $x = (y-3)^2 + 2$

$x + y = 5$

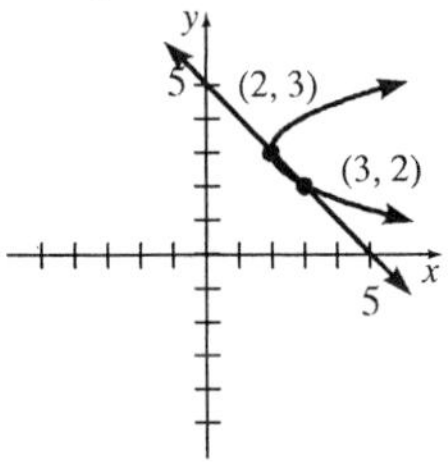


Check $(2, 3)$:

$2 = (3-3)^2 + 2$ $\quad$ $2 + 3 = 5$

$2 = 0 + 2$ $\quad$ $5 = 5$

$2 = 2$ $\quad$ true

true

Check $(3, 2)$:

$3 = (2-3)^2 + 2$ $\quad$ $3 + 2 = 5$

$3 = 1 + 2$ $\quad$ $5 = 5$

$3 = 3$ $\quad$ true

true

The solution set is $\{(2, 3), (3, 2)\}$.

**83.** $x = y^2 - 3$

$x = y^2 - 3y$

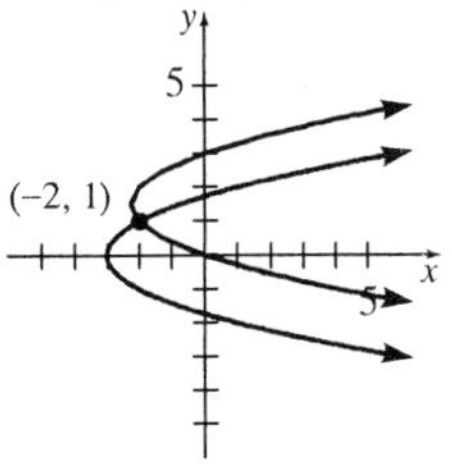


Check $(-2, 1)$:

$-2 = (1)^2 - 3$ $\quad$ $-2 = (1)^2 - 3(1)$

$-2 = 1 - 3$ $\quad$ $-2 = 1 - 3$

$-2 = -2$ true $\quad$ $-2 = -2$ true

The solution set is $\{(-2, 1)\}$.

**84.** $x = y^2 - 5$

$x^2 + y^2 = 25$

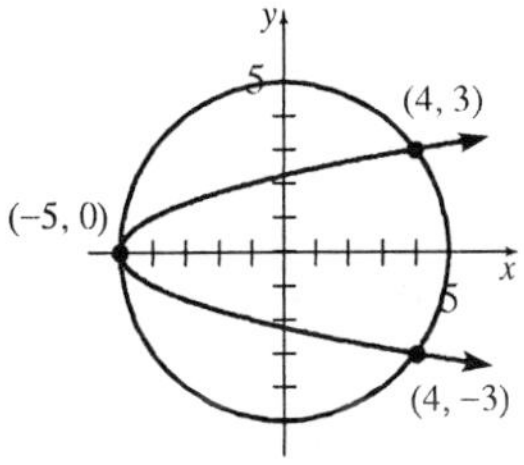


Check $(-5, 0)$:

$-5 = 0^2 - 5$

$-5 = 0 - 5$

$-5 = -5$ true

$(-5)^2 + 0^2 = 25$

$25 + 0 = 25$

$25 = 25$ true

Check $(4,-3)$:

$4 = (-3)^2 - 5$

$4 = 9 - 5$

$4 = 4$ true

$(4)^2 + (-3)^2 = 25$

$16 + 9 = 25$

$25 = 25$ true

Check $(4,3)$:

$4 = (3)^2 - 5 \qquad (4)^2 + (3)^2 = 25$

$4 = 9 - 5 \qquad 16 + 9 = 25$

$4 = 4$ true $\qquad 25 = 25$ true

The solution set is $\{(-5,0),(4,-3),(4,3)\}$.

**85.** $x = (y+2)^2 - 1$

$(x-2)^2 + (y+2)^2 = 1$

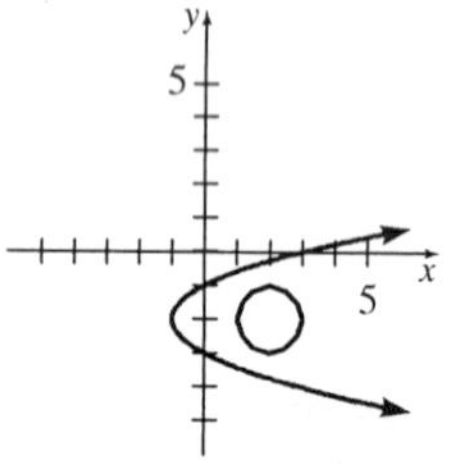


The two graphs do not cross. Therefore, the solution set is the empty set, $\{\ \}$ or $\varnothing$.

**86.** $x = 2y^2 + 4y + 5$

$(x+1)^2 + (y-2)^2 = 1$

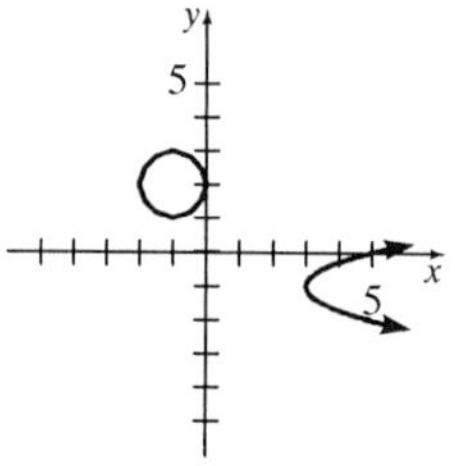


The two graphs do not cross. Therefore, the solution set is the empty set, $\{\ \}$ or $\varnothing$.

**87.** **a.** $y = ax^2$

$316 = a(1750)^2$

$316 = a(3062500)$

$\frac{316}{3062500} = a$

$0.0001032 = a$

The equation is $y = 0.0001032x^2$.

**b.** To find the height of the cable 1000 feet from the tower, find $y$ when $x = 1750 - 1000 = 750$.

$y = 0.0001032(750)^2$

$= 0.0001032(562,500) = 58.05$

The height of the cable is about 58 feet.

**89.** **a.** $y = ax^2$

$2 = a(6)^2$

$2 = a(36)$

$\frac{2}{36} = a$

$\frac{1}{18} = a$

The equation is $y = \frac{1}{18}x^2$.

**b.** $a = \dfrac{1}{4p}$

$\dfrac{1}{18} = \dfrac{1}{4p}$

$4p = 18$

$p = \dfrac{18}{4} = 4.5$

The receiver should be placed 4.5 feet from the base of the dish.

**91.** Answers will vary.

**93.** Answers will vary.

**95.** Answers will vary.

**97.** Answers will vary.

**99.**

$$y^2 + 2y - 6x + 13 = 0$$

$$y^2 + 2y + (-6x + 13) = 0$$

$a = 1 \qquad b = 2 \qquad c = -6x + 13$

$$y = \frac{-2 \pm \sqrt{2^2 - 4(1)(-6x+13)}}{2(1)}$$

$$= \frac{-2 \pm \sqrt{4 - 4(-6x+13)}}{2}$$

$$= \frac{-2 \pm \sqrt{4 + 24x - 52}}{2}$$

$$= \frac{-2 \pm \sqrt{24x - 48}}{2} = \frac{-2 \pm \sqrt{4(6x-12)}}{2}$$

$$= \frac{-2 \pm 2\sqrt{6x-12}}{2}$$

$$= -1 \pm \sqrt{6x-12}$$

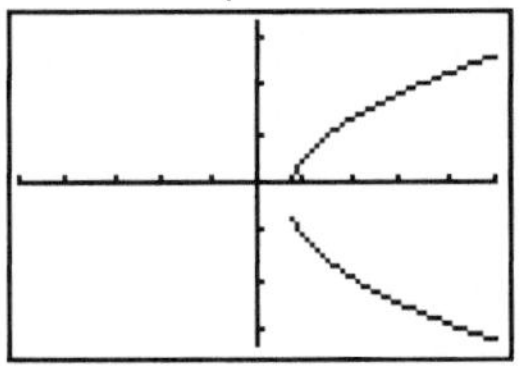

**101.** Answers will vary. For example, consider Exercise 19.

$$x = (y-2)^2 - 4$$

$$x + 4 = (y-2)^2$$

$$\pm\sqrt{x+4} = y - 2$$

$$2 \pm \sqrt{x+4} = y$$

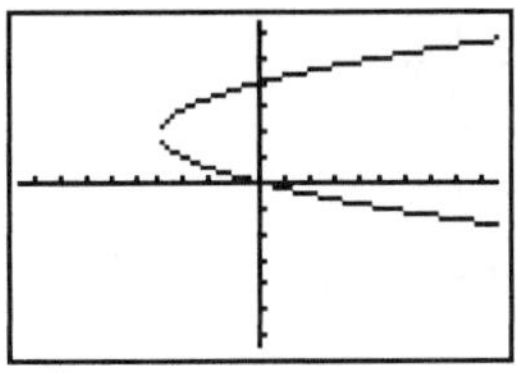

**103.** Answers will vary.

**105.** $f(x) = 2^{1-x}$

| $x$ | $f(x)$ |
|---|---|
| -2 | 8 |
| -1 | 4 |
| 0 | 2 |
| 1 | 1 |
| 2 | $\frac{1}{2}$ |

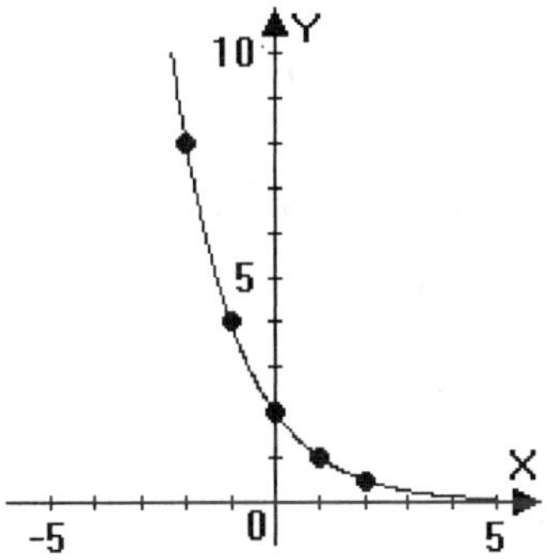

**106.** $f(x)=\frac{1}{3}x-5$

$y=\frac{1}{3}x-5$

Interchange $x$ and $y$ and solve for $y$.

$x=\frac{1}{3}y-5$

$x+5=\frac{1}{3}y$

$3x+15=y$

$f^{-1}(x)=3x+15$

**107.** $4x-3y=12$

$3x-4y=2$

Multiply the first equation by –3 and the second equation by 4.

$-12x+9y=-36$

$12x-16y=8$

$-7y=-28$

$y=4$

Back-substitute 4 for $y$ to find $x$.

$4x-3(4)=12$

$4x-12=12$

$4x=24$

$x=6$

The solution is $(6,4)$ and the solution set is $\{(6,4)\}$.

## 10.5 Exercise Set

**1.** $x+y=2$

$y=x^2-4$

Substitute $x^2+4$ for $y$ in the first equation and solve for $x$.

$x+(x^2-4)=2$

$x+x^2-4=2$

$x^2+x-6=0$

$(x+3)(x-2)=0$

Apply the zero product principle.

$x+3=0$ or $x-2=0$

$x=-3$ $\quad$ $x=2$

Substitute –3 and 2 for $x$ in the second equation to find $y$.

$x=-3$ or $x=2$

$y=(-3)^2-4$ $\quad$ $y=2^2-4$

$y=9-4$ $\quad$ $y=4-4$

$y=5$ $\quad$ $y=0$

The solutions are $(-3,5)$ and $(2,0)$ and the solution set is $\{(-3,5),\ (2,0)\}$.

**3.** $x+y=2$

$y=x^2-4x+4$

Substitute $x^2-4x+4$ for $y$ in the first equation and solve for $x$.

$x+x^2-4x+4=2$

$x^2-3x+4=2$

$x^2-3x+2=0$

$(x-2)(x-1)=0$

Apply the zero product principle.

$x-2=0$ or $x-1=0$

$x=2$ $\quad$ $x=1$

Substitute 1 and 2 for $x$ to find $y$.

$x = 2$ or $x = 1$

$x + y = 2$ $\quad$ $x + y = 2$

$2 + y = 2$ $\quad$ $1 + y = 2$

$y = 0$ $\quad$ $y = 1$

The solutions are $(2,0)$ and $(1,1)$ and the solution set is $\{(1,1),(2,0)\}$.

**5.** $y = x^2 - 4x - 10$

$y = -x^2 - 2x + 14$

Substitute $-x^2 - 2x + 14$ for $y$ in the first equation and solve for $x$.

$$-x^2 - 2x + 14 = x^2 - 4x - 10$$
$$0 = 2x^2 - 2x - 24$$
$$0 = x^2 - x - 12$$
$$0 = (x - 4)(x + 3)$$

Apply the zero product principle.

$x - 4 = 0$ or $x + 3 = 0$

$x = 4$ $\quad$ $x = -3$

Substitute 3 and 4 for $x$ to find $y$.

$x = 4$

$y = 4^2 - 4(4) - 10$

$= 16 - 16 - 10 = -10$

$x = -3$

$y = (-3)^2 - 4(-3) - 10$

$= 9 + 12 - 10 = 11$

The solutions are $(4,-10)$ and $(-3,11)$ and the solution set is $\{(-3,11),(4,-10)\}$.

**7.** $x^2 + y^2 = 25$

$x - y = 1$

Solve the second equation for $x$.

$x - y = 1$

$x = y + 1$

Substitute $y + 1$ for $x$ to find $y$.

$$x^2 + y^2 = 25$$
$$(y+1)^2 + y^2 = 25$$
$$y^2 + 2y + 1 + y^2 = 25$$
$$2y^2 + 2y + 1 = 25$$
$$2y^2 + 2y - 24 = 0$$
$$y^2 + y - 12 = 0$$
$$(y+4)(y-3) = 0$$

Apply the zero product principle.

$y + 4 = 0$ or $y - 3 = 0$

$y = -4$ $\quad$ $y = 3$

Substitute –4 and 3 for $y$ to find $x$.

$y = -4$ $\quad$ $y = 3$

$x = -4 + 1$ $\quad$ $x = 3 + 1$

$x = -3$ $\quad$ $x = 4$

The solutions are $(-3,-4)$ and $(4,3)$ and the solution set it $\{(-3,-4),$ $(4,3)\}$.

**9.** $xy = 6$

$2x - y = 1$

Solve the first equation for $y$.

$xy = 6$

$y = \frac{6}{x}$

Substitute $\frac{6}{x}$ for $y$ in the second equation and solve for $x$.

$$2x - \frac{6}{x} = 1$$

$$x\left(2x - \frac{6}{x}\right) = x(1)$$

$$2x^2 - 6 = x$$

$$2x^2 - x - 6 = 0$$

$$(2x+3)(x-2) = 0$$

Apply the zero product principle.

$x - 2 = 0$ or $2x + 3 = 0$

$x = 2$ $\quad 2x = -3$

$x = -\frac{3}{2}$

Substitute 2 and $-\frac{3}{2}$ for $x$ to find $y$.

$x = 2$ or $x = -\frac{3}{2}$

$2y = 6$

$y = 3$ $\quad -\frac{3}{2}y = 6$

$$-\frac{2}{3}\left(-\frac{3}{2}\right)y = \left(-\frac{2}{3}\right)6$$

$$y = -4$$

The solutions are $(2,3)$ and $\left(-\frac{3}{2},-4\right)$ and the solution set is $\left\{(2,3),\left(-\frac{3}{2},-4\right)\right\}$.

**11.** $y^2 = x^2 - 9$

$2y = x - 3$

Solve the second equation for $x$.

$2y = x - 3$

$2y + 3 = x$

Substitute $2y + 3$ for $x$ to find $y$.

$$y^2 = (2y+3)^2 - 9$$

$$y^2 = 4y^2 + 12y + 9 - 9$$

$$y^2 = 4y^2 + 12y$$

$$0 = 3y^2 + 12y$$

$$0 = 3y(y+4)$$

Apply the zero product principle.

$3y = 0$ or $y + 4 = 0$

$y = 0$ $\quad y = -4$

Substitute –4 and 0 for $y$ to find $x$.

$y = 0$ or $y = -4$

$2(0) + 3 = x$ $\quad 2(-4) + 3 = x$

$3 = x$ $\quad -8 + 3 = x$

$-5 = x$

The solutions are $(3,0)$ and $(-5,-4)$ and the solution set is $\{(-5,-4),(3,0)\}$.

**13.** $xy = 3$

$x^2 + y^2 = 10$

Solve the first equation for $y$.

$xy = 3$

$$y = \frac{3}{x}$$

Substitute $\frac{3}{x}$ for $y$ to find $x$.

$$x^2 + \left(\frac{3}{x}\right)^2 = 10$$

$$x^2 + \frac{9}{x^2} = 10$$

$$x^2\left(x^2 + \frac{9}{x^2}\right) = x^2(10)$$

$$x^4 + 9 = 10x^2$$

$$x^4 - 10x^2 + 9 = 0$$

$$(x^2 - 9)(x^2 - 1) = 0$$

$$(x+3)(x-3)(x+1)(x-1) = 0$$

Apply the zero product principle.

$x+3=0 \qquad x-3=0$

$x=-3 \qquad x=3$

$x+1=0 \qquad x-1=0$

$x=-1 \qquad x=1$

Substitute $\pm 1$ and $\pm 3$ for $x$ to find $y$.

$x=-3 \qquad x=3$

$y=\frac{3}{-3} \qquad y=\frac{3}{3}$

$y=-1 \qquad y=1$

$x=-1 \qquad x=1$

$y=\frac{3}{-1} \qquad y=\frac{3}{1}$

$y=-3 \qquad y=3$

The solutions are $(-3,-1),(-1,-3)$, $(1,3)$ and $(3,1)$ and the solution set is $\{(-3,-1),(-1,-3),(1,3),(3,1)\}$.

**15.** $x+y=1$

$x^2+xy-y^2=-5$

Solve the first equation for $y$.

$x+y=1$

$y=-x+1$

Substitute $-x+1$ for $y$ and solve for $x$.

$$x^2+x(-x+1)-(-x+1)^2=-5$$
$$x^2-x^2+x-(x^2-2x+1)=-5$$
$$x^2-x^2+x-x^2+2x-1=-5$$
$$-x^2+3x-1=-5$$
$$-x^2+3x+4=0$$
$$x^2-3x-4=0$$
$$(x-4)(x+1)=0$$

Apply the zero product principle.

$x-4=0$ or $x+1=0$

$x=4 \qquad x=-1$

Substitute $-1$ and 4 for $x$ to find $y$.

$x=4$ or $x=-1$

$y=-4+1 \qquad y=-(-1)+1$

$y=-3 \qquad y=1+1$

$y=2$

The solutions are $(4,-3)$ and $(-1,2)$ and the solution set is $\{(4,-3),(-1,2)\}$.

**17.** $x+y=1$

$(x-1)^2+(y+2)^2=10$

Solve the first equation for $y$.

$x+y=1$

$y=-x+1$

Substitute $-x+1$ for $y$ to find $x$.

$$(x-1)^2+((-x+1)+2)^2=10$$
$$(x-1)^2+(-x+1+2)^2=10$$
$$(x-1)^2+(-x+3)^2=10$$
$$x^2-2x+1+x^2-6x+9=10$$
$$2x^2-8x+10=10$$
$$2x^2-8x=0$$
$$2x(x-4)=0$$

Apply the zero product principle.

$2x=0$ or $x-4=0$

$x=0 \qquad x=4$

Substitute 0 and 4 for $x$ to find $y$.

$x=0$ or $x=4$

$y=-0+1 \qquad y=-4+1$

$y=1 \qquad y=-3$

The solutions are $(0,1)$ and $(4,-3)$ and the solution set is $\{(0,1),\ (4,-3)\}$.

**19.** Solve the system by addition.

$$\begin{aligned} x^2 + y^2 &= 13 \\ x^2 - y^2 &= 5 \\ \hline 2x^2 &= 18 \\ x^2 &= 9 \\ x &= \pm 3 \end{aligned}$$

Substitute $\pm 3$ for $x$ to find $y$.

$$\begin{aligned} x &= \pm 3 \\ (\pm 3)^2 + y^2 &= 13 \\ 9 + y^2 &= 13 \\ y^2 &= 4 \\ y &= \pm 2 \end{aligned}$$

The solutions are $(-3,-2),(-3,2)$, $(3,-2)$ and $(3,2)$ and the solution set is $\{(-3,-2),(-3,2),(3,-2),(3,2)\}$.

**21.**

$$\begin{aligned} x^2 - 4y^2 &= -7 \\ 3x^2 + y^2 &= 31 \end{aligned}$$

Multiply the first equation by –3 and add to the second equation.

$$\begin{aligned} -3x^2 + 12y^2 &= 21 \\ 3x^2 + y^2 &= 31 \\ \hline 13y^2 &= 52 \\ y^2 &= 4 \\ y &= \pm 2 \end{aligned}$$

Substitute –2 and 2 for $y$ to find $x$.

$$\begin{aligned} y &= \pm 2 \\ 3x^2 + (\pm 2)^2 &= 31 \\ 3x^2 + 4 &= 31 \\ 3x^2 &= 27 \\ x^2 &= 9 \\ x &= \pm 3 \end{aligned}$$

The solutions are $(-3,-2),(-3,2)$, $(3,-2)$ and $(3,2)$ and the solution set is $\{(-3,-2),(-3,2),(3,-2),(3,2)\}$.

**23.**

$$\begin{aligned} 3x^2 + 4y^2 - 16 &= 0 \\ 2x^2 - 3y^2 - 5 &= 0 \end{aligned}$$

Multiply the first equation by 3 and the second equation by 4 and solve by addition.

$$\begin{aligned} 9x^2 + 12y^2 - 48 &= 0 \\ 8x^2 - 12y^2 - 20 &= 0 \\ \hline 17x^2 - 68 &= 0 \\ 17x^2 &= 68 \\ x^2 &= 4 \\ x &= \pm 2 \end{aligned}$$

Substitute $\pm 2$ for $x$ to find $y$.

$$\begin{aligned} x &= \pm 2 \\ 2(\pm 2)^2 - 3y^2 - 5 &= 0 \\ 2(4) - 3y^2 - 5 &= 0 \\ 8 - 3y^2 - 5 &= 0 \\ 3 - 3y^2 &= 0 \\ 3 &= 3y^2 \\ 1 &= y^2 \\ \pm 1 &= y \end{aligned}$$

The solutions are $(-2,-1),(-2,1)$, $(2,-1)$ and $(2,1)$ and the solution set is $\{(-2,-1),(-2,1),(2,-1),(2,1)\}$.

**25.** $x^2 + y^2 = 25$

$(x-8)^2 + y^2 = 41$

Multiply the first equation by –1 and solve by addition.

$$\begin{array}{r} -x^2 \quad\quad - y^2 = -25 \\ (x-8)^2 + y^2 = \ \ 41 \\ \hline -x^2 + (x-8)^2 = 16 \end{array}$$

$-x^2 + x^2 - 16x + 64 = 16$

$-16x + 64 = 16$

$-16x = -48$

$x = 3$

Substitute 3 for $x$ to find $y$.

$x = 3$

$3^2 + y^2 = 25$

$6 + y^2 = 25$

$y^2 = 16$

$y = \pm 4$

The solutions are $(3,-4)$ and $(3,4)$ and the solution set is $\{(3,-4),\ (3,4)\}$.

**27.** $y^2 - x = 4$

$x^2 + y^2 \quad = 4$

Multiply the first equation by –1 and solve by addition.

$$\begin{array}{r} -y^2 + x = -4 \\ x^2 + y^2 \quad = \ \ 4 \\ \hline x^2 + x = 0 \end{array}$$

$x(x+1) = 0$

Apply the zero product principle.

$x = 0$ or $x + 1 = 0$

$x = -1$

Substitute –1 and 0 for $x$ to find $y$.

$x = 0$ or $x = -1$

$y^2 - 0 = 4$ $\quad$ $y^2 - (-1) = 4$

$y^2 = 4$ $\quad$ $y^2 + 1 = 4$

$y = \pm 2$ $\quad$ $y^2 = 3$

$y = \pm\sqrt{3}$

The solutions are $(0,-2),(0,2)$, $(-1,-\sqrt{3})$ and $(-1,\sqrt{3})$ and the solution set is $\{(-1,-\sqrt{3}),(-1,\sqrt{3}),(0,-2),(0,2)\}$.

**29.** $3x^2 + 4y^2 = 16$

$2x^2 - 3y^2 = 5$

Multiply the first equation by –2 and the second equation by 3 and solve by addition.

$$\begin{array}{r} -6x^2 - 8y^2 = -32 \\ 6x^2 - 9y^2 = \ \ 15 \\ \hline -17y^2 = -17 \end{array}$$

$y^2 = 1$

$y = \pm 1$

Substitute $\pm 1$ for $y$ to find $x$.

$y = \pm 1$

$3x^2 + 4(\pm 1)^2 = 16$

$3x^2 + 4(1) = 16$

$3x^2 + 4 = 16$

$3x^2 = 12$

$x^2 = 4$

$x = \pm 2$

The solutions are $(-2,1),(2,1)$, $(-2,-1)$ and $(2,-1)$ and the solution set is $\{(-2,-1),(-2,1),(2,-1),\ (2,1)\}$.

**31.** $2x^2 + y^2 = 18$

$xy = 4$

Solve the second equation for $y$.

$xy = 4 \rightarrow y = \frac{4}{x}$

Substitute $\frac{4}{x}$ for $y$ in the second equation and solve for $x$.

$$2x^2 + \left(\frac{4}{x}\right)^2 = 18$$

$$2x^2 + \frac{16}{x^2} = 18$$

$$x^2\left(2x^2 + \frac{16}{x^2}\right) = x^2(18)$$

$$2x^4 + 16 = 18x^2$$

$$2x^4 - 18x^2 + 16 = 0$$

$$x^4 - 9x^2 + 8 = 0$$

$$(x^2 - 8)(x^2 - 1) = 0$$

$$(x^2 - 8)(x+1)(x-1) = 0$$

Apply the zero product principle.

$x^2 - 8 = 0 \qquad x+1 = 0 \qquad x-1 = 0$

$x^2 = 8 \qquad x = -1 \qquad x = 1$

$x = \pm\sqrt{8}$

$x = \pm 2\sqrt{2}$

Substitute $\pm 2\sqrt{2}$ and for $x$ to find $y$.

$x = 1 \qquad x = -1$

$y = \frac{4}{1} \qquad y = \frac{4}{-1}$

$y = 4 \qquad y = -4$

$x = 2\sqrt{2} \qquad x = -2\sqrt{2}$

$y = \frac{4}{2\sqrt{2}} \qquad y = \frac{4}{-2\sqrt{2}}$

$y = \frac{2\sqrt{2}}{2} \qquad y = -\frac{2\sqrt{2}}{2}$

$y = \sqrt{2} \qquad y = -\sqrt{2}$

The solutions are $(2\sqrt{2}, \sqrt{2})$, $(-2\sqrt{2}, -\sqrt{2})$, $(1, 4)$ and $(-1, -4)$ and the solution set is $\{(-2\sqrt{2}, -\sqrt{2}), (-1, -4), (1, 4), (2\sqrt{2}, \sqrt{2})\}$.

**33.** $x^2 + 4y^2 = 20$

$x + 2y = 6$

Solve the second equation for $x$.

$x + 2y = 6$

$x = 6 - 2y$

Substitute $6 - 2y$ for $x$ to find $y$.

$$(6 - 2y)^2 + 4y^2 = 20$$

$$36 - 24y + 4y^2 + 4y^2 = 20$$

$$36 - 24y + 8y^2 = 20$$

$$8y^2 - 24y + 16 = 0$$

$$y^2 - 3y + 2 = 0$$

$$(y-2)(y-1) = 0$$

Apply the zero product principle.

$y - 2 = 0$ or $y - 1 = 0$

$y = 2 \qquad y = 1$

Substitute 1 and 2 for $y$ to find $x$.

$y = 2$ or $y = 1$

$x = 6 - 2(2) \qquad x = 6 - 2(1)$

$x = 6 - 4 \qquad x = 6 - 2$

$x = 2 \qquad x = 4$

The solutions are $(2, 2)$ and $(4, 1)$ and the solution set is $\{(2, 2), (4, 1)\}$.

**35.** Eliminate $y$ by adding the two equations.

$$x^3 + y = 0$$
$$\underline{x^2 - y = 0}$$
$$x^3 + x^2 = 0$$
$$x^2(x+1) = 0$$

Apply the zero product principle.

$x = 0$ or $x = -1$

Substitute –1 and 0 for $x$ to find $y$.

$$x = 0 \quad \text{or} \quad x = -1$$
$$0^2 - y = 0 \qquad (-1)^2 - y = 0$$
$$-y = 0 \qquad 1 - y = 0$$
$$y = 0 \qquad -y = -1$$
$$y = 1$$

The solutions are $(0,0)$ and $(-1,1)$ and the solution set is $\{(-1,1),\ (0,0)\}$.

**37.** $x^2 + (y-2)^2 = 4$

$$x^2 - 2y = 0$$

Solve the second equation for $x^2$.

$$x^2 - 2y = 0$$
$$x^2 = 2y$$

Substitute $2y$ for $x^2$ in the first equation and solve for $y$.

$$2y + (y-2)^2 = 4$$
$$2y + y^2 - 4y + 4 = 4$$
$$y^2 - 2y + 4 = 4$$
$$y^2 - 2y = 0$$
$$y(y-2) = 0$$

Apply the zero product principle.

$y = 0$ or $y - 2 = 0$

$$y = 2$$

Substitute 0 and $\frac{4}{5}$ for $y$ to find $x$.

$$y = 0 \quad \text{or} \quad y = 2$$
$$x^2 = 2(0) \qquad x^2 = 2(2)$$
$$x^2 = 0 \qquad x^2 = 4$$
$$x = 0 \qquad x = \pm 2$$

The solutions are $(0,0), (-2,2)$ and $(2,2)$ and the solution set is $\{(0,0), (-2,2), (2,2)\}$.

**39.** $y = (x+3)^2$

$$x + 2y = -2$$

Substitute $(x+3)^2$ for $y$ in the second equation.

$$x + 2(x+3)^2 = -2$$
$$x + 2(x^2 + 6x + 9) = -2$$
$$x + 2x^2 + 12x + 18 = -2$$
$$2x^2 + 13x + 18 = -2$$
$$2x^2 + 13x + 20 = 0$$
$$(2x+5)(x+4) = 0$$

Apply the zero product principle.

$2x + 5 = 0$ or $x + 4 = 0$

$$x = -\frac{5}{2} \qquad x = -4$$

Substitute $-\frac{5}{2}$ and –4 for $x$ to find $y$.

$$x = -\frac{5}{2} \quad \text{or} \quad x = -4$$
$$-\frac{5}{2} + 2y = -2 \qquad -4 + 2y = -2$$
$$-5 + 4y = -4 \qquad 2y = 2$$
$$4y = 1 \qquad y = 1$$
$$y = \frac{1}{4}$$

The solutions are $\left(-\frac{5}{2}, \frac{1}{4}\right)$ and $(-4,1)$ and the solution set is $\left\{(-4,1), \left(-\frac{5}{2}, \frac{1}{4}\right)\right\}$.

**41.** $x^2 + y^2 + 3y = 22$

$2x + y = -1$

Solve the second equation for $y$.

$2x + y = -1$

$y = -2x - 1$

Substitute $-2x - 1$ for $y$ to find $x$.

$$x^2 + (-2x-1)^2 + 3(-2x-1) = 22$$
$$x^2 + 4x^2 + 4x + 1 - 6x - 3 = 22$$
$$5x^2 - 2x - 2 = 22$$
$$5x^2 - 2x - 24 = 0$$
$$(5x-12)(x+2) = 0$$

Apply the zero product principle.

$5x - 12 = 0$ or $x + 2 = 0$

$5x = 12$    $x = -2$

$x = \frac{12}{5}$

Substitute –2 and $\frac{12}{5}$ for $x$ to find $y$.

$x = \frac{12}{5}$ or $x = -2$

$y = -2\left(\frac{12}{5}\right) - 1$    $y = -2(-2) - 1$

$y = -\frac{24}{5} - \frac{5}{5}$    $y = 4 - 1$

$y = -\frac{29}{5}$    $y = 3$

The solutions are $\left(\frac{12}{5}, -\frac{29}{5}\right)$ and $(-2, 3)$ and the solution set is $\left\{\left(\frac{12}{5}, -\frac{29}{5}\right), (-2, 3)\right\}$.

**43.** Let $x$ = one of the numbers

Let $y$ = the other number

$x + y = 10$

$xy = 24$

Solve the second equation for $y$.

$xy = 24$

$y = \frac{24}{x}$

Substitute $\frac{24}{x}$ for $y$ in the first equation and solve for $x$.

$$x + \frac{24}{x} = 10$$
$$x\left(x + \frac{24}{x}\right) = x(10)$$
$$x^2 + 24 = 10x$$
$$x^2 - 10x + 24 = 0$$
$$(x-6)(x-4) = 0$$

Apply the zero product principle.

$x - 6 = 0$ or $x - 4 = 0$

$x = 6$    $x = 4$

Substitute 6 and 4 for $x$ to find $y$.

$x = 6$    $x = 4$

$y = \frac{24}{6}$ or $y = \frac{24}{4}$

$y = 4$    $y = 6$

The numbers are 4 and 6.

**45.** Let $x$ = one of the numbers

Let $y$ = the other number

$$x^2 - y^2 = 3$$
$$\underline{2x^2 + y^2 = 9}$$
$$3x^2 = 12$$
$$x^2 = 4$$
$$x = \pm 2$$

Substitute $\pm 2$ for $x$ to find $y$.

$$x = \pm 2$$
$$(\pm 2)^2 - y^2 = 3$$
$$4 - y^2 = 3$$
$$-y^2 = -1$$
$$y^2 = 1$$
$$y = \pm 1$$

The numbers are either 2 and –1, 2 and 1, –2 and –1, or –2 and 1.

**47.** $2x^2 + xy = 6$

$x^2 + 2xy = 0$

Multiply the first equation by $-2$ and add the two equations.

$$-4x^2 - 2xy = -12$$
$$x^2 + 2xy = 0$$
$$\overline{\quad -3x^2 = -12 \quad}$$
$$x^2 = 4$$
$$x = \pm 2$$

Back-substitute these values for $x$ in the second equation and solve for $y$.

For $x = -2$: $(-2)^2 + 2(-2)y = 0$
$$4 - 4y = 0$$
$$-4y = -4$$
$$y = 1$$

For $x = 2$: $(2)^2 + 2(2)y = 0$
$$4 + 4y = 0$$
$$4y = -4$$
$$y = -1$$

The solution set is $\{(-2,1),(2,-1)\}$.

**48.** $4x^2 + xy = 30$

$x^2 + 3xy = -9$

Multiply the first equation by $-3$ and add the equations.

$$-12x^2 - 3xy = -90$$
$$x^2 + 3xy = -9$$
$$\overline{\quad -11x^2 = -99 \quad}$$
$$x^2 = 9$$
$$x = \pm 3$$

Back-substitute these values for $x$ in the second equation and solve for $y$.

For $x = -3$: $(-3)^2 + 3(-3)y = -9$
$$9 - 9y = -9$$
$$-9y = -18$$
$$y = 2$$

For $x = 3$: $(3)^2 + 3(3)y = 9$
$$9 + 9y = -9$$
$$9y = -18$$
$$y = -2$$

The solution set is $\{(-3,2),(3,-2)\}$.

**49.** $-4x + y = 12$

$y = x^3 + 3x^2$

Substitute $x^3 + 3x^2$ for $y$ in the first equation and solve for $x$.

$$-4x + (x^3 + 3x^2) = 12$$
$$x^3 + 3x^2 - 4x - 12 = 0$$
$$x^2(x+3) - 4(x+3) = 0$$
$$(x+3)(x^2 - 4) = 0$$
$$(x+3)(x-2)(x+2) = 0$$

$x = -3$, $x = 2$, or $x = -2$

Substitute these values for $x$ in the second equation and solve for $y$.

For $x=-3$: $y=(-3)^3+3(-3)^2$
$=-27+27$
$=0$

For $x=2$: $y=(2)^3+3(2)^2$
$=8+12$
$=20$

For $x=-2$: $y=(-2)^3+3(-2)^2$
$=-8+12$
$=4$

The solution set is
$\{(-3,0),(2,20),(-2,4)\}$.

**50.** $-9x+y=45$
$y=x^3+5x^2$

Substitute $x^3+5x^2$ for $y$ in the first equation and solve for $x$.

$$-9x+(x^3+5x^2)=45$$
$$x^3+5x^2-9x-45=0$$
$$x^2(x+5)-9(x+5)=0$$
$$(x+5)(x^2-9)=0$$
$$(x+5)(x-3)(x+3)=0$$
$x=-5$, $x=3$, or $x=-3$

Substitute these values for $x$ in the second equation and solve for $y$.

For $x=-5$: $y=(-5)^3+5(-5)^2$
$=-125+125=0$

For $x=3$: $y=(3)^3+5(3)^2$
$=27+45=72$

For $x=-3$: $y=(-3)^3+5(-3)^2$
$=-27+45=18$

The solution set
is $\{(-5,0),(3,72),(-3,18)\}$.

**51.** $\frac{3}{x^2}+\frac{1}{y^2}=7$
$\frac{5}{x^2}-\frac{2}{y^2}=-3$

Multiply the first equation by 2 and add the equations.

$$\frac{6}{x^2}+\frac{2}{y^2}=14$$
$$\frac{5}{x^2}-\frac{2}{y^2}=-3$$
$$\frac{11}{x^2}=11$$
$$x^2=1$$
$$x=\pm 1$$

Back-substitute these values for $x$ in the first equation and solve for $y$.

For $x=-1$:
$$\frac{3}{(-1)^2}+\frac{1}{y^2}=7$$
$$3+\frac{1}{y^2}=7$$
$$\frac{1}{y^2}=4$$
$$y^2=\frac{1}{4}$$
$$y=\pm\frac{1}{2}$$

For $x=1$:
$$\frac{3}{(1)^2}+\frac{1}{y^2}=7$$
$$3+\frac{1}{y^2}=7$$
$$\frac{1}{y^2}=4$$
$$y^2=\frac{1}{4}$$
$$y=\pm\frac{1}{2}$$

The solution set is
$\left\{\left(-1,-\frac{1}{2}\right),\left(-1,\frac{1}{2}\right),\left(1,-\frac{1}{2}\right),\left(1,\frac{1}{2}\right)\right\}$.

**52.** $\frac{2}{x^2}+\frac{1}{y^2}=11$

$\frac{4}{x^2}-\frac{2}{y^2}=-14$

Multiply the first equation by 2 and add the two equations.

$\frac{4}{x^2}+\frac{2}{y^2}=22$

$\frac{4}{x^2}-\frac{2}{y^2}=-14$

$\frac{8}{x^2}=8$

$x^2=1$

$x=\pm 1$

Back-substitute these values for $x$ in the first equation and solve for $y$.

For $x=-1$:

$\frac{2}{(-1)^2}+\frac{1}{y^2}=11$

$2+\frac{1}{y^2}=11$

$\frac{1}{y^2}=9$

$y^2=\frac{1}{9}$

$y=\pm\frac{1}{3}$

For $x=1$:

$\frac{2}{(1)^2}+\frac{1}{y^2}=11$

$2+\frac{1}{y^2}=11$

$\frac{1}{y^2}=9$

$y^2=\frac{1}{9}$

$y=\pm\frac{1}{3}$

The solution set is $\left\{\left(-1,-\frac{1}{3}\right),\left(-1,\frac{1}{3}\right),\left(1,-\frac{1}{3}\right),\left(1,\frac{1}{3}\right)\right\}$.

**53.** Answers will vary. One example:

Circle: $x^2+y^2=9$

Ellipse: $\frac{x^2}{9}+\frac{y^2}{49}=1$

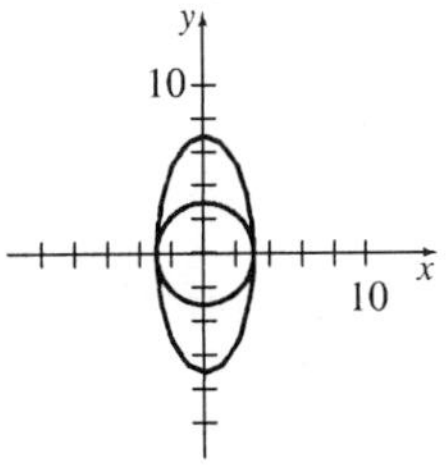


Solutions: $(-3,0)$ and $(3,0)$.

**54.** Answers will vary. One example follows:

Line: $x - y = 2$

Parabola: $x = y^2$

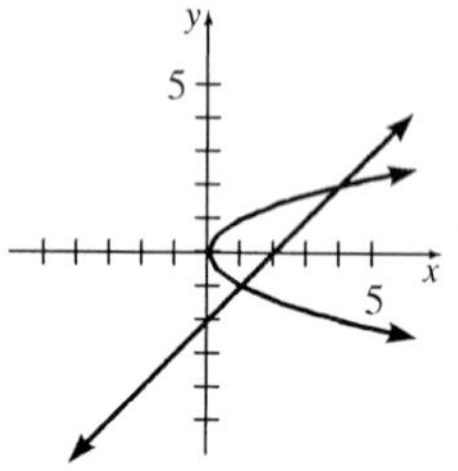


Solutions: $(1,-1)$ and $(4,2)$

**55.** $16x^2 + 4y^2 = 64$

$y = x^2 - 4$

Solve the second equation for $x^2$.

$y = x^2 - 4$

$y + 4 = x^2$

Substitute $x^2 - 4$ for $y$ in the first equation and solve for $x$.

$16(y+4) + 4y^2 = 64$

$16y + 64 + 4y^2 = 64$

$16y + 4y^2 = 0$

$4y(4 + y) = 0$

Apply the zero product principle.

$4y = 0$ or $4 + y = 0$

$y = 0$ $\quad$ $y = -4$

Substitute 0 and 4 for $y$ to find $x$.

$y = 0$ or $y = -4$

$0 = x^2 - 4$ $\quad$ $-4 = x^2 - 4$

$4 = x^2$ $\quad$ $0 = x^2$

$\pm 2 = x$ $\quad$ $0 = x$

The comet intersects the planet's orbit at the points $(2,0), (-2,0)$ and $(0,-4)$.

**57.** Let $x$ = the length of the rectangle

Let $y$ = the width of the rectangle

Perimeter: $2x + 2y = 36$

Area: $xy = 77$

Solve the second equation for $y$.

$xy = 77$

$y = \frac{77}{x}$

Substitute $\frac{77}{x}$ for $y$ in the first equation and solve for $x$.

$2x + 2\left(\frac{77}{x}\right) = 36$

$2x + \frac{154}{x} = 36$

$x\left(2x + \frac{154}{x}\right) = x(36)$

$2x^2 + 154 = 36x$

$2x^2 - 36x + 154 = 0$

$x^2 - 18x + 77 = 0$

$(x-7)(x-11) = 0$

Apply the zero product principle.

$x - 7 = 0$ or $x - 11 = 0$

$x = 7$ $\quad$ $x = 11$

Substitute 7 and 11 for $x$ to find $y$.

$x = 7$ or $x = 11$

$y = \frac{77}{7}$ $\quad$ $y = \frac{77}{11}$

$y = 11$ $\quad$ $y = 7$

The dimensions of the rectangle are 7 feet by 11 feet.

**59.** Let $x$ = the length of the screen
Let $y$ = the width of the screen

$$x^2 + y^2 = 10^2$$
$$xy = 48$$

Solve the second equation for $y$.

$$xy = 48$$
$$y = \frac{48}{x}$$

Substitute $\frac{48}{x}$ for $y$ to find $x$.

$$x^2 + \left(\frac{48}{x}\right)^2 = 10^2$$
$$x^2 + \frac{2304}{x^2} = 100$$
$$x^2\left(x^2 + \frac{2304}{x^2}\right) = x^2(100)$$
$$x^4 + 2304 = 100x^2$$
$$x^4 - 100x^2 + 2304 = 0$$
$$(x^2 - 64)(x^2 - 36) = 0$$
$$(x+8)(x-8)(x+6)(x-6) = 0$$

Apply the zero product principle.

$x + 8 = 0$ $\quad$ $x - 8 = 0$
$x = -8$ $\quad$ $x = 8$

$x + 6 = 0$ $\quad$ $x + 6 = 0$
$x = -6$ $\quad$ $x = -6$

We disregard –8 and –6 because we cannot have a negative length.
Substitute 8 and 6 for $x$ to find $y$.

$x = 8$ or $x = 6$

$y = \frac{48}{8}$ $\quad$ $y = \frac{48}{6}$

$y = 6$ $\quad$ $y = 8$

The dimensions of the screen are 8 inches by 6 inches.

**61.**
$$x^2 - y^2 = 21$$
$$4x + 2y = 24$$

Solve for y in the second equation.

$$4x + 2y = 24$$
$$2y = 24 - 4x$$
$$y = 12 - 2x$$

Substitute $12 - 2x$ for $y$ and solve for $x$.

$$x^2 - (12 - 2x)^2 = 21$$
$$x^2 - (144 - 48x + 4x^2) = 21$$
$$x^2 - 144 + 48x - 4x^2 = 21$$
$$-3x^2 + 48x - 144 = 21$$
$$-3x^2 + 48x - 165 = 0$$
$$x^2 - 16x + 55 = 0$$
$$(x-5)(x-11) = 0$$

Apply the zero product principle.

$x - 5 = 0$ or $x - 11 = 0$
$x = 5$ $\quad$ $x = 11$

Substitute 5 and 11 for $x$ to find $y$.

$x = 5$ or $x = 11$

$y = 12 - 2(5)$ $\quad$ $y = 12 - 2(11)$
$y = 12 - 10$ $\quad$ $y = 12 - 22$
$y = 2$ $\quad$ $y = -10$

We disregard –10 because we can't have a negative length measurement. The larger square is 5 meters by 5 meters and the smaller square to be cut out is 2 meters by 2 meters.

**63.** Answers will vary.

**65.** Answers will vary.

**67.** Answers will vary. For example, consider Exercise 1.

$x + y = 2 \qquad y = x^2 - 4$

$y = -x + 2$

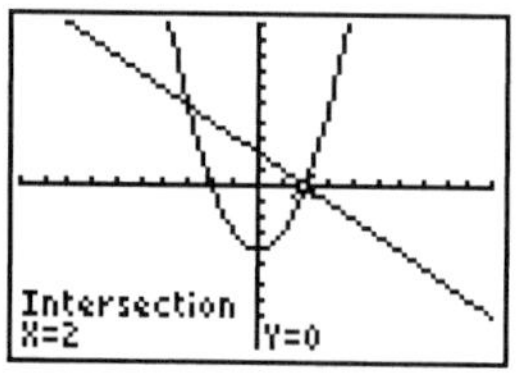


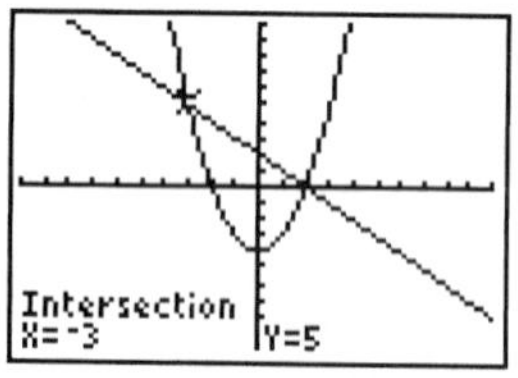


The solutions are $(2,0)$ and $(-3,5)$.

This is the same answer obtained in Exercise 1.

**69.** Statement **b.** is true. As shown in the graph below, a parabola and a circle can intersect in at most four points, and therefore, has at most four real solutions.

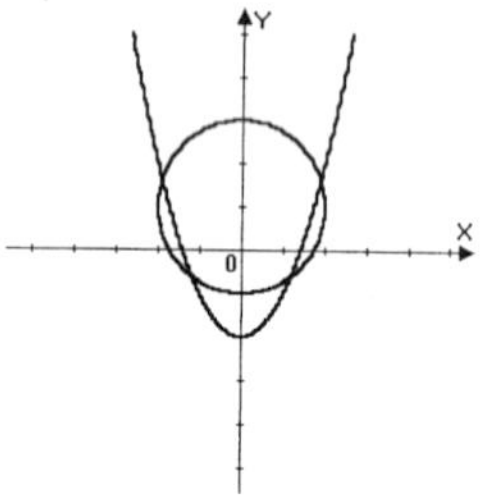

Statement **a.** is false. As shown in the graph below, a circle and a line can intersect in at most two points, and therefore has at most two real solutions.

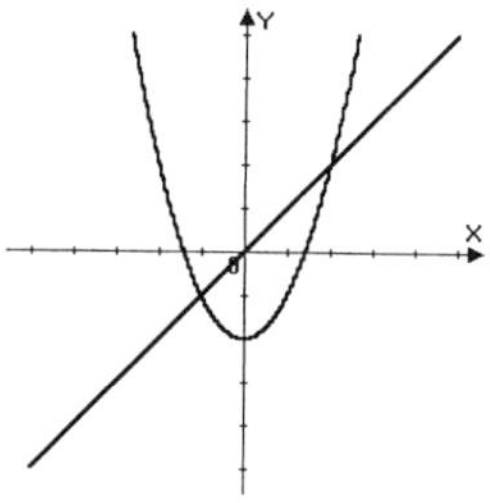

Statement **c.** is false. As shown in the graphs below, it is possible that a system of two equations in two variables whose graphs represent circles do not intersect, or intersect in a single point. This means that the system would have no solution, or a single solution, respectively.

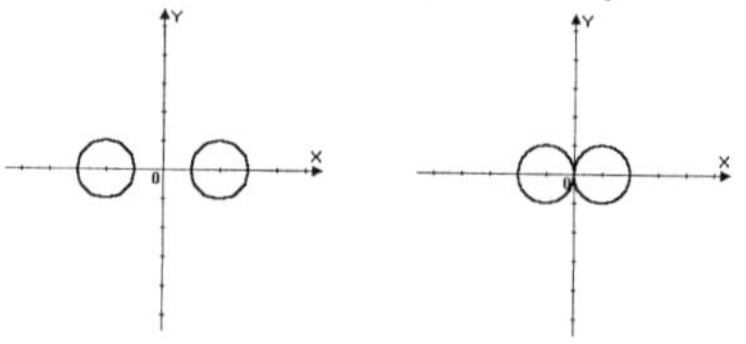

Statement **d.** is false. As shown in the graph below, a circle and a parabola can intersect in one point, and therefore have only one real solution.

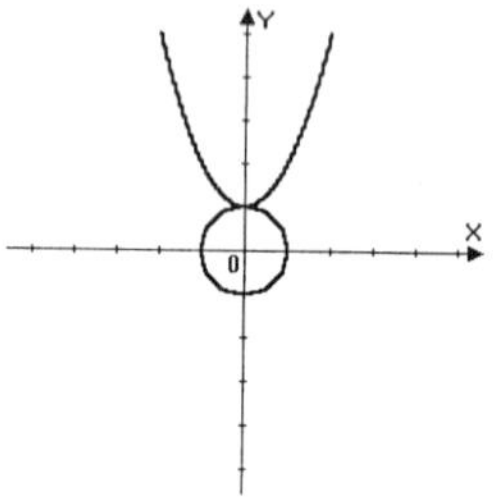

**71.** $\log_y x = 3$

$\log_y (4x) = 5$

Rewrite the equations.

$y^3 = x$

$y^5 = 4x$

Substitute $y^3$ for $x$ in the second equation and solve for $y$.

$$y^5 = 4y^3$$
$$y^5 - 4y^3 = 0$$
$$y^3\left(y^2 - 4\right) = 0$$
$$y^3\left(y+2\right)\left(y-2\right) = 0$$

Apply the zero product principle.

$y^3 = 0 \qquad y+2=0 \qquad y-2=0$

$y = 0 \qquad y=-2 \qquad y=2$

We disregard 0 and $-2$ because the base of a logarithm must be greater than zero.

Substitute 2 for $y$ to find $x$.

$$y^3 = x$$
$$2^3 = x$$
$$8 = x$$

The solution is $(8,2)$ and the solution set is $\{(8,2)\}$.

**73.** $3x - 2y \le 6$

First, find the intercepts to the equation $3x - 2y = 6$. Find the $x$–intercept by setting $y$ equal to zero.

$$3x - 2(0) = 6$$
$$3x = 6$$
$$x = 2$$

Find the $y$–intercept by setting $x$ equal to zero.

$$3(0) - 2y = 6$$
$$-2y = 6$$
$$y = -3$$

Next, use the origin as a test point.

$$3(0) - 2(0) \le 6$$
$$0 - 0 \le 6$$
$$0 \le 6$$

This is a true statement. This means that the origin will fall in the shaded half-plane.

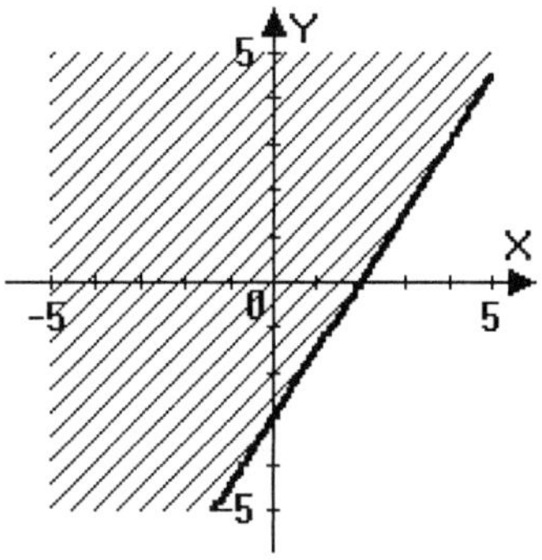


**74.** $m = \dfrac{y_2 - y_1}{x_2 - x_1} = \dfrac{5-(-3)}{1-(-2)} = \dfrac{5+3}{1+2} = \dfrac{8}{3}$

The slope is $\dfrac{8}{3}$.

**75.** $\left(3x-2\right)\left(2x^2-4x+3\right)$

$$= 3x\left(2x^2-4x+3\right) - 2\left(2x^2-4x+3\right)$$
$$= 6x^3 - 12x^2 + 9x - 4x^2 + 8x - 6$$
$$= 6x^3 - 16x^2 + 17x - 6$$

## Chapter 10 Review Exercises

**1.** $d = \sqrt{\left(3-(-2)\right)^2 + \left(9-(-3)\right)^2}$

$$= \sqrt{(3+2)^2 + (9+3)^2}$$
$$= \sqrt{5^2 + 12^2} = \sqrt{25+144}$$
$$= \sqrt{169} = 13$$

The distance between the points is 13 units.

**2.** $d = \sqrt{\left(-2-(-4)\right)^2 + (5-3)^2}$

$$= \sqrt{(-2+4)^2 + 2^2} = \sqrt{2^2 + 4}$$
$$= \sqrt{4+4} = \sqrt{8} = \sqrt{4\cdot 2} = 2\sqrt{2}$$

The distance between the points is $2\sqrt{2}$ or approximately 2.83 units.

**3.** $$\text{Midpoint} = \left(\frac{2+(-12)}{2}, \frac{6+4}{2}\right) = \left(\frac{-10}{2}, \frac{10}{2}\right) = (-5, 5)$$

The midpoint is $(-5,5)$.

**4.** $$\text{Midpoint} = \left(\frac{4+(-15)}{2}, \frac{-6+2}{2}\right) = \left(\frac{-11}{2}, \frac{-4}{2}\right) = \left(-\frac{11}{2}, -2\right)$$

The midpoint is $\left(-\frac{11}{2}, -2\right)$.

**5.** $$(x-0)^2 + (y-0)^2 = 3^2$$
$$x^2 + y^2 = 9$$

**6.** $$(x-(-2))^2 + (y-4)^2 = 6^2$$
$$(x+2)^2 + (y-4)^2 = 36$$

**7.** $$x^2 + y^2 = 1$$
$$(x-0)^2 + (y-0)^2 = 1^2$$

The center is $(0,0)$ and the radius is 1 units.

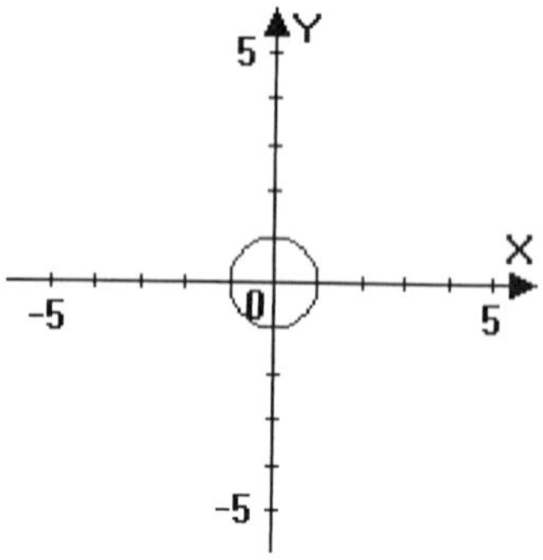


**8.** $$(x+2)^2 + (y-3)^2 = 9$$
$$(x-(-2))^2 + (y-3)^2 = 3^2$$

The center is $(-2,3)$ and the radius is 3 units.

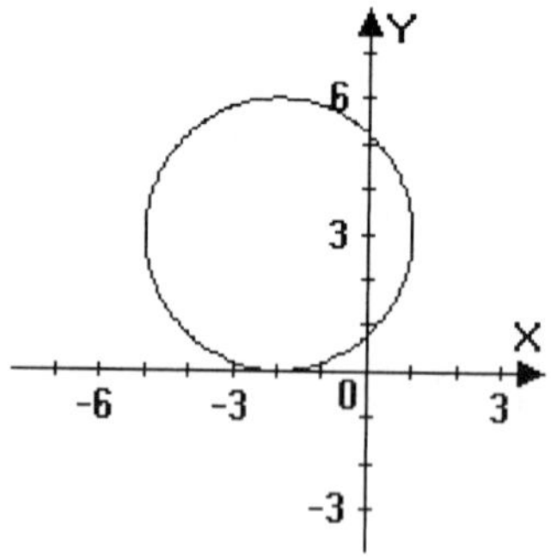


**9.** $$x^2 + y^2 - 4x + 2y - 4 = 0$$
$$(x^2 - 4x \quad) + (y^2 + 2y \quad) = 4$$

Complete the squares.

$$\left(\frac{b}{2}\right)^2 = \left(\frac{-4}{2}\right)^2 = (-4)^2 = 4$$
$$\left(\frac{b}{2}\right)^2 = \left(\frac{2}{2}\right)^2 = 1^2 = 1$$
$$(x^2 - 4x + 4) + (y^2 + 2y + 1) = 4 + 4 + 1$$
$$(x-2)^2 + (y+1)^2 = 9$$
$$(x-2)^2 + (y-(-1))^2 = 3^2$$

The center is $(2,-1)$ and the radius is 3 units.

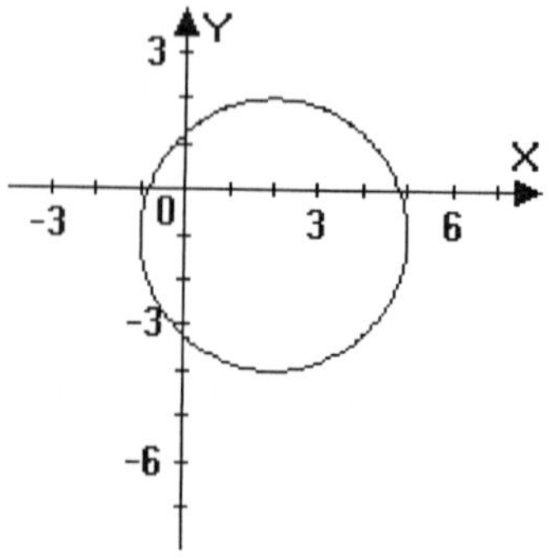

**10.** $x^2 + y^2 - 4y = 0$

$x^2 + (y^2 - 4y \quad ) = 0$

Complete the square.

$\left(\frac{b}{2}\right)^2 = \left(\frac{-4}{2}\right)^2 = (-4)^2 = 4$

$x^2 + (y^2 - 4y + 4) = 0 + 4$

$(x-0)^2 + (y-2)^2 = 4$

$(x-0)^2 + (y-2)^2 = 2^2$

The center is $(0,2)$ and the radius is 2 units.

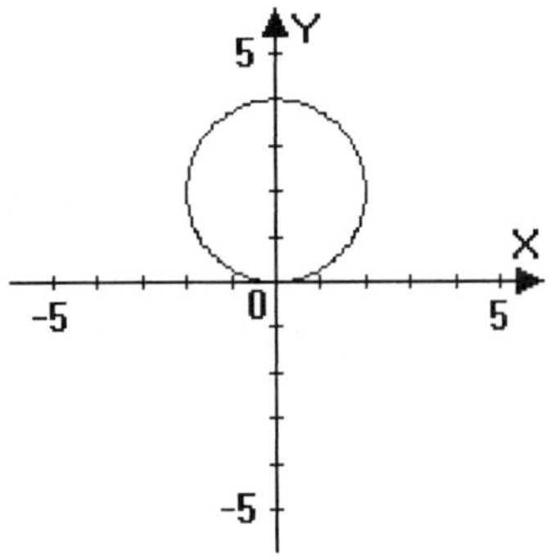


**11.** $\frac{x^2}{36} + \frac{y^2}{25} = 1$

Because the denominator of the $x^2$ – term is greater than the denominator of the $y^2$ – term, the major axis is horizontal. Since $a^2 = 36$, $a = 6$ and the vertices are $(-6,0)$ and $(6,0)$. Since $b^2 = 25$, $b = 5$ and endpoints of the minor axis are $(0,-5)$ and $(0,5)$.

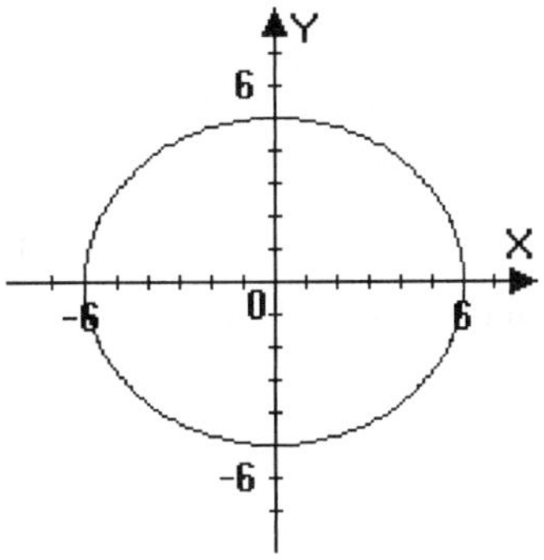


**12.** $\frac{x^2}{25} + \frac{y^2}{16} = 1$

Because the denominator of the $x^2$ – term is greater than the denominator of the $y^2$ – term, the major axis is horizontal. Since $a^2 = 25$, $a = 5$ and the vertices are $(-5,0)$ and $(5,0)$. Since $b^2 = 16$, $b = 4$ and endpoints of the minor axis are $(0,-4)$ and $(0,4)$.

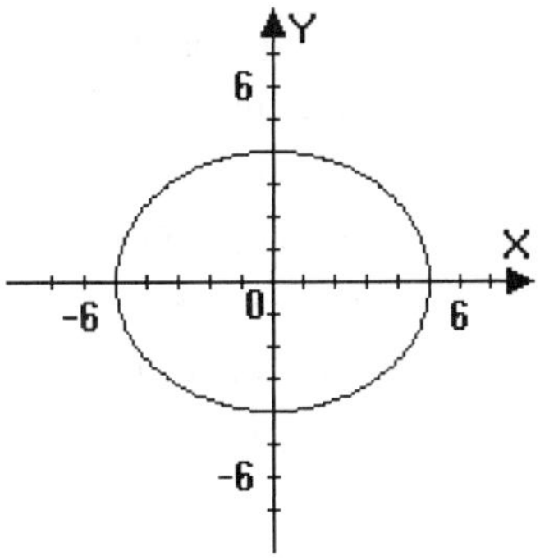


**13.** $4x^2 + y^2 = 16$

$\frac{4x^2}{16} + \frac{y^2}{16} = \frac{16}{16}$

$\frac{x^2}{4} + \frac{y^2}{16} = 1$

Because the denominator of the $y^2$ – term is greater than the denominator of the $x^2$ – term, the major axis is vertical. Since $a^2 = 16$, $a = 4$ and the vertices are $(0,-4)$ and $(0,4)$. Since $b^2 = 4$, $b = 2$ and endpoints of the minor axis are $(-2,0)$ and $(2,0)$.

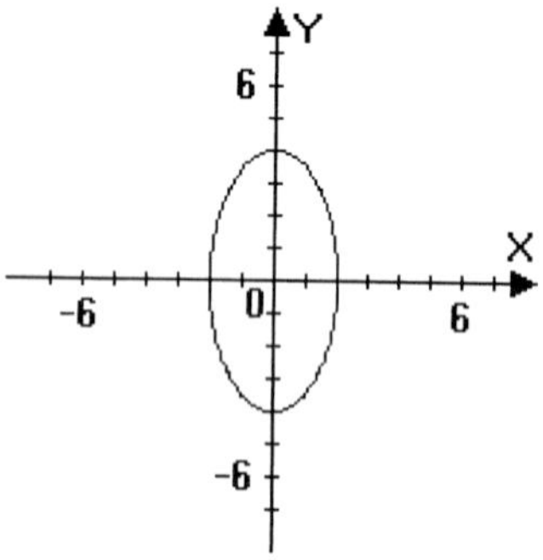


**14.** $4x^2 + 9y^2 = 36$

$$\frac{4x^2}{36} + \frac{9y^2}{36} = \frac{36}{36}$$

$$\frac{x^2}{9} + \frac{y^2}{4} = 1$$

Because the denominator of the $x^2$ – term is greater than the denominator of the $y^2$ – term, the major axis is horizontal. Since $a^2 = 9$, $a = 3$ and the vertices are $(-3,0)$ and $(3,0)$. Since $b^2 = 4$, $b = 2$ and endpoints of the minor axis are $(0,-2)$ and $(0,2)$.

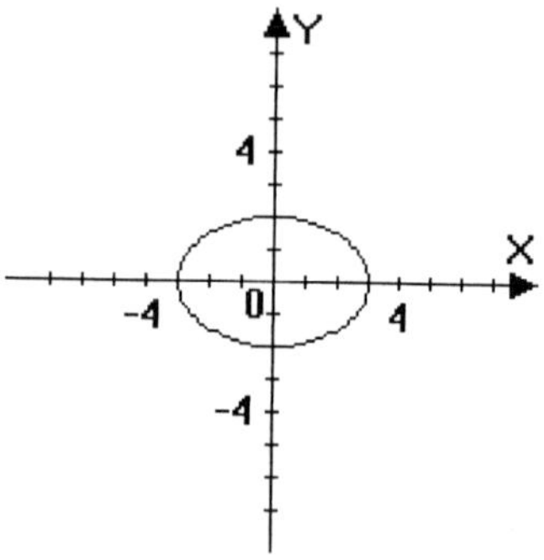


**15.** $\frac{(x-1)^2}{16} + \frac{(y+2)^2}{9} = 1$

The center of the ellipse is $(1,-2)$. Because the denominator of the $x^2$ – term is greater than the denominator of the $y^2$ – term, the major axis is horizontal. Since $a^2 = 16$, $a = 4$ and the vertices lie 4 units to the left and right of the center. Since $b^2 = 9$, $b = 3$ and endpoints of the minor axis lie 3 units above and below the center.

| Center | Vertices | Endpoints of Minor Axis |
|---|---|---|
| $(1,-2)$ | $(1-4,-2)$ $=(-3,-2)$ | $(1,-2-3)$ $=(1,-5)$ |
| | $(1+4,-2)$ $=(5,-2)$ | $(1,-2+3)$ $=(1,1)$ |

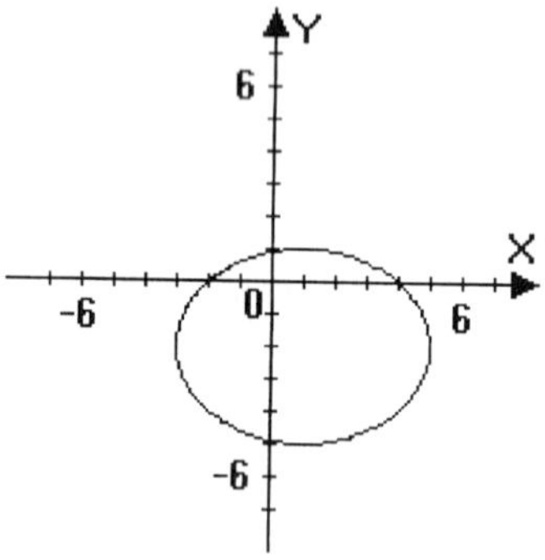


**16.** $\frac{(x+1)^2}{9} + \frac{(y-2)^2}{16} = 1$

The center of the ellipse is $(-1,2)$. Because the denominator of the $y^2$ – term is greater than the denominator of the $x^2$ – term the major axis is vertical. Since $a^2 = 16$, $a = 4$

and the vertices lie 4 units above and below the center. Since $b^2 = 9,\ b = 3$ and endpoints of the minor axis lie 3 units to the left and right of the center.

| Center | Vertices | Endpoints of Minor Axis |
|---|---|---|
| $(-1,2)$ | $(-1,2-4)$ $=(-1,-2)$ | $(-1-3,2)$ $=(-4,2)$ |
| | $(-1,2+4)$ $=(-1,6)$ | $(-1+3,2)$ $=(2,2)$ |

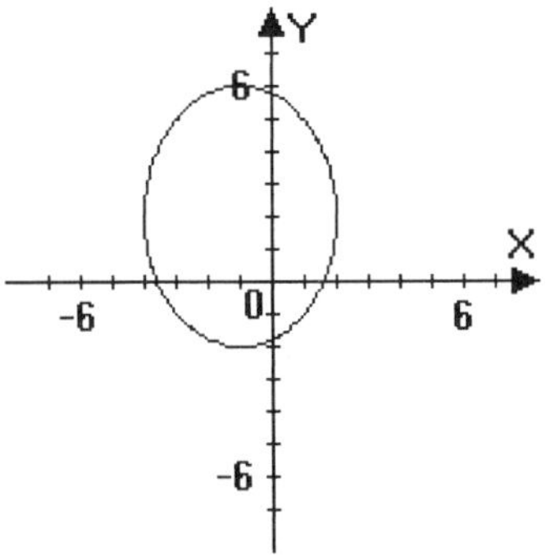


**17.** From the figure, we see that the major axis is horizontal with $a = 25$, and $b = 15$.

$$\frac{x^2}{25^2}+\frac{y^2}{15^2}=1$$

$$\frac{x^2}{625}+\frac{y^2}{225}=1$$

Since the truck is 14 feet wide, we need to determine the height of the archway at 14 feet to the right of center.

$$\frac{14^2}{625}+\frac{y^2}{225}=1$$

$$\frac{196}{625}+\frac{y^2}{225}=1$$

$$5625\left(\frac{196}{625}+\frac{y^2}{225}\right)=5625(1)$$

$$9(196)+25y^2=5625$$

$$1764+25y^2=5625$$

$$25y^2=3861$$

$$y^2=\frac{3861}{25}$$

$$y=\sqrt{\frac{3861}{25}}\approx 12.43$$

The height of the archway 14 feet from the center is approximately 12.43 feet. Since the truck is 12 feet high, the truck will clear the archway.

**18.** $$\frac{x^2}{16}-y^2=1$$

$$\frac{x^2}{16}-\frac{y^2}{1}=1$$

The equation is in the form $\frac{x^2}{a^2}-\frac{y^2}{b^2}=1$ with $a^2=16$, and $b^2=1$. We know the transverse axis lies on the $x$-axis and the vertices are $(-4,0)$ and $(4,0)$.

Because $a^2=16$ and $b^2=1$, $a=4$ and $b=1$. Construct a rectangle using –4 and 4 on the $x$–axis, and –1 and 1 on the $y$–axis. Draw extended diagonals to obtain the asymptotes. Graph the hyperbola.

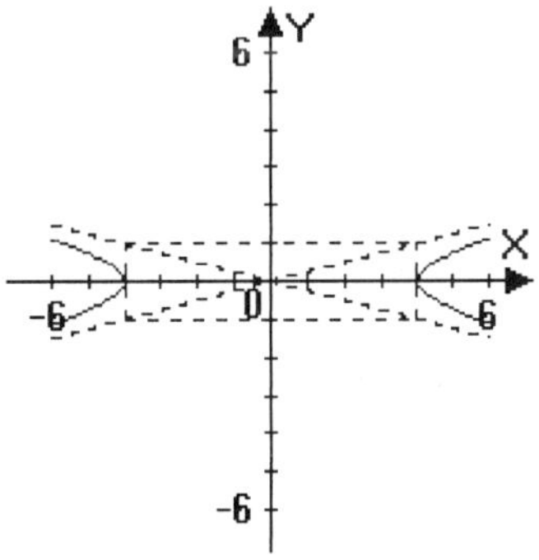

**19.** $\frac{y^2}{16} - x^2 = 1$

$\frac{y^2}{16} - \frac{x^2}{1} = 1$

The equation is in the form $\frac{y^2}{a^2} - \frac{x^2}{b^2} = 1$ with $a^2 = 16$, and $b^2 = 1$. We know the transverse axis lies on the $y$-axis and the vertices are $(0,-4)$ and $(0,4)$.

Because $a^2 = 16$ and $b^2 = 1$, $a = 4$ and $b = 1$. Construct a rectangle using –1 and 1 on the $x$–axis, and –4 and 4 on the $y$–axis. Draw extended diagonals to obtain the asymptotes. Graph the hyperbola.

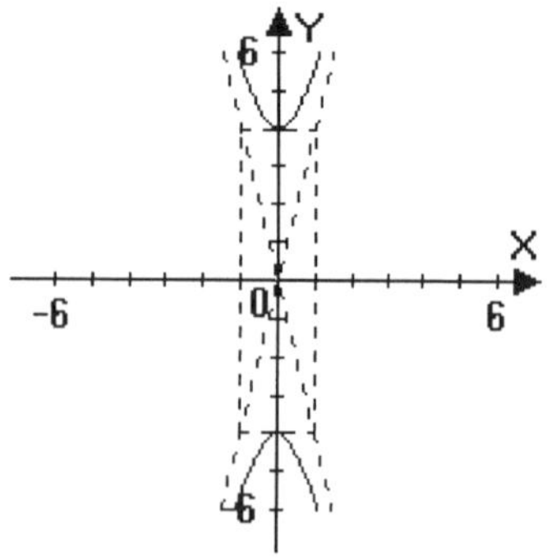


**20.** $9x^2 - 16y^2 = 144$

$\frac{9x^2}{144} - \frac{16y^2}{144} = \frac{144}{144}$

$\frac{x^2}{16} - \frac{y^2}{9} = 1$

The equation is in the form $\frac{x^2}{a^2} - \frac{y^2}{b^2} = 1$ with $a^2 = 16$, and $b^2 = 9$. We know the transverse axis lies on the $x$-axis and the vertices are $(-4,0)$ and $(4,0)$.

Because $a^2 = 16$ and $b^2 = 9$, $a = 4$ and $b = 3$. Construct a rectangle using –4 and 4 on the $x$–axis, and –3 and 3 on the $y$–axis. Draw extended diagonals to obtain the asymptotes. Graph the hyperbola.

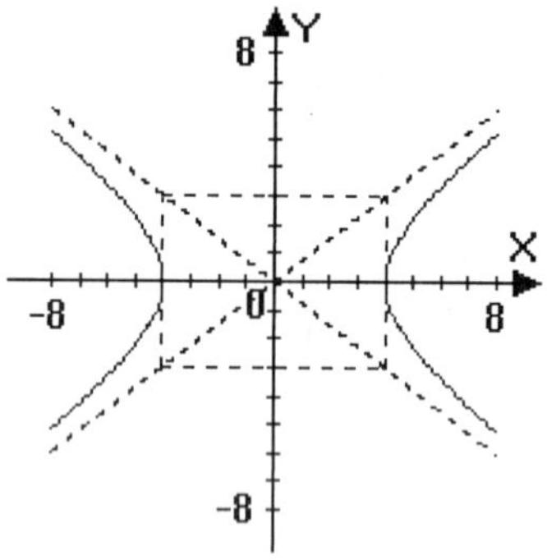


**21.** $4y^2 - x^2 = 16$

$\frac{4y^2}{16} - \frac{x^2}{16} = \frac{16}{16}$

$\frac{y^2}{4} - \frac{x^2}{16} = 1$

The equation is in the form $\frac{y^2}{a^2} - \frac{x^2}{b^2} = 1$ with $a^2 = 16$, and $b^2 = 4$. We know the transverse axis lies on the $y$-axis and the vertices are $(0,-4)$ and $(0,4)$.

Because $a^2 = 16$ and $b^2 = 4$, $a = 4$ and $b = 2$. Construct a rectangle using –2 and 2 on the $x$–axis, and –4 and 4 on the $y$–axis. Draw extended diagonals to obtain the asymptotes. Graph the hyperbola.

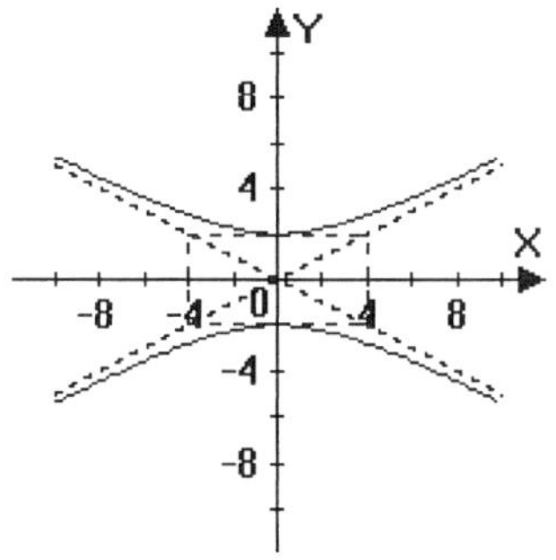


**22.** $x=(y-3)^2-4$

This is a parabola of the form $x=a(y-k)^2+h$. Since $a=1$ is positive, the parabola opens to the right. The vertex of the parabola is $(-4,3)$.

The axis of symmetry is $y=3$.

Replace $y$ with 0 to find the $x$–intercept.

$x=(0-3)^2-4=(-3)^2-4=9-4=5$

The $x$–intercept is 5. Replace $x$ with 0 to find the $y$–intercepts.

$0=(y-3)^2-4$

$0=y^2-6y+9-4$

$0=y^2-6y+5$

$0=(y-5)(y-1)$

Apply the zero product principle.

$y-5=0$ and $y-1=0$

$y=5$ $\qquad$ $y=1$

The $y$–intercepts are 1 and 5.

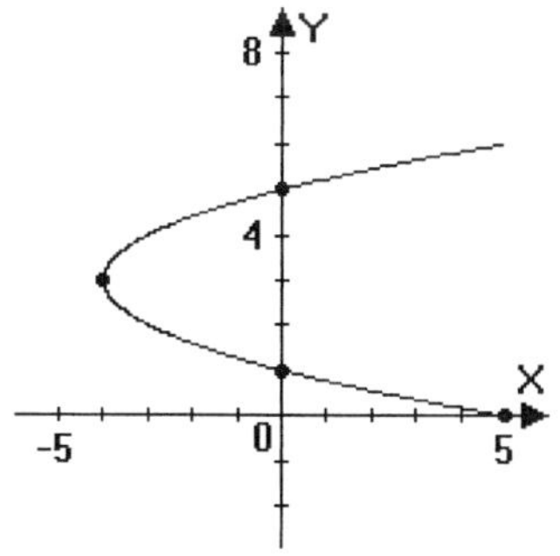


**23.** $x=-2(y+3)^2+2$

This is a parabola of the form $x=a(y-k)^2+h$. Since $a=-2$ is negative, the parabola opens to the left. The vertex of the parabola is $(2,-3)$.

The axis of symmetry is $y=-3$.

Replace $y$ with 0 to find the $x$–intercept.

$x=-2(0+3)^2+2=-2(3)^2+2$

$=-2(9)+2=-18+2=-16$

The $x$–intercept is -16. Replace $x$ with 0 to find the $y$–intercepts.

$0=-2(y+3)^2+2$

$0=-2(y^2+6y+9)+2$

$0=-2y^2-12y-18+2$

$0=-2y^2-12y-16$

$0=y^2+6y+8$

$0=(y+4)(y+2)$

Apply the zero product principle.

$y+4=0$ and $y+2=0$

$y=-4$ $\qquad$ $y=-2$

The $y$–intercepts are –4 and –2.

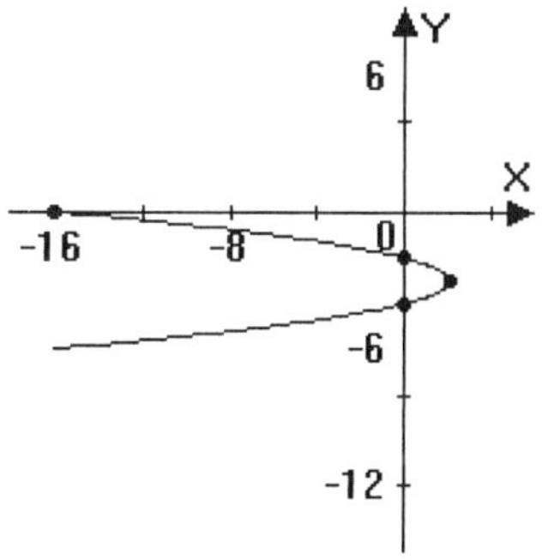

**24.** $x = y^2 - 8y + 12$

This is a parabola of the form $x = ay^2 + by + c$. Since $a = 1$ is positive, the parabola opens to the right. The $y$–coordinate of the vertex is $-\frac{b}{2a} = -\frac{-8}{2(1)} = -\frac{-8}{2} = 4$. The $x$–coordinate of the vertex is

$x = 4^2 - 8(4) + 12 = 16 - 32 + 12$

$= 16 - 32 + 12 = -4.$

The vertex of the parabola is $(-4, 4)$.

The axis of symmetry is $y = 4$.

Replace $y$ with 0 to find the $x$–intercept.

$x = 0^2 - 8(0) + 12 = 12$

The $x$–intercept is 12. Replace $x$ with 0 to find the $y$–intercepts.

$0 = y^2 - 8y + 12$

$0 = (y-6)(y-2)$

Apply the zero product principle.

$y - 6 = 0$ and $y - 2 = 0$

$y = 6$ $\quad$ $y = 2$

The $y$–intercepts are 2 and 6.

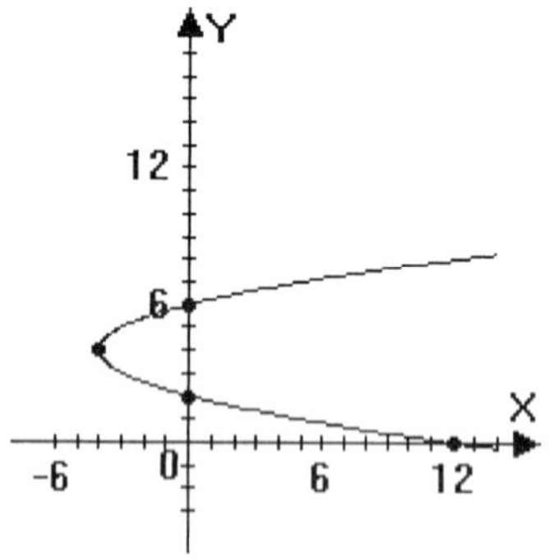


**25.** $x = -y^2 - 4y + 6$

This is a parabola of the form $x = ay^2 + by + c$. Since $a = -1$ is negative, the parabola opens to the left. The $y$–coordinate of the vertex is $-\frac{b}{2a} = -\frac{-4}{2(-1)} = -\frac{-4}{-2} = -2$. The $x$–coordinate of the vertex is

$x = -(-2)^2 - 4(-2) + 6$

$= -4 + 8 + 6 = 10.$

The vertex of the parabola is $(10, -2)$.

The axis of symmetry is $y = -2$.

Replace $y$ with 0 to find the $x$–intercept.

$x = -0^2 - 4(0) + 6 = 0^2 - 0 + 6 = 6$

The $x$–intercept is 6. Replace $x$ with 0 to find the $y$–intercepts.

$0 = -y^2 - 4y + 6$

Solve using the quadratic formula.

$$y = \frac{-b \pm \sqrt{b^2 - 4ac}}{2a}$$

$$= \frac{-(-4) \pm \sqrt{(-4)^2 - 4(-1)(6)}}{2(-1)}$$

$$= \frac{4 \pm \sqrt{16 + 24}}{-2} = \frac{4 \pm \sqrt{40}}{-2}$$

$$= \frac{4 \pm 2\sqrt{10}}{-2} = -2 \pm \sqrt{10}$$

The $y$–intercepts are $-2 \pm \sqrt{10}$.

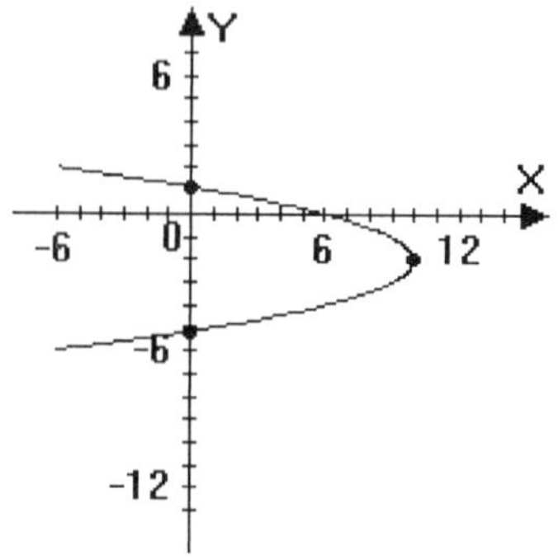

**26.** $x+8y=y^2+10$
Since only one variable is squared, the graph of the equation is a parabola.

**27.** $16x^2=32-y^2$

$16x^2+y^2=32$

Because $x^2$ and $y^2$ have different positive coefficients, the equation's graph is an ellipse.

**28.** $x^2=25+25y^2$

$x^2-25y^2=25$

Because $x^2$ and $y^2$ have opposite signs, the equation's graph is a hyperbola.

**29.** $x^2=4-y^2$

$x^2+y^2=4$

Because $x^2$ and $y^2$ have the same positive coefficient, the equation's graph is a circle.

**30.** $36y^2=576+16x^2$

$36y^2-16x^2=576$

Because $x^2$ and $y^2$ have opposite signs, the equation's graph is a hyperbola.

**31.** $\frac{(x+3)^2}{9}+\frac{(y-4)^2}{25}=1$

Because $x^2$ and $y^2$ have different positive coefficients, the equation's graph is an ellipse.

**32.** $y=x^2+6x+9$
Since only one variable is squared, the graph of the equation is a parabola.

**33.** $5x^2+5y^2=180$
Because $x^2$ and $y^2$ have the same positive coefficient, the equation's graph is a circle.
Divide both sides of the equation by 5.
$x^2+y^2=36$
The center is (0, 0) and the radius is 6 units.

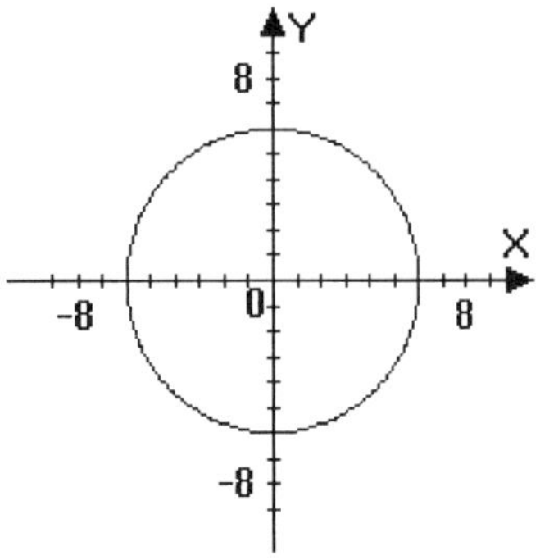


**34.** $4x^2+9y^2=36$

Because $x^2$ and $y^2$ have different positive coefficients, the equation's graph is an ellipse.

$\frac{4x^2}{36}+\frac{9y^2}{36}=\frac{36}{36}$

$\frac{x^2}{9}+\frac{y^2}{4}=1$

Because the denominator of the $x^2$ – term is greater than the denominator of the $y^2$ – term, the major axis is horizontal. Since $a^2=9$, $a=3$ and the vertices are $(-3,0)$ and $(3,0)$. Since $b^2=4$, $b=2$ and endpoints of the minor axis are $(0,-2)$ and $(0,2)$.

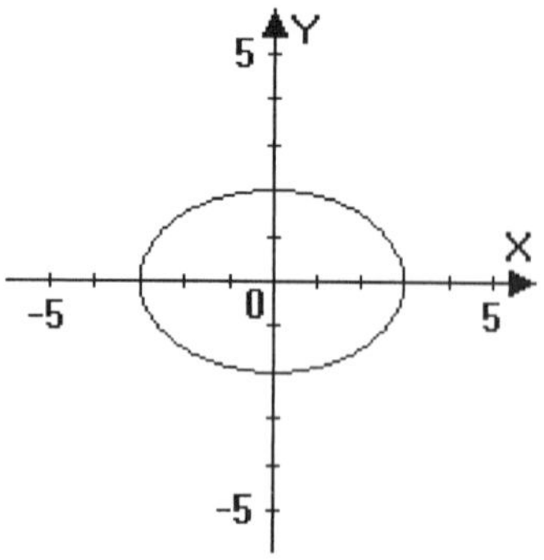


**35.** $4x^2 - 9y^2 = 36$

Because $x^2$ and $y^2$ have opposite signs, the equation's graph is a hyperbola.

$$\frac{4x^2}{36} - \frac{9y^2}{36} = \frac{36}{36}$$

$$\frac{x^2}{9} - \frac{y^2}{4} = 1$$

The equation is in the form $\frac{x^2}{a^2} - \frac{y^2}{b^2} = 1$ with $a^2 = 9$, and $b^2 = 4$. We know the transverse axis lies on the *x*-axis and the vertices are $(-3, 0)$ and $(3, 0)$.

Because $a^2 = 9$ and $b^2 = 4$, $a = 3$ and $b = 2$. Construct a rectangle using –3 and 3 on the *x*–axis, and –2 and 2 on the *y*–axis. Draw extended diagonals to obtain the asymptotes. Graph the hyperbola.

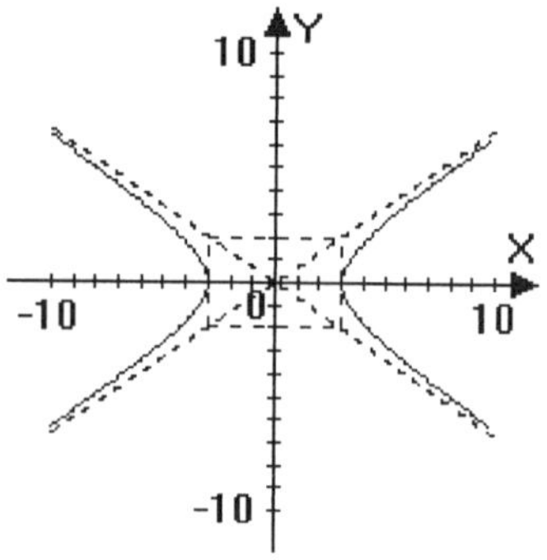


**36.** $\frac{x^2}{25} + \frac{y^2}{1} = 1$

Because $x^2$ and $y^2$ have different positive coefficients, the equation's graph is an ellipse.

Because the denominator of the $x^2$ – term is greater than the denominator of the $y^2$ – term, the major axis is horizontal. Since $a^2 = 25$, $a = 5$ and the vertices are $(-5, 0)$ and $(5, 0)$. Since $b^2 = 1$, $b = 1$ and endpoints of the minor axis are $(0, -1)$ and $(0, 1)$.

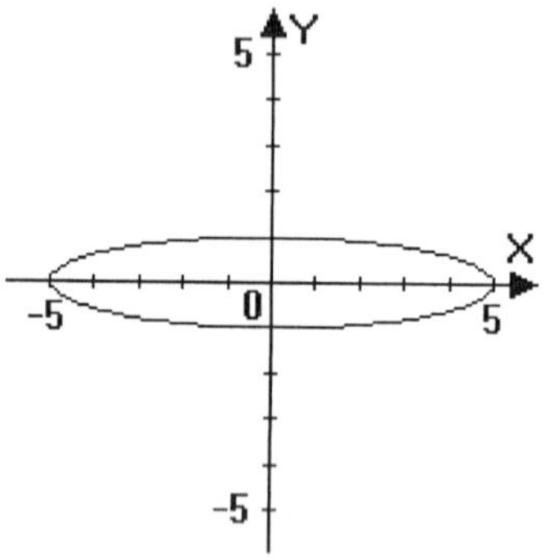


**37.** $x + 3 = -y^2 + 2y$

$$x = -y^2 + 2y - 3$$

Since only one variable is squared, the graph of the equation is a parabola. This is a parabola of the form $x = ay^2 + by + c$. Since $a = -1$ is negative, the parabola opens to the left. The *y*–coordinate of the vertex is $-\frac{b}{2a} = -\frac{2}{2(-1)} = -\frac{2}{-2} = 1$. The *x*–coordinate of the vertex is $x = -1^2 + 2(1) - 3 = -1 + 2 - 3 = -2$.

The vertex of the parabola is $(-2, 1)$.

Replace *y* with 0 to find the *x*–intercept.

$x = -0^2 + 2(0) - 3 = 0 + 0 - 3 = -3$

The $x$–intercept is –3. Replace $x$ with 0 to find the $y$–intercepts.

$0 = -y^2 + 2y - 3$

Solve using the quadratic formula.

$$y = \frac{-2 \pm \sqrt{2^2 - 4(-1)(-3)}}{2(-1)}$$

$$= \frac{-2 \pm \sqrt{4-12}}{-2} = \frac{-2 \pm \sqrt{-8}}{-2}$$

We do not need to simplify further. The solutions are complex and there are no $y$–intercepts.

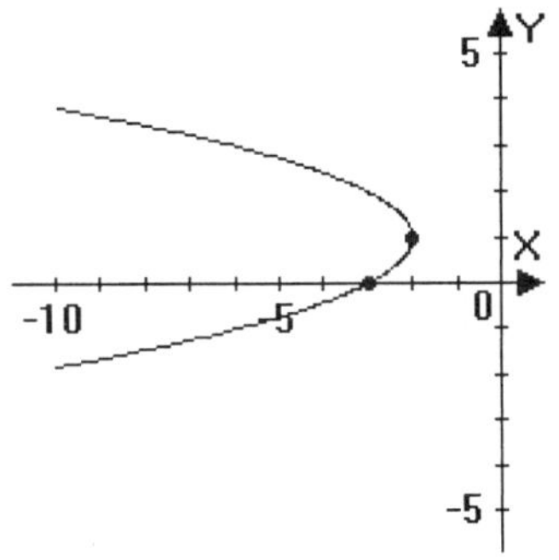


**38.** $y - 3 = x^2 - 2x$

$y = x^2 - 2x + 3$

Since only one variable is squared, the graph of the equation is a parabola. This is a parabola of the form $y = ax^2 + bx + c$. Since $a = 1$ is positive, the parabola opens to the right. The $x$–coordinate of the vertex is

$-\frac{b}{2a} = -\frac{-2}{2(1)} = -\frac{-2}{2} = 1$. The $y$–coordinate of the vertex is

$y = 1^2 - 2(1) + 3 = 1 - 2 + 3 = 2$.

The vertex of the parabola is $(1, 2)$.

Replace $x$ with 0 to find the $y$–intercept.

$y = 0^2 - 2(0) + 3 = 0 - 0 + 3 = 3$

The $y$–intercept is 3. Replace $y$ with 0 to find the $x$–intercepts.

$0 = x^2 - 2x + 3$

Solve using the quadratic formula.

$$x = \frac{-2 \pm \sqrt{2^2 - 4(-1)(-3)}}{2(-1)}$$

$$= \frac{-2 \pm \sqrt{4-12}}{-2} = \frac{-2 \pm \sqrt{-8}}{-2}$$

We do not need to simplify further. The solutions are complex and there are no $x$–intercepts.

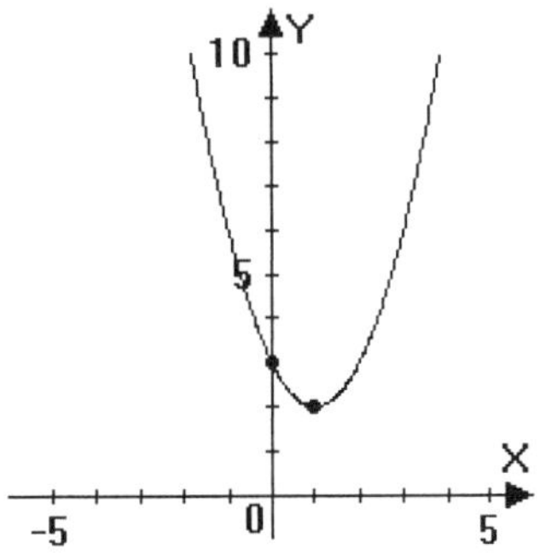


**39.** $\frac{(x+2)^2}{16} + \frac{(y-5)^2}{4} = 1$

Because $x^2$ and $y^2$ have different positive coefficients, the equation's graph is an ellipse.

The center of the ellipse is $(-2, 5)$.

Because the denominator of the $x^2$ – term is greater than the denominator of the $y^2$ – term, the major axis is horizontal. Since $a^2 = 16$, $a = 4$ and the vertices lie 4 units to the left and right of the center. Since $b^2 = 4$, $b = 2$ and endpoints of the minor axis lie two units above and below the center.

| Center | Vertices | Endpoints of Minor Axis |
|---|---|---|
| $(-2,5)$ | $(-2-4,5)$ $=(-6,5)$ | $(-2,5-2)$ $=(-2,3)$ |
| | $(-2+4,5)$ $=(2,5)$ | $(-2,5+2)$ $=(-2,7)$ |

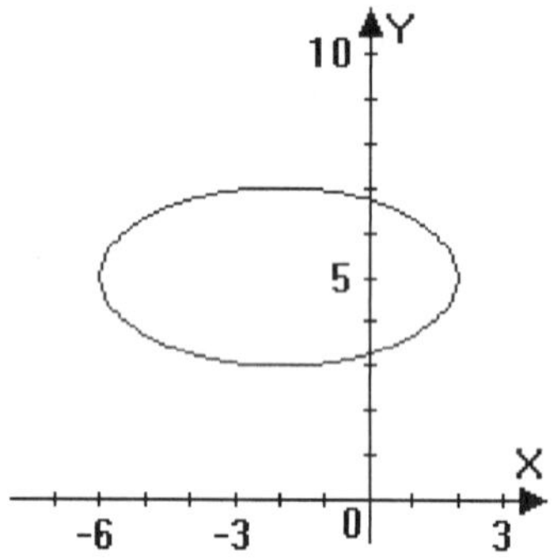


**40.** $(x-3)^2+(y+2)^2=4$

Because $x^2$ and $y^2$ have the same positive coefficient, the equation's graph is a circle.

$(x-3)^2+(y+2)^2=4$

The center is $(3,-2)$ and the radius is 2 units.

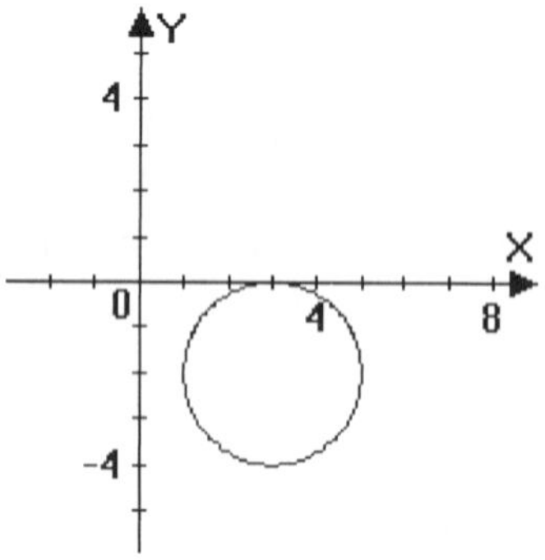


**41.** $$x^2+y^2+6x-2y+6=0$$

$(x^2+6x\quad)+(y^2-2y\quad)=-6$

Complete the squares.

$\left(\frac{b}{2}\right)^2=\left(\frac{6}{2}\right)^2=(3)^2=9$

$\left(\frac{b}{2}\right)^2=\left(\frac{-2}{2}\right)^2=(-1)^2=1$

$(x^2+6x+9)+(y^2-2y+1)=-6+9+1$

$$(x+3)^2+(y-1)^2=4$$

The center is $(-3,1)$ and the radius is 2 units.

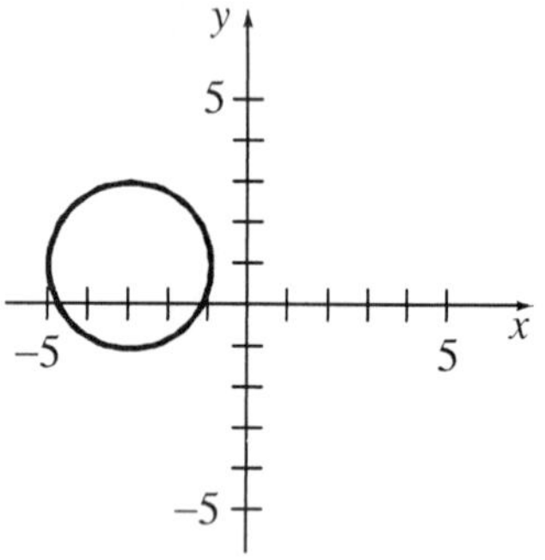


**42. a.** Using the point (6, 3), substitute for $x$ and $y$ to find $a$ in $y=ax^2$.

$3=a(6)^2$

$3=a(36)$

$a=\frac{3}{36}=\frac{1}{12}$

The equation for the parabola is $y=\frac{1}{12}x^2$.

**b.**
$$a = \frac{1}{4p}$$
$$\frac{1}{12} = \frac{1}{4p}$$
$$4p = 12$$
$$p = 3$$
The light source should be placed at the point (0, 3). This is the point 3 inches above the vertex.

**43.** $5y = x^2 - 1$
$x - y = 1$
Solve the second equation for *y*.
$x - y = 1$
$-y = -x + 1$
$y = x - 1$
Substitute $x - 1$ for $y$ in the first equation.
$5(x-1) = x^2 - 1$
$5x - 5 = x^2 - 1$
$0 = x^2 - 5x + 4$
$0 = (x-4)(x-1)$
Apply the zero product principle.
$x - 4 = 0$ and $x - 1 = 0$
$x = 4$ $\quad$ $x = 1$
Back-substitute 1 and 4 for *x* to find *y*.
$x = 4$ and $x = 1$
$y = x - 1$ $\quad$ $y = x - 1$
$y = 4 - 1$ $\quad$ $y = 1 - 1$
$y = 3$ $\quad$ $y = 0$
The solutions are $(1,0)$ and $(4,3)$ and the solution set is $\{(1,0),(4,3)\}$.

**44.** $y = x^2 + 2x + 1$
$x + y = 1$
Solve the second equation for *y*.
$x + y = 1$
$y = -x + 1$
Substitute $-x + 1$ for $y$ in the first equation.
$-x + 1 = x^2 + 2x + 1$
$0 = x^2 + 3x$
$0 = x(x+3)$
Apply the zero product principle.
$x = 0$ and $x + 3 = 0$
$x = -3$
Back-substitute $-3$ and 0 for *x* to find *y*.
$x = 0$ and $x = -3$
$y = -x + 1$ $\quad$ $y = -x + 1$
$y = -0 + 1$ $\quad$ $y = -(-3) + 1$
$y = 1$ $\quad$ $y = 3 + 1$
$y = 4$
The solutions are $(-3,4)$ and $(0,1)$ and the solution set is $\{(-3,4),(0,1)\}$.

**45.** $x^2 + y^2 = 2$
$x + y = 0$
Solve the second equation for *y*.
$x + y = 0$
$y = -x$
Substitute $-x$ for $y$ in the first equation.
$x^2 + (-x)^2 = 2$
$x^2 + x^2 = 2$
$2x^2 = 2$
$x^2 = 1$
$x = \pm 1$
Back-substitute $-1$ and 1 for *x* to find *y*.

$x=-1$ and $x=1$

$y=-x$ $\quad y=-x$

$y=-(-1)$ $\quad y=-1$

$y=1$

The solutions are $(-1,1)$ and $(1,-1)$ and the solution set is $\{(-1,1),\ (1,-1)\}$.

**46.** $2x^2+y^2=24$

$x^2+y^2=15$

Multiple the second equation by –1 and add to the first equation.

$$\begin{aligned} 2x^2+y^2 &= 24 \\ -x^2-y^2 &= -15 \\ \hline x^2 &= 9 \\ x &= \pm 3 \end{aligned}$$

Back-substitute –3 and 3 for $x$ to find $y$.

$$\begin{aligned} x &= \pm 3 \\ (\pm 3)^2+y^2 &= 15 \\ 9+y^2 &= 15 \\ y^2 &= 6 \\ y &= \pm\sqrt{6} \end{aligned}$$

The solutions are $(-3,-\sqrt{6})$, $(-3,\sqrt{6})$, $(3,-\sqrt{6})$ and $(3,\sqrt{6})$ and the solution set is $\{(-3,-\sqrt{6}),$ $(-3,\sqrt{6}),(3,-\sqrt{6}),(3,\sqrt{6})\}$.

**47.** $xy-4=0$

$y-x=0$

Solve the second equation for $y$.

$y-x=0$

$y=x$

Substitute $x$ for $y$ in the first equation and solve for $x$.

$$\begin{aligned} x(x)-4 &= 0 \\ x^2-4 &= 0 \\ (x+2)(x-2) &= 0 \end{aligned}$$

Apply the zero product principle.

$x+2=0$ and $x-2=0$

$x=-2$ $\quad x=2$

Back-substitute –2 and 2 for $x$ to find $y$.

$x=-2$ and $x=2$

$y=x$ $\quad y=x$

$y=-2$ $\quad y=2$

The solutions are $(-2,-2)$ and $(2,2)$ and the solution set is $\{(-2,-2),$ $(2,2)\}$.

**48.** $y^2=4x$

$x-2y+3=0$

Solve the second equation for $x$.

$x-2y+3=0$

$x=2y-3$

Substitute $2y-3$ for $x$ in the first equation and solve for $y$.

$$\begin{aligned} y^2 &= 4(2y-3) \\ y^2 &= 8y-12 \\ y^2-8y+12 &= 0 \\ (y-6)(y-2) &= 0 \end{aligned}$$

Apply the zero product principle.

$y-6=0$ and $y-2=0$

$y=6$ $\quad y=2$

Back-substitute 2 and 6 for $y$ to find $x$.

$y=6$ and $y=2$

$x=2y-3$ $\quad x=2y-3$

$x=2(6)-3$ $\quad x=2(2)-3$

$x=12-3$ $\quad x=4-3$

$x=9$ $\quad x=1$

The solutions are $(1,2)$ and $(9,6)$ and the solution set is $\{(1,2),(9,6)\}$.

**49.** $x^2 + y^2 = 10$

$y = x + 2$

Substitute $x + 2$ for $y$ in the first equation and solve for $x$.

$$x^2 + (x+2)^2 = 10$$
$$x^2 + x^2 + 4x + 4 = 10$$
$$2x^2 + 4x + 4 = 10$$
$$2x^2 + 4x - 6 = 0$$
$$x^2 + 2x - 3 = 0$$
$$(x+3)(x-1) = 0$$

Apply the zero product principle.

$x + 3 = 0$ and $x - 1 = 0$

$x = -3$ $\quad$ $x = 1$

Back-substitute $-3$ and 1 for $x$ to find $y$.

$x = -3$ and $x = 1$

$y = x + 2$ $\quad$ $y = x + 2$

$y = -3 + 2$ $\quad$ $y = 1 + 2$

$y = -1$ $\quad$ $y = 3$

The solutions are $(-3,-1)$ and $(1,3)$ and the solution set is $\{(-3,-1), (1,3)\}$.

**50.** $xy = 1$

$y = 2x + 1$

Substitute $2x + 1$ for $y$ in the first equation and solve for $x$.

$$x(2x+1) = 1$$
$$2x^2 + x = 1$$
$$2x^2 + x - 1 = 0$$
$$(2x-1)(x+1) = 0$$

Apply the zero product principle.

$2x - 1 = 0$ and $x + 1 = 0$

$2x = 1$ $\quad$ $x = -1$

$x = \frac{1}{2}$

Back-substitute $-1$ and $\frac{1}{2}$ for $x$ to find $y$.

$x = -1$ and $x = \frac{1}{2}$

$y = 2x + 1$ $\quad$ $y = 2x + 1$

$y = 2(-1) + 1$ $\quad$ $y = 2\left(\frac{1}{2}\right) + 1$

$y = -2 + 1$ $\quad$ $y = 1 + 1$

$y = -1$ $\quad$ $y = 2$

The solutions are $(-1,-1)$ and $\left(\frac{1}{2}, 2\right)$ and the solution set is $\left\{(-1,-1), \left(\frac{1}{2}, 2\right)\right\}$.

**51.** $x + y + 1 = 0$

$x^2 + y^2 + 6y - x = -5$

Solve for $y$ in the first equation.

$x + y + 1 = 0$

$y = -x - 1$

Substitute $-x - 1$ for $y$ in the second equation and solve for $x$.

$$x^2 + (-x-1)^2 + 6(-x-1) - x = -5$$
$$x^2 + x^2 + 2x + 1 - 6x - 6 - x = -5$$
$$2x^2 - 5x - 5 = -5$$
$$2x^2 - 5x = 0$$
$$x(2x-5) = 0$$

Apply the zero product principle.

$x = 0$ and $2x - 5 = 0$

$2x = 5$

$x = \frac{5}{2}$

Back-substitute 0 and $\frac{5}{2}$ for $x$ to find $y$.

$x = 0$ and $x = \frac{5}{2}$

$y = -x - 1$ $\quad$ $y = -x - 1$

$y = -0 - 1$ $\quad$ $y = -\frac{5}{2} - 1$

$y = -1$ $\quad$ $y = -\frac{7}{2}$

The solutions are $(0, -1)$ and $\left(\frac{5}{2}, -\frac{7}{2}\right)$ and the solution set is $\left\{(0, -1), \left(\frac{5}{2}, -\frac{7}{2}\right)\right\}$.

**52.** $x^2 + y^2 = 13$

$x^2 - y = 7$

Solve for $x^2$ in the second equation.

$x^2 - y = 7$

$x^2 = y + 7$

Substitute $y + 7$ for $x^2$ in the first equation and solve for $y$.

$(y + 7) + y^2 = 13$

$y^2 + y + 7 = 13$

$y^2 + y - 6 = 0$

$(y + 3)(y - 2) = 0$

Apply the zero product principle.

$y + 3 = 0$ and $y - 2 = 0$

$y = -3$ $\quad$ $y = 2$

Back-substitute –3 and 2 for $y$ to find $x$.

$y = -3$ and $y = 2$

$x^2 = y + 7$ $\quad$ $x^2 = y + 7$

$x^2 = -3 + 7$ $\quad$ $x^2 = 2 + 7$

$x^2 = 4$ $\quad$ $x^2 = 9$

$x = \pm 2$ $\quad$ $x = \pm 3$

The solutions are $(-3, 2), (-2, -3)$, $(2, -3)$ and $(3, 2)$ and the solution set is $\{(-3, 2), (-2, -3), (2, -3), (3, 2)\}$.

**53.** $2x^2 + 3y^2 = 21$

$3x^2 - 4y^2 = 23$

Multiply the first equation by 4 and the second equation by 3.

$8x^2 + 12y^2 = 84$

$9x^2 - 12y^2 = 69$

$17x^2 = 153$

$x^2 = 9$

$x = \pm 3$

Back-substitute $\pm 3$ for $x$ to find $y$.

$x = \pm 3$

$2(\pm 3)^2 + 3y^2 = 21$

$2(9) + 3y^2 = 21$

$18 + 3y^2 = 21$

$3y^2 = 3$

$y^2 = 1$

$y = \pm 1$

We have $x = \pm 3$ and $y = \pm 1$, the solutions are $(-3, -1), (-3, 1), (3, -1)$ and $(3, 1)$ and the solution set is $\{(-3, -1), (-3, 1), (3, -1), (3, 1)\}$.

**54.** Let $x$ = the length of the rectangle

Let $y$ = the width of the rectangle

$2x + 2y = 26$

$xy = 40$

Solve the first equation for $y$.

$2x + 2y = 26$

$x + y = 13$

$y = 13 - x$

Substitute $13 - x$ for $y$ in the second equation.

$x(13-x)=40$

$13x-x^2=40$

$0=x^2-13x+40$

$0=(x-8)(x-5)$

Apply the zero product principle.

$x-8=0$ and $x-5=0$

$x=8$ $\quad$ $x=5$

Back-substitute 5 and 8 for $x$ to find $y$.

$x=8$ and $x=5$

$y=13-8$ $\quad$ $y=13-5$

$y=5$ $\quad$ $y=8$

The solutions are the same. The dimensions are 8 meters by 5 meters.

**55.** $2x+y=8$

$xy=6$

Solve the first equation for $y$.

$2x+y=8$

$y=-2x+8$

Substitute $-2x+8$ for $y$ in the second equation.

$x(-2x+8)=6$

$-2x^2+8x=6$

$-2x^2+8x-6=0$

$x^2-4x+3=0$

$(x-3)(x-1)=0$

Apply the zero product principle.

$x-3=0$ and $x-1=0$

$x=3$ $\quad$ $x=1$

Back-substitute 1 and 3 for $x$ to find $y$.

$x=3$ and $x=1$

$y=-2x+8$ $\quad$ $y=-2x+8$

$y=-2(3)+8$ $\quad$ $y=-2(1)+8$

$y=-6+8$ $\quad$ $y=-2+8$

$y=2$ $\quad$ $y=6$

The solutions are the points $(1,6)$ and $(3,2)$.

**56.** Using the formula for the area, we have $x^2+y^2=2900$. Since there are 240 feet of fencing available, we have:

$x+(x+y)+y+y+(x-y)+x=240$

$x+x+y+y+y+x-y+x=240$

$4x+2y=240.$

The system of two variables in two equations is as follows.

$x^2+y^2=2900$

$4x+2y=240$

Solve the second equation for $y$.

$4x+2y=240$

$2y=-4x+240$

$y=-2x+120$

Substitute $-2x+120$ for $y$ to find $x$.

$x^2+(-2x+120)^2=2900$

$x^2+4x^2-480x+14400=2900$

$5x^2-480x+11500=0$

$x^2-96x+2300=0$

$(x-50)(x-46)=0$

Apply the zero product principle.

$x-50=0$ and $x-46=0$

$x=50$ $\quad$ $x=46$

Back-substitute 46 and 50 for $x$ to find $y$.

$x=50$ $\quad$ $x=46$

$y=-2x+120$ $\quad$ $y=-2x+120$

$y=-2(50)+120$ $\quad$ $y=-2(46)+120$

$y=-100+120$ $\quad$ $y=-92+120$

$y=20$ $\quad$ $y=28$

The solutions are $x=50$ feet and $y=20$ feet or $x=46$ feet and $y=28$ feet.

## Chapter 10 Test

**1.** $d=\sqrt{(2-(-1))^2+(-3-5)^2}$

$=\sqrt{(3)^2+(-8)^2}$

$=\sqrt{9+64}=\sqrt{73}\approx 8.54$

The distance between the points is $\sqrt{73}$ or 8.54 units.

**2.** $\text{Midpoint}=\left(\frac{-5+12}{2},\frac{-2+(-6)}{2}\right)$

$=\left(\frac{7}{2},\frac{-8}{2}\right)=\left(\frac{7}{2},-4\right)$

The midpoint is $\left(\frac{7}{2},-4\right)$.

**3.** $(x-3)^2+(y-(-2))^2=5^2$

$(x-3)^2+(y+2)^2=25$

**4.** $(x-5)^2+(y+3)^2=49$

$(x-5)^2+(y-(-3))^2=7^2$

The center is $(5,-3)$ and the radius is 7 units.

**5.** $x^2+y^2+4x-6y-3=0$

$(x^2+4x\quad)+(y^2-6y\quad)=3$

Complete the squares.

$\left(\frac{b}{2}\right)^2=\left(\frac{4}{2}\right)^2=(2)^2=4$

$\left(\frac{b}{2}\right)^2=\left(\frac{-6}{2}\right)^2=(-3)^2=9$

$(x^2+4x+4)+(y^2-6y+9)=3+4+9$

$(x+2)^2+(y-3)^2=16$

$(x-(-2))^2+(y-3)^2=4^2$

The center is $(-2,3)$ and the radius is 4 units.

**6.** $x=-2(y+3)^2+7$

$x=-2(y-(-3))^2+7$

The vertex of the parabola is $(7,-3)$.

**7.** $x=y^2+10y+23$

The $y$–coordinate of the vertex is

$-\frac{b}{2a}=-\frac{10}{2(1)}=-\frac{10}{2}=-5.$

The $x$–coordinate of the vertex is

$x=(-5)^2+10(-5)+23$

$=25-50+23$

$=25-50+23=-2.$

The vertex of the parabola is $(-2,-5)$.

**8.** $\frac{x^2}{4}-\frac{y^2}{9}=1$

Because $x^2$ and $y^2$ have opposite signs, the equation's graph is a hyperbola.

The equation is in the form $\frac{x^2}{a^2}-\frac{y^2}{b^2}=1$ with $a^2=4$, and $b^2=9$. We know the transverse axis lies on the $x$-axis and the vertices are $(-2,0)$ and $(2,0)$.

Because $a^2=4$ and $b^2=9$, $a=2$ and $b=3$. Construct a rectangle using –2 and 2 on the $x$–axis, and –3 and 3 on the $y$–axis. Draw extended diagonals to obtain the asymptotes. Graph the hyperbola.

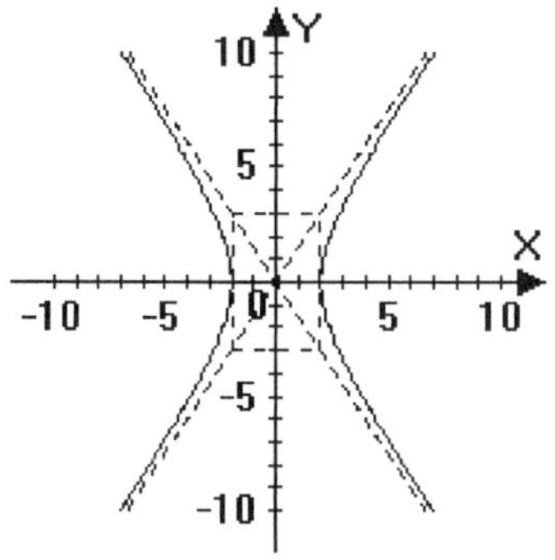


**9.** $4x^2 + 9y^2 = 36$

Because $x^2$ and $y^2$ have different positive coefficients, the equation's graph is an ellipse.

$$\frac{4x^2}{36} + \frac{9y^2}{36} = \frac{36}{36}$$

$$\frac{x^2}{9} + \frac{y^2}{4} = 1$$

Because the denominator of the $x^2$ – term is greater than the denominator of the $y^2$ – term, the major axis is horizontal. Since $a^2 = 9$, $a = 3$ and the vertices are $(-3,0)$ and $(3,0)$. Since $b^2 = 4$, $b = 2$ and endpoints of the minor axis are $(0,-2)$ and $(0,2)$.

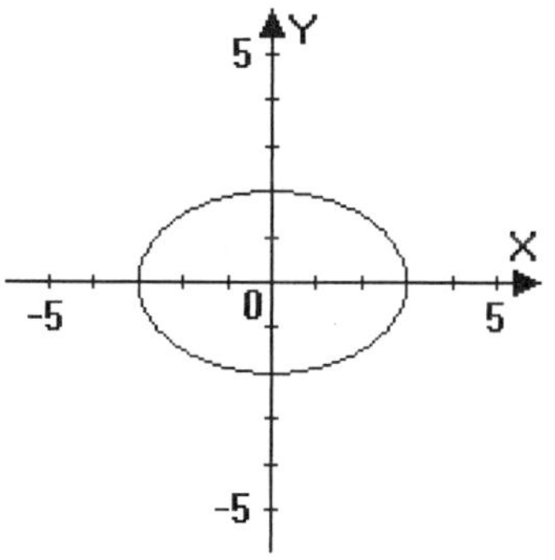


**10.** $x = (y+1)^2 - 4$

Since only one variable is squared, the graph of the equation is a parabola. This is a parabola of the form $x = a(y-k)^2 + h$. Since $a = 1$ is positive, the parabola opens to the right. The vertex of the parabola is $(4,-1)$.

Replace $y$ with 0 to find the $x$–intercept.

$x = (0+1)^2 - 4 = (1)^2 - 4 = 1 - 4 = -3.$

The $x$–intercept is 0. Replace $x$ with 0 to find the $y$–intercepts.

$0 = (y+1)^2 - 4$

$0 = y^2 + 2y + 1 - 4$

$0 = y^2 + 2y - 3$

$0 = (y+3)(y-1)$

Apply the zero product principle.

$y + 3 = 0$ and $y + 1 = 0$

$y = -3$ $\quad$ $y = -1$

The $y$–intercepts are –3 and –1.

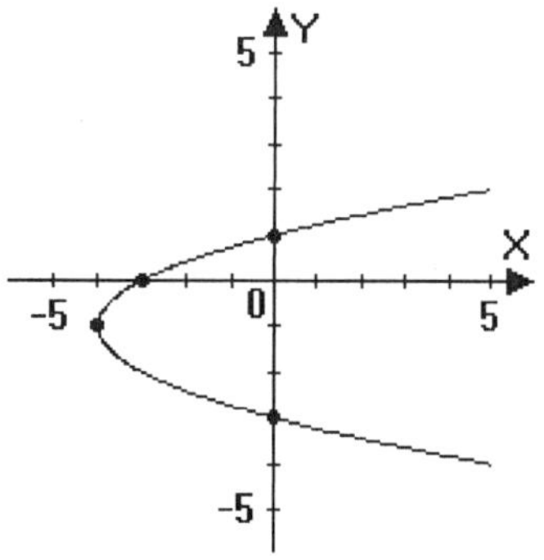


**11.** $16x^2 + y^2 = 16$

Because $x^2$ and $y^2$ have different positive coefficients, the equation's graph is an ellipse.

$$\frac{16x^2}{16} + \frac{y^2}{16} = \frac{16}{16}$$

$$\frac{x^2}{1} + \frac{y^2}{16} = 1$$

Because the denominator of the $y^2$ – term is greater than the denominator of the $x^2$ – term, the major axis is vertical. Since $a^2 = 16$, $a = 4$ and the vertices are $(0,-4)$ and $(0,4)$. Since $b^2 = 1$, $b = 1$ and endpoints of the minor axis are $(-1,0)$ and $(1,0)$.

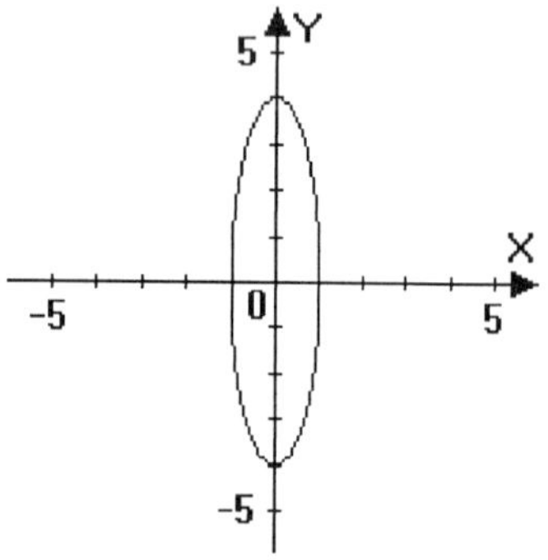


**12.** $25y^2 = 9x^2 + 225$

$25y^2 - 9x^2 = 225$

Because $x^2$ and $y^2$ have opposite signs, the equation's graph is a hyperbola.

$$\frac{25y^2}{225} - \frac{9x^2}{225} = \frac{225}{225}$$

$$\frac{y^2}{9} - \frac{x^2}{25} = 1$$

The equation is in the form $\frac{y^2}{a^2} - \frac{x^2}{b^2} = 1$ with $a^2 = 9$, and $b^2 = 25$. We know the transverse axis lies on the $y$-axis and the vertices are $(0,-3)$ and $(0,3)$.

Because $a^2 = 9$ and $b^2 = 25$, $a = 35$ and $b = 5$. Construct a rectangle using –5 and 5 on the $x$–axis, and –3 and 3 on the $y$–axis. Draw extended diagonals to obtain the asymptotes. Graph the hyperbola.

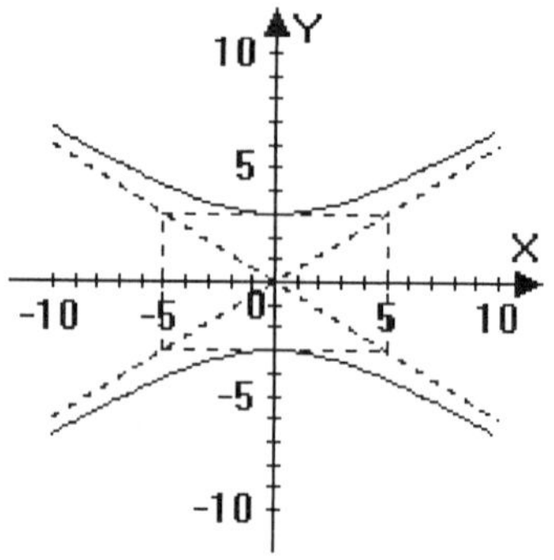


**13.** $x = -y^2 + 6y$

Since only one variable is squared, the graph of the equation is a parabola. This is a parabola of the form $x = ay^2 + by + c$. Since $a = 1$ is positive, the parabola opens to the right. The $y$–coordinate of the vertex is $-\frac{b}{2a} = -\frac{6}{2(-1)} = -\frac{6}{-2} = 3$. The $x$–coordinate of the vertex is $x = -3^2 + 6(3) = -9 + 18 = 9$.

The vertex of the parabola is $(9,3)$.

Replace $y$ with 0 to find the $x$–intercept.

$x = -0^2 + 6(0) = 0 + 0 = 0$

The $x$–intercept is 0. Replace $x$ with 0 to find the $y$–intercepts.

$0 = -y^2 + 6y$

$0 = -y(y-6)$

Apply the zero product principle.

$-y = 0$ and $y - 6 = 0$

$y = 0$ $\quad y = 6$

The $y$–intercepts are 0 and 6.

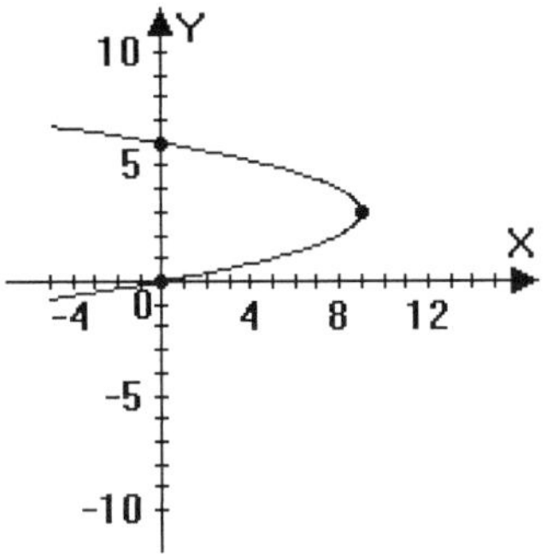


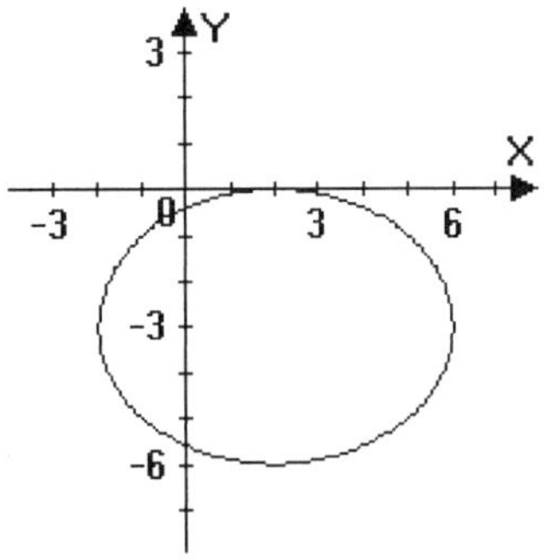


**14.** $\frac{(x-2)^2}{16}+\frac{(y+3)^2}{9}=1$

Because $x^2$ and $y^2$ have different positive coefficients, the equation's graph is an ellipse.
The center of the ellipse is $(2,-3)$.
Because the denominator of the $x^2-$term is greater than the denominator of the $y^2-$term, the major axis is horizontal. Since $a^2=16$, $a=4$ and the vertices lie 4 units to the left and right of the center. Since $b^2=9$, $b=3$ and endpoints of the minor axis lie 3 units above and below the center.

| Center | Vertices | Endpoints of Minor Axis |
|---|---|---|
| $(2,-3)$ | $(2-4,-3)$ $=(-2,-3)$ | $(2,-3-3)$ $=(2,-6)$ |
| | $(2+4,-3)$ $=(6,-3)$ | $(2,-3+3)$ $=(2,0)$ |

**15.** $(x+1)^2+(y+2)^2=9$

Because $x^2$ and $y^2$ have the same positive coefficient, the equation's graph is a circle.
The center of the circle is $(-1,-2)$ and the radius is 3.

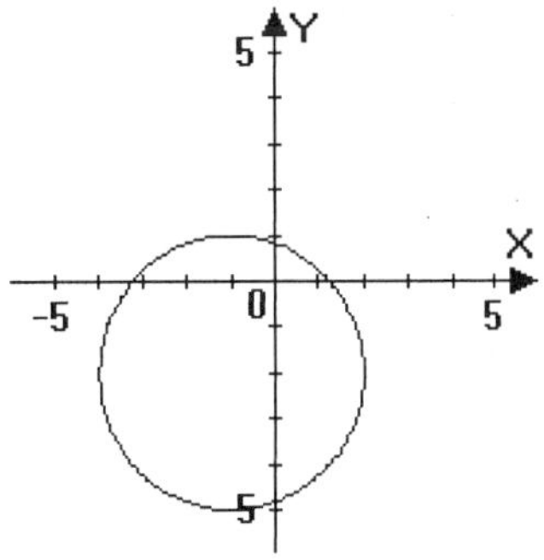


**16.**

$$\frac{x^2}{4}+\frac{y^2}{4}=1$$

$$4\left(\frac{x^2}{4}+\frac{y^2}{4}\right)=4(1)$$

$$x^2+y^2=4$$

Because $x^2$ and $y^2$ have the same positive coefficient, the equation's graph is a circle.
The center of the circle is $(0,0)$ and the radius is 2.

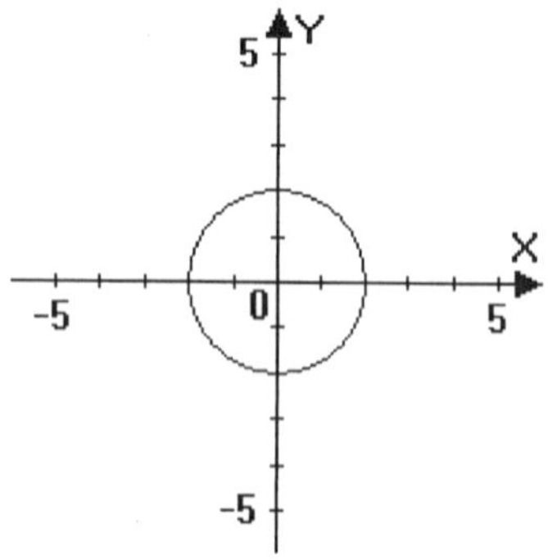


**17.** $x^2 + y^2 = 25$

$x + y = 1$

Solve the second equation for $y$.

$x + y = 1$

$y = -x + 1$

Substitute $-x + 1$ for $y$ to find $x$.

$$x^2 + (-x+1)^2 = 25$$
$$x^2 + x^2 - 2x + 1 = 25$$
$$2x^2 - 2x + 1 = 25$$
$$2x^2 - 2x - 24 = 0$$
$$x^2 - x - 12 = 0$$
$$(x-4)(x+3) = 0$$

Apply the zero product principle.

$x - 4 = 0$ and $x + 3 = 0$

$x = 4$ $\quad$ $x = -3$

Back-substitute $-3$ and $4$ for $x$ to find $y$.

| $x = 4$ | and $x = -3$ |
|---|---|
| $y = -x + 1$ | $y = -x + 1$ |
| $y = -4 + 1$ | $y = -(-3) + 1$ |
| $y = -3$ | $y = 3 + 1$ |
| | $y = 4$ |

The solutions are $(-3, 4)$ and $(4, -3)$ and the solution set is $\{(-3, 4), (4, -3)\}$.

**18.** $2x^2 - 5y^2 = -2$

$3x^2 + 2y^2 = 35$

Multiply the first equation by 2 and the second equation by 5.

$$4x^2 - 10y^2 = -4$$
$$\underline{15x^2 + 10y^2 = 175}$$
$$19x^2 = 171$$
$$x^2 = 9$$
$$x = \pm 3$$

In this case, we can back-substitute 9 for $x^2$ to find $y$.

$$x^2 = 9$$
$$2x^2 - 5y^2 = -2$$
$$2(9) - 5y^2 = -2$$
$$18 - 5y^2 = -2$$
$$-5y^2 = -20$$
$$y^2 = 4$$
$$y = \pm 2$$

We have $x = \pm 3$ and $y = \pm 2$, the solutions are $(-3, -2), (-3, 2), (3, -2)$ and $(3, 2)$ and the solution set is $\{(-3, -2), (-3, 2), (3, -2), (3, 2)\}$.

**19.** $2x + y = 39$

$xy = 180$

Solve the first equation for $y$.

$2x + y = 39$

$y = 39 - 2x$

Substitute $39 - 2x$ for $y$ to find $x$.

$$x(39 - 2x) = 180$$
$$39x - 2x^2 = 180$$
$$0 = 2x^2 - 39x + 180$$
$$0 = (2x - 15)(x - 12)$$

Apply the zero product principle.

$2x-15=0 \quad \text{and} \quad x-12=0$

$2x=15 \qquad\qquad x=12$

$x=\frac{15}{2}$

Back-substitute $\frac{15}{2}$ and 12 for $x$ to find $y$.

$x=\frac{15}{2} \quad \text{and} \quad x=12$

$y=39-2x \qquad y=39-2x$

$y=39-2\left(\frac{15}{2}\right) \qquad y=39-2(12)$ T

$y=39-15 \qquad y=39-24$

$y=24 \qquad y=15$

he dimensions are 15 feet by 12 feet or 24 feet by $\frac{15}{2}$ or 7.5 feet.

**20.** Let $x$ = the length of the rectangle
Let $y$ = the width of the rectangle

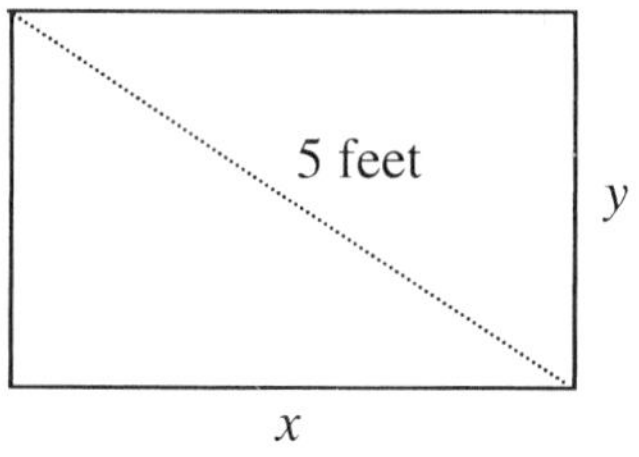


Using the Pythagorean Theorem, we obtain $x^2+y^2=5^2$. Since the perimeter is 14 feet, we have $2x+2y=14$.
The system of two equations in two variables is as follows.

$x^2+y^2=25$

$2x+2y=14$

Solve the second equation for $y$.

$2x+2y=14$

$2y=14-2x$

$y=7-x$

Substitute $7-x$ for $y$ to find $x$.

$x^2+(7-x)^2=25$

$x^2+49-14x+x^2=25$

$2x^2-14x+49=25$

$2x^2-14x+24=0$

$x^2-7x+12=0$

$(x-4)(x-3)=0$

Apply the zero product principle.

$x-4=0 \quad \text{and} \quad x-3=0$

$x=4 \qquad\qquad x=3$

Back-substitute 3 and 4 for $x$ to find $y$.

$x=4 \quad \text{and} \quad x=3$

$y=7-x \qquad y=7-x$

$y=7-4 \qquad y=7-3$

$y=3 \qquad y=4$

The solutions are the same. The dimensions are 4 feet by 3 feet.

### Cumulative Review Exercises (Chapters 1-10)

**1.** $3x+7>4 \quad \text{or} \quad 6-x<1$

$3x>-3 \qquad -x<-5$

$x>-1 \qquad x>5$

$x>-1 \quad \text{or} \quad x>5$

The solution set is $\{x|x>-1\}$ or $(-1,\infty)$.

**2.** $x(2x-7)=4$

$2x^2-7x=4$

$2x^2-7x-4=0$

$(2x+1)(x-4)=0$

Apply the zero product principle.

$$\begin{aligned} 2x+1&=0 \quad \text{and} \quad x-4=0 \\ 2x&=-1 \qquad\qquad x=4 \\ x&=-\frac{1}{2} \end{aligned}$$

The solutions are $-\frac{1}{2}$ and 4 and the solution set is $\left\{-\frac{1}{2}, 4\right\}$.

**3.** $$\frac{5}{x-3}=1+\frac{30}{x^2-9}$$
$$\frac{5}{x-3}=1+\frac{30}{(x+3)(x-3)}$$

Multiply both sides of the equation by the LCD, $(x+3)(x-3)$.

$$(x+3)(x-3)\left(\frac{5}{x-3}\right)=(x+3)(x-3)\left(1+\frac{30}{(x+3)(x-3)}\right)$$
$$(x+3)(5)=(x+3)(x-3)+30$$
$$5x+15=x^2-9+30$$
$$15=x^2-5x+21$$
$$0=x^2-5x+6$$
$$0=(x-3)(x-2)$$

Apply the zero product principle.

$$x-3=0 \quad \text{and} \quad x-2=0$$
$$x=3 \qquad\qquad x=2$$

We disregard 3 because it would make the denominator zero. The solution is 2 and the solution set is $\{2\}$.

**4.** $3x^2+8x+5<0$

Solve the related quadratic equation.

$$3x^2+8x+5=0$$
$$(3x+5)(x+1)=0$$

Apply the zero product principle.

$$3x+5=0 \quad \text{or} \quad x+1=0$$
$$3x=-5 \qquad x=-1$$
$$x=-\frac{5}{3}$$

The boundary points are $-\frac{5}{3}$ and $-1$.

| Test Interval | Test Number | Test | Conclusion |
|---|---|---|---|
| $\left(-\infty,-\frac{5}{3}\right)$ | -2 | $3(-2)^2+8(-2)+5<0$ <br> $1<0$, false | $\left(-\infty,-\frac{5}{3}\right)$ does not belong to the solution set. |
| $\left(-\frac{5}{3},-1\right)$ | $-\frac{4}{3}$ | $3\left(-\frac{4}{3}\right)^2+8\left(-\frac{4}{3}\right)+5<0$ <br> $-11<0$, true | $\left(-\frac{5}{3},-1\right)$ belongs to the solution set. |
| $(-1,\infty)$ | 0 | $3(0)^2+8(0)+5<0$ <br> $5<0$, false | $(-1,\infty)$ does not belong to the solution set. |

The solution set is $\left(-\frac{5}{3},-1\right)$ or $\left\{x\middle|-\frac{5}{3}<x<-1\right\}$.

**5.**

$$3^{2x-1}=81$$
$$3^{2x-1}=3^4$$
$$2x-1=4$$
$$2x=5$$
$$x=\frac{5}{2}$$

The solution is $\frac{5}{2}$ and the solution set is $\left\{\frac{5}{2}\right\}$.

**6.**

$$30e^{0.7x}=240$$
$$e^{0.7x}=80$$
$$\ln e^{0.7x}=\ln 8$$
$$0.7x=\ln 8$$
$$x=\frac{\ln 8}{0.7}=\frac{2.08}{0.7}\approx 2.97$$

The solution is $\frac{2.08}{0.7}\approx 2.97$ and the solution set is $\left\{\frac{2.08}{0.7}\approx 2.97\right\}$.

**7.** $3x^2 + 4y^2 = 39$

$5x^2 - 2y^2 = -13$

Multiply the second equation by 2 and add to the first equation.

$$\begin{aligned} 3x^2 + 4y^2 &= 39 \\ 10x^2 - 4y^2 &= -26 \\ \hline 13x^2 &= 13 \\ x^2 &= 1 \\ x &= \pm 1 \end{aligned}$$

In this case, we can back-substitute 9 for $x^2$ to find $y$.

$$\begin{aligned} x^2 &= 1 \\ 3x^2 + 4y^2 &= 39 \\ 3(1) + 4y^2 &= 39 \\ 3 + 4y^2 &= 39 \\ 4y^2 &= 36 \\ y^2 &= 9 \\ y &= \pm 3 \end{aligned}$$

We have $x = \pm 1$ and $y = \pm 3$, the solutions are $(-1,-3),(-1,3),(1,-3)$ and $(1,3)$ and the solution set is $\{(-1,-3),(-1,3),(1,-3),(1,3)\}$.

**8.** $f(x) = -\frac{2}{3}x + 4$

$y = -\frac{2}{3}x + 4$

The $y$–intercept is 4 and the slope is $-\frac{2}{3}$. We can write the slope as $m = \frac{-2}{3} = \frac{\text{rise}}{\text{run}}$ and use the intercept and the slope to graph the function.

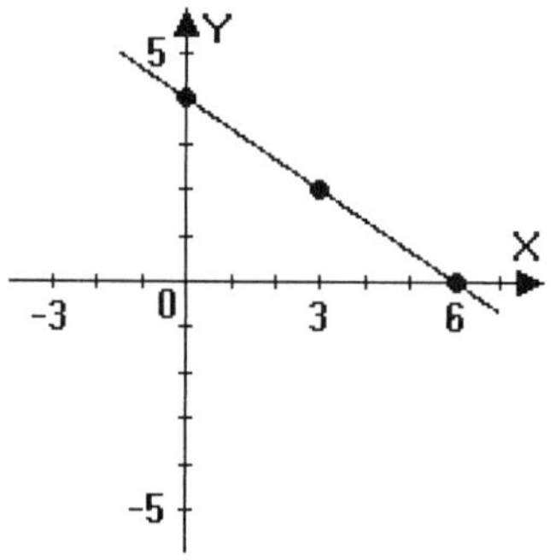


**9.** $3x - y > 6$

First, find the intercepts to the equation $3x - y = 6$.

Find the $x$–intercept by setting $y$ equal to zero.

$$\begin{aligned} 3x - 0 &= 6 \\ 3x &= 6 \\ x &= 2 \end{aligned}$$

Find the $y$–intercept by setting $x$ equal to zero.

$$\begin{aligned} 3(0) - y &= 6 \\ -y &= 6 \\ y &= -6 \end{aligned}$$

Next, use the origin as a test point.

$$\begin{aligned} 3(0) - 0 &> 6 \\ 0 - 0 &> 6 \\ 0 &> 6 \end{aligned}$$

This is a false statement. This means that the origin will not fall in the shaded half-plane.

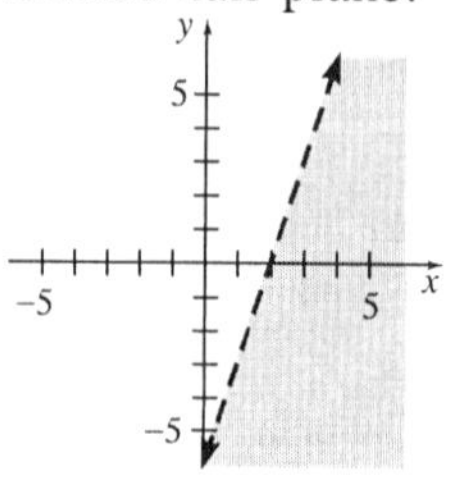

**10.** $x^2 + y^2 + 4x - 6y + 9 = 0$

Because $x^2$ and $y^2$ have the same positive coefficient, the equation's graph is a circle.

$(x^2 + 4x \quad) + (y^2 - 6y \quad) = -9$

Complete the squares.

$\left(\frac{b}{2}\right)^2 = \left(\frac{4}{2}\right)^2 = (2)^2 = 4$

$\left(\frac{b}{2}\right)^2 = \left(\frac{-6}{2}\right)^2 = (-3)^2 = 9$

$(x^2 + 4x + 4) + (y^2 - 6y + 9) = -9 + 4 + 9$

$(x+2)^2 + (y-3)^2 = 4$

The center of the circle is $(-2,3)$ and the radius is 2.

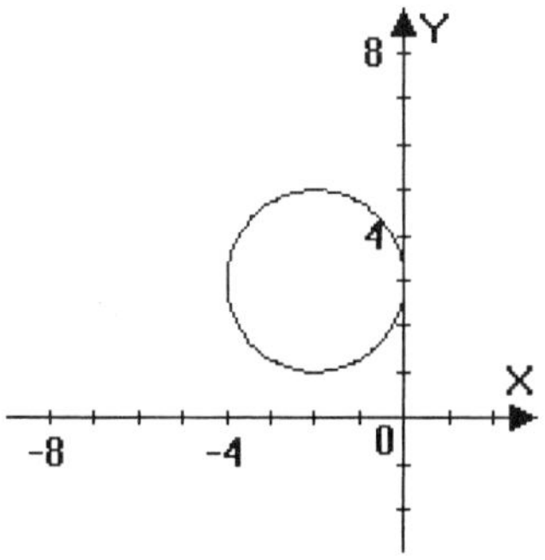


**11.** $9x^2 - 4y^2 = 36$

Because $x^2$ and $y^2$ have opposite signs, the equation's graph is a hyperbola.

$\frac{9x^2}{36} - \frac{4y^2}{36} = \frac{36}{36}$

$\frac{x^2}{4} - \frac{y^2}{9} = 1$

The equation is in the form $\frac{x^2}{a^2} - \frac{y^2}{b^2} = 1$ with $a^2 = 4$, and $b^2 = 9$.

We know the transverse axis lies on the $x$-axis and the vertices are $(-2,0)$ and $(2,0)$. Because $a^2 = 4$ and $b^2 = 9$, $a = 2$ and $b = 3$.

Construct a rectangle using –2 and 2 on the $x$–axis, and –3 and 3 on the $y$–axis. Draw extended diagonals to obtain the asymptotes. Graph the hyperbola.

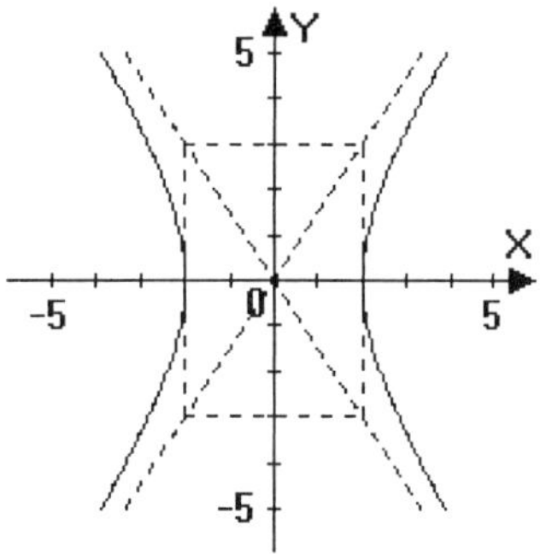


**12.** $-2(3^2 - 12)^3 - 45 \div 9 - 3$

$= -2(9-12)^3 - 45 \div 9 - 3$

$= -2(-3)^3 - 45 \div 9 - 3$

$= -2(-27) - 45 \div 9 - 3$

$= 54 - 5 - 3 = 46$

**13.** $(3x^3 - 19x^2 + 17x + 4) \div (3x - 4)$

Rewrite the polynomials in descending order and divide.

$$\begin{array}{r}
x^2 - 5x - 1 \\
3x - 4 \overline{)\, 3x^3 - 19x^2 + 17x + 4} \\
\underline{3x^3 - 4x^2} \qquad\qquad \\
-15x^2 + 17x \qquad \\
\underline{-15x^2 + 20x} \qquad \\
-3x + 4 \\
\underline{-3x + 4} \\
0
\end{array}$$

$\frac{3x^3 - 19x^2 + 17x + 4}{3x - 4} = x^2 - 5x - 1$

**14.** $\sqrt[3]{4x^2y^5}\cdot\sqrt[3]{4xy^2}$

$= \sqrt[3]{4x^2y^5 4xy^2} = \sqrt[3]{16x^3y^7}$

$= \sqrt[3]{8\cdot 2x^3y^6y} = 2xy^2\sqrt[3]{2y}$

**15.** $(2+3i)(4-i)$

$= 8-2i+12i-3i^2 = 8+10i-3(-1)$

$= 8+10i+3 = 11+10i$

**16.** $12x^3-36x^2+27x = 3x(4x^2-12x+9)$

$= 3x(2x-3)^2$

**17.** $x^3-2x^2-9x+18$

$= x^2(x-2)-9(x-2)$

$= (x-2)(x^2-9)$

$= (x-2)(x+3)(x-3)$

**18.** Since the radicand must be positive, the domain will exclude all values of $x$ which make the radicand less than zero.

$6-3x \geq 0$

$-3x \geq -6$

$x \leq 2$

The domain of $f$ = $\{x|x\leq 2\}$ or $(-\infty, 2]$.

**19.** $\dfrac{1-\sqrt{x}}{1+\sqrt{x}} = \dfrac{1-\sqrt{x}}{1+\sqrt{x}}\cdot\dfrac{1-\sqrt{x}}{1-\sqrt{x}}$

$= \dfrac{(1-\sqrt{x})^2}{1^2+(\sqrt{x})^2}$

$= \dfrac{(1-\sqrt{x})^2}{1+x}$ or $\dfrac{1-2\sqrt{x}+x}{1+x}$

**20.** $\dfrac{1}{3}\ln x + 7\ln y = \ln x^{\frac{1}{3}} + \ln y^7$

$= \ln\left(x^{\frac{1}{3}}y^7\right)$

**21.** $(3x^3-5x^2+2x-1)\div(x-2)$

$$\begin{array}{r|rrrr} 2 & 3 & -5 & 2 & -1 \\ & & 6 & 2 & 8 \\ \hline & 3 & 1 & 4 & 7 \end{array}$$

$(3x^3-5x^2+2x-1)\div(x-2)$

$= 3x^2+x+4+\dfrac{7}{x-2}$

**22.** $x = -2\sqrt{3}$ and $x = 2\sqrt{3}$

$x+2\sqrt{3} = 0$ $\quad$ $x-2\sqrt{3} = 0$

Multiply the factors to obtain the polynomial.

$(x+2\sqrt{3})(x-2\sqrt{3}) = 0$

$x^2-(2\sqrt{3})^2 = 0$

$x^2-4\cdot 3 = 0$

$x^2-12 = 0$

**23.** Let $x$ = the rate of the slower car

| | $r$ • | $t$ | $= d$ |
|---|---|---|---|
| Fast | $x + 10$ | 2 | $2(x + 10)$ |
| Slow | $x$ | 2 | $2x$ |

$2(x+10)+2x = 180$

$2x+20+2x = 180$

$4x+20 = 180$

$4x = 160$

$x = 40$

The rate of the slower car is 40 miles per hour and the rate of the faster car is 40 + 10 = 50 miles per hour.

**24.** Let $x$ = the number of miles driven in a day

$C_R = 39 + 0.16x$

$C_A = 25 + 0.24x$

Set the costs equal.

$$39 + 0.16x = 25 + 0.24x$$
$$39 = 25 + 0.08x$$
$$14 = 0.08x$$
$$\frac{14}{0.08} = x$$
$$x = 175$$

The cost is the same when renting from either company when 175 miles are driven in a day.

$C_R = 39 + 0.16(175) = 39 + 28 = 67$

When 175 miles are driven, the cost is \$67.

**25.** Let $x$ = the number of apples
Let $y$ = the number of bananas

$3x + 2y = 354$

$2x + 3y = 381$

Multiply the first equation by –3 and the second equation by 2 and solve by addition.

$$\begin{array}{r} -9x - 6y = -1062 \\ 4x + 6y = \quad 762 \\ \hline -5x = -300 \end{array}$$
$$x = 60$$

Back-substitute 60 for $x$ to find $y$.

$$3(60) + 2y = 354$$
$$180 + 2y = 354$$
$$2y = 174$$
$$y = 87$$

There are 60 calories in an apple and 87 calories in a banana.

# Chapter 11
# Sequences, Series, and the Binomial Theorem

## 11.1 Exercise Set

**1.** $a_n = 3n + 2$

$a_1 = 3(1) + 2 = 3 + 2 = 5$

$a_2 = 3(2) + 2 = 6 + 2 = 8$

$a_3 = 3(3) + 2 = 9 + 2 = 11$

$a_4 = 3(4) + 2 = 12 + 2 = 14$

The first four terms are 5, 8, 11, 14.

**3.** $a_n = 3^n$

$a_1 = 3^1 = 3$

$a_2 = 3^2 = 9$

$a_3 = 3^3 = 27$

$a_4 = 3^4 = 81$

The first four terms are 3, 9, 27, 81.

**5.** $a_n = (-3)^n$

$a_1 = (-3)^1 = -3$

$a_2 = (-3)^2 = 9$

$a_3 = (-3)^3 = -27$

$a_4 = (-3)^4 = 81$

The first four terms are –3, 9, –27, 81.

**7.** $a_n = (-1)^n (n+3)$

$a_1 = (-1)^1 (1+3) = -1(4) = -4$

$a_2 = (-1)^2 (2+3) = 1(5) = 5$

$a_3 = (-1)^3 (3+3) = -1(6) = -6$

$a_4 = (-1)^4 (4+3) = 1(7) = 7$

The first four terms are –4, 5, –6, 7.

**9.** $a_n = \frac{2n}{n+4}$

$a_1 = \frac{2(1)}{1+4} = \frac{2}{5}$

$a_2 = \frac{2(2)}{2+4} = \frac{4}{6} = \frac{2}{3}$

$a_3 = \frac{2(3)}{3+4} = \frac{6}{7}$

$a_4 = \frac{2(4)}{4+4} = \frac{8}{8} = 1$

The first four terms are $\frac{2}{5}, \frac{2}{3}, \frac{6}{7}, 1$.

**11.** $a_n = \frac{(-1)^{n+1}}{2^n - 1}$

$a_1 = \frac{(-1)^{1+1}}{2^1 - 1} = \frac{(-1)^2}{2-1} = \frac{1}{1} = 1$

$a_2 = \frac{(-1)^{2+1}}{2^2 - 1} = \frac{(-1)^3}{4-1} = \frac{-1}{3} = -\frac{1}{3}$

$a_3 = \frac{(-1)^{3+1}}{2^3 - 1} = \frac{(-1)^4}{8-1} = \frac{1}{7}$

$a_4 = \frac{(-1)^{4+1}}{2^4 - 1} = \frac{(-1)^5}{16-1} = \frac{-1}{15} = -\frac{1}{15}$

The first four terms are $1, -\frac{1}{3}, \frac{1}{7}, -\frac{1}{15}$.

**13.**
$$a_n = \frac{n^2}{n!}$$
$$a_1 = \frac{1^2}{1!} = \frac{1}{1} = 1$$
$$a_2 = \frac{2^2}{2!} = \frac{4}{2\cdot 1} = \frac{4}{2} = 2$$
$$a_3 = \frac{3^2}{3!} = \frac{9}{3\cdot 2\cdot 1} = \frac{3}{2}$$
$$a_4 = \frac{4^2}{4!} = \frac{16}{4\cdot 3\cdot 2\cdot 1} = \frac{2}{3}$$
The first four terms are $1, 2, \frac{3}{2}, \frac{2}{3}$.

**15.**
$$a_n = 2(n+1)!$$
$$a_1 = 2(1+1)! = 2\cdot 2! = 2\cdot 2\cdot 1 = 4$$
$$a_2 = 2(2+1)! = 2\cdot 3! = 2\cdot 3\cdot 2\cdot 1 = 12$$
$$a_3 = 2(3+1)! = 2\cdot 4! = 2\cdot 4\cdot 3\cdot 2\cdot 1 = 48$$
$$a_4 = 2(4+1)! = 2\cdot 5! = 2\cdot 5\cdot 4\cdot 3\cdot 2\cdot 1 = 240$$

**17.**
$$\sum_{i=1}^{6} 5i = 5(1)+5(2)+5(3)+5(4)+5(5)+5(6) = 5+10+15+20+25+30 = 105$$

**19.**
$$\sum_{i=1}^{4} 2i^2 = 2(1)^2+2(2)^2+2(3)^2+2(4)^2 = 2(1)+2(4)+2(9)+2(16)$$
$$= 2+8+18+32 = 60$$

**21.**
$$\sum_{k=1}^{5} k(k+4) = 1(1+4)+2(2+4)+3(3+4)+4(4+4)+5(5+4)$$
$$= 1(5)+2(6)+3(7)+4(8)+5(9) = 5+12+21+32+45 = 115$$

**23.**
$$\sum_{i=1}^{4}\left(-\frac{1}{2}\right)^i = \left(-\frac{1}{2}\right)^1+\left(-\frac{1}{2}\right)^2+\left(-\frac{1}{2}\right)^3+\left(-\frac{1}{2}\right)^4 = -\frac{1}{2}+\frac{1}{4}+\left(-\frac{1}{8}\right)+\frac{1}{16}$$
$$= -\frac{1}{2}\cdot\frac{8}{8}+\frac{1}{4}\cdot\frac{4}{4}+\left(-\frac{1}{8}\right)\frac{2}{2}+\frac{1}{16} = -\frac{8}{16}+\frac{4}{16}-\frac{2}{16}+\frac{1}{16}$$
$$= \frac{-8+4-2+1}{16} = -\frac{5}{16}$$

**25.**
$$\sum_{i=5}^{9} 11 = 11+11+11+11+11 = 55$$

**27.** $$\sum_{i=0}^{4}\frac{(-1)^i}{i!}=\frac{(-1)^0}{0!}+\frac{(-1)^1}{1!}+\frac{(-1)^2}{2!}+\frac{(-1)^3}{3!}+\frac{(-1)^4}{4!}=\frac{1}{1}+\frac{-1}{1}+\frac{1}{2\cdot1}+\frac{-1}{3\cdot2\cdot1}+\frac{1}{4\cdot3\cdot2\cdot1}$$
$$=1-1+\frac{1}{2}-\frac{1}{6}+\frac{1}{24}=\frac{1}{2}\cdot\frac{12}{12}-\frac{1}{6}\cdot\frac{4}{4}+\frac{1}{24}=\frac{12}{24}-\frac{4}{24}+\frac{1}{24}=\frac{12-4+1}{24}=\frac{9}{24}=\frac{3}{8}$$

**29.** $$\sum_{i=1}^{5}\frac{i!}{(i-1)!}=\frac{1!}{(1-1)!}+\frac{2!}{(2-1)!}+\frac{3!}{(3-1)!}+\frac{4!}{(4-1)!}+\frac{5!}{(5-1)!}=\frac{1!}{0!}+\frac{2!}{1!}+\frac{3!}{2!}+\frac{4!}{3!}+\frac{5!}{4!}$$
$$=\frac{1}{1}+\frac{2\cdot\cancel{1!}}{\cancel{1!}}+\frac{3\cdot\cancel{2!}}{\cancel{2!}}+\frac{4\cdot\cancel{3!}}{\cancel{3!}}+\frac{5\cdot\cancel{4!}}{\cancel{4!}}=1+2+3+4+5=15$$

**31.** $$1^2+2^2+3^2+\ldots+15^2=\sum_{i=1}^{15}i^2$$

**33.** $$2+2^2+2^3+\ldots+2^{11}=\sum_{i=1}^{11}2^i$$

**35.** $$1+2+3+\ldots+30=\sum_{i=1}^{30}i$$

**37.** $$\frac{1}{2}+\frac{2}{3}+\frac{3}{4}+\ldots+\frac{14}{14+1}=\sum_{i=1}^{14}\frac{i}{i+1}$$

**39.** $$4+\frac{4^2}{2}+\frac{4^3}{3}+\ldots+\frac{4^n}{n}=\sum_{i=1}^{n}\frac{4^i}{i}$$

**41.** $$1+3+5+\ldots+(2n-1)=\sum_{i=1}^{n}(2i-1)$$

**43.** $$5+7+9+11+\ldots+31=\sum_{k=2}^{15}(2k+1)\text{ or}$$
$$=\sum_{k=1}^{14}(2k+3)$$

**45.** $$a+ar+ar^2+\ldots+ar^{12}=\sum_{k=0}^{12}ar^k$$

**47.** $$a+(a+d)+(a+2d)+\ldots+a(a+nd)$$
$$=\sum_{k=0}^{n}(a+kd)$$

**49.** $$\sum_{i=1}^{5}(a_i^2+1)=\left((-4)^2+1\right)+\left((-2)^2+1\right)+\left((0)^2+1\right)+\left((2)^2+1\right)+\left((4)^2+1\right)$$
$$=17+5+1+5+17$$
$$=45$$

**50.** $\sum_{i=1}^{5}(b_i^2-1)=\left((4)^2-1\right)+\left((2)^2-1\right)+\left((0)^2-1\right)+\left((-2)^2-1\right)+\left((-4)^2-1\right)$

$=15+3+(-1)+3+15$

$=35$

**51.** $\sum_{i=1}^{5}(2a_i+b_i)=\left(2(-4)+4\right)+\left(2(-2)+2\right)+\left(2(0)+0\right)+\left(2(2)+(-2)\right)+\left(2(4)+(-4)\right)$

$=-4+(-2)+0+2+4=0$

**52.** $\sum_{i=1}^{5}(a_i+3b_i)=\left(-4+3(4)\right)+\left(-2+3(2)\right)+\left(0+3(0)\right)+\left(2+3(-2)\right)+\left(4+3(-4)\right)$

$=8+4+0-4-8=0$

**53.** $\sum_{i=4}^{5}\left(\frac{a_i}{b_i}\right)^2=\left(\frac{2}{-2}\right)^2+\left(\frac{4}{-4}\right)^2=(-1)^2+(-1)^2=1+1=2$

**54.** $\sum_{i=4}^{5}\left(\frac{a_i}{b_i}\right)^3=\left(\frac{2}{-2}\right)^3+\left(\frac{4}{-4}\right)^3=(-1)^3+(-1)^3=(-1)+(-1)=-2$

**55.** $\sum_{i=1}^{5}a_i^2+\sum_{i=1}^{5}b_i^2=\left((-4)^2+(-2)^2+0^2+2^2+4^2\right)+\left(4^2+2^2+0^2+(-2)^2+(-4)^2\right)$

$=\left(16+4+0+4+16\right)+\left(16+4+0+4+16\right)=80$

**56.** $\sum_{i=1}^{5}a_i^2-\sum_{i=3}^{5}b_i^2=\left((-4)^2+(-2)^2+0^2+2^2+4^2\right)-\left(0^2+(-2)^2+(-4)^2\right)$

$=\left(16+4+0+4+16\right)-\left(0+4+16\right)=40-20=20$

**57.** **a.** $\sum_{i=1}^{7}a_i=36.4+36.5+35.6+34.5+32.3+31.6+32.9=239.38$

**b.** $\frac{\sum_{i=1}^{7}a_i}{7}=\frac{239.8}{7}=34.3$

From 1995 through 2001, the average number of people living below poverty level each year was approximately 34.3 million.

**59. a.** $\frac{1}{8}\sum_{i=1}^{8} a_i = \frac{1}{8}(14.1+14.2+13.7+12.6+10.9+8.7+7.6+6.5) = \frac{1}{8}(88.3) = 11.0375$

From 1993 through 2000, the average number of welfare recipients each year was 11.0375 million.

**b.**

$a_1 = -1.23(1)+16.55 = 15.32$ $\quad a_5 = -1.23(5)+16.55 = 10.4$

$a_2 = -1.23(2)+16.55 = 14.09$ $\quad a_6 = -1.23(6)+16.55 = 9.17$

$a_3 = -1.23(3)+16.55 = 12.86$ $\quad a_7 = -1.23(7)+16.55 = 7.94$

$a_4 = -1.23(4)+16.55 = 11.63$ $\quad a_8 = -1.23(8)+16.55 = 6.71$

$$\frac{1}{8}\sum_{i=1}^{8} a_i = \frac{1}{8}(15.32+14.09+12.86+11.63+10.4+9.17+7.94+6.71)$$

$$= \frac{1}{8}(88.12) = 11.015$$

This is a reasonable model.

**61.** $a_{20} = 6000\left(1+\frac{0.06}{4}\right)^{20} = 6000(1+0.015)^{20} = 6000(1.015)^{20} = 8081.13$

The balance in the account after 5 years if \$8081.13.

**63. – 69.** Answers will vary.

**71.** Answers will vary. For example, consider Exercise 17.

$\sum_{i=1}^{6} 5i$

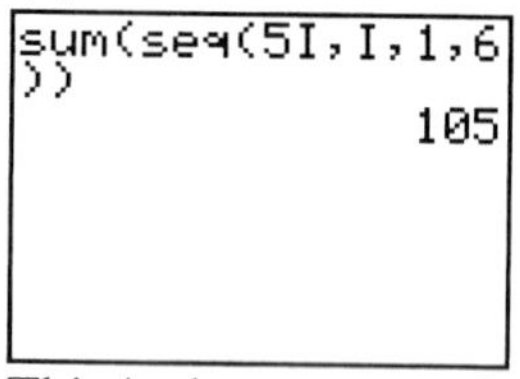


This is the same result obtained in Exercise 17.

**73.** $a_n = \frac{n}{n+1}$;

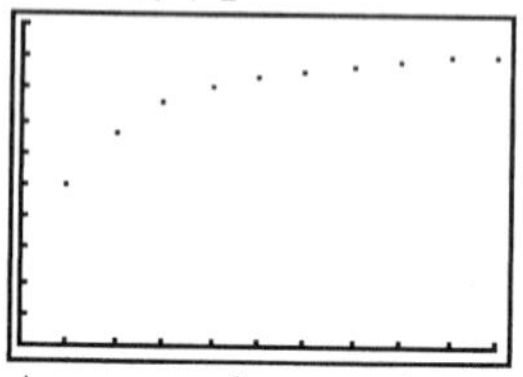

As $n$ gets larger, the terms get closer to 1.

**75.** $a_n = \frac{2n^2+5n-7}{n^3}$

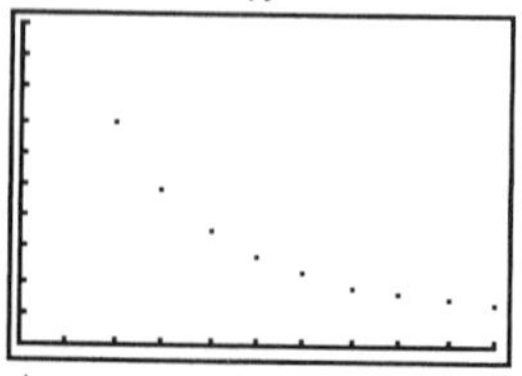

As $n$ gets larger, the terms get closer to 0.

**77.** Statement **c.** is true. $\sum_{i=1}^{4} 3i + \sum_{i=1}^{4} 4i = \sum_{i=1}^{4} 7i$

$$\sum_{i=1}^{4} 3i + \sum_{i=1}^{4} 4i = (3\cdot1+3\cdot2+3\cdot3+3\cdot4)+(4\cdot1+4\cdot2+4\cdot3+4\cdot4) = 3(1+2+3+4)+4(1+2+3+4)$$
$$= (1+2+3+4)(3+4) = (1+2+3+4)(7) = 7(1+2+3+4)$$
$$\sum_{i=1}^{4} 7i = 7\cdot1+7\cdot2+7\cdot3+7\cdot4 = 7(1+2+3+4)$$

Statement **a.** is false. $\sum_{i=1}^{2}(-1)^i 2^i = (-1)^1 2^1 + (-1)^2 2^2 = -1(2)+1(4) = -2+4 = 2$

Statement **b.** is false. $\sum_{i=1}^{2} a_i b_i \neq \sum_{i=1}^{2} a_i \sum_{i=1}^{2} b_i$

$$\sum_{i=1}^{2} a_i b_i = a_1 b_1 + a_2 b_2$$
$$\sum_{i=1}^{2} a_i \sum_{i=1}^{2} b_i = (a_1+a_2)(b_1+b_2) = a_1b_1 + a_1b_2 + a_2b_1 + a_2b_2$$

Statement **d.** is false. $\sum_{i=0}^{6}(-1)^i (i+1)^2 \neq \sum_{j=1}^{7}(-1)^j j^2$

$$\sum_{i=0}^{6}(-1)^i (i+1)^2$$
$$= (-1)^0(0+1)^2 + (-1)^1(1+1)^2 + (-1)^2(2+1)^2 + (-1)^3(3+1)^2 + (-1)^4(4+1)^2 + (-1)^5(5+1)^2 + (-1)^6(6+1)^2$$
$$= 1(1)^2 - 1(2)^2 + 1(3)^2 - 1(4)^2 + 1(5)^2 - 1(6)^2 + 1(7)^2 = 1(1)-1(4)+1(9)-1(16)+1(25)-1(36)+1(49)$$
$$= 1-4+9-16+25-36+49 = 28$$
$$\sum_{j=1}^{7}(-1)^j j^2 = (-1)^1 1^2 + (-1)^2 2^2 + (-1)^3 3^2 + (-1)^4 4^2 + (-1)^5 5^2 + (-1)^6 6^2 + (-1)^7 7^2$$
$$= -1(1)+1(4)-1(9)+1(16)-1(25)+1(36)-1(49)$$
$$= -1+4-9+16-25+36-49 = -28$$

**79.** $a_n = n^2$

**81.** $a_n = n(n+2)$

**83.** $a_n = 2n+3$

**85.** $a_n = (-2)^{n+1}$

**87.** $\dfrac{(n+4)!}{(n+2)!} == (n+3)(n+4) = n^2 + 7n + 12$

**89.**
$$\sum_{i=1}^{4} \log(2i)$$
$$= \log 2(1) + \log 2(2) + \log 2(3) + \log 2(4)$$
$$= \log 2 + \log 4 + \log 6 + \log 8 = \log 384$$

**91.** $a_n = a_{n-1} + 5$

$a_1 = 7$

$a_2 = a_{2-1} + 5 = a_1 + 5 = 7 + 5 = 12$

$a_3 = a_{3-1} + 5 = a_2 + 5 = 12 + 5 = 17$

$a_4 = a_{4-1} + 5 = a_3 + 5 = 17 + 5 = 22$

The first four terms are 7, 12, 17, 22.

**92.** $\sqrt[3]{40x^4y^7} = \sqrt[3]{8\cdot 5x^3xy^6y} = 2xy^2\sqrt[3]{5xy}$

**93.** $27x^3 - 8 = (3x-2)(9x^2+6x+4)$

**94.**
$$\frac{6}{x}+\frac{6}{x+2}=\frac{5}{2}$$
$$2x(x+2)\left(\frac{6}{x}+\frac{6}{x+2}\right)=2x(x+2)\left(\frac{5}{2}\right)$$
$$2(x+2)(6)+2x(6)=x(x+2)(5)$$
$$12(x+2)+12x=5x(x+2)$$
$$12x+24+12x=5x^2+10x$$
$$24x+24=5x^2+10x$$
$$0=5x^2-14x-24$$
$$0=(5x+6)(x-4)$$

Apply the zero product principle.

$5x+6=0$ or $x-4=0$

$5x=-6$ $\quad x=4$

$x=-\frac{6}{5}$

The solutions are $-\frac{6}{5}$ and 4 and the solution set is $\left\{-\frac{6}{5},4\right\}$.

## 11.2 Exercise Set

**1.** Since $6-2=4$, $d=4$.

**3.** Since $-2-(-7)=5$, $d=5$.

**5.** Since $711-714=-3$, $d=-3$.

**7.** $a_1=200$

$a_2=200+20=220$

$a_3=220+20=240$

$a_4=240+20=260$

$a_5=260+20=280$

$a_6=280+20=300$

The first six terms are 200, 220, 240, 260, 280, and 300.

**9.** $a_1=-7$

$a_2=-7+4=-3$

$a_3=-3+4=1$

$a_4=1+4=5$

$a_5=5+4=9$

$a_6=9+4=13$

The first six terms are –7, !3, 1, 5, 9, 13.

**11.** $a_1=300$

$a_2=300-90=210$

$a_3=210-90=120$

$a_4=120-90=30$

$a_5=30-90=-60$

$a_6=-60-90=-150$

The first six terms are 300, 210, 120, 30, $-60$, $-150$.

**13.** $a_1=\frac{5}{2}$

$a_2=\frac{5}{2}-\frac{1}{2}=\frac{4}{2}=2$

$a_3=\frac{4}{2}-\frac{1}{2}=\frac{3}{2}$

$a_4 = \frac{3}{2} - \frac{1}{2} = \frac{2}{2} = 1$

$a_5 = 1 - \frac{1}{2} = \frac{1}{2}$

$a_6 = \frac{1}{2} - \frac{1}{2} = 0$

The first six terms are $\frac{5}{2}, 2, \frac{3}{2}, 1, \frac{1}{2}, 0$.

**15.** $a_1 = -0.4$

$a_2 = -0.4 - 1.6 = -2$

$a_3 = -2 - 1.6 = -3.6$

$a_4 = -3.6 - 1.6 = -5.2$

$a_5 = -5.2 - 1.6 = -6.8$

$a_6 = -6.8 - 1.6 = -8.4$

The first six terms are $-0.4, -2, -3.6, -5.2, -6.8, -8.4$.

**17.** $a_6 = 13 + (6-1)4 = 13 + (5)4$
$= 13 + 20 = 33$

**19.** $a_{50} = 7 + (50-1)5 = 7 + (49)5$
$= 7 + 245 = 252$

**21.** $a_{200} = -40 + (200-1)5 = -40 + (199)5$
$= -40 + 995 = 955$

**23.** $a_{60} = 35 + (60-1)(-3) = 35 + (59)(-3)$
$= 35 + (-177) = -142$

**25.** $a_n = a_1 + (n-1)d = 1 + (n-1)4$
$= 1 + 4n - 4 = 4n - 3$

$a_{20} = 4(20) - 3 = 80 - 3 = 77$

**27.** $a_n = a_1 + (n-1)d = 7 + (n-1)(-4)$
$= 7 - 4n + 4 = 11 - 4n$
$a_{20} = 11 - 4(20) = 11 - 80 = -69$

**29.** $a_n = a_1 + (n-1)d = -20 + (n-1)(-4)$
$= -20 - 4n + 4 = -4n - 16$
$a_{20} = -4(20) - 16 = -80 - 16 = --96$

**31.** $a_n = a_1 + (n-1)d = -\frac{1}{3} + (n-1)\left(\frac{1}{3}\right)$
$= -\frac{1}{3} + \frac{1}{3}n - \frac{1}{3} = \frac{1}{3}n - \frac{2}{3}$
$a_{20} = \frac{1}{3}(20) - \frac{2}{3} = \frac{20}{3} - \frac{2}{3} = \frac{18}{3} = 6$

**33.** $a_n = a_1 + (n-1)d = 4 + (n-1)(-0.3)$
$= 4 - 0.3n + 0.3 = 4.3 - 0.3n$
$a_{20} = 4.3 - 0.3(20) = 4.3 - 6 = -1.7$

**35.** First find $a_{20}$.
$a_{20} = 4 + (20-1)6 = 4 + (19)6$
$= 4 + 114 = 118$
$S_{20} = \frac{20}{2}(4 + 118) = 10(122) = 1220$

**37.** First find $a_{50}$.
$a_{50} = -10 + (50-1)4 = -10 + (49)4$
$= -10 + 196 = 186$
$S_{50} = \frac{50}{2}(-10 + 186) = 25(176) = 4400$

**39.** First find $a_{100}$.
$a_{100} = 1 + (100-1)1 = 1 + (99)1$
$= 1 + 99 = 100$
$S_{100} = \frac{100}{2}(1 + 100) = 50(101) = 5050$

**41.** First find $a_{60}$.
$a_{60} = 2 + (60-1)2 = 2 + (59)2$
$= 2 + 118 = 120$
$S_{60} = \frac{60}{2}(2 + 120) = 30(122) = 3660$

**43.** Since there are 12 even integers between 21 and 45, find $a_{12}$.

$$a_{12} = 22 + (12-1)2 = 22 + (11)2$$
$$= 22 + 22 = 44$$
$$S_{12} = \frac{12}{2}(22+44) = 6(66) = 396$$

**45.**
$$\sum_{i=1}^{17}(5i+3) = (5(1)+3)+(5(2)+3)+(5(3)+3)+\ldots+(5(17)+3)$$
$$= (5+3)+(10+3)+(15+3)+\ldots+(85+3) = 8+13+18+\ldots+88$$
$$S_{17} = \frac{17}{2}(8+88) = \frac{17}{2}(96) = 17(48) = 816$$

**47.**
$$\sum_{i=1}^{30}(-3i+5) = (-3(1)+5)+(-3(2)+5)+(-3(3)+5)+\ldots+(-3(30)+5)$$
$$= (-3+5)+(-6+5)+(-9+5)+\ldots+(-90+5) = 2+(-1)+(-4)+\ldots+(-85)$$
$$S_{30} = \frac{30}{2}(2+(-85)) = 15(-83) = -1245$$

**49.**
$$\sum_{i=1}^{100}4i = 4(1)+4(2)+4(3)+\ldots+4(100) = 4+8+12+\ldots+400$$
$$S_{100} = \frac{100}{2}(4+400) = 50(404) = 20,200$$

**51.** First find $a_{14}$ and $b_{12}$:

$$a_{14} = a_1 + (n-1)d$$
$$= 1 + (14-1)(-3-1) = -51$$
$$b_{12} = b_1 + (n-1)d$$
$$= 3 + (12-1)(8-3) = 58$$

So, $a_{14} + b_{12} = -51 + 58 = 7$.

**52.** First find $a_{16}$ and $b_{18}$:

$$a_{16} = a_1 + (n-1)d$$
$$= 1 + (16-1)(-3-1) = -59$$
$$b_{18} = b_1 + (n-1)d$$
$$= 3 + (18-1)(8-3) = 88$$

So, $a_{16} + b_{18} = -59 + 88 = 29$.

**53.** $a_n = a_1 + (n-1)d$

$-83 = 1 + (n-1)(-3-1)$

$-83 = 1 + -4(n-1)$

$-84 = -4n + 4$

$-88 = -4n$

$n = 22$

There are 22 terms.

**54.** $b_n = b_1 + (n-1)d$

$93 = 3 + (n-1)(8-3)$

$93 = 3 + 5(n-1)$

$93 = 5n - 2$

$95 = 5n$

$n = 19$

There are 19 terms.

**55.** $S_n = \frac{n}{2}(a_1 + a_n)$

For $\{a_n\}$:

$S_{14} = \frac{14}{2}(a_1 + a_{14}) = 7(1 + (-51)) = -350$

For $\{b_n\}$:

$S_{14} = \frac{14}{2}(b_1 + b_{14}) = 7(3 + 68) = 497$

So $\sum_{n=1}^{14} b_n - \sum_{n=1}^{14} a_n = 497 - (-350) = 847$

**56.** First find $a_{15}$ and $b_{15}$:

$a_{15} = a_1 + (n-1)d$

$= 1 + (15-1)(-3-1) = -55$

$b_{15} = b_1 + (n-1)d$

$= 3 + (15-1)(8-3) = 73$

Using $S_n = \frac{n}{2}(a_1 + a_n)$ for $\{a_n\}$:

$S_{15} = \frac{15}{2}(a_1 + a_{15}) = 7.5(1 + (-55))$

$= -405$

And then for $\{b_n\}$:

$S_{15} = \frac{15}{2}(b_1 + b_{15}) = 7.5(3 + 73) = 570$

So

$\sum_{n=1}^{15} b_n - \sum_{n=1}^{15} a_n = 570 - (-405) = 975$

**57.** Two points on the graph are $(1, 1)$ and $(2, -3)$. Finding the slope of the line; $m = \frac{y_2 - y_1}{x_2 - x_2} = \frac{-3-1}{2-1} = \frac{-4}{1} = -4$

Using the point-slope form of an equation of a line;

$y - y_2 = m(x - x_2)$

$y - 1 = -4(x-1)$

$y - 1 = -4x + 4$

$y = -4x + 5$

Thus, $f(x) = -4x + 5$.

**58.** Two points on the graph are $(1, 3)$ and $(2, 8)$. Finding the slope of the line; $m = \dfrac{y_2 - y_1}{x_2 - x_2} = \dfrac{8-3}{2-1} = \dfrac{5}{1} = 5$

Using the point-slope form of an equation of a line;

$$y - y_2 = 5(x - x_2)$$
$$y - 3 = 5(x - 1)$$
$$y - 3 = 5x - 5$$
$$y = 5x - 2$$

Thus, $g(x) = 5x - 2$.

**59.** Using $a_n = a_1 + (n-1)d$ and $a_2 = 4$:

$$a_2 = a_1 + (2-1)d$$
$$4 = a_1 + d$$

And since $a_6 = 16$:

$$a_6 = a_1 + (6-1)d$$
$$16 = a_1 + 5d$$

The system of equations is

$$4 = a_1 + d$$
$$16 = a_1 + 5d$$

Solving the first equation for $a_1$:

$$a_1 = 4 - d$$

Substituting the value into the second equation and solving for *d*:

$$16 = (4 - d) + 5d$$
$$16 = 4 + 4d$$
$$12 = 4d$$
$$3 = d$$

Then $a_n = a_1 + (n-1)d$

$$a_n = 1 + (n-1)3$$
$$a_n = 1 + 3n - 3$$
$$a_n = 3n - 2$$

**60.** Using $a_n = a_1 + (n-1)d$ and $a_3 = 7$:

$$a_3 = a_1 + (3-1)d$$
$$7 = a_1 + 2d$$

And since $a_8 = 17$:

$$a_8 = a_1 + (8-1)d$$
$$17 = a_1 + 7d$$

The system of equations is

$$7 = a_1 + 2d$$
$$17 = a_1 + 7d$$

Solving the first equation for $a_1$:

$$a_1 = 7 - 2d$$

Substituting the value into the second equation and solving for *d*:

$$17 = (7 - 2d) + 7d$$
$$17 = 7 + 5d$$
$$10 = 5d$$
$$2 = d$$

So $a_1 = 7 - 2(2) = 7 - 4 = 3$.

Then $a_n = a_1 + (n-1)2$

$$a_n = 3 + (n-1)2$$
$$a_n = 3 + 2n - 2$$
$$a_n = 2n + 1$$

**61.** **a.**

$$a_n = a_1 + (n-1)d$$
$$a_n = 150 + (n-1)(1.7)$$
$$a_n = 150 + 1.7n - 1.7$$
$$a_n = 1.7n + 148.3$$

**b.**

$$a_{40} = 1.7(40) + 148.3$$
$$= 68 + 148.3$$
$$= 216.3 \text{ pounds}$$

**63.** Answers will vary.

**65.** Company *A*

$$\begin{aligned}a_n &= 24000+(n-1)1600\\ &= 24000+1600n-1600\\ &= 1600n+22400\end{aligned}$$

$$\begin{aligned}a_{10} &= 1600(10)+22400\\ &= 16000+22400=38400\end{aligned}$$

Company *B*

$$\begin{aligned}a_n &= 28000+(n-1)1000\\ &= 28000+1000n-1000\\ &= 1000n+27000\end{aligned}$$

$$\begin{aligned}a_{10} &= 1000(10)+27000\\ &= 10000+27000=37000\end{aligned}$$

Company *A* will pay \$1400 more in year 10.

**67.** **a.** Total cost:

$$\$7107+\$7310+\$7586+\$8046 = \$30,049$$

**b.**

$$a_1 = 309(1)+6739=7048$$
$$a_2 = 309(2)+6739=7357$$
$$a_3 = 309(3)+6739=7666$$
$$a_4 = 309(4)+6739=7975$$

Total cost:

$$7048+7357+7666+7975 = \$30,046$$

The model describes the actual sum very well.

**69.** Answers will vary.

**71.** Company *A*

$$\begin{aligned}a_n &= 19000+(n-1)2600\\ &= 19000+2600n-2600\\ &= 2600n+16400\end{aligned}$$

$$\begin{aligned}a_{10} &= 2600(10)+16400\\ &= 26000+16400=42400\end{aligned}$$

$$S_n = \frac{n}{2}(a_1+a_{10})$$

$$\begin{aligned}S_{10} &= \frac{10}{2}(19000+42400)\\ &= 5(61400)=\$307,000\end{aligned}$$

Company *B*

$$\begin{aligned}a_n &= 27000+(n-1)1200\\ &= 27000+1200n-1200\\ &= 1200n+25800\end{aligned}$$

$$\begin{aligned}a_{10} &= 1200(10)+25800\\ &= 12000+25800\\ &= 37800\end{aligned}$$

$$S_n = \frac{n}{2}(a_1+a_{10})$$

$$\begin{aligned}S_{10} &= \frac{10}{2}(27000+37800)\\ &= 5(64800)=\$324,000\end{aligned}$$

Company *B* pays the greater total amount.

**73.**

$$\begin{aligned}a_{38} &= a_1+(n-1)d\\ &= 20+(38-1)(3)\\ &= 20+37(3)=131\end{aligned}$$

$$S_{38} = \frac{n}{2}(a_1+a_{38})$$

$$\begin{aligned}&= \frac{38}{2}(20+131)\\ &= 19(151)\\ &= 2869\end{aligned}$$

There are 2869 seats in this section of the stadium.

**75. – 77.** Answers will vary.

**79.** Answers will vary. For example, consider Exercise 45.

$$\sum_{i=1}^{17}(5i+3)$$

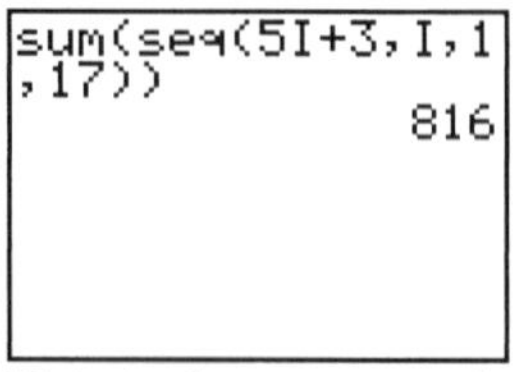


This is the same result obtained in Exercise 45.

**81.** From the sequence, we see that $a_1 = 21700$ and $d = 23172 - 21700 = 1472$.
We know that $a_n = a_1 + (n-1)d$. We can substitute what we know to find $n$.

$$314628 = 21700 + (n-1)1472$$
$$292928 = (n-1)1472$$
$$\frac{292928}{1472} = \frac{(n-1)1472}{1472}$$
$$199 = n-1$$
$$200 = n$$

314,628 is the 200th term of the sequence.

**83.** $1+3+5+\ldots+(2n-1)$

$$S_n = \frac{n}{2}(a_1 + a_n) = \frac{n}{2}(1+(2n-1))$$
$$= \frac{n}{2}(1+2n-1) = \frac{n}{2}(2n)$$
$$= n(n) = n^2$$

**84.** $\log(x^2 - 25) - \log(x+5) = 3$

$$\log\left(\frac{x^2-25}{x+5}\right) = 3$$
$$\log\left(\frac{(x-5)(x+5)}{x+5}\right) = 3$$
$$\log(x-5) = 3$$
$$x - 5 = 10^3$$
$$x - 5 = 1000$$
$$x = 1005$$

**85.** $x^2 + 3x \le 10$

Solve the related quadratic equation.

$$x^2 + 3x - 10 = 0$$
$$(x+5)(x-2) = 0$$

Apply the zero product principle.

$$x+5=0 \quad \text{or} \quad x-2=0$$
$$x=-5 \qquad\qquad x=2$$

The boundary points are $-5$ and 2.

| Test Interval | Test Number | Test | Conclusion |
|---|---|---|---|
| $(-\infty,-5]$ | -6 | $(-6)^2+3(-6)\le 10$ <br> $18\le 10$, false | $(-\infty,-5]$ does not belong to the solution set. |
| $[-5,2]$ | 0 | $0^2+3(0)\le 10$ <br> $0\le 10$, true | $[-5,2]$ belongs to the solution set. |
| $[2,\infty)$ | 3 | $3^2+3(3)\le 10$ <br> $18\le 10$, false | $[2,\infty)$ does not belong to the solution set. |

The solution set is $[-5,2]$ or $\{x \mid -5 \le x \le 2\}$.

**86.**

$$A = \frac{Pt}{P+t}$$
$$A(P+t) = Pt$$
$$AP + At = Pt$$
$$AP - Pt = -At$$
$$P(A-t) = -At$$
$$P = -\frac{At}{A-t} \text{ or } \frac{At}{t-A}$$

## 11.3 Exercise Set

**1.** $r = \frac{a_2}{a_1} = \frac{15}{5} = 3$

**3.** $r = \frac{a_2}{a_1} = \frac{30}{-15} = -2$

**5.** $r = \frac{a_2}{a_1} = \frac{\frac{9}{2}}{3} = \frac{9}{2}\cdot\frac{1}{3} = \frac{3}{2}$

**7.** $r = \frac{a_2}{a_1} = \frac{0.04}{-0.4} = -0.1$

**9.** The first term is 2.

The second term is $2 \cdot 3 = 6$.

The third term is $6 \cdot 3 = 18$.

The fourth term is $18 \cdot 3 = 54$.

The fifth term is $54 \cdot 3 = 162$.

**11.** The first term is 20.

The second term is $20 \cdot \frac{1}{2} = 10$.

The third term is $10 \cdot \frac{1}{2} = 5$.

The fourth term is $5 \cdot \frac{1}{2} = \frac{5}{2}$.

The fifth term is $\frac{5}{2} \cdot \frac{1}{2} = \frac{5}{4}$.

**13.** The first term is –4.

The second term is $-4(-10) = 40$.

The third term is $40(-10) = -400$.

The fourth term is $-400(-10) = 4000$.

The fifth term is

$4000(-10) = -40,000$.

**15.** The first term is $-\frac{1}{4}$.

The second term is $-\frac{1}{4}(-2) = \frac{1}{2}$.

The third term is $\frac{1}{2}(-2) = -1$.

The fourth term is $-1(-2) = 2$.

The fifth term is $2(-2) = -4$.

**17.** $a_8 = 6(2)^{8-1} = 6(2)^7 = 6(128) = 768$

**19.** $a_{12} = 5(-2)^{12-1} = 5(-2)^{11}$
$= 5(-2048) = -10,240$

**21.** $a_6 = 6400\left(-\frac{1}{2}\right)^{6-1} = 6400\left(-\frac{1}{2}\right)^5$
$= -200$

**23.** $a_8 = 1,000,000(0.1)^{8-1}$
$= 1,000,000(0.1)^7$
$= 1,000,000(0.0000001) = 0.1$

**25.** $r = \frac{a_2}{a_1} = \frac{12}{3} = 4$

$a_n = a_1 r^{n-1} = 3(4)^{n-1}$

$a_7 = 3(4)^{7-1} = 3(4)^6$
$= 3(4096) = 12,288$

**27.** $r = \frac{a_2}{a_1} = \frac{6}{18} = \frac{1}{3}$

$a_n = a_1 r^{n-1} = 18\left(\frac{1}{3}\right)^{n-1}$

$a_7 = 18\left(\frac{1}{3}\right)^{7-1} = 18\left(\frac{1}{3}\right)^6$
$= 18\left(\frac{1}{729}\right) = \frac{18}{729} = \frac{2}{81}$

**29.** $r = \frac{a_2}{a_1} = \frac{-3}{1.5} = -2$

$a_n = a_1 r^{n-1} = 1.5(-2)^{n-1}$

$a_7 = 1.5(-2)^{7-1} = 1.5(-2)^6$
$= 1.5(64) = 96$

**31.** $r = \frac{a_2}{a_1} = \frac{-0.004}{0.0004} = -10$

$a_n = a_1 r^{n-1} = 0.0004(-10)^{n-1}$

$a_7 = 0.0004(-10)^{7-1} = 0.0004(-10)^6$
$= 0.0004(1000000) = 400$

**33.** $r = \frac{a_2}{a_1} = \frac{6}{2} = 3$

$$S_{12} = \frac{2\left(1-3^{12}\right)}{1-3} = \frac{2(1-531441)}{-2}$$
$$= \frac{2(-531440)}{-2} = \frac{-1,062,880}{-2}$$
$$= 531,440$$

**35.** $r = \frac{a_2}{a_1} = \frac{-6}{3} = -2$

$$S_{11} = \frac{a_1\left(1-r^n\right)}{1-r} = \frac{3\left(1-(-2)^{11}\right)}{1-(-2)}$$
$$= \frac{\cancel{3}\left(1-(-2048)\right)}{\cancel{3}} = 2049$$

**37.** $r = \frac{a_2}{a_1} = \frac{3}{-\frac{3}{2}} = 3 \div \left(-\frac{3}{2}\right)$

$$= 3 \cdot \left(-\frac{2}{3}\right) = -2$$

$$S_{14} = \frac{a_1\left(1-r^n\right)}{1-r} = \frac{-\frac{3}{2}\left(1-(-2)^{14}\right)}{1-(-2)}$$
$$= \frac{-\frac{3}{2}\left(1-(16384)\right)}{3} = \frac{-\frac{3}{2}(-16385)}{3}$$
$$= -\frac{3}{2}(-16385) \div 3 = \frac{49155}{2} \cdot \frac{1}{3}$$
$$= \frac{16385}{2}$$

**39.** $\sum_{i=1}^{8} 3^i = \frac{3\left(1-3^8\right)}{1-3} = \frac{3(1-6561)}{-2}$

$$= \frac{3(-6560)}{-2} = \frac{-19680}{-2} = 9840$$

**41.** $\sum_{i=1}^{10} 5 \cdot 2^i = \frac{10\left(1-2^{10}\right)}{1-2} = \frac{10(1-1024)}{-1}$

$$= \frac{10(-1023)}{-1} = 10,230$$

**43.** $\sum_{i=1}^{6} \left(\frac{1}{2}\right)^{i+1} = \frac{\frac{1}{4}\left(1-\left(\frac{1}{2}\right)^6\right)}{1-\frac{1}{2}} = \frac{\frac{1}{4}\left(1-\frac{1}{64}\right)}{\frac{1}{2}}$

$$= \frac{\frac{1}{4}\left(\frac{64}{64}-\frac{1}{64}\right)}{\frac{1}{2}} = \frac{\frac{1}{4}\left(\frac{63}{64}\right)}{\frac{1}{2}}$$
$$= \frac{1}{4}\left(\frac{63}{64}\right) \div \frac{1}{2} = \frac{1}{4}\left(\frac{63}{64}\right) \cdot \frac{2}{1}$$
$$= \frac{63}{128}$$

**45.** $r = \frac{a_2}{a_1} = \frac{\frac{1}{3}}{1} = \frac{1}{3}$

$$S = \frac{a_1}{1-r} = \frac{1}{1-\frac{1}{3}} = \frac{1}{\frac{2}{3}} = 1 \div \frac{2}{3} = 1 \cdot \frac{3}{2} = \frac{3}{2}$$

**47.** $r = \frac{a_2}{a_1} = \frac{\frac{3}{4}}{3} = \frac{3}{4} \div 3 = \frac{3}{4} \cdot \frac{1}{3} = \frac{1}{4}$

$$S = \frac{a_1}{1-r} = \frac{3}{1-\frac{1}{4}} = \frac{3}{\frac{3}{4}} = 3 \div \frac{3}{4}$$
$$= 3 \cdot \frac{4}{3} = \frac{12}{3} = 4$$

**49.** $r = \dfrac{a_2}{a_1} = \dfrac{-\frac{1}{2}}{1} = -\dfrac{1}{2}$

$S = \dfrac{a_1}{1-r} = \dfrac{1}{1-\left(-\frac{1}{2}\right)} = \dfrac{1}{\frac{3}{2}} = 1 \div \dfrac{3}{2}$

$= 1 \cdot \dfrac{2}{3} = \dfrac{2}{3}$

**51.** $r = -0.3$

$a_1 = 26(-0.3)^{1-1} = 26(-0.3)^0$

$= 26(1) = 26$

$S = \dfrac{26}{1-(-0.3)} = \dfrac{26}{1.3} = 20$

**53.** $0.\overline{5} = \dfrac{a_1}{1-r} = \dfrac{\frac{5}{10}}{1-\frac{1}{10}} = \dfrac{\frac{5}{10}}{\frac{9}{10}} = \dfrac{5}{10} \div \dfrac{9}{10}$

$= \dfrac{5}{10} \cdot \dfrac{10}{9} = \dfrac{5}{9}$

**55.** $0.\overline{47} = \dfrac{a_1}{1-r} = \dfrac{\frac{47}{100}}{1-\frac{1}{100}} = \dfrac{\frac{47}{100}}{\frac{99}{100}}$

$= \dfrac{47}{100} \div \dfrac{99}{100} = \dfrac{47}{100} \cdot \dfrac{100}{99} = \dfrac{47}{99}$

**57.** $0.\overline{257} = \dfrac{a_1}{1-r} = \dfrac{\frac{257}{1000}}{1-\frac{1}{1000}} = \dfrac{\frac{257}{1000}}{\frac{999}{1000}}$

$= \dfrac{257}{1000} \div \dfrac{999}{1000} = \dfrac{257}{1000} \cdot \dfrac{1000}{999}$

$= \dfrac{257}{999}$

**59.** The sequence is arithmetic with common difference $d = 1$.

**61.** The sequence is geometric with common ratio $r = 2$.

**63.** The sequence is neither arithmetic nor geometric.

**65.** First find $a_{10}$ and $b_{10}$:

$a_{10} = a_1 r^{n-1}$

$= (-5)\left(\dfrac{10}{-5}\right)^{10-1} = (-5)(-2)^9$

$= 2560$

$b_{10} = b_1 + (n-1)d$

$= 10 + (10-1)(-5-10)$

$= 10 + (9)(-15) = -125$

So,

$a_{10} + b_{10} = 2560 + (-125) = 2435$.

**66.** First find $a_{11}$ and $b_{11}$:

$a_{11} = a_1 r^{n-1}$

$= (-5)\left(\dfrac{10}{-5}\right)^{11-1} = (-5)(-2)^{10}$

$= -5120$

$b_{11} = b_1 + (n-1)d$

$= 10 + (11-1)(-5-10)$

$= 10 + (10)(-15) = -140$

So,

$a_{11} + b_{11} = -5120 + (-140) = -5260$.

**67.** From Exercise 65, $a_{10} = 2560$ and $b_{10} = -125$.

For $\{a_n\}$, $r = \dfrac{10}{-5} = -2$ and:

$S_{10} = \dfrac{a_1(1-r^n)}{1-r} = \dfrac{(-5)\left(1-(-2)^{10}\right)}{1-(-2)}$

$= \dfrac{(-5)(-1023)}{3} = 1705$

For $\{b_n\}$,

$$S_n = \frac{n}{2}\left(b_1 + b_n\right) = \frac{10}{2}(10 + (-125))$$
$$= 5(-115) = -575$$

So,

$$\sum_{n=1}^{10} a_n - \sum_{n=1}^{10} b_n = 1705 - (-575) = 2280$$

**68.** For $\{a_n\}$, $r = \frac{10}{-5} = -2$ and :

$$S_{11} = \frac{a_1(1-r^n)}{1-r} = \frac{(-5)\left(1-(-2)^{11}\right)}{1-(-2)}$$
$$= \frac{(-5)(2049)}{3} = -3415$$

For $\{b_n\}$,

$$b_{11} = b_1 + (n-1)d$$
$$= 10 + (11-1)(-5-10)$$
$$= 10 + (10)(-15) = -140$$
$$S_{11} = \frac{n}{2}\left(b_1 + b_n\right) = \frac{11}{2}(10 + (-140))$$
$$= 5.5(-130) = -715$$

So, $\sum_{n=1}^{11} a_n - \sum_{n=1}^{11} b_n = -3415 - (-715)$

$$= -2700$$

**69.** For $\{a_n\}$,

$$S_6 = \frac{a_1(1-r^n)}{1-r} = \frac{(-5)\left(1-(-2)^{6}\right)}{1-(-2)}$$
$$= \frac{(-5)(-63)}{3} = 105$$

For $\{c_n\}$,

$$S = \frac{a_1}{1-r} = \frac{-2}{1-\frac{1}{-2}} = \frac{-2}{\frac{3}{2}} = -\frac{4}{3}$$

So, $S_6 \cdot S = 105\left(-\frac{4}{3}\right) = -140$

**70.** For $\{a_n\}$,

$$S_9 = \frac{a_1(1-r^n)}{1-r} = \frac{(-5)\left(1-(-2)^{9}\right)}{1-(-2)}$$
$$= \frac{(-5)(513)}{3} = -855$$

For $\{c_n\}$,

$$S = \frac{c_1}{1-r} = \frac{-2}{1-\frac{1}{-2}} = \frac{-2}{\frac{3}{2}} = -\frac{4}{3}$$

So, $S_9 \cdot S = -855\left(-\frac{4}{3}\right) = 1140$

**71.** It is given that $a_4 = 27$. Using the formula $a_n = a_1 r^{n-1}$ when $n = 4$ we have:

$$27 = 8r^{4-1}$$
$$\frac{27}{8} = r^3$$
$$r = \sqrt[3]{\frac{27}{8}} = \frac{3}{2}$$

Then

$$a_n = a_1 r^{n-1}$$
$$a_2 = 8\left(\frac{3}{2}\right)^{2-1} = 8\left(\frac{3}{2}\right) = 12$$
$$a_3 = 8\left(\frac{3}{2}\right)^{3-1} = 8\left(\frac{3}{2}\right)^2 = 8\left(\frac{9}{4}\right) = 18$$

**72.** It is given that $a_4 = -54$. Using the formula $a_n = a_1 r^{n-1}$ when $n = 4$ we have:

$$-54 = 2r^{4-1}$$
$$-27 = r^3$$
$$r = \sqrt[3]{-27} = -3$$

Then

$$a_n = a_1 r^{n-1}$$
$$a_2 = 2(-3)^{2-1} = 2(-3) = -6$$
$$a_3 = 2(-3)^{3-1} = 2(-3)^2 = 2(9) = 18$$

**73.** $r = \frac{a_2}{a_1} = \frac{2}{1} = 2$

$a_{15} = 1(2)^{15-1} = (2)^{14} = 16384$

On the fifteenth day, you will put aside \$16,384 for savings.

**75.** $r = 1.04$

$$\begin{aligned} a_7 &= 3,000,000(1.04)^{7-1} \\ &= 3,000,000(1.04)^6 \\ &= 3,000,000(1.265319) \\ &= 3,795,957 \end{aligned}$$

The athlete's salary for year 7 will be \$3,795,957.

**77. a.**

$$r_{1990 \text{ to } 1991} = \frac{30.15}{29.76} \approx 1.013$$

$$r_{1991 \text{ to } 1992} = \frac{30.54}{30.15} \approx 1.013$$

$$r_{1992 \text{ to } 1993} = \frac{30.94}{30.54} \approx 1.013$$

$$r_{1993 \text{ to } 1994} = \frac{31.34}{30.94} \approx 1.013$$

$$r_{1994 \text{ to } 1995} = \frac{31.75}{31.34} \approx 1.013$$

$$r_{1995 \text{ to } 1996} = \frac{32.16}{31.75} \approx 1.013$$

$$r_{1996 \text{ to } 1997} = \frac{32.58}{32.16} \approx 1.013$$

Since the population of each year can be found by multiplying the preceding year by 1.013, the population is increasing geometrically.

**b.** $a_n = a_1 r^{n-1} = 29.76(1.013)^{n-1}$

**c.** Since year 2000 is the 11th term, find $a_{11}$.

$$\begin{aligned} a_{11} &= 29.76(1.013)^{11-1} \\ &= 29.76(1.013)^{10} \approx 33.86 \end{aligned}$$

The population of California will be approximately 33.86 million in 2000. The geometric sequence models the actual population of 33.87 million very well.

**79.** $r = \frac{a_2}{a_1} = \frac{2}{1} = 2$

$$S_{15} = \frac{a_1(1-r^n)}{1-r} = \frac{1(1-(2)^{15})}{1-2}$$

$$= \frac{(1-32768)}{-1} = \frac{(-32767)}{-1} = 32767$$

Your savings will be \$32,767 over the 15 days.

**81.** $r = 1.05$

$$S_{20} = \frac{a_1(1-r^n)}{1-r} = \frac{24000(1-(1.05)^{20})}{1-1.05}$$

$$= \frac{24000(1-2.6533)}{-0.05}$$

$$= \frac{24000(-1.6533)}{-0.05} = 793583$$

The total lifetime salary over the 20 years is \$793,583.

**83.** $r = 0.9$

$$S_{10} = \frac{a_1(1-r^n)}{1-r} = \frac{20(1-(0.9)^{10})}{1-0.9}$$

$$= \frac{20(1-0.348678)}{0.1}$$

$$= \frac{20(0.651322)}{0.1} = 130.264$$

After 10 swings, the pendulum covers a distance of approximately 130.26 inches.

**85.**

$$A = P\frac{\left(1+\frac{r}{n}\right)^{nt}-1}{\frac{r}{n}}$$

$$= 2500\frac{\left(1+\frac{0.09}{1}\right)^{1(40)}-1}{\frac{0.09}{1}}$$

$$= 2500\frac{(1+0.09)^{40}-1}{0.09}$$

$$= 2500\frac{(1.09)^{40}-1}{0.09}$$

$$= 2500\frac{31.4094-1}{0.09}$$

$$= 2500\frac{30.4094}{0.09} = 844706$$

After 40 years, the value of the IRA will be $844,706.

**87.**

$$A = P\frac{\left(1+\frac{r}{n}\right)^{nt}-1}{\frac{r}{n}}$$

$$= 600\frac{\left(1+\frac{0.08}{4}\right)^{4(18)}-1}{\frac{0.08}{4}}$$

$$= 600\frac{(1+0.02)^{72}-1}{0.02}$$

$$= 600\frac{(1.02)^{72}-1}{0.02}$$

$$= 600\frac{4.16114-1}{0.02} = 600\frac{3.16114}{0.02}$$

$$= 94834.2$$

The value of the TSA after 18 years will be $94,834.20.

**89.** $r = 60\% = 0.6$

$a_1 = 6(.6) = 3.6$

$$S = \frac{3.6}{1-0.6} = \frac{3.6}{0.4} = 9$$

The total economic impact of the factory will be $9 million per year.

**91.** $r = \frac{1}{4}$

$$S = \frac{\frac{1}{4}}{1-\frac{1}{4}} = \frac{\frac{1}{4}}{\frac{3}{4}} = \frac{1}{4} \div \frac{3}{4} = \frac{1}{4}\cdot\frac{4}{3} = \frac{1}{3}$$

Eventually $\frac{1}{3}$ of the largest square will be shaded.

**93. – 99.** Answers will vary.

**101.** Answers will vary. For example, consider Exercise 25.

$a_n = 3(4)^{n-1}$

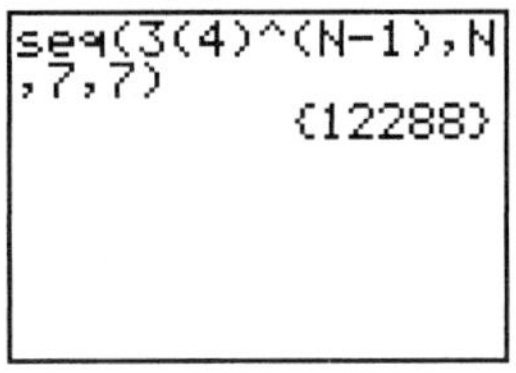


This matches the result obtained in Exercise 25.

**103.** $$f(x)=\frac{2\left[1-\left(\frac{1}{3}\right)^{x}\right]}{1-\frac{1}{3}}$$

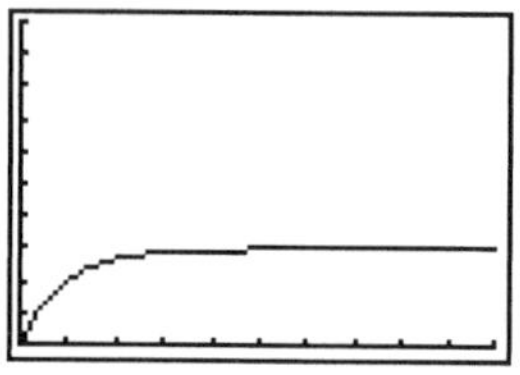

$$S=\frac{2}{1-\frac{1}{3}}=\frac{2}{\frac{2}{3}}=2\div\frac{2}{3}=2\cdot\frac{3}{2}=3$$

The sum of the series and the asymptote of the function are both 3.

**105.** Statement **d.** is true. The common ratio is $0.5=\frac{1}{2}$.

Statement **a.** is false. The sequence is not geometric. The fourth term would have to be $24\cdot 4=96$ for the sequence to be geometric.
Statement **b.** is false. We do not need to know the terms between $\frac{1}{8}$ and $\frac{1}{512}$, but we do need to know how many terms there are between $\frac{1}{8}$ and $\frac{1}{512}$.

Statement **c.** is false. The sum of the sequence is $\frac{10}{1-\left(-\frac{1}{2}\right)}$.

**107.** $$A=P\frac{\left(1+\frac{r}{n}\right)^{nt}-1}{\frac{r}{n}}$$

$$1{,}000{,}000=P\frac{\left(1+\frac{0.10}{12}\right)^{12(30)}-1}{\frac{0.10}{12}}$$

$$1{,}000{,}000=P\frac{\left(1+\frac{1}{120}\right)^{360}-1}{\frac{1}{120}}$$

$$1{,}000{,}000=P\frac{\left(1\frac{1}{120}\right)^{360}-1}{\frac{1}{120}}$$

$$1{,}000{,}000=P\frac{19.8374-1}{\frac{1}{120}}$$

$$\frac{1}{120}(1{,}000{,}000)=\frac{1}{120}\left(P\frac{18.8374}{\frac{1}{120}}\right)$$

$$\frac{25000}{3}=18.8374P$$

$$\frac{25000}{3(18.8374)}=P$$

$$442.382=P$$

You should deposit approximately \$442.38 per month.

**108.** $$\sqrt{28}-3\sqrt{7}+\sqrt{63}$$
$$=\sqrt{4\cdot 7}-3\sqrt{7}+\sqrt{9\cdot 7}$$
$$=2\sqrt{7}-3\sqrt{7}+3\sqrt{7}$$
$$=2\sqrt{7}$$

**109.** $2x^2 = 4 - x$

$2x^2 + x - 4 = 0$

$a = 2 \quad b = 1 \quad c = -4$

Solve using the quadratic formula.

$$x = \frac{-1 \pm \sqrt{1^2 - 4(2)(-4)}}{2(2)}$$

$$= \frac{-1 \pm \sqrt{1+32}}{4} = \frac{-1 \pm \sqrt{33}}{4}$$

The solutions are $\frac{-1 \pm \sqrt{33}}{4}$ and the solution set is $\left\{\frac{-1 \pm \sqrt{33}}{4}\right\}$.

**110.**

$$\frac{6}{\sqrt{3}-\sqrt{5}} = \frac{6}{\sqrt{3}-\sqrt{5}} \cdot \frac{\sqrt{3}+\sqrt{5}}{\sqrt{3}+\sqrt{5}}$$

$$= \frac{6\left(\sqrt{3}+\sqrt{5}\right)}{3-5}$$

$$= \frac{6\left(\sqrt{3}+\sqrt{5}\right)}{-2}$$

$$= -3\left(\sqrt{3}+\sqrt{5}\right)$$

## Mid-Chapter Check Point

**1.**

$$a_n = (-1)^{n+1} \frac{n}{(n-1)!}$$

$$a_1 = (-1)^{1+1} \frac{1}{(1-1)!} = (-1)^2 \frac{1}{0!} = 1 \cdot 1 = 1$$

$$a_2 = (-1)^{2+1} \frac{2}{(2-1)!} = (-1)^3 \frac{2}{1!} = (-1)(2) = -2$$

$$a_3 = (-1)^{3+1} \frac{3}{(3-1)!} = (-1)^4 \frac{3}{2!} = 1 \cdot \frac{3}{2} = \frac{3}{2}$$

$$a_4 = (-1)^{4+1} \frac{4}{(4-1)!} = (-1)^5 \frac{4}{3!} = (-1)\frac{4}{6} = -\frac{2}{3}$$

$$a_5 = (-1)^{5+1} \frac{5}{(5-1)!} = (-1)^6 \frac{5}{4!} = 1 \cdot \frac{5}{24} = \frac{5}{24}$$

**2.** Using $a_n = a_1 + (n-1)d$;

$a_1 = 5$

$a_2 = 5 + (2-1)(-3) = 5 + 1(-3) = 5 - 3 = 2$

$a_3 = 5 + (3-1)(-3) = 5 + 2(-3) = 5 - 6 = -1$

$a_4 = 5 + (4-1)(-3) = 5 + 3(-3) = 5 - 9 = -4$

$a_5 = 5 + (5-1)(-3) = 5 + 4(-3) = 5 - 12 = -7$

**3.** Using $a_n = a_1 r^{n-1}$;

$a_1 = 5$

$a_2 = 5(-3)^{2-1} = 5(-3)^1 = 5(-3) = -15$

$a_3 = 5(-3)^{3-1} = 5(-3)^2 = 5(9) = 45$

$a_4 = 5(-3)^{4-1} = 5(-3)^3 = 5(-27) = -135$

$a_5 = 5(-3)^{5-1} = 5(-3)^4 = 5(81) = 405$

**4.** $d = a_2 - a_1 = 6 - 2 = 4$

$$\begin{aligned} a_n &= a_1 + (n-1)d \\ &= 2 + (n-1)4 \\ &= 2 + 4n - 4 \\ &= 4n - 2 \end{aligned}$$

$a_{20} = 4(20) - 2 = 78$

**5.** $r = \dfrac{a_2}{a_1} = \dfrac{6}{3} = 2$

$$\begin{aligned} a_n &= a_1 r^{n-1} \\ &= 3(2)^{n-1} \end{aligned}$$

$$\begin{aligned} a_{10} &= 3(2)^{10-1} \\ &= 3(2)^9 \\ &= 1536 \end{aligned}$$

**6.** $d = a_2 - a_1 = 1 - \dfrac{3}{2} = -\dfrac{1}{2}$

$$\begin{aligned} a_n &= a_1 + (n-1)d \\ &= \frac{3}{2} + (n-1)\left(-\frac{1}{2}\right) \\ &= \frac{3}{2} - \frac{1}{2}n + \frac{1}{2} \\ &= -\frac{1}{2}n + 2 \end{aligned}$$

$$\begin{aligned} a_{30} &= -\frac{1}{2}(30) + 2 \\ &= -15 + 2 \\ &= -13 \end{aligned}$$

**7.** $S_n = \dfrac{a_1(1-r^n)}{1-r}$; $r = \dfrac{a_2}{a_1} = \dfrac{10}{5} = 2$

$$S_{10} = \frac{5(1-2^{10})}{1-2} = \frac{5(-1023)}{-1} = 5115$$

**8.** First find $a_{10}$;

$$d = a_2 - a_1 = 0 - (-2) = 2$$

$$a_{50} = a_1 + (n-1)d = -2 + (50-1)(2) = -2 + 49(2) = 96$$

$$S_{50} = \frac{n}{2}(a_1 + a_n) = \frac{50}{2}(-2+96) = 25(94) = 2350$$

**9.** First find $a_{10}$;

$$r = \frac{a_2}{a_1} = \frac{40}{-20} = -2$$

$$a_{10} = a_1 r^{n-1} = -20(2)^{10-1} = -20(2)^9 = -20(512) = -10240$$

$$S_{10} = \frac{a_1(1-r^n)}{1-r} = \frac{-20(-1-(-2)^{10})}{1-(-2)} = \frac{-20(-1023)}{3} = \frac{20460}{3} = 6820$$

**10.** First find $a_{100}$;

$$d = a_2 - a_1 = -2 - 4 = -6$$

$$a_{100} = a_1 + (n-1)d = 4 + (100-1)(-6) = 4 + 99(-6) = -590$$

$$S_{100} = \frac{n}{2}(a_1 + a_n) = \frac{100}{2}(4-590) = 50(-586) = -29,300$$

**11.**

$$\sum_{i=1}^{4}(i+4)(i-1) = (1+4)(1-1) + (2+4)(2-1) + (3+4)(3-1) + (4+4)(4-1)$$

$$= 5(0) + 6(1) + 7(2) + 8(3) = 0 + 6 + 14 + 24 = 44$$

**12.**

$$\sum_{i=1}^{50}(3i-2) = (3\cdot 1 - 2) + (3\cdot 2 - 2) + (3\cdot 3 - 3) + \ldots + (3\cdot 50 - 2)$$

$$= (3-2) + (6-2) + (9-3) + \ldots + (150-2)$$

$$= 1 + 4 + 6 + \ldots + 148$$

The sum of this arithmetic sequence is given by $S_n = \frac{n}{2}(a_1 + a_n)$;

$$S_{50} = \frac{50}{2}(1+148) = 25(149) = 3725$$

**13.**

$$\sum_{i=1}^{6}\left(\frac{3}{2}\right)^i = \left(\frac{3}{2}\right)^1 + \left(\frac{3}{2}\right)^2 + \left(\frac{3}{2}\right)^3 + \left(\frac{3}{2}\right)^4 + \left(\frac{3}{2}\right)^5 + \left(\frac{3}{2}\right)^6$$

$$= \frac{3}{2} + \frac{9}{4} + \frac{27}{8} + \frac{81}{16} + \frac{243}{32} + \frac{729}{64} = \frac{1995}{64}$$

**14.** $$\sum_{i=1}^{\infty}\left(-\frac{2}{5}\right)^{i-1}=\left(-\frac{2}{5}\right)^{1-1}+\left(-\frac{2}{5}\right)^{2-1}+\left(-\frac{2}{5}\right)^{3-1}+\ldots$$
$$=\left(-\frac{2}{5}\right)^{0}+\left(-\frac{2}{5}\right)^{1}+\left(-\frac{2}{5}\right)^{2}+\ldots$$
$$=1+\left(-\frac{2}{5}\right)+\frac{4}{25}+\ldots$$

This is an infinite geometric sequence with $r=\frac{a_2}{a_1}=\frac{-\frac{2}{5}}{1}=-\frac{2}{5}$.

Using $S=\frac{a_1}{1-r}=\frac{1}{1-\left(-\frac{2}{5}\right)}=\frac{1}{\frac{7}{5}}=\frac{5}{7}$

**15.** $$0.\overline{45}=\frac{a_1}{1-r}=\frac{\frac{45}{100}}{1-\frac{1}{100}}=\frac{\frac{45}{100}}{\frac{99}{100}}$$
$$=\frac{45}{100}\div\frac{99}{100}=\frac{45}{100}\cdot\frac{100}{99}=\frac{45}{99}=\frac{5}{11}$$

**16.** Answers will vary. An example is $\sum_{i=1}^{18}\frac{i}{i+2}$.

**17.** The arithmetic sequence is 16, 48, 80, 112, ....
First find $a_{15}$ where $d=a_2-a_1=48-16=32$.
$a_{15}=a_1+(n-1)d=16+(15-1)(32)=16+14(32)=16+448=464$.
The distance the skydiver falls during the $15^{\text{th}}$ second is 464 feet.
$S_{15}=\frac{n}{2}(a_1+a_n)=\frac{15}{2}(16+464)=7.5(480)=3600$
The total distance the skydiver falls in 15 seconds is 3600 feet.

**18.** $r=0.10$
$$A=P(1+r)^t$$
$$=120000(1+0.10)^{10}$$
$$\approx 311249$$
The value of the house after 10 years is \$311,249.

## 11.4 Exercise Set

**1.** $\binom{8}{3}=\frac{8!}{3!(8-3)!}$

$=\frac{8!}{3!5!}=\frac{8\cdot 7\cdot \cancel{6}\cdot \cancel{5!}}{\cancel{3\cdot 2}\cdot 1\cdot \cancel{5!}}$

$=56$

**3.** $\binom{12}{1}=\frac{12!}{1!(12-1)!}=\frac{12\cdot \cancel{11!}}{1\cdot \cancel{11!}}=12$

**5.** $\binom{6}{6}=\frac{\cancel{6!}}{\cancel{6!}(6-6)!}=\frac{1}{0!}=\frac{1}{1}=1$

**7.** $\binom{100}{2}=\frac{100!}{2!(100-2)!}=\frac{100\cdot 99\cdot \cancel{98!}}{2\cdot 1\cdot \cancel{98!}}$

$=4950$

**9.** Applying the Binomial Theorem to $(x+2)^3$, we have $a=x,\ b=2,$ and $n=3$.

$(x+2)^3=\binom{3}{0}x^3+\binom{3}{1}x^2(2)+\binom{3}{2}x(2)^2+\binom{3}{3}2^3$

$=\frac{3!}{0!(3-0)!}x^3+\frac{3!}{1!(3-1)!}2x^2+\frac{3!}{2!(3-2)!}4x+\frac{3!}{3!(3-3)!}8$

$=\frac{\cancel{3!}}{1\cdot\cancel{3!}}x^3+\frac{3\cdot\cancel{2!}}{1\cdot\cancel{2!}}2x^2+\frac{3\cdot\cancel{2!}}{\cancel{2!}1!}4x+\frac{\cancel{3!}}{\cancel{3!}0!}8=x^3+3\left(2x^2\right)+3(4x)+1(8)$

$=x^3+6x^2+12x+8$

**11.** Applying the Binomial Theorem to $(3x+y)^3$, we have $a=3x,\ b=y,$ and $n=3$.

$(3x+y)^3=\binom{3}{0}(3x)^3+\binom{3}{1}(3x)^2y+\binom{3}{2}(3x)y^2+\binom{3}{3}y^3$

$=\frac{3!}{0!(3-0)!}27x^3+\frac{3!}{1!(3-1)!}9x^2y+\frac{3!}{2!(3-2)!}3xy^2+\frac{3!}{3!(3-3)!}y^3$

$=\frac{\cancel{3!}}{1\cdot\cancel{3!}}27x^3+\frac{3\cdot\cancel{2!}}{1\cdot\cancel{2!}}9x^2y+\frac{3\cdot\cancel{2!}}{\cancel{2!}1!}3xy^2+\frac{\cancel{3!}}{\cancel{3!}0!}y^3=27x^3+3\left(9x^2y\right)+3\left(3xy^2\right)+1\left(y^3\right)$

$=27x^3+27x^2y+9xy^2+y^3$

**13.** Applying the Binomial Theorem to $(5x-1)^3$, we have $a=5x,\ b=-1,$ and $n=3$.

$$(5x-1)^3=\binom{3}{0}(5x)^3+\binom{3}{1}(5x)^2(-1)+\binom{3}{2}(5x)(-1)^2+\binom{3}{3}(-1)^3$$

$$=\frac{3!}{0!(3-0)!}125x^3-\frac{3!}{1!(3-1)!}25x^2+\frac{3!}{2!(3-2)!}5x(1)-\frac{3!}{3!(3-3)!}$$

$$=\frac{\cancel{3!}}{1\cdot\cancel{3!}}125x^3-\frac{3\cdot\cancel{2!}}{1\cdot\cancel{2!}}25x^2+\frac{3\cdot\cancel{2!}}{\cancel{2!}1!}5x-\frac{\cancel{3!}}{\cancel{3!}0!}=125x^3-3(25x^2)+3(5x)-1$$

$$=125x^3-75x^2+15x-1$$

**15.** Applying the Binomial Theorem to $(2x+1)^4$, we have $a=2x,\ b=1,$ and $n=4$.

$$(2x+1)^4=\binom{4}{0}(2x)^4+\binom{4}{1}(2x)^3+\binom{4}{2}(2x)^2+\binom{4}{3}2x+\binom{4}{4}$$

$$=\frac{4!}{0!(4-0)!}16x^4+\frac{4!}{1!(4-1)!}8x^3\cdot 1+\frac{4!}{2!(4-2)!}4x^2\cdot 1^2+\frac{4!}{3!(4-3)!}2x\cdot 1^3+\frac{4!}{4!(4-4)!}\cdot 1^4$$

$$=\frac{\cancel{4!}}{0!\cancel{4!}}16x^4+\frac{4!}{1!3!}8x^3\cdot 1+\frac{4!}{2!2!}4x^2\cdot 1+\frac{4!}{3!1!}2x\cdot 1+\frac{\cancel{4!}}{\cancel{4!}0!}\cdot 1$$

$$=1(16x^4)+\frac{4\cdot\cancel{3!}}{1\cdot\cancel{3!}}8x^3+\frac{4\cdot 3\cdot\cancel{2!}}{2\cdot 1\cdot\cancel{2!}}4x^2+\frac{4\cdot\cancel{3!}}{\cancel{3!}\cdot 1}2x+1=16x^4+4(8x^3)+6(4x^2)+4(2x)+1$$

$$=16x^4+32x^3+24x^2+8x+1$$

**17.** Applying the Binomial Theorem to $(x^2+2y)^4$, we have $a=x^2,\ b=2y,$ and $n=4$.

$$(x^2+2y)^4=\binom{4}{0}(x^2)^4+\binom{4}{1}(x^2)^3(2y)+\binom{4}{2}(x^2)^2(2y)^2+\binom{4}{3}x^2(2y)^3+\binom{4}{4}(2y)^4$$

$$=\frac{4!}{0!(4-0)!}x^8+\frac{4!}{1!(4-1)!}2x^6y+\frac{4!}{2!(4-2)!}4x^4y^2+\frac{4!}{3!(4-3)!}8x^2y^3+\frac{4!}{4!(4-4)!}16y^4$$

$$=\frac{\cancel{4!}}{0!\cancel{4!}}x^8+\frac{4!}{1!3!}2x^6y+\frac{4!}{2!2!}4x^4y^2+\frac{4!}{3!1!}8x^2y^3+\frac{\cancel{4!}}{\cancel{4!}0!}16y^4$$

$$=1(x^8)+\frac{4\cdot\cancel{3!}}{1\cdot\cancel{3!}}2x^6y+\frac{4\cdot 3\cdot\cancel{2!}}{2\cdot 1\cdot\cancel{2!}}4x^4y^2+\frac{4\cdot\cancel{3!}}{\cancel{3!}\cdot 1}8x^2y^3+16y^4$$

$$=x^8+4(2x^6y)+6(4x^4y^2)+4(8x^2y^3)+16y^4=x^8+8x^6y+24x^4y^2+32x^2y^3+16y^4$$

**19.** Applying the Binomial Theorem to $(y-3)^4$, we have $a=y,\ b=-3,$ and $n=4$.

$$(y-3)^4=\binom{4}{0}y^4+\binom{4}{1}y^3(-3)+\binom{4}{2}y^2(-3)^2+\binom{4}{3}y(-3)^3+\binom{4}{4}(-3)^4$$

$$=\frac{4!}{0!(4-0)!}y^4-\frac{4!}{1!(4-1)!}3y^3+\frac{4!}{2!(4-2)!}9y^2-\frac{4!}{3!(4-3)!}27y+\frac{4!}{4!(4-4)!}81$$

$$=\frac{\cancel{4!}}{0!\cancel{4!}}y^4-\frac{4!}{1!3!}3y^3+\frac{4!}{2!2!}9y^2-\frac{4!}{3!1!}27y+\frac{\cancel{4!}}{\cancel{4!}0!}81$$

$$=1\left(y^4\right)-\frac{4\cdot\cancel{3!}}{1\cdot\cancel{3!}}3y^3+\frac{4\cdot3\cdot\cancel{2!}}{2\cdot1\cdot\cancel{2!}}9y^2-\frac{4\cdot\cancel{3!}}{\cancel{3!}\cdot1}27y+81$$

$$=y^4-4\left(3y^3\right)+6\left(9y^2\right)-4(27y)+81=y^4-12y^3+54y^2-108y+81$$

**21.** Applying the Binomial Theorem to $\left(2x^3-1\right)^4$, we have $a=2x^3,\ b=-1,$ and $n=4$.

$$\left(2x^3-1\right)^4=\binom{4}{0}\left(2x^3\right)^4+\binom{4}{1}\left(2x^3\right)^3(-1)+\binom{4}{2}\left(2x^3\right)^2(-1)^2+\binom{4}{3}\left(2x^3\right)(-1)^3+\binom{4}{4}(-1)^4$$

$$=\frac{4!}{0!(4-0)!}16x^{12}-\frac{4!}{1!(4-1)!}8x^9+\frac{4!}{2!(4-2)!}4x^6-\frac{4!}{3!(4-3)!}2x^3+\frac{4!}{4!(4-4)!}$$

$$=\frac{\cancel{4!}}{0!\cancel{4!}}16x^{12}-\frac{4!}{1!3!}8x^9+\frac{4!}{2!2!}4x^6-\frac{4!}{3!1!}2x^3+\frac{\cancel{4!}}{\cancel{4!}0!}$$

$$=1\left(16x^{12}\right)-\frac{4\cdot\cancel{3!}}{1\cdot\cancel{3!}}8x^9+\frac{4\cdot3\cdot\cancel{2!}}{2\cdot1\cdot\cancel{2!}}4x^6-\frac{4\cdot\cancel{3!}}{\cancel{3!}\cdot1}2x^3+1$$

$$=16x^{12}-4\left(8x^9\right)+6\left(4x^6\right)-4\left(2x^3\right)+1=16x^{12}-32x^9+24x^6-8x^3+1$$

**23.** Applying the Binomial Theorem to $(c+2)^5$, we have $a=c,\ b=2,$ and $n=5$.

$$(c+2)^5=\binom{5}{0}c^5+\binom{5}{1}c^4(2)+\binom{5}{2}c^3(2)^2+\binom{5}{3}c^2(2)^3+\binom{5}{4}c(2)^4+\binom{5}{5}2^5$$

$$=\frac{\cancel{5!}}{0!\cancel{5!}}c^5+\frac{5!}{1!(5-1)!}2c^4+\frac{5!}{2!(5-2)!}4c^3+\frac{5!}{3!(5-3)!}8c^2+\frac{5!}{4!(5-4)!}16c+\frac{5!}{5!(5-5)!}32$$

$$=1c^5+\frac{5\cdot\cancel{4!}}{1\cdot\cancel{4!}}2c^4+\frac{5\cdot4\cdot\cancel{3!}}{2\cdot1\cdot\cancel{3!}}4c^3+\frac{5\cdot4\cdot\cancel{3!}}{\cancel{3!}2\cdot1}8c^2+\frac{5\cdot\cancel{4!}}{\cancel{4!}\cdot1}16c+\frac{\cancel{5!}}{\cancel{5!}0!}32$$

$$=c^5+5\left(2c^4\right)+10\left(4c^3\right)+10\left(8c^2\right)+5(16c)+1(32)=c^5+10c^4+40c^3+80c^2+80c+32$$

**25.** Applying the Binomial Theorem to $(x-1)^5$, we have $a=x,\ b=-1,$ and $n=5.$

$$(x-1)^5=\binom{5}{0}x^5+\binom{5}{1}x^4(-1)+\binom{5}{2}x^3(-1)^2+\binom{5}{3}x^2(-1)^3+\binom{5}{4}x(-1)^4+\binom{5}{5}(-1)^5$$

$$=\frac{\cancel{5!}}{0!\cancel{5!}}x^5-\frac{5!}{1!(5-1)!}x^4+\frac{5!}{2!(5-2)!}x^3-\frac{5!}{3!(5-3)!}x^2+\frac{5!}{4!(5-4)!}x-\frac{5!}{5!(5-5)!}$$

$$=1x^5-\frac{5\cdot\cancel{4!}}{1\cdot\cancel{4!}}x^4+\frac{5\cdot4\cdot\cancel{3!}}{2\cdot1\cdot\cancel{3!}}x^3-\frac{5\cdot4\cdot\cancel{3!}}{\cancel{3!}2\cdot1}x^2+\frac{5\cdot\cancel{4!}}{\cancel{4!}\cdot1}x-\frac{\cancel{5!}}{\cancel{5!}0!}$$

$$=x^5-5x^4+10x^3-10x^2+5x-1$$

**27.** Applying the Binomial Theorem to $(3x-y)^5$, we have $a=3x,\ b=-y,$ and $n=5.$

$$(3x-y)^5$$

$$=\binom{5}{0}(3x)^5+\binom{5}{1}(3x)^4(-y)+\binom{5}{2}(3x)^3(-y)^2+\binom{5}{3}(3x)^2(-y)^3+\binom{5}{4}3x(-y)^4+\binom{5}{5}(-y)^5$$

$$=\frac{\cancel{5!}}{0!\cancel{5!}}243x^5-\frac{5!}{1!(5-1)!}81x^4y+\frac{5!}{2!(5-2)!}27x^3y^2$$

$$-\frac{5!}{3!(5-3)!}9x^2y^3+\frac{5!}{4!(5-4)!}3xy^4-\frac{5!}{5!(5-5)!}y^5$$

$$=243x^5-\frac{5\cdot\cancel{4!}}{1\cdot\cancel{4!}}81x^4y+\frac{5\cdot4\cdot\cancel{3!}}{2\cdot1\cdot\cancel{3!}}27x^3y^2-\frac{5\cdot4\cdot\cancel{3!}}{\cancel{3!}2\cdot1}9x^2y^3+\frac{5\cdot\cancel{4!}}{\cancel{4!}\cdot1}3xy^4-\frac{\cancel{5!}}{\cancel{5!}0!}y^5$$

$$=243x^5-5\left(81x^4y\right)+10\left(27x^3y^2\right)-10\left(9x^2y^3\right)+5\left(3xy^4\right)-y^5$$

$$=243x^5-405x^4y+207x^3y^2-90x^2y^3+15xy^4-y^5$$

**29.** Applying the Binomial Theorem to $(2a+b)^6$, we have $a=2a,\ b=b,$ and $n=6$.

$(2a+b)^6$

$$=\binom{6}{0}(2a)^6+\binom{6}{1}(2a)^5 b+\binom{6}{2}(2a)^4 b^2+\binom{6}{3}(2a)^3 b^3+\binom{6}{4}(2a)^2 b^4+\binom{6}{5}2ab^5+\binom{6}{6}b^6$$

$$=\frac{6!}{0!(6-0)!}64a^6+\frac{6!}{1!(6-1)!}32a^5b+\frac{6!}{2!(6-2)!}16a^4b^2+\frac{6!}{3!(6-3)!}8a^3b^3$$

$$+\frac{6!}{4!(6-4)!}4a^2b^4+\frac{6!}{5!(6-5)!}2ab^5+\frac{6!}{6!(6-6)!}b^6$$

$$=\frac{\cancel{6!}}{1\cancel{6!}}64a^6+\frac{6\cdot\cancel{5!}}{1\cdot\cancel{5!}}32a^5b+\frac{6\cdot5\cdot\cancel{4!}}{2\cdot1\cdot\cancel{4!}}16a^4b^2+\frac{\cancel{6}\cdot5\cdot4\cdot\cancel{3!}}{\cancel{3}\cdot\cancel{2}\cdot1\cdot\cancel{3!}}8a^3b^3$$

$$+\frac{6\cdot5\cdot\cancel{4!}}{\cancel{4!}2\cdot1}4a^2b^4+\frac{6\cdot\cancel{5!}}{\cancel{5!}1}2ab^5+\frac{\cancel{6!}}{\cancel{6!}\cdot1}b^6$$

$$=64a^6+6(32a^5b)+15(16a^4b^2)+20(8a^3b^3)+15(4a^2b^4)+6(2ab^5)+1b^6$$

$$=64a^6+192a^5b+240a^4b^2+160a^3b^3+60a^2b^4+12ab^5+b^6$$

**31.** $(x+2)^8$

First Term $\quad \binom{n}{r-1}a^{n-r+1}b^{r-1}=\binom{8}{1-1}x^{8-1+1}2^{1-1}=\binom{8}{0}x^8 2^0$

$$=\frac{8!}{0!(8-0)!}x^8\cdot1=\frac{\cancel{8!}}{0!\cancel{8!}}x^8=x^8$$

Second Term $\quad \binom{n}{r-1}a^{n-r+1}b^{r-1}=\binom{8}{2-1}x^{8-2+1}2^{2-1}=\binom{8}{1}x^7 2^1=\frac{8!}{1!(8-1)!}2x^7$

$$=\frac{8\cdot\cancel{7!}}{1\cdot\cancel{7!}}2x^7=8\cdot2x^7=16x^7$$

Third Term $\quad \binom{n}{r-1}a^{n-r+1}b^{r-1}=\binom{8}{3-1}x^{8-3+1}2^{3-1}=\binom{8}{2}x^6 2^2=\frac{8!}{2!(8-2)!}4x^6$

$$=\frac{8\cdot7\cdot\cancel{6!}}{2\cdot1\cdot\cancel{6!}}4x^6=28\cdot4x^6=112x^6$$

**33.** $(x-2y)^{10}$

First Term $\binom{n}{r-1}a^{n-r+1}b^{r-1}=\binom{10}{1-1}x^{10-1+1}(-2y)^{1-1}=\binom{10}{0}x^{10}(-2y)^{0}$

$$=\frac{10!}{0!(10-0)!}x^{10}\cdot 1=\frac{\cancel{10!}}{0!\cancel{10!}}x^{10}=x^{10}$$

Second Term $\binom{n}{r-1}a^{n-r+1}b^{r-1}=\binom{10}{2-1}x^{10-2+1}(-2y)^{2-1}=\binom{10}{1}x^{9}(-2y)^{1}=-\frac{10!}{1!(10-1)!}2x^{9}y$

$$=-\frac{10\cdot\cancel{9!}}{1\cdot\cancel{9!}}2x^{9}y=-10\cdot 2x^{9}y=-20x^{9}y$$

Third Term $\binom{n}{r-1}a^{n-r+1}b^{r-1}=\binom{10}{3-1}x^{10-3+1}(-2y)^{3-1}=\binom{10}{2}x^{8}(-2y)^{2}=\frac{10!}{2!(10-2)!}4x^{8}y^{2}$

$$=\frac{10\cdot 9\cdot\cancel{8!}}{2\cdot 1\cdot\cancel{8!}}4x^{8}y^{2}=45\cdot 4x^{8}y^{2}=180x^{8}y^{2}$$

**35.** $(x^2+1)^{16}$

First Term $\binom{n}{r-1}a^{n-r+1}b^{r-1}=\binom{16}{1-1}(x^2)^{16-1+1}(1)^{1-1}=\binom{16}{0}(x^2)^{16}1^{0}$

$$=\frac{16!}{0!(16-0)!}x^{32}\cdot 1=\frac{\cancel{16!}}{0!\cancel{16!}}x^{32}=x^{32}$$

Second Term $\binom{n}{r-1}a^{n-r+1}b^{r-1}=\binom{16}{2-1}(x^2)^{16-2+1}(1)^{2-1}=\binom{16}{1}(x^2)^{15}1^{1}$

$$=\frac{16!}{1!(16-1)!}x^{30}\cdot 1=\frac{16\cdot\cancel{15!}}{1\cdot\cancel{15!}}x^{30}=16x^{30}$$

Third Term $\binom{n}{r-1}a^{n-r+1}b^{r-1}=\binom{16}{3-1}(x^2)^{16-3+1}(1)^{3-1}=\binom{16}{2}(x^2)^{14}1^{2}$

$$=\frac{16!}{2!(16-2)!}x^{28}\cdot 1=\frac{16\cdot 15\cdot\cancel{14!}}{2\cdot 1\cdot\cancel{14!}}x^{28}=120x^{28}$$

**37.** $(y^3-1)^{20}$

First Term $\binom{n}{r-1}a^{n-r+1}b^{r-1}=\binom{20}{1-1}(y^3)^{20-1+1}(-1)^{1-1}=\binom{20}{0}(y^3)^{20}(-1)^{0}$

$$=\frac{20!}{0!(20-0)!}y^{60}\cdot 1=\frac{\cancel{20!}}{0!\cancel{20!}}y^{60}=y^{60}$$

Second Term $\binom{n}{r-1}a^{n-r+1}b^{r-1}=\binom{20}{2-1}\left(y^3\right)^{20-2+1}(-1)^{2-1}=\binom{20}{1}\left(y^3\right)^{19}(-1)^1$

$$=\frac{20!}{1!(20-1)!}y^{57}\cdot(-1)=-\frac{20\cdot \cancel{19!}}{1\cdot \cancel{19!}}y^{57}=-20y^{57}$$

Third Term $\binom{n}{r-1}a^{n-r+1}b^{r-1}=\binom{20}{3-1}\left(y^3\right)^{20-3+1}(-1)^{3-1}=\binom{20}{2}\left(y^3\right)^{18}(-1)^2$

$$=\frac{20!}{2!(20-2)!}y^{54}\cdot 1=\frac{20\cdot 19\cdot \cancel{18!}}{2\cdot 1\cdot \cancel{18!}}y^{54}=190y^{54}$$

**39.** $(2x+y)^6$

Third Term $\binom{n}{r-1}a^{n-r+1}b^{r-1}=\binom{6}{3-1}(2x)^{6-3+1}y^{3-1}=\binom{6}{2}(2x)^4y^2=\frac{6!}{2!(6-2)!}16x^4y^2$

$$=\frac{6\cdot 5\cdot \cancel{4!}}{2\cdot 1\cdot \cancel{4!}}16x^4y^2=15\left(16x^4y^2\right)=240x^4y^2$$

**41.** $(x-1)^9$

Fifth Term $\binom{n}{r-1}a^{n-r+1}b^{r-1}=\binom{9}{5-1}x^{9-5+1}(-1)^{5-1}=\binom{9}{4}x^5(-1)^4=\frac{9!}{4!(9-4)!}x^5\cdot 1$

$$=\frac{9\cdot 8\cdot 7\cdot \cancel{6}\cdot \cancel{5!}}{4\cdot \cancel{3\cdot 2}\cdot 1\cdot \cancel{5!}}x^5=126x^5$$

**43.** $\left(x^2+y^3\right)^8$

Sixth Term $\binom{n}{r-1}a^{n-r+1}b^{r-1}=\binom{8}{6-1}\left(x^2\right)^{8-6+1}\left(y^3\right)^{6-1}=\binom{8}{5}\left(x^2\right)^3\left(y^3\right)^5=\frac{8!}{5!(8-5)!}x^6y^{15}$

$$=\frac{8\cdot 7\cdot \cancel{6}\cdot \cancel{5!}}{\cancel{5!}\cdot \cancel{3\cdot 2}\cdot 1}x^6y^{15}=56x^6y^{15}$$

**45.** $\left(x-\frac{1}{2}\right)^9$

Fourth Term $\binom{n}{r-1}a^{n-r+1}b^{r-1}=\binom{9}{4-1}x^{9-4+1}\left(-\frac{1}{2}\right)^{4-1}=\binom{9}{3}x^6\left(-\frac{1}{2}\right)^3=-\frac{9!}{3!(9-3)!}\cdot\frac{1}{8}x^6$

$$=-\frac{9\cdot \cancel{8}\cdot 7\cdot \cancel{6!}}{3\cdot 2\cdot 1\cdot \cancel{6!}}\cdot\frac{1}{\cancel{8}}x^6=-\frac{21}{2}x^6$$

**47.**

$$\left(x^3+x^{-2}\right)^4=\binom{4}{0}\left(x^3\right)^4+\binom{4}{1}\left(x^3\right)^3\left(x^{-2}\right)+\binom{4}{2}\left(x^3\right)^2\left(x^{-2}\right)^2+\binom{4}{3}\left(x^3\right)^1\left(x^{-2}\right)^3+\binom{4}{4}\left(x^{-2}\right)^4$$

$$=\frac{4!}{0!(4-0)!}x^{12}+\frac{4!}{1!(4-1)!}x^9x^{-2}+\frac{4!}{2!(4-2)!}x^6x^{-4}+\frac{4!}{3!(4-3)!}x^3x^{-6}+\frac{4!}{4!(4-4)!}x^{-8}$$

$$=\frac{\cancel{4!}}{0!\cdot\cancel{4!}}x^{12}+\frac{4\cdot\cancel{3!}}{1!\cdot\cancel{3!}}x^7+\frac{4\cdot3\cdot\cancel{2!}}{2\cdot1\cdot\cancel{2!}}x^2+\frac{4\cdot\cancel{3!}}{\cancel{3!}\cdot1!}x^{-3}+\frac{\cancel{4!}}{\cancel{4!}\cdot0!}x^{-8}$$

$$=x^{12}+4x^7+6x^2+\frac{4}{x^3}+\frac{1}{x^8}$$

**48.**

$$\left(x^2+x^{-3}\right)^4=\binom{4}{0}\left(x^2\right)^4+\binom{4}{1}\left(x^2\right)^3\left(x^{-3}\right)+\binom{4}{2}\left(x^2\right)^2\left(x^{-3}\right)^2+\binom{4}{3}\left(x^2\right)^1\left(x^{-3}\right)^3+\binom{4}{4}\left(x^{-3}\right)^4$$

$$=\frac{4!}{0!(4-0)!}x^8+\frac{4!}{1!(4-1)!}x^6x^{-3}+\frac{4!}{2!(4-2)!}x^4x^{-6}+\frac{4!}{3!(4-3)!}x^2x^{-9}+\frac{4!}{4!(4-4)!}x^{-12}$$

$$=\frac{\cancel{4!}}{0!\cdot\cancel{4!}}x^8+\frac{4\cdot\cancel{3!}}{1!\cdot\cancel{3!}}x^3+\frac{4\cdot3\cdot\cancel{2!}}{2\cdot1\cdot\cancel{2!}}x^{-2}+\frac{4\cdot\cancel{3!}}{\cancel{3!}\cdot1!}x^{-7}+\frac{\cancel{4!}}{\cancel{4!}\cdot0!}x^{-12}$$

$$=x^8+4x^3+\frac{6}{x^2}+\frac{4}{x^7}+\frac{1}{x^{12}}$$

**49.**

$$\left(x^{\frac{1}{3}}-x^{-\frac{1}{3}}\right)^3=\left(x^{\frac{1}{3}}+\left(-x^{-\frac{1}{3}}\right)\right)^3=\binom{3}{0}\left(x^{\frac{1}{3}}\right)^3+\binom{3}{1}\left(x^{\frac{1}{3}}\right)^2\left(-x^{-\frac{1}{3}}\right)+$$

$$+\binom{3}{2}\left(x^{\frac{1}{3}}\right)^1\left(-x^{-\frac{1}{3}}\right)^2+\binom{3}{3}\left(-x^{-\frac{1}{3}}\right)^3$$

$$=\frac{3!}{0!(3-0)!}x^1+\frac{3!}{1!(3-1)!}x^{\frac{2}{3}}\cdot -x^{-\frac{1}{3}}+\frac{3!}{2!(3-2)!}x^{\frac{1}{3}}x^{-\frac{2}{3}}+\frac{3!}{3!(3-3)!}\cdot -x^{-1}$$

$$=\frac{\cancel{3!}}{0!\cdot\cancel{3!}}x+\frac{3\cdot\cancel{2!}}{1!\cdot\cancel{2!}}\cdot -x^{\frac{1}{3}}+\frac{\cdot3\cdot\cancel{2!}}{\cancel{2!}\cdot1!}x^{-\frac{1}{3}}+\frac{\cancel{3!}}{\cancel{3!}\cdot0!}\cdot -x^{-1}$$

$$=x-3x^{\frac{1}{3}}+\frac{3}{x^{\frac{1}{3}}}-\frac{1}{x}$$

**50.**

$$\left(x^{\frac{2}{3}}-\frac{1}{\sqrt[3]{x}}\right)^3=\left(x^{\frac{2}{3}}+\left(-\frac{1}{x^{\frac{1}{3}}}\right)\right)^3=\left(x^{\frac{2}{3}}+\left(-x^{-\frac{1}{3}}\right)\right)^3=$$

$$=\left(x^{\frac{2}{3}}+\left(-x^{-\frac{1}{3}}\right)\right)^3=\binom{3}{0}\left(x^{\frac{2}{3}}\right)^3+\binom{3}{1}\left(x^{\frac{2}{3}}\right)^2\left(-x^{-\frac{1}{3}}\right)+$$

$$+\binom{3}{2}\left(x^{\frac{2}{3}}\right)^1\left(-x^{-\frac{1}{3}}\right)^2+\binom{3}{3}\left(-x^{-\frac{1}{3}}\right)^3$$

$$=\frac{3!}{0!(3-0)!}x^2+\frac{3!}{1!(3-1)!}x^{\frac{4}{3}}\cdot -x^{-\frac{1}{3}}+\frac{3!}{2!(3-2)!}x^{\frac{2}{3}}x^{-\frac{2}{3}}+\frac{3!}{3!(3-3)!}\cdot -x^{-1}$$

$$=\frac{\cancel{3!}}{0!\cdot\cancel{3!}}x^2+\frac{3\cdot\cancel{2!}}{1!\cdot\cancel{2!}}\cdot -x^1+\frac{\cdot 3\cdot\cancel{2!}}{\cancel{2!}\cdot 1!}x^0+\frac{\cancel{3!}}{\cancel{3!}\cdot 0!}\cdot -x^{-1}$$

$$=x^2-3x+3-\frac{1}{x}$$

**51.**

$$\left(-1+\sqrt{3i}\right)^3=\binom{3}{0}(-1)^3+\binom{3}{1}(-1)^2\left(\sqrt{3i}\right)+\binom{3}{2}(-1)^1\left(\sqrt{3i}\right)^2+\binom{3}{3}\left(\sqrt{3i}\right)^3$$

$$=\frac{3!}{0!(3-0)!}\cdot -1+\frac{3!}{1!(3-1)!}1\cdot\sqrt{3i}+\frac{3!}{2!(3-2)!}\cdot -1\cdot -3+\frac{3!}{3!(3-3)!}\cdot -3\sqrt{3i}$$

$$=\frac{\cancel{3!}}{0!\cancel{3!}}(-1)+\frac{3\cdot\cancel{2!}}{1!\cancel{2!}}\sqrt{3i}+\frac{3\cdot\cancel{2!}}{\cancel{2!}1!}\cdot 3+\frac{\cancel{3!}}{\cancel{3!}0!}\cdot -3\sqrt{3i}$$

$$=-1+3\sqrt{3i}+9-3\sqrt{3i}=8$$

**52.**

$$\left(-1-\sqrt{3i}\right)^3=\binom{3}{0}(-1)^3+\binom{3}{1}(-1)^2\left(-\sqrt{3i}\right)+\binom{3}{2}(-1)^1\left(-\sqrt{3i}\right)^2+\binom{3}{3}\left(-\sqrt{3i}\right)^3$$

$$=\frac{3!}{0!(3-0)!}\cdot -1+\frac{3!}{1!(3-1)!}1\cdot -\sqrt{3i}+\frac{3!}{2!(3-2)!}\cdot -1\cdot -3+\frac{3!}{3!(3-3)!}\cdot 3\sqrt{3i}$$

$$=-\frac{\cancel{3!}}{0!\cancel{3!}}-\frac{3\cdot\cancel{2!}}{1!\cancel{2!}}\sqrt{3i}+\frac{3\cdot\cancel{2!}}{\cancel{2!}1!}\cdot 3+\frac{\cancel{3!}}{\cancel{3!}0!}\cdot 3\sqrt{3i}$$

$$=-1-3\sqrt{3i}+9+3\sqrt{3i}=8$$

**53.** $f(x)=x^4+7$;

$$\frac{f(x+h)-f(x)}{h}$$

$$=\frac{(x+h)^4+7-\left(x^4+7\right)}{h}$$

$$=\frac{\binom{4}{0}x^4+\binom{4}{1}x^3h+\binom{4}{2}x^2h^2+\binom{4}{3}xh^3+\binom{4}{4}h^4+7-x^4-7}{h}$$

$$=\frac{\frac{4!}{0!(4-0)!}x^4+\frac{4!}{1!(4-1)!}x^3h+\frac{4!}{2!(4-2)!}x^2h^2+\frac{4!}{3!(4-3)!}xh^3+\frac{4!}{4!(4-4)!}h^4-x^4}{h}$$

$$=\frac{\frac{\cancel{4!}}{0!\cancel{4!}}x^4+\frac{4\cdot\cancel{3!}}{1!\cancel{3!}}x^3h+\frac{4\cdot3\cdot\cancel{2!}}{\cancel{2!}\cdot2\cdot1}x^2h^2+\frac{4\cdot\cancel{3!}}{\cancel{3!}1!}xh^3+\frac{\cancel{4!}}{\cancel{4!}0!}h^4-x^4}{h}$$

$$=\frac{\cancel{x^4}+4x^3h+6x^2h^2+4xh^3+h^4-\cancel{x^4}}{h}$$

$$=\frac{h(4x^3+6x^2h+4xh^2+h^3)}{h}$$

$$=4x^3+6x^2h+4xh^2+h^3$$

**54.** $f(x)=x^5+8$

$$\frac{f(x+h)-f(x)}{h}$$

$$=\frac{(x+h)^5+8-(x^5-8)}{h}$$

$$=\frac{\binom{5}{0}x^5+\binom{5}{1}x^4h+\binom{5}{2}x^3h^2+\binom{5}{3}x^2h^3+\binom{5}{4}xh^4+\binom{5}{5}h^5+8-x^5-8}{h}$$

$$=\frac{\frac{5!}{0!(5-0)!}x^5+\frac{5!}{1!(5-1)!}x^4h+\frac{5!}{2!(5-2)!}x^3h^2+\frac{5!}{3!(5-3)!}x^2h^3+\frac{5!}{4!(5-4)!}xh^4+\frac{5!}{5!(5-5)!}h^5-x^5}{h}$$

$$=\frac{\frac{\cancel{5!}}{0!\cancel{5!}}x^5+\frac{5\cdot\cancel{4!}}{1!\cancel{4!}}x^4h+\frac{5\cdot4\cdot\cancel{3!}}{2!\cancel{3!}}x^3h^2+\frac{5\cdot4\cdot\cancel{3!}}{\cancel{3!}2!}x^2h^3+\frac{5\cdot\cancel{4!}}{\cancel{4!}1!}xh^4+\frac{\cancel{5!}}{\cancel{5!}0!}h^5-x^5}{h}$$

$$= \frac{x^5 + 5x^4h + 10x^3h^2 + 10x^2h^3 + 5xh^4 + h^5 - x^5}{h}$$

$$= \frac{5x^4h + 10x^3h^2 + 10x^2h^3 + 5xh^4 + h^5}{h}$$

$$= \frac{h(5x^4 + 10x^3h + 10x^2h^2 + 5xh^3 + h^4)}{h}$$

$$= 5x^4 + 10x^3h + 10x^2h^2 + 5xh^3 + h^4$$

**55.** We want to find the $(5+1) = 6^{th}$ term.

$$\binom{n}{r} a^{n-r} b^r = \binom{10}{5} \left(\frac{3}{x}\right)^{10-5} \left(\frac{x}{3}\right)^5 = \frac{10!}{5!(10-5)!} \left(\frac{3}{x}\right)^5 \left(\frac{x}{3}\right)^5$$

$$= \frac{\cancel{10} \cdot 9 \cdot 8 \cdot 7 \cdot 6 \cdot \cancel{5!}}{\cancel{5!} \cdot \cancel{5} \cdot 4 \cdot 3 \cdot \cancel{2} \cdot 1} \left(\frac{3}{x}\right)^5 \left(\frac{x}{3}\right)^5 = 252 \cdot \frac{3^5}{x^5} \cdot \frac{x^5}{3^5} = 252$$

**56.** We want to find the $(6+1) = 7^{th}$ term.

$$\binom{n}{r} a^{n-r} b^r = \binom{12}{6} \left(\frac{1}{x}\right)^{12-6} \left(-x^2\right)^6 = \frac{12!}{6!(12-6)!} \left(\frac{1}{x}\right)^6 \left(-x^2\right)^6$$

$$= \frac{\cancel{12} \cdot 11 \cdot 10 \cdot 9 \cdot 8 \cdot 7 \cdot \cancel{6!}}{\cancel{6!} \cdot \cancel{6} \cdot 5 \cdot 4 \cdot 3 \cdot \cancel{2} \cdot 1} \left(\frac{1}{x}\right)^6 \left(-x^2\right)^6 = 924 \cdot \frac{1}{x^6} \cdot x^{12} = 924x^{12-6} = 924x^6$$

**57.** $g(x) = 0.12(x+3)^3 - (x+3)^2 + 3(x+3) + 15$

$$= 0.12\left[\binom{3}{0}x^3 + \binom{3}{1}x^2 \cdot 3 + \binom{3}{2}x \cdot 3^2 + \binom{3}{3} \cdot 3^3\right] - \left(x^2 + 6x + 9\right) + 3x + 9 + 15$$

$$= 0.12\left[\frac{3!}{0!(3-0)!}x^3 + \frac{3!}{1!(3-1)!}3x^2 + \frac{3!}{2!(3-2)!}9x + \frac{3!}{3!(3-3)!} \cdot 27\right] - x^2 - 6x - 9 + 3x + 9 + 15$$

$$= 0.12\left(x^3 + 9x^2 + 27x + 27\right) - x^2 - 3x + 15$$

$$= 0.12x^3 + 1.08x^2 + 3.24x + 3.24 - x^2 - 3x + 15$$

$$= 0.12x^3 + 0.08x^2 + 0.24x + 18.24$$

**59. – 67.** Answers will vary.

$f_1(x) = (x+2)^3$

**69.** $f_2(x) = x^3$

$f_3(x)=x^3+6x^2$

$f_4(x)=x^3+6x^2+12x$

$f_5(x)=x^3+6x^2+12x+8$

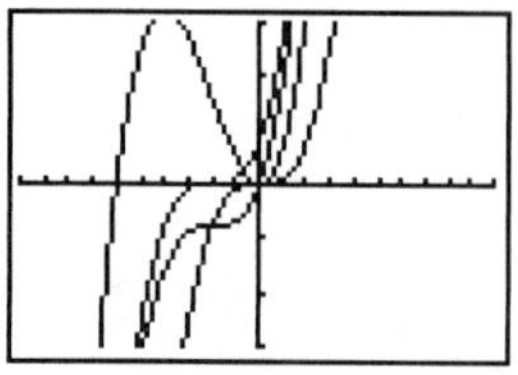

Graphs $f_1$ and $f_5$ are the same. This means that the functions are equivalent. Graphs $f_2$ through $f_4$ are increasingly similar to the graphs of $f_1$ and $f_5$.

**71.** Applying the Binomial Theorem to $(x-1)^3$, we have $a=x,\ b=-1,$ and $n=3$.

$$(x-1)^3=\binom{3}{0}x^3+\binom{3}{1}x^2(-1)+\binom{3}{2}x(-1)^2+\binom{3}{3}(-1)^3$$

$$=\frac{3!}{0!(3-0)!}x^3-\frac{3!}{1!(3-1)!}x^2+\frac{3!}{2!(3-2)!}x(1)-\frac{3!}{3!(3-3)!}$$

$$=\frac{\cancel{3!}}{1\cdot\cancel{3!}}x^3-\frac{3\cdot\cancel{2!}}{1\cdot\cancel{2!}}x^2+\frac{3\cdot\cancel{2!}}{\cancel{2!}1!}x-\frac{\cancel{3!}}{\cancel{3!}0!}=x^3-3x^2+3x-1$$

Graph using the method from Exercises 69 and 70.

$f_1(x)=(x-1)^3$ $\qquad f_2(x)=x^3$

$f_3(x)=x^3+3x^2$ $\qquad f_4(x)=x^3+3x^2+3x$

$f_5(x)=x^3-3x^2+3x-1$

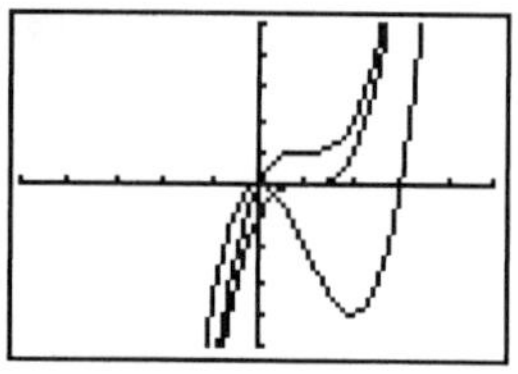

Graphs $f_1$ and $f_5$ are the same. This means that the functions are equivalent. Graphs $f_2$ through $f_4$ are increasingly similar to the graphs of $f_1$ and $f_5$.

**73.** Applying the Binomial Theorem to $(x+2)^6$, we have $a=x,\ b=2,$ and $n=6$.

$$(x+2)^6 = \binom{6}{0}x^6 + \binom{6}{1}x^5 2 + \binom{6}{2}x^4 2^2 + \binom{6}{3}x^3 2^3 + \binom{6}{4}x^2 2^4 + \binom{6}{5}x 2^5 + \binom{6}{6}2^6$$

$$= \frac{6!}{0!(6-0)!}x^6 + \frac{6!}{1!(6-1)!}2x^5 + \frac{6!}{2!(6-2)!}4x^4 + \frac{6!}{3!(6-3)!}8x^3 + \frac{6!}{4!(6-4)!}16x^2$$

$$+ \frac{6!}{5!(6-5)!}32x + \frac{6!}{6!(6-6)!}64$$

$$= \frac{\cancel{6!}}{1\cancel{6!}}x^6 + \frac{6\cdot\cancel{5!}}{1\cdot\cancel{5!}}2x^5 + \frac{6\cdot5\cdot\cancel{4!}}{2\cdot1\cdot\cancel{4!}}4x^4 + \frac{\cancel{6}\cdot5\cdot4\cdot\cancel{3!}}{\cancel{3}\cdot\cancel{2}\cdot1\cdot\cancel{3!}}8x^3 + \frac{6\cdot5\cdot\cancel{4!}}{\cancel{4!}2\cdot1}16x^2 + \frac{6\cdot\cancel{5!}}{\cancel{5!}1}32x + \frac{\cancel{6!}}{\cancel{6!}\cdot1}64$$

$$= x^6 + 6(2x^5) + 15(4x^4) + 20(8x^3) + 15(16x^2) + 6(32x) + 1(64)$$

$$= x^6 + 12x^5 + 60x^4 + 160x^3 + 240x^2 + 192x + 64$$

Graph using the method from Exercises 69 and 70.

$f_1(x) = (x+2)^6$

$f_2(x) = x^6$

$f_3(x) = x^6 + 12x^5$

$f_4(x) = x^6 + 12x^5 + 60x^4$

$f_5(x) = x^6 + 12x^5 + 60x^4 + 160x^3$ .

$f_6(x) = x^6 + 12x^5 + 60x^4 + 160x^3 + 240x^2$

$f_7(x) = x^6 + 12x^5 + 60x^4 + 160x^3 + 240x^2 + 192x$

$f_8(x) = x^6 + 12x^5 + 60x^4 + 160x^3 + 240x^2 + 192x + 64$

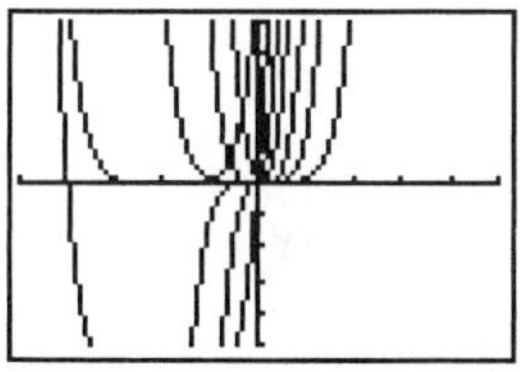

Graphs $f_1$ and $f_8$ are the same. This means that the functions are equivalent. Graphs $f_2$ through $f_7$ are increasingly similar to the graphs of $f_1$ and $f_8$.

**75.** Statement **b.** is true. The Binomial Theorem can be written in condensed form as $(a+b)^n = \sum_{r=0}^{n} \binom{n}{r} a^{n-r} b^r$

Statement **a.** is false. The binomial expansion for $(a+b)^n$ contains $n+1$ terms.

Statement **c.** is false. The sum of the binomial coefficients in $(a+b)^n$ is $2^n$.

Statement **d.** is false. There are values of $a$ and $b$ for which $(a+b)^4 = a^4 + b^4$. Consider $a = 0$ and $b = 1$.

$$(0+1)^4 = 0^4 + 1^4$$
$$(1)^4 = 0+1$$
$$1 = 1$$

**77.** In $(x^2 + y^2)^5$, the term containing $x^4$ is the term in which $a = x^2$ is squared. Applying the Binomial Theorem, the following pattern results. In the first term, $x^2$ is taken to the fifth power. In the second term, $x^2$ is taken to the fourth power. In the third term $x^2$ is taken to the third power. In the fourth term, $x^2$ is taken to the second power. This is the term we are looking for. Applying the Binomial Theorem to $(x^2 + y^2)^5$, we have $a = x^2$, $b = y^2$, and $n = 5$. We are looking for the $r^{th}$ term where $r = 4$.

$$\binom{n}{r-1} a^{n-r+1} b^{r-1}$$
$$= \binom{5}{4-1} (x^2)^{5-4+1} (y^2)^{4-1}$$
$$= \binom{5}{3} (x^2)^2 (y^2)^3$$
$$= \frac{5!}{3!(5-3)!} x^4 y^6$$
$$= \frac{5!}{3!2!} x^4 y^6$$
$$= \frac{5 \cdot 4 \cdot \cancel{3!}}{\cancel{3!} 2 \cdot 1} x^4 y^6$$
$$= 10x^4 y^6$$

**78.**
$$f(a+1) = (a+1)^2 + 2(a+1) + 3$$
$$= a^2 + 2a + 1 + 2a + 2 + 3$$
$$= a^2 + 4a + 6$$

**79.** $f(x) = x^2 + 5x \qquad g(x) = 2x - 3$

$$f(g(x)) = f(2x-3)$$
$$= (2x-3)^2 + 5(2x-3)$$
$$= 4x^2 - 12x + 9 + 10x - 15$$
$$= 4x^2 - 2x - 6$$

$$g(f(x)) = g(x^2 + 5x)$$
$$= 2(x^2 + 5x) - 3$$
$$= 2x^2 + 10x - 3$$

**80.** $\frac{x}{x+3}-\frac{x+1}{2x^2-2x-24}$

$=\frac{x}{x+3}-\frac{x+1}{2(x^2-x-12)}$

$=\frac{x}{x+3}-\frac{x+1}{2(x-4)(x+3)}$

$=\frac{x}{x+3}\cdot\frac{2(x-4)}{2(x-4)}-\frac{x+1}{2(x-4)(x+3)}$

$=\frac{x}{x+3}\cdot\frac{2x-8}{2(x-4)}-\frac{x+1}{2(x-4)(x+3)}$

$=\frac{2x^2-8x}{2(x-4)(x+3)}-\frac{x+1}{2(x-4)(x+3)}$

$=\frac{2x^2-8x-(x+1)}{2(x-4)(x+3)}$

$=\frac{2x^2-8x-x-1}{2(x-4)(x+3)}$

$=\frac{2x^2-9x-1}{2(x-4)(x+3)}$

## Review Exercises

**1.** $a_n=7n-4$

$a_1=7(1)-4=7-4=3$

$a_2=7(2)-4=14-4=10$

$a_3=7(3)-4=21-4=17$

$a_4=7(4)-4=28-4=24$

The first four terms are 3, 10, 17, 24.

**2.** $a_n=(-1)^n\frac{n+2}{n+1}$

$a_1=(-1)^1\frac{1+2}{1+1}=-\frac{3}{2}$

$a_2=(-1)^2\frac{2+2}{2+1}=\frac{4}{3}$

$a_3=(-1)^3\frac{3+2}{3+1}=-\frac{5}{4}$

$a_4=(-1)^4\frac{4+2}{4+1}=\frac{6}{5}$

The first four terms are $-\frac{3}{2},\frac{4}{3},-\frac{5}{4},\frac{6}{5}$.

**3.** $a_n=\frac{1}{(n-1)!}$

$a_1=\frac{1}{(1-1)!}=\frac{1}{0!}=\frac{1}{1}=1$

$a_2=\frac{1}{(2-1)!}=\frac{1}{1!}=\frac{1}{1}=1$

$a_3=\frac{1}{(3-1)!}=\frac{1}{2!}=\frac{1}{2\cdot 1}=\frac{1}{2}$

$a_4=\frac{1}{(4-1)!}=\frac{1}{3!}=\frac{1}{3\cdot 2\cdot 1}=\frac{1}{6}$

The first four terms are $1,1,\frac{1}{2},\frac{1}{6}$.

**4.** $a_n=\frac{(-1)^{n+1}}{2^n}$

$a_1=\frac{(-1)^{1+1}}{2^1}=\frac{(-1)^2}{2}=\frac{1}{2}$

$a_2=\frac{(-1)^{2+1}}{2^2}=\frac{(-1)^3}{4}=-\frac{1}{4}$

$a_3=\frac{(-1)^{3+1}}{2^3}=\frac{(-1)^4}{8}=\frac{1}{8}$

$a_4=\frac{(-1)^{4+1}}{2^4}=\frac{(-1)^5}{16}=-\frac{1}{16}$

The first four terms are $\frac{1}{2},-\frac{1}{4},\frac{1}{8},-\frac{1}{16}$.

**5.** $\sum_{i=1}^{5}(2i^2-3)$

$=(2(1)^2-3)+(2(2)^2-3)+(2(3)^2-3)+(2(4)^2-3)+(2(5)^2-3)+(2(6)^2-3)$

$=(2(1)-3)+(2(4)-3)+(2(9)-3)+(2(16)-3)+(2(25)-3)$

$=(2-3)+(8-3)+(18-3)+(32-3)+(50-3)=-1+5+15+29+47=95$

**6.** $\sum_{i=0}^{4}(-1)^{i+1}i!=(-1)^{0+1}0!+(-1)^{1+1}1!+(-1)^{2+1}2!+(-1)^{3+1}3!+(-1)^{4+1}4!$

$=(-1)^1 1+(-1)^2 1+(-1)^3 2\cdot 1+(-1)^4 3\cdot 2\cdot 1+(-1)^5 4\cdot 3\cdot 2\cdot 1$

$=-1+1-2+6-24=-20$

**7.** $\frac{1}{3}+\frac{2}{4}+\frac{3}{5}+\ldots+\frac{15}{17}=\sum_{i=1}^{15}\frac{i}{i+2}$

**8.** $4^3+5^3+6^3+\ldots+13^3=\sum_{i=4}^{13}i^3$

**9.** $a_1=7$

$a_2=7+4=11$

$a_3=11+4=15$

$a_4=15+4=19$

$a_5=19+4=23$

$a_6=23+4=27$

The first six terms are
7, 11, 15, 19, 23, 27

**10.** $a_1=-4$

$a_2=-4-5=-9$

$a_3=-9-5=-14$

$a_4=-14-5=-19$

$a_5=-19-5=-24$

$a_6=-24-5=-29$

The first six terms are
$-4,-9,-14,-19,-24,-29$.

**11.** $a_1=\frac{3}{2}$

$a_2=\frac{3}{2}-\frac{1}{2}=\frac{2}{2}=1$

$a_3=1-\frac{1}{2}=\frac{1}{2}$

$a_4=\frac{1}{2}-\frac{1}{2}=0$

$a_5=0-\frac{1}{2}=-\frac{1}{2}$

$a_6=-\frac{1}{2}-\frac{1}{2}=-\frac{2}{2}=-1$

The first six terms are
$\frac{3}{2},1,\frac{1}{2},0,-\frac{1}{2},-1.$

**12.** $a_6=5+(6-1)3=5+(5)3$

$=5+15=20$

**13.** $a_{12}=-8+(12-1)(-2)=-8+11(-2)$

$=-8+(-22)=-30$

**14.** $a_{14}=14+(14-1)(-4)=14+13(-4)$

$=14+(-52)=-38$

**15.** $d = -3-(-7) = 4$

$a_n = -7+(n-1)4 = -7+4n-4$

$= 4n-11$

$a_{20} = 4(20)-11 = 80-11 = 69$

**16.** $a_n = 200+(n-1)(-20)$

$= 200-20n+20$

$= 220-20n$

$a_{20} = 220-20(20)$

$= 220-400 = -180$

**17.** $a_n = -12+(n-1)\left(-\frac{1}{2}\right)$

$= -12-\frac{1}{2}n+\frac{1}{2}$

$= -\frac{24}{2}-\frac{1}{2}n+\frac{1}{2}$

$= -\frac{1}{2}n-\frac{23}{2}$

$a_{20} = -\frac{1}{2}(20)-\frac{23}{2}$

$= -\frac{20}{2}-\frac{23}{2} = -\frac{43}{2}$

**18.** $d = 8-15 = -7$

$a_n = 15+(n-1)(-7) = 15-7n+7$

$= 22-7n$

$a_{20} = 22-7(20) = 22-140 = -118$

**19.** First, find $d$.

$d = 12-5 = 7$

Next, find $a_{22}$.

$a_{22} = 5+(22-1)7 = 5+(21)7$

$= 5+147 = 152$

Now, find the sum.

$S_{22} = \frac{22}{2}(5+152) = 11(157) = 1727$

**20.** First, find $d$.

$d = -3-(-6) = 3$

Next, find $a_{15}$.

$a_{15} = -6+(15-1)3 = -6+(14)3$

$= -6+42 = 36$

Now, find the sum.

$S_{15} = \frac{15}{2}(-6+36) = \frac{15}{2}(30) = 225$

**21.** We are given that $a_1 = 300$.

$S_{100} = \frac{100}{2}(3+300)$

$= 50(303) = 15150$

**22.** $$\sum_{i=1}^{16}(3i+2) = (3(1)+2)+(3(2)+2)+(3(3)+2)+...+(3(16)+2)$$
$$= (3+2)+(6+2)+(9+2)+...+(48+2)$$
$$= 5+8+11+...+50$$
$$S_{16} = \frac{16}{2}(5+50) = 8(55) = 440$$

**23.** $$\sum_{i=1}^{25}(-2i+6) = (-2(1)+6)+(-2(2)+6)+(-2(3)+6)+...+(-2(25)+6)$$
$$= (-2+6)+(-4+6)+(-6+6)+...+(-50+6)$$
$$= 4+2+0+...+(-44)$$
$$S_{25} = \frac{25}{2}(4+(-44)) = \frac{25}{2}(-40) = -500$$

**24.** $$\sum_{i=1}^{30}(-5i) = (-5(1))+(-5(2))+(-5(3))+...+(-5(30))$$
$$= -5+(-10)+(-15)+...+(-150)$$
$$S_{30} = \frac{30}{2}(-5+(-150)) = 15(-155) = -2325$$

**25. a.** $$a_n = 20+(n-1)(0.52)$$
$$= 20+0.52n-0.52$$
$$= 0.52n+19.48$$

**b.** $$a_n = 0.52(111)+19.48$$
$$= 57.72+19.48$$
$$= 77.2$$
The percentage of white-collar works in the labor force by the year 2010 is approximately 77.2%.

**26.** $$a_{10} = 31500+(10-1)2300$$
$$= 31500+(9)2300$$
$$= 31500+20700 = 52200$$
$$S_{10} = \frac{10}{2}(31500+52200)$$
$$= 5(83700) = 418500$$
The total salary over a ten-year period is
\$418,500.

**27.** $$a_{35} = 25+(35-1)1 = 25+(34)1$$
$$= 25+34 = 59$$
$$S_{35} = \frac{35}{2}(25+59) = \frac{35}{2}(84) = 1470$$
There are 1470 seats in the theater.

**28.** The first term is 3.
The second term is $3 \cdot 2 = 6$.
The third term is $6 \cdot 2 = 12$.
The fourth term is $12 \cdot 2 = 24$.
The fifth term is $24 \cdot 2 = 48$.

**29.** The first term is $\frac{1}{2}$.
The second term is $\frac{1}{2} \cdot \frac{1}{2} = \frac{1}{4}$.
The third term is $\frac{1}{4} \cdot \frac{1}{2} = \frac{1}{8}$.
The fourth term is $\frac{1}{8} \cdot \frac{1}{2} = \frac{1}{16}$.
The fifth term is $\frac{1}{16} \cdot \frac{1}{2} = \frac{1}{32}$.

**30.** The first term is 16.
The second term is $16 \cdot -\frac{1}{4} = -4$.
The third term is $-4 \cdot -\frac{1}{4} = 1$.
The fourth term is $1 \cdot -\frac{1}{4} = -\frac{1}{4}$.
The fifth term is $-\frac{1}{4} \cdot -\frac{1}{4} = \frac{1}{16}$.

**31.** The first term is –5.
The second term is $-5 \cdot -1 = 5$.
The third term is $5 \cdot -1 = -5$.
The fourth term is $-5 \cdot -1 = 5$.
The fifth term is $5 \cdot -1 = -5$.

**32.** $a_7 = 2(3)^{7-1} = 2(3)^6 = 2(729) = 1458$

**33.**
$$a_6 = 16\left(\frac{1}{2}\right)^{6-1} = 16\left(\frac{1}{2}\right)^5$$
$$= 16\left(\frac{1}{32}\right) = \frac{1}{2}$$

**34.**
$$a_5 = -3(2)^{5-1} = -3(2)^4$$
$$= -3(16) = -48$$

**35.**
$$a_n = a_1 r^{n-1} = 1(2)^{n-1}$$
$$a_8 = 1(2)^{8-1} = 1(2)^7 = 1(128) = 128$$

**36.**
$$a_n = a_1 r^{n-1} = 100\left(\frac{1}{10}\right)^{n-1}$$
$$a_8 = 100\left(\frac{1}{10}\right)^{8-1} = 100\left(\frac{1}{10}\right)^7$$
$$= 100\left(\frac{1}{10000000}\right) = \frac{1}{100000}$$

**37.**
$$d = \frac{-4}{12} = -\frac{1}{3}$$
$$a_n = a_1 r^{n-1} = 12\left(-\frac{1}{3}\right)^{n-1}$$
$$a_8 = 12\left(-\frac{1}{3}\right)^{8-1} = 12\left(-\frac{1}{3}\right)^7$$
$$= 12\left(-\frac{1}{2187}\right) = -\frac{12}{2187} = -\frac{4}{729}$$

**38.**
$$r = \frac{a_2}{a_1} = \frac{-15}{5} = -3$$
$$S_{15} = \frac{5\left(1-(-3)^{15}\right)}{1-(-3)}$$
$$= \frac{5\left(1-(-14348907)\right)}{4}$$
$$= \frac{5(14348908)}{4} = \frac{71744540}{4}$$
$$= 17{,}936{,}135$$

**39.** $r = \frac{a_2}{a_1} = \frac{4}{8} = \frac{1}{2}$

$S_7 = \frac{8\left(1-\left(\frac{1}{2}\right)^7\right)}{1-\frac{1}{2}} = \frac{8\left(1-\frac{1}{128}\right)}{\frac{1}{2}}$

$= \frac{8\left(\frac{128}{128}-\frac{1}{128}\right)}{\frac{1}{2}} = \frac{8\left(\frac{127}{128}\right)}{\frac{1}{2}}$

$= \frac{8}{1}\left(\frac{127}{128}\right) \div \frac{1}{2} = \frac{8}{1}\left(\frac{127}{128}\right) \cdot \frac{2}{1}$

$= \frac{2032}{128} = \frac{127}{8} = 15.875$

**40.** $\sum_{i=1}^{6} 5^i = \frac{5(1-5^6)}{1-5} = \frac{5(1-15625)}{-4}$

$= \frac{5(-15624)}{-4} = 5(3906)$

$= 19,530$

**41.** $\sum_{i=1}^{7} 3(-2)^i = \frac{-6\left(1-(-2)^7\right)}{1-(-2)}$

$= \frac{-6(1-(-128))}{3}$

$= \frac{-6(129)}{3}$

$= -2(129) = -258$

**42.** $\sum_{i=1}^{5} 2\left(\frac{1}{4}\right)^{i-1} = \frac{2\left(1-\left(\frac{1}{4}\right)^5\right)}{1-\frac{1}{4}}$

$= \frac{2\left(1-\frac{1}{1024}\right)}{\frac{3}{4}} = \frac{2\left(\frac{1024}{1024}-\frac{1}{1024}\right)}{\frac{3}{4}}$

$= \frac{2\left(\frac{1023}{1024}\right)}{\frac{3}{4}} = \frac{\frac{2046}{1024}}{\frac{3}{4}} = \frac{2046}{1024} \div \frac{3}{4}$

$= \frac{2046}{1024} \cdot \frac{4}{3} = \frac{682}{256} = \frac{341}{128}$

**43.** $r = \frac{a_2}{a_1} = \frac{3}{9} = \frac{1}{3}$

$S = \frac{9}{1-\frac{1}{3}} = \frac{9}{\frac{2}{3}} = 9 \div \frac{2}{3} = 9 \cdot \frac{3}{2} = \frac{27}{2}$

**44.** $r = \frac{a_2}{a_1} = \frac{-1}{2} = -\frac{1}{2}$

$S = \frac{2}{1-\left(-\frac{1}{2}\right)} = \frac{2}{\frac{3}{2}} = 2 \div \frac{3}{2} = 2 \cdot \frac{2}{3} = \frac{4}{3}$

**45.** $r = \frac{a_2}{a_1} = \frac{4}{-6} = -\frac{2}{3}$

$S = \frac{-6}{1-\left(-\frac{2}{3}\right)} = \frac{-6}{\frac{5}{3}} = -6 \div \frac{5}{3}$

$= -6 \cdot \frac{3}{5} = -\frac{18}{5}$

**46.** $\sum_{i=1}^{\infty} 5(0.8)^i = \frac{4}{1-0.8} = \frac{4}{0.2} = 20$

**47.**

$$0.\overline{6} = \frac{a_1}{1-r} = \frac{\frac{6}{10}}{1-\frac{1}{10}} = \frac{\frac{6}{10}}{\frac{9}{10}} = \frac{6}{10} \div \frac{9}{10}$$

$$= \frac{6}{10} \cdot \frac{10}{9} = \frac{2}{3}$$

**48.**

$$0.\overline{47} = \frac{a_1}{1-r} = \frac{\frac{47}{100}}{1-\frac{1}{100}} = \frac{\frac{47}{100}}{\frac{99}{100}}$$

$$= \frac{47}{100} \div \frac{99}{100} = \frac{47}{100} \cdot \frac{100}{99} = \frac{47}{99}$$

**49.** **a.**

$$r_{1998-1999} = \frac{20.72}{19.96} \approx 1.038$$

$$r_{1999-2000} = \frac{21.51}{20.72} \approx 1.038$$

$$r_{2000-2001} = \frac{22.33}{21.51} \approx 1.038$$

The population is increasing geometrically since dividing the population for each year by the population for the preceding years gives approximately 1.038 for each division.

**b.** $a_n = a_1 r^{n-1} = 19.96(1.038)^{n-1}$

**c.** $a_{10} = 19.96(1.038)^{10-1} \approx 27.92$. In 2008, the population will be approximately 27.92 million.

**50.** $r = 1.06$

$$a_n = a_1 r^{n-1}$$

$$= 32000(1.06)^{n-1}$$

$$a_6 = 32000(1.06)^{6-1}$$

$$= 32000(1.06)^5$$

$$= 32000(1.338226)$$

$$= 42823.22$$

The salary in the sixth year is approximately \$42,823.22.

$$S_6 = \frac{a_1(1-r^n)}{1-r}$$

$$= \frac{32000\left(1-(1.06)^6\right)}{1-1.06}$$

$$= \frac{32000(1-1.418519)}{-0.06}$$

$$= \frac{32000(-0.418519)}{-0.06}$$

$$= 223210.13$$

The total salary over the six years is approximately \$223,210.13.

**51.**

$$A = P\frac{\left(1+\frac{r}{n}\right)^{nt}-1}{\frac{r}{n}}$$

$$= 200\frac{\left(1+\frac{0.10}{12}\right)^{12(18)}-1}{\frac{0.10}{12}}$$

$$= 200\frac{\left(1+\frac{1}{120}\right)^{216}-1}{\frac{1}{120}}$$

$$= 200\frac{\left(\frac{121}{120}\right)^{216}-1}{\frac{1}{120}} = 200\frac{6.004693-1}{\frac{1}{120}}$$

$$= 200\frac{5.004693}{\frac{1}{120}} = 120112.63$$

After 18 years, the value of the account will be approximately $120,112.63.

**52.** $r = 70\% = 0.7$

$a_1 = 4(.7) = 2.8$

$$S = \frac{2.8}{1-0.7} = \frac{2.8}{0.3} = 9.\overline{3}$$

The total spending in the town will be approximately $9.3 million each year.

**53.**
$$\binom{11}{8} = \frac{11!}{8!(11-8)!} = \frac{11\cdot 10\cdot 9\cdot \cancel{8!}}{\cancel{8!}\cdot 3\cdot 2\cdot 1} = 165$$

**54.**
$$\binom{90}{2} = \frac{90!}{2!(90-2)!} = \frac{90\cdot 89\cdot \cancel{88!}}{2\cdot 1\cdot \cancel{88!}}$$
$$= 4005$$

**55.** Applying the Binomial Theorem to $(2x+1)^3$, we have $a = 2x,\ b = 1,$ and $n = 3$.

$$(2x+1)^3 = \binom{3}{0}(2x)^3 + \binom{3}{1}(2x)^2\cdot 1 + \binom{3}{2}(2x)\cdot 1^2 + \binom{3}{3}1^3$$

$$= \frac{3!}{0!(3-0)!}8x^3 + \frac{3!}{1!(3-1)!}4x^2\cdot 1 + \frac{3!}{2!(3-2)!}2x\cdot 1 + \frac{3!}{3!(3-3)!}1$$

$$= \frac{\cancel{3!}}{1\cdot \cancel{3!}}8x^3 + \frac{3\cdot \cancel{2!}}{1\cdot \cancel{2!}}4x^2 + \frac{3\cdot \cancel{2!}}{\cancel{2!}1!}2x + \frac{\cancel{3!}}{\cancel{3!}0!}$$

$$= 8x^3 + 3(4x^2) + 3(2x) + 1 = 8x^3 + 12x^2 + 6x + 1$$

**56.** Applying the Binomial Theorem to $(x^2-1)^4$, we have $a=x^2$, $b=-1$, and $n=4$.

$$(x^2-1)^4=\binom{4}{0}(x^2)^4+\binom{4}{1}(x^2)^3(-1)+\binom{4}{2}(x^2)^2(-1)^2+\binom{4}{3}x^2(-1)^3+\binom{4}{4}(-1)^4$$

$$=\frac{4!}{0!(4-0)!}x^8-\frac{4!}{1!(4-1)!}x^6+\frac{4!}{2!(4-2)!}x^4-\frac{4!}{3!(4-3)!}x^2+\frac{4!}{4!(4-4)!}1$$

$$=\frac{\cancel{4!}}{0!\cancel{4!}}x^8-\frac{4!}{1!3!}x^6+\frac{4!}{2!2!}x^4-\frac{4!}{3!1!}x^2+\frac{\cancel{4!}}{\cancel{4!}0!}$$

$$=1(x^8)-\frac{4\cdot\cancel{3!}}{1\cdot\cancel{3!}}x^6+\frac{4\cdot3\cdot\cancel{2!}}{2\cdot1\cdot\cancel{2!}}x^4-\frac{4\cdot\cancel{3!}}{\cancel{3!}\cdot1}x^2+1$$

$$=x^8-4x^6+6x^4-4x^2+1$$

**57.** Applying the Binomial Theorem to $(x+2y)^5$, we have $a=x$, $b=2y$, and $n=5$.

$$(x+2y)^5=\binom{5}{0}x^5+\binom{5}{1}x^4(2y)+\binom{5}{2}x^3(2y)^2+\binom{5}{3}x^2(2y)^3+\binom{5}{4}x(2y)^4+\binom{5}{5}(2y)^5$$

$$=\frac{\cancel{5!}}{0!\cancel{5!}}x^5+\frac{5!}{1!(5-1)!}2x^4y+\frac{5!}{2!(5-2)!}4x^3y^2+\frac{5!}{3!(5-3)!}8x^2y^3$$

$$+\frac{5!}{4!(5-4)!}16xy^4+\frac{5!}{5!(5-5)!}32y^5$$

$$=1x^5+\frac{5\cdot\cancel{4!}}{1\cdot\cancel{4!}}2x^4y+\frac{5\cdot4\cdot\cancel{3!}}{2\cdot1\cdot\cancel{3!}}4x^3y^2+\frac{5\cdot4\cdot\cancel{3!}}{\cancel{3!}2\cdot1}8x^2y^3+\frac{5\cdot\cancel{4!}}{\cancel{4!}\cdot1}16xy^4+\frac{\cancel{5!}}{\cancel{5!}0!}32y^5$$

$$=x^5+5(2x^4y)+10(4x^3y^2)+10(8x^2y^3)+5(16xy^4)+1(32y^5)$$

$$=x^5+10x^4y+40x^3y^2+80x^2y^3+80xy^4+32y^5$$

**58.** Applying the Binomial Theorem to $(x-2)^6$, we have $a = x,\ b = -2,$ and $n = 6$.

$$\begin{aligned}
&(x-2)^6\\
&=\binom{6}{0}x^6+\binom{6}{1}x^5(-2)+\binom{6}{2}x^4(-2)^2+\binom{6}{3}x^3(-2)^3+\binom{6}{4}x^2(-2)^4+\binom{6}{5}x(-2)^5+\binom{6}{6}(-2)^6\\
&=\frac{6!}{0!(6-0)!}x^6+\frac{6!}{1!(6-1)!}x^5(-2)+\frac{6!}{2!(6-2)!}x^4(-2)^2+\frac{6!}{3!(6-3)!}x^3(-2)^3\\
&\qquad+\frac{6!}{4!(6-4)!}x^2(-2)^4+\frac{6!}{5!(6-5)!}x(-2)^5+\frac{6!}{6!(6-6)!}(-2)^6\\
&=\frac{\cancel{6!}}{1\cancel{6!}}x^6-\frac{6\cdot\cancel{5!}}{1\cdot\cancel{5!}}2x^5+\frac{6\cdot5\cdot\cancel{4!}}{2\cdot1\cdot\cancel{4!}}4x^4-\frac{\cancel{6}\cdot5\cdot4\cdot\cancel{3!}}{\cancel{3\cdot2}\cdot1\cdot\cancel{3!}}8x^3\\
&\qquad+\frac{6\cdot5\cdot\cancel{4!}}{\cancel{4!}2\cdot1}16x^2-\frac{6\cdot\cancel{5!}}{\cancel{5!}1}32x+\frac{\cancel{6!}}{\cancel{6!}\cdot1}64\\
&=x^6-6(2x^5)+15(4x^4)-20(8x^3)+15(16x^2)-6(32x)+1\cdot64\\
&=x^6-12x^5+60x^4-160x^3+240x^2-192x+64
\end{aligned}$$

**59.** $(x^2+3)^8$

First Term $\binom{n}{r-1}a^{n-r+1}b^{r-1}=\binom{8}{1-1}(x^2)^{8-1+1}3^{1-1}=\binom{8}{0}(x^2)^8 3^0$

$$=\frac{8!}{0!(8-0)!}x^{16}\cdot1=\frac{\cancel{8!}}{0!\cancel{8!}}x^{16}=x^{16}$$

Second Term $\binom{n}{r-1}a^{n-r+1}b^{r-1}=\binom{8}{2-1}(x^2)^{8-2+1}3^{2-1}=\binom{8}{1}(x^2)^7 3^1=\frac{8!}{1!(8-1)!}3x^{14}$

$$=\frac{8\cdot\cancel{7!}}{1\cdot\cancel{7!}}3x^{14}=8\cdot3x^{14}=24x^{14}$$

Third Term $\binom{n}{r-1}a^{n-r+1}b^{r-1}=\binom{8}{3-1}(x^2)^{8-3+1}3^{3-1}=\binom{8}{2}(x^2)^6 3^2=\frac{8!}{2!(8-2)!}9x^{12}$

$$=\frac{8\cdot7\cdot\cancel{6!}}{2\cdot1\cdot\cancel{6!}}9x^{12}=28\cdot9x^{12}=252x^{12}$$

**60.** $(x-3)^9$

First Term $\binom{n}{r-1}a^{n-r+1}b^{r-1}=\binom{9}{1-1}x^{9-1+1}(-3)^{1-1}=\binom{9}{0}x^9(-3)^0$

$$=\frac{9!}{0!(9-0)!}x^9\cdot 1=\frac{\cancel{9!}}{0!\cancel{9!}}x^9=x^9$$

Second Term $\binom{n}{r-1}a^{n-r+1}b^{r-1}=\binom{9}{2-1}x^{9-2+1}(-3)^{2-1}=\binom{9}{1}x^8(-3)^1=-\frac{9!}{1!(9-1)!}3x^8$

$$=-\frac{9\cdot\cancel{8!}}{1\cdot\cancel{8!}}3x^8=-9\cdot 3x^8=-27x^8$$

Third Term $\binom{n}{r-1}a^{n-r+1}b^{r-1}=\binom{9}{3-1}x^{9-3+1}(-3)^{3-1}=\binom{9}{2}x^7(-3)^2=\frac{9!}{2!(9-2)!}9x^7$

$$=\frac{9\cdot 8\cdot\cancel{7!}}{2\cdot 1\cdot\cancel{7!}}9x^7=36\cdot 9x^7=324x^7$$

**61.** $(x+2)^5$

Fourth Term $\binom{n}{r-1}a^{n-r+1}b^{r-1}=\binom{5}{4-1}x^{5-4+1}(2)^{4-1}=\binom{5}{3}x^2(2)^3=\frac{5!}{3!(5-3)!}8x^2$

$$=\frac{5!}{3!2!}8x^2=\frac{5\cdot 4\cdot\cancel{3!}}{\cancel{3!}\cdot 2\cdot 1}8x^2=(10)8x^2=80x^2$$

**62.** $(2x-3)^6$

Fifth Term $\binom{n}{r-1}a^{n-r+1}b^{r-1}=\binom{6}{5-1}(2x)^{6-5+1}(-3)^{5-1}=\binom{6}{4}(2x)^2(-3)^4$

$$=\frac{6!}{4!(6-4)!}4x^2(81)=\frac{6!}{4!2!}324x^2$$

$$=\frac{6\cdot 5\cdot\cancel{4!}}{\cancel{4!}\cdot 2\cdot 1}324x^2$$

$$=(15)324x^2=4860x^2$$

## Chapter 11 Test

**1.** $a_n = \dfrac{(-1)^{n+1}}{n^2}$

$a_1 = \dfrac{(-1)^{1+1}}{1^2} = \dfrac{(-1)^2}{1} = \dfrac{1}{1} = 1$

$a_2 = \dfrac{(-1)^{2+1}}{2^2} = \dfrac{(-1)^3}{4} = \dfrac{-1}{4} = -\dfrac{1}{4}$

$a_3 = \dfrac{(-1)^{3+1}}{3^2} = \dfrac{(-1)^4}{9} = \dfrac{1}{9}$

$a_4 = \dfrac{(-1)^{4+1}}{4^2} = \dfrac{(-1)^5}{16} = \dfrac{-1}{16} = -\dfrac{1}{16}$

$a_5 = \dfrac{(-1)^{5+1}}{5^2} = \dfrac{(-1)^6}{25} = \dfrac{1}{25}$

The first five terms are

$1, -\dfrac{1}{4}, \dfrac{1}{9}, -\dfrac{1}{16}, \dfrac{1}{25}$

**2.** $\sum_{i=1}^{5}(i^2+10)$

$= (1^2+10)+(2^2+10)+(3^2+10)+(4^2+10)+(5^2+10)$

$= (1+10)+(4+10)+(9+10)+(16+10)+(25+10)$

$= 11+14+19+26+35 = 105$

**3.** $\dfrac{2}{3}+\dfrac{3}{4}+\dfrac{4}{5}+\ldots+\dfrac{21}{22} = \sum_{i=2}^{21}\dfrac{i}{i+1}$

**4.** $d = 9-4 = 5$

$a_n = 4+(n-1)5 = 4+5n-5$

$= 5n-1$

$a_{12} = 5(12)-1 = 60-1 = 59$

**5.** $d = \dfrac{a_2}{a_1} = \dfrac{4}{16} = \dfrac{1}{4}$

$a_n = a_1 r^{n-1} = 16\left(\dfrac{1}{4}\right)^{n-1}$

$a_{12} = 16\left(\dfrac{1}{4}\right)^{12-1} = 16\left(\dfrac{1}{4}\right)^{11}$

$= 16\left(\dfrac{1}{4194304}\right) = \dfrac{16}{4194304}$

$= \dfrac{1}{262144}$

**6.** $d = -14-(-7) = -7$

$a_{10} = -7+(10-1)(-7)$

$= -7+(9)(-7)$

$= -7+(-63)$

$= -70$

$S_{10} = \dfrac{10}{2}(-7+(-70))$

$= 5(-77)$

$= -385$

**7.** $\sum_{i=1}^{20}(3i-4)$

$= (3(1)-4)+(3(2)-4)+(3(3)-4)+\ldots+(3(20)-4)$

$= (3-4)+(6-4)+(9-4)+\ldots+(60-4)$

$= -1+2+5+\ldots+56$

$S_{20} = \dfrac{20}{2}(-1+56)$

$= 10(55)$

$= 550$

**8.**

$$r = \frac{a_2}{a_1} = \frac{-14}{7} = -2$$

$$S_{10} = \frac{7\left(1-(-2)^{10}\right)}{1-(-2)} = \frac{7(1-1024)}{3} = \frac{7(-1023)}{3} = -2387$$

**9.**

$$\sum_{i=1}^{15}(-2)^i = \frac{-2\left(1-(-2)^{15}\right)}{1-(-2)} = \frac{-2(1-(-32768))}{3} = \frac{-2(32769)}{3} = -21{,}846$$

**10.**

$$r = \frac{1}{2}$$

$$S = \frac{4}{1-\frac{1}{2}} = \frac{4}{\frac{1}{2}} = 4 \div \frac{1}{2} = 4 \cdot \frac{2}{1} = 8$$

**11.**

$$0.\overline{73} = \frac{a_1}{1-r} = \frac{\frac{73}{100}}{1-\frac{1}{100}} = \frac{\frac{73}{100}}{\frac{99}{100}} = \frac{73}{100} \div \frac{99}{100} = \frac{73}{100} \cdot \frac{100}{99} = \frac{73}{99}$$

**12.** $r = 1.04$

$$S_8 = \frac{a_1\left(1-r^n\right)}{1-r} = \frac{30000\left(1-(1.04)^8\right)}{1-1.04} = \frac{30000(1-1.368569)}{-0.04} = \frac{30000(-0.368569)}{-0.04} = 276{,}426.75$$

The total salary over the eight years is approximately $276,426.75

**13.**

$$\binom{9}{2} = \frac{9!}{2!(9-2)!} = \frac{9!}{2!7!} = \frac{9 \cdot 8 \cdot \cancel{7!}}{2 \cdot 1 \cdot \cancel{7!}} = 36$$

**14.** Applying the Binomial Theorem to $\left(x^2-1\right)^5$, we have $a=x^2$, $b=-1$, and $n=5$.

$$\begin{aligned}
&\left(x^2-1\right)^5\\
&=\binom{5}{0}\left(x^2\right)^5+\binom{5}{1}\left(x^2\right)^4(-1)+\binom{5}{2}\left(x^2\right)^3(-1)^2+\binom{5}{3}\left(x^2\right)^2(-1)^3+\binom{5}{4}\left(x^2\right)(-1)^4+\binom{5}{5}(-1)^5\\
&=\frac{\cancel{5!}}{0!\cancel{5!}}x^{10}-\frac{5!}{1!(5-1)!}x^8+\frac{5!}{2!(5-2)!}x^6-\frac{5!}{3!(5-3)!}x^4+\frac{5!}{4!(5-4)!}x^2-\frac{5!}{5!(5-5)!}\\
&=1x^{10}-\frac{5\cdot\cancel{4!}}{1\cdot\cancel{4!}}x^8+\frac{5\cdot4\cdot\cancel{3!}}{2\cdot1\cdot\cancel{3!}}x^6-\frac{5\cdot4\cdot\cancel{3!}}{\cancel{3!}2\cdot1}x^4+\frac{5\cdot\cancel{4!}}{\cancel{4!}\cdot1}x^2-\frac{\cancel{5!}}{\cancel{5!}0!}\\
&=x^{10}-5x^8+10x^6-10x^4+5x^2-1
\end{aligned}$$

**15.** $\left(x+y^2\right)^8$

First Term $\binom{n}{r-1}a^{n-r+1}b^{r-1}=\binom{8}{1-1}x^{8-1+1}\left(y^2\right)^{1-1}=\binom{8}{0}x^8\left(y^2\right)^0$

$$=\frac{8!}{0!(8-0)!}x^8\cdot1=\frac{\cancel{8!}}{0!\cancel{8!}}x^8=x^8$$

Second Term $\binom{n}{r-1}a^{n-r+1}b^{r-1}=\binom{8}{2-1}x^{8-2+1}\left(y^2\right)^{2-1}=\binom{8}{1}x^7\left(y^2\right)^1$

$$=\frac{8!}{1!(8-1)!}x^7y^2=\frac{8\cdot\cancel{7!}}{1\cdot\cancel{7!}}x^7y^2=8x^7y^2$$

Third Term $\binom{n}{r-1}a^{n-r+1}b^{r-1}=\binom{8}{3-1}x^{8-3+1}\left(y^2\right)^{3-1}=\binom{8}{2}x^6\left(y^2\right)^2$

$$=\frac{8!}{2!(8-2)!}x^6y^4=\frac{8\cdot7\cdot\cancel{6!}}{2\cdot1\cdot\cancel{6!}}x^6y^4=28x^6y^4$$

## Cumulative Review Exercises (Chapters 1-11)

**1.** $\sqrt{2x+5}-\sqrt{x+3}=2$

$$\sqrt{x+3}=2+\sqrt{2x+5}$$
$$x+3=\left(2+\sqrt{2x+5}\right)^2$$
$$x+3=4+4\sqrt{2x+5}+2x+5$$
$$x+3=4\sqrt{2x+5}+2x+9$$
$$-x-6=4\sqrt{2x+5}$$
$$(-x-6)^2=16(2x+5)$$
$$x^2+12x+36=32x+80$$
$$x^2-20x-44=0$$
$$(x-22)(x+2)=0$$
$$x=22 \text{ or } x=-2$$

The solution is 22 and the solution set is $\{22\}$.

**2.** $(x-5)^2=-49$

$$x-5=\pm\sqrt{-49}$$
$$x=5\pm\sqrt{-49}$$
$$x=5\pm 7i$$

**3.** $x^2+x>6$

Solve the related quadratic equation.

$$x^2+x=6$$
$$x^2+x-6=0$$
$$(x+3)(x-2)=0$$

Apply the zero product principle.

$x+3=0$ or $x-2=0$

$x=-3$ $\quad$ $x=2$

The boundary points are 3 and 2.

| Test Interval | Test Number | Test | Conclusion |
|---|---|---|---|
| $(-\infty,-3)$ | $-4$ | $(-4)^2+(-4)>6$<br>$12>6$, true | $(-\infty,-3)$ does belong to the solution set. |
| $(-3,2)$ | 0 | $(0)^2+0>6$<br>$0>6$, false | $(-3,2)$ does not belong to the solution set. |
| $(2,\infty)$ | 3 | $(3)^2+3>6$<br>$12>6$, true | $(2,\infty)$ does belong to the solution set. |

The solution set is $(-\infty,-3)\cup(2,\infty)$ or $\{x \mid x<-3 \text{ or } x>2\}$

**4.** $6x-3(5x+2)=4(1-x)$

$$6x-15x-6=4-4x$$
$$-9x-6=4-4x$$
$$-5x=10$$
$$x=-2$$

The solution is $-2$ or $\{-2\}$

**5.**

$$\frac{4}{x-3}-\frac{6}{x+3}=\frac{24}{x^2-9}$$
$$\frac{4}{x-3}-\frac{6}{x+3}=\frac{24}{(x-3)(x+3)}$$
$$(x-3)(x+3)\left(\frac{4}{x-3}-\frac{6}{x+3}\right)=(x-3)(x+3)\left(\frac{24}{(x-3)(x+3)}\right)$$
$$4(x+3)-6(x-3)=24$$
$$4x+12-6x+18=24$$
$$-2x+30=24$$
$$-2x=-6$$
$$x=3$$

Since 3 would make one or more of the denominators in the original equation zero, we disregard it and conclude that there is no solution. The solution set is $\varnothing$ or $\{\ \ \}$.

**6.** $3x+2<4$ and $4-x>1$

$$3x<2 \qquad -x>-3$$
$$x<\frac{2}{3} \qquad x<3$$

The solution set is $\left\{x \middle| x<\frac{2}{3}\right\}$ or $\left(-\infty,\frac{2}{3}\right)$.

**7.**

$$3x-2y+z=7$$
$$2x+3y-z=13$$
$$x-y+2z=-6$$

Multiply the second equation by 2 and add to the third equation.

$$4x+6y-2z=26$$
$$x-y+2z=-6$$
$$\overline{5x+5y=20}$$

Add the first equation to the second equation.

$$\begin{aligned} 3x-2y+z&=7\\ 2x+3y-z&=13\\ \hline 5x+y&=20 \end{aligned}$$

We now have a system of two equations in two variables.

$$5x+5y=20$$
$$5x+y=20$$

Multiply the second equation by $-1$ and add to the first equation.

$$\begin{aligned} 5x+5y&=20\\ -5x-y&=-20\\ \hline 4y&=0\\ y&=0 \end{aligned}$$

Back-substitute 0 for $y$ to find $x$.

$$\begin{aligned} 5x+y&=20\\ 5x+0&=20\\ 5x&=20\\ x&=4 \end{aligned}$$

Back-substitute 4 for $x$ and 0 for $y$ to find $z$.

$$\begin{aligned} 3x-2y+z&=7\\ 3(4)-2(0)+z&=7\\ 12-0+z&=7\\ 12+z&=7\\ z&=-5 \end{aligned}$$

The solution is $(4, 0, -5)$ and the solution set is $\{(4, 0, -5)\}$.

**8.**

$$\begin{aligned} \log_9 x+\log_9(x-8)&=1\\ \log_9(x(x-8))&=1\\ x(x-8)&=9^1\\ x^2-8x&=9\\ x^2-8x-9&=0\\ (x-9)(x+1)&=0 \end{aligned}$$

Apply the zero product principle.

$x-9=0$ or $x+1=0$

$x=9$ $\quad$ $x=-1$

Since we cannot take a log of a negative number, we disregard $-1$ and conclude that the solution is 9 and the solution set is $\{9\}$.

**9.** $2x^2 - 3y^2 = 5$

$3x^2 - 4y^2 = 16$

Multiply the first equation by –3 and the second equation by 2 and solve by addition.

$$-6x^2 + 9y^2 = -15$$
$$\underline{6x^2 + 8y^2 = \ \ 32}$$
$$17y^2 = 17$$
$$y^2 = 1$$
$$y = \pm 1$$

Back-substitute $\pm 1$ for $y$ to find $x$.

$$y = \pm 1$$
$$2x^2 - 3(\pm 1)^2 = 5$$
$$2x^2 - 3(1) = 5$$
$$2x^2 - 3 = 5$$
$$2x^2 = 8$$
$$x^2 = 4$$
$$x = \pm 2$$

The solutions are $(-2,-1), (-2,1)$, $(2,-1)$ and $(2,1)$ and the solution set is $\{(-2,-1),(-2,1),(2,-1),(2,1)\}$.

**10.** $2x^2 - y^2 = -8$

$x - y = 6$

Solve the second equation for $x$.

$$x - y = 6$$
$$x = y + 6$$

Substitute $y + 6$ for $x$.

$$2(y+6)^2 - y^2 = -8$$
$$2(y^2 + 12x + 36) - y^2 = -8$$
$$2y^2 + 24x + 72 - y^2 = -8$$
$$y^2 + 24x + 72 = -8$$
$$y^2 + 24x + 80 = 0$$
$$(y+20)(y+4) = 0$$

Apply the zero product principle.

$y + 20 = 0$ or $y + 4 = 0$

$y = -20$ $\quad$ $y = -4$

Back-substitute –4 and –20 for $y$ to find $x$.

$y=-20$ or $y=-4$

$x=-20+6$  $x=-4+6$

$x=-14$  $x=2$

**11.** $f(x)=(x+2)^2-4$

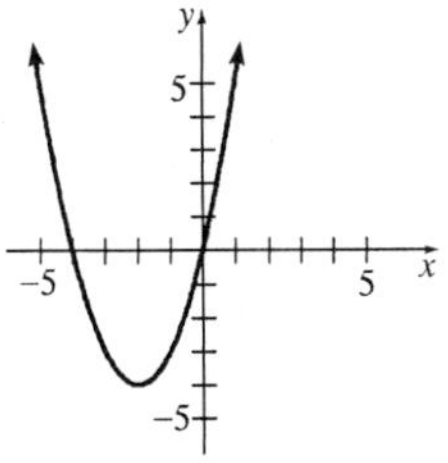

**12.** $y<-3x+5$

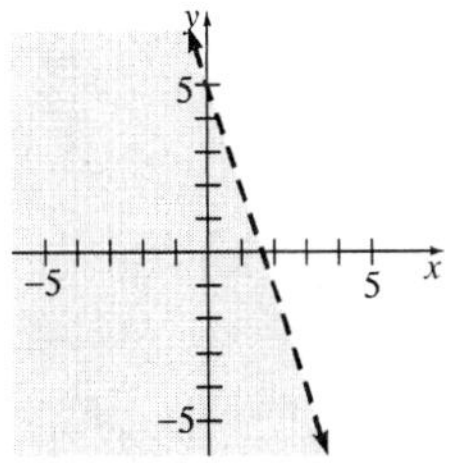

**13.** $f(x)=3^{x-2}$

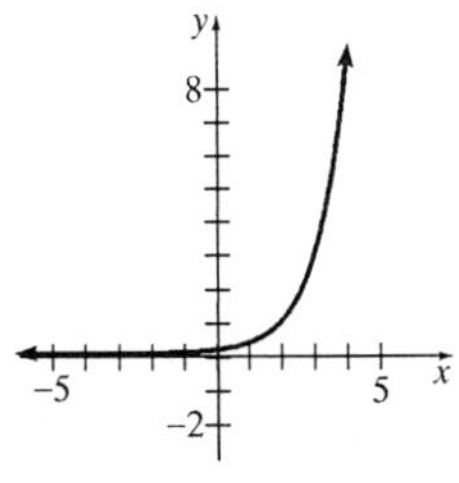

**14.** $\dfrac{x^2}{16}+\dfrac{y^2}{4}=1$

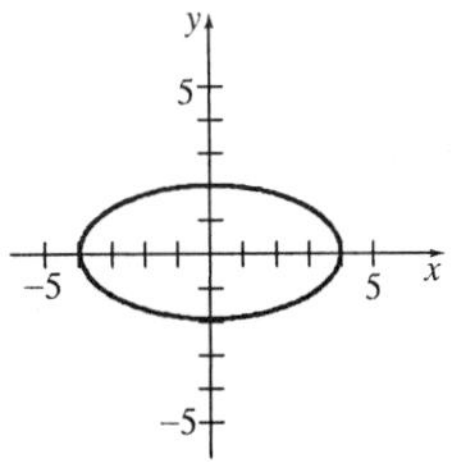

**15.** $x^2-y^2=9$

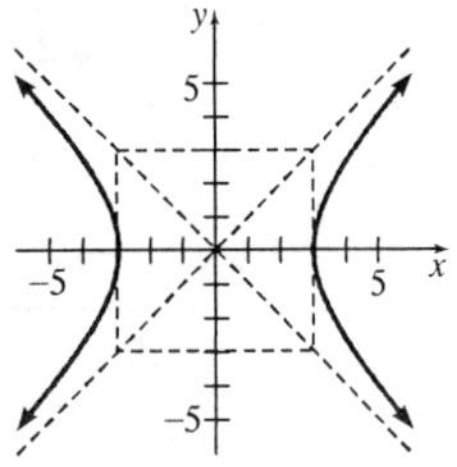

**16.** $\dfrac{2x+1}{x-5}-\dfrac{4}{x^2-3x-10}$

$=\dfrac{2x+1}{x-5}-\dfrac{4}{(x-5)(x+2)}$

$=\dfrac{2x+1}{x-5}\cdot\dfrac{x+2}{x+2}-\dfrac{4}{(x-5)(x+2)}$

$=\dfrac{(2x+1)(x+2)-4}{(x-5)(x+2)}$

$=\dfrac{2x^2+5x+2-4}{(x-5)(x+2)}$

$=\dfrac{2x^2+5x-2}{(x-5)(x+2)}$

**17.** $$\frac{\frac{1}{x-1}+1}{\frac{1}{x+1}-1}=\frac{\frac{1}{x-1}+\frac{x-1}{x-1}}{\frac{1}{x+1}-\frac{x+1}{x+1}}$$

$$=\frac{\frac{1+(x-1)}{x-1}}{\frac{1-(x+1)}{x+1}}$$

$$=\frac{\frac{x}{x-1}}{\frac{-x}{x+1}}=\frac{x}{x-1}\cdot\frac{x+1}{-x}$$

$$=-\frac{x+1}{x-1}$$

**18.** $$\frac{6}{\sqrt{5}-\sqrt{2}}\cdot\frac{\sqrt{5}+\sqrt{2}}{\sqrt{5}+\sqrt{2}}=\frac{6\left(\sqrt{5}+\sqrt{2}\right)}{5-2}$$

$$=\frac{6\left(\sqrt{5}+\sqrt{2}\right)}{3}$$

$$=2\left(\sqrt{5}+\sqrt{2}\right)$$

$$=2\sqrt{5}+2\sqrt{2}$$

**19.** $$8\sqrt{45}+2\sqrt{5}-7\sqrt{20}$$

$$=8\sqrt{9\cdot 5}+2\sqrt{5}-7\sqrt{4\cdot 5}$$

$$=8\cdot 3\sqrt{5}+2\sqrt{5}-7\cdot 2\sqrt{5}$$

$$=24\sqrt{5}+2\sqrt{5}-14\sqrt{5}$$

$$=12\sqrt{5}$$

**20.** $$\frac{5}{\sqrt[3]{2x^2y}}=\frac{5}{2^{\frac{1}{3}}x^{\frac{2}{3}}y^{\frac{1}{3}}}\cdot\frac{2^{\frac{2}{3}}x^{\frac{1}{3}}y^{\frac{2}{3}}}{2^{\frac{2}{3}}x^{\frac{1}{3}}y^{\frac{2}{3}}}$$

$$=\frac{5\sqrt[3]{4xy^2}}{2xy}$$

**21.** $$5ax+5ay-4bx-4by$$

$$=5a(x+y)-4b(x+y)$$

$$=(x+y)(5a-4b)$$

**22.** $$5\log x-\frac{1}{2}\log y=\log x^5-\log y^{\frac{1}{2}}$$

$$=\log\left(\frac{x^5}{y^{\frac{1}{2}}}\right)$$

$$=\log\left(\frac{x^5}{\sqrt{y}}\right)$$

**23.** $$\frac{1}{p}+\frac{1}{q}=\frac{1}{f};$$

$$\frac{1}{p}=\frac{1}{f}-\frac{1}{q}$$

$$\frac{1}{p}=\frac{1}{f}\cdot\frac{q}{q}-\frac{1}{q}\cdot\frac{f}{f}$$

$$\frac{1}{p}=\frac{q}{qf}-\frac{f}{qf}$$

$$\frac{1}{p}=\frac{q-f}{qf}$$

$$p(q-f)=qf$$

$$p=\frac{qf}{q-f}$$

**24.** $$d=\sqrt{\left(6-(-3)\right)^2+\left(-1-(-4)\right)^2}$$

$$=\sqrt{9^2+3^2}=\sqrt{81+9}=\sqrt{90}$$

$$=\sqrt{9\cdot 10}=3\sqrt{10}$$

The distance is $3\sqrt{10}$ units.

**25.** 
$$\sum_{i=2}^{5}(i^3-4)$$
$$=(2^3-4)+(3^3-4)+(4^3-4)+(5^3-4)$$
$$=(8-4)+(27-4)+(64-4)+(125-4)$$
$$=4+23+60+121=208$$

**26.** First, find $d$.

$d=6-2=4$

Next, find $a_{30}$.

$$a_{30}=2+(30-1)4=2+(29)4$$
$$=2+116=118$$

Now, find the sum.

$$S_{30}=\frac{30}{2}(2+118)=15(120)=1800$$

**27.**
$$0.\bar{3}=\frac{a_1}{1-r}=\frac{\frac{3}{10}}{1-\frac{1}{10}}=\frac{\frac{3}{10}}{\frac{9}{10}}=\frac{3}{10}\div\frac{9}{10}$$
$$=\frac{3}{10}\cdot\frac{10}{9}=\frac{1}{3}$$

**28.** Applying the Binomial Theorem to $(2x-y^3)^4$, we have $a=2x,\ b=-y^3,$ and $n=4$.

$$(2x-y^3)^4$$
$$=\binom{4}{0}(2x)^4+\binom{4}{1}(2x)^3(-y^3)+\binom{4}{2}(2x)^2(-y^3)^2+\binom{4}{3}2x(-y^3)^3+\binom{4}{4}(-y^3)^4$$
$$=\frac{4!}{0!(4-0)!}16x^4-\frac{4!}{1!(4-1)!}8x^3y^3+\frac{4!}{2!(4-2)!}4x^2y^6-\frac{4!}{3!(4-3)!}2xy^9+\frac{4!}{4!(4-4)!}y^{12}$$
$$=\frac{\cancel{4!}}{0!\cancel{4!}}16x^4-\frac{4!}{1!3!}8x^3y^3+\frac{4!}{2!2!}4x^2y^6-\frac{4!}{3!1!}2xy^9+\frac{\cancel{4!}}{\cancel{4!}0!}y^{12}$$
$$=1(16x^4)-\frac{4\cdot\cancel{3!}}{1\cdot\cancel{3!}}8x^3y^3+\frac{4\cdot3\cdot\cancel{2!}}{2\cdot1\cdot\cancel{2!}}4x^2y^6-\frac{4\cdot\cancel{3!}}{\cancel{3!}\cdot1}2xy^9+y^{12}$$
$$=16x^4-4(8x^3y^3)+6(4x^2y^6)-4(2xy^9)+y^{12}$$
$$=16x^4-32x^3y^3+24x^2y^6-8xy^9+y^{12}$$

**29.** $f(x) = \dfrac{2}{x^2 + 2x - 15}$

Set the denominator equal to 0 to find the domain:

$x^2 + 2x - 15 = 0$

$(x+5)(x-3) = 0$

$x = -5,\ x = 3$

So, $\{x \mid x$ is a real number and $x \neq -5$ and $x \neq 3\}$

**30.** $f(x) = \sqrt{2x-6}$;

We can not take the square root of a negative number.

$2x - 6 \geq 0$

$2x \geq 6$

$x \geq 3$

So, $\{x \mid x \geq 3\}$ or $[3, \infty)$.

**31.** $f(x) = \ln(1-x)$

We can only take the ln of positive numbers.

$1 - x > 0$

$-x > -1$

$x < 1$

So, $\{x \mid x < 1\}$ or $(-\infty, 1)$.

**32.** Let $w =$ width of the rectangle.

Then $l = 2w + 2$.

The perimeter of a rectangle is given by $P = 2w + 2l$.

Then $P = 2w + 2l$

$22 = 2w + 2(2w + 2)$

$22 = 2w + 4w + 4$

$22 = 6w + 4$

$18 = 6w$

$w = 3$

Then $l = 2w + 2 = 2(3) + 2 = 8$

The dimension of the rectangle is 8 feet by 3 feet.

**33.** $A = P(1+r)^t$

$19610 = P(1+0.06)^1$

$19610 = 1.06P$

$P = 18500$

Your salary before the raise is \$18,500.

**34.** $F(t) = 1 - k\ln(t+1)$

$\dfrac{1}{2} = 1 - k\ln(3+1)$

$\dfrac{1}{2} = 1 - k\ln 4$

$k\ln 4 = \dfrac{1}{2}$

$k = \dfrac{1}{2\ln 4} \approx 0.3607$

Then $F(t) \approx 1 - 0.3607\ln(t+1)$

$F(6) \approx 1 - 0.3607\ln(6+1)$

$\approx 1 - 0.3607\ln(7)$

$\approx 1 - 0.7019$

$\approx 0.298$ or $\dfrac{298}{1000}$

# Chapter 11
# More on Polynomial and Rational Functions

## 11.1 Exercise Set

**1.** $f(x) = 5x^3 + 7x^2 - x + 9$

Since $f$ is an odd-degree polynomial and since the leading coefficient, 5, is positive, the graph falls to the left and rises to the right.

**3.** $f(x) = 5x^4 + 7x^2 - x + 9$

Since $f$ is an even-degree polynomial and since the leading coefficient, 5, is positive, the graph rises to the left and rises to the right.

**5.** $f(x) = -5x^4 + 7x^2 - x + 9$

Since $f$ is an even-degree polynomial and since the leading coefficient, $-5$, is negative, the graph falls to the left and falls to the right.

**7.** $f(x) = 2(x-5)(x+4)^2$

$2(x-5)(x+4)^2 = 0$

$(x-5)(x+4)^2 = 0$

Apply the zero-product principle:

$x - 5 = 0$ or $(x+4)^2 = 0$

$x = 5$ $\quad x + 4 = 0$

$x = -4$

The zeros of $f$ are 5 and $-4$. Since the factor $x-1$ only occurs once (exponent 1), the zero 5 has multiplicity 1, Since the multiplicity is odd, the graph crosses the $x$-axis at 5. Since the factor $x+4$ occurs twice (exponent 2), the zero $-4$ has multiplicity 2. Since the multiplicity is even, the graph touches the $x$-axis and turns around at 2.

**9.** $f(x) = 4(x-3)(x+6)^3$

$4(x-3)(x+6)^3 = 0$

$(x-3)(x+6)^3 = 0$

Apply the zero-product principle:

$x - 3 = 0$ or $(x+6)^2 = 0$

$x = 3$ $\quad x + 6 = 0$

$x = -6$

The zeros of $f$ are 3 and $-6$. Since the factor $x-3$ only occurs once (exponent 1), the zero 3 has multiplicity 1. Since the multiplicity is odd, the graph crosses the $x$-axis at 3. Since the factor $x+6$ occurs three times (exponent 3), the zero $-6$ has multiplicity 3. Since the multiplicity is odd, the graph crosses the $x$-axis at $-6$.

**11.** $f(x) = x^3 - 2x^2 + x$

$x^3 - 2x^2 + x = 0$

$x(x^2 - 2x + 1) = 0$

$x(x-1)^2 = 0$

Apply the zero-product principle:

$x = 0$ or $(x-1)^2 = 0$

$x - 1 = 0$

$x = 1$

The zeros of $f$ are 0 and 1. Since the factor $x$ only occurs once (exponent 1), the zero 0 has multiplicity 1, Since the multiplicity is odd, the graph crosses the $x$-axis at 0. Since the factor $x-1$ occurs twice (exponent 2), the zero 1 has multiplicity 2. Since the multiplicity is even, the graph touches the $x$-axis and turns around at 1.

**13.** $f(x)=x^3+7x^2-4x-28$

$$x^3+7x^2-4x-28=0$$
$$x^2(x+7)-4(x+7)=0$$
$$(x^2-4)(x+7)=0$$
$$(x+2)(x-2)(x+7)=0$$

Apply the zero-product principle:

$x+2=0$ or $x-2=0$ or $x+7=0$

$x=-2$ $\quad$ $x=2$ $\quad$ $x=-7$

The zeros of *f* are $-7$, $-2$, and 2. Since the each factor occurs only once (exponent 1), all three zeros have multiplicity 1, and the graph crosses the *x*-axis at $-7$, $-2$, and 2.

**15.** $f(x)=x^3+x$

$$f(-x)=(-x)^3+(-x)$$
$$=-x^3-x=-(x^3+x)$$

Since $f(-x)=-f(x)$, the function is odd.

**17.** $g(x)=x^2+x$

$$g(-x)=(-x)^2+(-x)=x^2-x$$

Since $f(-x)\neq f(x)$ and since $f(-x)\neq -f(x)$, the function is neither even nor odd.

**19.** $h(x)=x^2-x^4$

$$h(-x)=(-x)^2-(-x)^4=x^2-x^4$$

Since $h(-x)=h(x)$, the function is even.

**21.** $f(x)=x^2-x^4+1$

$$f(-x)=(-x)^2-(-x)^4+1$$
$$=x^2-x^4+1$$

Since $f(-x)=f(x)$, the function is even.

**23.** $f(x)=\frac{1}{5}x^6-3x^2$

$$f(-x)=\frac{1}{5}(-x)^6-3(-x)^2$$
$$=\frac{1}{5}x^6-3x^2$$

Since $f(-x)=f(x)$, the function is even.

**25.** $f(x)=x(1-x^2)$

$$f(-x)=(-x)\left[1-(-x)^2\right]=-x(1-x^2)$$

Since $f(-x)=-f(x)$, the function is odd.

**27.** **a.** This function cannot be a polynomial function. We can tell because the end behavior of the graph does not behave like that of polynomial functions.

**b.** The graph is symmetric with respect to the *y*-axis, so the function is even.

**29.** **a.** This function could be a polynomial function.

**b.** The graph is symmetric with respect to the origin, so the function is odd.

**31.** $f(x)=x^3+2x^2-x-2$

**a.** Since *f* is an odd-degree polynomial and since the leading coefficient, 1, is positive, the graph falls to the left and rises to the right.

**b.**
$$x^3+2x^2-x-2=0$$
$$x^2(x+2)-1(x+2)=0$$
$$(x^2-1)(x+2)=0$$
$$(x+1)(x-1)(x+2)=0$$

Apply the zero-product principle:

$x+1=0$ or $x-1=0$ or $x+2=0$

$x=-1$ $\quad$ $x=1$ $\quad$ $x=-2$

The zeros of *f* are $-2$, $-1$, and 1. The graph crosses the *x*-axis at $-2$, $-1$, and 1 since all three have multiplicity 1.

**c.** $f(0) = 0^3 + 2\cdot 0^2 - 0 - 2 = -2$

**d.** $f(-x) = (-x)^3 + 2(-x)^2 - (-x) - 2$

$= -x^3 + 2x^2 + x - 2$

Since $f(-x) \neq f(x)$ and since $f(-x) \neq -f(x)$, the function is neither even nor odd, and the graph is neither symmetric with respect to the *y*-axis nor the origin.

**e.**

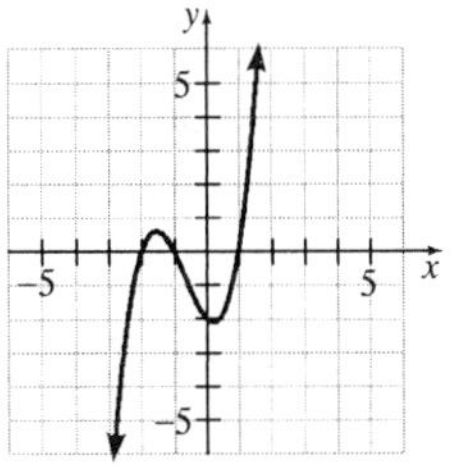


**33.** $f(x) = x^4 - 9x^2$

**a.** Since *f* is an even-degree polynomial and since the leading coefficient, 1, is positive, the graph rises to the left and rises to the right.

**b.**
$$x^4 - 9x^2 = 0$$
$$x^2(x^2 - 9) = 0$$
$$x^2(x+3)(x-3) = 0$$

Apply the zero-product principle:

$x^2 = 0$ or $x + 3 = 0$ or $x - 3 = 0$

$x = 0$ $\quad x = -3$ $\quad x = 3$

The zeros of *f* are $-3$, 0, and 3. The graph crosses the *x*-axis at $-3$ and 3 since both have multiplicity 1. The graph touches the *x*-axis and turns around at 0 since it has multiplicity 2.

**c.** $f(0) = 0^4 - 9\cdot 0^2 = 0$

**d.** $f(-x) = (-x)^4 - 9(-x)^2 = x^4 - 9x^2$

Since $f(-x) = f(x)$, the graph is even and is symmetric with respect to the *y*-axis.

**e.**

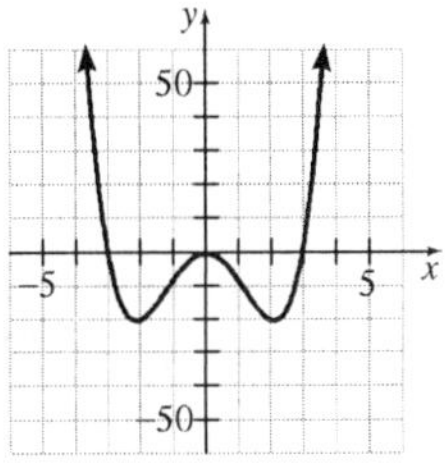


**35.** $f(x) = -x^4 + 16x^2$

**a.** Since *f* is an even-degree polynomial and since the leading coefficient, $-1$, is negative, the graph falls to the left and falls to the right.

**b.**
$$-x^4 + 16x^2 = 0$$
$$x^4 - 16x^2 = 0$$
$$x^2(x^2 - 16) = 0$$
$$x^2(x+4)(x-4) = 0$$

Apply the zero-product principle:

$x^2 = 0$ or $x + 4 = 0$ or $x - 4 = 0$

$x = 0$ $\quad x = -4$ $\quad x = 4$

The zeros of *f* are $-4$, 0, and 4. The graph crosses the *x*-axis at $-4$ and 4 since they both have multiplicity 1. The graph touches the *x*-axis and turns around at 0 since it has multiplicity 2.

**c.** $f(0) = -0^4 + 16\cdot 0^2 = 0$

**d.** $f(-x) = -(-x)^4 + 16(-x)^2$

$= -x^4 + 16x^2$

Since $f(-x) = f(x)$, the graph is even and is symmetric with respect to the *y*-axis.

**e.**

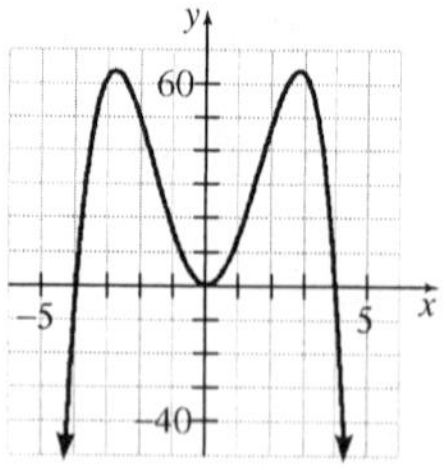


**37.** $f(x) = x^4 - 2x^3 + x^2$

**a.** Since *f* is an even-degree polynomial and since the leading coefficient, 1, is positive, the graph rises to the left and rises to the right.

**b.**
$$x^4 - 2x^3 + x^2 = 0$$
$$x^2(x^2 - 2x + 1) = 0$$
$$x^2(x-1)^2 = 0$$

Apply the zero-product principle:

$x^2 = 0$ or $(x-1)^2 = 0$

$x = 0$ $\qquad x - 1 = 0$

$\qquad x = 1$

The zeros of *f* are 0 and 1. The graph touches the *x*-axis and turns around at both 0 and 1 since both have multiplicity 2.

**c.** $f(0) = 0^4 - 2 \cdot 0^3 + 0^2 = 0$

**d.** $f(-x) = (-x)^4 - 2(-x)^3 + (-x)^2$
$= x^4 + 2x^3 + x^2$

Since $f(-x) \neq f(x)$ and since $f(-x) \neq -f(x)$, the function is neither even nor odd, and the graph is neither symmetric with respect to the *y*-axis nor the origin.

**e.**

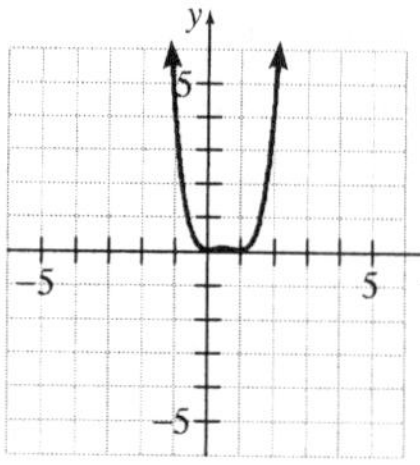


**39.** $f(x) = -2x^4 + 4x^3$

**a.** Since *f* is an even-degree polynomial and since the leading coefficient, $-2$, is negative, the graph falls to the left and falls to the right.

**b.**
$$-2x^4 + 4x^3 = 0$$
$$-2x^3(x-2) = 0$$

Apply the zero-product principle:

$-2x^3 = 0$ or $x - 2 = 0$

$x^3 = 0$ $\qquad x = 2$

$x = 0$

The zeros of *f* are 0 and 2. The graph crosses the *x*-axis at both 0 and 2 since they have multiplicities 3 and 1, respectively.

**c.** $f(0) = -2 \cdot 0^4 + 4 \cdot 0^3 = 0$

**d.** $f(-x) = -2(-x)^4 + 4(-x)^3$
$= -2x^4 - 4x^3$

Since $f(-x) \neq f(x)$ and since $f(-x) \neq -f(x)$, the function is neither even nor odd, and the graph is neither symmetric with respect to the *y*-axis nor the origin.

**e.**

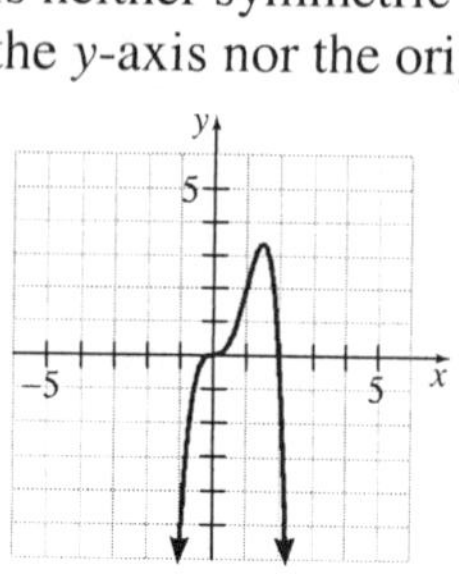

**41.** $f(x) = 6x^3 - 9x - x^5 = -x^5 + 6x^3 - 9x$

**a.** Since $f$ is an odd-degree polynomial and since the leading coefficient, $-1$, is negative, the graph rises to the left and falls to the right.

**b.**
$$-x^5 + 6x^3 - 9x = 0$$
$$-x(x^4 - 6x^2 + 9) = 0$$
$$-x(x^2 - 3)^2 = 0$$
$$-x\left[(x+\sqrt{3})(x-\sqrt{3})\right]^2 = 0$$
$$-x(x+\sqrt{3})^2(x-\sqrt{3})^2 = 0$$

Apply the zero-product principle:

$-x = 0$ or $x + \sqrt{3} = 0$ or $x - \sqrt{3} = 0$

$x = 0$ $\quad x = -\sqrt{3}$ $\quad x = \sqrt{3}$

The zeros of $f$ are $-\sqrt{3}$, 0, and $\sqrt{3}$. The graph crosses the $x$-axis at 0 since it has multiplicity 1. The graph touches the $x$-axis and turns around at both $-\sqrt{3}$ and $\sqrt{3}$ since both have multiplicity 2.

**c.** $f(0) = -0^5 + 6 \cdot 0^3 - 9 \cdot 0 = 0$

**d.**
$$f(-x) = -(-x)^5 + 6(-x)^3 - 9(-x)$$
$$= x^5 - 6x^3 + 9x$$
$$= -(-x^5 + 6x^3 - 9x)$$

Since $f(-x) = -f(x)$, the function is odd, and the graph is symmetric with respect to the origin.

**e.**

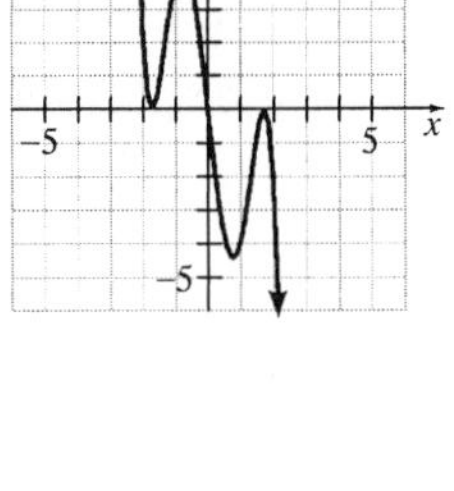


**43.** $f(x) = 3x^2 - x^3 = -x^3 + 3x^2$

**a.** Since $f$ is an odd-degree polynomial and since the leading coefficient, $-1$, is negative, the graph rises to the left and falls to the right.

**b.**
$$-x^3 + 3x^2 = 0$$
$$x^3 - 3x^2 = 0$$
$$x^2(x-3) = 0$$

Apply the zero-product principle:

$x^2 = 0$ or $x - 3 = 0$

$x = 0$ $\quad x = 3$

The zeros of $f$ are 0 and 3. The graph crosses the $x$-axis at 3 since it has multiplicity 1. The graph touches the $x$-axis and turns around at 0 since it has multiplicity 2.

**c.** $f(0) = -0^3 + 3 \cdot 0^2 = 0$

**d.** $f(x) = 3(-x)^2 - (-x)^3 = 3x^2 + x^3$

Since $f(-x) \neq f(x)$ and since $f(-x) \neq -f(x)$, the function is neither even nor odd, and the graph is neither symmetric with respect to the $y$-axis nor the origin.

**e.**

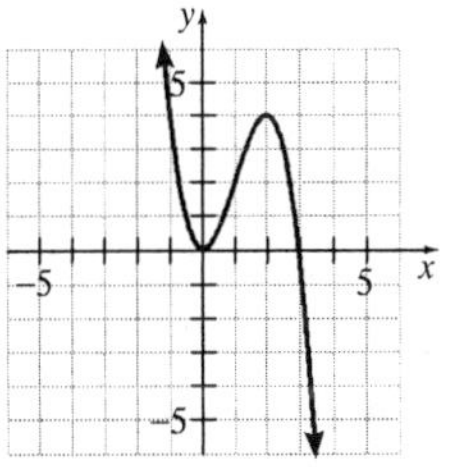


**45.** $f(x) = -3(x-1)^2(x^2 - 4)$

**a.** Since $f$ is an even-degree polynomial ($4^{th}$ degree) and since the leading coefficient, $-3$, is negative, the graph falls to the left and falls to the right.

**b.** $-3(x-1)^2(x^2-4)=0$

$(x-1)^2(x^2-4)=0$

$(x-1)^2(x+2)(x-2)=0$

Apply the zero-product principle:

$(x-1)^2=0$ or $x+2=0$ or $x-2=0$

$x-1=0$ $\quad x=-2$ $\quad x=2$

$x=1$

The zeros of *f* are $-2$, 1, and 2. The graph crosses the *x*-axis at both $-2$ and 2 since both have multiplicity 1. The graph touches the *x*-axis and turns around at 1 since it has multiplicity 2.

**c.** $f(0)=-3(0-1)^2(0^2-4)=12$

**d.** Since $f(-x)\neq f(x)$ and since $f(-x)\neq -f(x)$, the function is neither even nor odd, and the graph is neither symmetric with respect to the *y*-axis nor the origin.

**e.**

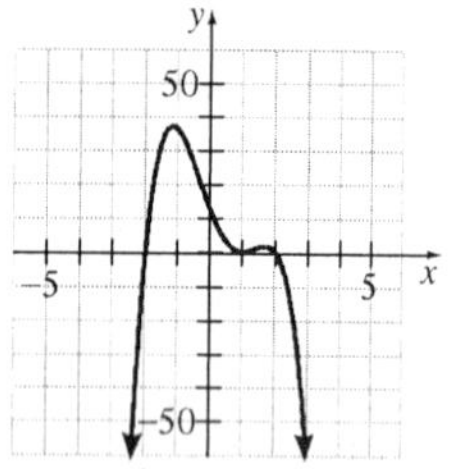


**47. a.** The *x*-intercepts of the graph are $-2$, 1, and 4, so they are the zeros. Since the graph actually crosses the *x*-axis at all three places, all three have odd multiplicity.

**b.** Since the graph has two turning points, the function must be at least of degree 3. Since $-2$, 1, and 4 are the zeros, $x+2$, $x-1$, and $x-4$ are factors of the function. The lowest odd multiplicity is 1. From the end behavior, we can tell that the leading coefficient must be positive. Thus, the function is

$f(x)=(x+2)(x-1)(x-4)$.

**c.** $f(0)=(0+2)(0-1)(0-4)=8$

**48. a.** The *x*-intercepts of the graph are $-3$, 2, and 5, so they are the zeros. Since the graph actually crosses the *x*-axis at all three places, all three have odd multiplicity.

**b.** Since the graph has two turning points, the function must be at least of degree 3. Since $-3$, 2, and 5 are the zeros, $x+3$, $x-2$, and $x-5$ are factors of the function. The lowest odd multiplicity is 1. From the end behavior, we can tell that the leading coefficient must be positive. Thus, the function is

$f(x)=(x+3)(x-2)(x-5)$.

**c.** $f(0)=(0+3)(0-2)(0-5)=30$

**49. a.** The *x*-intercepts of the graph are $-1$ and 3, so they are the zeros. Since the graph crosses the *x*-axis at $-1$, it has odd multiplicity. Since the graph touches the *x*-axis and turns around at 3, it has even multiplicity.

**b.** Since the graph has two turning points, the function must be at least of degree 3. Since $-1$ and 3 are the zeros, $x+1$ and $x-3$ are factors of the function. The lowest odd multiplicity is 1, and the lowest even multiplicity is 2. From the end behavior, we can tell that the leading coefficient must be positive. Thus, the function is

$f(x)=(x+1)(x-3)^2$.

**c.** $f(0)=(0+1)(0-3)^2=9$

**50.** **a.** The $x$-intercepts of the graph are $-2$ and 1, so they are the zeros. Since the graph crosses the $x$-axis at $-2$, it has odd multiplicity. Since the graph touches the $x$-axis and turns around at 1, it has even multiplicity.

**b.** Since the graph has two turning points, the function must be at least of degree 3. Since $-2$ and 1 are the zeros, $x+2$ and $x-1$ are factors of the function. The lowest odd multiplicity is 1, and the lowest even multiplicity is 2. From the end behavior, we can tell that the leading coefficient must be positive. Thus, the function is

$$f(x)=(x+2)(x-1)^2.$$

**c.** $f(0)=(0+2)(0-1)^2=2$

**51.** **a.** The $x$-intercepts of the graph are $-3$ and 2, so they are the zeros. Since the graph touches the $x$-axis and turns around at both $-3$ and 2, both have even multiplicity.

**b.** Since the graph has three turning points, the function must be at least of degree 4. Since $-3$ and 2 are the zeros, $x+3$ and $x-2$ are factors of the function. The lowest even multiplicity is 2. From the end behavior, we can tell that the leading coefficient must be negative. Thus, the function is

$$f(x)=-(x+3)^2(x-2)^2.$$

**c.** $f(0)=-(0+3)^2(0-2)^2=-36$

**52.** **a.** The $x$-intercepts of the graph are $-1$ and 4, so they are the zeros. Since the graph touches the $x$-axis and turns around at both $-1$ and 4, both have even multiplicity.

**b.** Since the graph has two turning points, the function must be at least of degree 3. Since $-1$ and 4 are the zeros, $x+1$ and $x-4$ are factors of the function. The lowest even multiplicity is 2. From the end behavior, we can tell that the leading coefficient must be negative. Thus, the function is

$$f(x)=-(x+1)^2(x-4)^2.$$

**c.** $f(0)=-(0+1)^2(0-4)^2=-16$

**53.** **a.** The $x$-intercepts of the graph are $-2$, $-1$, and 1, so they are the zeros. Since the graph crosses the $x$-axis at $-1$ and 1, they both have odd multiplicity. Since the graph touches the $x$-axis and turns around at $-2$, it has even multiplicity.

**b.** Since the graph has five turning points, the function must be at least of degree 6. Since $-2$, $-1$, and 1 are the zeros, $x+2$, $x+1$, and $x-1$ are factors of the function. The lowest even multiplicity is 2, and the lowest odd multiplicity is 1. However, to reach degree 6, one of the odd multiplicities must be 3. From the end behavior, we can tell that the leading coefficient must be positive. The function is

$$f(x)=(x+2)^2(x+1)(x-1)^3.$$

**c.** $f(0)=(0+2)^2(0+1)(0-1)^3=-4$

**54.** **a.** The $x$-intercepts of the graph are $-2$, $-1$, and 1, so they are the zeros. Since the graph crosses the $x$-axis at $-2$ and 1, they both have odd multiplicity. Since the graph touches the $x$-axis and turns around at $-1$, it has even multiplicity.

**b.** Since the graph has five turning points, the function must be at least of degree 6. Since $-2$, $-1$, and 1 are the zeros, $x+2$, $x+1$, and $x-1$ are factors of the function. The lowest even multiplicity is 2, and the lowest odd multiplicity is 1. However, to reach degree 6, one of the odd multiplicities must be 3. From the end behavior, we can tell that the leading coefficient must be positive. The function is

$$f(x)=(x+2)(x+1)^2(x-1)^3.$$

**c.** $f(0)=(0+2)(0+1)^2(0-1)^3=-2$

**55. a.** Marital satisfaction for wives was decreasing between stages I and IV and between stages VI and VII.

**b.** Marital satisfaction for wives was increasing between stages IV and VI and between stages VII and VIII.

**c.** Three turning points are shown in the graph for wives.

**d.** Since there are three turning points, the degree of the polynomial function of best fit would be 4.

**e.** The leading coefficient would be positive because the graph rises to the left and rises to the right.

**57. – 67.** Answers will vary.

**69.** Answers will vary depending on which exercises are selected.

**71.** $f(x)=x^3-7x^2-4x+28$

$$x^3-7x^2-4x+28=0$$
$$x^2(x-7)-4(x-7)=0$$
$$(x^2-4)(x-7)=0$$
$$(x+2)(x-2)(x-7)=0$$

Apply the zero-product principle:

$$x+2=0 \quad \text{or} \quad x-2=0 \quad \text{or} \quad x+7=0$$
$$x=-2 \qquad x=2 \qquad x=-7$$

The zeros of $f$ are $-2$, 2, and 7. All three have multiplicity 1.

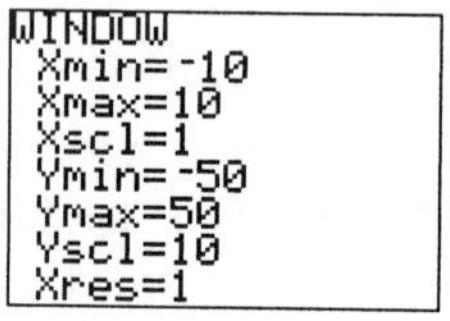


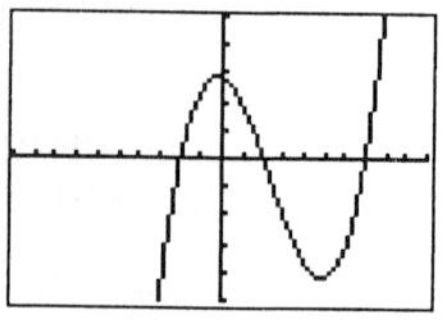

The graph does in fact cross the $x$-axis at $-2$, 2, and 7.

**73.** Statement **c.** is true.

For part **a.**, the function given will rise to the left and fall to the right.

For part **b.**, the model should only be used to describe phenomena for limited periods of time.

For part **d.**, the graph of a function with origin symmetry must rise to one side and fall to the other.

**75.** Answer will vary. An example is $f(x)=x^2(x-2)=x^3-2x^2$

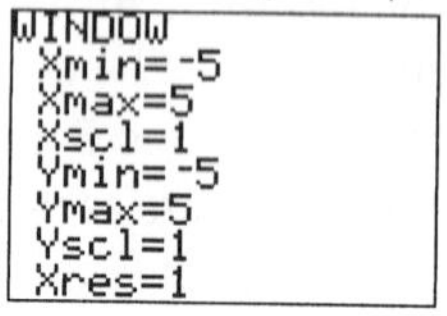


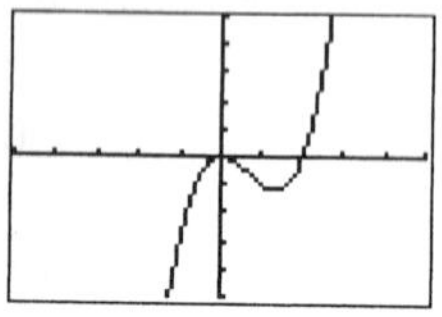

**76.** $$\frac{1}{4x}-\frac{3}{4}=\frac{7}{x}$$

$$4x\left(\frac{1}{4x}-\frac{3}{4}\right)=4x\left(\frac{7}{x}\right)$$
$$1-3x=28$$
$$-3x=27$$
$$x=-9$$

The solution is $-9$, and the solution set is $\{-9\}$.

**77.** $8\sqrt{45}-3\sqrt{80}=8\sqrt{9\cdot 5}-3\sqrt{16\cdot 5}$
$=8\cdot 3\sqrt{5}-3\cdot 4\sqrt{5}$
$=24\sqrt{5}-12\sqrt{5}=12\sqrt{5}$

**78.** $x^2+2x=2$

$x^2+2x-2=0$

Apply the quadratic formula:
$a=1 \quad b=2 \quad c=-2$

$$x=\frac{-2\pm\sqrt{2^2-4(1)(-2)}}{2(1)}$$
$$=\frac{-2\pm\sqrt{12}}{2}$$
$$=\frac{-2\pm\sqrt{4\cdot 3}}{2}$$
$$=\frac{-2\pm 2\sqrt{3}}{2}=-1\pm\sqrt{3}$$

The solutions are $-1\pm\sqrt{3}$, and the solution set is $\left\{-1\pm\sqrt{3}\right\}$.

## 11.2 Exercise Set

**1.** $f(x)=x^3-9x^2+26x-24$

Since 3 is a zero of $f(x)$, we know $f(3)=0$. By the Factor Theorem, $x-3$ is a factor of $f(x)$. We can use synthetic division to divide $f(x)$ by $x-3$.

$$\begin{array}{r|rrrr} 3 & 1 & -9 & 26 & -24 \\ & & 3 & -18 & 24 \\ \hline & 1 & -6 & 8 & 0 \end{array}$$

Thus, $x^3-9x^2+26x-24=0$
$(x-3)(x^2-6x+8)=0$
$(x-3)(x-4)(x-2)=0$

Apply the zero-product property:
$x-3=0$ or $x-4=0$ or $x-2=0$
$x=3 \qquad x=4 \qquad x=2$

The solutions are 2, 3, and 4, and the solution set is $\{2, 3, 4\}$.

**3.** $f(x)=2x^3+3x^2-8x+3$

Since $-3$ is a zero of $f(x)$, we know $f(-3)=0$. By the Factor Theorem, $x+3$ is a factor of $f(x)$. We can use synthetic division to divide $f(x)$ by $x+3$.

$$\begin{array}{r|rrrr} -3 & 2 & 3 & -8 & 3 \\ & & -6 & 9 & -3 \\ \hline & 2 & -3 & 1 & 0 \end{array}$$

Thus, $2x^3+3x^2-8x+3=0$
$(x+3)(2x^2-3x+1)=0$
$(x+3)(2x-1)(x-1)=0$

Apply the zero-product property:
$x+3=0$ or $2x-1=0$ or $x-1=0$
$x=-3 \qquad 2x=1 \qquad x=1$
$x=\frac{1}{2}$

The solutions are $-3$, $\frac{1}{2}$, and 1, and the solution set is $\left\{-3,\frac{1}{2},1\right\}$.

**5.** $f(x)=12x^3+16x^2-5x-3$

Since $-\frac{3}{2}$ is a zero of $f(x)$, we know $f\left(-\frac{3}{2}\right)=0$. By the Factor Theorem, $x+\frac{3}{2}$ is a factor of $f(x)$. We can use synthetic division to divide $f(x)$ by $x+\frac{3}{2}$.

$$\begin{array}{r|rrrr} -\frac{3}{2} & 12 & 16 & -5 & -3 \\ & & -18 & 3 & 3 \\ \hline & 12 & -2 & -2 & 0 \end{array}$$

Thus, $12x^3+16x^2-5x-3=0$

$$\left(x+\frac{3}{2}\right)\left(12x^2-2x-2\right)=0$$
$$2\left(x+\frac{3}{2}\right)\left(6x^2-x-1\right)=0$$
$$(2x+3)\left(6x^2-x-1\right)=0$$
$$(2x+3)(3x+1)(2x-1)=0$$

Apply the zero-product property:

$2x+3=0$ or $3x+1=0$ or $2x-1=0$

$2x=-3 \qquad 3x=-1 \qquad 2x=1$

$x=-\frac{3}{2} \qquad x=-\frac{1}{3} \qquad x=\frac{1}{2}$

The solutions are $-\frac{3}{2}$, $-\frac{1}{3}$, and $-\frac{1}{2}$, and the solution set is $\left\{-\frac{3}{2},-\frac{1}{3},\frac{1}{2}\right\}$.

**7.** $f(x)=2x^3+3x^2-8x-12$

Since 2 is a zero of $f(x)$, we know $f(2)=0$. By the Factor Theorem, $x-2$ is a factor of $f(x)$. We can use synthetic division to divide $f(x)$ by $x-2$.

$$\begin{array}{r|rrrr} 2 & 2 & 3 & -8 & -12 \\ & & 4 & 14 & 12 \\ \hline & 2 & 7 & 6 & 0 \end{array}$$

Thus, $2x^3+3x^2-8x-12=0$

$$(x-2)\left(2x^2+7x+6\right)=0$$
$$(x-2)(2x+3)(x+2)=0$$

Apply the zero-product property:

$x-2=0$ or $2x+3=0$ or $x+2=0$

$x=2 \qquad 2x=-3 \qquad x=-2$

$x=-\frac{3}{2}$

The solutions are $-2$, $-\frac{3}{2}$, and 2, and the solution set is $\left\{-2,-\frac{3}{2},2\right\}$.

**9.** $f(x)=x^3+x^2-4x-4$

List all factors of the constant term $-4$:
$\pm1,\ \pm2,\ \pm4$

List all factors of the leading coefficient 1:
$\pm1$

The possible rational zeros are:

$$\frac{\text{Factors of } -4}{\text{Factors of } 1}=\frac{\pm1,\ \pm2,\ \pm4}{\pm1}$$
$=\pm1,\ \pm2,\ \pm4$

**11.** $f(x)=3x^4-11x^3-x^2+19x+6$

List all factors of the constant term 6:
$\pm1,\ \pm2,\ \pm3,\ \pm6$

List all factors of the leading coefficient 3:
$\pm1,\ \pm3$

The possible rational zeros are:

$$\frac{\text{Factors of } 6}{\text{Factors of } 3}=\frac{\pm1,\ \pm2,\ \pm3,\ \pm6}{\pm1,\ \pm3}$$
$=\pm1,\ \pm2,\ \pm3,\ \pm6,\ \pm\frac{1}{3},\ \pm\frac{2}{3}$

**13.** $f(x) = 4x^4 - x^3 + 5x^2 - 2x - 6$

List all factors of the constant term $-6$: $\pm 1,\ \pm 2,\ \pm 3,\ \pm 6$

List all factors of the leading coefficient 4: $\pm 1,\ \pm 2,\ \pm 4$

The possible rational zeros are:

$$\frac{\text{Factors of } -6}{\text{Factors of } 4} = \frac{\pm 1,\ \pm 2,\ \pm 3,\ \pm 6}{\pm 1,\ \pm 2,\ \pm 4}$$

$$= \pm 1, \pm 2, \pm 3, \pm 6, \pm\frac{1}{2}, \pm\frac{3}{2}, \pm\frac{1}{4}, \pm\frac{3}{4}$$

**15.** $f(x) = x^5 - x^4 - 7x^3 + 7x^2 - 12x - 12$

List all factors of the constant term $-12$: $\pm 1,\ \pm 2,\ \pm 3,\ \pm 4,\ \pm 6,\ \pm 12$

List all factors of the leading coefficient 1: $\pm 1$

The possible rational zeros are:

$$\frac{\text{Factors of } -12}{\text{Factors of } 1}$$

$$= \frac{\pm 1,\ \pm 2,\ \pm 3,\ \pm 4,\ \pm 6,\ \pm 12}{\pm 1}$$

$$= \pm 1,\ \pm 2,\ \pm 3,\ \pm 4,\ \pm 6,\ \pm 12$$

**17.** $f(x) = x^3 + x^2 - 4x - 4$

**a.** List all factors of the constant term $-4$: $\pm 1,\ \pm 2,\ \pm 4$

List all factors of the leading coefficient 1: $\pm 1$

The possible rational zeros are:

$$\frac{\text{Factors of } -4}{\text{Factors of } 1} = \frac{\pm 1,\ \pm 2,\ \pm 4}{\pm 1}$$

$$= \pm 1,\ \pm 2,\ \pm 4$$

**b.** We begin by testing 1:

| 1 | 1 | 1 | −4 | −4 |
|---|---|---|---|---|
| | | 1 | 1 | −2 |
| | 1 | 2 | −2 | −6 |

The remainder is $-6$, not 0, so 1 is not a zero of *f*.

Next, test $-1$:

| −1 | 1 | 1 | −4 | −4 |
|---|---|---|---|---|
| | | −1 | 0 | 4 |
| | 1 | 0 | −4 | 0 |

The remainder is 0, so $-1$ *is* a zero of *f*.

**c.** From part (b) and the Factor Theorem, we know that $x+1$ is a factor of *f*. Thus,

$$x^3 + x^2 - 4x - 4 = 0$$
$$(x+1)(x^2 - 4) = 0$$
$$(x+1)(x+2)(x-2) = 0$$
$$x+1=0 \quad \text{or} \quad x+2=0 \quad \text{or} \quad x-2=0$$
$$x=-1 \qquad x=-2 \qquad x=2$$

The rational zeros of *f* are $-2$, $-1$, and 2.

**19.** $f(x) = 2x^3 - 3x^2 - 11x + 6$

**a.** List all factors of the constant term 6: $\pm 1,\ \pm 2,\ \pm 3,\ \pm 6$

List all factors of the leading coefficient 2: $\pm 1,\ \pm 2$

The possible rational zeros are:

$$\frac{\text{Factors of } 6}{\text{Factors of } 2} = \frac{\pm 1,\ \pm 2,\ \pm 3,\ \pm 6}{\pm 1,\ \pm 2}$$

$$= \pm 1,\ \pm 2,\ \pm 3,\ \pm 6,\ \pm\frac{1}{2},\ \pm\frac{3}{2}$$

**b.** We begin by testing 1:

| 1 | 2 | −3 | −11 | 6 |
|---|---|---|---|---|
| | | 2 | −1 | −12 |
| | 2 | −1 | −12 | −6 |

The remainder is $-6$, not 0, so 1 is not a zero of *f*.

Next, test $-1$:

$$\begin{array}{r|rrrr} -1 & 2 & -3 & -11 & 6 \\ & & -2 & 5 & 6 \\ \hline & 2 & -5 & -6 & 12 \end{array}$$

The remainder is 12, not 0, so $-1$ is not a zero of *f*.

Test 2:

$$\begin{array}{r|rrrr} 2 & 2 & -3 & -11 & 6 \\ & & 4 & 2 & -18 \\ \hline & 2 & 1 & -9 & -12 \end{array}$$

The remainder is $-12$, not 0, so 2 is not a zero of *f*.

Test $-2$:

$$\begin{array}{r|rrrr} -2 & 2 & -3 & -11 & 6 \\ & & -4 & 14 & -6 \\ \hline & 2 & -7 & 3 & 0 \end{array}$$

The remainder is 0, so $-2$ *is* a zero of *f*.

**c.** From part (b) and the Factor Theorem, we know that $x+2$ is a factor of *f*. Thus,

$$2x^3-3x^2-11x+6=0$$
$$(x+2)(2x^2-7x+3)=0$$
$$(x+2)(2x-1)(x-3)=0$$
$$x+2=0 \quad \text{or} \quad 2x-1=0 \quad \text{or} \quad x-3=0$$
$$x=-2 \qquad x=\frac{1}{2} \qquad x=3$$

The rational zeros of *f* are $-2$, $\frac{1}{2}$, and 3.

**21.** $f(x)=3x^3+7x^2-22x-8$

**a.** List all factors of the constant term $-8$: $\pm1,\ \pm2,\ \pm4,\ \pm8$
List all factors of the leading coefficient 3: $\pm1,\ \pm3$

The possible rational zeros are:

$$\frac{\text{Factors of } -8}{\text{Factors of } 3}=\frac{\pm1,\ \pm2,\ \pm4,\ \pm8}{\pm1,\ \pm3}$$
$$=\pm1,\pm2,\pm4,\pm8,\pm\frac{1}{3},\pm\frac{2}{3},\pm\frac{4}{3},\pm\frac{8}{3}$$

**b.** We begin by testing 1:

$$\begin{array}{r|rrrr} 1 & 3 & 7 & -22 & -8 \\ & & 3 & 10 & -12 \\ \hline & 3 & 10 & -12 & -20 \end{array}$$

The remainder is $-20$, not 0, so 1 is not a zero of *f*.

Next, test $-1$:

$$\begin{array}{r|rrrr} -1 & 3 & 7 & -22 & -8 \\ & & -3 & -4 & 26 \\ \hline & 3 & 4 & -26 & 18 \end{array}$$

The remainder is 18, not 0, so $-1$ is not a zero of *f*.

Test 2:

$$\begin{array}{r|rrrr} 2 & 3 & 7 & -22 & -8 \\ & & 6 & 26 & 8 \\ \hline & 3 & 13 & 4 & 0 \end{array}$$

The remainder is 0, so 2 *is* a zero of *f*.

**c.** From part (b) and the Factor Theorem, we know that $x-2$ is a factor of *f*. Thus,

$$3x^3+7x^2-22x-8=0$$
$$(x-2)(3x^2+13x+4)=0$$
$$(x-2)(3x+1)(x+4)=0$$
$$x-2=0 \quad \text{or} \quad 3x+1=0 \quad \text{or} \quad x+4=0$$
$$x=2 \qquad x=-\frac{1}{3} \qquad x=-4$$

The rational zeros of *f* are $-4$, $-\frac{1}{3}$, and 2.

**23.** $x^3 - 2x^2 - 11x + 12 = 0$

**a.** List all factors of the constant term 12: $\pm1, \pm2, \pm3, \pm4, \pm6, \pm12$

List all factors of the leading coefficient 1: $\pm1$

The possible rational zeros are:

$$\frac{\text{Factors of 12}}{\text{Factors of 1}} = \frac{\pm1, \pm2, \pm3, \pm4, \pm6, \pm12}{\pm1} = \pm1, \pm2, \pm3, \pm4, \pm6, \pm12$$

**b.** We begin by testing 1:

$$\begin{array}{r|rrrr} 1 & 1 & -2 & -11 & 12 \\ & & 1 & -1 & -12 \\ \hline & 1 & -1 & -12 & 0 \end{array}$$

The remainder is 0, so 1 *is* a root.

**c.** From part (b) and the Factor Theorem, we know that $x-1$ is a factor of the polynomial. Thus,

$$x^3 - 2x^2 - 11x + 12 = 0$$
$$(x-1)(x^2 - x - 12) = 0$$
$$(x-1)(x-4)(x+3) = 0$$
$$x-1=0 \quad \text{or} \quad x-4=0 \quad \text{or} \quad x+3=0$$
$$x=1 \qquad x=4 \qquad x=-3$$

The solutions are $-3$, 1 and 4, and the solution set is $\{-3, 1, 4\}$.

**25.** $x^3 - 10x - 12 = 0$

**a.** List all factors of the constant term $-12$: $\pm1, \pm2, \pm3, \pm4, \pm6, \pm12$

List all factors of the leading coefficient 1: $\pm1$

The possible rational zeros are:

$$\frac{\text{Factors of 12}}{\text{Factors of 1}} = \frac{\pm1, \pm2, \pm3, \pm4, \pm6, \pm12}{\pm1} = \pm1, \pm2, \pm3, \pm4, \pm6, \pm12$$

**b.** We begin by testing 1:

$$\begin{array}{r|rrrr} 1 & 1 & 0 & -10 & -12 \\ & & 1 & 1 & -9 \\ \hline & 1 & 1 & -9 & -21 \end{array}$$

The remainder is $-21$, not 0, so 1 is not a root.

Next, test $-1$:

$$\begin{array}{r|rrrr} -1 & 1 & 0 & -10 & -12 \\ & & -1 & 1 & 9 \\ \hline & 1 & -1 & -9 & -3 \end{array}$$

The remainder is $-3$, not 0, so $-1$ is not a root.

Test 2:

$$\begin{array}{r|rrrr} 2 & 1 & 0 & -10 & -12 \\ & & 2 & 4 & -12 \\ \hline & 1 & 2 & -6 & -24 \end{array}$$

The remainder is $-24$, not 0, so 2 is not a root.

Test $-2$:

$$\begin{array}{r|rrrr} -2 & 1 & 0 & -10 & -12 \\ & & -2 & 4 & 12 \\ \hline & 1 & -2 & -6 & 0 \end{array}$$

The remainder is 0, so $-2$ *is* a root.

**c.** From part (b) and the Factor Theorem, we know that $x+2$ is a factor of the polynomial. Thus,

$$x^3 - 10x - 12 = 0$$
$$(x+2)(x^2 - 2x - 6) = 0$$

Note that $x^2 - 2x - 6$ does not factor, so we use the quadratic formula:

$x+2=0 \quad \text{or} \quad x^2-2x-6=0$

$x=-2 \qquad a=1,\ b=-2,\ c=-6$

$$x=\frac{-(-2)\pm\sqrt{(-2)^2-4(1)(-6)}}{2(1)}$$

$$=\frac{2\pm\sqrt{28}}{2}=\frac{2\pm2\sqrt{7}}{2}=1\pm\sqrt{7}$$

The solutions are $-2$ and $1\pm\sqrt{7}$, and the solution set is $\{-2,\ 1\pm\sqrt{7}\}$.

**27.** $6x^3+25x^2-24x+5=0$

**a.** List all factors of the constant term 5: $\pm1,\ \pm5$

List all factors of the leading coefficient 6: $\pm1,\ \pm2,\ \pm3,\ \pm6$

The possible rational zeros are:

$$\frac{\text{Factors of 5}}{\text{Factors of 6}}=\frac{\pm1,\ \pm5}{\pm1,\ \pm2,\ \pm3,\ \pm6}$$

$$=\pm1,\pm5,\pm\frac{1}{2},\pm\frac{5}{2},\pm\frac{1}{3},\pm\frac{5}{3},\pm\frac{1}{5},\pm\frac{5}{6}$$

**b.** We begin by testing 1:

| 1 | 6 | 25 | −24 | 5 |
|---|---|---|---|---|
| | | 6 | 31 | 7 |
| | 6 | 31 | 7 | 12 |

The remainder is 12, not 0, so 1 is not a root.

Next, test $-1$:

| −1 | 6 | 25 | −24 | 5 |
|---|---|---|---|---|
| | | −6 | −19 | 43 |
| | 6 | 19 | −43 | 48 |

The remainder is 48, not 0, so $-1$ is not a root.

Test 5:

| 5 | 6 | 25 | −24 | 5 |
|---|---|---|---|---|
| | | 30 | 275 | 1255 |
| | 6 | 55 | 251 | 1260 |

The remainder is 1260, not 0, so 5 is not a root.

Test $-5$:

| −5 | 6 | 25 | −24 | 5 |
|---|---|---|---|---|
| | | −30 | 25 | −5 |
| | 6 | −5 | 1 | 0 |

The remainder is 0, so $-5$ *is* a root.

**c.** From part (b) and the Factor Theorem, we know that $x+5$ is a factor of the polynomial. Thus,

$$6x^3+25x^2-24x+5=0$$

$$(x+5)(6x^2-5x+1)=0$$

$$(x+5)(3x-1)(2x-1)=0$$

$x+5=0 \quad \text{or} \quad 3x-1=0 \quad \text{or} \quad 2x-1=0$

$x=-5 \qquad x=\frac{1}{3} \qquad x=\frac{1}{2}$

The solutions are $-5$, $\frac{1}{3}$, and $\frac{1}{2}$, and the solution set is $\left\{-5,\ \frac{1}{3},\ \frac{1}{2}\right\}$.

**29.** $x^4-2x^3-5x^2+8x+4=0$

**a.** List all factors of the constant term 4: $\pm1,\ \pm2,\ \pm4$

List all factors of the leading coefficient 1: $\pm1$

The possible rational zeros are:

$$\frac{\text{Factors of 4}}{\text{Factors of 1}}=\frac{\pm1,\ \pm2,\ \pm4}{\pm1}$$

$$=\pm1,\ \pm2,\ \pm4$$

**b.** We begin by testing 1:

$$\begin{array}{r|rrrrr} 1 & 1 & -2 & -5 & 8 & 4 \\ & & 1 & -1 & -6 & 2 \\ \hline & 1 & -1 & -6 & 2 & 6 \end{array}$$

The remainder is 6, not 0, so 1 is not a root.

Next, test $-1$:

$$\begin{array}{r|rrrrr} -1 & 1 & -2 & -5 & 8 & 4 \\ & & -1 & 3 & 2 & -10 \\ \hline & 1 & -3 & -2 & 10 & -6 \end{array}$$

The remainder is $-6$, not 0, so $-1$ is not a root.

Test 2:

$$\begin{array}{r|rrrrr} 2 & 1 & -2 & -5 & 8 & 4 \\ & & 2 & 0 & -10 & -4 \\ \hline & 1 & 0 & -5 & -2 & 0 \end{array}$$

The remainder is 0, so 2 *is* a root.

**c.** From part (b) and the Factor Theorem, we know that $x-2$ is a factor of the polynomial. Thus,

$$x^4-2x^3-5x^2+8x+4=0$$
$$(x-2)(x^3-5x-2)=0$$

To solve the equation above, we need to factor $x^3-5x-2$. We continue testing potential roots:

Test $-2$:

$$\begin{array}{r|rrrr} -2 & 1 & 0 & -5 & -2 \\ & & -2 & 4 & 2 \\ \hline & 1 & -2 & -1 & 0 \end{array}$$

The remainder is 0, so $-2$ *is* a root.

Summarizing our findings so far, we have

$$x^4-2x^3-5x^2+8x+4=0$$
$$(x-2)(x^3-5x-2)=0$$
$$(x-2)(x+2)(x^2-2x-1)=0$$

We know that 2 and $-2$ are solutions to the equation. Note that $x^2-2x-1$ does not factor, so we use the quadratic formula:

$$x^2-2x-1=0$$
$$a=1,\ b=-2,\ c=-1$$
$$x=\frac{-(-2)\pm\sqrt{(-2)^2-4(1)(-1)}}{2(1)}$$
$$=\frac{2\pm\sqrt{8}}{2}=\frac{2\pm2\sqrt{2}}{2}=1\pm\sqrt{2}$$

The solutions are $-2$, 2, and $1\pm\sqrt{2}$, and the solution set is $\{-2,2,1\pm\sqrt{2}\}$.

**31.** $f(x)=x^3+2x^2+5x+4$

Because all the coefficients are positive, there are no sign changes. Thus, $f$ has no positive real zeros.

$$f(-x)=(-x)^3+2(-x)^2+5(-x)+4$$
$$=-x^3+2x^2-5x+4$$

There are three sign changes. Thus, $f$ has either 3 or 1 negative real zeros.

**33.** $f(x)=5x^3-3x^2+3x-1$

There are three sign changes. Thus, $f$ has either 3 or 1 positive real zeros.

$$f(-x)=5(-x)^3-3(-x)^2+3(-x)-1$$
$$=-5x^3-3x^2-3x-1$$

Because all the coefficients are negative, there are no sign changes. Thus, $f$ has no negative real zeros.

**35.** $f(x)=2x^4-5x^3-x^2-6x+4$

There are two sign changes. Thus, $f$ has either 2 or 0 positive real zeros.

$$f(-x)=2(-x)^4-5(-x)^3-(-x)^2-6(-x)+4$$
$$=2x^4+5x^3-x^2+6x+4$$

There are two sign changes. Thus, $f$ has either 2 or 0 negative real zeros.

**37.** $f(x) = x^3 - 4x^2 - 7x + 10$

We begin by using the Rational Zero Theorem to determine possible rational zeros.

Factors of the constant term 10: $\pm 1,\ \pm 2,\ \pm 5,\ \pm 10$

Factors of the leading coefficient 1: $\pm 1$

The possible rational zeros are:

$$\frac{\text{Factors of 10}}{\text{Factors of 1}} = \frac{\pm 1,\ \pm 2,\ \pm 5,\ \pm 10}{\pm 1}$$
$$= \pm 1,\ \pm 2,\ \pm 5,\ \pm 10$$

Next, we use Descartes's Rule of Signs. Since $f(x)$ has two sign changes, it will have either 2 or 0 positive real zeros.

$$f(-x) = (-x)^3 - 4(-x)^2 - 7(-x) + 10$$
$$= -x^3 - 4x^2 + 7x + 10$$

Since $f(-x)$ has one sign change, $f$ will have exactly 1 negative real zero.

We test values from above until we find a zero. One possibility is shown next:

Test $-2$:

| | | | | |
|---|---|---|---|---|
| $-2$ | 1 | $-4$ | $-7$ | 10 |
| | | $-2$ | 12 | $-10$ |
| | 1 | $-6$ | 5 | 0 |

The remainder is 0, so $-2$ *is* a zero of $f$. Using the Factor Theorem, we know that $x + 2$ is a factor of $f$. Thus,

$$x^3 - 4x^2 - 7x + 10 = 0$$
$$(x+2)(x^2 - 6x + 5) = 0$$
$$(x+2)(x-1)(x-5) = 0$$
$$x + 2 = 0 \quad \text{or} \quad x - 1 = 0 \quad \text{or} \quad x - 5 = 0$$
$$x = -2 \qquad x = 1 \qquad x = 5$$

The zeros of $f$ are $-2$, 1, and 5.

**39.** $2x^3 - x^2 - 9x - 4 = 0$

We begin by using the Rational Zero Theorem to determine possible rational zeros.

Factors of the constant term $-4$: $\pm 1,\ \pm 2,\ \pm 4$

Factors of the leading coefficient 2: $\pm 1,\ \pm 2$

The possible rational zeros are:

$$\frac{\text{Factors of } -4}{\text{Factors of 2}} = \frac{\pm 1,\ \pm 2,\ \pm 4}{\pm 1,\ \pm 2}$$
$$= \pm 1,\ \pm 2,\ \pm 4,\ \pm \frac{1}{2}$$

Next, we let $f(x) = 2x^3 - x^2 - 9x - 4$ and use Descartes's Rule of Signs. Since $f(x)$ has one sign change, it will have exactly 1 positive real zero.

$$f(-x) = 2(-x)^3 - (-x)^2 - 9(-x) - 4$$
$$= -2x^3 - x^2 + 9x - 4$$

Since $f(-x)$ has two sign changes, $f$ will have either 2 or 0 negative real zeros.

We test values from above until we find a zero. One possibility is shown next:

Test $-\frac{1}{2}$:

| | | | | |
|---|---|---|---|---|
| $-\frac{1}{2}$ | 2 | $-1$ | $-9$ | $-4$ |
| | | $-1$ | 1 | 4 |
| | 2 | $-2$ | $-8$ | 0 |

The remainder is 0, so $-\frac{1}{2}$ *is* a zero of $f$.

Using the Factor Theorem, we know that $x + \frac{1}{2}$ is a factor of $f$. Thus,

$$2x^3 - x^2 - 9x - 4 = 0$$
$$\left(x + \frac{1}{2}\right)\left(2x^2 - 2x - 8\right) = 0$$
$$2\left(x + \frac{1}{2}\right)\left(x^2 - x - 4\right) = 0$$
$$(2x+1)\left(x^2 - x - 4\right) = 0$$

Note that $x^2 - x - 4$ does not factor, so we use the quadratic formula:

$2x + 1 = 0$ or $x^2 - x - 4 = 0$

$x = -\frac{1}{2}$ $\quad a = 1,\ b = -1,\ c = -4$

$$x = \frac{-(-1) \pm \sqrt{(-1)^2 - 4(1)(-4)}}{2(1)}$$
$$= \frac{1 \pm \sqrt{17}}{2}$$

The solutions are $-\frac{1}{2}$ and $\frac{1 \pm \sqrt{17}}{2}$, and the solution set is $\left\{-\frac{1}{2}, \frac{1 \pm \sqrt{17}}{2}\right\}$.

**41.** $x^4 - 3x^3 - 20x^2 - 24x - 8 = 0$

We begin by using the Rational Zero Theorem to determine possible rational zeros.

Factors of the constant term $-8$: $\pm 1,\ \pm 2,\ \pm 4,\ \pm 8$

Factors of the leading coefficient 1: $\pm 1$

The possible rational zeros are:

$$\frac{\text{Factors of } -8}{\text{Factors of } 1} = \frac{\pm 1,\ \pm 2,\ \pm 4,\ \pm 8}{\pm 1}$$
$$= \pm 1,\ \pm 2,\ \pm 4,\ \pm 8$$

Next, we let $f(x) = x^4 - 3x^3 - 20x^2 - 24x - 8$ and use Descartes's Rule of Signs.

Since $f(x)$ has one sign change, it will have exactly 1 positive real zero.

$$f(-x) = (-x)^4 - 3(-x)^3 - 20(-x)^2 - 24(-x) - 8$$
$$= x^4 + 3x^3 - 20x^2 + 24x - 8$$

Since $f(-x)$ has three sign changes, $f$ will have either 3 or 1 negative real zeros.

We test values from above until we find a zero. One possibility is shown next:

Test $-1$:

| | | | | | |
|---|---|---|---|---|---|
| $-1$ | 1 | $-3$ | $-20$ | $-24$ | $-8$ |
| | | $-1$ | 4 | 16 | 8 |
| | 1 | $-4$ | $-16$ | $-8$ | 0 |

The remainder is 0, so $-1$ *is* a zero of $f$. Using the Factor Theorem, we know that $x + 1$ is a factor of $f$. Thus,

$$x^4 - 3x^3 - 20x^2 - 24x - 8 = 0$$
$$(x+1)\left(x^3 - 4x^2 - 16x - 8\right) = 0$$

To solve the equation above, we need to factor $x^3 - 4x^2 - 16x - 8$. We continue testing potential roots:

Test $-2$:

| | | | | |
|---|---|---|---|---|
| $-2$ | 1 | $-4$ | $-16$ | $-8$ |
| | | $-2$ | 12 | 8 |
| | 1 | $-6$ | $-4$ | 0 |

The remainder is 0, so $-2$ *is* a root and $x + 2$ is a factor.

Summarizing our findings so far, we have

$$x^4 - 3x^3 - 20x^2 - 24x - 8 = 0$$
$$(x+1)\left(x^3 - 4x^2 - 16x - 8\right) = 0$$
$$(x+1)(x+2)\left(x^2 - 6x - 4\right) = 0$$

We know that $-1$ and $-2$ are solutions to the equation. Note that $x^2 - 6x - 2$ does not factor, so we use the quadratic formula: $x^2 - 6x - 2 = 0$

$a = 1,\ b = -6,\ c = -4$

$$x=\frac{-(-6)\pm\sqrt{(-6)^2-4(1)(-4)}}{2(1)}$$

$$=\frac{6\pm\sqrt{52}}{2}=\frac{6\pm2\sqrt{13}}{2}=3\pm\sqrt{13}$$

The solutions are $-2$, $-1$, and $3\pm\sqrt{13}$, and the solution set is $\{-2,\ -1,\ 3\pm\sqrt{13}\}$.

**43.** $f(x)=3x^4-11x^3-x^2+19x+6$

We begin by using the Rational Zero Theorem to determine possible rational zeros.

Factors of the constant term 6:
$\pm1,\ \pm2,\ \pm3,\ \pm6$

Factors of the leading coefficient 3:
$\pm1,\ \pm3$

The possible rational zeros are:

$$\frac{\text{Factors of 6}}{\text{Factors of 3}}=\frac{\pm1,\ \pm2,\ \pm3,\ \pm6}{\pm1,\ \pm3}$$

$$=\pm1,\ \pm2,\ \pm3,\ \pm6,\ \pm\frac{1}{3},\ \pm\frac{2}{3}$$

Next, we use Descartes's Rule of Signs.

Since $f(x)$ has two sign changes, it will have either 2 or 0 positive real zero.

$$f(-x)=3(-x)^4-11(-x)^3-(-x)^2+19(-x)+6$$

$$=3x^4+11x^3-x^2-19x+6$$

Since $f(-x)$ has two sign changes, $f$ will have either 2 or 0 negative real zeros.

We test values from above until we find a zero. One possibility is shown next:

Test $-1$:

| | | | | | |
|---|---|---|---|---|---|
| $-1$ | 3 | $-11$ | $-1$ | 19 | 6 |
| | | $-3$ | 14 | $-13$ | $-6$ |
| | 3 | $-14$ | 13 | 6 | 0 |

The remainder is 0, so $-1$ *is* a zero of *f*. Using the Factor Theorem, we know that $x+1$ is a factor of *f*. Thus,

$$3x^4-11x^3-x^2+19x+6=0$$

$$(x+1)(3x^3-14x^2+13x+6)=0$$

To solve the equation above, we need to factor $3x^3-14x^2+13x+6$. We continue testing potential roots:

Test 2:

| | | | | |
|---|---|---|---|---|
| 2 | 3 | $-14$ | 13 | 6 |
| | | 6 | $-16$ | $-6$ |
| | 3 | $-8$ | $-3$ | 0 |

The remainder is 0, so 2 *is* a zero and $x-2$ is a factor.

Summarizing our findings so far, we have

$$3x^4-11x^3-x^2+19x+6=0$$

$$(x+1)(3x^3-14x^2+13x+6)=0$$

$$(x+1)(x-2)(3x^2-8x-3)=0$$

$$(x+1)(x-2)(3x+1)(x-3)=0$$

We know that $-1$ and 2 are zeros of the function. To find the remaining zeros, we set the remaining two factors equal to zero:

$3x+1=0$ or $x-3=0$

$x=-\frac{1}{3}$ $\qquad$ $x=3$

The zeros are $-1$, $-\frac{1}{3}$, 2, and 3.

**45.** $4x^4 - x^3 + 5x^2 - 2x - 6 = 0$

We begin by using the Rational Zero Theorem to determine possible rational zeros.

Factors of the constant term $-6$: $\pm1,\ \pm2,\ \pm3,\ \pm6$

Factors of the leading coefficient 4: $\pm1,\ \pm2,\ \pm4$

The possible rational zeros are:

$$\frac{\text{Factors of } -6}{\text{Factors of } 4} = \frac{\pm1,\ \pm2,\ \pm3,\ \pm6}{\pm1,\ \pm2,\ \pm4}$$

$$= \pm1, \pm2, \pm3, \pm6, \pm\frac{1}{2}, \pm\frac{3}{2}, \pm\frac{1}{4}, \pm\frac{3}{4}$$

Next, we let $f(x) = 3x^4 - 11x^3 - 3x^2 - 6x + 8$ and use Descartes's Rule of Signs.

Since *f*(*x*) has two sign changes, it will have either 2 or 0 positive real zeros.

$$f(-x) = 3(-x)^4 - 11(-x)^3 - 3(-x)^2 - 6(-x) + 8$$
$$= 3x^4 + 11x^3 - 3x^2 + 6x + 8$$

Since $f(-x)$ has two sign changes, *f* will have either 2 or 0 negative real zeros.

We test values from above until we find a zero. One possibility is shown next:

Test 1:

| | | | | | |
|---|---|---|---|---|---|
| 1 | 4 | −1 | 5 | −2 | −6 |
| | | 4 | 3 | 8 | 6 |
| | 4 | 3 | 8 | 6 | 0 |

The remainder is 0, so 1 *is* a zero of *f*. Using the Factor Theorem, we know that $x - 1$ is a factor of *f*. Thus,

$$3x^4 - 11x^3 - 3x^2 + 6x + 8 = 0$$
$$(x-1)(4x^3 + 3x^2 + 8x + 6) = 0$$

To solve the equation above, we need to factor $4x^3 + 3x^2 + 8x + 6$. We continue testing potential roots:

Test $-\frac{3}{4}$:

| | | | | |
|---|---|---|---|---|
| $-\frac{3}{4}$ | 4 | 3 | 8 | 6 |
| | | −3 | 0 | −6 |
| | 4 | 0 | 8 | 0 |

The remainder is 0, so $-\frac{3}{4}$ *is* a zero and $x + \frac{3}{4}$ is a factor.

Summarizing our findings so far, we have

$$3x^4 - 11x^3 - 3x^2 + 6x + 8 = 0$$
$$(x-1)(4x^3 + 3x^2 + 8x + 6) = 0$$
$$(x-1)\left(x + \frac{3}{4}\right)(4x^2 + 8) = 0$$
$$4(x-1)\left(x + \frac{3}{4}\right)(x^2 + 2) = 0$$

We know that 1 and $-\frac{3}{4}$ are roots of the equation. Note that $x^2 + 2$ does not factor, so we use the square-root principle: $x^2 + 2 = 0$

$$x^2 = -2$$
$$x = \pm\sqrt{-2} = \pm\sqrt{2}i$$

The roots are $-\frac{3}{4}$, 1, and $\pm\sqrt{2}i$, and the solution set is $\left\{-\frac{3}{4},\ 1,\ \pm\sqrt{2}i\right\}$.

**47.** $2x^5+7x^4-18x^2-8x+8=0$
We begin by using the Rational Zero Theorem to determine possible rational zeros.

Factors of the constant term 8: $\pm1,\ \pm2,\ \pm4,\ \pm8$

Factors of the leading coefficient 2: $\pm1,\ \pm2$

The possible rational zeros are:

$$\frac{\text{Factors of 8}}{\text{Factors of 2}}=\frac{\pm1,\ \pm2,\ \pm4,\ \pm8}{\pm1,\ \pm2}$$

$$=\pm1,\pm2,\pm4,\pm8,\pm\frac{1}{2}$$

Next, we let $f(x)=2x^5+7x^4-18x^2-8x+8$ and use Descartes's Rule of Signs.

Since *f*(*x*) has two sign changes, it will have either 2 or 0 positive real zeros.

$$f(-x)=2(-x)^5+7(-x)^4-18(-x)^2-8(-x)+8$$
$$=-2x^5+7x^4-18x^2+8x+8$$

Since $f(-x)$ has three sign changes, *f* will have either 3 or 1 negative real zeros.

We test values from above until we find a zero. One possibility is shown next:

Test $-2$:

| | | | | | | |
|---|---|---|---|---|---|---|
| $-2$ | 2 | 7 | 0 | −18 | −8 | 8 |
| | | −4 | −6 | 12 | 12 | −8 |
| | 2 | 3 | −6 | −6 | 4 | 0 |

The remainder is 0, so $-2$ *is* a zero of *f*. Using the Factor Theorem, we know that $x+2$ is a factor of *f*. Thus,

$$2x^5+7x^4-18x^2-8x+8=0$$
$$(x+2)(2x^4+3x^3-6x^2-6x+4)=0$$

To solve the equation above, we need to factor $2x^4+3x^3-6x^2-6x+4$. We continue testing potential roots:

Test $-2$:

| | | | | | |
|---|---|---|---|---|---|
| $-2$ | 2 | 3 | −6 | −6 | 4 |
| | | −4 | 2 | 8 | −4 |
| | 2 | −1 | −4 | 2 | 0 |

The remainder is 0, so $-2$ *is* a zero of *f* with multiplicity 2. Using the Factor Theorem, we know that $x+2$ is a factor of *f*. Thus,

$$2x^5+7x^4-18x^2-8x+8=0$$
$$(x+2)(2x^4+3x^3-6x^2-6x+4)=0$$
$$(x+2)^2(2x^3-x^2-4x+2)=0$$

We need to factor the equation $2x^3-x^2-4x+2=0$:

Test $\frac{1}{2}$:

| | | | | |
|---|---|---|---|---|
| $\frac{1}{2}$ | 2 | −1 | −4 | 2 |
| | | 1 | 0 | −2 |
| | 2 | 0 | −4 | 0 |

The remainder is 0, so $\frac{1}{2}$ *is* a zero and $x-\frac{1}{2}$ is a factor.

Summarizing our findings so far, we have

$$2x^5+7x^4-18x^2-8x+8=0$$
$$(x+2)(2x^4+3x^3-6x^2-6x+4)=0$$
$$(x+2)^2(2x^3-x^2-4x+2)=0$$
$$(x+2)^2\left(x+\frac{1}{2}\right)(2x^2-4)=0$$
$$2(x+2)^2\left(x+\frac{1}{2}\right)(x^2-2)=0$$

We know that $-2$ and $\frac{1}{2}$ are roots of the equation. Note that $x^2-2$ does not factor, so we use the square-root principle: $x^2-2=0$

$$x^2=2$$

$$x=\pm\sqrt{2}$$

The roots are $-2$, $\frac{1}{2}$, and $\pm\sqrt{2}$, and the solution set is $\left\{-2, \frac{1}{2}, \pm\sqrt{2}\right\}$. Again, note that $-2$ has multiplicity 2.

**49.** $f(x)=-x^3+x^2+16x-16$

**a.** From the graph provided, we can see that $-4$ is an $x$-intercept and is thus a zero of the function. We verify this below:

$$\begin{array}{r|rrrr} \underline{-4} & -1 & 1 & 16 & -16 \\ & & 4 & -20 & 16 \\ \hline & -1 & 5 & -4 & 0 \end{array}$$

Thus,

$$-x^3+x^2+16x-16=0$$
$$(x+4)(-x^2+5x-4)=0$$
$$-(x+4)(x^2-5x+4)=0$$
$$-(x+4)(x-1)(x-4)=0$$

$x+4=0$ or $x-1=0$ or $x-4=0$

$x=-4$ $\quad x=1$ $\quad x=4$

The zeros are $-4$, 1, and 4.

**b.**

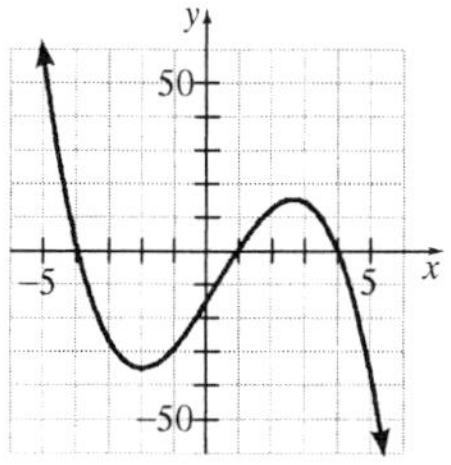


**50.** $f(x)=-x^3+3x^2-4$

**a.** From the graph provided, we can see that $-1$ is an $x$-intercept and is thus a zero of the function. We verify this below:

$$\begin{array}{r|rrrr} \underline{-1} & -1 & 3 & 0 & -4 \\ & & 1 & -4 & 4 \\ \hline & -1 & 4 & -4 & 0 \end{array}$$

Thus,

$$-x^3+3x^2-4=0$$
$$(x+1)(-x^2+4x-4)=0$$
$$-(x+1)(x^2-4x+4)=0$$
$$-(x+1)(x-2)^2=0$$

$x+1=0$ or $(x-2)^2=0$

$x=-1$ $\quad x-2=0$

$x=2$

The zeros are $-1$ and 2.

**b.**

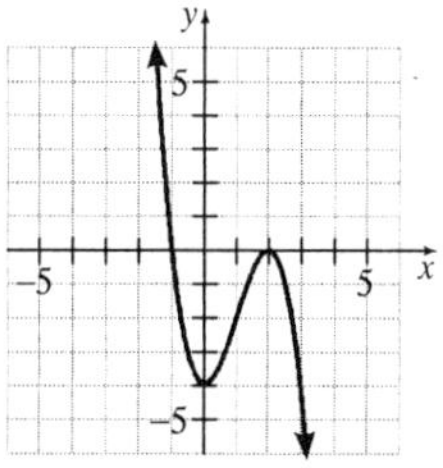


**51.** $f(x)=4x^3-8x^2-3x+9$

**a.** From the graph provided, we can see that $-1$ is an $x$-intercept and is thus a zero of the function. We verify this below:

$$\begin{array}{r|rrrr} \underline{-1} & 4 & -8 & -3 & 9 \\ & & -4 & 12 & -9 \\ \hline & 4 & -12 & 9 & 0 \end{array}$$

Thus,

$$4x^3-8x^2-3x+9=0$$
$$(x+1)(4x^2-12x+9)=0$$
$$(x+1)(2x-3)^2=0$$

$$x+1=0 \quad \text{or} \quad (2x-3)^2=0$$
$$x=-1 \qquad 2x-3=0$$
$$2x=3$$
$$x=\frac{3}{2}$$

The zeros are $-1$ and $\frac{3}{2}$.

**b.**

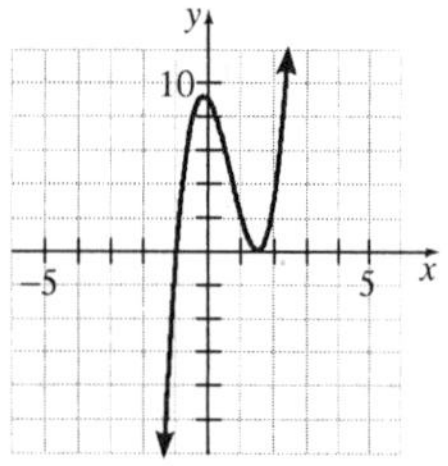


**52.** $f(x)=3x^3+2x^2+2x-1$

**a.** From the graph provided, we can see that $\frac{1}{3}$ is an $x$-intercept and is thus a zero of the function. We verify this below:

| $\frac{1}{3}$ | 3 | 2 | 2 | −1 |
|---|---|---|---|---|
| | | 1 | 1 | 1 |
| | 3 | 3 | 3 | 0 |

Thus, $3x^3+2x^2+2x-1=0$

$$\left(x-\frac{1}{3}\right)(3x^2+3x+3)=0$$

$$3\left(x-\frac{1}{3}\right)(x^2+x+1)=0$$

Note that $x^2+x+1$ will not factor, so we use the quadratic formula:

$$x-\frac{1}{3}=0 \quad \text{or} \quad x^2+x+1=0$$
$$x=\frac{1}{3} \qquad a=1 \quad b=1 \quad c=1$$

$$x=\frac{-1\pm\sqrt{1^2-4(1)(1)}}{2(1)}$$
$$=\frac{-1\pm\sqrt{-3}}{2}$$
$$=\frac{-1\pm\sqrt{3}i}{2}=-\frac{1}{2}\pm\frac{\sqrt{3}}{2}i$$

The zeros are $\frac{1}{3}$ and $-\frac{1}{2}\pm\frac{\sqrt{3}}{2}i$.

**b.**

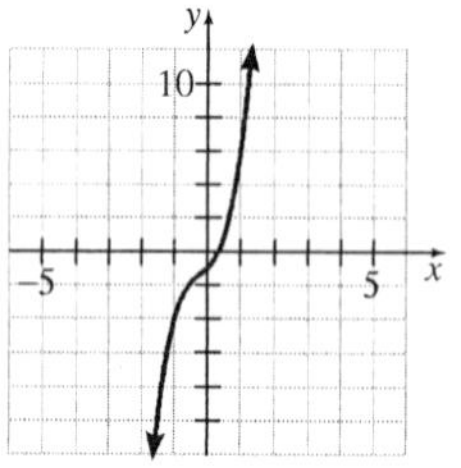


**53.** $f(x)=2x^4-3x^3-7x^2-8x+6$

**a.** From the graph provided, we can see that $\frac{1}{2}$ is an $x$-intercept and is thus a zero of the function. We verify this below:

| $\frac{1}{2}$ | 2 | −3 | −7 | −8 | 6 |
|---|---|---|---|---|---|
| | | 1 | −1 | −4 | −6 |
| | 2 | −2 | −8 | −12 | 0 |

Thus, $2x^4-3x^3-7x^2-8x+6=0$

$$\left(x-\frac{1}{2}\right)(2x^3-2x^2-8x-12)=0$$

$$2\left(x-\frac{1}{2}\right)(x^3-x^2-4x-6)=0$$

To factor $x^3-x^2-4x-6$, we use the Rational Zero Theorem to determine possible rational zeros.

Factors of the constant term $-6$: $\pm1,\ \pm2,\ \pm3,\ \pm6$

Factors of the leading coefficient 1: $\pm1$

The possible rational zeros are:

$$\frac{\text{Factors of } -6}{\text{Factors of } 1} = \frac{\pm1,\ \pm2,\ \pm3,\ \pm6}{\pm1}$$
$= \pm1,\ \pm2,\ \pm3,\ \pm6$

We test values from above until we find a zero. One possibility is shown next:

Test 3:

| | | | | |
|---|---|---|---|---|
| 3⌋ | 1 | −1 | −4 | −6 |
| | | 3 | 6 | 6 |
| | 1 | 2 | 2 | 0 |

The remainder is 0, so 3 is a zero of *f*.

$$2x^4 - 3x^3 - 7x^2 - 8x + 6 = 0$$
$$\left(x - \frac{1}{2}\right)\left(2x^3 - 2x^2 - 8x - 12\right) = 0$$
$$2\left(x - \frac{1}{2}\right)\left(x^3 - x^2 - 4x - 6\right) = 0$$
$$2\left(x - \frac{1}{2}\right)(x-3)\left(x^2 + 2x + 2\right) = 0$$

Note that $x^2 + x + 1$ will not factor, so we use the quadratic formula:

$a = 1 \quad b = 2 \quad c = 2$

$$x = \frac{-2 \pm \sqrt{2^2 - 4(1)(2)}}{2(1)}$$
$$= \frac{-2 \pm \sqrt{-4}}{2} = \frac{-2 \pm 2i}{2} = -1 \pm i$$

The zeros are $\frac{1}{2}$, 3, and $-1 \pm i$.

**b.**

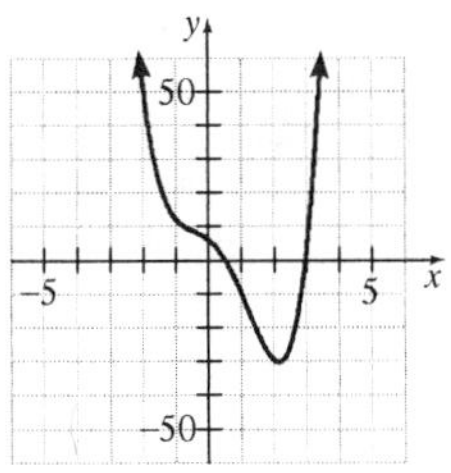


**54.** $f(x) = 2x^4 + 2x^3 - 22x^2 - 18x + 36$

**a.** From the graph provided, we can see that 1 and 3 are *x*-intercepts and are thus zeros of the function. We verify this below:

| | | | | | |
|---|---|---|---|---|---|
| 1⌋ | 2 | 2 | −22 | −18 | 36 |
| | | 2 | 4 | −18 | −36 |
| | 2 | 4 | −18 | −36 | 0 |

Thus, $2x^4 + 2x^3 - 22x^2 - 18x + 36$
$= (x-1)\left(2x^3 + 4x^2 - 18x - 36\right)$

| | | | | |
|---|---|---|---|---|
| 3⌋ | 2 | 4 | −18 | −36 |
| | | 6 | 30 | 36 |
| | 2 | 10 | 12 | 0 |

Thus,

$$2x^4 + 2x^3 - 22x^2 - 18x + 36 = 0$$
$$(x-1)(x-3)\left(2x^2 + 10x + 12\right) = 0$$
$$2(x-1)(x-3)\left(x^2 + 5x + 6\right) = 0$$
$$2(x-1)(x-3)(x+3)(x+2) = 0$$
$$x = 1,\ x = 3,\ x = -3,\ x = -2$$

The zeros are −3, −2, 1, and 3.

**b.**

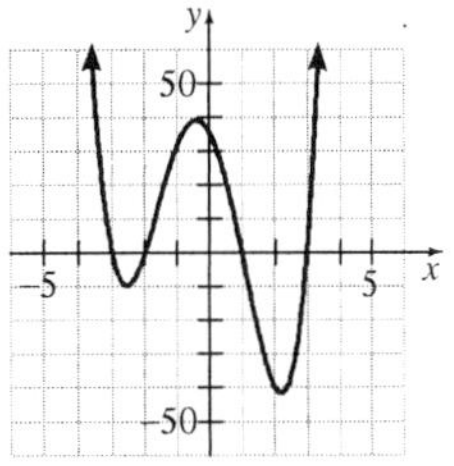


**55.** $f(x) = 3x^5 + 2x^4 - 15x^3 - 10x^2 + 12x + 8$

**a.** From the graph provided, we can see that 1 and 2 are *x*-intercepts and are thus zeros of the function. We verify this below:

| | | | | | | |
|---|---|---|---|---|---|---|
| 1⌋ | 3 | 2 | −15 | −10 | 12 | 8 |
| | | 3 | 5 | −10 | −20 | −8 |
| | 3 | 5 | −10 | −20 | −8 | 0 |

Thus,
$3x^5+2x^4-15x^3-10x^2+12x+8$
$=(x-1)(3x^4+5x^3-10x^2-20x-8)$

| | | | | | |
|---|---|---|---|---|---|
| 2⌋ | 3 | 5 | −10 | −20 | −8 |
| | | 6 | 22 | 24 | 8 |
| | 3 | 11 | 12 | 4 | 0 |

Thus,
$3x^5+2x^4-15x^3-10x^2+12x+8$
$=(x-1)(3x^4+5x^3-10x^2-20x-8)$
$=(x-1)(x-2)(3x^3+11x^2+12x+4)$

To factor $3x^3+11x^2+12x+4$, we use the Rational Zero Theorem to determine possible rational zeros.

Factors of the constant term 4:
$\pm 1,\ \pm 2,\ \pm 4$

Factors of the leading coefficient 3:
$\pm 1,\ \pm 3$

The possible rational zeros are:

$$\frac{\text{Factors of } 4}{\text{Factors of } 3}=\frac{\pm 1,\ \pm 2,\ \pm 4}{\pm 1,\ \pm 3}$$
$$=\pm 1,\ \pm 2,\ \pm 4,\ \pm\frac{1}{3},\ \pm\frac{2}{3},\ \pm\frac{4}{3}$$

We test values from above until we find a zero. One possibility is shown next:

Test $-1$:

| | | | | |
|---|---|---|---|---|
| −1⌋ | 3 | 11 | 12 | 4 |
| | | −3 | −8 | −4 |
| | 3 | 8 | 4 | 0 |

The remainder is 0, so $-1$ is a zero of $f$. We can now finish the factoring:

$$3x^5+2x^4-15x^3-10x^2+12x+8=0$$
$$(x-1)(3x^4+5x^3-10x^2-20x-8)=0$$
$$(x-1)(x-2)(3x^3+11x^2+12x+4)=0$$
$$(x-1)(x-2)(x+1)(3x^2+8x+4)=0$$
$$(x-1)(x-2)(x+1)(3x+2)(x+2)=0$$
$$x=1,\ x=2,\ x=-1,\ x=-\frac{2}{3},\ x=-2$$

The zeros are $-2$, $-1$, $-\frac{2}{3}$, 1 and 2.

**b.**

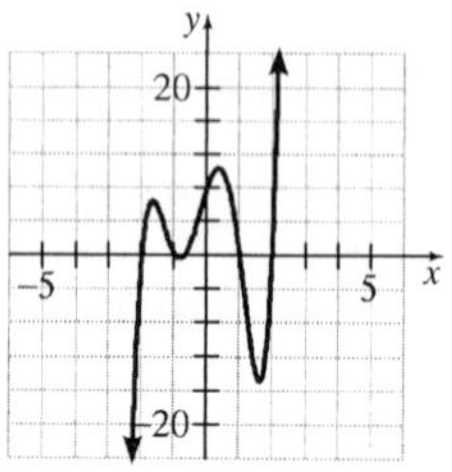


**56.** $f(x)=-5x^4+4x^3-19x^2+16x+4$

**a.** From the graph provided, we can see that 1 is an $x$-intercept and is thus a zero of the function. We verify this below:

| | | | | | |
|---|---|---|---|---|---|
| 1⌋ | −5 | 4 | −19 | 16 | 4 |
| | | −5 | −1 | −20 | −4 |
| | −5 | −1 | −20 | −4 | 0 |

Thus, $-5x^4+4x^3-19x^2+16x+4=0$
$$(x-1)(-5x^3-x^2-20x-4)=0$$
$$-(x-1)(5x^3+x^2+20x+4)=0$$

To factor $5x^3+x^2+20x+4$, we use the Rational Zero Theorem to determine possible rational zeros.

Factors of the constant term 4:
$\pm 1,\ \pm 2,\ \pm 4$

Factors of the leading coefficient 5:
$\pm 1,\ \pm 5$

The possible rational zeros are:

$$\frac{\text{Factors of } 4}{\text{Factors of } 5} = \frac{\pm 1,\ \pm 2,\ \pm 4}{\pm 1,\ \pm 5}$$

$$= \pm 1,\ \pm 2,\ \pm 4,\ \pm\frac{1}{5},\ \pm\frac{2}{5},\ \pm\frac{4}{5}$$

We test values from above until we find a zero. One possibility is shown next:

Test $-\frac{1}{5}$:

$$\begin{array}{r|rrrr} -\frac{1}{5} & 5 & 1 & 20 & 4 \\ & & -1 & 0 & -4 \\ \hline & 5 & 0 & 20 & 0 \end{array}$$

The remainder is 0, so $-\frac{1}{5}$ is a zero of *f*.

$$-5x^4 + 4x^3 - 19x^2 + 16x + 4 = 0$$
$$(x-1)(-5x^3 - x^2 - 20x - 4) = 0$$
$$-(x-1)(5x^3 + x^2 + 20x + 4) = 0$$
$$-(x-1)\left(x+\frac{1}{5}\right)(5x^2+20) = 0$$
$$-5(x-1)\left(x+\frac{1}{5}\right)(x^2+4) = 0$$
$$-5(x-1)\left(x+\frac{1}{5}\right)(x+2i)(x-2i) = 0$$
$$x = 1,\ x = -\frac{1}{5},\ x = -2i,\ x = 2i$$

The zeros are $-\frac{1}{5}$, 1, and $\pm 2i$.

**b.**

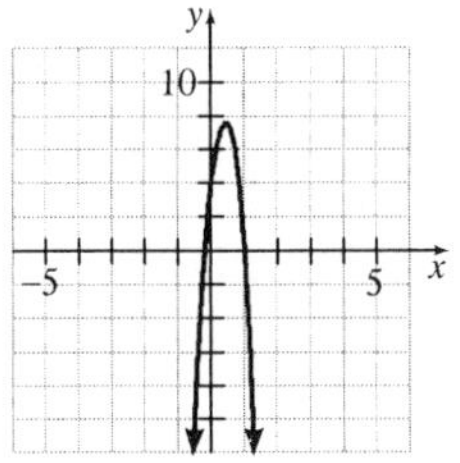


**57. a.** From the graph, it appears that $f(x) = 27$ when $x = 40\text{s}$. This means that artists completed 27% of their professional work in their 40s.

**b.** From the graph, it appears that the degree must be even and that the leading coefficient must be negative.

**59.** $V = lwh$

$$48 = (8-2x)(10-2x)x$$
$$48 = 2(4-x)\cdot 2(5-x)\cdot x$$
$$12 = (4-x)(5-x)x$$
$$12 = 20x - 9x^2 + x^3$$
$$x^3 - 9x^2 + 20x - 12 = 0$$

To factor $x^3 - 9x^2 + 20x - 12$, we use the Rational Zero Theorem to determine possible rational roots.

Factors of the constant term $-12$:
$\pm 1,\ \pm 2,\ \pm 3,\ \pm 4,\ \pm 6,\ \pm 12$

Factors of the leading coefficient 1: $\pm 1$

The possible rational roots are:

$$\frac{\text{Factors of } -12}{\text{Factors of } 1} = \frac{\pm 1,\ \pm 2,\ \pm 3,\ \pm 4,\ \pm 6,\ \pm 12}{\pm 1}$$
$$= \pm 1,\ \pm 2,\ \pm 3,\ \pm 4,\ \pm 6,\ \pm 12$$

We test values from above until we find a root. One possibility is shown next:

Test 1:

$$\begin{array}{r|rrrr} 1 & 1 & -9 & 20 & -12 \\ & & 1 & -8 & 12 \\ \hline & 1 & -8 & 12 & 0 \end{array}$$

The remainder is 0, so 1 is a root. Thus,

$$x^3 - 9x^2 + 20x - 12 = 0$$
$$(x-1)(x^2 - 8x + 12) = 0$$
$$(x-1)(x-2)(x-6) = 0$$

$(x-1)(x-2)(x-6)=0$

$x=1,\ x=2,\ x=6$

We disregard 6 because if $x=6$ would result in a negative length and width of the box, which is impossible. Thus, either a 1-inch square or a 2-inch square can be cut.

**61. – 69.** Answers will vary.

**71.** $2x^3-15x^2+22x+15=0$

Factors of the constant term 15:
$\pm1,\ \pm3,\ \pm5,\ \pm15$

Factors of the leading coefficient 2:
$\pm1,\ \pm2$

The possible rational zeros are:

$$\frac{\text{Factors of 15}}{\text{Factors of 2}}=\frac{\pm1,\ \pm3,\ \pm5,\ \pm15}{\pm1,\ \pm2}$$

$$=\pm1,\pm3,\pm5,\pm15,\pm\frac{1}{2},\pm\frac{3}{2},\pm\frac{5}{2},\pm\frac{15}{2}$$

**71.** $6x^3-19x^2+16x-4=0$

Factors of the constant term $-4$:
$\pm1,\ \pm2,\ \pm4$

Factors of the leading coefficient 3:
$\pm1,\ \pm2,\ \pm3,\ \pm6$

The possible rational zeros are:

$$\frac{\text{Factors of }-4}{\text{Factors of 6}}=\frac{\pm1,\ \pm2,\ \pm4}{\pm1,\ \pm2,\ \pm3,\ \pm6}$$

$$=\pm1,\pm2,\pm4,\pm\frac{1}{2},\pm\frac{1}{3},\pm\frac{2}{3},\pm\frac{4}{3},\pm\frac{1}{6}$$

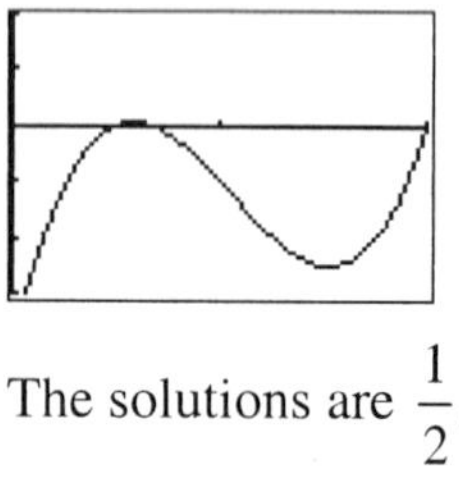

The solutions are $\frac{1}{2}$, $\frac{2}{3}$, and 2.

**73.** $4x^4+4x^3+7x^2-x-2=0$

Factors of the constant term $-2$:
$\pm1,\ \pm2$

Factors of the leading coefficient 3:
$\pm1,\ \pm2,\ \pm4$

The possible rational zeros are:

$$\frac{\text{Factors of }-2}{\text{Factors of 4}}=\frac{\pm1,\ \pm2}{\pm1,\ \pm2,\ \pm4}$$

$$=\pm1,\ \pm2,\ \pm\frac{1}{2},\ \pm\frac{1}{4}$$

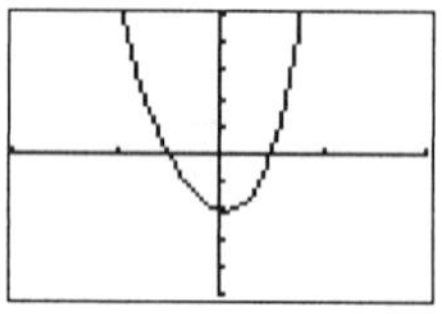

The solutions are $-\frac{1}{2}$ and $\frac{1}{2}$.

**75.** $f(x)=x^5-x^4+x^3-x^2+x-8$

Since $f(x)$ has five sign changes, it will have 5, 3, or 1 positive real zeros.

$$f(-x)=(-x)^5-(-x)^4+(-x)^3-(-x)^2+(-x)-8$$
$$=-x^5-x^4-x^3-x^2-x-8$$

Since $f(-x)$ has no sign changes, $f$ will have 0 negative real zeros.

Graph the function using the following window setting.

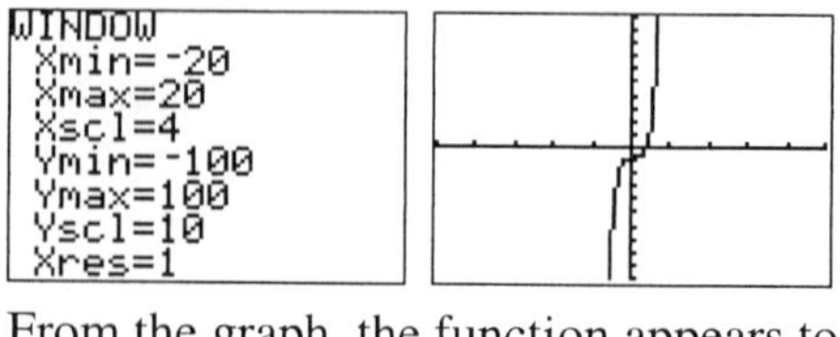


From the graph, the function appears to have 1 positive real solution, which agrees with our result.

**77.** Statement **c.** is true.

For part **a.,** the equation has no sign changes. Therefore by Descartes's Rule of Signs, it has no positive real roots.

For part **b.**, Descartes's Rule of Signs does not generally give the exact number of positive and negative real roots for polynomial equation, though it may in specific instances do so.

Since part **c**. is true, part **d.** is not.

**79.** $(2x+1)(x+5)(x+2)-3x(x+5)=208$

$$2x^3+15x^2+27x+10-3x^2-15x=208$$
$$2x^3+12x^2+12x-198=0$$
$$x^3+6x^2+6x-99=0$$

To factor $x^3+6x^2+6x-99=0$, we use the Rational Zero Theorem to determine possible rational roots.

Factors of the constant term $-99$:
$\pm1,\ \pm3,\ \pm9,\ \pm11,\ \pm33,\ \pm99$

Factors of the leading coefficient 1: $\pm1$

The possible rational roots are:

$$\frac{\text{Factors of } -99}{\text{Factors of } 1} = \frac{\pm1,\ \pm3,\ \pm9,\ \pm11,\ \pm33,\ \pm99}{\pm1}$$
$$=\pm1,\ \pm3,\ \pm9,\ \pm11,\ \pm33,\ \pm99$$

Using Descartes's Rule of signs, since $x^3+6x^2+6x-99$ has one sign change, it must have exactly one positive real root. Now, since $x$ represents a length, we are only interested in finding this one positive real root.

We test positive values from above until we find the root:

Test 3:

$$\begin{array}{r|rrrr} 3 & 1 & 6 & 6 & -99 \\ & & 3 & 27 & 99 \\ \hline & 1 & 9 & 33 & 0 \end{array}$$

The remainder is 0, so 3 is a root. In fact, it is the one positive real we need. Thus, the measure of the side designated by $x$ is 3 inches.

**81.** $x^2+2x+3>11$

$x^2+2x+3>11$

Solve the related quadratic equation.

$$x^2+2x+3=11$$
$$x^2+2x-8=0$$
$$(x+4)(x-2)=0$$

Apply the zero product principle.

$x+4=0$ or $x-2=0$

$x=-4$ $\quad$ $x=2$

The boundary points are $-4$ and 2.

| Test Interval | Test No. | Test | Conclusion |
|---|---|---|---|
| $(-\infty,-4)$ | $-5$ | $(-5)^2+2(-5)+3>11$<br>$17>11$,<br>True | $(-\infty,-4)$ belongs to the solution set. |
| $(-4,2)$ | 0 | $0^2+2(0)+3>11$<br>$3>11$,<br>False | $(-4,2)$ does not belong to the solution set. |
| $(2,\infty)$ | 3 | $3^2+2(3)+3>11$<br>$18>11$,<br>True | $\left(\frac{3}{2},\infty\right)$ belongs to the solution set. |

The solution set is $\{x\,|\,x<-4 \text{ or } x>2\}$ or $(-\infty,-4)\cup(2,\infty)$.

**82.** $16x^2 + 9y^2 = 144$

$$\frac{16x^2 + 9y^2}{144} = \frac{144}{144}$$

$$\frac{x^2}{9} + \frac{y^2}{16} = 1$$

$$\frac{x^2}{3^2} + \frac{y^2}{4^2} = 1$$

The graph of the equation will be an ellipse centered at $(0,0)$ with vertices $(0,4)$ and $(0,-4)$. The $x$-intercepts will be $(3,0)$ and $(-3,0)$.

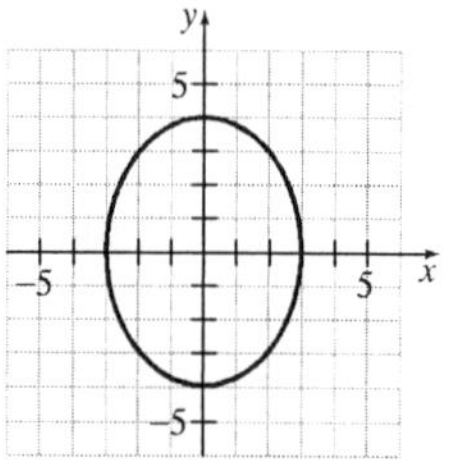


**83.** $x^3 - 125y^3 = x^3 - (5y)^3$

$= (x-5y)\left(x^2 + x(5y) + (5y)^2\right)$

$= (x-5y)\left(x^2 + 5xy + 25y^2\right)$

## Mid-Chapter Check Point – Chapter 11

1. $f(x) = (x-2)^2(x+1)^3$

   $(x-2)^2(x+1)^3 = 0$

   Apply the zero-product principle:

   $(x-2)^2 = 0$ or $(x+1)^3 = 0$

   $x - 2 = 0$ $\quad$ $x + 1 = 0$

   $x = 2$ $\quad$ $x = -1$

   The zeros are $-1$ and 2.

   The graph of *f* crosses the $x$-axis at $-1$, since the zero has multiplicity 3. The graph touches the $x$-axis and turns around at 2 since the zero has multiplicity 2.

   Since *f* is an odd-degree polynomial, degree 5, and since the leading coefficient, 1, is positive, the graph falls to the left and rises to the right.

   Plot additional points as necessary and construct the graph.

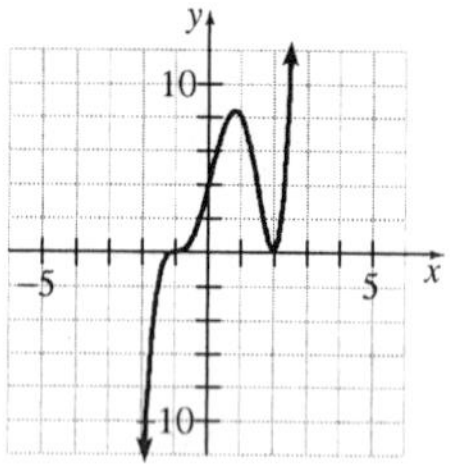


2. $f(x) = -(x-2)^2(x+1)^2$

   $-(x-2)^2(x+1)^2 = 0$

   Apply the zero-product principle:

   $(x-2)^2 = 0$ or $(x+1)^2 = 0$

   $x - 2 = 0$ $\quad$ $x + 1 = 0$

   $x = 2$ $\quad$ $x = -1$

   The zeros are $-1$ and 2.

   The graph touches the $x$-axis and turns around both at $-1$ and 2 since both zeros have multiplicity 2.

   Since *f* is an even-degree polynomial, degree 4, and since the leading coefficient, $-1$, is negative, the graph falls to the left and falls to the right.

   Plot additional points as necessary and construct the graph.

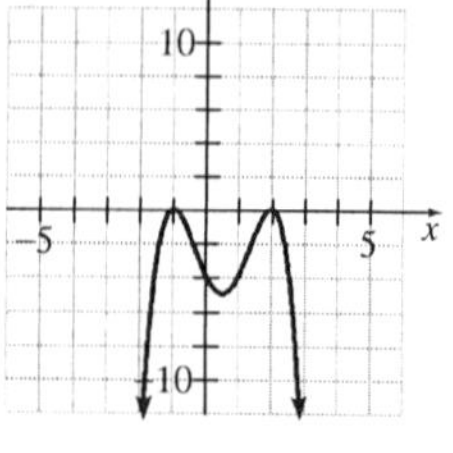

**3.** $f(x) = x^3 - x^2 - 4x + 4$

$$x^3 - x^2 - 4x + 4 = 0$$
$$x^2(x-1) - 4(x-1) = 0$$
$$(x^2 - 4)(x-1) = 0$$
$$(x+2)(x-2)(x-1) = 0$$

Apply the zero-product principle:

$x+2=0$ or $x-2=0$ or $x-1=0$

$x=-2$  $x=2$  $x=1$

The zeros are $-2$, 1, and 2.

The graph of *f* crosses the *x*-axis at all three zeros, $-2$, 1, and 2, since all have multiplicity 1.

Since *f* is an odd-degree polynomial, degree 3, and since the leading coefficient, 1, is positive, the graph falls to the left and rises to the right.

Plot additional points as necessary and construct the graph.

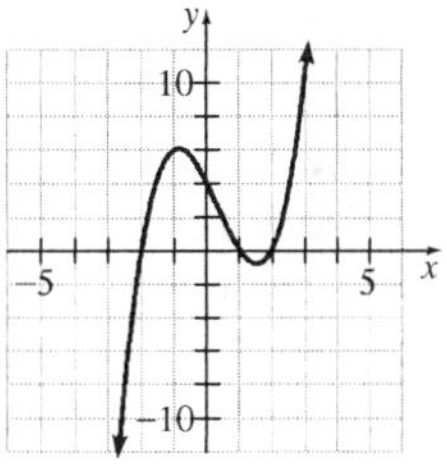


**4.** $f(x) = x^4 - 5x^2 + 4$

$$x^4 - 5x^2 + 4 = 0$$
$$(x^2 - 4)(x^2 - 1) = 0$$
$$(x+2)(x-2)(x+1)(x-1) = 0$$

Apply the zero-product principle,

$x=-2,\ x=2,\ x=-1,\ x=1$

The zeros are $-2$, $-1$, 1, and 2.

The graph crosses the *x*-axis at all four zeros, $-2$, $-1$, 1, and 2., since all have multiplicity 1.

Since *f* is an even-degree polynomial, degree 4, and since the leading coefficient, 1, is positive, the graph rises to the left and rises to the right.

Plot additional points as necessary and construct the graph.

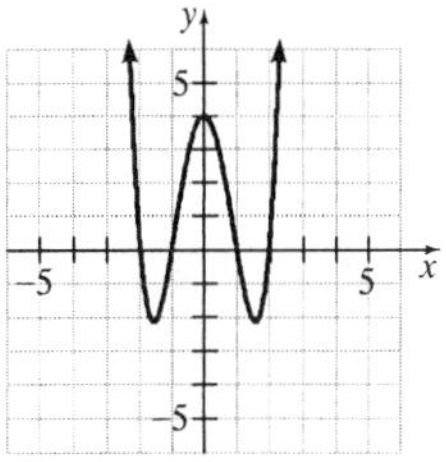


**5.** $f(x) = -(x+1)^6$

$$-(x+1)^6 = 0$$
$$(x+1)^6 = 0$$
$$x+1 = 0$$
$$x = -1$$

The zero is are $-1$.

The graph touches the *x*-axis and turns around at $-1$ since the zero has multiplicity 6.

Since *f* is an even-degree polynomial, degree 6, and since the leading coefficient, $-1$, is negative, the graph falls to the left and falls to the right.

Plot additional points as necessary and construct the graph.

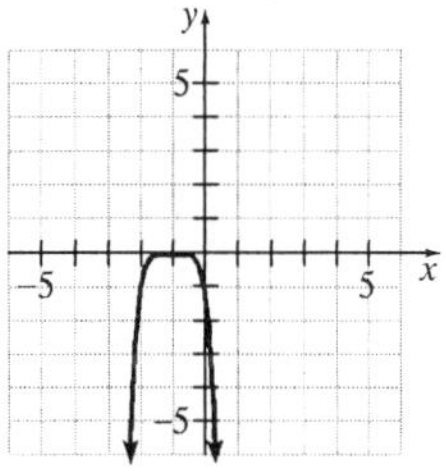

**6.** $f(x) = -6x^3 + 7x^2 - 1$

To find the zeros, we use the Rational Zero Theorem:

List all factors of the constant term $-1$: $\pm 1$

List all factors of the leading coefficient $-6$: $\pm 1,\ \pm 2,\ \pm 3,\ \pm 6$

The possible rational zeros are:

$$\frac{\text{Factors of } -1}{\text{Factors of } -6} = \frac{\pm 1}{\pm 1,\ \pm 2,\ \pm 3,\ \pm 6}$$

$$= \pm 1,\ \pm \frac{1}{2},\ \pm \frac{1}{3},\ \pm \frac{1}{6}$$

We test values from the above list until we find a zero. One is shown next:

Test 1:

$$\begin{array}{r|rrrr} 1 & -6 & 7 & 0 & -1 \\ & & -6 & 1 & 1 \\ \hline & -6 & 1 & 1 & 0 \end{array}$$

The remainder is 0, so 1 is a zero. Thus,

$$-6x^3 + 7x^2 - 1 = 0$$

$$(x-1)(-6x^2 + x + 1) = 0$$

$$-(x-1)(6x^2 - x - 1) = 0$$

$$-(x-1)(3x+1)(2x-1) = 0$$

Apply the zero-product property:

$$x = 1, \quad x = -\frac{1}{3}, \quad x = \frac{1}{2}$$

The zeros are $-\frac{1}{3}$, $\frac{1}{2}$, and 1.

The graph of $f$ crosses the $x$-axis at all three zeros, $-\frac{1}{3}$, $\frac{1}{2}$, and 1, since all have multiplicity 1.

Since $f$ is an odd-degree polynomial, degree 3, and since the leading coefficient, $-6$, is negative, the graph rises to the left and falls to the right.

Plot additional points as necessary and construct the graph.

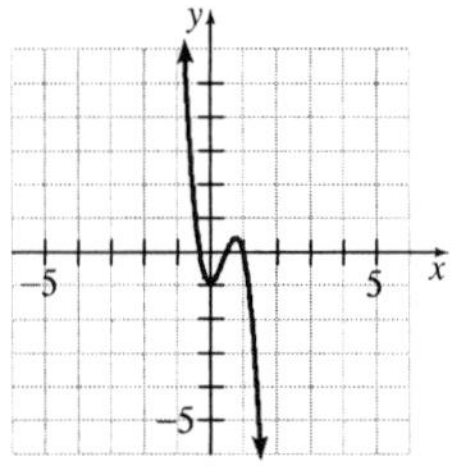

**7.** $f(x) = 2x^3 - 2x$

$$2x^3 - 2x = 0$$

$$2x(x^2 - 1) = 0$$

$$2x(x+1)(x-1) = 0$$

Apply the zero-product principle:

$x = 0, \quad x = -1, \quad x = 1$

The zeros are $-1$, 0, and 1.

The graph of $f$ crosses the $x$-axis at all three zeros, $-1$, 0, and 1, since all have multiplicity 1.

Since $f$ is an odd-degree polynomial, degree 3, and since the leading coefficient, 2, is positive, the graph falls to the left and rises to the right.

Plot additional points as necessary and construct the graph.

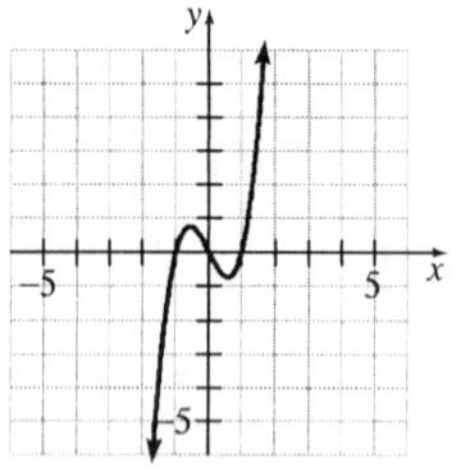

**8.** $f(x)=x^3-2x^2+26x$

$x^3-2x^2+26x=0$

$x(x^2-2x+26)=0$

Note that $x^2-2x+26$ does not factor, so we use the quadratic formula:

$x=0$ or $x^2-2x+26=0$

$a=1,\ b=-2,\ c=26$

$$x=\frac{-(-2)\pm\sqrt{(-2)^2-4(1)(26)}}{2(1)}$$

$$=\frac{2\pm\sqrt{-100}}{2}=\frac{2\pm10i}{2}=1\pm5i$$

The zeros are 0 and $1\pm5i$.

The graph of *f* crosses the *x*-axis at 0 (the only real zero), since it has multiplicity 1.

Since *f* is an odd-degree polynomial, degree 3, and since the leading coefficient, 1, is positive, the graph falls to the left and rises to the right.

Plot additional points as necessary and construct the graph.

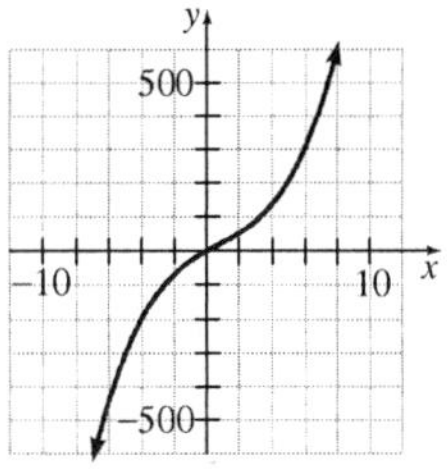


**9.** $f(x)=-x^3+5x^2-5x-3$

To find the zeros, we use the Rational Zero Theorem:

List all factors of the constant term $-3$: $\pm1,\ \pm3$

List all factors of the leading coefficient $-1$: $\pm1$

The possible rational zeros are:

$$\frac{\text{Factors of }-3}{\text{Factors of }-1}=\frac{\pm1,\ \pm3}{\pm1}=\pm1,\ \pm3$$

We test values from the previous list until we find a zero. One is shown next:

Test 3:

| | | | | |
|---|---|---|---|---|
| 3 | −1 | 5 | −5 | −3 |
| | | −3 | 6 | 3 |
| | −1 | 2 | 1 | 0 |

The remainder is 0, so 3 is a zero. Thus,

$-x^3+5x^2-5x-3=0$

$(x-3)(-x^2+2x+1)=0$

$-(x-3)(x^2-2x-1)=0$

Note that $x^2-2x-1$ does not factor, so we use the quadratic formula:

$x-3=0$ or $x^2-2x-1=0$

$x=3$ $\quad a=1,\ b=-2,\ c=-1$

$$x=\frac{-(-2)\pm\sqrt{(-2)^2-4(1)(-1)}}{2(1)}$$

$$=\frac{2\pm\sqrt{8}}{2}=\frac{2\pm2\sqrt{2}}{2}=1\pm\sqrt{2}$$

The zeros are 3 and $1\pm\sqrt{2}$.

The graph of *f* crosses the *x*-axis at all three zeros, 3 and $1\pm\sqrt{2}$, since all have multiplicity 1.

Since *f* is an odd-degree polynomial, degree 3, and since the leading coefficient, $-1$, is negative, the graph rises to the left and falls to the right.

Plot additional points as necessary and construct the graph.

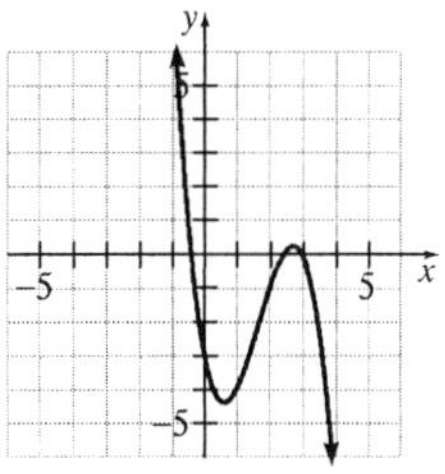

**10.** $x^3 - 3x + 2 = 0$

We begin by using the Rational Zero Theorem to determine possible rational roots.

Factors of the constant term 2: $\pm 1, \ \pm 2$

Factors of the leading coefficient 1: $\pm 1$

The possible rational zeros are:

$$\frac{\text{Factors of } 2}{\text{Factors of } 1} = \frac{\pm 1, \ \pm 2}{\pm 1} = \pm 1, \ \pm 2$$

We test values from above until we find a root. One is shown next:

Test 1:

$$\begin{array}{r|rrrr} 1 & 1 & 0 & -3 & 2 \\ & & 1 & 1 & -2 \\ \hline & 1 & 1 & -2 & 0 \end{array}$$

The remainder is 0, so 1 is a root of the equation. Thus,

$$x^3 - 3x + 2 = 0$$
$$(x-1)(x^2 + x - 2) = 0$$
$$(x-1)(x+2)(x-1) = 0$$
$$(x-1)^2(x+2) = 0$$

Apply the zero-product property:

$$(x-1)^2 = 0 \quad \text{or} \quad x + 2 = 0$$
$$x - 1 = 0 \qquad\qquad x = -2$$
$$x = 1$$

The solutions are $-2$ and 1, and the solution set is $\{-2, 1\}$.

**11.** $6x^3 - 11x^2 + 6x - 1 = 0$

We begin by using the Rational Zero Theorem to determine possible rational roots.

Factors of the constant term $-1$: $\pm 1$

Factors of the leading coefficient 6: $\pm 1, \ \pm 2, \ \pm 3, \ \pm 6$

The possible rational zeros are:

$$\frac{\text{Factors of } -1}{\text{Factors of } 6} = \frac{\pm 1}{\pm 1, \ \pm 2, \ \pm 3, \ \pm 6}$$
$$= \pm 1, \ \pm \frac{1}{2}, \ \pm \frac{1}{3}, \ \pm \frac{1}{6}$$

We test values from above until we find a root. One is shown next:

Test 1:

$$\begin{array}{r|rrrr} 1 & 6 & -11 & 6 & -1 \\ & & 6 & -5 & 1 \\ \hline & 6 & -5 & 1 & 0 \end{array}$$

The remainder is 0, so 1 is a root of the equation. Thus,

$$6x^3 - 11x^2 + 6x - 1 = 0$$
$$(x-1)(6x^2 - 5x + 1) = 0$$
$$(x-1)(3x-1)(2x-1) = 0$$

Apply the zero-product property:

$$x - 1 = 0 \quad \text{or} \quad 3x - 1 = 0 \quad \text{or} \quad 2x - 1 = 0$$
$$x = 1 \qquad\qquad x = \frac{1}{3} \qquad\qquad x = \frac{1}{2}$$

The solutions are $\frac{1}{3}, \frac{1}{2}$ and 1, and the solution set is $\left\{\frac{1}{3}, \frac{1}{2}, 1\right\}$.

**12.** $(2x+1)(3x-2)^3(2x-7) = 0$

Apply the zero-product property:

$$2x + 1 = 0 \quad \text{or} \quad (3x-2)^3 = 0 \quad \text{or} \quad 2x - 7 = 0$$
$$x = -\frac{1}{2} \qquad\qquad 3x - 2 = 0 \qquad\qquad x = \frac{7}{2}$$
$$x = \frac{2}{3}$$

The solutions are $-\frac{1}{2}, \frac{2}{3}$ and $\frac{7}{2}$, and the solution set is $\left\{-\frac{1}{2}, \frac{2}{3}, \frac{7}{2}\right\}$.

**13.** $2x^3+5x^2-200x-500=0$

We begin by using the Rational Zero Theorem to determine possible rational roots.

Factors of the constant term $-500$:
$\pm1,\ \pm2,\ \pm4,\ \pm5,\ \pm10,\ \pm20,\ \pm25,$
$\pm50,\ \pm100,\ \pm125,\ \pm250,\ \pm500$

Factors of the leading coefficient 2:
$\pm1,\ \pm2$

The possible rational zeros are:

$$\frac{\text{Factors of 500}}{\text{Factors of 2}}=\pm1,\ \pm2,\ \pm4,\ \pm5,$$
$$\pm10,\ \pm20,\ \pm25,\ \pm50,\ \pm100,\ \pm125,$$
$$\pm250,\ \pm500,\ \pm\frac{1}{2},\ \pm\frac{5}{2},\ \pm\frac{25}{2},\ \pm\frac{125}{2}$$

We test values from above until we find a root. One is shown next:

Test 10:

| | | | | |
|---|---|---|---|---|
| 10 | 2 | 5 | −200 | −500 |
| | | 20 | 250 | 500 |
| | 2 | 25 | 50 | 0 |

The remainder is 0, so 10 is a root of the equation. Thus,

$$2x^3+5x^2-200x-500=0$$
$$(x-10)(2x^2+25x+50)=0$$
$$(x-10)(2x+5)(x+10)=0$$

Apply the zero-product property:

$x-10=0$ or $2x+5=0$ or $x+10=0$

$x=10$ $\qquad x=-\frac{5}{2}$ $\qquad x=-10$

The solutions are $-10,\ -\frac{5}{2}$, and $10$, and the solution set is $\left\{-10,\ -\frac{5}{2},\ 10\right\}$.

**14.** $x^4-x^3-11x^2=x+12$

$x^4-x^3-11x^2-x-12=0$

We begin by using the Rational Zero Theorem to determine possible rational roots.

Factors of the constant term $-12$:
$\pm1,\ \pm2,\ \pm3,\ \pm4,\ \pm6,\ \pm12$

Factors of the leading coefficient 1: $\pm1$

The possible rational zeros are:

$$\frac{\text{Factors of }-12}{\text{Factors of 1}}$$
$$=\frac{\pm1,\ \pm2,\ \pm3,\ \pm4,\ \pm6,\ \pm12}{\pm1}$$
$$=\pm1,\ \pm2,\ \pm3,\ \pm4,\ \pm6,\ \pm12$$

We test values from this list we find a root. One possibility is shown next:

Test $-3$:

| | | | | | |
|---|---|---|---|---|---|
| −3 | 1 | −1 | −11 | −1 | −12 |
| | | −3 | 12 | −3 | 12 |
| | 1 | −4 | 1 | −4 | 0 |

The remainder is 0, so $-3$ is a root of the equation. Using the Factor Theorem, we know that $x-1$ is a factor. Thus,

$$x^4-x^3-11x^2-x-12=0$$
$$(x+3)(x^3-4x^2+x-4)=0$$
$$(x+3)\left[x^2(x-4)+1(x-4)\right]=0$$
$$(x+3)(x-4)(x^2+1)=0$$

As this point we know that $-3$ and 4 are roots of the equation. Note that $x^2+1$ does not factor, so we use the square-root principle: $x^2+1=0$

$$x^2=-1$$
$$x=\pm\sqrt{-1}=\pm i$$

The roots are $-3$, 4, and $\pm i$, and the solution set is $\{-3,\ 4,\ \pm i\}$.

**15.** $2x^4 + x^3 - 17x^2 - 4x + 6 = 0$

We begin by using the Rational Zero Theorem to determine possible rational roots.

Factors of the constant term 6: $\pm 1,\ \pm 2,\ \pm 3,\ \pm 6$

Factors of the leading coefficient 4: $\pm 1,\ \pm 2$

The possible rational roots are:

$$\frac{\text{Factors of } 6}{\text{Factors of } 2} = \frac{\pm 1,\ \pm 2,\ \pm 3,\ \pm 6}{\pm 1,\ \pm 2}$$

$$= \pm 1, \pm 2, \pm 3, \pm 6, \pm\frac{1}{2}, \pm\frac{3}{2}$$

We test values from above until we find a root. One possibility is shown next:

Test $-3$:

$$\begin{array}{r|rrrrr} -3 & 2 & 1 & -17 & -4 & 6 \\ & & -6 & 15 & 6 & -6 \\ \hline & 2 & -5 & -2 & 2 & 0 \end{array}$$

The remainder is 0, so $-3$ is a root. Using the Factor Theorem, we know that $x+3$ is a factor of the polynomial. Thus,

$$2x^4 + x^3 - 17x^2 - 4x + 6 = 0$$
$$(x+3)(2x^3 - 5x^2 - 2x + 2) = 0$$

To solve the equation above, we need to factor $2x^3 - 5x^2 - 2x + 2$. We continue testing potential roots:

Test $\frac{1}{2}$:

$$\begin{array}{r|rrrr} \frac{1}{2} & 2 & -5 & -2 & 2 \\ & & 1 & -2 & -2 \\ \hline & 2 & -4 & -4 & 0 \end{array}$$

The remainder is 0, so $\frac{1}{2}$ is a zero and $x - \frac{1}{2}$ is a factor.

Summarizing our findings so far, we have

$$2x^4 + x^3 - 17x^2 - 4x + 6 = 0$$
$$(x+3)(2x^3 - 5x^2 - 2x + 2) = 0$$
$$(x+3)\left(x - \frac{1}{2}\right)(2x^2 - 4x - 4) = 0$$
$$2(x+3)\left(x - \frac{1}{2}\right)(x^2 - 2x - 2) = 0$$

At this point, we know that $-3$ and $\frac{1}{2}$ are roots of the equation. Note that $x^2 - 2x - 2$ does not factor, so we use the quadratic formula:

$x^2 - 2x - 2 = 0$

$a = 1,\ \ b = -2,\ \ c = -2$

$$x = \frac{-(-2) \pm \sqrt{(-2)^2 - 4(1)(-2)}}{2(1)}$$
$$= \frac{2 \pm \sqrt{12}}{2} = \frac{2 \pm 2\sqrt{3}}{2} = 1 \pm \sqrt{3}$$

$$x = \frac{-(-1) \pm \sqrt{(-1)^2 - 4(1)(-4)}}{2(1)}$$
$$= \frac{1 \pm \sqrt{17}}{2}$$

The solutions are $-3$, $\frac{1}{2}$, and $1 \pm \sqrt{3}$, and the solution set is $\left\{-3,\ \frac{1}{2},\ 1 \pm \sqrt{3}\right\}$.

## 11.3 Exercise Set

**1.** $f(x)=\dfrac{x}{x+4}$

There are no common factors in the numerator and the denominator. The only zero of the denominator is $-4$. Thus, the line $x=-4$ is the only vertical asymptote for the graph of *f*.

**3.** $g(x)=\dfrac{x+3}{x(x+4)}$

There are no common factors in the numerator and the denominator. The zeros of the denominator are $-4$ and 0. Thus, the lines $x=-4$ and $x=0$ are vertical asymptotes for the graph of $g$.

**5.** $h(x)=\dfrac{x}{x(x+4)}$

The numerator and denominator have a factor in common. Therefore, we simplify *h*.

$h(x)=\dfrac{x}{x(x+4)}=\dfrac{1}{x+4},\ x\neq 0.$

There only zeros of the denominator of the simplified function is $-4$. Thus, the line $x=-4$ is a vertical asymptote for the graph of *h*. (Note: the graph of *h* would have a hole at $x=0$.)

**7.** $r(x)=\dfrac{x}{x^2+4}$

The denominator cannot be factored. No real numbers make the denominator 0 (i.e., the denominator has no real zeros). Thus, the graph of *r* has no vertical asymptotes.

**9.** $f(x)=\dfrac{12x}{3x^2+1}$

The degree of the numerator, 1, is less than the degree of the denominator, 2. Thus, the graph of *f* has the line $y=0$ (i.e., the *x*-axis) as a horizontal asymptote.

**11.** $g(x)=\dfrac{12x^2}{3x^2+1}$

The degree of the numerator, 2, is equal to the degree of the denominator, 2. The leading coefficients of the numerator and denominator are 12 and 3, respectively. Thus, the equation of the horizontal asymptote is $y=\dfrac{12}{3}$ or $y=4$.

**13.** $h(x)=\dfrac{12x^3}{3x^2+1}$

The degree of the numerator, 3, is greater than the degree of the denominator, 2. Thus, the graph of *h* has no horizontal asymptote.

**15.** $f(x)=\dfrac{-2x+1}{3x+5}$

The degree of the numerator, 1, is equal to the degree of the denominator, 1. The leading coefficients of the numerator and denominator are $-2$ and 3, respectively. Thus, the equation of the horizontal asymptote is $y=\dfrac{-2}{3}$ or $y=-\dfrac{2}{3}$.

**17.** $f(x)=\dfrac{4x}{x-2}$

Step 1: $f(-x)=\dfrac{4(-x)}{(-x)-2}$

$$=\frac{-4x}{-x-2}=\frac{4x}{x+2}$$

Because $f(x)$ does not equal $f(x)$ or $-f(x)$, the graph has neither $y$-axis symmetry nor origin symmetry.

Step 2: $f(0)=\dfrac{4\cdot 0}{0-2}=0$

The $y$-intercept is 0, so the graph passes through the origin.

Step 3: $4x=0$

$$x=0$$

There is only one $x$-intercept. This verifies that the graph passes through the origin.

Step 4: $x-2=0$

$$x=2$$

The line $x=2$ is the only vertical asymptote for the graph of $f$.

Step 5: The numerator and denominator have the same degree, 1. The leading coefficients of the numerator and denominator are 4 and 1, respectively. Thus, the equation of the horizontal asymptote is $y=\dfrac{4}{1}$ or $y=4$.

Step 6: Plot some points other than the intercepts on each side of the vertical asymptote:

| $x$ | $-1$ | 1 | 3 | 4 |
|---|---|---|---|---|
| $f(x)$ | $\frac{4}{3}$ | $-4$ | 12 | 8 |

Step 7: Use the preceding information to graph the function.

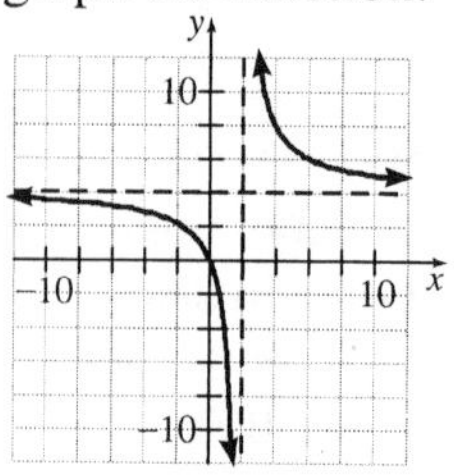


**19.** $f(x)=\dfrac{2x}{x^2-4}$

Step 1: $f(-x)=\dfrac{2(-x)}{(-x)^2-4}$

$$=\frac{-2x}{x^2-4}=-f(x)$$

Because $f(-x)=-f(x)$, the graph has origin symmetry.

Step 2: $f(0)=\dfrac{2\cdot 0}{0^2-4}=0$

The $y$-intercept is 0, so the graph passes through the origin.

Step 3: $2x=0$

$$x=0$$

There is only one $x$-intercept. This verifies that the graph passes through the origin.

Step 4: $x^2-4=0$

$$x^2=4$$

$$x=\pm 2$$

The lines $x=-2$ and $x=2$ are the vertical asymptotes for the graph of $f$.

Step 5: The degree of the numerator, 1, is less than the degree of the denominator, 2, so the graph will have the line $y=0$ ($x$-axis) as a horizontal asymptote.

Step 6: Plot some points other than the intercepts on each side of the verticals asymptotes:

| $x$ | $-3$ | $-1$ | $1$ | $3$ |
|---|---|---|---|---|
| $f(x)$ | $-\frac{6}{5}$ | $\frac{2}{3}$ | $-\frac{2}{3}$ | $\frac{6}{5}$ |

Step 7: Use the preceding information to graph the function.

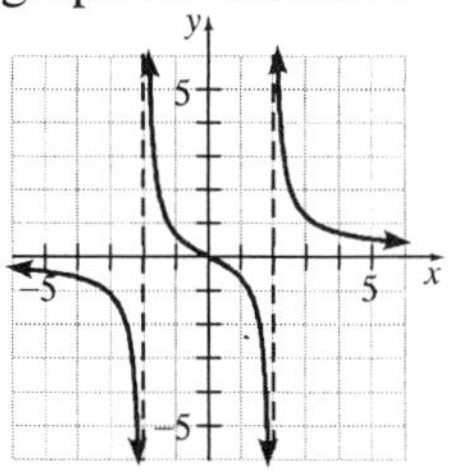

**21.** $f(x)=\dfrac{2x^2}{x^2-1}$

Step 1: $f(-x)=\dfrac{2(-x)^2}{(-x)^2-1}$

$=\dfrac{2x^2}{x^2-1}=f(x)$

Because $f(-x)=f(x)$, the graph has $y$-axis symmetry.

Step 2: $f(0)=\dfrac{2\cdot 0^2}{0^2-1}=0$

The $y$-intercept is 0, so the graph passes through the origin.

Step 3: $2x^2=0$

$x^2=0$

$x=0$

There is only one $x$-intercept. This verifies that the graph passes through the origin.

Step 4: $x^2-1=0$

$x^2=1$

$x=\pm 1$

The lines $x=-1$ and $x=1$ are the vertical asymptotes for the graph of $f$.

Step 5: The numerator and denominator have the same degree, 2. The leading coefficients of the numerator and denominator are 2 and 1, respectively. Thus, the equation of the horizontal asymptote is $y=\frac{2}{1}$ or $y=2$.

Step 6: Plot some points other than the intercepts on each side of the verticals asymptotes:

| $x$ | $-2$ | $-\frac{1}{2}$ | $\frac{1}{2}$ | $2$ |
|---|---|---|---|---|
| $f(x)$ | $\frac{8}{3}$ | $-\frac{2}{3}$ | $-\frac{2}{3}$ | $\frac{8}{3}$ |

Step 7: Use the preceding information to graph the function.

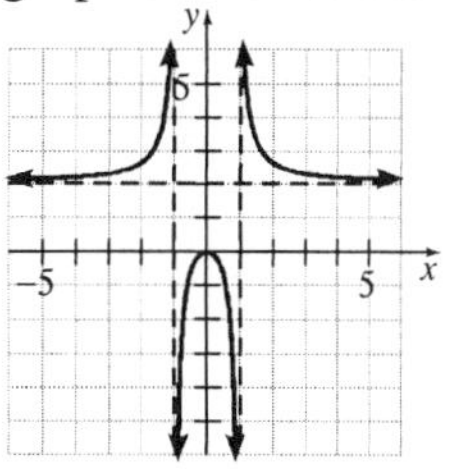

**23.** $f(x)=\dfrac{-x}{x+1}$

Step 1: $f(-x)=\dfrac{-(-x)}{(-x)+1}=\dfrac{x}{-x+1}$

Because $f(x)$ does not equal $f(x)$ or $-f(x)$, the graph has neither $y$-axis symmetry nor origin symmetry.

Step 2: $f(0)=\dfrac{-0}{0+1}=0$

The $y$-intercept is 0, so the graph passes through the origin.

Step 3: $-x=0$

$x=0$

There is only one $x$-intercept. This verifies that the graph passes through the origin.

Step 4: $x+1=0$

$$x=-1$$

The line $x=-1$ is the only vertical asymptote for the graph of *f*.

Step 5: The numerator and denominator have the same degree, 1. The leading coefficients of the numerator and denominator are $-1$ and 1, respectively. Thus, the equation of the horizontal asymptote is $y=\frac{-1}{1}$ or $y=-1$.

Step 6: Plot some points other than the intercepts on each side of the vertical asymptote:

| $x$ | $-3$ | $-2$ | $1$ | $2$ |
|---|---|---|---|---|
| $f(x)$ | $-\frac{3}{2}$ | $-2$ | $-\frac{1}{2}$ | $-\frac{2}{3}$ |

Step 7: Use the preceding information to graph the function.

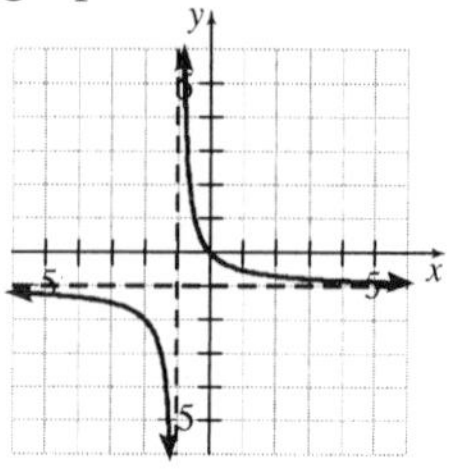


**25.** $f(x)=-\frac{1}{x^2-4}$

Step 1: $f(-x)=-\frac{1}{(-x)^2-4}$

$$=-\frac{1}{x^2-4}=f(x)$$

Because $f(-x)=f(x)$, the graph has *y*-axis symmetry.

Step 2: $f(0)=-\frac{1}{0^2-4}=\frac{1}{4}$

The *y*-intercept is $\frac{1}{4}$.

Step 3: Because the numerator is a constant, it is not possible for the numerator to equal 0. Thus, the graph has no *x*-intercepts.

Step 4: $x^2-4=0$

$$x^2=4$$

$$x=\pm 2$$

The lines $x=-2$ and $x=2$ are the vertical asymptotes.

Step 5: The degree of the numerator, 0, is less than the degree of the denominator, 2, so the graph will have the line $y=0$ (*x*-axis) as a horizontal asymptote.

Step 6: Plot some points other than the intercepts on each side of the verticals asymptotes:

| $x$ | $-3$ | $-1$ | $1$ | $3$ |
|---|---|---|---|---|
| $f(x)$ | $-\frac{1}{5}$ | $\frac{1}{3}$ | $\frac{1}{3}$ | $-\frac{1}{5}$ |

Step 7: Use the preceding information to graph the function.

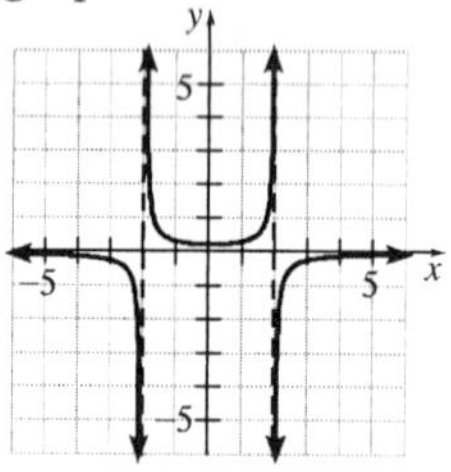


**27.** $f(x)=\frac{2}{x^2+x-2}$

Step 1: $f(-x)=\frac{2}{(-x)^2+(-x)-2}$

$$=\frac{2}{x^2-x-2}$$

Because $f(x)$ does not equal $f(x)$ or $-f(x)$, the graph has neither *y*-axis symmetry nor origin symmetry.

Step 2: $f(0)=\dfrac{2}{0^2+0-2}=-1$

The $y$-intercept is $-1$.

Step 3: Because the numerator is a constant, it is not possible for the numerator to equal 0. Thus, the graph has no $x$-intercepts.

Step 4: $x^2+x-2=0$

$(x+2)(x-1)=0$

$x=-2$ or $x=1$

The lines $x=-2$ and $x=1$ are the vertical asymptotes.

Step 5: The degree of the numerator, 0, is less than the degree of the denominator, 2, so the graph will have the line $y=0$ ($x$-axis) as a horizontal asymptote.

Step 6: Plot some points other than the intercepts on each side of the verticals asymptotes:

| $x$ | $-3$ | $-1$ | $\frac{1}{2}$ | $2$ |
|---|---|---|---|---|
| $f(x)$ | $\frac{1}{2}$ | $-1$ | $-\frac{8}{5}$ | $\frac{1}{2}$ |

Step 7: Use the preceding information to graph the function.

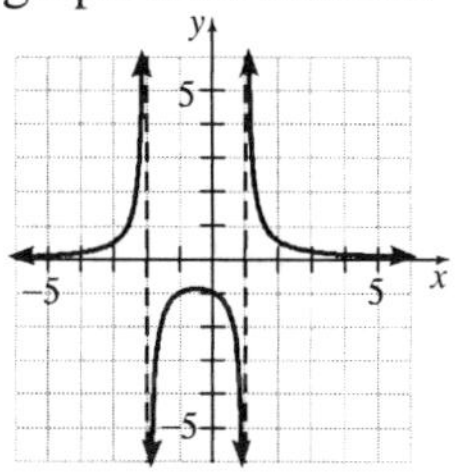


**29.** $f(x)=\dfrac{2x^2}{x^2+4}$

Step 1: $f(-x)=\dfrac{2(-x)^2}{(-x)^2+4}$

$=\dfrac{2x^2}{x^2+4}=f(x)$

Because $f(-x)=f(x)$, the graph has $y$-axis symmetry.

Step 2: $f(0)=\dfrac{2\cdot 0^2}{0^2+4}=0$

The $y$-intercept is 0, so the graph passes through the origin.

Step 3: $2x^2=0$

$x^2=0$

$x=0$

There is only one $x$-intercept. This verifies that the graph passes through the origin.

Step 4: $x^2+4=0$

$x^2=-4$

$x=\pm 2i$

Since these solutions are not real, the graph of $f$ will not have any vertical asymptotes.

Step 5: The numerator and denominator have the same degree, 2. The leading coefficients of the numerator and denominator are 2 and 1, respectively. Thus, the equation of the horizontal asymptote is $y=\dfrac{2}{1}$ or $y=2$.

Step 6: Plot some points other than the intercepts:

| $x$ | $-2$ | $-1$ | $1$ | $2$ |
|---|---|---|---|---|
| $f(x)$ | $1$ | $\frac{2}{5}$ | $\frac{2}{5}$ | $1$ |

Step 7: Use the preceding information to graph the function.

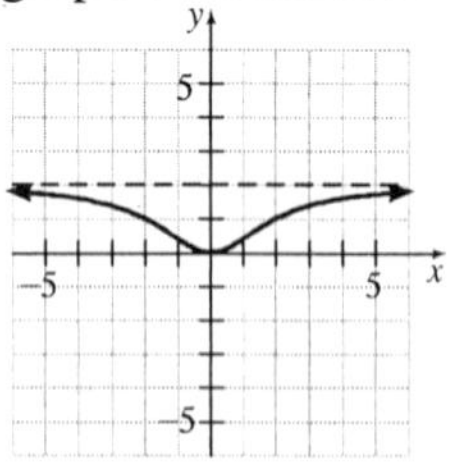


**31.** $f(x)=\dfrac{x+2}{x^2+x-6}$

Step 1: $f(-x)=\dfrac{(-x)+2}{(-x)^2+(-x)-6}$

$=\dfrac{-x+2}{x^2-x-6}$

Because $f(x)$ does not equal $f(x)$ or $-f(x)$, the graph has neither *y*-axis symmetry nor origin symmetry.

Step 2: $f(0)=\dfrac{0+2}{0^2+0-6}=-\dfrac{1}{3}$

The *y*-intercept is $-\dfrac{1}{3}$.

Step 3: $x+2=0$

$x=-2$

There is only one *x*-intercept, $-2$.

Step 4: $x^2+x-6=0$

$(x+3)(x-2)=0$

$x=-3$ or $x=2$

The lines $x=-3$ and $x=2$ are the vertical asymptotes for the graph of *f*.

Step 5: The degree of the numerator, 1, is less than the degree of the denominator, 2, so the graph will have the line $y=0$ (*x*-axis) as a horizontal asymptote.

Step 6: Plot some points other than the intercepts on each side of the verticals asymptotes:

| $x$ | $-4$ | $-1$ | $1$ | $3$ |
|---|---|---|---|---|
| $f(x)$ | $-\frac{1}{3}$ | $-\frac{1}{6}$ | $-\frac{3}{4}$ | $\frac{5}{6}$ |

Step 7: Use the preceding information to graph the function.

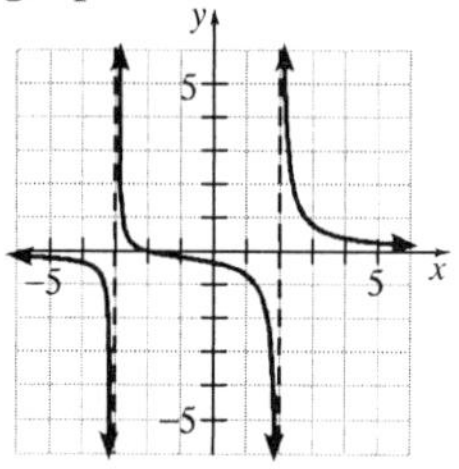


**33.** $f(x)=\dfrac{x^4}{x^2+2}$

Step 1: $f(-x)=\dfrac{(-x)^4}{(-x)^2+2}$

$=\dfrac{x^4}{x^2+2}=f(x)$

Because $f(-x)=f(x)$, the graph has *y*-axis symmetry.

Step 2: $f(0)=\dfrac{0^4}{0^2+2}=0$

The *y*-intercept is 0, so the graph passes through the origin.

Step 3: $x^4=0$

$x=0$

There is only one *x*-intercept. This verifies that the graph passes through the origin.

Step 4: $x^2+2=0$

$x^2=-2$

$x=\pm\sqrt{2}i$

Since these solutions are not real, the graph of *f* will not have any vertical asymptotes.

Step 5: The degree of the numerator, 4, is greater than the degree of the denominator, 2, so the graph will not have a horizontal asymptote.

Step 6: Plot some points other than the intercepts:

| $x$ | $-2$ | $-1$ | $1$ | $2$ |
|---|---|---|---|---|
| $f(x)$ | $\frac{8}{3}$ | $\frac{1}{3}$ | $\frac{8}{3}$ | $\frac{1}{3}$ |

Step 7: Use the preceding information to graph the function.

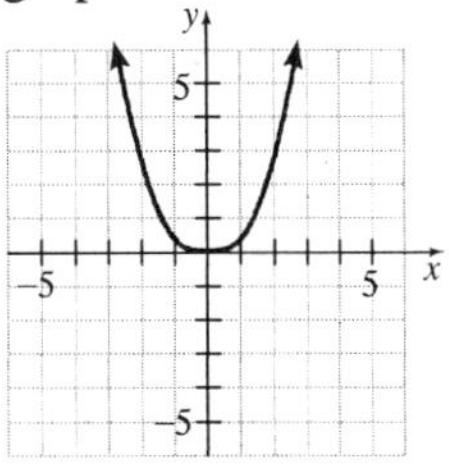


**35.** $f(x)=\dfrac{x^2+x-12}{x^2-4}$

Step 1: $f(-x)=\dfrac{(-x)^2+(-x)-12}{(-x)^2-4}$.

$=\dfrac{x^2-x-12}{x^2-4}$

Because $f(x)$ does not equal $f(x)$ or $-f(x)$, the graph has neither $y$-axis symmetry nor origin symmetry.

Step 2: $f(0)=\dfrac{0^2+0-12}{0^2-4}=3$

The $y$-intercept is 3.

Step 3: $x^2+x-12=0$

$(x+4)(x-3)=0$

$x=-4$ or $x=3$

The $x$-intercepts are $-4$ and 3.

Step 4: $x^2-4=0$

$x^2=4$

$x=\pm 2$

The lines $x=-2$ and $x=2$ are the vertical asymptotes for the graph of $f$.

Step 5: The numerator and denominator have the same degree, 2. The leading coefficients of the numerator and denominator are both 1. Thus, the equation of the horizontal asymptote is $y=\dfrac{1}{1}$ or $y=1$.

Step 6: Plot some points other than the intercepts on each side of the verticals asymptotes:

| $x$ | $-3$ | $-1$ | $1$ | $4$ |
|---|---|---|---|---|
| $f(x)$ | $-\frac{6}{5}$ | $4$ | $\frac{10}{3}$ | $\frac{2}{3}$ |

Step 7: Use the preceding information to graph the function.

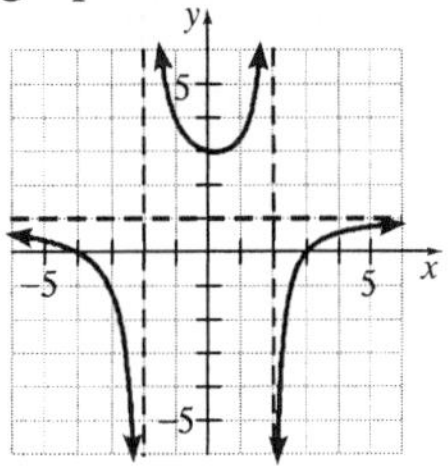


**37.** $f(x)=\dfrac{3x^2+x-4}{2x^2-5x}$

Step 1: $f(-x)=\dfrac{3(-x)^2+(-x)-4}{2(-x)^2-5(-x)}$

$=\dfrac{3x^2-x-4}{2x^2+5x}$

Because $f(x)$ does not equal $f(x)$ or $-f(x)$, the graph has neither $y$-axis symmetry nor origin symmetry.

Step 2: $f(0)=\dfrac{3\cdot 0^2+0-4}{2\cdot 0^2-5\cdot 0}=\text{undefined}$

The function does not have a $y$-intercept.

Step 3:
$$3x^2+x-4=0$$
$$(3x+4)(x-1)=0$$
$$x=-\frac{4}{3}\text{ or }x=1$$

The $x$-intercepts are $-\dfrac{4}{3}$ and 1.

Step 4:
$$2x^2-5x=0$$
$$x(2x-5)=0$$
$$x=0\text{ or }x=\frac{5}{2}$$

The lines $x=0$ ($y$-axis) and $x=\dfrac{5}{2}$ are the vertical asymptotes.

Step 5: The numerator and denominator have the same degree, 2. The leading coefficients of the numerator and denominator are 3 and 2, respectively. Thus, the equation of the horizontal asymptote is $y=\dfrac{3}{2}$.

Step 6: Plot some points other than the intercepts on each side of the verticals asymptotes:

| $x$ | $-2$ | $-\frac{1}{2}$ | 2 | 4 |
|---|---|---|---|---|
| $f(x)$ | $\frac{1}{3}$ | $\frac{11}{8}$ | $-5$ | 4 |

Step 7: Use the preceding information to graph the function.

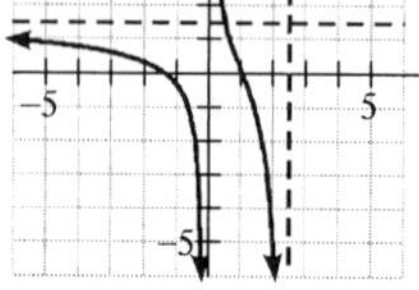


**39.** $\dfrac{5x^2}{x^2-4}\cdot\dfrac{x^2+4x+4}{10x^3}$

$$=\frac{5x^2}{(x+2)(x-2)}\cdot\frac{(x+2)^2}{10x^3}$$

$$=\frac{x+2}{2x(x-2)}$$

So, $f(x)=\dfrac{x+2}{2x(x-2)}$

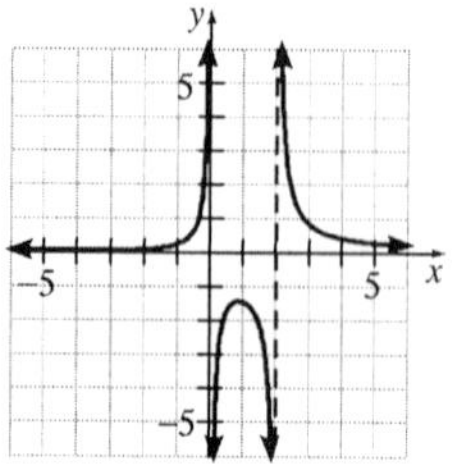


**40.** $\dfrac{x-5}{10x-2}\div\dfrac{x^2-10x+25}{25x^2-1}$

$$=\frac{x-5}{10x-2}\cdot\frac{25x^2-1}{x^2-10x+25}$$

$$=\frac{x-5}{2(5x-1)}\cdot\frac{(5x+1)(5x-1)}{(x-5)^2}$$

$$=\frac{5x+1}{2(x-5)}$$

So, $f(x)=\dfrac{5x+1}{2(x-5)}$

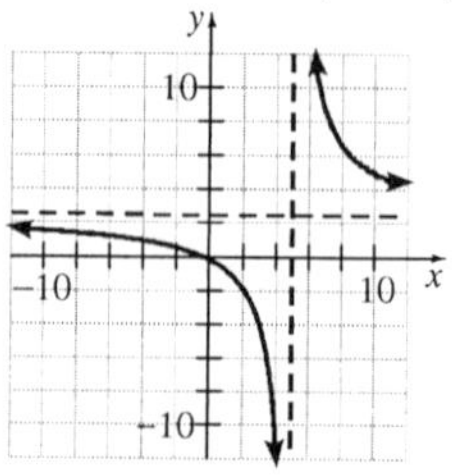

**41.** $\dfrac{x}{2x+6}-\dfrac{9}{x^2-9}$

$\dfrac{x}{2x+6}-\dfrac{9}{x^2-9}$

$=\dfrac{x}{2(x+3)}-\dfrac{9}{(x+3)(x-3)}$

$=\dfrac{x(x-3)-9(2)}{2(x+3)(x-3)}$

$=\dfrac{x^2-3x-18}{2(x+3)(x-3)}$

$=\dfrac{(x-6)\cancel{(x+3)}}{2\cancel{(x+3)}(x-3)}=\dfrac{x-6}{2(x-3)}$

So, $f(x)=\dfrac{x-6}{2(x-3)}$

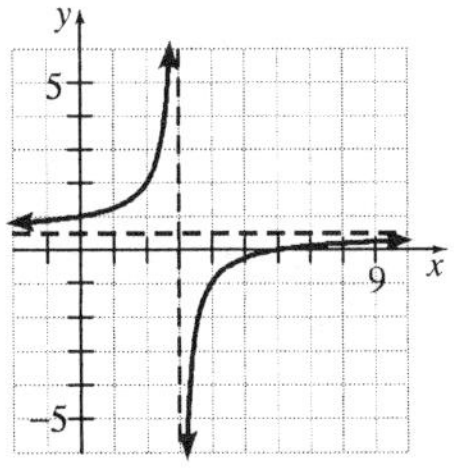


**42.** $\dfrac{2}{x^2+3x+2}-\dfrac{4}{x^2+4x+3}$

$=\dfrac{2}{(x+2)(x+1)}-\dfrac{4}{(x+3)(x+1)}$

$=\dfrac{2(x+3)-4(x+2)}{(x+2)(x+1)(x+3)}$

$=\dfrac{2x+6-4x-8}{(x+2)(x+1)(x+3)}$

$=\dfrac{-2x-2}{(x+2)(x+1)(x+3)}$

$=\dfrac{-2\cancel{(x+1)}}{(x+2)\cancel{(x+1)}(x+3)}=\dfrac{-2}{(x+2)(x+3)}$

So, $f(x)=\dfrac{-2}{(x+2)(x+3)}$

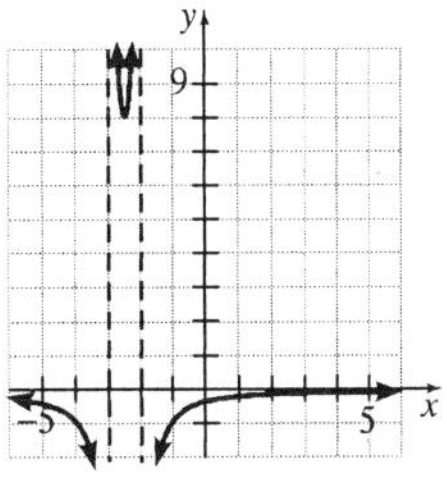


**43.** $\dfrac{1-\dfrac{3}{x+2}}{1+\dfrac{1}{x-2}}=\dfrac{1-\dfrac{3}{x+2}}{1+\dfrac{1}{x-2}}\cdot\dfrac{(x+2)(x-2)}{(x+2)(x-2)}$

$=\dfrac{(x+2)(x-2)-3(x-2)}{(x+2)(x-2)+(x+2)}$

$=\dfrac{x^2-4-3x+6}{x^2-4+x+2}$

$=\dfrac{x^2-3x+2}{x^2+x-2}$

$=\dfrac{(x-2)\cancel{(x-1)}}{(x+2)\cancel{(x-1)}}=\dfrac{x-2}{x+2}$

So, $f(x)=\dfrac{x-2}{x+2}$

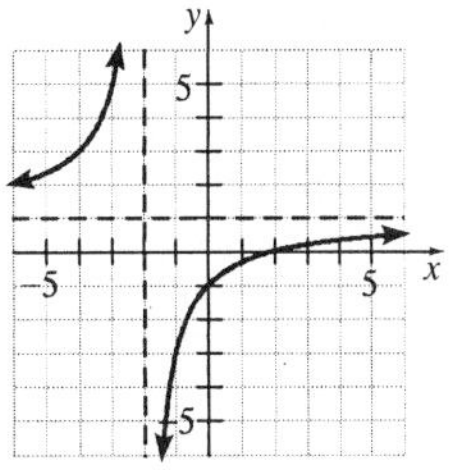

**44.** $\dfrac{x-\dfrac{1}{x}}{x+\dfrac{1}{x}}\cdot\dfrac{x}{x}=\dfrac{x^2-1}{x^2+1}=\dfrac{(x-1)(x+1)}{x^2+1}$

So, $f(x)=\dfrac{(x-1)(x+1)}{x^2+1}$

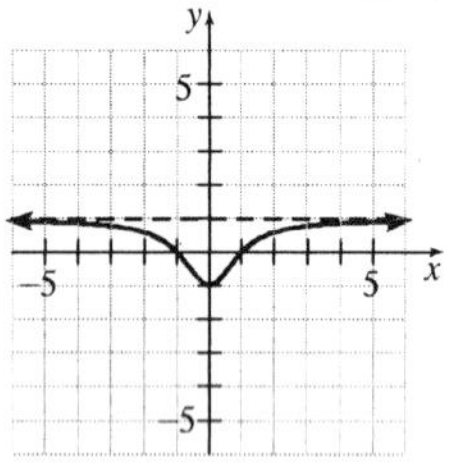


**45. a.** $f(5)=\dfrac{70(5)+100}{0.02(5)+1}=\dfrac{450}{1.1}\approx 409$

About 409 bass will be in the lake after 5 months.

**b.** $y=\dfrac{70}{0.02}$ or $y=3500$

Over time, the number of bass in he lake will approach 3500.

**47. a.** From the graph the pH level of the human mouth 42 minutes after a person eats food containing sugar will be about 6.0.

**b.** From the graph, the pH level is lowest after about 6 minutes.

$$f(6)=\frac{6.5(6)^2-20.4(6)+234}{6^2+36}$$
$$=4.8$$

The pH level after 6 minutes (i.e. the lowest pH level) is 4.8.

**c.** From the graph, the pH level appears to approach 6.5 as time goes by. Therefore, the normal pH level must be 6.5.

**d.** $y=6.5$

Over time, the pH level rises back to the normal level.

**e.** During the first hour, the pH level drops quickly below normal, and then slowly begins to approach the normal level.

**49. a.** $\dfrac{128.3}{134.5}\approx 0.954$

In 1995, there were 954 males per 1000 females.

**b.** $\dfrac{141.7}{146.7}\approx 0.966$

In 2002, there were 966 males per 1000 females.

**c.** $f(x)=\dfrac{1.256x+74.2}{1.324x+76.71}$

**d.** 1995 is 45 years after 1950, so we find

$$f(45)=\frac{1.256(45)+74.2}{1.324(45)+76.71}\approx 0.959$$

The model predicts 959 males per 1000 females in 1995. The result from the function modeled the actual number fairly well.

**d.** 2002 is 52 years after 1950, so we find

$$f(52)=\frac{1.256(52)+74.2}{1.324(52)+76.71}\approx 0.958$$

The model predicts 958 males per 1000 females in 2002. The result from the function modeled the actual number fairly well.

**e.** $y=\dfrac{1.256}{1.324}$ or $y=0.949$

This means that over time, the number of males per 1000 females will approach 949.

**51.** **a.** $F(0)=\dfrac{80}{0^2+4(0)+1}=80$

The temperature of the dessert is $80°F$ when it is initially placed in the freezer.

**b.** 1 hour: $F(1)=\dfrac{80}{1^2+4(1)+1}\approx 13.3°F$

2 hours: $F(2)=\dfrac{80}{2^2+4(2)+1}\approx 6.2°F$

3 hours: $F(3)=\dfrac{80}{3^2+4(3)+1}\approx 3.6°F$

4 hours: $F(4)=\dfrac{80}{4^2+4(4)+1}\approx 2.4°F$

5 hours: $F(5)=\dfrac{80}{5^2+4(5)+1}\approx 1.7°F$

**c.** $y=0$

Over time, the desert's temperature will approach $0°F$.

**d.**

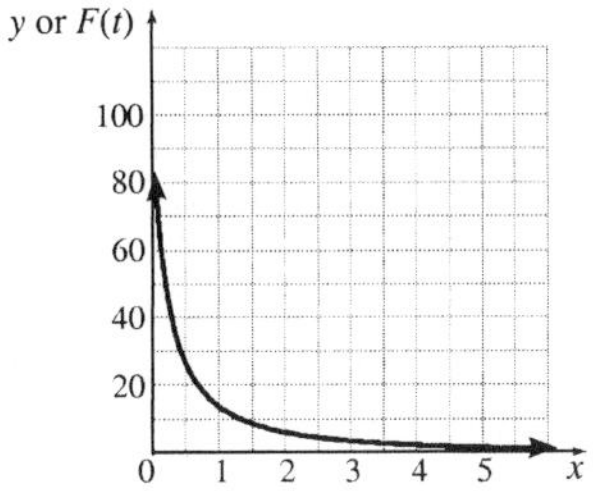


**53. – 57.** Answers will vary.

**59.** Answers will vary depending on which exercises are selected.

**61.** Statement **d.** is true.

For part **a.**, a rational function *can* have both a vertical and horizontal asymptote. For example, see exercise 17.

For part **b.**, it is possible to have a rational function whose graph has no *y*-intercept. For example, see exercise 37.

For part **c.**, a rational function can only have either one or zero horizontal asymptotes.

**63.** Answers may vary. One possibility is $f(x)=\dfrac{-3}{x-3}$.

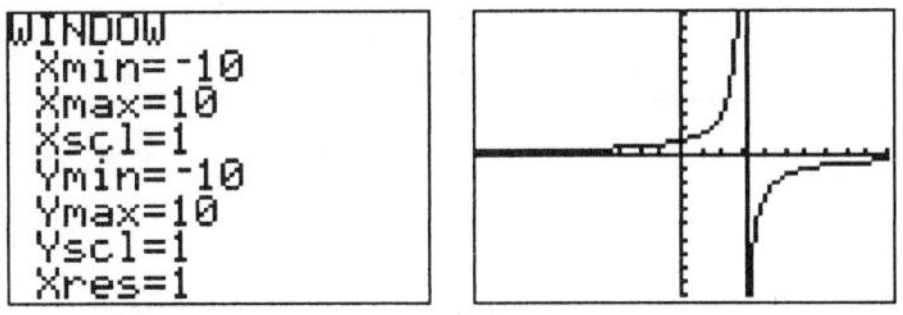


**65.**
$$\log_3 x+\log_3(x-2)=1$$
$$\log_3[x(x-2)]=1$$
$$x(x-2)=3^1$$
$$x^2-2x-3=0$$
$$(x-3)(x+1)=0$$
$$x=3 \text{ or } x=-1$$

We disregard $-1$ because we cannot take logarithms of negative numbers. Thus, the solution is 3, and the solution set is $\{3\}$.

**66.** $|4x-5|=15$

$4x-5=-15$ or $4x-5=15$

$4x=-10$    $4x=20$

$x=-\dfrac{10}{4}=-\dfrac{5}{2}$    $x=5$

The solutions are $-\dfrac{5}{2}$ and 5, and the solution set is $\left\{-\dfrac{5}{2},5\right\}$.

**67.**
$$\sqrt{2xy}\cdot\sqrt{10xy^2}=\sqrt{20x^2y^3}$$
$$=\sqrt{4\cdot 5\cdot x^2\cdot y^2\cdot y}$$
$$=2xy\sqrt{5y}$$

## Chapter 11 Review Exercises

**1.** $f(x) = -2(x-1)(x+2)^2(x+5)^3$

$-2(x-1)(x+2)^2(x+5)^3 = 0$

$(x-1)(x+2)^2(x+5)^3 = 0$

Apply the zero-product principle:

$x-1=0$ or $(x+2)^2=0$ or $(x+5)^2=0$

$x=1$ $\quad x+2=0$ $\quad x+5=0$

$x=-2$ $\quad x=-5$

The zeros of *f* are $-5$, $-2$, and 1. Since the factor $x-1$ only occurs once (exponent 1), the zero 1 has multiplicity 1. Since the multiplicity is odd, the graph crosses the *x*-axis at 1. Since the factors of $x+2$ occurs twice (exponent 2), the zero $-2$ has multiplicity 2. Since the multiplicity is even, the graph touches the *x*-axis and turns around at $-2$. Since the factor $x+5$ occurs three times (exponent 3), the zero $-5$ has multiplicity 3. Since the multiplicity is odd, the graph crosses the *x*-axis at $-5$.

**2.** $f(x) = x^3 - 5x^2 - 25x + 125$

$x^3 - 5x^2 - 25x + 125 = 0$

$x^2(x-5) - 25(x-5) = 0$

$(x^2-25)(x-5) = 0$

$(x+5)(x-5)(x-5) = 0$

$(x+5)(x-5)^2 = 0$

Apply the zero-product principle:

$x+5=0$ or $(x-5)^2=0$

$x=-5$ $\quad x-5=0$

$x=5$

The zeros of *f* are $-5$ and 5. Since the factor $x+5$ occurs only once (exponent 1), the zero $-5$ has multiplicity 1, and the graph crosses the *x*-axis at $-5$. Since the factor $x-5$ occurs twice (exponent 2), the zero 5 has multiplicity 2, and the graph touches the *x*-axis and turns around at 5.

**3.** $f(x) = x^3 - 5x$

$f(-x) = (-x)^3 - 5(-x)$

$= -x^3 + 5x = -(x^3 - 5x)$

Since $f(-x) = -f(x)$, the function is odd. Therefore, the graph of *f* is symmetric with respect to the origin.

**4.** $f(x) = x^4 - 2x^2 + 1$

$f(-x) = (-x)^4 - 2(-x)^2 + 1$

$= x^4 - 2x^2 + 1$

Since $f(-x) = f(x)$, the function is even. Therefore, the graph of *f* is symmetric with respect to the *y*-axis.

**5.** $f(x) = 2x\sqrt{1-x^2}$

$f(-x) = 2(-x)\sqrt{1-(-x)^2}$

$= -2x\sqrt{1-x^2}$

Since $f(-x) = -f(x)$, the function is odd. Therefore, the graph of *f* is symmetric with respect to the origin.

**6.** $f(x) = x^3 - 2x^2 - 9x + 9$

**a.** Since *f* is an odd-degree polynomial and since the leading coefficient, 1, is positive, the graph falls to the left and rises to the right.

**b.** $f(-x) = (-x)^3 - (-x)^2 - 9(-x) + 9$

$= -x^3 - x^2 + 9x + 2$

Since $f(-x) \neq f(x)$ and since $f(-x) \neq -f(x)$, the function is neither even nor odd, and the graph is neither symmetric with respect to the *y*-axis nor the origin.

**c.**

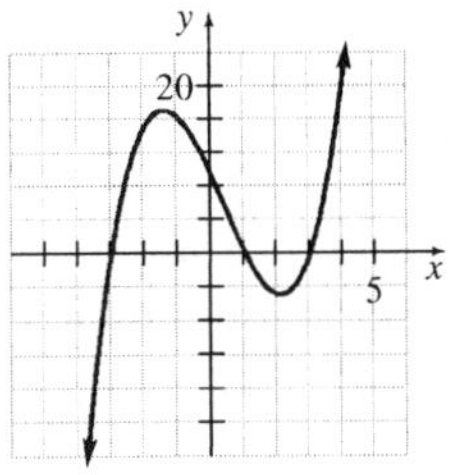


**7.** $f(x) = 4x - x^3 = -x^3 + 4x$

**a.** Since $f$ is an odd-degree polynomial and since the leading coefficient, $-1$, is negative, the graph rises to the left and falls to the right.

**b.** $f(-x) = 4(-x) - (-x)^3$

$= -4x + x^3 = -(4x - x^3)$

Since $f(-x) = -f(x)$, the function is odd, and the graph is symmetric with respect to the origin.

**c.**

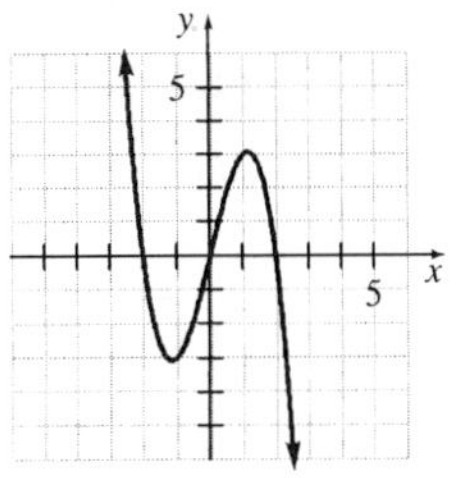


**8.** $f(x) = 2x^3 + 3x^2 - 8x - 12$

**a.** Since $f$ is an odd-degree polynomial and since the leading coefficient, 2, is positive, the graph falls to the left and rises to the right.

**b.** $f(-x) = 2(-x)^3 + 3(-x)^2 - 8(-x) - 12$

$= -2x^3 + 3x^2 + 8x - 12$

Since $f(-x) \neq f(x)$ and since $f(-x) \neq -f(x)$, the function is neither even nor odd, and the graph is neither symmetric with respect to the $y$-axis nor the origin.

**c.**

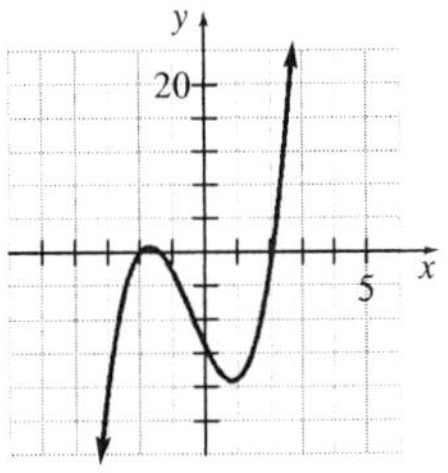


**9.** $f(x) = -x^4 + 25x^2$

**a.** Since $f$ is an even-degree polynomial and since the leading coefficient, $-1$, is negative, the graph falls to the left and falls to the right.

**b.** $f(-x) = -(-x)^4 + 25(-x)^2$

$= -x^4 + 25x^2$

Since $f(-x) = f(x)$, the graph is even and is symmetric with respect to the $y$-axis.

**c.**

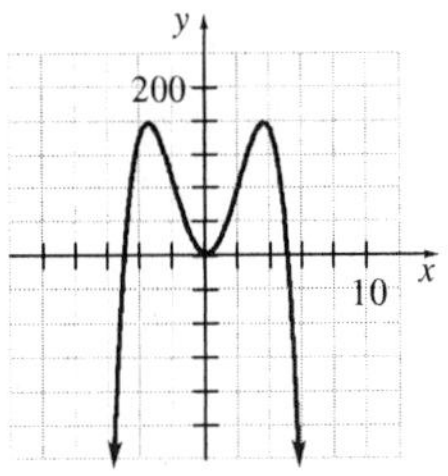


**10.** $f(x) = -x^4 + 6x^3 - 9x^2$

**a.** Since $f$ is an even-degree polynomial and since the leading coefficient, $-1$, is negative, the graph falls to the left and falls to the right.

**b.** $f(-x) = -(-x)^4 + 6(-x)^3 - 9(-x)^2$

$= -x^4 - 6x^3 - 9x^2$

Since $f(-x) \neq f(x)$ and since $f(-x) \neq -f(x)$, the function is neither even nor odd, and the graph is neither symmetric with respect to the $y$-axis nor the origin.

**c.**

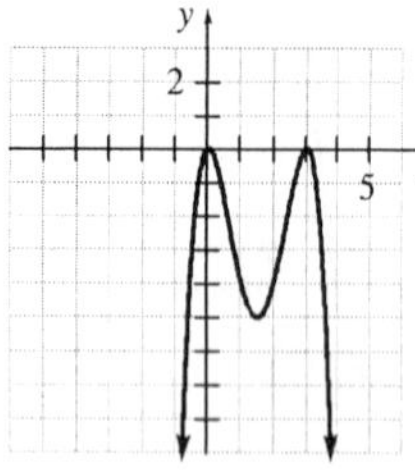


**11.** $f(x)=3x^4-15x^3$

**a.** Since *f* is an even-degree polynomial and since the leading coefficient, 3, is positive, the graph rises to the left and rises to the right.

**b.** $f(-x)=3(-x)^4-15(-x)^3$

$=3x^4+15x^3$

Since $f(-x)\neq f(x)$ and since $f(-x)\neq -f(x)$, the function is neither even nor odd, and the graph is neither symmetric with respect to the *y*-axis nor the origin.

**c.**

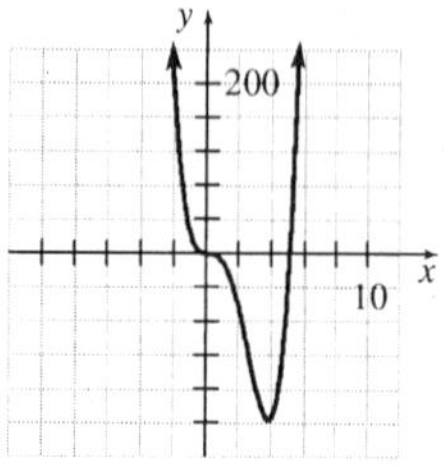


**12.** $f(x)=x^3-17x+4$

Since 4 is a zero of $f(x)$, we know $f(4)=0$. By the Factor Theorem, $x-4$ is a factor of $f(x)$. We can use synthetic division to divide $f(x)$ by $x-4$.

| | | | | |
|---|---|---|---|---|
| 4 | 1 | 0 | −17 | 4 |
| | | 4 | 16 | −4 |
| | 1 | 4 | −1 | 0 |

Thus, $x^3-17x+4=0$

$(x-4)(x^2+4x-1)=0$

Note that $x^2+4x-1$ does not factor, so we use the quadratic formula:

$x-4=0$ or $x^2+4x-1=0$

$x=4$ $\quad a=1,\ b=4,\ c=-1$

$$x=\frac{-4\pm\sqrt{4^2-4(1)(-1)}}{2(1)}$$

$$=\frac{-4\pm\sqrt{20}}{2}=\frac{-4\pm2\sqrt{5}}{2}=-2\pm\sqrt{5}$$

The solutions are 4 and $-2\pm\sqrt{5}$, and the solution set is $\{4,\ -2\pm\sqrt{5}\}$.

**13.** Let $f(x)=3x^3+10x^2-x-12$.

Since −3 is a root of the equation, we know $f(-3)=0$. By the Factor Theorem, $x+3$ is a factor of $f(x)$. We can use synthetic division to divide $f(x)$ by $x+3$.

| | | | | |
|---|---|---|---|---|
| −3 | 3 | 10 | −1 | −12 |
| | | −9 | −3 | 12 |
| | 3 | 1 | −4 | 0 |

Thus, $3x^3+10x^2-x-12=0$

$(x+3)(3x^2+x-4)=0$

$(x+3)(3x+4)(x-1)=0$

Apply the zero-product property:

$x+3=0$ or $3x+4=0$ or $x-1=0$

$x=-3$ $\quad 3x=-4$ $\quad x=1$

$x=-\frac{4}{3}$

The solutions are $-3$, $-\frac{4}{3}$, and 1, and the solution set is $\left\{-3,-\frac{4}{3},1\right\}$.

**14.** $f(x)=x^4-6x^3+14x^2-14x+5$
List all factors of the constant term 5: $\pm1,\ \pm5$
List all factors of the leading coefficient 1: $\pm1$
The possible rational zeros are:
$$\frac{\text{Factors of 5}}{\text{Factors of 1}}=\frac{\pm1,\ \pm5}{\pm1}=\pm1,\ \pm5$$

**15.** $f(x)=3x^5-2x^4-15x^3+10x^2+12x-8$
List all factors of the constant term $-8$: $\pm1,\ \pm2,\ \pm4,\ \pm8$
List all factors of the leading coefficient 3: $\pm1,\ \pm3$
The possible rational zeros are:
$$\frac{\text{Factors of }-8}{\text{Factors of 3}}=\frac{\pm1,\ \pm2,\ \pm4,\ \pm8}{\pm1,\ \pm3}$$
$$=\pm1,\pm2,\pm4,\pm8,\pm\frac{1}{3},\pm\frac{2}{3},\pm\frac{4}{3},\pm\frac{8}{3}$$

**16.** $f(x)=3x^4-2x^3-8x+5$
There are two sign changes. Thus, $f$ has either 2 or 0 positive real zeros.
$$f(-x)=3(-x)^4-2(-x)^3-8(-x)+5$$
$$=3x^4+2x^3+8x+5$$
There are no sign changes. Thus, $f$ has no negative real zeros.

**17.** $f(x)=2x^5-3x^3-5x^2+3x-1$
There are three sign changes. Thus, $f$ has either 3 or 1 positive real zeros.
$$f(-x)=2(-x)^5-3(-x)^3$$
$$-5(-x)^2+3(-x)-1$$
$$=-2x^5+3x^3-5x^2-3x-1$$
There are two sign changes. Thus, $f$ has either 2 or 0 negative real zeros.

**18.** Let $f(x)=2x^4+6x^2+8$.
There are no sign changes, so $f$ has no positive real zeros.
$$f(-x)=2(-x)^4+6(-x)^2+8$$
$$=2x^4+6x^2+8$$
Again, there are no sign changes, so $f$ has no negative real zeros. Thus, there are no real zeros at all.

**19.** $f(x)=x^3+3x^2-4$

**a.** List all factors of the constant term $-4$: $\pm1,\ \pm2,\ \pm4$
List all factors of the leading coefficient 1: $\pm1$
The possible rational zeros are:
$$\frac{\text{Factors of }-4}{\text{Factors of 1}}=\frac{\pm1,\ \pm2,\ \pm4}{\pm1}$$
$$=\pm1,\ \pm2,\ \pm4$$

**b.** $f(x)=x^3+3x^2-4$
There is one sign change. Thus, $f$ has exactly 1 positive real zeros.
$$f(-x)=(-x)^3+3(-x)^2-4$$
$$=-x^3+3x^2-4$$
There are two sign changes. Thus, $f$ has either 2 or 0 negative real zeros.

**c.** We begin by testing 1:

$$\begin{array}{r|rrrr} 1 & 1 & 3 & 0 & -4 \\ & & 1 & 4 & 4 \\ \hline & 1 & 4 & 4 & 0 \end{array}$$

The remainder is 0, so 1 is a root.

**d.** From part (c) and the Factor Theorem, we know that $x-1$ is a factor of the polynomial. Thus,

$$x^3+3x^2-4=0$$
$$(x-1)(x^2+4x+4)=0$$
$$(x-1)(x+2)^2=0$$
$$x-1=0 \quad \text{or} \quad (x+2)^2=0$$
$$x=1 \qquad x+2=0$$
$$x=-2$$

The solutions are $-2$ and 1, and the solution set is $\{-2,\ 1\}$.

**20.** $f(x)=6x^3+x^2-4x+1$

**a.** List all factors of the constant term 1: $\pm 1$

List all factors of the leading coefficient 6: $\pm 1,\ \pm 2,\ \pm 3,\ \pm 6$

The possible rational zeros are:

$$\frac{\text{Factors of }1}{\text{Factors of }6}=\frac{\pm 1}{\pm 1,\ \pm 2,\ \pm 3,\ \pm 6}$$
$$=\pm 1,\ \pm\frac{1}{2},\ \pm\frac{1}{3},\ \pm\frac{1}{6}$$

**b.** $f(x)=6x^3+x^2-4x+1$

There are two sign changes. Thus, $f$ has either 2 or 0 positive real zeros.

$$f(-x)=6(-x)^3+(-x)^2-4(-x)+1$$
$$=-6x^3+x^2+4x+1$$

There is one sign change. Thus, $f$ has exactly 1 negative real zeros.

**c.** We begin by testing $-1$:

| | | | | |
|---|---|---|---|---|
| $-1$ | 6 | 1 | $-4$ | 1 |
| | | $-6$ | 5 | $-1$ |
| | 6 | $-5$ | 1 | 0 |

The remainder is 0, so $-1$ is a root.

**d.** From part (c) and the Factor Theorem, we know that $x+1$ is a factor of the polynomial. Thus,

$$6x^3+x^2-4x+1=0$$
$$(x+1)(6x^2-5x+1)=0$$
$$(x-1)(3x-1)(2x-1)=0$$
$$x+1=0; \quad 3x-1=0; \quad 2x-1=0$$
$$x=-1 \qquad x=\frac{1}{3} \qquad x=\frac{1}{2}$$

The solutions are $-1$, $\frac{1}{3}$, and $\frac{1}{2}$, and the solution set is $\left\{-2,\ \frac{1}{3},\ \frac{1}{2}\right\}$.

**21.** $8x^3-36x^2+46x-15=0$

**a.** List all factors of the constant term $-15$: $\pm 1,\ \pm 3,\ \pm 5,\ \pm 15$

List all factors of the leading coefficient 8: $\pm 1,\ \pm 2,\ \pm 4$

The possible rational zeros are:

$$\frac{\text{Factors of }-15}{\text{Factors of }8}=\frac{\pm 1,\ \pm 3,\ \pm 5,\ \pm 15}{\pm 1,\ \pm 2,\ \pm 4}$$
$$=\pm 1,\pm 3,\pm 5,\pm 15,\pm\frac{1}{2},\pm\frac{3}{2},\pm\frac{5}{2},\pm\frac{15}{2},$$
$$\pm\frac{1}{4},\pm\frac{3}{4},\pm\frac{5}{4},\pm\frac{15}{4}$$

**b.** $f(x)=8x^3-36x^2+46x-15$

There are three sign changes. Thus, $f$ has either 3 or 1 positive real zeros.

$$f(-x)=8(-x)^3-36(-x)^2+46(-x)-15$$
$$=-8x^3-36x^2-46x-15$$

There are no sign changes. Thus, $f$ has no negative real zeros.

**c.** We test values from above until we find a root. One possibility is shown next:

Test $\frac{1}{2}$:

$$\begin{array}{r|rrrr} \frac{1}{2} & 8 & -36 & 46 & -15 \\ & & 4 & -16 & 15 \\ \hline & 8 & -32 & 30 & 0 \end{array}$$

The remainder is 0, so $-1$ is a root.

**d.** From part (c) and the Factor Theorem, we know that $x-\frac{1}{2}$ is a factor of the polynomial. Thus,

$$8x^3-36x^2+46x-15=0$$

$$\left(x-\frac{1}{2}\right)\left(8x^2-32x+30\right)=0$$

$$2\left(x-\frac{1}{2}\right)\left(4x^2-16x+15\right)=0$$

$$2\left(x-\frac{1}{2}\right)(2x-3)(2x-5)=0$$

$$x-\frac{1}{2}=0; \quad 2x-3=0; \quad 2x-5=0$$

$$x=\frac{1}{2} \qquad x=\frac{3}{2} \qquad x=\frac{5}{2}$$

The solutions are $\frac{1}{2}$, $\frac{3}{2}$, and $\frac{5}{2}$, and the solution set is $\left\{\frac{1}{2},\frac{3}{2},\frac{5}{2}\right\}$.

**22.** $x^4-x^3-7x^2+x+6=0$

**a.** List all factors of the constant term 6: $\pm1,\ \pm2,\ \pm3,\ \pm6$

List all factors of the leading coefficient 1: $\pm1$

The possible rational zeros are:

$$\frac{\text{Factors of 6}}{\text{Factors of 1}}=\frac{\pm1,\ \pm2,\ \pm3,\ \pm6}{\pm1}$$

$$=\pm1,\ \pm2,\ \pm3,\ \pm6$$

**b.** $f(x)=x^4-x^3-7x^2+x+6$

There are two sign changes. Thus, $f$ has either 2 or 0 positive real zeros.

$$f(-x)=(-x)^4-(-x)^3-7(-x)^2+(-x)+6$$
$$=x^4+x^3-7x^2-x+6$$

There are three sign changes. Thus, $f$ has either 3 or 1 negative real zeros.

**c.** We test values from above until we find a root. One possibility is shown next:

Test 1:

$$\begin{array}{r|rrrrr} 1 & 1 & -1 & -7 & 1 & 6 \\ & & 1 & 0 & -7 & -6 \\ \hline & 1 & 0 & -7 & -6 & 0 \end{array}$$

The remainder is 0, so 1 is a root.

**d.** From part (c) and the Factor Theorem, we know that $x-1$ is a factor of the polynomial. Thus,

$$x^4-x^3-7x^2+x+6=0$$

$$(x-1)\left(x^3-7x-6\right)=0$$

To solve the equation above, we need to factor $x^3-7x-6$. We continue testing potential roots:

Test $-1$:

$$\begin{array}{r|rrrr} -1 & 1 & 0 & -7 & -6 \\ & & -1 & 1 & 6 \\ \hline & 1 & -1 & -6 & 0 \end{array}$$

The remainder is 0, so $-1$ is a zero and $x+1$ is a factor.

Summarizing our findings so far, we have

$$x^4-x^3-7x^2+x+6=0$$

$$(x-1)\left(x^3-7x-6\right)=0$$

$$(x-1)(x+1)\left(x^2-x-6\right)=0$$

Factor the remaining binomial to solve:

$(x-1)(x+1)(x-3)(x+2)=0$

$x=1;\ x=-1;\ x=3;\ x=-2$

The solutions are $-2$, $-1$, 2, and 3, and the solution set is $\{-2,-1,2,3\}$.

**23.** $4x^4+7x^3-2=0$

**a.** List all factors of the constant term $-2$: $\pm1,\ \pm2$

List all factors of the leading coefficient 4: $\pm1,\ \pm2,\ \pm4$

The possible rational zeros are:

$$\frac{\text{Factors of } -2}{\text{Factors of } 4}=\frac{\pm1,\ \pm2}{\pm1,\ \pm2,\ \pm4}$$

$$=\pm1,\ \pm2,\ \pm\frac{1}{2},\ \pm\frac{1}{4}$$

**b.** $f(x)=4x^4+7x^2-2$

There is one sign change. Thus, $f$ has exactly 1 positive real zero.

$$f(-x)=4(-x)^4+7(-x)^2-2$$
$$=4x^4+7x^2-2$$

There is one sign change. Thus, $f$ has exactly 1 negative real zero.

**c.** We test values from above until we find a root. One possibility is shown next:

Test $\frac{1}{2}$:

| $\frac{1}{2}$ | 4 | 0 | 7 | 0 | −2 |
|---|---|---|---|---|---|
| | | 2 | 1 | 4 | 2 |
| | 4 | 2 | 8 | 4 | 0 |

The remainder is 0, so $\frac{1}{2}$ is a root.

**d.** From part (c) and the Factor Theorem, we know that $x-\frac{1}{2}$ is a factor of the polynomial. Thus,

$$4x^4+7x^2-2=0$$
$$\left(x-\frac{1}{2}\right)\left(4x^3+2x^2+8x+4\right)=0$$

To solve the equation above, we need to factor $4x^3+2x^2+8x+4$. We continue testing potential roots:

Test $-\frac{1}{2}$:

| $-\frac{1}{2}$ | 4 | 2 | 8 | 4 |
|---|---|---|---|---|
| | | −2 | 0 | −4 |
| | 4 | 0 | 8 | 0 |

The remainder is 0, so $-\frac{1}{2}$ is a zero and $x+\frac{1}{2}$ is a factor.

Thus, so far, we have:

$$4x^4+7x^2-2=0$$
$$\left(x-\frac{1}{2}\right)\left(4x^3+2x^2+8x+4\right)=0$$
$$\left(x-\frac{1}{2}\right)\left(x+\frac{1}{2}\right)\left(4x^2+8\right)=0$$

We know that $-\frac{1}{2}$ and $\frac{1}{2}$ are the two real roots. We now so $4x^2+8=0$ to find the remaining roots:

$$4x^2=-8$$
$$x^2=-2$$
$$x=\pm\sqrt{-2}=\pm\sqrt{2}i$$

The solutions are $-\frac{1}{2}$, $\frac{1}{2}$, and $\pm\sqrt{2}i$, and the solution set is $\left\{-\frac{1}{2},\frac{1}{2},\pm\sqrt{2}i\right\}$.

**24.** $f(x)=2x^4+x^3-9x^2-4x+4$

**a.** List all factors of the constant term 4: $\pm1,\ \pm2,\ \pm4$

List all factors of the leading coefficient 2: $\pm1,\ \pm2$

The possible rational zeros are:

$$\frac{\text{Factors of 4}}{\text{Factors of 2}}=\frac{\pm1,\ \pm2,\ \pm4}{\pm1,\ \pm2}$$

$$=\pm1,\ \pm2,\ \pm4,\ \pm\frac{1}{2}$$

**b.** $f(x)=2x^4+x^3-9x^2-4x+4$

There are two sign changes. Thus, $f$ has either 2 or 0 positive real zeros.

$$f(-x)=2(-x)^4+(-x)^3-9(-x)^2-4(-x)+4$$
$$=2x^4-x^3-9x^2+4x+4$$

There are two sign changes. Thus, $f$ has either 2 or 0 real zeros.

**c.** We test values from above until we find a root. One possibility is shown next:

Test $-1$:

| | | | | | |
|---|---|---|---|---|---|
| $-1$ | 2 | 1 | −9 | −4 | 4 |
| | | −2 | 1 | 8 | −4 |
| | 2 | −1 | −8 | 4 | 0 |

The remainder is 0, so $-1$ is a zero.

**d.** From part (c) and the Factor Theorem, we know that $x+1$ is a factor of the polynomial. Thus,

$$2x^4+x^3-9x^2-4x+4=0$$
$$(x+1)(2x^3-x^2-8x+4)=0$$

To solve the equation above, we need to factor $2x^3-x^2-8x+4$. We continue testing potential roots:

Test $-2$:

| | | | | |
|---|---|---|---|---|
| $-2$ | 2 | −1 | −8 | 4 |
| | | −4 | 10 | −4 |
| | 2 | −5 | 2 | 0 |

The remainder is 0, so $-2$ is a zero and $x+2$ is a factor.

Thus, so far, we have:

$$2x^4+x^3-9x^2-4x+4=0$$
$$(x+1)(2x^3-x^2-8x+4)=0$$
$$(x+1)(x+2)(2x^2-5x+2)=0$$

Factor the remaining binomial to solve:

$$(x+1)(x+2)(2x-1)(x-2)=0$$
$$x=-1;\ x=-2,\ x=\frac{1}{2},\ x=2$$

The solutions are $-2,\ -1,\ \frac{1}{2}$, and 2, and the solution set is $\left\{-2,-1,\frac{1}{2},2\right\}$.

**25.** $f(x)=\dfrac{2x}{x^2-9}$

$$x^2-9=0$$
$$x^2=9$$
$$x=\pm3$$

The lines $x=-3$ and $x=3$ are the vertical asymptotes for the graph of $f$.

The degree of the numerator, 1, is less than the degree of the denominator, 2, so the horizontal asymptote is the line $y=0$ ($x$-axis).

Plot additional points as needed, and graph the function.

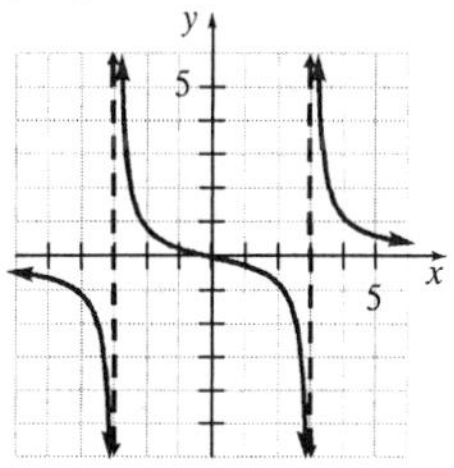

**26.** $f(x)=\dfrac{2x-4}{x+3}$

$x+3=0$

$x=-3$

The line $x=-3$ is the only vertical asymptote of *f*.

The numerator and denominator have the same degree, 1. The leading coefficients of the numerator and denominator are 2 and 1, respectively. Thus, the equation of the horizontal asymptote is $y=\dfrac{2}{1}$ or $y=2$.

Plot additional points as needed, and graph the function.

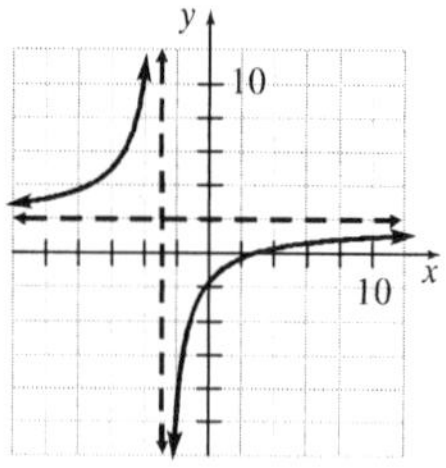

**27.** $f(x)=\dfrac{4x^2}{x^2-1}$

$x^2-1=0$

$x^2=1$

$x=\pm 1$

The lines $x=-1$ and $x=1$ are the vertical asymptotes for the graph of *f*.

The numerator and denominator have the same degree, 2. The leading coefficients of the numerator and denominator are 4 and 1, respectively. Thus, the equation of the horizontal asymptote is $y=\dfrac{4}{1}$ or $y=4$.

Plot additional points as needed, and graph the function.

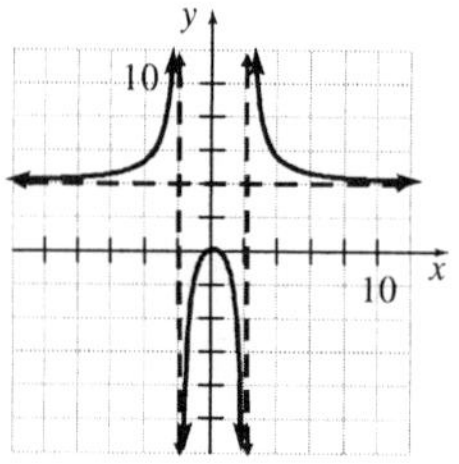

**28.** $f(x)=\dfrac{x^2}{x^2+1}$

$x^2+1=0$

$x^2=-1$

$x=\pm i$

Since these solutions are not real, the graph of *f* will not have any vertical asymptotes.

The numerator and denominator have the same degree, 2. The leading coefficients of the numerator and denominator are both 1. Thus, the equation of the horizontal asymptote is $y=\dfrac{1}{1}$ or $y=1$.

Plot additional points as needed, and graph the function.

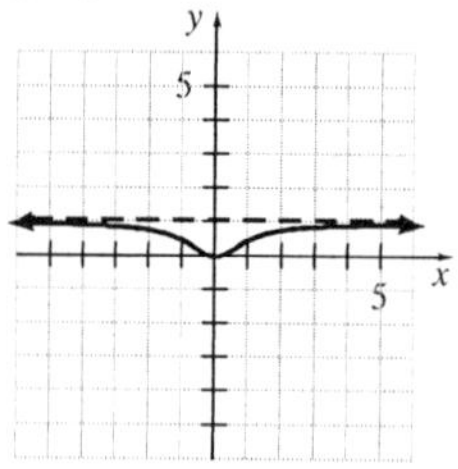

**29.** $f(x)=\dfrac{x^4}{x^2+2}$

$x^2+2=0$

$x^2=-2$

$x=\pm\sqrt{2}i$

Since these solutions are not real, the graph of *f* will not have any vertical asymptotes.

The degree of the numerator, 4, is greater than the degree of the denominator, 2, so the graph will not have a horizontal asymptote.

Plot points as needed, and graph the function.

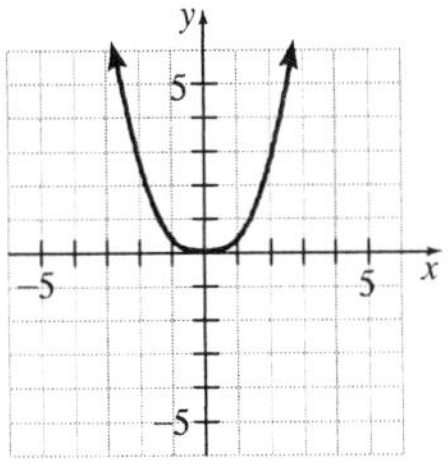


**30.** $f(x)=\dfrac{x^2-3x-4}{x^2-x-6}$

$$x^2-x-6=0$$
$$(x-3)(x+2)=0$$
$$x-3=0 \quad \text{or} \quad x+2=0$$
$$x=3 \qquad x=-2$$

The lines $x=-2$ and $x=3$ are the vertical asymptotes for the graph of *f*.

The numerator and denominator have the same degree, 2. The leading coefficients of the numerator and denominator are both 1. Thus, the equation of the horizontal asymptote is $y=\frac{1}{1}$ or $y=1$.

Plot points as needed, and graph the function.

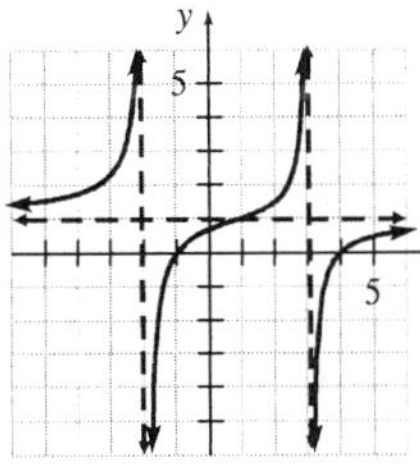


**31.** $f(x)=\dfrac{x^2+4x+3}{(x+2)^2}$

$$(x+2)^2=0$$
$$x+2=0$$
$$x=-2$$

The line $x=-2$ is the only vertical asymptotes for the graph of *f*.

The numerator and denominator have the same degree, 2. The leading coefficients of the numerator and denominator are both 1. Thus, the equation of the horizontal asymptote is $y=\frac{1}{1}$ or $y=1$.

Plot points as needed, and graph the function.

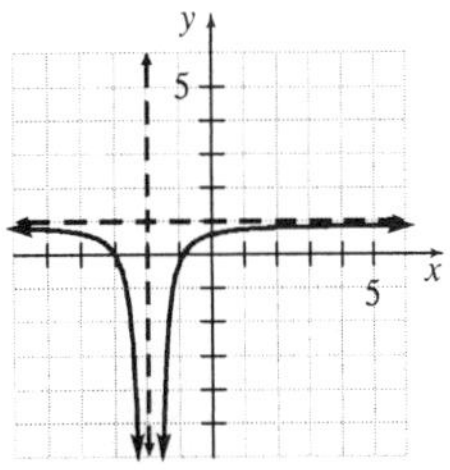


**32.** The numerator and denominator have the same degree, 1. The leading coefficient of the numerator and denominator are 150 and 0.05, respectively. Thus, the equation of the horizontal asymptote is $y=\dfrac{150}{0.05}$ or $y=3000$. This means that, over time, the number of bass in the lake will approach 3000.

**33.** The degree of the numerator, 0, is less than the degree of the denominator, 2, so the horizontal asymptote is the line $y=0$ (*x*-axis). This means that, as the number of years of education increases, the percentage of people unemployed approaches 0.

## Chapter 11 Test

**1.** $f(x) = x^3 - 5x^2 - 4x + 20$

**a.** $x^3 - 5x^2 - 4x + 20 = 0$

$x^2(x-5) - 4(x-5) = 0$

$(x^2 - 4)(x-5) = 0$

$(x+2)(x-2)(x-5) = 0$

$x+2=0; \quad x-2=0; \quad x-5=0$

$x=-2 \quad x=2 \quad x=5$

The zeros of *f* are $-2$, 2, and 5.

**b.** Since *f* is an odd-degree polynomial and since the leading coefficient, 1, is positive, the graph falls to the left and rises to the right.

Plot additional points as needed, and graph the function.

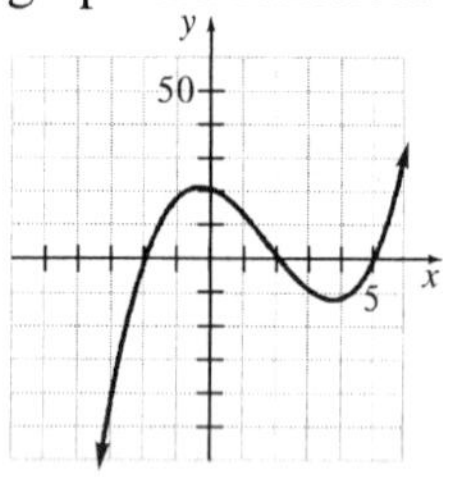


**2.** $f(x) = x^4 - x^2$

$f(-x) = (-x)^4 - (-x)^2 = x^4 - x^2$

Since $f(-x) = f(x)$, the function is even. Therefore, the graph of *f* is symmetric with respect to the *y*-axis. However, the graph shown is symmetric with respect to the origin, not with respect to the *y*-axis. This means that the graph shown cannot be that of *f*.

**3.** $f(x) = 6x^3 - 19x^2 + 16x - 4$

**a.** The graph appears to have an integer *x*-intercept at 2. This means that 2 is a zero of *f*.

**b.** Since 2 is a zero of *f*, we know that $x-2$ is a factor.

| | | | | |
|---|---|---|---|---|
| 2 | 6 | −19 | 16 | −4 |
| | | 12 | −14 | 4 |
| | 6 | −7 | 2 | 0 |

Thus, $6x^3 - 19x^2 + 16x - 4 = 0$

$(x-2)(6x^2 - 7x + 2) = 0$

$(x-2)(3x-2)(2x-1) = 0$

$x-2=0 \quad 3x-2=0 \quad 2x-1=0$

$x=2 \quad x=\frac{2}{3} \quad x=\frac{1}{2}$

The other two zeros are $\frac{1}{2}$ and $\frac{2}{3}$.

**4.** $f(x) = 2x^3 + 11x^2 - 7x - 6$

List all factors of the constant term $-6$: $\pm1, \pm2, \pm3, \pm6$

List all factors of the leading coefficient 2: $\pm1, \pm2$

The possible rational zeros are:

$$\frac{\text{Factors of } -6}{\text{Factors of } 2} = \frac{\pm1, \pm2, \pm3, \pm6}{\pm1, \pm2}$$

$$= \pm1, \pm2, \pm3, \pm6, \pm\frac{1}{2}, \pm\frac{3}{2}$$

**5.** $f(x) = 3x^5 - 2x^4 - 2x^2 + x - 1$

There are three sign changes. Thus, *f* has either 3 or 1 positive real zeros.

$f(-x) = 3(-x)^5 - 2(-x)^4 - 2(-x)^2 + (-x) - 1$

$= -3x^5 - 2x^3 - 2x^2 - x - 1$

There are no sign changes. Thus, *f* has no negative real zeros.

**6.** $x^3+6x^2-x-30=0$

List all factors of the constant term $-30$: $\pm1,\pm2,\pm3,\pm5,\pm6,\pm10,\pm15,\pm30$

List all factors of the leading coefficient 1: $\pm1$

The possible rational roots are:

$$\frac{\text{Factors of } -30}{\text{Factors of } 1} = \frac{\pm1,\pm2,\pm3,\pm5,\pm6,\pm10,\pm15,\pm30}{\pm1}$$
$$=\pm1,\pm2,\pm3,\pm5,\pm6,\pm10,\pm15,\pm30$$

We test values from above until we find a root. One possibility is shown next:

Test 2:

$$\begin{array}{r|rrrr} 2 & 1 & 6 & -1 & -30 \\ & & 2 & 16 & 30 \\ \hline & 1 & 8 & 15 & 0 \end{array}$$

The remainder is 0, so 2 is a root. By the Factor Theorem, we know that $x-2$ is a factor of the polynomial. Thus,

$$x^3+6x^2-x-30=0$$
$$(x-2)(x^2+8x+15)=0$$
$$(x-2)(x+5)(x+3)=0$$
$$x-2=0 \qquad x+5=0 \qquad x+3=0$$
$$x=2 \qquad x=-5 \qquad x=-3$$

The solutions are $-5$, $-3$, and 2, and the solution set is $\{-5,-3,2\}$.

**7. a.** $f(x)=2x^4-x^3-13x^2+5x+15$

List all factors of the constant term $-15$: $\pm1,\ \pm3,\ \pm5,\ \pm15$

List all factors of the leading coefficient 2: $\pm1,\ \pm2$

The possible rational zeros are:

$$\frac{\text{Factors of } -15}{\text{Factors of } 2} = \frac{\pm1,\ \pm3,\ \pm5,\ \pm15}{\pm1,\ \pm2}$$
$$=\pm1,\pm2,\pm5,\pm15,\pm\frac{1}{2},\pm\frac{3}{2},\pm\frac{5}{2},\pm\frac{15}{2}$$

**b.** From the graph, it appears that $-1$ and $\frac{3}{2}$ are zeros. We verify this next:

$$\begin{array}{r|rrrrr} -1 & 2 & -1 & -13 & 5 & 15 \\ & & -2 & 3 & 10 & -15 \\ \hline & 2 & -3 & -10 & 15 & 0 \end{array}$$

Thus, $-1$ is a zero, and the polynomial factors as follows:

$$2x^4-x^3-13x^2+5x+15=0$$
$$(x+1)(2x^3-3x^2-10x+15)=0$$

$$\begin{array}{r|rrrr} \frac{3}{2} & 2 & -3 & -10 & 15 \\ & & 3 & 0 & -15 \\ \hline & 2 & 0 & -10 & 0 \end{array}$$

Thus, $\frac{3}{2}$ is a zero, and the polynomial factors further:

$$(x+1)\left(x-\frac{3}{2}\right)(2x^2-10)=0$$

So far, we have verified that $-1$ and $\frac{3}{2}$ are zeros of *f*. We find the remaining zeros by solving:

$$2x^2-10=0$$
$$2x^2=10$$
$$x^2=5$$
$$x=\pm\sqrt{5}$$

The zeros are $-1$, $\frac{3}{2}$, and $\pm\sqrt{5}$.

**8.** $f(x)=\dfrac{4x-2}{x-1}$

$x-1=0$

$x=1$

The line $x=1$ is the only vertical asymptote of *f*.

The numerator and denominator have the same degree, 1. The leading coefficients of the numerator and denominator are 4 and 1, respectively. Thus, the equation of the horizontal asymptote is $y=\frac{4}{1}$ or $y=4$.

Plot additional points as needed, and graph the function.

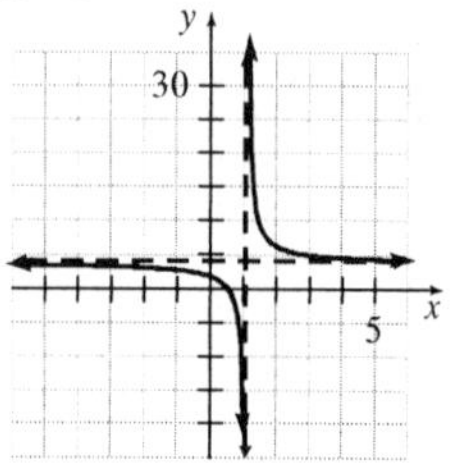


**9.** $f(x)=\dfrac{x^2-1}{x^2-4}$

$x^2-4=0$

$x^2=4$

$x=\pm 2$

The lines $x=-2$ and $x=2$ are the vertical asymptotes for the graph of *f*.

The numerator and denominator have the same degree, 2. The leading coefficients of the numerator and denominator are both 1. Thus, the equation of the horizontal asymptote is $y=\frac{1}{1}$ or $y=1$.

Plot points as needed, and graph the function.

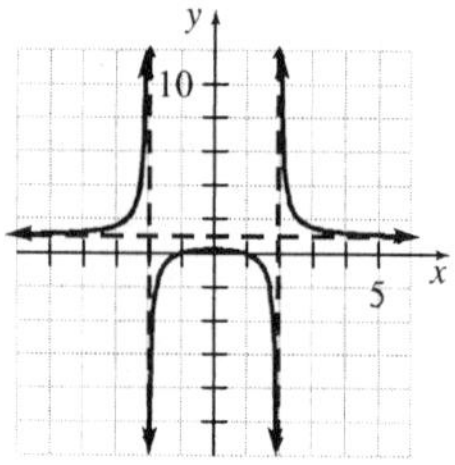


**10.** $f(x)=\dfrac{4x^2}{x^2+3}$

$x^2+3=0$

$x^2=-3$

$x=\pm\sqrt{3}i$

Since these solutions are not real, the graph of *f* will not have any vertical asymptotes.

The numerator and denominator have the same degree, 2. The leading coefficients of the numerator and denominator are 4 and 1, respectively. Thus, the equation of the horizontal asymptote is $y=\frac{4}{1}$ or $y=4$.

Plot points as needed, and graph the function.

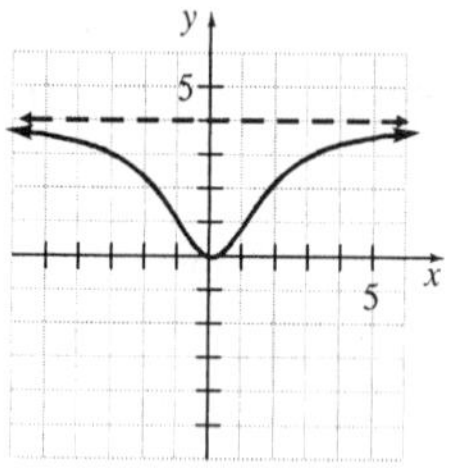

## Cumulative Review Exercises (Chapters 1 – 11)

**1.** $x^3-4x^2-10x+4=0$

Using the rational zeros theorem, we see that the potential rational zeros are $\pm1,\pm2,\pm4$. Let

$f(x)=x^3-4x^2-10x+4$ and use the factor and remainder theorems to find a rational zero.

$$f(-1)=(-1)^3-4(-1)^2-10(-1)+4$$
$$=9$$
$$f(-2)=(-2)^3-4(-2)^2-10(-2)+4$$
$$=0$$

Since $f(-2)=0$, we know that $(x+2)$ must be a factor. Use synthetic division to find a reduced polynomial.

| | | | | |
|---|---|---|---|---|
| $-2$ | 1 | $-4$ | $-10$ | 4 |
| | | $-2$ | 12 | $-4$ |
| | 1 | $-6$ | 2 | 0 |

Therefore, we have

$$(x+2)(x^2-6x+2)=0$$

The second factor is an irreducible quadratic. Set each factor equal to 0 and solve.

$$x+2=0$$
$$x=-2$$

or

$$x^2-6x+2=0$$
$$a=1, b=-6, c=2$$
$$x=\frac{-(-6)\pm\sqrt{(-6)^2-4(1)(2)}}{2(1)}$$
$$x=\frac{6\pm\sqrt{28}}{2}=\frac{6\pm2\sqrt{7}}{2}=3\pm\sqrt{7}$$

The solution set is

$\{-2,3+\sqrt{7},3-\sqrt{7}\}$.

**2.** $2x+y=3$

$$y=2x^2-1$$

Substitute $2x^2-1$ for $y$ in the first equation and solve for $x$.

$$2x+(2x^2-1)=3$$
$$2x^2+2x-1=3$$
$$2x^2+2x-4=0$$
$$x^2+x-2=0$$
$$(x+2)(x-1)=0$$
$$x+2=0 \quad \text{or} \quad x-1=0$$
$$x=-2 \qquad\qquad x=1$$

Substitute each of these values for $x$ back into the second equation to solve for $y$.

For $x=-2$: $y=2(-2)^2-1=7$

For $x=1$: $y=2(1)^2-1=1$

The solution set is $\{(-2,7),(1,1)\}$.

**3.** $\sqrt{x}+6=x$

$$\sqrt{x}=x-6$$
$$(\sqrt{x})^2=(x-6)^2$$
$$x=x^2-12x+36$$
$$0=x^2-13x+36$$
$$0=(x-9)(x-4)$$
$$x-9=0 \quad \text{or} \quad x-4=0$$
$$x=9 \qquad\qquad x=4$$

Check $x=9$: $\sqrt{9}+6=9$

$$3+6=9$$
$$9=9 \text{ true}$$

Check $x=4$: $\sqrt{4}+6=4$

$$2+6=4$$
$$8=4 \text{ false}$$

The solution set is $\{9\}$.

**4.** $x^{2/3}+3x^{1/3}-18=0$

Let $u=x^{1/3}$.

$u^2+3u-18=0$

$(u+6)(u-3)=0$

$u+6=0$ or $u-3=0$

$u=-6$ $\quad u=3$

Therefore,

$x^{1/3}=-6$ or $x^{1/3}=3$

$x=(-6)^3$ $\quad x=3^3$

$=-216$ $\quad x=27$

The solution set is $\{-216, 27\}$

**5.** $\log_2 x+\log_2 3=1$

$\log_2(3x)=1$ (product rule)

$2^1=3x$ (change form)

$\frac{2}{3}=x$

The solution set is $\left\{\frac{2}{3}\right\}$.

**6.** $x+y-z=-2$

$3x-2y-5z=7$

$2x-3y+4z=17$

Start by writing an augmented matrix for the system.

$$\left[\begin{array}{ccc|c} 1 & 1 & -1 & -2 \\ 3 & -2 & -5 & 7 \\ 2 & -3 & 4 & 17 \end{array}\right]$$

Now use row operations to get the matrix into row-echelon form.

$$\left[\begin{array}{ccc|c} 1 & 1 & -1 & -2 \\ 3 & -2 & -5 & 7 \\ 2 & -3 & 4 & 17 \end{array}\right] \begin{array}{l} \\ -3R_1+R_2 \\ -2R_1+R_3 \end{array}$$

$$\left[\begin{array}{ccc|c} 1 & 1 & -1 & -2 \\ 0 & -5 & -2 & 13 \\ 0 & -5 & 6 & 21 \end{array}\right] \begin{array}{l} \\ \\ -R_2+R_3 \end{array}$$

$$\left[\begin{array}{ccc|c} 1 & 1 & -1 & -2 \\ 0 & -5 & -2 & 13 \\ 0 & 0 & 8 & 8 \end{array}\right] \begin{array}{l} \\ -\frac{1}{5}R_2 \\ \frac{1}{18}R_3 \end{array}$$

$$\left[\begin{array}{ccc|c} 1 & 1 & -1 & -2 \\ 0 & 1 & \frac{2}{5} & -\frac{13}{5} \\ 0 & 0 & 1 & 1 \end{array}\right]$$

The equivalent system of equations is

$x+y-z=-2$

$y+\frac{2}{5}z=-\frac{13}{5}$

$z=1$

Back-substitute 1 for $z$ in the second equation and solve for $y$.

$y+\frac{2}{5}(1)=-\frac{13}{5}$

$y+\frac{2}{5}=-\frac{13}{5}$

$y=-\frac{15}{5}=-3$

Back-substitute $-3$ for $y$ and 1 for $z$ in the first equation.

$x-3-1=-2$

$x=2$

The solution is $(2,-3,1)$ or $\{(2,-3,1)\}$.

**7.** $\frac{x-4}{x+6}>0$

The left side of the inequality is 0 when $x=4$ and undefined when $x=-6$. Use these values to divide up the real number line into sub-intervals. Then pick a test point in each sub-interval to determine the sign of the left side on that interval.

| | $(-\infty,-6)$ | $(-6,4)$ | $(4,\infty)$ |
|---|---|---|---|
| Test | $x=-7$ | $x=0$ | $x=5$ |
| Left side | 11 | $-\frac{2}{3}$ | $\frac{1}{11}$ |
| Conclusion | $+++$ | $---$ | $+++$ |

The left side is positive on the intervals $(-\infty,-6)$ and $(4,\infty)$. Therefore, the solution set is $\{x\,|\,x<-6 \text{ or } x>4\}$ or $(-\infty,-6)\cup(4,\infty)$.

**8.** $\dfrac{x^2}{16}-\dfrac{y^2}{9}=1$

The center of the ellipse is $(0,0)$. Because the denominator of the $x^2$ – term is greater than the denominator of the $y^2$ – term, the major axis is horizontal. Since $a^2=16$, $a=4$ and the vertices lie 5 units to the right and left of the center. Since $b^2=9$, $b=3$ and endpoints of the minor axis lie 3 units above and below the center. Draw a rectangle using the vertices and minor axis endpoints, then extend the diagonals to obtain the graphs of the asymptotes.

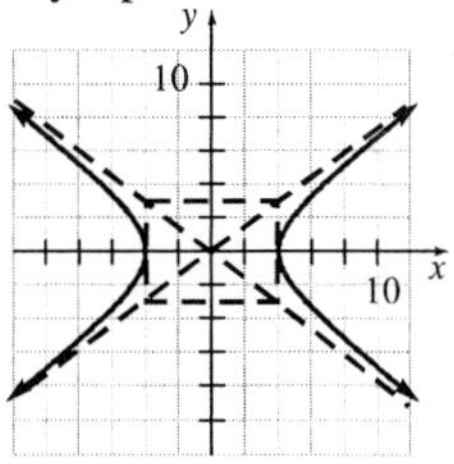


**9.** $3x-2y\le -6$

First, graph the equation $3x-2y=-6$. Rewrite the equation in slope-intercept form by solving for *y*.

$$3x-2y=-6$$
$$-2y=-3x-6$$
$$y=\frac{3}{2}x+3$$

y!intercept = 3

slope = $\dfrac{3}{2}$

Next, use the origin as a test point.

$$3x-2y\le -6$$
$$3(0)-2(0)\le -6$$
$$0\le -6$$

This is a false statement. This means that the point $(0,0)$ will fall in the half-plane that is not shaded.

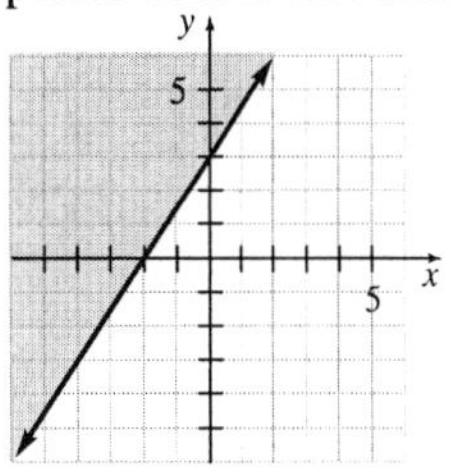


**10.** $f(x)=\dfrac{2x}{x+1}$

This is a rational function with domain $\{x\,|\,x \text{ is a real number and } x\neq -1\}$.

As $x\to\infty$, $\dfrac{2x}{x+1}\to\dfrac{2x}{x}=2$. So the function has a horizontal asymptote of $y=2$. The denominator equals 0 when $x=-1$, but the numerator is non-zero. Therefore, there is a vertical asymptote at $x=-1$.

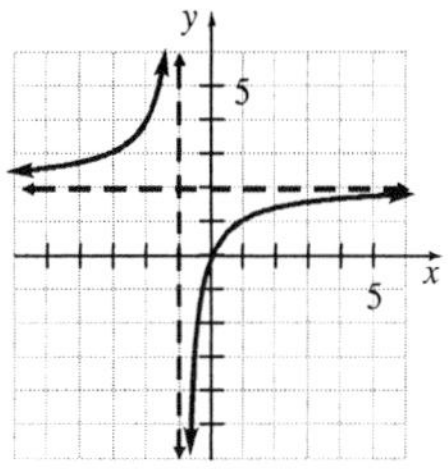


**11.** $(5x-7)^2=(5x)^2-2(5x)(-7)+(7)^2$
$$=25x^2+70x+49$$

**12.** $(2x-3)(4x^2+6x+9)$
$$=2x(4x^2+6x+9)-3(4x^2+6x+9)$$
$$=8x^3+12x^2+18x-12x^2-18x-27$$
$$=8x^3-27$$

**13.** $\frac{3}{x^2-x}-\frac{2x}{3x-3}=\frac{3}{x(x-1)}-\frac{2x}{3(x-1)}$

$=\frac{3\cdot 3}{3x(x-1)}-\frac{2x\cdot x}{3x(x-1)}$

$=\frac{9}{3x(x-1)}-\frac{2x^2}{3x(x-1)}$

$=\frac{9-2x^2}{3x(x-1)}$

**14.** $\sqrt{54}-\sqrt{150}=\sqrt{9\cdot 6}-\sqrt{25\cdot 6}$

$=\sqrt{9}\sqrt{6}-\sqrt{25}\sqrt{6}$

$=3\sqrt{6}-5\sqrt{6}$

$=(3-5)\sqrt{6}$

$=-2\sqrt{6}$

**15.** $(1+2i)(6-3i)$

$=1\cdot 6-1\cdot 3i+2i\cdot 6-2i\cdot 3i$

$=6-3i+12i-6i^2$

$=6+9i-6(-1)$

$=12+9i$

**16.** $3x^3+6x^2-3x-6$

$=3(x^3+6x^2-x-2)$

$=3(x^2(x+2)-1(x+2))$

$=3(x+2)(x^2-1)$

$=3(x+2)(x-1)(x+1)$

**17.** $25x^2-20xy+4y^2$

$=(5x)^2-2(5x)(2y)+(2y)^2$

$=(5x-2y)^2$

**18.** Since parallel lines have the same slope, we can obtain our slope by putting the given equation in slope-intercept form.

$4x+y=7$

$y=-4x+7$

Therefore, the slope of our line is $m=-4$ and the line must pass through the point $(2,-3)$. Using the point-slope form of a line, we get

$y-y_1=m(x-x_1)$

$y-(-3)=-4(x-2)$

$y+3=-4x+8$

$y=-4x+5$

The equation of our line is $y=-4x+5$.

**19.** $\begin{vmatrix} 4 & -3 \\ 5 & 7 \end{vmatrix}=(4)(7)-(5)(-3)$

$=28+15=43$

**20.** $f(x)=x^2-4x+7$

$g(x)=x+3$

$f(g(x))=f(x+3)$

$=(x+3)^2-4(x+3)+7$

$=x^2+6x+9-4x-12+7$

$=x^2+2x+4$

**21.** $\frac{1}{2}\log x+3\log y=\log x^{1/2}+\log y^3$

$=\log\left(x^{1/2}y^3\right)$

$=\log\left(\sqrt{x}\cdot y^3\right)$

**22.** $\dfrac{9x}{\sqrt[3]{3x^2}} = \dfrac{9x}{\sqrt[3]{3x^2}} \cdot \dfrac{\sqrt[3]{9x}}{\sqrt[3]{9x}}$

$= \dfrac{9x\sqrt[3]{9x}}{\sqrt[3]{27x^3}} = \dfrac{9x\sqrt[3]{9x}}{3x}$

$= 3\sqrt[3]{9x}$

**23.** Let $x$ = the number of cups of coffee and $y$ = the number of hours you sleep.

$y = \dfrac{k}{x^2}$

$y_2 = \dfrac{k}{(2x)^2} = \dfrac{k}{4x^2} = \dfrac{1}{4}\left(\dfrac{k}{x^2}\right)$

Doubling the number of cups of coffee will reduce the number of hours of sleep to one-fourth of the original amount. So, if you originally got 8 hours of sleep, you would only get $\dfrac{1}{4}(8) = 2$ hours of sleep if you doubled the number of cups of coffee.

**24.** Let $x$ = the cost of a bath towel, and $y$ = the cost of a hand towel.
From the problem statement, we obtain the following system of equations:

$3x + 2y = 31$

$2x + 5y = 39$

Multiply the first equation by $-2$ and the second equation by 3, then add the equations.

$-6x - 4y = -62$

$\underline{6x + 15y = 117}$

$11y = 55$

$y = 5$

Let $y = 5$ in the first equation and solve for $x$.

$3x + 2y = 31$

$3x + 2(5) = 31$

$3x + 10 = 31$

$3x = 21$

$x = 7$

Each bath towel costs \$7 and each hand towel costs \$5.

**25.** Consider the diagram below:

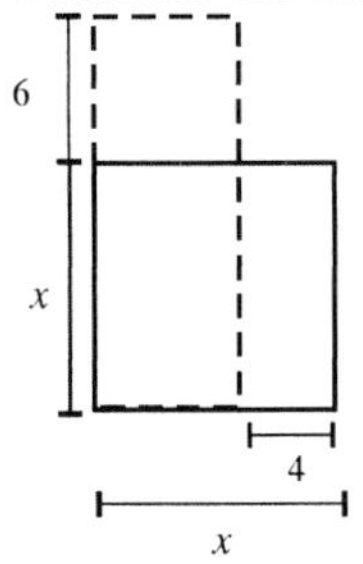


The width of the rectangle is $(x-4)$ and the length is $(x+6)$. The areas of the two figures are to be the same, so we have the equation

$x^2 = (x+6)(x-4)$

We can solve this equation for $x$ and then determine the dimensions of each figure.

$x^2 = (x+6)(x-4)$

$x^2 = x^2 - 4x + 6x - 24$

$0 = 2x - 24$

$0 = x - 12$

$x = 12$

Therefore, the original square measured 12 centimeters on each side and the new rectangle measures 8 centimeters by 18 centimeters.

# Chapter 12
# Sequences, Induction, and Probability

## 12.1 Exercise Set

**1.** $a_n = 3n + 2$

$a_1 = 3(1) + 2 = 3 + 2 = 5$

$a_2 = 3(2) + 2 = 6 + 2 = 8$

$a_3 = 3(3) + 2 = 9 + 2 = 11$

$a_4 = 3(4) + 2 = 12 + 2 = 14$

The first four terms are 5, 8, 11, 14.

**3.** $a_n = 3^n$

$a_1 = 3^1 = 3$

$a_2 = 3^2 = 9$

$a_3 = 3^3 = 27$

$a_4 = 3^4 = 81$

The first four terms are 3, 9, 27, 81.

**5.** $a_n = (-3)^n$

$a_1 = (-3)^1 = -3$

$a_2 = (-3)^2 = 9$

$a_3 = (-3)^3 = -27$

$a_4 = (-3)^4 = 81$

The first four terms are –3, 9, –27, 81.

**7.** $a_n = (-1)^n (n+3)$

$a_1 = (-1)^1 (1+3) = -1(4) = -4$

$a_2 = (-1)^2 (2+3) = 1(5) = 5$

$a_3 = (-1)^3 (3+3) = -1(6) = -6$

$a_4 = (-1)^4 (4+3) = 1(7) = 7$

The first four terms are –4, 5, –6, 7.

**9.** $a_n = \frac{2n}{n+4}$

$a_1 = \frac{2(1)}{1+4} = \frac{2}{5}$

$a_2 = \frac{2(2)}{2+4} = \frac{4}{6} = \frac{2}{3}$

$a_3 = \frac{2(3)}{3+4} = \frac{6}{7}$

$a_4 = \frac{2(4)}{4+4} = \frac{8}{8} = 1$

The first four terms are $\frac{2}{5}, \frac{2}{3}, \frac{6}{7}, 1$.

**11.** $a_n = \frac{(-1)^{n+1}}{2^n - 1}$

$a_1 = \frac{(-1)^{1+1}}{2^1 - 1} = \frac{(-1)^2}{2-1} = \frac{1}{1} = 1$

$a_2 = \frac{(-1)^{2+1}}{2^2 - 1} = \frac{(-1)^3}{4-1} = \frac{-1}{3} = -\frac{1}{3}$

$a_3 = \frac{(-1)^{3+1}}{2^3 - 1} = \frac{(-1)^4}{8-1} = \frac{1}{7}$

$a_4 = \frac{(-1)^{4+1}}{2^4 - 1} = \frac{(-1)^5}{16-1} = \frac{-1}{15} = -\frac{1}{15}$

The first four terms are $1, -\frac{1}{3}, \frac{1}{7}, -\frac{1}{15}$.

**13.** $a_n = \frac{n^2}{n!}$

$a_1 = \frac{1^2}{1!} = \frac{1}{1} = 1$

$a_2 = \frac{2^2}{2!} = \frac{4}{2 \cdot 1} = \frac{4}{2} = 2$

$a_3 = \frac{3^2}{3!} = \frac{9}{3 \cdot 2 \cdot 1} = \frac{3}{2}$

$a_4 = \frac{4^2}{4!} = \frac{16}{4 \cdot 3 \cdot 2 \cdot 1} = \frac{2}{3}$

The first four terms are $1, 2, \frac{3}{2}, \frac{2}{3}$.

**15.** $a_n = 2(n+1)!$

$a_1 = 2(1+1)! = 2 \cdot 2! = 2 \cdot 2 \cdot 1 = 4$

$a_2 = 2(2+1)! = 2 \cdot 3! = 2 \cdot 3 \cdot 2 \cdot 1 = 12$

$a_3 = 2(3+1)! = 2 \cdot 4! = 2 \cdot 4 \cdot 3 \cdot 2 \cdot 1 = 48$

$a_4 = 2(4+1)! = 2 \cdot 5! = 2 \cdot 5 \cdot 4 \cdot 3 \cdot 2 \cdot 1 = 240$

**17.** $\sum_{i=1}^{6} 5i = 5(1) + 5(2) + 5(3) + 5(4) + 5(5) + 5(6) = 5 + 10 + 15 + 20 + 25 + 30 = 105$

**19.** $\sum_{i=1}^{4} 2i^2 = 2(1)^2 + 2(2)^2 + 2(3)^2 + 2(4)^2 = 2(1) + 2(4) + 2(9) + 2(16)$

$= 2 + 8 + 18 + 32 = 60$

**21.** $\sum_{k=1}^{5} k(k+4) = 1(1+4) + 2(2+4) + 3(3+4) + 4(4+4) + 5(5+4)$

$= 1(5) + 2(6) + 3(7) + 4(8) + 5(9) = 5 + 12 + 21 + 32 + 45 = 115$

**23.** $\sum_{i=1}^{4} \left(-\frac{1}{2}\right)^i = \left(-\frac{1}{2}\right)^1 + \left(-\frac{1}{2}\right)^2 + \left(-\frac{1}{2}\right)^3 + \left(-\frac{1}{2}\right)^4 = -\frac{1}{2} + \frac{1}{4} + \left(-\frac{1}{8}\right) + \frac{1}{16}$

$= -\frac{1}{2} \cdot \frac{8}{8} + \frac{1}{4} \cdot \frac{4}{4} + \left(-\frac{1}{8}\right)\frac{2}{2} + \frac{1}{16} = -\frac{8}{16} + \frac{4}{16} - \frac{2}{16} + \frac{1}{16}$

$= \frac{-8 + 4 - 2 + 1}{16} = -\frac{5}{16}$

**25.** $\sum_{i=5}^{9} 11 = 11 + 11 + 11 + 11 + 11 = 55$

**27.** $\sum_{i=0}^{4} \frac{(-1)^i}{i!} = \frac{(-1)^0}{0!} + \frac{(-1)^1}{1!} + \frac{(-1)^2}{2!} + \frac{(-1)^3}{3!} + \frac{(-1)^4}{4!} = \frac{1}{1} + \frac{-1}{1} + \frac{1}{2 \cdot 1} + \frac{-1}{3 \cdot 2 \cdot 1} + \frac{1}{4 \cdot 3 \cdot 2 \cdot 1}$

$= 1 - 1 + \frac{1}{2} - \frac{1}{6} + \frac{1}{24} = \frac{1}{2} \cdot \frac{12}{12} - \frac{1}{6} \cdot \frac{4}{4} + \frac{1}{24} = \frac{12}{24} - \frac{4}{24} + \frac{1}{24} = \frac{12 - 4 + 1}{24} = \frac{9}{24} = \frac{3}{8}$

**29.** $\sum_{i=1}^{5} \frac{i!}{(i-1)!} = \frac{1!}{(1-1)!} + \frac{2!}{(2-1)!} + \frac{3!}{(3-1)!} + \frac{4!}{(4-1)!} + \frac{5!}{(5-1)!} = \frac{1!}{0!} + \frac{2!}{1!} + \frac{3!}{2!} + \frac{4!}{3!} + \frac{5!}{4!}$

$= \frac{1}{1} + \frac{2 \cdot \cancel{1!}}{\cancel{1!}} + \frac{3 \cdot \cancel{2!}}{\cancel{2!}} + \frac{4 \cdot \cancel{3!}}{\cancel{3!}} + \frac{5 \cdot \cancel{4!}}{\cancel{4!}} = 1 + 2 + 3 + 4 + 5 = 15$

**31.** $1^2+2^2+3^2+\ldots+15^2=\sum_{i=1}^{15} i^2$

**33.** $2+2^2+2^3+\ldots+2^{11}=\sum_{i=1}^{11} 2^i$

**35.** $1+2+3+\ldots+30=\sum_{i=1}^{30} i$

**37.** $\frac{1}{2}+\frac{2}{3}+\frac{3}{4}+\ldots+\frac{14}{14+1}=\sum_{i=1}^{14}\frac{i}{i+1}$

**39.** $4+\frac{4^2}{2}+\frac{4^3}{3}+\ldots+\frac{4^n}{n}=\sum_{i=1}^{n}\frac{4^i}{i}$

**41.** $1+3+5+\ldots+(2n-1)=\sum_{i=1}^{n}(2i-1)$

**43.** $5+7+9+11+\ldots+31=\sum_{k=2}^{15}(2k+1)$ or

$=\sum_{k=1}^{14}(2k+3)$

**45.** $a+ar+ar^2+\ldots+ar^{12}=\sum_{k=0}^{12} ar^k$

**47.** $a+(a+d)+(a+2d)+\ldots+a(a+nd)$

$=\sum_{k=0}^{n}(a+kd)$

**49.** $\sum_{i=1}^{5}(a_i^2+1)=((-4)^2+1)+((-2)^2+1)+((0)^2+1)+((2)^2+1)+((4)^2+1)$

$=17+5+1+5+17$

$=45$

**50.** $\sum_{i=1}^{5}(b_i^2-1)=((4)^2-1)+((2)^2-1)+((0)^2-1)+((-2)^2-1)+((-4)^2-1)$

$=15+3+(-1)+3+15$

$=35$

**51.** $\sum_{i=1}^{5}(2a_i+b_i)=(2(-4)+4)+(2(-2)+2)+(2(0)+0)+(2(2)+(-2))+(2(4)+(-4))$

$=-4+(-2)+0+2+4=0$

**52.** $\sum_{i=1}^{5}(a_i+3b_i)=(-4+3(4))+(-2+3(2))+(0+3(0))+(2+3(-2))+(4+3(-4))$

$=8+4+0-4-8=0$

**53.** $\sum_{i=4}^{5}\left(\frac{a_i}{b_i}\right)^2 = \left(\frac{2}{-2}\right)^2 + \left(\frac{4}{-4}\right)^2 = (-1)^2 + (-1)^2 = 1 + 1 = 2$

**54.** $\sum_{i=4}^{5}\left(\frac{a_i}{b_i}\right)^3 = \left(\frac{2}{-2}\right)^3 + \left(\frac{4}{-4}\right)^3 = (-1)^3 + (-1)^3 = (-1) + (-1) = -2$

**55.** $\sum_{i=1}^{5} a_i^2 + \sum_{i=1}^{5} b_i^2 = \left((-4)^2 + (-2)^2 + 0^2 + 2^2 + 4^2\right) + \left(4^2 + 2^2 + 0^2 + (-2)^2 + (-4)^2\right)$

$= (16 + 4 + 0 + 4 + 16) + (16 + 4 + 0 + 4 + 16) = 80$

**56.** $\sum_{i=1}^{5} a_i^2 - \sum_{i=3}^{5} b_i^2 = \left((-4)^2 + (-2)^2 + 0^2 + 2^2 + 4^2\right) - \left(0^2 + (-2)^2 + (-4)^2\right)$

$= (16 + 4 + 0 + 4 + 16) - (0 + 4 + 16) = 40 - 20 = 20$

**57. a.** $\sum_{i=1}^{7} a_i = 36.4 + 36.5 + 35.6 + 34.5 + 32.3 + 31.6 + 32.9 = 239.38$

**b.** $\frac{\sum_{i=1}^{7} a_i}{7} = \frac{239.8}{7} = 34.3$

From 1995 through 2001, the average number of people living below poverty level each year was approximately 34.3 million.

**59. a.** $\frac{1}{8}\sum_{i=1}^{8} a_i = \frac{1}{8}\left(14.1 + 14.2 + 13.7 + 12.6 + 10.9 + 8.7 + 7.6 + 6.5\right) = \frac{1}{8}(88.3) = 11.0375$

From 1993 through 2000, the average number of welfare recipients each year was 11.0375 million.

**b.** $a_1 = -1.23(1) + 16.55 = 15.32$ $\quad a_5 = -1.23(5) + 16.55 = 10.4$

$a_2 = -1.23(2) + 16.55 = 14.09$ $\quad a_6 = -1.23(6) + 16.55 = 9.17$

$a_3 = -1.23(3) + 16.55 = 12.86$ $\quad a_7 = -1.23(7) + 16.55 = 7.94$

$a_4 = -1.23(4) + 16.55 = 11.63$ $\quad a_8 = -1.23(8) + 16.55 = 6.71$

$\frac{1}{8}\sum_{i=1}^{8} a_i = \frac{1}{8}\left(15.32 + 14.09 + 12.86 + 11.63 + 10.4 + 9.17 + 7.94 + 6.71\right)$

$= \frac{1}{8}(88.12) = 11.015$

This is a reasonable model.

**61.**
$$a_{20} = 6000\left(1+\frac{0.06}{4}\right)^{20} = 6000(1+0.015)^{20} = 6000(1.015)^{20} = 8081.13$$
The balance in the account after 5 years if \$8081.13.

**63. – 69.** Answers will vary.

**71.** Answers will vary. For example, consider Exercise 17.
$$\sum_{i=1}^{6} 5i$$
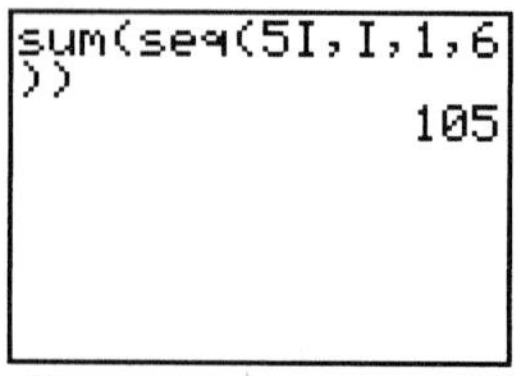


This is the same result obtained in Exercise 17.

**73.**
$$a_n = \frac{n}{n+1};$$
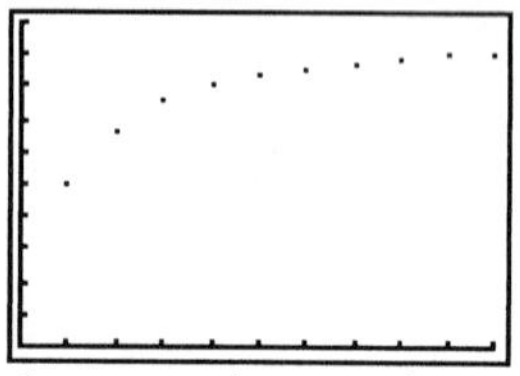

As $n$ gets larger, the terms get closer to 1.

**75.**
$$a_n = \frac{2n^2+5n-7}{n^3}$$
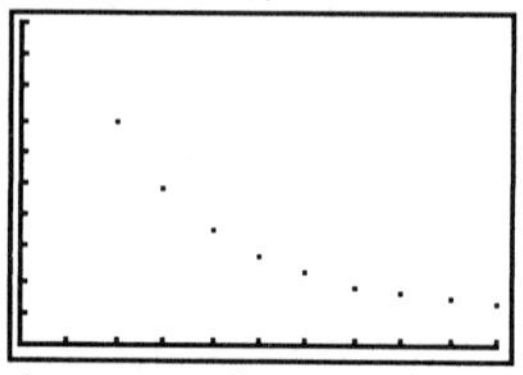

As $n$ gets larger, the terms get closer to 0.

**77.** Statement **c.** is true. $\sum_{i=1}^{4} 3i + \sum_{i=1}^{4} 4i = \sum_{i=1}^{4} 7i$
$$\sum_{i=1}^{4} 3i + \sum_{i=1}^{4} 4i = (3\cdot1+3\cdot2+3\cdot3+3\cdot4)+(4\cdot1+4\cdot2+4\cdot3+4\cdot4) = 3(1+2+3+4)+4(1+2+3+4)$$
$$= (1+2+3+4)(3+4) = (1+2+3+4)(7) = 7(1+2+3+4)$$
$$\sum_{i=1}^{4} 7i = 7\cdot1+7\cdot2+7\cdot3+7\cdot4 = 7(1+2+3+4)$$

Statement **a.** is false. $\sum_{i=1}^{2} (-1)^i 2^i = (-1)^1 2^1 + (-1)^2 2^2 = -1(2)+1(4) = -2+4 = 2$

Statement **b.** is false. $\sum_{i=1}^{2} a_i b_i \neq \sum_{i=1}^{2} a_i \sum_{i=1}^{2} b_i$

$\sum_{i=1}^{2} a_i b_i = a_1 b_1 + a_2 b_2$

$\sum_{i=1}^{2} a_i \sum_{i=1}^{2} b_i = (a_1 + a_2)(b_1 + b_2) = a_1 b_1 + a_1 b_2 + a_2 b_1 + a_2 b_2$

Statement **d.** is false. $\sum_{i=0}^{6} (-1)^i (i+1)^2 \neq \sum_{j=1}^{7} (-1)^j j^2$

$\sum_{i=0}^{6} (-1)^i (i+1)^2$

$= (-1)^0(0+1)^2 + (-1)^1(1+1)^2 + (-1)^2(2+1)^2 + (-1)^3(3+1)^2 + (-1)^4(4+1)^2 + (-1)^5(5+1)^2 + (-1)^6(6+1)^2$

$= 1(1)^2 - 1(2)^2 + 1(3)^2 - 1(4)^2 + 1(5)^2 - 1(6)^2 + 1(7)^2 = 1(1) - 1(4) + 1(9) - 1(16) + 1(25) - 1(36) + 1(49)$

$= 1 - 4 + 9 - 16 + 25 - 36 + 49 = 28$

$\sum_{j=1}^{7} (-1)^j j^2 = (-1)^1 1^2 + (-1)^2 2^2 + (-1)^3 3^2 + (-1)^4 4^2 + (-1)^5 5^2 + (-1)^6 6^2 + (-1)^7 7^2$

$= -1(1) + 1(4) - 1(9) + 1(16) - 1(25) + 1(36) - 1(49)$

$= -1 + 4 - 9 + 16 - 25 + 36 - 49 = -28$

**79.** $a_n = n^2$

**81.** $a_n = n(n+2)$

**83.** $a_n = 2n + 3$

**85.** $a_n = (-2)^{n+1}$

**87.** $\frac{(n+4)!}{(n+2)!} == (n+3)(n+4) = n^2 + 7n + 12$

**89.** $\sum_{i=1}^{4} \log(2i)$

$= \log 2(1) + \log 2(2) + \log 2(3) + \log 2(4)$

$= \log 2 + \log 4 + \log 6 + \log 8 = \log 384$

**91.** $a_n = a_{n-1} + 5$

$a_1 = 7$

$a_2 = a_{2-1} + 5 = a_1 + 5 = 7 + 5 = 12$

$a_3 = a_{3-1} + 5 = a_2 + 5 = 12 + 5 = 17$

$a_4 = a_{4-1} + 5 = a_3 + 5 = 17 + 5 = 22$

The first four terms are 7, 12, 17, 22.

**92.** $\sqrt[3]{40x^4y^7} = \sqrt[3]{8 \cdot 5x^3xy^6y} = 2xy^2\sqrt[3]{5xy}$

**93.** $27x^3 - 8 = (3x-2)(9x^2 + 6x + 4)$

**94.**

$$\frac{6}{x} + \frac{6}{x+2} = \frac{5}{2}$$

$$2x(x+2)\left(\frac{6}{x} + \frac{6}{x+2}\right) = 2x(x+2)\left(\frac{5}{2}\right)$$

$$2(x+2)(6) + 2x(6) = x(x+2)(5)$$

$$12(x+2) + 12x = 5x(x+2)$$

$$12x + 24 + 12x = 5x^2 + 10x$$

$$24x + 24 = 5x^2 + 10x$$

$$0 = 5x^2 - 14x - 24$$

$$0 = (5x+6)(x-4)$$

Apply the zero product principle.

$5x + 6 = 0$ or $x - 4 = 0$

$5x = -6$ $\quad$ $x = 4$

$x = -\frac{6}{5}$

The solutions are $-\frac{6}{5}$ and 4 and the solution set is $\left\{-\frac{6}{5}, 4\right\}$.

## 12.2 Exercise Set

**1.** Since $6-2=4,\ d=4.$

**3.** Since $-2-(-7)=5,\ d=5.$

**5.** Since $711-714=-3,\ d=-3.$

**7.**
$a_1=200$
$a_2=200+20=220$
$a_3=220+20=240$
$a_4=240+20=260$
$a_5=260+20=280$
$a_6=280+20=300$
The first six terms are 200, 220, 240, 260, 280, and 300.

**9.**
$a_1=-7$
$a_2=-7+4=-3$
$a_3=-3+4=1$
$a_4=1+4=5$
$a_5=5+4=9$
$a_6=9+4=13$
The first six terms are –7, !3, 1, 5, 9, 13.

**11.**
$a_1=300$
$a_2=300-90=210$
$a_3=210-90=120$
$a_4=120-90=30$
$a_5=30-90=-60$
$a_6=-60-90=-150$
The first six terms are 300, 210, 120, 30, $-60$, $-150$.

**13.**
$a_1=\frac{5}{2}$
$a_2=\frac{5}{2}-\frac{1}{2}=\frac{4}{2}=2$
$a_3=\frac{4}{2}-\frac{1}{2}=\frac{3}{2}$
$a_4=\frac{3}{2}-\frac{1}{2}=\frac{2}{2}=1$
$a_5=1-\frac{1}{2}=\frac{1}{2}$
$a_6=\frac{1}{2}-\frac{1}{2}=0$
The first six terms are $\frac{5}{2},2,\frac{3}{2},1,\frac{1}{2},0.$

**15.**
$a_1=-0.4$
$a_2=-0.4-1.6=-2$
$a_3=-2-1.6=-3.6$
$a_4=-3.6-1.6=-5.2$
$a_5=-5.2-1.6=-6.8$
$a_6=-6.8-1.6=-8.4$

The first six terms are $-0.4,-2,-3.6,$ $-5.2,-6.8,-8.4.$

**17.**
$a_6=13+(6-1)4=13+(5)4$
$=13+20=33$

**19.**
$a_{50}=7+(50-1)5=7+(49)5$
$=7+245=252$

**21.**
$a_{200}=-40+(200-1)5=-40+(199)5$
$=-40+995=955$

**23.**
$a_{60}=35+(60-1)(-3)=35+(59)(-3)$
$=35+(-177)=-142$

**25.**
$a_n=a_1+(n-1)d=1+(n-1)4$
$=1+4n-4=4n-3$
$a_{20}=4(20)-3=80-3=77$

**27.** $a_n = a_1 + (n-1)d = 7 + (n-1)(-4)$
$= 7 - 4n + 4 = 11 - 4n$
$a_{20} = 11 - 4(20) = 11 - 80 = -69$

**29.** $a_n = a_1 + (n-1)d = -20 + (n-1)(-4)$
$= -20 - 4n + 4 = -4n - 16$
$a_{20} = -4(20) - 16 = -80 - 16 = -96$

**31.** $a_n = a_1 + (n-1)d = -\frac{1}{3} + (n-1)\left(\frac{1}{3}\right)$
$= -\frac{1}{3} + \frac{1}{3}n - \frac{1}{3} = \frac{1}{3}n - \frac{2}{3}$
$a_{20} = \frac{1}{3}(20) - \frac{2}{3} = \frac{20}{3} - \frac{2}{3} = \frac{18}{3} = 6$

**33.** $a_n = a_1 + (n-1)d = 4 + (n-1)(-0.3)$
$= 4 - 0.3n + 0.3 = 4.3 - 0.3n$
$a_{20} = 4.3 - 0.3(20) = 4.3 - 6 = -1.7$

**35.** First find $a_{20}$.
$a_{20} = 4 + (20-1)6 = 4 + (19)6$
$= 4 + 114 = 118$
$S_{20} = \frac{20}{2}(4 + 118) = 10(122) = 1220$

**37.** First find $a_{50}$.
$a_{50} = -10 + (50-1)4 = -10 + (49)4$
$= -10 + 196 = 186$
$S_{50} = \frac{50}{2}(-10 + 186) = 25(176) = 4400$

**39.** First find $a_{100}$.
$a_{100} = 1 + (100-1)1 = 1 + (99)1$
$= 1 + 99 = 100$
$S_{100} = \frac{100}{2}(1 + 100) = 50(101) = 5050$

**41.** First find $a_{60}$.
$a_{60} = 2 + (60-1)2 = 2 + (59)2$
$= 2 + 118 = 120$
$S_{60} = \frac{60}{2}(2 + 120) = 30(122) = 3660$

**43.** Since there are 12 even integers between 21 and 45, find $a_{12}$.
$a_{12} = 22 + (12-1)2 = 22 + (11)2$
$= 22 + 22 = 44$
$S_{12} = \frac{12}{2}(22 + 44) = 6(66) = 396$

**45.** $\sum_{i=1}^{17}(5i+3) = (5(1)+3) + (5(2)+3) + (5(3)+3) + ... + (5(17)+3)$
$= (5+3) + (10+3) + (15+3) + ... + (85+3) = 8 + 13 + 18 + ... + 88$
$S_{17} = \frac{17}{2}(8 + 88) = \frac{17}{2}(96) = 17(48) = 816$

**47.** $\sum_{i=1}^{30}(-3i+5) = (-3(1)+5) + (-3(2)+5) + (-3(3)+5) + ... + (-3(30)+5)$
$= (-3+5) + (-6+5) + (-9+5) + ... + (-90+5) = 2 + (-1) + (-4) + ... + (-85)$
$S_{30} = \frac{30}{2}(2 + (-85)) = 15(-83) = -1245$

**49.** $\sum_{i=1}^{100} 4i = 4(1)+4(2)+4(3)+...+4(100) = 4+8+12+...+400$

$S_{100} = \frac{100}{2}(4+400) = 50(404) = 20,200$

**51.** First find $a_{14}$ and $b_{12}$:

$a_{14} = a_1 + (n-1)d$

$= 1 + (14-1)(-3-1) = -51$

$b_{12} = b_1 + (n-1)d$

$= 3 + (12-1)(8-3) = 58$

So, $a_{14} + b_{12} = -51 + 58 = 7$.

**52.** First find $a_{16}$ and $b_{18}$:

$a_{16} = a_1 + (n-1)d$

$= 1 + (16-1)(-3-1) = -59$

$b_{18} = b_1 + (n-1)d$

$= 3 + (18-1)(8-3) = 88$

So, $a_{16} + b_{18} = -59 + 88 = 29$.

**53.** $a_n = a_1 + (n-1)d$

$-83 = 1 + (n-1)(-3-1)$

$-83 = 1 + -4(n-1)$

$-84 = -4n + 4$

$-88 = -4n$

$n = 22$

There are 22 terms.

**54.** $b_n = b_1 + (n-1)d$

$93 = 3 + (n-1)(8-3)$

$93 = 3 + 5(n-1)$

$93 = 5n - 2$

$95 = 5n$

$n = 19$

There are 19 terms.

**55.** $S_n = \frac{n}{2}(a_1 + a_n)$

For $\{a_n\}$:

$S_{14} = \frac{14}{2}(a_1 + a_{14}) = 7(1 + (-51)) = -350$

For $\{b_n\}$:

$S_{14} = \frac{14}{2}(b_1 + b_{14}) = 7(3 + 68) = 497$

So $\sum_{n=1}^{14} b_n - \sum_{n=1}^{14} a_n = 497 - (-350) = 847$

**56.** First find $a_{15}$ and $b_{15}$:

$a_{15} = a_1 + (n-1)d$

$= 1 + (15-1)(-3-1) = -55$

$b_{15} = b_1 + (n-1)d$

$= 3 + (15-1)(8-3) = 73$

Using $S_n = \frac{n}{2}(a_1 + a_n)$ for $\{a_n\}$:

$S_{15} = \frac{15}{2}(a_1 + a_{15}) = 7.5(1 + (-55))$

$= -405$

And then for $\{b_n\}$:

$S_{15} = \frac{15}{2}(b_1 + b_{15}) = 7.5(3 + 73) = 570$

So

$\sum_{n=1}^{15} b_n - \sum_{n=1}^{15} a_n = 570 - (-405) = 975$

**57.** Two points on the graph are $(1, 1)$ and $(2, -3)$. Finding the slope of the line; $m = \dfrac{y_2 - y_1}{x_2 - x_2} = \dfrac{-3-1}{2-1} = \dfrac{-4}{1} = -4$

Using the point-slope form of an equation of a line;

$$y - y_2 = m(x - x_2)$$
$$y - 1 = -4(x-1)$$
$$y - 1 = -4x + 4$$
$$y = -4x + 5$$

Thus, $f(x) = -4x + 5$.

**58.** Two points on the graph are $(1, 3)$ and $(2, 8)$. Finding the slope of the line; $m = \dfrac{y_2 - y_1}{x_2 - x_2} = \dfrac{8-3}{2-1} = \dfrac{5}{1} = 5$

Using the point-slope form of an equation of a line;

$$y - y_2 = 5(x - x_2)$$
$$y - 3 = 5(x-1)$$
$$y - 3 = 5x - 5$$
$$y = 5x - 2$$

Thus, $g(x) = 5x - 2$.

**59.** Using $a_n = a_1 + (n-1)d$ and $a_2 = 4$:

$$a_2 = a_1 + (2-1)d$$
$$4 = a_1 + d$$

And since $a_6 = 16$:

$$a_6 = a_1 + (6-1)d$$
$$16 = a_1 + 5d$$

The system of equations is

$$4 = a_1 + d$$
$$16 = a_1 + 5d$$

Solving the first equation for $a_1$:

$$a_1 = 4 - d$$

Substituting the value into the second equation and solving for *d*:

$$16 = (4-d) + 5d$$
$$16 = 4 + 4d$$
$$12 = 4d$$
$$3 = d$$

Then $a_n = a_1 + (n-1)d$

$$a_n = 1 + (n-1)3$$
$$a_n = 1 + 3n - 3$$
$$a_n = 3n - 2$$

**60.** Using $a_n = a_1 + (n-1)d$ and $a_3 = 7$:

$a_3 = a_1 + (3-1)d$

$7 = a_1 + 2d$

And since $a_8 = 17$:

$a_8 = a_1 + (8-1)d$

$17 = a_1 + 7d$

The system of equations is

$7 = a_1 + 2d$

$17 = a_1 + 7d$

Solving the first equation for $a_1$:

$a_1 = 7 - 2d$

Substituting the value into the second equation and solving for *d*:

$17 = (7-2d) + 7d$

$17 = 7 + 5d$

$10 = 5d$

$2 = d$

So $a_1 = 7 - 2(2) = 7 - 4 = 3$.

Then $a_n = a_1 + (n-1)2$

$a_n = 3 + (n-1)2$

$a_n = 3 + 2n - 2$

$a_n = 2n + 1$

**61.** **a.** $a_n = a_1 + (n-1)d$

$a_n = 150 + (n-1)(1.7)$

$a_n = 150 + 1.7n - 1.7$

$a_n = 1.7n + 148.3$

**b.** $a_{40} = 1.7(40) + 148.3$

$= 68 + 148.3$

$= 216.3$ pounds

**63.** Answers will vary.

**65.** Company *A*

$a_n = 24000 + (n-1)1600$

$= 24000 + 1600n - 1600$

$= 1600n + 22400$

$a_{10} = 1600(10) + 22400$

$= 16000 + 22400 = 38400$

Company *B*

$a_n = 28000 + (n-1)1000$

$= 28000 + 1000n - 1000$

$= 1000n + 27000$

$a_{10} = 1000(10) + 27000$

$= 10000 + 27000 = 37000$

Company *A* will pay \$1400 more in year 10.

**67.** **a.** Total cost:

$\$7107 + \$7310 + \$7586 + \$8046$

$= \$30,049$

**b.** $a_1 = 309(1) + 6739 = 7048$

$a_2 = 309(2) + 6739 = 7357$

$a_3 = 309(3) + 6739 = 7666$

$a_4 = 309(4) + 6739 = 7975$

Total cost:

$7048 + 7357 + 7666 + 7975$

$= \$30,046$

The model describes the actual sum very well.

**69.** Answers will vary.

**71.** Company *A*

$a_n = 19000 + (n-1)2600$

$= 19000 + 2600n - 2600$

$= 2600n + 16400$

$a_{10} = 2600(10) + 16400$

$= 26000 + 16400 = 42400$

$$S_n = \frac{n}{2}(a_1 + a_{10})$$

$$S_{10} = \frac{10}{2}(19000 + 42400)$$

$$= 5(61400) = \$307,000$$

Company $B$

$$a_n = 27000 + (n-1)1200$$

$$= 27000 + 1200n - 1200$$

$$= 1200n + 25800$$

$$a_{10} = 1200(10) + 25800$$

$$= 12000 + 25800$$

$$= 37800$$

$$S_n = \frac{n}{2}(a_1 + a_{10})$$

$$S_{10} = \frac{10}{2}(27000 + 37800)$$

$$= 5(64800) = \$324,000$$

Company $B$ pays the greater total amount.

**73.** $a_{38} = a_1 + (n-1)d$

$$= 20 + (38-1)(3)$$

$$= 20 + 37(3) = 131$$

$$S_{38} = \frac{n}{2}(a_1 + a_{38})$$

$$= \frac{38}{2}(20 + 131)$$

$$= 19(151)$$

$$= 2869$$

There are 2869 seats in this section of the stadium.

**75. – 77.** Answers will vary.

**79.** Answers will vary. For example, consider Exercise 45.

$$\sum_{i=1}^{17}(5i+3)$$

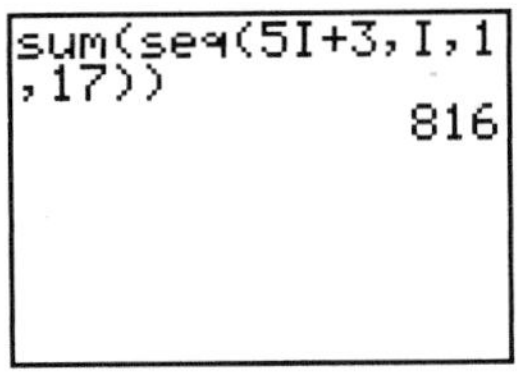


This is the same result obtained in Exercise 45.

**81.** From the sequence, we see that $a_1 = 21700$ and $d = 23172 - 21700 = 1472$.

We know that $a_n = a_1 + (n-1)d$. We can substitute what we know to find $n$.

$$314628 = 21700 + (n-1)1472$$

$$292928 = (n-1)1472$$

$$\frac{292928}{1472} = \frac{(n-1)1472}{1472}$$

$$199 = n - 1$$

$$200 = n$$

314,628 is the 200th term of the sequence.

**83.** $1+3+5+...+(2n-1)$

$$S_n = \frac{n}{2}(a_1 + a_n)$$
$$= \frac{n}{2}(1+(2n-1))$$
$$= \frac{n}{2}(1+2n-1)$$
$$= \frac{n}{2}(2n)$$
$$= n(n) = n^2$$

**84.** $\log(x^2-25)-\log(x+5)=3$

$$\log\left(\frac{x^2-25}{x+5}\right)=3$$
$$\log\left(\frac{(x-5)(x+5)}{x+5}\right)=3$$
$$\log(x-5)=3$$
$$x-5=10^3$$
$$x-5=1000$$
$$x=1005$$

**85.** $x^2+3x\le 10$

Solve the related quadratic equation.

$$x^2+3x-10=0$$
$$(x+5)(x-2)=0$$

Apply the zero product principle.

$x+5=0$ or $x-2=0$

$x=-5$ $\quad$ $x=2$

The boundary points are $-5$ and 2.

| Test Interval | Test Number | Test | Conclusion |
|---|---|---|---|
| $(-\infty,-5]$ | -6 | $(-6)^2+3(-6)\le 10$<br>$18\le 10$, false | $(-\infty,-5]$ does not belong to the solution set. |
| $[-5,2]$ | 0 | $0^2+3(0)\le 10$<br>$0\le 10$, true | $[-5,2]$ belongs to the solution set. |
| $[2,\infty)$ | 3 | $3^2+3(3)\le 10$<br>$18\le 10$, false | $[2,\infty)$ does not belong to the solution set. |

The solution set is $[-5,2]$ or $\{x \mid -5\le x\le 2\}$.

**86.**
$$A = \frac{Pt}{P+t}$$
$$A(P+t) = Pt$$
$$AP + At = Pt$$
$$AP - Pt = -At$$
$$P(A-t) = -At$$
$$P = -\frac{At}{A-t} \text{ or } \frac{At}{t-A}$$

## 12.3 Exercise Set

**1.** $r = \frac{a_2}{a_1} = \frac{15}{5} = 3$

**3.** $r = \frac{a_2}{a_1} = \frac{30}{-15} = -2$

**5.** $r = \frac{a_2}{a_1} = \frac{\frac{9}{2}}{3} = \frac{9}{2} \cdot \frac{1}{3} = \frac{3}{2}$

**7.** $r = \frac{a_2}{a_1} = \frac{0.04}{-0.4} = -0.1$

**9.** The first term is 2.
The second term is $2 \cdot 3 = 6$.
The third term is $6 \cdot 3 = 18$.
The fourth term is $18 \cdot 3 = 54$.
The fifth term is $54 \cdot 3 = 162$.

**11.** The first term is 20.
The second term is $20 \cdot \frac{1}{2} = 10$.
The third term is $10 \cdot \frac{1}{2} = 5$.
The fourth term is $5 \cdot \frac{1}{2} = \frac{5}{2}$.
The fifth term is $\frac{5}{2} \cdot \frac{1}{2} = \frac{5}{4}$.

**13.** The first term is –4.
The second term is $-4(-10) = 40$.
The third term is $40(-10) = -400$.
The fourth term is $-400(-10) = 4000$.
The fifth term is $4000(-10) = -40,000$.

**15.** The first term is $-\frac{1}{4}$.
The second term is $-\frac{1}{4}(-2) = \frac{1}{2}$.
The third term is $\frac{1}{2}(-2) = -1$.
The fourth term is $-1(-2) = 2$.
The fifth term is $2(-2) = -4$.

**17.** $a_8 = 6(2)^{8-1} = 6(2)^7 = 6(128) = 768$

**19.** $a_{12} = 5(-2)^{12-1} = 5(-2)^{11}$
$= 5(-2048) = -10,240$

**21.** $a_6 = 6400\left(-\frac{1}{2}\right)^{6-1} = 6400\left(-\frac{1}{2}\right)^5$
$= -200$

**23.** $a_8 = 1{,}000{,}000(0.1)^{8-1}$
$= 1{,}000{,}000(0.1)^7$
$= 1{,}000{,}000(0.0000001) = 0.1$

**25.** $r = \dfrac{a_2}{a_1} = \dfrac{12}{3} = 4$
$a_n = a_1 r^{n-1} = 3(4)^{n-1}$
$a_7 = 3(4)^{7-1} = 3(4)^6$
$= 3(4096) = 12{,}288$

**27.** $r = \dfrac{a_2}{a_1} = \dfrac{6}{18} = \dfrac{1}{3}$
$a_n = a_1 r^{n-1} = 18\left(\dfrac{1}{3}\right)^{n-1}$
$a_7 = 18\left(\dfrac{1}{3}\right)^{7-1} = 18\left(\dfrac{1}{3}\right)^6$
$= 18\left(\dfrac{1}{729}\right) = \dfrac{18}{729} = \dfrac{2}{81}$

**29.** $r = \dfrac{a_2}{a_1} = \dfrac{-3}{1.5} = -2$
$a_n = a_1 r^{n-1} = 1.5(-2)^{n-1}$
$a_7 = 1.5(-2)^{7-1} = 1.5(-2)^6$
$= 1.5(64) = 96$

**31.** $r = \dfrac{a_2}{a_1} = \dfrac{-0.004}{0.0004} = -10$
$a_n = a_1 r^{n-1} = 0.0004(-10)^{n-1}$
$a_7 = 0.0004(-10)^{7-1} = 0.0004(-10)^6$
$= 0.0004(1000000) = 400$

**33.** $r = \dfrac{a_2}{a_1} = \dfrac{6}{2} = 3$
$S_{12} = \dfrac{2(1-3^{12})}{1-3} = \dfrac{2(1-531441)}{-2}$
$= \dfrac{2(-531440)}{-2} = \dfrac{-1{,}062{,}880}{-2}$
$= 531{,}440$

**35.** $r = \dfrac{a_2}{a_1} = \dfrac{-6}{3} = -2$
$S_{11} = \dfrac{a_1(1-r^n)}{1-r} = \dfrac{3(1-(-2)^{11})}{1-(-2)}$
$= \dfrac{\cancel{3}(1-(-2048))}{\cancel{3}} = 2049$

**37.** $r = \dfrac{a_2}{a_1} = \dfrac{3}{-\frac{3}{2}} = 3 \div \left(-\dfrac{3}{2}\right)$
$= 3 \cdot \left(-\dfrac{2}{3}\right) = -2$
$S_{14} = \dfrac{a_1(1-r^n)}{1-r} = \dfrac{-\frac{3}{2}(1-(-2)^{14})}{1-(-2)}$
$= \dfrac{-\frac{3}{2}(1-(-16384))}{3} = \dfrac{-\frac{3}{2}(16385)}{3}$
$= -\dfrac{3}{2}(16385) \div 3 = -\dfrac{49155}{2} \cdot \dfrac{1}{3}$
$= -\dfrac{16385}{2}$

**39.** $\displaystyle\sum_{i=1}^{8} 3^i = \dfrac{3(1-3^8)}{1-3} = \dfrac{3(1-6561)}{-2}$
$= \dfrac{3(-6560)}{-2} = \dfrac{-19680}{-2} = 9840$

**41.** $$\sum_{i=1}^{10} 5\cdot 2^i = \frac{10\left(1-2^{10}\right)}{1-2} = \frac{10(1-1024)}{-1}$$
$$= \frac{10(-1023)}{-1} = 10,230$$

**43.** $$\sum_{i=1}^{6}\left(\frac{1}{2}\right)^{i+1} = \frac{\frac{1}{4}\left(1-\left(\frac{1}{2}\right)^6\right)}{1-\frac{1}{2}} = \frac{\frac{1}{4}\left(1-\frac{1}{64}\right)}{\frac{1}{2}}$$
$$= \frac{\frac{1}{4}\left(\frac{64}{64}-\frac{1}{64}\right)}{\frac{1}{2}} = \frac{\frac{1}{4}\left(\frac{63}{64}\right)}{\frac{1}{2}}$$
$$= \frac{1}{4}\left(\frac{63}{64}\right)\div\frac{1}{2} = \frac{1}{4}\left(\frac{63}{64}\right)\cdot\frac{2}{1}$$
$$= \frac{63}{128}$$

**45.** $$r = \frac{a_2}{a_1} = \frac{\frac{1}{3}}{1} = \frac{1}{3}$$
$$S = \frac{a_1}{1-r} = \frac{1}{1-\frac{1}{3}} = \frac{1}{\frac{2}{3}} = 1\div\frac{2}{3} = 1\cdot\frac{3}{2} = \frac{3}{2}$$

**47.** $$r = \frac{a_2}{a_1} = \frac{\frac{3}{4}}{3} = \frac{3}{4}\div 3 = \frac{3}{4}\cdot\frac{1}{3} = \frac{1}{4}$$
$$S = \frac{a_1}{1-r} = \frac{3}{1-\frac{1}{4}}$$
$$= \frac{3}{\frac{3}{4}} = 3\div\frac{3}{4}$$
$$= 3\cdot\frac{4}{3} = \frac{12}{3} = 4$$

**49.** $$r = \frac{a_2}{a_1} = \frac{-\frac{1}{2}}{1} = -\frac{1}{2}$$
$$S = \frac{a_1}{1-r} = \frac{1}{1-\left(-\frac{1}{2}\right)} = \frac{1}{\frac{3}{2}} = 1\div\frac{3}{2}$$
$$= 1\cdot\frac{2}{3} = \frac{2}{3}$$

**51.** $r = -0.3$
$$a_1 = 26(-0.3)^{1-1} = 26(-0.3)^0$$
$$= 26(1) = 26$$
$$S = \frac{26}{1-(-0.3)} = \frac{26}{1.3} = 20$$

**53.** $$0.\overline{5} = \frac{a_1}{1-r} = \frac{\frac{5}{10}}{1-\frac{1}{10}} = \frac{\frac{5}{10}}{\frac{9}{10}} = \frac{5}{10}\div\frac{9}{10}$$
$$= \frac{5}{10}\cdot\frac{10}{9} = \frac{5}{9}$$

**55.** $$0.\overline{47} = \frac{a_1}{1-r} = \frac{\frac{47}{100}}{1-\frac{1}{100}} = \frac{\frac{47}{100}}{\frac{99}{100}}$$
$$= \frac{47}{100}\div\frac{99}{100} = \frac{47}{100}\cdot\frac{100}{99} = \frac{47}{99}$$

**57.** $$0.\overline{257} = \frac{a_1}{1-r} = \frac{\frac{257}{1000}}{1-\frac{1}{1000}} = \frac{\frac{257}{1000}}{\frac{999}{1000}}$$
$$= \frac{257}{1000}\div\frac{999}{1000} = \frac{257}{1000}\cdot\frac{1000}{999}$$
$$= \frac{257}{999}$$

**59.** The sequence is arithmetic with common difference $d = 1$.

**61.** The sequence is geometric with common ratio $r = 2$.

**63.** The sequence is neither arithmetic nor geometric.

**65.** First find $a_{10}$ and $b_{10}$:

$$a_{10} = a_1 r^{n-1}$$
$$= (-5)\left(\frac{10}{-5}\right)^{10-1} = (-5)(-2)^9$$
$$= 2560$$
$$b_{10} = b_1 + (n-1)d$$
$$= 10 + (10-1)(-5-10)$$
$$= 10 + (9)(-15) = -125$$

So,

$$a_{10} + b_{10} = 2560 + (-125) = 2435.$$

**66.** First find $a_{11}$ and $b_{11}$:

$$a_{11} = a_1 r^{n-1}$$
$$= (-5)\left(\frac{10}{-5}\right)^{11-1} = (-5)(-2)^{10}$$
$$= -5120$$
$$b_{11} = b_1 + (n-1)d$$
$$= 10 + (11-1)(-5-10)$$
$$= 10 + (10)(-15) = -140$$

So,

$$a_{11} + b_{11} = -5120 + (-140) = -5260.$$

**67.** From Exercise 65, $a_{10} = 2560$ and $b_{10} = -125$.

For $\{a_n\}$, $r = \frac{10}{-5} = -2$ and :

$$S_{10} = \frac{a_1(1-r^n)}{1-r} = \frac{(-5)\left(1-(-2)^{10}\right)}{1-(-2)}$$
$$= \frac{(-5)(-1023)}{3} = 1705$$

For $\{b_n\}$,

$$S_n = \frac{n}{2}(b_1 + b_n) = \frac{10}{2}(10 + (-125))$$
$$= 5(-115) = -575$$

So,

$$\sum_{n=1}^{10} a_n - \sum_{n=1}^{10} b_n = 1705 - (-575) = 2280$$

**68.** For $\{a_n\}$, $r = \frac{10}{-5} = -2$ and :

$$S_{11} = \frac{a_1(1-r^n)}{1-r} = \frac{(-5)\left(1-(-2)^{11}\right)}{1-(-2)}$$
$$= \frac{(-5)(2049)}{3} = -3415$$

For $\{b_n\}$,

$$b_{11} = b_1 + (n-1)d$$
$$= 10 + (11-1)(-5-10)$$
$$= 10 + (10)(-15) = -140$$
$$S_{11} = \frac{n}{2}(b_1 + b_n) = \frac{11}{2}(10 + (-140))$$
$$= 5.5(-130) = -715$$

So, $$\sum_{n=1}^{11} a_n - \sum_{n=1}^{11} b_n = -3415 - (-715)$$
$$= -2700$$

**69.** For $\{a_n\}$,

$$S_6 = \frac{a_1(1-r^n)}{1-r} = \frac{(-5)\left(1-(-2)^6\right)}{1-(-2)}$$
$$= \frac{(-5)(-63)}{3} = 105$$

For $\{c_n\}$,

$$S = \frac{a_1}{1-r} = \frac{-2}{1-\frac{1}{-2}} = \frac{-2}{\frac{3}{2}} = -\frac{4}{3}$$

So, $S_6 \cdot S = 105\left(-\frac{4}{3}\right) = -140$

**70.** For $\{a_n\}$,

$$S_9 = \frac{a_1(1-r^n)}{1-r} = \frac{(-5)\left(1-(-2)^9\right)}{1-(-2)}$$
$$= \frac{(-5)(513)}{3} = -855$$

For $\{c_n\}$,

$$S = \frac{a_1}{1-r} = \frac{-2}{1-\frac{1}{-2}} = \frac{-2}{\frac{3}{2}} = -\frac{4}{3}$$

So, $S_9 \cdot S = -855\left(-\frac{4}{3}\right) = 1140$

**71.** It is given that $a_4 = 27$. Using the formula $a_n = a_1r^{n-1}$ when $n = 4$ we have:

$$27 = 8r^{4-1}$$
$$\frac{27}{8} = r^3$$
$$r = \sqrt[3]{\frac{27}{8}} = \frac{3}{2}$$

Then

$$a_n = a_1r^{n-1}$$
$$a_2 = 8\left(\frac{3}{2}\right)^{2-1} = 8\left(\frac{3}{2}\right) = 12$$
$$a_3 = 8\left(\frac{3}{2}\right)^{3-1} = 8\left(\frac{3}{2}\right)^2 = 8\left(\frac{9}{4}\right) = 18$$

**72.** It is given that $a_4 = -54$. Using the formula $a_n = a_1r^{n-1}$ when $n = 4$ we have:

$$-54 = 2r^{4-1}$$
$$-27 = r^3$$
$$r = \sqrt[3]{-27}$$
$$= -3$$

Then

$$a_n = a_1r^{n-1}$$
$$a_2 = 2(-3)^{2-1}$$
$$= 2(-3)$$
$$= -6$$
$$a_3 = 2(-3)^{3-1}$$
$$= 2(-3)^2$$
$$= 2(9)$$
$$= 18$$

**73.**

$$r = \frac{a_2}{a_1} = \frac{2}{1} = 2$$
$$a_{15} = 1(2)^{15-1}$$
$$= (2)^{14}$$
$$= 16384$$

On the fifteenth day, you will put aside \$16,384 for savings.

**75.** $r = 1.04$

$$a_7 = 3{,}000{,}000(1.04)^{7-1}$$
$$= 3{,}000{,}000(1.04)^6$$
$$= 3{,}000{,}000(1.265319)$$
$$= 3{,}795{,}957$$

The athlete's salary for year 7 will be \$3,795,957.

**77.** **a.**

$$r_{1990 \text{ to } 1991} = \frac{30.15}{29.76} \approx 1.013$$

$$r_{1991 \text{ to } 1992} = \frac{30.54}{30.15} \approx 1.013$$

$$r_{1992 \text{ to } 1993} = \frac{30.94}{30.54} \approx 1.013$$

$$r_{1993 \text{ to } 1994} = \frac{31.34}{30.94} \approx 1.013$$

$$r_{1994 \text{ to } 1995} = \frac{31.75}{31.34} \approx 1.013$$

$$r_{1995 \text{ to } 1996} = \frac{32.16}{31.75} \approx 1.013$$

$$r_{1996 \text{ to } 1997} = \frac{32.58}{32.16} \approx 1.013$$

Since the population of each year can be found by multiplying the preceding year by 1.013, the population is increasing geometrically.

**b.** $a_n = a_1 r^{n-1} = 29.76(1.013)^{n-1}$

**c.** Since year 2000 is the 11th term, find $a_{11}$.

$$a_{11} = 29.76(1.013)^{11-1}$$
$$= 29.76(1.013)^{10} \approx 33.86$$

The population of California will be approximately 33.86 million in 2000. The geometric sequence models the actual population of 33.87 million very well.

**79.** $r = \frac{a_2}{a_1} = \frac{2}{1} = 2$

$$S_{15} = \frac{a_1(1-r^n)}{1-r} = \frac{1(1-(2)^{15})}{1-2}$$
$$= \frac{(1-32768)}{-1} = \frac{(-32767)}{-1} = 32767$$

Your savings will be \$32,767 over the 15 days.

**81.** $r = 1.05$

$$S_{20} = \frac{a_1(1-r^n)}{1-r}$$
$$= \frac{24000(1-(1.05)^{20})}{1-1.05}$$
$$= \frac{24000(1-2.6533)}{-0.05}$$
$$= \frac{24000(-1.6533)}{-0.05}$$
$$= 793583$$

The total lifetime salary over the 20 years is \$793,583.

**83.** $r = 0.9$

$$S_{10} = \frac{a_1(1-r^n)}{1-r}$$
$$= \frac{20(1-(0.9)^{10})}{1-0.9}$$
$$= \frac{20(1-0.348678)}{0.1}$$
$$= \frac{20(0.651322)}{0.1}$$
$$= 130.264$$

After 10 swings, the pendulum covers a distance of approximately 130.26 inches.

**85.**

$$A = P\frac{\left(1+\frac{r}{n}\right)^{nt}-1}{\frac{r}{n}}$$

$$= 2500\frac{\left(1+\frac{0.09}{1}\right)^{1(40)}-1}{\frac{0.09}{1}}$$

$$= 2500\frac{(1+0.09)^{40}-1}{0.09}$$

$$= 2500\frac{(1.09)^{40}-1}{0.09}$$

$$= 2500\frac{31.4094-1}{0.09}$$

$$= 2500\frac{30.4094}{0.09} = 844706$$

After 40 years, the value of the IRA will be \$844,706.

**87.**

$$A = P\frac{\left(1+\frac{r}{n}\right)^{nt}-1}{\frac{r}{n}}$$

$$= 600\frac{\left(1+\frac{0.08}{4}\right)^{4(18)}-1}{\frac{0.08}{4}}$$

$$= 600\frac{(1+0.02)^{72}-1}{0.02}$$

$$= 600\frac{(1.02)^{72}-1}{0.02}$$

$$= 600\frac{4.16114-1}{0.02} = 600\frac{3.16114}{0.02}$$

$$= 94834.2$$

The value of the TSA after 18 years will be \$94,834.20.

**89.** $r = 60\% = 0.6$

$a_1 = 6(.6) = 3.6$

$$S = \frac{3.6}{1-0.6} = \frac{3.6}{0.4} = 9$$

The total economic impact of the factory will be \$9 million per year.

**91.** $r = \frac{1}{4}$

$$S = \frac{\frac{1}{4}}{1-\frac{1}{4}} = \frac{\frac{1}{4}}{\frac{3}{4}}$$

$$= \frac{1}{4} \div \frac{3}{4} = \frac{1}{4}\cdot\frac{4}{3}$$

$$= \frac{1}{3}$$

Eventually $\frac{1}{3}$ of the largest square will be shaded.

**93. – 99.** Answers will vary.

**101.** Answers will vary. For example, consider Exercise 25.

$a_n = 3(4)^{n-1}$

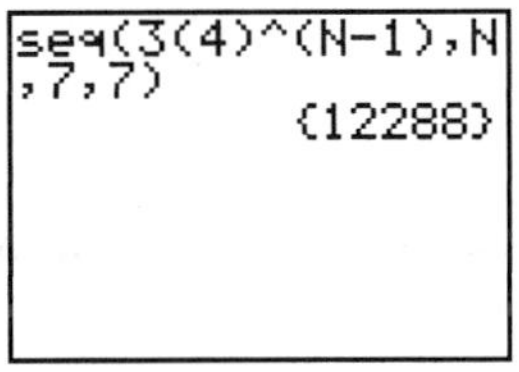


This matches the result obtained in Exercise 25.

**103.**

$$f(x)=\frac{2\left[1-\left(\frac{1}{3}\right)^{x}\right]}{1-\frac{1}{3}}$$

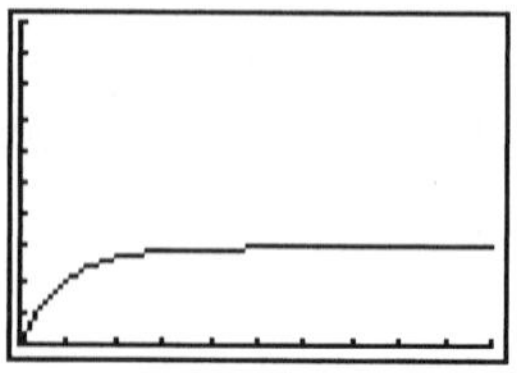

$$S=\frac{2}{1-\frac{1}{3}}=\frac{2}{\frac{2}{3}}=2\div\frac{2}{3}=2\cdot\frac{3}{2}=3$$

The sum of the series and the asymptote of the function are both 3.

**105.** Statement **d.** is true. The common ratio is $0.5=\frac{1}{2}$.

Statement **a.** is false. The sequence is not geometric. The fourth term would have to be $24\cdot 4=96$ for the sequence to be geometric.

Statement **b.** is false. We do not need to know the terms between $\frac{1}{8}$ and $\frac{1}{512}$, but we do need to know how many terms there are between $\frac{1}{8}$ and $\frac{1}{512}$.

Statement **c.** is false. The sum of the sequence is $\frac{10}{1-\left(-\frac{1}{2}\right)}$.

**107.**

$$A=P\frac{\left(1+\frac{r}{n}\right)^{nt}-1}{\frac{r}{n}}$$

$$1,000,000=P\frac{\left(1+\frac{0.10}{12}\right)^{12(30)}-1}{\frac{0.10}{12}}$$

$$1,000,000=P\frac{\left(1+\frac{1}{120}\right)^{360}-1}{\frac{1}{120}}$$

$$1,000,000=P\frac{\left(1\frac{1}{120}\right)^{360}-1}{\frac{1}{120}}$$

$$1,000,000=P\frac{19.8374-1}{\frac{1}{120}}$$

$$\frac{1}{120}(1,000,000)=\frac{1}{120}\left(P\frac{18.8374}{\frac{1}{120}}\right)$$

$$\frac{25000}{3}=18.8374P$$

$$\frac{25000}{3(18.8374)}=P$$

$$442.382=P$$

You should deposit approximately \$442.38 per month.

**108.**

$$\sqrt{28}-3\sqrt{7}+\sqrt{63}$$
$$=\sqrt{4\cdot 7}-3\sqrt{7}+\sqrt{9\cdot 7}$$
$$=2\sqrt{7}-3\sqrt{7}+3\sqrt{7}$$
$$=2\sqrt{7}$$

**109.**

$$2x^2 = 4 - x$$
$$2x^2 + x - 4 = 0$$
$$a = 2 \quad b = 1 \quad c = -4$$

Solve using the quadratic formula.

$$x = \frac{-1 \pm \sqrt{1^2 - 4(2)(-4)}}{2(2)}$$
$$= \frac{-1 \pm \sqrt{1+32}}{4} = \frac{-1 \pm \sqrt{33}}{4}$$

The solutions are $\frac{-1 \pm \sqrt{33}}{4}$ and the solution set is $\left\{\frac{-1 \pm \sqrt{33}}{4}\right\}$.

**110.**

$$\frac{6}{\sqrt{3}-\sqrt{5}} = \frac{6}{\sqrt{3}-\sqrt{5}} \cdot \frac{\sqrt{3}+\sqrt{5}}{\sqrt{3}+\sqrt{5}}$$
$$= \frac{6\left(\sqrt{3}+\sqrt{5}\right)}{3-5}$$
$$= \frac{6\left(\sqrt{3}+\sqrt{5}\right)}{-2}$$
$$= -3\left(\sqrt{3}+\sqrt{5}\right)$$

## Mid-Chapter Check Point

**1.**

$$a_n = (-1)^{n+1}\frac{n}{(n-1)!}$$
$$a_1 = (-1)^{1+1}\frac{1}{(1-1)!} = (-1)^2\frac{1}{0!} = 1 \cdot 1 = 1$$
$$a_2 = (-1)^{2+1}\frac{2}{(2-1)!} = (-1)^3\frac{2}{1!} = (-1)(2) = -2$$
$$a_3 = (-1)^{3+1}\frac{3}{(3-1)!} = (-1)^4\frac{3}{2!} = 1 \cdot \frac{3}{2} = \frac{3}{2}$$
$$a_4 = (-1)^{4+1}\frac{4}{(4-1)!} = (-1)^5\frac{4}{3!} = (-1)\frac{4}{6} = -\frac{2}{3}$$
$$a_5 = (-1)^{5+1}\frac{5}{(5-1)!} = (-1)^6\frac{5}{4!} = 1 \cdot \frac{5}{24} = \frac{5}{24}$$

**2.** Using $a_n = a_1 + (n-1)d$;

$a_1 = 5$

$a_2 = 5 + (2-1)(-3) = 5 + 1(-3) = 5 - 3 = 2$

$a_3 = 5 + (3-1)(-3) = 5 + 2(-3) = 5 - 6 = -1$

$a_4 = 5 + (4-1)(-3) = 5 + 3(-3) = 5 - 9 = -4$

$a_5 = 5 + (5-1)(-3) = 5 + 4(-3) = 5 - 12 = -7$

**3.** Using $a_n = a_1 r^{n-1}$;

$a_1 = 5$

$a_2 = 5(-3)^{2-1} = 5(-3)^1 = 5(-3) = -15$

$a_3 = 5(-3)^{3-1} = 5(-3)^2 = 5(9) = 45$

$a_4 = 5(-3)^{4-1} = 5(-3)^3 = 5(-27) = -135$

$a_5 = 5(-3)^{5-1} = 5(-3)^4 = 5(81) = 405$

**4.** $d = a_2 - a_1 = 6 - 2 = 4$

$$\begin{aligned} a_n &= a_1 + (n-1)d \\ &= 2 + (n-1)4 \\ &= 2 + 4n - 4 \\ &= 4n - 2 \end{aligned}$$

$a_{20} = 4(20) - 2 = 78$

**5.** $r = \frac{a_2}{a_1} = \frac{6}{3} = 2$

$$\begin{aligned} a_n &= a_1 r^{n-1} \\ &= 3(2)^{n-1} \end{aligned}$$

$a_{10} = 3(2)^{10-1} = 3(2)^9 = 1536$

**6.** $d = a_2 - a_1 = 1 - \frac{3}{2} = -\frac{1}{2}$

$$\begin{aligned} a_n &= a_1 + (n-1)d \\ &= \frac{3}{2} + (n-1)\left(-\frac{1}{2}\right) \\ &= \frac{3}{2} - \frac{1}{2}n + \frac{1}{2} \\ &= -\frac{1}{2}n + 2 \end{aligned}$$

$a_{30} = -\frac{1}{2}(30) + 2 = -15 + 2 = -13$

**7.** $S_n = \frac{a_1(1-r^n)}{1-r}; r = \frac{a_2}{a_1} = \frac{10}{5} = 2$

$S_{10} = \frac{5(1-2^{10})}{1-2} = \frac{5(-1023)}{-1} = 5115$

**8.** First find $a_{10}$;

$$d = a_2 - a_1 = 0 - (-2) = 2$$

$$a_{50} = a_1 + (n-1)d = -2 + (50-1)(2) = -2 + 49(2) = 96$$

$$S_{50} = \frac{n}{2}(a_1 + a_n) = \frac{50}{2}(-2+96) = 25(94) = 2350$$

**9.** First find $a_{10}$;

$$r = \frac{a_2}{a_1} = \frac{40}{-20} = -2$$

$$a_{10} = a_1 r^{n-1} = -20(2)^{10-1} = -20(2)^9 = -20(512) = -10240$$

$$S_{10} = \frac{a_1(1-r^n)}{1-r} = \frac{-20(-1-(-2)^{10})}{1-(-2)} = \frac{-20(-1023)}{3} = \frac{20460}{3} = 6820$$

**10.** First find $a_{100}$;

$$d = a_2 - a_1 = -2 - 4 = -6$$

$$a_{100} = a_1 + (n-1)d = 4 + (100-1)(-6) = 4 + 99(-6) = -590$$

$$S_{100} = \frac{n}{2}(a_1 + a_n) = \frac{100}{2}(4-590) = 50(-586) = -29,300$$

**11.**
$$\sum_{i=1}^{4}(i+4)(i-1) = (1+4)(1-1) + (2+4)(2-1) + (3+4)(3-1) + (4+4)(4-1)$$
$$= 5(0) + 6(1) + 7(2) + 8(3) = 0 + 6 + 14 + 24 = 44$$

**12.**
$$\sum_{i=1}^{50}(3i-2) = (3\cdot1-2) + (3\cdot2-2) + (3\cdot3-3) + \ldots + (3\cdot50-2)$$
$$= (3-2) + (6-2) + (9-3) + \ldots + (150-2)$$
$$= 1 + 4 + 6 + \ldots + 148$$

The sum of this arithmetic sequence is given by $S_n = \frac{n}{2}(a_1 + a_n)$;

$$S_{50} = \frac{50}{2}(1+148) = 25(149) = 3725$$

**13.**
$$\sum_{i=1}^{6}\left(\frac{3}{2}\right)^i = \left(\frac{3}{2}\right)^1 + \left(\frac{3}{2}\right)^2 + \left(\frac{3}{2}\right)^3 + \left(\frac{3}{2}\right)^4 + \left(\frac{3}{2}\right)^5 + \left(\frac{3}{2}\right)^6$$
$$= \frac{3}{2} + \frac{9}{4} + \frac{27}{8} + \frac{81}{16} + \frac{243}{32} + \frac{729}{64} = \frac{1995}{64}$$

**14.**
$$\sum_{i=1}^{\infty}\left(-\frac{2}{5}\right)^{i-1}=\left(-\frac{2}{5}\right)^{1-1}+\left(-\frac{2}{5}\right)^{2-1}+\left(-\frac{2}{5}\right)^{3-1}+\ldots$$
$$=\left(-\frac{2}{5}\right)^{0}+\left(-\frac{2}{5}\right)^{1}+\left(-\frac{2}{5}\right)^{2}+\ldots$$
$$=1+\left(-\frac{2}{5}\right)+\frac{4}{25}+\ldots$$

This is an infinite geometric sequence with $r=\frac{a_2}{a_1}=\frac{-\frac{2}{5}}{1}=-\frac{2}{5}$.

Using $S=\frac{a_1}{1-r}=\frac{1}{1-\left(-\frac{2}{5}\right)}=\frac{1}{\frac{7}{5}}=\frac{5}{7}$

**15.**
$$0.\overline{45}=\frac{a_1}{1-r}=\frac{\frac{45}{100}}{1-\frac{1}{100}}=\frac{\frac{45}{100}}{\frac{99}{100}}$$
$$=\frac{45}{100}\div\frac{99}{100}=\frac{45}{100}\cdot\frac{100}{99}=\frac{45}{99}=\frac{5}{11}$$

**16.** Answers will vary. An example is $\sum_{i=1}^{18}\frac{i}{i+2}$.

**17.** The arithmetic sequence is 16, 48, 80, 112, ….
First find $a_{15}$ where $d=a_2-a_1=48-16=32$.
$a_{15}=a_1+(n-1)d=16+(15-1)(32)=16+14(32)=16+448=464$.
The distance the skydiver falls during the 15th second is 464 feet.
$S_{15}=\frac{n}{2}(a_1+a_n)=\frac{15}{2}(16+464)=7.5(480)=3600$
The total distance the skydiver falls in 15 seconds is 3600 feet.

**18.** $r=0.10$
$A=P(1+r)^t$
$=120000(1+0.10)^{10}$
$\approx 311249$
The value of the house after 10 years is $311,249.

## 12.4 Exercise Set

**1.** $\binom{8}{3}=\frac{8!}{3!(8-3)!}$

$=\frac{8!}{3!5!}=\frac{8\cdot 7\cdot \cancel{6}\cdot \cancel{5!}}{\cancel{3\cdot 2}\cdot 1\cdot \cancel{5!}}=56$

**3.** $\binom{12}{1}=\frac{12!}{1!(12-1)!}=\frac{12\cdot \cancel{11!}}{1\cdot \cancel{11!}}=12$

**5.** $\binom{6}{6}=\frac{\cancel{6!}}{\cancel{6!}(6-6)!}=\frac{1}{0!}=\frac{1}{1}=1$

**7.** $\binom{100}{2}=\frac{100!}{2!(100-2)!}=\frac{100\cdot 99\cdot \cancel{98!}}{2\cdot 1\cdot \cancel{98!}}$

$=4950$

**9.** Applying the Binomial Theorem to $(x+2)^3$, we have $a=x,\ b=2,$ and $n=3$.

$$(x+2)^3=\binom{3}{0}x^3+\binom{3}{1}x^2(2)+\binom{3}{2}x(2)^2+\binom{3}{3}2^3$$

$$=\frac{3!}{0!(3-0)!}x^3+\frac{3!}{1!(3-1)!}2x^2+\frac{3!}{2!(3-2)!}4x+\frac{3!}{3!(3-3)!}8$$

$$=\frac{\cancel{3!}}{1\cdot \cancel{3!}}x^3+\frac{3\cdot \cancel{2!}}{1\cdot \cancel{2!}}2x^2+\frac{3\cdot \cancel{2!}}{\cancel{2!}1!}4x+\frac{\cancel{3!}}{\cancel{3!}0!}8=x^3+3(2x^2)+3(4x)+1(8)$$

$$=x^3+6x^2+12x+8$$

**11.** Applying the Binomial Theorem to $(3x+y)^3$, we have $a=3x,\ b=y,$ and $n=3$.

$$(3x+y)^3=\binom{3}{0}(3x)^3+\binom{3}{1}(3x)^2y+\binom{3}{2}(3x)y^2+\binom{3}{3}y^3$$

$$=\frac{3!}{0!(3-0)!}27x^3+\frac{3!}{1!(3-1)!}9x^2y+\frac{3!}{2!(3-2)!}3xy^2+\frac{3!}{3!(3-3)!}y^3$$

$$=\frac{\cancel{3!}}{1\cdot \cancel{3!}}27x^3+\frac{3\cdot \cancel{2!}}{1\cdot \cancel{2!}}9x^2y+\frac{3\cdot \cancel{2!}}{\cancel{2!}1!}3xy^2+\frac{\cancel{3!}}{\cancel{3!}0!}y^3=27x^3+3(9x^2y)+3(3xy^2)+1(y^3)$$

$$=27x^3+27x^2y+9xy^2+y^3$$

**13.** Applying the Binomial Theorem to $(5x-1)^3$, we have $a=5x,\ b=-1,$ and $n=3$.

$$(5x-1)^3=\binom{3}{0}(5x)^3+\binom{3}{1}(5x)^2(-1)+\binom{3}{2}(5x)(-1)^2+\binom{3}{3}(-1)^3$$

$$=\frac{3!}{0!(3-0)!}125x^3-\frac{3!}{1!(3-1)!}25x^2+\frac{3!}{2!(3-2)!}5x(1)-\frac{3!}{3!(3-3)!}$$

$$=\frac{\cancel{3!}}{1\cdot\cancel{3!}}125x^3-\frac{3\cdot\cancel{2!}}{1\cdot\cancel{2!}}25x^2+\frac{3\cdot\cancel{2!}}{\cancel{2!}1!}5x-\frac{\cancel{3!}}{\cancel{3!}0!}=125x^3-3\left(25x^2\right)+3(5x)-1$$

$$=125x^3-75x^2+15x-1$$

**15.** Applying the Binomial Theorem to $(2x+1)^4$, we have $a=2x,\ b=1,$ and $n=4$.

$$(2x+1)^4=\binom{4}{0}(2x)^4+\binom{4}{1}(2x)^3+\binom{4}{2}(2x)^2+\binom{4}{3}2x+\binom{4}{4}$$

$$=\frac{4!}{0!(4-0)!}16x^4+\frac{4!}{1!(4-1)!}8x^3\cdot 1+\frac{4!}{2!(4-2)!}4x^2\cdot 1^2+\frac{4!}{3!(4-3)!}2x\cdot 1^3+\frac{4!}{4!(4-4)!}\cdot 1^4$$

$$=\frac{\cancel{4!}}{0!\cancel{4!}}16x^4+\frac{4!}{1!3!}8x^3\cdot 1+\frac{4!}{2!2!}4x^2\cdot 1+\frac{4!}{3!1!}2x\cdot 1+\frac{\cancel{4!}}{\cancel{4!}0!}\cdot 1$$

$$=1\left(16x^4\right)+\frac{4\cdot\cancel{3!}}{1\cdot\cancel{3!}}8x^3+\frac{4\cdot 3\cdot\cancel{2!}}{2\cdot 1\cdot\cancel{2!}}4x^2+\frac{4\cdot\cancel{3!}}{\cancel{3!}\cdot 1}2x+1=16x^4+4\left(8x^3\right)+6\left(4x^2\right)+4(2x)+1$$

$$=16x^4+32x^3+24x^2+8x+1$$

**17.** Applying the Binomial Theorem to $\left(x^2+2y\right)^4$, we have $a=x^2,\ b=2y,$ and $n=4$.

$$\left(x^2+2y\right)^4=\binom{4}{0}\left(x^2\right)^4+\binom{4}{1}\left(x^2\right)^3(2y)+\binom{4}{2}\left(x^2\right)^2(2y)^2+\binom{4}{3}x^2(2y)^3+\binom{4}{4}(2y)^4$$

$$=\frac{4!}{0!(4-0)!}x^8+\frac{4!}{1!(4-1)!}2x^6y+\frac{4!}{2!(4-2)!}4x^4y^2+\frac{4!}{3!(4-3)!}8x^2y^3+\frac{4!}{4!(4-4)!}16y^4$$

$$=\frac{\cancel{4!}}{0!\cancel{4!}}x^8+\frac{4!}{1!3!}2x^6y+\frac{4!}{2!2!}4x^4y^2+\frac{4!}{3!1!}8x^2y^3+\frac{\cancel{4!}}{\cancel{4!}0!}16y^4$$

$$=1\left(x^8\right)+\frac{4\cdot\cancel{3!}}{1\cdot\cancel{3!}}2x^6y+\frac{4\cdot 3\cdot\cancel{2!}}{2\cdot 1\cdot\cancel{2!}}4x^4y^2+\frac{4\cdot\cancel{3!}}{\cancel{3!}\cdot 1}8x^2y^3+16y^4$$

$$=x^8+4\left(2x^6y\right)+6\left(4x^4y^2\right)+4\left(8x^2y^3\right)+16y^4=x^8+8x^6y+24x^4y^2+32x^2y^3+16y^4$$

**19.** Applying the Binomial Theorem to $(y-3)^4$, we have $a=y,\ b=-3,$ and $n=4$.

$$(y-3)^4=\binom{4}{0}y^4+\binom{4}{1}y^3(-3)+\binom{4}{2}y^2(-3)^2+\binom{4}{3}y(-3)^3+\binom{4}{4}(-3)^4$$

$$=\frac{4!}{0!(4-0)!}y^4-\frac{4!}{1!(4-1)!}3y^3+\frac{4!}{2!(4-2)!}9y^2-\frac{4!}{3!(4-3)!}27y+\frac{4!}{4!(4-4)!}81$$

$$=\frac{\cancel{4!}}{0!\cancel{4!}}y^4-\frac{4!}{1!3!}3y^3+\frac{4!}{2!2!}9y^2-\frac{4!}{3!1!}27y+\frac{\cancel{4!}}{\cancel{4!}0!}81$$

$$=1(y^4)-\frac{4\cdot\cancel{3!}}{1\cdot\cancel{3!}}3y^3+\frac{4\cdot3\cdot\cancel{2!}}{2\cdot1\cdot\cancel{2!}}9y^2-\frac{4\cdot\cancel{3!}}{\cancel{3!}\cdot1}27y+81$$

$$=y^4-4(3y^3)+6(9y^2)-4(27y)+81=y^4-12y^3+54y^2-108y+81$$

**21.** Applying the Binomial Theorem to $(2x^3-1)^4$, we have $a=2x^3,\ b=-1,$ and $n=4$.

$$(2x^3-1)^4=\binom{4}{0}(2x^3)^4+\binom{4}{1}(2x^3)^3(-1)+\binom{4}{2}(2x^3)^2(-1)^2+\binom{4}{3}(2x^3)(-1)^3+\binom{4}{4}(-1)^4$$

$$=\frac{4!}{0!(4-0)!}16x^{12}-\frac{4!}{1!(4-1)!}8x^9+\frac{4!}{2!(4-2)!}4x^6-\frac{4!}{3!(4-3)!}2x^3+\frac{4!}{4!(4-4)!}$$

$$=\frac{\cancel{4!}}{0!\cancel{4!}}16x^{12}-\frac{4!}{1!3!}8x^9+\frac{4!}{2!2!}4x^6-\frac{4!}{3!1!}2x^3+\frac{\cancel{4!}}{\cancel{4!}0!}$$

$$=1(16x^{12})-\frac{4\cdot\cancel{3!}}{1\cdot\cancel{3!}}8x^9+\frac{4\cdot3\cdot\cancel{2!}}{2\cdot1\cdot\cancel{2!}}4x^6-\frac{4\cdot\cancel{3!}}{\cancel{3!}\cdot1}2x^3+1$$

$$=16x^{12}-4(8x^9)+6(4x^6)-4(2x^3)+1=16x^{12}-32x^9+24x^6-8x^3+1$$

**23.** Applying the Binomial Theorem to $(c+2)^5$, we have $a=c,\ b=2,$ and $n=5$.

$$(c+2)^5=\binom{5}{0}c^5+\binom{5}{1}c^4(2)+\binom{5}{2}c^3(2)^2+\binom{5}{3}c^2(2)^3+\binom{5}{4}c(2)^4+\binom{5}{5}2^5$$

$$=\frac{\cancel{5!}}{0!\cancel{5!}}c^5+\frac{5!}{1!(5-1)!}2c^4+\frac{5!}{2!(5-2)!}4c^3+\frac{5!}{3!(5-3)!}8c^2+\frac{5!}{4!(5-4)!}16c+\frac{5!}{5!(5-5)!}32$$

$$=1c^5+\frac{5\cdot\cancel{4!}}{1\cdot\cancel{4!}}2c^4+\frac{5\cdot4\cdot\cancel{3!}}{2\cdot1\cdot\cancel{3!}}4c^3+\frac{5\cdot4\cdot\cancel{3!}}{\cancel{3!}2\cdot1}8c^2+\frac{5\cdot\cancel{4!}}{\cancel{4!}\cdot1}16c+\frac{\cancel{5!}}{\cancel{5!}0!}32$$

$$=c^5+5(2c^4)+10(4c^3)+10(8c^2)+5(16c)+1(32)=c^5+10c^4+40c^3+80c^2+80c+32$$

**25.** Applying the Binomial Theorem to $(x-1)^5$, we have $a = x,\ b = -1,$ and $n = 5$.

$$(x-1)^5 = \binom{5}{0}x^5 + \binom{5}{1}x^4(-1) + \binom{5}{2}x^3(-1)^2 + \binom{5}{3}x^2(-1)^3 + \binom{5}{4}x(-1)^4 + \binom{5}{5}(-1)^5$$

$$= \frac{\cancel{5!}}{0!\cancel{5!}}x^5 - \frac{5!}{1!(5-1)!}x^4 + \frac{5!}{2!(5-2)!}x^3 - \frac{5!}{3!(5-3)!}x^2 + \frac{5!}{4!(5-4)!}x - \frac{5!}{5!(5-5)!}$$

$$= 1x^5 - \frac{5\cdot\cancel{4!}}{1\cdot\cancel{4!}}x^4 + \frac{5\cdot 4\cdot\cancel{3!}}{2\cdot 1\cdot\cancel{3!}}x^3 - \frac{5\cdot 4\cdot\cancel{3!}}{\cancel{3!}2\cdot 1}x^2 + \frac{5\cdot\cancel{4!}}{\cancel{4!}\cdot 1}x - \frac{\cancel{5!}}{\cancel{5!}0!}$$

$$= x^5 - 5x^4 + 10x^3 - 10x^2 + 5x - 1$$

**27.** Applying the Binomial Theorem to $(3x-y)^5$, we have $a = 3x,\ b = -y,$ and $n = 5$.

$$(3x-y)^5$$

$$= \binom{5}{0}(3x)^5 + \binom{5}{1}(3x)^4(-y) + \binom{5}{2}(3x)^3(-y)^2 + \binom{5}{3}(3x)^2(-y)^3 + \binom{5}{4}3x(-y)^4 + \binom{5}{5}(-y)^5$$

$$= \frac{\cancel{5!}}{0!\cancel{5!}}243x^5 - \frac{5!}{1!(5-1)!}81x^4y + \frac{5!}{2!(5-2)!}27x^3y^2$$

$$- \frac{5!}{3!(5-3)!}9x^2y^3 + \frac{5!}{4!(5-4)!}3xy^4 - \frac{5!}{5!(5-5)!}y^5$$

$$= 243x^5 - \frac{5\cdot\cancel{4!}}{1\cdot\cancel{4!}}81x^4y + \frac{5\cdot 4\cdot\cancel{3!}}{2\cdot 1\cdot\cancel{3!}}27x^3y^2 - \frac{5\cdot 4\cdot\cancel{3!}}{\cancel{3!}2\cdot 1}9x^2y^3 + \frac{5\cdot\cancel{4!}}{\cancel{4!}\cdot 1}3xy^4 - \frac{\cancel{5!}}{\cancel{5!}0!}y^5$$

$$= 243x^5 - 5\left(81x^4y\right) + 10\left(27x^3y^2\right) - 10\left(9x^2y^3\right) + 5\left(3xy^4\right) - y^5$$

$$= 243x^5 - 405x^4y + 207x^3y^2 - 90x^2y^3 + 15xy^4 - y^5$$

**29.** Applying the Binomial Theorem to $(2a+b)^6$, we have $a=2a,\ b=b,$ and $n=6.$

$(2a+b)^6$

$$=\binom{6}{0}(2a)^6+\binom{6}{1}(2a)^5b+\binom{6}{2}(2a)^4b^2+\binom{6}{3}(2a)^3b^3+\binom{6}{4}(2a)^2b^4+\binom{6}{5}2ab^5+\binom{6}{6}b^6$$

$$=\frac{6!}{0!(6-0)!}64a^6+\frac{6!}{1!(6-1)!}32a^5b+\frac{6!}{2!(6-2)!}16a^4b^2+\frac{6!}{3!(6-3)!}8a^3b^3$$

$$+\frac{6!}{4!(6-4)!}4a^2b^4+\frac{6!}{5!(6-5)!}2ab^5+\frac{6!}{6!(6-6)!}b^6$$

$$=\frac{\cancel{6!}}{1\cancel{6!}}64a^6+\frac{6\cdot\cancel{5!}}{1\cdot\cancel{5!}}32a^5b+\frac{6\cdot5\cdot\cancel{4!}}{2\cdot1\cdot\cancel{4!}}16a^4b^2+\frac{\cancel{6}\cdot5\cdot4\cdot\cancel{3!}}{\cancel{3\cdot2}\cdot1\cdot\cancel{3!}}8a^3b^3$$

$$+\frac{6\cdot5\cdot\cancel{4!}}{\cancel{4!}2\cdot1}4a^2b^4+\frac{6\cdot\cancel{5!}}{\cancel{5!}1}2ab^5+\frac{\cancel{6!}}{\cancel{6!}\cdot1}b^6$$

$$=64a^6+6\left(32a^5b\right)+15\left(16a^4b^2\right)+20\left(8a^3b^3\right)+15\left(4a^2b^4\right)+6\left(2ab^5\right)+1b^6$$

$$=64a^6+192a^5b+240a^4b^2+160a^3b^3+60a^2b^4+12ab^5+b^6$$

**31.** $(x+2)^8$

First Term $\binom{n}{r-1}a^{n-r+1}b^{r-1}=\binom{8}{1-1}x^{8-1+1}2^{1-1}=\binom{8}{0}x^8 2^0$

$$=\frac{8!}{0!(8-0)!}x^8\cdot1=\frac{\cancel{8!}}{0!\cancel{8!}}x^8=x^8$$

Second Term $\binom{n}{r-1}a^{n-r+1}b^{r-1}=\binom{8}{2-1}x^{8-2+1}2^{2-1}=\binom{8}{1}x^7 2^1=\frac{8!}{1!(8-1)!}2x^7$

$$=\frac{8\cdot\cancel{7!}}{1\cdot\cancel{7!}}2x^7=8\cdot2x^7=16x^7$$

Third Term $\binom{n}{r-1}a^{n-r+1}b^{r-1}=\binom{8}{3-1}x^{8-3+1}2^{3-1}=\binom{8}{2}x^6 2^2=\frac{8!}{2!(8-2)!}4x^6$

$$=\frac{8\cdot7\cdot\cancel{6!}}{2\cdot1\cdot\cancel{6!}}4x^6=28\cdot4x^6=112x^6$$

**33.** $(x-2y)^{10}$

First Term $\binom{n}{r-1}a^{n-r+1}b^{r-1}=\binom{10}{1-1}x^{10-1+1}(-2y)^{1-1}=\binom{10}{0}x^{10}(-2y)^0$

$$=\frac{10!}{0!(10-0)!}x^{10}\cdot 1=\frac{\cancel{10!}}{0!\cancel{10!}}x^{10}=x^{10}$$

Second Term $\binom{n}{r-1}a^{n-r+1}b^{r-1}=\binom{10}{2-1}x^{10-2+1}(-2y)^{2-1}=\binom{10}{1}x^9(-2y)^1=-\frac{10!}{1!(10-1)!}2x^9y$

$$=-\frac{10\cdot\cancel{9!}}{1\cdot\cancel{9!}}2x^9y=-10\cdot 2x^9y=-20x^9y$$

Third Term $\binom{n}{r-1}a^{n-r+1}b^{r-1}=\binom{10}{3-1}x^{10-3+1}(-2y)^{3-1}=\binom{10}{2}x^8(-2y)^2=\frac{10!}{2!(10-2)!}4x^8y^2$

$$=\frac{10\cdot 9\cdot\cancel{8!}}{2\cdot 1\cdot\cancel{8!}}4x^8y^2=45\cdot 4x^8y^2=180x^8y^2$$

**35.** $(x^2+1)^{16}$

First Term $\binom{n}{r-1}a^{n-r+1}b^{r-1}=\binom{16}{1-1}(x^2)^{16-1+1}(1)^{1-1}=\binom{16}{0}(x^2)^{16}1^0$

$$=\frac{16!}{0!(16-0)!}x^{32}\cdot 1=\frac{\cancel{16!}}{0!\cancel{16!}}x^{32}=x^{32}$$

Second Term $\binom{n}{r-1}a^{n-r+1}b^{r-1}=\binom{16}{2-1}(x^2)^{16-2+1}(1)^{2-1}=\binom{16}{1}(x^2)^{15}1^1$

$$=\frac{16!}{1!(16-1)!}x^{30}\cdot 1=\frac{16\cdot\cancel{15!}}{1\cdot\cancel{15!}}x^{30}=16x^{30}$$

Third Term $\binom{n}{r-1}a^{n-r+1}b^{r-1}=\binom{16}{3-1}(x^2)^{16-3+1}(1)^{3-1}=\binom{16}{2}(x^2)^{14}1^2$

$$=\frac{16!}{2!(16-2)!}x^{28}\cdot 1=\frac{16\cdot 15\cdot\cancel{14!}}{2\cdot 1\cdot\cancel{14!}}x^{28}=120x^{28}$$

**37.** $(y^3-1)^{20}$

First Term $\binom{n}{r-1}a^{n-r+1}b^{r-1}=\binom{20}{1-1}(y^3)^{20-1+1}(-1)^{1-1}=\binom{20}{0}(y^3)^{20}(-1)^0$

$$=\frac{20!}{0!(20-0)!}y^{60}\cdot 1=\frac{\cancel{20!}}{0!\cancel{20!}}y^{60}=y^{60}$$

Second Term $\binom{n}{r-1}a^{n-r+1}b^{r-1}=\binom{20}{2-1}\left(y^3\right)^{20-2+1}(-1)^{2-1}=\binom{20}{1}\left(y^3\right)^{19}(-1)^1$

$$=\frac{20!}{1!(20-1)!}y^{57}\cdot(-1)=-\frac{20\cdot\cancel{19!}}{1\cdot\cancel{19!}}y^{57}=-20y^{57}$$

Third Term $\binom{n}{r-1}a^{n-r+1}b^{r-1}=\binom{20}{3-1}\left(y^3\right)^{20-3+1}(-1)^{3-1}=\binom{20}{2}\left(y^3\right)^{18}(-1)^2$

$$=\frac{20!}{2!(20-2)!}y^{54}\cdot 1=\frac{20\cdot 19\cdot\cancel{18!}}{2\cdot 1\cdot\cancel{18!}}y^{54}=190y^{54}$$

**39.** $(2x+y)^6$

Third Term $\binom{n}{r-1}a^{n-r+1}b^{r-1}=\binom{6}{3-1}(2x)^{6-3+1}y^{3-1}=\binom{6}{2}(2x)^4y^2=\frac{6!}{2!(6-2)!}16x^4y^2$

$$=\frac{6\cdot 5\cdot\cancel{4!}}{2\cdot 1\cdot\cancel{4!}}16x^4y^2=15\left(16x^4y^2\right)=240x^4y^2$$

**41.** $(x-1)^9$

Fifth Term $\binom{n}{r-1}a^{n-r+1}b^{r-1}=\binom{9}{5-1}x^{9-5+1}(-1)^{5-1}=\binom{9}{4}x^5(-1)^4=\frac{9!}{4!(9-4)!}x^5\cdot 1$

$$=\frac{9\cdot 8\cdot 7\cdot\cancel{6}\cdot\cancel{5!}}{4\cdot\cancel{3\cdot 2}\cdot 1\cdot\cancel{5!}}x^5=126x^5$$

**43.** $\left(x^2+y^3\right)^8$

Sixth Term $\binom{n}{r-1}a^{n-r+1}b^{r-1}=\binom{8}{6-1}\left(x^2\right)^{8-6+1}\left(y^3\right)^{6-1}=\binom{8}{5}\left(x^2\right)^3\left(y^3\right)^5=\frac{8!}{5!(8-5)!}x^6y^{15}$

$$=\frac{8\cdot 7\cdot\cancel{6}\cdot\cancel{5!}}{\cancel{5!}\cdot\cancel{3\cdot 2}\cdot 1}x^6y^{15}=56x^6y^{15}$$

**45.** $\left(x-\frac{1}{2}\right)^9$

Fourth Term $\binom{n}{r-1}a^{n-r+1}b^{r-1}=\binom{9}{4-1}x^{9-4+1}\left(-\frac{1}{2}\right)^{4-1}=\binom{9}{3}x^6\left(-\frac{1}{2}\right)^3=-\frac{9!}{3!(9-3)!}\cdot\frac{1}{8}x^6$

$$=-\frac{9\cdot\cancel{8}\cdot 7\cdot\cancel{6!}}{3\cdot 2\cdot 1\cdot\cancel{6!}}\cdot\frac{1}{\cancel{8}}x^6=-\frac{21}{2}x^6$$

**47.** $$\left(x^3+x^{-2}\right)^4=\binom{4}{0}\left(x^3\right)^4+\binom{4}{1}\left(x^3\right)^3\left(x^{-2}\right)+\binom{4}{2}\left(x^3\right)^2\left(x^{-2}\right)^2+\binom{4}{3}\left(x^3\right)^1\left(x^{-2}\right)^3+\binom{4}{4}\left(x^{-2}\right)^4$$

$$=\frac{4!}{0!(4-0)!}x^{12}+\frac{4!}{1!(4-1)!}x^9x^{-2}+\frac{4!}{2!(4-2)!}x^6x^{-4}+\frac{4!}{3!(4-3)!}x^3x^{-6}+\frac{4!}{4!(4-4)!}x^{-8}$$

$$=\frac{\cancel{4!}}{0!\cdot\cancel{4!}}x^{12}+\frac{4\cdot\cancel{3!}}{1!\cdot\cancel{3!}}x^7+\frac{4\cdot3\cdot\cancel{2!}}{2\cdot1\cdot\cancel{2!}}x^2+\frac{4\cdot\cancel{3!}}{\cancel{3!}\cdot1!}x^{-3}+\frac{\cancel{4!}}{\cancel{4!}\cdot0!}x^{-8}$$

$$=x^{12}+4x^7+6x^2+\frac{4}{x^3}+\frac{1}{x^8}$$

**48.** $$\left(x^2+x^{-3}\right)^4=\binom{4}{0}\left(x^2\right)^4+\binom{4}{1}\left(x^2\right)^3\left(x^{-3}\right)+\binom{4}{2}\left(x^2\right)^2\left(x^{-3}\right)^2+\binom{4}{3}\left(x^2\right)^1\left(x^{-3}\right)^3+\binom{4}{4}\left(x^{-3}\right)^4$$

$$=\frac{4!}{0!(4-0)!}x^8+\frac{4!}{1!(4-1)!}x^6x^{-3}+\frac{4!}{2!(4-2)!}x^4x^{-6}+\frac{4!}{3!(4-3)!}x^2x^{-9}+\frac{4!}{4!(4-4)!}x^{-12}$$

$$=\frac{\cancel{4!}}{0!\cdot\cancel{4!}}x^8+\frac{4\cdot\cancel{3!}}{1!\cdot\cancel{3!}}x^3+\frac{4\cdot3\cdot\cancel{2!}}{2\cdot1\cdot\cancel{2!}}x^{-2}+\frac{4\cdot\cancel{3!}}{\cancel{3!}\cdot1!}x^{-7}+\frac{\cancel{4!}}{\cancel{4!}\cdot0!}x^{-12}$$

$$=x^8+4x^3+\frac{6}{x^2}+\frac{4}{x^7}+\frac{1}{x^{12}}$$

**49.** $$\left(x^{\frac{1}{3}}-x^{-\frac{1}{3}}\right)^3=\left(x^{\frac{1}{3}}+\left(-x^{-\frac{1}{3}}\right)\right)^3=\binom{3}{0}\left(x^{\frac{1}{3}}\right)^3+\binom{3}{1}\left(x^{\frac{1}{3}}\right)^2\left(-x^{-\frac{1}{3}}\right)+$$

$$+\binom{3}{2}\left(x^{\frac{1}{3}}\right)^1\left(-x^{-\frac{1}{3}}\right)^2+\binom{3}{3}\left(-x^{-\frac{1}{3}}\right)^3$$

$$=\frac{3!}{0!(3-0)!}x^1+\frac{3!}{1!(3-1)!}x^{\frac{2}{3}}\cdot-x^{-\frac{1}{3}}+\frac{3!}{2!(3-2)!}x^{\frac{1}{3}}x^{-\frac{2}{3}}+\frac{3!}{3!(3-3)!}\cdot-x^{-1}$$

$$=\frac{\cancel{3!}}{0!\cdot\cancel{3!}}x+\frac{3\cdot\cancel{2!}}{1!\cdot\cancel{2!}}\cdot-x^{\frac{1}{3}}+\frac{\cdot3\cdot\cancel{2!}}{\cancel{2!}\cdot1!}x^{-\frac{1}{3}}+\frac{\cancel{3!}}{\cancel{3!}\cdot0!}\cdot-x^{-1}$$

$$=x-3x^{\frac{1}{3}}+\frac{3}{x^{\frac{1}{3}}}-\frac{1}{x}$$

**50.** $$\left(x^{\frac{2}{3}}-\frac{1}{\sqrt[3]{x}}\right)^3=\left(x^{\frac{2}{3}}+\left(-\frac{1}{x^{\frac{1}{3}}}\right)\right)^3=\left(x^{\frac{2}{3}}+\left(-x^{-\frac{1}{3}}\right)\right)^3=$$

$$=\left(x^{\frac{2}{3}}+\left(-x^{-\frac{1}{3}}\right)\right)^3=\binom{3}{0}\left(x^{\frac{2}{3}}\right)^3+\binom{3}{1}\left(x^{\frac{2}{3}}\right)^2\left(-x^{-\frac{1}{3}}\right)+$$

$$+\binom{3}{2}\left(x^{\frac{2}{3}}\right)^1\left(-x^{-\frac{1}{3}}\right)^2+\binom{3}{3}\left(-x^{-\frac{1}{3}}\right)^3$$

$$=\frac{3!}{0!(3-0)!}x^2+\frac{3!}{1!(3-1)!}x^{\frac{4}{3}}\cdot -x^{-\frac{1}{3}}+\frac{3!}{2!(3-2)!}x^{\frac{2}{3}}x^{-\frac{2}{3}}+\frac{3!}{3!(3-3)!}\cdot -x^{-1}$$

$$=\frac{\cancel{3!}}{0!\cdot\cancel{3!}}x^2+\frac{3\cdot\cancel{2!}}{1!\cdot\cancel{2!}}\cdot -x^1+\frac{\cdot 3\cdot\cancel{2!}}{\cancel{2!}\cdot 1!}x^0+\frac{\cancel{3!}}{\cancel{3!}\cdot 0!}\cdot -x^{-1}$$

$$=x^2-3x+3-\frac{1}{x}$$

**51.** $$\left(-1+\sqrt{3i}\right)^3=\binom{3}{0}(-1)^3+\binom{3}{1}(-1)^2\left(\sqrt{3i}\right)+\binom{3}{2}(-1)^1\left(\sqrt{3i}\right)^2+\binom{3}{3}\left(\sqrt{3i}\right)^3$$

$$=\frac{3!}{0!(3-0)!}\cdot -1+\frac{3!}{1!(3-1)!}1\cdot\sqrt{3i}+\frac{3!}{2!(3-2)!}\cdot -1\cdot -3+\frac{3!}{3!(3-3)!}\cdot -3\sqrt{3i}$$

$$=\frac{\cancel{3!}}{0!\cancel{3!}}(-1)+\frac{3\cdot\cancel{2!}}{1!\cancel{2!}}\sqrt{3i}+\frac{3\cdot\cancel{2!}}{\cancel{2!}1!}\cdot 3+\frac{\cancel{3!}}{\cancel{3!}0!}\cdot -3\sqrt{3i}$$

$$=-1+3\sqrt{3i}+9-3\sqrt{3i}=8$$

**52.** $$\left(-1-\sqrt{3i}\right)^3=\binom{3}{0}(-1)^3+\binom{3}{1}(-1)^2\left(-\sqrt{3i}\right)+\binom{3}{2}(-1)^1\left(-\sqrt{3i}\right)^2+\binom{3}{3}\left(-\sqrt{3i}\right)^3$$

$$=\frac{3!}{0!(3-0)!}\cdot -1+\frac{3!}{1!(3-1)!}1\cdot -\sqrt{3i}+\frac{3!}{2!(3-2)!}\cdot -1\cdot -3+\frac{3!}{3!(3-3)!}\cdot 3\sqrt{3i}$$

$$=-\frac{\cancel{3!}}{0!\cancel{3!}}-\frac{3\cdot\cancel{2!}}{1!\cancel{2!}}\sqrt{3i}+\frac{3\cdot\cancel{2!}}{\cancel{2!}1!}\cdot 3+\frac{\cancel{3!}}{\cancel{3!}0!}\cdot 3\sqrt{3i}$$

$$=-1-3\sqrt{3i}+9+3\sqrt{3i}=8$$

**53.** $f(x) = x^4 + 7$;

$$\frac{f(x+h) - f(x)}{h}$$

$$= \frac{(x+h)^4 + 7 - \left(x^4 + 7\right)}{h}$$

$$= \frac{\binom{4}{0}x^4 + \binom{4}{1}x^3h + \binom{4}{2}x^2h^2 + \binom{4}{3}xh^3 + \binom{4}{4}h^4 + 7 - x^4 - 7}{h}$$

$$= \frac{\frac{4!}{0!(4-0)!}x^4 + \frac{4!}{1!(4-1)!}x^3h + \frac{4!}{2!(4-2)!}x^2h^2 + \frac{4!}{3!(4-3)!}xh^3 + \frac{4!}{4!(4-4)!}h^4 - x^4}{h}$$

$$= \frac{\frac{\cancel{4!}}{0!\cancel{4!}}x^4 + \frac{4\cdot\cancel{3!}}{1!\cancel{3!}}x^3h + \frac{4\cdot 3\cdot\cancel{2!}}{\cancel{2!}\cdot 2\cdot 1}x^2h^2 + \frac{4\cdot\cancel{3!}}{\cancel{3!}1!}xh^3 + \frac{\cancel{4!}}{\cancel{4!}0!}h^4 - x^4}{h}$$

$$= \frac{\cancel{x^4} + 4x^3h + 6x^2h^2 + 4xh^3 + h^4 - \cancel{x^4}}{h}$$

$$= \frac{h(4x^3 + 6x^2h + 4xh^2 + h^3)}{h}$$

$$= 4x^3 + 6x^2h + 4xh^2 + h^3$$

**54.** $f(x) = x^5 + 8$

$$\frac{f(x+h) - f(x)}{h}$$

$$= \frac{(x+h)^5 + 8 - (x^5 - 8)}{h}$$

$$= \frac{\binom{5}{0}x^5 + \binom{5}{1}x^4h + \binom{5}{2}x^3h^2 + \binom{5}{3}x^2h^3 + \binom{5}{4}xh^4 + \binom{5}{5}h^5 + 8 - x^5 - 8}{h}$$

$$= \frac{\frac{5!}{0!(5-0)!}x^5 + \frac{5!}{1!(5-1)!}x^4h + \frac{5!}{2!(5-2)!}x^3h^2 + \frac{5!}{3!(5-3)!}x^2h^3 + \frac{5!}{4!(5-4)!}xh^4 + \frac{5!}{5!(5-5)!}h^5 - x^5}{h}$$

$$= \frac{\frac{\cancel{5!}}{0!\cancel{5!}}x^5 + \frac{5\cdot\cancel{4!}}{1!\cancel{4!}}x^4h + \frac{5\cdot 4\cdot\cancel{3!}}{2!\cancel{3!}}x^3h^2 + \frac{5\cdot 4\cdot\cancel{3!}}{\cancel{3!}2!}x^2h^3 + \frac{5\cdot\cancel{4!}}{\cancel{4!}1!}xh^4 + \frac{\cancel{5!}}{\cancel{5!}0!}h^5 - x^5}{h}$$

$$= \frac{x^5 + 5x^4h + 10x^3h^2 + 10x^2h^3 + 5xh^4 + h^5 - x^5}{h}$$
$$= \frac{5x^4h + 10x^3h^2 + 10x^2h^3 + 5xh^4 + h^5}{h}$$
$$= \frac{h(5x^4 + 10x^3h + 10x^2h^2 + 5xh^3 + h^4)}{h}$$
$$= 5x^4 + 10x^3h + 10x^2h^2 + 5xh^3 + h^4$$

**55.** We want to find the $(5+1) = 6^{th}$ term.

$$\binom{n}{r} a^{n-r} b^r = \binom{10}{5}\left(\frac{3}{x}\right)^{10-5}\left(\frac{x}{3}\right)^5 = \frac{10!}{5!(10-5)!}\left(\frac{3}{x}\right)^5\left(\frac{x}{3}\right)^5$$
$$= \frac{\cancel{10}\cdot 9\cdot 8\cdot 7\cdot 6\cdot \cancel{5!}}{\cancel{5!}\cdot \cancel{5}\cdot 4\cdot 3\cdot \cancel{2}\cdot 1}\left(\frac{3}{x}\right)^5\left(\frac{x}{3}\right)^5 = 252\cdot\frac{3^5}{x^5}\cdot\frac{x^5}{3^5} = 252$$

**56.** We want to find the $(6+1) = 7^{th}$ term.

$$\binom{n}{r} a^{n-r} b^r = \binom{12}{6}\left(\frac{1}{x}\right)^{12-6}\left(-x^2\right)^6 = \frac{12!}{6!(12-6)!}\left(\frac{1}{x}\right)^6\left(-x^2\right)^6$$
$$= \frac{\cancel{12}\cdot 11\cdot 10\cdot 9\cdot 8\cdot 7\cdot \cancel{6!}}{\cancel{6!}\cdot \cancel{6}\cdot 5\cdot 4\cdot 3\cdot \cancel{2}\cdot 1}\left(\frac{1}{x}\right)^6\left(-x^2\right)^6 = 924\cdot\frac{1}{x^6}\cdot x^{12} = 924x^{12-6} = 924x^6$$

**57.** $g(x) = 0.12(x+3)^3 - (x+3)^2 + 3(x+3) + 15$

$$= 0.12\left[\binom{3}{0}x^3 + \binom{3}{1}x^2\cdot 3 + \binom{3}{2}x\cdot 3^2 + \binom{3}{3}\cdot 3^3\right] - \left(x^2 + 6x + 9\right) + 3x + 9 + 15$$
$$= 0.12\left[\frac{3!}{0!(3-0)!}x^3 + \frac{3!}{1!(3-1)!}3x^2 + \frac{3!}{2!(3-2)!}9x + \frac{3!}{3!(3-3)!}\cdot 27\right] - x^2 - 6x - 9 + 3x + 9 + 15$$
$$= 0.12\left(x^3 + 9x^2 + 27x + 27\right) - x^2 - 3x + 15$$
$$= 0.12x^3 + 1.08x^2 + 3.24x + 3.24 - x^2 - 3x + 15$$
$$= 0.12x^3 + 0.08x^2 0.24x + 18.24$$

**59. – 67.** Answers will vary.

**69.** $f_1(x) = (x+2)^3$

$f_2(x) = x^3$

$f_3(x) = x^3 + 6x^2$

$f_4(x) = x^3 + 6x^2 + 12x$

$f_5(x) = x^3 + 6x^2 + 12x + 8$

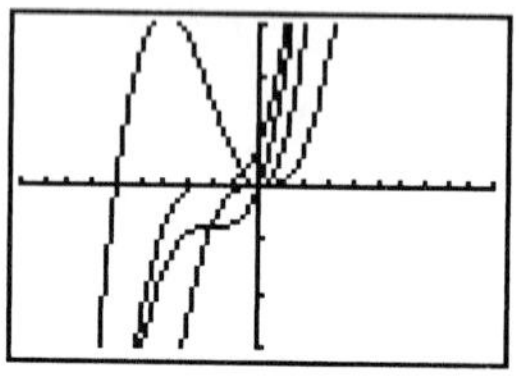

Graphs $f_1$ and $f_5$ are the same. This means that the functions are equivalent. Graphs $f_2$ through $f_4$ are increasingly similar to the graphs of $f_1$ and $f_5$.

**71.** Applying the Binomial Theorem to $(x-1)^3$, we have $a = x,\ b = -1,$ and $n = 3$.

$$(x-1)^3 = \binom{3}{0}x^3 + \binom{3}{1}x^2(-1) + \binom{3}{2}x(-1)^2 + \binom{3}{3}(-1)^3$$

$$= \frac{3!}{0!(3-0)!}x^3 - \frac{3!}{1!(3-1)!}x^2 + \frac{3!}{2!(3-2)!}x(1) - \frac{3!}{3!(3-3)!}$$

$$= \frac{\cancel{3!}}{1\cdot\cancel{3!}}x^3 - \frac{3\cdot\cancel{2!}}{1\cdot\cancel{2!}}x^2 + \frac{3\cdot\cancel{2!}}{\cancel{2!}1!}x - \frac{\cancel{3!}}{\cancel{3!}0!} = x^3 - 3x^2 + 3x - 1$$

Graph using the method from Exercises 69 and 70.

$f_1(x) = (x-1)^3$ $\qquad f_2(x) = x^3$

$f_3(x) = x^3 + 3x^2$ $\qquad f_4(x) = x^3 + 3x^2 + 3x$

$f_5(x) = x^3 - 3x^2 + 3x - 1$

Graphs $f_1$ and $f_5$ are the same. This means that the functions are equivalent. Graphs $f_2$ through $f_4$ are increasingly similar to the graphs of $f_1$ and $f_5$.

**73.** Applying the Binomial Theorem to $(x+2)^6$, we have $a=x,\ b=2,$ and $n=6$.

$$(x+2)^6=\binom{6}{0}x^6+\binom{6}{1}x^5 2+\binom{6}{2}x^4 2^2+\binom{6}{3}x^3 2^3+\binom{6}{4}x^2 2^4+\binom{6}{5}x 2^5+\binom{6}{6}2^6$$

$$=\frac{6!}{0!(6-0)!}x^6+\frac{6!}{1!(6-1)!}2x^5+\frac{6!}{2!(6-2)!}4x^4+\frac{6!}{3!(6-3)!}8x^3+\frac{6!}{4!(6-4)!}16x^2$$

$$+\frac{6!}{5!(6-5)!}32x+\frac{6!}{6!(6-6)!}64$$

$$=\frac{\cancel{6!}}{1\cancel{6!}}x^6+\frac{6\cdot\cancel{5!}}{1\cdot\cancel{5!}}2x^5+\frac{6\cdot5\cdot\cancel{4!}}{2\cdot1\cdot\cancel{4!}}4x^4+\frac{\cancel{6}\cdot5\cdot4\cdot\cancel{3!}}{\cancel{3\cdot2}\cdot1\cdot\cancel{3!}}8x^3+\frac{6\cdot5\cdot\cancel{4!}}{\cancel{4!}2\cdot1}16x^2+\frac{6\cdot\cancel{5!}}{\cancel{5!}1}32x+\frac{\cancel{6!}}{\cancel{6!}\cdot1}64$$

$$=x^6+6(2x^5)+15(4x^4)+20(8x^3)+15(16x^2)+6(32x)+1(64)$$

$$=x^6+12x^5+60x^4+160x^3+240x^2+192x+64$$

Graph using the method from Exercises 69 and 70.

$f_1(x)=(x+2)^6$

$f_2(x)=x^6$

$f_3(x)=x^6+12x^5$

$f_4(x)=x^6+12x^5+60x^4$

$f_5(x)=x^6+12x^5+60x^4+160x^3$

$f_6(x)=x^6+12x^5+60x^4+160x^3+240x^2$

$f_7(x)=x^6+12x^5+60x^4+160x^3+240x^2+192x$

$f_8(x)=x^6+12x^5+60x^4+160x^3+240x^2+192x+64$

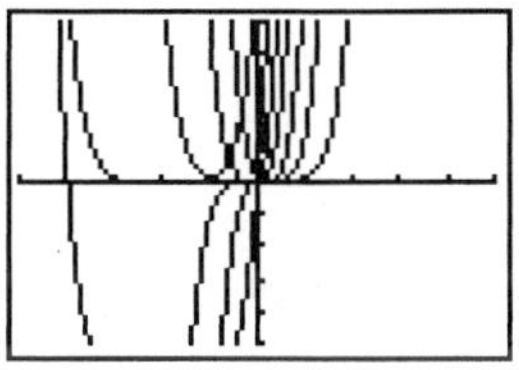

Graphs $f_1$ and $f_8$ are the same. This means that the functions are equivalent. Graphs $f_2$ through $f_7$ are increasingly similar to the graphs of $f_1$ and $f_8$.

**75.** Statement **b.** is true. The Binomial Theorem can be written in condensed form as $(a+b)^n = \sum_{r=0}^{n} \binom{n}{r} a^{n-r} b^r$

Statement **a.** is false. The binomial expansion for $(a+b)^n$ contains $n+1$ terms.

Statement **c.** is false. The sum of the binomial coefficients in $(a+b)^n$ is $2^n$.

Statement **d.** is false. There are values of $a$ and $b$ for which $(a+b)^4 = a^4 + b^4$. Consider $a = 0$ and $b = 1$.

$$(0+1)^4 = 0^4 + 1^4$$
$$(1)^4 = 0+1$$
$$1 = 1$$

**77.** In $(x^2+y^2)^5$, the term containing $x^4$ is the term in which $a = x^2$ is squared. Applying the Binomial Theorem, the following pattern results. In the first term, $x^2$ is taken to the fifth power. In the second term, $x^2$ is taken to the fourth power. In the third term $x^2$ is taken to the third power. In the fourth term, $x^2$ is taken to the second power. This is the term we are looking for. Applying the Binomial Theorem to $(x^2+y^2)^5$, we have $a = x^2$, $b = y^2$, and $n = 5$. We are looking for the $r^{th}$ term where $r = 4$.

$$\binom{n}{r-1} a^{n-r+1} b^{r-1}$$
$$= \binom{5}{4-1} (x^2)^{5-4+1} (y^2)^{4-1}$$
$$= \binom{5}{3} (x^2)^2 (y^2)^3 = \frac{5!}{3!(5-3)!} x^4 y^6$$
$$= \frac{5!}{3!2!} x^4 y^6 = \frac{5 \cdot 4 \cdot \cancel{3!}}{\cancel{3!}2 \cdot 1} x^4 y^6 = 10x^4 y^6$$

**78.**
$$f(a+1) = (a+1)^2 + 2(a+1) + 3$$
$$= a^2 + 2a + 1 + 2a + 2 + 3$$
$$= a^2 + 4a + 6$$

**79.**
$$f(x) = x^2 + 5x \qquad g(x) = 2x - 3$$
$$f(g(x)) = f(2x-3)$$
$$= (2x-3)^2 + 5(2x-3)$$
$$= 4x^2 - 12x + 9 + 10x - 15$$
$$= 4x^2 - 2x - 6$$

$$g(f(x)) = g(x^2 + 5x)$$
$$= 2(x^2 + 5x) - 3$$
$$= 2x^2 + 10x - 3$$

**80.**
$$\frac{x}{x+3} - \frac{x+1}{2x^2 - 2x - 24}$$
$$= \frac{x}{x+3} - \frac{x+1}{2(x^2 - x - 12)}$$
$$= \frac{x}{x+3} - \frac{x+1}{2(x-4)(x+3)}$$
$$= \frac{x}{x+3} \cdot \frac{2(x-4)}{2(x-4)} - \frac{x+1}{2(x-4)(x+3)}$$
$$= \frac{x}{x+3} \cdot \frac{2x-8}{2(x-4)} - \frac{x+1}{2(x-4)(x+3)}$$

$$= \frac{2x^2 - 8x}{2(x-4)(x+3)} - \frac{x+1}{2(x-4)(x+3)}$$

$$= \frac{2x^2 - 8x - (x+1)}{2(x-4)(x+3)}$$

$$= \frac{2x^2 - 8x - x - 1}{2(x-4)(x+3)}$$

$$= \frac{2x^2 - 9x - 1}{2(x-4)(x+3)}$$

## 12.4 Exercise Set

**1.** $S_n : 1 + 3 + 5 + (2n-1) = n^2$

$S_1 : 1 = 1^2$

$1 = 1$, True

$S_2 : 1 + 3 = 2^2$

$4 = 4$, True

$S_3 : 1 + 3 + 5 = 3^2$

$9 = 9$, True

**3.** $S_n$ : 2 is a factor of $n^2 - n$.

$S_1$ : 2 is a factor of $1^2 - 1$

Since $1^2 - 1 = 0 = 2 \cdot 0$, $S_1$ is true.

$S_2$ : 2 is a factor of $2^2 - 2$

Since $2^2 - 2 = 2 = 2 \cdot 1$, $S_2$ is true

$S_3$ : 2 is a factor of $3^2 - 3$

Since $3^2 - 3 = 6 = 2 \cdot 3$, $S_3$ is true

**5.** $S_n : 4 + 8 + 12 + \cdots + 4n = 2n(n+1)$

$S_k : 4 + 8 + 12 + \cdots + 4k = 2k(k+1)$

$S_{k+1} : 4 + 8 + 12 + \cdots + 4(k+1) = 2(k+1)[(k+1)+1] = 2(k+1)(k+2)$

**7.** $S_n : 3 + 7 + 11 + \cdots + (4n-1) = n(2n+1)$

$S_k : 3 + 7 + 11 + \cdots + (4k-1) = k(2k+1)$

$S_{k+1} : 3 + 7 + 11 + \cdots + [4(k+1)-1] = (k+1)[2(k+1)+1] = (k+1)(2k+3)$

**9.** $S_n$ : 2 is a factor of $n^2 - n + 2$

$S_k$ : 2 is a factor of $k^2 - k + 2$

$S_{k+1}$ : 2 is a factor of $(k+1)^2 - (k+1) + 2 = k^2 + 2k + 1 - k - 1 + 2 = k^2 + k + 2$

**11.** $4+8+12+\cdots+4n=2n(n+1)$

*Show that $S_1$ is true:* $4=2\cdot 1(1+1)$

$4=2\cdot 1\cdot 2$

$4=4$, True

*Show that if $S_k$ is true, then $S_{k+1}$ is true:*

Assume $S_k: 4+8+12+\cdots+4k=2k(k+1)$ is true. Then,

$$4+8+12+\cdots+4k+4(k+1)=2k(k+1)+4(k+1)$$
$$4+8+12+\cdots+4k+4(k+1)=2k^2+2k+4k+4$$
$$4+8+12+\cdots+4k+4(k+1)=2k^2+6k+4$$
$$4+8+12+\cdots+4k+4(k+1)=2(k^2+3k+2)$$
$$4+8+12+\cdots+4k+4(k+1)=2(k+1)(k+2)$$
$$4+8+12+\cdots+4k+4(k+1)=2(k+1)[(k+1)+1]$$

The final statement is $S_{k+1}$. Thus, by mathematical induction, we have proven that $4+8+12+\cdots+4n=2n(n+1)$.

**13.** $1+3+5+\cdots+(2n-1)=n^2$

*Show that $S_1$ is true:* $1=1^2$

$1=1$, True

*Show that if $S_k$ is true, then $S_{k+1}$ is true:*

Assume $S_k: 1+3+5+\cdots+(2k-1)=k^2$ is true. Then,

$$1+3+5+\cdots+(2k-1)+[2(k+1)-1]=k^2+[2(k+1)-1]$$
$$1+3+5+\cdots+(2k-1)+[2(k+1)-1]=k^2+2k+2-1$$
$$1+3+5+\cdots+(2k-1)+[2(k+1)-1]=k^2+2k+1$$
$$1+3+5+\cdots+(2k-1)+[2(k+1)-1]=(k+1)^2$$

The final statement is $S_{k+1}$. Thus, by mathematical induction, we have proven that $1+3+5+\cdots+(2n-1)=n^2$.

**15.** $3+7+11+\cdots+(4n-1)=n(2n+1)$

*Show that* $S_1$ *is true*: $3=1(2\cdot 1+1)$

$3=1(2+1)$

$3=1(3)$

$3=3$, True

*Show that if* $S_k$ *is true, then* $S_{k+1}$ *is true*:

Assume $S_k: 3+7+11+\cdots+(4k-1)=k(2k+1)$ is true. Then,

$$3+7+11+\cdots+(4k-1)+[4(k+1)-1]=k(2k+1)+[4(k+1)-1]$$

$$3+7+11+\cdots+(4k-1)+[4(k+1)-1]=2k^2+k+4k+4-1$$

$$3+7+11+\cdots+(4k-1)+[4(k+1)-1]=2k^2+5k+3$$

$$3+7+11+\cdots+(4k-1)+[4(k+1)-1]=(k+1)(2k+3)$$

$$3+7+11+\cdots+(4k-1)+[4(k+1)-1]=(k+1)[2(k+1)+1]$$

The final statement is $S_{k+1}$. Thus, by mathematical induction, we have proven that $3+7+11+\cdots+(4n-1)=n(2n+1)$.

**17.** $1+2+2^2+\cdots+2^{n-1}=2^n-1$

*Show that* $S_1$ *is true*: $1=2^1-1$

$1=2-1$

$1=1$, True

*Show that if* $S_k$ *is true, then* $S_{k+1}$ *is true*:

Assume $S_k: 1+2+2^2+\cdots+2^{k-1}=2^k-1$ is true. Then,

$$1+2+2^2+\cdots+2^{k-1}+2^{(k+1)-1}=2^k-1+2^{(k+1)-1}$$

$$1+2+2^2+\cdots+2^{k-1}+2^{(k+1)-1}=2^k-1+2^k$$

$$1+2+2^2+\cdots+2^{k-1}+2^{(k+1)-1}=2\cdot 2^k-1$$

$$1+2+2^2+\cdots+2^{k-1}+2^{(k+1)-1}=2^{k+1}-1$$

The final statement is $S_{k+1}$. Thus, by mathematical induction, we have proven that $1+2+2^2+\cdots+2^{n-1}=2^n-1$.

**19.** $2+4+8+\cdots+2^n=2^{n+1}-2$

*Show that* $S_1$ *is true*: $2=2^{1+1}-2$

$2=2^2-2$

$2=4-2$

$2=2$, True

*Show that if $S_k$ is true, then $S_{k+1}$ is true*:

Assume $S_k: 2+4+8+\cdots+2^k = 2^{k+1}-2$ is true. Then,

$$2+4+8+\cdots+2^k+2^{k+1} = 2^{k+1}-2+2^{k+1}$$

$$2+4+8+\cdots+2^k+2^{k+1} = 2\cdot 2^{k+1}-2$$

$$2+4+8+\cdots+2^k+2^{k+1} = 2^{(k+1)+1}-2$$

The final statement is $S_{k+1}$. Thus, by mathematical induction, we have proven that $2+4+8+\cdots+2^n = 2^{n+1}-2$.

**21.** $$1\cdot 2+2\cdot 3+3\cdot 4+\cdots+n(n+1) = \frac{n(n+1)(n+2)}{3}$$

*Show that $S_1$ is true*: $1\cdot 2 = \frac{1(1+1)(1+2)}{3}$

$$2 = \frac{1(2)(3)}{3}$$

$$2 = 2, \text{ True}$$

*Show that if $S_k$ is true, then $S_{k+1}$ is true*:

Assume $S_k: 1\cdot 2+2\cdot 3+3\cdot 4+\cdots+k(k+1) = \frac{k(k+1)(k+2)}{3}$ is true. Then,

$$1\cdot 2+2\cdot 3+3\cdot 4+\cdots+k(k+1)+(k+1)[(k+1)+1] = \frac{k(k+1)(k+2)}{3}+(k+1)[(k+1)+1]$$

$$1\cdot 2+2\cdot 3+3\cdot 4+\cdots+k(k+1)+(k+1)[(k+1)+1] = \frac{k(k+1)(k+2)}{3}+(k+1)(k+2)$$

$$1\cdot 2+2\cdot 3+3\cdot 4+\cdots+k(k+1)+(k+1)[(k+1)+1] = \frac{k(k+1)(k+2)+3(k+1)(k+2)}{3}$$

$$1\cdot 2+2\cdot 3+3\cdot 4+\cdots+k(k+1)+(k+1)[(k+1)+1] = \frac{(k+1)(k+2)(k+3)}{3}$$

$$1\cdot 2+2\cdot 3+3\cdot 4+\cdots+k(k+1)+(k+1)[(k+1)+1] = \frac{(k+1)[(k+1)+1][(k+1)+2]}{3}$$

The final statement is $S_{k+1}$. Thus, by mathematical induction, we have proven that

$$1\cdot 2+2\cdot 3+3\cdot 4+\cdots+n(n+1) = \frac{n(n+1)(n+2)}{3}.$$

**23.** $\frac{1}{1\cdot 2}+\frac{1}{2\cdot 3}+\frac{1}{3\cdot 4}+\cdots+\frac{1}{n(n+1)}=\frac{n}{n+1}$

*Show that* $S_1$ *is true*: $\frac{1}{1\cdot 2}=\frac{1}{1+1}$

$\frac{1}{2}=\frac{1}{2}$, True

*Show that if* $S_k$ *is true, then* $S_{k+1}$ *is true*:

Assume $S_k:\frac{1}{1\cdot 2}+\frac{1}{2\cdot 3}+\frac{1}{3\cdot 4}+\cdots+\frac{1}{k(k+1)}=\frac{k}{k+1}$ is true. Then,

$$\frac{1}{1\cdot 2}+\frac{1}{2\cdot 3}+\frac{1}{3\cdot 4}+\cdots+\frac{1}{k(k+1)}+\frac{1}{(k+1)[(k+1)+1]}=\frac{k}{k+1}+\frac{1}{(k+1)[(k+1)+1]}$$

$$\frac{1}{1\cdot 2}+\frac{1}{2\cdot 3}+\frac{1}{3\cdot 4}+\cdots+\frac{1}{k(k+1)}+\frac{1}{(k+1)[(k+1)+1]}=\frac{k}{k+1}+\frac{1}{(k+1)(k+2)}$$

$$\frac{1}{1\cdot 2}+\frac{1}{2\cdot 3}+\frac{1}{3\cdot 4}+\cdots+\frac{1}{k(k+1)}+\frac{1}{(k+1)[(k+1)+1]}=\frac{k(k+2)}{(k+1)(k+2)}+\frac{1}{(k+1)(k+2)}$$

$$\frac{1}{1\cdot 2}+\frac{1}{2\cdot 3}+\frac{1}{3\cdot 4}+\cdots+\frac{1}{k(k+1)}+\frac{1}{(k+1)[(k+1)+1]}=\frac{k(k+2)+1}{(k+1)(k+2)}$$

$$\frac{1}{1\cdot 2}+\frac{1}{2\cdot 3}+\frac{1}{3\cdot 4}+\cdots+\frac{1}{k(k+1)}+\frac{1}{(k+1)[(k+1)+1]}=\frac{k^2+2k+1}{(k+1)(k+2)}$$

$$\frac{1}{1\cdot 2}+\frac{1}{2\cdot 3}+\frac{1}{3\cdot 4}+\cdots+\frac{1}{k(k+1)}+\frac{1}{(k+1)[(k+1)+1]}=\frac{(k+1)^2}{(k+1)(k+2)}$$

$$\frac{1}{1\cdot 2}+\frac{1}{2\cdot 3}+\frac{1}{3\cdot 4}+\cdots+\frac{1}{k(k+1)}+\frac{1}{(k+1)[(k+1)+1]}=\frac{k+1}{k+2}$$

$$\frac{1}{1\cdot 2}+\frac{1}{2\cdot 3}+\frac{1}{3\cdot 4}+\cdots+\frac{1}{k(k+1)}+\frac{1}{(k+1)[(k+1)+1]}=\frac{k+1}{(k+1)+1}$$

The final statement is $S_{k+1}$. Thus, by mathematical induction, we have proven that

$\frac{1}{1\cdot 2}+\frac{1}{2\cdot 3}+\frac{1}{3\cdot 4}+\cdots+\frac{1}{n(n+1)}=\frac{n}{n+1}$.

**25.** $\sum_{i=1}^{n} 5\cdot 6^i = 6(6^n - 1)$

*Show that $S_1$ is true:*

$\sum_{i=1}^{1} 5\cdot 6^i = 6(6^1 - 1)$

$5\cdot 6^1 = 6(6-1)$

$5\cdot 6 = 6\cdot 5$, True

*Show that if $S_k$ is true, then $S_{k+1}$ is true:*

Assume $S_k : \sum_{i=1}^{k} 5\cdot 6^i = 6(6^k - 1)$ is true.

Then,

$\sum_{i=1}^{k} 5\cdot 6^i + 5\cdot 6^{k+1} = 6(6^k - 1) + 5\cdot 6^{k+1}$

$\sum_{i=1}^{k+1} 5\cdot 6^i = 6^{k+1} - 6 + 5\cdot 6^{k+1}$

$\sum_{i=1}^{k+1} 5\cdot 6^i = 6\cdot 6^{k+1} - 6$

$\sum_{i=1}^{k+1} 5\cdot 6^i = 6(6^{k+1} - 1)$

The final statement is $S_{k+1}$. Thus, by mathematical induction, we have proven that $\sum_{i=1}^{n} 5\cdot 6^i = 6(6^n - 1)$.

**26.** $\sum_{i=1}^{n} 7\cdot 8^i = 8(8^n - 1)$

*Show that $S_1$ is true:*

$\sum_{i=1}^{1} 7\cdot 8^i = 8(8^1 - 1)$

$7\cdot 8^1 = 8(8-1)$

$7\cdot 8 = 8\cdot 7$, True

*Show that if $S_k$ is true, then $S_{k+1}$ is true:*

Assume $S_k : \sum_{i=1}^{k} 7\cdot 8^i = 8(8^k - 1)$ is true.

Then,

$\sum_{i=1}^{k} 7\cdot 8^i + 7\cdot 8^{k+1} = 8(8^k - 1) + 7\cdot 8^{k+1}$

$\sum_{i=1}^{k+1} 7\cdot 8^i = 8^{k+1} - 8 + 7\cdot 8^{k+1}$

$\sum_{i=1}^{k+1} 7\cdot 8^i = 8\cdot 8^{k+1} - 8$

$\sum_{i=1}^{k+1} 7\cdot 8^i = 8(8^{k+1} - 1)$

The final statement is $S_{k+1}$. Thus, by mathematical induction, we have proven that $\sum_{i=1}^{n} 7\cdot 8^i = 8(8^n - 1)$.

**27.** $n + 2 > n$

*Show that $S_1$ is true:* $1 + 2 > 1$

$3 > 1$, True

*Show that if $S_k$ is true, then $S_{k+1}$ is true:*

Assume $S_k : k + 2 > k$ is true. Then,

$k + 2 + 1 > k + 1$

$(k+1) + 2 > k + 1$

The final statement is $S_{k+1}$. Thus, by mathematical induction, we have proven that $n + 2 > n$.

**28.** If $0 < x < 1$, then $0 < x^n < 1$.

*Show that $S_1$ is true:* $0 < x^1 < 1$ is true because it is equivalent to the if statement $0 < x < 1$.

*Show that if $S_k$ is true, then $S_{k+1}$ is true:*

Assume $S_k : 0 < x^k < 1$ is true. Then,

$0\cdot x < x^k \cdot x < 1\cdot x$

$0 < x^{k+1} < x$

Now we know that $x < 1$, so $0 < x^{k+1} < 1$ is true.

The final statement is $S_{k+1}$. Thus, by mathematical induction, we have proven that $0 < x^n < 1$.

**29.** $(ab)^n = a^n b^n$

*Show that* $S_1$ *is true*: $(ab)^1 = a^1 b^1$

$$ab = ab$$

*Show that if* $S_k$ *is true, then* $S_{k+1}$ *is true*:

Assume $S_k : (ab)^k = a^k b^k$ is true. Then,

$$(ab)^k \cdot ab = a^k b^k \cdot ab$$
$$(ab)^{k+1} = a^k \cdot a \cdot b^k \cdot b$$
$$(ab)^{k+1} = a^{k+1} b^{k+1}$$

The final statement is $S_{k+1}$. Thus, by mathematical induction, we have proven that $(ab)^n = a^n b^n$.

**30.** $\left(\frac{a}{b}\right)^n = \frac{a^n}{b^n}$

*Show that* $S_1$ *is true*: $\left(\frac{a}{b}\right)^1 = \frac{a^1}{b^1}$

$$\frac{a}{b} = \frac{a}{b},$$

*Show that if* $S_k$ *is true, then* $S_{k+1}$ *is true*:

Assume $S_k : \left(\frac{a}{b}\right)^k = \frac{a^k}{b^k}$ is true. Then,

$$\left(\frac{a}{b}\right)^k \cdot \frac{a}{b} = \frac{a^k}{b^k} \cdot \frac{a}{b}$$
$$\left(\frac{a}{b}\right)^{k+1} = \frac{a^k \cdot a}{b^k \cdot b}$$
$$\left(\frac{a}{b}\right)^{k+1} = \frac{a^{k+1}}{b^{k+1}}$$

The final statement is $S_{k+1}$. Thus, by mathematical induction, we have proven that $\left(\frac{a}{b}\right)^n = \frac{a^n}{b^n}$.

**31.** $n^2 + n$ is divisible by 2.

*Show that* $S_1$ *is true*: $1^2 + 1 = 2$, which is divisible by 2.

*Show that if* $S_k$ *is true, then* $S_{k+1}$ *is true*:

Assume $S_k$ is true. Then, $k^2 + k$ is divisible by 2. Now, since $2(k+1)$ is a multiple of 2, it must be divisible by 2. Therefore, $k^2 + k + 2(k+1)$ is divisible by 2 also. Simplifying, we obtain:

$$k^2 + k + 2(k+1) = k^2 + k + 2k + 2$$
$$= k^2 + 2k + 1 + k + 1$$
$$= (k+1)^2 + (k+1)$$

Thus, $(k+1)^2 + (k+1)$ is divisible by 2. The final statement is $S_{k+1}$. Thus, by mathematical induction, we have proven that $n^2 + n$ is divisible by 2.

**32.** $n^2 + 3n$ is divisible by 2.

*Show that* $S_1$ *is true*: $1^2 + 3 \cdot 1 = 4$, which is divisible by 2.

*Show that if* $S_k$ *is true, then* $S_{k+1}$ *is true*:

Assume $S_k$ is true. Then, $k^2 + 3k$ is divisible by 2. Now, since $2(k+2)$ is a multiple of 2, it must be divisible by 2. Therefore, $k^2 + 3k + 2(k+2)$ is divisible by 2 also. Simplifying, we obtain:

$$k^2 + 3k + 2(k+2) = k^2 + 3k + 2k + 4$$
$$= k^2 + 2k + 1 + 3k + 3$$
$$= (k+1)^2 + 3(k+1)$$

Thus, $(k+1)^2 + 3(k+1)$ is divisible by 2. The final statement is $S_{k+1}$. Thus, by mathematical induction, we have proven that $n^2 + 3n$ is divisible by 2.

**33. – 35.** Answers will vary.

**37.** $S_n: \left(1-\frac{1}{2}\right)\left(1-\frac{1}{3}\right)\left(1-\frac{1}{4}\right)\cdots\left(1-\frac{1}{n+1}\right)=?$

$S_1: \left(1-\frac{1}{2}\right)=\frac{1}{2}$

$S_2: \left(1-\frac{1}{2}\right)\left(1-\frac{1}{3}\right)=\frac{1}{3}$

$S_3: \left(1-\frac{1}{2}\right)\left(1-\frac{1}{3}\right)\left(1-\frac{1}{4}\right)=\frac{1}{4}$

$S_4: \left(1-\frac{1}{2}\right)\left(1-\frac{1}{3}\right)\left(1-\frac{1}{4}\right)\left(1-\frac{1}{5}\right)=\frac{1}{5}$

$S_5: \left(1-\frac{1}{2}\right)\left(1-\frac{1}{3}\right)\left(1-\frac{1}{4}\right)\left(1-\frac{1}{5}\right)\left(1-\frac{1}{6}\right)=\frac{1}{6}$

Conjecture: $S_n: \left(1-\frac{1}{2}\right)\left(1-\frac{1}{3}\right)\left(1-\frac{1}{4}\right)\cdots\left(1-\frac{1}{n+1}\right)=\frac{1}{n+1}$

Proof: We have already shown that $S_1$ is true above.

*Show that if $S_k$ is true, then $S_{k+1}$ is true*:

Assume $S_k: \left(1-\frac{1}{2}\right)\left(1-\frac{1}{3}\right)\left(1-\frac{1}{4}\right)\cdots\left(1-\frac{1}{k+1}\right)=\frac{1}{k+1}$ is true. Then,

$$\left(1-\frac{1}{2}\right)\left(1-\frac{1}{3}\right)\left(1-\frac{1}{4}\right)\cdots\left(1-\frac{1}{k+1}\right)\left(1-\frac{1}{(k+1)+1}\right)=\frac{1}{k+1}\left(1-\frac{1}{(k+1)+1}\right)$$

$$\left(1-\frac{1}{2}\right)\left(1-\frac{1}{3}\right)\left(1-\frac{1}{4}\right)\cdots\left(1-\frac{1}{k+1}\right)\left(1-\frac{1}{(k+1)+1}\right)=\frac{1}{k+1}\left(1-\frac{1}{k+2}\right)$$

$$\left(1-\frac{1}{2}\right)\left(1-\frac{1}{3}\right)\left(1-\frac{1}{4}\right)\cdots\left(1-\frac{1}{k+1}\right)\left(1-\frac{1}{(k+1)+1}\right)=\frac{1}{k+1}\left(\frac{k+2}{k+2}-\frac{1}{k+2}\right)$$

$$\left(1-\frac{1}{2}\right)\left(1-\frac{1}{3}\right)\left(1-\frac{1}{4}\right)\cdots\left(1-\frac{1}{k+1}\right)\left(1-\frac{1}{(k+1)+1}\right)=\frac{1}{k+1}\left(\frac{k+1}{k+2}\right)$$

$$\left(1-\frac{1}{2}\right)\left(1-\frac{1}{3}\right)\left(1-\frac{1}{4}\right)\cdots\left(1-\frac{1}{k+1}\right)\left(1-\frac{1}{(k+1)+1}\right)=\frac{1}{k+2}$$

$$\left(1-\frac{1}{2}\right)\left(1-\frac{1}{3}\right)\left(1-\frac{1}{4}\right)\cdots\left(1-\frac{1}{k+1}\right)\left(1-\frac{1}{(k+1)+1}\right)=\frac{1}{(k+1)+1}$$

The final statement is $S_{k+1}$. Thus, by mathematical induction, we have proven that

$S_n: \left(1-\frac{1}{2}\right)\left(1-\frac{1}{3}\right)\left(1-\frac{1}{4}\right)\cdots\left(1-\frac{1}{n+1}\right)=\frac{1}{n+1}$.

**39.**

$$V = C(1-t)$$
$$V = C - Ct$$
$$Ct + V = C$$
$$Ct = C - V$$
$$t = \frac{C-V}{C}$$

**40.** $x^3 + 2x^2 - 5x - 6 = 0$

We begin by using the Rational Zero Theorem to determine possible rational roots.

Factors of the constant term $-6$: $\pm 1,\ \pm 2,\ \pm 3,\ \pm 6$

Factors of the leading coefficient 1: $\pm 1$

The possible rational zeros are:

$$\frac{\text{Factors of } -6}{\text{Factors of } 1} = \frac{\pm 1,\ \pm 2,\ \pm 3,\ \pm 6}{\pm 1}$$
$$= \pm 1,\ \pm 2,\ \pm 3,\ \pm 6$$

We test values from above until we find a root. One is shown next:

Test $-1$:

| | | | | |
|---|---|---|---|---|
| $-1$ | 1 | 2 | $-5$ | $-6$ |
| | | $-1$ | $-1$ | 6 |
| | 1 | 1 | $-6$ | 0 |

The remainder is 0, so $-1$ is a root of the equation. Thus,

$$x^3 + 2x^2 - 5x - 6 = 0$$
$$(x+1)(x^2 + x - 6) = 0$$
$$(x+1)(x+3)(x-2) = 0$$

Apply the zero-product property:

$x + 1 = 0$ or $x + 3 = 0$ or $x - 2 = 0$

$x = -1$ $\quad x = -3$ $\quad x = 2$

The solutions are $-3$, $-1$ and 2, and the solution set is $\{-3,\ -1,\ 2\}$.

**41.**

$$x^2 + y^2 - 2x + 4y - 4 = 0$$
$$x^2 - 2x \quad + y^2 + 4y \quad = 4$$
$$(x^2 - 2x + 1) + (y^2 + 4y + 4) = 4 + 1 + 4$$
$$(x-1)^2 + (y+2)^2 = 9$$
$$(x-1)^2 + (y+2)^2 = 3^2$$

The center of the circle is $(1,-2)$, and the radius is 3.

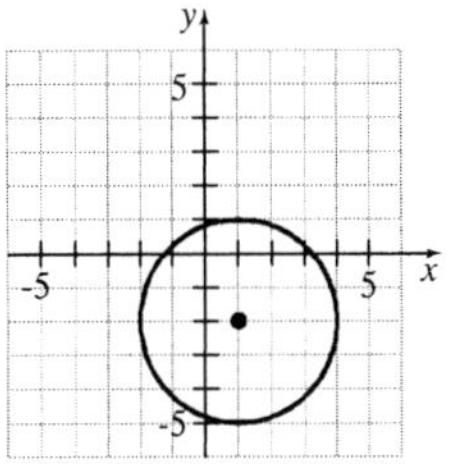


## 12.6 Exercise Set

**1.** $${}_9P_4 = \frac{9!}{(9-4)!} = \frac{9!}{5!} = \frac{9\cdot 8\cdot 7\cdot 6\cdot \cancel{5!}}{\cancel{5!}}$$
$$= 9\cdot 8\cdot 7\cdot 6 = 3024$$

**3.** $${}_8P_5 = \frac{8!}{(8-5)!} = \frac{8!}{3!} = \frac{8\cdot 7\cdot 6\cdot 5\cdot 4\cdot \cancel{3!}}{\cancel{3!}}$$
$$= 8\cdot 7\cdot 6\cdot 5\cdot 4 = 6720$$

**5.** $${}_6P_6 = \frac{6!}{(6-6)!} = \frac{6!}{0!} = \frac{6\cdot 5\cdot 4\cdot 3\cdot 2\cdot 1}{1}$$
$$= 720$$

**7.** $${}_8P_0 = \frac{8!}{(8-0)!} = \frac{8!}{8!} = 1$$

**9.** $${}_9C_5 = \frac{9!}{(9-5)!5!} = \frac{9!}{4!5!} = \frac{9\cdot 8\cdot 7\cdot 6\cdot \cancel{5!}}{4\cdot 3\cdot 2\cdot 1\cdot \cancel{5!}}$$
$$= 126$$

**11.** $${}_{11}C_4 = \frac{11!}{(11-4)!4!} = \frac{11!}{7!4!}$$
$$= \frac{11 \cdot 10 \cdot 9 \cdot 8 \cdot \cancel{7!}}{\cancel{7!} \cdot 4 \cdot 3 \cdot 2 \cdot 1} = 330$$

**13.** $${}_7C_7 = \frac{7!}{(7-7)!7!} = \frac{7!}{0!7!} = 1$$

**15.** $${}_5C_0 = \frac{5!}{(5-0)!0!} = \frac{5!}{5!0!} = 1$$

**17.** Combinations; the order in which the 6 people selected does not matter.

**19.** Permutations; the order in which the letters are selected does matter.

**21.** $$\frac{{}_7P_3}{3!} - {}_7C_3 = \frac{\frac{7!}{(7-3)!}}{3!} - \frac{7!}{(7-3)!3!}$$
$$= \frac{\frac{7!}{4!}}{3!} - \frac{7!}{4!3!}$$
$$= \frac{7!}{4!3!} - \frac{7!}{4!3!} = 0$$

**23.** $$1 - \frac{{}_3P_2}{{}_4P_3} = 1 - \frac{\frac{3!}{(3-2)!}}{\frac{4!}{(4-3)!}} = \frac{\frac{3!}{1!}}{\frac{4!}{1!}} = 1 - \frac{3!}{4!}$$
$$= 1 - \frac{3!}{4 \cdot 3!} = 1 - \frac{1}{4} = \frac{3}{4}$$

**25.** $$\frac{{}_7C_3}{{}_5C_4} - \frac{98!}{96!} = \frac{\frac{7!}{(7-3)!3!}}{\frac{5!}{(5-4)!4!}} - \frac{98 \cdot 97 \cdot \cancel{96!}}{\cancel{96!}}$$
$$= \frac{\frac{7!}{4!3!}}{\frac{5!}{1!4!}} - 95067$$
$$= \frac{\frac{7 \cdot 6 \cdot 5 \cdot \cancel{4!}}{\cancel{4!}3 \cdot 2 \cdot 1}}{\frac{5 \cdot \cancel{4!}}{1!\cancel{4!}}} - 9506$$
$$= \frac{35}{5} - 9506$$
$$= 7 - 9506 = -9499$$

**27.** $$\frac{{}_4C_2 \cdot {}_6C_1}{{}_{18}C_3} = \frac{\frac{4!}{(4-2)!2!} \cdot \frac{6!}{(6-1)!1!}}{\frac{18!}{(18-3)!3!}}$$
$$= \frac{\frac{4!}{2!2!} \cdot \frac{6!}{5!1!}}{\frac{18!}{15!3!}}$$
$$= \frac{\frac{4 \cdot 3 \cdot \cancel{2!}}{\cancel{2!}2 \cdot 1} \cdot \frac{6 \cdot \cancel{5!}}{\cancel{5!}1!}}{\frac{18 \cdot 17 \cdot 16 \cdot \cancel{15!}}{\cancel{15!}3 \cdot 2 \cdot 1}}$$
$$= \frac{36}{816} = \frac{3}{68}$$

**29.** $9 \cdot 3 = 27$
There are 27 ways you can order the car.

**31.** $2 \cdot 4 \cdot 5 = 40$
There are 40 ways to order a drink.

**33.** $3 \cdot 3 \cdot 3 \cdot 3 \cdot 3 = 3^4 = 243$
There are 243 ways to answer the questions.

**35.** $8 \cdot 2 \cdot 9 = 144$
There are 144 area codes possible.

**37.** Find the number of ways the remaining five performers can by arranged:
$5 \cdot 4 \cdot 3 \cdot 2 \cdot 1 = 120$
There are 120 ways to schedule the appearances.

**39.** Find the number of arrangements for the remaining three paragraphs:
$3 \cdot 2 \cdot 1 = 6$
6 five-sentence paragraphs can be formed.

**41.** $${}_{10}P_3 = \frac{10!}{(10-3)!} = \frac{10!}{7!} = 720$$
There are 720 ways the offices can be filled.

**43.** $${}_{13}P_7 = \frac{13!}{(13-7)!} = \frac{13!}{6!} = 8648640$$
There are 8,648,640 ways to arrange the program for this segment.

**45.** $${}_{6}P_3 = \frac{6!}{(6-3)!} = \frac{6!}{3!} = 120$$
There are 120 ways the first three finishers can come in.

**47.** $${}_{9}P_5 = \frac{9!}{(9-5)!} = \frac{9!}{4!} = 15120$$
There are 15,120 lineups possible.

**49.** $${}_{6}C_2 = \frac{6!}{(6-2)!2!} = \frac{6!}{4!2!} = 20$$
There are 20 ways to select the three city commissioners.

**51.** $${}_{12}C_4 = \frac{12!}{(12-4)!4!} = \frac{12!}{8!4!}495$$
You can take 495 different collections of 4 books.

**53.** $${}_{17}C_8 = \frac{17!}{(17-8)!8!} = \frac{17!}{9!8!} = 24310$$
You can drive 24,310 different groups of 8 children.

**55.** $${}_{53}C_6 = \frac{53!}{(53-6)!6!} = \frac{53!}{47!6!} = 22957480$$
There are 22,957,480 selections possible.

**57.** $6 \cdot 5 \cdot 4 \cdot 3 = 360$
There are 360 ways the first four finishers can come in.

**59.** $${}_{13}C_6 = \frac{13!}{(13-6)!6!} = \frac{13!}{7!6!} = 1716$$
There are 1716 ways 6 people can be selected.

**61.** $${}_{20}C_3 = \frac{20!}{(20-3)!3!} = \frac{20!}{17!3!} = 1140$$
There are 1140 ways to select 3 members.

**63.** $${}_{7}P_4 = \frac{7!}{(7-4)!} = \frac{7!}{3!} = 840$$
840 four-letter passwords can be formed.

**65.** $${}_{15}P_3 = \frac{15!}{(15-3)!} = \frac{15!}{12!} = 2730$$
2730 cones can be created.

**67. – 73.** Answers will vary..

**75.** For example, consider Exercise 1.

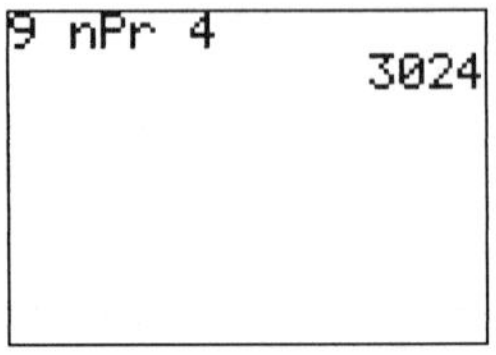


This matches the result obtained in Exercise 1.

**77.** Statement **c** is true.

$$3!\,{}_7C_3 = \cancel{3!}\cdot\frac{7!}{(7-3)!\cancel{3!}} = \frac{7!}{(7-3)!} = {}_7P_3$$

Statement **a** is false. The number of ways to choose four questions out of ten questions is ${}_{10}C_4$.

Statement **b** is false. If $r > 1$, ${}_nP_r$ is greater than ${}_nC_r$.

Statement **d** is false. The number of ways to pick a winner and first runner-up is ${}_{20}C_2$

**79.** The first digit must be either 2 or 4. The second and third digits can be 2, 4, 6, 7, 8, or 9. The fourth digit must be either 7 or 9.
$2\cdot 6\cdot 6\cdot 2 = 144$
There are 144 four-digit odd numbers less than 6000 that can be formed.

**81.**
$$\begin{aligned}(f\circ g)(x) &= f\left(g(x)\right) = f\left(4x-1\right)\\ &= (4x-1)^2 + 2(4x-1) - 5\\ &= 16x^2 - 8x + 1 + 8x - 2 - 5\\ &= 16x^2 - 6\end{aligned}$$

**82.** $|2x-5| > 3$

$$\begin{aligned} 2x-5 &> 3 \quad \text{or} \quad 2x-5 < -3\\ 2x &> 8 \qquad\qquad 2x < 2\\ x &> 4 \qquad\qquad\ \ x < 1\end{aligned}$$

$\{x \mid x < 1 \text{ or } x > 4\}$ or $(-\infty, 1)\cup(4, \infty)$

**83** $f(x) = (x-2)^2(x+1)$

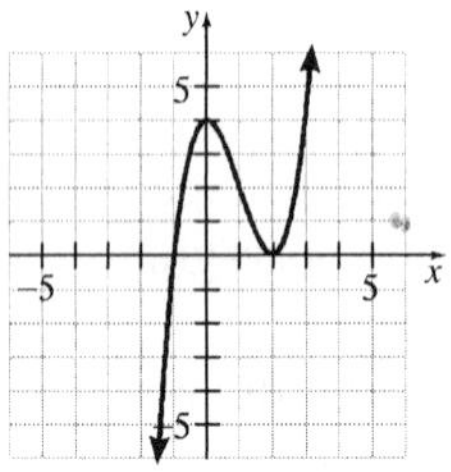


## 12.7 Exercise Set

**1.** $P(\textit{female}) = \dfrac{23}{89}$

**3.** $P(\textit{in the Army}) = \dfrac{36}{89}$

**5.** $P(\textit{a woman in the Air Force}) = \dfrac{6}{89}$

**7.** $P(\textit{woman, among Air Force}) = \dfrac{6}{18} = \dfrac{1}{3}$

**9.** $P(\textit{woman in Air Force}) = \dfrac{6}{23}$

**11.** $P(4) = \dfrac{1}{6}$

**13.** $P(\textit{odd number}) = \dfrac{3}{6} = \dfrac{1}{2}$

**15** $P(\textit{greater than } 4) = \dfrac{2}{6} = \dfrac{1}{3}$

**17.** $P(queen) = \frac{4}{52} = \frac{1}{13}$

**19.** $P(picture\ card) = \frac{12}{52} = \frac{3}{13}$

**21.** $P(two\ heads) = P(HH) = \frac{1}{4}$

**23.** $P(at\ least\ one\ male\ child) = \frac{7}{8}$

**25.** $P(sum\ is\ 4) = \frac{3}{36} = \frac{1}{12}$

**27.** First find the number of combinations of 6 out of 51 numbers:

$${}_{51}C_6 = \frac{51!}{(51-6)!6!} = \frac{51!}{45!6!} = 18009460$$

The probability that a person with one combinations of six numbers will win is $\frac{1}{18{,}009{,}460} \approx 0.0000000555$.

If 100 different tickets are purchased, the probability of winning is

$$100 \cdot \frac{1}{18{,}009{,}460} = \frac{5}{900{,}473}$$
$$= 0.00000555$$

**29.** **a.** ${}_{52}C_5 = \frac{52!}{(52-5)!5!}$

$$= \frac{52!}{47!5!} = 2{,}598{,}960$$

**b.** ${}_{12}C_5 = \frac{12!}{(12-5)!5!} = \frac{12!}{7!5!} = 1287$

**c.** $P(diamond\ flush) = \frac{1287}{2{,}598{,}960}$

$$\approx 0.000495$$

**31.** $P(not\ completed\ 4\ years\ of\ college)$

$= 1 - P(completed\ 4\ years\ of\ college)$

$= 1 - \frac{45}{174} = \frac{43}{58}$

**33.** $P(completed\ H.S.\ or$
$less\ than\ 4\ yrs\ college)$

$= P(completed\ H.S) +$
$P(less\ than\ 4\ yrs\ college)$

$= \frac{56}{174} + \frac{44}{174} = \frac{100}{174} = \frac{50}{87}$

**35.** $P(completed\ 4\ yrs\ H.S.\ or\ man)$

$= P(completed\ 4\ yrs\ H.S.) + P(man)$
$- P(man\ who\ completed\ 4\ yrs\ H.S)$

$= \frac{56}{174} + \frac{82}{174} - \frac{25}{174} = \frac{113}{174}$

**37.** $P(not\ king) = 1 - P(king) = 1 - \frac{4}{52}$

$$= 1 - \frac{1}{13} = \frac{12}{13}$$

**39.** $P(2\ or\ 3) = P(2) + P(3)$

$$= \frac{4}{52} + \frac{4}{52}$$
$$= \frac{8}{52} = \frac{2}{13}$$

**41.** $P(7\ or\ red\ card)$

$= P(7) + P(red\ card) - P(7\ and\ red)$

$= \frac{4}{52} + \frac{26}{52} - \frac{2}{52} = \frac{28}{52} = \frac{7}{13}$

**43** $P(odd\ or\ less\ than\ 6)$

$= P(odd) + P(less\ than\ 6)$

$- P(odd\ \#\ less\ than\ 6)$

$= \frac{4}{8} + \frac{5}{8} - \frac{3}{8} = \frac{6}{8} = \frac{3}{4}$

**45.** $P(professor\ or\ male) = P(professor)$ -

$= \frac{19}{40} + \frac{22}{40} - \frac{8}{40} = \frac{33}{40}$

**47.** $P(2\ \text{and}\ 3) = P(2) \cdot P(3) = \frac{1}{6} \cdot \frac{1}{6} = \frac{1}{36}$

**49.** $P(even\ and\ greater\ than\ 2)$

$= P(even) \cdot P\left(greater\ than\ 2\right)$

$= \frac{3}{6} \cdot \frac{4}{6} = \frac{1}{3}$

**51.** $P(all\ heads) = \frac{1}{2} \cdot \frac{1}{2} \cdot \frac{1}{2} \cdot \frac{1}{2} \cdot \frac{1}{2} \cdot \frac{1}{2}$

$= \left(\frac{1}{2}\right)^6 = \frac{1}{64}$

**53.** **a.** $P(hit\ 2\ yrs\ in\ a\ row)$

$= P(hit\ in\ 1^{st}\ year\ and\ 2^{nd}\ year)$

$= P(hit\ 1^{st}\ year) \cdot P(hit\ 2^{nd}\ year)$

$= \frac{1}{16} \cdot \frac{1}{16} = \frac{1}{256}$

**b.** $P(hit\ 3\ yrs\ in\ a\ row) = \frac{1}{16} \cdot \frac{1}{16} \cdot \frac{1}{16}$

$= \frac{1}{4096}$

**c.** First find the probability that South Florida will not be hit by a major hurricane in a single year:

$P(not\ hit) = 1 - \frac{1}{16} = \frac{15}{16}$

$P(not\ hit\ in\ next\ 10\ years) = \left(\frac{15}{16}\right)^{10}$

$\approx 0.524$

**d.** $P(hit\ at\ least\ once)$

$= 1 - P(hit\ none)$

$= 1 - \left(\frac{15}{16}\right)^{10} \approx 0.476$

**55. – 61.** Answers will vary.

**63.** $P$(someone who tests positive for cocaine uses cocaine)

$$= \frac{\text{the number of employees who test positive and are cocaine users}}{\text{number of employees who test positive}}$$

$$= \frac{90\% \text{ of } 1\% \text{ of } 10{,}000}{\text{\# who test positive who actually use cocaine plus \# who test positive who do not use cocaine}}$$

$$= \frac{(.90)(0.01)(10,000)}{(0.90)(0.01)(10,000)+(0.10)(0.99)(10,000)}$$

$$= \frac{90}{90+990} = \frac{90}{1080} = \frac{1}{12};$$

Answers will vary.

**65.** Answers will vary.

**67.** **a.** $P$(Democrat who is not a business major)

$$= \frac{\text{\# of students who are Democrats but not business majors}}{\text{\# of students}}$$

$$= \frac{29-5}{50} = \frac{24}{50} = \frac{12}{25}$$

**b.** $P$(neither Democrat nor business major)

$= 1 - P$(Democrat or business major)

$= 1 - \left(P(\text{Democrat}) + P(\text{business major}) - P(\text{Democrat and business major})\right)$

$$= 1 - \left(\frac{29}{50} + \frac{11}{50} - \frac{5}{50}\right) = 1 - \frac{35}{50} = \frac{15}{50} = \frac{3}{10}$$

**69.** $$f(x) = \frac{x^2-1}{x^2-4}$$

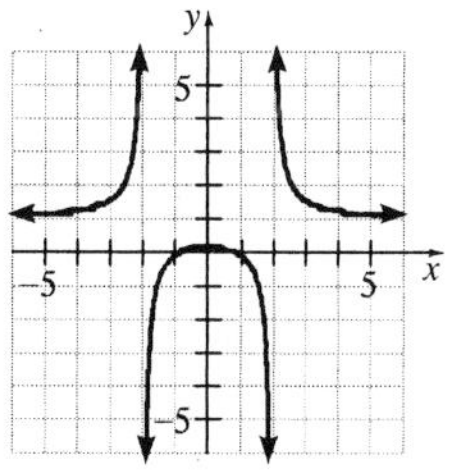

**70.** $\log_2(x+5)+\log_2(x-1)=4$

$$\log_2(x+5)(x-1)=4$$
$$\log_2\left(x^2+4x-5\right)=4$$
$$x^2+4x-5=2^4$$
$$x^2+4x-5=16$$
$$x^2+4x-21=0$$
$$(x+7)(x-3)=0$$

Using the zero product rule,
$x=-7$ or $x=3$

Omit $x=-7$ since
$\log_2(x+5)=\log_2(-7+5)=\log_2(-2)$
and we cannot take the log of a negative number.
So, $x=3$ or $\{3\}$.

**71.** Using synthetic division:

| | | | | |
|---|---|---|---|---|
| $-2\rfloor$ | 1 | 5 | 3 | −10 |
| | | −2 | −6 | 6 |
| | 1 | 3 | −3 | −4 |

The quotient is $x^2+3x-3x$ and the remainder is $-4$.

Then $\dfrac{x^3+5x^2+3x-10}{x+2}$

$$=x^2+3x-3-\frac{4}{x+2}$$

## Chapter 12 Review Exercises

**1.** $a_n=7n-4$

$$a_1=7(1)-4=7-4=3$$
$$a_2=7(2)-4=14-4=10$$
$$a_3=7(3)-4=21-4=17$$
$$a_4=7(4)-4=28-4=24$$

The first four terms are 3, 10, 17, 24.

**2.**

$$a_n=(-1)^n\frac{n+2}{n+1}$$
$$a_1=(-1)^1\frac{1+2}{1+1}=-\frac{3}{2}$$
$$a_2=(-1)^2\frac{2+2}{2+1}=\frac{4}{3}$$
$$a_3=(-1)^3\frac{3+2}{3+1}=-\frac{5}{4}$$
$$a_4=(-1)^4\frac{4+2}{4+1}=\frac{6}{5}$$

The first four terms are $-\frac{3}{2},\frac{4}{3},-\frac{5}{4},\frac{6}{5}$.

**3.**

$$a_n=\frac{1}{(n-1)!}$$
$$a_1=\frac{1}{(1-1)!}=\frac{1}{0!}=\frac{1}{1}=1$$
$$a_2=\frac{1}{(2-1)!}=\frac{1}{1!}=\frac{1}{1}=1$$
$$a_3=\frac{1}{(3-1)!}=\frac{1}{2!}=\frac{1}{2\cdot 1}=\frac{1}{2}$$
$$a_4=\frac{1}{(4-1)!}=\frac{1}{3!}=\frac{1}{3\cdot 2\cdot 1}=\frac{1}{6}$$

The first four terms are $1,1,\frac{1}{2},\frac{1}{6}$.

**4.**

$$a_n=\frac{(-1)^{n+1}}{2^n}$$
$$a_1=\frac{(-1)^{1+1}}{2^1}=\frac{(-1)^2}{2}=\frac{1}{2}$$
$$a_2=\frac{(-1)^{2+1}}{2^2}=\frac{(-1)^3}{4}=-\frac{1}{4}$$
$$a_3=\frac{(-1)^{3+1}}{2^3}=\frac{(-1)^4}{8}=\frac{1}{8}$$
$$a_4=\frac{(-1)^{4+1}}{2^4}=\frac{(-1)^5}{16}=-\frac{1}{16}$$

The first four terms are $\frac{1}{2},-\frac{1}{4},\frac{1}{8},-\frac{1}{16}$.

**5.** $\sum_{i=1}^{5}(2i^2-3)$

$=(2(1)^2-3)+(2(2)^2-3)+(2(3)^2-3)+(2(4)^2-3)+(2(5)^2-3)+(2(6)^2-3)$

$=(2(1)-3)+(2(4)-3)+(2(9)-3)+(2(16)-3)+(2(25)-3)$

$=(2-3)+(8-3)+(18-3)+(32-3)+(50-3)=-1+5+15+29+47=95$

**6.** $\sum_{i=0}^{4}(-1)^{i+1}i!=(-1)^{0+1}0!+(-1)^{1+1}1!+(-1)^{2+1}2!+(-1)^{3+1}3!+(-1)^{4+1}4!$

$=(-1)^1 1+(-1)^2 1+(-1)^3 2\cdot1+(-1)^4 3\cdot2\cdot1+(-1)^5 4\cdot3\cdot2\cdot1$

$=-1+1-2+6-24=-20$

**7.** $\frac{1}{3}+\frac{2}{4}+\frac{3}{5}+\ldots+\frac{15}{17}=\sum_{i=1}^{15}\frac{i}{i+2}$

**8.** $4^3+5^3+6^3+\ldots+13^3=\sum_{i=4}^{13}i^3$

**9.** $a_1=7$

$a_2=7+4=11$

$a_3=11+4=15$

$a_4=15+4=19$

$a_5=19+4=23$

$a_6=23+4=27$

The first six terms are
7, 11, 15, 19, 23, 27

**10.** $a_1=-4$

$a_2=-4-5=-9$

$a_3=-9-5=-14$

$a_4=-14-5=-19$

$a_5=-19-5=-24$

$a_6=-24-5=-29$

The first six terms are
$-4,-9,-14,-19,-24,-29$.

**11.** $a_1=\frac{3}{2}$

$a_2=\frac{3}{2}-\frac{1}{2}=\frac{2}{2}=1$

$a_3=1-\frac{1}{2}=\frac{1}{2}$

$a_4=\frac{1}{2}-\frac{1}{2}=0$

$a_5=0-\frac{1}{2}=-\frac{1}{2}$

$a_6=-\frac{1}{2}-\frac{1}{2}=-\frac{2}{2}=-1$

The first six terms are $\frac{3}{2},1,\frac{1}{2},0,-\frac{1}{2},-1$.

**12.** $a_6=5+(6-1)3=5+(5)3$

$=5+15=20$

**13.** $a_{12}=-8+(12-1)(-2)=-8+11(-2)$

$=-8+(-22)=-30$

**14.** $a_{14}=14+(14-1)(-4)=14+13(-4)$

$=14+(-52)=-38$

**15.** $d = -3-(-7) = 4$

$a_n = -7+(n-1)4 = -7+4n-4$

$= 4n-11$

$a_{20} = 4(20)-11 = 80-11 = 69$

**16.** $a_n = 200+(n-1)(-20)$

$= 200-20n+20$

$= 220-20n$

$a_{20} = 220-20(20)$

$= 220-400 = -180$

**17.** $a_n = -12+(n-1)\left(-\frac{1}{2}\right)$

$= -12-\frac{1}{2}n+\frac{1}{2}$

$= -\frac{24}{2}-\frac{1}{2}n+\frac{1}{2}$

$= -\frac{1}{2}n-\frac{23}{2}$

$a_{20} = -\frac{1}{2}(20)-\frac{23}{2}$

$= -\frac{20}{2}-\frac{23}{2} = -\frac{43}{2}$

**18.** $d = 8-15 = -7$

$a_n = 15+(n-1)(-7) = 15-7n+7$

$= 22-7n$

$a_{20} = 22-7(20) = 22-140 = -118$

**19.** First, find $d$.

$d = 12-5 = 7$

Next, find $a_{22}$.

$a_{22} = 5+(22-1)7 = 5+(21)7$

$= 5+147 = 152$

Now, find the sum.

$S_{22} = \frac{22}{2}(5+152) = 11(157) = 1727$

**20.** First, find $d$.

$d = -3-(-6) = 3$

Next, find $a_{15}$.

$a_{15} = -6+(15-1)3 = -6+(14)3$

$= -6+42 = 36$

Now, find the sum.

$S_{15} = \frac{15}{2}(-6+36) = \frac{15}{2}(30) = 225$

**21.** We are given that $a_1 = 300$.

$S_{100} = \frac{100}{2}(3+300)$

$= 50(303) = 15150$

**22.** $\sum_{i=1}^{16}(3i+2) = (3(1)+2)+(3(2)+2)+(3(3)+2)+\ldots+(3(16)+2)$

$= (3+2)+(6+2)+(9+2)+\ldots+(48+2)$

$= 5+8+11+\ldots+50$

$S_{16} = \frac{16}{2}(5+50) = 8(55) = 440$

**23.** $$\sum_{i=1}^{25}(-2i+6) = (-2(1)+6)+(-2(2)+6)+(-2(3)+6)+\ldots+(-2(25)+6)$$
$$= (-2+6)+(-4+6)+(-6+6)+\ldots+(-50+6)$$
$$= 4+2+0+\ldots+(-44)$$
$$S_{25} = \frac{25}{2}(4+(-44)) = \frac{25}{2}(-40) = -500$$

**24.** $$\sum_{i=1}^{30}(-5i) = (-5(1))+(-5(2))+(-5(3))+\ldots+(-5(30))$$
$$= -5+(-10)+(-15)+\ldots+(-150)$$
$$S_{30} = \frac{30}{2}(-5+(-150)) = 15(-155) = -2325$$

**25.** **a.** $$a_n = 20+(n-1)(0.52)$$
$$= 20+0.52n-0.52$$
$$= 0.52n+19.48$$

**b.** $$a_n = 0.52(111)+19.48$$
$$= 57.72+19.48$$
$$= 77.2$$
The percentage of white-collar works in the labor force by the year 2010 is approximately 77.2%.

**26.** $$a_{10} = 31500+(10-1)2300$$
$$= 31500+(9)2300$$
$$= 31500+20700 = 52200$$
$$S_{10} = \frac{10}{2}(31500+52200)$$
$$= 5(83700) = 418500$$
The total salary over a ten-year period is $418,500.

**27.** $$a_{35} = 25+(35-1)1 = 25+(34)1$$
$$= 25+34 = 59$$
$$S_{35} = \frac{35}{2}(25+59) = \frac{35}{2}(84) = 1470$$
There are 1470 seats in the theater.

**28.** The first term is 3.
The second term is $3 \cdot 2 = 6$.
The third term is $6 \cdot 2 = 12$.
The fourth term is $12 \cdot 2 = 24$.
The fifth term is $24 \cdot 2 = 48$.

**29.** The first term is $\frac{1}{2}$.
The second term is $\frac{1}{2} \cdot \frac{1}{2} = \frac{1}{4}$.
The third term is $\frac{1}{4} \cdot \frac{1}{2} = \frac{1}{8}$.
The fourth term is $\frac{1}{8} \cdot \frac{1}{2} = \frac{1}{16}$.
The fifth term is $\frac{1}{16} \cdot \frac{1}{2} = \frac{1}{32}$.

**30.** The first term is 16.

The second term is $16 \cdot -\frac{1}{4} = -4$.

The third term is $-4 \cdot -\frac{1}{4} = 1$.

The fourth term is $1 \cdot -\frac{1}{4} = -\frac{1}{4}$.

The fifth term is $-\frac{1}{4} \cdot -\frac{1}{4} = \frac{1}{16}$.

**31.** The first term is –5.
The second term is $-5 \cdot -1 = 5$.
The third term is $5 \cdot -1 = -5$.
The fourth term is $-5 \cdot -1 = 5$.
The fifth term is $5 \cdot -1 = -5$.

**32.** $a_7 = 2(3)^{7-1} = 2(3)^6 = 2(729) = 1458$

**33.**
$$a_6 = 16\left(\frac{1}{2}\right)^{6-1} = 16\left(\frac{1}{2}\right)^5$$
$$= 16\left(\frac{1}{32}\right) = \frac{1}{2}$$

**34.**
$$a_5 = -3(2)^{5-1} = -3(2)^4$$
$$= -3(16) = -48$$

**35.**
$$a_n = a_1 r^{n-1} = 1(2)^{n-1}$$
$$a_8 = 1(2)^{8-1} = 1(2)^7 = 1(128) = 128$$

**36.**
$$a_n = a_1 r^{n-1} = 100\left(\frac{1}{10}\right)^{n-1}$$
$$a_8 = 100\left(\frac{1}{10}\right)^{8-1} = 100\left(\frac{1}{10}\right)^7$$
$$= 100\left(\frac{1}{10000000}\right) = \frac{1}{100000}$$

**37.**
$$d = \frac{-4}{12} = -\frac{1}{3}$$
$$a_n = a_1 r^{n-1} = 12\left(-\frac{1}{3}\right)^{n-1}$$
$$a_8 = 12\left(-\frac{1}{3}\right)^{8-1} = 12\left(-\frac{1}{3}\right)^7$$
$$= 12\left(-\frac{1}{2187}\right) = -\frac{12}{2187} = -\frac{4}{729}$$

**38.**
$$r = \frac{a_2}{a_1} = \frac{-15}{5} = -3$$
$$S_{15} = \frac{5\left(1-(-3)^{15}\right)}{1-(-3)}$$
$$= \frac{5(1-(-14348907))}{4}$$
$$= \frac{5(14348908)}{4} = \frac{71744540}{4}$$
$$= 17{,}936{,}135$$

**39.**
$$r = \frac{a_2}{a_1} = \frac{4}{8} = \frac{1}{2}$$
$$S_7 = \frac{8\left(1-\left(\frac{1}{2}\right)^7\right)}{1-\frac{1}{2}} = \frac{8\left(1-\frac{1}{128}\right)}{\frac{1}{2}}$$
$$= \frac{8\left(\frac{128}{128}-\frac{1}{128}\right)}{\frac{1}{2}} = \frac{8\left(\frac{127}{128}\right)}{\frac{1}{2}}$$
$$= \frac{8}{1}\left(\frac{127}{128}\right) \div \frac{1}{2} = \frac{8}{1}\left(\frac{127}{128}\right) \cdot \frac{2}{1}$$
$$= \frac{2032}{128} = \frac{127}{8} = 15.875$$

**40.**
$$\sum_{i=1}^{6} 5^i = \frac{5\left(1-5^6\right)}{1-5} = \frac{5(1-15625)}{-4}$$
$$= \frac{5(-15624)}{-4} = 5(3906)$$
$$= 19,530$$

**41.**
$$\sum_{i=1}^{7} 3(-2)^i = \frac{-6\left(1-(-2)^7\right)}{1-(-2)}$$
$$= \frac{-6(1-(-128))}{3}$$
$$= \frac{-6(129)}{3}$$
$$= -2(129) = -258$$

**42.**
$$\sum_{i=1}^{5} 2\left(\frac{1}{4}\right)^{i-1} = \frac{2\left(1-\left(\frac{1}{4}\right)^5\right)}{1-\frac{1}{4}}$$
$$= \frac{2\left(1-\frac{1}{1024}\right)}{\frac{3}{4}} = \frac{2\left(\frac{1024}{1024}-\frac{1}{1024}\right)}{\frac{3}{4}}$$
$$= \frac{2\left(\frac{1023}{1024}\right)}{\frac{3}{4}} = \frac{\frac{2046}{1024}}{\frac{3}{4}} = \frac{2046}{1024} \div \frac{3}{4}$$
$$= \frac{2046}{1024} \cdot \frac{4}{3} = \frac{682}{256} = \frac{341}{128}$$

**43.**
$$r = \frac{a_2}{a_1} = \frac{3}{9} = \frac{1}{3}$$
$$S = \frac{9}{1-\frac{1}{3}} = \frac{9}{\frac{2}{3}} = 9 \div \frac{2}{3} = 9 \cdot \frac{3}{2} = \frac{27}{2}$$

**44.**
$$r = \frac{a_2}{a_1} = \frac{-1}{2} = -\frac{1}{2}$$
$$S = \frac{2}{1-\left(-\frac{1}{2}\right)} = \frac{2}{\frac{3}{2}} = 2 \div \frac{3}{2} = 2 \cdot \frac{2}{3} = \frac{4}{3}$$

**45.**
$$r = \frac{a_2}{a_1} = \frac{4}{-6} = -\frac{2}{3}$$
$$S = \frac{-6}{1-\left(-\frac{2}{3}\right)} = \frac{-6}{\frac{5}{3}} = -6 \div \frac{5}{3}$$
$$= -6 \cdot \frac{3}{5} = -\frac{18}{5}$$

**46.**
$$\sum_{i=1}^{\infty} 5(0.8)^i = \frac{4}{1-0.8} = \frac{4}{0.2} = 20$$

**47.**
$$0.\overline{6} = \frac{a_1}{1-r} = \frac{\frac{6}{10}}{1-\frac{1}{10}} = \frac{\frac{6}{10}}{\frac{9}{10}} = \frac{6}{10} \div \frac{9}{10}$$
$$= \frac{6}{10} \cdot \frac{10}{9} = \frac{2}{3}$$

**48.**
$$0.\overline{47} = \frac{a_1}{1-r} = \frac{\frac{47}{100}}{1-\frac{1}{100}} = \frac{\frac{47}{100}}{\frac{99}{100}}$$
$$= \frac{47}{100} \div \frac{99}{100} = \frac{47}{100} \cdot \frac{100}{99} = \frac{47}{99}$$

**49.** **a.**

$$r_{1998-1999} = \frac{20.72}{19.96} \approx 1.038$$

$$r_{1999-2000} = \frac{21.51}{20.72} \approx 1.038$$

$$r_{2000-2001} = \frac{22.33}{21.51} \approx 1.038$$

The population is increasing geometrically since dividing the population for each year by the population for the preceding years gives approximately 1.038 for each division.

**b.** $a_n = a_1 r^{n-1} = 19.96(1.038)^{n-1}$

**c.** $a_{10} = 19.96(1.038)^{10-1} \approx 27.92$.

In 2008, the population will be approximately 27.92 million.

**50.** $r = 1.06$

$$a_n = a_1 r^{n-1} = 32000(1.06)^{n-1}$$

$$a_6 = 32000(1.06)^{6-1} = 32000(1.06)^5$$

$$= 32000(1.338226) = 42823.22$$

The salary in the sixth year is approximately $42,823.22.

$$S_6 = \frac{a_1(1-r^n)}{1-r} = \frac{32000\left(1-(1.06)^6\right)}{1-1.06}$$

$$= \frac{32000(1-1.418519)}{-0.06}$$

$$= \frac{32000(-0.418519)}{-0.06} = 223210.13$$

The total salary over the six years is approximately $223,210.13.

**51.**

$$A = P\frac{\left(1+\frac{r}{n}\right)^{nt} - 1}{\frac{r}{n}}$$

$$= 200\frac{\left(1+\frac{0.10}{12}\right)^{12(18)} - 1}{\frac{0.10}{12}}$$

$$= 200\frac{\left(1+\frac{1}{120}\right)^{216} - 1}{\frac{1}{120}}$$

$$= 200\frac{\left(\frac{121}{120}\right)^{216} - 1}{\frac{1}{120}} = 200\frac{6.004693-1}{\frac{1}{120}}$$

$$= 200\frac{5.004693}{\frac{1}{120}} = 120112.63$$

After 18 years, the value of the account will be approximately $120,112.63.

**52.** $r = 70\% = 0.7$

$$a_1 = 4(.7) = 2.8$$

$$S = \frac{2.8}{1-0.7} = \frac{2.8}{0.3} = 9.\overline{3}$$

The total spending in the town will be approximately $9.3 million each year.

**53.**

$$\binom{11}{8} = \frac{11!}{8!(11-8)!}$$

$$= \frac{11\cdot 10\cdot 9\cdot \cancel{8!}}{\cancel{8!}\cdot 3\cdot 2\cdot 1} = 165$$

**54.** $\binom{90}{2} = \dfrac{90!}{2!(90-2)!} = \dfrac{90 \cdot 89 \cdot \cancel{88!}}{2 \cdot 1 \cdot \cancel{88!}} = 4005$

**55.** Applying the Binomial Theorem to $(2x+1)^3$, we have $a = 2x,\ b = 1,$ and $n = 3.$

$$\begin{aligned}(2x+1)^3 &= \binom{3}{0}(2x)^3 + \binom{3}{1}(2x)^2 \cdot 1 + \binom{3}{2}(2x) \cdot 1^2 + \binom{3}{3}1^3 \\ &= \frac{3!}{0!(3-0)!}8x^3 + \frac{3!}{1!(3-1)!}4x^2 \cdot 1 + \frac{3!}{2!(3-2)!}2x \cdot 1 + \frac{3!}{3!(3-3)!}1 \\ &= \frac{\cancel{3!}}{1 \cdot \cancel{3!}}8x^3 + \frac{3 \cdot \cancel{2!}}{1 \cdot \cancel{2!}}4x^2 + \frac{3 \cdot \cancel{2!}}{\cancel{2!}1!}2x + \frac{\cancel{3!}}{\cancel{3!}0!} \\ &= 8x^3 + 3(4x^2) + 3(2x) + 1 = 8x^3 + 12x^2 + 6x + 1\end{aligned}$$

**56.** Applying the Binomial Theorem to $(x^2-1)^4$, we have $a = x^2,\ b = -1,$ and $n = 4.$

$$\begin{aligned}(x^2-1)^4 &= \binom{4}{0}(x^2)^4 + \binom{4}{1}(x^2)^3(-1) + \binom{4}{2}(x^2)^2(-1)^2 + \binom{4}{3}x^2(-1)^3 + \binom{4}{4}(-1)^4 \\ &= \frac{4!}{0!(4-0)!}x^8 - \frac{4!}{1!(4-1)!}x^6 + \frac{4!}{2!(4-2)!}x^4 - \frac{4!}{3!(4-3)!}x^2 + \frac{4!}{4!(4-4)!}1 \\ &= \frac{\cancel{4!}}{0!\cancel{4!}}x^8 - \frac{4!}{1!3!}x^6 + \frac{4!}{2!2!}x^4 - \frac{4!}{3!1!}x^2 + \frac{\cancel{4!}}{\cancel{4!}0!} \\ &= 1(x^8) - \frac{4 \cdot \cancel{3!}}{1 \cdot \cancel{3!}}x^6 + \frac{4 \cdot 3 \cdot \cancel{2!}}{2 \cdot 1 \cdot \cancel{2!}}x^4 - \frac{4 \cdot \cancel{3!}}{\cancel{3!} \cdot 1}x^2 + 1 \\ &= x^8 - 4x^6 + 6x^4 - 4x^2 + 1\end{aligned}$$

**57.** Applying the Binomial Theorem to $(x+2y)^5$, we have $a = x,\ b = 2y,$ and $n = 5.$

$$\begin{aligned}(x+2y)^5 &= \binom{5}{0}x^5 + \binom{5}{1}x^4(2y) + \binom{5}{2}x^3(2y)^2 + \binom{5}{3}x^2(2y)^3 + \binom{5}{4}x(2y)^4 + \binom{5}{5}(2y)^5 \\ &= \frac{\cancel{5!}}{0!\cancel{5!}}x^5 + \frac{5!}{1!(5-1)!}2x^4y + \frac{5!}{2!(5-2)!}4x^3y^2 + \frac{5!}{3!(5-3)!}8x^2y^3 \\ &\quad + \frac{5!}{4!(5-4)!}16xy^4 + \frac{5!}{5!(5-5)!}32y^5 \\ &= 1x^5 + \frac{5 \cdot \cancel{4!}}{1 \cdot \cancel{4!}}2x^4y + \frac{5 \cdot 4 \cdot \cancel{3!}}{2 \cdot 1 \cdot \cancel{3!}}4x^3y^2 + \frac{5 \cdot 4 \cdot \cancel{3!}}{\cancel{3!}2 \cdot 1}8x^2y^3 + \frac{5 \cdot \cancel{4!}}{\cancel{4!} \cdot 1}16xy^4 + \frac{\cancel{5!}}{\cancel{5!}0!}32y^5 \\ &= x^5 + 5(2x^4y) + 10(4x^3y^2) + 10(8x^2y^3) + 5(16xy^4) + 1(32y^5) \\ &= x^5 + 10x^4y + 40x^3y^2 + 80x^2y^3 + 80xy^4 + 32y^5\end{aligned}$$

**58.** Applying the Binomial Theorem to $(x-2)^6$, we have $a = x,\ b = -2,$ and $n = 6$.

$$(x-2)^6$$
$$= \binom{6}{0}x^6 + \binom{6}{1}x^5(-2) + \binom{6}{2}x^4(-2)^2 + \binom{6}{3}x^3(-2)^3 + \binom{6}{4}x^2(-2)^4 + \binom{6}{5}x(-2)^5 + \binom{6}{6}(-2)^6$$
$$= \frac{6!}{0!(6-0)!}x^6 + \frac{6!}{1!(6-1)!}x^5(-2) + \frac{6!}{2!(6-2)!}x^4(-2)^2 + \frac{6!}{3!(6-3)!}x^3(-2)^3$$
$$+ \frac{6!}{4!(6-4)!}x^2(-2)^4 + \frac{6!}{5!(6-5)!}x(-2)^5 + \frac{6!}{6!(6-6)!}(-2)^6$$
$$= \frac{\cancel{6!}}{1\cancel{6!}}x^6 - \frac{6\cdot\cancel{5!}}{1\cdot\cancel{5!}}2x^5 + \frac{6\cdot5\cdot\cancel{4!}}{2\cdot1\cdot\cancel{4!}}4x^4 - \frac{\cancel{6}\cdot5\cdot4\cdot\cancel{3!}}{\cancel{3\cdot2}\cdot1\cdot\cancel{3!}}8x^3$$
$$+ \frac{6\cdot5\cdot\cancel{4!}}{\cancel{4!}2\cdot1}16x^2 - \frac{6\cdot\cancel{5!}}{\cancel{5!}1}32x + \frac{\cancel{6!}}{\cancel{6!}\cdot1}64$$
$$= x^6 - 6(2x^5) + 15(4x^4) - 20(8x^3) + 15(16x^2) - 6(32x) + 1\cdot64$$
$$= x^6 - 12x^5 + 60x^4 - 160x^3 + 240x^2 - 192x + 64$$

**59.** $(x^2+3)^8$

First Term $\binom{n}{r-1}a^{n-r+1}b^{r-1} = \binom{8}{1-1}(x^2)^{8-1+1}3^{1-1} = \binom{8}{0}(x^2)^8 3^0$

$$= \frac{8!}{0!(8-0)!}x^{16}\cdot1 = \frac{\cancel{8!}}{0!\cancel{8!}}x^{16} = x^{16}$$

Second Term $\binom{n}{r-1}a^{n-r+1}b^{r-1} = \binom{8}{2-1}(x^2)^{8-2+1}3^{2-1} = \binom{8}{1}(x^2)^7 3^1 = \frac{8!}{1!(8-1)!}3x^{14}$

$$= \frac{8\cdot\cancel{7!}}{1\cdot\cancel{7!}}3x^{14} = 8\cdot3x^{14} = 24x^{14}$$

Third Term $\binom{n}{r-1}a^{n-r+1}b^{r-1} = \binom{8}{3-1}(x^2)^{8-3+1}3^{3-1} = \binom{8}{2}(x^2)^6 3^2 = \frac{8!}{2!(8-2)!}9x^{12}$

$$= \frac{8\cdot7\cdot\cancel{6!}}{2\cdot1\cdot\cancel{6!}}9x^{12} = 28\cdot9x^{12} = 252x^{12}$$

**60.** $(x-3)^9$

First Term $\binom{n}{r-1}a^{n-r+1}b^{r-1} = \binom{9}{1-1}x^{9-1+1}(-3)^{1-1} = \binom{9}{0}x^9(-3)^0$

$$= \frac{9!}{0!(9-0)!}x^9 \cdot 1 = \frac{\cancel{9!}}{0!\cancel{9!}}x^9 = x^9$$

Second Term $\binom{n}{r-1}a^{n-r+1}b^{r-1} = \binom{9}{2-1}x^{9-2+1}(-3)^{2-1} = \binom{9}{1}x^8(-3)^1 = -\frac{9!}{1!(9-1)!}3x^8$

$$= -\frac{9\cdot\cancel{8!}}{1\cdot\cancel{8!}}3x^8 = -9\cdot 3x^8 = -27x^8$$

Third Term $\binom{n}{r-1}a^{n-r+1}b^{r-1} = \binom{9}{3-1}x^{9-3+1}(-3)^{3-1} = \binom{9}{2}x^7(-3)^2 = \frac{9!}{2!(9-2)!}9x^7$

$$= \frac{9\cdot 8\cdot\cancel{7!}}{2\cdot 1\cdot\cancel{7!}}9x^7 = 36\cdot 9x^7 = 324x^7$$

**61.** $(x+2)^5$

Fourth Term $\binom{n}{r-1}a^{n-r+1}b^{r-1} = \binom{5}{4-1}x^{5-4+1}(2)^{4-1} = \binom{5}{3}x^2(2)^3 = \frac{5!}{3!(5-3)!}8x^2$

$$= \frac{5!}{3!2!}8x^2 = \frac{5\cdot 4\cdot\cancel{3!}}{\cancel{3!}\cdot 2\cdot 1}8x^2 = (10)8x^2 = 80x^2$$

**62.** $(2x-3)^6$

Fifth Term $\binom{n}{r-1}a^{n-r+1}b^{r-1} = \binom{6}{5-1}(2x)^{6-5+1}(-3)^{5-1} = \binom{6}{4}(2x)^2(-3)^4$

$$= \frac{6!}{4!(6-4)!}4x^2(81) = \frac{6!}{4!2!}324x^2 = \frac{6\cdot 5\cdot\cancel{4!}}{\cancel{4!}\cdot 2\cdot 1}324x^2$$

$$= (15)324x^2 = 4860x^2$$

**63.** $5+10+15+\cdots+5n=\dfrac{5n(n+1)}{2}$

*Show that $S_1$ is true:* $5=\dfrac{5\cdot 1(1+1)}{2}$

$$5=\frac{5\cdot 1\cdot 2}{2}$$

$$5=5,\ \text{True}$$

*Show that if $S_k$ is true, then $S_{k+1}$ is true:*

Assume $S_k: 5+10+15+\cdots+5k=\dfrac{5k(k+1)}{2}$ is true. Then,

$$5+10+15+\cdots+5k+5(k+1)=\frac{5k(k+1)}{2}+5(k+1)$$

$$5+10+15+\cdots+5k+5(k+1)=\frac{5k(k+1)}{2}+\frac{10(k+1)}{2}$$

$$5+10+15+\cdots+5k+5(k+1)=\frac{5k^2+5k+10k+10}{2}$$

$$5+10+15+\cdots+5k+5(k+1)=\frac{5k^2+15k+10}{2}$$

$$5+10+15+\cdots+5k+5(k+1)=\frac{5(k^2+3k+2)}{2}$$

$$5+10+15+\cdots+5k+5(k+1)=\frac{5(k+1)(k+2)}{2}$$

$$5+10+15+\cdots+5k+5(k+1)=\frac{5(k+1)[(k+1)+1]}{2}$$

The final statement is $S_{k+1}$. Thus, by mathematical induction, we have proven that

$5+10+15+\cdots+5n=\dfrac{5n(n+1)}{2}$.

**64.** $1+4+4^2+\cdots+4^{n-1}=\dfrac{4^n-1}{3}$

*Show that $S_1$ is true:* $1=\dfrac{4^1-1}{3}$

$$1=\frac{3}{3}$$

$$1=1,\ \text{True}$$

*Show that if* $S_k$ *is true, then* $S_{k+1}$ *is true*:

Assume $S_k: 1+4+4^2+\cdots+4^{k-1}=\dfrac{4^k-1}{3}$ is true. Then,

$$1+4+4^2+\cdots+4^{k-1}+4^{(k+1)-1}=\frac{4^k-1}{3}+4^{(k+1)-1}$$

$$1+4+4^2+\cdots+4^{k-1}+4^{(k+1)-1}=\frac{4^k-1}{3}+4^k$$

$$1+4+4^2+\cdots+4^{k-1}+4^{(k+1)-1}=\frac{4^k-1}{3}+\frac{3\cdot 4^k}{3}$$

$$1+4+4^2+\cdots+4^{k-1}+4^{(k+1)-1}=\frac{4\cdot 4^k-1}{3}$$

$$1+4+4^2+\cdots+4^{k-1}+4^{(k+1)-1}=\frac{4^{k+1}-1}{3}$$

The final statement is $S_{k+1}$. Thus, by mathematical induction, we have proven that

$1+4+4^2+\cdots+4^{n-1}=\dfrac{4^n-1}{3}$.

**65.** $2+6+10+\cdots+(4n-2)=2n^2$

*Show that* $S_1$ *is true*: $2=2\cdot 1^2$

$2=2$, True

*Show that if* $S_k$ *is true, then* $S_{k+1}$ *is true*:

Assume $S_k: 2+6+10+\cdots+(4k-2)=2k^2$ is true. Then,

$$2+6+10+\cdots+(4k-2)+[4(k+1)-2]=2k^2+[4(k+1)-2]$$

$$2+6+10+\cdots+(4k-2)+[4(k+1)-2]=2k^2+4k+4-2$$

$$2+6+10+\cdots+(4k-2)+[4(k+1)-2]=2k^2+4k+2$$

$$2+6+10+\cdots+(4k-2)+[4(k+1)-2]=2(k^2+2k+1)$$

$$2+6+10+\cdots+(4k-2)+[4(k+1)-2]=2(k+1)^2$$

The final statement is $S_{k+1}$. Thus, by mathematical induction, we have proven that

$2+6+10+\cdots+(4n-2)=2n^2$.

**66.** $1\cdot 3+2\cdot 4+3\cdot 5+\cdots+n(n+2)=\frac{n(n+1)(2n+7)}{6}$

*Show that $S_1$ is true:* $1\cdot 3=\frac{1(1+1)(2\cdot 1+7)}{6}$

$$3=\frac{1(2)(9)}{6}$$

$$3=\frac{18}{6}$$

$$3=3, \text{ True}$$

*Show that if $S_k$ is true, then $S_{k+1}$ is true:*

Assume $S_k: 1\cdot 3+2\cdot 4+3\cdot 5+\cdots+k(k+2)=\frac{k(k+1)(2k+7)}{6}$ is true. Then,

$$1\cdot 3+2\cdot 4+3\cdot 5+\cdots+k(k+2)+(k+1)[(k+2)+1]=\frac{k(k+1)(2k+7)}{6}+(k+1)[(k+2)+1]$$

$$1\cdot 3+2\cdot 4+3\cdot 5+\cdots+k(k+2)+(k+1)[(k+2)+1]=\frac{k(k+1)(2k+7)}{6}+(k+1)(k+3)$$

$$1\cdot 3+2\cdot 4+3\cdot 5+\cdots+k(k+2)+(k+1)[(k+2)+1]=\frac{k(k+1)(2k+7)+6(k+1)(k+3)}{6}$$

$$1\cdot 3+2\cdot 4+3\cdot 5+\cdots+k(k+2)+(k+1)[(k+2)+1]=\frac{(k+1)[k(2k+7)+6(k+3)]}{6}$$

$$1\cdot 3+2\cdot 4+3\cdot 5+\cdots+k(k+2)+(k+1)[(k+2)+1]=\frac{(k+1)[2k^2+7k+6k+18]}{6}$$

$$1\cdot 3+2\cdot 4+3\cdot 5+\cdots+k(k+2)+(k+1)[(k+2)+1]=\frac{(k+1)(2k^2+13k+18)}{6}$$

$$1\cdot 3+2\cdot 4+3\cdot 5+\cdots+k(k+2)+(k+1)[(k+2)+1]=\frac{(k+1)(k+2)(2k+9)}{6}$$

$$1\cdot 3+2\cdot 4+3\cdot 5+\cdots+k(k+2)+(k+1)[(k+2)+1]=\frac{(k+1)[(k+1)+1][2(k+1)+7]}{6}$$

The final statement is $S_{k+1}$. Thus, by mathematical induction, we have proven that

$$1\cdot 3+2\cdot 4+3\cdot 5+\cdots+n(n+2)=\frac{n(n+1)(2n+7)}{6}.$$

**67.** $_8P_3 = \frac{8!}{(8-3)!} = \frac{8\cdot 7\cdot 6\cdot \cancel{5!}}{\cancel{5!}} = 336$

**68.** $_9P_5 = \frac{9!}{(9-5)!} = \frac{9\cdot 8\cdot 7\cdot 6\cdot 5\cdot \cancel{4!}}{\cancel{4!}} 15{,}120$

**69.** $_8C_3 = \frac{8!}{(8-3)!3!} = \frac{8\cdot 7\cdot 6\cdot \cancel{5!}}{\cancel{5!}3\cdot 2\cdot 1} = 56$

**70.** $_{13}C_{11} = \frac{13!}{(13-11)!11!} = \frac{13\cdot 12\cdot \cancel{11!}}{2\cdot 1\cdot \cancel{11!}} = 78$

**71.** $4\cdot 5 = 20$
You have 20 choices with this brand of pens.

**72.** $3\cdot 3\cdot 3\cdot 3\cdot 3 = 3^5 = 243$
There are 243 possibilities.

**73.** $_{15}P_4 = \frac{15!}{(15-4)!} = \frac{15!}{11!} = 32760$
There are 32,760 ways to fill the offices.

**74.** $_{20}C_4 = \frac{20!}{(20-4)!4!} = \frac{20!}{16!4!} = 4845$
There are 4845 ways to select the 4 actors.

**75.** $_{20}C_3 = \frac{20!}{(20-3)!3!} = \frac{20!}{17!3!} = 1140$
You can take 1140 sets of 3 CDs.

**76.** $_{20}P_4 = \frac{20!}{(20-4)!} = \frac{20!}{16!} = 116280$
The director has 116,280 ways to select actors for the roles.

**77.** $5! = 120$
There are 120 ways to line up the five airplanes.

**78.** $P(liberal) = \frac{17}{100}$

**79.** $P(not\ conservative)$
$= 1 - P(conservative)$
$= 1 - \frac{33}{100} = \frac{67}{100}$

**80.** $P(\text{moderate or conservative})$
$= P(\text{moderate}) + P(\text{conservative})$
$= \frac{50}{100} + \frac{33}{100} = \frac{83}{100}$

**81.** $P(\text{conservative or attended college})$
$= P(\text{conservative}) + P(\text{attended college})$
$- P(\text{conservative and attended college})$
$= \frac{33}{100} + \frac{45}{100} - \frac{20}{100} = \frac{58}{100} = \frac{29}{50}$

**82.** $P(\text{high school only}) = \frac{13}{33}$

**83.** $P(\text{liberal}) = \frac{10}{45} = \frac{2}{9}$

**84.** $P(\text{less than 5}) = P(\text{rolling a 1, 2, 3, 4})$
$= \frac{4}{6} = \frac{2}{3}$

**85.** $P(\text{less than 3 or greater than 4})$
$= P(\text{less than 3}) + P(\text{greater than 4})$
$= P(\text{rolling a 1, 2}) + P(\text{rolling a 5, 6})$
$= \frac{2}{6} + \frac{2}{6} = \frac{4}{6} = \frac{2}{3}$

**86.** $P(ace\ or\ king) = P(ace) + P(king)$
$= \frac{4}{52} + \frac{4}{52} = \frac{8}{52} = \frac{2}{13}$

**87.** $P(\text{queen or red})$

$= P(\text{queen}) + P(\text{red}) - P(\text{red queen})$

$= \frac{4}{52} + \frac{26}{52} - \frac{2}{52} = \frac{28}{52} = \frac{7}{13}$

**88.** $P(\text{not yellow}) = \frac{5}{6}$

**89.** $P(\text{red or greater than } 3)$

$= P(\text{red}) + P(\text{greater than } 3)$

$\quad - P(\text{red and greater than } 3)$

$= \frac{3}{6} + \frac{3}{6} - \frac{1}{6} = \frac{5}{6}$

**90.** $P(\text{green and less than } 4)$

$= P(\text{green}) \cdot P(\text{less than } 4)$

$= \frac{2}{6} \cdot \frac{3}{6} = \frac{1}{6}$

**91.** **a.** There are a total of ${}_{20}C_5 = 15,504$ combinations of 5 numbers from 1 to 20 that can be chosen.

$P(\text{winning}) = \frac{1}{15504} \approx 0.0000645$

**b.** $P(\text{winning}) = 100\left(\frac{1}{15504}\right)$

$= \frac{25}{3876} \approx 0.00645$

**92.** $P(\text{five boys}) = \left(\frac{1}{2}\right)^5 = \frac{1}{32}$

**93.** **a.** $P(\text{flood 2 yrs in a row})$

$= (0.2)(0.2)$

$= 0.04$

The probability of a flood two years in a row is 0.04.

**b.** $P(\text{flood 3 yrs in a row})$

$= (0.2)(0.2)(0.2)$

$= 0.008$

The probability of a flood for three consecutive years is 0.008.

**c.** $P(\text{no flooding for 4 yrs}) = (0.8)^4$

$= 0.4096$

The probability of no flood for four consecutive years in 0.4096.

## Chapter 12 Test

**1.** $a_n = \frac{(-1)^{n+1}}{n^2}$

$a_1 = \frac{(-1)^{1+1}}{1^2} = \frac{(-1)^2}{1} = \frac{1}{1} = 1$

$a_2 = \frac{(-1)^{2+1}}{2^2} = \frac{(-1)^3}{4} = \frac{-1}{4} = -\frac{1}{4}$

$a_3 = \frac{(-1)^{3+1}}{3^2} = \frac{(-1)^4}{9} = \frac{1}{9}$

$a_4 = \frac{(-1)^{4+1}}{4^2} = \frac{(-1)^5}{16} = \frac{-1}{16} = -\frac{1}{16}$

$a_5 = \frac{(-1)^{5+1}}{5^2} = \frac{(-1)^6}{25} = \frac{1}{25}$

The first five terms are

$1, -\frac{1}{4}, \frac{1}{9}, -\frac{1}{16}, \frac{1}{25}$.

**2.** $\sum_{i=1}^{5}(i^2 + 10)$

$= (1^2 + 10) + (2^2 + 10) + (3^2 + 10)$

$\quad + (4^2 + 10) + (5^2 + 10)$

$= (1 + 10) + (4 + 10) + (9 + 10)$

$\quad + (16 + 10) + (25 + 10)$

$= 11 + 14 + 19 + 26 + 35 = 105$

**3.** $\frac{2}{3}+\frac{3}{4}+\frac{4}{5}+\ldots+\frac{21}{22}=\sum_{i=2}^{21}\frac{i}{i+1}$

**4.**
$$d=9-4=5$$
$$a_n=4+(n-1)5=4+5n-5$$
$$=5n-1$$
$$a_{12}=5(12)-1=60-1=59$$

**5.**
$$d=\frac{a_2}{a_1}=\frac{4}{16}=\frac{1}{4}$$
$$a_n=a_1r^{n-1}=16\left(\frac{1}{4}\right)^{n-1}$$
$$a_{12}=16\left(\frac{1}{4}\right)^{12-1}=16\left(\frac{1}{4}\right)^{11}$$
$$=16\left(\frac{1}{4194304}\right)=\frac{16}{4194304}$$
$$=\frac{1}{262144}$$

**6.**
$$d=-14-(-7)=-7$$
$$a_{10}=-7+(10-1)(-7)$$
$$=-7+(9)(-7)$$
$$=-7+(-63)=-70$$
$$S_{10}=\frac{10}{2}(-7+(-70))$$
$$=5(-77)=-385$$

**7.**
$$\sum_{i=1}^{20}(3i-4)$$
$$=(3(1)-4)+(3(2)-4)+(3(3)-4)+\ldots+(3(20)-4)$$
$$=(3-4)+(6-4)+(9-4)+\ldots+(60-4)$$
$$=-1+2+5+\ldots+56$$
$$S_{20}=\frac{20}{2}(-1+56)=10(55)=550$$

**8.**
$$r=\frac{a_2}{a_1}=\frac{-14}{7}=-2$$
$$S_{10}=\frac{7\left(1-(-2)^{10}\right)}{1-(-2)}=\frac{7(1-1024)}{3}$$
$$=\frac{7(-1023)}{3}=-2387$$

**9.**
$$\sum_{i=1}^{15}(-2)^i=\frac{-2\left(1-(-2)^{15}\right)}{1-(-2)}$$
$$=\frac{-2(1-(-32768))}{3}$$
$$=\frac{-2(32769)}{3}=-21{,}846$$

**10.**
$$r=\frac{1}{2}$$
$$S=\frac{4}{1-\frac{1}{2}}=\frac{4}{\frac{1}{2}}=4\div\frac{1}{2}=4\cdot\frac{2}{1}=8$$

**11.**
$$0.\overline{73}=\frac{a_1}{1-r}=\frac{\frac{73}{100}}{1-\frac{1}{100}}=\frac{\frac{73}{100}}{\frac{99}{100}}$$
$$=\frac{73}{100}\div\frac{99}{100}=\frac{73}{100}\cdot\frac{100}{99}=\frac{73}{99}$$

**12.** $r=1.04$
$$S_8=\frac{a_1\left(1-r^n\right)}{1-r}=\frac{30000\left(1-(1.04)^8\right)}{1-1.04}$$
$$=\frac{30000(1-1.368569)}{-0.04}$$
$$=\frac{30000(-0.368569)}{-0.04}$$
$$=276{,}426.75$$

The total salary over the eight years is approximately \$276,426.75

**13.** $\binom{9}{2}=\frac{9!}{2!(9-2)!}=\frac{9!}{2!7!}=\frac{9\cdot 8\cdot \cancel{7!}}{2\cdot 1\cdot \cancel{7!}}=36$

**14.** Applying the Binomial Theorem to $\left(x^2-1\right)^5$, we have $a=x^2$, $b=-1$, and $n=5$.

$$\begin{aligned}
&\left(x^2-1\right)^5\\
&=\binom{5}{0}\left(x^2\right)^5+\binom{5}{1}\left(x^2\right)^4(-1)+\binom{5}{2}\left(x^2\right)^3(-1)^2+\binom{5}{3}\left(x^2\right)^2(-1)^3+\binom{5}{4}\left(x^2\right)(-1)^4+\binom{5}{5}(-1)^5\\
&=\frac{\cancel{5!}}{0!\cancel{5!}}x^{10}-\frac{5!}{1!(5-1)!}x^8+\frac{5!}{2!(5-2)!}x^6-\frac{5!}{3!(5-3)!}x^4+\frac{5!}{4!(5-4)!}x^2-\frac{5!}{5!(5-5)!}\\
&=1x^{10}-\frac{5\cdot \cancel{4!}}{1\cdot \cancel{4!}}x^8+\frac{5\cdot 4\cdot \cancel{3!}}{2\cdot 1\cdot \cancel{3!}}x^6-\frac{5\cdot 4\cdot \cancel{3!}}{\cancel{3!}2\cdot 1}x^4+\frac{5\cdot \cancel{4!}}{\cancel{4!}\cdot 1}x^2-\frac{\cancel{5!}}{\cancel{5!}0!}\\
&=x^{10}-5x^8+10x^6-10x^4+5x^2-1
\end{aligned}$$

**15.** $\left(x+y^2\right)^8$

First Term $\binom{n}{r-1}a^{n-r+1}b^{r-1}=\binom{8}{1-1}x^{8-1+1}\left(y^2\right)^{1-1}=\binom{8}{0}x^8\left(y^2\right)^0$

$$=\frac{8!}{0!(8-0)!}x^8\cdot 1=\frac{\cancel{8!}}{0!\cancel{8!}}x^8=x^8$$

Second Term $\binom{n}{r-1}a^{n-r+1}b^{r-1}=\binom{8}{2-1}x^{8-2+1}\left(y^2\right)^{2-1}=\binom{8}{1}x^7\left(y^2\right)^1=\frac{8!}{1!(8-1)!}x^7y^2$

$$=\frac{8\cdot \cancel{7!}}{1\cdot \cancel{7!}}x^7y^2=8x^7y^2$$

Third Term $\binom{n}{r-1}a^{n-r+1}b^{r-1}=\binom{8}{3-1}x^{8-3+1}\left(y^2\right)^{3-1}=\binom{8}{2}x^6\left(y^2\right)^2=\frac{8!}{2!(8-2)!}x^6y^4$

$$=\frac{8\cdot 7\cdot \cancel{6!}}{2\cdot 1\cdot \cancel{6!}}x^6y^4=28x^6y^4$$

**16.** $1+4+7+\cdots+(3n-2)=\dfrac{n(3n-1)}{2}$

*Show that* $S_1$ *is true*: $1=\dfrac{1(3\cdot 1-1)}{2}$

$$1=\frac{1(2)}{2}$$

$$1=1,\ \text{True}$$

*Show that if* $S_k$ *is true, then* $S_{k+1}$ *is true*:

Assume $S_k: 1+4+7+\cdots+(3k-2)=\dfrac{k(3k-1)}{2}$ is true. Then,

$$1+4+7+\cdots+(3k-2)+\left[3(k+1)-2\right]=\frac{k(3k-1)}{2}+\left[3(k+1)-2\right]$$

$$1+4+7+\cdots+(3k-2)+\left[3(k+1)-2\right]=\frac{k(3k-1)}{2}+2k+3-2$$

$$1+4+7+\cdots+(3k-2)+\left[3(k+1)-2\right]=\frac{k(3k-1)}{2}+2k+1$$

$$1+4+7+\cdots+(3k-2)+\left[3(k+1)-2\right]=\frac{k(3k-1)+2(2k+1)}{2}$$

$$1+4+7+\cdots+(3k-2)+\left[3(k+1)-2\right]=\frac{3k^2-k+4k+2}{2}$$

$$1+4+7+\cdots+(3k-2)+\left[3(k+1)-2\right]=\frac{3k^2+3k+2}{2}$$

$$1+4+7+\cdots+(3k-2)+\left[3(k+1)-2\right]=\frac{(k+1)(3k+2)}{2}$$

$$1+4+7+\cdots+(3k-2)+\left[3(k+1)-2\right]=\frac{(k+1)\left[3(k+1)-1\right]}{2}$$

The final statement is $S_{k+1}$. Thus, by mathematical induction, we have proven that $1+4+7+\cdots+(3n-2)=\dfrac{n(3n-1)}{2}$.

**17.** ${}_{11}P_3=\dfrac{11!}{(11-3)!}=\dfrac{11!}{8!}=990$

There are 990 ways to fill the three positions.

**18.** ${}_{10}C_4=\dfrac{10!}{(10-4)!4!}=\dfrac{10!}{6!4!}=210$

You can take 210 different sets of four books.

**19.** $10 \cdot 10 \cdot 10 \cdot 10 = 10000$
10,000 telephone numbers can be formed.

**20.** $P(\text{not brown eyes}) = 1 - P(\text{brown eyes})$

$$= 1 - \frac{40}{50}$$

$$= \frac{60}{100} = \frac{3}{5}$$

**21.** $P(\text{brown eyes or blue eyes})$

$$= P(\text{brown eyes}) + P(\text{blue eyes})$$

$$= \frac{40}{100} + \frac{38}{100} = \frac{78}{100} = \frac{39}{50}$$

**22.** $P(\text{female or green eyes})$

$$= P(\text{female}) + P(\text{green eyes}) - P(\text{female with green eyes})$$

$$= \frac{50}{100} + \frac{22}{100} - \frac{12}{100} = \frac{60}{100} = \frac{3}{5}$$

**23.** $P(\text{male, given blue eyes}) = \frac{18}{38} = \frac{9}{19}$

**24.** The number of ways to choose the six different numbers is

$${}_{15}C_6 = \frac{15!}{(15-6)!6!} = \frac{15!}{9!6!} = 5005.$$

$$P(\text{winning}) = 50 \cdot \frac{1}{5005} = \frac{10}{1001}$$

**25.** $P(\text{black or picture})$

$$= P(\text{black}) + P(\text{picture}) - P(\text{black picture card})$$

$$= \frac{26}{52} + \frac{12}{52} - \frac{6}{52} = \frac{32}{52} = \frac{8}{13}$$

**26.** $P(\text{freshman or female})$

$$= P(\text{fresh.}) + P(\text{female}) - P(\text{female fresh.})$$

$$= \frac{25}{50} + \frac{20}{50} - \frac{15}{50} = \frac{30}{50} = \frac{3}{5}$$

**27.** $P(\text{all questions correct}) = \left(\frac{1}{4}\right)^4 = \frac{1}{256}$

**28.** $P(\text{red and blue}) = \frac{2}{8} \cdot \frac{2}{8} = \frac{4}{64} = \frac{1}{16}$

## Cumulative Review Exercises (Chapters 1-12)

**1.**

$$\sqrt{2x+5} - \sqrt{x+3} = 2$$

$$\sqrt{x+3} = 2 + \sqrt{2x+5}$$

$$x+3 = \left(2 + \sqrt{2x+5}\right)^2$$

$$x+3 = 4 + 4\sqrt{2x+5} + 2x + 5$$

$$x+3 = 4\sqrt{2x+5} + 2x + 9$$

$$-x-6 = 4\sqrt{2x+5}$$

$$(-x-6)^2 = 16(2x+5)$$

$$x^2 + 12x + 36 = 32x + 80$$

$$x^2 - 20x - 44 = 0$$

$$(x-22)(x+2) = 0$$

$$x = 22 \text{ or } x = -2$$

The solution is 22, and the solution set is $\{22\}$.

**2.**

$$(x-5)^2 = -49$$

$$x - 5 = \pm\sqrt{-49}$$

$$x = 5 \pm \sqrt{-49}$$

$$x = 5 \pm 7i$$

**3.** $x^2 + x > 6$

Solve the related quadratic equation.

$$x^2 + x = 6$$

$$x^2 + x - 6 = 0$$

$$(x+3)(x-2) = 0$$

Apply the zero product principle.

$$x+3=0 \quad \text{or} \quad x-2=0$$

$$x=-3 \qquad\qquad x=2$$

The boundary points are 3 and 2.

| Test Interval | Test Number | Test | Conclusion |
|---|---|---|---|
| $(-\infty, -3)$ | $-4$ | $(-4)^2 + (-4) > 6$<br>$12 > 6$, true | $(-\infty, -3)$ does belong to the solution set. |
| $(-3, 2)$ | 0 | $(0)^2 + 0 > 6$<br>$0 > 6$, false | $(-3, 2)$ does not belong to the solution set. |
| $(2, \infty)$ | 3 | $(3)^2 + 3 > 6$<br>$12 > 6$, true | $(2, \infty)$ does not belong to the solution set. |

The solution set is $(-\infty, -3) \cup (2, \infty)$ or $\{x \mid x < -3 \text{ or } x > 2\}$

**4.** $6x - 3(5x+2) = 4(1-x)$

$$6x - 15x - 6 = 4 - 4x$$

$$-9x - 6 = 4 - 4x$$

$$-5x = 10$$

$$x = -2$$

The solution is $-2$, and the solution set is $\{-2\}$.

**5.**

$$\frac{4}{x-3}-\frac{6}{x+3}=\frac{24}{x^2-9}$$

$$\frac{4}{x-3}-\frac{6}{x+3}=\frac{24}{(x-3)(x+3)}$$

$$(x-3)(x+3)\left(\frac{4}{x-3}-\frac{6}{x+3}\right)=(x-3)(x+3)\left(\frac{24}{(x-3)(x+3)}\right)$$

$$4(x+3)-6(x-3)=24$$

$$4x+12-6x+18=24$$

$$-2x+30=24$$

$$-2x=-6$$

$$x=3$$

Since 3 would make one or more of the denominators in the original equation zero, we disregard it and conclude that there is no solution. The solution set is $\varnothing$ or $\{\ \ \}$.

**6.** $3x+2<4$ and $4-x>1$

$3x<2 \qquad -x>-3$

$x<\frac{2}{3} \qquad x<3$

The solution set is

$\left\{x \middle| x<\frac{2}{3}\right\}$ or $\left(-\infty, \frac{2}{3}\right)$.

**7.**

$$3x-2y+z=7$$
$$2x+3y-z=13$$
$$x-y+2z=-6$$

Multiply the second equation by 2 and add to the third equation.

$$4x+6y-2z=26$$
$$x-y+2z=-6$$
$$\overline{5x+5y=20}$$

Add the first equation to the second equation.

$$3x-2y+z=7$$
$$2x+3y-z=13$$
$$\overline{5x+y=20}$$

We now have a system of two equations in two variables.

$$5x+5y=20$$
$$5x+y=20$$

Multiply the second equation by $-1$ and add to the first equation.

$$5x+5y=20$$
$$-5x-y=-20$$
$$\overline{4y=0}$$
$$y=0$$

Back-substitute 0 for $y$ to find $x$.

$$5x+y=20$$
$$5x+0=20$$
$$5x=20$$
$$x=4$$

Back-substitute 4 for $x$ and 0 for $y$ to find $z$.

$$3x-2y+z=7$$
$$3(4)-2(0)+z=7$$
$$12-0+z=7$$
$$12+z=7$$
$$z=-5$$

The solution is $(4, 0, -5)$ and the solution set is $\{(4, 0, -5)\}$.

**8.** $\log_9 x + \log_9 (x-8) = 1$

$$\log_9 (x(x-8)) = 1$$
$$x(x-8) = 9^1$$
$$x^2 - 8x = 9$$
$$x^2 - 8x - 9 = 0$$
$$(x-9)(x+1) = 0$$

Apply the zero product principle.

$x - 9 = 0$ or $x + 1 = 0$

$x = 9$ $\qquad x = -1$

Since we cannot take a log of a negative number, we disregard –1 and conclude that the solution is 9 and the solution set is $\{9\}$.

**9.** $2x^2 - 3y^2 = 5$

$3x^2 - 4y^2 = 16$

Multiply the first equation by –3 and the second equation by 2 and solve by addition.

$$-6x^2 + 9y^2 = -15$$
$$6x^2 + 8y^2 = 32$$
$$17y^2 = 17$$
$$y^2 = 1$$
$$y = \pm 1$$

Back-substitute $\pm 1$ for $y$ to find $x$.

$$y = \pm 1$$
$$2x^2 - 3(\pm 1)^2 = 5$$
$$2x^2 - 3(1) = 5$$
$$2x^2 - 3 = 5$$
$$2x^2 = 8$$
$$x^2 = 4$$
$$x = \pm 2$$

The solutions are $(-2,-1), (-2,1)$, $(2,-1)$ and $(2,1)$ and the solution set is $\{(-2,-1),(-2,1),(2,-1),(2,1)\}$.

**10.** $2x^2 - y^2 = -8$

$x - y = 6$

Solve the second equation for $x$.

$x - y = 6$

$x = y + 6$

Substitute $y+6$ for $x$.

$$2(y+6)^2 - y^2 = -8$$
$$2(y^2 + 12x + 36) - y^2 = -8$$
$$2y^2 + 24x + 72 - y^2 = -8$$
$$y^2 + 24x + 72 = -8$$
$$y^2 + 24x + 80 = 0$$
$$(y+20)(y+4) = 0$$

Apply the zero product principle.

$y + 20 = 0$ or $y + 4 = 0$

$y = -20$ $\qquad y = -4$

Back-substitute –4 and –20 for $y$ to find $x$.

$y = -20$ or $y = -4$

$x = -20 + 6$ $\qquad x = -4 + 6$

$x = -14$ $\qquad x = 2$

**11.** $3x^3 + 4x^2 - 7x + 2 = 0$

Form the list $p$, all factors of 2:

$p: \pm 1, \pm 2$

Form the list $q$, all factors of 3:

$q: \pm 1, \pm 3$

Forming all ratios:

$\frac{p}{q}: \pm\frac{1}{3}, \pm\frac{2}{3}, \pm 1, \pm 2$

Using synthetic division to test $\frac{2}{3}$:

| $\frac{2}{3}$ | 3 | 4 | –7 | 2 |
|---|---|---|---|---|
| | | 2 | 4 | –2 |
| | 3 | 6 | –3 | 0 |

Since the remainder is 0, then $\frac{2}{3}$ is a factor. Rewrite the equation:

$$\left(x-\frac{2}{3}\right)\left(3x^2+6x-3\right)=0$$

$$3\left(x-\frac{2}{3}\right)(x^2+2x-1)=0$$

The zero product principle,

$$x^2+2x-1=0$$

$$x=\frac{-2\pm\sqrt{2^2-4(1)(-1)}}{2(1)}$$

$$=\frac{-2\pm\sqrt{4+4}}{2}$$

$$=\frac{-2\pm\sqrt{8}}{2}=\frac{-2\pm2\sqrt{2}}{2}$$

$$=-2\pm\sqrt{2}$$

The solution set is $\left\{\frac{2}{3},-1\pm\sqrt{2}\right\}$.

**12.** $f(x)=(x+2)^2-4$

Graph $y=x^2$ and shift left 2 units and then down 4 units.

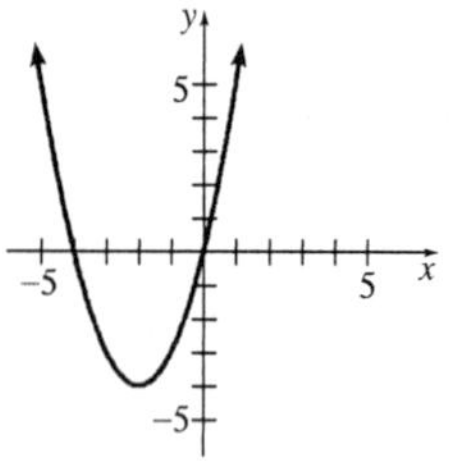


**13.** $y<-3x+5$

Replace the > by = and graph the line $y=-3x+5$ with a dashed line.

Choosing a test point of (0, 0) on the left of the line:

$0<-3(0)+5$

$0<0+5$

$0<5$, true

Shade the half-plane containing the test point.

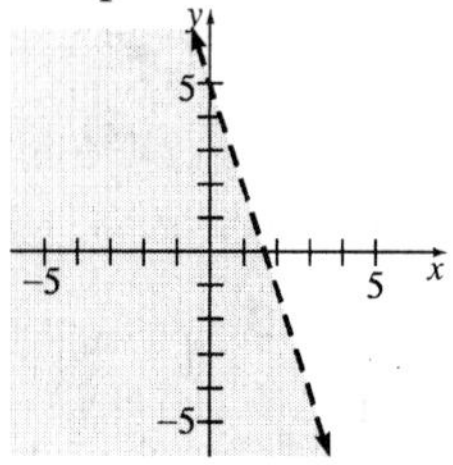


**14.** $f(x)=3^{x-2}$

Graph $y=3^x$ and shift right 2 units.

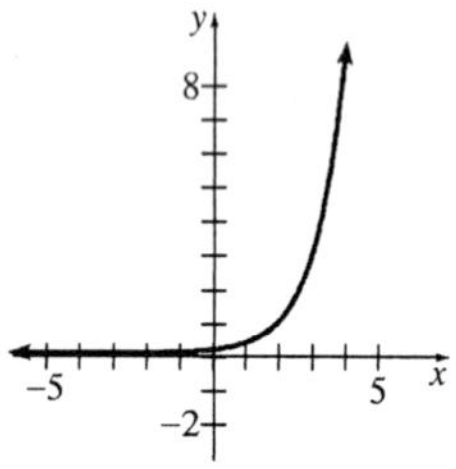


**15.** $\frac{x^2}{16}+\frac{y^2}{4}=1$

Because $x^2$ and $y^2$ have different positive coefficients, the equation's graph is an ellipse. Since the denominator of the $x^2$ – term is greater than the denominator of the $y^2$ – term, the major axis is horizontal. Since $a^2=16$, $a=4$ and the vertices are $(-4,0)$ and $(4,0)$. Since $b^2=4$, $b=2$ and endpoints of the minor axis are $(0,-2)$ and $(0,2)$.

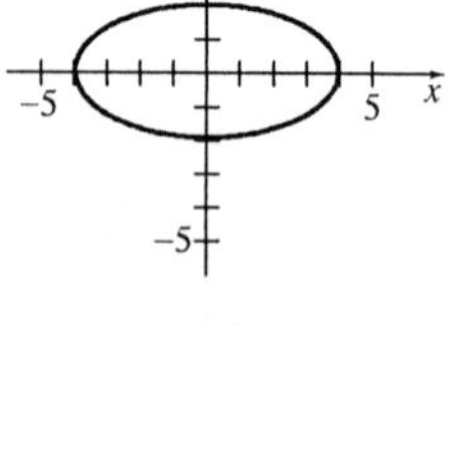

**16.** $x^2 - y^2 = 9$

$\frac{x^2}{9} - \frac{y^2}{9} = 1$

The equation is in the form

$\frac{x^2}{a^2} - \frac{y^2}{b^2} = 1$ with $a^2 = 9$, and $b^2 = 9$.

We know the transverse axis lies on the $x$-axis and the vertices are $(-3,0)$ and $(3,0)$. Because $a^2 = 3$ and $b^2 = 3$, $a = 3$ and $b = 3$. Construct a rectangle using –3 and 3 on the $x$–axis, and –3 and 3 on the $y$–axis. Draw extended diagonals to obtain the asymptotes. Graph the hyperbola.

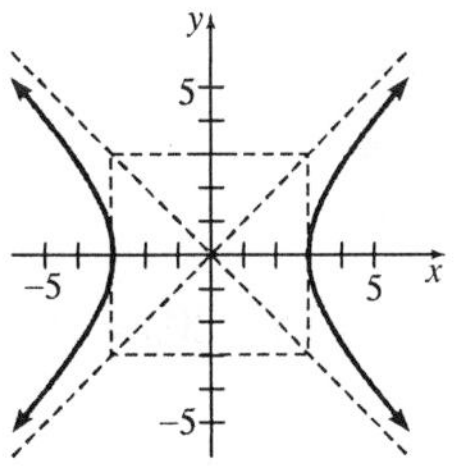


**17.** $f(x) = \frac{x-1}{x-2}$

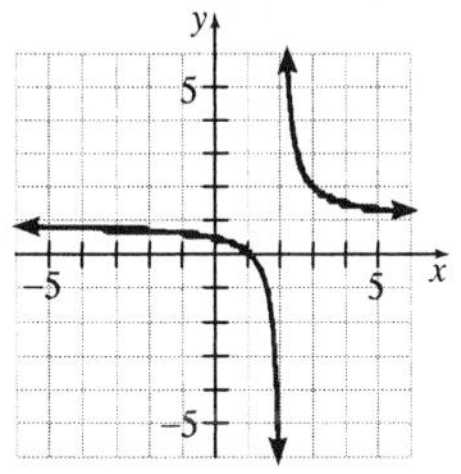


**18.** $\frac{2x+1}{x-5} - \frac{4}{x^2-3x-10}$

$= \frac{2x+1}{x-5} - \frac{4}{(x-5)(x+2)}$

$= \frac{2x+1}{x-5} \cdot \frac{x+2}{x+2} - \frac{4}{(x-5)(x+2)}$

$= \frac{(2x+1)(x+2)-4}{(x-5)(x+2)}$

$= \frac{2x^2+5x+2-4}{(x-5)(x+2)}$

$= \frac{2x^2+5x-2}{(x-5)(x+2)}$

**19.** $\frac{\frac{1}{x-1}+1}{\frac{1}{x+1}-1} = \frac{\frac{1}{x-1}+\frac{x-1}{x-1}}{\frac{1}{x+1}-\frac{x+1}{x+1}}$

$= \frac{\frac{1+(x-1)}{x-1}}{\frac{1-(x+1)}{x+1}}$

$= \frac{\frac{x}{x-1}}{\frac{-x}{x+1}}$

$= \frac{x}{x-1} \cdot \frac{x+1}{-x}$

$= -\frac{x+1}{x-1}$

**20.** $\frac{6}{\sqrt{5}-\sqrt{2}} \cdot \frac{\sqrt{5}+\sqrt{2}}{\sqrt{5}+\sqrt{2}} = \frac{6\left(\sqrt{5}+\sqrt{2}\right)}{5-2}$

$= \frac{6\left(\sqrt{5}+\sqrt{2}\right)}{3}$

$= 2\left(\sqrt{5}+\sqrt{2}\right)$

$= 2\sqrt{5}+2\sqrt{2}$

**21.** $8\sqrt{45}+2\sqrt{5}-7\sqrt{20}$
$=8\sqrt{9\cdot 5}+2\sqrt{5}-7\sqrt{4\cdot 5}$
$=8\cdot 3\sqrt{5}+2\sqrt{5}-7\cdot 2\sqrt{5}$
$=24\sqrt{5}+2\sqrt{5}-14\sqrt{5}$
$=12\sqrt{5}$

**22.** $\dfrac{5}{\sqrt[3]{2x^2y}}=\dfrac{5}{2^{\frac{1}{3}}x^{\frac{2}{3}}y^{\frac{1}{3}}}\cdot\dfrac{2^{\frac{2}{3}}x^{\frac{1}{3}}y^{\frac{2}{3}}}{2^{\frac{2}{3}}x^{\frac{1}{3}}y^{\frac{2}{3}}}=\dfrac{5\sqrt[3]{4xy^2}}{2xy}$

**23.** $5ax+5ay-4bx-4by$
$=5a(x+y)-4b(x+y)$
$=(x+y)(5a-4b)$

**24.** $5\log x-\dfrac{1}{2}\log y=\log x^5-\log y^{\frac{1}{2}}$
$=\log\left(\dfrac{x^5}{y^{\frac{1}{2}}}\right)$
$=\log\left(\dfrac{x^5}{\sqrt{y}}\right)$

**25.** $\dfrac{1}{p}+\dfrac{1}{q}=\dfrac{1}{f}$
$\dfrac{1}{p}=\dfrac{1}{f}-\dfrac{1}{q}$
$\dfrac{1}{p}=\dfrac{1}{f}\cdot\dfrac{q}{q}-\dfrac{1}{q}\cdot\dfrac{f}{f}$
$\dfrac{1}{p}=\dfrac{q}{qf}-\dfrac{f}{qf}$
$\dfrac{1}{p}=\dfrac{q-f}{qf}$
$p(q-f)=qf$
$p=\dfrac{qf}{q-f}$

**26.** $d=\sqrt{(6-(-3))^2+(-1-(-4))^2}$
$=\sqrt{9^2+3^2}=\sqrt{81+9}=\sqrt{90}$
$=\sqrt{9\cdot 10}=3\sqrt{10}$
The distance is $3\sqrt{10}$ units.

**27.** $\sum_{i=2}^{5}(i^3-4)$
$=(2^3-4)+(3^3-4)+(4^3-4)+(5^3-4)$
$=(8-4)+(27-4)+(64-4)+(125-4)$
$=4+23+60+121=208$

**28.** First, find $d$.
$d=6-2=4$
Next, find $a_{30}$.
$a_{30}=2+(30-1)4=2+(29)4$
$=2+116=118$
Now, find the sum.
$S_{30}=\dfrac{30}{2}(2+118)=15(120)=1800$

**29.** $0.\bar{3}=\dfrac{a_1}{1-r}=\dfrac{\frac{3}{10}}{1-\frac{1}{10}}=\dfrac{\frac{3}{10}}{\frac{9}{10}}=\dfrac{3}{10}\div\dfrac{9}{10}$
$=\dfrac{3}{10}\cdot\dfrac{10}{9}=\dfrac{1}{3}$

**30.** Applying the Binomial Theorem to $\left(2x-y^3\right)^4$, we have $a=2x$, $b=-y^3$, and $n=4$.

$$\left(2x-y^3\right)^4$$
$$=\binom{4}{0}(2x)^4+\binom{4}{1}(2x)^3\left(-y^3\right)+\binom{4}{2}(2x)^2\left(-y^3\right)^2+\binom{4}{3}2x\left(-y^3\right)^3+\binom{4}{4}\left(-y^3\right)^4$$
$$=\frac{4!}{0!(4-0)!}16x^4-\frac{4!}{1!(4-1)!}8x^3y^3+\frac{4!}{2!(4-2)!}4x^2y^6-\frac{4!}{3!(4-3)!}2xy^9+\frac{4!}{4!(4-4)!}y^{12}$$
$$=\frac{\cancel{4!}}{0!\cancel{4!}}16x^4-\frac{4!}{1!3!}8x^3y^3+\frac{4!}{2!2!}4x^2y^6-\frac{4!}{3!1!}2xy^9+\frac{\cancel{4!}}{\cancel{4!}0!}y^{12}$$
$$=1\left(16x^4\right)-\frac{4\cdot\cancel{3!}}{1\cdot\cancel{3!}}8x^3y^3+\frac{4\cdot3\cdot\cancel{2!}}{2\cdot1\cdot\cancel{2!}}4x^2y^6-\frac{4\cdot\cancel{3!}}{\cancel{3!}\cdot1}2xy^9+y^{12}$$
$$=16x^4-4\left(8x^3y^3\right)+6\left(4x^2y^6\right)-4\left(2xy^9\right)+y^{12}$$
$$=16x^4-32x^3y^3+24x^2y^6-8xy^9+y^{12}$$

**31.** $f(x)=\dfrac{2}{x^2+2x-15}$

Set the denominator equal to 0 to find the domain:

$x^2+2x-15=0$

$(x+5)(x-3)=0$

$x=-5,\ x=3$

So, $\{x\mid x$ is a real number and $x\neq-5$ and $x\neq3\}$

**32.** $f(x)=\sqrt{2x-6}$;

We can not take the square root of a negative number.

$2x-6\geq0$

$2x\geq6$

$x\geq3$

So, $\{x|x\geq3\}$ or $[3,\infty)$.

**33.** $f(x)=\ln(1-x)$

We can only take the ln of positive numbers.

$1-x>0$

$-x>-1$

$x<1$

So, $\{x|x<1\}$ or $(-\infty,1)$.

**34.** Let $w=$ width of the rectangle.

Then $l=2w+2$.

The perimeter of a rectangle is given by $P=2w+2l$.

Then $P=2w+2l$

$22=2w+2(2w+2)$

$22=2w+4w+4$

$22=6w+4$

$18=6w$

$w=3$

Then $l=2w+2=2(3)+2=8$

The dimension of the rectangle is 8 feet by 3 feet.

**35.**
$$A = P(1+r)^t$$
$$19610 = P(1+0.06)^1$$
$$19610 = 1.06P$$
$$P = 18500$$

Your salary before the raise is \$18,500.

**36.**
$$F(t) = 1 - k\ln(t+1)$$
$$\frac{1}{2} = 1 - k\ln(3+1)$$
$$\frac{1}{2} = 1 - k\ln 4$$
$$k\ln 4 = \frac{1}{2}$$
$$k = \frac{1}{2\ln 4} \approx 0.3607$$

Then
$$F(t) \approx 1 - 0.3607\ln(t+1)$$
$$F(6) \approx 1 - 0.3607\ln(6+1)$$
$$\approx 1 - 0.3607\ln(7)$$
$$\approx 1 - 0.7019$$
$$\approx 0.298 \text{ or } \frac{298}{100}$$

# Appendix A
# Distance and Midpoint Formulas; Circles

**1.**
$$
\begin{aligned}
d &= \sqrt{(x_2 - x_1)^2 + (y_2 - y_1)^2} \\
&= \sqrt{(14-2)^2 + (8-3)^2} \\
&= \sqrt{12^2 + 5^2} \\
&= \sqrt{144 + 25} \\
&= \sqrt{169} \\
&= 13 \text{ units}
\end{aligned}
$$

**3.**
$$
\begin{aligned}
d &= \sqrt{(x_2 - x_1)^2 + (y_2 - y_1)^2} \\
&= \sqrt{(6-4)^2 + (3-1)^2} \\
&= \sqrt{2^2 + 2^2} \\
&= \sqrt{4 + 4} \\
&= \sqrt{8} \\
&= 2\sqrt{2} \text{ or } 2.83 \text{ units}
\end{aligned}
$$

**5.**
$$
\begin{aligned}
d &= \sqrt{(x_2 - x_1)^2 + (y_2 - y_1)^2} \\
&= \sqrt{(-3-0)^2 + (4-0)^2} \\
&= \sqrt{(-3)^2 + 4^2} \\
&= \sqrt{9 + 16} \\
&= \sqrt{25} \\
&= 5 \text{ units}
\end{aligned}
$$

**7.**
$$
\begin{aligned}
d &= \sqrt{(x_2 - x_1)^2 + (y_2 - y_1)^2} \\
&= \sqrt{(3-(-2))^2 + (-4-(-6))^2} \\
&= \sqrt{5^2 + 2^2} \\
&= \sqrt{25 + 4} \\
&= \sqrt{29} \text{ or } 5.39 \text{ units}
\end{aligned}
$$

**9.**
$$
\begin{aligned}
d &= \sqrt{(x_2 - x_1)^2 + (y_2 - y_1)^2} \\
&= \sqrt{(4-0)^2 + (1-(-3))^2} \\
&= \sqrt{4^2 + 4^2} \\
&= \sqrt{16 + 16} \\
&= \sqrt{32} \\
&= 4\sqrt{2} \text{ or } 5.66 \text{ units}
\end{aligned}
$$

**11.**
$$
\begin{aligned}
d &= \sqrt{(x_2 - x_1)^2 + (y_2 - y_1)^2} \\
&= \sqrt{(-0.5-3.5)^2 + (6.2-8.2)^2} \\
&= \sqrt{(-4)^2 + (-2)^2} \\
&= \sqrt{16 + 4} \\
&= \sqrt{20} \\
&= 2\sqrt{5} \text{ or } 4.47 \text{ units}
\end{aligned}
$$

**13.**
$$
\begin{aligned}
d &= \sqrt{(x_2 - x_1)^2 + (y_2 - y_1)^2} \\
&= \sqrt{(\sqrt{5}-0)^2 + (0+\sqrt{3})^2} \\
&= \sqrt{\sqrt{5}^2 + \sqrt{3}^2} \\
&= \sqrt{5 + 3} \\
&= \sqrt{8} \\
&= 2\sqrt{2} \text{ or } 2.83 \text{ units}
\end{aligned}
$$

**15.**
$$
\begin{aligned}
d &= \sqrt{(x_2 - x_1)^2 + (y_2 - y_1)^2} \\
&= \sqrt{(-\sqrt{3}-3\sqrt{3})^2 + (4\sqrt{5}-\sqrt{5})^2} \\
&= \sqrt{\left(-4\sqrt{3}\right)^2 + \left(3\sqrt{5}\right)^2} \\
&= \sqrt{48 + 45} \\
&= \sqrt{93} \text{ or } 9.64 \text{ units}
\end{aligned}
$$

**17.**
$$\begin{aligned} d &= \sqrt{(x_2 - x_1)^2 + (y_2 - y_1)^2} \\ &= \sqrt{\left(\frac{1}{3} - \frac{7}{3}\right)^2 + \left(\frac{6}{5} - \frac{1}{5}\right)^2} \\ &= \sqrt{(-2)^2 + 1^2} \\ &= \sqrt{4+1} \\ &= \sqrt{5} \text{ or } 2.24 \text{ units} \end{aligned}$$

**19.**
$$\begin{aligned} \text{Midpoint} &= \left(\frac{6+2}{2}, \frac{8+4}{2}\right) \\ &= \left(\frac{8}{2}, \frac{12}{2}\right) \\ &= (4, 6) \end{aligned}$$

**21.**
$$\begin{aligned} \text{Midpoint} &= \left(\frac{-2+(-6)}{2}, \frac{-8+(-2)}{2}\right) \\ &= \left(\frac{-8}{2}, \frac{-10}{2}\right) \\ &= (-4, -5) \end{aligned}$$

**23.**
$$\begin{aligned} \text{Midpoint} &= \left(\frac{-3+6}{2}, \frac{-4+(-8)}{2}\right) \\ &= \left(\frac{3}{2}, \frac{-12}{2}\right) \\ &= \left(\frac{3}{2}, -6\right) \end{aligned}$$

**25.**
$$\begin{aligned} \text{Midpoint} &= \left(\frac{-\frac{7}{2}+\left(-\frac{5}{2}\right)}{2}, \frac{\frac{3}{2}+\left(-\frac{11}{2}\right)}{2}\right) \\ &= \left(\frac{-6}{2}, \frac{-4}{2}\right) \\ &= (-3, -2) \end{aligned}$$

**27.**
$$\begin{aligned} \text{Midpoint} &= \left(\frac{8+(-6)}{2}, \frac{3\sqrt{5}+7\sqrt{5}}{2}\right) \\ &= \left(\frac{2}{2}, \frac{10\sqrt{5}}{2}\right) \\ &= \left(1, 5\sqrt{5}\right) \end{aligned}$$

**29.**
$$\begin{aligned} \text{Midpoint} &= \left(\frac{\sqrt{18}+\sqrt{2}}{2}, \frac{-4+4}{2}\right) \\ &= \left(\frac{3\sqrt{2}+\sqrt{2}}{2}, \frac{0}{2}\right) \\ &= \left(\frac{4\sqrt{2}}{2}, 0\right) \\ &= \left(2\sqrt{2}, 0\right) \end{aligned}$$

**31.**
$$\begin{aligned} (x-0)^2 + (y-0)^2 &= 7^2 \\ x^2 + y^2 &= 49 \end{aligned}$$

**33.**
$$\begin{aligned} (x-3)^2 + (y-2)^2 &= 5^2 \\ (x-3)^2 + (y-2)^2 &= 25 \end{aligned}$$

**35.**
$$\begin{aligned} (x-(-1))^2 + (y-4)^2 &= 2^2 \\ (x+1)^2 + (y-4)^2 &= 4 \end{aligned}$$

**37.**
$$\begin{aligned} (x-(-3))^2 + (y-(-1))^2 &= \left(\sqrt{3}\right)^2 \\ (x+3)^2 + (y+1)^2 &= 3 \end{aligned}$$

**39.**
$$\begin{aligned} (x-(-4))^2 + (y-0)^2 &= 10^2 \\ (x+4)^2 + y^2 &= 100 \end{aligned}$$

**41.** $$x^2+y^2=16$$
$$(x-0)^2+(y-0)^2=4^2$$
Center: $(0, 0)$, $r=4$

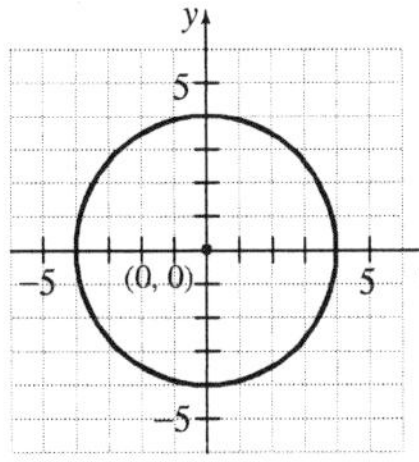


**43.** $$(x-3)^2+(y-1)^2=36$$
$$(x-3)^2+(y-1)^2=6^2$$
Center: $(3, 1)$, $r=6$

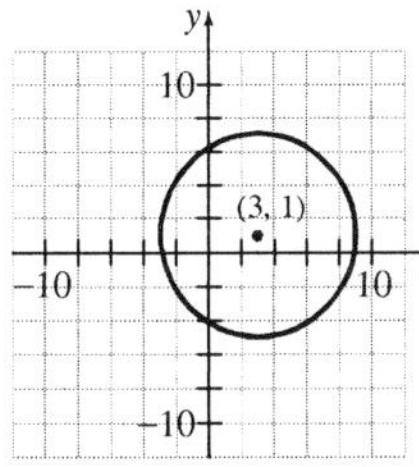


**45.** $$(x+3)^2+(y-2)^2=4$$
$$\left(x-(-3)\right)^2+(y-2)^2=2^2$$
Center: $(-3, 2)$, $r=2$

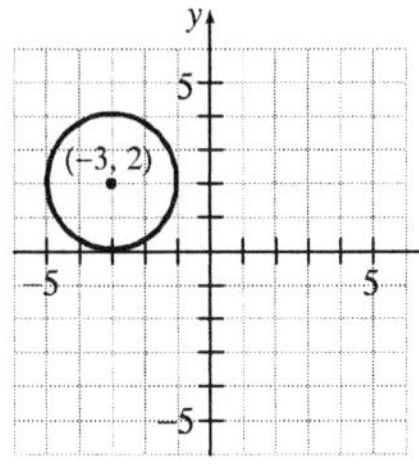


**47.** $$(x+2)^2+(y+2)^2=4$$
$$\left(x-(-2)\right)^2+\left(y-(-2)\right)^2=2^2$$
Center: $(-2, -2)$, $r=2$

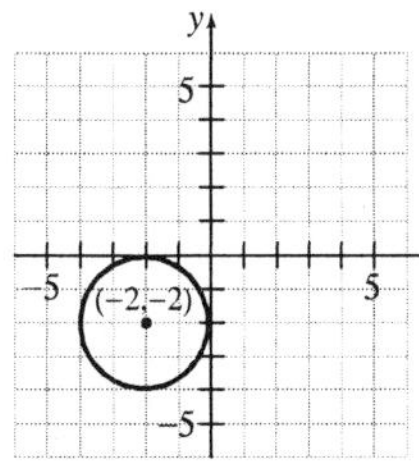


**49.** $$x^2+y^2+6x+2y+6=0$$
$$x^2+6x+y^2+2y=-6$$
$$(x^2+6x+9)+(y^2+2y+1)=-6+9+1$$
$$(x+3)^2+(y+1)^2=4$$
Rewrite to find center and radius:
$$(x+3)^2+(y+1)^2=4$$
$$\left(x-(-3)\right)^2+\left(y-(-1)\right)=2^2$$
Center: $(-3, -1)$, $r=2$

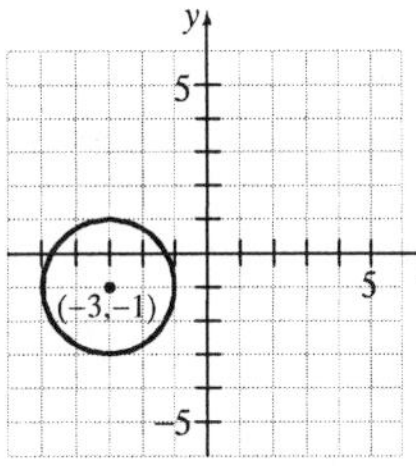


**51.** $$x^2+y^2-10x-6y-30=0$$
$$x^2-10x+y^2-6y=30$$
$$(x^2-10x+25)+(y^2-6y+9)=30+25+9$$
$$(x-5)^2+(y-3)^2=64$$
Rewrite to find center and radius:
$$(x-5)^2+(y-3)^2=8^2$$
Center: $(5, 3)$, $r=8$

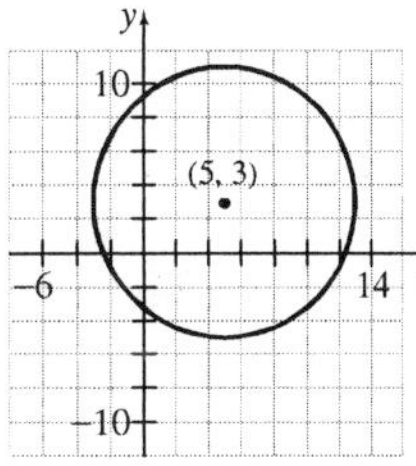

**53.**

$$x^2 + y^2 + 8x - 2y - 8 = 0$$
$$x^2 + 8x + y^2 - 2y = 8$$
$$(x^2 + 8x + 16) + (y^2 - 2y + 1) = 8 + 16 + 1$$
$$(x+4)^2 + (y-1)^2 = 25$$

Rewrite to find center and radius:

$(x-(-4))^2 + (y-1)^2 = 5^2$:

Center: $(-4, 1)$, $r = 5$

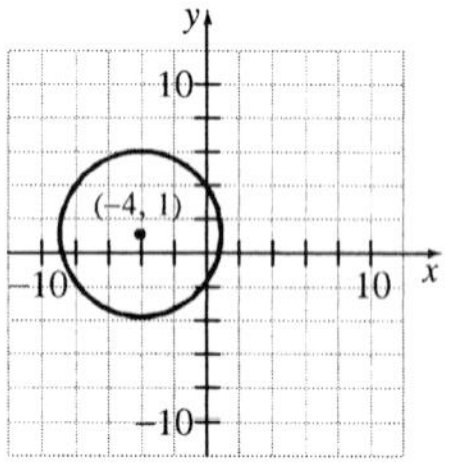


**55.** $x^2 - 2x + y^2 - 15 = 0$

$$x^2 - 2x + y^2 = 15$$
$$(x^2 - 2x + 1) + y^2 = 15 + 1$$
$$(x-1)^2 + y^2 = 16$$

Rewrite to find center and radius:

$(x-1)^2 + (y-0)^2 = 4^2$

Center: $(1, 0)$, $r = 4$

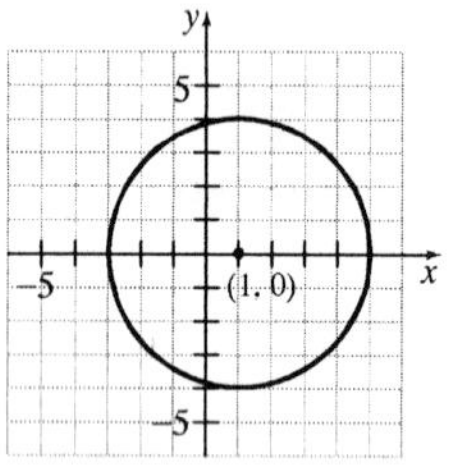

# Appendix B
## Summation Notation and the Binomial Theorem

**1.** $\sum_{i=1}^{6} 5i = 5(1)+5(2)+5(3)+5(4)+5(5)+5(6) = 5+10+15+20+25+30 = 105$

**3.** $\sum_{i=1}^{4} 2i^2 = 2(1)^2+2(2)^2+2(3)^2+2(4)^2 = 2(1)+2(4)+2(9)+2(16) = 2+8+18+32 = 60$

**5.** $$\sum_{k=1}^{5} k(k+4) = 1(1+4)+2(2+4)+3(3+4)+4(4+4)+5(5+4)$$
$$= 1(5)+2(6)+3(7)+4(8)+5(9) = 5+12+21+32+45 = 115$$

**7.** $$\sum_{i=1}^{4}\left(-\frac{1}{2}\right)^i = \left(-\frac{1}{2}\right)^1+\left(-\frac{1}{2}\right)^2+\left(-\frac{1}{2}\right)^3+\left(-\frac{1}{2}\right)^4 = -\frac{1}{2}+\frac{1}{4}+\left(-\frac{1}{8}\right)+\frac{1}{16}$$
$$= -\frac{1}{2}\cdot\frac{8}{8}+\frac{1}{4}\cdot\frac{4}{4}+\left(-\frac{1}{8}\right)\frac{2}{2}+\frac{1}{16} = -\frac{8}{16}+\frac{4}{16}-\frac{2}{16}+\frac{1}{16}$$
$$= \frac{-8+4-2+1}{16} = -\frac{5}{16}$$

**9.** $\sum_{i=5}^{9} 11 = 11+11+11+11+11 = 55$

**11.** $$\sum_{i=0}^{4}\frac{(-1)^i}{i!} = \frac{(-1)^0}{0!}+\frac{(-1)^1}{1!}+\frac{(-1)^2}{2!}+\frac{(-1)^3}{3!}+\frac{(-1)^4}{4!} = \frac{1}{1}+\frac{-1}{1}+\frac{1}{2\cdot1}+\frac{-1}{3\cdot2\cdot1}+\frac{1}{4\cdot3\cdot2\cdot1}$$
$$= 1-1+\frac{1}{2}-\frac{1}{6}+\frac{1}{24} = \frac{1}{2}\cdot\frac{12}{12}-\frac{1}{6}\cdot\frac{4}{4}+\frac{1}{24} = \frac{12}{24}-\frac{4}{24}+\frac{1}{24} = \frac{12-4+1}{24} = \frac{9}{24} = \frac{3}{8}$$

**13.** $$\sum_{i=1}^{5}\frac{i!}{(i-1)!} = \frac{1!}{(1-1)!}+\frac{2!}{(2-1)!}+\frac{3!}{(3-1)!}+\frac{4!}{(4-1)!}+\frac{5!}{(5-1)!} = \frac{1!}{0!}+\frac{2!}{1!}+\frac{3!}{2!}+\frac{4!}{3!}+\frac{5!}{4!}$$
$$= \frac{1}{1}+\frac{2\cdot\cancel{1!}}{\cancel{1!}}+\frac{3\cdot\cancel{2!}}{\cancel{2!}}+\frac{4\cdot\cancel{3!}}{\cancel{3!}}+\frac{5\cdot\cancel{4!}}{\cancel{4!}} = 1+2+3+4+5 = 15$$

**15.** $$\binom{8}{3} = \frac{8!}{3!(8-3)!} = \frac{8!}{3!5!} = \frac{8\cdot7\cdot\cancel{6}\cdot\cancel{5!}}{\cancel{3\cdot2}\cdot1\cdot\cancel{5!}} = 56$$

**17.** $\binom{12}{1} = \dfrac{12!}{1!(12-1)!} = \dfrac{12 \cdot \cancel{11!}}{1 \cdot \cancel{11!}} = 12$

**19.** $\binom{6}{6} = \dfrac{\cancel{6!}}{\cancel{6!}(6-6)!} = \dfrac{1}{0!} = \dfrac{1}{1} = 1$

**21.** $\binom{100}{2} = \dfrac{100!}{2!(100-2)!} = \dfrac{100 \cdot 99 \cdot \cancel{98!}}{2 \cdot 1 \cdot \cancel{98!}} = 4950$

**23.**
$$(x+2)^3 = \binom{3}{0}x^3 + \binom{3}{1}x^2(2) + \binom{3}{2}x(2)^2 + \binom{3}{3}2^3$$
$$= \frac{3!}{0!(3-0)!}x^3 + \frac{3!}{1!(3-1)!}2x^2 + \frac{3!}{2!(3-2)!}4x + \frac{3!}{3!(3-3)!}8$$
$$= \frac{\cancel{3!}}{1 \cdot \cancel{3!}}x^3 + \frac{3 \cdot \cancel{2!}}{1 \cdot \cancel{2!}}2x^2 + \frac{3 \cdot \cancel{2!}}{\cancel{2!}1!}4x + \frac{\cancel{3!}}{\cancel{3!}0!}8 = x^3 + 3(2x^2) + 3(4x) + 1(8)$$
$$= x^3 + 6x^2 + 12x + 8$$

**25.**
$$(3x+y)^3 = \binom{3}{0}(3x)^3 + \binom{3}{1}(3x)^2 y + \binom{3}{2}(3x)y^2 + \binom{3}{3}y^3$$
$$= \frac{3!}{0!(3-0)!}27x^3 + \frac{3!}{1!(3-1)!}9x^2y + \frac{3!}{2!(3-2)!}3xy^2 + \frac{3!}{3!(3-3)!}y^3$$
$$= \frac{\cancel{3!}}{1 \cdot \cancel{3!}}27x^3 + \frac{3 \cdot \cancel{2!}}{1 \cdot \cancel{2!}}9x^2y + \frac{3 \cdot \cancel{2!}}{\cancel{2!}1!}3xy^2 + \frac{\cancel{3!}}{\cancel{3!}0!}y^3 = 27x^3 + 3(9x^2y) + 3(3xy^2) + 1(y^3)$$
$$= 27x^3 + 27x^2y + 9xy^2 + y^3$$

**27.**
$$(5x-1)^3 = \binom{3}{0}(5x)^3 + \binom{3}{1}(5x)^2(-1) + \binom{3}{2}(5x)(-1)^2 + \binom{3}{3}(-1)^3$$
$$= \frac{3!}{0!(3-0)!}125x^3 - \frac{3!}{1!(3-1)!}25x^2 + \frac{3!}{2!(3-2)!}5x(1) - \frac{3!}{3!(3-3)!}$$
$$= \frac{\cancel{3!}}{1 \cdot \cancel{3!}}125x^3 - \frac{3 \cdot \cancel{2!}}{1 \cdot \cancel{2!}}25x^2 + \frac{3 \cdot \cancel{2!}}{\cancel{2!}1!}5x - \frac{\cancel{3!}}{\cancel{3!}0!} = 125x^3 - 3(25x^2) + 3(5x) - 1$$
$$= 125x^3 - 75x^2 + 15x - 1$$

**29.**
$$(2x+1)^4 = \binom{4}{0}(2x)^4 + \binom{4}{1}(2x)^3 + \binom{4}{2}(2x)^2 + \binom{4}{3}2x + \binom{4}{4}$$
$$= \frac{4!}{0!(4-0)!}16x^4 + \frac{4!}{1!(4-1)!}8x^3 \cdot 1 + \frac{4!}{2!(4-2)!}4x^2 \cdot 1^2 + \frac{4!}{3!(4-3)!}2x \cdot 1^3 + \frac{4!}{4!(4-4)!} \cdot 1^4$$
$$= \frac{\cancel{4!}}{0!\cancel{4!}}16x^4 + \frac{4!}{1!3!}8x^3 \cdot 1 + \frac{4!}{2!2!}4x^2 \cdot 1 + \frac{4!}{3!1!}2x \cdot 1 + \frac{\cancel{4!}}{\cancel{4!}0!} \cdot 1$$
$$= 1\left(16x^4\right) + \frac{4 \cdot \cancel{3!}}{1 \cdot \cancel{3!}}8x^3 + \frac{4 \cdot 3 \cdot \cancel{2!}}{2 \cdot 1 \cdot \cancel{2!}}4x^2 + \frac{4 \cdot \cancel{3!}}{\cancel{3!} \cdot 1}2x + 1$$
$$= 16x^4 + 4\left(8x^3\right) + 6\left(4x^2\right) + 4(2x) + 1$$
$$= 16x^4 + 32x^3 + 24x^2 + 8x + 1$$

**31.**
$$\left(x^2+2y\right)^4 = \binom{4}{0}\left(x^2\right)^4 + \binom{4}{1}\left(x^2\right)^3(2y) + \binom{4}{2}\left(x^2\right)^2(2y)^2 + \binom{4}{3}x^2(2y)^3 + \binom{4}{4}(2y)^4$$
$$= \frac{4!}{0!(4-0)!}x^8 + \frac{4!}{1!(4-1)!}2x^6y + \frac{4!}{2!(4-2)!}4x^4y^2 + \frac{4!}{3!(4-3)!}8x^2y^3 + \frac{4!}{4!(4-4)!}16y^4$$
$$= \frac{\cancel{4!}}{0!\cancel{4!}}x^8 + \frac{4!}{1!3!}2x^6y + \frac{4!}{2!2!}4x^4y^2 + \frac{4!}{3!1!}8x^2y^3 + \frac{\cancel{4!}}{\cancel{4!}0!}16y^4$$
$$= 1\left(x^8\right) + \frac{4 \cdot \cancel{3!}}{1 \cdot \cancel{3!}}2x^6y + \frac{4 \cdot 3 \cdot \cancel{2!}}{2 \cdot 1 \cdot \cancel{2!}}4x^4y^2 + \frac{4 \cdot \cancel{3!}}{\cancel{3!} \cdot 1}8x^2y^3 + 16y^4$$
$$= x^8 + 4\left(2x^6y\right) + 6\left(4x^4y^2\right) + 4\left(8x^2y^3\right) + 16y^4$$
$$= x^8 + 8x^6y + 24x^4y^2 + 32x^2y^3 + 16y^4$$

**33.**
$$(y-3)^4 = \binom{4}{0}y^4 + \binom{4}{1}y^3(-3) + \binom{4}{2}y^2(-3)^2 + \binom{4}{3}y(-3)^3 + \binom{4}{4}(-3)^4$$
$$= \frac{4!}{0!(4-0)!}y^4 - \frac{4!}{1!(4-1)!}3y^3 + \frac{4!}{2!(4-2)!}9y^2 - \frac{4!}{3!(4-3)!}27y + \frac{4!}{4!(4-4)!}81$$
$$= \frac{\cancel{4!}}{0!\cancel{4!}}y^4 - \frac{4!}{1!3!}3y^3 + \frac{4!}{2!2!}9y^2 - \frac{4!}{3!1!}27y + \frac{\cancel{4!}}{\cancel{4!}0!}81$$
$$= 1\left(y^4\right) - \frac{4 \cdot \cancel{3!}}{1 \cdot \cancel{3!}}3y^3 + \frac{4 \cdot 3 \cdot \cancel{2!}}{2 \cdot 1 \cdot \cancel{2!}}9y^2 - \frac{4 \cdot \cancel{3!}}{\cancel{3!} \cdot 1}27y + 81$$
$$= y^4 - 4\left(3y^3\right) + 6\left(9y^2\right) - 4(27y) + 81 = y^4 - 12y^3 + 54y^2 - 108y + 81$$

**35.**
$$(2x^3-1)^4 = \binom{4}{0}(2x^3)^4 + \binom{4}{1}(2x^3)^3(-1) + \binom{4}{2}(2x^3)^2(-1)^2 + \binom{4}{3}(2x^3)(-1)^3 + \binom{4}{4}(-1)^4$$
$$= \frac{4!}{0!(4-0)!}16x^{12} - \frac{4!}{1!(4-1)!}8x^9 + \frac{4!}{2!(4-2)!}4x^6 - \frac{4!}{3!(4-3)!}2x^3 + \frac{4!}{4!(4-4)!}$$
$$= \frac{\cancel{4!}}{0!\cancel{4!}}16x^{12} - \frac{4!}{1!3!}8x^9 + \frac{4!}{2!2!}4x^6 - \frac{4!}{3!1!}2x^3 + \frac{\cancel{4!}}{\cancel{4!}0!}$$
$$= 1(16x^{12}) - \frac{4\cdot\cancel{3!}}{1\cdot\cancel{3!}}8x^9 + \frac{4\cdot3\cdot\cancel{2!}}{2\cdot1\cdot\cancel{2!}}4x^6 - \frac{4\cdot\cancel{3!}}{\cancel{3!}\cdot1}2x^3 + 1$$
$$= 16x^{12} - 4(8x^9) + 6(4x^6) - 4(2x^3) + 1 = 16x^{12} - 32x^9 + 24x^6 - 8x^3 + 1$$

**37.**
$$(c+2)^5 = \binom{5}{0}c^5 + \binom{5}{1}c^4(2) + \binom{5}{2}c^3(2)^2 + \binom{5}{3}c^2(2)^3 + \binom{5}{4}c(2)^4 + \binom{5}{5}2^5$$
$$= \frac{\cancel{5!}}{0!\cancel{5!}}c^5 + \frac{5!}{1!(5-1)!}2c^4 + \frac{5!}{2!(5-2)!}4c^3 + \frac{5!}{3!(5-3)!}8c^2 + \frac{5!}{4!(5-4)!}16c + \frac{5!}{5!(5-5)!}32$$
$$= 1c^5 + \frac{5\cdot\cancel{4!}}{1\cdot\cancel{4!}}2c^4 + \frac{5\cdot4\cdot\cancel{3!}}{2\cdot1\cdot\cancel{3!}}4c^3 + \frac{5\cdot4\cdot\cancel{3!}}{\cancel{3!}2\cdot1}8c^2 + \frac{5\cdot\cancel{4!}}{\cancel{4!}\cdot1}16c + \frac{\cancel{5!}}{\cancel{5!}0!}32$$
$$= c^5 + 5(2c^4) + 10(4c^3) + 10(8c^2) + 5(16c) + 1(32) = c^5 + 10c^4 + 40c^3 + 80c^2 + 80c + 32$$

**39.**
$$(x-1)^5 = \binom{5}{0}x^5 + \binom{5}{1}x^4(-1) + \binom{5}{2}x^3(-1)^2 + \binom{5}{3}x^2(-1)^3 + \binom{5}{4}x(-1)^4 + \binom{5}{5}(-1)^5$$
$$= \frac{\cancel{5!}}{0!\cancel{5!}}x^5 - \frac{5!}{1!(5-1)!}x^4 + \frac{5!}{2!(5-2)!}x^3 - \frac{5!}{3!(5-3)!}x^2 + \frac{5!}{4!(5-4)!}x - \frac{5!}{5!(5-5)!}$$
$$= 1x^5 - \frac{5\cdot\cancel{4!}}{1\cdot\cancel{4!}}x^4 + \frac{5\cdot4\cdot\cancel{3!}}{2\cdot1\cdot\cancel{3!}}x^3 - \frac{5\cdot4\cdot\cancel{3!}}{\cancel{3!}2\cdot1}x^2 + \frac{5\cdot\cancel{4!}}{\cancel{4!}\cdot1}x - \frac{\cancel{5!}}{\cancel{5!}0!}$$
$$= x^5 - 5x^4 + 10x^3 - 10x^2 + 5x - 1$$

**41.** $(3x-y)^5$

$$=\binom{5}{0}(3x)^5+\binom{5}{1}(3x)^4(-y)+\binom{5}{2}(3x)^3(-y)^2+\binom{5}{3}(3x)^2(-y)^3+\binom{5}{4}3x(-y)^4+\binom{5}{5}(-y)^5$$

$$=\frac{\cancel{5!}}{0!\cancel{5!}}243x^5-\frac{5!}{1!(5-1)!}81x^4y+\frac{5!}{2!(5-2)!}27x^3y^2$$

$$-\frac{5!}{3!(5-3)!}9x^2y^3+\frac{5!}{4!(5-4)!}3xy^4-\frac{5!}{5!(5-5)!}y^5$$

$$=243x^5-\frac{5\cdot\cancel{4!}}{1\cdot\cancel{4!}}81x^4y+\frac{5\cdot4\cdot\cancel{3!}}{2\cdot1\cdot\cancel{3!}}27x^3y^2-\frac{5\cdot4\cdot\cancel{3!}}{\cancel{3!}2\cdot1}9x^2y^3+\frac{5\cdot\cancel{4!}}{\cancel{4!}\cdot1}3xy^4-\frac{\cancel{5!}}{\cancel{5!}0!}y^5$$

$$=243x^5-5\left(81x^4y\right)+10\left(27x^3y^2\right)-10\left(9x^2y^3\right)+5\left(3xy^4\right)-y^5$$

$$=243x^5-405x^4y+207x^3y^2-90x^2y^3+15xy^4-y^5$$

**43.** $(2a+b)^6$

$$=\binom{6}{0}(2a)^6+\binom{6}{1}(2a)^5b+\binom{6}{2}(2a)^4b^2+\binom{6}{3}(2a)^3b^3+\binom{6}{4}(2a)^2b^4+\binom{6}{5}2ab^5+\binom{6}{6}b^6$$

$$=\frac{6!}{0!(6-0)!}64a^6+\frac{6!}{1!(6-1)!}32a^5b+\frac{6!}{2!(6-2)!}16a^4b^2+\frac{6!}{3!(6-3)!}8a^3b^3$$

$$+\frac{6!}{4!(6-4)!}4a^2b^4+\frac{6!}{5!(6-5)!}2ab^5+\frac{6!}{6!(6-6)!}b^6$$

$$=\frac{\cancel{6!}}{1\cancel{6!}}64a^6+\frac{6\cdot\cancel{5!}}{1\cdot\cancel{5!}}32a^5b+\frac{6\cdot5\cdot\cancel{4!}}{2\cdot1\cdot\cancel{4!}}16a^4b^2+\frac{\cancel{6}\cdot5\cdot4\cdot\cancel{3!}}{\cancel{3\cdot2}\cdot1\cdot\cancel{3!}}8a^3b^3$$

$$+\frac{6\cdot5\cdot\cancel{4!}}{\cancel{4!}2\cdot1}4a^2b^4+\frac{6\cdot\cancel{5!}}{\cancel{5!}1}2ab^5+\frac{\cancel{6!}}{\cancel{6!}\cdot1}b^6$$

$$=64a^6+6\left(32a^5b\right)+15\left(16a^4b^2\right)+20\left(8a^3b^3\right)+15\left(4a^2b^4\right)+6\left(2ab^5\right)+1b^6$$

$$=64a^6+192a^5b+240a^4b^2+160a^3b^3+60a^2b^4+12ab^5+b^6$$